T0136232

FLORA *of the* GRAN DESIERTO AND RÍO COLORADO OF NORTHWESTERN MEXICO

THE SOUTHWEST CENTER SERIES

Ignaz Pfefferkorn, *Sonora: A Description of the Province*

Carl Lumholtz, *New Trails in Mexico*

Buford Pickens, *The Missions of Northern Sonora: A 1935 Field Documentation*

Gary Paul Nabhan, editor, *Counting Sheep: Twenty Ways of Seeing Desert Bighorn*

Eileen Oktavek, *Answered Prayers: Miracles and Milagros along the Border*

Curtis M. Hinsley and David R. Wilcox, editors, *Frank Hamilton Cushing and the Hemenway Southwestern Archaeological Expedition, 1886–1889,* volume 1: *The Southwest in the American Imagination: The Writings of Sylvester Baxter, 1881–1899*

Lawrence J. Taylor and Maeve Hickey, *The Road to Mexico*

Donna J. Guy and Thomas E. Sheridan, editors, *Contested Ground: Comparative Frontiers on the Northern and Southern Edges of the Spanish Empire*

Julian D. Hayden, *The Sierra Pinacate*

Paul S. Martin, David Yetman, Mark Fishbein, Phil Jenkins, Thomas R. Van Devender, and Rebecca K. Wilson, editors, *Gentry's Río Mayo Plants: The Tropical Deciduous Forest and Environs of Northwest Mexico.*

W J McGee, *Trails to Tiburón: The 1894 and 1895 Field Diaries of W J McGee,* transcribed by Hazel McFeely Fontana, annotated and with an introduction by Bernard L. Fontana

Richard Stephen Felger, *Flora of the Gran Desierto and Río Colorado of Northwestern Mexico*

FLORA

of the

GRAN DESIERTO AND RÍO COLORADO OF NORTHWESTERN MEXICO

BY RICHARD STEPHEN FELGER

The University of Arizona Press

∞ This book is printed on acid-free, archival-quality paper.
Manufactured in the United States of America.

05 04 03 02 01 00 6 5 4 3 2 1

Library of Congress Cataloging-in-Publication Data
Felger, Richard Stephen
Flora of the Gran Desierto and Río Colorado of northwestern Mexico / Richard S. Felger
p. cm. — (The Southwest Center series)
Includes bibliographical references (p.) and index.
ISBN 0-8165-2044-5 (alk. paper)
1. Desert plants—Mexico—Sonora (State) 2. Desert plants—Sonoran Desert. I. Title. II. Series.
QK211.F385 1999
581.7'54'097217—dc21

99-6732

CIP

British Library Cataloguing-in-Publication Data
A catalogue record for this book is available from the British Library.

For Jocelyn M. Wallace

& H. B. Wallace

Contents

Many friends, colleagues, and organizations helped make this book happen. This project was realized by support from the Wallace Research Foundation and earlier from the Wallace Genetic Foundation. I also acknowledge support from the Sunny Knickerbocker Foundation, the Southwest Parks and Monuments Association, and the U.S. Fish and Wildlife Service. Jane Ivancovich contributed during the early part of the work. Publication funds were provided by the Wallace Research Foundation, Sunny Knickerbocker Foundation, Southwest Center of the University of Arizona, James Aronson, Gil Gillenwater, Jean Russell, James Henrickson, Esther Capin, Council Rock Land and Cattle, Dan Felger, Peggy Hitchcock, Jim Hills, Duncan Laurie, Todd Gillenwater, and Leonard Scheff.

A significant experience that set me on the path leading to this volume was gained during a six week period, more than twenty-five years ago, when I camped near Tinaja Tule on the western flank of the Sierra Pinacate with Ann Woodin, her family, and various friends. I am grateful for that privilege. Since then many friends and colleagues have contributed additional help and companionship in the field, valuable information, and reviews.

I thank all who have helped with the project, especially the following: Dan Austin (Convolvulaceae); Marc Baker (Cactaceae); Theodore M. Barkley (Asteraceae); Tom Bowen (field work); Bill Broyles (field work, geography, history, and sustained encouragement); Katie Bucher (field work); Sterling Bunnell (field work); Alberto Búrquez (field work, ecology, geography, and biological, historical, scientific, and technical information as well as sustained encouragement); Fernando Chiang-Cabrera (*Lycium*, Solanaceae, and botanical data); Kim Cliffton (field work); Jane Cole (bibliography); Charles Conner (field work and technical information); Kevin Dahl (field work and manuscript assistance); Tom Daniel (Acanthaceae); Mark A. Dimmitt (field work and botanical and ecological information); Luke Evans (field work and geography); Exequiel Ezcurra (field work, ecological and geographical information, and inspiration); Wayne Ferren (*Suaeda*, Chenopodiaceae); Mark Fishbein (Apocynaceae, Asclepiadaceae, and other botanical information); Paul Fryxell (Malvaceae); Linnea Gentry (editorial assistance); Judy Gibson (botanical and archival data); Gil Gillenwater, Todd Gillenwater, and Troy Gillenwater (field work); Edward Glenn (field work, ecology of the Río Colorado delta, and halophytes); Lucretia Brezeale Hamilton (illustrations); Mary Frances Hamilton (archival assistance); Julie Hawkins (*Parkinsonia*, Fabaceae); Trica Oshant Hawkins (field work); Julian Hayden (field work, geography, history, and encouragement); James Henrickson (Aizoaceae, Amaranthaceae, Chenopodiaceae, Portulacaceae, and many other taxa, manuscript reviews, and sustained inspiration); Wendy Hodgson (Agavaceae); Colin Hughes (Fabaceae); Kevin Horstman (maps); Phil Jenkins (Solanaceae, review of the manuscript, and a great deal of botanical and technical expertise); Matthew B. Johnson (Fabaceae, and illustrations); Peter S. Johnson (map coordinates); Gene Joseph (field work); Dennis Kearns (Cucurbitaceae); Peter Kresan (geology, photographs); Meredith A. Lane (Asteraceae); Linda Leigh (field work and extensive help with the manuscript); Cathy Moser Marlett (illustrations); Angelina Martínez Yrízar (ecology, geography, and technical assistance); Michael McClure (field work); Lucinda McDade (Acanthaceae, extensive manuscript review and botanical information); Reid Moran (Crassulaceae); James D. Morefield (Asteraceae); Gary P. Nabhan (field work, O'odham names); Nancy Nicholson (illustrations); James Norris (field work); David Ortíz Reyna (field work); Bruce Parfitt (Cactaceae and other taxa); William D. Peachey (geology and cactus ecology); Donald Pinkava (Cactaceae and manuscript review);

Barry Prigge (Loasaceae); Adrianne Rankin (field work, ecological and geographical information); Jon Rebman (Cactaceae); Charlotte Reeder (Poaceae, bibliographic and botanical information); John Reeder (Poaceae, bibliographic and botanical information, and manuscript review); Frances Runyan (illustrations); Sue Rutman (field work and technical information); Dean Saxton (O'odham names); Silke Schneider (field work, encouragement, and tolerance); John Semple (Asteraceae); Richard Spellenberg (Nyctaginaceae); Victor Steinmann (field work, Euphorbiaceae, and botanical information); John Strother (Asteraceae); Scott Sundberg (Asteraceae); Barbara Tellman (field work, photographs); Laurence Toolin (manuscript review); Dale S. Turner (field work and manuscript assistance); Billie Turner (Asteraceae); Raymond M. Turner (ecological and geographical information, and sharing illustrations); Tom Van Devender (manuscript review, ecological and paleo-botanical information); Warren L. Wagner (Onagraceae); Michael Windham (ferns and fern relatives); James J. White (archival illustrations from Hunt Institute for Botanical Documentation); Anita Williams de Alvarez (field work, history, and geography); George Yatskievych (ferns and fern relatives, Lennoaceae, Orobanchaceae, and extensive botanical information and manuscript review); Allan Zimmerman (field work, Cactaceae and other taxa).

I especially thank the curators and staff of the various herbaria who provided facilities, hospitality, information, and loans of specimens. In this regard I thank researchers and staff of the herbaria at the Arizona State University, California Academy of Sciences, Desert Botanical Garden, Jepson Herbarium, Instituto de Biología de la Universidad Nacional Autónoma de México, Missouri Botanical Garden, Rancho Santa Ana Botanic Garden, San Diego Museum of Natural History, University of Texas, and University of California, Berkeley. A major part of my work was carried out at the University of Arizona herbarium, and I sincerely thank Phil Jenkins, Lucinda McDade and other staff members, volunteers, and herbarium scholars for their generous help.

I am grateful to the institutions and presses that have granted permission for use of illustrations, including the Hunt Institute for Botanical Documentation, the Sherman Foundation, Stanford University Press, University of Arizona Press, University of California Press, and University of Washington Press. William K. Hartmann and Fisher Books gave permission to use a photograph (© 1989 Fisher Books) from *Desert Heart.* The New York Botanical Garden granted permission to reprint illustrations from the *Intermountain Flora: Vascular Plants of the Intermountain West, U.S.A.*, vol. 3 part A, © 1977; vol. 4, © 1984; vol. 3 part B, © 1989; vol. 5, © 1994; vol. 6, © 1997, The New York Botanical Garden. The University of California Press, Berkeley, and the Jepson Herbarium granted permission to use illustrations from *The Jepson manual: higher plants of California*, J. Hickman (ed.), University of California, Berkeley, © 1993, the Regents of the University California. Stanford University Press gave permission to use illustrations from the *Illustrated Flora of the Pacific States*, 4 vols., by LeRoy Abrams and Roxana Stinchfield Ferris, © 1960 by the Board of Trustees of the Leland Stanford Junior University.

James Henrickson has given unwavering encouragement and an enormous wealth of botanical information, shared numerous illustrations by Bobbi Angell and Felicia Bond, and lent his monumental and extremely useful unpublished *Flora of the Chihuahuan Desert Region* (Henrickson & Johnston n.d.). Over the years Charlotte Reeder and John Reeder have been a source of inspiration and knowledge—I have been privileged to gain a deep appreciation of the grasses from these intrepid agrostologists. Michael F. Wilson has been involved in many tasks in the making of this book, and I am grateful for his efforts and dedication. I thank Kirsteen E. Anderson for copy editing, Bill Benoit for expert design and production, and attention to detail, and the many artists who contributed illustrations. Bill Broyles and Joseph C. Wilder of the Southwest Center of the University of Arizona made this book a reality.

Abbreviations

The talent of many illustrators and photographers and the cooperation of several institutions have contributed to the illustrations for this book. Each is identified in individual credit lines below the illustrations as follows:

ILLUSTRATORS

AB—Alfonso Barbosa
AC—Agnes Chase
AE—Amy Eisenberg
AS—Anthony Salazar
BA—Bobbi Angell
BT—Barbara Tellman
CLV—unable to identify
CMM—Cathy Moser Marlett
DTM—Daniel Trembly MacDougal
EDC—E. D. Church
ER—Emily Reed
FB—Felicia Bond
FLS—Frank Lamson-Scribner
FR—Frances Runyan
ILW—Ira L. Wiggins
IS—Isaac Sprague
JRJ—Jeanne R. Janish
KCH—Kevin C. Horstman
KD—Karin Douthit
KK—Karen Klitz
KS—Kristina Schierenbeck
LAV—Linda Ann Vorobik
LBH—Lucretia Brezeale Hamilton
LH—Leta Hughey
LTD—Lauramay T. Dempster
MAD—Mark A. Dimmitt
MB—Marlo D. Buchmann
MBJ—Matthew B. Johnston
MBP—Mary Barnas Pomeroy
MD—Maggie Day
MDB—M. D. Baker (Miss)
MFW—Michael F. Wilson
ME—Matilda Essig
MK—Margaret Kurzius
MWG—Mary Wright Gill
NE—Nancy Evans Weaver
NLN—Nancy L. Nicholson
PB—Phyllis Brick
PLK—Peter L. Kresan

ILLUSTRATORS (cont.)

PM—Paul Mirocha
PR—P. Roetter
RAJ—Robin A. Jess
RHP—Robert H. Peebles
RJP—Ruth J. Powell
RMT—Raymond M. Turner
RSF—Richard S. Felger
TH—Thomas Holm
VK—Valloo Kapadia
WKH—William K. Hartmann
WM—William S. Moye III

ARCHIVES AND HERBARIA

ARIZ—University of Arizona, Tucson
ASU—Arizona State University, Tempe
CAS—California Academy of Sciences, San Francisco
COTECOCA (HERMOSILLO)—Comisión Técnico Consultiva para la Determinación Regional de los Coeficientes de Agrostadero
COTECOCA (MEXICO CITY)—Comisión Técnico Consultiva para la Determinación Regional de los Coeficientes de Agrostadero
DAV—University of California, Davis
DES—Desert Botanical Garden, Phoenix
DS—Dudley Herbarium of Stanford University, San Francisco
ENCB—Escuela Nacional de Ciencias Biológicas, Instituto Politécnico Nacional, Mexico City
F—Field Museum of Natural History, Chicago
HUNT—Hunt Institute for Botanical Documentation, Carnegie Mellon University. Pittsburgh
HUNT/H-US—Hitchcock-Chase Collection of Grass Drawings, Hunt Institute for Botanical Documentation, Carnegie Mellon University, Pittsburgh, indefinite loan from Smithsonian Institution
HUNT/US—Hunt Institute for Botanical Documentation, Carnegie Mellon University, Pittsburgh, indefinite loan from Smithsonian Institution
HUNT/USDA—United States Department of Agriculture Forest Service Collection, Hunt Institute for Botanical Documentation, Carnegie Mellon University, Pittsburgh
JEPS—Jepson Herbarium, Berkeley
MEXU—Instituto de Biología, Universidad Nacional Autónoma de México, Mexico City
MO—Missouri Botanical Garden, Saint Louis
ORPI—Organ Pipe Cactus National Monument, Arizona
POM—Pomona College, Claremont
RSA—Rancho Santa Ana Botanic Garden, Claremont
SD—San Diego Society of Natural History, San Diego
SFC—Sherman Foundation Collection, Corona del Mar
TEX—University of Texas, Austin
UC—University of California, Berkeley
UCR—University of California, Riverside
UNM—University of New Mexico, Albuquerque
US—United States National Herbarium, Smithsonian Institution, Washington, D.C.

Throughout this book, an asterisk (*) denotes a non-native or introduced, exotic plant.

† indicates a plant that is expected in the Gran Desierto region given its known range, but that has not actually been collected in the region.

FLORA *of the* GRAN DESIERTO AND RÍO COLORADO OF NORTHWESTERN MEXICO

Part 1
The Environment and Human Interactions

On a warm spring day the desert breeze is alive with buzzing insects and the smell of yellow palo verde flowers. It has been a long day's hike from our camp at Tinaja del Tule to Pinacate Peak, long because of the distance and so many things to see and record. From the cinder-cone summit I see lava fields, great craters, dunes, distant sawtooth desert mountains, the sea, and the Sierra San Pedro Mártir more than a hundred miles across the vanquished delta. I imagine a delta still teeming with life. I see a quiet, primordial empire.

Less than three months later, during late May, I return to find most of the waterholes dry—even Tinaja del Tule. There are animal bones where the last water disappeared. The heat—the searing heat—the dust, and the dryness penetrate everything. The wildflowers are gone. Even creosote bush leaves are shriveled, brown, and brittle.

The northwestern corner of the state of Sonora, Mexico, makes up the area encompassed in this flora, referred to in the text as "the flora area"; it is a logistically convenient eco-political unit (see Felger & Broyles 1997 and maps). The U.S. border provides the northern boundary, the delta of the Río Colorado and the Mexican portion of the river the western edge, and the Gulf of California the southern limit. The 112°50′ W–113°30′ W meridians, which roughly coincide with Mexico Highway 8 from Sonoyta to Puerto Peñasco, make up the approximate eastern edge of the study area. The area is approximately 15,000 km^2. It is a substantial portion of the extremely arid center, or heart, of the Sonoran Desert. Within this region there are expansive dune fields, maritime strands, a small river, a once-great river and its delta, tidal wetlands, desert plains, steep granitic mountains, desert oases, and an enormous black and red volcanic field featuring its own mountain, lava flows, cinder cones, and formidable craters. Also included is the Quitobaquito oasis, along the international border but mostly on the Arizona side.

Despite this aridity the vegetation is surprisingly diverse (Ezcurra 1984; Ezcurra, Equihua, & López-Portillo 1987; Felger 1980). As would be expected of an arid region, however, the flora is relatively small in relation to area (McLaughlin & Bowers 1999). The flora includes 589 species in 327 genera and 85 families. Among these are 79 species not native to the region, although the majority are urban and agricultural weeds that have not become established in undisturbed natural desert habitats.

Paleoclimate

Wonderfully detailed reconstructions of the vegetation and climate of many desert areas in northern Mexico and the southwestern United States are available for the past 30,000 to 45,000 years (Betancourt, Van Devender, & Martin 1990). These reconstructions are based on analysis of plant contents from fossil packrat middens. The vegetation history of the past several million years is dominated by the advent and conclusion of glacial periods. Past glacial periods have lasted five to ten times as long as interglacials (Porter 1989; Winograd et al. 1997). Thus, climate and vegetation similar to what is in the flora area today would have been present for perhaps only 5% to 10% of the last 1.8 million years. In the flora area ice-age climates have favored the expansion of woodland and chaparral species from higher to lower elevations and from the north to the south. Likewise, species with subtropical affini-

Northwestern Sonora and adjacent southwestern Arizona. PM.

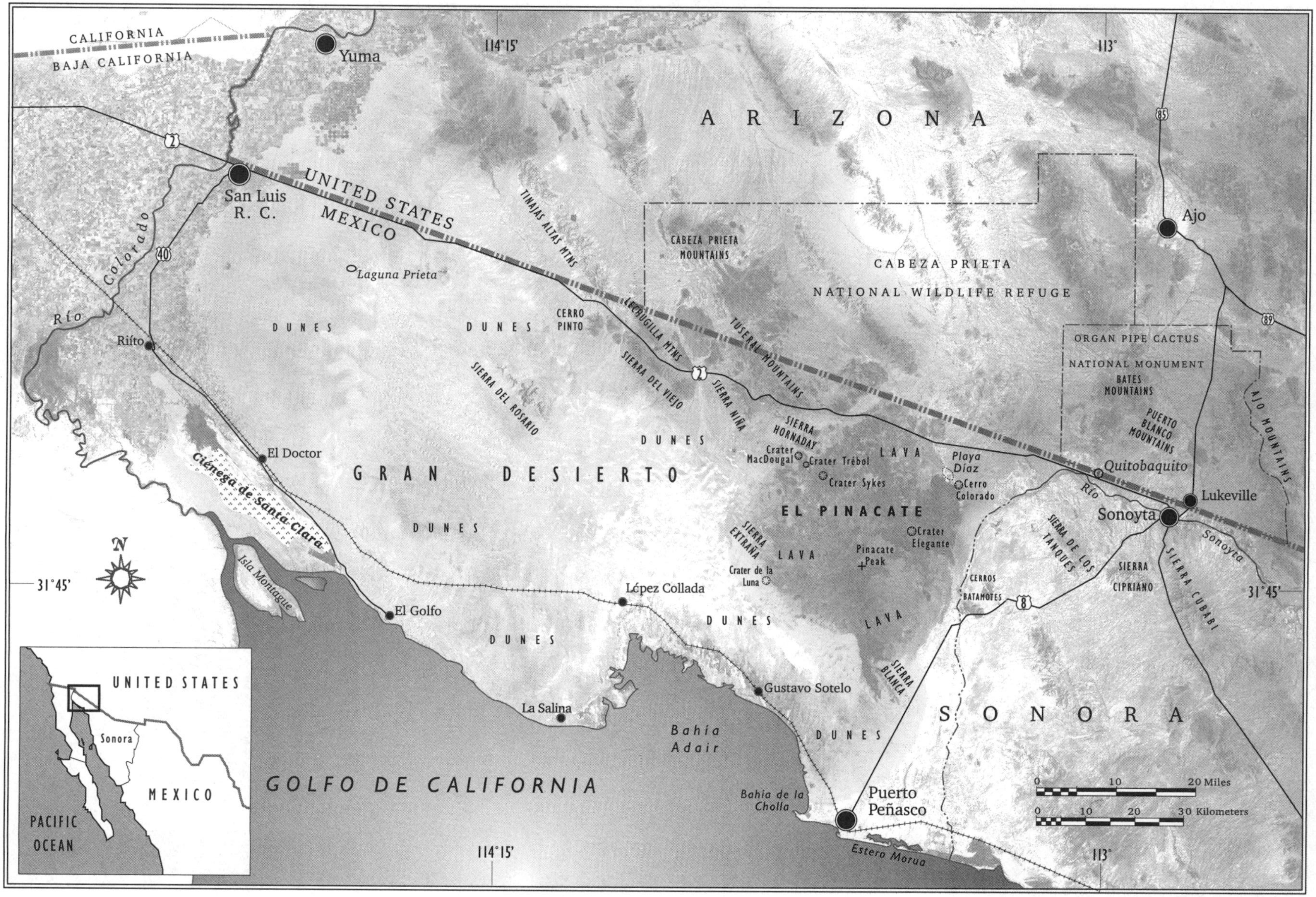

ties have shifted to the south and into lower elevations during glacial periods. Successional stages from ice-age woodlands to subtropical deserts likely occurred during each of the 15 to 20 glacial-interglacial cycles during the Pleistocene (Imbrie & Imbrie 1979).

The most recent ice age, the Wisconsin, which ended about 11,000 years ago, produced profound changes in climate and vegetation of the Sonoran Desert region. During this ice age, pinyon-juniper woodlands were widespread in the Arizona Upland subdivision of the Sonoran Desert at elevations above 550 m. A drier woodland with Joshua tree (*Yucca brevifolia*) occurred at lower elevations in the Lower Colorado River valley. Desertscrub communities apparently persisted below 300 m during the entire Pleistocene (Van Devender 1990).

The development of modern vegetation soon after 11,000 years ago is recorded in packrat middens from the Hornaday Mountains (Van Devender et al. 1990), Puerto Blanco Mountains (Van Devender 1987), Tinajas Altas Mountains (Van Devender 1990), and Picacho Peak in southeastern California (Cole 1986). Saguaro (*Carnegiea gigantea*) and brittlebush (*Encelia farinosa*) had migrated north from warmer regions in Mexico by that time, although dry woodland plants were still present in the region. Sonoran Desert vegetation began to form about 9000 years ago as ice-age woodland species finally retreated upslope. The modern climatic regime and plant communities became established about 4500 years ago with the arrivals of ironwood (*Olneya tesota*), foothill palo verde (*Parkinsonia microphylla*), and organpipe cactus (*Stenocereus thurberi*). The current vegetation is apparently the most arid association with the lowest species richness that has occurred in this area during the present interglacial period (Van Devender et al. 1990).

Successional changes have left populations of some species "stranded" in refugia (areas with special conditions). Certain 'sky island' plants found to the north or at higher elevations occur on the somewhat less arid north slopes of Pinacate Peak and the Sierra del Viejo, where they are separated from their nearest populations by substantial distances of inhospitable lowland desert (see discussion in "Higher Elevations of Sierra Pinacate").

PRESENT CLIMATE

Today the Gran Desierto is one of the most arid regions of North America. Frequent winds accentuate the aridity. Rainfall is biseasonal. The usually gentle winter-spring rains, called *las equipatas,* are produced by storms originating in the Pacific Ocean. These migratory cyclones, or Pacific frontal storms, can deliver widespread light rains or drizzle that may last many hours. However, the spring storms can sometimes be violent and torrential, especially during an El Niño year. El Niño years bring maximum winter-spring rains, which generally result in spectacular displays of spring wildflowers. El Niño years occurred during the winters of 1972–73, 1976–77, 1982–83, 1986–87, 1991–92 (Diaz & Markgraf 1992), early 1993, 1993–94, and 1997–98.

Summer rains, called *las aguas,* result from an abbreviated monsoon of southern or tropical origin. These rains are derived from moist, maritime air—essentially from the Bermuda High—which moves over the continent following summer warming of the land. The results are usually convective and often violent and highly localized thunderstorms with spectacular lightning and heavy rainfall of brief duration. These sporadic, monsoon-spawned thunderstorms may commence about a month after the summer solstice and continue into September. Much rain may fall in a very short time, sometimes leading to flash floods, while nearby areas may receive no rain all summer. In some years the summer rains may be very scant or lacking. The high temperatures and seasonal drought of late spring and early summer is severely limiting for plant life.

The September-October, late summer–early fall tropical storms, known as *chubascos,* sometimes bring large amounts of rain. These occasional hurricane-fringe storms can result in spectacular development of many perennials adapted to warm-weather growth, especially some of the larger shrubs such as *Bursera, Condalia, Fouquieria,* and *Jatropha.* Indeed, it is at such times that most stem growth may occur.

Annual average rainfall in the Sonoyta region—which falls within the Arizona Upland subdivision of the Sonoran Desert (Shreve 1951)—is roughly 200 to 265 mm (table 1). Precipitation declines sharply

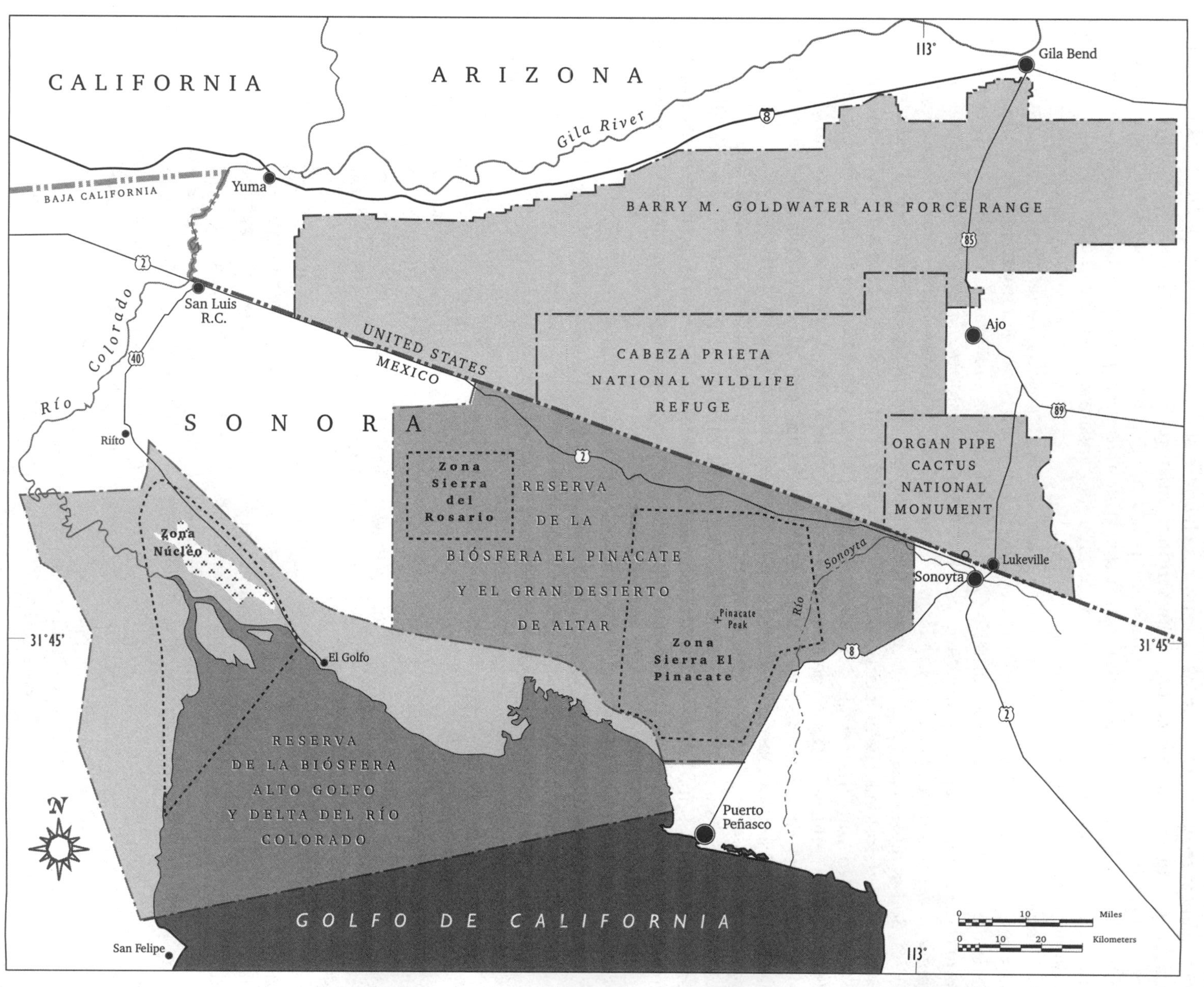

Reserves and protected areas in northwestern Sonora and adjacent southwestern Arizona. PM.

to the south and west over most of the western and southern part of the flora area (within the Lower Colorado Valley subdivision) the average annual rainfall is less than 100 mm. However, average annual values can be misleading in attempting to explain the distribution of plants in this desert region.

The data present several notable examples of the extreme variability of rainfall in the region. The record annual lows for Puerto Peñasco, San Luis, and Riito were all set in 1956, and the winter-spring drought of 1998–99 was likewise extreme. The year after the 1956 drought each station received far

TABLE 1. SUMMARY OF WEATHER DATA FOR NORTHWESTERN SONORA.

STATION	ELEVATION (m)	TEMPERATURE					PRECIPITATION			
		mean			extreme		mean		extreme	
		annual (°C)	January (°C)	July (°C)	minimum (°C)	maximum (°C)	annual (mm)	warm season (%)	minimum (mm)	maximum (mm)
Sonoyta	300	21.2	11.2	31.9			195.6	62	31.9	373.0
Aguajita/ Quitobaquito	230	20.7	10.0	30.1	-9.5	44.5	267.0	54		
Pinacate Peak	1190								133	178
Hornaday Mountains	230						120	61		
Puerto Penasco	5	20.1	11.4	29.3			86.4	48	6.0	241.0
Sierra del Rosario		25.3	12.8	36.7	-2.2	48.9			0	76
San Luis	25	23.1	13.0	34.0			55.3	42	0.1	152.0
El Riito	5	21.9	11.8	32.5			40.2	36	0.8	127.0

NOTES: Warm-season percentage is based on rainfall during July, August, September, and October.

DATA ARE COMPILED from Durrenberger & Murrieta (1978), Hastings (1964), Hastings & Humphrey (1969), May (1973), and unpublished data from the National Park Service. The data are of varying quality. Data from Organ Pipe Cactus National Monument—Aguajita Wash and Quitobaquito (two sites within 3 km of each other)—are highly reliable but cover a limited time span. Aguajita data cover January 1988–October 1991; Quitobaquito data are for rainfall only and cover 1982–1990. Aguajita data are broken into 13 reporting periods, so the January mean covers January 1–28, and the July mean covers June 16–August 10 (two periods).

DATA FROM MAY (1973) were used for Pinacate Peak (station on the summit), Hornaday Mountains, and Sierra del Rosario. They cover the period January 1971–December 1972. May gave only limited data, but his are the only data available for this part of the region.

THE OTHER DATA presented cover much greater time spans: Sonoyta 1949–1967, Puerto Peñasco 1949–1967, San Luis 1927–1967, El Riito 1950–1967. The mean data come from Hastings (1964) and Hastings & Humphrey (1969). Additional data for precipitation extremes come from Durrenberger & Murrieta (1978) and include the years 1968-1977. Durrenberger & Murrieta's data were not incorporated in the rest because there are inconsistencies for the years that overlap and the means have not been calculated.

more than average rainfall. Although months or years without significant rainfall are common, rain may sometimes be torrential, and much of the year's total precipitation may occur during a period of a few hours. Sonoyta recorded 177 mm in 24 hours during September 1970, nearly three-quarters of that year's total for the station.

The amount of rainfall in one season or year versus the next is highly variable and unpredictable (Ezcurra & Rodrígues 1986). In many respects the unpredictability of rainfall is of greater significance to plant life than are yearly averages. The vegetation is dynamic and its present appearance, especially in the Pinacate region, is largely a result of extremely dry years in the mid-twentieth century (R. M. Turner 1990).

The gradient of total rainfall declines from east to west, but there is an important seasonal effect in rainfall patterns. Summer rainfall is generally greater toward the east whereas winter rainfall seems to be greater toward the west. There is also a significant increase in rainfall and decrease in temperatures at peak elevations in the Sierra Pinacate. The highest precipitation in the region has been recorded from the Sonoyta region (e.g., Sonoyta and Aguajita/Quitobaquito) and on Pinacate Peak (table 1). At the other extreme there was a 34-month period in the Sierra del Rosario during which no measurable precipitation was recorded (May 1973).

Summers are long and very hot. Average maximum daily temperature exceeds 38°C (100°F) during June, July, and August, and temperatures exceeding 38°C are common from late April to early October (Felger et al. 1992). In general, temperatures are slightly lower toward the coast and peak elevations. The highest temperature reported for the Sonoran Desert was 56.7°C (134°F) in the Sierra Blanca recorded by Larry May (1973) in late June 1971.

Winter daytime temperatures are commonly 15.6 to 23.8°C (60–75°F), and on a few nights during colder months temperatures may dip several degrees below freezing. A minimum of −8.3°C (17.0°F) was recorded at a station 1.6 km south of Los Vidrios in November 1972 (May 1973), and similar minimum temperatures can be expected atop Sierra Pinacate. During the winter of 1988–89, a weather station in Aguajita Wash, near Quitobaquito, recorded a total of 22 days with overnight temperatures below 0°C. Temperatures remained below freezing for more than 11 hours on two of those days. On rare occasions light snowfall blankets the Pinacate peaks, but it usually remains for only a few hours.

Although mild winter freezing is commonplace, certain habitats or microhabitats can be nearly or perhaps entirely frost-free, e.g., crater walls, south- or west-facing rock slopes (especially near the coast), and areas beneath large shrubs or small trees with dense canopies. The lack of severe frost is a critical factor that permits a number of frost-sensitive species with subtropical affinities to survive in the region. However, for certain highly frost-sensitive species even these moderate freezing temperatures are severely limiting. Among the many species that appear to be limited by freezing weather are *Bursera microphylla, Hyptis emoryi, Jatropha cinerea, J. cuneata, Lophocereus schottii, Sebastiania bilocularis, Solanum hindsianum,* and *Stenocereus thurberi.*

Major Habitats

Northwestern Sonora lies in the southwestern corner of the southern Basin and Range Physiographic Province. The region is at the western margin of the North American plate, adjacent to the complex boundary between the North American and Pacific plates (Kresan 1997). Within the flora area of northwestern Sonora are several readily recognizable major habitats. These habitats may correspond to geomorphic, edaphic, hydrologic, climatic, and ethnologic boundaries (e.g., land-use history or political divisions). The distributions of plants in these major habitats are summarized in Appendix A. As in any classification of nature, the boundaries are often indistinct and further refinements might be made.

Available soil moisture, the overriding factor shaping the distribution of desert plants, differs significantly between the major habitats. Especially in arid regions there is a soil-texture effect on plants. Coarser soils generally provide more available water for plants than do finer soils. The available runoff water is another obvious factor choreographing the distributions of desert plants (e.g., Nobel 1978). Freezing temperatures are also important in restricting the local distributions of many plants. Other factors to consider are salt and alkali accumulation, human influences, and environmental degradation. Variations in these factors among habitats may help explain local distributional patterns (Búrquez et al. 1999; McAuliffe 1999; Shreve 1951).

Sonoyta Region

The international border crossing at Lukeville and the town of Sonoyta form the northeastern corner of the flora area. Major features include sandy valley-plains and low granitic mountains, hills, and bajadas. The Río Sonoyta courses westward then turns south toward the Gulf of California. The desert-floor elevations in the Sonoyta region range from about 325 m at its southwestern margins to about 400 m at Sonoyta. The granitic Sierra de los Tanques, with a peak of 590 m, lies within the region. Also included is the western bajada and base of Sierra Cipriano, as well as some smaller hills of rhyolitic volcanics.

The Sonoyta region has a surprisingly distinctive and generally richer floristic makeup than the rest of northwestern Sonora. The Sonoyta region forms part of the western border of Arizona Upland, one of the seven vegetational subdivisions of the Sonoran Desert delineated by Forrest Shreve (1951). The rest of northwestern Sonora is part of Shreve's Lower Colorado Valley phytogeographic region. The Sonoyta region experiences slightly higher and more dependable rainfall than do the lowlands of the rest of the flora area in northwestern Sonora. The flora of the Sonoyta region contains at least 314 species (Appendix A).

As the desert floor gradually dips below approximately 325 m to the south and west, there is a general decline in the number of plant species as well as geographic replacement of plant taxa (Appendix A). For example, at the approximate boundary of the Arizona Upland and Lower Colorado Valley, the hedgehog cactus *Echinocereus engelmannii* var. *acicularis* is replaced by var. *chrysocentrus*. Velvet mesquite (*Prosopis velutina*) shows a similar boundary with western honey mesquite (*P. glandulosa* var. *torreyana*) as do two zygophylls, *Fagonia californica* subsp. *longipes* and subsp. *californica*. *Opuntia acanthocarpa* and *O. echinocarpa* are chollas with similar growth forms; the former occurs in the Sonoyta region and is abruptly replaced by the latter in the adjacent Lower Colorado Valley.

A number of species occur in the Sonoyta region and at higher elevations in the Sierra Pinacate but are absent from the intervening lowlands; among them are *Daucus pusillus, Dichelostemma pulchellum, Digitaria californica, Jatropha cinerea, Malacothrix sonorae, Muhlenbergia porteri,* and *Pectocarya recurvata.*

The most common, or conspicuous, larger plants of the Sonoyta region include *Ambrosia deltoidea, A. dumosa, Carnegiea gigantea, Jatropha cuneata, Larrea divaricata, Lophocereus schottii, Olneya tesota, Opuntia acanthocarpa, Parkinsonia microphylla, Prosopis velutina, Sebastiania bilocularis,* and *Stenocereus thurberi*. Creosote bush (*Larrea*) and triangle-leaf bursage (*Ambrosia deltoidea*) are among the most common perennials across valley floors. The riverine habitat of the Río Sonoyta is discussed under "Wetland Habitats," and urban and agricultural impacts are considered under "Human Influences."

Pinacate Volcanic Field

Pinacate is the local Mexican name for the 'stink bug', large black beetles in the genus *Eleodes*. They are common in the Sonoran Desert and especially in the Pinacate region. When threatened, the pinacate puts its head down, the other end up, and emits a foul-smelling secretion.

The Pinacate volcanic field, or El Pinacate, covering approximately 2000 km^2, is one of the youngest and most spectacular lava fields in North America (Hartmann 1989; Hayden 1998; Kresan 1997; Lynch 1981). The geology of the Pinacate region has received considerable attention and has been compared with the surface of Mars (e.g., Hartmann 1989). This volcanic field is mostly in Sonora, but the northernmost cones and flows are in Arizona. The field includes Volcán Santa Clara, also known as Sierra Pinacate or Sierra Santa Clara, numerous cinder cones, extensive lava flows, and some of the world's most spectacular steam blast maar craters. As Kresan (1997:583) describes it, "the volcanic complex consists of a small shield-type volcano with some 500 eruptive centers, including cinder cones and fissure eruptions, along its flanks. Although presently dormant, volcanic activity has been continuous but episodic during the last 1.7 million years and consists of two distinct periods of volcanism. First, the shield volcano, often referred to as Volcán Santa Clara, was built by successive eruptions from a central summit vent complex, active from 1.7 to 1.1 million years ago. Then, smaller, individual basaltic cones and lava flows erupted on the slopes of Santa Clara and extended out onto the desert surrounding the central vent complex. I anxiously await the next eruption."

Northwestern Sonora and adjacent southwestern Arizona. Composite Landsat 5 Thematic Mapper scenes taken on various dates. KCH. © Kevin C. Horstman 1999.

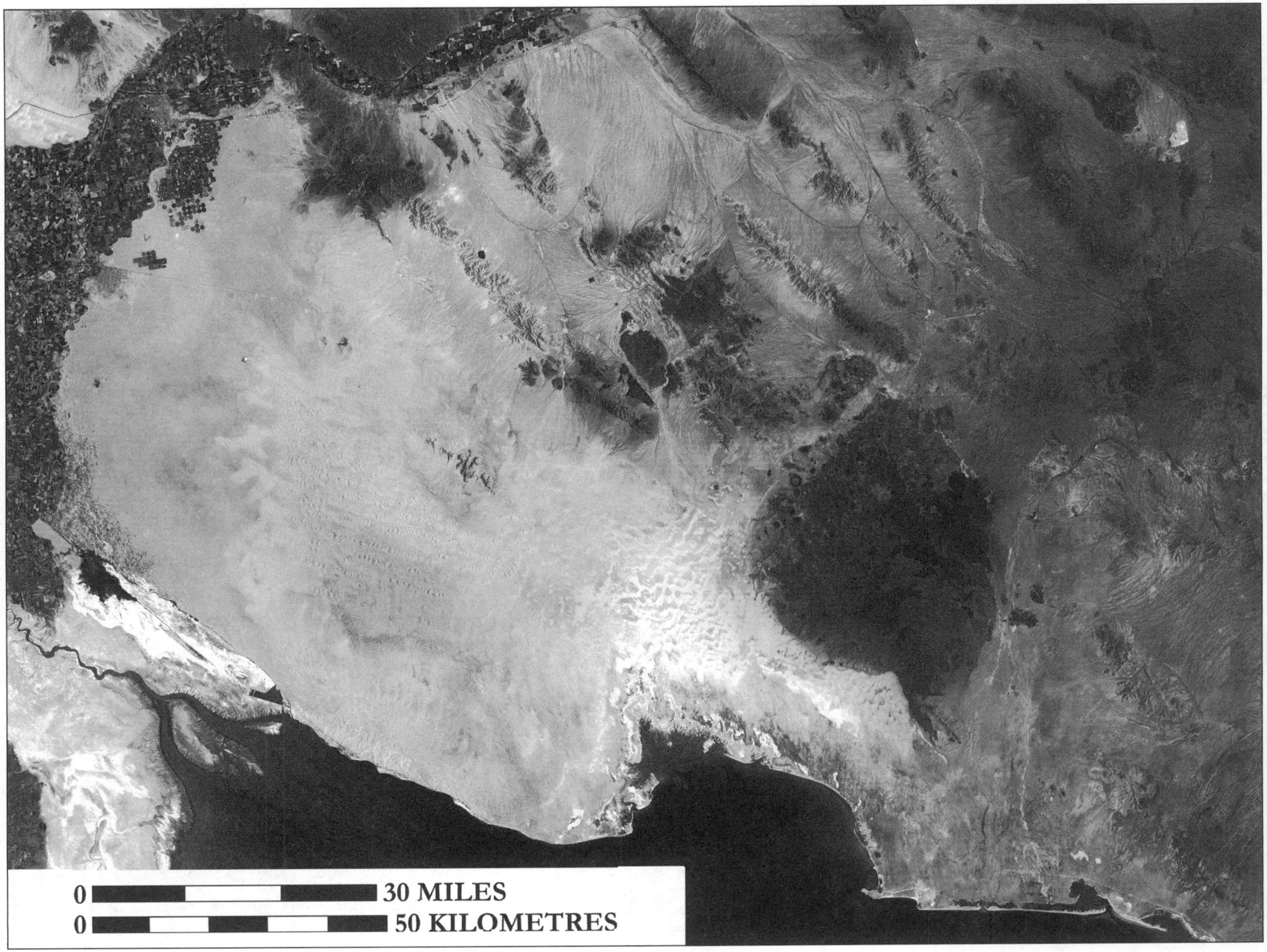

0 30 MILES
0 50 KILOMETRES

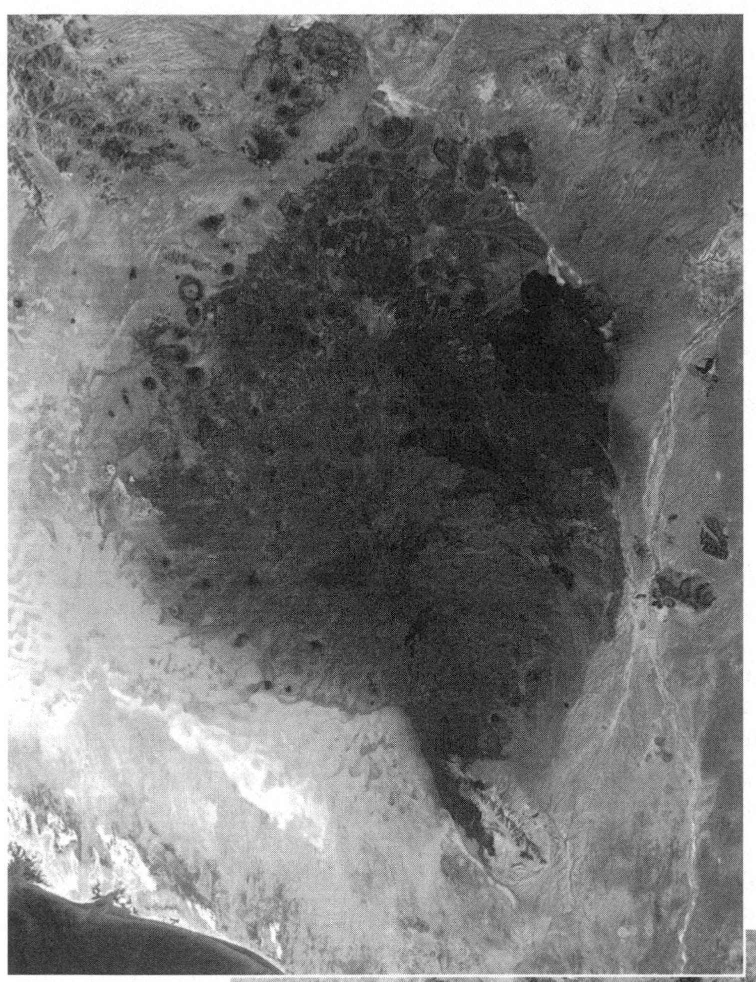

Pinacate region. Note MacDougal Crater upper left, Mexico Highway 2 across north end of lava field, and road south from Pinacate Junction to Tezontle cinder mine and Elegante Crater. Composite Landsat 5 Thematic Mosaic scenes taken on various dates. KCH. © Kevin C. Horstman 1999.

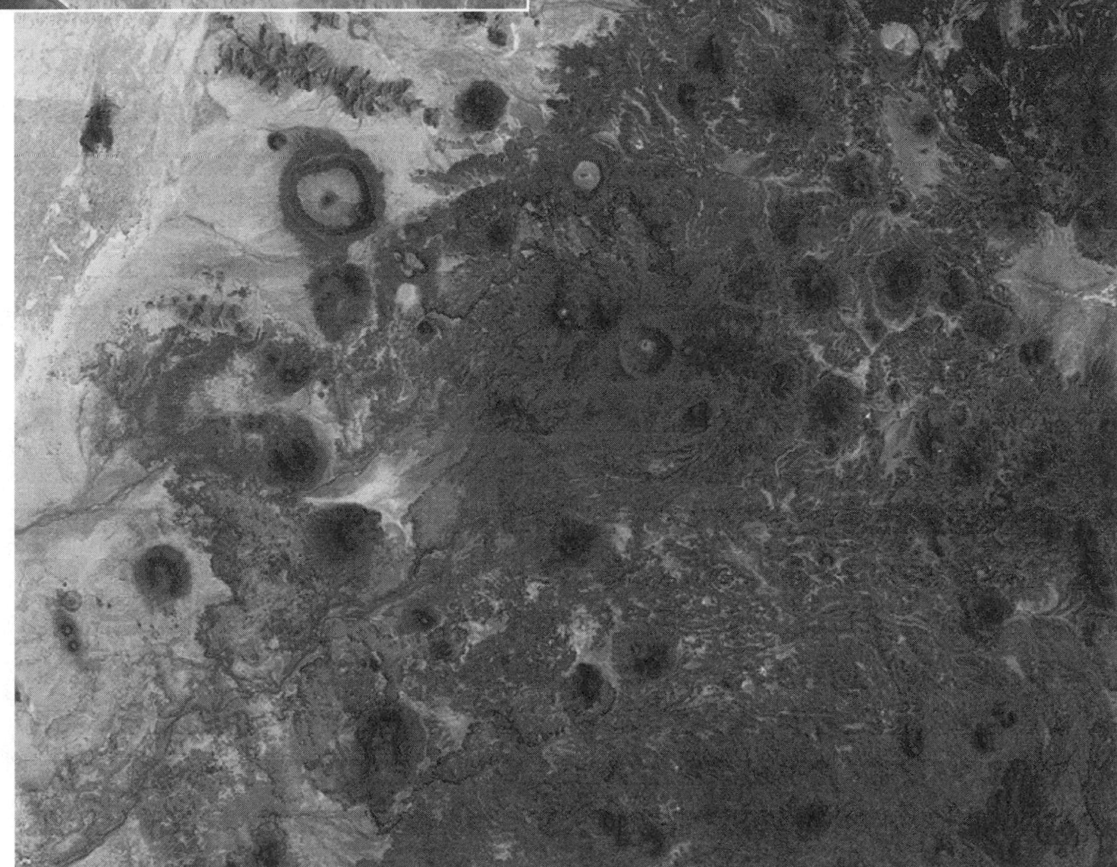

Northwestern portion of Pinacate region. Note MacDougal Crater upper left, Hornaday Mountains north of MacDougal Crater, and Sykes Crater slightly above center of image. Composite Landsat 5 Thematic Mapper Mosaic scenes taken on various dates. KCH. © Kevin C. Horstman 1999.

The meager soils of the volcanic region are largely wind-deposited sands and silts derived from redeposited Río Colorado delta sediments and airborne dust. Some soil pockets have developed weak stratification, perhaps due in part to organic buildup from plant detritus, especially from ephemerals (desert annuals), and from the activity of soil algae and lichens. The playas within the boundaries of the volcanic field have deep strata of fine-textured silty/sandy soil. The entire Pinacate volcanic field, from lowest to highest elevation, supports a flora of at least 309 species. The 267 plant species documented at the lower elevations, mostly below approximately 600 m, are listed in Appendix A under the Pinacate volcanic field.

Peripheral Lava Flows

The volcanic flanks of El Pinacate are complex, with many different habitats and corresponding vegetational and floristic variation. Among the most common and widespread large perennials are brittlebush (*Encelia farinosa*), triangle-leaf bursage (*Ambrosia deltoidea*), white bursage (*A. dumosa*), creosote bush (*Larrea divaricata*), saguaro (*Carnegiea gigantea*), teddybear cholla (*Opuntia bigelovii*), and palo verde (*Parkinsonia microphylla*). Other widespread, common large shrubs and trees include ocotillo (*Fouquieria splendens*), ironwood (*Olneya tesota*), and limberbush (*Jatropha cuneata*). Elephant tree (*Bursera microphylla*), barrel cactus (*Ferocactus cylindraceus*), pygmy cedar (*Peucephyllum schottii*), rhatany (*Krameria grayi*), galleta grass (*Pleuraphis rigida*), globemallow (*Sphaeralcea ambigua*), and sweetbush (*Bebbia juncea*) are often conspicuous.

Along the larger drainageways one almost invariably finds the arborescent legume trio of ironwood, mesquite (*Prosopis glandulosa*), and blue palo verde (*Parkinsonia florida*), plus an array of smaller perennials and ephemerals. In more arid environments, the distributions of many of the shrubs, trees and other perennials are limited to the drainageways in a familiar desert pattern called *xeroriparian*.

Several extreme habitats stand out as being virtually devoid of perennials or having very low perennial diversity and density. Hot, south- and west-facing slopes of many cinder cones are especially harsh places for perennials. On some cinder cones only one or two truly perennial species are present, usually *Ambrosia dumosa* and *Larrea divaricata*, or only *A. dumosa*. Other perennials that may be present tend to be relatively short-lived, e.g., scattered individuals of *Ditaxis neomexicana* and *Sphaeralcea ambigua* subsp. *ambigua*. One such extreme habitat is Pyramid Peak, a cinder cone surrounded by dunes 3.5 km northwest of MacDougal Crater. Over most of this hill, especially on the south- and west-facing slopes and most of the ridges, the perennial cover ranges from 1% to 5%. *Ambrosia dumosa* is thinly scattered across the northwest side of the cinder cone, and *Ephedra aspera* is present but rare. The north-facing slopes support several additional perennials, although their density is low: *Encelia farinosa, Eriogonum inflatum, Fouquieria splendens, Larrea divaricata,* and *Peucephyllum schottii.*

Following times of favorable winter-spring rains these otherwise nearly barren slopes support many ephemerals, which undoubtedly make up the majority of local biomass production. Common among these ephemerals are *Allionia incarnata, Aristida adscensionis, Chaenactis carphoclinia, Cryptantha angustifolia, C. maritima, Eriogonum* spp., *Geraea canescens, Perityle emoryi, Phacelia ambigua,* and *Plantago ovata.* These plants mostly grow in sand

Lava hills at Campo Rojo, northwest side of Pinacate region; sinita (*Lophocereus schottii*) with mostly brittlebush (*Encelia farinosa*) in foreground and on slopes. MAD.

and silt (loess) blown in from adjacent dunes and washed into pockets between the lava rocks. In washes at the edge of Pyramid Peak and other such cinder hills and dunes are *Ambrosia dumosa, Encelia farinosa, Fouquieria splendens, Larrea divaricata,* and a few plants of *Atriplex canescens, Opuntia fulgida,* and *Pleuraphis rigida.*

Ash flats can also be inhospitable habitats for perennials. These flats are riddled with rodent burrows, which may prevent establishment of perennials. The large apron of ash flat surrounding Cerro Colorado is virtually devoid of perennials except for the deep-rooted and apparently non-palatable *Euphorbia polycarpa* and *Tiquilia palmeri.*

The Craters

It was my first trip to El Pinacate. I parked the Jeep, walked up a nondescript hillside, and was suddenly standing at the rim of a great crater. Before me was a mile-wide gaping hole in the earth like nothing I had ever seen—surprising and mysterious because there was no warning. Until you learn to pick out the nearly featureless outer rim, there is nothing special to warn you of the presence of one of these great craters that seem to exhale an eerie silence. The descent into MacDougal Crater was fairly easy, scrambling down steep, crumbly tuff breccia, carefully climbing down a partial break in the cliffs that ring the crater and finally down the rocky talus slope to the flat crater floor. The air was still, there was no wind, there was no view except crater walls and sky. I was reminded of the words of Captain Juan

MacDougal Crater and Hornaday Mountains, looking southeast. Molino Crater beyond MacDougal Crater; Pinacate volcanic field and peaks in background; note shield aspect to peak, upper right. PLK. © Peter L. Kresan 1999.

Playa at bottom of MacDougal Crater, with crater walls in background; dense stand of spring annuals with prickly poppy (*Argemone gracilenta*). WKH.

Molina Crater showing the three parts of the crater. WKH. (Hartmann 1989).

Mateo Manje when he discovered Elegante Crater on March 20, 1701. He described the crater as "a big hole of such profundity that it caused terror and fear" (Burrus 1971:476).

Ten major craters are scattered across the northern and western flanks of the lava shield. The six largest are Elegante, Sykes, MacDougal, Cerro Colorado, Molina, and Moon (Jahns 1959). Although their origins are probably varied, most are maar, or collapse, craters. According to current theory, these craters were formed when molten lava encountered ground water, causing it to boil immediately and generate a great steam explosion. The explosion was followed by deposition of a tuff ring and its subsequent collapse into the chamber from which the explosion occurred.

The western craters—such as MacDougal, Molina, and especially Moon—are partially filled with sand, which continues to blow in from the dunes of the Gran Desierto. Indeed there may be undiscovered craters buried in the dunes. Cerro Colorado, the most distinctive of the large craters, lies slightly apart from the main volcanic shield. The two deepest craters, Elegante and Sykes, farthest from moving dunes, are the least sand filled. Large fissures, or cracks, in the playa at the bottom of Elegante Crater, some more than 30 cm across, are probably due to continuing subsidence. Perhaps as recently as about 20,000 years ago there was a freshwater lake in Elegante Crater, as evidenced by remains of lake sediments and freshwater snail shells.

The vegetation and flora of the various craters are relatively rich and remarkably similar, especially among craters of comparable size and depth. The majority of species present in the entire Pinacate region can be found in just the several larger craters. During a one-day survey I located nearly 100 plant species in Sykes Crater (Appendix B), which probably represents about 90% of the total flora of that crater. This rich assemblage of species undoubtedly is related to the relatively high habitat diversity. The usual lack of wind would reduce transpiration, but this factor might be offset by the generally higher temperatures inside the craters. On a summer afternoon the dark lava on west-facing crater walls can be too hot to touch—yet little herbs such as *Euphorbia polycarpa* grow out of crevices in these rocks.

Small playas, mostly about 200 to 300 m across in the larger craters, have formed in the lowest points of the craters. The playa flora and vegetation are remarkably predictable from crater to crater, and are also similar to those of playas elsewhere in the Pinacate region. Crater playas may develop 100% vegetation coverage (mostly ephemerals) during wet periods and remain green weeks after the surrounding vegetation becomes desiccated. There is even vegetational zonation, correlated with soil moisture, in concentric bands within and around the playas. Mesquite (*Prosopis glandulosa*) is the most conspicuous and predictable larger perennial. Other common perennials include ragweed (*Ambrosia confertiflora*), spiny poppy (*Argemone gracilenta*), the spurge *Euphorbia albomarginata*, blue palo verde (*Parkinsonia florida*), and devil's claw (*Proboscidea althaeafolia*).

In springtime the crater playas are often carpeted orange with annual globemallow (*Sphaeralcea coulteri*), which may reach 1 m or more in height. Other ephemerals are likewise abundant. Summer rains often bring head-height, 100% cover of another set of ephemerals dominated by *Amaranthus palmeri, Gaura parviflora,* and *Ambrosia confertiflora.* Walking through this amaranth jungle in the hot, still air of a sweaty summer day can produce nose-running clouds of allergenic pollen. Other common summer ephemerals of the crater playas are *Boerhavia* spp., *Bouteloua aristidoides, B. barbata, Cuscuta* spp., *Dalea mollis, Kallstroemia grandiflora, Leptochloa panicea,* and *Pectis papposa.*

Bighorn sheep, coyotes, and other large mammals—including humans—come and go at will from the craters, but desert tortoises living in the bottom of MacDougal Crater are isolated. Indeed, how a population of tortoises became established inside MacDougal Crater is difficult to conjure, unless they were put there by earlier people. Because livestock and vehicles do not reach the deeper craters, the natural vegetation is protected. Thus, these craters provide important natural laboratories. Long-term vegetation studies in MacDougal Crater by Ray Turner (1990) demonstrate the dynamic nature of desert vegetation. Stands of mesquites and creosote bushes have grown and perished in cycles responding to climatic change and weather events. Although the crater floors are relatively free of non-native weedy plants, I was surprised and indignant to find a few weedy tamarisk shrubs (*Tamarix ramosissima*) established in the bottom of Elegante Crater.

During summer-fall thunderstorms the nearly vertical drainage channels on the crater walls can briefly turn into waterfalls. In an early September storm at MacDougal Crater I witnessed torrential waterfalls spilling into flash floods ripping across the crater floor. The water flow lasted about ten minutes and the storm clouds quickly passed. Descending into the hot, steamy crater I found freshly cut channels near the crater walls where as much as 1 m of sandy soil had been washed away.

Higher Elevations of the Sierra Pinacate

On October 9, 1698, the intrepid Jesuit explorer-missionary Padre Eusebio Kino and his party were the first Europeans to ascend El Pinacate. On November 5, 1706, Kino again climbed the mountain. From that lofty vantage point he reported seeing the head of the Gulf of California and Baja California: "We

Pinacate volcanic region, north of main peaks, ca. 800 m elevation. *Opuntia chlorotica*, right foreground. MAD.

North side of Pinacate Peak, ca. 900 m. RSF.

saw very plainly the connection of this our land with that to the west" (Burrus 1971:156; also see Ives 1966). Kino used this information to support his claim that Baja California was a peninsula and not an island (see Hartmann 1989). The Pinacate uplands, mostly 600 to 900 m in elevation, are surmounted by the steep, twin cinder cones of Carnegie Peak and the slightly higher Pinacate Peak, which rises to 1290 m. In places where the cinder soils are relatively stable, there is often a rich development of both perennials and ephemerals. This cinder mulch overlays fine-textured soils built up in part by fine dust filtering through the cinders.

Temperatures are a little cooler and the climate is slightly wetter than in the surrounding desert. Winter freezing is certainly a factor in excluding or limiting the development of some lowland perennials. For example, plants of *Jatropha cinerea* on the slopes of Pinacate Peak are repeatedly frozen to ground level.

The mountaintop of El Pinacate is a biological sky island. At least 165 species occur at higher elevations in the Sierra Pinacate, mostly above about 650 to 800 m; they are listed in Appendix A. Sky-island plants isolated at higher elevations are often on north-facing slopes, and some are restricted to north-facing cliffs and slopes on and near Pinacate Peak itself (Van Devender et al. 1990). *Senecio pinacatensis,* known only from the higher elevations, is the only species endemic to the Pinacate volcanic field (Felger 1991). At least 23 sky-island plants do not occur elsewhere in the flora area, and some are not known anywhere else in mainland Mexico: *Artemisia ludoviciana, Astrolepis cochisensis, A. sinuata, Berberis haematocarpa, Bothriochloa barbinodis, Bromus berterianus, Calycoseris parryi, Cleome isomeris, Gilia minor, Glandularia gooddingii, Keckiella antirrhinoides, Mentzelia veatchiana, Opuntia chlorotica, Pellaea mucronata, Pholistoma auritum, Rafinesquia californica, Rhus aromatica, Salvia mohavensis, Senecio pinacatensis, Stipa speciosa, Teucrium glandulosum,* and *Zephyranthes longifolia.*

Cliff faces and steep slopes on the north side of the Pinacate Peak cinder cone support an extraordinary pocket of chaparral-like plants, largely comprised of sky-island plants. Yet the desert nature of the nearby vegetation is clearly demonstrated by the presence of *Larrea* to peak elevation on the more arid, south- and west-facing slopes. Several other species in the Pinacate uplands also occur along the northern edge of northwestern Sonora (near the international boundary) but are absent from the intervening lowland desert: *Digitaria californica, Jatropha cinerea, Muhlenbergia porteri, Pectocarya recurvata,* and *Penstemon pseudospectabilis.* Many sky-island plants have biogeographic affinities with the California or Mojave flora, and appear to be relicts of more widespread distributions during Pleistocene times. Indeed, some have been recovered from fossil packrat middens in nearby low elevations (e.g., *Bothriochloa* and *Stipa;* Van Devender et al. 1990).

Granitic Ranges

Mountains referred to here as the granitic ranges are linear fault-block ranges characterized by light-colored intrusive igneous rocks. They have very steep slopes and sawtooth ridges, and are generally oriented in a northwest-southeast direction, parallel to the main local faults (Cortés et al. 1976). These ranges formed as a result of faulting during a period of tectonic extension in the late Tertiary. They consist of pre-Tertiary intrusive and metamorphic rocks of different ages pre-dating the Pinacate eruptions (Merriam 1972). Most abundant are Paleozoic gneiss and Cretaceous granite. They are separated by deep basins filled with tertiary to recent gravels. Tertiary volcanic rocks perched on the granitic ranges are mostly basalt and tuff erupted before the Pinacate volcanic field was formed (Kresan 1997).

Most of the granitic ranges within the flora area lie to the northwest of the main mass of the Pinacate volcanic field. The most notable is a series along Mexico 2 comprised of the Sierra Nina, Sierra del Viejo, and Cerro Pinto, which for the sake of convenience are termed the northwest series. The Sierra del Viejo, rising to 790 m, is highest in this series. Closer to the Pinacate lava are the Sierra Hornaday, Sierra Extraña, and Sierra Enterrada, so-called because it is partially buried by dune sand. Sierra Blanca is separated from the northwest series by the intervening Pinacate lava. South of the Pinacate lava are small granitic ranges such as the one at Bahía de la Cholla. Sierra del Rosario, another granitic range, is located to the southwest of the northwest series and is surrounded by dunes of the Gran Desierto. This range is treated separately in a following section and in Appendix A in order to emphasize the effects of its great aridity and isolation.

The granitic ranges support a flora of at least 173 species (Appendix A). A number of these species do not extend into the Pinacate volcanic field: *Agave deserti, Camissonia arenaria, Crossosoma bigelovii, Cryptantha racemosa, Dudleya arizonica, Erigeron oxyphyllus, Gymnosperma glutinosum, Hesperoyucca whipplei, Hoffmannseggia microphylla, Lotus rigidus, Mentzelia puberula, Nolina bigelovii, Opuntia basilaris, Rhus kearneyi, Thamnosma montana.*

The three agavoid plants in the flora area—*Agave deserti, Hesperoyucca whipplei,* and *Nolina bigelovii*—occur intermixed in the northwest series of granitic mountains. Only *Nolina* extends to the extremely arid Sierra del Rosario. *Hesperoyucca* has the narrowest range and occurs only on Sierra del Viejo, nested well within the range of the others. The *Agave* and the much smaller rosette succulent *Dudleya arizonica* occur in the northwestern granitic mountains and Sierra Blanca, but are absent from the intervening Pinacate volcanic field. In fact there are no agavoids or other succulent rosette plants in the Pinacate field. Their absence from the volcanic region is probably due to edaphic factors. Basalt does not hold water as readily as granitic rock. There are vast differences in the soils that form in the crevices of the different kinds of rock. Granite weathers into crystalline, coarse-textured soils and allows for soil build-up even in crevices. In contrast, the meager soils between the volcanic rocks are largely the result of wind deposition and tend to be fine-textured. In addition, granitic rock weathers more rapidly, even in a dry climate, than does volcanic rock.

Sierra del Rosario

This isolated range is even more arid than the other, larger granitic ranges. Approximately 78 km^2 in area, it supports a flora of at least 111 species (Felger 1980; Appendix A). Surrounded by dunes of the Gran Desierto, the Sierra del Rosario consists of a series of peaks with their bases buried by dunes and alluvium. Elevation of the highest peak is 562 m. The northeastern portion of the range is partially buried by drifting dunes, but elsewhere the mountains are generally flanked by sand flats about 2 to 5 km in width, which in turn are surrounded by dunes. At the base of some peaks are limited alluvial deposits forming nearly level terrain with desert pavement surfaces dissected by erosional gullies.

Morning fog over Sierra del Rosario, southern outlier peak, looking southwest, November 1977. PLK, © Peter L. Kresan 1977 (Felger 1980).

Dunes and Sierra del Rosario, looking south-southwest. Semistabilized dunes in foreground, and star dune field at center. PLK, © Peter L. Kresan 1977 (Felger 1980).

Canyons in these mountains are very steep, short, and rugged. Carl Lumholtz (1912:311) described the range as follows:

> It runs in the usual direction and consists of several parts, some of them single mountains, stretched out for about fourteen miles. There are two main bodies of the sierra, each little range being perhaps four miles long; then follow mountains at both ends, more or less connected, some of them, especially those toward the north-west, being half submerged in huge sand dunes. The shape of the mountains is the usual one, the crests resembling the teeth of a saw. This succession of hilltops is conspicuous even at a great distance; hence the name of *rosario* (rosary), which has been proposed, is appropriate and it should be called Sierra del Rosario.

There is no surface water and no evidence of sustained human occupation, although it was visited by western O'odham. Other O'odham people crossed the region on their annual trek to the shores of the Gulf of California to collect salt and seashells (Hayden 1967, 1972; Lumholtz 1912). In March 1916 Charles Sheldon (1979:79–80) made the following entries in his diary: "An old Indian trail, well preserved and clear, passes along the east side of the mountain. This trail goes straight through and does not branch off toward the mountain as would be usual if there was water. . . . The nearest fresh water is Laguna Prieta 20 miles away, and Tinajas Altas 25 miles or more."

The flora of the northwest granitic mountains is richer than that of the nearby Sierra del Rosario, about 20 to 25 km distant. More than two dozen species in the northwest ranges are absent from the Sierra del Rosario. The isolated Sierra del Rosario, of lower elevation and smaller mountain mass, is clearly an even harsher environment than the other ranges. Perhaps most notable is the absence of *Opuntia bigelovii,* present on virtually all other mountains in northwestern Sonora. In contrast, four Sierra del Rosario plants have not been found in the northwestern series: *Cistanthe ambigua, Fagonia densa, Mammillaria tetrancistra,* and *Mirabilis tenuiloba.*

In 1975 Kim Cliffton, Gary Nabhan, and I surveyed the vegetation of the Sierra del Rosario (Felger 1980). At that time there had been minimal woodcutting and the vegetation had suffered minimal disturbance. Perennial vegetation along a major wash covered 84% of the ground, whereas nearby desert pavement and sandy-gravel flats supported less than 10% perennial plant cover. No non-native plants have been found in the Sierra del Rosario, although several non-natives occur in the nearby dunes.

Dunes

The dunes of the Gran Desierto of northwestern Sonora form the major part of the largest sand sea, or area of moving dunes, in North America (Davis 1990; Lancaster, Greeley, & Christensen 1987). This sand desert covers about 5000 km^2. Most of this system is made up of active crescentic and star dunes, with relatively few linear dunes. The highest dunes, to the west of Sierra del Rosario, have a relief, from trough to dune crest, of 80 to 150 m.

Occasionally one encounters the underlying desert floor, or hardpan, at the bottom of a blowout. Apparently water does not penetrate this hardpan and no plants grow on it. The inland dunes originate largely from sands and silts blown eastward from the delta of the Río Colorado, beginning at least 10,000 years ago (Kresan 1997).

The enormous Gran Desierto dune system has a flora of only 85 species (Felger 1980; Appendix A), amazingly few species for such an immense area. The relatively low species richness probably can be attributed to the low habitat diversity, aridity, and harsh conditions brought about by moving sand.

High, unstable dune ridges and their moving, advancing slopes are barren. However, most of the dune areas support at least some vegetation. Zonation is evident, with gradual shifts in species composition from the highest, least stable, shifting dunes to low stabilized dunes and sand flats. Vegetation is present on dune surfaces with slopes up to 40%, and a few scattered plants occur on slopes up to 60%.

During years of favorable winter-spring rainfall, especially El Niño years, the dunes support spectacular displays of wildflowers. Among the most conspicuous are sand verbena (*Abronia villosa*), *Baileya pauciradiata,* desert cryptantha (*Cryptantha angustifolia*), spectacle-pod (*Dimorphocarpa pinnatifida*), Ari-

Crest of star dune near north side of Sierra del Rosario, looking southward with complex dune field in background. PLK. © Peter L. Kresan 1977.

Low, stabilized dunes with big galleta grass (*Pleuraphis rigida*) and Gran Desierto camphor weed (*Heterotheca thiniicola*). Looking northward, Sierra Blanca with Sierra Pinacate in background. BT.

Cactus flat, about 5 km E of Pinacate Junction. Saguaro (*Carnegiea gigantea*), creosote bush (*Larrea divaricata*), and desert club cholla (*Opuntia kunzei*) in foreground. BT.

Tinajas de los Pápagos. DTM, 1907 (courtesy William Hartmann).

zona lupine (*Lupinus arizonicus*), evening primrose (*Oenothera deltoides*), and Spanish needles (*Palafoxia arida*). Some of the dune-adapted plants show a trend toward gigantism (taller and more robust plants) and a strict growth habit (straight and erect, not at all spreading); e.g., *Croton wigginsii, Helianthus niveus, Larrea divaricata* var. *arenaria, Palafoxia arida* var. *gigantea,* and *Petalonyx thurberi.* The upright growth form and lack of lower branches seems to represent adaptation to wind shear caused by moving sand near the ground surface, as well as shifting sands that may partially bury or expose these plants. Notably prominent are plants with dense whitish- or silvery-haired leaves and stems, e.g., *Ambrosia dumosa, Croton wigginsii, Dicoria canescens, Eriogonum deserticola,* and *Psorothamnus emoryi.* Deep roots are likewise a common feature, enabling plants to utilize moisture deep in the sand.

The floristic makeup of the dune communities, particularly among perennial species, is highly predictable, with relatively little change in species composition across the dune fields. Among the most common and conspicuous plants on higher, shifting (moving or unstable) dunes are the following perennials: *Ambrosia dumosa, Asclepias subulata, Croton wigginsii, Ephedra trifurca, Hesperocallis undulata, Petalonyx thurberi, Psorothamnus emoryi, Tiquilia plicata;* and several ephemerals: *Abronia villosa, Dicoria canescens, Dimorphocarpa pinnatifida, Drymaria viscosa, Helianthus niveus, Lupinus arizonicus, Oenothera deltoides.* In addition, several others are invariably encountered on lower, yet unstable and moving dunes, including *Atriplex canescens, Euphorbia platysperma, Larrea divaricata,* and *Prosopis glandulosa.*

Characteristic common plants on low, partially stabilized dunes as well as other sandy habitats include the following perennials: *Ambrosia dumosa, Aristida californica, Hesperocallis undulata, Larrea divaricata, Pleuraphis rigida, Prosopis glandulosa, Tiquilia palmeri, Triteleiopsis palmeri;* and ephemerals: *Abronia villosa, Amaranthus fimbriatus, Eriogonum trichopes, Langloisia setosissima, Oenothera primiveris, Pectis papposa, Stephanomeria schottii.*

Ephemerals comprise about 65% of the dune flora, the majority of which are winter-spring (cool-season) species. Shreve (1951:127) pointed out that "sand . . . is a particularly favorable soil for the ephemerals of the cool season, because of the deep infiltration of moisture and the rapid warming of the surface early in the season and early in the day."

Twenty percent of the dune flora, or 15 taxa, are regional dune endemics at the species or infraspecific levels, or show unique, apparently ecotypic variation. Endemic species are *Croton wigginsii, Dimorphocarpa pinnatifida, Eriogonum deserticola, Euphorbia platysperma, Heterotheca thiniicola, Pholisma sonorae,* and *Stephanomeria schottii.* Differentiation at the infraspecific level is seen among *Astragalus magdalenae* var. *peirsonii, Camissonia claviformis* subsp. *yumae, Palafoxia arida* var. *gigantea,* and *Larrea divaricata* var. *arenaria.* Ecotypic variation is discernable among *Helianthus niveus, Petalonyx thurberi, Plantago ovata,* and *Wislizenia refracta* subsp. *palmeri.* Some of these plants may represent dune endemics worthy of taxonomic recognition. *Heterotheca thiniicola* is narrowly restricted to a small area of dunes northeast of Puerto Peñasco. The other dune endemics listed here variously extend into adjacent areas of southeast California, northeast Baja California, and/or southwest Arizona, within the same general sand sea system.

Only four non-native species, all ephemerals, occur among the dune flora: *Brassica tournefortii* and *Schismus arabicus* are common and well established. *Sonchus asper* is relatively rare on the dunes, and *Mollugo cerviana* might actually be native.

Desert Plains or Flats

Lowland plains with predominantly sandy soil cover large areas of northwestern Sonora. Although these areas do not comprise a homogenous region or habitat, it is convenient to consider them together. Included are lowland desert regions that do not fall into other major habitats. Due to the relative lack of relief and habitat diversity, the local species richness of perennials may be relatively low. Over vast areas of desert plains the most common larger perennials are creosote bush (*Larrea divaricata*) and bursage (*Ambrosia deltoidea* and/or *A. dumosa*), and ephemerals are seasonally numerous. There is, however, considerable floristic and vegetational variation between different habitats within this region. Although many species occur nearly throughout the lowland desert plains, others are more restricted in distribution and are found mostly to the north, east, south, or west of the Pinacate volcanic field (Appendix A).

The distinction between lower dunes and desert plains is blurred. By and large the dunes are distinguished by moving sands, and sand flats by relatively stable soil surfaces characteristically held in place by a cryptogamic crust of blue-green and green soil algae, bacteria, diatoms, fungi, and lichens. There are correlated floristic distinctions; for example, *Oenothera primiveris* and *Tiquilia palmeri* occur on low stabilized dunes and sand flats, whereas *Oenothera deltoides* and *Tiquilia plicata* are on moving dunes even if they are low in relief. In addition, many areas of desert plains are held in place by desert pavements of pebbles or small rocks.

There are a number of playas or dry lakebeds, which are low-lying areas without drainage outlets, characterized by fine-textured clayish-silty soils. Typical members of the playa communities include *Amaranthus palmeri, Ambrosia confertiflora, Argemone gracilenta, Euphorbia albomarginata,* and *Malvella sagittifolia.*

The most extensive regional playas are Playa Díaz north of Cerro Colorado and the low-lying areas around Rancho los Vidrios. Playa Díaz and the large alluvial fan leading into it may collect substantial runoff from a large watershed in the northern part of the Pinacate volcanic field. In August 1986 there was sufficient rainfall upstream to produce an enormous flash flood that covered the upper two-thirds of the playa but did not reach the lower one-third. By mid-September this alluvial fan and the previously flooded upper two-thirds of Playa Díaz were carpeted in a great green meadow. Most of this vegetation was comprised of ephemeral grasses and dicots and herbaceous perennials. At such times each square meter may contain more than 1000 individual plants. The September 1986 Playa Díaz flora included *Amaranthus palmeri, Ambrosia confertiflora, Boerhavia* spp., *Bouteloua aristidoides, B. barbata, Cyperus* sp., *Dalea mollis, *Eragrostis cilianensis, E. pectinacea, Eriochloa aristata, Euphorbia abramsiana, E. albomarginata, *E. prostrata, E. trachysperma, Kallstroemia* spp., *Leptochloa panicea, L. viscida, Malvella leprosa, Muhlenbergia microsperma, Panicum* spp. *Pectis* spp., *Physalis lobata,* and *Tidestromia lanuginosa.*

At the same time the lower third of the playa and the surrounding terrain remained dry and devoid of green growth. The boundary between lush green meadow and parched desert was amazingly abrupt. By late October the meadow had dried up, and most of the plants had withered and disappeared, leaving virtually no evidence of the former biomass.

Coastal Habitats

The coastal zone of northwestern Sonora includes *esteros* (desert estuaries), bays, beach dunes, a few granitic and volcanic hills, salt flats, and the delta of the Río Colorado. Some of the near-shore area of Bahía Adair is buried in extensive shell middens left by earlier peoples. The salt flats, extensive in some areas, are generally devoid of plants. Plants of the delta and esteros are discussed under "Wetland Habitats."

Plant species restricted to the coastal margins are relatively few but their distributions clearly demonstrate the effects of the sea (Appendix A). Although rainfall is especially low along the coast, maritime dew is often heavy, as anyone who has camped near the shore well knows. During winter and spring the plants are often wet at dawn from dew that condensed during the night. For example, the dwarf coastal shrub *Frankenia palmeri* excretes salt, which is hygroscopic, causing water to condense, drip down the stems, and soak into the soil.

An intriguing distributional pattern is seen in *Eriogonum fasciculatum.* Variety *fasciculatum* has been found only on the rocky hills at Puerto Peñasco and along the Pacific coast of the Californias. It is distinguished from the more widespread desert taxon, var. *polifolium,* by sparser pubescence on stems, leaves, inflorescences, bracts, and floral parts. Another interesting disjunct is *Ditaxis serrata* var. *californica,* distinguished from the densely pubescent var. *serrata* by being glabrous. Variety *californica* occurs in Baja California and California and is otherwise known only from Puerto Peñasco.

In the southwestern corner of the flora area, mostly between El Golfo and La Salina, to the east of the Río Colorado delta, is an extremely arid region, the Mesa de Sonora. In this area tectonic activity has resulted in uplift and exposure of Pleistocene deposits and nearly continuous beach cliffs of highly eroded badlands (Ortlieb & Roldan 1981). Heavy maritime dew is common during the cooler months. Many of the larger, more conspicuous and widespread desert plants are absent. There is a noticeable absence of cacti except rare, small colonies of teddy-bear cholla (*Opuntia bigelovii*).

Dryland plants restricted to the coastal margin of northwestern Sonora include *Abronia maritima, Astragalus magdalenae* var. *magdalenae, Atriplex barclayana, Ditaxis serrata* var. *californica, Eriogonum fasciculatum* var. *fasciculatum, Euphorbia petrina,* and *Frankenia palmeri.*

Wetland Habitats

Although northwestern Sonora is an extremely arid region, its wetlands are diverse and some were once extensive, although they are by no means evenly distributed across the region. The wetland flora is listed in Appendix A. The freshwater habitats were critical for native people and travelers during historic times. The tidal wetlands provide critical habitat and nutrients for fisheries and other marine life. Wetland habitats are among the most fragile and vulnerable habitats in the Sonoran Desert, and most in the region have suffered great destruction from human onslaught.

Tidal Wetlands

Halophytic plant communities of relatively low species richness and highly consistent makeup occur in protected embayments and desert estuaries known as *esteros* (Felger & Lowe 1976). Unlike brackish-water estuaries, the water in these esteros or tidal lagoons is hypersaline due to high evaporation and negligible freshwater influx from the desert. In contrast, the vast tidal marshes fringing the delta of the Río Colorado formerly were washed with brackish and even fresh water.

Halophytes characteristically form 100% plant cover in the saline wet mud in esteros inundated by seawater at high tide. This vegetation seldom exceeds 50 or 60 cm in height, and only a few members of the community reach 1 m in height. This low, saltscrub vegetation is made up largely of succulents and salt-excreting perennials. Succulent species are *Allenrolfea occidentalis, Batis maritima, Salicornia bigelovii, S. subterminalis, S. virginica, Suaeda esteroa,* and *S. puertopenascoa. Atriplex linearis* is essentially non-succulent, *A. barclayana* is semisucculent, and the older leaves of *Cressa truxillensis* often become succulent. *Frankenia salina* and the two saltgrasses, *Distichlis palmeri* and *Monanthochloë littoralis,* have salt-excreting organs in their leaves. Most of the estero plants are perennials. *Salicornia bigelovii* is the only strictly annual species present, and *Suaeda esteroa* is an annual or short-lived perennial. Biomass productivity in the esteros is very high. The estero and tidal-flat vegetation is crucial to many marine organisms and fisheries as well as the numerous shore birds.

Tidal wetlands and salt flats at west end of Bahía Adair. Composite Landsat 5 Thematic Mapper scenes taken in spring, 1993. KCH. © Kevin C. Horstman © 1999.

Whereas most marine wetland species have very large geographic ranges, a few are regional endemics. *Distichlis palmeri,* restricted to the shores of the Gulf of California, is the only species of grass entirely endemic to the Sonoran Desert (see Gould & Moran 1981; Reeder & Felger 1989). *Suaeda puertopenascoa* is known only from esteros in northwestern Sonora (Watson & Ferren 1991).

The major esteros in the eastern part of the flora area, from east to west, are Morúa, Peñasco, Bahía de la Cholla, Cerro Prieto, and La Pinta. Conversion of the estero at Puerto Peñasco into a harbor has destroyed the original vegetation, Bahía de la Cholla is seriously degraded by off-road vehicles and refuse dumping, and resort development and urbanization threatens much of the remaining coastal area. The great tidal sloughs and inlets on the western side of Bahía Adair support the most extensive halophyte communities in the region and at the time of this writing remain in primordial condition. Extensive tidal wetland vegetation also occurs in sloughs, inlets, and islands at the delta of the Río Colorado.

Oases

Widely scattered desert oases of highly varying size occur in the region. The most important of these are described below.

Bahía Adair Pozos. A series of small artesian springs scattered along the edge of expansive salt flats (coastal *sabkhas* in the terminology of the geomorphologist) bordering the western part of Bahía Adair provides unique habitats (Ezcurra et al. 1988). These small freshwater oases occur in the midst of highly saline flats, in areas of thick salt crusts devoid of vegetation. Known locally as *pozos,* the term for 'hole' or 'well', these island-like waterholes support a vegetation and flora entirely different from that of the surrounding desert. The pozos provide essential fresh water for the bird fauna and some of the mammals, and were also utilized by people.

The Gran Desierto aquifer appears to consist of sand and gravel deposited in ancient riverbeds and subsequently overlain by dunes. Toward the coast the alluvial aquifer becomes confined, or buried, beneath the relatively impermeable clays of the salt flats. These clays act as a barrier that causes artesian pressure to develop within the underlying aquifer. Pozos appear to develop where permeability of the clay is increased, possibly by cracking upon desiccation or by flocculation due to ion exchange. A buried fluvial system may explain the occurrence of clusters of pozos in some salt flats and their absence in many others: pozos occur only in salt flats with an underlying waterway. May (1973) noted that pozos can form in a relatively short time. Particularly after a rainy period, a partial dissolution of the salt crust can often be seen in distinct patches along the salt flats. Most of these wet patches are surrounded by small hillocks where alkali weed (*Nitrophila occidentalis*) grows. Smaller pozos may be dug by coyotes. In many cases fresh water upwells from the freshly dug hole, which is then maintained by the drinking and digging activities of birds and mammals (Lumholtz 1912; May 1973).

Alkali weed is the first plant to colonize places where the aquifer has broken through the overlying clays and reaches near or to the surface. Saltgrass (*Distichlis spicata*) is the second plant to colonize a pozo, and larger pozos (oases) are colonized by a more diverse flora.

The open waterhole in the center of vegetated pozos may range from less than 30 cm wide in the youngest pozos to several meters across in the larger pozos. Each waterhole is surrounded by a sandy hummock formed by accumulation of undecomposed roots (mostly from *Distichlis spicata*) and blown sand. In some instances the hummocks reach heights of 1.5 m. The highest hummocks occur at La Salina and show an organic matter content of 36% to 45%. This value contrasts with the surrounding salt flats, which have from less than 0.1% to 0.5% organic matter. The peaty hummocks of the pozos provide an interesting substrate for potential pollen and carbon-dating studies.

The flora of the pozos is markedly different from that of the surrounding desert. The smaller pozos have only a few species of halophytes. The largest pozos are at La Salina, with a flora made up of characteristic wetland plants. These pozos are isolated plant community relicts of the delta of the Río Colorado. Because the original delta ecosystem is now virtually destroyed, the local extinction of any wetland species in the pozo flora will most probably be followed not by recolonization of the same flora, but by establishment of introduced weed species such as salt cedar (*Tamarix ramosissima*).

Whereas the smallest pozos may support only one or two species of vascular plants, the larger individual pozos at La Salina may have as many as 18 species in a dense ring of vegetation. The more than two dozen individual waterholes (pozos) around the margin of the La Salina salt flat support a total flora of 25 species. Only *Sarcobatus* and *Machaeranthera carnosa* do not occur at La Salina—they occur at some of the smaller pozos toward the eastern end of the Bahía Adair pozo field. The total wetland flora of Bahía Adair pozos consists of 27 species: *Allenrolfea occidentalis, Anemopsis californica, Apocynum cannabinum, Atriplex barclayana, Baccharis emoryi, Cyperus laevigatus, Distichlis spicata, Eleocharis rostellata, Heliotropium curassavicum, Juncus acutus, J. cooperi, Lythrum californicum, Machaeranthera carnosa, Nitrophila occidentalis, Phragmites australis, Pluchea odorata, P. sericea,* **Polypogon monspeliensis, Prosopis pubescens, Ruppia maritima, Salix exigua, Sarcobatus vermiculatus, Scirpus americanus, S. maritimus, Sporobolus airoides,* **Tamarix ramosissima,* and *Typha domingensis.*

Laguna Prieta. This small, usually dry lake or *salina,* about 25 km southeast of San Luis, lies hidden in a depression surrounded by low dunes. It was well known to original inhabitants and a famous watering place for travelers and cattle drives crossing the Gran Desierto in the nineteenth and early twentieth centuries (Lumholtz 1912). In the mid-twentieth century it was a shallow, briny lake (Hayden 1997).

During the latter part of the twentieth century the water level dropped, apparently due to pumping of ground water in nearby areas. White salt crust covers most of the surface. Water seeps and pozolike waterholes at the margins of the salina support locally dense pockets of wetland vegetation. A single immature cottonwood (*Populus fremontii*) and other hydrophytes stand out bright green along the west side of the lakebed, but wetland vegetation is much sparser on the east side.

In the early 1990s there was an abandoned house with a few eucalyptus trees near the northwest edge of the lakebed, and at the south end an unoccupied ranchito with decaying corrals. A hole dug in the salt pan showed the water level to be about 35 cm below the surface. A large salt cedar tree (*Tamarix aphylla*), a large palo verde (*Parkinsonia aculeata*), and several small, dying oleanders (*Nerium oleander*) were found. A grove of about 50 rather drought-stunted date palms (*Phoenix dactylifera*) near the south edge of the lakebed survived with no apparent irrigation. In addition there were some screwbean (*Prosopis pubescens*) and mesquite trees (*Prosopis glandulosa*). For the most part the wetland vegetation remained intact, although there was some seasonal cattle grazing. Wetland plants near the edge of the lakebed and the adjacent low, alkaline, wet flat are *Allenrolfea occidentalis, Anemopsis californica, Baccharis emoryi, *Cynodon dactylon, Distichlis spicata, Juncus cooperi, Nitrophila occidentalis, Phragmites australis, Pluchea odorata, P. sericea, Populus fremontii, Prosopis pubescens, Scirpus americana, *Tamarix ramosissima,* and *Typha domingensis.*

Quitobaquito. Quitobaquito is a shaded oasis in the desert, a legendary place with a dependable supply of fresh water. It has been a crossroads of cultural activity as well as a center of biological dynamism and diversity. Except for the roar of trucks and buses along the nearby Mexico 2, a casual visitor might think that Quitobaquito is a pristine habitat. In fact, the area has a long and varied history of land use and modification by people of diverse ethnic backgrounds. Changes continue to occur owing to current management practices, or lack thereof, and colonization by non-native weedy plants from nearby agricultural and urban areas (Felger et al. 1992).

The history of settlement at Quitobaquito is one of coming and going. The area was probably always peripheral to the more extensive agricultural areas and settlements along the nearby Río Sonoyta (Ives 1950a). 'A'al Waippia, 'little springs' or 'little wells', is the Hia-ceḍ O'odham name for Quitobaquito. Historically, the Hia-ceḍ O'odham (the Western or "Sand" Papago) had a major village in the vicinity of Quitobaquito. In 1698 and 1699 Padre Eusebio Kino visited the settlement of 'A'al Waippia and called it San Sergio. O'odham influences on the vegetation and flora included periodic burning, brush clearing, plowing, transplanting wild and cultivated plants, livestock grazing, irrigating, and harvesting wild plants (Felger et al. 1992). O'odham occupation ended in 1957 with the forced sale of Jim Orozco's holdings to the U.S. National Park Service.

A series of springs lies along a fault on the south side of the Quitobaquito Hills in Organ Pipe Cactus National Monument, in western Pima County, Arizona. These springs are, from large to small, Quitobaquito, Williams (Rincón), Aguajita, and Burro; there are also a few smaller springs and seeps. Natural springs are rare in the region, and to have four major ones in a cluster is unique. Although Quitobaquito is the best known of the springs in the region, the other associated springs are also biotically important. The springs and the artificial pond at Quitobaquito support a diversity of wetland plants and animal life not found in the surrounding desert (Cole & Whiteside 1965; Felger et al. 1992; Huey 1942; Johnson, Brown, & Goldwasser 1983; Kingsley & Bailowitz 1987; Kingsley, Bailowitz, & Smith 1987). An endemic subspecies of desert pupfish (*Cyprinodon macularius eremus*) known only from Quitobaquito is not found in the nearby Río Sonoyta (Miller & Fuiman 1987).

The pond at Quitobaquito, covering 0.22 ha and about 1 m deep, is supplied with water from springs on the hillside north of the pond. There are many seeps along this hillside but only two main springs. From these springs water flowed about 100 m through open ditches to the pond. In 1989 the small ditch leading from the spring to the pond was reconstructed and lined with ferro-cement, and the two main springs were encased in cement and covered with locked metal gates. The soil around the springs and pond is moist and alkaline. Although conditions vary somewhat from spring to spring, the wetland habitat at each is basically similar.

Although Quitobaquito is politically on the Arizona side of the fence, it is culturally and biologically linked to northwestern Sonora. This oasis nearly straddles the international border, and the few species not known from the Sonora side most likely occurred there in earlier times. Quitobaquito lies at the boundary of the Arizona Upland and Lower Colorado Valley subdivisions. The total flora of the Quitobaquito region includes 261 species (Felger et al. 1992); 34 are wetland plants: *Anemopsis californica, Baccharis salicifolia, Centaurium calycosum, *Cynodon dactylon, Cyperus laevigatus, C. squarrosus, Distichlis spicata, *Eclipta prostrata, Eleocharis geniculata, E. rostellata, Eustoma exaltatum, Heliotropium curassavicum, Juncus articus, J. bufonius, J. cooperi, Myosurus minimus, Najas marina, Nitrophila occidentalis, Phragmites australis, Pluchea odorata, P. sericea, *Poa annua, *Polypogon monspeliensis, *P. viridis, Populus fremontii, Potamogeton pectinatus, Prosopis pubescens, Salix gooddingii, Scirpus americanus, Sporobolus airoides, *Tamarix ramosissima, Typha domingensis, Veronica peregrina, Zannichellia palustris.*

Tinajas. There are about a dozen clusters of major water-holding bedrock depressions, or tinajas, in the Pinacate volcanic field (Broyles 1996a; Broyles et al. 1997). They occur singly or strung out like a chain of beads along major arroyos or canyons, mostly at low to middle elevations. These tinajas may hold fresh water for many months and a few usually have water year-round. However, during extreme drought years even normally perennial tinajas are known to dry up (May 1973; Julian Hayden, personal communication 1985).

There are no significant tinajas in the granitic mountains within the area covered by the present study, or at least none with "wetland" plants. However, there are a number of substantial tinajas in granitic mountains in nearby southern Arizona, such as the Tinajas Altas and Cabeza Prieta Tanks (Broyles et al. 1997). Only a few species of wetland plants occur in soil pockets at various tinajas. Solid rock walls and the paucity and unstable nature of permanently wet soil preclude substantial wetland vegetation. In addition, the open water attracts herbivores. Wetland species in this locally restricted habitat include **Cynodon dactylon, Cyperus esculentus, C. squarrosus, Erigeron lobatus, Gnaphalium palustre, Nama stenocarpum,* and **Sonchus oleraceus.*

Rivers

Río Sonoyta. The Río Sonoyta is a small river coursing westward from Sonoyta and then southward into the east side of the Pinacate region. Near Sonoyta there is still a meager, intermittent surface flow. This water and its associated riparian vegetation are diminishing due to extensive ground-water pumping in the Sonoyta Valley, especially since the 1970s. A flood on the night of August 6, 1891, initiated disastrous arroyo cutting, resulting in lowering of the water table. Soon after, the series of ciénegas (marshes) at Sonoyta dried up and the village was relocated to its present location (Ives 1989; Lumholtz 1912). Woodcutting and cattle grazing probably contributed to the erosion and demise of the ciénegas. We can only imagine the wetland plants that once flourished there.

The present-day river runs through Sonoyta in a deeply eroded, refuse-strewn channel. Downstream it appears little changed from early-twentieth-century photographs except that salt cedar (*Tamarix ramosissima*) has formed impenetrable thickets, apparently largely replacing mesquite, and the riparian vegetation is generally diminished.

In the late seventeenth and early eighteenth centuries Kino described a place called Los Carrizales, undoubtedly named for *carrizo* or giant reed (*Phragmites australis*). In the 1990s carrizo was limited to a local population several kilometers upriver from Los Carrizales.

Common riparian or semiriparian plants growing along the Río Sonoyta are *Ambrosia ambrosioides, Anemopsis californica, Atriplex lentiformis, Baccharis salicifolia, *Cynodon dactylon, Cyperus laevigatus, Heliotropium curassavicum, Hymenoclea monogyra, Juncus articus, Leptochloa fusca, Machaeranthera arida, M. carnosa, *Melilotus indica, Nitrophila occidentalis, *Phalaris minor, Pluchea sericea, *Polygonum hydro-*

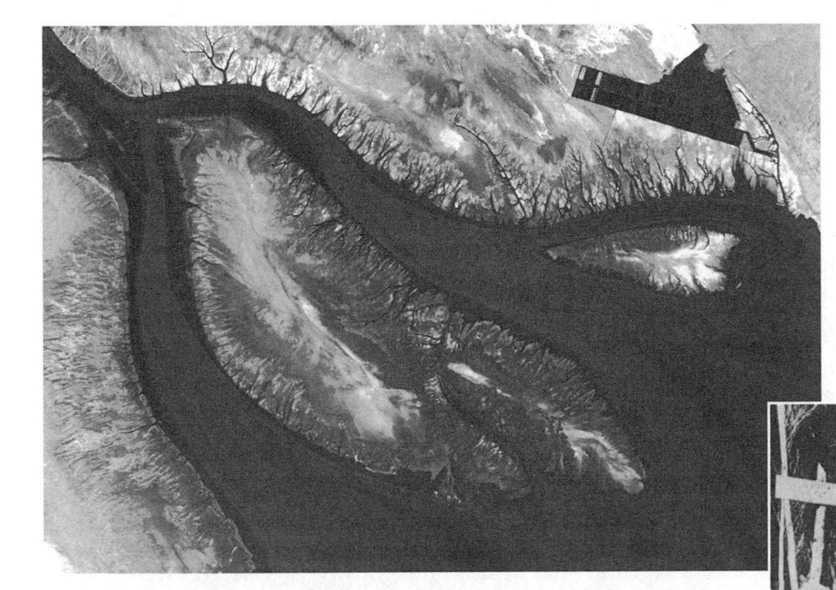

Mouth of the Río Colorado showing tidal wetlands. Isla Montague on left, the much smaller Isla Pelícano on right, and a shrimp farm. Composite Landsat 5 Thematic Mapper scenes taken spring, 1993. KCH.

Steamship and woodlot among cottonwood and willow forest in the Río Colorado delta, about 1900. These appear to be young trees of about the same age class, which have grown rapidly. SFC.

*piperoides, *Polypogon monspeliensis, Potamogeton pectinatus, Prosopis pubescens, Rumex inconspicuus, Salix gooddingii, Scirpus americanus, Spergularia salina, Sporobolus airoides, Suaeda moquinii, Symphyotrichum subulatum, *Tamarix ramosissima, Typha domingensis, Zannichellia palustris.*

Río Colorado. The Río Colorado delta originally embraced an area of 2000 to 3000 km^2, much of it comprised of abundant wetland and aquatic vegetation (Sykes 1937). Like the Nile, the delta and lower portion of the Río Colorado were replenished with nutrients from annual floods. Twentieth-century upriver dams and diversions killed the river and destroyed most of the original delta ecosystem (see "Human Influences"). The eastern portion of the delta, politically in Sonora, is included in this study but not the western portion in the state of Baja California which includes the Río Hardy, a western branch of the river.

The mouth of the delta is washed with tidal seawater that floods and drains the low, muddy Montague and Gore islands and the tidal sloughs and flats of the opposite mainland delta. Extensive stands of Palmer grass (*Distichlis palmeri*) and other halophytes grow within reach of daily tidal fluctuations that can attain 10 m. These dense, prickly leaved meadows stabilize the mud flats, sloughs, and delta islands. Palmer grass meadows once extended upriver about as far as the upper reaches of the full force of the tidal bore (see Lingenfelter 1978; Sykes 1937; Vasey 1889). Grain from this saltgrass served as a major food resource for the Cocopa and today shows promise as a modern, salt-tolerant grain crop (see Part II). The Palmer grass stands seem to be losing ground, and recent developments in the delta threaten many of the remaining stands.

Beyond the influence of the tidal seawater, the river and its channels sustained gallery forests of cottonwood and willow, abundant wildlife, and a rich agricultural lifestyle (e.g., Kelly 1977; Leopold 1949; Sykes 1937). Backwater lagoons teemed with waterfowl and other aquatic life. The Green Lagoons described so elegantly by Aldo Leopold (1949) were off the northeast side of Mesa de Andrade, just upstream from the present-day Ciénega de Santa Clara (see Meine 1988).

The dense, verdant riverine habitats of the Río Colorado stood in sharp contrast to the adjacent, extremely sparse desert vegetation of the Mesa de Sonora. This contrast can still be seen in the vicinity of El Doctor, where the almost barren badlands of Mesa de Sonora abut the seeps and marshes of the riverbed margin. Cattails, reedgrass, rushes, sedges, and other emergent and submerged hydrophytes cloak this wetland with a brilliant green mantle. This flora is similar to that surrounding the Ciénega de Santa Clara and is more or less a continuation of that wetland system.

Nineteenth- and early-twentieth-century photographs show dense, tall forests of cottonwood and willow all along the river channels. These are fast-growing trees, and with all that water, the heat, and a rich supply of nutrients, their growth must have been phenomenal. On slightly higher ground the cottonwood and willow forests were fringed with mesquite and screwbean trees interlaced with an understory of wetland species. The floristic makeup of this extinct forest will probably never be well known.

This vast riverine system formed myriad meandering and anastomosing channels that writhed across the deltaic flatlands, often changing course from flood to flood or year to year. Intervening areas appeared high and dry but had a very high water table. The soil surfaces were undoubtedly highly alkaline. The vegetation may have resembled the present-day semiriparian and mostly halophytic vegetation of the Ciénega de Santa Clara and El Doctor wetlands. In 1922 Aldo Leopold recorded that *cachanilla* (*Pluchea sericea*) "grew in dense impenetrable groves throughout the delta" (Meine 1988:207).

The delta wetland supported an estimated 200 to 400 species of plants (Ezcurra et al. 1988). The delta populations of most of these species met local extinction with the destruction of the wetland ecosystem during the first half of the twentieth century. This flora was scarcely studied before the delta dried up and salinized. The cottonwood and willow forests are gone from the Mexican side of the river, and only tiny vestiges remain north of the international border. The remnant pockets of wetland vegetation in the delta region deserve vigorous protection, and a surprising degree of restoration is possible (e.g., Glenn et al. 1996). Some idea of the former diversity of the riverine flora can be gleaned from early herbarium and ethnobotanical records near Yuma (e.g., *Ammannia robusta* and *Centaurium calycosum*) and the present-day riverbank flora north of the Mexico border (e.g., Castetter & Bell 1951; McLaughlin, Bowers, & Hall 1987).

During the late 1980s and 1990s young riparian communities have become re-established in areas just below the limits of agriculture. For example, cottonwood trees to about 10 m tall, saplings, and seedlings as well as an understory of willows and arrowweed (*Pluchea sericea*) have become established along the river channel and its meanders above the confluence with the Río Hardy. In these areas the river channel is not entrenched, and the water salinity is 1000–2000 ppm (Ed Glenn, personal communication 1997). Farther downstream, where the water salinity is about 4000 ppm, there are extensive expanses of *Tamarix ramosissima* and *Phragmites australis*.

Ciénega de Santa Clara. The Ciénega de Santa Clara is the largest and richest wetland in the northern part of the Gulf of California and anywhere in the Sonoran Desert (Glenn et al. 1992; Glenn et al. 1996). It is located in the southeastern part of the delta region. The ciénega shows on LANDSAT maps as varying year to year from approximately 12,500 to 20,000 ha of wetland. This vast brackish-water marsh probably has more than 80% coverage of dense wetland vegetation. There are numerous small fish, e.g., mullet and the endangered desert pupfish, *Cyprinodon macularius* (Zengel & Glenn 1996). The ciénega supports an "extremely diverse avifauna" (Eddlemann 1989) including the endangered Yuma clapper rail, usually invisible in the cattails. The region is a well-known major stop along the flyway for migratory birds.

Historically this part of the delta waxed and waned with the meandering of the river (see Sykes 1937). It was a vast marsh fed by overflow from the Río Colorado, and late-nineteenth-century maps show the ciénega covering about the same extent as it does in the 1990s. Sykes (1937) mapped it as the Salado Rillito. Due to upriver diversions and dams, by 1974 the marsh had shrunk to upper and lower portions of approximately 30 and 180 ha respectively, separated by salt flats (see Glenn et al. 1992). The two marshes were fed by agricultural runoff and brackish-water seeps. Despite its small size, the upper marsh was a critical habitat for wetland plants and associated animal species, including the desert pupfish and Yuma clapper rail.

Since 1977 the ciénega has been fed by water pumped into the Wellton-Mohawk drainage canal (Canal de Descarga R. Sánchez Toboada), which carries semi-saline irrigation water drained from fields in the lower Gila River valley. This water is too salty for ordinary agriculture but serves for wildlife and wetland vegetation. As part of the United States government's attempt to meet treaty agreements with Mexico concerning Colorado River water, the Yuma Desalting Plant was built to desalinate Colorado River water being delivered to Mexico. The Mohawk canal water flowing into the Ciénega de Santa Clara was initially meant to be a temporary disposal of wastewater that later would be desalinated. The flow into the ciénega was scheduled to be reduced to just one-third of the present amount, and the water would be two to three times more saline (Glenn et al. 1992; Glenn et al. 1996; Zengel et al. 1995). However, the desalinization plant is not in operation, and it appears unlikely that this will happen.

In the late 1980s and 1990s the flooded area was 36 km long. The southern portion is shallow, hypersaline, and devoid of vascular plants, and floods with seawater during the highest tides. The northernmost one-third, covering roughly 4000 to 6000 ha, is thickly vegetated with cattail (*Typha domingensis*) interspersed with extensive stands of common reed (*Phragmites australis*) and bulrush (*Scirpus americanus*). Large quantities of submerged aquatic plants, largely holly-leaved water nymph (*Najas marina*) and some widgeon grass (*Ruppia maritima*), provide important food for waterfowl. The backwaters and permanently wet soils support dense growth of halophytic or semi-halophytic wetland species, including *Allenrolfea occidentalis, Anemopsis californica, Atriplex canescens, A. lentiformis, Cressa truxillensis, Distichlis palmeri, D. spicata, Heliotropium curassavicum, Juncus cooperi, Leptochloa fusca, Najas marina, Phragmites australis, Pluchea sericea, *Polypogon monspeliensis, Prosopis pubescens, Rumex inconspicuus, Ruppia maritima, Salicornia subterminalis, Scirpus americanus, S. maritimus, Sesuvium verrucosum, Sporobolus airoides, Suaeda moquinii, *Tamarix ramosissima,* and *Typha domingensis*.

Xeroriparian Habitats

The many larger arroyos, or major desert washes, which in the Old World would be known as *wadis,* can carry substantial amounts of water. These usually dry streamways may have water for only a few hours during the occasional brief times of flash flooding. Also known as xeroriparian habitats, they support denser vegetation than the surrounding desert.

Desert legume trees, notably mesquite (*Prosopis glandulosa* and *P. velutina*), ironwood (*Olneya tesota*), and palo verde (*Parkinsonia florida*), form the usual xeroriparian trio. In the largest drainageways these trees and their associated flora crowd the margins in the manner of a gallery forest, with the watercourse itself often broad and barren. Gaps between these trees may be densely filled with shrubs such as desert wolfberry (*Lycium* spp.), hummingbird bush (*Justicia californica*), desert lavender (*Hyptis emoryi*), and many others. In the driest habitats even saguaros (*Carnegiea gigantea*) and other plants generally characteristic of non-riparian habitats become restricted to dry streamways. In other extremely arid sandy areas the essentially leafless smoketree (*Psorothamnus spinosus*) is the common xeroriparian leguminous tree.

History and Human Influences

Human impact in northwestern Sonora ranges from practically nil to extreme. Towns and cities, agriculture, industrialization, roads, woodcutting, off-road vehicles, tourism, smuggling, hunting, and mining have added to the economy and taken their toll on the ecosystem. Most of the biota of the Río Colorado and its great delta have been destroyed, and the Río Sonoyta is withering.

Although areas near the highways, towns, and rivers are seriously modified by human activities, the vast majority of the desert remains as close to pristine condition as nearly any place at the end of the twentieth century. The result is that the impact on the native flora and vegetation is highly variable.

The Hohokam and earlier people are known only by their artifacts (Hayden 1967, 1969, 1972, 1976, 1982). These people left basalt tools including the unusual gyratory crushers (see *Prosopis*), pottery, projectile points, shrines, trails, petroglyphs, sleeping circles, bedrock mortars, sacred items in caves, and vast shell middens along the coast.

The region was long occupied by western groups of Uto-Aztecan speaking Piman people, the Hia-ceḍ O'odham. To the outside world they were known as the Sand Papago or Pápagos Areñero or Pinacateño (Bell, Anderson, & Stewart 1980; Lumholtz 1912; Thomas 1991). Their lives necessarily centered around waterholes, oases, and the Río Sonoyta. In the better-watered places hunting and gathering was combined with agriculture and they managed local habitats (Felger et al. 1992; Nabhan et al. 1982). Kino estimated that 1000 people lived in the Sonoyta valley area in the late 1600s and early 1700s (Burrus 1971). During the mid-1800s Mexican settlers began moving into the area, and in 1850 an estimated 250 acres were under cultivation in the Río Sonoyta valley.

The people who lived in the desert expanses between the rivers subsisted mostly by hunting and gathering. Between the mid- and late-nineteenth century the Hia-ceḍ O'odham disappeared from much of the Pinacate region (Ives 1964; Lumholtz 1912), leaving the desert essentially unoccupied. Some were killed in hostilities, and by the early 1900s the remaining Areñero and Pinacateño people had simply left a region where the living was hard (Lumholtz 1912). They amalgamated with other O'odham people in such places as Sonoyta, Quitobaquito, and Quitovac, or found employment in the mines at Ajo (e.g., Bell, Anderson, and Stewart 1980; Nabhan et al. 1982).

The western margin of the flora area was occupied by the Cocopa, a Hokan-speaking or Yuman people. Living in the delta region and lower reaches of the Río Colorado, they had a rich agricultural tradition and access to the delta's vast resources (e.g., Alvarez de Williams 1983; Castetter & Bell 1951; Kelly 1977). Brief descriptions of these resources begin with the records of early Spanish explorers, notably Hernando de Alarcón, who sailed up the Río Colorado in 1540 (Bolton 1936; Hammond & Rey 1940).

The intrepid Jesuit priest-explorer Eusebio Kino visited the Pinacate region several times between 1698 and 1706. He originally named the highest peak in the volcanic field Santa Brígida, but changed the name to Santa Clara on the map he made after his second visit. Kino and other members of his expeditions, Padre Salvatierra and Captain Juan Mateo Manje, compared the lavas to those of Etna and Vesuvius. Thus these became the first "extinct" volcanoes to be recognized and recorded in the written history in North America (Burrus 1971; Hartmann 1989; Ives 1989).

The outside world had little recorded contact with the Pinacate region until the mid-1800s, when the Gold Rush drew people across the infamous Camino del Diablo—the old border route from the present-day Organ Pipe Cactus National Monument to Yuma. Sonoyta was a thriving village, and boat traffic and non-Indian settlement along the Río Colorado steadily increased.

Following the California Gold Rush of 1849, settlements proliferated at the Colorado River crossing in the vicinity of Yuma. A military fort was established as more settlers arrived and commerce expanded. From 1854 until 1877 when the railroad reached Yuma, steamboats regularly plied the river from its mouth to well above Yuma. There was a regular steamship connection from San Francisco, California, via the Gulf of California to Puerto Isabel at the delta, where passengers and freight were transferred to river steamers.

Enormous quantities of wood were cut to fuel the river-going steamboats during the last half of the nineteenth century. The hard wood of mesquite would have been more desirable than the relatively soft wood of cottonwood and willow, which were more readily available. Whatever effect this woodcutting may have had on the riparian forests is not recorded, but there are no reports of wood shortages. These riparian forests survived until upriver diversions and dams devastated the river.

During the golden age of steamboat traffic on the river, a series of woodyards was set up to supply the ships. The woodyards were about 50 km apart—the distance most steamboats could travel upriver per day. Because of sandbars and other dangers, the steamers did not travel at night, and the woodyards were set up to refuel the ships while they were tied up for the night. The woodyards were owned and run by Yankees who employed Cocopas to cut, transport, and load the wood. Traveling upriver from the delta, the woodyards were Port Famine, below Lerdo, the Gridiron, Ogden's Landing, Hualapai Smith's, and Pedrick's below Yuma (then called Arizona City). Apart from Colonia Lerdo, the only other settlements along the river were primarily the scattered Cocopa rancherías.

The river and its flora remained essentially intact until the early twentieth century. The critical events were diversion of the river into the Alamo Channel (Imperial Canal) from 1905 to 1907, which flooded the Salton Sink; completion of Laguna Dam near Yuma in 1909; and finally the great upriver dams beginning with Boulder (Hoover) Dam, completed in 1935. The closing of the gates of Glenn Canyon Dam in 1963 delivered the coup de grace (e.g., Reisner 1986). The resurrection of the Ciénega de Santa Clara and new interest in management of wetland habitats offer a glimmer of hope for partial restoration of the delta ecosystems.

The fishing village of Puerto Peñasco was linked to Arizona by paved road during World War II, and the railroad across the Gran Desierto was completed in 1949 (Barrios Matrecito 1977; Ives 1950b). Until about the mid-twentieth century the desert between Sonoyta and the Río Colorado was isolated and at times treacherous to those naive to its ways. In the 1950s a paved highway, Mexico 2, was completed across the northern edge of the Gran Desierto from Sonoyta to San Luis. Since then the populations of Puerto Peñasco, Sonoyta, and San Luis have grown dramatically.

During the latter half of the twentieth century a number of small ranching and farming *ejidos* (agricultural cooperatives) were founded near the highways on the fringes of the Pinacate volcanic field; most of them have been abandoned. Many ejido lands were sold to developers following changes to the Mexican Constitution in 1991 that allowed privatization of ejido properties and direct foreign investment. Development in the forms of new highways and roads, binational *maquiladora* factories, and urbanization is escalating.

Conservation

Until nearly the end of the twentieth century the remoteness of the Pinacate region and the Gran Desierto provided protection from the onslaught of modern civilization. However, the need for greater environmental protection had long been recognized. In the 1990s Mexico established two spectacular national biosphere reserves encompassing a large part of northwestern Sonora. In addition, the vast majority of the Arizona side of the shared international border region is encompassed by three major protected areas that comprise the proposed Sonoran Desert National Park: Organ Pipe Cactus National Monument, Cabeza Prieta National Wildlife Reserve, and the Barry M. Goldwater Air Force Range. This chain of reserves forms the largest zone of contiguous protected desert anywhere in the Americas. These reserves contain a diversity of habitats supporting a flora of at least 782 species, subspecies, and varieties in 412 genera and 100 families (Felger et al. 1997).

The Reserva de la Biósfera El Pinacate y El Gran Desierto de Altar was created by Mexican presidential proclamation in 1993. It covers 714,656 ha of the Gran Desierto in northwestern Sonora (Búrquez 1998; Búrquez & Castillo 1994; Búrquez & Martínez-Yrízar 1997). There are two designated zonas núcleos or protected (core) areas. The larger core area is the Zona Sierra El Pinacate covering 228,113 ha. Also included are granitic sierra-type ranges, such as the Sierra Blanca and the Sierra Hornaday. The smaller core area, Zona Sierra del Rosario, covering 41,392 ha, includes most of the granitic sierra and extensive surrounding dune fields. Surrounding the two nuclear areas are 445,151 ha of the zona de amortiguamiento, a managed-use buffer area. The buffer area incorporates major sections of the eolian dunes, and Mexico Highway 2 runs through its northern edge.

Cattle grazing and associated human activities, mostly at the northern margins of the region, have damaged local areas, especially some of the waterholes such as Tinajas de los Pápagos. Other significant threats to the natural environment include off-road vehicle traffic and other recreation activities, woodcutting, cinder mining, chemical pollution from agriculture, proposed *maquiladora* industrial plants, and ground-water pumping to supply the growing urban areas of Sonoyta, Puerto Peñasco, and

San Luis. An illegal road from Campo Rojo to the saddle between Pinacate and Carnegie peaks is a potentially dangerous means for introduction of exotic plants and wildlife poaching and harassment. A planned coastal highway, probably near the existing railroad along the boundary with the Alto Golfo reserve, will have devastating environmental consequences.

Reserva de la Biósfera Alto Golfo y Delta del Río Colorado was likewise created by presidential proclamation in 1993 (Búrquez & Martínez-Yrízar 1997). The 942,270 ha reserve overlays the delta of the Río Colorado, the uppermost reaches of the Gulf of California, and adjacent coastal areas. About 60% of the reserve area is ocean, spanning the gulf north of a line between San Felipe, Baja California, and Puerto Peñasco, Sonora. The zona núcleo, or core area, covers 160,620 ha and includes the delta with Islas Montague and Gore, the Ciénega de Santa Clara, and the nearby El Doctor wetlands. The surrounding buffer zone (zona de amortiguamiento) includes the saline mud flats and extreme desert north of San Felipe, extensive areas of the Mesa de Sonora, the La Salina pozos, tidal wetlands of Bahía Adair, and the desert between the Pinacate reserve and Bahía Adair.

Of special concern are the endangered fish, totoaba (*Totoaba macdonaldi*), the Gulf of California harbor porpoise, or vaquita (*Phocoena sinus*), and large populations of pupfish (*Cyprinodon macularius*) in the Ciénega de Santa Clara. Salinized water, pollution, over-fishing, and agriculture continue to threaten the remaining delta ecosystem. Despite the tremendous damage, remnants of a rich ecosystem persist; for example in the great Ciénega de Santa Clara and associated El Doctor wetlands (Glenn et al. 1992; Zengel et al. 1995). Vast but apparently diminishing stands of Palmer grass (*Distichlis palmeri*) cover much of Islas Montague and Gore and other tidal wetlands.

There is still considerable subsurface and on occasion surface water flowing into the delta. With improved management this water could be used to revive substantial parts of the delta wetlands. Further restoration might be accomplished through reduced use of toxic pesticides, improved recycling of agricultural and urban wastewater, and improved quality of water discharged from upriver users.

Agriculture and Horticulture

The Cocopa farmed along the lower Río Colorado and delta region. From pre-Spanish times until the early twentieth century they made extensive use of indigenous strains of the ubiquitous native American trinity: maize (*Zea mays*), beans (*Phaseolus acutifolius*), and squash (*Cucurbita argyrosperma*). In 1701 Padre Kino and his companions found O'Odham people cultivating maize (corn) along the river at Sonoyta (Burrus 1971). Sonoran panic grass (*Panicum hirticaule* var. *miliaceum*), and other small grains (e.g., barnyard grass, *Echinochloa crusgalli*) were harvested in summertime. Seeds of these crops were commonly planted along the riverbanks in the receding annual spring floodwaters, usually in May and June. In addition, in about one out of two years there was sufficient fall floodwater to allow planting of winter wheat (Castetter & Bell 1951).

Hernando de Alarcón brought wheat to the lower Río Colorado in 1540 (Bolton 1936; Hammond & Rey 1940) but that early introduction was apparently lost, and it remained for Padre Eusebio Kino to reintroduce this and other crops in the late seventeenth and early eighteenth centuries (Bolton 1936). The missionaries also introduced cattle, which thrived on the native grasses and other forage.

Mearns (1907) and Lumholtz (1912) mention crops being irrigated with water from the Río Sonoyta and small-scale agriculture at Quitobaquito irrigated with spring-fed water. Crops grown along the Río Sonoyta included wheat, barley, oats, maize (corn), sugarcane, beans (probably mostly teparies), and orchard trees. Maize, tepary beans, and other crops have been grown at several small dryland, runoff agricultural fields among the Pinacate volcanic complex; examples are the floodwater field at Suvuk and one near Tinajas de los Pápagos (Bell, Anderson, & Stewart 1980; Childs 1954; Lumholtz 1912; Nabhan 1985). Around 1980 about 0.5 km^2 of wheat was dry-farmed in a playa on the southwest side of Cerros Batamote (ca. 31°42′ N, 113°16′ W). These fields represent the limits of arid agriculture in North America.

During the mid-twentieth century extensive industrial agriculture was developed in the Río Colorado valley south and west of San Luis. In the 1970s and 1980s modern agriculture was extended along the Sonoyta Valley. Both regions are irrigated with rapidly diminishing reserves of fossil ground water pumped from deep wells. Fields in the San Luis Valley are also irrigated with water from the Río Colorado. The abundant supply of irrigation water in the San Luis Valley has fostered a thriving agriculture.

Commercial agricultural crops in the Sonoyta region include alfalfa, common beans, cotton, cucumber, melons, rye, sesame, sorghum, squash, watermelon, and wheat. Most of the produce goes to California. The more important cultivated crops include alfalfa, Brussels sprouts, cabbage, carrot, cauliflower, cilantro, chilies, cotton, melons (cantaloupe, honeydew, and others), onion (especially green onions), radish, squash (various kinds), sorghum, watermelon, and wheat. Other commercial crops include green beans, grapes, lettuce, and maize (corn).

Where people have settled and water has been available they have planted trees and created oases. In 1894 Edgar Mearns (1907:116) recorded that "fig trees thrive in a half-wild state, cottonwoods and willows border the fields and acequias, and a luxuriant thicket of young mesquite edges the [Río] Sonoyta, where it has not been dug away to make place for fields and gardens." At Santo Domingo, downstream from Sonoyta, he collected specimens of elderberry (*Sambucus mexicana*) from a large, vigorous tree that was undoubtedly planted. Approaching Sonoyta from the west, Carl Lumholtz (1912:287–288) noted: "It was toward the end of February and spring-time greeted us in the oasis. Already at a distance the attractive light green color of the cotton-wood trees . . . and willows were evident, the new leaves being half-grown; peach, apple, and almond trees in bloom, and the verdure of the wheat-fields among the Indian houses was pleasing to the eye."

When supplied with additional water, many subtropical or even temperate plants can be grown in the region. Winter freezing is one of the major limiting factors for the cultivation of plants of tropical affinity. Freezing can be more severe in Sonoyta than in San Luis, whereas Puerto Peñasco has nearly frost-free winters. For example, *guaje* (*Leucaena leucocephala*) and *guamúchil* (*Pithecellobium dulce*) are often severely damaged by winter freezing in Sonoyta, but usually sustain only slight freeze damage at San Luis. *Arbol del fuego* or royal poinciana (*Delonix regia*) is grown in San Luis and Puerto Peñasco but is too tender for Sonoyta. Trees and shrubs commonly cultivated in the region are listed in Appendix C. Except for the cottonwood (*Populus fremontii*) these are not native to the region.

Non-native Plants

Eighty-eight of the 561 species comprising the flora for northwestern Sonora are not native to the region. These are introduced, naturalized plants, and/or urban and agricultural weeds (Appendix D).

The opportunity for human-influenced plant introductions is not new. Occasional visitors, beginning with Kino in 1701, traveled via horses or mules. Weed seeds can be transported in animal hair, deposited through their dung, or transported with animal fodder. Lumholtz (1912) and others brought along feed for their animals, and sometimes cattle were driven across the Gran Desierto to the Colorado River (Lumholtz 1912). In 1992 I found small numbers of three weed species at Campo Rojo that were obviously associated with horse manure, i.e., *Eragrostis lehmanniana, Hordeum murale,* and *Sisymbrium irio*. These plants apparently did not become established (see Part II).

The majority of the non-natives are colonizing species, established or occurring only on disturbed habitats (see Mooney & Drake 1986). Two categories of colonizing plants are distinguished: (1) ruderals, recorded from artificial habitats, such as irrigated fields and urban areas; and (2) disturbed habitat plants recorded only in non-irrigated or non-watered but human-altered habitats such as along roadsides. Most of the colonizing plants do not survive in the natural desert habitats of the region. Some (e.g., *Brassica nigra, Sesamum orientale,* and *Triticum aestivum*), are not established in the flora area as reproducing populations. These plants are sometimes encountered along highways and probably result from spillage from trucks.

Only 23 non-native species, or 4% of the total flora, are present as well-established, reproducing populations in undisturbed (natural) habitats in northwestern Sonora (i.e., invading species, see Mooney & Drake 1986). These 23 and the disturbed habitat species (in disturbed but non-irrigated habitats), 65 species in all, represent 11% of the total flora (539 species) of the "natural" and non-irrigated "disturbed" habitats.

As the area continues to be altered by human activities additional non-native, weedy species will enter the region. Conversely, local extinctions can be expected as ground-water pumping lowers the water tables in the Sonoyta and San Luis Valleys. Particularly critical is the drying and destruction of remaining wetland habitats.

Growth Forms

The usual growth form of each species as it occurs in the flora area is given in Appendix A. Shreve (1951) pointed out that diversity in growth forms is characteristic of deserts, especially the Sonoran Desert. The life-form, or growth-form spectrum of the flora of northwestern Sonora is characteristic of a very arid, desert region, with herbaceous plants, especially annuals (ephemerals), out-numbering all others (see Felger 1980; Inouye 1991; Raunkiaer 1934; Shmida & Burgess 1988; Shreve 1951; Venable & Pake 1999).

Annuals (called therophytes by Raunkiaer) are plants that pass the most unfavorable season(s) as seeds, and indeed the severity of a habitat, especially an arid one, is reflected by the percentage of its flora comprised of annuals. The distinction between ephemeral and annual life-forms was emphasized by Shreve (1951), because the classical definition of an annual is not appropriate in deserts such as the Sonoran Desert. Annuals, in the strict sense, germinate after the last killing freeze in spring and produce seeds and die with the first killing freeze in fall or early winter. Desert ephemerals are annuals that can complete their life cycle within a single season.

In the Sonoran Desert there is a large and diverse group of ephemerals that responds exclusively to the cool-season rains (winter-spring); and another, less diverse group that responds only to hot-weather rains (summer or summer-fall). Most cool-season ephemerals utilize the C_3 photosynthetic pathway, whereas hot-season ephemerals tend to use the C_4 pathway (Kemp 1983; Mulroy & Rundel 1977). A third major category, made up of species that can respond to rainfall at essentially any time of year, are the non-seasonal ephemerals. In addition, a case can be made for a very minor fourth group, which I am calling long-season ephemerals. These are spring ephemerals that may survive through the summer.

"True" annuals, characteristic of temperate climates, are scarce in the Sonoran Desert. Species in the flora area that sometimes exhibit these characteristics include only a few weeds in urban-agricultural habitats or wetland plants. These plants tend to have temperate origins; e.g., *Atriplex wrightii*, certain sedges (*Cyperus* spp.), and perhaps *Conyza canadensis, Lactuca serriola,* and *Xanthium strumarium.*

The total ephemeral/annual flora of northwestern Sonora is made up of 248 species, or 46% of the total flora. The percentage and seasonal distribution of the ephemeral flora are comparable to the general pattern of the Sonoran Desert, where roughly 50% of the species are annuals with 60% to 80% of these being winter annuals (Venable & Pake 1999). Except in extremely dry years the ephemeral vegetation probably accounts for the majority of biomass production in the region. Certain habitats, such as playas and sandy-soil flats or plains, may support far less than 5% perennial plant cover, but with sufficient rainfall ephemerals may form meadowlike expanses of 100% plant cover. At such times there may be more than 1000 individual plants per meter square, mostly comprised of grasses, or the individual plants may be large and few per meter square (e.g., *Lupinus arizonicus*).

One of the most striking features of desert ephemerals is their extreme plasticity in size (Inouye 1991). Many ephemerals can germinate with minimal rainfall and develop into tiny plants that produce seeds in one or a few fruits before perishing. With sufficient soil moisture individuals of these same species can become extraordinarily large and produce many flowers and fruits. During El Niño years the attractive, white-flowered evening primrose (*Oenothera deltoides*) may develop shrublike plants more than 1–1.5 m across with hundreds of flowers, but during a spring of low rainfall the plants may reproduce as small, essentially stemless rosette plants with only a few flowers and fruits. The common summer ephemeral *Amaranthus palmeri* may grow to 3 m or more in height, although on occasion I have found drought-stunted seed-producing plants less than 10 cm tall. Individual plants of *Kallstroemia grandiflora* may reach 3 m across on sandy soil following favorable summer and early fall rains, although plants with stems 15–30 cm are commonplace.

Reproductive as well as vegetative parts may vary with soil moisture. For example, *Pectis papposa,* a common hot-weather ephemeral, shows a tendency to produce awnless achenes during drought stress. Likewise, under extreme drought conditions the ubiquitous non-seasonal ephemeral three-awn grass *Aristida adscensionis* may produce spikelets with greatly reduced awns or even no awns.

Winter-spring or cool-season ephemerals, represented by 173 species, make up 70% of the total ephemeral flora. Many cool-season ephemerals begin life with a basal rosette of leaves (Mulroy & Rundel 1977), which seems to be an adaptation to take advantage of the warmer temperatures at ground level and potential water conservation (i.e., reduced effects of wind). Perennials that also respond only to the winter-spring growing season may show similar strategies (e.g., *Triteleiopsis palmeri*).

Winter-spring ephemerals tend to have temperate or northern affinities (Shreve 1951). Often their closest biogeographic affinities are with the Mojave Desert or Californian region, or desert regions of the Californias, western Arizona, southern Nevada, and sometimes southwestern Utah. Many winter-spring ephemerals are at their southern limits in the Pinacate–Gran Desierto region, at least in mainland Mexico, e.g., *Ambrosia acanthicarpa, Astragalus lentiginosus, Bromus berterianus, Calycoseris parryi, Camissonia arenaria, C. boothii, Cryptantha costata, C. ganderi, Geraea canescens, Prenanthella exigua, Psathyrotes ramosissima,* and *Rafinesquia californica.*

Rainfall patterns vary considerably from year to year, and each year may bring a different vegetational and floristic spectrum of winter-spring ephemerals or wildflowers. There is a general progression of early to late cool-season ephemeral species. The mustards (Brassicaceae), borages (Boraginaceae), *Brandegea bigelovii* (Cucurbitaceae), and certain composites (e.g., *Perityle emoryi*) are among the earliest cool-season ephemerals to germinate and flower. They can appear as early as mid to late October—as soon as the period of high soil temperature has passed. Some early germinating, fast-growing winter-spring weedy ephemerals of Old World origin, such as *Brassica tournefortii, Bromus rubens,* and *Schismus,* seem "to get a jump" on some of the natives and out-compete them. For example, the spreading rosette leaves of *Brassica tournefortii* and leafy stems of *Schismus* often cover up and inhibit growth or germination of the seemingly slower growing, or perhaps later germinating natives.

Among cool-season ephemerals the early, or rosette, leaves are almost always larger than the stem leaves, which in some instances are represented by scales or may be absent. As temperatures rise and the soil dries, the rosette leaves usually wither or fall away by the time the plants begin flowering and fruiting. Some spring ephemerals, such as *Camissonia californica, Chorizanthe brevicornu,* various *Eriogonum* species, and *Stephanomeria schottii,* are essentially leafless at flowering or fruiting time and resemble miniature palo verdes.

The ephemeral desert buckwheat, *Eriogonum deflexum,* and to a lesser extent *E. trichopes* and the annual chenopods *Atriplex elegans* and *A. wrightii,* sometimes continue to flower well into late spring or even early summer. Some plants may germinate late in spring and mature in early summer. Occasionally they survive through the summer to produce flowers and seeds in fall. Drought limits their life span, and I have not found such plants in the flora area surviving beyond one year.

Summer, or hot-weather, ephemerals are represented by 50 species, or 20% of the total ephemeral flora. These plants tend to have southern or tropical affinities. As with cool-season ephemerals, they appear to be derived from perennial ancestors. In contrast to the cool-season ephemerals, hot-weather ephemerals do not form basal rosettes and there is little variation in species composition from early to late summer. Among the most numerous and ubiquitous, or predictable, members of the summer ephemeral flora are *Amaranthus* spp., *Boerhavia* spp., *Bouteloua aristidoides, B. barbata, Kallstroemia* spp., *Pectis papposa,* and several non-seasonal species such as *Aristida adscensionis* and *Euphorbia* spp.

Non-seasonal ephemerals, comprising 25 species, or 10% of the total ephemeral flora, are capable of germinating and maturing at any season. These plants are primarily limited by soil moisture rather than temperature. Some are frost sensitive; they may disappear or be damaged by freezing weather but can quickly recover or re-appear when conditions are favorable (e.g., *Euphorbia setiloba*). Among the most common non-seasonal ephemerals are small euphorbs (*Ditaxis neomexicana, D. serrata, Euphorbia micromera,* and *E. polycarpa*), papilionoid legumes (*Dalea mollis, Marina parryi,* and *Phaseolus filiformis*), *Datura discolor,* and certain grasses (*Aristida adscensionis, Cenchrus palmeri,* and *Muhlenbergia microsperma*).

Facultative perennials, or annual/perennial species ("AP" plants in Appendix A), include 18 species or 3% of the total flora. Two closely related borages, *Cryptantha holoptera* and *C. racemosa,* and two closely related evening primroses, *Camissonia arenaria* and *C. cardiophylla,* germinate only with winter-spring rains (as do their strictly ephemeral congeneric species), but may survive more than one year as short-lived perennials. The most common and widespread annual/perennial plants include two closely related desert spurges *Euphorbia micromera* and *E. polycarpa,* the composite *Machaeranthera coulteri,* and the weedy grass *Pennisetum ciliare.*

Shrubs and woody perennials with perennial growth above ground level include the nanophanerophytes and chamaephytes distinguished only by their height, or size (Appendix A). Most are dicotyledons. Two monocotyledons, galleta grass (*Pleuraphis rigida*) and bush muhly (*Muhlenbergia porteri*), are unusual among Sonoran Desert grasses by taking on the growth-form of a small shrub.

Woody perennials and shrubs form a diverse group accounting for the majority of the perennial plant cover in the non-wetland desert habitats. Most are drought deciduous, and some exhibit drought-induced dieback of portions or even all of their stems. Even species usually perceived as evergreen, such as creosote bush (*Larrea divaricata*) lose most or occasionally all of their leaves during the most severe droughts. Desert sumac (*Rhus kearneyi*) and pygmy cedar (*Peucephyllum schottii*) are evergreen but may lose many of their leaves in extreme drought.

Tree species are few in deserts (Felger, Johnson, & Wilson 2000). The largest trees in the flora area are the cottonwoods (*Populus fremontii*) and willows (*Salix gooddingii*). These trees are winter-deciduous, reflecting their temperate origins. The largest non-riparian tree is the introduced tamarisk tree (*Tamarix aphylla*). The most widespread tree-sized plants in the natural landscape are species of legumes and cacti. The arborescent legume flora is comprised of three species of *Prosopis, Olneya tesota,* two species of *Parkinsonia,* and *Psorothamnus spinosus.* In addition, *Condalia globosa* (Rhamnaceae) occasionally becomes arborescent. For the most part trees in northwestern Sonora are concentrated along riparian or xeroriparian drainageways (washes, arroyos, or riverbeds), although *Parkinsonia microphylla* is common on slopes and cinder-soil habitats.

The mesquites and screwbean (*Prosopis*) are winter-deciduous and leaf out in spring after the last frost; they are deeply rooted and largely independent of rainfall. The palo verdes (*Parkinsonia*) and most other leaf-bearing woody plants and perennials are drought deciduous, and generally produce leaves in response to soil moisture at almost any time of the year. Among the perennial species of tropical and subtropical affinities—such as the larger legumes, elephant tree (*Bursera microphylla*), and ocotillo (*Fouquieria splendens*)—leaves also may fall or cease to be produced during the colder weeks or months of the year.

Plants with succulent stems and/or leaves form a very conspicuous part of the landscape across most of northwestern Sonora. However, succulent plants are rare or even absent from the harshest, driest regions toward the western margin of the Gran Desierto such as the dunes and Mesa de Sonora.

Succulents span a wide range of growth forms. The saguaro (*Carnegiea gigantea*) is the tallest organism across much of the desert landscape. Other large cacti that sometimes exceed shrub size are *sinita* (*Lophocereus schottii*), organpipe (*Stenocereus thurberi*), and jumping cholla (*Opuntia fulgida*). Perennial, rosette-forming leaf succulents—*Agave deserti, Hesperoyucca whipplei,* and the much smaller *Dudleya arizonica*—are restricted to granitic rock slopes and soils.

Among succulent ephemerals there is no dominant pattern of seasonality. Cool-season species include *Cistanthe* spp. (Portulacaceae), *Crassula connata* (Crassulaceae), and *Mesembryanthemum crystallinum* (Aizoaceae). Among hot-season species are *Sesuvium verrucosum* (Aizoaceae) and *Portulaca* spp. (Portulacaceae). Succulent halophytes, which include perennial as well as some annual or ephemeral species, are *Abronia maritima* (Nyctaginaceae), *Batis maritima* (Bataceae), *Heliotropium curassavicum* (Boraginaceae), and eight species of Chenopodiaceae.

It is well known that vining plants are few in deserts. The only truly common or widespread perennial vines in northwestern Sonora are the milkweed *Sarcostemma cynanchoides* and the malpigh *Janusia gracilis. Brandegea bigelovii* (Cucurbitaceae), a cool-season ephemeral, is the only other truly common vining species. Eleven other species, which include hot- and cool-season ephemerals and perennials, complete the list of vining plants: *Antirrhinum filipes, Aristolochia watsonii, Cucurbita digitata, Cuscuta* spp., *Ipomoea* spp., *Metastelma arizonicum, Phaseolus filiformis,* and *Tragia nepetifolia.*

There are only six species of geophytes in northwestern Sonora: four "liliaceous" species that produce bulbs or corms, yellow devil's claw (*Proboscidea althaeafolia,* Martyniaceae), and coyote gourd (*Cucurbita digitata,* Cucurbitaceae). These plants are widespread across the region and show active growth and flowering at various seasons.

While studying the plants of the region I was impressed by the number of species possessing dissimilar dispersal units (heteromorphic propagules). These are plants that bear seeds or other dispersal units of more than one shape or morphology, size, weight, or dispersal mechanism. Heteromorphic propagules offer more than one strategy for dispersal and survival, and would seem to be a significant adaptation in an arid environment with highly variable rainfall and environmental conditions (Venable & Pake 1999).

In some cases seeds or seedlike organs (such as nutlets or achenes) within the same fruit or flower head differ from one another in shape, size, or morphology and disperse at different times (e.g., certain *Cryptantha* and *Filago*). In other cases, such as among certain *Portulaca* species and *Brassica tournefortii,* the seeds themselves may be similar but one or more seeds are retained within the fruit. The fruit (e.g., capsule lid or beak) and its retained seed(s) weigh more than the free seeds. Among the cryptanthas the "odd nutlet" is tenaciously held within the calyx and ultimately disperses with it long after its siblings have been dispersed. Other *Cryptantha* have smooth and rough nutlets produced by the same flower. Among several grasses (*Enneapogon desvauxii, Leptochloa dubia,* and *Muhlenbergia microsperma*) and the borage *Pectocarya heterocarpa* the basal flowers are cleistogamous whereas the upper flowers open normally. The heavier or more persistent propagules are more likely to remain closer to the parent plant (where the parent was successful), whereas the lighter or non-persistent ones are more likely to colonize new areas.

At least 33 species, or approximately 6% of the total flora, exhibit heteromorphic propagules and/or dispersal mechanisms: *Brassica tournefortii, Chorizanthe* spp., *Cryptantha* spp., *Enneapogon desvauxii, Eucrypta chrysanthemifolia, Filago* spp., *Horsfordia newberryi, Lappula occidentalis, Leptochloa dubia, Lotus strigosus, Machaeranthera coulteri, Mentzelia albicaulis* complex, *Muhlenbergia microsperma, Pectis papposa, Pectocarya heterocarpa, Perityle emoryi, Portulaca oleracea, Prenanthella exigua, Proboscidea* spp., *Sphaeralcea* spp., *Suaeda esteroa,* and *Trianthema portulacastrum.* These are herbaceous plants and most are ephemerals.

Botanical History

The earliest collections in the region are those of Arthur Schott and others members of the boundary survey of 1855 (Torrey 1858). The only significant early collections from the lower Río Colorado were made by the intrepid Edward Palmer in October 1869, December 1884–January 1885, and April 1889 (McVaugh 1956; Vasey & Rose 1890). Apart from studies of the Río Colorado (see Sykes 1937), the earliest biological survey of the region was conducted by Edgar Mearns, a physician attached to the International Boundary Survey of 1892–94 (Mearns 1907). Mearns and his assistants made a few botanical collections near Sonoyta and Quitobaquito, and along the Río Colorado (see Hitchcock 1913).

The next exploration to be documented in print was the Carnegie Institution expedition in 1907 (Hornaday 1908). This account is "a whizzing good tale of turn-of-the-century 'gentlemen' hunters bent on shooting anything that moved" (Lynch 1981:7). They named many of the area's geologic features. Daniel T. MacDougal and Glenton Sykes, members of the Carnegie expedition, made the first extensive plant collection from the Pinacate region. In order to set his altimeter, Sykes made an incredible solo march from Tinaja del Tule to Bahía Adair and back in 13 hours (Hornaday 1908:238–240). He brought back not only marine specimens but several plants, although Hornaday made no mention of Sykes' collections (see *Croton wigginsii* and *Heliotropium convolvulaceum*). The MacDougal-Sykes collections, described by Rose & Standley (1912), resulted in descriptions of a number of new species, many of which have been reduced to synonymy. During his 1909–10 expedition Carl

Lumholtz (1912) collected a limited number of herbarium specimens (see *Pholismas sonorae*). He provided an eloquent description of the land, people, plants, and animals, and named and described many geologic features.

During the 1930s and 1940s Forrest Shreve, Ira Wiggins, and other pioneer desert botanists collected in the northern part of the Pinacate region, along the road to Puerto Peñasco, and in immediately adjacent areas in southern Arizona (Shreve 1951; Wiggins 1964). Prominent post-World War II collectors in the region included Lyman Benson, Frank Gould, David Keck, and many others. Toward the latter part of the twentieth century, in addition to my work, numerous investigators contributed to the botanical knowledge of the region, e.g., Janice Bowers, Tony Burgess, Alberto Búrquez, Miguel Equihua, Exequiel Ezcurra, Jorge López-Portillo, Gary P. Nabhan, Raymond M. Turner, Thomas Van Devender, Grady Webster, and many others.

Norman Simmons' (1966) *Flora of the Cabeza Prieta Game Range* includes many species also occurring in adjacent northwestern Sonora, as does the *Flora of Organ Pipe Cactus National Monument* (Bowers 1980). My *Vegetation and Flora of the Gran Desierto, Sonora, Mexico* (Felger 1980), although covering more than half the land surface of northwestern Sonora, includes only about one-fifth of the total flora of the region. Other floristic listings include the flora and ethnobotany of the Quitobaquito area (Felger et al. 1992), Felger (1992), and Felger et al. (1997).

Researchers from the Instituto de Ecología in Mexico City conducted intensive studies of the region in the late 1970s and the 1980s. They produced a vegetation map and analyzed the perennial vegetation of the Gran Desierto (Ezcurra 1984). Ezcurra et al. (1988) described the hydrology, flora, and biogeography of the Bahía Adair pozos (oases). Ray Turner's (1990) long-term study in MacDougal Crater documented dynamic change in desert vegetation. Based on studies of plant remains from fossil packrat middens, Tom Van Devender reconstructed floristic and climatic change in the region since the early Pleistocene (Van Devender 1987; Van Devender et al. 1990; Van Devender, Toolin, & Burgess 1990).

Flora of the Gran Desierto and Río Colorado is the first comprehensive treatment of plant life focusing on the center of the Sonoran Desert. It adds to current knowledge of the flora of the Sonoran Desert with new identification keys, original descriptions, distributional information, illustrations, and natural history information. I hope this work will be useful for continued conservation of the great natural reserves on both sides of the international boundary.

Part 2 The Flora

The species accounts are organized in pteridophytes, then the one gymnosperm genus, then the dicotyledons and monocotyledons. Within these divisions, all entries are listed alphabetically by family, genus, and species. (See table 2 for a list of major plant groups and the ten largest families and genera.) Identification keys for the various taxa (e.g., families, genera, and species) are based on plants from the flora area and are not necessarily applicable to other regions. The keys are artificial and not meant to be phylogenetic. The same genus or species may appear more than once within a key.

The descriptions and notes on the families and genera pertain to those taxa more or less as they occur throughout their entire geographic ranges. In contrast, descriptions and measurements for the various species and infraspecific taxa pertain only to plants and populations from northwestern Sonora and immediately adjacent areas unless otherwise stated. This enables readers to compare or contrast plants from this extremely arid desert with their relatives across their entire distributions, whether it be locally in the Sonoran Desert or globally. Very often there is little agreement on the size of a genus, family, and other taxa, and for this reason I have not repeated the qualification *about* or *approximately* every time numbers of taxa are cited.

Largely for the sake of convenience, a conservative approach is used for circumscription of plant families, generally following Cronquist (1981) and Mabberley (1997). Recent molecular studies are showing that many traditional families are not monophyletic and new alignments are needed, e.g., the artificial but convenient grouping of the Liliaceae and relationships among the Orobanchaceae, Plantaginaceae, Scrophulariaceae, etc.

In the descriptions I have emphasized characters that seem important to understanding the variation and adaptations of plants in this arid environment. For this reason there is sometimes more emphasis on vegetative characters and less on certain characters emphasized in other floras. More detailed description, discussion, and interpretation are offered for taxa not well covered elsewhere, as well as those of special interest, uniqueness, or significance in the flora area.

The accepted names are in bold-face font. Throughout the text non-native plants are marked with an asterisk (*). Species expected or reported but not actually documented for the flora area are indicated with a dagger (†). Selected pertinent synonyms are listed, especially those not appearing in the *Flora of the Sonoran Desert* (Wiggins 1964). The first synonym listed is the basionym. Synonyms, when provided, are in square brackets following the accepted scientific name. Authors of scientific names follow the listings of Brummitt & Powell (1992). Place and date of publication are provided for taxa based on type specimens from the flora area. Common names, when available, follow the scientific name(s), with the Mexican common names italicized and listed first, followed by the American common names. Selected O'odham names, largely from Felger et al. (1997), follow the other vernacular names and are also italicized. Selected references are provided for most families and genera.

The presentation and format of the infraspecific taxa (subspecies or variety) depend largely on my judgment of the significance of the differentiation; after all, variation and evolution are not consistent. When an infraspecific taxon seems worthy of recognition, then it is given in the heading and may be included in the keys. When the biological significance seems questionable, then I often merge the infraspecific taxa into the discussion for the species.

Measurements for length or height precede those for width, and the terms *length, long,* or *height* are omitted unless needed to avoid confusion. Unless otherwise noted, it is assumed that the flowers are bisexual (perfect). Furthermore, unless otherwise specified, the color given for a flower or other structure is the most conspicuous or dominant color. The term *radial* refers to flowers that are radially symmetrical, and *bilateral* refers to those that are bilaterally symmetrical.

I measured a number of specimens of each taxon (generally 10 to 20+ measurements depending upon variation encountered), provided enough specimens were available. In addition to the usual or commonly encountered range of variation, the extreme or uncommon variation is often provided in parentheses, e.g., (5) 10–15 (20). Sizes of larger plants, such as shrubs and trees, were usually only roughly estimated, and sizes for herbaceous plants were often rounded off.

I have emphasized both distributions of taxa restricted to the Sonoran Desert and relationships within this region. Special attention is given to details and refinements of geographic ranges not covered in other, readily available regional floras. For example, the local distributions in Sonora, especially in western Sonora, are often poorly known but are significant to the understanding of relationships among taxa in the flora area. Usual flowering times follow the descriptions.

Specific herbarium specimens are cited (in italics) in order to give information on distribution and habitat, as well as to provide verification. I have inspected all specimens cited except those noted as "not seen." Localities from labels have been standardized to conform to local usage and the gazetteer of place-names in this volume (see also Broyles et al. 1997). For example "Rocky Point" is cited as "Puerto Peñasco," "Cholla Bay" as "Bahía de la Cholla," and "4.7 miles east of Los Vidrios on Mex 2" as "Pinacate Junction."

For my herbarium records I have substituted *F* for *Felger*. In most cases the first two sets of my collections are deposited at ARIZ and MEXU or ENCB; and additional specimens are mostly at ASU, CAS, RSA, SD, TEX, UC, and UCR. All specimens cited are at ARIZ unless otherwise indicated by the abbreviations for herbaria given here and in Holmgren, Holmgren, & Barnett (1990). If a specimen is at ARIZ, I do not generally cite a duplicate at another herbarium. In cases where more than one collector is listed on the label, generally only the first collector's name is given. If no collection number is given on the label, the specimen is identified by the date of collection; for example, *Ezcurra 30 Oct 1982*. When the date of collection is significant, such as collections of historic interest or type collections, both the collection number and date are given. In a few cases the herbarium accession number follows the herbarium abbreviation; this is provided to avoid confusion, especially in the case of multiple specimens of type collections.

Key to the Major Plant Groups

1. Plants reproducing by spores, lacking flowers or seeds: herbaceous ferns, spikemoss, and water clover. ____ **Pteridophytes—Ferns and Fern Allies (p. 42)**

1′ Plants reproducing by seeds; flowers present or absent.

  2. Twiggy shrubs with slender greenish stems and scale leaves; true flowers absent, the plants producing either stamens in cone-like clusters or 1 seed surrounded by bracts. ____ **Gymnosperms—Ephedraceae (p. 49)**

  2′ Plants producing true flowers and seeds within fruits. ____ **Angiosperms (p. 51)**

    3. Leaves various, usually net-veined or leaves absent; flowers various, often 4- or 5-merous; plants terrestrial. ____ **Dicotyledons (p. 51)**

    3′ Leaves narrow with single or parallel veins; flowers 2- or 3-merous (except submerged aquatics): agavoids, arrowhead, cattails, lily relatives, grasses, sedges, rushes, and aquatic plants. ____ **Monocotyledons (p. 469)**

Table 2. Summary of the major plant groups of northwestern Sonora, including the ten largest families and genera with the most species (families and genera listed in descending order by number of species).

	Families	Genera	Species	Non-native Species
Pteridophytes	3	6	8	0
Gymnosperms	1	1	2	0
Dicotyledons	70	269	478	46
Monocotyledons	11	51	101	33
Total	**85**	**327**	**589**	**79**
Families				
Asteraceae		59	90	12
Poaceae		39	75	32
Fabaceae		17	35	3
Euphorbiaceae		8	29	3
Cactaceae		10	26	0
Chenopodiaceae		10	24	6
Boraginaceae		6	19	0
Solanaceae		7	21	4
Malvaceae		8	18	2
Brassicaceae		13	4	4
Genera				
Euphorbia			15	3
Opuntia			11	0
Cryptantha			9	0
Atriplex			8	1
Eriogonum			8	0
Mentzelia			8	0
Lycium			8	0
Ambrosia			6	0
Astragalus			6	0
Camissonia			6	0
Phacelia			6	0

PTERIDOPHYTES

FERNS AND FERN ALLIES

Mostly perennials, minute to tree-sized; characteristically reproducing by shedding spores that germinate and grow into small (to 15 mm) gametophyte plants producing sex cells that require water for fertilization. (Some ferns circumvent sexual reproduction, a strategy that can be advantageous in arid regions.) Few in deserts; mostly in wet, especially tropical and subtropical, regions of the world; 10,000+ species.

Three distinct groups of pteridophytes occur in northwestern Sonora. These plants require moist conditions for active growth. The cheilanthoid ferns and spike-moss (*Selaginella*) are perennial, drought-avoiding resurrection plants—their leaves, leaflets, or stems curl up during drought and rather quickly expand when wet. These are restricted to rocky places, especially at higher elevations. *Marsilea,* restricted in the flora area to low-lying habitats of the desert plain, has taken on a desert ephemeral-like way of life that is uncommon among pteridophytes.

Ferns and fern allies make up only 1.5% of the flora of northwestern Sonora as compared to nearly 4% of the world flora of vascular plant species. However, only *Notholaena californica* is really common and widespread in northwestern Sonora. References: Davidse, Sousa, & Knapp 1995; Flora of North America Editorial Committee 1993; Lellinger 1985; Mickel 1979a; Tryon & Tryon 1982; Yatskievych & Windham 1986.

Key to the Families

1. Leaves less than 4 mm, simple, somewhat scalelike and sessile. ______ **Selaginellaceae**

1′ Leaves more than 10 mm, compound or deeply divided, not scalelike, the petioles conspicuous.

  2. Leaves with 4 leaflets at apex of petiole, resembling a 4-leaf clover, the leaflet margins entire to apically notched. ______ **Marsileaceae**

  2′ Leaves deeply divided (pinnatifid) or with more than 4 leaflets, not resembling a cloverleaf. ______ **Pteridaceae**

MARSILEACEAE WATER-CLOVER FAMILY

Aquatic or amphibious herbs with rhizomes, long petioles, and sporocarps; 3 genera, 50 species.

MARSILEA WATER-CLOVER

Annual and perennial herbs with short creeping to long rhizomes. Aquatic, often in temporary ponds, in which case usually with floating leaves; when semiaquatic or terrestrial on wet or muddy soils, then with erect-ascending leaves. Leaves resembling a 4-leaf clover, with 4 closely set mostly fan-shaped leaflets at the apex of the petiole; land forms with functional pulvini that orient the leaflets during the day and cause them to fold together at night. Sporangia enclosed in nutlike, often drought-resistant, bony, stalked sporocarps that "split open and, through the absorption of water, a gelatinous, thread-like, sporangium-bearing structure is extruded" (Lellinger 1985:304).

Cosmopolitan, largely Old World, mostly tropical to temperate regions, very few in deserts; 45 species. When subjected to freezing weather, many of the species overwinter as sporocarps, thus qualifying *Marsilea* as one of the few fern genera with functionally annual species.

Marsilea shares ". . . many characteristics with those of weedy seed plants, such as long propagule dormancies, self-compatibility, rapid growth, and occurrence in early successional habitats. . . ." It further differs from most ferns ". . . in its use of biotic rather than abiotic dispersal agents" (Johnson 1986:27). References: Bloom 1955; Johnson 1985, 1986; Malone & Proctor 1965.

Marsilea vestita Hooker & Greville [*M. mucronata* A. Braun] HAIRY WATER CLOVER. Non-seasonal ephemerals in northwestern Sonora, producing long, slender rhizomes or stolons with 1 leaf per node; more often remaining small, stunted, and producing only a dense tuft of nodes, leaves, and sporocarps. Leaves densely hairy, less so with age and size (aquatic forms, unknown in the flora area, are glabrous); petioles 2–20 cm. Sporocarps on unbranched stalks from near leaf bases, the surfaces olive-green to brown or blackish, densely hairy, with a short, superior tooth or tubercle.

Sometimes locally common in playa-like depressions or other poorly drained temporary pools in low-lying areas in the northern part of the Pinacate region; often emergent from very shallow water on dried, cracked mud. The plants are generally visible for about 6–8 weeks after the water dries up. During dry seasons there is no sign of the plants.

Western 2/3 of North America, from Canada to central Mexico and western Peru; nearly throughout Sonora. Introduced elsewhere and sometimes weedy. This species "occupies the largest continuous range of any New World *Marsilea,* and exhibits regional variations in characters" (Johnson 1986:67).

This is the only Sonoran Desert pteridophyte to be functionally ephemeral and the only one found away from rocky hills or mountains. Sporocarps taken from 77-year-old herbarium specimens originally collected in California produced viable plants, and century-old sporocarps from a herbarium specimen of another species, *M. oligospora,* also produced plants (Johnson 1985). Birds are likely agents of dispersal because the sporocarps survive undamaged after passing through the digestive tract of certain waterfowl that feed on the plants (Malone & Proctor 1965).

2 km N and 3.6 km E of Los Vidrios, *Burgess 6793.* Pinacate Junction, *F 85-991.* 2 mi S of Tinajas de los Pápagos, *F 86-495.*

PTERIDACEAE MAIDENHAIR FERN FAMILY

Herbaceous perennials, the stems short, creeping to erect, bearing hairs and/or scales. Leaves petioled, usually coiled as fiddle-heads (circinate vernation) in bud; blades mostly 1- to 4-times divided; lower surfaces especially often showing adaptations for dry habitats, e.g., glandular secretions, hairs, scales, or waxes. Fertile and sterile leaves similar (ours) or dissimilar. Sporangia often in sori on lower leaf surfaces at margins or along veins; leaf blade or leaflet margins often rolled under, protecting the sporangia as false indusia.

Worldwide, best developed in dry tropical to subtropical or mild temperate regions, relatively few in deserts; 40 genera, 1000 species.

The 4 genera in the flora area are cheilanthoid ferns. In dry climates these ferns occupy moist, often wet niches and mostly grow in association with rocks and rocky soils. Nobel (1978) demonstrated the importance of rainfall channelling by rocks, which provides a favorable niche for ferns in desert habitats. Although ours do not occur in moist or wet niches, they are found only among rocks and often on north-facing exposures. They have leaves that curl up during drought and expand when wet.

Some cheilanthoid "species or races are triploid or tetraploid. These can spread in the driest deserts because their gametophytes do not require any water for fertilization, and so can develop new sporophyte plants with viable spores under conditions impossible for sexual species" (Lellinger 1985:131). At least half of the cheilanthoids in the flora area belong to taxa reported to be apogamous (reproducing asexually). References: Cronquist et al. 1972; Gastony & Rollo 1998; Tryon & Tryon 1973.

1. Leaves once pinnate, the blades more than 5 times as long as wide, the petioles about 1/8 as long as the blade. ____ **Astrolepis**

1′ Leaves 2–3 times pinnate or deeply pinnatifid, the blades less than twice as long as wide, the petioles usually at least half as long as the blade.

  2. Leaf blades (portion above petiole) mostly more than 20 cm. ____ **Pellaea**

  2′ Leaf blades mostly less than 15 (18) cm.

    3. Leaflets densely woolly, the hairs obscuring the surfaces and extending beyond the margins. ____ **Cheilanthes**

    3′ Leaflets not densely woolly, the upper surfaces green and not obscured by hairs, the hairs not extending beyond the margins. ____ **Notholaena**

ASTROLEPIS STAR-SCALED CLOAK FERN

Small to medium-sized densely tufted ferns with short-creeping, scaly rhizomes. Leaves linear, once pinnate, the leaf-axis scaly; leaflets somewhat thickened, entire to shallowly lobed or divided, the margins not modified and not recurved, the upper surfaces often having deciduous stellate-pectinate scales (star-shaped and comblike), the lower surfaces densely covered with overlapping ciliate scales and an underlying layer of stellate scales. Sporangia along veins near leaflet margins and partially hidden by the scales. Southwestern United States to South America and West Indies; 8 species. Reference: Benham & Windham 1993.

1. Leaves usually 10–20 (28) × 1–1.5 cm; leaflets usually 4–8 mm; scales of lower leaf surface 0.5–0.8 mm. ____ **A. cochisensis**

1′ Leaves often 20–37 × 3–3.4 cm; leaflets usually more than 15 mm; scales of lower leaf surface 1.6–2.1 mm. ____ **A. sinuata**

Astrolepis cochisensis (Goodding) D.M. Benham & Windham subsp. **cochisensis** [*Notholaena cochisensis* Goodding. *N. sinuata* var. *cochisensis* (Goodding) Weatherby. *Cheilanthes cochisensis* (Goodding) Mickel] Rhizome scales 6–7.5 mm. Leaves (6) 10–28 × 1–1.5 cm; petioles (1.5) 2–4.5 + cm. Larger leaflets 4.5–8 mm, shallowly lobed; upper surfaces dark green with few to many scales; lower surfaces obscured by dense, overlapping scales 0.5–0.8 mm (to 1.0 mm including cilia at tip and base), ovate to narrowly lance-attenuate, brown at middle with irregularly ciliate-fringed membranous margins.

Higher elevations on the north slopes of Sierra Pinacate. Two small populations located: One at 960 m, 1.2 km N of Pinacate Peak (*F 87-43*), with highly localized small colonies among lava rocks in an open, sunny habitat with *Notholaena californica* and *Pellaea mucronata.* The other population at 1100 m, among rock rubble at the northeast base of Pinacate Peak (*F 92-89*), where it is scarce and grows among the locally much more common *A. sinuata* and *Notholaena californica.*

Primarily on calcareous substrates in the Sonoran, Mojave, and western Chihuahuan Deserts; Coahuila, Chihuahua, Sonora, and Baja California, and California to Oklahoma and Texas. The species, with 3 subspecies, ranges southward to the state of Aguascalientes. This subspecies is an apogamous triploid.

Distinguished from *A. sinuata* by its smaller plants, leaves, and leaflets, with leaflets only shallowly lobed. The rhizome scales are similar, but those on the lower leaflet surfaces are larger for *A. sinuata. Astrolepis cochisensis* is known to be toxic to livestock whereas *A. sinuata* is not.

Astrolepis sinuata (Lagasca ex Swartz) D.M. Benham & Windham subsp. **sinuata** [*Acrostichum sinuatum* Lagasca ex Swartz. *Cheilanthes sinuata* (Lagasca ex Swartz) Domin. *Notholaena sinuata* (Lagasca ex Swartz) Kaulfuss] Rhizome scales very narrowly linear-attenuate, mostly 5–7.5 mm, chestnut brown with narrow, membranous (whitish), and minutely toothed margins (darkening with age). Leaves often 20–37 × 3–3.4 + cm; petioles 2–9 + cm. Larger leaflets (15) 18–30 mm, with 3 or 4 conspicuous lobes on each side; upper surfaces olive-green with few to many scales; lower surfaces obscured by dense, overlapping scales 1.6–2.1 mm, narrowly lance-attenuate, brown at midbase with white ciliate-fringed broad membranous margins.

North and northeast side of Pinacate Peak, 1050–1150 m; winter-shaded and locally common in soil pockets on cliffs and narrow ledges, often among brushy, "chaparral-like" vegetation, and in rock rubble at the cliff base, often with *Notholaena californica* and *Pellaea mucronata* (*F 86-449, F 92-90*).

Subsp. *sinuata* ranges from Arizona to Texas to South America, and in the Baja California Peninsula, Georgia, and the West Indies; it is an apogamous triploid. Subspecies *mexicana* D.M. Benham, a sexually reproducing diploid, ranges from western Texas and southeastern New Mexico to Central America.

CHEILANTHES LIP FERN

Small to medium-sized ferns; rhizomes short- or long-creeping and scaly (rhizomes short and thick in ours, with scales that darken with age). Leaves mostly 2- or 3- (5-) pinnate; hairy, scaly or waxy, or

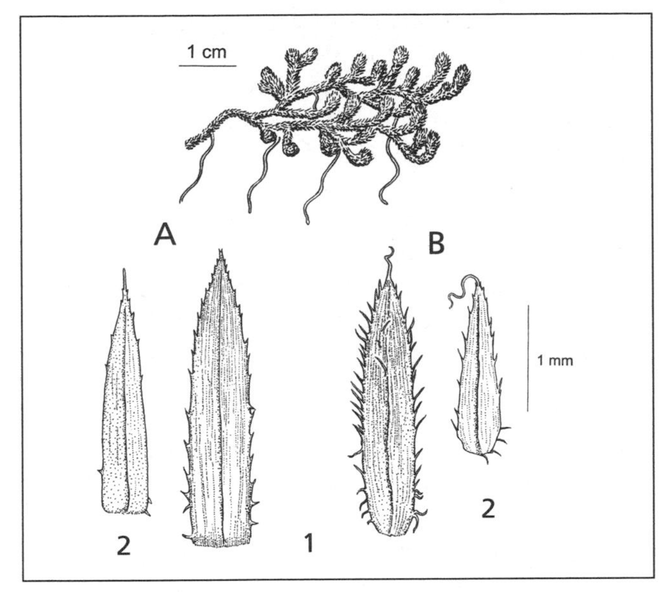

Selaginella:* *(A) S. eremophila*; *(B) S. arizonica. Ventral (adaxial) surfaces of leaves: ***(1)*** from upper side of stem ***(2)*** from lower side of stem. MFW.

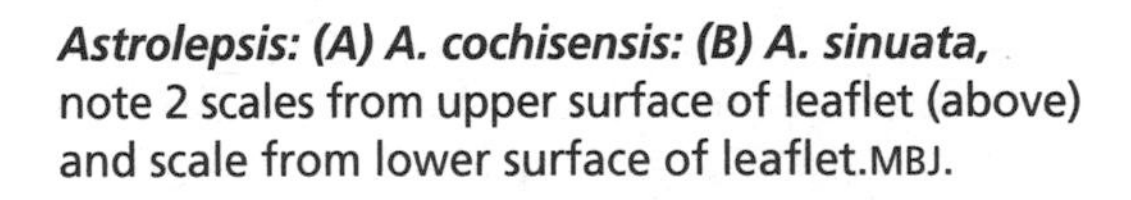

Astrolepsis:* *(A) A. cochisensis:* *(B) A. sinuata, note 2 scales from upper surface of leaflet (above) and scale from lower surface of leaflet. MBJ.

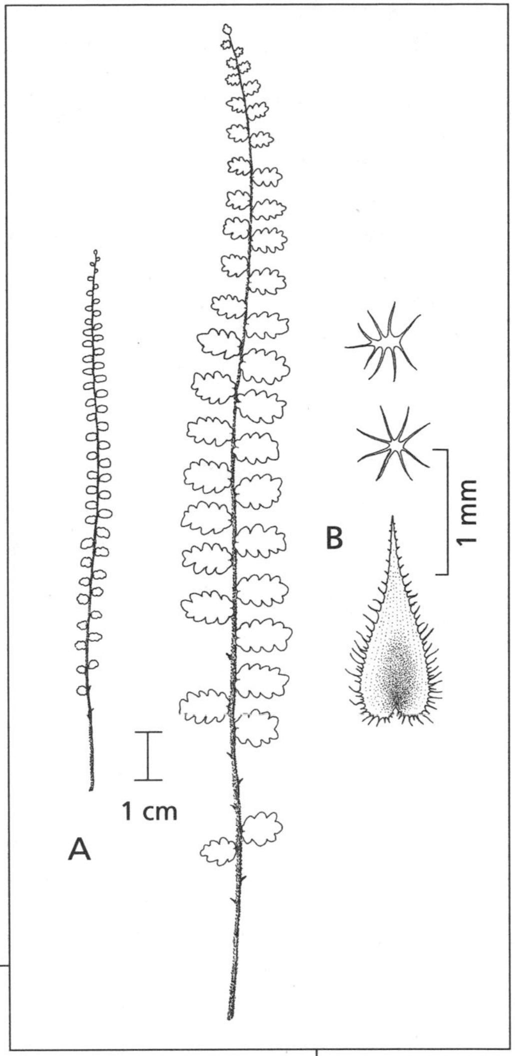

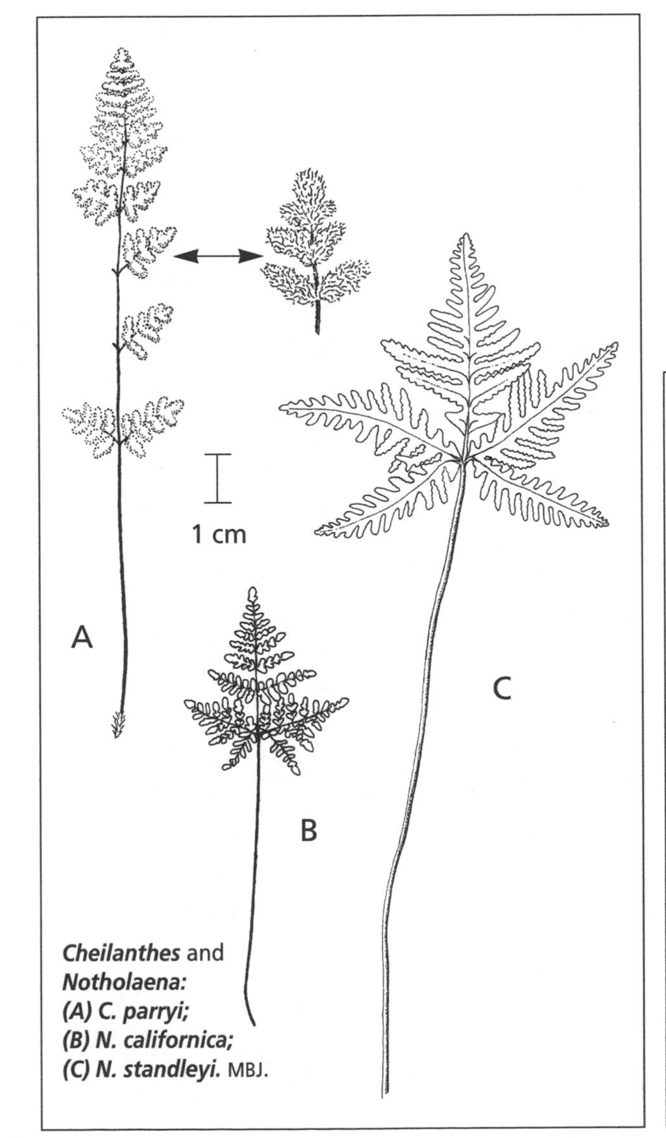

Cheilanthes and ***Notholaena:***
(A) C. parryi;
(B) N. californica;
(C) N. standleyi. MBJ.

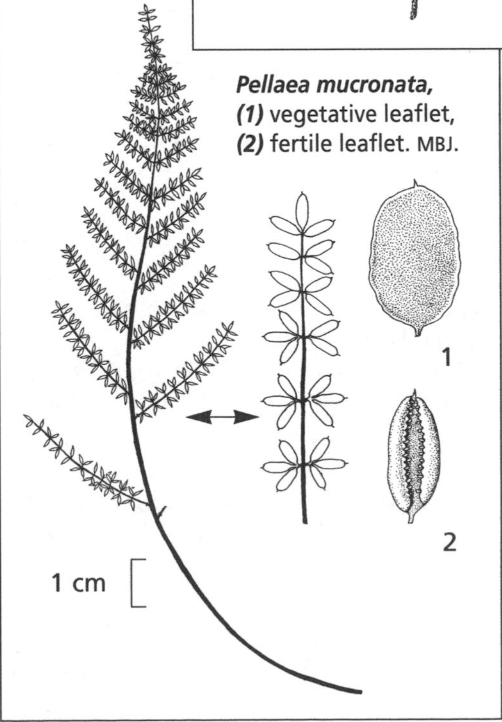

Pellaea mucronata,
(1) vegetative leaflet,
(2) fertile leaflet. MBJ.

glandular, especially on lower surfaces. Sporangia in sori near margins of leaf segments or leaflets, the margins sometimes folded over the sori.

Mostly warm, seasonally dry or semiarid regions, mostly in the Americas, some in deserts; 150+ species. The separation of *Cheilanthes* and *Notholaena* is based on technical but seemingly significant biological characters. References: Mickel 1979b; Windham & Rabe 1993.

Cheilanthes parryi (D.C. Eaton) Domin [*Notholaena parryi* D.C. Eaton] PARRY'S LIP FERN. Rhizome scales with slender blackish centers and orange margins. Leaves (5) 8–15 × 1.8–3.6 cm (fully expanded); petioles slender, dark brown to blackish with sparse white hairs, becoming glabrate with age; blades oblong-lanceolate, twice pinnate, densely woolly; leaflets with hairs obscuring the surfaces and extending beyond the margins, white on the upper surfaces, and white to brown and denser on the lower surfaces.

Locally common on north-facing rocky slopes, mostly at higher elevations, in the Sierra Pinacate, Sierra del Viejo, and relatively rare in the Sierra del Rosario. Probably also in other granitic mountains in northwestern Sonora; not known elsewhere in Sonora. Also desert and semiarid regions in Arizona, southern Nevada, southwestern Utah, southern California, and Baja California.

Sierra Pinacate, N slope, *F 18680.* Elegante Crater, *Webster 22363* (DAV). Sierra del Viejo, *F 85-727.* Sierra del Rosario, *Búrquez 14 Mar 1991.*

NOTHOLAENA CLOAK FERN

Often similar to *Cheilanthes.* Lower leaf surfaces with white or yellow farina (glandular exudate). Sporangia in more or less continuous bands on leaflet margins, the unmodified leaflet margins recurved and partially covering the sporangia.

Mostly in warm, seasonally dry regions, some in deserts. Mostly Mexico and southwestern United States, also Central America, South America, and the Caribbean; 25 species. References: Tryon 1956; Windham 1993a.

1. Leaf segments divided into separate small leaflets; rhizome scales with toothed margins. ____ **N. californica**
1′ Leaves deeply dissected but not divided into separate leaflets; rhizome scales with entire to slightly ragged margins but not toothed. ____ **N. standleyi**

Notholaena californica D.C. Eaton subsp. **californica** [*Cheilanthes deserti* Mickel] CALIFORNIA CLOAK FERN. Rhizome scales dark red-brown, with membranous (translucent) to orange-brown, toothed margins, the center turning blackish with age. Petioles (2.5) 4–12.5 cm, dark brown, glabrous, or with some felt-like hairs when young. Leaf blades triangular-ovate to pentagonal, nearly as wide at base as long, 2–5.5 × 1.5–4.8 cm, 3- or 4-times divided, the pinnae divided into separate small leaflets; upper surfaces bright green to olive-green, dotted with small glands, the lower surfaces often obscured by pale to bright golden yellow glandular exudate (farina).

This is the most common and widespread fern in the region, apparently in all local mountains, especially at higher elevations. Among rocks, usually on north- and east-facing slopes. Especially common on the east and north sides of Sierra Pinacate. Occasionally with *N. standleyi* in granitic mountains in the Sonoyta region but mostly at elevations below the zone of *N. standleyi.* Southern California to the Cape Region of Baja California Sur, southern Arizona, and Sonora south to the vicinity of Bahía Kino. Probably apogamous.

There are 2 chemical races, or chemotypes, which are treated as subspecies. They are distinguished by chemical composition and color of the farina. The second subspecies occurs in California and Baja California, and has white farina.

Sierra de los Tanques, 380 m, *F 89-18.* Sierra Cipriano, near base of mountain, *F 91-20.* Sierra Blanca, *F 87-25.* Tinajas de Emilia, *F 87-38.* N of Pinacate Peak, 960 m, *F 87-45A.* Elegante Crater, *F 19689.* Hornaday Mts, *Burgess 6845.* Sierra del Viejo, *F 16897.* Sierra del Rosario, *F 20668.*

Notholaena standleyi Maxon [*Cheilanthes standleyi* (Maxon) Mickel] STAR CLOAK FERN. Rhizome scales with a dark red-brown center, the margins membranous to orange-brown and entire to slightly ragged, the center turning blackish with age. Petioles (3) 5–13 cm, dark brown, glabrous or with some felt-like hairs when young. Leaf blades pentagonal in outline, more or less as wide as long, 3–7 cm wide, divided into 5 major, deeply cleft pinnate pinnae or segments; upper surfaces olive-green, the lower surfaces often obscured by golden yellow glands. Seigler & Wollenweber (1983) found 3 geographic chemical races of this species. Ours belong to the western "golden race," so called because of the golden-colored exudate or farina on the lower surface of the leaf blades.

Common in granitic *cerros* in the Sonoyta region among Arizona Upland vegetation; mostly on north-facing slopes from the base of the mountains to the summit in the Sierra de los Tanques and in the Sierra Cipriano. At lower elevations sometimes growing with *N. californica.* Also near the northeastern border of the flora area in Organ Pipe Monument and the Sierra Cubabi, and southward in western Sonora to the Guaymas region. Southwestern United States from western Oklahoma and Texas to Colorado and Arizona, and northern Mexico including Baja California and southeastward to Tamaulipas, Durango, and Puebla. The Sonoran Desert plants of this species are usually smaller than those in more mesic regions. Probably sexual diploids.

Sierra Cipriano, *F 91-11.* Sierra de los Tanques, *F 89-17.* Quitobaquito Hills, *Darrow 2421.*

PELLAEA CLIFF-BRAKE

Mostly medium-sized ferns with short- to long-creeping rhizomes. Leaves 1–3 pinnate, mostly glabrous. Leaflets or leaf segments mostly stalked, the margins often folded in drought and/or rolled under to form continuous, poorly defined false indusia protecting the sporangia and extending the length of the segment.

Worldwide, largely temperate to seasonally dry subtropical rocky regions; 40 species. The North American species are in section *Pellaea,* centered in Mexico and southwestern United States. References: Tryon 1957; Windham 1993b.

Pellaea mucronata (D.C. Eaton) D.C. Eaton [*Allosorus mucronatus* D.C. Eaton] BIRD'S-FOOT CLIFF-BRAKE. Rhizomes short and stout, the scales with a blackish midstripe and light brown margins. Leaves 20–30 cm, twice pinnate, or 3-pinnate with some to most of the pinnules (especially the lower ones) on short "branchlets" with 3 leaflets; petioles dark brown, glabrous, terete (among immature leaves sometimes folded, or grooved upon drying). Leaflets often relatively thick and fold longitudinally to enclose the sporangia and the lower surfaces during dry periods (most of the year).

Among rocks and on cliffs at higher elevations (950–1100 m) on the north side of Sierra Pinacate (*F 87-42, 87-51*). Not common and generally growing with *Astrolepis sinuata* and *Notholaena californica.* Western Nevada, California, and Baja California; not known elsewhere in mainland Mexico. The nearest populations are in the Sierra San Pedro Mártir of Baja California and above the Mojave Desert in mountains of southeastern California.

Pellaea mucronata is largely a Californian chaparral element. It is closely related to *P. truncata* Goodding (*P. longimucronata* Hooker), of southwestern United States and northwestern Mexico, including the Sierra San Pedro Mártir in Baja California and the Ajo Mountains of Organ Pipe Monument. *P. mucronata* differs in having smaller, narrower (presumably more xeromorphic) leaflets and pinnae with some trifoliolate pinnules—the latter character found only occasionally in *P. truncata.*

There are 2 weakly differentiated varieties of *P. mucronata.* Among the Pinacate plants, characteristics of both varieties are sometimes present on the same individual, but overall the plants more closely resemble var. *californica* (Lemmon) Windham.

SELAGINELLACEAE SPIKE-MOSS FAMILY

The family consists of a single genus.

SELAGINELLA SPIKE-MOSS

Perennial herbs, usually small, mosslike, and creeping. Leaves small, scalelike, with a single vein, and often ending in a bristle (seta). Sporangia in axils of special leaves, usually (ours) organized into terminal 4-sided miniature conelike structures.

Largely tropical, extending into temperate, arctic, and alpine climates; some in semiarid or even desert regions; 700+ species. Many tropical species are epiphytes.

Only 2 species, *S. arizonica* and *S. eremophila,* truly extend into the core of the Sonoran Desert, mostly at intermediate to higher elevations with northern exposures. These species are in subgenus *Tetragonostachys,* which includes 50 species nearly worldwide. These plants "occupy a unique place in xeric ecology for they are vascular plants adapted to being completely desiccated and reviving a few hours after moisture becomes available" (Tryon 1955:1). "They are mostly plants of dry habitats and they usually do not thrive in the presence of other plants" (Tryon 1955:15). *S. arizonica* and *S. eremophila* are in the series *Eremophilae,* which includes 5 species in the southwestern United States through much of Mexico and Peru to Argentina. In this series the dorsal and ventral leaves are different from each other. This group seems to be relatively advanced among the *Tetragonostachys*. References: Tryon 1955; Wiggins 1971.

Selaginella eremophila Maxon. *FLOR DE PIEDRA*; DESERT SPIKE-MOSS. Plants firm and wiry when dry, softer when wet, visibly green only during brief wet periods. Stems curling up during dry weather and rapidly unfolding when wet (even dry, dead plants open up when wet); rooting along prostrate to semiprostrate stems, forming dense mats close to the ground, mostly 1–3 cm tall, sometimes to 5 cm in shaded areas. Leaves 0.6–2.2 mm (not including the bristle tips); lower surface with a prominent midrib; margins minutely serrated; leaf tips with a slender, squiggly, quickly deciduous bristle.

"The tortuous, delicate, and early deciduous setae set this species off from all others. The setae are attached only on the young leaves in the apical bud and are so inconspicuous that they had been overlooked until C. Weatherby noted them in 1943" (Tryon 1955:80). The setae tend to break off even more readily when the plants have been dried and rehydrated.

Among rocks on locally open, sparsely vegetated granitic slopes in the Sonoyta region. Small, dense but isolated populations occur on north-facing slopes in the Sierra Cipriano and the Sierra de los Tanques to their peaks. The Sierra Cipriano population is large and locally extensive.

Southeastern California in the Colorado Desert, northeastern Baja California, southwestern Arizona in Yuma and western Pima Counties from the Tinajas Altas Mountains to Organ Pipe Monument, and northwestern Sonora south to the Sierra Seri (vicinity 29°17′ N, 112°09′ W).

This species is one of the few pteridophytes endemic to the Sonoran Desert. Mike Windham (personal communication 1998) has discovered that *S. eremophila* actually may be comprised of 2 closely related species.

Selaginella eremophila resembles *S. arizonica* Maxon; the latter can be distinguished by its straight leaf-tip bristle (sometimes once-bent but otherwise straight, stiff, thickish, and persistent, or breaking off above a persistent base). The leaf hairs and leaf shape also differs between the two species, and plants of *S. eremophila* tend to be more compact. *S. arizonica* occurs in the Agua Dulce Mountains in Cabeza Prieta Refuge, Organ Pipe Monument, and the Sierra Cubabi south of Sonoyta. Specimens keying to *S. eremophila* grow intermixed with *S. arizonica* in the Sierra Seri in western Sonora and Alamo Canyon in the Ajo Mountains.

Sierra Cipriano, *F 91-10*. Sierra de los Tanques, *F 89-28*. Quitobaquito, *F 88-113*.

GYMNOSPERMS
CONE-BEARING PLANTS

EPHEDRACEAE JOINT-FIR FAMILY

The family has a single genus.

EPHEDRA

Mostly shrubs, the younger stems photosynthetic and jointed. Leaves scalelike, opposite or whorled. Mostly dioecious (ours); bearing 1 to several small cones (strobili) at the nodes of young branches, the seed (ovulate or female) cones different from the pollen (staminate or male) cones.

Desert and semiarid regions of Eurasia, northern Africa, and the Americas; 60 species, 15 in North America, mostly in the Southwest. Often in dry, temperate regions with moderate to heavy winter freezing and winter-spring rainfall. In Sonora only in the northern part of the state north of 30°30′ N. The Asiatic species *E. sinica* is the original source of the drug ephedrine. Other species in North America and elsewhere have likewise served as important medicinal plants. Reference: Stevenson 1993.

1. Scale leaves 2 per node, about as long as wide; rocky soils. ________ **E. aspera**
1′ Scale leaves 3 per node, more than twice as long as wide; sandy soils. ________ **E. trifurca**

Ephedra aspera Engelmann ex S. Watson [*E. nevadensis* S. Watson var. *aspera* (Engelmann ex S. Watson) L.D. Benson] *TEPOPOTE*; BOUNDARY EPHEDRA. Shrubs 1–1.3 (1.8) m, as broad as tall, the lower limbs woody and sometimes very thick. Branches and twigs usually straight; twigs either all yellow-green or all blue-green (these sometimes becoming yellow-green with age). Leaves 2 per node, 1.5–2.5 (3) mm, fused at base, obtuse, soon fraying and deciduous. Seed cones 2–5 (8) per node, sessile, 1-seeded. "Flowering" March, the seeds ripe in April.

Rocky slopes, mostly toward higher elevations and on north- and east-facing slopes; Sonoyta region, Pinacate volcanic field, and granitic ranges including Sierra del Rosario. Especially common at higher elevations in the Sierra Pinacate. Southward in Sonora to the Sierra del Viejo southwest of Caborca. Also Baja California, southeastern California to Utah and southwestern Texas, and southward to Zacatecas.

Sierra del Rosario, *F 20705.* Sierra del Viejo, *F 85-724.* Sierra Extraña, *F 19059.* Hornaday Mts, *Burgess 6834.* NE slopes of Cerro del Pinacate, *Sanders 5694.* 10 km SW of Sonoyta, *F 88-176.*

Ephedra trifurca Torrey ex S. Watson. *CANUTILLO, TEPOPOTE*; LONG-LEAVED JOINT-FIR. Sprawling shrubs 1–2 (2.5) m, reaching 3–4 m across, usually with several well-developed trunks. Lower branches often decumbent or spreading near the ground, thick and woody, the twigs bright green, long, slender, and arching. Leaves 3 per node, 5–10 mm, fused to ca. 2/3 of their length, narrowly acute, semipersistent, splitting and fraying with age. Cones 1 to several per node, sessile, 1-seeded. "Flowering" late February–March, the seeds ripe in April.

Sandy flats, hills, and low to high dunes; capping sand hummocks and small to medium-sized dunes. Across most of the flora area except the Sonoyta region. One of the few common shrubs on the moving dunes and often the only green plant in sight. Sonora in the flora area and the northeastern corner of the state in Chihuahuan Desert vegetation. Southeastern California and northeastern Baja California to western Texas, northern Chihuahua, and Coahuila.

El Golfo, *F 75-8.* 11 mi E of San Luis, *F 16683.* Dunes near Sierra del Rosario, *Lumholtz 24* (GH). S of Moon Crater, *Webster 24259.* NE of Puerto Peñasco, *F 88-232.* Strand at Estero Morúa, *F 87-20.*

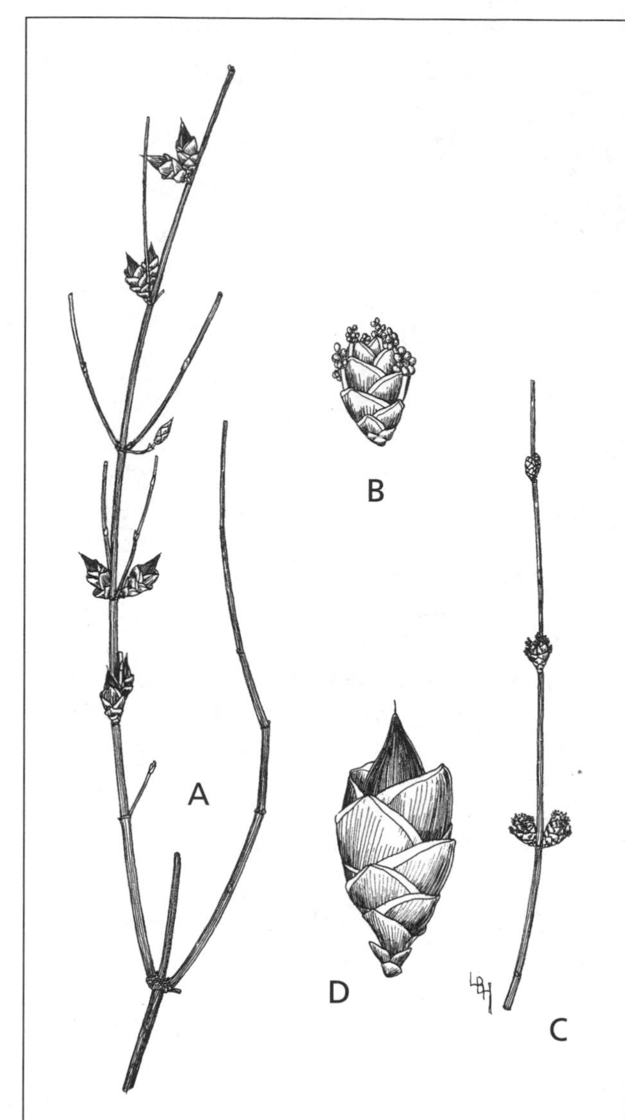

Ephedra aspera: ***(A)*** branch with seed cones; ***(B)*** pollen cone; ***(C)*** branch with pollen cones; ***(D)*** seed cone showing the tip of the single seed. LBH (Benson & Darrow 1945).

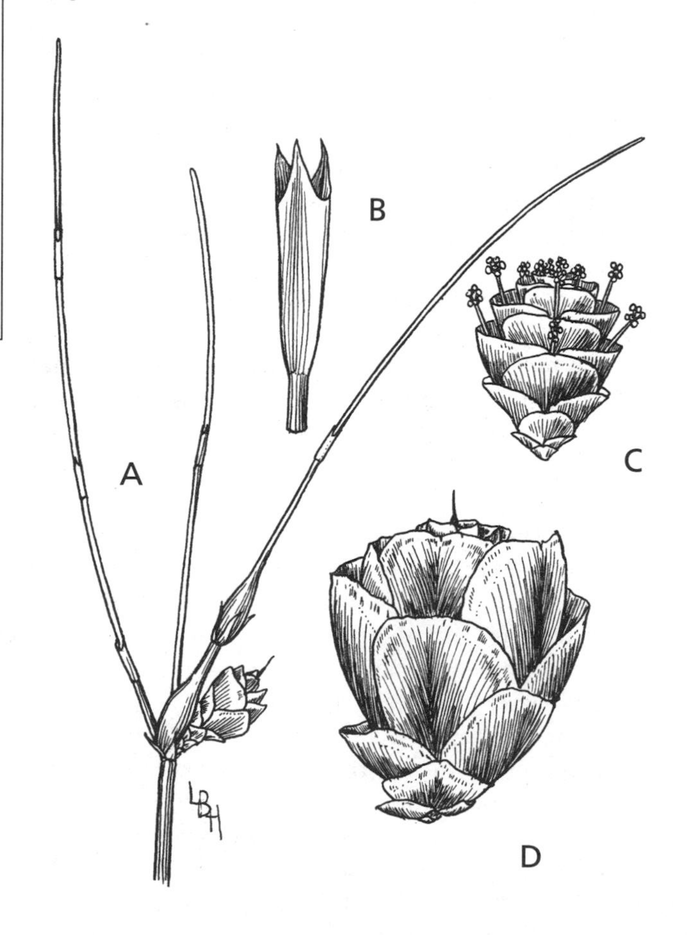

Ephedra trifurca: ***(A)*** branch with a seed cone, the swellings on the stem are insect galls; ***(B)*** a whorl of three fused leaves; ***(C)*** pollen cone; ***(D)*** seed cone. LBH (Benson & Darrow 1945).

ANGIOSPERMS

FLOWERING PLANTS: DICOTYLEDONS

KEY TO THE FAMILIES

1. Plants strictly parasitic; without chlorophyll, or with chlorophyll and the parasitic attachment above ground, conspicuous, and without roots. ______ **Key 1**

1′ Plants not parasitic; with chlorophyll (if partially parasitic then the parasitic attachment below ground and not conspicuous, e.g., *Krameriaceae*).

2. Plants leafless or essentially so, or with scale leaves or with few soon-deciduous and reduced leaves (includes trees, shrubs, and succulents but not annuals or other plants that were leafy at another time—check for leaf scars). ______ **Key 2**

2′ Plants leafy, at least the young growth leafy (when plants are dormant, check for leaf scars or dried leaves).

3. Composites; individual flowers small and borne in a head resembling a single large flower, the head surrounded by a series of somewhat sepal-like bracts forming an involucre (*Stylocline* an exception), the heads often sunflower- or daisy-like, with a central group of tubular disk flowers often surrounded by a ring of ray flowers, or the heads dandelion-like and all flowers with a strap-shaped corolla; ovary inferior, the fruit a cypsela (*achene*), topped by scales or bristles forming the pappus, or the pappus sometimes none: Asteraceae and composite-like plants. ______ **Key 3**

3′ Inflorescence and flowers various but not grouped into heads subtended by a series of bracts forming an involucre (as in 3); usually with green sepals instead of a pappus; (non-composites with composite-like flower heads key in both leads).

4. Vines, the stems twining on other plants or sprawling across the ground and with milky sap or tendrils. ______ **Key 4**

4′ Not vines, without tendrils.

5. Plants conspicuously woody; shrubs or trees.

6. Leaves compound. ______ **Key 5**

6′ Leaves simple.

7. Leaves opposite. ______ **Key 6**

7′ Leaves alternate.

8. Leaves entire. ______ **Key 7**

8′ Leaves toothed to lobed or divided. ______ **Key 8**

5′ Plants herbaceous or only moderately woody at base.

9. Plants with conspicuously succulent leaves and stems. ______ **Key 9**

9′ Plants not conspicuously succulent.

10. Leaves compound. ______ **Key 10**

10′ Leaves simple.

11. Larger leaves basal or in a conspicuous basal rosette (basal rosette leaves may be deciduous), the flowering stems leafless or the stem leaves substantially reduced in size and number. (Note: Borderline cases key out in either lead.) ______ **Key 11**

11′ Midstem leaves well developed; if a basal rosette present, then the stem leaves conspicuous and usually at least 1/4 to 1/3 as large as the basal leaves.

12. Midstem leaves opposite or whorled. ______ **Key 12**

12′ Midstem leaves alternate. ______ **Key 13**

Key 1. Parasitic Plants

1. Stems vining, threadlike, and uniformly yellow or orange; flowers white. ____________ **Convolvulaceae (*Cuscuta*)**
1′ Plants not vining, the stems not threadlike, not yellow or orange; flowers not white.
  2. Plants attached to shrub or tree branches above ground.
    3. Only on indigo bush (*Psorothamnus emoryi*), appearing stemless, the visible parts of the plant (flowers and capsules) less than 5 mm long and purple-brown. ____________ **Rafflesiaceae**
    3′ On various larger shrubs and trees, the stems conspicuous, green or greenish, more than 10 cm long. ____________ **Viscaceae**
  2′ Plants attached to roots underground.
    4. Inflorescence stems widest at apex, button- or mushroom-shaped at surface of sand; flowers radial. ____________ **Lennoaceae**
    4′ Inflorescence stems widest at base, elongated, gradually tapering to apex well above soil surface; flowers bilateral. ____________ **Orobanchaceae**

Key 2. Plants Leafless or Essentially So, or with Scale Leaves

(Note: Generally excluded are plants leafy at another season.)

1. Plants armed with spines, thorns, or spinescent twigs.
  2. Cacti; stems succulent, with areoles bearing spines; leaves either none or succulent and soon deciduous. ____________ **Cactaceae**
  2′ Not cacti; stems not noticeably succulent, without areoles, the nodes spineless or with only 1 spine; leaves none or not succulent.
    3. Ocotillo; tall, wandlike mostly unbranched stems from a common base; nodes with a single stout spine or some without spines. ____________ **Fouquieriaceae**
    3′ Stems not as in 3; nodes unarmed, the plants with spinescent-tipped twigs, irregularly spaced thorns, or both.
      4. Smoke tree; stems dotted with conspicuous lenslike glands. ____________ **Fabaceae (*Psorothamnus spinosus*)**
      4′ Stems not dotted with lenslike glands.
        5. Fruits star-shaped with woody segments (carpels), in dense clusters persisting for several years or more. ____________ **Simaroubaceae**
        5′ Fruits not woody, not persisting.
          6. Composites; flowers stalkless in heads surrounded by multiple bracts (involucre), with white ray corollas and yellow disk corollas; fruits small achenes. ____________ **Asteraceae (*Chloracantha*)**
          6′ Flowers not in heads as in 6, each flower separate on a conspicuous stalk; fruits not small achenes.
            7. Scarcely woody shrubs usually less than 1 m tall; fruits burlike and spiny. ____________ **Krameriaceae**
            7′ Woody shrubs usually more than 1 m tall; fruits not burlike, without spines.
              8. Flowers and fruits without a basal disk, the pedicels not persistent; fruits rounded berries drying capsule-like with 2 seeds; rare in northwestern Sonora (known only from Playa Díaz). ____________ **Koeberliniaceae**
              8′ Flowers and fruits with an expanded basal disk, the pedicels and disk semipersistent; fruit fleshy, 1-seeded. ____________ **Rhamnaceae (*Ziziphus*)**
1′ Plants unarmed.
  9. Stems succulent.
    10. Sap milky; stems not jointed; non-saline desert habitats. ____________ **Asclepiadaceae (*Asclepias*)**
    10′ Sap not milky; stems jointed; halophytes, coastal and alkaline-saline soils. ____________ **Chenopodiaceae (*Allenrolfea, Salicornia*)**
  9′ Stems not succulent.
    11. Nodes with semipersistent scale leaves.
      12. Scale leaves opposite or in 3s. ____________ **Gymnosperms: Ephedraceae**
      12′ Scale leaves alternate. ____________ **Tamaricaceae**

11′ Leaves few, quickly deciduous, not scalelike (plants seasonally leafless or nearly so and thus placed in this key).
  13. Stems conspicuously dotted with large lenslike glands; granitic slopes. ____ **Rutaceae**
  13′ Stems not dotted with large lenslike glands; various habitats.
    14. Composites; flowers stalkless in heads surrounded by multiple sepal-like bracts (involucre), the individual fruits 1-seeded. ____ **Asteraceae** **(*Baccharis sarothroides, Bebbia, Porophyllum, Stephanomeria pauciflora*)**
    14′ Flowers stalked, not surrounded by bracts, the fruits multiple-seeded. ____ **Fabaceae (*Lotus rigidus*)**

Key 3. Composites (Asteraceae) and Composite-Like Plants

1. Perennials with thick rootstocks and above-ground stolons; leaves more than 8 cm; wetland habitats. ____ **Saururaceae**
1′ Ephemerals to perennials, the rootstocks usually not as in 1, without above-ground stolons; leaves various; wetland or dryland habitats.
  2. Individual flowers (florets) producing a single 1-seeded fruit or sometimes some florets sterile; flowers in a single flower head sometimes very different (ray and disk florets); some florets in a flower head or the entire flower head sometimes unisexual; sepals modified into scales, bristles, or hairs, or absent. ____ **Asteraceae**
  2′ Individual flowers producing multiple seeds or two 1-seeded segments; flowers all bisexual; sepals not as in 2.
    3. Leaves linear, sessile; corollas blue. ____ **Polemoniaceae (*Eriastrum*)**
    3′ Leaves not linear, with a conspicuous stalk or petiole and a well-differentiated blade; corollas not blue.
      4. Leaves and bracts with spinescent-tipped lobes. ____ **Apiaceae (*Eryngium*)**
      4′ Leaves and bracts not spinescent. ____ **Polygonaceae (*Eriogonum*)**

Key 4. Vines

1. Plants with stinging hairs or tendrils; flowers unisexual.
  2. Plants with stinging hairs and without tendrils; leaves 1.5–4 cm. ____ **Euphorbiaceae (*Tragia*)**
  2′ Plants without stinging hairs and with tendrils; leaves usually 3.5–20 cm. ____ **Cucurbitaceae**
1′ Plants with neither stinging hairs nor tendrils; flowers bisexual.
  3. Ovary inferior; perianth 3.5–5 cm, of a single, bilateral segment, toothlike above and funnel-like below; leaves arrow-shaped, the sap not milky. ____ **Aristolochiaceae**
  3′ Ovary superior; perianth not as in 3, both calyx and corolla present; leaves not arrow-shaped, or if arrow-shaped then the sap milky.
    4. Ephemerals, the flowers bilateral.
      5. Non-seasonal ephemerals; leaves compound with 3 leaflets, the flower stalks not vining, the flowers pealike with pink corollas. ____ **Fabaceae (*Phaseolus*)**
      5′ Winter-spring ephemerals; leaves simple, the flower stalks long and vining, the flowers snapdragon-like with yellow corollas. ____ **Scrophulariaceae (*Antirrhinum filipes*)**
    4′ Perennials; the flowers radial or nearly so.
      6. Hairs attached at middle with 2 branches extending in opposite directions; flowers bright yellow; fruits with 2 or 3 papery wings. ____ **Malpighiaceae**
      6′ Glabrous or with simple hairs; flowers not bright yellow; fruits not winged.
        7. Sap milky; leaf blades more than twice as long as wide, linear to arrow-shaped; flowers less than 1.4 cm wide, greenish or maroon and white; fruits more than 4 cm, more than twice as long as wide. ____ **Asclepiadaceae (*Cynanchum, Sarcostemma*)**
        7′ Sap not milky; leaf blades about as long as wide, heart-shaped and entire to deeply lobed; flowers 1.5 cm or more in width, blue or pink; fruits less than 1.5 cm, about as long as wide. ____ **Convolvulaceae (*Ipomoea*)**

Key 5. Shrubs or Trees with Compound Leaves

1. Leaves opposite; flowers solitary in leaf axils; fruit a 5-lobed capsule or breaking into 5 one-seeded segments. ____ **Zygophyllaceae (*Fagonia, Larrea*)**
1′ Leaves alternate.
  2. Leaves with 3 leaflets.
    3. Leaflet margins wavy; flowers small, white to pink, the sepals and petals 5 each; inflorescences mostly shorter than the leaves. ____ **Anacardiaceae**
    3′ Leaflet margins entire; flowers showy, yellow, the sepals and petals 4 each; inflorescences much longer than the leaves. ____ **Capparaceae (*Cleome isomeris, Wislizenia*)**
  2′ Most or all leaves with more than 3 leaflets.
    4. Leaflet margins with spine-tipped teeth. ____ **Berberidaceae**
    4′ Leaflet margins mostly entire, not spinescent.
      5. Plants unarmed; trunks and limbs thick, pithy, soft, and semisucculent; leaves highly aromatic, once pinnate, odd pinnate, without pulvini, stipules, and leafstalk glands; fruits 1-seeded drupes. ____ **Burseraceae**
      5′ Plants either armed or unarmed; wood not pithy and semisucculent; leaves not aromatic, once or twice pinnate, odd or even pinnate, often with pulvini on leaf and leaflet stalks or bases; stipules often well developed, the leafstalks sometimes with a prominent gland; fruits pods, mostly multiple-seeded. ____ **Fabaceae**

Key 6. Shrubs or Trees with Simple, Opposite Leaves

1. Leaf margins serrated to toothed or the leaves notched at tip.
  2. Herbage not gummy-resinous; leaf tips acute to obtuse, the leaf margins wavy to toothed. ____ **Lamiaceae (*Hyptis, Salvia mohavensis*)**
  2′ Herbage gummy-resinous; leaves deeply notched at apex (actually 2 fused leaflets with entire margins that appear to be a simple leaf). ____ **Zygophyllaceae (*Larrea*)**
1′ Leaves entire.
  3. Leaves mostly 2.5–3.5 mm, the margins deeply inrolled; coastal habitats. ____ **Frankeniaceae (*Frankenia palmeri*)**
  3′ Leaves mostly more than 5 mm, the margins not inrolled; inland habitats.
    4. Stems square in cross section; corollas dark blue; fruiting calyx enlarged and inflated like a bag. ____ **Lamiaceae (*Salazaria*)**
    4′ Stems rounded in cross section; corollas yellow or red-orange; calyx not inflated.
      5. Leaves all opposite; corollas red-orange; fruits club-shaped, more than twice as long as wide. ____ **Acanthaceae (*Justicia*)**
      5′ Leaves opposite near base, often alternate on upper stems; corollas yellow; fruits ovoid to globose, less than twice as long as wide.
        6. Corollas radial; calyx of 7–10+ linear lobes; capsules of 2 papery spheres opening around middle; Sonoyta region. ____ **Oleaceae**
        6′ Corollas bilateral, snapdragon-like; calyx of 5 lanceolate lobes; capsules solitary, firm, ovoid-acuminate (tapering at apex), opening at apex; higher elevations in the Sierra Pinacate. ____ **Scrophulariaceae (*Keckiella*)**

Key 7. Shrubs or Trees with Simple, Alternate, Entire Leaves

1. Plants armed with spines or thorns.
  2. Ocotillo; essentially trunkless with long, wandlike spiny stems, the spines regularly spaced and similar in size, or some stems without spines; flowers red-orange in terminal inflorescences. ____ **Fouquieriaceae**
  2′ Plants otherwise; flowers not red or orange.
    3. Fruits fleshy or semifleshy (berries or drupes), smooth-surfaced but sometimes constricted near middle; flowers radial, petals present..
      4. Leaf veins conspicuous, usually 3 diverging from leaf base, the pubescence of simple hairs; fruits 1-seeded. ____ **Rhamnaceae**

4′ Leaf veins not readily visible, or leaf veins prominent and pinnate and the herbage densely pubescent with star-shaped hairs; fruits 2- to many-seeded. ______ **Solanaceae (*Lycium, Solanum hindsianum*)**

3′ Fruits dry, winged capsules or bristly burs; flowers bilateral or petals absent.

5. Flowers rather inconspicuous, unisexual, without petals; fruits winged capsules, not spiny; leaves 1–2.5 cm; coastal habitats. ______ **Chenopodiaceae (*Sarcobatus*)**

5′ Flowers bisexual, with showy, purplish sepals and petals; fruits bristly-spiny, globose, not winged; leaves 0.3–1.0 (1.5) cm; widespread. ______ **Krameriaceae**

1′ Plants unarmed.

6. Leaves scalelike, less than 3 mm, clasping or sheathing the stem, mostly with salt-excreting glands. ______ **Tamaricaceae**

6′ Leaves not scalelike, usually more than 3 mm, lacking salt-excreting glands.

7. Plants dotted with yellowish lens-shaped (lenticular) glands; leaves very sparse; flowers dark blue-purple. ______ **Rutaceae**

7′ Plants not dotted with glands; stems (seasonally) leafy; flowers not blue-purple.

8. Plants, or at least the leaves, glabrous or glabrate.

9. Stems relatively thick and very flexible, often with knobby short shoots; sap copious, watery or bloodlike; flowers unisexual. ______ **Euphorbiaceae (*Jatropha*)**

9′ Stems not unusually thick, not especially flexible; sap not especially watery, not bloodlike; flowers bisexual.

10. Leaf blades narrow, mostly narrowly oblong to lanceolate or oblanceolate, more than twice as long as wide.

11. Herbage smooth, not viscid-sticky; flowers on racemes longer than the leaves; fruits comprised of 2 hard, nutlet-like halves, not winged. ______ **Capparaceae (*Wislizenia*)**

11′ Herbage viscid-sticky; flowers in clusters much shorter than the leaves; fruits with 3 papery wings. ______ **Sapindaceae**

10′ Leaf blades broad, mostly ovate to elliptic, less than twice as long as wide.

12. Leaves mostly less than 15 mm, essentially stalkless or stalks less than 1 mm; flowers with a hypanthium and several (rarely 1) separate pistils. ______ **Crossomataceae**

12′ Leaves mostly more than 15 mm, the stalks mostly much more than 1 mm; flowers without a hypanthium and with a single pistil.

13. Leaves (4) 5.5–16 cm; flowers tubular, more than twice as long as wide, yellow and green; seeds tiny and numerous. ______ **Solanaceae (*Nicotiana glauca*)**

13′ Leaves 1.5–4.5 cm; flowers not tubular, wider than long, the petals white; seeds 5 or fewer, not minute. ______ **Phytolaccaceae**

8′ Plants with hairs or scales.

14. Leaves shiny, dark green above, dull gray to brownish below with silvery to brownish scales. ______ **Capparaceae (*Capparis*)**

14′ Upper and lower leaf surfaces similar.

15. Flowers unisexual; leaves and sepals semisucculent, or the leaves scurfy whitish gray with saclike hairs and the female flowers enclosed in winged or scalelike bracts. ______ **Chenopodiaceae (*Atriplex, Suaeda*)**

15′ Flowers bisexual; leaves not as in 15.

16. Hairs either star-shaped or rough and sandpaper-like (plants scabrous).

17. Hairs star-shaped, not rough.

18. Flowers inconspicuous, the corollas white or absent, the anthers small, without pores; fruits 2- or 3-seeded. ______ **Euphorbiaceae (*Croton*)**

18′ Flowers large and showy, the corollas lavender (rarely white), the anthers large (7–10 mm) and with apical pores; fruits many-seeded. ______ **Solanaceae (*Solanum hindsianum*)**

17′ Hairs rough (plants scabrous), simple or branched (dendritic), not star-shaped.

19. Hairs branched (dendritic, like miniature trees); flowers white; calyx of 5 sepals; fruits solitary, cylindrical, indehiscent, 1-seeded, enclosed in bracts. ______ **Loasaceae (*Petalonyx linearis*)**

19′ Hairs simple; flowers yellow; calyx of 7–10+ linear lobes; fruits of 2 papery, 4-seeded hemispheres, not enclosed in bracts. ______ **Oleaceae**

16′ Hairs simple, soft or relatively firm but not rough.

20. Plants with copious milky sap, the stems rather thick, gnarly, and flexible; actual flowers minute, unisexual, aggregated into a flower-like structure predominantly yellow-green, with 5 yellow-green glands bearing white to yellow-white petal-like appendages. ______ **Euphorbiaceae (*Euphorbia misera*)**

20′ Sap not milky; stems brittle or firm, not unusually thick; flowers not as in 20.

21. Flowers often in loose clusters but not surrounded by bracts, purple, bilateral, ca. 15 mm wide; fruits conspicuous spiny burs. ______ **Krameriaceae**

21′ Flowers several, in dense clusters surrounded by bracts, white to pink, the individual flowers radial, length and width ca. 5 mm or less; fruits small, inconspicuous nutlets, not bristly or spiny. ______ **Polygonaceae (*Eriogonum* in part)**

Key 8. Shrubs or Trees with Simple, Alternate, Toothed to Lobed or Divided Leaves

1. Plants armed with spines, prickles, or spinescent twigs.

2. Hairs star-shaped; leaf midribs and stem internodes often with single spines or prickles; flowers large and showy, the corollas lavender (rarely white). ______ **Solanaceae (*Solanum hindsianum*)**

2′ Hairs simple; leaf midribs and stem internodes unarmed, the nodes with single or paired spines, and/or the twigs spinescent or thornlike; flowers small and inconspicuous, the corollas yellowish green or absent.

3. Twigs spine-tipped, the plants otherwise without spines; leaf blades with soft hairs, the margins mostly entire; fruits blackish blue to dark purplish; widespread but seldom common in the flora area. ______ **Rhamnaceae (*Ziziphus*)**

3′ Many twigs with single or paired spines at nodes; leaf blades scabrous; fruits orange; Sonoyta region, highly localized. ______ **Ulmaceae**

1′ Plants unarmed.

4. Plants with branched or star-shaped hairs.

5. Hairs branched (dendritic), white and firm, with whorls of tiny grappling hooks, the herbage like sandpaper in texture and adhering like Velcro; leaves sessile. ______ **Loasaceae (*Petalonyx thurberi*)**

5′ Hairs branched from the base, 2-branched or star-shaped, relatively soft or glassy, the herbage not like sandpaper and not adhering; leaves petioled. (Leaf margins sometimes with inconspicuously toothed or wavy margins, or some leaves entire.)

6. Flowers small, unisexual; fruits 3-lobed, 3-seeded; young herbage with sparse, 2-armed, glassy hairs. ______ **Euphorbiaceae (*Ditaxis brandegeei*)**

6′ Flowers large or small, bisexual; fruits not 3-lobed, with more than 3 seeds; hairs star-shaped, relatively soft, the herbage often densely hairy.

7. Fruits fleshy (a berry), globose, and smooth; corollas lavender, the petals fused; anthers large (7–10 mm) with terminal pores. ______ **Solanaceae (*Solanum hindsianum*)**

7′ Fruits dry (capsules or schizocarps), with hairs or bristles; corollas various colors, the petals separate or united only at base; anthers small, without pores.

8. Fruits more than 5 mm long or wide, globose or not, not studded with blunt spines; flowers showy and relatively large (more than 10 mm wide) with conspicuous petals, variously colored but not maroon, the stamens numerous; widespread and common. ______ **Malvaceae (*Abutilon, Hibiscus, Horsfordia, Sphaeralcea*)**

8′ Fruits ca. 5 mm wide, globose, studded with blunt spines; flowers minute (ca. 5–6 mm wide), the perianth maroon, the stamens 5; localized on rocky slopes, Sonoyta region and higher western granitic mountains. ______ **Sterculiaceae**

4′ Plants with simple hairs or glabrous or glabrate.
9. Leaf blades smooth and relatively firm; flowers bisexual, the petals pink; fruits orange, symmetric drupes, densely glandular; north-facing slopes of the northwestern granitic mountains. **Anacardiaceae (*Rhus kearneyi*)**
9′ Leaf blades not firm; flowers unisexual, the male and females very different; fruits not as in 9; various habitats.
10. Desert habitats; monoecious; seeds not minute, lacking long hairs. **Euphorbiaceae (*Acalypha, Jatropha, Sebastiania*)**
10′ Willows and cottonwoods; wetland habitats with permanent water; dioecious; seeds minute, each with a tuft of long, silky hairs. **Salicaceae**

Key 9. Herbaceous Plants with Succulent Leaves and Stems

1. Perennials.
2. Leaves in a basal rosette; on slopes of granitic rock. **Crassulaceae (*Dudleya*)**
2′ Leaves not in a basal rosette; non-rocky, often saline or alkaline habitats.
3. Leaves alternate.
4. Flowers in terminal, spikelike inflorescences coiled at the tip, the corollas white, small but conspicuous. **Boraginaceae (*Heliotropium curassavicum*)**
4′ Inflorescences not coiled, the flowers inconspicuous in small axillary clusters among the leaves or leafy bracts, lacking corollas. **Chenopodiaceae (*Atriplex barclayana, Suaeda*)**
3′ Leaves opposite.
5. Leaves linear, of equal width nearly throughout. **Chenopodiaceae (*Nitrophila, Suaeda*)**
5′ Leaves linear-oblanceolate to oval, widest at or above middle.
6. Leaves conspicuously stalked, the blades less than twice as long as wide; flowers in umbellate clusters, the perianth bright purple-magenta **Nyctaginaceae (*Abronia maritima*)**
6′ Leaves stalkless or nearly so, at least twice as long as wide; flowers solitary or in fleshy, conelike spikes, the perianth green, partially pink, or none.
7. Leaves moderately flattened, widest at about the middle; flowers bisexual, conspicuous, solitary in leaf axils, the perianth (calyx) lobes 5, pink inside. **Aizoaceae (*Sesuvium*)**
7′ Leaves rounded in cross section, widest well above middle; flowers unisexual, minute, green, several in succulent, conelike axillary spikes, the perianth none or inconspicuous. **Bataceae**
1′ Ephemerals or annuals.
8. Leaves opposite.
9. Plants diminutive, the leaves beadlike, less than 4 mm. **Crassulaceae (*Crassula*)**
9′ Plants not diminutive, the leaves more than (9) 10 mm.
10. Flowers in umbellate clusters, the perianth showy, pinkish; fruits winged, aggregated into a rounded cluster on a prominent stalk. **Nyctaginaceae (*Abronia villosa*)**
10′ Flowers inconspicuous to moderately showy, not in umbellate clusters; fruits not as in 10.
11. Perianth or perianth-like segments pink to white, at least somewhat petal-like; fruits multiple-seeded. **Aizoaceae**
11′ Perianth minute and inconspicuous, green or reddish green, not petal-like; fruits 1-seeded. **Chenopodiaceae (*Monolepis, Suaeda*)**
8′ Leaves alternate.
12. Flowers in spikelike inflorescences coiled at the tip; corollas fused well above the middle to form a conspicuous tube. **Boraginaceae (*Heliotropium curassavicum*)**
12′ Inflorescences not coiled (or moderately coiled in *Monolepis*); perianth segments distinct or fused basally, lacking a conspicuous tube.
13. Perianth inconspicuous, with sepals only; fruits 1-seeded. **Chenopodiaceae (*Monolepis, Suaeda*)**
13′ Perianth often small but conspicuous, with sepals and petals; fruits multiple-seeded. **Portulacaceae**

Key 10. Herbaceous Plants with Compound Leaves

1. Leaves opposite.
  2. Leaflet margins toothed to deeply cut; flowers bilateral, in spikelike to racemose inflorescences. __**Verbenaceae**
  2′ Leaflets entire or sometimes with a spinescent tip; flowers radial, solitary in leaf axils. ________**Zygophyllaceae (*Fagonia, Kallstroemia, Tribulus*)**
1′ Leaves alternate or in a basal rosette.
  3. Flowers in umbellate or dense headlike clusters; fruits comprised of two 1-seeded segments.__________**Apiaceae**
  3′ Flowers not in umbellate or headlike clusters; fruits multiple-seeded, or two 1-seeded segments in *Wislizenia* (Capparaceae).
    4. Sepals 2 or 4, separate or united at base in *Wislizenia*; petals 4 or more; stamens 6 to many.
      5. Sepals 2, falling as the flower opens; flowers and fruits with a ringlike disk (hypanthium). ________**Papaveraceae (*Eschscholzia*)**
      5′ Sepals 4, not falling as the flower opens; flowers and fruits without a hypanthium.
        6. Leaves pinnately much dissected to appear compound. ________**Brassicaceae**
        6′ Leaves palmately or digitately compound with 3 or 5 leaflets.________**Capparaceae (*Cleome, Wislizenia refracta*** subsp. ***refracta*)**
    4′ Sepals 5, united; petals or corolla lobes 5; stamens 5 or 10.
      7. Stipules none; flowers radial, the petals united, the styles 2 or 2-branched; stamens 5.________**Hydrophyllaceae** (in part)
      7′ Stipules usually present; flowers bilateral, the petals separate, the style and stigma unbranched; stamens 10. ________**Fabaceae** (in part)

Key 11. Herbaceous Plants with Simple Basal Leaves

(Note: Leaves mostly in basal rosettes, the stem leaves absent or reduced.)

1. Perennials from a woody rootstock or thickened roots.
  2. Plants with long stolons and creeping, woody rootstocks; leaf blades entire; flowers aggregated into a conelike structure surrounded at base by white, petal-like bracts; wetland habitats. ___________**Saururaceae**
  2′ Plants without stolons, the rootstocks not creeping; leaves deeply cleft; flowers in elongated racemes, dark blue; dryland habitats in the Sonoyta region. ________**Ranunculaceae**
1′ Annuals or ephemerals (winter-spring except *Mollugo*).
  3. Flowers minute, solitary in rigid bracts (involucres) with 3 or 6 apical-spined teeth. ____________**Polygonaceae (*Chorizanthe*)**
  3′ Flowers minute or not, not surrounded by rigid involucres as in 3.
    4. Ovary inferior.
      5. Sepals minute, inconspicuous or spine-tipped, the flowers in umbellate clusters, or grouped 2–6, or in a dense headlike cluster; fruit comprised of two 1-seeded segments. ______________**Apiaceae**
      5′ Sepals conspicuous although sometimes small, not spine-tipped, the flowers not in umbellate clusters or as in 5; fruits multiple-seeded.
        6. Flowers 4-merous (sepals and petals each 4, the stamens 8). ________________**Onagraceae** (in part)
        6′ Flowers 5-merous (sepals, petals, and stamens 5 or multiples of 5) or the stamens and petals many.
          7. Plants glabrous or with few, simple, soft hairs at base. ______________________**Campanulaceae**
          7′ Plants with rough, branched hairs adhering like Velcro. ________________**Loasaceae (*Mentzelia*)**
    4′ Ovary superior.
      8. Diminutive, emergent herbs; flowers single on a slender stalk, inconspicuous, the fruiting receptacle elongated (cylindrical) and spike-like (like a mouse tail), with numerous pistils (the single flower somewhat resembles the many-flowered spikes of *Plantago*); sepals petal-like, spurred; apparently extirpated from the flora area.____________________**Ranunculaceae (*Myosurus*)**
      8′ Flower receptacle not elongated, not resembling a mouse tail; pistils not numerous.

9. Perianth parts 6, in 2 separate whorls; fruits 1-seeded. ____ **Polygonaceae (*Eriogonum*** in part, ***Nemacaulis, Rumex*)**

9′ Flowers not as in 9; fruits with 2 or more seeds, or 1-seeded in some *Cryptantha* (Boraginaceae) and *Achyronychia* (Caryophyllaceae).

10. Sepals separate.

11. Stamens 8 to many; sepals 2, falling as the flower opens; flowers and fruits borne on a ringlike disk. ____ **Papaveraceae (*Eschscholzia*)**

11′ Stamens 10 or fewer; sepals 4 or 5, not falling as the flower opens; flowers and fruits not borne on a ringlike disk.

12. Sepals 4; petals 4; stamens 6 (or fewer in *Lepidium*). ____ **Brassicaceae**

12′ Flowers 5-merous.

13. Winter-spring ephemerals; plants hairy; leaves more than 2 cm, deeply incised or divided; petals 5, pink or violet. ____ **Geraniaceae**

13′ Warm-weather ephemerals; plants glabrous; leaves less than 1.2 cm, entire; petals none. ____ **Molluginaceae**

10′ Sepals united at least near their bases.

14. Corollas bilateral.

15. Flowers in woolly heads with blue corollas, or the corollas pink and the leaves with bristle-tipped lobes. ____ **Polemoniaceae (*Eriastrum, Langloisia setosissima*)**

15′ Flowers not in woolly heads, the corollas white or white and red, the leaves entire or the lobes not bristle-tipped.

16. Flowering stems zigzag, the individual flower stalks longer than the flowers; desert habitats. ____ **Campanulaceae**

16′ Stems straight, the individual flower stalks inconspicuous, shorter than the flowers; wetland habitats or at least on temporarily wet soils. ____ **Scrophulariaceae (*Veronica*)**

14′ Corollas radial.

17. Plants glabrous; stipules and sepals white and papery. ____ **Caryophyllaceae (*Achyronychia*)**

17′ Plants hairy; stipules none, the sepals not white and papery.

18. Flowers 4-merous, the perianth membranous or papery when dry; fruits (capsules) opening around the middle. ____ **Plantaginaceae**

18′ Flowers 5-merous, the perianth not papery; fruits not opening around the middle.

19. Herbage with coarse, often glassy hairs; fruits separating into 4 or fewer 1-seeded nutlets. ____ **Boraginaceae** (in part)

19′ Hairs not coarse or glassy; fruits few- to many-seeded capsules, not separating into nutlets.

20. Style and stigma unbranched; leaves entire, mostly more than 5 cm; corollas white, opening at night. ____ **Solanaceae (*Nicotiana clevelandii*)**

20′ Style branches or stigmas 2 or 3; leaves not as in 20; corollas blue, lavender, or pink (if white then tinged with lavender or pink), open in daytime.

21. Styles 2 or 2-branched. ____ **Hydrophyllaceae (*Eucrypta, Nama, Phacelia*)**

21′ Styles 3-branched. ____ **Polemoniaceae (*Eriastrum, Gilia, Langloisia schottii*)**

Key 12. Herbaceous Plants with Midstem Leaves Simple and Opposite or Whorled

1. Flowers unisexual, inconspicuous.
   2. Sap not milky; fruits 1-seeded, not 3-lobed; female flowers and fruits enclosed in a pair of sepal-like bracts; leaves scurfy gray or whitish due to inflated hairs that collapse upon drying. __________**Chenopodiaceae (*Atriplex*** in part**)**
   2′ Sap milky; fruits 3-lobed, each lobe with a single seed; female flowers and fruits not enclosed in bracts; leaves glabrous or hairy but not scurfy gray. ________**Euphorbiaceae (*Euphorbia, Stillingia spinulosa*)**

1′ Flowers bisexual, inconspicuous or not.
   3. Plants densely white woolly, the hairs branched (dendritic); flowers yellow and minute, without petals (some leaves alternate).__________**Amaranthaceae (*Tidestromia*)**
   3′ Plants glabrous or hairy but not densely woolly, the hairs not branched.
      4. Fruits more than 2 cm.
         5. Sap not milky; fruits 2–5 cm (less than 5 mm wide when more than 3.5 cm long); leaves conspicuously petioled.
            6. Winter-spring ephemerals; fruits 2.5–5 cm, less than 5 mm wide, tapered to a slender beak, at maturity separating into segments and twisting spirally; widespread. __________**Geraniaceae**
            6′ Hot-weather ephemerals; capsules 2–3.2 cm long, 7 mm wide, oblong, not tapered; roadsides, rare. __________**Pedaliaceae**
         5′ Sap milky; fruits at least 5 cm long and more than 5 mm wide; leaves sessile or with very short petioles.
            7. Leaf margins entire and not thickened; wetland habitats. __________**Apocynaceae**
            7′ Leaf margins thickened and ragged (erose); sandy desert habitats. __________**Asclepiadaceae (*Asclepias erosa*)**
      4′ Fruits less than 1.5 cm.
         8. Leaves appearing whorled.
            9. Perennial subshrubs; stems square in cross section with white-margined corners, the herbage rough-scabrous, not glandular. __________**Rubiaceae**
            9′ Ephemerals; stems round in cross section, the herbage glabrous or glandular, not rough-scabrous.
               10. Herbage glandular. __________**Caryophyllaceae (*Drymaria*)**
               10′ Herbage glabrous.
                  11. Warm-weather ephemerals 3–14 cm tall; leaves 3.5–11 mm; stems threadlike; ovary and capsule not open at apex. __________**Molluginaceae**
                  11′ Non-seasonal ephemerals 5–40 cm tall; leaves at least 15–40 mm; stems slender but not threadlike; ovary and capsule open and gaping at apex. __________**Resedaceae**
         8′ Leaves opposite.
            12. Leaf margins toothed, lobed, or deeply cut.
               13. Stems delicate or weak and trailing; leaves alternate above; corollas radial; fruits of delicate, thin-walled capsules.__________**Hydrophyllaceae (*Eucrypta, Pholistoma*)**
               13′ Stems often not as in 13; leaves all opposite; corollas bilateral, although sometimes not conspicuously so; fruits of firm-walled capsules or separating into nutlets.
                  14. Stems rounded in cross section; fruits of capsules with numerous seeds. __**Scrophulariaceae (*Penstemon pseudospectabilis*)**
                  14′ Stems square in cross section; fruits separating into 2 or 4 one-seeded nutlets.
                     15. Styles entire. __________**Verbenaceae**
                     15′ Styles 2-cleft or 2-lobed. __________**Lamiaceae (*Salvia columbariae, Teucrium*)**
            12′ Leaf margins entire or nearly so (sometimes shallowly crenulate or wavy, but not regularly and conspicuously lobed or toothed) or inrolled.
               16. Calyx of 7–10+ linear lobes; capsules of 2 papery spheres opening around middle; lower leaves opposite, the upper ones alternate; Sonoyta region. __________**Oleaceae**
               16′ Calyx of 4 or 5 sepals or lobes; fruits not as in 16; leaves all opposite.

17. Flowers radial.
  18. Leaf margins inrolled; tidal flats. ______________ **Frankeniaceae (*Frankenia salina*)**
  18′ Leaf margins not inrolled; desert or coastal habitats but not in tidal flats.
    19. Plants glabrous; flowers mostly 1–4 cm; fruits 1–1.5 cm, many-seeded; wetland habitats (Quitobaquito and perhaps Río Colorado). ________ **Gentianaceae**
    19′ Plants glabrous or not; flowers less than 1 cm or united into a slender tube more than 10 cm; fruits less than 1 cm, 1- to many-seeded; various habitats.
      20. Leaves narrow, 3 or more times longer than wide, mostly less than 5 (8) mm wide.
        21. Stems and leaves succulent. ______________ **Aizoaceae (*Sesuvium*)**
        21′ Stems and leaves not succulent.
          22. Wetland habitats.
            23. Plants sparsely to densely glandular pubescent; stamens 5. ______________ **Caryophyllaceae (*Spergularia*)**
            23′ Plants glabrous; stamens 4.______________ **Lythraceae (*Ammannia*)**
          22′ Dry, desert habitats.
            24. Flowers diurnal; sepals or petals (if present) notched or deeply cut (fringed) at the apex. ______________ **Caryophyllaceae**
            24′ Flowers nocturnal; calyx and corolla lobes entire. __**Polemoniaceae (*Linanthus*)**
      20′ Leaves not especially narrow, mostly less than twice as long as wide, mostly more than 6 mm wide.
        25. Stipules present; flowers sessile, solitary, and partly enclosed by sheathing stipule and petiole bases; fruits several-seeded, circumscissile capsules. ______________ **Aizoaceae (*Trianthema*)**
        25′ Stipules none; flowers pedicelled and mostly several or more per inflorescence; fruits 1-seeded and indehiscent. ____________ **Nyctaginaceae (except *Allionia*)**

17′ Flowers bilateral.
  26. Flowers lavender, in 3s with 3 closely clustered flowers resembling a single flower; fruit 1-seeded. ______________ **Nyctaginaceae (*Allionia*)**
  26′ Flowers not lavender, not in 3s as in 26; fruits with 4 or more seeds.
    27. Capsules 4-seeded, abruptly narrowed below to a slender stalk (stipe). ______________ **Acanthaceae (Carlowrightia)**
    27′ Capsules many-seeded, not abruptly narrowed below. __________ **Scrophulariaceae (*Mimulus, Penstemon parryi, Veronica*)**

Key 13. Herbaceous Plants with Midstem Leaves Simple and Alternate

1. Fruits 3-lobed (sometimes 1 or 2 lobes or carpels not forming due to abortion), each lobe with a single seed; flowers unisexual, inconspicuous, the stigma 3-branched. (Note: In *Euphorbia eriantha* the flowers are in a flowerlike cyathium or cluster; this species and *Stillingia* differ from all other plants in Key 13 in having milky sap.) ______________ **Euphorbiaceae (*Ditaxis, Croton, Euphorbia eriantha, Stillingia*)**

1′ Fruits other than in 1.
  2. Plants with star-shaped or branched hairs.
    3. Hairs harsh, branched, minutely hooked, the leaves and capsules adhering like Velcro. __________ **Loasaceae**
    3′ Hairs not harsh, not hooked and do not stick.
      4. Hairs branched above base (dendritic); leaves opposite and alternate on the same plant. ______________ **Amaranthaceae (*Tidestromia*)**
      4′ Hairs star-shaped (stellate), branched from base; leaves all alternate.

5. Stamens 6 or more.
   6. Stamens 6; sepals and petals each 4, the sepals separate. ____ **Brassicaceae (*Dimorphocarpa, Dithyrea, Lesquerella, Lyrocarpa*)**
   6′ Stamens many; sepal lobes and petals each 5, the sepals united below. ____ **Malvaceae**

5′ Stamens 5 or fewer.
   7. Corollas conspicuous, white or violet; anthers with terminal pores; fruits fleshy, globose berries. ____ **Solanaceae (*Solanum*)**
   7′ Corollas minute or none; anthers without terminal pores; fruits dry.
      8. Fruits with 2 globose, nearly smooth segments 2–2.5 mm wide; winter-spring ephemerals with very weak herbaceous stems. ____ **Apiaceae (*Bowlesia*)**
      8′ Fruits solitary, globose capsules ca. 5 mm wide, studded with blunt spines; sepals maroon, the petals none; annuals to perennials with tough woody or semiwoody stems. ____ **Sterculiaceae**

2′ Plants glabrous or with simple (unbranched) hairs.
   9. Ovary inferior (or appearing inferior).
      10. Perennials from a carrot-shaped tuberous root; stems prostrate to vining, the leaves arrow-shaped; perianth bilateral with a single toothlike segment 3.5–5 cm. ____ **Aristolochiaceae**
      10′ Plants and leaves not as in 10; perianth radial, with 1 or 2 whorls of 4 or 5 segments.
         11. Fruits with 2 indehiscent 1-seeded segments; perianth 5-merous, often inconspicuous; inflorescences umbellate or aggregated into a conelike or headlike structure. ____ **Apiaceae** (in part)
         11′ Fruits multiple-seeded capsules; perianth conspicuous although sometimes small, the flowers 4- to 6-merous; inflorescences not umbellate or headlike.
            12. Flowers mostly 5- or 6-merous, the sepals ca. 1 mm or less in length, the petals bright purple; wetland habitats at La Salina and perhaps the Río Colorado. ____ **Lythraceae**
            12′ Flowers 4-merous, the sepals more than 2 mm, the petals various colors but not bright purple; widespread. ____ **Onagraceae**

   9′ Ovary superior.
      13. Flowers small, inconspicuous, the perianth none or of only 1 whorl; fruits 1-seeded.
         14. Fruits opening around middle. ____ **Amaranthaceae (*Amaranthus*** in part**)**
         14′ Fruits not opening around middle.
            15. Inflorescence branches and flower stalks relatively thick, sinuous, and corky. ____ **Amaranthaceae (*Amaranthus crassipes*)**
            15′ Inflorescence branches and flowering stalks not as in 15.
               16. Sepals 4; leaves delicate and soft, the stems weak and semisucculent; winter-spring ephemerals. ____ **Urticaceae**
               16′ Sepals 1–5 but not 4; leaves not delicate and soft, the stems not weak and succulent (except sometimes in *Monolepis*, Chenopodiaceae); ephemerals or perennials.
                  17. Stipules absent. ____ **Chenopodiaceae (*Atriplex, Bassia, Chenopodium, Monolepis, Salsola*)**
                  17′ Stipules present, fused, forming sheaths around the stem above the usually swollen nodes. ____ **Polygonaceae (*Polygonum*)**

      13′ Perianth present, with both an outer and an inner series (whorl), often conspicuous; fruit 1- to many-seeded.
         18. Flowers small, the perianth segments 6, all similar, often petal-like; fruits 1-seeded. ____ **Polygonaceae (*Chorizanthe, Eriogonum, Nemacaulis*)**
         18′ Flowers usually conspicuous, with both sepals and petals, the sepals different from the petals; fruits with 2 or more seeds (except in some *Cryptantha*, Boraginaceae).
            19. Flowers, at least the corollas, bilateral; fruits capsules.
               20. Fruits at least 2 cm; herbage mucilaginous.
                  21. Fruits more than 4 cm, woody, tapered at both ends, the apex curved into a long beak splitting into 2 long hooks or claws; corollas lavender-pink or yellow. ____ **Martyniaceae**

21′ Fruits 2–3.2 cm, not woody, cylindrical, opening at apex, not beaked, not curved; corollas white or tinged with pink. ____ **Pedaliaceae**

20′ Fruits less than 1.5 cm; herbage glabrous or hairy, but not mucilaginous (viscid in *Antirrhinum cyathiferum* and *Mohavea*, Scrophulariaceae).

22. Flowers in woolly heads; corolla lobes blue; stamens 5. ____ **Polemoniaceae (*Eriastrum*)**

22′ Flowers not in woolly heads; corollas yellow, lavender, or pink; stamens 2–4.

23. Plants glabrous; leaves linear-filiform; ovary gaping open at apex; petals minute and white; stamens 3. ____ **Resedaceae**

23′ Plants glabrous or glandular hairy; leaves broader, not linear-filiform; ovary not gaping open; petals showy and brightly colored, not white; stamens 2 or 4. ____ **Scrophulariaceae (*Antirrhinum, Mohavea*)**

19′ Flowers radial; fruits capsules or not.

24. Course, robust annual to perennial herbs, the herbage, sepals, and fruits densely prickly-spiny; petals white or yellow, 2.5 cm or more. ____ **Papaveraceae (*Argemone*)**

24′ Plants not as in 24, not densely spiny.

25. Roots thick, blackish, and deep; perennials; stems prostrate or matlike; leaf veins conspicuously incised; corollas lavender. ____ **Boraginaceae (*Tiquilia*)**

25′ Roots not as in 25; ephemerals or annuals; stems mostly not prostrate or matted; leaf veins not conspicuously incised; corollas lavender or not.

26. Sepals and petals all separate; stamens 6 or more (except *Lepidium*, Brassicaceae), more numerous than the petals.

27. Stamens 6 (or fewer in *Lepidium*); sepals and petals each 4, the sepals not falling as the flower opens. ____ **Brassicaceae**

27′ Stamens 8 to many; petals usually 5, the sepals 2 or 3, falling as the flower opens. ____ **Papaveraceae**

26′ Sepals and/or petals fused basally; stamens 2 or 5, as many or fewer than the corolla lobes.

28. Calyx parted into 6–10+ narrow segments; stamens 2; fruits with 2 papery hemispheres. ____ **Oleaceae**

28′ Calyx 5-lobed; stamens 5; fruits not as in 28.

29. Inflorescence branches coiled at tips.

30. Seeds 4 or fewer per fruit; style and stigma unbranched. ____ **Boraginaceae** (in part)

30′ Seeds more than 4 per fruit; stigma 2-branched. ____ **Hydrophyllaceae** (in part)

29′ Inflorescence branches not coiled.

31. Style undivided, the stigma 3-cleft. ____ **Polemoniaceae**

31′ Styles and stigmas unbranched or bifid (2-branched or 2-cleft).

32. Styles and stigmas unbranched.

33. Seeds 4 or fewer per fruit. ____ **Boraginaceae**

33′ Seeds many per fruit. ____ **Solanaceae (*Calibrachoa, Nicotiana, Physalis, Datura, Solanum*)**

32′ Styles or stigmas 2 or 2-branched (bifid).

34. Leaves petioled, the blades not entire (margins sometimes merely wavy). ____ **Hydrophyllaceae (*Eucrypta, Phacelia* in part, *Pholistoma*)**

34′ Leaves sessile or with short petioles, the blades entire. ____ **Convolvulaceae (*Cressa, Evolvulus*)**

ACANTHACEAE ACANTHUS FAMILY

Herbs, shrubs, sometimes vines, rarely trees. Leaves usually opposite (ours) and decussate, simple, mostly entire; stipules none. Flowers subtended by bracts, radial to bilateral (ours), the petals and sepals united below. Stamens often 2 (ours). Ovary superior. Fruit a capsule, usually stalked (stipitate), 2-chambered, elastically dehiscent; mostly (ours) with retinacula (hook-like structures subtending the seeds). Seeds 2 to several.

Worldwide, mostly in tropics and subtropics, few in temperate regions; 230 genera, 4000 species. About 35 species occur in southeastern Sonora and southwestern Chihuahua (Gentry 1942; Martin et al. 1998) but only 2 extend into northwestern Sonora. Major factors limiting their distribution in northwestern Sonora seem to be the paucity of summer rains and freezing weather. Reference: Daniel 1984.

1. Subshrubs usually less than 1 m; flowers white with purple and yellow markings, the petals spreading; capsules 1 cm. ____ **Carlowrightia**
1′ Shrubs usually more than 1.5 m; flowers red-orange, tubular; capsules 1.5–2 cm. ____ **Justicia**

CARLOWRIGHTIA

Subshrubs and perennial herbs with woody bases. Flowers relatively small. Calyx 5-lobed. Corollas strongly bilateral and 2-lipped (ours) to nearly radial. Capsules flattened, club-shaped, conspicuously narrowed at base. Seeds flat, disk-shaped, 4 or fewer.

Central America to southwestern United States; 12 species, mostly in Mexico, 6 in the Sonoran Desert. At least half of the species are self-compatible, and if the weather is not hot enough they can produce cleistogamous flowers. References: Bourell & Daniel 1988; Daniel 1983, 1988.

Carlowrightia arizonica A. Gray [*C. californica* Brandegee. *C. cordifolia* A. Gray] *LEMILLA*. Perennial subshrubs often 15–30 cm, or to 1 m in the protection of a spiny shrub, much-branched, often leafless. Stems slender and brittle; herbage densely pubescent with minute hairs and an inconspicuous understory of minute glands. Leaves quickly drought deciduous, sessile to petioled, 0.6 (dry seasons)–5 cm (hot, wet weather); blades mostly lanceolate. Floral bracts persistent. Corollas ca. 1 cm in diameter, somewhat pealike, falling as a unit, white with a yellow eye and purple guide lines on upper lip (formed by 2 fused petals). Capsules ca. 1 cm, glabrous; seeds 4, ovate- to cordate-discoid, 3 mm wide, the surfaces conspicuously papillose. Mostly late spring and summer rainy season.

Sonoyta region, Pinacate volcanic complex, and granitic ranges. Mostly along large washes, canyons, and on rocky slopes, especially north-facing; absent from the more arid, exposed habitats. Costa Rica to west Texas to southern Arizona, disjunct in Anza-Borrego in San Diego County, California. The geographic range essentially spans that of the entire genus.

In open places it is almost always grazed by rabbits, rodents, and especially chuckwallas, which reduce the plant to a mass of short, stubby stems. The corollas snap open about sunrise during hot weather and fall by late morning or with the heat of the day. It is a "taxonomically complex species with numerous diverse morphological forms" (Daniel 1984:167).

N side of Sierra del Viejo, *F 85-723*. Tinajas de los Pápagos, *F 86-485A*. Hill at Pinacate Junction, *F 9689*. Aguajita, *F 88-427*.

JUSTICIA

Perennial herbs or shrubs. Herbage often with cystoliths (silica deposits) in epidermal and interior tissue cells. Leaves petioled; blades lanceolate to ovate. Calyx 4- or 5-lobed. Corollas 2-lipped, the tube expanded (ampliate) above, with a furrow on the upper (dorsal) portion in which the style rests. Anther sacs often unevenly attached to the filament. Seeds 2 or 4. Worldwide, mostly tropical; 600 species. This is the largest genus in the family.

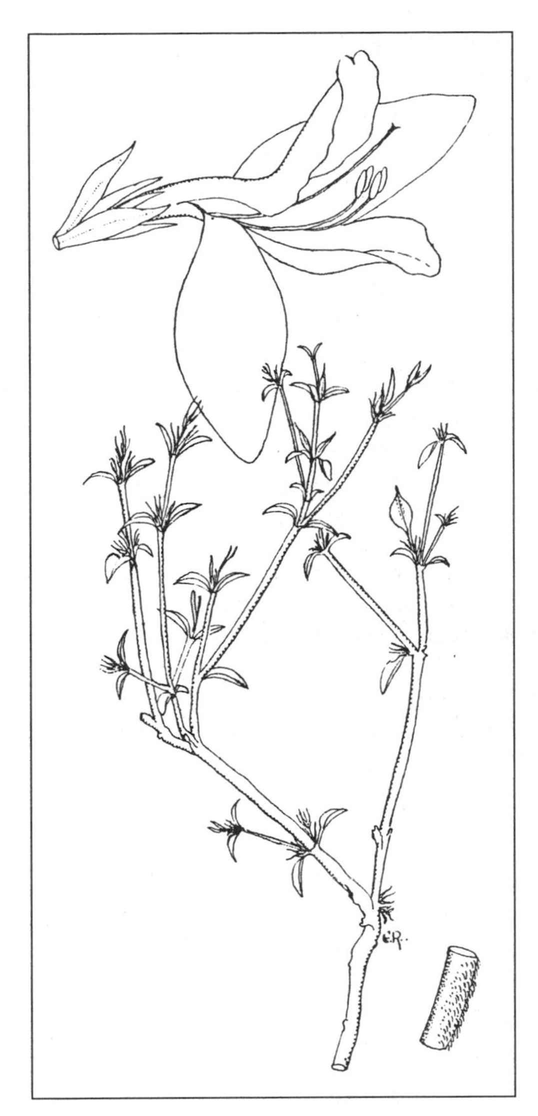

Carlowrightia arizonica FR. © James Henrickson 1999.

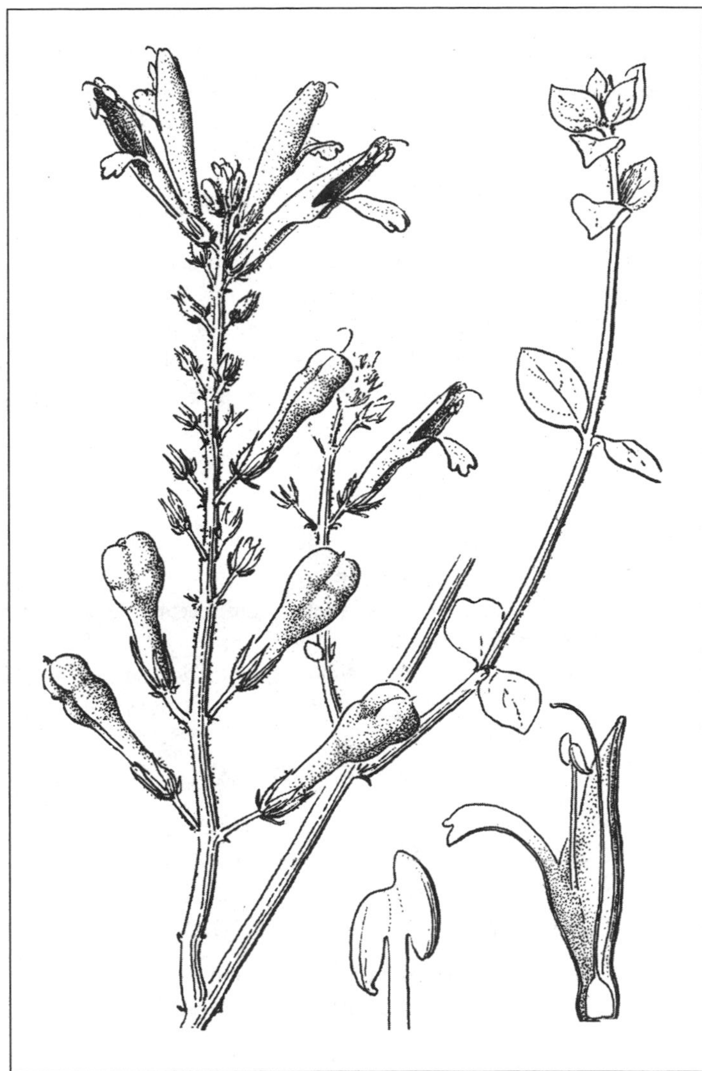

Justicia californica. (Abrams & Ferris 1960).

Justicia californica (Bentham) D.N. Gibson [*Beloperone californica* Bentham] *CHUPARROSA;* HUMMINGBIRD BUSH; *WIPISIMAL*. Sprawling shrubs often 1–2.5 m, forming mounds of loosely interlacing stems. Stems brittle, pale blue-green, densely pubescent with short hairs, essentially leafless much of the year. Leaves highly variable in size and number and very quickly drought deciduous, 2–8 cm; petioles prominent; blades ovate to deltate, often moderately thick, the margins entire or nearly so. Corollas 2.5–4.0 cm, orange-red, with a slender tube flaring above. Capsules 1.5–2.0 cm, club-shaped, pubescent. Seeds 4, rounded, 3.0–3.3 mm in diameter, glabrous. Various seasons; mass flowering often March and sometimes November-December.

Across much of the flora area except dunes; Sonoyta region, Pinacate volcanic complex, and granitic ranges. Mostly along washes and arroyos or canyons including their slopes, and on cinder flats and slopes at higher elevations in the Sierra Pinacate. Especially common along major gravelly watercourses with palo verde (*Parkinsonia florida*), desert lavender (*Hyptis*), and ironwood (*Olneya*), where mornings during spring flowering turn into hummingbird battlefields. The flowers attract Costa's, Rufous, and Allen's hummingbirds, as well as honeybees and butterflies. *Chuparrosa,* derived from *chupar* 'to suck', is used in Mexico for various hummingbird-adapted flowers (see Felger & Moser 1985). Carpenter bees (*Xylocopa*) and sometimes hummingbirds slit the floral tube and consume nectar, perhaps without providing pollination.

Thornscrub in the Cape Region of Baja California Sur and northwestern Sinaloa and northward in the Sonoran Desert to southeastern California and southwestern Arizona.

Base of Sierra del Rosario, *F 20738*. 12 mi SW of Sonoyta, *F 9824*. Arroyo Tule, *F 19228F*. 0.5 km S of Carnegie Peak, *F 19881*. W of Campo Rojo, *Soule 30 Jan 1983*.

AIZOACEAE AIZOON FAMILY

Annual or perennial herbs and shrubs; mostly succulents, some halophytes. Leaves opposite (in the flowal area) or alternate, without stipules or sometimes stipule-like structures (here called stipules). Flowers radial. Sepals mostly 5. Petals none but numerous petal-like staminodes often present. Stamens 5–many. Ovary superior to inferior. Fruit usually a dry capsule. Seeds mostly kidney- to pear-shaped.

Mainly dry subtropics and tropics, especially the Southern Hemisphere, mostly Old World; 130 genera, 2000 species. Great diversity in dry regions of southern Africa with myriad bizarre, often very small "living rock" succulents such as *Conophytum* and *Lithops* and their relatives. Several trailing ice-plants, including *Carpobrotus edulis,* are cultivated at Puerto Peñasco. Reference: Bogle 1970.

1. Winter-spring ephemerals, extremely succulent; flowers with numerous slender petal-like staminodes; capsules opening at top by valves. ______ **Mesembryanthemum**

1′ Hot-weather ephemerals or perennials, semisucculent to succulent; staminodes none; capsules opening around the middle.

  2. Leaves subsessile or short petioled, the blades more than twice as long as wide; stipules none. ______ **Sesuvium**

  2′ Leaves conspicuously petioled, the blades about as wide as long; stipules present. ______ **Trianthema**

MESEMBRYANTHEMUM *HIELITOS,* ICEPLANT

Diverse low, succulent annual or perennial herbs. Leaves highly succulent; stipules none. Flowers solitary in leaf axils or in few-flowered inflorescences. Staminodes petal-like, slender, numerous, often showy; fertile stamens many. Ovary half-inferior with 5 carpels. Fruits often fleshy at first, becoming dry capsules with valves at the flattened top opening when moist. Old World; 70 species.

1. Leaves flat, broader than thick, the blade and petiole evident. ______ **M. crystallinum**

1′ Leaves cylindric, sessile (petiole and blade not distinguishable). ______ **M. nodiflorum**

***Mesembryanthemum crystallinum** Linnaeus [*Cryophytum crystallinum* N.E. Brown. *Gasoul crystallinum* (Linnaeus) Rothmaler] *HIELITOS;* CRYSTAL ICEPLANT. Winter-spring ephemerals; spectacularly fleshy with large, watery, crystal-like vesicles, green or green and red, highly variable in size, usually less than 15 cm tall. Stems often 5–20 (30) cm, forming spreading mats. Leaves often 2–10 cm, narrowed basally or petiolated; blades usually obovate to broadly spatulate, or somewhat triangular. Flowers ca. 1 cm wide, the staminodes white. Capsules ca. 1 cm wide, rounded like a clenched fist when dry, quickly gaping open by slitlike valves when wet (even on years-old dry specimens), expanding to ca. 1.5 cm to reveal the seeds and closing again when dry. Seeds 0.85–1.1 mm, numerous, light brown, slightly roughened.

Widely scattered, often well-established local populations in sandy, mostly disturbed habitats including roadsides and upper beaches; sometimes in natural areas. Near the Río Colorado (e.g., near El Golfo, Riito, and San Luis), vicinity of Puerto Peñasco, and in spring 1995 first recorded near Sonoyta. Elsewhere in Sonora at Las Guasimas (south of Guaymas) and near El Desemboque San Ignacio. Not recorded in Sonora prior to the 1970s, and in Arizona known from a few widely scattered places (Pinkava et al. 1978). Native of South Africa, extensively naturalized along the Pacific coast of the Californias and South America, as well as the Mediterranean region.

5 km E of Puerto Peñasco, 19 May 1973, *F 20865*. San Luis, *F 93-211*. La Abra Plain, near boundary marker 170, *Rutman 11 May 1995*.

***Mesembryanthemum nodiflorum** Linnaeus. SLENDER-LEAF ICEPLANT. Winter-spring ephemerals; extremely succulent. Stems to 25+ cm. Leaves 10–28 mm, cylindric, sessile. Flowers ca. 5 mm wide, the staminodes white, fading yellow. Capsules obovoid, ca. 5 mm; seeds brown, 0.7–0.8 mm.

First recorded for Sonora in spring 1995, west of Sonoyta. During this El Niño year Sue Rutman found many thousands of these plants together with *M. crystallinum* spreading across the international border from the nearby roadside along Mex 2. Native to South Africa, naturalized along the Pacific coast of the Californias and South America, as well as in Australia. Occasional elsewhere in southern Arizona.

La Abra Plain, near boundary marker 170, *Rutman 11 May 1995.*

SESUVIUM SEA PURSLANE

Annual or perennial herbs or subshrubs; succulent and glabrous. Stipules none. Flowers terminal but appearing axillary, single or clustered, sessile to pedicelled. Calyx fleshy, united below, the lobes 5, pink to purple inside, with membranous margins and a fleshy horn on back near apex. Stamens 5 to many. Ovary superior, mostly 3 carpelled. Capsules membranous, enclosed by persistent sepals, circumscissile. Seeds numerous, often rounded to kidney-shaped, mostly smooth and shiny, rarely rough, enclosed by a thin, persistent, membranous aril, each seed on a tiny persistent stalk (funiculus).

Tropics to warm-temperate regions, mostly New World; 8 species. Mainly coastal or inland on alkaline or saline soils.

1. Perennials; stems much-branched, rooting at some nodes of trailing or buried stems; maritime wetlands or beaches; flowers pedicelled; seeds 1.1–1.3 mm, the surfaces usually sculptured. ____________ **S. portulacastrum**

1′ Annuals or perennials; stems radiating from the single rootstock; inland wetland, alkaline, or agricultural habitats; flowers sessile or nearly so; seeds 0.8–0.9 mm, the surfaces usually smooth. ____________ **S. verrucosum**

†**Sesuvium portulacastrum** (Linnaeus) Linnaeus [*Portulaca portulacastrum* Linnaeus. *Sesuvium sessile* Persoon] SEA PURSLANE. Tidally wet, saline soils along the shores of the Gulf of California and Pacific coast of Baja California Sur. Sonora northward at least to the estero at Puerto Lobos (*F 20310*) and expected farther north. The maritime, long-lived *Sesuvium* from the Gulf of California region differs from other *S. portulacastrum* in having sculptured rather than smooth seeds and may represent an undescribed taxon (James Henrickson, personal communication 1995). *S. portulacastrum* is primarily a coastal species from Florida to Texas and the Gulf of California to South America, the Caribbean, and in the Old World tropics.

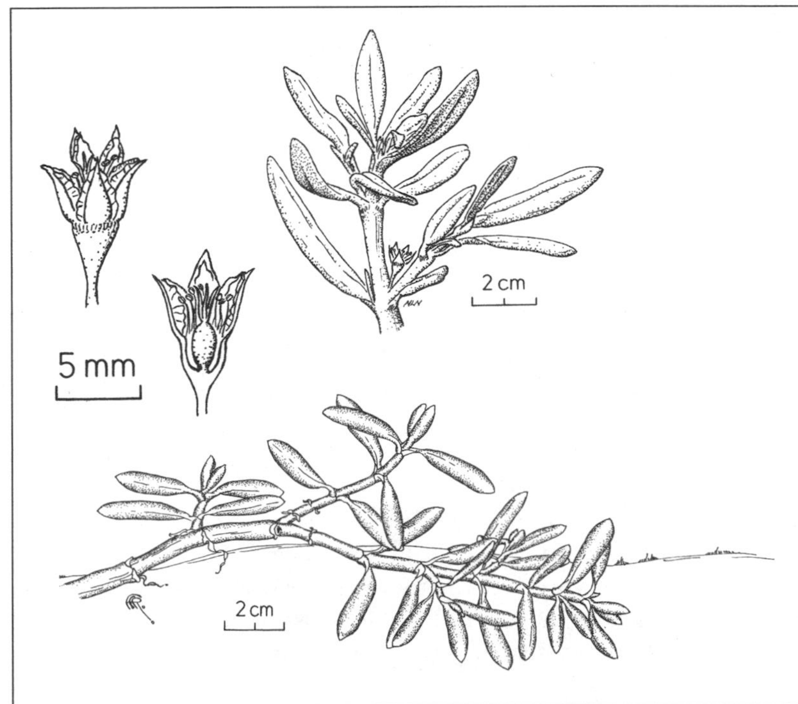

Sesuvium portulacastrum.
NLN and FR (Felger & Moser 1985).

Sesuvium verrucosum Rafinesque. WESTERN SEA PURSLANE. Warm-weather ephemerals or short-lived perennials. Stems radiating from a weak taproot on ephemeral or first-season plants, and from a basal caudex on perennials. Herbage often red-green. Leaves narrowly to broadly spatulate, succulent and thick but conspicuously flattened, the larger leaves 1.5–5.0 cm, subsessile to short petioled, the leaf bases with broadly winged, membranous margins. Flowers sessile or essentially so, 8–11 mm. Sepals thick and succulent, the lobes pink and petal-like inside. Stamens many, the filaments pink and fleshy (appearing flat or winged when dry), the anthers often bright magenta. Seeds 0.8–0.9 × 0.65–0.7 mm, obovoid, markedly broader above the middle, not conspicuously notched, the surfaces dark red-brown, smooth and shiny; funiculus persistent.

Alkaline and semi-saline soils in wetland and agricultural areas of the Ciénega de Santa Clara margins and elsewhere in the lower Río Colorado valley to the Imperial Valley. Northern Mexico and southern United States; mostly an inland species on alkaline or saline soils. In Sonora sometimes near the coast but not in tidally inundated or marine habitats. For example, it is common near the mouth of the Río Yaqui but not on wetland soils bordering the mangrove esteros where one finds *S. portulacastrum.* At least in Sonora the stems and leaves of *S. verrucosum* are generally not as thick and succulent as those of *S. portulacastrum.*

Lerdo, *Price 3 Dec 1898* (DS). Ciénega de Santa Clara, *F 90-200.* Farmland S of San Luis, *F 85-1041.*

TRIANTHEMA

Annual or perennial herbs. Leaves opposite, those of each pair unequal; stipules present or absent. Sepals 5, united below, often petal-like, the lobes often with a fleshy horn on back near apex. Stamens 5–many; staminodes none. Ovary superior, 1-carpelled. Fruit a circumscissile capsule, the seeds 2 to several; capsule lid hard and retaining 1 or more seeds, the other seed(s) remaining in the basal portion (thus two strategies for dispersal). Tropics and warm-temperate regions; 20 in the Old World and 1 also in the New World.

Trianthema portulacastrum Linnaeus. *VERDOLAGO DE COCHI;* HORSE PURSLANE; *KAṢVAÑ.* Hot-weather ephemerals; semisucculent, usually reddish green. Stems first ascending then spreading, relatively weak and prostrate, often 15–60 cm. Leaves often 3.0–4.5 (9) cm on robust, young plants, the older plants with smaller and usually thicker leaves to ca. 2 cm; petioles prominent; blades obovate to orbicular. Stipules and expanded leaf bases forming a membranous sheath around the stem. Flowers solitary, sessile, partly enclosed by sheathing leaf bases. Calyx lobes petal-like, 2.3–12.5 mm, pink with a green horn. Anthers pink-violet. Style with spreading transparent hairs or papillae, thus appearing bearded. Capsules several-seeded; cap truncate or more often 2-horned, retaining 1 or 2 seeds. Seeds 2 mm, kidney-shaped, blackish brown, dull, and rough.

Common and widespread in the lowland desert including the Sonoyta region, and in and near the volcanic and granitic ranges. Especially common on low, semi-saline or sandy soils, and also gravelly and rocky soils; often abundant in disturbed habitats. Widespread and often weedy in the Americas, possibly introduced from the Old World in post-Columbian times.

Two forms are evident. One is larger and more robust, with larger organs including leaves and flowers, the sepals often 10.0–12.5 mm. The other and more common form is smaller with parts all or mostly smaller, the sepals as small as 2.5 mm. Both forms often occur intermixed in the flora area as well as elsewhere in Sonora and nearby Arizona. Intermediate forms appear to be absent.

Sonoyta, *F 85-942.* Aguajita: *F 88-431* ("giant" form); *F 88-440* ("normal" form).

AMARANTHACEAE AMARANTH FAMILY

Mostly herbs, some shrubs, rarely trees or vines. Leaves alternate or opposite, simple, often entire; stipules none. Bracts usually subtending the flowers, these usually dry, chaffy scales and sometimes colorful. Flowers small, bisexual or sometimes unisexual, usually radial. Sepals usually 4 or 5, often dry, membranaceous and not colorful; petals none. Ovary superior. Fruit 1-seeded, usually an achene, small nut, or circumscissile capsule. Seeds usually shiny.

Worldwide, mostly tropical and subtropical, some in temperate and arid regions; 71 genera, 750 species. References: Robinson 1981; Standley 1917.

1. Leaves alternate, green or reddish green, glabrous or with short simple hairs, the surfaces clearly visible; flowers unisexual. ______ **Amaranthus**

1′ Leaves opposite or alternate, densely white woolly with branched hairs obscuring the surfaces; flowers bisexual. ______ **Tidestromia**

AMARANTHUS

Annuals with well-developed taproots. Leaves alternate, petioled. Dioecious or monoecious. Fruits small, indehiscent or circumscissile (splitting around the middle. Seeds lens-shaped and compressed, erect in the fruit, smooth, red-brown to blackish among wild amaranths, usually white or light colored among domesticated grain amaranths.

Worldwide, mostly native to the New World, often weedy; 60 species. The dioecious amaranths, including *A. palmeri* and *A. watsonii,* seem to have expanded into new weedy habitats. The grain amaranths, long ago domesticated in North America, are derived from the dioecious amaranths. Castetter & Bell (1951:189) reported that a grain amaranth, which they called *A. caudatus* Linnaeus, was cultivated along the lower Colorado River by the Cocopa. It is more likely to have been *A. hypochondriacus* Linnaeus (see Sauer 1967). Seeds of most wild amaranths were important food resources, and the young shoots were utilized as greens (e.g., Castetter & Bell 1951; Felger & Moser 1985). Ours are hot-weather ephemerals or annuals. References: Sauer 1955, 1967.

1. Fruits indehiscent, scaly papillose to warty near tip; inflorescence branches and pedicels sinuous, becoming thickened and corky; stems ascending, sprawling, or prostrate, the plants mostly small, the longer stems weak. ______ **A. crassipes**

1′ Fruits dehiscent (circumscissile), smooth-surfaced; inflorescence branches and pedicels mostly straight, not thickened and corky; stems usually erect or ascending, the plants often robust, the larger stems not weak.

  2. Calyx not urceolate (urn-shaped), the pistillate sepals not spatulate, not narrowed into a claw at the base; flowers in axillary clusters. ______ **A. albus**

  2′ Calyx urceolate, the pistillate sepals spatulate, narrowed into a claw just above the base; flowers in both axillary clusters and terminal spikes.

    3. Monoecious; inflorescences "soft," the bracts not spiny or stiff; pistillate sepals conspicuously fringed; stamens 3. ______ **A. fimbriatus**

    3′ Dioecious; inflorescence bracts (not sepals) stiff and often sharp; pistillate sepals not fringed; stamens 5.

      4. Herbage glabrous; sepals conspicuously spinescent. ______ **A. palmeri**

      4′ Herbage glandular pubescent; sepals not spinescent (or not conspicuously so). ______ **A. watsonii**

Amaranthus albus Linnaeus [*A. graecizans* of authors. Not *A. graecizans* Linnaeus] PIGWEED, TUMBLEWEED PIGWEED. Robust, weedy annuals, often becoming tumbleweeds at maturity, often 0.8–1.0 m. Stems erect to ascending, becoming yellow. Monoecious. Inflorescences comprised of axillary clusters. Bracts stiff, sharp, more than twice as long as the sepals. Flowers all with 3 sepals, the pistillate sepals nearly equal in size and not narrowed basally. Stamens 3. Fruits circumscissile. Seeds 0.8 mm diameter, lens-shaped, dark red-brown.

Common agricultural weed in the San Luis and Sonoyta regions, and disturbed habitats in the Río Colorado delta. Known from southern Arizona since the late 1800s. Native to tropical America; weedy in many parts of the world. There are 2 varieties; ours are var. *albus,* characterized in part by finely pubescent or glabrous stems.

8 km W of Sonoyta, N of Río Sonoyta, *F 87-306.* Ciénega de Santa Clara, *F 92-527.*

Amaranthus crassipes Schlechtendal var. **crassipes.** Glabrous, sometimes semisucculent, usually branching from the base upward, the stems and branches at first erect to ascending, often less than 15

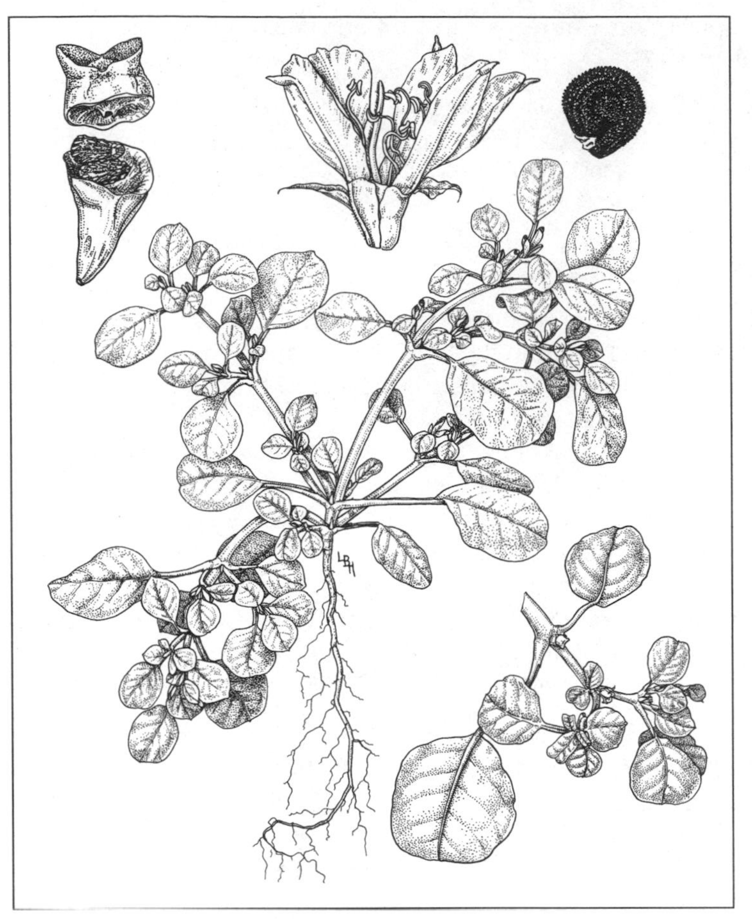

Trianthema portulacastrum. LBH (Parker 1958).

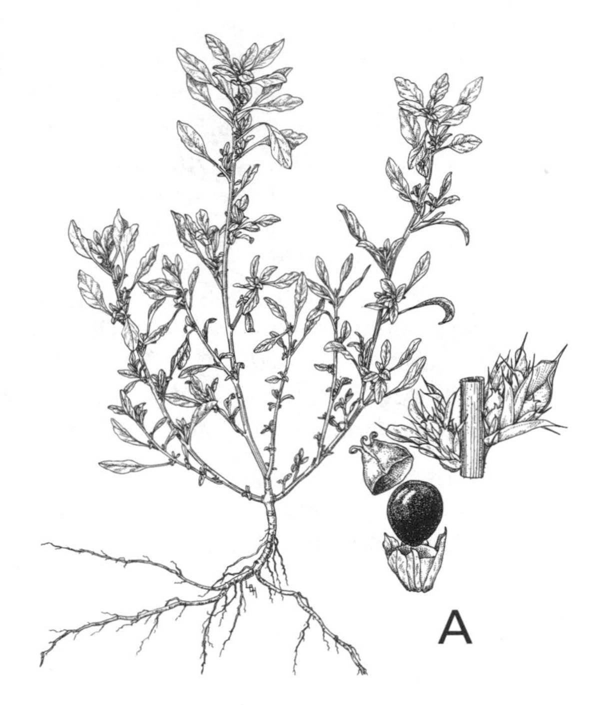

Amaranthus: (A) A. albus; (B) A. fimbriatus; (C) A. palmeri, (1) male flower subtended by bracts, (2) female flower subtended by bracts. LBH (Parker 1958).

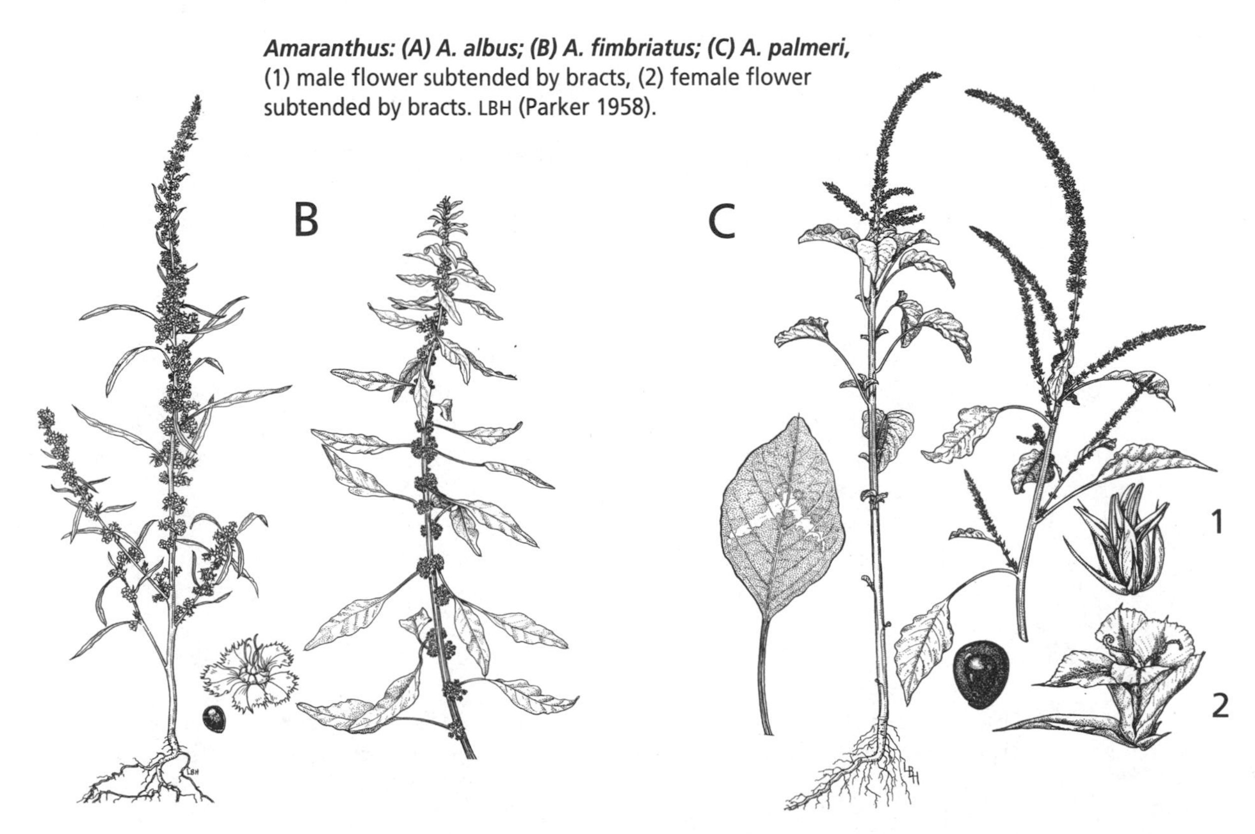

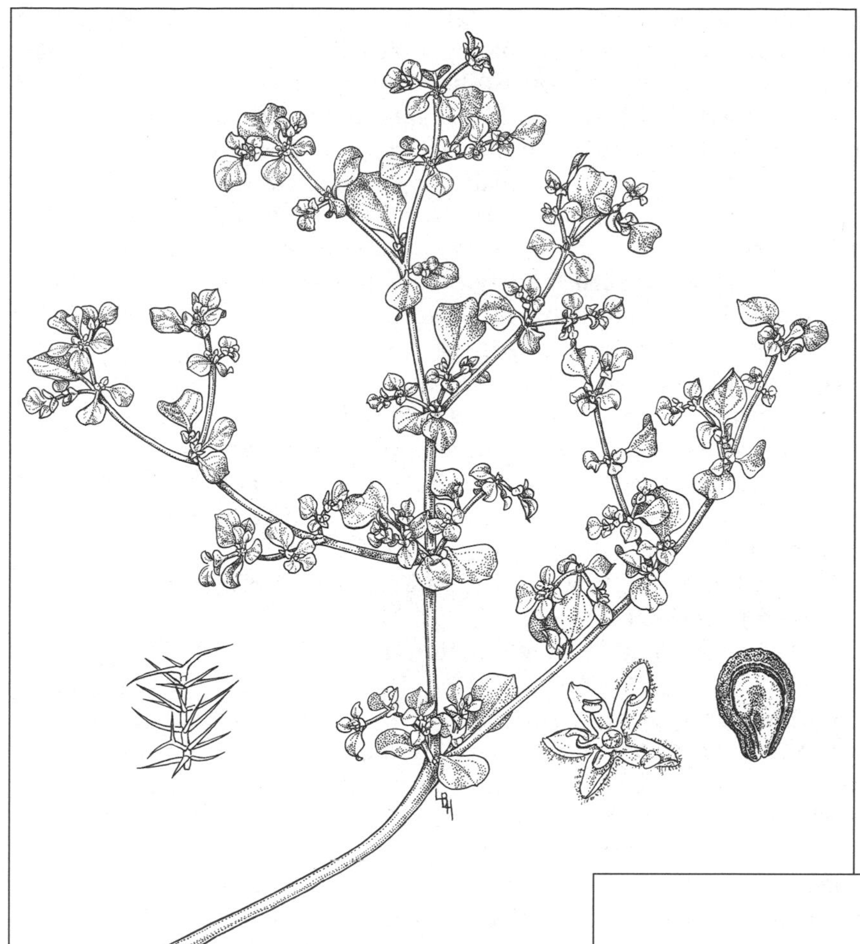

Tidestromia lanuginosa. Note branched hair. LBH (Parker 1958).

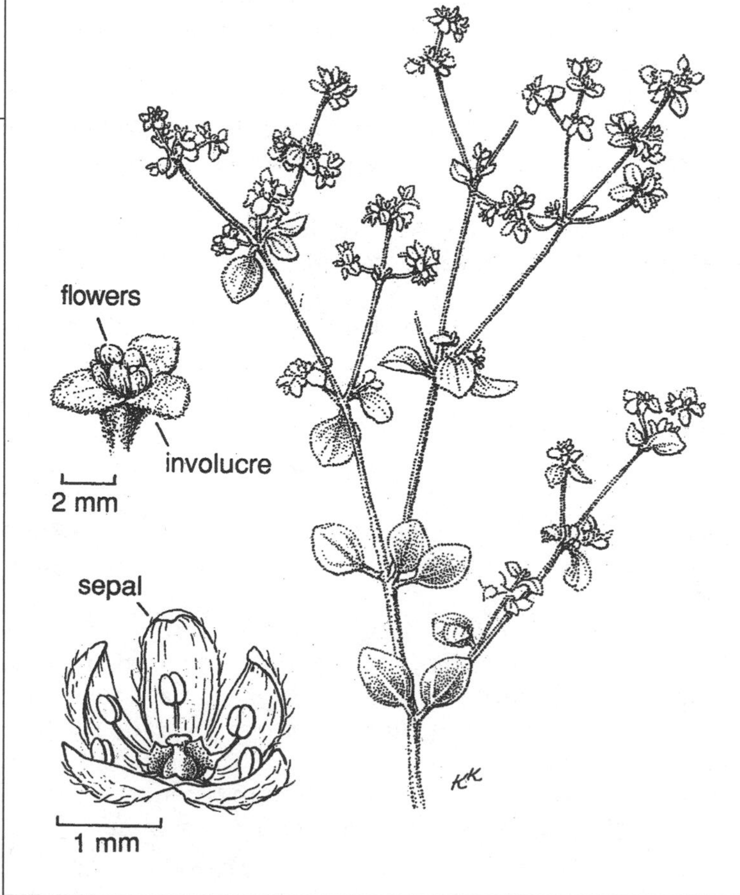

Tidestromia oblongifolia. KK (Hickman 1993).

cm, perishing or soon spreading to prostrate, sometimes to 45 cm. Leaves 2.5–7.0 cm; petioles and blades about equal in length; blades obovate to broadly elliptic or nearly orbicular.

Monoecious. Inflorescences comprised of few to many short, stubby axillary clusters; flowering from lowermost nodes upward. (Even young, rosette-like plants only 1 cm tall can produce dense basal flower clusters.) Inflorescence branches and pedicels sinuous, becoming thickened and corky, the bracts also thickened. Pistillate bracts minute, keeled. Pistillate sepals 5, spatulate, clawed, about as long as the ovary, of 2 sizes, persistent, 0.8–1.4 mm; staminate sepals 4, nearly equal, 1.2–1.5 mm. Styles 2 or sometimes 3, prominent and persistent but the tips breaking off, their bases expanded and compressed. Stamens 3. Utricles 1.8–2.0 mm (including persistent styles), indehiscent, conspicuously compressed, scaly papillose, becoming warty with age, especially above the middle. Some fruits persistent, some falling with their sepals; inflorescence often remaining intact (probably as an adaptation for water dispersal). Seeds 1.0 × 0.9 mm, obovoid-lenticular, moderately compressed, dark red-brown.

Low-lying, poorly drained, fine-textured vertisol soils of playas, or sandy soils of adjacent alluvial plains on the east of the Pinacate region. Var. *crassipes* occurs in southern Texas, southern Alabama, and Key West, Florida, through the West Indies, northern South America, and southern and eastern Mexico. It is disjunct and perhaps not native in the flora area of northwestern Sonora and southwestern Arizona in western Pima County, where it is known from only a few collections. Var. *warnockii* (I.M. Johnston) Henrickson [*A. warnockii* I.M. Johnston], from coastal and southern Texas, has glaucous green and narrower (oblanceolate) leaves (Henrickson 1999).

Rancho los Vidrios, *Equihua 26 Aug 1982.* 0.5 km S of Pinacate Junction, *F 86-361.*

Amaranthus fimbriatus (Torrey) Bentham [*Serratea berlandieri* (Moquin) Uline var. *fimbriata* Torrey] *BLEDO, QUELITILLO;* FRINGED AMARANTH; *CUHUKKIA 'I:WAGĬ.* Plants sometimes persisting until December or even spring, usually less than 1 m, glabrous, upright with mostly straight ascending branches, pale green or often red especially late in the season. Leaves narrowly lanceolate. Monoecious. Inflorescences terminal and axillary; bracts herbaceous and not prickly. Pistillate flowers urn-shaped, their sepals 5, fringed (rarely nearly entire), white with green veins. Stamens 3. Fruits circumscissile. Seeds 0.85–1.0 × 0.75–0.9 mm, obovoid-lenticular, red-brown to blackish when fully ripe.

Widespread and common, mostly on sandy-gravelly soils; arroyos and washes, interdune troughs, cinder slopes to peak elevation in the Sierra Pinacate, less common on exposed desert flats and rocky slopes. Sinaloa and Baja California Sur to southwestern United States. The tender shoots were one of the few greens relished by the Seri and were probably likewise used by people in the flora area and elsewhere.

Bahía de la Cholla, *F 16818.* Dunes N of Sierra del Rosario, *F 20766.* MacDougal Crater, *F 10714.* N slope of Pinacate Peak, *F 86-438.* Mex 2 west of rd to Rancho Guadalupe Victoria, *F 10814.*

Amaranthus palmeri S. Watson. *BLEDO, QUELITE;* CARELESS WEED; *CUHUGIA.* Plants sometimes persisting through fall but killed by the first wave of near-freezing temperatures; highly variable in size depending upon soil moisture, (0.1) 0.2–2.6 (3.1) m, usually erect with a well-developed main axis, glabrous or sparsely pubescent but not glandular. Leaves mostly lanceolate, highly variable in size, the lower-stem leaves largest and soon deciduous, often with petioles 1.5–7.5 cm and blades 3–12 cm. Dioecious. Inflorescences terminal, indeterminate, long and slender, the tips usually drooping, and also with short, axillary clusters. Floral bracts and sepals stiff and spinescent. Pistillate sepals 5. Stamens 5. Fruits circumscissile. Seeds 1.1–1.25 × 0.9–1.0 mm, obovoid-lenticular, red-brown to blackish when fully ripe.

Widespread during favorable years, mostly in the lowlands; Sonoyta region, Pinacate volcanic complex, granitic ranges, open desert, and as urban and agricultural weeds. Apparently absent from dunes. Common on sandy-silty or gravelly soils; abundant in crater-floor playas; low-lying, silty-clay flats and depressions; floodplains; arroyo beds; and roadside depressions. In such places, but especially in the playas at the center of the large craters, plants 1–3 m tall seasonally can form 100% coverage. Walking through this allergenic forest in the intense summer heat can stir a yellow rain of pollen. The dead stalks often persist until the following summer.

The primary distribution extends from southern California to Oklahoma and through much of Mexico. It is weedy through much of its range and has spread to many other parts of North America and elsewhere in the world.

W of Sonoyta, *F 86-317*. Suvuk, *Nabhan 381*. Tinajas de los Pápagos, *Turner 59-32*. MacDougal Crater, *F 10439*. S of San Luis, *F 85-1031*.

Amaranthus watsonii Standley. *BLEDO, QUELITE;* CARELESS WEED. Warm-weather ephemerals in the flora area, resembling *A. palmeri* but differing primarily in having glandular-pubescent herbage. Sauer (1955) emphasized the often broader, thicker leaves with retuse (notched) leaf tips, broader (obtuse or retuse) pistillate outer tepals, and shorter pistillate and staminate floral bracts with shorter excurrent, spinescent midribs. Pistillate sepals 5. Stamens 5.

Endemic to the Gulf of California region. Uncontroversial, conspicuously glandular *A. watsonii* is common along the coast of Sonora at least as far north as Puerto Lobos (30°15′ N, 112°15′ W) and in Baja California as far north as San Felipe. It seems to have evolved from a common ancestor of the more widespread *A. palmeri,* and some specimens from the Río Colorado delta region appear intermediate between *A. watsonii* and *A. palmeri.* "The Colorado River and associated irrigation works have apparently provided the pathway by which this species, often mixed with *A. palmeri,* has moved into southern California as a weed of irrigated fields and citrus groves" (Sauer 1955:38). As in many Gulf of California–Baja California endemics, the presence of glandular hairs is the primary distinction from related, "mainland" taxa. Over most of its range in the Gulf of California region *A. watsonii* is a common winter-spring as well as a summer-fall ephemeral.

NE margin of Ciénega de Santa Clara, *F 92-511* (close to the Lerdo site). Colonia Lerdo, 1889, *Palmer 953* & *958* (GH, US).

TIDESTROMIA

Heat-loving annual or perennial herbs, some slightly woody at base, with branched hairs (an uncommon character in the family and unique among the family for the Sonoran Desert). Leaves opposite or alternate, or congested above and appearing whorled. Flowers bisexual, each produced in a little cup-shaped involucre growing around the flower, this involucre becoming hardened. New involucres are produced from the margin of the cups of older involucres; this proliferation results in highly clustered inflorescences, which continue through the season as long as moisture and warm weather hold out. Seeds globose. Mexico and Southwest United States; 6 species. The flowers attract flies but produce little nectar (James Henrickson, personal communication 1995).

1. Annuals; leaves of lower stem usually less hairy than leaves of flowering branches; widespread. ______ **T. lanuginosa**

1′ Perennials; leaves of lower stems and flowering branches similarly and densely hairy; Cerro Pinto (south of Mex 2). ______ **T. oblongifolia**

Tidestromia lanuginosa (Nuttall) Standley [*Achyranthes lanuginosa* Nuttall] *HIERBA CENIZA, HIERBA LANUDA;* HONEYSWEET. Summer-fall ephemerals, sometimes persisting until December but perishing with the first near-freezing temperatures; low and spreading, with reddish stems and densely pubescent scurfy-whitish foliage. Leaves mostly opposite, occasionally alternate, petioled, the blades broadly obovate to orbicular, reduced and congested upwards and appearing whorled, the lower leaves often 1.5–3.5 (4.5) cm, soon withering. Flowers 2.0–2.5 mm wide, the sepals translucent yellow, the filaments greenish yellow, the anthers bright yellow.

Common and widespread nearly throughout the region, including the volcanic and granitic areas, desert plains, and Sonoyta region. Silty, sandy, and gravelly soils of desert flats and arroyo and canyon bottoms; sometimes on desert pavements; also along roadsides and an urban and agricultural weed. Less common on rocky slopes and interdune troughs; not on moving dunes. Baja California Peninsula, and Sinaloa and Zacatecas to Utah and Kansas; weedy through much of its range. Rare in desert mountains of southern California, where it has been collected recently in San Bernadino County by Andy Sanders (UCR). However, the report of it at "Otlay Lake, San Diego County" (Munz 1974:63, 64) is based on a misidentified specimen of *Glinus lotoides* (James Henrickson, personal communication 1999).

N of Puerto Peñasco, *F 85-890*. Suvuk, *Nabhan 366*. N of Cerro Colorado, *F 10413*. MacDougal Crater, *F 10506*. Sierra del Rosario, *F 20760*.

Tidestromia oblongifolia (S. Watson) Standley [*Cladothrix oblongifolia* S. Watson. *Tidestromia oblongifolia* subsp. *cryptantha* (S. Watson) Wiggins] Herbaceous perennials, extremely xerophytic, dying back to hard, woody bases during extreme drought; densely velvety woolly, the hairs obscuring leaf surfaces. Leaves alternate or opposite, petioled, the blades ovate to orbicular, the lower leaves often 2.0–3.5 cm, the upper leaves reduced.

In Sonora known only from Cerro Pinto (south of Mex 2), where it is common among granitic rocks on the south side of the mountain (*F 75-41*). Otherwise in the Mojave and Colorado Deserts of California, northern Baja California, southern Nevada, and western Arizona.

The leaves are noticeably thicker than those of *T. lanuginosa*. Reports of this species in western Texas (Correll & Johnston 1970) are in error; it does not occur there (James Henrickson, personal communication 1987; Standley 1917). Subsp. *cryptantha* seems to be based on nothing more than small-leaved, dry-season plants (James Henrickson, personal communication 1986; Standley 1917).

ANACARDIACEAE Sumac or Cashew Family

Woody shrubs, trees, and vines, often containing toxic substances. Leaves alternate, simple to pinnate; stipules none. Inflorescences paniculate. Flowers bisexual or unisexual, small, with a ring-shaped nectary disk between the stamens and ovary. Sepals mostly 5, united below; petals as many as the sepals or rarely none. Ovary superior, the fruit usually a drupe.

Worldwide, especially tropics and subtropics; 75 genera, 600 species. Includes cashew, mango, pistachio, the Baja California elephant tree (*Pachycormus discolor*), and poison oak and poison ivy (*Toxicodendron*). *Rhus* is the largest genus. The Peruvian pepper tree (*Schinus molle*) is cultivated as a fast-growing shade tree in the Sonoran Desert but is susceptible to Texas root rot.

RHUS Sumac

Shrubs or small trees. Leaves pinnate, trifoliolate, or sometimes unifoliolate (simple). Mostly dioecious with a few bisexual flowers. Sepals mostly 5, overlapping (imbricate), the petals as many as sepals and also overlapping. Fertile stamens 5. Fruit a small drupe.

Worldwide, temperate and subtropical, with highest diversity in South America; 150 species, 8 in Sonora. *Rhus* is somewhat of an exception in the family because of its mostly extra-tropical distribution, with centers of distribution in both the Northern and Southern Hemispheres. The North American species are in 2 subgenera: *Rhus* (*Sumac*) and *Lobadium* (*Schmaltzia*). Ours are in subgenus *Lobadium,* having sessile or subsessile flowers and fruits with both non-glandular and reddish glandular hairs.

Sumacs are important members of chaparral vegetation in Mexico and the United States. Except for a small population of *R. terebinthifolia* (*R. hartmanii*) in the mountains north of Guaymas, no other *Rhus* are known from western Sonora. The South African *R. lancea* is widely cultivated in the Sonoran Desert as an ornamental tree. References: Barkley 1937, 1940; Brizicky 1963; Moran 1969; Young 1975.

1. Leaves with 3 leaflets, winter deciduous; Pinacate Peak. ____ **R. aromatica**

1′ Leaves simple (or rarely a few leaves on a shrub with 3 leaflets), essentially evergreen; granitic slopes in the northwestern part of the flora area. ____ **R. kearneyi**

Rhus aromatica Aiton var. **simplicifolia** (Greene) Cronquist [*R. canadensis* var. *simplicifolia* Greene. *Schmaltzia simplicifolia* (Greene) Greene] Skunkbush. Woody shrubs 1.8–2.3 m, the branches and twigs slender and flexible. Young twigs, petioles, and inflorescences moderately hairy, the leaf blades sparsely short pubescent or glabrate. Leaves winter deciduous, (1.4) 2–3.5 (4) cm, with 3, thin, dull, bicolored leaflets, the margins shallowly lobed. Inflorescences terminal and axillary, with spike-like branches developing with the foliage in fall, overwintering, and producing small, white flowers late February-March before the leaves appear.

Rhus aromatica var. ***simplicifolia.***
LBH (Humphrey 1960).

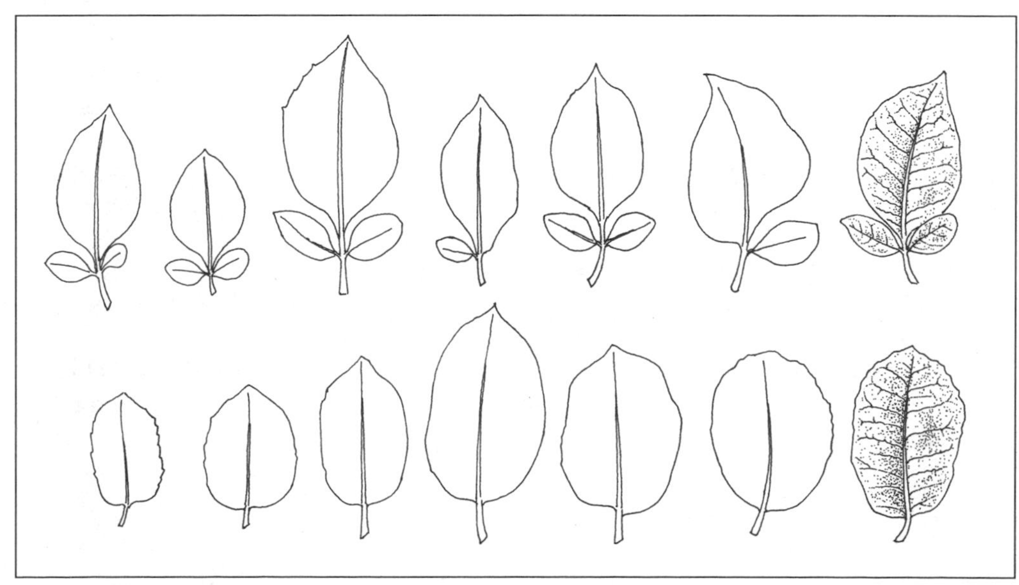

Rhus kearneyi
subsp. ***kearneyi.***
Leaves from a
single shrub
(F 89-47). MBJ.

Probably no more than several dozen shrubs on the north-facing cliffs of Pinacate Peak at 950–1080 m (*F 87-439*). This *Rhus* grows at elevations above the desert in northern and eastern Sonora and is virtually absent from the Sonoran Desert except on scattered mountaintops. *R. aromatica* ranges across much of North America from Canada to southern Mexico; highly variable and taxonomically confusing with about 9 varieties. Var. *simplicifolia* occurs from Mexico to southwestern United States including parts of Utah. To the north and east it is replaced by var. *trilobata* (Torrey & A. Gray) A. Gray.

Rhus kearneyi F.A. Barkley subsp. **kearneyi** [*Schmaltzia kearneyi* (F.A. Barkley) F.A. Barkley] DESERT SUMAC. Hardwood shrubs often 1.5–2 (3) m, usually broader than tall, with dense foliage and stout, tough, rigid branches. Youngest stems with minute simple hairs, soon glabrate. Leaves 3.5–8.5 × 2.0–3.8 cm, simple or occasionally 3-lobed or with 3 leaflets on long shoots, somewhat leathery, shiny green to somewhat bluish glaucous, bicolored, glabrous or with sparse glandular hairs; margins entire or sometimes irregularly undulate or serrate even on the same plant. Evergreen to tardily deciduous during extended drought. (During severe freezing weather in January 1987, cultivated plants in Tucson defoliated but rapidly recovered.)

Inflorescences terminal, spikelike, or paniculate, compact and densely flowered, mostly 1.5–3 cm. Flowers white to pink; sepals 5, green and pink, 2.5–3.5 mm, broadly ovate-orbicular, obtuse, ciliate with simple, non-glandular hairs; petals white, fading pink, 4.5–5.0 mm, broadly elliptic to ovate or obovate, narrowed basally to a claw. Nectary disk of young flowers glistening yellow, with age turning dark yellow-orange and then maroon-red. Ovary greenish white, the style white, the stigma white, fading dark pink. Fruits ca. 1 cm. November-March.

Fairly common at higher elevations in steep canyons on precipitous north-facing slopes in the Sierra del Viejo and nearby Sierra Nina. Also in the Gila and Tinajas Altas Mountains in adjacent Arizona, where it was first discovered. Disjunct populations of this subspecies occur in the Sierra del Viejo southwest of Caborca and Sierra San Pedro Mártir in Baja California.

This species is in section *Styphonia,* subsection *Styphonia.* Characterized by terminal inflorescences and thick, coriaceous, and usually simple evergreen leaves, this subsection is considered to be the most advanced in its subgenus. *R. kearneyi* appears to be a relict of a more widespread Pleistocene distribution (Turner, Bowers, & Burgess 1995). The size, shape, and general appearance of this shrub, especially the foliage, are unusual for the flora area. Two other subspecies occur in mountains of Baja California and Baja California Sur. *R. kearneyi* seems most closely related to *R. integrifolia* of the Pacific coast of the Californias; the two species differ in leaf shape and pubescence.

Sierra Nina, *F 89-47.* Sierra del Viejo (along Mex 2), *Bezy 376.*

APIACEAE (UMBELLIFERAE) CARROT FAMILY

Aromatic perennial herbs or less often annuals, rarely softwood shrubs or trees. Leaves usually alternate, sometimes in basal rosettes, often compound or dissected; petioles usually sheathing the stems. Inflorescences usually umbellate, often with bracts, sometimes dense and headlike, sometimes having morphological differences among individual flowers as in the composites (Asteraceae). Flowers mostly insect-pollinated and self-fertile, relatively uniform, usually small and radial. Perianth 5-merous; calyx reduced (often represented by small teeth on top of the ovary) or absent; petals separate, usually incurved at apex, or rarely absent. Stamens 5. Ovary inferior; styles 2, often swollen at base to form the unique stylopodium, an often colorful organ confluent with a nectary disk. Fruits dry, with 2 one-seeded segments (mericarps) almost always separating at maturity.

Worldwide, mostly temperate; 445 genera, 3540 species. Eighteen species are listed for the Sonoran Desert; some are non-native and others occur only at the margins of the desert. Ours and most other Sonoran Desert species are cool-season ephemerals that begin life with a basal rosette. The family includes food and spice plants such as carrot, celery, coriander, dill, parsley, and parsnip, as well as poisonous plants including hemlock. Reference: Mathias & Constance 1944–45.

1. Stems weak and trailing; leaves simple, palmately lobed to orbicular and peltate; umbels simple or flowers in an interrupted spike.
   2. Winter-spring ephemerals with stellate hairs; desert habitats. ____ **Bowlesia**
   2′ Glabrous perennials; wetlands. ____ **Hydrocotyle**
1′ Stems erect to spreading, not weak; leaves pinnately deeply lobed or toothed to divided; umbels compound.
   3. Plants or fruits not spinescent; disturbed, well-watered habitats.
      4. Fruits 4–5 mm, not prominently ribbed; calyx teeth 0.5–1.0 mm; petals unequal, 1.0–3.5 mm. ____ **Coriandrum**
      4′ Fruits 1.5–2.0 mm, prominently ribbed; calyx absent; petals equal in size, less than 0.5 mm. ____ **Cyclospermum**
   3′ Leaves, bracts, or fruits spinescent; natural habitats.
      5. Leaves finely divided; bracts soft, not spinescent (fruits with spines); flowers and fruits clearly pedicelled in rather open umbels. ____ **Daucus**
      5′ Leaves coarsely toothed; bracts stiff and spinescent at maturity; flowers and fruits sessile in dense headlike or cone-shaped inflorescences. ____ **Eryngium**

BOWLESIA

Annual or perennial herbs with slender leafy stems and stellate hairs, or glabrate. Leaves small, simple. Inflorescences of unbranched umbels, stalked or nearly sessile. South America with 13 species; 2 species also in North America, perhaps adventive. Reference: Mathias & Constance 1965.

Bowlesia incana Ruiz & Pavón. Delicate winter-spring ephemerals with stellate hairs throughout. Stems 4–45 cm, slender, weak, and reclining. Leaves petioled, the blades wider than long, 10–23+ mm wide, with (3) 5 or 7 broad lobes. Sepals and petals scalelike, 0.5 mm. Peduncles 2–6-flowered, shorter than petioles, sometimes vestigial. Fruits sessile or nearly so, the 2 segments each ovoid-globose, 2.0–2.4 mm.

Shaded places, especially beneath shrubs on north-facing arroyo banks; often with *Parietaria floridana* or beneath mesquites among mesquite-leaf litter in silty-clayish soil. Arizona Upland in the northeastern corner of the flora area westward to Represo Cipriano along the Arizona border. Also eastward in northern Sonora and southward in western Sonora at least to the vicinity of El Desemboque San Ignacio. North America, mostly in southwestern United States, and northwestern Mexico; also in southern South America. This species is questionably adventive in North America. It was collected in Texas as early as 1828 by Jean Louis Berlandier and in California in 1833 by David Douglas (Matthias & Constance 1965).

Río Sonoyta, Sonoyta, *F 86-88.* 10 km SW of Sonoyta, *F 88-168.* Represo Cipriano, *F 89-41.*

CORIANDRUM

Glabrous annuals. Leaves 2–3-times pinnate; leaflets 2-lobed, often broadest near apex, the leaflets of upper leaves linear. Petals unequal in size. Fruits nearly round. Europe and the Near East; 2 species.

***Coriandrum sativum** Linnaeus. CILANTRO; CORIANDER. Winter-spring ephemerals, aromatic. Lower leaves in a basal rosette, the leaflets ovate to obovate with deeply cut margins. Flowering stems erect, the stem leaves divided into linear-filiform segments. Sepals mostly 0.5–0.7 mm, to 1 mm in fruit. Petals white to slightly pink, ranging 1.0–3.5 mm even in the same flower. Fruits 4 mm.

An important culinary herb commonly cultivated in Sonoran gardens. Occasional, stunted plants, probably waifs from nearby gardens, in Sonoyta along the banks of the Río Sonoyta (*F 86-92*). Native to southern Europe.

CYCLOSPERMUM

Aromatic annuals. Leaves dissected or divided into narrow segments. Sepals inconspicuous or absent. Inflorescences of simple or compound umbels. Fruits globose, conspicuously ribbed. Native to South

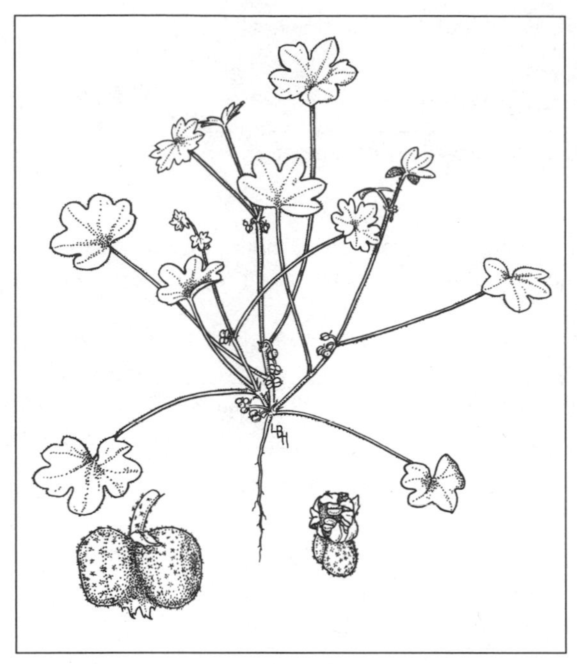

Bowlesia incana. LBH (Parker 1958).

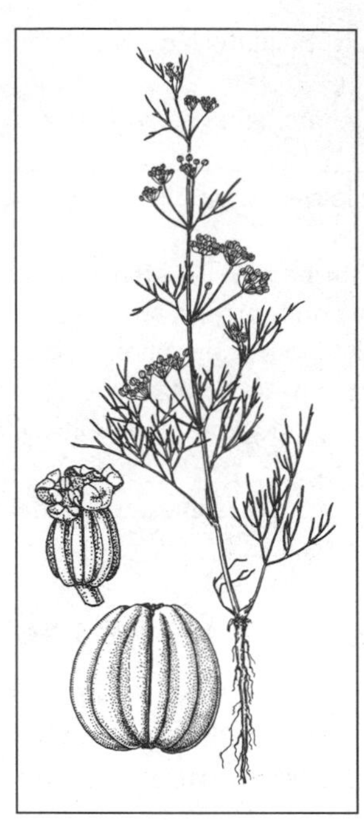

Cyclospermum leptophyllum var. ***leptophyllum.*** LBH (Parker 1958).

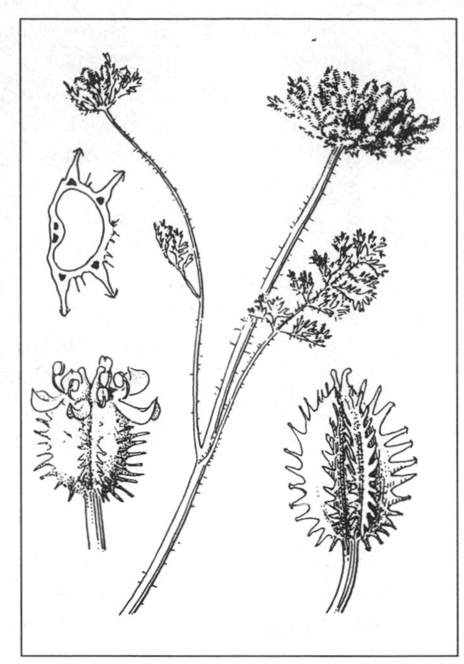

Daucus pusillus. JRJ (Abrams 1951).

America; 4 species. *Cyclospermum* (chromosome number: $n = 7$) is sometimes united with *Apium* ($n = 11$). References: Burtt 1991; Constance 1990.

***Cyclospermum leptophyllum** (Persoon) Sprague ex Britton & Wilson var. **leptophyllum** [*Pimpinella leptophylla* Persoon. *Apium leptophyllum* (Persoon) F. Mueller ex Bentham] SLENDER CELERY. Stems to 30 cm, rather coarse, grooved, erect to spreading (or somewhat prostrate when mowed in a lawn). Leaves 3–8 cm, pinnately 3 or 4 times divided into linear-filiform segments. Petals 0.3–0.4 mm, white, often fading pink. Fruits 1.5–2.0 mm, broadly ellipsoid to globose, the mericarps each with 5 prominent, narrow ribs.

Weed in Sonoyta and expected elsewhere; it thrives in mowed lawns, becoming knotty or spreading. Weedy in warm regions worldwide; probably native to Brazil, where there are 2 well-developed varieties.

Sonoyta, lawn at plaza, *F 91-3.*

DAUCUS *ZANAHORIA,* CARROT

Annuals or biennials with taproots and erect, leafy stems. Leaves dissected into small, narrow segments. Stylopodium absent. Sepals present or absent. Flowers pedicelled. Worldwide; 22 species. Only 1 species native to North America. The cultivated carrot is *D. carota* subsp. *sativa.*

Daucus pusillus Michaux. *ZANAHORIA SILVESTRE;* AMERICAN WILD CARROT. Winter-spring ephemerals with stiff white (hispid) hairs sometimes papilla-based on stems and inflorescence branches. Stems slender, 7–50 cm. Leaves highly dissected. Umbels densely flowered, on stout peduncles 3.5–27 cm, the bracts leafy. Sepals absent; petals 0.6 mm, pale yellow. Fruits burlike, the body dark colored, 3 mm, intricately sculptured with yellow barb-tipped spines.

Common along the Río Sonoyta and at higher elevations in the Sierra Pinacate. Southern United States, northern Mexico, and South America; often in disturbed habitats. In moister regions, such as parts of California, the plants are often more robust and the fruits can be larger.

Sonoyta, *F 86-81.* El Papalote, *F 86-118.* SE of Carnegie Peak, 850 m, *F 19921.*

Eryngium

Usually biennial or perennial herbs, with a stout rootstock or taproot. Usually glabrous, often spiny and thistle-like (an unusual feature in the family). Leaves often rather firm, entire to toothed or cleft, the teeth often spinose. Flowers and fruits sessile or subsessile in dense heads (aggregations of 1-flowered, minute *umbels*) subtended by conspicuous and often spiny bracts. Stylopodium absent. Temperate and warm parts of the world, very few extend into deserts; 230 species. Includes various edible and ornamental garden plants.

Eryngium nasturtiifolium Jussieu ex F. Delaroche. Winter-spring ephemerals in the flora area (often biennial or perennial outside the desert), glabrous, 3–12 cm across. (In better-watered, semiriparian habitats south of Guaymas it often forms trailing stems to 1 meter with flower heads scattered along the stems.) First leaves in a basal rosette, 1.5–6+ cm, relatively thin, green, soon withering. Leaves coarsely pinnatifid-toothed, the teeth mucronate to spine-tipped; thickness of the leaves and harshness of the spines vary with soil moisture. Inflorescences cone-shaped; floral bracts 5–15 mm, stiff, entire, spinescent-tipped, narrowly lanceolate, the larger ones often intergrading with the leaves. Sepals 1.6–2.5 mm, green with white margins and midrib, lanceolate, sharp-pointed, otherwise entire and resembling the bracts. Petals 0.5 mm, white, incurved, soon deciduous. Stamens quickly deciduous. Ovary and fruit with white hairs or scales, the upper 2 rows enlarged as inflated scales 0.8–1.6+ mm, ovate, cellular, the apex acuminate to short-awned and minutely barbellate.

Temporarily wet clay soil at the margin of Playa Díaz to the playa at Pinacate Junction, often beneath or near mesquites. Also in similar habitats in nearby Cabeza Prieta Refuge, the only known locality in Arizona; Baja California Sur, Río Grande Plains of Texas and southern Sonora through much of Mexico and Cuba.

Specimens from farther south and east in Mexico and in Texas have broader sepals with white, membranous margins erose or denticulate toward the apex; the differences seem to be clinal. *E. cervantesii* Delaroche is the only other Mexican species in the genus that has a double row of scales on the fruits.

Pinacate Junction, *F 19652*. Ciénega 25 mi W of Sonoyta, 17 Mar 1936, *Shreve 7604*.

Hydrocotyle

Small herbaceous perennials with slender, creeping rootstocks or stems, rooting at nodes. Leaves simple, the blades peltate or not, entire to deeply parted. Flowers very small, on a simple umbel, interrupted spike, or proliferous (bearing vegetative propagules). Calyx lobes minute or absent, the petals green to white or yellow. Fruits ovoid to ellipsoid, strongly flattened. Worldwide, mostly tropical and temperate regions of the Southern Hemisphere; 130 species.

Hydrocotyle verticillata Thunberg var. **verticillata.** WATER PENNYWORT. Glabrous; petioles 3–10+ cm, the blades 18–35 mm wide, peltate, shallowly lobed. Inflorescences of interrupted spikes shorter than the leaves, sometimes 2-branched above, the flowers and fruits sessile to short-stalked. Flowers greenish white, the petals 0.5–0.8 mm. Fruits ribbed, 2 mm wide, 1.5 mm high.

Recorded in the flora area only from the El Doctor wetlands at the margin of the Río Colorado riverbed (*F 93-251*), in boglike "floating," peaty vegetation. Also on banks of canals and the Colorado River in Imperial County, California; perhaps formerly widespread in the delta. North and South America, and in the Old World.

The North American varieties, including var. *triradiata* Fernald (distinguished by longer peduncles and pedicels), are probably not worth recognizing. There is, however, at least one distinct variety in South America (Lincoln Constance, personal communication 1993.)

APOCYNACEAE DOGBANE FAMILY

Trees, shrubs, herbs, and vines, usually with milky sap; many with toxic compounds. Leaves simple, usually opposite, entire; mostly without stipules. Flowers radial, often fragrant, large and showy. Calyx and corolla each 5-lobed, the corollas twisted in bud. Fruits usually paired, with 2 carpels, fleshy or dry, indehiscent or dehiscent. Seeds often with a terminal tuft of hairs.

Worldwide, tropical and subtropical, a few in temperate and some in arid regions; 165 genera, 1900 species. The tropical tree *Rauvolfia* is the source of tranquilizing drugs, and many others have medicinal qualities. Also included are *laurel* or oleander (*Nerium*), extensively cultivated in the Sonoran Desert, fragipani (*Plumeria*), and periwinkle (*Vinca*). Reference: Rosatti 1989.

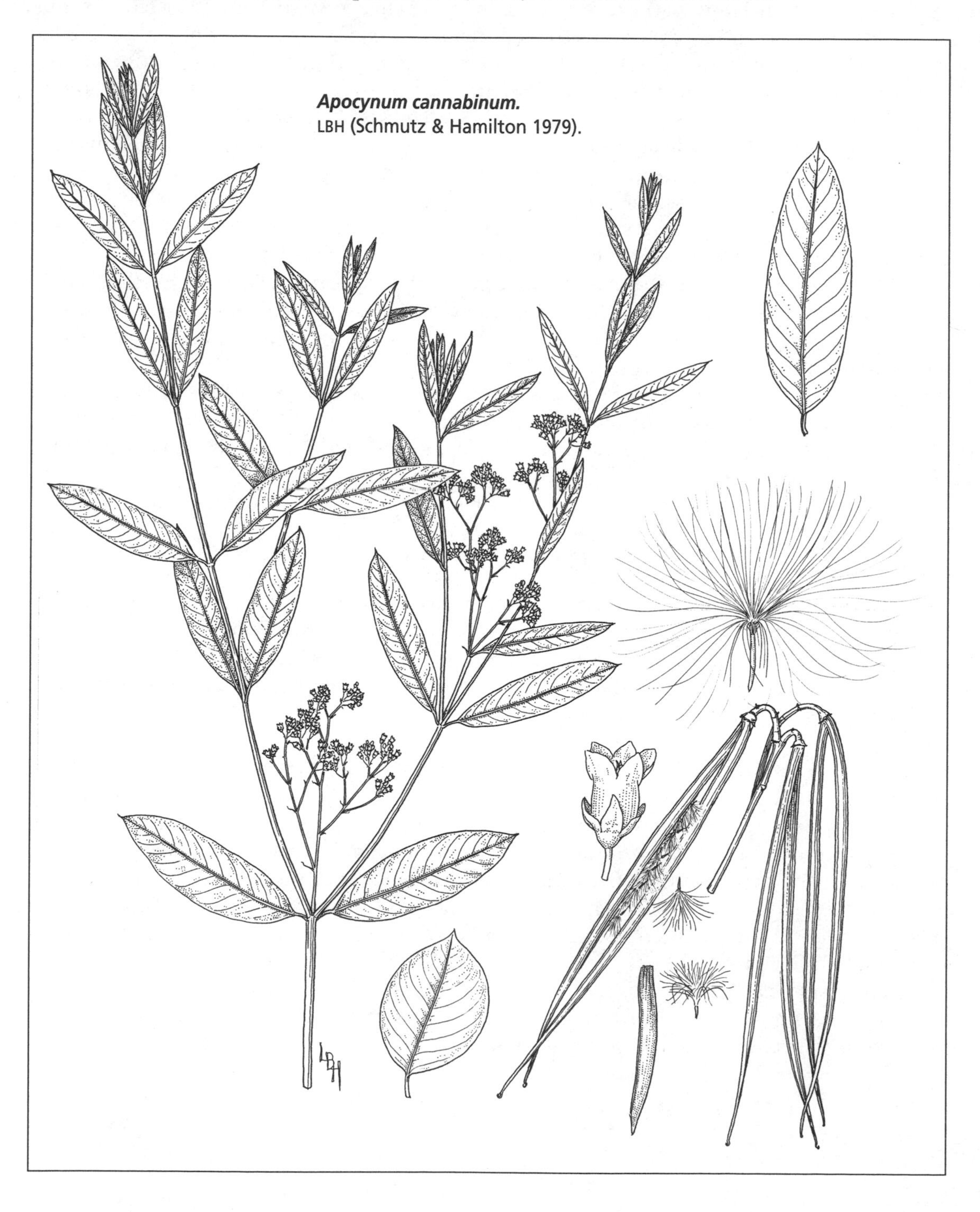

Apocynum cannabinum.
LBH (Schmutz & Hamilton 1979).

APOCYNUM

Herbaceous perennials with fibrous rhizomes. Stems usually rather long, slender, flexible, and leafy. Flowers relatively small among the family. Interior of corolla bearing 5 small triangular spurs alternating with 5 nectaries at base of ovary. Fruits dry and dehiscent. Seeds with tufts of long, silky hairs.

Extreme northern Mexico to Canada; 7 species with few sharp morphological boundaries between them, often weedy. The genus is unusual in the family in being almost entirely temperate. Native Americans made cordage from the stems. References: Hartman 1986; Uphof 1968; Woodson 1930.

Apocynum cannabinum Linnaeus [*A. sibiricum* Jacquin. *A. sibiricum* var. *salignum* (Greene) Fernald] INDIAN HEMP. Large herbs, dying back to rootstock in winter; glabrous (elsewhere often pubescent). Stems often 0.5–1.5 (2) m. Larger leaves ca. 6–10 cm, lanceolate, the petioles 2–3 mm, the leaves reduced upwards. Flowers white, apparently not fragrant. Fruits slender, at least 4–10 cm.

Well established in permanently wet soil at several La Salina pozos (*F 86-555*), growing with and partially shaded by large wetland perennials such as reedgrass (*Phragmites*) and sandbar willow (*Salix exigua*). This is the only record for *Apocynum* well within the Sonoran Desert. There are no specimens from the Río Colorado delta, although there used to be large areas of suitable habitat. Canada, much of the United States, and the northern parts of the northernmost states in Mexico; to 2300 m in Arizona.

Many different Native American peoples made strong, silklike cordage and thread from the pliable fibers of the stem, and it was esteemed for making nets (e.g., Bean & Saubel 1972). However, the "wild hemp" used by the Cocopa (cf. Alvarez de Williams 1987) was almost certainly the legume herb *Sesbania* rather than *Apocynum*.

ARISTOLOCHIACEAE PIPEVINE FAMILY

Perennials, mostly vines, woody or sometimes herbaceous. Leaves alternate, usually simple. Flowers often bizarre, the calyx enlarged. Fruits many-seeded capsules. Worldwide, mostly tropical, a few in temperate regions and very few in deserts; 5 genera, 600 species. Reference: Barringer 1997.

ARISTOLOCHIA PIPEVINE, BIRTHWORT

Mostly climbing shrubs or vines. Flowers bilateral, smelling of carrion. Calyx forming a tube, bent or curved with a petal-like, variously shaped and often intricate 1- or 3-lobed limb, mostly green- to purple-brown; petals none. Ovary inferior. Worldwide, mostly tropical; 300 species. Many of the species, including ours, are important medicinal herbs (e.g., Felger & Moser 1985). Reference: Pfeifer 1970.

Aristolochia watsonii Wooton & Standley [*A. brevipes* Bentham var. *acuminata* S. Watson. *A. porphyrophylla* Pfeifer] *HIERBA DEL INDIO;* INDIAN ROOT. Perennials from a single thickened, carrot-shaped root; dying back to root during drought and with freezing weather. Stems slender, herbaceous, trailing, often less than 30 cm or vining to 1.0–1.5 m in shaded, moist habitats. Larger leaves (2.5) 3.5–12.0 cm, the blades narrowly triangular-hastate, or the lower leaves often broadly triangular to triangular-hastate under favorable conditions, the lobes ("ears") as long as or longer than the petioles. Flowers 3.5–5.0 cm, solitary in leaf axils. Calyx tube slightly inflated surrounding style and stamens (just above the ovary) and narrowed at the throat, the limb somewhat tooth-shaped (1-lobed), yellow-green with brown-purple spots mostly along 5 prominent veins, the margin and tip dark maroon. Unopened capsules ovoid, 1.6–2.5 cm, with a narrow ridge or wing along the midrib of each of 5 valves. Seeds flattened, blackish. Flowering during warm to hot weather.

Sonoyta region along the international border westward to Quitobaquito; locally common in gravelly soils along larger arroyos and brushy floodplains and in Organ Pipe Monument. Northwestern Nayarit to southwestern New Mexico and southern 2/3 of Arizona; Baja California Sur, and southern Baja California. This is one of the smallest of the aristolochias. The herbage is often consumed by the red-orange caterpillars of the pipevine swallowtail butterfly. The caterpillars sequester aristolochic acid from the plants, which renders them unpalatable to birds.

2.6 mi W of Sonoyta, *F 86-406.* El Papalote, *F 86-325.*

ASCLEPIADACEAE MILKWEED FAMILY

Herbs, vines, shrubs, some trees, and diverse succulents; mostly with latex sap. Leaves mostly opposite; stipules absent or small and soon deciduous. Inflorescence usually umbellate. Flowers radial, 5-merous except pistils, mostly complex and highly specialized. Sepals and petals both united below. Filaments often fused to form the column; anthers fused to the enlarged stigma head, the combined structure (stamens, style, and stigma) called the gynostegium. A 5-lobed structure (believed to be an outgrowth of the filament bases), termed a crown (corona), usually present between the corolla and gynostegium. Pollen massed into pollinia (paired, waxy, sac-like structures) joined by wishbone-shaped translator arms to a corpusculum, the entire structure called the pollinarium. (The corpusculum may be engaged by the leg or tongue of an insect, which in its search for nectar carries the pollinarium to another flower. Pollination is achieved when an insect visits another flower and inadvertently inserts the pollinarium into the slit between adjacent anthers. Orchids are the only other plants with similarly complex pollinaria.) Ovaries superior, 2, free at base but joined at apex by a common stigma. Fruit a follicle (dry, dehiscent, 1-carpelled fruit usually with more than 1 seed and opening at the ventral suture), usually ovoid to linear, usually only 1 of the paired ovaries develops. Seeds usually many, flattened and often winged, generally with a tuft of soft hairs at apex.

Worldwide, mostly tropical and subtropical, a few in temperate regions; 250 genera, 2500 species. Apart from *Asclepias,* nearly all North American members are vines. A profusion of Old World arid-land genera are succulents that approach certain cacti in growth form. Some asclepiads, such as *Cryptostegia grandiflora* (naturalized in southern Sonora) and *Asclepias erosa,* are sources of latex for rubber. References: Rosatti 1989; Woodson 1954.

1. Stems erect, not vining. ______ **Asclepias**
1′ Stems vining or trailing, not supporting themselves.
    2. Stems with a longitudinal line of usually curved hairs between the nodes; flowers 4 mm, the inner corolla lobe surfaces obscured by hairs. ______ **Metastelma**
    2′ Stems glabrate or sparsely pubescent with scattered, straight hairs; flowers more than 5 mm, the corolla surfaces readily visible. ______ **Sarcostemma**

ASCLEPIAS MILKWEED

Perennial herbs or rarely shrubs, erect or decumbent, never twining. Leaves usually opposite. Flowers in umbellate clusters. Corolla lobes often reflexed and obscuring the calyx. Corona of 5 separate folded, hoodlike structures (the hood), each often with an internal, elongated horn; corona particularly prominent because of the reflexed corolla lobes. Fruits linear to broadly ovoid, generally erect. Seeds many, flat, lightweight, usually with silky hairs at apex.

New World, 150 species; perhaps as many as 200 if the genus is broadly interpreted to include African species. Only diploids ($2n = 22$) are known among the North American species. References: Fishbein 1996; Wyatt & Broyles 1990.

1. Stems leafy, not reedlike; leaves elliptic to ovate, usually more than 5 cm wide. ______ **A. erosa**
1′ Stems reedlike, leafless or leaves few, linear-filiform, less than 1 mm wide.
    2. Stems usually relatively few, usually more than 1.5 m; hoods 2.0–2.5 mm, not exceeding the anther head. ______ **A. albicans**
    2′ Stems many, usually reaching 1 m; hoods 6–10 mm, longer than the anther head. ______ **A. subulata**

Asclepias albicans S. Watson. *YAMATE;* WHITE-STEM MILKWEED, WAX MILKWEED. Stems reaching 3 m, few, mostly branching near the base, erect, reedlike, whitish to glaucous bluish white and waxy. Leaves opposite or sometimes in 3s, narrowly linear, 1.5–5.0 cm × ca. 0.5 mm, few and very quickly decidu-

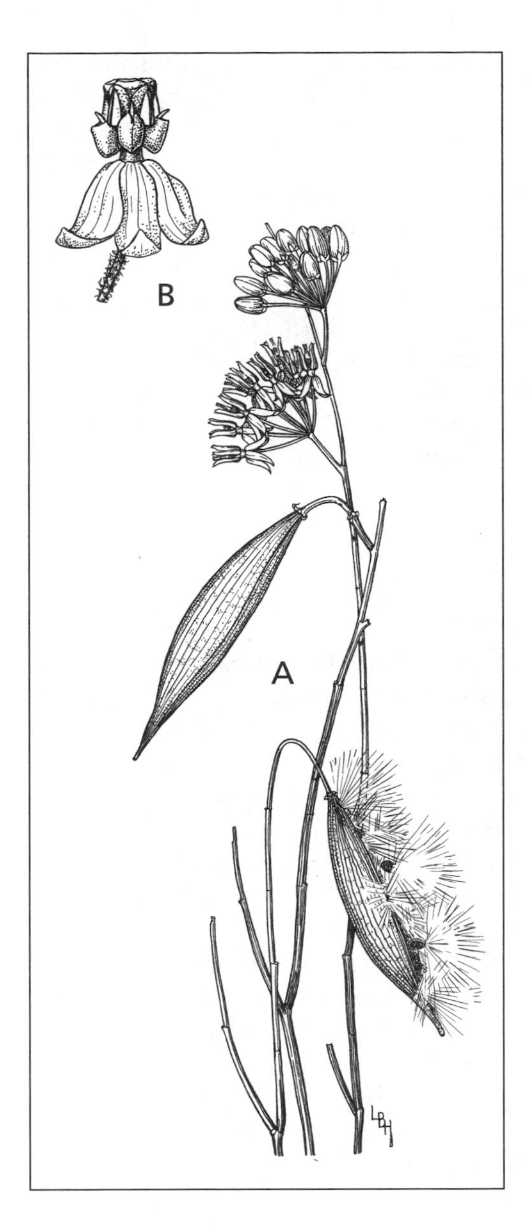

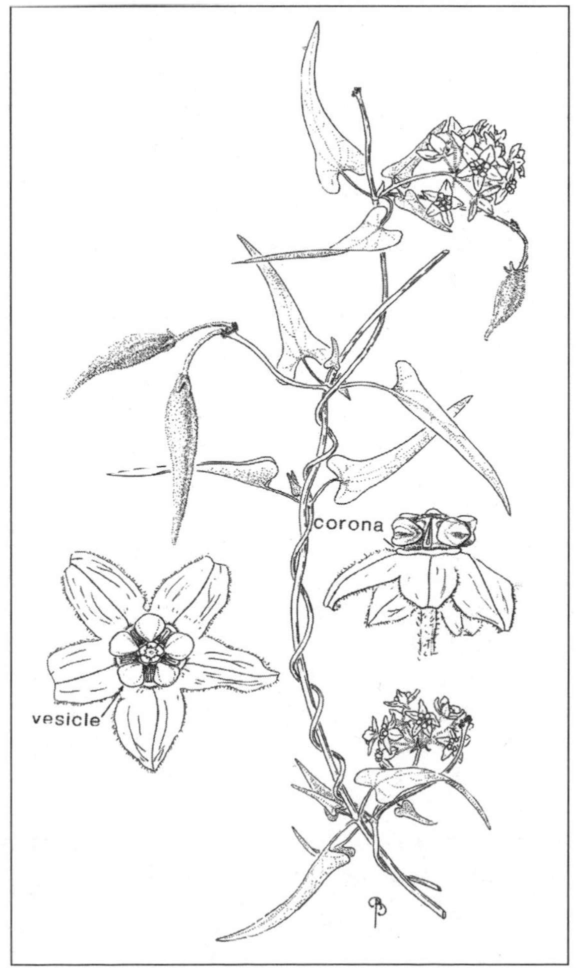

Above:
Asclepias: (A) A. subulata;
(B) A. albicans.
LBH (Benson & Darrow 1981).

Upper right:
Asclepias erosa.
BA (Cronquist et al. 1984).

Lower right:
Sarcostemma cynanchoides
subsp. ***hartwegii.***
BA (Cronquist et al. 1984).

ous. Umbels 1 to several per stem, terminal and from the upper nodes. Flowers cream-white often suffused with pink, fading pale yellow; corolla lobes 6–7 mm, the hoods 2.0–2.5 mm. Fruits 7.5–15 cm. Seeds 6–7 mm. Flowering various seasons.

Rocky slopes and canyons, especially in the granitic ranges, where it is most common toward higher elevations and on north-facing slopes. In the Pinacate volcanic complex it generally occurs on the more arid, exposed slopes such as cliff walls of the large craters. Rarely along washes, on low dunes, and on sand flats in the western part of the flora area. Southeastern California, both Baja California states, western Arizona, and northwestern Sonora south to Cerro Tepopa (29°21′ N) and Islas San Esteban and Tiburón. The report of *A. albicans* in northwestern Sinaloa (Wiggins 1964:1107) is incorrect and based on *A. subaphylla* Woodson, a reed-stem milkweed species that occurs south of the Sonoran Desert.

Asclepias albicans usually has fewer and sparser branches than does *A. subulata,* but young plants are often many-stemmed like those of *A. subulata.* Although Woodson (1954) considered *A. albicans* and *A. subulata* as convergent in growth form and not especially closely related, it appears that they are indeed close relatives (Fishbein 1996).

Sierra del Rosario, *F 20669.* Bahía de la Cholla, granitic hill, *F 13160.* Moon Crater, *F 10587.* 58 mi E of San Luis, *F 5769.*

Asclepias erosa Torrey. *HIERBA DEL CUERVO;* GIANT SAND-MILKWEED. Robust herbs often 1.2–1.8 m, the stems single or multiple from a perennial caudex with a stout, fleshy taproot, or perhaps sometimes annuals. Herbage and inflorescences white woolly when young, glabrate with age. Leaves mostly 13–23 × 9.5–13 cm, broadly ovate below to ovate-lanceolate above, nearly sessile, rather thick and firm, the midrib and lateral veins prominent, the margins thickened and ragged (erose). Sepals green and pubescent; corollas 8–9 mm, green-white; hoods 4–5 mm, white. Fruits 5–8 cm. Seeds 11–12.5 mm. Growing from winter to early summer; flowering mostly late spring and sometimes early summer.

Common on sandy plains and low stabilized dunes between San Luis and the granitic ranges in the northwestern part of the flora area; not known elsewhere in Sonora. Also Baja California, southeastern California, southern Nevada, southwestern Utah, and western Arizona.

The large leaves are unusual among the generally microphyllous flora of the region, all the more so because this species flourishes in late spring and early summer when virtually all other herbaceous plants have shut down due to drought. Aside from *Agave deserti* and perhaps *Nicotiana glauca* this is the largest-leaved non-riparian plant in the flora area. The flowers are visited by large orange and black spider wasps (*Pepsis* and *Hemipepsis*), large yellow-banded tiphiid wasps, and hairstreak butterflies. The wasps avidly drink nectar from the flowers in the hot sun. Many orange and black milkweed bugs (*Lygaeus* and *Oncopeltus*) are invariably present, often sucking from unripe fruits. The flower clusters are often visited by large numbers of ants tending yellow aphids.

18 and 35 mi E of San Luis, *F 77-19, F 90-174.*

Asclepias subulata Decaisne. *JUMETE, MATA CANDELILLA;* REED-STEM MILKWEED. Stems many, slender, and erect, mostly 0.8–1 m. Leaves narrowly linear, (1.5) 4+ cm × 0.5 mm, few and very quickly deciduous. Umbels 1 to several per stem, terminal and from upper nodes. Flowers waxy, cream- and yellow-white, the corolla lobes 9–10 mm, the hood 7–10 mm. Fruits 8.5–15 cm. Seeds 7.2–8.8 mm. Flowering profusely at various seasons; often one of the few (or the only) plants flowering in late May and June, especially in roadside habitats; flowers frequented by the large orange-winged tarantula hawk wasp (*Hemipepsis* sp.) and at least one other spider wasp (*Pepsis thisbe*).

Lowlands, generally on sandy and fine-textured soils; playas, sand flats, dunes to the limits of vegetation on 40% to 50% slopes, and disturbed habitats including roadsides. Widespread but seldom common across most of the flora area, except sometimes fairly common in sand flats in the delta of the Río Colorado and Mesa de Sonora between El Golfo and La Salina. Plants in northwestern Sonora seem more robust, taller, and with thicker stems than those farther south. Sinaloa, western Sonora, and the Cape Region of Baja California Sur to southeastern California, western Arizona, and southern Nevada.

25 mi N of Puerto Peñasco, *F 13205.* NW of Puerto Peñasco, *Reichenbacher 903.* Moon Crater, *F 18639.* High dunes W of Sierra del Rosario, *F 75-31.*

Metastelma

Perennial vines of small stature. Leaves opposite, narrowed to truncate at base (not cordate). Flowers minute, the corona of ligulate (straplike, or awl-shaped) segments. Fruits narrow with small seeds. Tropics and subtropics of North and South America and the Caribbean; 100 species. Reference: Sundell 1981.

Metastelma arizonicum A. Gray [*Cynanchum arizonicum* (A. Gray) Shinners. *Metastelma watsonianum* Standley. *M. albiflorum* S. Watson. Not *M. albiflorum* Grisebach] Small vines with slender stems twining in shrubs. Stems with a longitudinal line of hairs between the nodes, the hairs short, white, and often recurved, and also with a similar line of hairs along the upper leaf midrib. Leaves 1.2–3.0 cm, short petioled, the blades linear to linear-elliptic, thickish, dark green above, the margins usually inrolled during drought. Flowers 4 mm, in small axillary clusters, green except for very dense pubescence of white hairs on the inner surface of the corolla lobes. (These hairs point downward toward the center of the flower. A small insect attracted to the flower would be directed toward the center of the flower, and the downward-pointing hairs might prevent its access elsewhere.) Calyx lobes shorter than the corolla lobes, both narrowly lanceolate. Fruits 4.5–6.0 cm, narrowly fusiform, smooth. Warmer months.

Barely entering the flora area in the Sierra Cipriano (*F 91-33*), where it is locally common in canyons and brushy, mostly north-facing slopes. Southeastern Sonora to southern Arizona.

Sarcostemma

Perennial vines; some in dry habitats in the Old World (these in a different subgenus than the New World species) with succulent stems, reduced leaves, and sometimes shrubby. Leaves opposite, the New World species usually with cordate to hastate leaves. Corona a ring of 5 separate, inflated vesicles. Fruits elongated (fusiform) and smooth. Dry tropics to deserts and warm-temperate regions; New World, and Africa to Australia. Liede (1996) placed the 15 New World species in *Funastrum*, "but her results do not argue against recognition of a broad interpretation of *Sarcostemma* with perhaps as many as 100 species" (Mark Fishbein, personal communication, 1999). References: Holm 1950; Liede 1996.

Sarcostemma cynanchoides Decaisne subsp. **hartwegii** (Vail) R.W. Holm [*Philibertela hartwegii* Vail. *Funastrum hartwegii* (Vail) Schlechtendal] *GÜIROTE;* CLIMBING MILKWEED; *WI'IBGAM.* Often robust, trailing or twining vines with woody bases and rootstocks. Stems slender, often dying back severely, sometimes entirely, during drought. Herbage glabrate or sparsely pubescent with small hairs, the peduncles, pedicels, and calyx with small, short white hairs. Leaves gradually drought deciduous, linear to linear-oblong or narrowly triangular, often hastate at base; upper, or dry-season, leaves often 3–8 cm, the lower, or wet-season, leaves often 11–15.5 × 2.5–4.8 cm at base; usually with one or more yellow-brown conical glands (drying brown) 0.2–0.4 mm at the base of the midrib on upper surface of the blade. Peduncles and pedicels slender. Calyx 1.5–2.8 mm, green to green and maroon; corollas 5.5–7.5 mm diameter, maroon with white, the margins ciliate. Corona and vesicles white, the vesicles longer than the gynostegium and surrounding it. Fruits 6.5–8.5 cm, fusiform and smooth. Various seasons, especially spring.

Mostly in riparian or semiriparian habitats such as major arroyos, sometimes festooning trees and shrubs. Also widely scattered in open desert and sometimes growing on saguaros or sprawling across the ground in dry arroyos or washes, in soil pockets in lava, and in rock outcrops in the Sierra Pinacate. Near sea level to peak elevations, generally not on dunes; one of the few common vines in the region. The flowers are edible and have an onion-like flavor (Felger & Moser 1985).

Central Mexico and Baja California Sur to southern California, southwestern Utah, and western Texas. Replaced by subsp. *cynanchoides*—distinguished in part by broader, shorter leaves and white flowers—from eastern Arizona to Arkansas and northeastern Mexico.

Carnegie Peak, *F 19826*. Sierra Extraña, *F 19086*. S of Tinajas de los Pápagos, *F 10555*. Cerro Colorado, *F 10792*. Sonoyta, *F 86-91*. 24 mi SW of Sonoyta, *Shreve 7588*.

ASTERACEAE (COMPOSITAE) ASTER OR DAISY FAMILY

Annual and perennial herbs and shrubs (ours), sometimes vines and small trees. Leaves highly variable, usually simple but often deeply lobed to highly dissected; stipules none (sometimes with stipule-like leafy appendages at leaf bases, e.g., *Senecio pinacatensis* and *Verbesina encelioides*). Flowers (florets) usually small, highly variable; bisexual, staminate, or pistillate; pistillate florets either fertile (producing fruit) or sterile (asexual, or neutral, not producing fruit). Florets organized into a head (capitulum) surrounded by bracts (phyllaries) forming the involucre (or bracts rarely absent, e.g., *Stylocline*). Heads usually on a peduncle, usually with several to numerous florets, rarely reduced to a single floret. Composites in the flora area with 4 major types of florets:

1. **Disk florets:** Corollas tubular, (3- or 4-) 5-lobed, radial or rarely slightly bilateral; usually bisexual or sometimes functionally staminate (the stigma present but no fruit produced).
2. **Ray or pistillate florets:** Corollas with a short tube below a limb, the limb (ray or ligule) strap-shaped (extended on one side, the floret thus bilateral) and usually 3-toothed or 3-veined, or rarely the limb greatly reduced or absent (the florets eligulate, e.g., *Conyza*). Ray florets pistillate and fertile (producing fruit) or sterile. The ray florets surrounding the disk florets (unless the plants monoecious or dioecious, and the pistillate and staminate florets on different heads).
3. **Ligulate florets:** Corollas tubular, bilateral, with a 5-lobed limb (ligule). Bisexual, all florets in the head of the same type, although the inner ones often smaller, the florets usually numerous in each head.
4. **Bilabiate florets:** Corollas bilateral, 2-lipped, the inner (toward center) lip 2-lobed, the outer lip 3-lobed; all florets in the head of the same type.

Calyx modified into the pappus or absent, the pappus diverse, represented by bristles, awns, or scales, simple or feathery (plumose), etc. Stamens inserted on the corolla, the anthers usually united around the style, the filaments not fused; or stamens absent from unisexual pistillate flowers. Ovary inferior, 1-chambered, the ovule solitary, the style solitary, the stigma 2-branched. Fruit 1-seeded, referred to here as an achene but technically a cypsela.

Asteraceae is the largest family of vascular plants, with 1530 genera, 22,750 species, or about 9% of the world's flora. Mexico probably contains more than 10% of the world's species. The composite family is especially well developed and diverse in semiarid regions of the world. It is the largest plant family in the flora area, with 88 species in 55 genera, or 15% of the total flora. Thirteen species are non-native and 47 genera have only native species. Their having evolved efficient chemical defenses against herbivores is believed to be a major factor in the worldwide success of the family (Cronquist 1981). Indeed, while studying Sonoran Desert composites one is impressed by the prevalence and diversity of glands on the stems, leaves, and especially on exposed surfaces of phyllaries and corollas. These glands often occur in combination with a great diversity of hairs.

Several small composite shrubs form important components of the local desert vegetation; e.g., *Ambrosia deltoidea, A. dumosa, Encelia farinosa,* and *Isocoma acradenia.* Species with showy yellow flower heads or corollas are especially common; e.g., *Encelia farinosa, Geraea canescens,* and *Pectis papposa.* A minority of the species have nocturnal or crepuscular, generally white to pink flowers, e.g., *Stephanomeria* spp. There are no red-flowered composites in the region, and blue flowers occur only among *Erigeron.* Most are insect-pollinated. The Ambrosiinae, e.g., *Ambrosia* and *Hymenoclea,* are wind-pollinated, with corollas absent from pistillate flowers and reduced on staminate flowers. Others, such as *Filago* and *Stylocline,* have reduced corollas and are presumably selfing. Species with annual or ephemeral growth forms make up 50% of the composite flora of northwestern Sonora. Worldwide 24.7% are annuals, 2.3% biennial herbs, 51.2% perennial herbs, 20.2% shrubs, 1.5% trees, and 0.1% other growth forms such as vines (John Strother, personal communication 1998). References: Cronquist 1994; McVaugh 1984; Rzedowski 1978; Turner 1996, 1997.

1. Sap milky; florets all conspicuous, similar in shape (inner florets often smaller), ligulate (ligules 5-lobed and "raylike"), and bisexual.
   2. Pappus of lanceolate, papery (paleaceous) scales, cleft at apex with the midrib extending into a slender bristle, the bristle not plumose. ______ **Uropappus**
   2′ Pappus of slender, feathery or threadlike bristles (if expanded and scalelike at base, then the apex plumose).
      3. Pappus bristles plumose.
         4. Phyllaries (13) 17–22 mm; pappus bristles (6) 9.5–13 mm; ligules of larger (outer) florets usually 15–30 mm; achenes tapering into a slender beak. ______ **Rafinesquia**
         4′ Phyllaries 6–10.5 mm; pappus bristles 2.2–8 mm; ligules 6–12 mm; achenes columnar (not tapering), ending abruptly (truncate). ______ **Stephanomeria**
      3′ Pappus bristles threadlike (capillary), smooth to barbellate but not plumose.
         5. Achenes beaked, the beak slender like a wire and about as long as or longer than the body of the achene. ______ **Lactuca**
         5′ Achenes not beaked (sometimes narrowed to a neck but the neck not slender like a wire and much shorter than the achene body).
            6. Achenes flattened, rounded at apex; stems leafy, at least below; leaf margins mostly spinose-toothed. ______ **Sonchus**
            6′ Achenes cylindrical, truncate at apex; stem leaves absent, few, or much reduced; leaf margins not spinose-toothed.
               7. Florets 3 or 4 per head, the phyllaries 4–5 mm. ______ **Prenanthella**
               7′ Florets ca. 10 or more, the phyllaries more than 7 mm.
                  8. Upper part of plants, including involucres, with conspicuous tack-shaped glands, otherwise glabrous or glabrate; achenes 7 mm, narrowed to a slender neck just below pappus. ______ **Calycoseris**
                  8′ Plants without tack-shaped hairs, the involucres and new growth moderately woolly; achenes 1.8–2.4 mm, cylindrical (without a neck). ______ **Malacothrix**

1′ Sap not milky; florets not ligulate; heads with (1) ray and disk florets, the rays sterile or pistillate, usually 3-toothed or 3-lobed, or (2) disk or disklike florets only, the corollas showy to reduced or lacking, or (3) bilabiate (2-lipped) florets only.
   9. Vegetative parts (herbage and phyllaries) conspicuously resinous-glutinous (sticky) and aromatic (these plants may also key out elsewhere).
      10. Monoecious, florets of each flower head of a single sex, the pistillate florets enclosed in burs or nutlike structures.
         11. Leaves and leaf segments not filiform (threadlike), more than 4 mm wide; burs with straight or hooked spines widest at base. ______ **Ambrosia**
         11′ Leaves and leaf segments filiform, less than 2 mm wide; burs with flat wings narrowed basally. ______ **Hymenoclea**
      10′ Dioecious, or flower heads with staminate and/or pistillate and/or bisexual florets, flowers not in burs or nutlike structures.
         12. Ephemerals or herbaceous perennials with vegetative parts that die back to ground level at end of season.
            13. Leaf margins toothed or lobed less than halfway to midrib; flowers dull whitish; achenes 1 mm. ______ **Conyza**
            13′ Leaves pinnately 2 or 3 times dissected nearly to midrib; flowers bright yellow; achenes at least 2.5 mm. ______ **Hymenothrix**
         12′ Bushy perennials or shrubs without seasonal dieback or only partially dying back at end of season.
            14. Leaves filiform, terete or nearly so, not lobed, less than 1.5 mm wide.
               15. Subshrubs with many slender stems branching from base, scarcely woody; flower heads on peduncles, in dense clusters on upper stems, 4–5 mm. ______ **Gutierrezia**
               15′ Woody shrubs usually 1 m or more tall with a thick, woody trunk or several main woody branches; flower heads solitary and sessile at stem tips, at least 1 cm. ______ **Peucephyllum**

14′ Leaves not filiform, at least 2.0 mm wide or if very narrow, then at least some leaves toothed or lobed; blades flattened or at least not terete.

16. Pappus none; lower leaf surface with a prominent and raised midrib. __________ **Gymnosperma**

16′ Pappus conspicuous.

17. Plants often 1–2 m tall; flowers unisexual, dull white. __________ **Baccharis**

17′ Plants often 1 m or less in height; flowers bisexual, bright yellow. __________ **Isocoma**

9′ Vegetative parts not conspicuously resinous-glutinous and sticky.

18. Heads consisting of bilabiate florets only; achenes expanded at apex into a disk bearing numerous pappus bristles.

19. Perennial herbs; leaves stiff, spinescent-toothed, largely semipersistent, not losing their shape when dry. __________ **Acourtia**

19′ Shrubs; leaves thin and soft, not spinescent-toothed, shrivelling when dry. __________ **Trixis**

18′ Heads with ray and disk florets, or only disk or disklike florets, these not bilabiate; achenes various.

20. Heads with disk and ray florets, the rays usually obvious (taxa with small, inconspicuous or early-deciduous rays will key out in either choice; if in doubt go to 20′).

21. Pappus absent on all achenes (caution: refers to absence of pappus bristles surrounding the corolla at top of achene; do not confuse hairs on sides of achenes with pappus).

22. Leaves opposite; rays white, minute, and in several rows (disk yellow); wetland habitats. __________ **Eclipta**

22′ Leaves alternate or basal; rays yellow (disk yellow or brownish); dryland habitats.

23. Leaves glabrate (minutely scabrous); stems slender and woody; heads ca. 1.5 mm wide. __________ **Gymnosperma**

23′ Leaves hairy; herbaceous or if slightly woody then stems not slender; heads more than 10 mm wide.

24. Larger leaves basal or near ground; ephemerals, rarely annuals; achenes more or less cylindrical and ribbed. __________ **Baileya**

24′ Leaves terminal or along stems; ephemerals to shrubs; achenes either flattened or 4-angled, not ribbed.

25. Rounded, perennial bushes; leaves mostly crowded (close together) at stem tips; achenes flattened, the margins outlined with white hairs. __________ **Encelia farinosa**

25′ Ephemerals to perennials; leaves scattered along stems; achenes angular or only slightly compressed, the margins undifferentiated (pappus deciduous, absent from older achenes). __________ **Helianthus**

21′ Pappus present, at least on disk achenes.

26. Rays deeply cleft into 3 (or 4) large, conspicuous, and moderately spreading lobes; achenes densely covered with ascending golden brown hairs and pappus of broad, nearly transparent scales. __________ **Gaillardia**

26′ Rays not deeply cleft into large lobes; both achenes and pappus not as in 26.

27. Leaves deeply divided (at least the larger, lower leaves), parted at least to middle of leaf.

28. Plants glabrous; leaves and phyllaries with conspicuous dark-colored oil glands, these oval or round and within tissue of the organ. __________ **Thymophylla**

28′ Plants with hairs; leaves and phyllaries often with glands but these not as in 28.

29. Pappus of awn-tipped scales (flattened).

30. Heads numerous, crowded, in clusters touching one another or nearly so; achenes 2.5–4 mm, the pappus of 10–15 scales. __________ **Hymenothrix**

30′ Heads solitary or several on a branch, usually widely spaced; achenes 2 mm, the pappus of 5 or 6 scales. __________ **Hymenoxys**

29′ Pappus of slender bristles or hairs (not at all flattened).
31. Mostly ephemerals to annuals, plants mostly less than 50 cm; phyllaries graduated; pappus of rather stiff, light brown bristles; lowlands to higher elevations. __________ **Machaeranthera**
31′ Perennials and flowering in first season; semishrubby to shrubby, usually 1 m or more; phyllaries in 2 series, an inner series of larger phyllaries and an outer series of smaller phyllaries; pappus of soft, white hairs; Sierra Pinacate above ca. 650 m. __________ **Senecio pinacatensis**

27′ Leaves entire or lobed less than halfway to midrib.
32. Heads usually (2) 3–5 cm wide including rays; receptacles with chaffy bracts subtending disk florets.
33. Phyllaries with striking ciliate margins of long white hairs; winter-spring ephemerals. __________ **Geraea**
33′ Phyllaries not ciliate with long white hairs; non-seasonal ephemerals or shrubs.
34. Upper surface of leaves greener than lower surface; ray florets fertile (producing achenes). __________ **Verbesina**
34′ Both leaf surfaces of same color; ray florets sterile (not producing achenes).
35. Sunflowers, herbaceous annuals to short-lived perennials; if more than 1 m tall, then relatively few-branched with only 1 to several main stems, the stems slender or stout; heads (3.5) 4–12` cm wide; pappus deciduous. __________ **Helianthus**
35′ Much-branched hemispherical shrubs, the stems slender; heads 2.6–3.5 cm wide; pappus persistent. __________ **Viguiera**
32′ Heads mostly less than 2 cm wide; receptacles naked, disk florets without subtending bracts.
36. Perennials: shrubby and woody, or semiwoody with rhizomes, and/or evidence of last year's stems.
37. Herbage densely white hairy; flowers yellow.
38. Stems, leaves, and bracts with firm, appressed hairs; pappus of more than 2 dozen slender bristles; dunes. __________ **Heterotheca**
38′ Stems and to a lesser extent leaves and bracts with soft white-woolly hairs; pappus of 6 or fewer scales; mountains. __________ **Psilostrophe**
37′ Herbage not densely white hairy, the stems and leaves glabrate or glabrous, glaucous, or hairs small and not woolly; flowers yellow, white, or blue.
39. Herbage either somewhat glutinous or gland-dotted; ray and disk florets not markedly different, all florets yellow to yellow-orange.
40. Herbage not glutinous, the glands (e.g., in phyllaries) elliptic, 1–2 mm, bright maroon; phyllaries 11–13.5 mm; Sonoyta region. __________ **Adenophyllum**
40′ Herbage glutinous, the glands rounded, ca. 0.1 mm wide and same color as herbage; phyllaries 3.5–4.5 mm; Pinacate Peak. __________ **Gutierrezia**
39′ Herbage neither glutinous nor gland-dotted; ray and disk florets well differentiated, the rays white to pale bluish, the disk yellow.
41. Spiny shrubs or subshrubs; glabrate (phyllaries minutely ciliate); rays white; riverbeds and nearby wetland habitats or agricultural areas. __________ **Chloracantha**
41′ Herbaceous perennials, not spiny; young herbage and phyllaries sparsely to moderately pubescent; rays bluish to whitish; granitic slopes. __________ **Erigeron oxyphyllus**

36′ Ephemerals or annuals.
42. Plants glabrous, dotted with conspicuous oil glands, pungently aromatic; leaves entire with prominent bristles at base; rays yellow. ____**Pectis**
42′ Plants sparsely to densely hairy, at least on new growth—or if essentially glabrous then blades not entire; oil glands lacking or at least not conspicuous, not pungent; leaf bases not bristly; rays white, lavender, or pinkish.
43. Plants white woolly; rays white, the disk yellow. ____**Eriophyllum**
43′ Plants glabrous or hairy but not woolly.
44. Plants mostly low and spreading, less than 10 cm; hairs stiff and white; peduncles mostly shorter than flower heads; leaves entire (margins have bristly hairs). ____**Monoptilon**
44′ Plants usually taller than wide, commonly more than 10 cm; glabrous or with soft hairs; peduncles often longer than heads; leaves entire to toothed or lobed.
45. Ray and disk corollas uniformly bright yellow. ____**Senecio**
45′ Rays white or lavender, the disk corollas white or yellow.
46. Plants glabrous; leaves sessile and entire or minutely toothed. ____**Symphyotrichum**
46′ Plants hairy, glabrate, or essentially glabrous; at least the lower leaves conspicuously toothed to lobed.
47. Plants with spreading soft, white hairs; leaves pinnatifid to pinnately lobed or toothed; rays lavender, filiform, numerous. ____**Erigeron divergens**
47′ Plants with short, non-spreading hairs to glabrate or essentially glabrous; leaves palmately toothed to lobed; rays white, linear but not filiform. ____**Perityle**

20′ Heads with disk florets only; outer florets without an obvious ligule or ray, or if ray florets present then inconspicuous or reduced, or lacking a well-developed ligule—eligulate (if in doubt about presence of rays, then take this choice).
48. Heads unisexual, the pistillate florets enclosed in a bur or woody, winged involucre or nutlike structure.
49. Leaves and leaf segments linear-filiform; burs winged but not spiny. ____**Hymenoclea**
49′ Leaves and leaf segments not linear-filiform; burs spiny.
50. Perennials or if annuals, then burs with straight spines (not hooked); burs 15 mm or less. ____**Ambrosia**
50′ Annuals, the burs with hooked spines; burs (18) 25–35 mm. ____**Xanthium**
48′ Heads not unisexual (except some heads in *Dicoria*), the pistillate florets not enclosed in burs or as in 48.
51. Annuals; heads and leaves spinescent (rarely not spiny in some cultivars of *Carthamus*, safflower).
52. Stems white, not winged; heads more than 4 cm wide, the corollas bright orange (or yellow-orange); pappus none. ____**Carthamus**
52′ Stems green, winged with decurrent leaf bases; heads 2.5–2.8 cm wide, the corollas bright yellow; pappus bristles present. ____**Centaurea**
51′ Annuals or perennials; heads and leaves not spinescent (or if somewhat spinescent, then the plants shrubs).
53. Plants tomentose or white woolly, at least the lower leaf surfaces and stems.
54. Perennials; higher elevations of Sierra Pinacate. ____**Artemisia**
54′ Winter-spring ephemerals or annuals (occasionally summer-fall in *Trichoptilium*); mostly lower elevations.

55. Stems thick or not, but not threadlike; leaves petioled, the blades shallowly to coarsely toothed; individual flowers small but readily visible, corollas bright yellow; achenes more than 2 mm.
  56. Leaf blades as wide as or wider than long, the marginal teeth blunt; pappus with more than 100 bristles. ________ **Psathyrotes**
  56′ Leaf blades longer than wide, the marginal teeth often sharply pointed; pappus with 5 broad scales divided above into many slender bristles. ________ **Trichoptilium**
55′ Stems slender, mostly threadlike; leaves sessile or nearly so, the margins entire or nearly so; individual flowers minute, inconspicuous, and dull colored (not bright yellow; rarely reddish), scarcely visible except under magnification; achenes 1 mm or less ("fuzzy little composites" including filaginoids).
  57. Majority of bracts on head not directly associated with florets; all florets with a pappus; wet soil at waterholes. ________ **Gnaphalium**
  57′ Majority of bracts on head partially or completely enclosing a floret; outer several florets without a pappus; widespread, desert habitats and dry watercourses (true filaginoids).
    58. Disk florets bisexual, the disk achenes usually developing and with a copious pappus (averaging more than 12 bristles/floret); receptacle often tack-shaped. ________ **Filago**
    58′ Disk florets staminate, the disk achenes not developing and their pappus none or vestigial (averaging fewer than 12 bristles); receptacle often conical or cylindrical. ________ **Stylocline**

53′ Plants glabrous or hairy but not woolly.
  59. Leaves opposite; ephemerals or annuals.
    60. Peduncles of individual flower heads readily evident; florets numerous, the outer florets white, the central (disk) florets yellow. ________ **Eclipta**
    60′ Heads sessile in dense clusters; actual individual heads 1- (2)-flowered, florets yellow-green. ________ **Flaveria**
  59′ Leaves alternate, or in basal or near-basal rosettes, or if opposite, then plants shrubby perennials or only first leaves opposite and subsequent leaves alternate.
    61. Florets subtended by chaffy bracts of receptacle, the bracts folded around the achenes and falling with them; shrubby or bushy perennials.
      62. Leaves alternate; achenes 7–10 mm, the margins with long white hairs; pappus none. ________ **Encelia frutescens**
      62′ Leaves opposite or the upper ones alternate; achenes 2.8–4 mm, the margins not different from the body; pappus present (rays present but often falling early). ________ **Viguiera**
    61′ Receptacle naked, without chaffy bracts.
      63. Plants glaucous, the herbage bluish green.
        64. Plants not aromatic, lacking oil glands; flowers bright yellow. ________ **Machaeranthera carnosa**
        64′ Plants pungently aromatic, the leaves and bracts with conspicuous elongate-elliptic oil glands; flowers whitish to pinkish. ________ **Porophyllum**
      63′ Plants not glaucous, the herbage not bluish green.
        65. Pappus of plumose bristles; many-stemmed perennial bushes. ________ **Bebbia**
        65′ Pappus not plumose; annuals and perennials.
          66. Annuals or ephemerals, or if perennials, then the vegetative parts dying back and renewed each year (or season).
            67. Leaves 1–3-times pinnatisect (pinnately divided to midrib).

68. Spring ephemerals; flowers white to pale pinkish; inflorescences open, with few heads, the heads usually well separated; phyllaries 5–10 mm. ______ **Chaenactis**

68′ Annuals or perennials; flowers bright yellow; inflorescences with dense clusters of heads, the heads mostly touching; phyllaries 2.7–4 mm. __ **Hymenothrix**

67′ Leaves entire or margins lobed or parted halfway to midrib or less.

69. Leaf surfaces mostly grayish or grayish green with coarse, stiff grayish or white hairs; achenes at least 4 mm.

70. Flowers unisexual; achenes flat, 4–5.5 mm, the margins broad, yellow and papery; pappus none. ______ **Dicoria**

70′ Flowers bisexual; achenes narrowly columnar and 4-angled, 7.5–14 mm, the margins not differentiated; pappus well developed. ______ **Palafoxia**

69′ Leaf surfaces usually green, glabrous or the hairs not as in 69; achenes to 3.2 mm.

71. Leaves very thin, almost membranous; flowers yellow; phyllaries 6–7.5 mm; achenes 2.8–3.2 mm; pappus of more than 25 soft, white capillary bristles. ______ **Senecio mohavensis**

71′ Leaves often thin but not membranous; flowers white to lavender; phyllaries 2.5–5.5 mm; achenes 0.8–1.3 mm; pappus bristles about 20 or fewer.

72. Annuals; phyllaries 2.5–3.5 mm; flowers whitish; weeds mostly in disturbed habitats. ______ **Conyza**

72′ Perennials and also flowering in first year; longer phyllaries 4–5.5 mm; phyllaries and flowers rose-lavender; wetland, mostly natural habitats. ______ **Pluchea odorata**

66′ Shrubby or bushy perennials, vegetative parts mostly persistent all year.

73. Leaves linear and densely crowded at stem tips, like a miniature fir tree; heads sessile, mostly solitary at stem tips; inner series of phyllaries all alike, plus an outer series of few, narrower, and sometimes shorter phyllaries; flowers yellow. ______ **Peucephyllum**

73′ Leaves not linear and crowded at stem tips; heads on peduncles, usually not solitary; phyllaries conspicuously graduated; flowers whitish, pinkish, or lavender.

74. Broom- or willowlike shrubs usually at least (1) 1.5–2 m tall, the branches erect to ascending (at least the larger ones); leaves sessile or the blade gradually narrowed to an indistinct petiole less than 1/6 length of leaf, the blade entire to toothed but not spinose-toothed.

75. Herbage yellow-green to glaucous, glabrate or with short, inconspicuous hairs; heads unisexual; flowers whitish. ______ **Baccharis**

75′ Herbage densely silvery hairy; heads with both staminate and pistillate florets; flowers pinkish. ______ **Pluchea sericea**

74′ Plants not broom- or willowlike, mostly 1–1.5 m or less in height, the branches mostly spreading; junction of blade and petiole abrupt and well marked, the petiole usually more than 1/3 as long as blade (at least among the lower leaves), or the petioles shorter and blades spinose-toothed.

76. Petioles shorter than or as long as leaf blades; pappus bristles uniform, slender (capillary), not barbellate, the margins not differentiated. ______ **Brickellia**

76′ Petioles mostly longer than leaf blades; pappus of membranous-margined scales and slender, barbellate bristles. ______ **Pleurocoronis**

ACOURTIA

Perennial herbs with a caudex or knotty base bearing tufts of coarse brown wool. Leaves basal or alternate, usually firm, the margins spiny-toothed. Phyllaries stiff, graduated. Florets bilabiate, all alike, bisexual and fertile. Achenes elongated (linear-cylindric to fusiform), often glandular, apex often expanded into a disk bearing a pappus of numerous barbellate bristles.

Acourtia and *Perezia* are treated as distinct genera or sections of the same genus. *Perezia* is Andean. *Acourtia* ranges from Central America to southwestern United States with 41 species, only 5 of which occur north of Mexico. References: McVaugh 1984; Simpson 1978.

Acourtia wrightii (A. Gray) Reveal & R.M. King [*Perezia wrightii* A. Gray] BROWNFOOT. Stems to 1.5 m, leafy. Leaves mostly 8–12.5 × 3.0–5.5 cm, thin and dry to the touch, sessile, ovate to lanceolate, coarsely and irregularly toothed, the bases clasping the stem. Inflorescences with many-headed clusters of pink flowers. Phyllaries oblong to elliptic with obtuse tips, thin, green to pink-tinged, the margins often membranous-ciliate, the larger phyllaries 6.5–7.5 mm. Achenes 5.5–6.5 mm, densely glandular; pappus 7.5–9.5 mm. Mostly March-April and September-November.

Arizona Upland near Sonoyta and the Sierra Cipriano; granitic rocky or gravelly soils along arroyos and washes dissecting bajadas. Central mainland Mexico to southwestern United States but not in California.

10 km SW of Sonoyta, *F 88-161.*

ADENOPHYLLUM

Mostly robust annual or perennial herbs, sometimes shrubs; strong scented with conspicuous oil glands. Leaves and branches opposite below, often alternate above. Leaves mostly pinnate or pinnatifid, the segments or teeth usually bristle-tipped. Phyllaries weakly united below and separating with age, equal, usually subtended by a series of smaller, accessory bracts. Rays few, fertile, yellow, orange, or red; disk florets usually numerous, bisexual and fertile. Achenes obpyramidal; pappus variable. Southeastern California to Texas and southward to Nicaragua; 10 species, segregated from *Dyssodia* sensu lato. References: Strother 1969, 1986.

Adenophyllum porophylloides (A. Gray) Strother [*Dyssodia porophylloides* A. Gray] Suffrutescent perennials, often 50–80 cm, glabrous, moderately glaucous. Leaves mostly alternate, the lower ones pinnately parted into 3 or 5 segments, reaching 3–4 cm but often smaller, the segments slender and entire to lanceolate and toothed; upper leaves often entire. Phyllaries distinct to base, 10.0–13.5 mm, yellow-green, turning light brown with age, the oil glands maroon, elliptic, often 1–2 mm; accessory bracts conspicuous. Disk florets yellow with red-purple tips, the rays orange-yellow, either about as

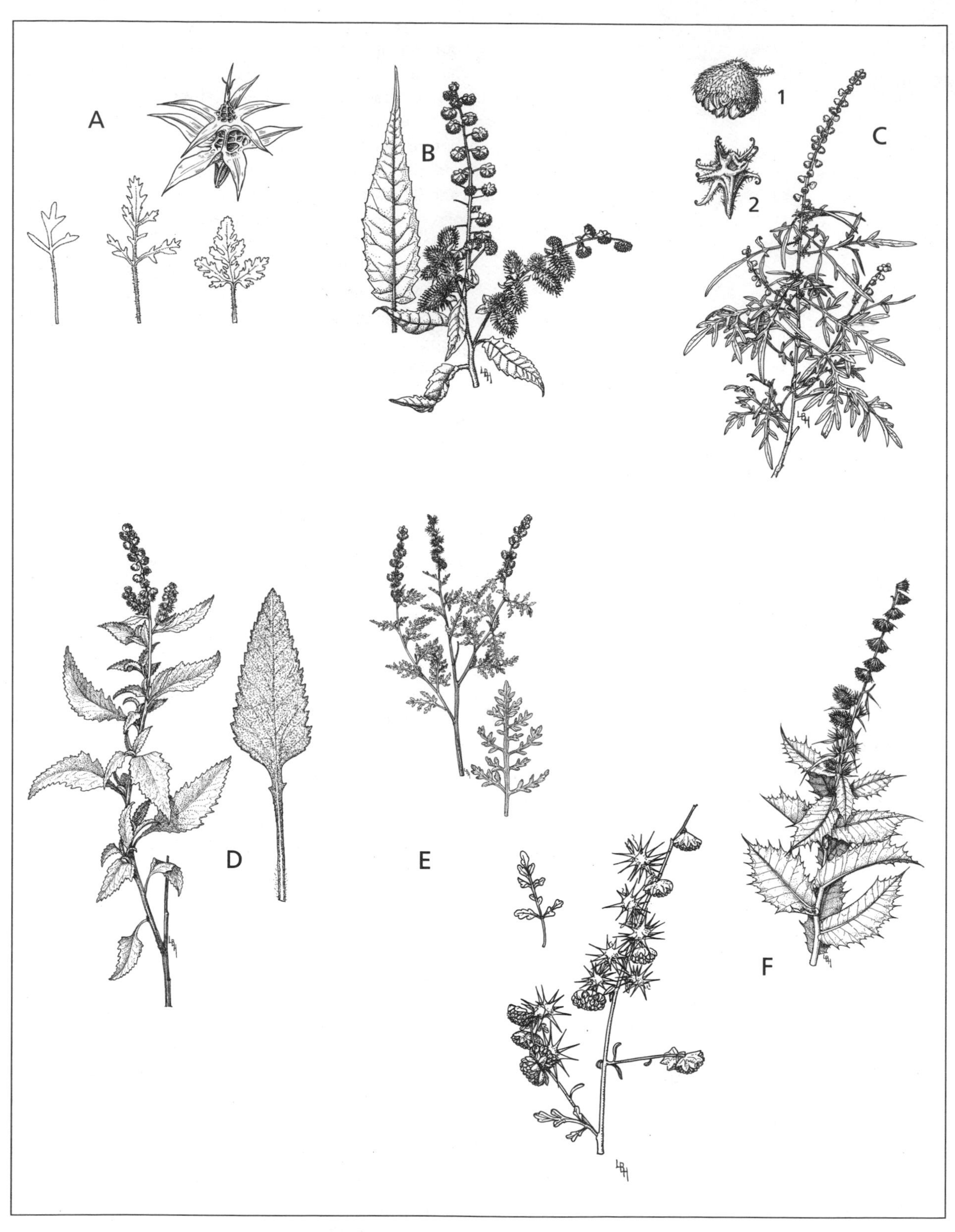

Ambrosia:* (A) *A. acanthicarpa;* (B) *A. ambrosioides;* (C) *A. confertiflora, (1) flower head with male flowers, (2) bur; **(D) *A. deltoidea;* (E) *A. dumosa;* (F) *A. ilicifolia.*** LBH (Benson & Darrow 1945, 1981; Humphrey 1960; Parker 1958).

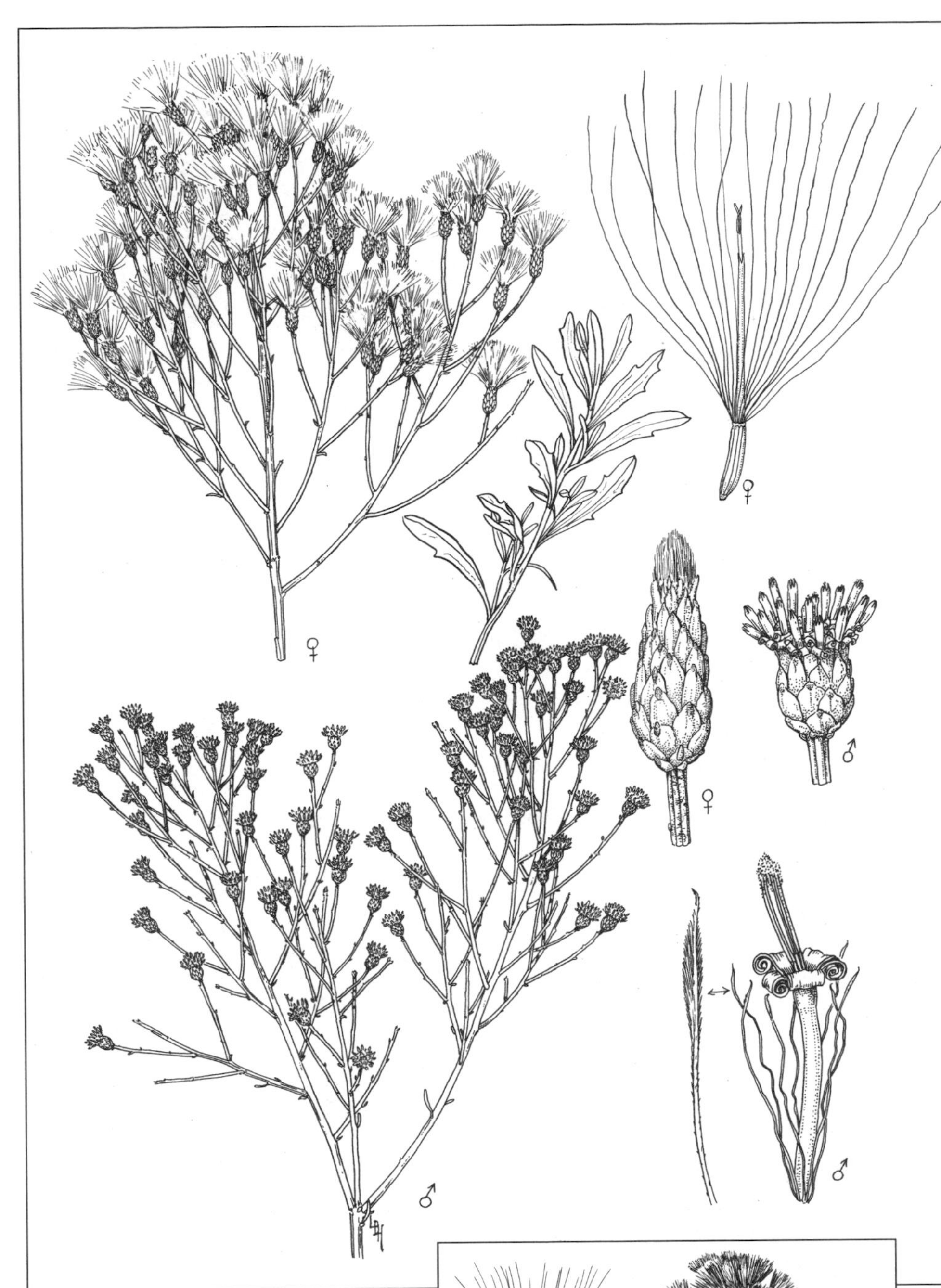

Baccharis sarothroides. LBH (Benson & Darrow 1981).

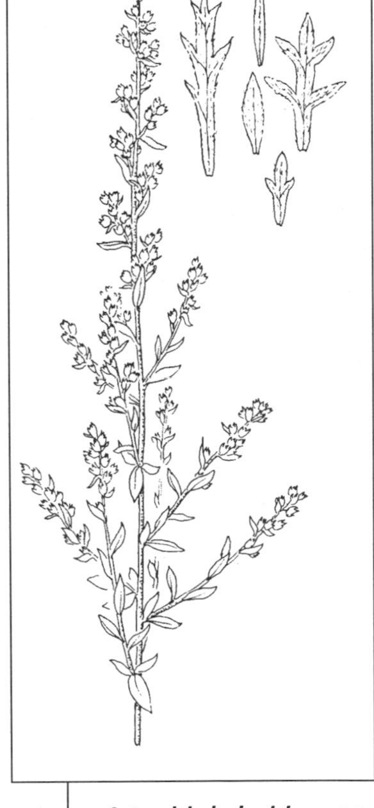

Artemisia ludoviciana. BA

Baccharis salicifolia. LBH (Benson & Darrow 1981; Parker 1958).

Baileya multiradiata. LBH (Schmutz, Freeman, & Reed 1968).

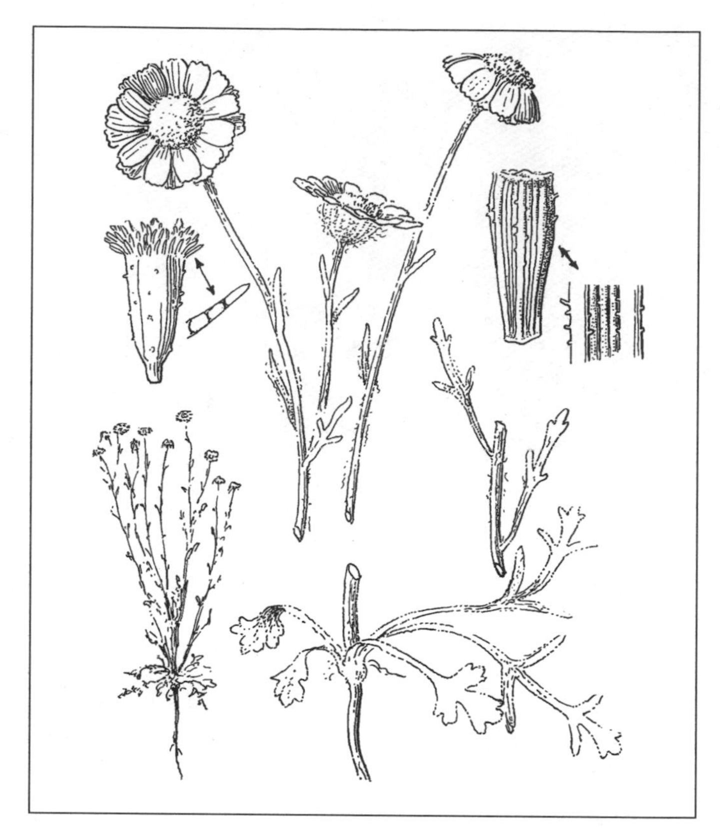

Baileya pleniradiata. JRJ (Abrams & Ferris 1960).

Brickellia coulteri var. ***coulteri.*** BA

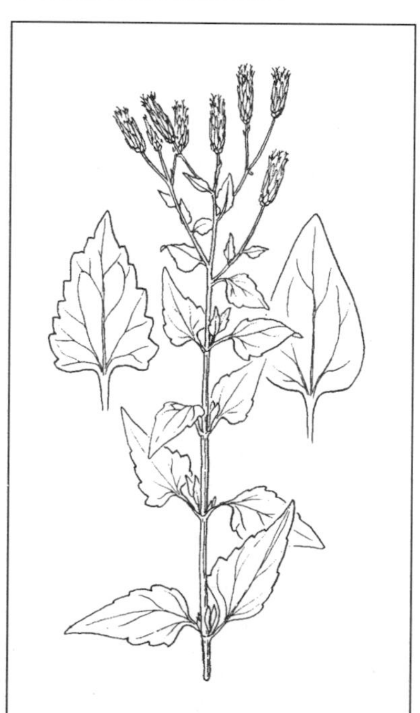

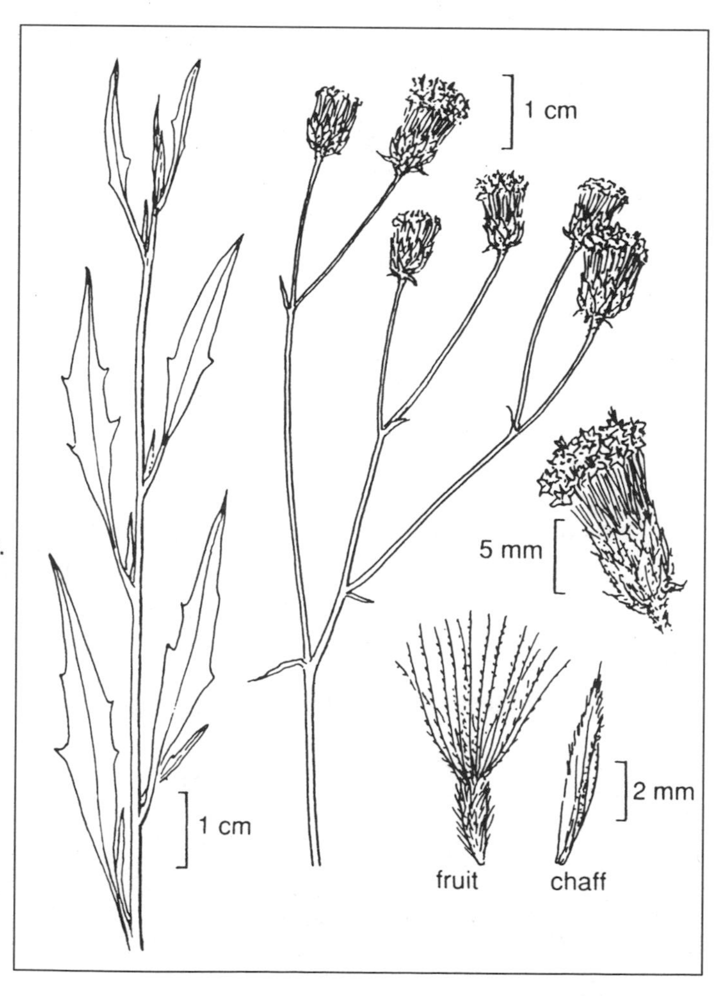

Bebbia juncea var. ***aspera.*** ER (Hickman 1993).

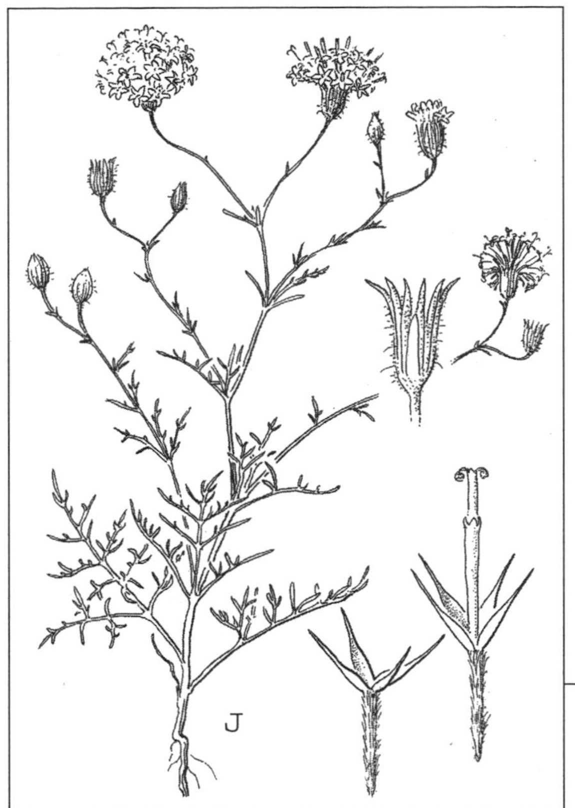

Chaenactis carphoclinia var. ***carphoclinia.*** JRJ (Abrams & Ferris 1960).

Centaurea melitensis. LBH (Parker 1972).

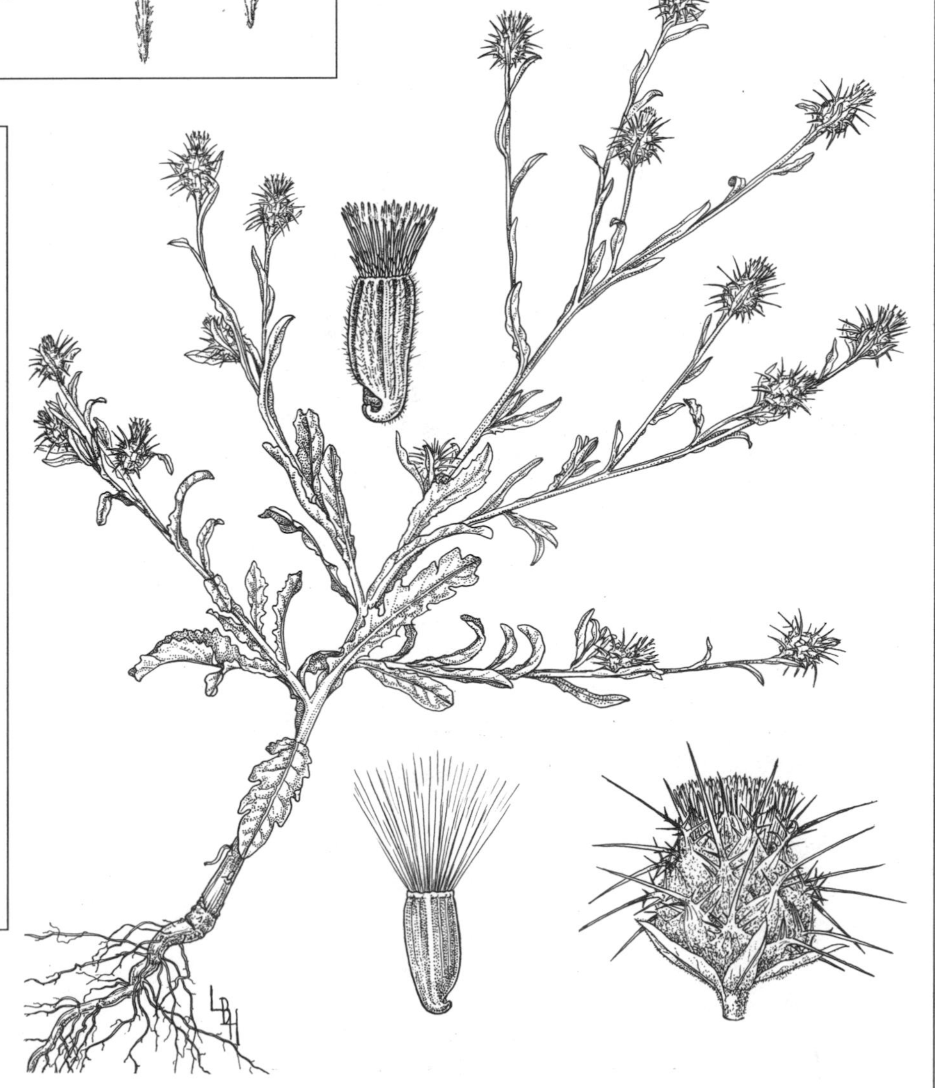

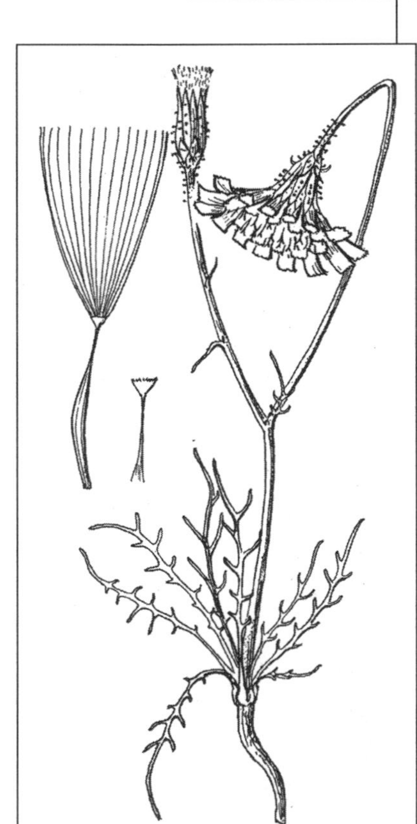

Calycoseris parryi. (Abrams & Ferris 1960).

Chloracantha spinosa var. ***spinosa.*** LBH (Parker 1958).

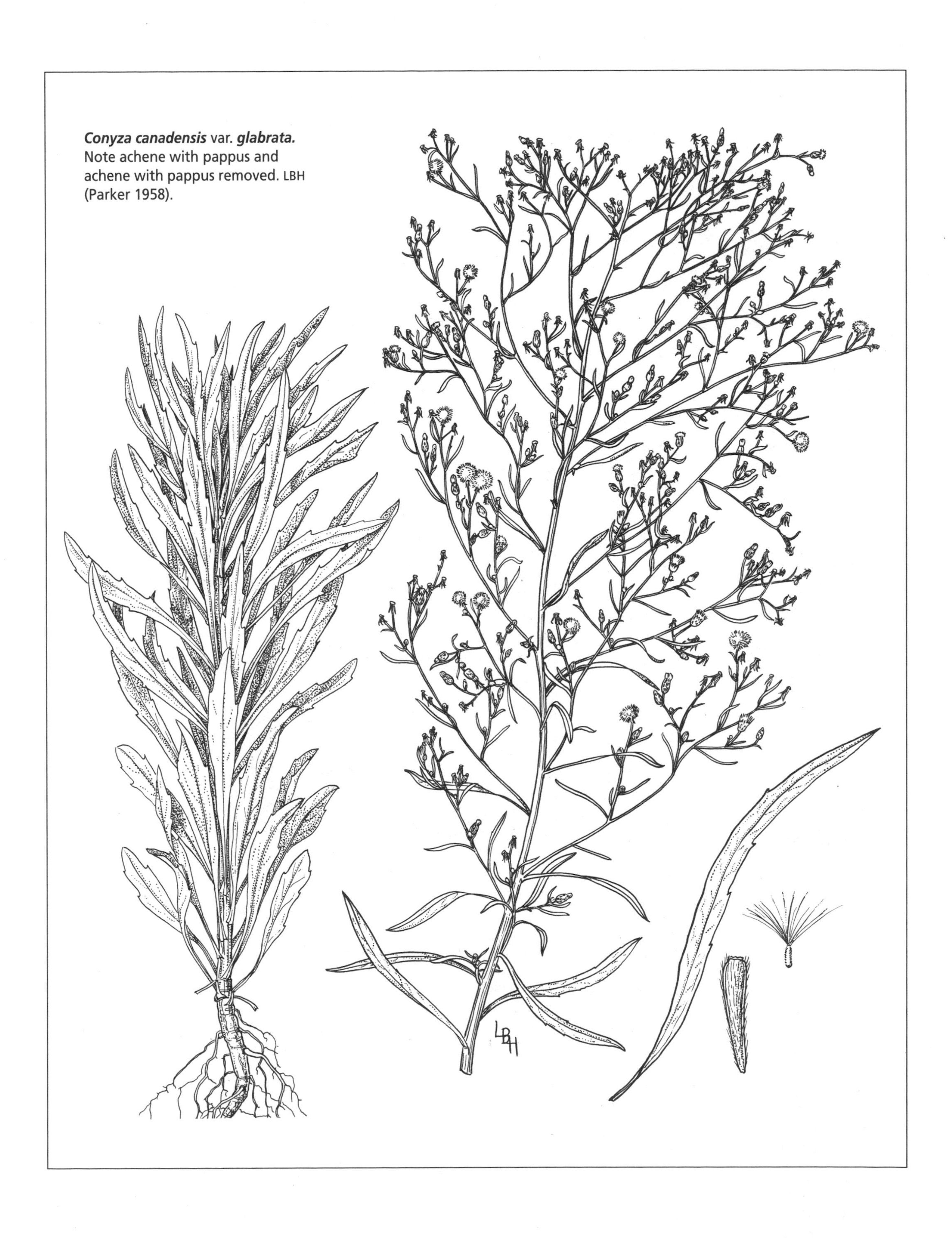

Conyza canadensis var. ***glabrata.*** Note achene with pappus and achene with pappus removed. LBH (Parker 1958).

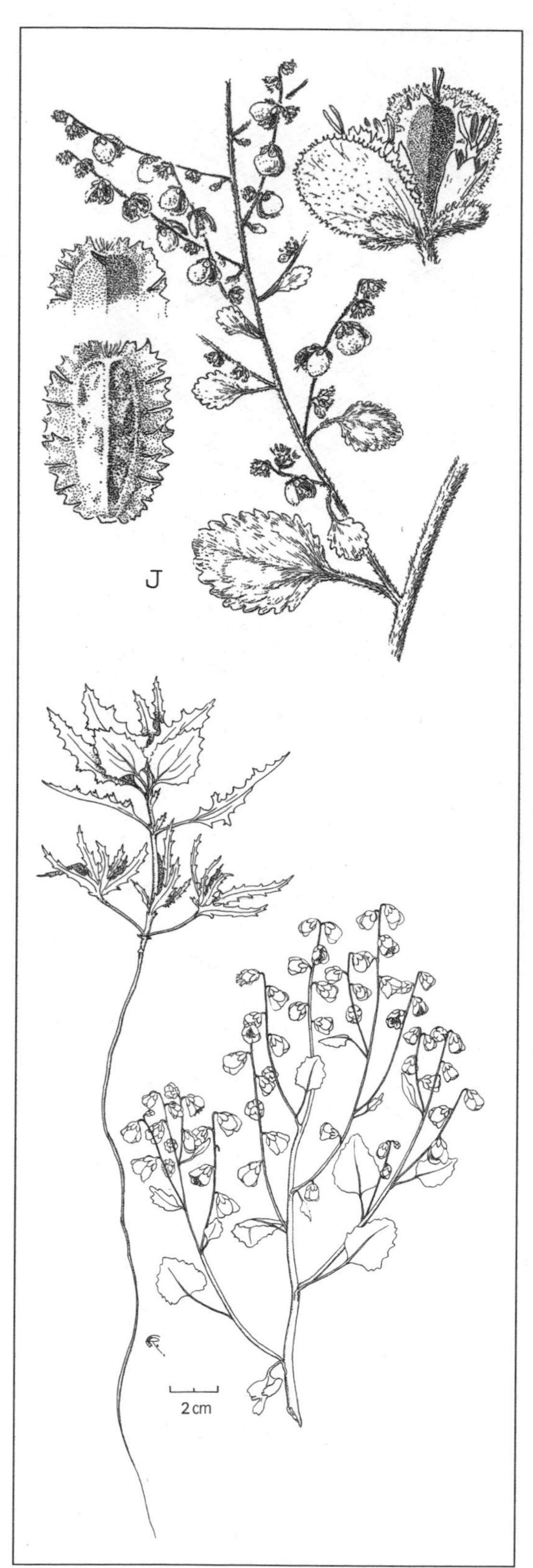

Dicoria canescens. JRJ (Abrams & Ferris 1960, above) and FR (Felger 1980).

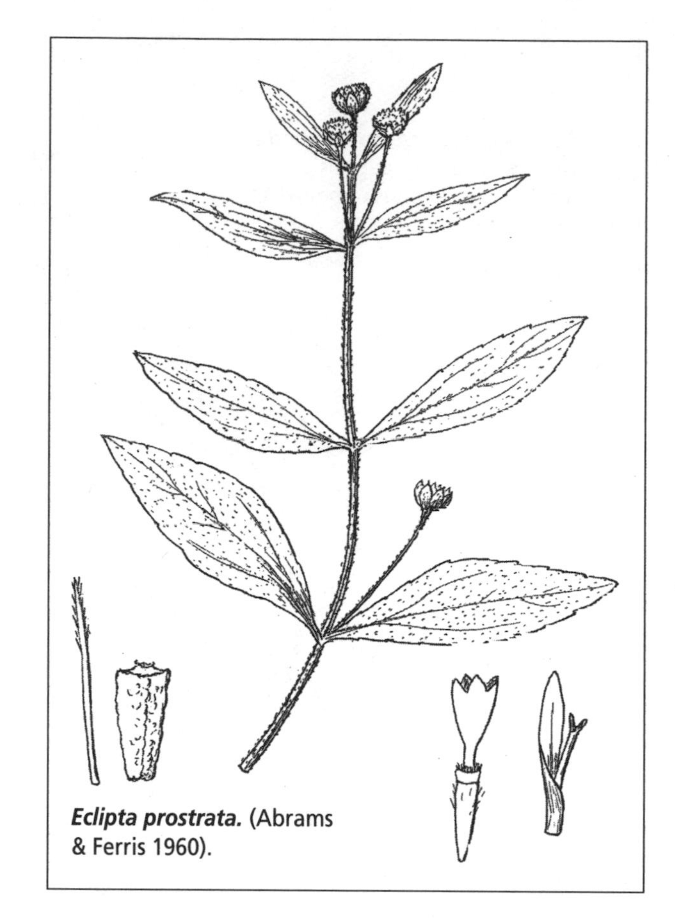

Eclipta prostrata. (Abrams & Ferris 1960).

Encelia farinosa var. ***farinosa.*** LBH (Benson & Darrow 1945).

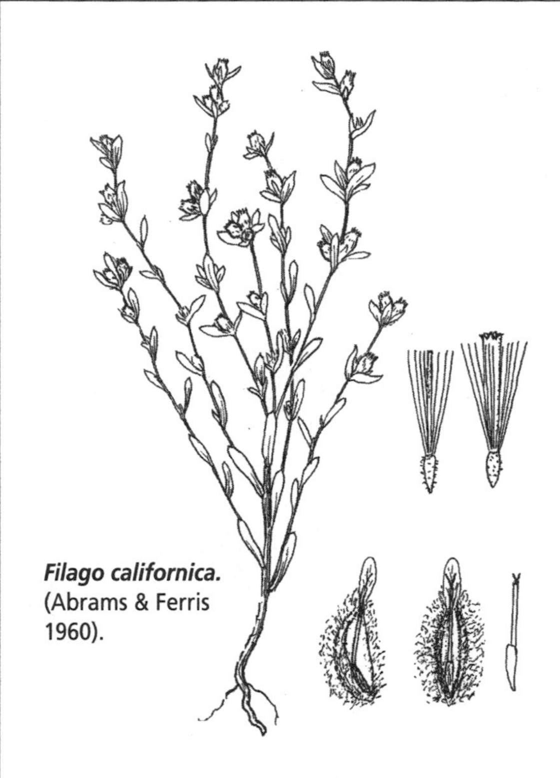

Filago californica. (Abrams & Ferris 1960).

Filaginoid and gnaphaloid flower heads. Diagrams showing relative positions of bracts, pistillate flowers (solid circles), and disk florets (open circles) comprising bisexual flowers (open circle with dot) and staminate flowers (open circle without dot): ***(A) Filago; (B) Stylocline; (C) Gnaphalium.*** MBJ.

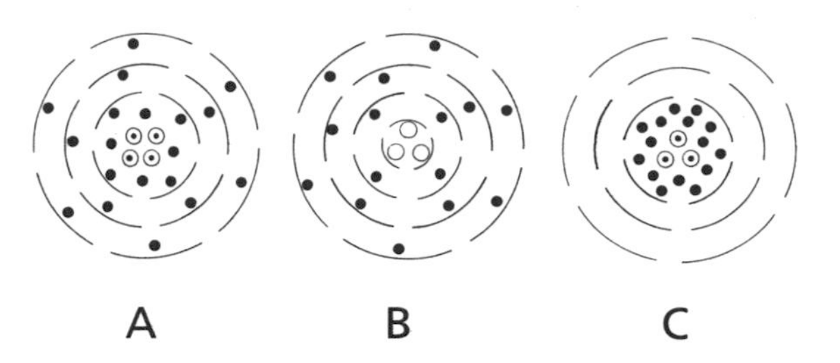

Filago achenes. Upper row, achenes of outermost pistillate florets; lower row, achenes of bisexual inner florets: ***(A) F. arizonica; (B) F. california; (C) F. depressa.*** MBJ (from scanning electron microscope photos by James D. Morefield).

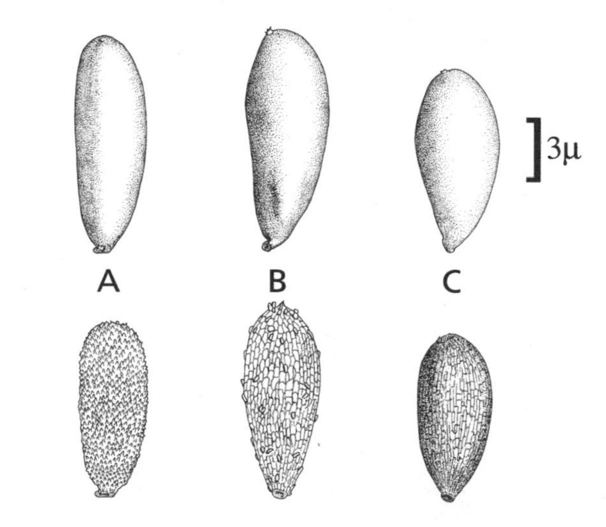

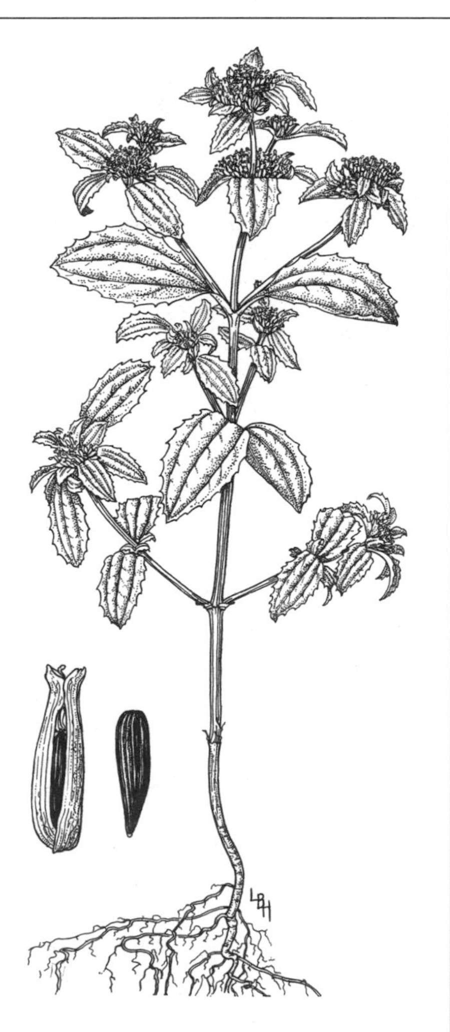

Flaveria trinervia. Note two phyllaries enclosing an achene. LBH.

Gaillardia arizonica. RAJ (Cronquist 1994).

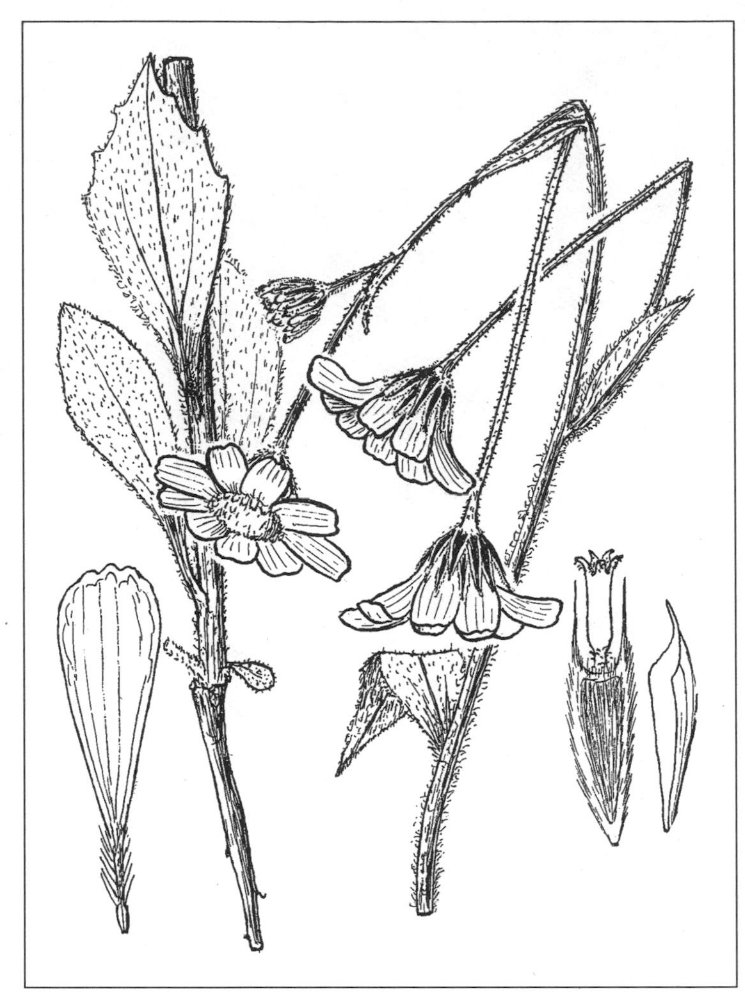

Geraea canescens. (Abrams & Ferris 1960).

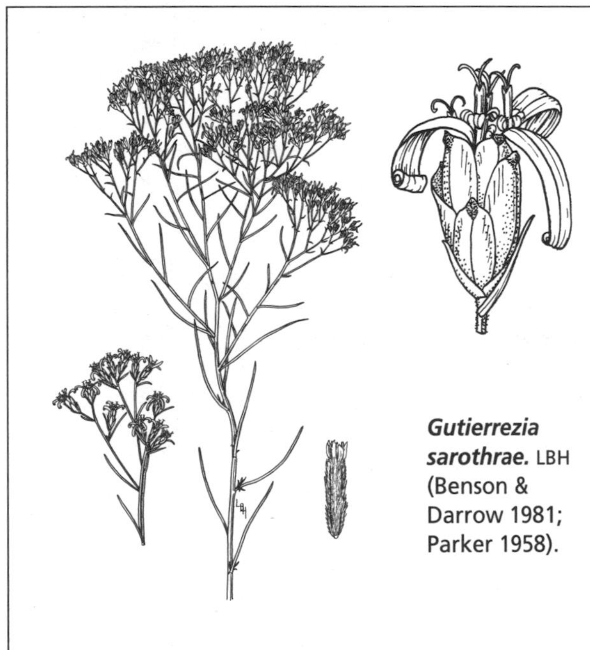

Gutierrezia sarothrae. LBH (Benson & Darrow 1981; Parker 1958).

Gymnosperma glutinosum. BA. © James Henrickson 1999.

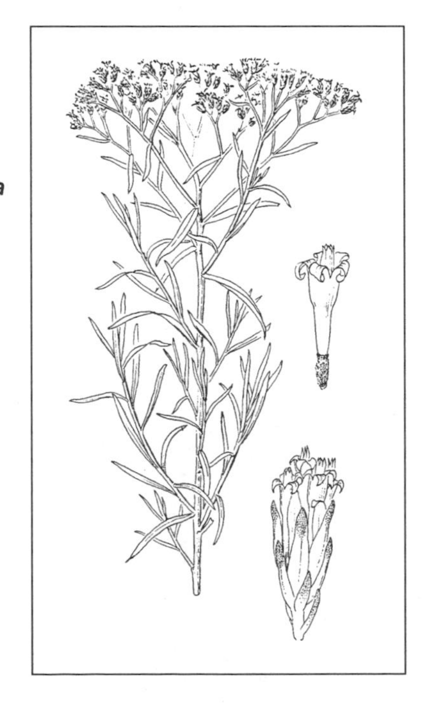

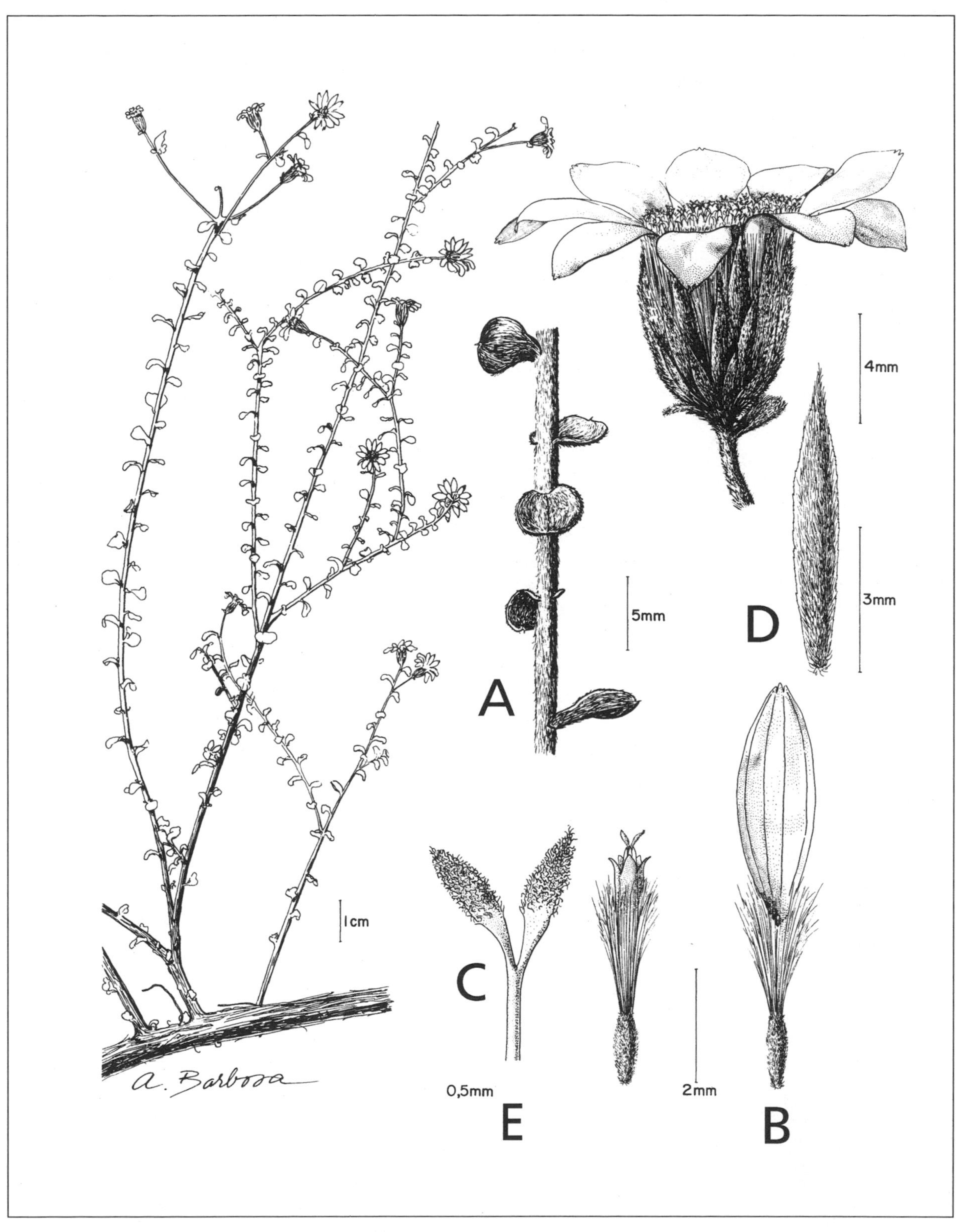

Heterotheca thiniicola: (A) stem with leaves; (B) ray floret; (C) disk floret; (D) inner phyllary; (E) style branches of a disk floret. AB (Rzedowski & Ezcurra 1986).

Hymenoclea monogyra. LBH (Benson & Darrow 1981).

Hymenoclea salsola var. ***pentalepis.*** JRJ (Abrams & Ferris 1960).

Hymenoxys odorata. LBH (Parker 1958).

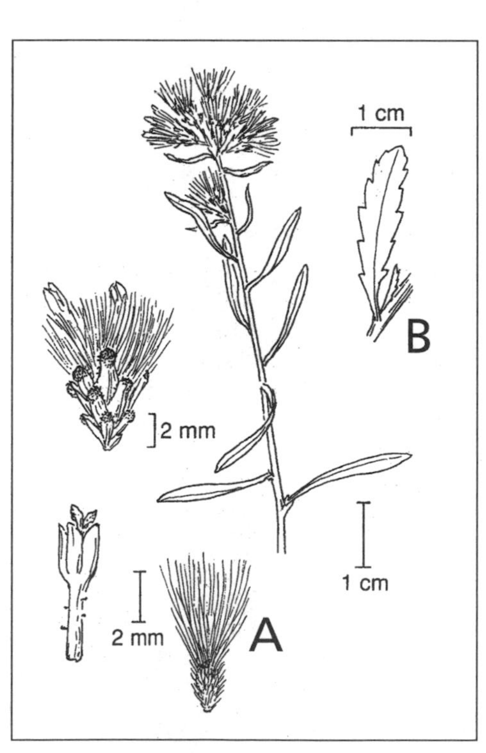

Isocoma acradenia: (A) var. ***acradenia; (B)*** var. ***eremophila.*** ER(Hickman 1993).

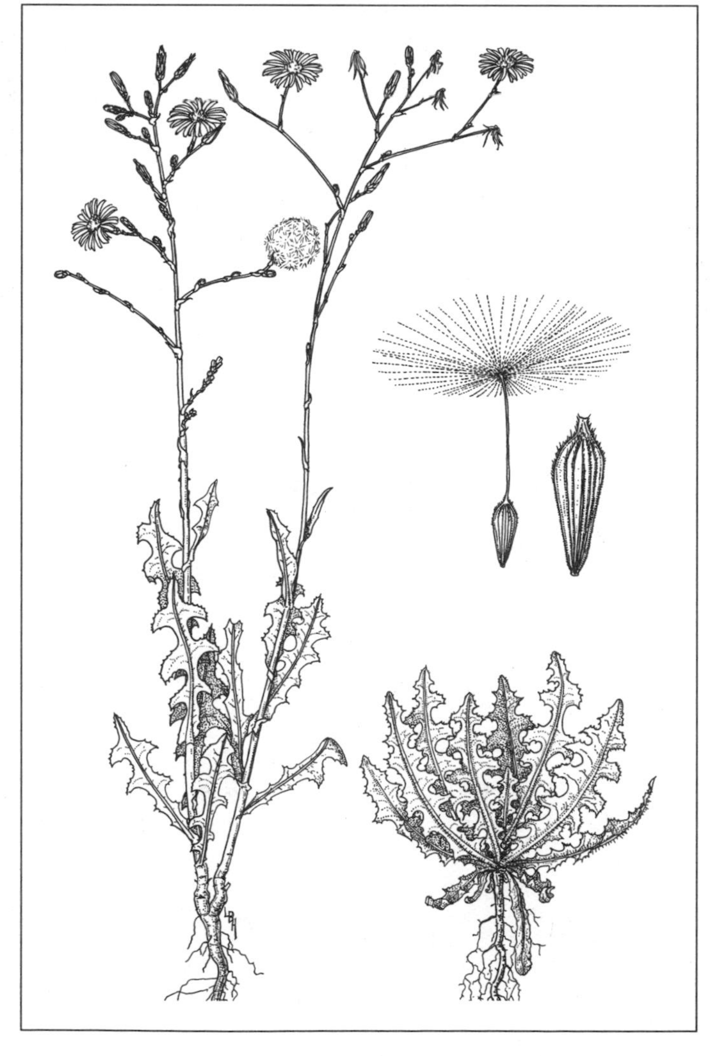

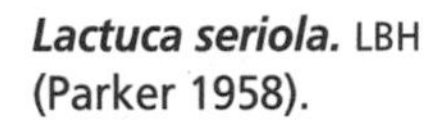

Lactuca seriola. LBH (Parker 1958).

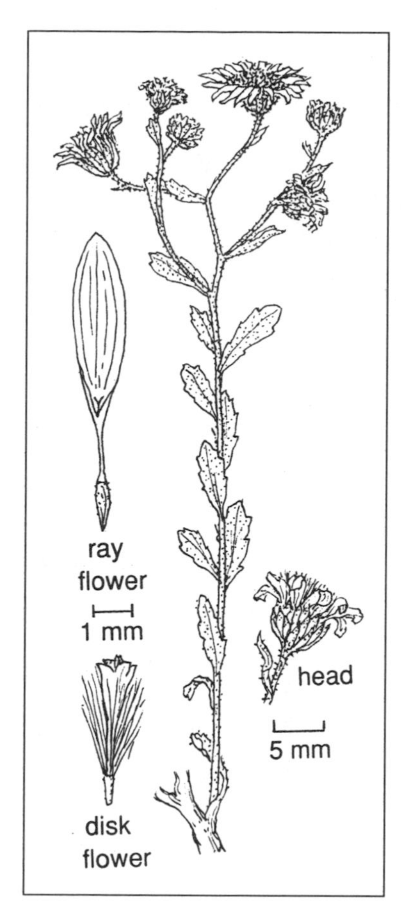

Machaeranthera coulteri var. ***arida.*** ER (Hickman 1993).

Malacothrix glabrata. RAJ (Cronquist 1994).

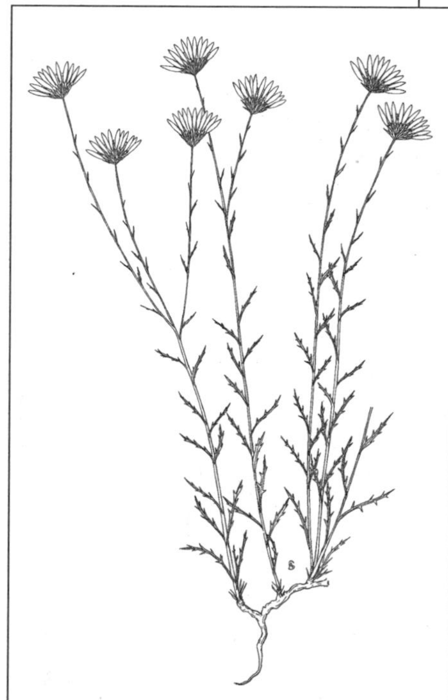

Machaeranthera pinnatifida subsp. ***gooddingii.*** AS (Cronquist 1994).

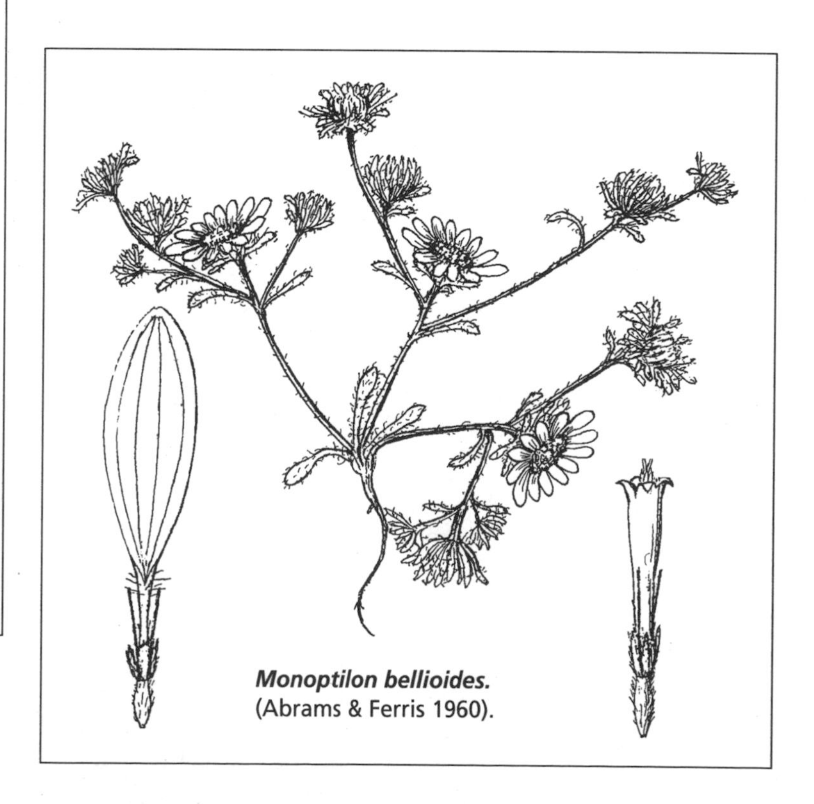

Monoptilon bellioides. (Abrams & Ferris 1960).

Palafoxia arida var. ***arida*** FR (Felger 1980).

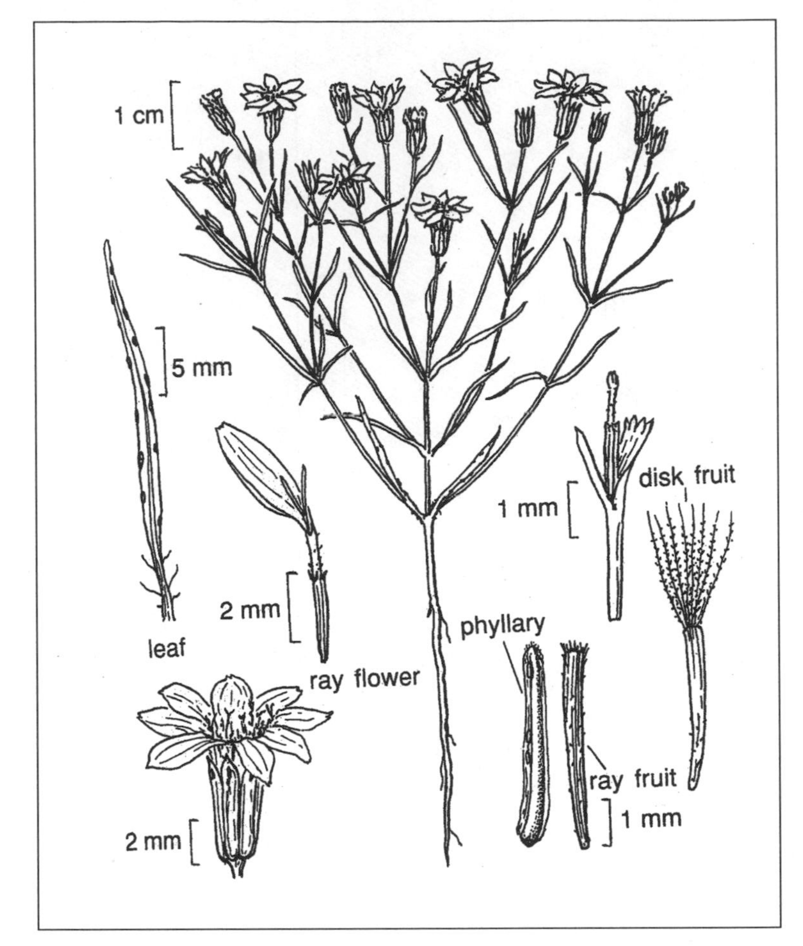

Pectis papposa var. ***papposa.*** ER (Hickman 1993).

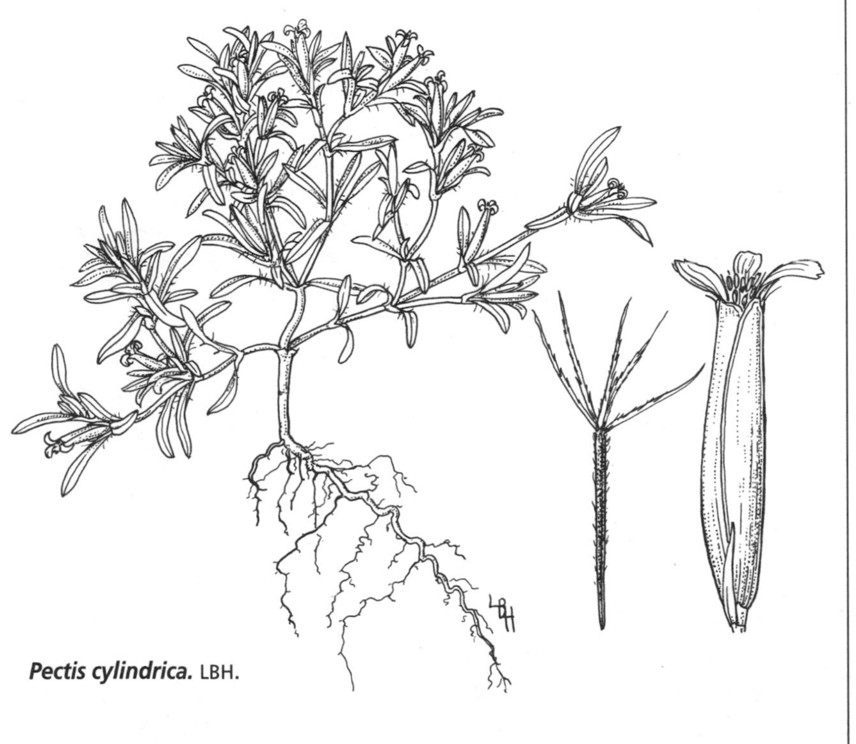

Pectis cylindrica. LBH.

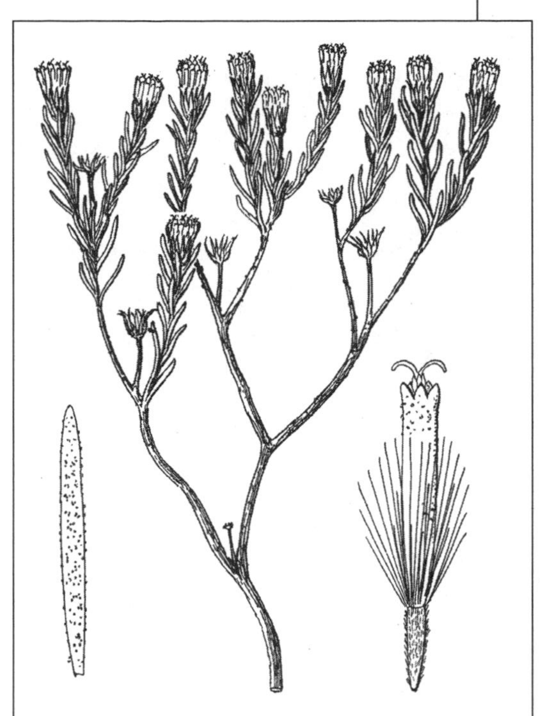

Peucephyllum schottii.
(Abrams & Ferris 1960).

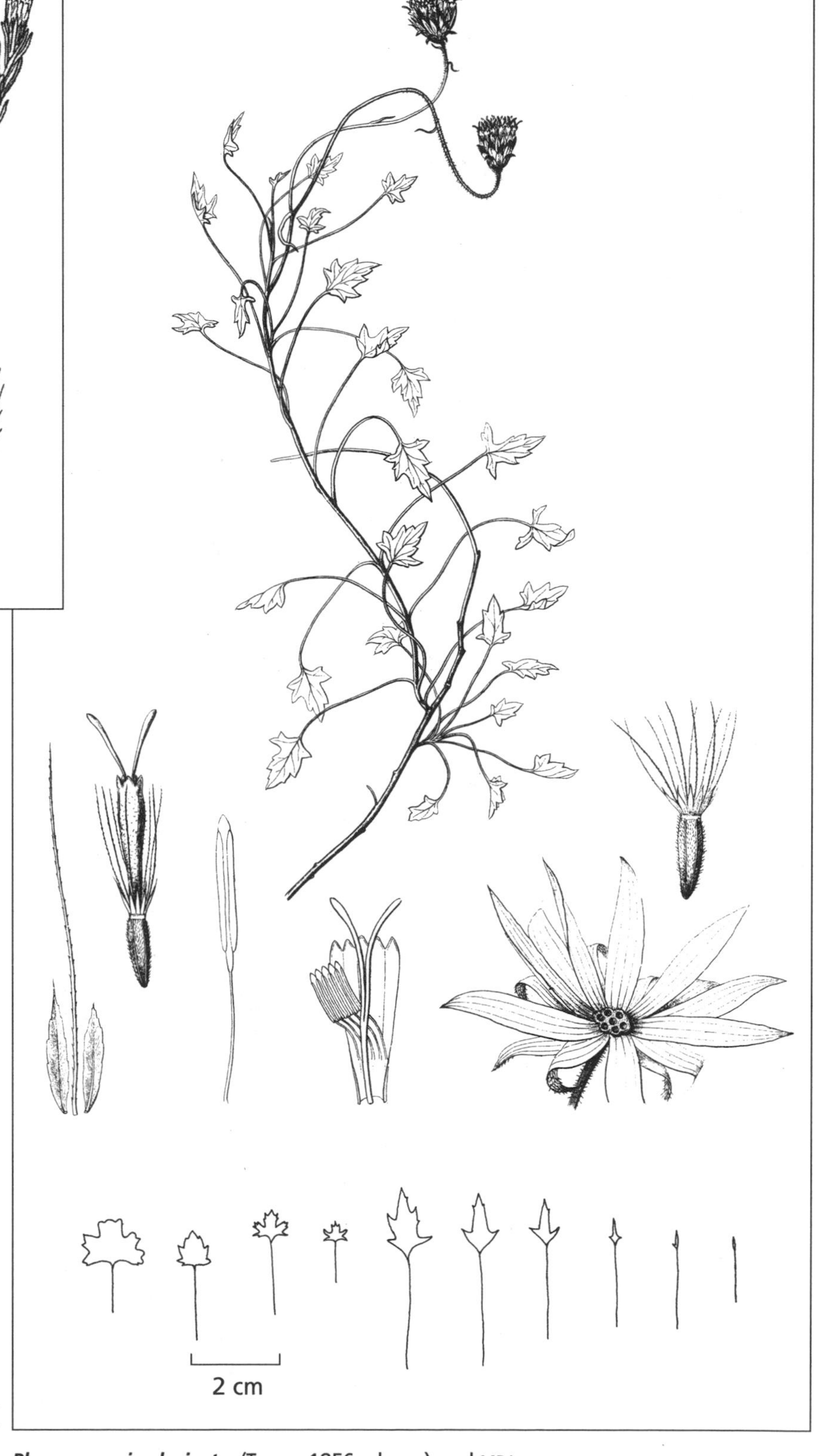

Pleurocoronis pluriseta. (Torrey 1856, above) and MBJ.

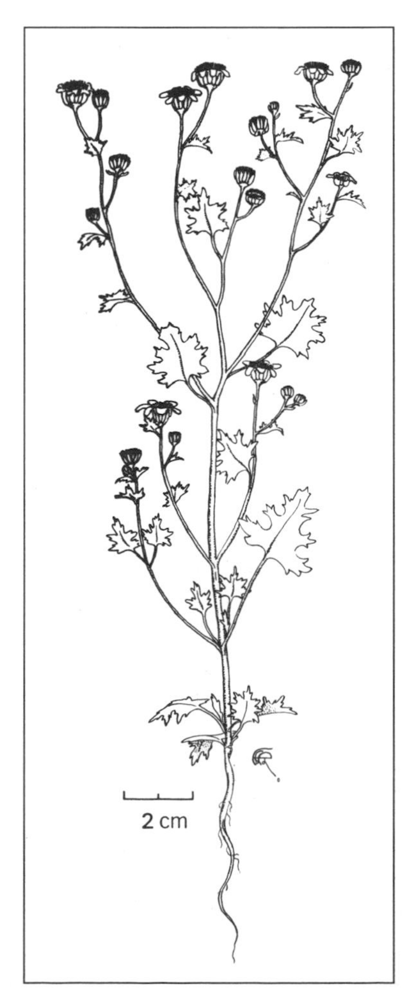

Perityle emoryi. fr (FELGER & MOSER 1985).

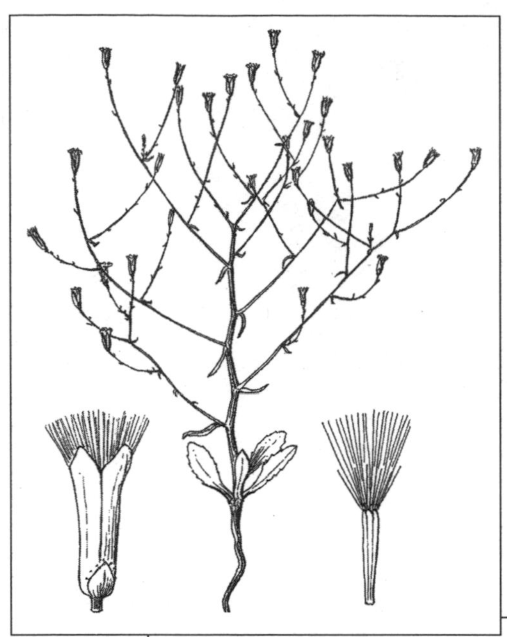

Prenanthella exigua. (Abrams & Ferris 1960).

Porophyllum gracile. RAJ (Cronquist 1994).

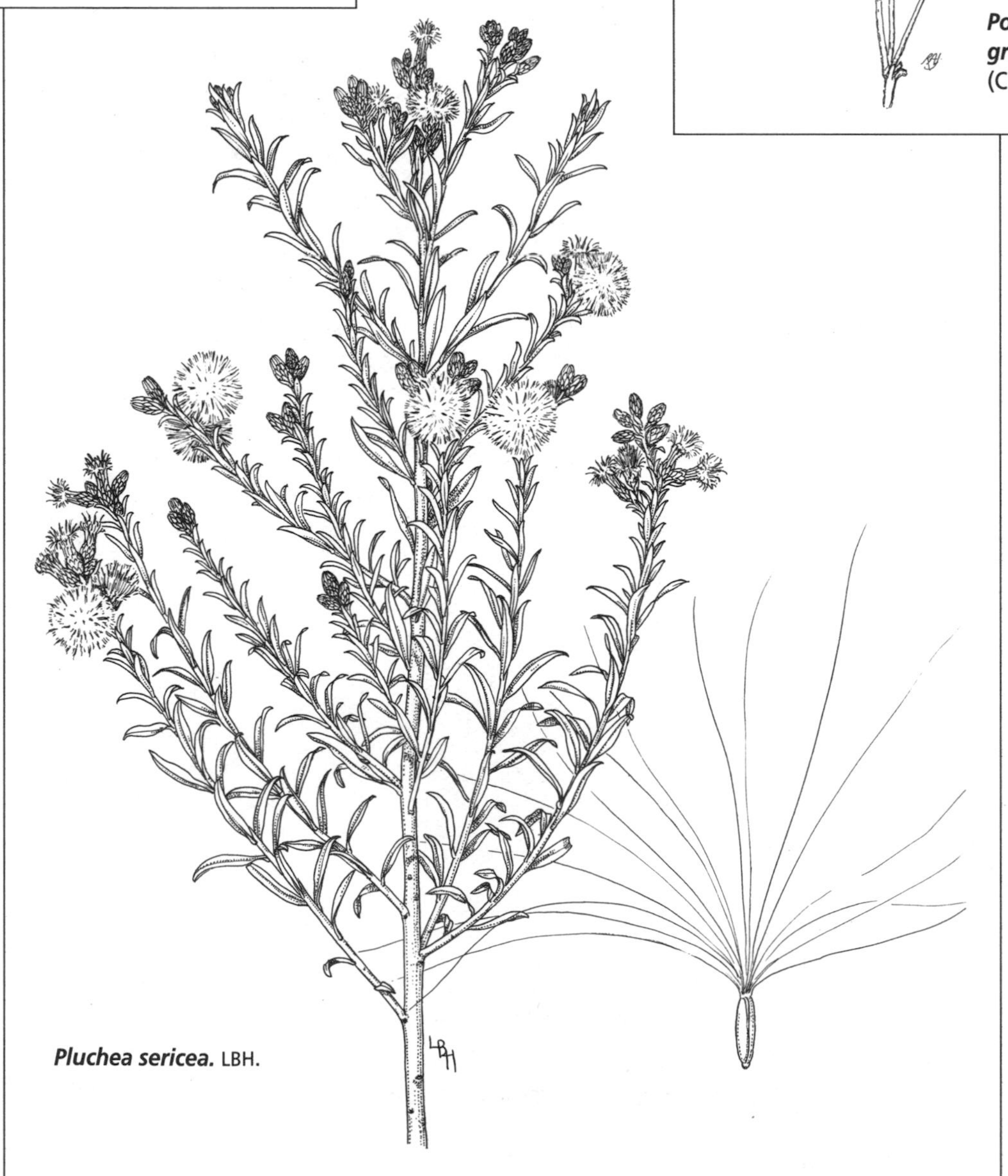

Pluchea sericea. LBH.

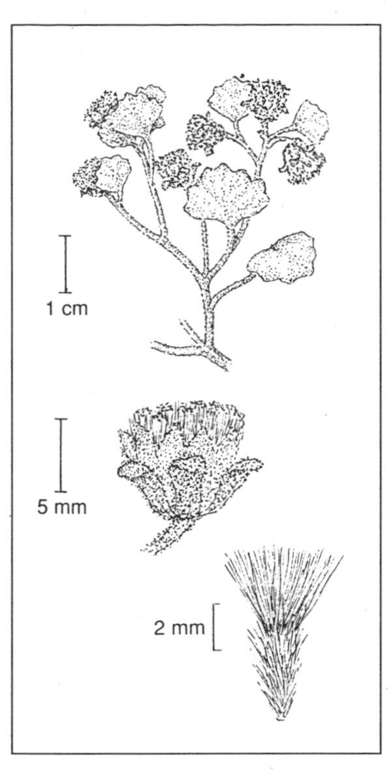

Psathyrotes ramosissima. ER (Hickman 1993).

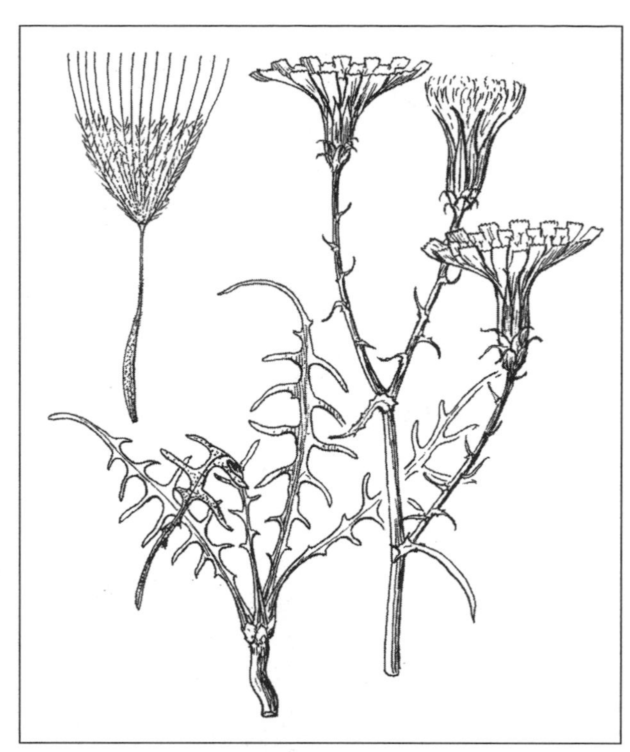

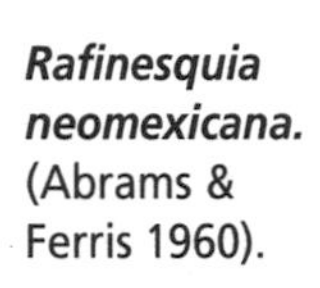

Rafinesquia neomexicana. (Abrams & Ferris 1960).

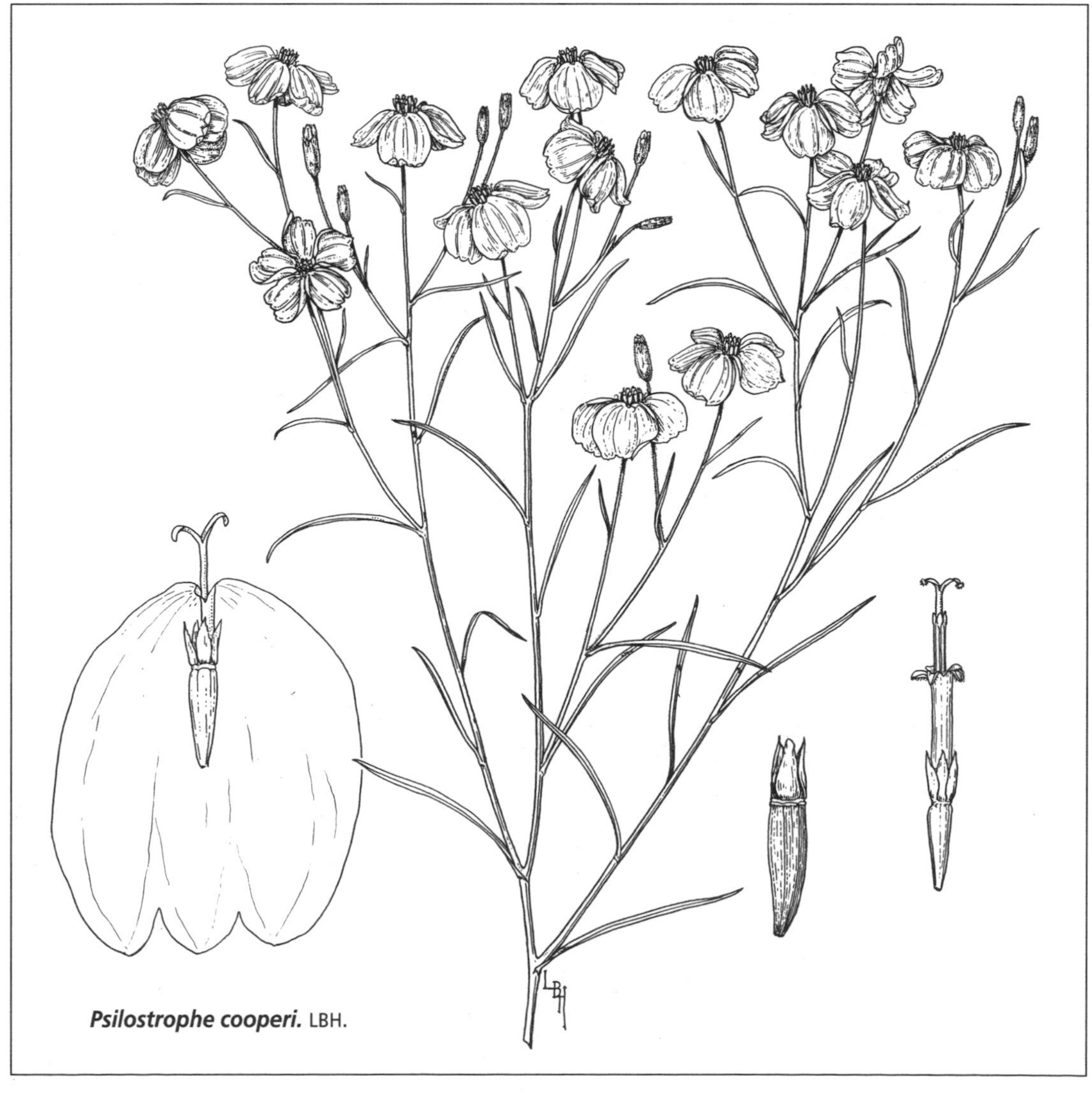

Psilostrophe cooperi. LBH.

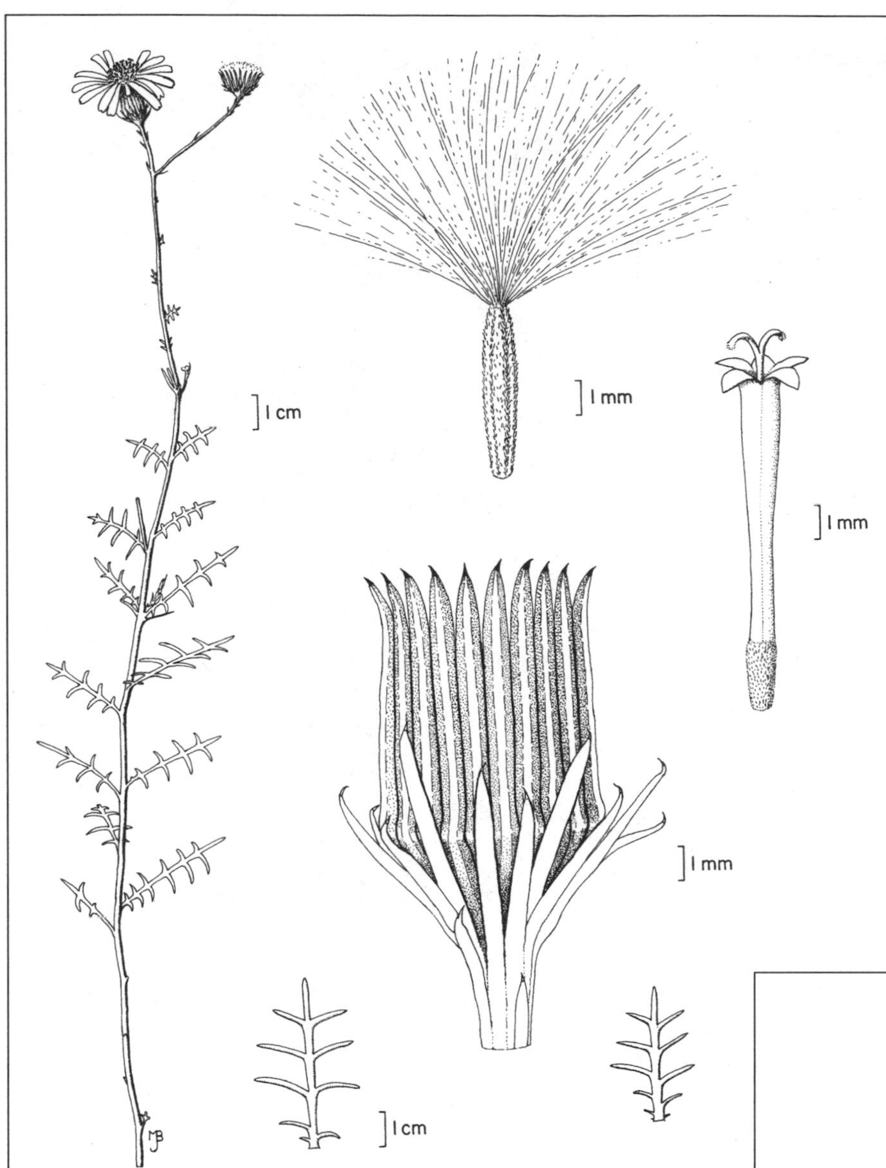

Senecio pinacatensis. MBJ (Felger 1991).

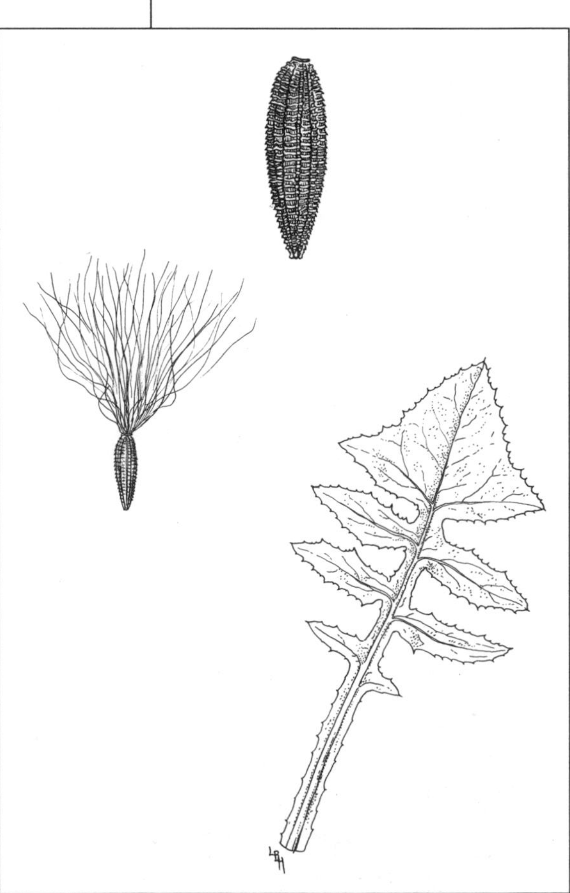

Sonchus oleraceus. LBH (Parker 1972).

Sonchus asper subsp. ***asper.*** LBH (Parker 1958).

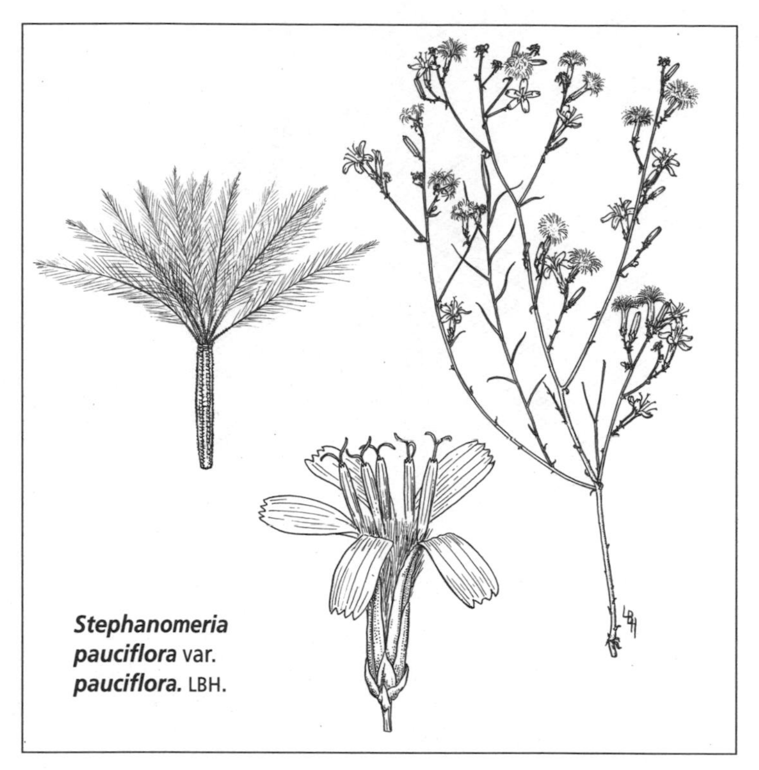
***Stephanomeria pauciflora* var. *pauciflora*.** LBH.

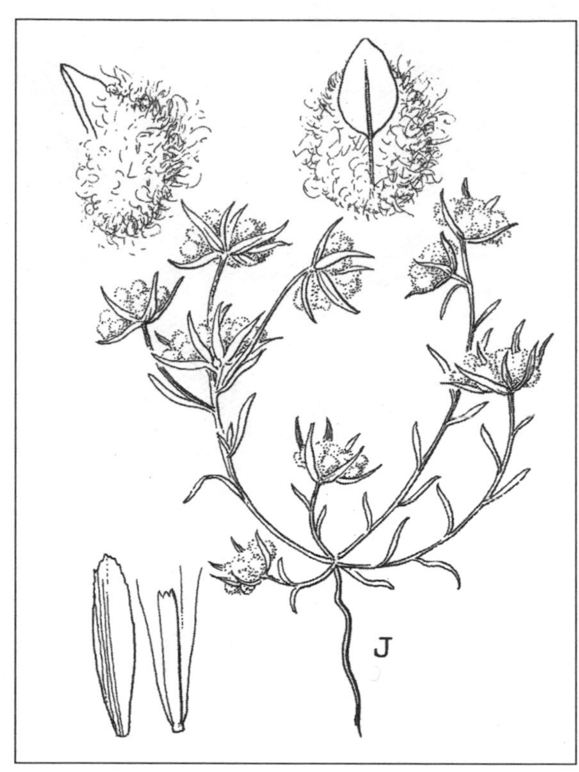

***Stylocline micropoides*.** JRJ (Abrams & Ferris 1960).

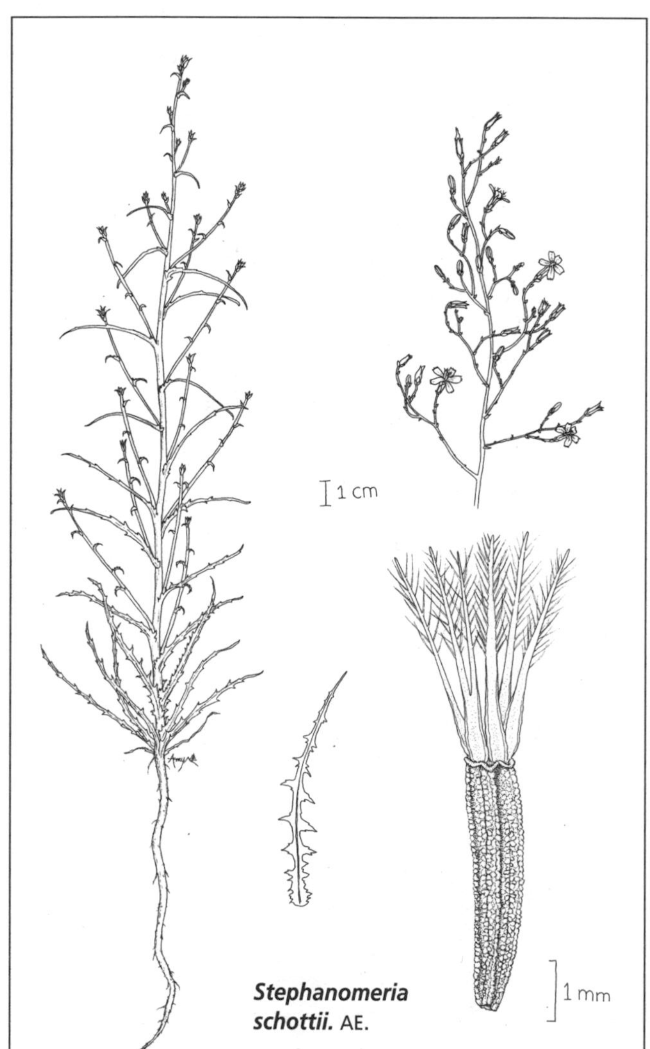

***Stephanomeria schottii*.** AE.

***Trichoptilium incisum*.** (Abrams & Ferris 1960).

Trixis californica var. ***californica.***
(Abrams & Ferris 1960).

Verbesina encelioides. LBH
(Parker 1972).

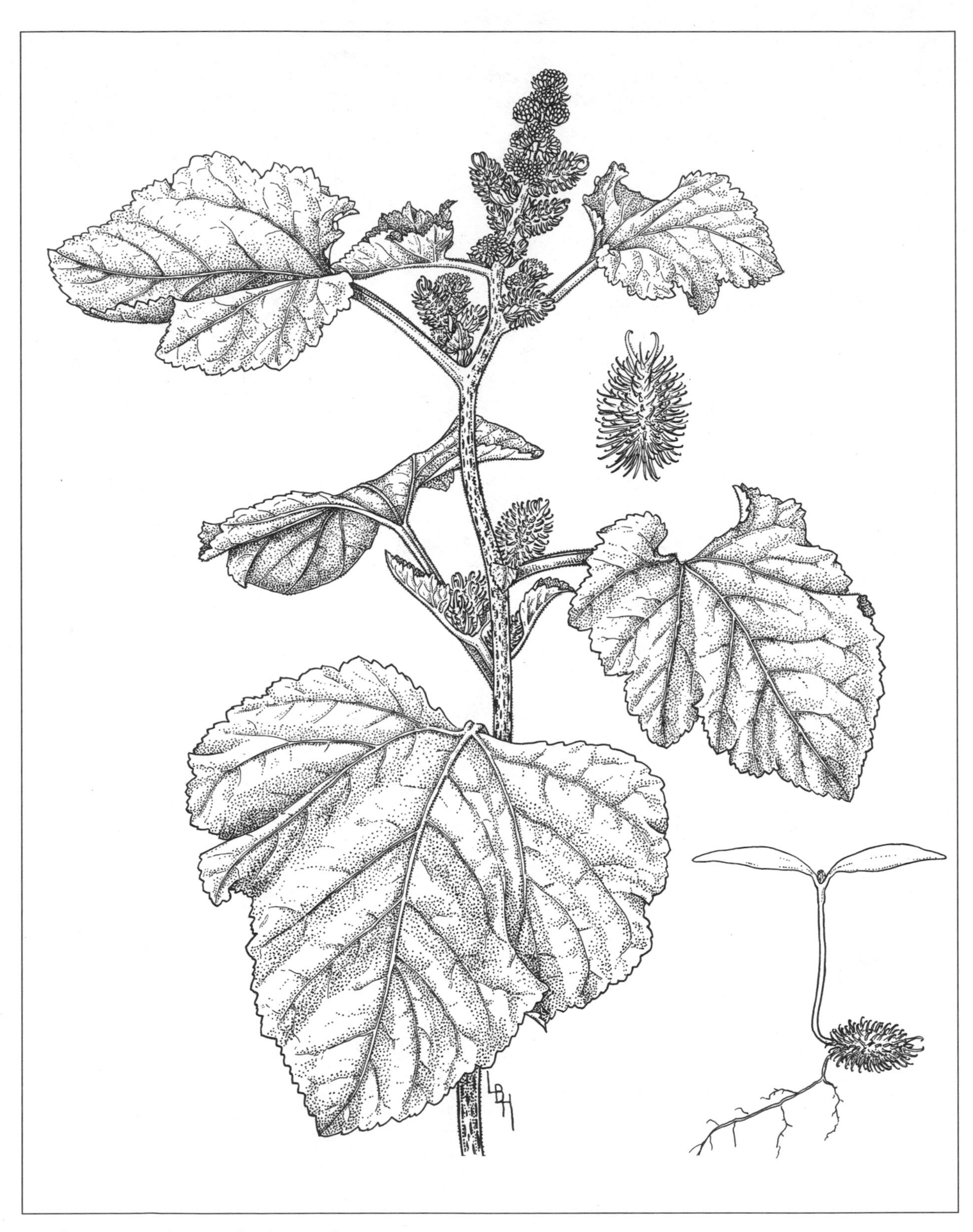

Xanthium strumarium. LBH (Parker 1958).

long as disk florets or absent. Achenes ca. 5 mm, blackish gray, with short, mostly curved, appressed hairs on slender ribs; pappus of 8–12 basally thickened scales divided into coarse, tawny brown bristles about twice as long as the achenes. March-April, sometimes October-December.

Sonoyta region; granitic soils of upper bajadas, mostly along small drainageways, lower slopes of mountains or larger hills, and more common on north-facing canyon and mountain slopes. More widespread at slightly higher elevations eastward in Sonora and in Organ Pipe Monument. Southeastern California, western and southern Arizona, both Baja California states, and southward in western Sonora to the vicinity of Puerto Libertad.

10 km SW of Sonoyta, *F 88-158.* Sierra de los Tanques, *F 89-30.*

AMBROSIA RAGWEED, BURSAGE

Annuals and herbaceous or shrubby perennials with sessile or stalked glands. Leaves alternate (except in *A. acanthicarpa*). Monoecious. Inflorescences spicate to racemose, the heads unisexual with disk florets, the staminate heads above pistillate heads (or intermixed in *A. dumosa*); wind-pollinated, the pollen hay-fever causing. Staminate heads with cup- or plate-shaped involucres, the phyllaries united at least basally; bearing chaffy bracts each with 1 or more florets; pappus none. Pistillate heads with 1 to several florets and beaks, each beak representing a floret, the fruiting involucral bracts (phyllaries) hard, developing into a spiny bur or nutlike structure (measurements for burs include the spines; each spine representing the distal portion of a phyllary); corollas and pappus none. Seeds germinating within the bur.

Two shrubby bursages, *A. deltoidea* and *A. dumosa,* are among the most widespread and abundant perennials in the regional vegetation. Warm regions of the world, mostly southwestern North America; 43 species. *Ambrosia* is closely allied to *Hymenoclea* and *Xanthium* (Karis 1995). Reference: Payne 1964.

1. Herbaceous, not woody.
    2. Annuals; spines of bur straight, flattened, and often channeled. ________ **A. acanthicarpa**
    2′ Perennials, stems dying back to the ground; most spines of bur hooked and terete. ________ **A. confertiflora**

1′ Shrubby, stems woody at least at base.
    3. Leaves pinnately to tripinnately deeply dissected. ________ **A. dumosa**
    3′ Leaf margins toothed or rarely nearly entire, not deeply dissected.
        4. Leaves sessile, firm, with spine-tipped teeth. ________ **A. ilicifolia**
        4′ Leaves petioled, "soft," and flexible; marginal teeth, if present, not spine-tipped.
            5. Shrubs with erect, few-branched stems usually more than 1 m tall; leaf blades more than 7 cm; burs with hooked spines. ________ **A. ambrosioides**
            5′ Subshrubs usually less than 0.8 m tall with much-branched, spreading stems; leaf blades less than 5 (6.5) cm; burs with straight spines (rarely a few hooked spines near apex). ________ **A. deltoidea**

Ambrosia acanthicarpa Hooker [*Franseria acanthicarpa* (Hooker) Coville] ANNUAL BURSAGE. Annuals, apparently non-seasonal, often 0.5–1 m, forming a deep taproot, extremely variable—with a single main stem to dense, much-branched, globose, and bushy; gray-green with dense, coarse, white hairs (strigose-hispid). Leaves mostly 2.5–8.0 cm, opposite below, alternate above, petioled, the blades lobed to bipinnate, the segments narrowly linear to broadly ovate. Burs with several or more sharp spines, the lower spines larger, 4.5–8.0 mm, linear to lanceolate or narrowly deltate, straight, and flat to channeled.

Sandy and gravelly soils; roadsides, washes, sand flats, and low dunes. Often one of the most abundant summer weeds in the vicinity of Puerto Peñasco and in some years northward as a roadside weed nearly to Sonoyta; also on relatively undisturbed dunes near the coast in the vicinity of Puerto Peñasco. Not known elsewhere in Sonora or in nearby parts of Arizona. Western Canada and western United States to Chihuahua and Sonora; many of the populations are geographically disjunct.

This species is unique for the genus in having opposite as well as alternate leaves. Plants from gravelly, roadside soils near Sonoyta have highly dissected leaves comparable to specimens from Arizona and California. Plants on dunes and unstable sand near the coast tend to be more robust and larger, nearly globose, and almost like a tumbleweed when mature; they have stouter stems, seem to be more densely white pubescent, and have larger, thicker leaf blades with larger, broader lobes. The dune and non-dune plants appear to intergrade. Payne (1964) pointed out extreme variation in leaf morphology in other regions.

Puerto Peñasco, *F 85-63.* 24 mi SW of Sonoyta, *F 13212.* 20 mi S[W] of Sonoyta, *Bowers 2588.*

Ambrosia ambrosioides (Cavanilles) W.W. Payne [*Franseria ambrosioides* Cavanilles] *CHICURA;* CANYON RAGWEED; *ÑUNUWI JE:J*. Shrubs mostly 1.2–2+ m, the stems scarcely woody, relatively slender, dark colored and few-branched. Herbage viscid resinous-glandular (especially when young), and with coarse, mostly spreading, white hairs. Petioles 2–4 cm; leaf blades lanceolate, often 7–25 cm, rough surfaced, commonly studded with small insect galls, the margins ragged-toothed. Tardily drought deciduous, the dry, dead leaves persisting for 1 or 2 seasons; leaves and young stems frost killed, but the plants quickly recovering. Burs like a cocklebur (*Xanthium*), ca. 15 mm, ellipsoid, with many hooked spines and sessile and stalked greenish golden, glistening glands (visible with 10× magnification). March-May; fruiting in the same season.

Common beneath or near desert trees (especially ironwood, mesquite, and palo verde) in washes, arroyos, and canyon bottoms. This is one of the larger leaved plants in the flora area. Sonoyta region, especially the Río Sonoyta valley, lower elevations in the Pinacate volcanic field, larger drainageways in the open desert, and near the northwestern granitic hills and mountains westward to the alluvial valley between Sierra del Viejo and Cerro Pinto. Durango and Sinaloa to the southern half of Arizona, and Baja California Sur and the southern portion of Baja California.

W of Mina del Desierto, *F 88-98.* Tinajas de los Pápagos, *Turner 59-31.* MacDougal Crater, *F 10458.* 12 mi S of Sonoyta, *F 9825.* E side of Pinacate region, 31°45′ N, 113°22′ W, *F 87-31.*

Ambrosia confertiflora de Candolle [*Franseria confertiflora* (de Candolle) Rydberg] *ESTAFIATE;* SLIMLEAF RAGWEED; *MO'OTADK*. Herbaceous perennials from a hard, knotty base, with stout, deeply buried, woody taproots; stems often 40–75 cm, erect, and leafy with white, mostly appressed hairs. Leaves green, often 6–17 cm, 2 or 3 times pinnately divided. Burs 3–4 mm with small, terete, hooked spines. Winter dormant, the new shoots appearing in spring and/or following summer rains; tardily drought deciduous and/or stems dying back to the ground. Flowering and fruiting May-June and mostly September-December.

Locally common in the Sonoyta and Pinacate regions, near the granitic mountains, and on desert flats; clayish soil flats, playas, and other poorly drained low-lying places such as roadside ditches, where it is often associated with *Argemone gracillenta, Euphorbia albomarginata,* and *Malvella* spp.; an extensive population stretches across Playa Díaz. Also often in sandy-gravelly washes and arroyo beds with ironwood, mesquite, and palo verde; common in disturbed habitats in towns and farmland, often in heavily grazed habitats (not palatable to cattle). Southwestern United States to Zacatecas and the Baja California Peninsula; often weedy.

1 km SSW of Tinajas de los Pápagos, *F 10606.* Molina Crater, *F 10643.* Playa Díaz, *F 10810.* Puerto Peñasco, *F 85-794.*

Ambrosia deltoidea (Torrey) W.W. Payne [*Franseria deltoidea* Torrey] *CHAMIZO FORRAJERO;* TRIANGLE-LEAF BURSAGE; *TAḌSAḌ, WA:GITA*. Bushy subshrubs often 25–100 cm, with slender, leafy, few-branched stems; tardily drought deciduous. Leaves gray-green to olive-green, densely short woolly, especially during drier conditions; with age becoming sticky with glandular exudate, especially on the upper surfaces, matting the hairs so that the leaves appear glabrous; new leaves greener and less hairy during relatively more favorable times. Leaf blades (8) 12–50 (65) mm, roughly triangular-ovate, often highly variable within a population; petioles (5) 6–15 (34) mm. Burs 6.5–11.5 mm wide and about as long, densely glandular with sessile and short-stalked glands, ranging from very sparse pubescence of white hairs to densely woolly (often even on the same branch) especially in the axils of the spines; spines few to many, the lower spines larger, straight, usually flattened, and sometimes channelled and/or winged, the upper spines smaller, narrower, sometimes nearly terete, rarely hooked. Flowering and fruiting winter and spring.

During drought the stems and leaves become covered with amber-colored resinous-sticky exudate. In extreme drought the plants become leafless except for a few small, budlike terminal leaves encased with resin. Growth ceases during the long, hot summer and the plants become dormant—even during rainy times. If temperatures drop, however, as during unusually cloudy or rainy days in summer, some new leaves may be produced in a very short time. Luxuriant growth may be produced as early as mid- to late September. It is an important nurse plant for seedlings of many of the larger perennials such as cacti, e.g., chollas and saguaro, and the host plant for broomrape (*Orobanche cooperi*).

One of the most widespread and abundant perennials in the flora area; commonly associated with *Larrea* and *Ambrosia dumosa,* but reaching maximum development and density in slightly less xeric habitats than does *A. dumosa.* Maximum densities often along small drainageways or arroyos and at the edges of xeroriparian galleries of ironwood, mesquite, and palo verde along major arroyos. Also rocky slopes, playas of crater floors, clay soil, sandy creosote bush flats, to peak elevation in the Sierra Pinacate, and on low stabilized dunes at the eastern edge of the Gran Desierto dune fields. Common in the granitic mountains in canyons and arroyos and on rocky slopes, especially the gentler slopes; absent from the more arid, exposed slopes. Not on moving dunes and not in the Sierra del Rosario. It seems strange that it ranges into such an arid part of the Gran Desierto but goes no farther west—it is absent from California and most of Baja California. Central and southwestern Arizona, Sonora south to about Tastiota and Bahía Colorado (approximately 28°18′ N), Baja California Sur, and extreme southern Baja California.

Apparent hybrids between *A. deltoidea* and *A. dumosa* occasionally are encountered. Leaves of the putative hybrids are intermediate in appearance; they have not been found in flower or fruit.

Ambrosia deltoidea is closely related to *A. chenopodifolia* (Bentham) W.W. Payne, which is distinguished primarily by its more densely woolly burs with often terete spines and a weak trend towards relatively broader leaf blades. Plants of intermediate character are commonplace, especially at the geographic boundaries. Wiggins (1964) claims that *A. chenopodifolia* is further distinguished by the absence of glands on the bur, but this is not a reliable characteristic. *A. chenopodifolia* more or less replaces *A. deltoidea* in western Sonora near Puerto Libertad and through much of Baja California and San Diego County in southern California.

Dunes N of Gustavo Sotelo, *Ezcurra 14 Apr 1981* (MEXU). Bahía de la Cholla, *F 16850.* S of Sonoyta, *F 9809.* Pinacate Junction, *F 86-354.* Moon Crater, *F 10584.* 11 mi E of San Luis, *F 16688.* Putative hybrid, *A. deltoidea* × *A. dumosa:* 3.2 km SSW of Quitobaquito, *F 89-15.*

Ambrosia dumosa (A. Gray) W.W. Payne [*Franseria dumosa* A. Gray] *CHAMIZO;* WHITE BURSAGE; *TAḌSAḌ.* Dwarf hemispherical shrubs, much-branched, often 30–120 cm, the branches spreading. Twigs and leaves strigose with dense white hairs, the twigs sometimes spinescent-tipped. Leaves (4) 10–32 (40+) mm, the size depending upon soil moisture and season; very tardily drought deciduous and ultimately leafless in extreme drought; petioles prominent, often winged—more so with favorable moisture conditions; leaf blades 1–3 times pinnately dissected into small "rounded" segments variable in shape depending upon moisture conditions, dull green becoming whitish with drier conditions. Male and female heads often intermixed (unique within the genus), each female flower head below a male head. Burs 7.0–9.5 mm wide, glandular, sometimes with sparse, slender white hairs, the spines straight and flattened, the lower ones larger and often channeled. Summer dormant; flowering and fruiting winter and spring.

White bursage is the single most widespread and common perennial in the region, accounting for a major portion of the perennial biomass. Desert flats of rocky and sandy soils, stable and moving dunes, and rocky slopes to peak elevations in the granitic and volcanic ranges on all slope exposures. Frequently associated with creosote bush (*Larrea*) and often on rugged, extremely arid volcanic slopes beyond the local distribution and tolerance of creosote bush—on these cinder slopes it may be the only common perennial present. *A. dumosa* extends into much more xeric habitats than does *A. deltoidea*. For example, *A. dumosa* is abundant and widespread in the Sierra del Rosario, but *A. deltoidea* does not occur there.

Ambrosia dumosa is one of the most abundant perennials of the Sonoran and Mojave Deserts in southern California; southern Nevada and extreme southwestern Utah southward to Baja California Sur and Sonora in the vicinity of Bahía Kino. Raven et al. (1968) reported sympatric diploid, tetraploid, and hexaploid chromosome levels for the species, with hexaploids restricted to California deserts, and diploids and tetraploids in the flora area.

El Golfo, *Burgess 701*. 30 mi E of San Luis, *F 5765*. Dunes NE of Sierra del Rosario, *F 20416*. Bahía de la Cholla, *Gould 4136*. Campo Rojo, *Starr 720*. NW slope of Pinacate Peak, *F 86-456*. MacDougal Crater, *F 9941*.

Ambrosia ilicifolia (A. Gray) W.W. Payne [*Franseria ilicifolia* A. Gray] HOLLY-LEAF BURSAGE. Shrubs, broadly spreading, often 0.5–1.0 m tall, sometimes to 3 m across. Stems thick but scarcely woody. Leaves very tardily drought-killed to partially evergreen, 3.5–10 cm, sessile, dry and rough to the touch, relatively stiff, and glandular hairy; margins with coarse, spine-tipped teeth; dead leaves persistent and turning white. Burs (10) 15–18 mm, densely glandular with stalked glandular hairs, the spines many, curved and hooked with grooves above; burs resembling a cocklebur (*Xanthium*). Flowering and fruiting winter and spring.

In and among the granitic ranges west of the Pinacate volcanic field; most numerous along broad, dry washes leading out of the larger mountains; also on granitic slopes and pediments, especially on steep, north- and east-facing slopes and canyons. Arid regions surrounding the northern part of the Gulf of California and the lower Colorado River; in the flora area of Sonora, southwestern Arizona, southeastern California, northeastern Baja California, and Gulf of California islands: Tiburón, San Esteban, San Lorenzo, and Angel de la Guarda.

The dry, dead leaves rustling in the wind sound startlingly like a rattlesnake. (Sidewinders are sometimes coiled beneath the dense foliage.) During a bighorn sheep count in June at Cabeza Prieta Refuge, while concealed in a blind, Luke Evans (personal communication 1998) saw a bighorn put its mouth around a raceme of dry burs, strip them off with its teeth, and chew them up.

Sierra del Rosario, *F 20675*. Cerro Pinto, *F 89-61*. Sierra del Viejo, *F 5727*. Sierra Nina, *F 89-46*.

ARTEMISIA WORMWOOD, SAGE

Annual to perennial herbs or shrubs, usually aromatic and bitter-flavored. Leaves alternate, entire to dissected. Heads small, with (1) bisexual disk florets only; or (2) the outer florets pistillate, disklike with a greatly reduced corolla, and the inner, disk florets either bisexual or staminate. Phyllaries overlapping in 2 to several series, dry and membranous. Receptacle usually naked. Achenes usually glabrous; pappus usually none, occasionally of short scales.

Mostly dry, temperate regions, often with cold winters, but some extending into humid and even tropical regions: nearly worldwide, mostly Eurasia, and secondarily in North America; 350 species. *Artemisia* is absent from most of the Sonoran Desert and is the only member of the Anthemideae in the flora area, although 13% of the composite species of the world belong to this tribe. Typically wind-pollinated and often causing hay fever. Sagebrush (*A. tridentata*) is one of the major perennials of the Great Basin Desert. More than two dozen species are used for seasoning, medicinal purposes, and tea; e.g., absinthe, tarragon, and wormwood; others are used as livestock feed. References: Cronquist 1994; Hall & Clements 1923; Keck 1946; Ling 1995.

Artemisia ludoviciana Nuttall subsp. **albula** (Wooton) D.D. Keck [*A. albula* Wooton. *A. ludoviciana* subsp. *mexicana* (Willdenow ex Sprengel) var. *albula* (Wooton) Shinners. *A. microcephala* Wooton 1889. Not *A. microcephala* Hillebrand 1888] *CHAMIZO CENIZO;* WESTERN MUGWEED, WHITE SAGE. Herbaceous perennials. Stems arching or nearly upright, 0.4–1.0 m, leafy, slender, brittle, white woolly, and dying back during drought and severe freezing. Leaves mostly 3.0–4.5 cm (lowermost early-season leaves to 7 cm) with a few large, coarse teeth, woolly on both surfaces, grayish to whitish green, bicolored, and often entire and reduced upwards. Heads 3–3.5 mm, in elongated leafy or bracteate terminal panicles. Phyllaries thin, woolly outside, persistent, outer ones smaller and greener, inner ones 2.5–3.5 mm, mostly obovate to oblanceolate, green with broad, transparent-membranous margins; receptacle naked. Outer florets pistillate (lacking stamens) and fertile; corollas ca. 1 mm, conical, resembling a sheath around the style, the lobes extremely reduced, appressed to the style column. Inner florets bisexual and fertile; corollas 1.8–2.0 mm, 5-lobed. Corollas of both kinds of florets dotted with golden, globose glands. Achenes 0.9–1.3 mm, ellipsoid-obovoid, light brown, faintly ribbed.

Higher elevations in the Sierra Pinacate; along small washes or arroyos and canyons, and on north- and east-facing slopes and cliffs, mostly in slightly protected places. I know of no other records for this genus in western Sonora; elsewhere its intrusion into the Sonoran Desert is mostly at the upper elevational limits of the desert.

This species, as broadly interpreted, ranges from Alaska and Canada through much of the United States to northern Mexico; it is highly variable with a confusing array of subspecies and varieties. Subsp. *albula* occurs in southwestern United States and northwestern Mexico including Baja California and northern Sonora, mostly at elevations above the desert, and eastward to Chihuahua and Coahuila. This taxon is distinguished by its generally smaller leaves and open inflorescences. Cronquist (1994) classified it as *A. ludoviciana* subsp. *mexicana* var. *albula,* whereas Ling (1995) listed *A. albula* as a distinct species.

S of Pinacate Peak, 875 m, *F 19317*. Pinacate Peak, *F 86-444.*

ASTER

Mostly fibrous-rooted, perennial herbs, sometimes annual or biennial, rarely shrubby. Leaves simple, alternate, mostly with toothed margins. Heads with fertile ray and disk florets, often showy, or ray florets absent; disk golden yellow. Phyllaries graduated. Pappus bristles usually barbellate.

Traditionally encompassing about 250 species, mostly in temperate regions of North America, also in the Old World and South America, few in the American tropics and fewer in deserts. There has been a trend to split the genus asunder into smaller, presumably monophyletic units and to restrict *Aster* to the Old World. Nesom (1994a, b) distributed 181 New World species in 14 genera. Traditional treatment gives 3 *Aster* species in northwestern Sonora: *A. carnosus* (*Machaeranthera carnosa*), *A. spinosus* (*Chloracantha spinosa*), and *A. subulatus* (*Symphyotrichum subulatum*).

BACCHARIS

Shrubs or sometimes subshrubs or perennial herbs; usually resinous. Leaves alternate. Dioecious (ours) or rarely monoecious. Heads small to medium-sized, of disk florets; flowers often white. Phyllaries graduated. Pistillate pappus of numerous soft capillary bristles; staminate pappus of fewer, firmer, bristles. Achenes small, light colored, 5–10-ribbed (veined).

New World, greatest diversity in South America, mostly tropical, subtropical, and warm, dry regions; 400 species. References: Blake 1926; Boldt 1989; Cuatrecasas 1968a.

1. Leaves 5–12 cm; achenes 5-ribbed. ____ **B. salicifolia**

1′ Leaves 5 cm or less; achenes 10-ribbed.

  2. Stems leafy; herbage usually bluish green (slightly glaucous); leaves narrowly oblanceolate to obovate, entire or the larger leaves with a few coarse teeth; pappus bristles 13–15 mm; Río Colorado, La Salina, and Laguna Prieta. ____ **B. emoryi**

  2′ Stems usually nearly leafless or the leaves few and reduced; herbage yellow-green, not glaucous; leaves mostly linear, the larger leaves often minutely toothed (dentate); pappus bristles 7–11 mm; in the flora area not west of the Pinacate region. ____ **B. sarothroides**

Baccharis emoryi A. Gray. Broomlike, much-branched, woody shrubs 2.0–3.5 m, resembling *B. sarothroides* in growth form but leafier. Branches and twigs mostly ascending, the twigs striate-ridged. Herbage glabrous, usually slightly glaucous, sometimes with a slight red-purple cast in winter. Leaves mostly 1.5–5.0 cm, narrowly elliptic to oblanceolate or obovate, drought deciduous, mostly entire but the larger, lower leaves sometimes with few, low, broad teeth. Young herbage glandular-punctate and glutinous. Phyllaries ovate to lance-ovate in staminate heads, ovate to linear in pistillate heads, the margins membranous; inner phyllaries 8.0–9.0 mm. Achenes 2.0–2.5 mm, 10-ribbed, the pappus 13–15 mm. Pappus bristles of staminate flowers slender but flattened, barbellate to subplumose toward apex (seen at 30× magnification). Midwinter.

Damp or wet semisaline soil. Rare in the delta of the Río Colorado near El Doctor and at Laguna Prieta, and perhaps 100–150 shrubs at several of the larger pozos at La Salina. Also the Baja California side of the delta and northward along the river in western Arizona and southeastern California, to higher elevations in Arizona, New Mexico, Nevada, and Utah. Perhaps once more common in the Colorado delta.

The northern populations of *B. emoryi* (e.g., northwestern Arizona and Utah) have larger, broader, often thinner, and more conspicuously toothed leaves, and the plants are notably smaller. No other species of *Baccharis* occurs in the Río Colorado delta region including La Salina and Laguna Prieta. This shrub might be confused with *B. sarothroides* or *B. sergiloides* A. Gray. Both *B. emoryi* and *B. sarothroides* have pistillate pappus bristles 7–15 mm, whereas in *B. sergiloides* the pistillate pappus is 4–5 mm. *B. sergiloides* occurs in Baja California, California, Arizona, and Utah.

La Salina, *F 84-18.* Laguna Prieta, *F 85-755.* 1 km SW of El Doctor, *F 92-1000.*

Baccharis salicifolia (Ruiz & Pavón) Persoon [*Molina salicifolia* Ruiz & Pavón. *Baccharis glutinosa* Persoon. *B. viminea* de Candolle] BATAMOTE; SEEP WILLOW; *ṢUṢK KUAGSIG.* Shrubs ca. 2 m, glabrous, glandular-punctate, conspicuously glutinous-sticky and aromatic, nearly evergreen. Stems erect-ascending, many, and leafy. Leaves 5–8 (12) cm, lanceolate, willowlike, usually toothed. Phyllary margins erose-membranous, the larger (inner) phyllaries 2.0–2.8 mm. Achenes 1.0–1.1 mm, 5-ribbed; pappus bristles 4.0–5.5 mm. Staminate pappus bristles barbellate or subplumose near tip. Warmer months.

Mostly riparian in wet, sandy, gravelly, or poorly drained clayish soils. Large populations along the Río Sonoyta, where Mearns (1892–93) recorded it "in creek bottoms forming an almost impenetrable growth." Occasional at some of the Pinacate waterholes, roadside depressions with poor drainage, and playas. Sometimes a common agricultural weed along irrigation canals. Wetlands across much of Sonora and Arizona.

South America to Central America south of Nicaragua, Mexico north of the Isthmus of Tehuantepec including the Baja California Peninsula, and southwestern United States. Apparently more variable in South America than in North America. Cuatrecasas (1968b) recognized 3 weakly differentiated varieties and considered the plants north of South America as var. *longifolia* (de Candolle) Cuatrecasas.

Nested within the geographic distribution of typical dioecious *B. salicifolia* is a large, allopatric population of monoecious plants from Oaxaca to Nicaragua described as *B. monoica* G.L. Nesom (Nesom 1988). Other than the monoecious character, a striking feature in the genus, the characters distinguishing *B. monoica* from *B. salicifolia* seem weak.

Río Sonoyta at Sonoyta, *F 85-701.* SE side of Colinas Batamote, *Ezcurra 20 Sep 1980* (MEXU). Rancho los Vidrios, *Ezcurra 20 Sep 1980* (MEXU). Pinacate Junction, *F 19606.*

Baccharis sarothroides A. Gray. *ESCOBA AMARGA, ROMERILLO;* DESERT BROOM; *ṢUṢK KUAGĬ, ṢUṢK KUAGSIG.* Woody shrubs often 2.0–2.5 m, with broomlike green branches, often nearly leafless. Twigs angled or striate-ridged. Twigs and leaves densely dotted with short-stalked, glandular hairs producing copious sticky-glutinous exudate that soon covers the hairs and coats the leaf surfaces. New growth with quickly deciduous, linear to linear-lanceolate leaves reaching 1–3 (4) cm, the larger leaves often minutely toothed (dentate); most leaves much smaller or reduced to scales. Phyllary margins erose to ciliate-membranous, the outer phyllaries broadly ovate, the inner ones linear, 6.5–7.5 mm. Achenes 1.5–2.7 mm, 10-ribbed, the pappus 7–11 mm. Staminate pappus bristles slightly expanded (flattened) toward apex. Mostly flowering October-November.

Sandy-gravelly soils of washes in scattered localities in the Pinacate volcanic region, sometimes locally common in roadside ditches along Mex 2 and near Sonoyta; absent across most of the rest of the flora area. Sinaloa to Arizona, Nevada, New Mexico, California, and Baja California.

Sonoyta, 13 Jan 1894, *Mearns 2716* (US). 33 mi S of Sonoyta on rd to Puerto Peñasco, *Keck 4193* (DS). Quitobaquito, 5 Feb 1894, "Romerio," *Mearns 2775* (US).

BAILEYA *TECOMBLATE,* DESERT MARIGOLD

Annual or perennial herbs with a well-developed taproot; densely white woolly and with glistening orange glands even on the corollas and achenes. Leaves in basal rosettes and alternate on stems, pinnatifid below, generally reduced and often entire above. Phyllaries essentially equal, linear-lanceolate, green beneath the woolly hair. Heads on long peduncles or few-leaved stems, showy, with yellow ray and disk florets; ray florets fertile, large and showy, persistent and somewhat papery, bent down in age; disk florets many, fertile. Achenes papillose-hispid, conspicuously ribbed, clavate-cylindrical to moderately compressed laterally, truncate at the apex; pappus none.

Desert and semiarid southwestern United States and northern Mexico, mostly Mojave, Sonoran, and Chihuahuan Deserts; 3 species. Common through most of the lowlands of the flora area, especially on sandy soils; generally absent from moving dunes. Winter-spring ephemerals in the flora area, and at least *B. multiradiata* and probably *B. pleniradiata* also may respond to summer-fall rains or persist for more than one season. Reference: Turner 1993.

1. Ray florets 5–7; heads ca. 6 mm wide not including rays, mostly loosely cymose. ________ **B. pauciradiata**

1′ Ray florets 20–50; heads 10 mm or more in width not including rays, mostly single on an elongate peduncle.

  2. Stems usually leafy only at base or below the middle; heads usually 3.5–4.0 cm wide including rays; achene ribs all nearly similar; Sonoyta region. ________ **B. multiradiata**

  2′ Stems leafy to the middle or above; heads usually 2.4–3.2 cm wide including rays; achene ribs conspicuously unequal, larger at the angles; widespread but apparently not in the Sonoyta region. ________ **B. pleniradiata**

Baileya multiradiata Harvey & A. Gray ex Torrey [*B. thurberi* Rydberg. *B. multiradiata* var. *thurberi* (Rydberg) Kittell] MANY-FLOWERED DESERT MARIGOLD; *GI:KO*. Non-seasonal ephemerals, occasionally annuals or perhaps short-lived perennials. Leaves mostly in apparent basal rosettes and below the middle of the stems, the larger leaves 5.5–12.0 cm. Flowering stems ("peduncles") mostly 20–30 cm, with reduced leaves below and leafless above, with 1 head per stem. Phyllaries 5.5–6.5 mm, linear-lanceolate. Flower heads often 3.5–5.3 cm wide including rays; rays many, bright yellow, 15–20 × 5–8 mm, the apex conspicuously 3-toothed. Style branches truncate to slightly rounded at tips. Achenes 3.5–3.8 mm, the ribs essentially all alike and of relatively low relief.

Roadsides and sandy-gravelly soils in the Sonoyta region to about 30 km southwest of Sonoyta. Common in adjacent Organ Pipe Monument and eastward across much of northern Sonora to Coahuila, Durango, and Aguascalientes, as well as southeastern California, southern Nevada, and southwestern Utah to western Texas. *Baileya multiradiata* is not always readily distinguished from *B. pleniradiata,* especially among plants sampled late in the season and in fall; however the shape of the style is diagnostic.

8 km W of Sonoyta, *F 92-144*. 2.6 km SW of Sonoyta, *F 88-143*.

Baileya pauciradiata Harvey & A. Gray ex A. Gray. FEW-FLOWERED DESERT MARIGOLD. Drought-stunted plants ca. 10–15 cm tall with a single stem, or sometimes many-stemmed and bushy to 50 cm in seasons of exceptional rainfall. Larger leaves (3) 5–12 cm. Stems usually several-branched, with several to many heads, the peduncles 1.0–3.5 cm. Phyllaries 3.5–4.5 mm. Heads 1.5–2.0 cm across including rays, the rays 5–7 in number, 6–10 × 3.5–7.5 mm; flowers pale yellow (paler than those of *B. pleniradiata*). Achenes 3.5–4.5 (5.0) mm, the ribs conspicuous and more or less equal.

Sandy soils of desert plains across much of the flora area, including lower dunes, and especially common south and west of the Pinacate volcanic field; sometimes intermixed with *B. pleniradiata*. Not known elsewhere in Sonora or from the Sonoyta region. Also southwestern Arizona, southeastern California, and Baja California.

Mex 8 just S of km marker 83 [N of Puerto Peñasco], *Lockwood 98*. S of Moon Crater, *F 19035A*. NNE of Sierra del Rosario, *Burgess 6876*. Beach dunes E of El Golfo, *F 75-81*.

Baileya pleniradiata Harvey & A. Gray ex A. Gray. WOOLLY DESERT MARIGOLD. Mostly 10–45 cm, sometimes 50–60 × 100–110 cm during El Niño years. Larger leaves mostly 3.5–10 cm (sometimes 15–30 cm on especially robust plants). Stems several-branched, mostly with several heads, the flowering stems ("peduncles") often 2–14 cm. Heads usually 2.4–3.2 cm across including rays, or much smaller on drought-stunted plants. Phyllaries 3.5–4.5 mm. Rays many, pale yellow, 7.5–10 × 3.6–8.5 mm, 3-toothed to rounded-truncate at apex. Style branches pointed (acute). Achenes 2.5–3.5 (3.8) mm, the ribs unequal, those of the angles conspicuously thicker.

Especially common on sandy soils of flats and lower, stabilized dunes, interdune troughs, and crater floors; less common along gravelly washes; also highway roadsides and clay flats or playas. Lowlands throughout most of northwestern Sonora but replaced by *B. multiradiata* in the Sonoyta region. Also in Sonora from the vicinity of Sierra Cubabi southward near the coast to El Desemboque Río Concepción, and in Arizona, southeastern California, Baja California, Nevada, and southwestern Utah. Essentially a species of the Sonoran and Mojave Deserts.

Often persisting longer into the late-spring drought than most cool-season ephemerals, perhaps in part because of the deep taproot. It is the most common and widespread *Baileya* in the flora area.

Puerto Peñasco, *F 16853*. Gustavo Sotelo, *Ezcurra 14 Apr 1981* (MEXU). MacDougal Crater, *F 9901*. SSE of Cerro Colorado, *Ezcurra 17 Apr 1981* (MEXU). Sierra del Rosario, *F 20714*.

BEBBIA

Aromatic shrubby perennials, the receptacles with chaffy bracts. Southwestern United States and northwestern Mexico; 2 species. Reference: Whalen 1977.

Bebbia juncea (Bentham) Greene var. **aspera** Greene [*B. aspera* (Greene) A. Nelson] CHUCKWALLA DELIGHT, SWEETBUSH; HAUK 'U'US. Small shrubs, rounded, many stemmed, intricately branched, probably often short-lived, often 0.7–1.5 m. Stems slender and brittle. Branches and leaves opposite, or alternate above. Herbage and phyllaries scabrous, the hairs curved with hard, swollen bases like those of certain *Cryptantha* species. Leaves sparse, often semisucculent during favorable times, reaching 3–5.5+ cm but mostly smaller, and reduced upwards, linear to linear-oblanceolate, usually entire or the larger leaves with 1 to several lobes, mostly sessile; quickly drought deciduous, leafless or nearly so most of the year. Heads on long stems, yellow and fragrant, with disk florets only. Phyllaries green, lanceolate, the inner ones 4.5–6.0 mm, intergrading with the chaffy bracts. Chaffy bracts 5.5–7.5 mm, linear-lanceolate, orange-veined, the tips ciliate. Achenes 2.3–3.0 mm, compressed, 3-angled, asymmetric, and club-shaped, blackish when mature; pappus bristles plumose, 5–8 mm. Flowering almost any time of the year.

Widespread on rocky slopes, canyons, arroyos, washes, and cinder flats and slopes; Sonoyta region, Pinacate volcanic field to higher elevations, and granitic ranges. Often severely browsed by animals including chuckwallas (*Sauromalus obesus*).

Baja California and northwestern Sinaloa to southern California, Nevada, southwestern Utah, New Mexico, and extreme western Texas. Var. *aspera,* the most widespread of the two varieties, is replaced by the often leafier, less xeromorphic var. *juncea* in southern Baja California and Baja California Sur.

Tinajas de Emilia, *F 19718*. Cerro Colorado, *F 10801*. Lava hill W of MacDougal Crater, *F 9931*. Bahía de la Cholla, granitic hill, *F 16827*.

BRICKELLIA BRICKELLBUSH

Annual or perennial herbs or shrubs. Leaves mostly opposite, sometimes alternate, rarely whorled. Heads with disk florets only, usually not showy. Phyllaries in several series, prominently striped or ribbed. Achenes linear-columnar, narrowed toward the base, usually 10-ribbed, usually hairy; pappus of slender, barbellate to subplumose bristles. Americas, mostly dry regions, and concentrated in western Mexico and southwestern United States; 100 species. References: King 1987; Robinson 1917; B. L. Turner 1978, 1990.

1. Bark shredding in thin strips; leaves mostly alternate, the petiole less than 1/4 as long as blade; blades stiff, thickish, with spinescent teeth. ____ **B. atractyloides**

1′ Bark not shredding; leaves mostly opposite, the petiole ca. 1/3 as long as blade; blades not stiff, relatively thin, toothed but not spinescent. ____ **B. coulteri**

Brickellia atractyloides A. Gray var. **atractyloides.** Shrubs usually less than 1 m. Stems slender and brittle; bark shredding in thin strips. Herbage, peduncles, and phyllaries short glandular-pubescent or glabrate or glabrous, especially with age. Leaves 1.5–2.5 cm, short petioled, usually alternate, the blades mostly broadly ovate to nearly triangular, relatively stiff and dry to the touch, prominently veined, with several stout, spinose teeth along margins and at the apex; dead leaves moderately persistent. Heads many flowered, solitary, on stout peduncles 2–5 cm. Outer phyllaries 3.5–5 mm wide, leaflike, broadly ovate, the inner phyllaries 12–15 mm long, linear, the narrower ones 0.6–1.0 mm wide. Flowers pale yellow and purplish. Achenes 4–6 mm, blackish, with short white hairs; pappus 6–10 mm. Warmer months, especially late spring.

In Sonora known only from the Sierra del Viejo on north-facing canyon slopes nearly to the summit. Also western and central Arizona, southeastern California, southern Nevada, and southern Utah. Replaced to the west in southern California and Baja California by var. *odontolepis* B.L. Robinson and var. *arguta* (B.L. Robinson) Jepson; the varieties are sometimes treated as distinct species.

Sierra del Viejo [61 mi E of San Luis on Mex 2], *F 16879.* Tinajas Altas, *Van Devender 5 Mar 1983.*

Brickellia coulteri A. Gray var. **coulteri.** Bushy perennials or small shrubs, often reaching ca. 1 (1.5) m; stems slender and brittle; gradually drought deciduous. Herbage and outer phyllaries with crinkly white hairs and often glandular. Leaves mostly opposite, the petioles well developed; leaf blades relatively thin, ovate to triangular, coarsely toothed, green or occasionally purplish green, often 3–6+ cm, those of upper branches often 1.5–2.0 cm. Heads in loose, open panicles; heads often with 13–15 florets. Peduncles 1–2 cm or much shorter when drought stressed. Phyllaries appearing graduated, linear-attenuate, conspicuously striate-veined, the longer phyllaries 8–11 mm. Flowers yellow-green with purple-brown tips. Achenes 2.5–4.5 mm, blackish with short white hairs; pappus 5.5–7.0 mm. Various seasons.

Sonoyta region, Pinacate volcanic field, and granitic ranges. Mostly in protected places such as canyons and arroyos, and beneath desert trees including palo verde or ironwood along larger washes; not in open desert. In the granitic mountains generally in larger canyons on north- and east-facing slopes; seldom common in the Pinacate mountains.

Var. *coulteri* largely in the Sonoran Desert region in Arizona, most of Sonora, and the Baja California Peninsula. Two other varieties, southern Sonora to Aguascalientes and Michoacán, and eastward to Texas and San Luis Potosí.

2.6 mi W of Sonoyta, *F 86-405.* Sierra de los Tanques, *F 89-23.* W slope of Sierra Pinacate, ca. 750 m, *F 19286.*

CALYCOSERIS

Annuals (spring ephemerals in the flora area) with milky sap, branched from the base or near the base, and with conspicuous tack-shaped glandular hairs above, otherwise glabrous or glabrate. Leaves pinnately parted into narrow segments, the early leaves in a basal rosette, the stem leaves alternate and reduced upwards. Heads with ligulate (raylike) florets only; phyllaries in 2 series, those of the inner series nearly equal, those of the outer series much shorter and unequal. Achenes narrowed above into a neck or beak. Pappus bristles white, threadlike, readily deciduous. Southwestern United States, northern Sonora, and Baja California, mostly in deserts; 2 species.

"*Calycoseris* is very closely related to *Malacothrix,* the two genera differing essentially in pubescence and achene characters" (Tomb 1974:213). *Calycoseris* has glandular hairs not found elsewhere in the subtribe Stephanomeriinae (includes *Malacothrix, Rafinesquia, Stephanomeria*).

1. Corollas pale yellow; higher elevations of Sierra Pinacate. ____ **C. parryi**
1′ Corollas white; low elevations. ____ **C. wrightii**

Calycoseris parryi A. Gray. YELLOW TACKSTEM. Often 5–15 cm with few to many short, ascending and spreading branches. Leaves mostly basal, often 4–9 cm, often withering by flowering time. Inner phyllaries 10–12 mm with the midrib prominently thickened (keeled) below. Florets (rays) often 15+ mm, pale lemon-yellow, the outer florets with a rose-colored midstripe on the lower surface. Achenes 5 mm, deeply ribbed, otherwise relatively smooth.

Sierra Pinacate from about 550 m to peak elevation (*F 19433, F 92-446, F 92-485*); not known elsewhere in Sonora. Southern and western Arizona, southeastern California, Nevada, and Utah. The achenes differ from those of *C. wrightii* in the longer beak, deeper ribs or grooves (sulcae), and smoother surfaces.

Calycoseris wrightii A. Gray. WHITE TACKSTEM. Mostly 6–40 cm, usually few-branched. Leaves mostly basal, 3–10+ cm. Inner phyllaries 11–13 mm, with a green midstripe and membranous margins. Florets (rays) reaching 10–20 mm, white with pinkish brown streaks (drying purple-brown) on the lower surfaces. Achenes 7 mm, ribbed and with small bumps.

Sandy, gravelly, or rocky-gravelly soils of floodplains and bajadas, seldom common; northwestern granitic mountains and in the Sonoyta region. Southeastern California and Baja California eastward across most of northern Sonora and Arizona to western Texas, and northward to Nevada and Utah.

10 km SW of Sonoyta, *F 88-159.* Sonoyta, *F 86-98A.* 3.9 mi E of La Joyita, *Prigge 7248* (RSA).

CARTHAMUS

Annuals. Leaves alternate; leaves and phyllaries spinose except in some cultivars of *C. tinctorius.* Phyllaries with long, leafy blades or appendages. Heads often large and showy, with disk florets only, the corolla expanded above the tube. Achenes mostly 4-angled; pappus of scales or absent. Native to the Old World; 14 species. Reference: Ashri & Knowles 1959.

***Carthamus tinctorius** Linnaeus. *CÁRTAMO;* SAFFLOWER. Spring annuals in the flora area; glabrous, usually with a single main axis, the stems whitish. Leaves often 5–12+ cm, firm, obovate to elliptic; leaf and phyllary blade margins with spine-tipped teeth. Heads thistlelike, 4–5 cm across, the receptacle with numerous flat, papery, white scales. Flowers bright orange-yellow. Achenes 7.0–7.5 mm, white and plump; pappus none.

Common oil-seed crop in Sonora; usually in sandy soils, seldom reproducing, mostly along roadsides from passing trucks carrying safflower harvests.

Mex 2 at rd to Rancho Guadalupe Victoria, *F 20405.* Quitobaquito, *Bowers 1717.*

CENTAUREA STAR THISTLE

Annual, biennial, or perennial herbs. Leaves alternate, often pinnatifid, reduced upwards. Heads thistlelike, with disk florets only but marginal florets often raylike; receptacle with bristles. Flower color highly variable. Achenes attached obliquely or laterally to the receptacle; pappus present or absent.

Mostly Mediterranean, a few native to the Americas and Australia; 450 species. None native to the Sonoran Desert. Bachelor's button, a garden plant, is *C. cyanus.*

***Centaurea melitensis** Linnaeus. MALTA STAR THISTLE. Annuals, deep-rooted, thistlelike, often 50–60 cm, to 1 m or more in well-watered, partially shaded habitats. Herbage and phyllaries glandular, grayish with tangled woolly hairs at least when young. Lower leaves pinnatifid, 5–15+ cm, the stem leaves smaller, with decurrent bases forming narrow wings on the stems. Heads 1.5–2+ × 2.2–2.8 cm including phyllary spines, or 1.2–1.5 cm wide not including spines. Phyllaries with a stout, straw-colored terminal spine 6–8+ mm with 2 or 3 pairs of smaller, lateral spines, the spine tips purplish. Flowers bright yellow. Achenes 3 mm, thick but moderately compressed, the base with a notchlike hook formed by the receptacle attachment, the surface shiny brown with longitudinal white lines and sparse slender hairs; pappus bristles many, white, flattened, unequal in length, reaching 2–3 mm. Late spring-early summer.

In the Sonoyta region, mostly in low, temporarily wet habitats and roadsides. Well established in the mesquite grove at Quitobaquito (*F 89-251*) and occasional along Mex 2 (4.5 km W of Sonoyta, *F 86-149*). Localized and probably a fairly recent arrival in the flora area, although in Tucson at least since 1901 (Felger 1990). Scattered, mostly disturbed habitats across the Sonoran Desert. Widespread and weedy in the Americas; native to southern Europe and North Africa.

CHAENACTIS

Annuals, biennial herbs, or low perennials. Leaves alternate, entire to pinnate or dissected. Heads basically with disk florets only, but the marginal florets often enlarged and appearing intermediate between a disk and a ray floret; without receptacular bristles (except in *C. carphoclinia*). Achenes club-shaped, tapering to the base, generally terete with stiff hairs; pappus of fringed scales or absent. References: Morefield 1993a; Stockwell 1940.

Western United States, adjacent Canada, and northern Sonora and Baja California; 18 species. The 9 Sonoran Desert species are spring ephemerals.

1. Herbage with short white hairs, not cobwebby; phyllaries with an elongated, slender (terete) tip; heads with stout receptacular bristles among the florets; pappus scales 1.1–3 mm. ____ **C. carphoclinia**

1′ Herbage with cobwebby woolly hairs; phyllaries blunt, not with an elongated tip; without receptacular bristles; pappus scales 3.5–6.5 mm. ____ **C. stevioides**

Chaenactis carphoclinia A. Gray var. **carphoclinia** [*C. carphoclinia* var. *attenuata* (A. Gray) M.E. Jones] PEBBLY PINCUSHION. Stems wiry, zigzag, solitary or with several or more branches above, mostly 9–25 cm. Herbage with short white hairs, sometimes scarcely white woolly; herbage and phyllaries minutely glandular pubescent, or the leaves sometimes only sparsely glandular. Leaves pinnatisect with linear-filiform segments, the larger leaves 3–7 cm, quickly drought deciduous. Heads rounded, 10–14 mm, the florets white to pink. Phyllaries 5–8 mm, linear to narrowly lanceolate or oblanceolate, extending into a slender, terete point. Receptacular bristles interspersed among the florets, rather stiff, about as long or longer than the phyllaries, tan with curved to curled red tips. Achenes 3–4 mm, columnar, blackish; pappus scales 4, papery-membranous, 1.1–3.0 mm.

Sonoyta region, Pinacate volcanic field to at least 750 m, granitic ranges, and desert plains; sandy flats, desert pavements, rocky flats, slopes, and washes; absent from dunes. One of the most abundant and widespread cool-season wildflowers in the flora area. Southwestern Utah and southern Nevada to northeastern Baja California and northwestern Sonora southward at least to the Sierra Bacha (south of Puerto Libertad). A second taxon, var. *peirsonii* (Jepson) Munz, occurs in California.

El Papalote, *F 86-112*. Elegante Crater, *F 19946*. MacDougal Crater, *F 9913*. Moon Crater, *F 19285*. Sierra del Rosario, *F 20701*. Granitic mountains 28 mi W of Los Vidrios, *Bezy 447*.

Chaenactis stevioides Hooker & Arnott. DESERT PINCUSHION. Often 15–35 cm, the herbage, especially when young, sparsely to moderately white woolly; herbage, inflorescences, and phyllaries moderately to densely and minutely glandular pubescent. Leaves pinnatisect with linear segments, the larger leaves 3–10 cm, quickly drought deciduous. Flower heads rounded, 14–17 mm, the florets white or cream, the corollas pink in bud. Phyllaries 7–10 mm, linear to oblanceolate, the tips acute to obtuse and blunt-ended. Achenes 5.5–6 mm, columnar, blackish; pappus scales 4 (5), papery-membranous, 3.5–6.5 mm.

Widespread, mostly on sandy-gravelly soils, sandy plains, crater floors, washes, and desert pavements; absent from dunes. Southeastern Oregon to Colorado, Baja California, northern Sonora, and western New Mexico; in western Sonora south to about 29°20′ N (30 km S of El Desemboque San Ignacio).

Both *C. stevioides* and *C. carphoclinia* have similar southern limits in western Sonora and often occur intermixed, although *C. stevioides* ranges much farther eastward in northern Sonora. In the field *C. stevioides* usually can be distinguished by its more robust habit, slightly larger heads, and white or cream rather than pinkish florets.

41 mi SW of Sonoyta, *Shreve 7593.* Cerro Colorado, *Webster 22348.* MacDougal Crater, *F 9750.* Moon Crater, *F 19279.*

CHLORACANTHA SPINY ASTER

This genus of a single species has been allied with *Aster* or *Erigeron,* but its affinity seems closer to *Boltonia* and possibly *Heterotheca.* References: Nesom et al. 1991; Sundberg 1991.

Chloracantha spinosa (Bentham) G.L. Nesom var. **spinosa** [*Aster spinosus* Bentham var. *spinosus*] Rhizome-producing perennials, often 1–2.2 m with several to many nearly leafless, stiff, broomlike green stems branched above, and usually spiny (modified branches) at least near the base; glabrous. Branches and leaves alternate; leaves sparse, mostly linear, very quickly deciduous, sometimes to 3 cm on new growth, the upper leaves scalelike. Involucres 7–11 mm wide, outer phyllaries broader, inner phyllaries 3.8–5.3 mm, linear-lanceolate, thin, green with membranous margins. Heads showy with white rays and pale yellow disk florets. Achenes 1.6–2.0 mm, pale brown, oblong-columnar, 5-ribbed, glabrous; pappus bristles 5.5–7.0 mm. Flowering during warmer months.

Wet, usually alkaline soils along the Río Sonoyta and roadsides near Sonoyta (*F 86-395*), mostly in trashy, disturbed habitats with *Tamarix ramosissima,* and roadside and irrigation ditches among agricultural fields south of San Luis (*F 85-1028*). Also in the Imperial Valley and the lower Colorado River valley including Yuma. Baja California and California to Utah, Louisiana, and the central plateau of Mexico. This species includes 4 geographically segregated varieties, ranging from Panama to the United States. Var. *spinosa* is the most widespread and weedy variety, and the only one in the United States and northernmost Mexico.

Some specimens from the San Luis Valley, like those from the Imperial Valley and the Colorado River north of the border "are unusual in having fleshy stems, larger heads, and short branchlets in the capitulescence. . . . Their growth form may be partly due to high salinity" (Sundberg 1991:386).

CONYZA

Annual or perennial herbs with a taproot. Leaves alternate. Heads of numerous small white to purplish florets; ray (pistillate) florets eligulate (ray portion of corolla absent or small and not extending beyond the phyllaries) and resembling the disk florets; disk florets bisexual and fewer than ray florets. Pappus bristles fragile, threadlike, barbellate, usually elongated beyond the involucre at maturity.

Worldwide, mostly tropical and subtropical regions with some cosmopolitan weeds; 50+ species. Closely allied to *Erigeron,* a largely temperate genus. References: Cronquist 1943; Nesom 1990a, b.

1. Leaves linear, entire or with a few shallow teeth; herbage not glandular; phyllaries graduated, glabrous. ____ **C. canadensis**

1′ Leaves oblanceolate to obovate, conspicuously toothed to lobed; herbage glandular; phyllaries mostly equal, glandular. ____ **C. coulteri**

***Conyza canadensis** (Linnaeus) Cronquist var. **glabrata** (A. Gray) Cronquist [*Erigeron canadense* Linnaeus var. *glabratum* A. Gray] *COLA DE CABALLO;* HORSEWEED. Mostly summer–fall ephemerals (annuals elsewhere), slender, erect weeds often 1–2 m and unbranched except the terminal, much-branched flowering portion. Herbage sparsely pubescent, the leaves hispid with coarse white hairs mostly along margins. Leaves linear, reduced upwards; larger leaves often 3–8 cm × 2–6 mm, sessile but narrowed below, the margins entire or with a few shallow teeth, ciliate and strigose all around. Flower heads 2.5–4.0 mm wide. Phyllaries graduated, the longer ones 2.5–3.0 mm, glabrous or some sparsely pubescent with coarse white hairs, the midveins prominent, orange, and resinous, and margins membranous. Ray florets minute and numerous, white or white with pink tips. Achenes 1 mm, the pappus slightly more than twice as long as the achenes, minutely barbellate (scarcely visible at 30× magnification).

Agricultural weeds, also sometimes along roadsides and other disturbed habitats, mostly near Sonoyta and south of San Luis; probably not native in the flora area. A worldwide weed, native to North America. Three weakly differentiated varieties, each with a large geographic range over a portion of North America; var. *glabrata* is the common variety in western North America.

5 km W of Sonoyta, *F 85-959*. S of Pinacate Junction, *F 86-383*.

Conyza coulteri A. Gray [*Laennecia coulteri* (A. Gray) G.L. Nesom] Summer-fall ephemerals or weeds (annuals elsewhere), often 50–60 cm with a single main axis below and much-branched above. Herbage and phyllaries sticky with sessile to stalked glands and hispid with non-glandular white hairs. Lower leaves 4–8+ × 1.3–3 cm, sessile, oblanceolate, regularly lobed to coarsely toothed, the base usually clasping the stem; uppermost leaves smaller and often entire. Flower heads 3–4 mm wide. Phyllaries nearly equal, 3.0–4.0 mm; outer phyllaries with a broad, densely glandular hairy green midregion or midstripe and narrow membranous margins; inner phyllaries narrower, with a narrow midvein and broad, membranous margins. Ray florets minute and numerous, the corollas greatly reduced. Achenes 1.0–1.1 mm, the pappus 3 times as long as the achenes, conspicuously barbellate (seen at 30× magnification).

Agricultural weed in the Río Sonoyta valley, often in abandoned fields (*F 86-168*); occasional in relatively natural vegetation such as the Río Sonoyta floodplain and at Playa Díaz (*F 86-368*); rarely elsewhere in low, temporarily wet, disturbed places. Not native to the flora area. Widespread and often weedy in many life zones in southwestern United States and much of Mexico.

DICORIA

Annuals with coarse white hairs of various sizes. Lower leaves opposite, upper ones alternate. Heads of disk florets only, each with 5–20 staminate and 1–4 pistillate florets, or only staminate florets; staminate florets with a small corolla, the pistillate florets lacking a corolla. Phyllaries dimorphic, the outer ones surrounding female florets small and reflexed, the inner ones much larger, convex, loosely enclosing the achenes in an involucre like a paper lantern. Achenes flat with intricately sculptured margins, seemingly well adapted to wind dispersal; pappus none.

Dunes in northwestern Mexico and southwestern United States; 5 species. The generic name derives from the Greek *dis-* 'twice' and *koris* 'a bug', because of the appearance of the 2 achenes of the first described species. References: Cronquist 1994; Rydberg 1922; Strother 1982.

Dicoria canescens A. Gray [*D. calliptera* Rose & Standley, *Contr. U.S. Natl. Herb.* 16:18, pl. 12, 1912] DESERT DICORIA, BUGSEED. Robust annuals, spring and summer, often 0.5–1 m, globose and bushy, branching from the base upward, with gray-green herbage; canescent with appressed and spreading coarse white hairs. Seedlings rapidly developing a deep taproot; the juvenile leaves narrowly lanceolate and irregularly coarsely toothed, the subsequent leaves often 3.5–6 cm, broadly ovate to sub-orbicular and shallowly toothed to nearly entire. Achenes 4.0–5.5 mm, the body blackish and oblong, the margin yellow-brown, broad and papery, with toothed, often distinct lobes.

Germinating with winter-spring rains. After the root has penetrated far below the surface sand, the subsequent leaves become much broader. When the seeds ripen the plant breaks off at the base and becomes a tumbleweed. This is one of the few native plants in the region with a tumbleweed growth form (see also *Eriogonum trichopes*).

Often the most abundant plant on shifting dunes. Sonora southward on coastal dunes to the vicinity of Tastiota; western Arizona, southeastern California, southern Nevada, southwestern Utah. Two subspecies and several varieties of questionable significance in the northern part of the range; ours are subsp. *canescens*.

7 km N of Puerto Peñasco, *F 85-891*. Sand hills near Adair Bay, 20 Nov 1907, *Sykes 63* (type of *D. calliptera*, US 574268). El Golfo, *F 86-1061*. S of Moon Crater, *F 19107B*. NE of Sierra del Rosario, *F 20418*. Near Monument 197 on Mex 2, *Simmons 893*.

ECLIPTA

Annual or perennial herbs. Leaves opposite. Heads small, the receptacles with slender chaffy bracts, the ray and disk florets minute and numerous. Pappus absent or represented by rudimentary projections. Warm regions worldwide; 3 or 4, or perhaps only 1 species. References: Holm et al. 1977; McVaugh 1984.

***Eclipta prostrata** (Linnaeus) Linnaeus [*Verbesina prostrata* Linnaeus. *Eclipta alba* (Linnaeus) Hasskarl. *E. erecta* Linnaeus] *CHILE DE AGUA;* FALSE DAISY. Delicate ephemerals or annuals in the Sonoran Desert (often biennial or perennial in nondesert regions), quickly wilting when cut or uprooted; flowering more or less continuously during warm weather and winter dormant or freeze killed. Leaves 5–14 cm, sessile, mostly narrowly elliptic; margins with widely spaced, forward-pointing small teeth or serrations. Heads often 4–7 mm wide, the phyllaries 4–6 mm, the chaffy bracts of barbellate bristles; rays white, the disk florets yellow; fruiting heads ca. 7–10 mm, the phyllaries spreading slightly and enlarging to 5–7 mm. Achenes 2.6–2.9 mm, mostly moderately compressed, obconical, brown, tuberculate.

Wetlands in the Sonoyta region and San Luis Valley, and probably formerly along the Río Colorado; irrigation ditches, pond margins, and springs. Widely scattered in lowland Sonora in natural and artificial wetlands. Worldwide in warm regions; often a serious agricultural weed in regions of high rainfall; uncertain if native to the Old World or to the New World.

Sonoyta, Presa Derivadora, *F 86-301*. Quitobaquito, *Niles 724*. S of San Luis, *F 85-1025*. Yuma, *Wilkinson 20 Aug 1905*.

ENCELIA

Small shrubs and 1 perennial herb, usually pubescent. Leaves alternate. Ray florets sterile or absent; disk florets yellow or brown, bisexual. Disk florets and achenes enclosed by chaffy bracts. Achenes flat, the margins narrow, white and long-haired. Pappus none or of 2 slender awns.

Southwestern North America and South America; mostly in arid regions and along seacoasts, with greatest diversity in northwestern Mexico; 15 species. At least most of the species are self-incompatible. All species are interfertile in cultivation, producing fertile offspring in all studied species, and natural hybrids are common where the species occur together. References: Blake 1913; Clark 1998.

1. Leaves and stem tips commonly woolly with soft hairs; leaves mostly 3–10 × 1.5–3.5 cm; heads usually several or more in broad panicles, the peduncles glabrate; rays present. ____ **E. farinosa**

1′ Leaves and stems scabrous with rough hairs; leaves mostly 1.7–3.2 × 0.35–0.6 (1.0) cm; heads solitary, the peduncles hairy; rays absent. ____ **E. frutescens**

Encelia farinosa A. Gray var. **farinosa.** *INCIENSO, HIERBA DEL VASO;* BRITTLEBUSH; *TOHAWES*. Small shrubs, not long-lived, aromatic, mostly 0.5–1.6 m (excluding inflorescences); older stems scarcely woody, with a usually rounded or hemispherical, often dense crown. Vegetative stems relatively thick, brittle, white woolly, glabrate with age. Leaves drought deciduous and highly variable with soil moisture, semipersistent, 3–10 cm including petiole, 1.4–3.6 cm wide; blades mostly ovate, entire or nearly so, often white woolly (conspicuously greener and thinner when produced during wet periods, whiter and thicker when produced during dry times).

Flowering branches of slender, usually few-branched panicles (8) 20–30 cm, usually raised well above the foliage; peduncles glabrate or sparsely woolly, especially near the flower heads. Phyllaries graduated, green, lanceolate to ovate, sparsely hairy, often ciliate to moderately woolly, the longer phyllaries 3–5 mm. Flowers showy, the rays bright yellow, 12–18 mm, the disk florets yellow or maroon-brown. Achenes 3.5–5.0 mm, gray to blackish, obovate, shallowly notched at apex, the body and margins with long white hairs; chaffy bracts glandular, entire to toothed, slightly longer than the enclosed achene and falling with it; pappus none. Flowering mostly spring, and with summer-fall rains.

One of the most abundant and conspicuous perennials in the region; producing massive displays of showy yellow, daisy-like flowers during favorable seasons. Frost sensitive and sometimes severely nipped by freezing weather; fast growing and often quickly recovering from drought or moderate freeze damage. At maturity the heads turn downward and the achenes fall away.

Across much of the flora area including the Sonoyta region, granitic and volcanic ranges to peak elevations and on all exposures, desert plains, bajadas, hills, and arroyos or washes; generally not in creosote bush flats and not on moving dunes. Often a pioneer species where few or no other major perennials are present, such as on slopes of rugged lava hills and cinder cones. Desert and semiarid regions nearly throughout the western half of Sonora, northwestern Sinaloa, both states of Baja California, southeastern California, southern and western Arizona, southern Nevada, and extreme southwestern Utah.

Forma *farinosa* with all yellow heads, and forma *phenicodonta* S.F. Blake [var. *phenicodonta* (S.F. Blake) I.M. Johnston] with maroon-brown disk florets are common and freely intermixed in many parts of northwestern Sonora. The *farinosa* form is more common in many parts of the Pinacate volcanic field, whereas the *phenicodonta* form is often more common in the southwestern part of the field. The color differences are reported to be controlled by a single gene (Kyhos 1971). Another variety occurs in Baja California Sur.

A yellowish resin often oozes from wounds in the stems. When heated this resin becomes plastic and was used by Sonoran Desert people as a glue or sealant, for medicinal purposes, and as incense (see Felger & Moser 1985; Uphof 1968).

Forma **farinosa:** 25 mi E of San Luis, *F 16710*. Sierra del Rosario, *F 75-3*. Pinacate Junction, *F 20603*. Tinaja del Tule, *F 18785B*. Sykes Crater, *F 18924*. S slope Pinacate Peak, *F 19394*. Puerto Peñasco, *Dennis 14-16*. Forma **phenicodonta:** 45 mi W of Sonoyta on Mex 2, *Turner 59-8*. Sierra del Rosario, *F 75-2*. Pinacate Junction, *F 20604*. Tinaja del Tule, *F 18785A*. Sykes Crater, *F 18925*. SW of Sonoyta, *F 9829*.

Encelia frutescens (A. Gray) A. Gray subsp. **frutescens** [*Simsia frutescens* A. Gray] RAYLESS ENCELIA. Rounded shrubs often 0.9–1.5 m. Stems brittle and slender, at first densely hairy, glabrate and whitish with age. Herbage rough hairy (scabrous), the hairs white and often with expanded bases. Leaves 17–32 × 3.5–6 (12) mm, sparsely hairy, petioled, the blades green, narrowly ovate to oblong. Heads solitary on hairy peduncles mostly 3–9 cm; heads 1.3–1.6 cm wide, more or less round, with disk florets only. Phyllaries lanceolate to ovate, glandular, with thick white hairs, the larger phyllaries 7–9 mm. Chaffy bracts 8–12 mm, nearly transparent, glandular, with hairs smaller than those of the phyllaries, the outer ones especially fold tightly over the enclosed achene and fall with it. Achenes 7–10 mm, dark gray, the apex truncate (not notched), the body with sparse hairs, the margins with silky white hairs 2.5–3.0 mm; pappus none. (Caution: Do not confuse marginal hairs for pappus; in other regions there is sometimes a pappus of short weak awns.) Spring and late summer-fall.

Sandy soils, washes, sand flats, and stabilized or low, shifting dunes; near the Río Colorado to Puerto Peñasco and to the northwest and southeast margins of the Pinacate volcanic region; often locally common. Southeastern California, southwestern Arizona, northeastern Baja California, and northwestern Sonora south to the vicinity of Hermosillo. Another subspecies occurs in Baja California.

NE of Puerto Peñasco, *F 88-231*. 14 km W of Los Vidrios, *Burgess 5606*. S base of Hornaday Mts, *Burgess 6853*. 35 mi E of San Luis, *F 90-177*. NNE of El Golfo, *F 92-213*.

ERIGERON FLEABANE

Annual, biennial, or mostly perennial herbs, sometimes shrubs. Leaves alternate, sometimes in basal rosettes, usually entire. Phyllaries narrow, nearly equal to graduated. Ray florets fertile, often numerous with delicate narrow rays usually white, pink, blue, or purple, or sometimes the ray portion of the corolla absent. Disk florets tiny, numerous, almost always yellow. Achenes flattened to cylindric, 2-ribbed. Pappus of slender and often fragile bristles.

Americas and Eurasia; mostly temperate to boreal and montane in tropical America; at least 375 species. References: Cronquist 1947; Nesom 1982, 1989a, 1992.

1. Ephemerals, often 15–30 cm. ______ **E. divergens**
1′ Perennials from a woody rootstock, often 50–100 cm. ______ **E. oxyphyllus**

Erigeron divergens Torrey & A. Gray var. **divergens** [*E. lobatus* A. Nelson] DESERT FLEABANE. Winter-spring ephemerals (elsewhere often biennial or short-lived perennial herbs), often 13–30 cm; taproot well developed on larger plants. Stems slender and flexible. Leaves, stems, and phyllaries densely hairy with minute stalked glands and larger non-glandular, spreading white hairs 0.5–1.0 (2.0) mm. Leaves soft and pale green; lower leaves pinnately lobed, upper leaves reduced, entire, and narrowly oblanceolate. Phyllaries many, more or less in 2 whorls, mostly 4.0–4.5 mm, the inner ones broader with membranous margins; larger heads 1.7–2.0 cm wide (including the rays); rays delicate, thin, pale violet. Achenes 1.3 mm; pappus of slender (capillary) bristles 2.0–2.5 mm as well as shorter, broader, fringed bristles.

Best developed on wet soils in riparian, semiriparian, or temporarily waterlogged habitats; sometimes seasonally common in the Pinacate region on playas and flats with hard-packed clayish soils. Also sandy-gravelly soils along washes and arroyo beds, at waterholes, and less often in open desert along the Río Sonoyta near the Arizona border. Northwestern Mexico, interior western United States and adjacent Canada; a second variety in Arizona, Nevada, and Utah.

Erigeron lobatus is said to occur in southeastern Californica, southern and western Arizona, southern Nevada, and northwestern Sonora eastward to the vicinity of Magdalena, and to be usually distinguishable from *E. divergens* by leaves with rounded lobes, as well as the presence of both stalked glandular hairs and longer and spreading non-glandular hairs. Characteristics of both taxa can be found in the same local population and in plants sampled at different seasons and years.

Pinacate Junction, *F 20610*. Tinajas de Emilia, *F 87-33*. S of Tinajas de los Pápagos, *F 18725*. Arroyo Tule, *F 19547*. Sonoyta, *F 86-87*. El Papalote, *F 86-129*.

Erigeron oxyphyllus Greene. Herbaceous perennials from a branched, woody caudex (rootstock) 2–15 cm long, or the leafy stems arising directly from the rootstock; stems 0.5–1.0 m, erect to arching, slender and brittle, glabrous or glabrate, striate. Young shoots, leaves, and phyllaries sparsely to moderately pubescent with short, white, often appressed hairs. Leaves linear, 3–5 cm, drought deciduous. Involucres 8–10 mm wide, the phyllaries strongly graduated, the inner (larger) phyllaries 3.2–4.8 mm, minutely ciliate. Rays white to pale blue, not curled; disk florets with corollas 4.0–5.5 mm, pale yellow. Achenes 1.8–2.0 mm, flattened, pale yellow-brown. Pappus of slender, pale yellow, barbellate bristles. Flowering at least in spring.

Higher elevations on steep north-facing granitic canyons and slopes of the Sierra del Viejo south of Mex 2 (*F 16873, F 85-733*) and the Sierra Cubabi south of Sonoyta (*F 20545*). Otherwise known only from arid slopes in western Arizona.

ERIOPHYLLUM

Annual and perennial herbs, or sometimes suffrutescent. Herbage woolly; leaves of seedlings and young shoots often opposite, otherwise alternate. Phyllaries in 1 whorl; ray florets pistillate and fertile, or sometimes absent, the disk florets bisexual and fertile. Pappus usually present, with scales and bristles without a midrib. Western North America; 14 species. References: Carlquist 1956; Constance 1937.

Eriophyllum lanosum (A. Gray) A. Gray [*Burrielia lanosum* A. Gray] WOOLLY DAISY. Diminutive spring ephemerals; conspicuously white woolly. Stems slender, erect to ascending and spreading or decumbent, often 2–12 cm. Leaves linear to narrowly oblanceolate, entire, the larger ones 7–15 mm. Heads 5–8 mm, showy, solitary on slender peduncles, the phyllaries 5–6.5 mm, densely woolly; rays 5–7 mm, white with red candy-stripe lines below, the disk florets yellow. Achenes 2.5 (3) mm, slender, blackish with appressed white hairs; pappus white, of flattened outer scales and longer, awnlike inner bristles.

Sandy and granitic gravelly soils of river floodplains, valley plains, and bajadas in the Sonoyta region; apparently not common. Common in nearby Arizona and expected eastward in northern Sonora but specimens lacking. Also Baja California, southeastern California, southern and western Arizona, southern Nevada, and southwestern Utah.

Sonoyta, river floodplain, *F 86-98C*. 18 km SW of Sonoyta, *F 88-211*. 4 mi S of Sonoyta, *F 20545*.

FILAGO

BY JAMES D. MOREFIELD AND RICHARD FELGER

Annuals, usually small, often with slender stems, usually woolly, sometimes canescent. Leaves alternate or seemingly whorled (*F. arizonica*), entire, sessile or petioles obscure. Heads tiny, sessile, or peduncles very short, generally grouped in clusters, the actual phyllaries none or vestigial and grading into the chaffy bracts. Receptacle elongated, generally tack-shaped (withering below, the tip expanded). Corollas reduced and dull in color, or sometimes pink or purple at tips.

Ray florets unisexual, modified as slender, tubular (filiform), pistillate florets with no anthers, the ligule very reduced. (The tiny ray florets are easily mistaken for disk florets, and some floras describe the heads as discoid.) Pistillate florets in several series, spirally arranged, each floret in all or outer few series subtended by a chaffy bract, at least a few florets of the inner series usually without chaffy bracts. Outer chaffy bracts phyllary-like (not phyllaries because each directly subtends an individual floret), the body densely woolly outside, the margins thin and membranous, generally (in ours) with a winglike membranous appendage at the tip (an extension of the margin of the body), each of these chaffy bracts enclosing the floret and achene and generally falling with the achene. Innermost chaffy bracts whorled, concave, open and boat-shaped, rigid, partly or fully spreading at maturity (giving the head a star-shaped appearance), usually longer than the outermost chaffy bracts.

Disk florets usually all bisexual, innermost in position, without chaffy bracts (except the outermost in *F. arizonica*), the corollas slightly expanded above and 4- or 5-lobed.

Achenes minute (ours 0.6–1.0 mm). Achenes of outermost pistillate florets smooth and shiny (ours) or at least relatively smoother and shinier, without a pappus; achenes of innermost chaffy-bracted and chaffless (if present) pistillate florets like the disk achenes. Disk achenes slightly smaller, rougher (the surface with a cellular-patterned sculpturing) and duller or pubescent with unicellular trichomes (papillae), with a pappus of 12–30 deciduous white bristles minutely barbed above and subplumose below.

Native to temperate and warm-temperate Europe, southwest Asia, North Africa, and southwest North America; 24 species. Eight species in North America, 3 native, 4 introduced in western North America plus 1 introduced in eastern North America. The 3 native North American species (in subgenus *Oglifa*) are common winter-spring ephemerals in the Sonoran Desert.

The rather confusing differences in the achenes and their subtending bracts, or lack thereof, may be the result of adaptive strategies for different kinds of dispersal mechanisms (see *Cryptantha*). The outer pistillate florets, enclosed in bracts, lack a pappus and their achenes are smooth-surfaced without trichomes (papillae); here the dispersal unit is the achene and its enclosing bract. Achenes of the inner florets (inner pistillate florets and disk florets—the ones not enclosed in bracts) are papillate or at least rougher surfaced, and the disk florets have a pappus; here the dispersal unit is the achene plus the pappus, although the pappus is immediately deciduous. Perhaps the roughened surfaces of the nonenclosed achenes assist in dispersal.

As with *Stylocline,* the only other filaginoid genus in the flora area, these fuzzy little composites apparently are self-fertilizing. (They are, however, not strictly autogamous, because pollen does have to cross from the florets with anthers to those without, but this can happen within the same head. The bisexual florets in *Filago* probably are truly autogamous. *Stylocline* has unisexual florets.) In keeping with this habit the flowers are minute, dull colored, and with greatly reduced corollas. The winged appendages of the outer chaffy bracts apparently play an important role in fertilization. As the developing stigmas grow up against the incurved bracts, they are guided inward, over the stamens of the disk florets.

The various filaginoids (*Filago* and *Stylocline*) often grow intermixed in the flora area as well as elsewhere in the Sonoran Desert; sometimes 3 species occur at a single site. *Filago californica* seems to be the most widespread, or eurytopic (an ecological generalist), and *Stylocline micropoides* the next most widespread and common. *F. arizonica* is also common but more narrowly restricted in habitat. Although each of the other 2 species (*F. depressa* and *S. gnaphaloides*) is known from only a single collection in the flora area, there is no reason to think that they are rare here—they were discovered on herbarium sheets intermixed with other species.

As with certain other so-called "difficult" groups, e.g., *Cryptantha* and *Euphorbia* subgenus *Chamaesyce,* once you have mastered the subtleties of the technical key characters and correlated these with the gross appearance of the plant, there is seldom any problem in identifying them in the field, even without microscopic examination. Among the filaginoids, drought-stunted plants may present identification problems, and the growth form, general habit, or even achene texture can vary from the perceived norm. Although general growth habit usually serves for quick identification, characters of the disk florets are more reliable and definitive. An unanswered riddle is why there is not more speciation, or eco-geographic variation, especially in light of the probable self-fertilizing nature of the plants.

On casual inspection plants of *Filago* could be confused with those of *Stylocline* or *Evax;* the latter two genera are readily distinguished by the near lack of pappus on the inner florets (*Stylocline* usually has a vestigial pappus). *Evax* has not been found in the flora area. The low, spreading habit of *F. depressa* might be confused with that of *Stylocline*. The filaginoids might also be confused with *Gnaphalium palustre,* but the latter has no epappose (without a pappus) florets, has no true chaffy bracts on the receptacle, and is a plant of wet places. In the flora area *Filago* usually has a broad receptacle whereas *Stylocline* has an elongated, cylindrical receptacle. However, a young, fresh (not withered) flower head of *Filago* may appear to have a cylindrical receptacle. *Filago* has disk achenes, but in *Stylocline* the disk achenes do not develop (or are reduced or vestigial). The broader receptacle in *Filago* may function as a nutrient base for its disk florets. References: Arriagada 1998; Morefield 1993b; Wagenitz 1976.

1. Lower heads and leaves mostly restricted to the branch nodes (forks); leaves nearly linear, usually much longer than the heads; florets inside innermost chaffy bracts 4–12 in number, the minority (0–2) pistillate. ______ **F. arizonica**

1′ Lower heads and leaves more evenly distributed; leaves oblanceolate to oblong, usually not much longer than the heads; florets inside innermost chaffy bracts 12–40 in number, the minority (3–7) bisexual.

  2. Well-developed plants erect, taller than wide, usually branched from a dominant central stem; leaves mostly oblanceolate, acute; body of outer chaffy bracts firm, thickened; innermost (disk) achenes mostly sparsely papillate, with pappus of 17–23 bristles falling away in complete or partial rings; disk corolla lobes mostly 4, usually red-tipped. ______ **F. californica**

  2′ Well-developed plants mostly spreading, wider than tall, branched from base without a dominant central stem; leaves mostly oblong to obovate, obtuse; body of outer chaffy bracts soft, membranous; innermost (disk) achenes mostly smooth, with pappus of 11–15 bristles falling away singly or in 2s; disk corolla lobes mostly 5, usually yellowish. ______ **F. depressa**

Filago arizonica A. Gray. ARIZONA FLUFFWEED. Stems 3.0–9.5 cm, 1 to several from base, none clearly dominant, at first erect or ascending, often spreading with age, almost always leafless between the clusters of flower heads or with 1 leaf between clusters. Branching pattern tends to be proliferate (pseudo-dichotomous, usually with 2 side branches from beneath a cluster of flowering heads). Stem distance between clusters of flower heads on larger plants ca. 1–5 cm, the stems tend to be straight, often dark and woolly, the hairs often falling away with age. Leaves just below flower heads usually conspicuously longer than heads. Inner, chaffless florets (innermost pistillate florets, if present, plus disk florets) 4 –12; disk florets 4–10, the corolla 5-lobed. Disk achenes minutely and densely papillose (occasionally coarsely and sparsely papillate as in *F. californica,* but if so, then the "background" surface minutely and densely papillose), the pappus bristles falling away in complete or partial rings.

Often on fine-textured soils (silty alluvium or clayish to sandy and gravelly) or in low places along watercourses or other places where water temporarily accumulates in washes, playas (including those in crater floors), and other depressions. Sonoyta region (vicinity of Río Sonoyta and Quitobaquito), the Pinacate region to 900 m, and perhaps more widespread. Southward in Sonora at least to the vicinity of El Desemboque San Ignacio (*F 17430*), southwestern California including the Channel Islands, northern Baja California, and southern and central Arizona.

10 km SW of Sonoyta, *F 88-164.* Río Sonoyta, S of El Papalote, *F 86-167.* Quitobaquito, *F 86-177.* 1 km SW of Tezontle cinder mine, *F 88-257.* Hornaday Mts, *Burgess 6866-A.* Cráter Trébol [Molina], *F 92-328.*

Filago californica Nuttall. CALIFORNIA FLUFFWEED. Characteristically taller than wide, (2) 3–25 cm, slender and erect with a dominant main axis, unbranched or with few branches arising at various places along a main axis, or larger plants branching and sometimes spreading with age; stems more or less evenly leafy. Inner, chaffless florets (innermost pistillate plus disk florets) total 18–30. Disk florets with a pappus of 17–23 bristles, these falling away in complete or partial rings, the corollas 4-lobed, usually reddish. Inner (chaffless floret) achenes usually coarsely and sparsely papillose (surfaces with short, blunt trichomes or bumps).

Widespread and often common in sandy-gravelly to rocky soils; lowland flats, dry streamways, bajadas and terraces, and rocky slopes. At least in the Sonoyta region through the Pinacate volcanic field, including higher elevations. Northern California to southwestern Utah, and southward to Baja California, Arizona, western Texas, and Sonora southward to the Guaymas region.

Drought-stunted plants might be confused with *F. depressa. F. californica* can be distinguished by pappus bristles that separate in complete or partial rings and the 4, usually reddish corolla lobes. In other regions the plants sometimes reach 60 cm in height.

Between Tinajas de Emilia and Carnegie Peak, 750 m, *F 19803.* Hornaday Mts, *Burgess 6866-C.* Quitobaquito, *F 86-186B.* 10 km SW of Sonoyta, *F 88-167.*

Filago depressa A. Gray. Characteristically low, semiprostrate and spreading with age, densely white woolly, the internodes usually very short; generally broader than tall and reaching no more than 5 cm in height. Leaves mostly broad and blunt. Disk florets with pappus of 11–15 bristles falling away singly or in 2s, their corolla lobes 5, usually yellowish. Inner achenes usually smooth but dull, the outer achenes smooth and shiny.

Granitic sandy bajada on the south side of Sierra Cipriano (*F 88-197B*) in the Sonoyta region, and perhaps more widespread. Eastward in northern Sonora and southward in western Sonora at least to Bahía Kino. Also southern and western Arizona, southern Nevada, southeastern California, and Baja California.

FLAVERIA

Annual or perennial herbs or small trees. Leaves opposite and decussate. Heads numerous, densely crowded at the stem tips, the flowers yellow, each head with 1 (or 2) ray(s) or rays lacking, the disk florets 1–15; pappus present or absent.

Mostly in warm regions of North America, few in South America, 1 in Australia; 21 species. Some are weedy and adventive. Often on alkaline or gypseous soils and in disturbed habitats. Reference: Powell 1978.

***Flaveria trinervia** (Sprengel) C. Mohr [*Oedera trinervia* Sprengel] Hot-weather annuals, mostly less than 45 cm. Leaves 4–12 cm, bright, shiny green, lanceolate to broadly ovate, often elliptic, the margins toothed. Numerous tiny yellow-green flower heads in dense clusters simulating a single "normal" multi-flower composite head and nearly hidden in leafy stem tips; individual heads minute, 1-flowered (rarely 2), enclosed by 2 membranous bractlike phyllaries. Achenes 2.0–2.5 mm, blackish, club-shaped, with low but conspicuous ribs; pappus none.

Agricultural weed south of San Luis (*F 85-1026*); roadsides, irrigation ditches, and fields; not native to the Sonoran Desert. Widespread in the New World and adventive in the Old World.

GAILLARDIA BLANKET FLOWER

Taprooted annuals or herbaceous perennials. Leaves basal or alternate, entire to toothed or pinnatifid. Heads showy and solitary at tips of branches. Receptacle with some bristles among the disk florets. Rays sterile, yellow to purple- or red-brown; disk florets bisexual. Achenes hairy; pappus usually of awned scales. North and South America; 28 species. Some are cultivated as ornamental garden flowers.

Gaillardia arizonica A. Gray [*G. arizonica* var. *pringlei* (Rydberg) S.F. Blake] Spring ephemerals, with sessile, globular glands and spreading to appressed, often crinkled white hairs to more than 2 mm. Leaves mostly basal, 3–10 cm, several times longer than wide, mostly deeply pinnatisect, often pinnately lobed or merely toothed on smaller plants; petioles prominent and usually winged. Peduncles (scapes) slender, 8–45 cm. Heads showy, 3.5–4.0 cm wide; phyllaries separate, green and leafy, narrowly to sometimes broadly lanceolate, the outer (larger) phyllaries 6–14 mm. Rays and disk florets bright yellow, the outer corolla surfaces glandular and with sparse, short hairs; rays 12–28 mm, the tips parted into 3 (or 4) large lobes. Achenes 2.5–3.0 mm, obscured by dense, ascending hairs at first white, becoming golden brown; pappus scales broad and membranous, the midrib sometimes extended into an awn.

In the flora area known from 2 collections in sandy soils, one to the north and the other to the east of the Pinacate region. Fairly common in nearby Organ Pipe Monument and Cabeza Prieta Refuge, and probably elsewhere in northernmost Sonora. Also Arizona, Nevada, and southwestern Utah. Var. *pringlei,* distinguished only by awned pappus scales, does not seem worthy of taxonomic recognition; awned and unawned pappus scales often occur in the same flower head, plant, or population.

24 mi SW of Sonoyta, *Shreve 7589.* 1 km SW of Tezontle cinder mine, *F 88-259.*

GERAEA

Annual or short-lived perennial herbs. Leaves alternate. Phyllaries green, the margins ciliate with conspicuous white hairs. Receptacles with membranous chaffy bracts clasping the disk achenes and falling with them. Achenes very flat, blackish with whitish margins continuous with 2 slender, persistent awns. *Geraea viscida* (A. Gray) S.F. Blake from California and Baja California, the only other species in the genus, lacks ray florets. *Geraea* is allied to *Encelia* and natural but mostly sterile hybrids are known. References: Blake 1913; Clark 1998; Kyhos 1967.

Geraea canescens Torrey & A. Gray. DESERT GOLD, DESERT SUNFLOWER. Winter-spring ephemerals 15–100 cm, erect, often robust, usually with 1 to several ascending branches, or sometimes much-branched above with age. Upper stems and outer phyllaries densely glandular. Leaves (2.5) 5–13 cm, oblanceolate to obovate or elliptic, coarsely toothed, especially the upper half, or smaller leaves sometimes entire, rough to the touch, with coarse white hairs; petioles prominent and winged or the leaves sessile; upper stem leaves reduced. Peduncles long, or short on much-branched late-season plants; heads sunflower-like, showy, closing at night, opening after sunrise, the flowers bright yellow. Phyllaries graduated, the inner and longer ones 5–9 mm, dark green, linear-lanceolate, strikingly outlined with dense white hairs (ciliate). Rays sterile, (0.8) 1.5–3.2 cm. Achenes 5–7 mm including the broad white margin at the apex that joins the 2 awns, obpyramidal, with silky, white, ascending hairs often 1–2+ mm. December-April.

One of the most abundant and conspicuous wildflowers in the region; notable stands often occur on the north side of the Pinacate region. Sonoyta region, desert plains, Pinacate volcanic field, and granitic ranges. Sandy or rocky soils; desert pavements, rocky slopes, sand flats, and low stabilized dunes; not on higher shifting dunes and not known elsewhere in Sonora. Also Baja California, southwestern Arizona, southeastern California, southern Nevada, and southwestern Utah. This is the only annual species in the *Encelia* alliance (Clark 1998).

29 km SW of Sonoyta, *Burgess 4759*. W of Sonoyta, *F 7596*. E of Palo Verde Camp [Campo Rojo], *F 19950*. MacDougal Crater, *F 9921*. SE of Moon Crater, *Ezcurra 19 Apr 1981* (MEXU). Sierra del Rosario, *F 20758*. E of San Luis, *F 16692*.

GNAPHALIUM CUDWEED, EVERLASTING

Annual or perennial woolly herbs or small shrubs. Leaves alternate and entire. Heads appearing discoid, phyllaries overlapping and membranous (scarious); receptacles flat and without chaffy bracts. Florets tubular, the inner ones bisexual, the outer ones pistillate. Achenes oblong, pappus a single row of capillary bristles.

Worldwide, well developed in temperate and semiarid North America; 150+ species, often weedy. Ours might be confused with filaginoids (*Filago* and *Stylocline*) but can be distinguished by their flat true phyllaries (not chaffy bracts because they do not individually subtend florets), absence of epappose florets, and lack of chaffy bracts. References: Arriagada 1998; Espinosa-Garcia 1985.

Gnaphalium palustre Nuttall [*Filaginella palustris* (Nuttall) Holub] LOWLAND CUDWEED. Small winter-spring ephemerals, densely woolly, often much-branched from the base, sometimes low, spreading, and nearly prostrate; stems 2.5–12.5 cm. Leaves 1–3 cm, sessile or the lower ones with narrowly winged petioles, often oblanceolate, the margins sometimes wavy. Phyllaries flat, with a broad green midstripe and broad, dry, membranous margins. Pappus bristles falling separately or in groups (not as a unit united into a ring at the base). Achenes 0.45–0.55 mm, papillate.

Localized in wet, fine-textured, silty-sandy soil, often among rocks, at the edge of several tinajas at the southwestern side of the Pinacate volcanic field. Occasional in western Sonora south to the Guaymas region at the edges of muddy pools. Western North America, Canada to Baja California and Sonora, often in vernal pools or low wet areas.

Tinaja de los Chivos, *F 18787*. Tinaja del Tule, *F 19193, 20149*.

GUTIERREZIA SNAKEWEED, MATCHWEED

Annuals to herbaceous or woody perennials. Leaves alternate, often narrow. Herbage and phyllaries glandular-punctate and resinous. Flower heads small, the phyllaries graduated, persistent, with yellow or white ray and disk florets (1 species has disk florets only). Pappus present or occasionally reduced.

Ten species in western North America, central Mexico to western Canada, mostly in dry places, and 15 in South America. References: Lane 1982, 1985; Suh & Simpson 1990.

Gutierrezia sarothrae (Pursh) Britton & Rusby [*Solidago sarothrae* Pursh] *HIERBA DE LA VÍBORA;* BROOM SNAKEWEED; *SIW TAḌṢAGĬ*. Globose subshrubs to ca. 50 cm; stems woody below, slender, numerous, crowded, mostly erect to ascending. Leaves mostly 2–5 cm × 0.5–1.3 mm, narrowly linear with moderately sparse, short white hairs. Flower heads 4–5 mm, in dense, much-branched terminal clusters; involucres 3.5–4.5 × 2.5–3.0 mm, persistent, the phyllaries rather thick with a resin pocket near the tip; flowers bright yellow. Achenes 1.3–1.5 mm, the surface obscured by short, white, ascending hairs; pappus scales membranous, about as long as achenes on disk florets, shorter on ray achenes. September-October.

North-facing cliffs of Pinacate Peak (*F 87-48*) and rock ledges in the Sierra Cipriano (*F 91-14*). Also in the nearby Sierra Cubabi and eastward in northern Sonora, but not known southward in coastal Sonora. Central Mexico to western Canada, including Arizona and Baja California.

This is the most wide-ranging and variable species in the genus, as might be surmised from a listing of 26 synonyms (Lane 1985). Three additional species occur near the flora area: *G. arizonica* (A. Gray) M.A. Lane, *G. californica* (de Candolle) A. Gray, and *G. microcephala* (de Candolle) A. Gray.

GYMNOSPERMA GUMHEAD

This genus has a single species. It is closely related to *Gutierrezia* (basic chromosome number: $x = 4$) and *Amphiachyris* ($x = 4$ or 5), with 2 grassland species of Texas and central United States. In *Gymnosperma* $x = 8$. References: Lane 1982; Solbrig 1961.

Gymnosperma glutinosum (Sprengel) Lessing [*Selloa glutinosa* Sprengel. *G. corymbosa* de Candolle] Suffrutescent perennials or small shrubs, often 0.6–1 m and becoming woody at base; glabrate (minutely scabrous). Stem tips, leaves, and flower heads glistening with viscid, glandular exudate in dry seasons. (This exudate is apparently water soluble because the surfaces are green and virtually devoid of visible exudate after rainy periods.) Leaves alternate, sessile, tardily drought deciduous, dark olive-green, densely gland-dotted, linear-lanceolate to linear-oblanceolate, 2–8 cm × 2.0–4.5 mm, the midrib on lower surface prominently keeled. Flower heads mostly 1.5 mm wide, in dense terminal clusters, persistent; phyllaries graduated, membranous with a green midrib and resin pocket near tip, the larger phyllaries 3–4 mm. Flowers bright yellow; ray florets small and inconspicuous with reduced corollas and not exceeding the phyllaries. Ray and disk florets fertile; achenes 1.3–2.0 mm, cylindrical, with minute hairs; pappus reduced to an almost microscopic ring. Warmer months.

Higher elevations on north-facing slopes of the northwestern granitic mountains. Also in adjacent Arizona and widely scattered mountains southward in western Sonora. Southern Arizona to southern Texas and Guatemala. Apparently non-palatable to cattle and abundant in the Chihuahuan Desert region where heavy grazing is a factor (B. L. Turner, personal communication 1986).

Sierra Nina, *F 89-50.* Sierra del Viejo, *F 85-720.* Cerro Pinto, *F 89-53.*

HELIANTHUS GIRASOL, SUNFLOWER

Large annual or perennial herbs, some shrubs in South America; often with coarse hairs. Leaves opposite below, sometimes alternate above. Heads usually large, showy, and on prominent peduncles. Ray florets sterile, large, usually yellow, or rays rarely absent. Disk florets fertile, yellow or red-brown or red-purple. Achenes thick, mostly laterally compressed, enclosed by prominent chaffy bracts; pappus usually a pair of deciduous awned scales.

North and South America, greatest diversity in western North America north of Mexico; 67 species. Taxonomy of the sunflowers is difficult due to a perplexing lack of morphological boundaries. Cultivars of *H. annuus* are major oil-seed and forage crops, and were grown in North America in pre-Columbian times. The Jerusalem artichoke is *H. tuberosum.*

The sunflowers in the flora area are in section *Helianthus* (= section *Annui*), which includes 14 species in North America. These are generally annuals with most of the leaves alternate, and petioled often with ovate blades. Ours have bright yellow rays and a red-purple disk. *Helianthus* is closely allied to *Viguiera,* especially among South American species. References: González-Elizondo & Gómez-Sánchez 1992; Heiser 1961, 1963; Heiser et al. 1969; Rogers, Thompson, & Seiler 1982; Rieseberg et al. 1991; Schilling & Heiser 1981; Schilling et al. 1998.

1. Phyllaries at least 4 mm wide, the attenuate tip at least 4 mm; leaves green, sparsely hairy; disturbed habitats. ____ **H. annuus**

1′ Phyllaries 1.8–3.5 mm wide, the attenuate tip 1 mm or less or absent; leaves gray-green, moderately to densely hairy; natural and disturbed habitats. ____ **H. niveus**

***Helianthus annuus** Linnaeus [*H. annuus* subsp. *lenticularis* (Douglas ex Lindley) Cockerell] *GIRASOL, MIRASOL;* COMMON SUNFLOWER. Annuals, hispid with stiff, coarse hairs and several or more branches above, each terminating in a typical sunflower head, or small, stunted plants sometimes unbranched. Leaves reaching ca. 7–8 cm wide, the blades narrowly to broadly ovate; margins minutely serrate to coarsely toothed. Heads 6.5–8.0 cm wide including the rays, the rays often 2–3 cm. Larger phyllaries 5–6 mm wide, the attenuate tips 4–5 mm. Growing and flowering all year except during the several cooler months.

Well established as an agricultural weed, and occasionally along the Río Sonoyta, larger washes, or roadsides near Sonoyta. This is the common wild, weedy, annual sunflower of western Canada, the United States, and northwestern Mexico.

Sonoyta, *F 86-930*. Suvuk, *Nabhan 374*.

Occasional stunted plants of the giant, single-headed cultivated sunflower [var. *macrocephalus* (de Candolle) Cockerell] are found in larger washes near settlements and agriculture, such as near Sonoyta (*F 86-124*). Stems stout, unbranched; leaves more than 10 cm wide; heads including rays more than 12 cm wide, the rays usually at least 4 cm, the phyllaries at least 1 cm wide, the attenuate tips often at least 1 cm.

Helianthus niveus (Bentham) Brandegee [*Encelia nivea* Bentham. *Viguiera sonorae* Rose & Standley, *Contr. U.S. Natl. Herb.* 16:20, pl. 16, 1912] *GIRASOL;* DUNE SUNFLOWER; *HI:WAI.* Ephemeral to short-lived perennial herbs, 0.5–1 + m or occasionally to 3 m and shrubby with a woody trunk 5–8 cm in diameter. Herbage and phyllaries hairy, green to whitish, villous to canescent, and with sessile glands beneath the hairs. Leaves mostly 4–12 cm, the petioles prominent, the blades lanceolate to broadly ovate, the margins entire to serrated. Heads including rays (3.5) 4–9 cm wide; phyllaries graduated, lanceolate to sometimes ovate, green, the hairs often obscuring the surface or the green surface visible, the larger phyllaries 6.5–11 × 1.8–3.5 mm. Rays mostly 2–3 cm, sparsely glandular and with few short hairs on lower surfaces. Inner, larger chaffy bracts 8–9 mm, the tip papillose to bearded with short, white hairs. Achenes 4–5 mm, thick (rhombic in cross section), and mottled black and pale tan, with long, silky, forward-pointing hairs (or the achenes sometimes flatter, blackish and without hairs, these perhaps immature); pappus deciduous, of several shorter scales and 2 larger, awn-tipped scales. Primarily spring, also summer-fall with rains.

Abundant during favorable seasons or years, on sandy soils nearly throughout the flora area, including plains, roadsides, and dunes. Southeastern California to western Texas, and adjacent northern Mexico including the Baja California Peninsula and the Sonora coast south at least to the Guaymas region.

Populations on the shifting dunes of the Gran Desierto tend toward gigantism. This dune sunflower is distinguished by its more robust habit (often reaching 2–3 m) including larger leaves, well-developed woody trunk, erect habit, and generally larger heads with larger phyllaries 3.0–3.5 mm wide. At least some of these plants have unusually dense pubescence of white hairs on the new growth. This "gigas" trend increases toward the eastern side of the Gran Desierto: plants on the dunes south of the Pinacate lava shield are conspicuously larger than those near the Sierra del Rosario. Rainfall generally increases eastward across the Gran Desierto, which might explain the observed size variation. Similar large, shrubby dune sunflowers do not occur elsewhere in the Sonoran Desert. The giant dune sunflowers may deserve taxonomic recognition and seem to fit the pattern of gigantism seen in other Gran Desierto dune plants, e.g., *Croton wigginsii, Palafoxia arida* var. *gigantea,* and *Petalonyx thurberi.*

Two subspecies are recognized by González-Elizondo & Gómez-Sánchez (1992); ours are subsp. *tephrodes* (A. Gray) Heiser et al. Var. *niveus* is endemic to Baja California.

Papago Tanks, 20 Nov 1908 [*sic.* 1907], *MacDougal 57* (holotype of *Viguiera sonorae,* US). 5 km N of Puerto Peñasco, *F 85-781.* 21 mi SW of Sonoyta, *Bowers 2595.* Hornaday Mts, *Burgess 6846.* W of Los Vidrios, *F 16860.* 20 mi E of San Luis, *Bezy 463.* N of Sierra del Rosario, *F 20792.* Giant dune form: S of Moon Crater, *F 19005.* N of Sierra del Rosario, *F 75-44.*

Heterotheca Camphor Weed, Goldenaster
By Richard Felger and John Semple

Annual or perennial herbs from a woody taproot, rarely persisting to become shrubby; densely hairy and glandular, often aromatic. Leaves alternate, the margins mostly entire or sometimes undulate (wavy), the lower leaves petioled to sessile, the upper leaves sessile and smaller. Phyllaries numerous, graduated in size, the outer ones smallest. Receptacle alveolate, without chaffy bracts. Heads with ray and disk florets, both fertile (ray florets absent in 1 species); corollas yellow. Achenes obconic, often 3-angled in ray florets, laterally compressed in disk florets. Pappus double with an inner series of numerous firm, barbellate bristles and an outer series of shorter bristles or scales, or pappus of ray florets sometimes reduced or absent.

North America; 28 species in 3 sections (sensu Semple 1996). Section *Phyllotheca,* the Western Goldenasters, with 20 species, Mexico to Canada, mostly in western 2/3 of North America, includes the 1 species in the flora area and the western North American species of *Chrysopsis*. Many members of the genus are weedy. Within this taxonomically difficult genus one often sees remarkable variation within a single species and a striking lack of boundaries between many of the taxa. References: Nesom 1997; Rzedowski & Ezcurra 1986; Semple 1996.

Heterotheca thiniicola (Rzedowski & Ezcurra) B.L. Turner, *Phytologia* 63:127–128, 1987 [*Haplopappus thiniicola* Rzedowski & Ezcurra, *Ciencia Interamericana* 26:16–18, 1986. *Heterotheca sessiliflora* (Nuttall) Shinners var. *thiniicola* (Rzedowski & Ezcurra) G.L. Nesom, *Phytologia* 83:17] Gran Desierto camphor weed. Plants much-branched, subshrubby, sometimes flowering in the first year or season, or more often shrubs 90–130 cm, often broader than tall, woody below, semiwoody to herbaceous above, at first erect, ultimately spreading under weight of the branches. The shrubs with woody trunks often reaching 3–4.5+ cm in diameter, often formed by several stems twisting and growing together; bark brown, becoming fissured, fibrous, shredding with a ropelike texture and fraying strands, and accumulating sand in the fissures.

Herbage smelling sweet and spicy when crushed, densely strigose with coarse, white hairs; hairs on stems appressed to spreading, the larger ones ca. 1.0–1.5 mm; hairs on leaves mostly appressed, 0.5–1.0 (1.5) mm, the larger, lower leaves with longer hairs and the upper, smaller leaves with shorter hairs. Herbage and phyllaries with many small, globular, golden-colored glands. Leaves sessile, variable with moisture conditions. Leaves of primary stems or new ("wet season") growth obovate or oblong to mostly oblanceolate 11–32 × 4–12 mm, the margins conspicuously wavy (undulate), the tip pointed and mucronate; these leaves generally on widely spaced nodes, relatively thinner, and soon deciduous with drought. Leaves of upper branches, or late- or dry-season growth mostly 3–10 × 3–5 mm, nearly orbicular or broadly ovate to spatulate, the apex conspicuously notched to truncate or mucronate, the base cordate, clasping the stem, and rather firm, the margins cartilaginous and entire; these leaves on closely set nodes, and especially the smaller ones tardily drought deciduous.

Heads solitary at the ends of short branches during drier conditions, or several to many on panicle-like upper branches with wetter conditions; peduncles sometimes 3–5 cm but usually much shorter to subsessile, and with a few leafy bracts. Involucres 6–7 mm, campanulate; phyllaries ca. 40, gray-green, linear-oblong, entire, acute; inner phyllaries 4.5–5.7 × 0.7–1.0 mm, the outer phyllaries 2.0–2.6 mm long, firm and narrower than the inner ones. Receptacle alveolate. Corollas yellow, with a few hairs to essentially glabrous. Ray florets 7–13 in number, the ray (limb) ca. 6 mm, oblong. Disk florets 20–30 in number, ca. 4 mm. Style branches subulate, ca. 0.5 mm, the stigmatic portion densely pubescent, about as long as the style branch. Achenes 2.5–3.5 mm, densely pubescent with white to tawny hairs. Pappus 3.0–3.5 mm, persistent, the inner whorl of about 50 or more slender, tawny, barbellate bristles, the outer whorl of few, much shorter, and rather cryptic slender bristles. Flowering with soil moisture, at least in fall and early winter.

Known from only 2 localities about 15 km apart in the dunes south of the Pinacate volcanic field, about 10–15 km inland from the coast. *H. thiniicola* has been found only in a narrow zone of semistabilized parabolic dunes at the edge of lower stabilized dunes, where it grows with other characteristic dune plants. Although locally common, it is potentially vulnerable to extinction due to its very narrowly defined habitat and highly restricted geographic range. The growth form, size, and general architecture are extremely variable depending on moisture. Flowers and even foliage are generally not present during drought years. It is the only taxon in *Heterotheca* for which there is not even one chromosome count.

Semple (1996) indicates a close relationship with other goldenasters occurring in the Colorado River drainage, especially *H. sessiliflora* subsp. *fastigiata* (Greene) Semple in the mountains of southern California bordering the desert. It has a similar pappus and undulate-margined leaves. Although this goldenaster generally occurs well above the desert, Semple (1996:38) reports it as "rarely adventive in desert arroyos below mountains." Nesom (1997) submerged *H. thiniicola* as a variety of *H. sessiliflora,* a position that seems weak. *H. thiniicola* is among the most unique of all the western goldenasters. Although there are no *Heterotheca* known from the delta region of the Río Colorado, we can only dream about what once existed among the extirpated native flora prior to the big upriver dams and agricultural development (see Ezcurra et al. 1988).

8 km al NE de la Estación Gustavo Sotelo, 30 m, dunas de arena parabólicas, semi-móvilas, 17 Dec 1984, *Ezcurra 84001* (holotype, ENCB; isotype, TEX). Ca. 10 km NE de la Estación Gustavo Sotelo: *Ezcurra 16 Sep 1980, Ezcurra 10 Mar 1990* (probably same place as the type collection). 13.8 mi NE of Puerto Peñasco on Mex 8, then 7.6 mi by rd NW to base of large shifting dunes: *F 88-227; Reichenbacher 1655.*

HYMENOCLEA BURROBRUSH

Shrubs with aromatic, resinous herbage. Leaves alternate, linear-filiform, some with linear-filiform segments, the margins involute. Monoecious, or occasionally dioecious. Heads with disk florets only, the pistillate heads below staminate heads; flowers wind-pollinated and inconspicuous, but the membranous, spreading bracts of the fruiting heads distinctive and conspicuous. Staminate heads several flowered; pistillate heads modified as a bur with a single flower enclosed in woody bracts fused at their bases, the bracts winged, persistent, papery-membranous.

Southwestern United States and northern Mexico; 2 species. The genus is allied to *Ambrosia* and *Xanthium.* References: Karis 1995; Peterson 1974; Peterson & Payne 1973.

1. Shrubs usually taller than wide; stems erect; wings of fruiting bracts longer than wide; fall-flowering. ______ **H. monogyra**
1′ Shrubs as wide as tall; stems spreading; wings of fruiting bracts about as wide as long; spring-flowering. ______ **H. salsola**

Hymenoclea monogyra Torrey & A. Gray. *JÉCOTA;* SLENDER BURROBRUSH, CHEESE BUSH; *'I:WADHOḌ.* Slender shrubs to 2.5 m, with multiple, slender, mostly erect major stems mostly branching above. Leaves sparse, tardily drought deciduous, mostly (1) 2–7+ cm; leaves of young, vigorous, vegetative shoots often pinnately divided into several segments, the upper leaves reduced and mostly entire; leaves or segments 0.5 mm wide, grooved above (involute), the grooves filled with short, white, elongate-conical hairs. Fruiting burs 3.5–4 mm wide, the bract wings in a single whorl, the wings 0.8–1.4 mm wide, longer than wide. Flowering and fruiting in fall.

Locally common in large, broad gravelly washes south of the Quitobaquito region, between Aguajita Spring on the Arizona border (El Papalote, *F 86-121*) and the Río Sonoyta. Probably formerly along the lower Río Colorado. Southwestern United States to northern Jalisco; mostly along large washes, streamways, and rivers.

Hymenoclea salsola Torrey & A. Gray var. **pentalepis** (Rydberg) L.D. Benson [*H. pentalepis* Rydberg] WHITE BURROBUSH; *'I:WADHOḌ.* Globose shrubs, 0.4–1.5 m, densely branched, the branches often spreading. Stems striate. Leaves sparse, 1–7 cm, the larger ones often with 1–4 filiform segments and falling as the new shoots mature, the upper leaves reduced and entire, the leaves or segments 0.5–0.7 mm wide, grooved above with hairs as in *H. monogyra.* Staminate heads 2.5 mm high. Ovary and style of pistillate flowers densely glandular. Fruiting burs 5–7 mm wide, the bracts with wings in 1-several whorls at the middle of the bur, wings mostly 2.7–5.0 mm wide, about as long as wide, spreading like tiny airplane propellers. March-April, fruiting in the same season; flowers foul smelling, like a dead animal, especially when wet.

Common and widespread in large, gravelly washes, floodplains, sand flats, alluvial fans (bajadas), and cinder flats; Sonoyta region, Pinacate volcanic field, granitic ranges, and desert plains mostly near mountains; often especially conspicuous in washes at the bases of granitic ranges, upper bajadas, and cinder flats.

Two varieties, Sinaloa to Utah and Nevada. Var. *pentalepis* is the southern and only variety in Sinaloa and across most of Sonora, and with minor exceptions the only variety known from the Baja California Peninsula. This species thrives in disturbed, especially overgrazed, habitats.

30 mi E of San Luis, *F 5763*. Sierra del Rosario, *F 20740*. S of Moon Crater, *F 19088*. NE of Elegante Crater, *Ezcurra 13 Apr 1981* (MEXU). 24 mi SW of Sonoyta, *Shreve 7590*.

HYMENOTHRIX

Annual, biennial, or sometimes perennial herbs, with a taproot and erect stems. Leaves crowded at base and alternate on stems, usually much divided into elongated segments. Flowers yellow; rays present in only 1 species (ours), the disk florets bisexual. Achenes obpyramidal, 4- (5)-angled, blackish, pappus of 7–20 slender scales with transparent margins, the midrib extending into an awn. Southwestern United States and Mexico southward to the state of México; 5 species. Reference: Turner 1962.

Hymenothrix wislizeni A. Gray. Annuals or perennials, sometimes propagating by short rhizomes. Densely pubescent with glandular hairs and short, coarse, white non-glandular hairs; young growth or basal part of plant sticky-glutinous, the subsequent herbage less so. Stems erect, mostly 50–70 cm. Leaves pinnately 2 or 3 times dissected, the larger leaves 6–10+ cm, early leaves in a basal rosette, the stem leaves with fewer segments. Inflorescences much-branched, flat topped, the heads crowded in dense terminal clusters. Heads 8–10 mm, the phyllaries green, 2.7–4.0 mm. Ray and disk florets bright yellow, bisexual and fertile. Achenes 2.5–4.0 mm, moderately angled; pappus scales 10–13 in number. April and September-December.

Near Sonoyta (*F 86-316, F 86-505*); infrequent, usually small colonies in sandy-gravelly soils along roadsides and arroyos. Southern and central Arizona, northern Sonora, southwestern New Mexico, and northwestern Chihuahua.

HYMENOXYS BITTERWEED

Annual or perennial aromatic herbs dotted with microscopic globules of resinlike exudate. Leaves alternate or basal. Heads often showy, the ray and disk florets yellow and fertile, the rays turning down at maturity and persistent; phyllaries in 2 or 3 rows, densely villous. Achenes obpyramidal, hairy, mostly 5-angled, the pappus of 5–7 scales with the midrib often extending into an awn. Mostly North America, a few in non-tropical South America; 28 species. Reference: Bierner & Jansen 1998.

Hymenoxys odorata de Candolle. Spring ephemerals, 20–45 cm, with a taproot; sparsely pubescent, the herbage glandular-punctate. Stems few to many. Basal rosette leaves withering as the leafy stems develop; leaves pinnately dissected into thickish, blunt-ended, slender segments. Heads 5–10 mm wide (not including the rays); outer phyllaries rigid, strongly arched or broadly keeled, conspicuously thickened and united at base; inner phyllaries separate, longer than outer ones. Flowers bright yellow, the rays 7–10 mm, 8–10 in number. Achenes 2 mm, densely silvery-silky haired; pappus scales 5 or 6, acuminate to awn-tipped.

Fine-textured, poorly drained clay soils in the vicinity of Playa Díaz; apparently not common. Kansas to California south to Sonora, Chihuahua, Coahuila, Nuevo León, and Tamaulipas. In Sonora at least as far south as the Guaymas region in widely scattered local populations on poorly drained clay-silt soils.

Playa [Díaz] W of Cerro Colorado, *Ezcurra 1 Apr 1983* (MEXU). Edge of ciénega, 25 mi W of Sonoyta, 17 Mar 1936, *Shreve 7601*. Pinacate Junction, *F 93-215*.

ISOCOMA GOLDENBUSH

Small shrubs and subshrubs, often resinous-glutinous, sparsely hairy or glabrous. Leaves alternate, entire to toothed or pinnatisect, without obvious veins. Heads usually aggregated (clustered), of disk florets only, the florets yellow, abruptly expanded (ampliate) above the tube. Phyllaries conspicuously graduated, firm. Achenes ribbed, moderately to densely pubescent (sericeous); pappus of many coarse, persistent barbellate bristles of uneven length. Northern Mexico and southwestern United States; 10 species. Reference: Nesom 1991.

1. Phyllaries acute, apices herbaceous; stems grayish white, foliage gray-green; leaves lobed or toothed. ____ **I. tenuisecta**

1′ Phyllaries rounded, apices with a wartlike, thin-walled resin pocket; stems gleaming white; foliage light yellow-green; leaves entire or with 4–5 teeth per side. ____ **I. acradenia**

  2. Leaf margins entire to few toothed. ____ **I. acradenia** var. **acradenia**

  2′ Leaf margins mostly pinnately and coarsely toothed. ____ **I. acradenia** var. **eremophila**

Isocoma acradenia (Greene) Greene [*Bigelovia acradenia* Greene. *Haplopappus acradenius* (Greene) S.F. Blake] ALKALI GOLDENBUSH. Densely branched shrubs often 0.8–1.2 m with many mostly erect to ascending, slender, brittle, woody stems. Herbage glandular punctate, copiously resinous-glutinous from the glands; young herbage with sparse, short, white hairs soon mostly covered in resin, the older herbage appearing glabrate. Involucre obconical; phyllaries linear-oblong with a herbaceous wartlike resin pocket near the tip, the margins narrowly transparent-membranous and erose-ciliate at tip. Corollas, anthers, and stigmas bright yellow, longer than the pappus.

The wartlike resin pocket near the tip of the phyllaries is a characteristic that purportedly sets this species apart from all other isocomas. However, the phyllaries in *I. tenuisecta* may appear very similar. Three varieties, 1 only in California.

Isocoma acradenia var. **acradenia** [*Isocoma limitanea* Rose & Standley, *Contr. U.S. Natl. Herb.* 16:18, 1912] Leaves (1) 1.5–4.6 cm, narrowly oblanceolate to lanceolate, the margins entire or few toothed. Involucres (4.0) 4.5–6.5 mm. Achenes 2.0–3.5 mm. Spring and late summer-fall.

Sonoyta region in alkaline soils at edges of playas, sandy-silty flats, depressions, and urban and agricultural areas. Also southward sporadically along the Río Sonoyta to within ca. 20 km of the coast. Sonora eastward to the vicinity of Altar and southward to the vicinity of Guaymas; also deserts in southern Nevada, southern California, and western Arizona.

Quitobaquito, *F 5724*. 10 mi W of Sonoyta, *F 86-410*. Village of Sonoyta, 14 Nov 1907, *MacDougal 14* (holotype of *I. limitanea,* US). Río Sonoyta, 20 km NE of Puerto Peñasco, *Ezcurra 16 Oct 1980*.

Isocoma acradenia var. **eremophila** (Greene) G.L. Nesom [*I. eremophila* Greene. *I. acradenia* subsp. *eremophila* (Greene) Beauchamp. *Haplopappus acradenius* subsp. *eremophilus* (Greene) H.M. Hall] Leaves mostly 1.0–3.5 cm, linear to narrowly oblanceolate, pinnately and coarsely toothed, or sometimes entire to nearly entire on drought-stressed plants. Involucre 5–7 mm. Achenes 2.5–3.0 mm. October-December.

Coastal sand flats and dunes, sea bluffs, and alkali flats from the vicinity of Puerto Peñasco to the delta region of the Río Colorado, and northward in sand or sandy-silty and often alkaline soil flats near the river to the vicinity of San Luis; not known elsewhere in Sonora. Also Baja California, southern California, western Arizona, southern Nevada, and southwestern Utah.

3 km E of Puerto Peñasco, *F 85-766*. E of Gustavo Sotelo, ca. 1 km inland, *F 86-518*. 20 mi along beach E of El Golfo, *F 86-559*. NE of El Golfo, *F 85-1058*. Laguna Prieta, *F 85-756*.

†**Isocoma tenuisecta** Greene [*Haploppapus tenuisecta* (Greene) S.F. Blake ex L.D. Benson. *Isocoma fruticosa* Rose & Standley, *Contr. U.S. Natl. Herb.* 16:18, pl. 13, 1913] The report of this species in the flora area is based on a single, rather poor specimen: MacDougal Pass, *MacDougal 14 Nov 1907* (holotype of *Isocoma fruticosa,* US 574278). *I. acradenia* var. *acradenia* and *I. tenuisecta* are closely related, and specimens with intermediate characteristics are known "particularly in the area of Organ Pipe Cactus National Monument" (Nesom 1991). The MacDougal specimen has leaves seemingly more similar to those of *I. tenuisecta* than of *I. acradenia* var. *acradenia.* I was unable to locate any *Isocoma* in the vicinity of MacDougal Pass or any other record for *I. tenuisecta* in the flora area.

LACTUCA LETTUCE

Annual, biennial, or perennial herbs, sometimes low shrubs or rarely small trees, the sap milky. Early leaves sometimes in a basal rosette, stem leaves alternate. Heads small, of ligulate florets only; involucre cylindric, often broadening at the base in fruit (ours), the phyllaries unequal. Achenes compressed, ribbed, often with a long, slender beak or neck, the apex usually expanded; pappus persistent or deciduous, usually of silky white hairs.

Eurasia, North America, and Africa; 75 species. Cultivated lettuce, *L. sativa,* is grown as a winter vegetable in the Sonoran Desert.

***Lactuca serriola** Linnaeus [*L. scariola* Linnaeus] PRICKLY LETTUCE, COMPASS PLANT. Annuals, often germinating in late winter or spring, flowering in late spring and summer, and often persisting through the summer; 0.5–1.0 m, glabrous. Stems shiny white, mostly unbranched below; herbage often with stiff bristles, especially on the midrib of the lower leaf surface. First leaves in a basal rosette, these and lower stem leaves often 10–22 cm, deeply pinnatifid, the margins often with short, broad teeth tipped with white spines, the petioles winged; upper leaves reduced, sessile, and clasping the stem; stem leaves turned at base to hold the leaf edgewise and upright in a north-south plane (hence "compass plant"). Inflorescence a terminal, open panicle. Involucres enlarging after flowering to 11–15 mm, the phyllaries green, the larger ones conspicuously broadened at base, the tips with small tufts of hair. Corollas inconspicuous, pale yellow, fading purple-brown. Achene body 2.8–3.0 mm with short bristles near apex, the beak extremely slender, as long as to much longer than the body.

Urban and agricultural weed in the Sonoyta and San Luis regions and disturbed habitats in the Colorado delta. Native to Europe, now weedy in many parts of the world. The seeds of this species are the source of Egyptian lettuce-seed oil, used in food for its pleasant flavor and as a semi-drying oil.

Sonoyta, *F 86-306.* San Luis, *F 92-509.* Ciénega de Santa Clara, *F 92-519.*

MACHAERANTHERA GOLDENWEED

Taprooted annual, biennial, or perennial herbs or small shrubs. Leaves alternate, mostly sessile, entire to pinnately or bipinnately dissected or lobed, often spiny-toothed. Heads with fertile disk and ray florets or rays sometimes absent; ray corollas yellow, white, blue, lavender, or reddish; disk florets yellow; phyllaries bristle-tipped. Achenes compressed, mostly several ribbed or veined; pappus of slender, persistent bristles, or sometimes absent from ray florets.

Temperate and arid North America; 30 species. Many aspects of the relationships of the species within this genus and with other genera remain unclear (Morgan 1997). References: Cronquist & Keck 1957; Hartman 1990; Morgan 1997; Turner 1986.

1. Heads with disk florets only. ____ **M. carnosa**

1′ Heads with ray and disk florets.

  2. Ray corollas lavender, the pappus of ray florets reduced or absent; phyllaries not spinose-tipped. ____ **M. coulteri**

  2′ Ray and disk florets yellow, ray florets with a well-developed pappus; phyllaries usually spinose-tipped, at least the outer ones. ____ **M. pinnatifida**

Machaeranthera carnosa (A. Gray) G.L. Nesom var. **carnosa** [*Linosyris carnosa* A. Gray. *Aster carnosus* (A. Gray) A. Gray ex Hemsley. *Leucosyris carnosa* (A. Gray) Greene] ALKALI ASTER. Perennials, 0.5–0.8 m, sometimes dying back to ground level during drought, spreading by rhizomes. Stems often ascending and becoming partly decumbent, with few, often spreading branches. Leaves and young stems semisucculent and glaucous. Bracts and phyllaries minutely ciliate, the plants otherwise glabrous. Leaves few, often widely spaced, linear, awl-shaped and flattened above, entire, quickly deciduous, reaching 1.4–3.6 cm on new growth, but mostly less than 1 cm × 1.0–1.5 (2.0) mm, the upper ones greatly reduced. Heads of bright yellow disk florets; ray florets none. Phyllaries linear, green, fading brown, with papery-membranous margins, inner phyllaries 6–7 mm. Achenes 3–4 mm, linear, ribbed, the surface obscured by appressed white hairs; pappus bristles 6–7 mm. Warmer months.

La Soda at Bahía Adair (*F 86-521*), localized along the Río Sonoyta (near El Huérfano, *F 92-971*) and nearby at Quitobaquito (*F 86-219*); in Sonora otherwise known only from Quitovac. At these oases it grows on alkaline or alkaline-saline soils; at La Soda it is common at senescent pozos and on immediately adjacent dunes. It thrives on nearly barren ground where few or no other plants grow. Var. *carnosa* also occurs in southern Arizona. Var. *intricata* (A. Gray) G.L. Nesom [*Aster intricatus* (A. Gray) S.F. Blake] occupies the northern part of the range of the species in California and Nevada.

This unusual species is placed in the monotypic genus *Leucosyris* (Cronquist 1994; Sundberg 1986), or in a subgroup of 3 glabrate species with mostly entire and succulent leaves of the *Arida* group of *Machaeranthera* (Nesom 1989b). The *Arida* group is restricted to southwestern United States and northwestern Mexico. On the basis of morphological evidence, this species "is an oddball and as such may be best pulled out [of *Machaeranthera*] into its own genus" (Scott Sundberg, personal communication, 1999).

Machaeranthera coulteri (A. Gray) B.L. Turner & D.B. Horne var. **arida** (B.L. Turner & D.B. Horne) B.L. Turner [*M. arida* B.L. Turner & D.B. Horne. *M. arizonica* R.C. Jackson & R.R. Johnson, *Rhodora* 69:476, 1967] Ephemerals to short-lived perennials depending on soil moisture; taproot well developed. Plants (8) 15–40 cm, with a single stem or more often much-branched to sometimes bushy; glandular pubescent and often also with some non-glandular hairs. First leaves 4–6 cm, deeply pinnatifid, the segments often toothed or again divided and laciniate, usually quickly withering; subsequent leaves resembling mere segments of the first leaves, 6–16 × (1.5) 3–6 mm, sessile, more or less ovate, the base clasping the stem, irregularly toothed with several pairs of large, broad teeth.

Heads 1.5–3.5 cm wide including rays. Phyllaries graduated, mostly green, persistent, lanceolate to oblanceolate, green with membranous margins, the larger ones 4.0–5.5 mm, innermost phyllaries partially enclosing the ray achenes and often retaining them long after the disk achenes are shed. Rays 4.5–12 mm, lavender, slender. Disk florets yellow. Disk achenes (1.2) 1.5–1.8 mm, light brown, with several veins and small, sparse hairs, the pappus 2.0–3.0 (4.0) mm, the bristles many. Ray achenes sometimes to 2.2 mm, the pappus very reduced or more often absent. Flowering with even minimal soil moisture during warm weather, especially late spring-early summer.

Lowlands through much of northwestern Sonora on sandy and often alkali soils, sandy-gravelly washes, arroyos, riverbeds, sandy flats, lower stabilized dunes, and disturbed habitats such as roadsides and agricultural and urban areas. Northwestern Sonora south to the vicinity of Puerto Libertad, southwestern and western Arizona, southeastern California, and southern Nevada.

Machaeranthera coulteri var. *arida* is replaced on the Sonora coast between the vicinity of El Desemboque San Ignacio and Bahía Kino by plants approaching *M. crispa* (Brandegee) B.L. Turner & D.B. Horne. (*M. crispa* previously has been known only from the gulf coast of Baja California Sur.) *M. crispa* is in turn replaced southward along the Sonora coast by *M. coulteri* var. *coulteri*, from about the vicinity of Bahía Kino to the vicinity of Guaymas. These 3 taxa are narrowly distinguished: *M. coulteri* var. *coulteri* by its glabrous or glabrate herbage; var. *arida* by glandular pubescent herbage; and *M. crispa* by slightly larger heads and more densely glandular pubescent herbage.

San Luis, *Turner 1829*. S of Mex 2 [rd to Rancho Guadalupe Victoria], *Mason 1829*. 21 mi SW of Sonoyta, *Bowers 2590*. W of Estación Gustavo Sotelo, *F 84-3*. Quitobaquito, *n* = 5, 31 Mar 1962, *Johnson 3043-2* (isotype of *M. arizonica*).

Machaeranthera pinnatifida (Hooker) Shinners subsp. **gooddingii** (A. Nelson) B.L. Turner & R.L. Hartman [*Sideranthus gooddingii* A. Nelson. *S. viridis* Rose & Standley, *Contr. U.S. Natl. Herb.* 16:19, pl. 15, 1912. *Haplopappus spinulosus* (Pursh) de Candolle subsp. *gooddingii* (A. Nelson) H.M. Hall] SPINY GOLDENWEED. Herbaceous, short-lived perennials, flowering in first season, forming a short taproot; often 25–50+ cm, with glandular and non-glandular hairs; size highly variable depending on soil moisture. (Plants in cinder soils at higher elevations of the Sierra Pinacate are often exceptionally robust with large, showy flower heads.) Lowermost or first leaves pinnatifid, often 3–7.5 cm; upper and late- or dry-season leaves often 1–4 cm, linear, often with several pairs of laciniate, linear, or short segments, each segment tipped with a short white bristle. Flowering heads 1.7–4.6 cm wide including rays. Involucres 6–7 mm, the outer and often the inner (larger) phyllaries with a white spinose tip or projection (seta), the phyllaries numerous, linear-oblong with a green midstripe, the larger ones 4.0–7.0 mm. Ray and disk florets bright yellow, both with a pappus of many slender white bristles 3.5–5.0 mm; rays often 10–16 mm. Achenes 1.7–2.5 mm, sparsely to densely hairy, with 2 or more low ribs. Flowering response non-seasonal.

Widespread in the Sonoyta region, Pinacate volcanic field including higher elevations, and granitic ranges. Often in small rocky arroyos at lower elevations, and more common and widespread toward higher elevations in the Sierra Pinacate and the surrounding granitic mountains. Often most common on north- and east-facing rocky slopes.

A highly polymorphic species; central Mexico and the Cape Region of Baja California Sur to Nevada, Colorado, Montana, North Dakota, and adjacent Canada. Turner & Hartman (1976) divided the species into 2 subspecies and 7 varieties. Their subsp. *gooddingii* ranges from southern Nevada and southern Utah southward through most of the Baja California Peninsula and northwestern Sonora. They divide this subspecies into 4 varieties; ours are var. *gooddingii,* which occurs in southern Nevada and southern Utah to northwestern Sonora and Baja California Sur. This variety is distinguished by "elongated, leafy 'peduncles,' the leaves being progressively much-reduced and subspinose; plants decidedly sub-shrubby, up to 100 cm tall" (Turner & Hartman 1976:309).

Cerro Pinto, *Burgess 5611.* Sierra del Rosario, *F 75-12.* Hornaday Mts, *Burgess 6855.* W of Los Vidrios, *F 85-718.* Sonoyta, *F 85-966.* Campo Rojo, *Webster 24272.* S of Pinacate Peak, ca. 875 m, *F 19330.* Pinacate Mountains, *MacDougal 21 Nov 1907* (holotype of *Sideranthus viridis,* US 574279).

MALACOTHRIX DESERT DANDELION

Annual or perennial herbs, rarely woody at base, with milky sap. Leaves mostly basal, the stem leaves alternate and usually reduced. Heads often nodding in bud. Phyllaries strongly graduated and overlapping, grading into the accessory bracts below. Florets all ligulate, yellow, white, or rarely pink. Achenes columnar but short, beakless, the apex truncate; pappus bristles silky, white, often united near base and readily deciduous. Western North America; 13 species. References: Davis & Raven 1962; Williams 1957.

1. Leaves pinnatifid with slender threadlike segments; achenes 2.4 mm. ____ **M. glabrata**
1′ Leaves with broad, coarse teeth; achenes 1.8–2.0 mm. ____ **M. sonorae**

Malacothrix glabrata (A. Gray ex D.C. Eaton) A. Gray [*M. californica* de Candolle var. *glabrata* A. Gray ex D.C. Eaton] SMOOTH DESERT DANDELION. Spring ephemerals with a taproot. Herbage glabrate, slightly woolly when young. Plants, leaves, and flower heads highly variable in size depending on soil moisture. Leaves mostly in a basal rosette, 6–18 cm, pinnatifid with linear to threadlike segments, usually withering before flowering or fruiting. Flowering stems simple or few or sometimes many-branched, 10–45+ cm, mostly with a few leaves below. Heads 1–5 or more on long peduncles, nodding in bud. Involucres 2.5–3.0 cm wide, phyllaries densely to sparsely woolly on outer surfaces, the larger phyllaries ca. 20+ in number, (8) 9–14 mm, green with white to membranous margins. Ligules 12–16 (25) mm, cream-white to pale yellow, with yellow at junction of ligule and tube to form a yellow eye in the flower head, the outer ligules often with a pale red midstripe below; flower heads reaching 5 cm in diameter, closing at night, opening at ca. 7 A.M. (in April). Achenes 2.4 mm, brown, with 15 ribs; pappus bristles mostly separate, barbellate, the upper barbelles minute, the lower ones larger (Cronquist 1994: 434).

Widespread; Sonoyta region, desert plains, dune fields, and the Pinacate volcanic field to ca. 550 m; sandy or sometimes rocky soils of bajadas, low hills, cinder flats and slopes, sand flats, and low dunes, often with creosote bush. Northwestern Sonora and Baja California Sur to Oregon, Idaho, Utah, and Arizona.

Malacothrix glabrata is replaced by *M. californica* in nondesert regions in California and Baja California; plants of intermediate character occur at the western edge of the Sonoran Desert. The 2 taxa are sometimes treated as conspecific varieties.

MacDougal Crater, *Turner 86-20.* SW of Sierra Extraña, *F 88-72.* Sierra Pinacate, 550 m, *F 92-486.* NE of Elegante Crater, *F 92-155.* SW of Sonoyta, *Prigge 7286* (RSA). NE of Puerto Peñasco, *F 88-236.*

Malacothrix sonorae W.S. Davis & P.H. Raven. SONORAN DESERT DANDELION. Winter-spring ephemerals, often 10–20+ cm. Basal rosette leaves 2.5–5 cm, lanceolate to oblanceolate, with broad, coarse teeth or toothed segments; stem leaves reduced. Phyllaries 8–14 or more, 7–9 mm, the accessory bracts ca. 8–10 in number, less than 1/3 as long as the longer phyllaries. Florets 10–20+ (Davis & Raven [1962] report 30–61 florets), the ligules yellow. Achenes 1.8–2.0 mm, with 15 ribs, the uppermost part of the achene smooth, the apex with small teeth; pappus bristles similar to those of *M. glabrata.*

Eastern margin of the Sonoyta region in sandy, granitic bajadas and hills, as well as cinder soils at higher elevations in the Sierra Pinacate. Southern Arizona, from the Ajo Mountains to eastern Pima County, and northwestern Sonora to the vicinity of Hermosillo. Often growing through small shrubs such as *Ambrosia dumosa.* This species is a member of the *M. clevelandii* complex, which is widespread in California, Baja California, and Arizona and extends into neighboring Nevada and northern Sonora.

S side of Sierra Cipriano, *F 88-215.* 10 km SW of Sonoyta, *F 88-166.* N side of Sierra Pinacate, 950 m, *F 92-496.*

MONOPTILON DESERT STAR

Small, low-growing, winter-spring annuals. Phyllaries linear, all similar; heads upturned, showy, solitary on stem tips, with ray and disk florets. Achenes slightly compressed and hairy. Two closely related species, mostly in the Mojave and Sonoran Deserts.

Monoptilon bellioides (A. Gray) H.M. Hall [*Eremiastrum bellioides* A. Gray] Spring ephemerals, mostly low and spreading, 1.5–25 cm wide, with coarse, white hairs and sessile glands, especially on the phyllaries. Early leaves in a basal rosette, the stem leaves alternate; leaves linear-spatulate, often 7–12 mm, closely spaced below the heads, and few and widely spaced along stems, reaching 20–35 mm on robust plants. Heads showy, closing at night, opening after sunrise, mostly 1.2–1.7 (2) cm wide, 1 to several or more per plant, often obscuring the leaves; phyllaries 5–6 mm, narrowly lanceolate or elliptic with membranous margins and red, acuminate tips, the inner phyllaries partially enclosing the ray achenes. Rays 6–10 mm, white, fading to lavender, the disk florets yellow. Ray and disk florets fertile, the achenes 1.5 mm, obovate, brown; pappus of shorter, outer, white bristly scales, and inner, more numerous straw-colored bristles 2.0–2.5 mm.

Rocky and sandy soils of slopes and level terrain nearly throughout the region; especially conspicuous on desert pavements. Absent from dunes. Southward in Sonora to the vicinity of Puerto Libertad and eastward to the vicinity of Imuris. Also northeastern Baja California, western and southern Arizona, southeastern California, and southern Nevada.

Sierra del Rosario, *F 20691.* Hornaday Mts, *Burgess 6821.* Moon Crater, *F 19267.* S of Mex 2 [rd to Rancho Guadalupe Victoria], *Mason 1824.* Trail 0.5–1 mi above Campo Rojo, *Quirk 2 Mar 1985* (DES).

PALAFOXIA

Taprooted annuals and some herbaceous perennials. Stems erect to ascending. Leaves entire, usually at first opposite, otherwise alternate. Phyllaries in a single series. Rays present or absent; ray and disk florets white to purple or pink. Achenes 4-angled, obpyramidal; pappus of scales to linear bristles, with membranous margins and a prominent midrib. Mexico and southwestern to midwestern United States and Florida; 12 species. References: Turner & Morris 1975, 1976.

Palafoxia arida B.L. Turner & M.I. Morris var. **arida.** SPANISH NEEDLES. Winter-spring ephemerals, sometimes also growing with summer-fall rains, often ca. 30–90 cm; rarely surviving as annuals or short-lived perennials; erect, often with a single main axis and several or more branches from above middle, or sometimes many-branched and bushy when well watered; taproot well developed. Leaves with coarse, forward-pointing hairs, the stems scabrous with shorter, stiff hairs, the flowering stems and phyllaries with glandular hairs. Leaves 3–9 cm, linear to linear-lanceolate, sometimes semisucculent, the apex acute to obtuse. Heads with disk florets only, cylindric to narrowly turbinate (shaped like a top), (18) 20–25 (28) mm including florets; phyllaries linear, often with a narrow membranous margin. Corollas white to pink; anthers purple. Achenes 7.5–12.3 mm, narrowly obpyramidal, blackish, mostly with dense, coarse, short white hairs or sometimes glabrate.

Widespread in the lowlands nearly throughout the flora area. Sandy and gravelly soils, relatively uncommon on rockier soils; washes, floodplains, bajadas, crater floors, sand flats, and dunes. Sandy soils in the Mojave and Sonoran Deserts; southwestern United States, Baja California, northern Baja California Sur, and western Sonora.

Puerto Peñasco, *F 85-780.* El Golfo, *Goldberg 15 Feb 1975.* 18 mi E of San Luis, *F 77-14.* Base of Hornaday Mts, *Burgess 6802.* Sykes Crater, *F 18970.* S of Moon Crater, *F 19039.*

In central Sonora, at about Hermosillo and Bahía Kino, *P. arida* grades into *P. linearis* (Cavanilles) Lagascea (sensu Turner & Morris), which extends southward at least to Altata, Sinaloa, and a similar transition occurs on the Baja California Peninsula. Turner & Morris (1975:79) differentiate *P. linearis* as "sprawling shrublets having linear leaves with round or obtuse apices" and restrict it to "coastal sand dunes of southern Baja California." Plants from that region appear indistinguishable from those of coastal Sinaloa and southwestern Sonora. The southern populations are ephemerals, annuals, or short-lived perennials: their life span is facultative, depending upon soil moisture.

Palafoxia arida var. **gigantea** (M.E. Jones) B.L. Turner & M.I. Morris [*P. linearis* var. *gigantea* M.E. Jones] Robust spring and perhaps summer ephemerals, 0.7–1.5 m. Flower heads (25) 28–30 mm, the achenes 11.8–15 mm. Plants closely approaching or assignable to this variety occur on high, shifting dunes north and west of Sierra del Rosario. This variety, known otherwise only from dunes west of Yuma in southeastern California, is distinguished by its substantially larger, more robust habit, larger and greener leaves, larger flower heads, and longer achenes.

N and W of Sierra del Rosario, *F 75-32, F 20347, F 20786.*

PECTIS CHINCHWEED

Annual or perennial herbs, low in stature, mostly pungently aromatic. Leaves opposite, dotted with few to many relatively large and conspicuous oil glands embedded in the leaf tissue, and almost always with several pairs of marginal bristles fringing the leaf base. Heads with fertile ray and disk florets. Phyllaries with large oil glands like those of the leaves, equal in number to the ray florets and partially enclosing the ray achenes, each ray floret actually attached to the base of its subtending phyllary. Achenes linear, club-shaped; pappus variable, sometimes reduced.

New World; 100 species. Readily recognized by the generally pungent aroma, leaf bristles, and arrangement of the rays and phyllaries. *Pectis* is unusual among the Tageteae in having C_4 rather than C_3 photosynthesis, which undoubtedly plays an important role in its adaptation to hot climates. References: Keil 1975, 1977, 1996.

1. Heads sessile, not showy, almost hidden among upper leaves; achene hairs straight, the tips not enlarged; clay soils of playas and depressions. ______ **P. cylindrica**
1′ Heads on conspicuous peduncles, showy, not hidden; achene hairs curled and bulbous at tip; widespread. ______ **P. papposa**

Pectis cylindrica (Fernald) Rydberg [*P. prostrata* Cavanilles var. *cylindrica* Fernald] HIERBA DE CHINCHE. Non-aromatic summer-fall ephemerals; mostly low and decumbent, 1–10 cm tall, 1–20 cm across. Leaves (0.8) 1–2 (2.4) cm, narrowly oblanceolate, narrowed to a winged petiole-like base with prominent bristles; midrib and margins of lower leaf surfaces thickened and light colored, the margins minutely toothed. Heads with 3 ray florets and 8–13 or fewer disk florets; florets relatively inconspicuous, scarcely protruding from the subcylindrical involucre. Phyllaries equal, 7–9 mm, green, oblong. Rays bright yellow but relatively small and not obviously differentiated from the disk florets. Achenes 4.0–4.5 mm, columnar, slender, blackish, with straight to slightly sinuous white hairs. Pappus of ray and disk florets similar, of 3 or 4 larger and several smaller bristles, these linear-lanceolate, flattened, translucent, and fringed; bristles on disk florets slightly longer than the corollas.

Seasonally common at Playa Díaz and at least a few other, smaller playas. The nearest known populations are from Topawa and Menager's Dam on the Tohono O'odham Reservation in southern Arizona. Southwestern Texas to Arizona and northern Mexico from Coahuila and Tamaulipas to Durango, Sinaloa, Sonora, and the Baja California Peninsula.

Playa Díaz, *F 86-367*. 2 mi S of Tinajas de los Pápagos, *F 86-494*.

Pectis papposa Harvey & A. Gray var. **papposa.** *MANSANILLA DEL COYOTE;* DESERT CHINCHWEED; *BAN MANSANI:YA*. Summer-fall ephemerals; pungently aromatic; low and compact to "bushy" and spreading, highly variable in size, sometimes reaching 30 cm across, usually much smaller and occasionally reproducing when only 1.5 cm tall. Leaves 1.5–4.0 cm, linear with prominent bristles on a winged petiole-like base. Heads showy, flowers bright yellow. Phyllaries 3.5–4.0 mm, oblong, green with a firm midrib. Rays conspicuous, 6–8; ray achene pappus absent or reduced to a crown of minute scales. Disk pappus of many plumose bristles, conspicuous but shorter than the corollas, or on some plants the disk pappus also reduced to a crown of minute scales or absent. Achenes 2.8–3.5 mm, columnar, blackish, with appressed, white (to brownish) hairs, the tip curled inward and bulbous (occasionally the hairs few and straight).

Widespread throughout the flora area including volcanic and granitic areas. Best developed on sandy soils of washes, flats, and lower dunes; generally less common on rocky soils of pavements and slopes, although sometimes abundant on cinder slopes and flats. Extending to the summit of Sierra Pinacate and in all of the craters.

This is one of the most conspicuous and widespread hot-weather ephemerals in the Sonoran Desert. It appears after the first soaking rains of summer, and successive generations may occur with additional rains later in the summer and fall. The plants sometimes persist and continue flowering into December but are killed by the first freezing (or near-freezing?) weather. On rare occasions some plants may survive through winter or germinate in spring. Late-season or drought-stressed plants tend to produce disk achenes without pappus bristles, which would reduce energy expenditure and reduce dispersal distance.

This species occurs in all 4 North American deserts and often colors the landscape yellow. There are 2 well-marked geographic varieties. Var. *papposa* occurs primarily in the Great Basin, Mojave, and Sonoran Deserts, and also in thornscrub and tropical deciduous forest, from southeastern California and southwestern Utah to southern Baja California Sur, New Mexico, and central Sinaloa. It is widespread in Sonora. Var. *grandis* D.J. Keil is mostly in the Chihuahuan Desert. *P. papposa* is in section *Pectothrix* of subgenus *Pectidopsis,* a characteristically showy, yellow-flowered group of one dozen species found mainly in northern Mexico and southwestern United States, and disjunct in Peru.

Bradley & Haagen-Smit (1949) reported that cumin oil comprises the major portion of the essential oils of *P. papposa* and suggested it might be grown as a commercial crop for this oil, but their suggestion has not been pursued (Keil 1977). It is very fast growing, quickly maturing, prolific, thrives in very hot weather, and should be highly resistant to insect pests.

Another Sonoran Desert composite, *Amauria brandegeana* (Rose) Rydberg, endemic to the Baja California Peninsula, also has similar strange, inwardly curled, white hairs on blackish achenes. Other shared characteristics include size and general shape of the achenes and ephemeral habit, but otherwise the plants are scarcely similar (see *Perityle emoryi*).

E of San Luis, *F 16696*. Cerro Colorado, *F 10803*. MacDougal Pass, *F 10634*. Moon Crater, *F 10598*. Suvuk, *Burgess 6339*. N of Puerto Peñasco, *F 85-782*.

PERITYLE

Annual or perennial herbs or subshrubs. Leaves petioled, mostly alternate. Rays fertile, white or yellow, rarely absent; disk florets bisexual, fertile, numerous, yellow. Achenes oblong to elliptic, flattened, often blackish, with a thickened white margin, and ciliate; pappus with scales and often also 1 or 2 often unequal, minutely barbed awns.

Mexico, mostly in the northern states, and southwestern United States, plus one species disjunct in Chile and Peru; 60 species. References: Everly 1947; Powell 1974.

Perityle emoryi Torrey. DESERT ROCK DAISY. Cool-weather ephemerals, 5–60 cm, often semisucculent when well watered; bearing sessile and stalked glands and sparse non-glandular white hairs, glabrate with age. Leaves 1.5–6 (8+) cm, the blades about as wide as long, coarsely toothed to palmately lobed, the lobes also toothed. Phyllaries 3.5–5.5 mm, green, oblong-ovate to obovate, ribbed, the margins ciliate. Ray corollas often 4.0–4.5 mm, white, the disk florets yellow; tube of disk and ray corollas densely glandular. Receptacle pitted after achenes fall, its surface pale yellow-green when fresh, and smooth, shiny, wavy, and cellular like hardened foam. Achenes 2.5–3.1 mm, the faces blackish, the margins thin (not calloused), with a border of short, white hairs; disk achenes very flat, the faces glabrous or glabrate; ray achenes similarly flat to somewhat thickened on one face, the faces with whitish hairs and minute, sessile, golden glands. Pappus of small scales and a single bristle barbellate on its upper 1/4–1/3, or the bristle absent. (About 1/3 of the specimens from the flora area lack pappus bristles; there is no discernable pattern to the variation.)

One of the most abundant and widespread wildflowers in the region; in nearly every non-saline habitat and to peak elevations but usually absent from the higher dunes. Often one of the first ephemerals to appear after fall rains and one of the last to flourish with late spring rains.

Southwestern North America, often weedy: Sonora south to the Guaymas region, Baja California Sur to southern California, southern Nevada, southwestern Utah, and western Arizona. Also Peru and Chile.

Another Sonoran Desert composite, *Amauria brandegeana,* is strikingly similar to *Perityle emoryi* (Powell 1972). At first glance you would not even suspect them to be different species. Similarities include their yellow disks, white rays, blackish achenes, general appearance and size of the plants, heads and flowers, and ephemeral habit. *A. brandegeana,* endemic to the Baja California Peninsula, is distinguished by its angled achenes with strange curled hairs like those of *Pectis papposa.*

Puerto Peñasco, *Burgess 4751*. Tinajas de Emilia, *F 19766*. Cerro Colorado, *F 19640*. High elevation, Sierra Pinacate, *F 19291*. Sykes Crater, *F 20009*. Sierra Extraña, *F 19064*. Sierra del Rosario, *F 20736*.

PEUCEPHYLLUM

This taxonomically isolated genus of 1 species shares similarities with *Dyscritothamnus* of Hidalgo. Reference: Strother 1978.

Peucephyllum schottii A. Gray. *ROMERO;* DESERT FIR, PYGMY CEDAR. Much-branched shrub resembling a dwarf conifer; often 1–1.8 (2) m with a well-developed stout, woody trunk, the bark twisted and shredding. Conspicuously resinous; glands crowded, glistening and golden when fresh, sessile and dotlike on stems and leaves, and stalked on peduncles and sometimes on stem tips, phyllaries, and toward apex of corollas. Evergreen; leaves often 7–30 × 0.8–1.0 mm, alternate, crowded at ends of twigs, sessile, linear-filiform, terete or nearly so, bright green, rarely 1- or 2-lobed on a few upper leaves.

Heads 10–15 mm, solitary at stem tips. Phyllaries 9.0–10.5 mm, mostly linear-lanceolate, somewhat variable, in 2 whorls, and sometimes with highly variable outer bracts. Flowers fragrant, having disk florets only, at first pale yellow-green, the corolla lobes and upper tube becoming red-purple. Achenes 2.5–3.0 mm, dark brown to blackish, obconic with a knoblike yellow callus at the base, faintly ribbed, the surface nearly to wholly obscured by dense white hairs. Pappus of many bristles and scales, with slender white bristles to the outside, and longer flat bristles or narrow scales, 5–7 mm long, to the inside. January-May, and probably in fall.

Widespread on rocky soils and rock slopes, cliffs, and crater walls; volcanic and granite ranges. Sometimes on all slope exposures but elsewhere absent from south- and west-facing slopes; often more common at higher elevations. Arid regions of the Sonoran and Mojave Deserts of southwestern Utah, southern Nevada, southeastern California, western Arizona, the gulf side of Baja California and northern Baja California Sur, islands in the Gulf of California including Isla San Esteban, and the flora area of northwestern Sonora.

Sierra del Rosario, *F 20399*. Cerro Pinto, *F 89-59*. Sierra Extraña, *F 19042*. MacDougal Crater, *Kamb 2001*. NNW of Pinacate Peak, 800 m, *F 19487*. NE of Elegante Crater, *Johnson 13*. Puerto Peñasco, *Weibe 6 Jun 1962*.

PLEUROCORONIS

Small shrubs or subshrubs. Leaves opposite below, alternate above, petioled, the blades entire to dissected. Inflorescences of loose, terminal clusters of several heads; heads many flowered, with disk florets only. Involucre of an inner series of larger phyllaries and an outer series of smaller, graduated phyllaries. Corolla tube long and slender, the lobes minute; style branches club-shaped and conspicuously exserted. Achenes narrowly obpyramidal and glandular, pappus of barbellate bristles and membranous scales. Endemic to the Sonoran Desert; 3 closely related species. References: Johnston 1924a; King 1987.

Pleurocoronis pluriseta (A. Gray) R.M. King & H. Robinson [*Hofmeisteria pluriseta* A. Gray] ARROW-LEAF. Shrubs or subshrubs, often globose, 0.8–1.0 m across, much-branched, the stems slender and brittle. Herbage densely pubescent with stalked, glandular hairs or relatively less pubescent when well watered or shaded. Leaves tardily drought deciduous; petioles 6–38 mm, the blades mostly 5–15 (18) × 3.5–14 mm, ovate to lanceolate, usually with 1–3 pairs of large teeth, these sometimes again divided; leaf size and shape highly variable with moisture and shade. (On hot exposed cliffs the leaves may consist of little more than the petioles, with the blades reduced to narrow spear-shaped thickenings.)

Inner phyllaries 8–11 mm, linear, thin, often green or purplish, with 3 prominent veins and thin, membranous white margins, the tip acuminate. Corollas, styles, stigmas, and anthers white to pale yellow, or the style branches and stigmas purple. Achenes 2.5–3.0 mm, 4-angled, the surfaces minutely hairy with appressed hairs that spread when wet; pappus bristles 5.0–6.5 mm, intergrading with outer small membranous scales, or the scales sometimes absent. Apparently growing and flowering at any time of the year.

Steep rock slopes and cliffs, often on canyon walls, mostly with north- and east-facing exposures and especially common at higher elevations; Sonoyta region and the volcanic and granitic ranges. Northwestern Sonora, southwestern Arizona, Baja California, southeastern California, and southern Nevada.

There is a gradual (clinal) increase in leaf blade size southward along the coasts of Sonora and Baja California, leading into *P. laphamioides* (Rose) R.M. King & H. Robinson, distinguished by its larger, broader leaf blades and supposedly shorter petioles. Although the ratio of petiole to blade length is larger for the northern populations, the petioles of *P. pluriseta* are not necessarily longer. In both taxa moisture conditions greatly influence length of petioles and size of leaf blades.

A weaker but similar clinal trend is seen from west to east across the flora area, from regions in the Lower Colorado Valley to the Arizona Upland (e.g., the Sonoyta region). Plants in the Sierra Cipriano and the Ajo Mountains in nearby Arizona tend to have leaf blades about as broad as long. Although these plants might be considered *P. laphamioides* because of their broad leaves, their leaves are always appreciably smaller than those of undisputed *P. laphamioides* from farther south in Sonora and Baja California. In several instances larger-leaved plants from the Ajo Mountains, apparently collected during times of higher soil moisture, were identified by King (1967) as *P. laphamioides,* whereas smaller-leaved dry-season collections from the same place were identified as *P. pluriseta*. Nevertheless, even during drought stress the more southerly *P. laphamioides* populations do not produce the minute leaf blades often seen farther north among *P. pluriseta* populations. Johnston (1924a) treated these taxa as varieties, which might be a more realistic solution.

Bahía de la Cholla, *F 16830*. Cerro Colorado, *F 10779*. Arroyo Tule, *F 18820*. Sierra del Rosario, *F 20672*. Sierra del Viejo, *F 16872*. Sierra Cipriano, *F 91-13*.

PLUCHEA

Annual or perennial herbs or shrubs, usually pungently aromatic. Leaves alternate, simple, petioled to sessile and clasping. Rays absent; outer florets pistillate, numerous, in several series, the inner florets with sterile ovaries. Achenes small, often ca. 1 mm; pappus of slender, minutely barbed bristles, sometimes the bristles widened at tip. Warm regions of the world; 40+ species, often in brackish marshes and other saline or alkaline habitats. References: Arriagada 1998; Gillis 1977; Godfrey 1952; Nesom 1989c; Robinson & Cuatrecasas 1973.

1. Herbaceous annuals or perennials, usually 1.5 m or less in height, the branches not willowlike; herbage green, sticky, and stinky. ____ **P. odorata**

1′ Woody shrubs, usually more than 1.5 m, branches willowlike; herbage silvery green and neither sticky nor stinky. ____ **P. sericea**

Pluchea odorata (Linnaeus) Cassini var. **odorata,** 1826 [*Conyza odorata* Linnaeus, 1759. *C. purpurascens* Swartz, 1788. *Pluchea purpurascens* (Swartz) de Candolle, 1836. *P. camphorata* of authors. Not *P. camphorata* (Linnaeus) de Candolle, 1826] *JARA;* MARSH FLEABANE, ALKALI CAMPHOR WEED. Annuals or perennials from a thick, semifleshy root, sometimes with rhizomes. New growth generally emerging in early summer, maturing and flowering in fall, the stems dying in late fall or early winter; herbage frost sensitive. Stems often reaching 1–1.5 m, leafy, the lower leaves withering as the shoots mature. Fresh green herbage with sticky (viscid) as well as soft hairs, dotted with glands, and stinking. Leaves mostly 4–15 cm, variable, ovate to elliptic or lanceolate, toothed to entire or nearly so; lower leaves petioled, the upper leaves reduced and sessile. Phyllaries graduated, larger ones 4–5.5 mm. Corollas and phyllaries pale to bright rose-lavender. Achenes 0.8–1.2 mm, brown, columnar.

Seeps and marshes along the lower Río Colorado, and wet soils at pozos and permanent ponds at La Salina, Laguna Prieta, and Quitobaquito. Widespread in North American wetlands, often on alkaline soils. A second variety occurs in northeastern United States.

Pluchea camphorata (Linnaeus) de Candolle of southeastern United States and northeastern Mexico is closely related to *P. odorata* and is distinguished from it largely by its elongated rather than flat-topped inflorescences. The tangled taxonomic trail of *P. odorata* is summarized by Cronquist (1994), McVaugh (1984), and Kahn & Jarvis (1989). The plant called *P. odorata* by Wiggins (1964) and others, and now known as *P. symphytifolia* (Miller) Gillis, is a tropical American woody shrub that extends into southeastern Sonora and Baja California Sur.

S of El Doctor, Río Colorado riverbed, *F 85-1048*. La Salina *F 84-30*. Laguna Prieta, *Roth 12*. Quitobaquito, *Bowers 902*.

Pluchea sericea (Nuttall) Coville [*Polypappus sericeus* Nuttall. *Tessaria sericea* (Nuttall) Shinners] *CACHANILLA;* ARROWWEED; *KO:MAGI 'JU'US* Woody shrubs often 1.5–3+ m. Branches willowlike, long, leafy and upright; herbage densely silvery hairy. Leaves (1) 1.5–4.5+ cm, entire, sessile, mostly narrowly elliptic, lanceolate, or oblanceolate. Phyllaries graduated, the outer ones conspicuously broader, to 3 mm, the longer inner ones 5–6 mm. Flowers pink. Achenes 1.0–1.3 mm; pappus bristles dilated at apex, especially on central florets. Flowering at least March-June.

Sandy, silty-clayish, often alkaline soils, mostly in wet places or where water accumulates, occasionally on low dunes. It was especially abundant along the lower Río Colorado: in 1922 Aldo Leopold recorded "cachinilla" in dense, impenetrable groves throughout the delta (Meine 1988). It is now common along the riverbank and adjacent sandy mesas, at the Ciénega de Santa Clara, and as a weed in agricultural areas. Also along the Río Sonoyta and at Quitobaquito, Laguna Prieta, and La Salina. Also at Quitovac but otherwise scarcely extending southward in western Sonora. Southeastern California to southern Utah and western Texas, Chihuahua, Baja California, and northern Sonora.

The Cocopa esteemed the long, straight stems for house construction, roofing, and arrow shafts. A prized lac, yellow to dark red-brown, was collected from the stems and used as an all-purpose plastic adhesive and sealant (Euler & Jones 1956). This lac is common on the extensive stands along the margins of the lower Río Colorado and occasional elsewhere in the flora area.

Río Sonoyta 21 km W of Sonoyta, *F 85-971.* Pinacate Junction, *F 20607.* 11 mi NW of Estación Gustavo Sotelo along RR, *Webster 24242.* La Salina, *F 86-553.* Laguna Prieta, *F 85-753.* S of Riito, *F 85-1045.*

PorophyLLUM

Annual or perennial herbs or shrubs, usually glabrous and glaucous, with translucent oil glands and a strong, pungent odor somewhat like that of marigolds (*Tagetes*), to which *Porophyllum* is related. Leaves entire, opposite below, alternate above. Heads with disk florets only. Achenes linear, tapering at both ends, often blackish, with short ascending hairs; pappus of many slender, barbellate bristles.

Southwestern United States to Peru and Argentina; 30 species, 8 in the Sonoran Desert region. Two annual species cultivated in central and southern Mexico, known as *papalo* and *papaloquelite,* are consumed in great quantity as condiments or green vegetables Reference: Johnson 1969.

Porophyllum gracile Bentham. *HIERBA DEL VENADO.* Perennials, suffrutescent or subshrubby, pungently aromatic, glabrous and conspicuously glaucous, sometimes purple tinged. Stems straight, very brittle, striate. Leaves 1.5–6 cm, linear to threadlike, sparse and widely spaced, quickly drought deciduous, especially the larger ones. Leaves and phyllaries with prominent, dark purplish, elongated oil glands. Flower heads cylindrical, solitary at ends of branches or branchlets, the phyllaries 10–17 mm, linear-oblong, the margins membranous and rose-purple. Corollas white to pale purple with dark purple longitudinal lines. Achenes 8–13 mm, narrowly cylindrical, narrowed at both ends, blackish at maturity, moderately to densely covered with short, stiff, appressed hairs; pappus bristles pale brown. Receptacles often persistent after achenes mature and fall, leaving a "stump" of brainlike convoluted, polished thickenings. Flowering at various seasons.

Rocky or sandy-gravelly soils, especially along arroyos; most common in canyons and steep north- and east-facing slopes. Sonoyta region and granitic and volcanic ranges to peak or near peak elevations. Southeastern California, southern Nevada, and southwestern Utah to western Texas, Baja California Sur, and much of Sonora to northwestern Sinaloa. This is one of the most variable and wide-ranging species in the genus and occurs in drier habitats than any other *Porophyllum.* It is an important medicinal plant in Sonora (Felger & Moser 1985).

Cerro Pinto, *F 13218.* Sierra del Rosario, *F 20380.* Hornaday Mts, *Burgess 6863.* NNE of Pinacate Peak, 950 m, *F 86-423.* SW of Sonoyta, *F 9814.* Bahía de la Cholla, granitic hills, *F 13155.*

PRENANTHELLA

"The monotypic genus *Prenanthella* is the most reduced member of the Cichorieae in North America . . . [and] appears to be derived from near the genus *Malacothrix*" (Tomb 1974:214). Cronquist (1994) included *Prenanthella* in *Lygodesmia.* References: Tomb 1972, 1974.

Prenanthella exigua (A. Gray) Rydberg [*Prenanthes exigua* A. Gray. *Lygodesmia exigua* (A. Gray) A. Gray] Taprooted winter–spring ephemerals, 10–28 cm in height, with a well-developed main axis and paniculately branched above. Leaves and stems, and especially the branches, sprinkled with minute tack-shaped glandular hairs. Early leaves in a basal rosette or closely spaced along lower stem, oblanceolate, 1.7–6.0 cm, pinnately and coarsely toothed to incised, often turning red-green and withering as the upper branches develop; stem leaves reduced upward to subulate scales.

Heads of ligulate (raylike) florets only, 3- or 4-flowered, the phyllaries 4–5 mm, 3 or 4 plus 1 or 2 smaller accessory bracts. Corollas 5 mm, white with violet teeth, the ligules 2.5–3.0 mm; anthers white shading to violet at tips; stigmas violet with dark purple papillae. Achenes 3.0–3.7 mm, columnar with a truncate apex, light brown, with 5 narrow grooves; pappus bristles in a dense tuft, persistent, white, minutely barbellate, fused at their bases, unequal in length, the longer ones 2.5–3.0 mm, or pappus bristles sometimes absent from 1 or more achenes in a head, the epappose achenes more persistent and slightly longer than the pappus-bearing achenes.

Sandy-gravelly arroyos, roadsides, and rocky granitic and lava slopes; widely scattered in the Sonoyta and Pinacate volcanic regions; not known elsewhere in Mexico. Adjacent Arizona to Utah, Colorado, Nevada, southeastern California, and disjunct in western Texas (Johnston 1990).

Sierra Extraña, *F 19058.* 2.8 km NW of Cerro Colorado, *F 19629.* Arroyo Tule, *F 19204.* 22 mi W of Sonoyta, *F 7629.* Quitobaquito, *Gould 2990.*

PSATHYROTES

Conspicuously woolly-scurfy annuals or perennials with a turpentine-like odor. Leaves alternate, petioled, the blades more or less rounded in outline. Heads with disk florets only. Achenes columnar to obpyramidal; pappus of coarse to fine bristles. Strother & Pilz (1975:24) claim "five species of humble herbs, two in the Chihuahuan Desert and three in the Sonoran Desert and southwestern Great Basin." In contrast, Cronquist (1994) placed the 2 Chihuahuan species in a different genus.

Psathyrotes ramosissima (Torrey) A. Gray [*Tetradymia ramosissima* Torrey] DESERT VELVET. Spring ephemerals, compact, much-branched and rounded, often 3.5–10 cm high, densely leafy and woolly-scurfy, with a well-developed taproot. (Strother & Pilz [1975] reported that crushed fresh plants have a turpentine-like odor, but fresh plants sampled from the flora area did not have this odor.) Leaves 1.7–4 cm, thickish, velvety, gray-green, the blades broadly ovate to kidney-shaped, often 1.2–2.3+ cm wide, usually wider than long and coarsely toothed, with deeply incised veins. Heads 8–9 mm; peduncles well developed; corollas yellow, some with tips turning red. Achenes hidden by dense "furry" hair, the achenes including pappus 4.5–6.5 mm, the pappus of more than 100 bristles; achene hairs and pappus bristles bright iridescent copper color (dull yellow when immature or old and faded).

Seasonally common on otherwise barren or partially barren cinder flats and desert pavement with creosote bush at the northeast side of the Pinacate region, from the vicinity of Elegante Crater northward for about 15 km. Also common in desert pavement or pavement-like sandy soils adjacent to the Río Colorado. Not known elsewhere in Sonora; also northeastern Baja California, southeastern California, western Arizona, southern Nevada, and southwestern Utah.

1 mi S of Pinacate Junction, *F 20617.* Mesa de Andrade, *F 93-255.* 1 km N of El Doctor, *F 93-254.*

PSILOSTROPHE PAPER FLOWER

Perennial herbs or low shrubs, rarely biennial or annual; usually glandular nearly throughout. Leaves alternate. Heads with fertile ray florets and bisexual, fertile disk florets; ray corollas yellow or orange, papery, and persistent. Achenes small, glabrous or hairy; pappus of 4–6 scales.

Semiarid southwestern United States and northwestern Mexico; 7 species. References: Brown 1978; Heiser 1944.

Psilostrophe cooperi (A. Gray) Greene [*Riddellia cooperi* A. Gray] PAPER FLOWER. Much-branched mound-shaped perennial bushes ca. 0.5–0.8 m, probably not long-lived and sometimes flowering in first season. Stems, leaves, and phyllaries densely white woolly, the stems and leaves less woolly to glabrate with age. Stems leafy; leaves often 2.5–5+ cm, linear, gray-green, quickly drought deciduous. Heads on peduncles mostly 2–8 cm; phyllaries mostly 6–7 mm, green beneath the woolly hair, lanceolate, thickened and calluslike basally and along the midrib. Flowers bright yellow; rays 1.0–1.7 cm and about as wide, 3-lobed at tip, turning downward at maturity. Achenes 3.0 mm, glabrous, light colored, truncate at apex. At least March-April, and October.

Entering the flora area on the lower slopes and bajada of the Sierra Cipriano in the Sonoyta region, and locally disjunct in the Sierra Pinacate in scattered but well-established colonies on cinder slopes and rugged lava above ca. 800 m. Eastward in northwestern Sonora, across much of Arizona, southwestern Utah, southern Nevada, southeastern California, and Baja California.

N of Pinacate Peak, *F 86-431*. 11 mi SW of Sonoyta, *F 90-153*.

RAFINESQUIA

Taprooted annuals with milky sap. Stems hollow. First leaves in a basal rosette, the stem leaves alternate, pinnatifid, the stem leaves sessile, reduced upwards, their bases usually clasping the stem. Heads showy, medium to large, with ligulate florets only, the florets progressively smaller inward; inner phyllaries equal, lanceolate, thin, green with a prominent midrib and membranous margins; outer (accessory) phyllaries smaller, fewer, and unequal. Flowers mostly white. Achenes slender, dark pigmented, sculptured with rows of low bumps, glabrous or with tufts of short, thick, white hairs, tapering to a slender beak (neck) below the pappus; pappus bristles slender and plumose.

Southwestern United States and northwestern Mexico near the international border; 2 species, winter-spring ephemerals in the flora area.

1. Achenes 9–12 mm; pappus bristles plumose to tip, plumose hairs of pappus (the tiny branchlets) straight; higher elevations of Sierra Pinacate. ______ **R. californica**

1′ Achenes 12–14 mm; uppermost ca. 1/5 or 1/4 of pappus bristles not plumose (lacking branchlets), the plumose hairs often cobwebby; widespread. ______ **R. neomexicana**

Rafinesquia californica Nuttall. CALIFORNIA CHICORY. Often 15–30+ cm with a single or few-branched, self-supporting main stem. Leaves variously pinnately cleft or dissected; lower stem leaves 5–8 (15?) cm, withering as the stem develops. Heads relatively large, the size highly variable depending on plant size and soil moisture. Inner phyllaries 13–21 mm. Larger (outer) floret ligules ca. 1 cm, white with a broad rose-purple midstripe, the innermost florets white with yellow at the summit of the throat forming a yellow center or eye. Achenes 9–12 mm; pappus bristles 10–12 mm, plumose to the tips.

Differing from *R. neomexicana* by the self-supporting often more robust habit; single- to few-branched and often larger main stems; broader leaf segments; generally smaller flower heads that do not open widely; thicker, more sculptured, glabrous, and shorter achenes; more slender achene beaks; and the different pappus structure. The two often grow intermixed.

Higher elevations in Sierra Pinacate (sometimes as low as 550 m), often on cinder slopes, flats, rock outcrops, and small washes, growing in the open and not necessarily in the protection of a shrub (*F 19832, F 92-497*). Not known elsewhere in Sonora but expected eastward in the northern part of the state. The nearest known population is in the Ajo Mountains. Otherwise in California, Baja California, and Nevada to southwestern Utah and central and southern Arizona. Mostly occurring beyond the desert but entering the upper elevations and margins of the Sonoran Desert.

Rafinesquia neomexicana A. Gray. DESERT CHICORY. Mostly 15–60 cm. Stems usually zigzag, often stout although weak and herbaceous. Early leaves 4–10 cm, in a basal rosette, quickly withering as the stem develops, the blades thin, the segments broad when the plants well watered, threadlike under dry conditions. Heads relatively large, the size highly variable depending on plant size and soil moisture. Inner phyllaries (13) 17–22 mm. Outer (larger) florets pure white except for a pale rose-purple midstripe on the lower (outer) surface, the innermost florets white with yellow at the summit of the throat (tube); outer ligules (15) 20–30 mm. Achenes 12–14 mm; pappus bristles (6) 9.5–14 mm, plumose except at tips, dull white, often cobwebby below.

Apparently highly palatable to animals, it often grows in the protection of small shrubs, the flowering stems overtopping the "nurse" shrub, which is often *Ambrosia deltoidea* or, especially in drier years, a spiny shrub. Widely scattered across the lowlands of the flora area except the dunes, and in the Pinacate volcanic field including the higher elevations; sandy soils of washes and open desert plains, rocky soils in both the granitic and volcanic mountains, and especially common on cinder soils. In Sonora known only from the flora area but expected eastward across the northern part of the state. Also Baja California and southeastern California to western Texas, Chihuahua, Nevada, and southwestern Utah.

SW of Sonoyta, *F 92-400.* NE of Elegante Crater, *F 92-156.* Sierra Pinacate, 560 m, *F 92-483.* MacDougal Pass, *F 92-319.* 28 mi W of Los Vidrios on Mex 2, granitic mountains, *Bezy 454.* Sierra del Rosario, *F 92-228.*

SENECIO GROUNDSEL

Annual or perennial herbs, vines, shrubs, or small trees (these diverse growth forms include tropical trees to desert succulents and arctic herbs). Leaves alternate or basal, of diverse forms. Heads with ray and disk florets, or rays sometimes absent; rays in a single row, their number related to the number of phyllaries. Involucres usually cylindrical-campanulate, the phyllaries equal or nearly so, and often also with an outer series of smaller accessory bracts. Flowers usually yellow. Achenes usually columnar; pappus usually white, of numerous fine, soft capillary hairs and often barbellate.

One of the largest genera of flowering plants in the world, with 3000 species in the broadest sense, or 1300 after exclusion of currently proposed segregates (Barkley 1999). Worldwide, mostly in middle elevations in tropical and subtropical regions; at least 180 species in Mexico but poorly represented in the Sonoran Desert. The generic name derives from *senex* 'old man', referring to the many white hairs of the pappus, said to resemble the beard of an old man. References: Barkley 1978, 1999; Barkley, Clark, & Funston 1996; Ediger 1970; Felger 1991; Turner & Barkley 1990.

1. Coarse perennials, suffrutescent to shrubby (sometimes flowering in first year), usually more than 1 m; at least young herbage with white hairs; leaves semisucculent, pinnatisect with widely and regularly spaced, narrowly linear segments spreading at right angles. ____ **S. pinacatensis**

1′ Delicate winter-spring ephemerals (sometimes weakly perennial?), usually less than 0.5 m; glabrous except leaf axils and bases; leaves thin, toothed to irregularly lobed, the lobes not linear.
   2. Stems and leaves green, with woolly tufts at leaf bases and axils; rays conspicuous, ca. 10 mm. ____ **S. lemmonii**
   2′ Stems and lower leaves usually purple-green, glabrous; rays absent, or sometimes inconspicuous and several mm long. ____ **S. mohavensis**

Senecio lemmonii A. Gray. LEMMON GROUNDSEL. Delicate winter-spring ephemerals, often 25–65+ cm, glabrate except woolly tufts in leaf axils and moderately woolly leaf bases. Stems leafy, the leaves 4–14 cm, green, petioled below, sessile above and clasping the stems, the blades thin, mostly lanceolate to elliptic, toothed to sometimes entire. Heads 10–13 mm, the phyllaries 6–7.5 mm with dark tips. Flowers bright yellow. Rays well developed, ca. 8–13 in number, ca. 10 mm. Achenes 3.5–3.7 mm with short white hairs.

Northeastern margin of the Sonoyta region in granitic soils at mountain bases. Western and southern Arizona, northern Sonora southward sporadically to the Guaymas region, and Baja California. The few collections from the flora area are ephemerals; elsewhere, including the nearby Ajo Mountains, the plants are often short-lived perennials with slender, moderately woody bases. In Baja California it intergrades with *S. californicus* de Candolle (Ted Barkley, personal communication 1991).

Bajada on S side of Sierra Cipriano, 390 m, *F 88-210.*

Senecio mohavensis A. Gray. MOJAVE GROUNDSEL. Delicate, glabrous spring ephemerals 12–30 (40) cm. Stems leafy, solitary to well branched above. Stems and lower leaves usually purple-green. Leaves mostly 2–9 cm, more or less ovate to obovate, irregularly toothed to coarsely lobed and toothed, thin and almost membranous; lower leaves with winged petioles, the upper leaves sessile and broadly clasping the stem. Heads 8–13.5 mm; corollas yellow, with disk florets only or rays occasionally present and inconspicuous. Phyllaries (6) 7–8.5 mm. Achenes 2.8–3.2 mm, cylindrical, with short white hairs.

Recorded from the Sonoyta region and in the Pinacate volcanic field at Cerro Colorado and MacDougal Crater; seasonally common on the Arizona side of the border in the Tinajas Altas Mountains and Organ Pipe Monument. These small, delicate plants are often localized in favorable microhabitats such as along small arroyos or washes and on shaded, north-facing, rocky slopes, especially beneath cliffs, shrubs, or small trees. Mojave and Sonoran Deserts in southeastern California, Baja California, western Arizona, southern Nevada, and northwestern Sonora south to the vicinity of El Desemboque San Ignacio.

Liston, Rieseberg, & Elias (1989) elucidated the remarkable similarity and intercontinental disjunction between *S. mohavensis* and *S. flavus* (Decaisne) Schultz-Bipontinus of the Sahara-Arabian and Namibian Deserts. The plants of both species are self-compatible, and *S. mohavensis* seems to be an obligate selfer (self-fertilizing). The seeds retain viability for at least 15 years, and the achenes become mucilaginous when wet. *S. mohavensis* seems to be derived from *S. flavus* subsp. *breviflorus* Kadereit of southwest Asia and to have reached the New World through relatively recent long-distance bird dispersal.

Bajada on S side of Sierra Cipriano, *F 88-210.* Cerro Colorado, *Webster 22315.* MacDougal Crater, *Fishbein 913.* Quitobaquito, *F 88-114.* S of El Desemboque San Ignacio, vicinity 29°30′N, *F 17171.*

Senecio pinacatensis Felger, *Phytologia* 71:326–327, 1991. PINACATE GROUNDSEL. Short-lived perennials to 1.5 m, shrubby or semishrubby, often flowering in the first season or year. Mostly sparsely branched, the stems straight, brittle, striate, and leafy. Herbage densely white woolly to sometimes sparsely pubescent with appressed to spreading soft white hairs or, occasionally, glabrate. Leaves semi-fleshy, sessile, 3–6 cm, reduced near inflorescences, pectinately pinnatisect with 3–7 pairs of widely spaced linear segments, each ending in a conical callus 0.2–0.7 mm, the longer segments of larger leaves mostly 10–20 mm, plus 1 or 2 very reduced basal pairs of stipule-like segments, the leaf rachis narrowly winged.

Heads 1 to several at stem tips, the peduncles often with several bracts. Involucres cylindrical to cylindrical-campanulate and often somewhat urceolate, 7.5–11.0 mm wide at top, about as wide to slightly wider at base. Phyllaries 5.5–9.0 mm, stiff, thick, succulent, green, and callus-tipped, the margins overlapping (in a tongue-and-groove pattern). Accessory bracts about as many as the phyllaries. Corollas bright yellow. Ray florets 11–14, the rays 7.5–15+ × 3–4 mm, minutely 3-lobed at tip; rays apparently largest in the fall season and smaller in the cooler, spring season. Disk florets often (59) 70–78+. Achenes 4–6 mm, light brown, nearly cylindrical, with appressed hairs in longitudinal rows on low ridges; pappus bristles 10–12 mm, very numerous. February-May, and October-November. A diploid taxon, $2n = 40$.

Endemic to the higher elevations of the Sierra Pinacate, mostly above 750 or 800 m, in scattered pockets to near peak elevation in cinder soils, often on steep slopes and among rocks; mostly on the north and northeast side of the mountain, especially common near the bases of Carnegie and Pinacate Peaks. During favorable years occasionally to 650 m or lower.

Senecio pinacatensis is the only known endemic among the plants of the Pinacate volcanic field. It is a member of the *Suffruticosi* group, with 6 species including the *S. flaccidus* Lessing complex (= *S. douglasii* de Candolle complex). No other members of the *Suffruticosi* group are found in the flora area. Other members of this group usually occur on sandy soils of outwashes, alluvial fans, steam bottoms, and valley floors.

In *S. pinacatensis* the appearance of the leaves—in terms of texture, width of the segments, color, and pubescence—most closely resembles that of some specimens of the *S. flaccidus* complex (see

Turner & Barkley 1990). Noteworthy distinctions in *S. pinacatensis* are the relatively persistent woolly pubescence, the pectinate leaves, the rigid straight stems, and open sparsely branched shrubs that branch at near-right angles. Unlike members of the *S. flaccidus* group, the pattern of stem branching in *S. pinacatensis* is not one of arching upward from the base. After much consideration I chose to describe it at the species level rather than infraspecific rank because it is morphologically and geographically isolated, the differences being as strong as between other species of the *Suffruticosi* complex.

Senecio flaccidus ranges from southwestern United States to Puebla and Veracruz, Mexico. It includes var. *douglasii* (de Candolle) B.L. Turner & T. Barkley and var. *monoensis* (Greene) B.L. Turner & T. Barkley. Var. *monoensis,* largely a Sonoran Desert plant, occurs on the desert floor to the east and south of the flora area in Sonora. It is often a winter-spring ephemeral, in contrast to the shrubby var. *flaccidus* and var. *douglasii.*

1.1 km N of Pinacate Peak, 31°47′10″ N, 113°29′18″ W, 935 m, *F 86-426* (holotype, ARIZ). Sierra Pinacate, 780–1055 m, *F 19943, F 86-435.*

SONCHUS SOW THISTLE

Mostly perennial herbs, some annuals, rarely shrubs, the sap milky. Leaves alternate. Heads with ligulate florets only, the flowers yellow. With age the phyllaries often become swollen or thickened and calluslike at base. Achenes usually compressed, beakless, truncate at apex; pappus of numerous fine, soft hairs plus a few deciduous scales.

An Old World genus of Mediterranean origin; 50 species. The annual species with taproots, such as the 2 naturalized in the Sonoran Desert, are considered to be more advanced than the perennials. Reference: Boulos 1972–74.

1. Plants usually conspicuously spiny, the stems sometimes more than 1 m tall; achenes smooth between ribs, the margins thin and winglike. ______ **S. asper**

1′ Plants not conspicuously spiny, the stems seldom reaching 1 m tall; mature achenes wrinkled-roughened between the ribs (caution: may be difficult to see if achenes not mature), the margins not thin and winglike. ______ **S. oleraceus**

***Sonchus asper** (Linnaeus) Hill subsp. **asper** [*S. oleraceus* var. *asper* Linnaeus] *CHINITA;* PRICKLY SOW THISTLE; *HOI'IDKAM 'I:WAKĬ.* Winter-spring ephemerals; mostly glabrous except stalked glands reaching 1.0–1.3 mm in upper part of plant near flower heads. Larger plants spiny-prickly, often very robust, sometimes reaching 1.8 m. Stems hollow and relatively watery. Leaves pinnatifid with many, often spinescent-tipped teeth; early leaves 6–30+ cm, in a rosette, pinnatifid, the rosette and lower stem leaves with winged petioles and an enlarged terminal segment, the upper leaves sessile and clasping the stem; lowermost leaf segments (basal auricles) of stem leaves rounded (on giant, robust plants the basal auricle can be deeply cut with many spinescent teeth). Achenes 2.8–3.0 mm, flat, oval to oval-obovate, light brown, with 3 prominent ridges (ribs) on each side (face), the surface otherwise plain and smooth between the ridges, the margins of very thin wings; pappus more or less deciduous.

Widely scattered, usually disturbed or riparian habitats including waterholes and occasionally in interdune troughs. The Quitobaquito and Aguajita–El Papalote plants can become gigantic and strikingly robust, with spiny, thistlelike leaves. Native of Europe, naturalized and weedy worldwide.

Tinaja Huarache, *F 19556.* El Papalote, *F 86-100A.*

***Sonchus oleraceus** Linnaeus. *CHINITA;* COMMON SOW THISTLE; *HA:WĬ HEHEWO.* Winter-spring ephemerals similar to *S. asper* but in the flora area usually not nearly as large and robust, the lowermost leaf segments (basal auricles) of stem leaves narrow-angled (acute), the achenes generally narrower, roughened between the ribs, the margins not thin and winged, and the pappus tending to be persistent. However, the 2 species are not always easily distinguished. Achenes 2.5–3.4 mm, narrowly elliptic to oblanceolate.

Occasionally encountered in natural, usually xeroriparian or riparian habitats such as canyon bottoms, but mostly weedy in disturbed urban and agricultural habitats. Now nearly worldwide and a serious agricultural weed.

N of Puerto Peñasco, cleared sand flat near highway, *F 85-792.* Arroyo Tule, *F 19202.* El Papalote, *F 86-100B.*

STEPHANOMERIA WIRE-LETTUCE

Annual or perennial herbs with milky sap. Leaves alternate, often scalelike and reduced upwards. Involucres cylindrical, often narrow, the phyllaries about 5, in a single series and nearly equal, plus smaller, accessory bracts. Heads few flowered, the flowers ligulate and equal in size; corollas white to pink or rose. Achenes columnar, beakless, 5-angled or 5-ribbed, usually lightly pigmented; pappus bristles plumose. Western North America; 24 species. References: Gottlieb 1972; Gray 1876; Lehto 1979.

1. Perennials (also flowering in first year), often bushy and much-branched; ligules (rays) 10–12 mm; pappus bristles 6–8 mm, slender and not flattened. ______ **S. pauciflora**

1′ Spring ephemerals with 1 to several erect main stems, branched above; ligules 6.0–6.5 mm; pappus bristles 2.2–3.0 mm, flattened. ______ **S. schottii**

Stephanomeria pauciflora (Torrey) A. Nelson var. **pauciflora** [*Prenanthes pauciflora* Torrey] DESERT STRAW. Perennials, often flowering in first year; globose or mound-shaped, much-branched bushes often 1 m; glabrous (see following). Leaves variable, quickly drought deciduous, pinnatifid with few, spreading narrow segments, the early leaves of well-watered plants sometimes 10–15 cm but usually much shorter, the segments usually less than 5–8 mm; first leaves (of first-season plants) in a basal rosette. Heads 5-flowered, involucres 6.8–10.2 mm. Ligules (rays) ca. 10–12 mm, pale pinkish. Styles and stigmas purple, the style and lower (outer) surface of stigma branches papillate. Pappus bristles 15–17 in number, 6–8 mm, white to pale brown, slender (capillary), and not flattened; plumose except the lowermost ca. 1.5 mm, the plumose branches extremely slender and hairlike. Flowering any season, mostly spring and fall.

Gravelly-sandy washes, rocky slopes, cinder soils, sand flats, and lower dunes; generally absent from higher dunes. Widely scattered through much of the flora area including higher elevations in the granitic and volcanic ranges. California to Utah and Kansas, and southward to northern Mexico: Baja California to Chihuahua and Coahuila; southward in western Sonora to the vicinity of Bahía Kino. This is the most widespread species in the subtribe Stephanomeriinae.

All specimens I have seen from Sonora and southern Arizona are glabrous except 1 collection from Organ Pipe Monument (*Bowers 910*). This specimen has densely pubescent herbage with short, white hairs; in every other respect it compares with var. *pauciflora*. Var. *parishii* (Jepson) Munz, distinguished by densely white, tomentose herbage, is known from deserts in California and Nevada.

25 mi N of Puerto Peñasco, *F 13209.* Cerro Colorado, *F 10797.* S of Carnegie Peak, *F 19913.* 1 km SW of Tezontle cinder mine, *F 88-256.* Aguajita Spring, *F 86-293.*

Stephanomeria schottii (A. Gray) A. Gray [*Hemiptilium schottii* A. Gray] SCHOTT'S WIRE-LETTUCE. Spring ephemerals with a stout taproot; glabrous, mostly 30–60 cm, with a single or sometimes several major erect stems, branched above, the branches usually straight and ending in flower heads. Stems shiny silvery white except when very young. Early leaves thin, in a quickly withering basal rosette, 2.8–12 cm, linear with few widely spaced alternate teeth or pinnate segments 1–3 cm; stem leaves reduced upwards, the plants generally nearly leafless at flowering time.

Involucres 6.0–7.5 mm, cylindrical in bud, narrowly funnelform at flowering time to often campanulate as achenes mature; phyllaries 5 (or 6), subequal, plus a few short basal bracts, often green in bud, becoming purplish with white, membranous margins. Ligules 6.0–7.0 mm, 5-toothed at apex, white, flushed with violet, fading to cream with violet at the tip. Anthers white with dark violet-purple toward apex. Style pale violet with dark purple papillae on the exserted portion, the stigma intense pale purple or violet, papillose outside but not inside (above). Achenes 3.2–4.0 mm, columnar, light tan, tuberculate, sharply 5-angled, each face with a slitlike longitudinal groove. Pappus bristles 5, or sometimes splitting to form 6–8 bristles, rigid, persistent, somewhat unequal in length, the larger bristles 2.2–3.0 mm, flattened, narrowly elongated-triangular and gradually tapering (attenuate), with a shiny copper-colored midstripe; margins transparent white, often wavy (crisped) below, and with tiny plumose hairs above, the hairs not much longer than the width of the pappus bristle at its base. March-May. Basic chromosome number: $n = 8$.

Widespread and often common on sandy soils including sand flats, crater floors, interdune troughs, and barchan and transverse dunes; most common on low dunes and along sandy-gravelly washes. Western and southern margins of the Pinacate lava shield nearly to the gulf and the Río Colorado. Known otherwise only from similar habitats in adjacent Yuma County in southwestern Arizona, from the Mohawk Dunes to the Pinta Sands of Cabeza Prieta Refuge, as well as the type collection by Arthur Schott in May 1855, at "Camp Miller, Valley of the Gila" (Torrey 1858). After its initial discovery there was no report of this plant for more than 100 years (Lehto 1979). The principal part of its distribution is the flora area in Sonora.

Mature plants tend to become foul smelling. At dawn the flowers can produce an almost sickeningly sweet fragrance. Lehto (1979) says that the flowers are nocturnal; however, I have seen them closed at dusk and open from dawn through midmorning on warm days, and suspect that they remain open longer on cooler days. Other members of the genus in Arizona have pink, diurnal flowers.

Stephanomeria schottii seems most closely related to *S. exigua* Nuttall, a polymorphic species ranging from Oregon to Wyoming and south to Baja California, northern Sonora and New Mexico. *S. exigua* var. *pentachaeta* (D.C. Eaton) H.M. Hall, known from near Yuma and elsewhere in the Sonoran Desert, seems to approach *S. schottii* in that it has fewer (5), distinct, and basally dilated but uniformly white or light brown pappus bristles. In contrast, var. *exigua* may have many capillary pappus bristles.

13 mi NE of Puerto Peñasco on Mex 8, then 7.6 mi by rd NW, *F 92-433.* Dunes 10 km N of Gustavo Sotelo, *Ezcurra 14 Apr 1981* (MEXU). S of Moon Crater, *F 19023.* N of Sierra del Rosario, *F 20755.* 20 mi E of San Luis, *Bezy 471.*

STYLOCLINE NEST STRAW

BY JAMES D. MOREFIELD AND RICHARD FELGER

Annuals; plants small and white woolly. Leaves alternate or appearing whorled when subtending flower heads, the leaves entire, sessile, or petioles obscure, the lower leaves soon withering. Flower heads small, rounded, single or usually in clusters (glomerate) of 2–10, the clusters subtended by several leaves in a whorl-like arrangement; these leaves often longer than the cluster of flower heads and seemingly functioning like phyllaries. Actual phyllaries of individual flower heads usually absent (ours). Receptacles longer than wide, cylindrical to club-shaped or linear.

Florets unisexual. Pistillate florets (modified ray florets) in 2 to several outer series or rows, spirally arranged, lacking a pappus and without stamens; pistillate florets all subtended by phyllary-like chaffy bracts, at least the inner of these pouchlike, very woolly outside, and with an expanded membranous tip or margin; the bract and achene falling as a unit (chaffy bracts of innermost pistillate florets may be reduced). Pistillate achenes smooth and shiny, tiny, without a pappus. Disk florets 2–6, in 1 series (spiral), staminate, the corolla lobes usually 5, the chaffy bracts absent or small. Disk achenes vestigial or aborting, the pappus of 0–12 barbed bristles (see *Filago* for terminology).

Southwestern North America; 7 species. *Stylocline* appears most closely related to *Filago* subgenus *Oglifa,* especially *F. depressa*. The elongate receptacle is a conspicuous characteristic and often one easy to see except in immature plants, but it is not necessarily diagnostic. Reference: Morefield 1993c.

1. Membranous (hyaline) wing of achene-bearing chaffy bracts broadest near or below middle of whole bract, cordate (shaped like an inverted heart) or rounded. ________ **S. gnaphaloides**

1′ Wing of achene-bearing chaffy bracts broadest well above middle of whole bract. ________ **S. micropoides**

Stylocline gnaphaloides Nuttall [*S. arizonica* Coville] DESERT NEST STRAW. Spring ephemerals, resembling *S. micropoides* but generally smaller in stature. Larger leaves to 20 mm, broadly linear to oblong, generally obtuse. Membranous (hyaline) wing of fruiting chaffy bract ovate (broadest near base), often heart-shaped (cordate) at base, extending over full length of bract (chaff body). Pistillate achenes 0.8–1.0 mm. Disk achenes vestigial, generally with 1–4 pappus bristles.

Known in northwestern Sonora only from the Sonoyta region. Baja California and California to southeastern Arizona and northeastern Sonora. This species is primarily cismontane Californian, extending into similar habitats in northern Baja California and with a secondary center of distribution in semiarid central Arizona (originally described there as *S. arizonica*) and the adjacent Arizona Upland in northern Sonora. There is apparently a distributional gap in the intervening, extremely arid desert.

Distinguished from *S. micropoides* by the often smaller, relatively broader, and blunter leaves, usually fewer pappus bristles, and most readily by the broad, inverted heart-shaped wing of the chaffy bracts. The two species often grow intermixed.

18 km SW of Sonoyta, granitic sandy bajada on S side of Sierra Cipriano, 350 m, with *S. micropoides, F 88-212A.*

Stylocline micropoides A. Gray. Spring ephemerals; diminutive, woolly plants, mostly much less than 15 cm tall, erect with a single main axis to low, spreading, and much-branched from near base and lacking a main axis. Larger leaves often to 20 mm, oblanceolate to lanceolate or awl-like, acute to acuminate. Clusters of flower heads (glomerules) subtended by leaves longer than the flower heads. Individual heads globose. Receptacles whitish, cylindrical, 2.5–3.5 × 0.5–0.6 mm, persistent long after the florets fall, each floret leaving a minute pit at its point of attachment. Largest chaffy bracts 3.4–4.5 mm, enclosing achenes, each falling with an achene as a unit, the wing broadly lanceolate or oblanceolate to ovate or obovate. Pistillate achenes 1.0–1.3 mm, smooth, shiny. Disk achenes vestigial, generally with 3–8 pappus bristles.

Often on fine-textured soils or fine-textured soils mixed with rock; sometimes common in soil pockets on lava rock; flats and slopes. Seasonally common from the international border at Organ Pipe Monument southward through most or perhaps all of the Pinacate volcanic field to peak elevation, and in at least some of the granitic mountains. The plants disintegrate soon after they mature and the soil dries.

Southeastern California to southern Nevada; southwestern Utah southward to northern Baja California; western and southern Arizona to western Texas and adjacent northern Mexico including Chihuahua and Sonora southward at least to the central part of the state.

S of Pinacate Peak, ca. 850 m, *F 19926.* N base of Hornaday Mts, *Burgess 6866-B.* 1 km SW of Tezontle cinder mine, *F 88-258.* El Papalote, *F 86-116.* Bajada on S side of Sierra Cipriano, *F 88-212b.*

SYMPHYOTRICHUM

BY RICHARD FELGER, JOHN SEMPLE, AND SCOTT SUNDBERG

Herbs, usually perennial. Herbage glabrous or variously pubescent (generally strigose), glandular or not. Leaves alternate; upper stem leaves somewhat to greatly reduced, similar to the basal leaves or becoming narrower, sometimes clasping. Heads few to numerous in panicles, with fertile ray and disk florets. Phyllaries graduated, lanceolate, sometimes leafy, acute, the midvein not raised and not keeled. Rays white to blue-violet; disk florets yellow, rarely pink, sometimes becoming pink to red with age. Achenes obconic, often compressed, with 2–9 ribs, glabrate to densely strigose; pappus of a single series of barbellate bristles.

North America with a few in West Indies, Central America, and South America, and several naturalized in Europe; 90 species. Reference: Nesom 1994a, b; Semple, Heard, & Xiang 1996; Sundberg 1986.

Symphyotrichum subulatum (Michaux) G.L. Nesom [*Aster subulatus* Michaux] Glabrous, non-glandular annuals with a stout taproot (sometimes appearing to persist for a second year, but most likely annual). Stems 1 to several, mostly erect, leafy, to 60–80 cm when not grazed, often red-purple below. Leaves mostly narrowly lanceolate, entire to minutely toothed, the lower (larger) leaves often 5–9+ × 0.5–1.0 cm, sessile or nearly so, usually gone at flowering time. Heads *Conyza*-like. Phyllaries with attenuate, semifleshy reddish tips, the inner phyllaries 4.0–5.5 mm, thin, linear to linear-oblanceolate, green with membranous margins. Rays white to pink or pinkish lavender, 0.2–0.4 mm wide when dry, the disk florets yellow. Achenes 1.3–2.3 mm, linear-obconate, 5- (6)-angled and ribbed, sparsely pubescent with minute hairs; pappus 2.3–4.2 mm. Flowering at least September-October.

Wet soil along the banks of the Río Sonoyta and other wetland habitats in the vicinity of Sonoyta including irrigation canals. Also wetlands along the lower Colorado River in Arizona and California, and probably formerly along the Mexico portion of the Río Colorado and perhaps presently weedy along irrigation canals. The plants are often severely grazed, especially by cattle, the stubs producing stunted flowering branches.

This species is widespread in the warmer regions of the Americas, largely in riparian and other wetland habitats. Sundberg (1986) recognized 5 varieties of *A. subulatus,* whereas Nesom (1994b) treated them as 5 separate species in *Symphyotrichum.* If variants are recognized at the species level, the appropriate name is *S. expansum* (Poeppig ex Sprengel) G.L. Nesom.

The western Sonoran populations represent a subspecies of *S. subulatum* distinct from the typical subspecies of the Atlantic Coast of North America (Sundberg will be making new combinations for these subspecies). This subspecies is widespread in the southwestern United States and Mexico, and is widely distributed in Sonora including the major river deltas. *Aster exilis* Elliot is sometimes given as a synonym of *A. subulatus,* but there is no known type specimen and it probably represents a different species.

Sonoyta, *F 86-402.* S of Cerro el Huérfano, *F 92-984.* Presa Derivadora, *F 86-304.* Arizona: Yuma, irrigation ditches, *Thornber 25 Sep 1912.* California: Colorado River just E of Ft. Yuma, 11 Sep 1985, *McLaughlin 3050.*

THYMOPHYLLA

Annual or perennial herbs or subshrubs, strong scented with conspicuous oil glands. Leaves and branches opposite below, often alternate above. Phyllaries united at least 2/3 of their length and with mostly round glands, and also with a series of smaller, accessory bracts. Rays few, fertile, or sometimes reduced or absent, usually yellow; disk florets usually numerous, bisexual and fertile. Achenes obconic; pappus highly variable.

Southeastern California to Kansas and southward to Oaxaca and in Argentina; 13 species. References: Strother 1969, 1986.

Thymophylla concinna (A. Gray) Strother [*Hymenatherum concinnum* A. Gray. *Dyssodia concinna* (A. Gray) B.L. Robinson] *MANSANILLA DEL COYOTE; BAN MANSANI:YA.* Winter-spring ephemerals, mostly 3.5–10 cm tall, the stems becoming semiprostrate, sometimes reaching 20 cm; glabrous or sparsely hairy, pungently aromatic, dotted with small oil glands. Leaves 7–16 mm, pinnate with slender segments, alternate above. Heads often clustered at ends of leafy stems, showy, the rays white, 4–5 mm, the disk yellow. Phyllaries 4.5–6.0 mm, united nearly to apex, the accessory bracts few and inconspicuous or absent. Achenes 2.8–3.0 mm, slender, blackish; pappus bristles white, about as long as the achenes.

Mostly on sandy or gravelly soils; seasonally common. Arizona Upland at the edge of the flora area in the vicinity of Sonoyta and Quitobaquito (*F 7662, F 86-130*). Sonoran Desert in southern Arizona and central and western Sonora to the Guaymas region.

Trichoptilium

The genus has a single species.

Trichoptilium incisum (A. Gray) A. Gray [*Psathyrotes incisa* A. Gray] YELLOW HEAD. Winter-spring ephemerals, sometimes persisting through summer or rarely weakly perennial; erect to spreading, often compact, with several dichotomous branches and a taproot, the stems 6–8 cm (not including peduncles). Herbage white woolly and aromatic. Leaves clustered near base of plant, alternate or nearly opposite, (1) 1.5–5.5 cm, the blades gradually narrowed to a winged petiole, oblanceolate, shallowly lobed to sharply toothed with 3–7 (9+) coarse teeth.

Heads with disk florets only, raised well above the leaves on slender, glandular pubescent and sometimes woolly peduncles 2.5–8.5 cm tall; flowers bright yellow and showy. Phyllaries in outer and inner whorls, 6.0–7.5 mm, green, moderately woolly, lanceolate to elliptic, the inner phyllaries narrow, with a 0.5–0.8 mm orange to red glandular tip. Achenes 2.3–2.8 mm, obpyramidal, densely hairy; pappus of 5 broad scales divided into many, uneven, slender white to golden bristles, the scales thickened basally into a yellow callosity. Mostly winter-spring, November–late April or May, occasionally flowering with summer–fall rains if the plants survive early summer drought.

Sonoyta region (Quitobaquito), Pinacate volcanic field, and most of the granitic ranges; not known elsewhere in Sonora. Common and widespread on rocky soils of pavements, upper bajadas, and slopes, and sometimes along sandy-gravelly washes and roadsides. Also northern Baja California Sur to southeastern California, western Arizona, and southern Nevada: Mojave and Sonoran Deserts.

MacDougal Crater, *F 92-353*. Sierra Extraña, *F 19074*. Moon Crater, *Fishbein 911*. Cerro Pinto, *F 85-1018*. Sierra del Viejo, *F 16884*.

Trixis

Mostly shrubs. Leaves alternate, pinnately veined; leaf bases prominent and persistent on stems after leaves fall. Stomata usually on lower leaf surfaces only. Heads terminal; florets yellow, conspicuously 2-lipped (bilabiate), all bisexual and fertile. Achenes elongated with a short, slender neck, the apex expanded into a disk bearing a pappus of numerous soft, barbellate bristles.

Southwestern United States to Argentina and Chile; 50 or more species, 29 in North America. Centers of diversity in southwestern Mexico and southeastern South America. Reference: Anderson 1972.

Trixis californica Kellogg var. **californica.** Shrubs often to 1 m with erect-ascending, slender, brittle branches; new growth glandular and often densely pubescent with brown hairs. Leaves and flowers appearing at various seasons, especially spring; fast-growing under favorable conditions. Larger leaves quickly wither during drought, the plants ultimately leafless and sometimes the branches die back; new growth frost sensitive. Leaves mostly (2.5) 3–8 cm, mostly upright (ascending), the blades relatively thin, lanceolate, with minute hairs to sometimes glabrate, densely glandular below and sometimes above but not as densely so, the margins toothed to nearly entire; sessile or petioles mostly 1–2 mm, usually winged; dead leaves semipersistent. Inner phyllaries green, 10–15 mm, oblong, with thickened yellow-brown bases extending into a midrib. Flowers yellow and attractive, the florets ca. 1 cm. Achenes 8–10 mm, slender, with short, thick, often glandular hairs, the neck prominent and especially glandular; pappus about as long as achenes. Various seasons.

Sonoyta region, Pinacate volcanic field and granitic ranges to peak elevations, and desert plains mostly near mountains and hills; many habitats including bajadas, slopes, and crater floors, often along gravelly washes, arroyos, and canyons and beneath larger shrubs and trees or among rocks; absent from dunes and the most arid desert flats.

Southeastern California to western Texas and south to Baja California Sur and Sinaloa to San Luis Potosí, Zacatecas, and Nuevo León; through most of Sonora at elevations below the oak zones.

Variety *californica* is unique at least within the North American members of *Trixis* in having stomata on both surfaces of the leaves. This feature seems to be part of the character set involving ascending rather than spreading leaves, which is probably an adaptation to an arid or semiarid environment. A second variety, var. *peninsularis* (S.F. Blake) C. Anderson, is endemic to the Cape Region of Baja California Sur.

Sierra del Rosario, *F 20678*. Sierra del Viejo, *F 5742*. Tinajas de los Pápagos, *Turner 59-27*. Tinaja de los Chivos, *F 18814*. Campo Rojo, *Soule 19 Mar 1983*.

UROPAPPUS

The genus has a single species (following Jansen et al. 1991), although Cronquist (1994) retained it in *Microseris,* a genus of 20 species mostly in western North America with 2 in Australia and New Zealand.

Uropappus lindleyi (de Candolle) Nuttall [*Calais lindleyi* de Candolle. *Microseris lindleyi* (de Candolle) A. Gray. *M. linearifolia* (de Candolle) Schultz Bipontinus] SILVER PUFFS. Spring ephemerals with milky sap. Leaves in a basal rosette, mostly 10–15 cm, linear or linear-lanceolate or pinnate with few slender segments, glabrate or moderately pubescent with crinkled white hairs near the base. Stems 12–30 cm, leafless, erect, with small glands near the flower head, each stem bearing a single, erect flower head. Phyllaries graduated, overlapping, the inner ones 15–30 mm, broadly lanceolate; accessory bracts none. Florets all ligulate, many, pale yellow. Achenes 8.7–10.0 mm, blackish, linear-cylindric, and slightly tapered at each end, the apex slightly flared; pappus with 5 papery, silvery, linear-lanceolate scales 9–10 mm, these deeply notched at the apex with a long, slender awn from the notch.

Cinder soils at higher elevations in the Sierra Pinacate (*F 92-469*). The nearest known populations are in Cabeza Prieta Refuge and the Puerto Blanco and Ajo Mountains. Washington and Idaho to Baja California, northern Sonora, and Texas.

VERBESINA

Annual or perennial herbs, shrubs, or small trees. Leaves often rough surfaced, opposite or sometimes alternate, often decurrent at the base to form winged stems. Flowers mostly yellow; ray florets fertile, sterile, or absent, the disk florets bisexual and fertile, subtended by concave chaffy bracts. Disk achenes strongly flattened laterally, often winged on each edge and with 2 pappus awns or the awns absent.

Americas, largely tropical and subtropical; estimated 150 species, half in Mexico. Several species in the southern part of the Sonoran Desert.

***Verbesina encelioides** (Cavanilles) Bentham & Hooker ex A. Gray [*Ximensia encelioides* Cavanilles] GOLDEN CROWNBEARD. Coarse, foul-smelling annuals; non-seasonal but mostly during warmer times of year; often reaching 0.5–1.0 m, with a stout taproot. Herbage and phyllaries densely pubescent with coarse white hairs. Leaves opposite below, alternate above, gray-green to whitish, bicolored. Petioles prominent, partially to fully winged; larger leaves with stipule-like leafy appendages near the petiole base. Leaf blades mostly 3–7 cm, more or less ovate to triangular, coarsely toothed. Phyllaries green, lanceolate, the larger ones 10–13 mm. Heads 3.5–5.0 cm wide, showy, daisy-like, the flowers yellow; rays ca. 1.5–2.0 cm, 2- or 3-toothed at apex. Ray achenes often failing to develop. Chaffy bracts membranous and enclosing disk achenes; achenes 5.0–6.5 mm, with appressed white hairs, the body dark colored with a white to yellow, corky wing 1.0–1.5 mm wide except at the notched apex; pappus awns as long as or longer than the wings, or 1 or both awns absent.

Disturbed, often trashy places, usually in sandy-silty soils. Abundant at Puerto Peñasco (*F 85-796*) and probably elsewhere as an urban and agricultural weed. Not native to the flora area; weedy in western United States, much of Mexico, the Caribbean, and South America, and naturalized in the Old World.

VIGUIERA

Perennial herbs or shrubs, sometimes annuals, usually coarsely pubescent. Leaves opposite, at least below. Heads with ray and disk florets, or rays sometimes absent, the flowers usually yellow. Ray florets usually sterile. Disk florets subtended by chaffy bracts, these enclosing the achenes and falling with them. Achenes with appressed hairs and a pappus of 2 bristles and smaller scales, or glabrous and the pappus absent.

More than 150 species; about half in South America and the rest in Central America to southwestern United States; greatest diversity in Mexico from Durango to Oaxaca. Closely related to and often difficult to separate from *Helianthus*. Reference: Schilling 1990a.

Viguiera parishii Greene [*V. deltoidea* A. Gray var. *parishii* (Greene) Vasey & Rose] *ARIOSA;* PARISH GOLDENEYE. Globose shrubs mostly 1.0–1.5 m with many slender, brittle stems. Herbage with coarse, stiff hairs and subsessile glands; leaf blades and phyllaries especially scabrous with rough hairs often with conical bases. Leaves opposite or the upper ones alternate, (2) 3–5 (7) cm, petioled; blades variable, mostly ovate to broadly triangular-ovate, the margins sub-entire (especially smaller leaves) to coarsely toothed. Phyllaries graduated, ovate to lanceolate and somewhat hardened basally, the larger, inner phyllaries 5.5–10.0 mm and usually abruptly narrowed above. Heads 2.6–3.5 cm wide, daisy-like, solitary to few on each inflorescence, often widely spaced on long slender peduncles. Rays, disk corollas, anthers, and stigmas bright yellow, the older anthers often purple-brown, the chaffy bracts herbaceous, usually not glandular, often purple-brown and yellow; rays often 1.2–1.5 cm, falling quickly (sometimes absent?). Achenes (2.8) 3.4–4.0 mm, moderately compressed, with ascending to appressed white hairs, the margins ciliate but otherwise not differentiated from the body; pappus bristles stout. Flowering after rains, at least March-May and October.

Sonoyta region, Pinacate volcanic field, and granitic ranges, often on north-facing slopes and along washes or arroyos. Especially common at higher elevations in the Sierra Pinacate. Northwestern Sonora southward to the vicinity of Guaymas, northeastern Baja California, western and central Arizona, southern Nevada, and southeastern California. *V. parishii* is a member of the subgenus *Bahiopsis,* which includes 12 species, some of them very closely related, in southwestern United States and northwestern Mexico, with greatest diversity in the Baja California Peninsula.

Pinacate Peak, *F 86-440.* Tinajas de Emilia, *F 19720.* Tinaja de los Chivos, *F 18811.* W of Los Vidrios, *F 16862A.* Sierra del Viejo, *F 85-732.*

XANTHIUM COCKLEBUR

Coarse annuals with a stout taproot. Herbage glandular; leaves alternate. Monoecious. Heads unisexual with disk florets only, the staminate heads above the pistillate ones. Staminate heads with chaffy bracts subtending the many florets. Pistillate heads with 2 florets tightly enclosed in a bur with hooked spines, the bur formed of fused phyllaries; corolla and pappus none, the stigmas protruding from the 2 beaks. Cotyledons relatively large, dark green, and fleshy, the seed germinating from within the bur.

New World and Europe, 3 species, now worldwide weeds. The burs cling to clothing and fur. The plants generally have a short day flowering response. References: Cronquist 1994; Löve & Dansereau 1959; McVaugh 1984.

***Xanthium strumarium** Linnaeus. COCKLEBUR; *CARDILLO; WAIWEL.* Often growing in summer, or with warm weather in spring, and maturing in early fall, sometimes reaching ca. 1 m. Herbage gland-dotted and rough. Petioles often 3–10 cm, as long as or longer than the blades; blades mostly deltate-ovate, often 3-lobed, the margins toothed. Mature burs (1.8) 2.5–3.5 cm.

Disturbed habitats at least in the Sonoyta region as an urban and agricultural weed, along the Río Sonoyta riverbed and roadsides (*F 85-936*); expected in the San Luis Valley. Worldwide weed in warm-temperate regions.

BATACEAE SALTWORT FAMILY

One genus of 2 species of maritime halophytes; the second species in New Guinea and Australia.

Batis maritima Linnaeus. SALTWORT. Succulent, glabrous perennials, mostly forming extensive, dense mats of trailing and scrambling stems often rooting at the nodes. Leaves opposite, often 1.3–4 cm, linear-oblong to narrowly spatulate, thickly terete, yellow-green and extremely succulent. Dioecious. Inflorescences of fleshy cone-like axillary spikes or catkins often 7.5–15 mm, the flowers greatly reduced and inconspicuous.

Saline wet mud and sand of bays, esteros, and tidal flats, often inundated by high tides and locally abundant; not on beaches facing the open sea. Coastal southern California to the Cape Region of Baja California Sur, the Gulf of California; Pacific and Atlantic coasts of tropical and subtropical America, Galapagos Islands, and Hawaii. The herbage is edible but salty (Uphof 1968) and the Seri used the roots to sweeten coffee (Felger & Moser 1985).

Punta Peñasco, in *Distichlis* marsh, *Shreve 7598a*. Estero Cerro Prieto, *F 90-161*.

BERBERIDACEAE BARBERRY FAMILY

Mostly woody shrubs, some trees and perennial herbs. Leaves alternate, opposite, or whorled; stipules present or absent. Flowers radial, the perianth usually of several whorls, stamens 6. North temperate regions and South America; 15 genera, 650 species. Reference: Whetstone, Atkinson, & Spaulding 1997.

BERBERIS BARBERRY

Mostly shrubs, some trees, with yellow wood and inner bark. Leaves odd pinnate or unifoliolate, the margins usually with spine-tipped teeth. Flowers yellow, the stamens tactile (closing inward when touched). Fruit a few-seeded berry.

North temperate regions of both hemispheres and Andean South America, few in deserts; 500 species. In the Sonoran Desert only at its upper elevational limits; *B. harrisoniana* Kearney & Peebles is endemic to the Ajo and Kofa Mountains in southwestern Arizona. Some of the species harbor the black stem rust of wheat (*Puccinia graminis*), and are illegal to sell or transport in the United States and elsewhere. References: Lafèrriere 1992; Whittemore 1997a.

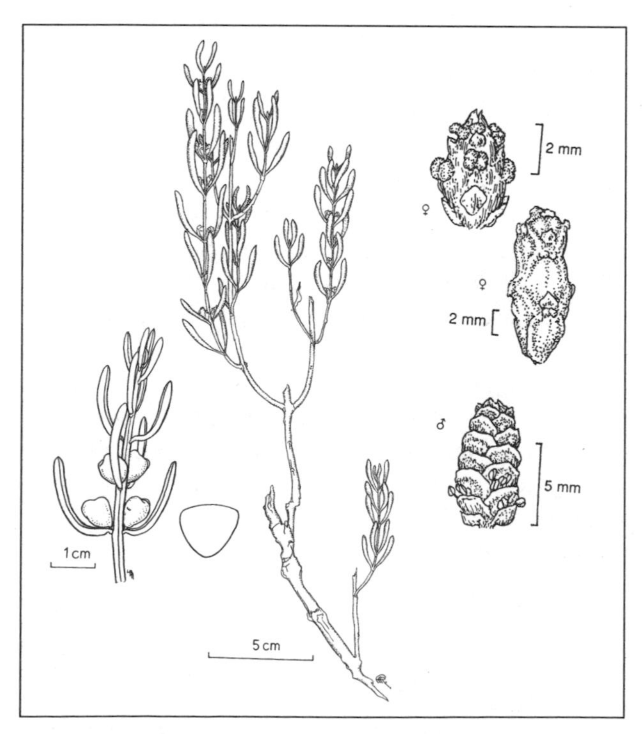

Batis maritima. FR (Felger & Moser 1985), LBH, and ER (Hickman 1993).

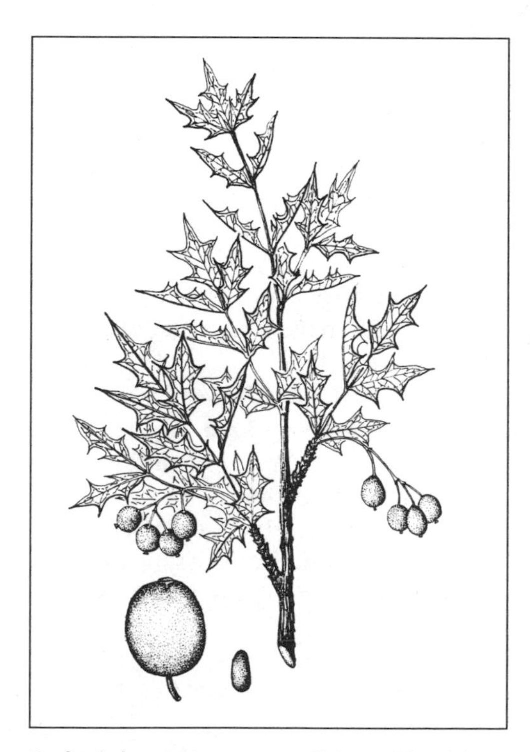

Berberis haematocarpa. FB. © James Henrickson 1999.

Berberis haematocarpa Wooton [*Mahonia haematocarpa* (Wooton) Fedde] RED BARBERRY. Woody shrubs mostly 1–2.5 m, rarely to 4 m with a trunk 15 cm in diameter; bark rough and fissured. Stems of long and short shoots. Leaves mostly evergreen, 4–10 cm; leaflets 5 or 7, bluish glaucous, firm with spine-tipped teeth, the terminal leaflet longest and often stalked. Inflorescences of small, few-flowered racemes. Berries 6–8 mm diameter, juicy, red-purple, edible. Winter and early spring, the fruits ripe in April.

Sierra Pinacate above 800 m; small, aggregated colonies mostly on north-facing slopes and cliffs, often in small steep arroyos or drainageways, or depressions in cinder soils (*F 86-451, F 92-492*). The nearest population is in the Ajo Mountains. Arizona to western Texas, southeastern California, southern Nevada, Colorado, and northern Sonora and Chihuahua; often in oak woodland and chaparral. No other *Berberis* occur in western Sonora.

BORAGINACEAE BORAGE FAMILY

Mostly herbs, some shrubs or trees, rarely vines, often with stiff, calcified or silicified hairs. Leaves simple, mostly entire and alternate, sometimes opposite; stipules none. Inflorescence branches often helicoid (curled at the tip like a scorpion tail: in the shape of a flattened spiral, straightening as they grow, the flowers on one side). Flowers usually 5-merous, mostly bisexual, radial, the sepals separate or united below, the corolla sympetalous. Ovary superior; style in ours undivided or bifid in *Tiquilia*.

Worldwide, especially diverse in western North America and the Mediterranean region east into Asia; 100 genera, 2300 species.

Although many tropical and subtropical borages have fleshy fruits, most temperate and aridland borages (ours) have dry fruits that split into 4 or fewer nutlets, each with 1 seed. Among our borages, those with fewer than 4 nutlets arrive at that condition by abortion of the developing fruit segments or parts. This is undoubtedly an evolutionarily advanced condition. Because the seeds do not separate from the hard surrounding case, the nutlet is functionally like a seed in a fashion similar to the grain (caryopsis) in the grasses or the achene (cypsela) in the composites. References: Cronquist et al. 1984; Higgins 1979.

1. Roots thick, blackish, and deep; perennials, prostrate or matlike; leaf veins conspicuously impressed; corollas lavender; stigmas 2, distinct. ____ **Tiquilia**

1′ Roots not thick and blackish; ephemerals or perennials, prostrate to taller than wide; leaf veins not conspicuously impressed; corollas white, yellow, or pale blue (*Lappula*); stigmas one.

 2. Perennials; glabrous, succulent, and halophytic. ____ **Heliotropium curassavicum**

 2′ Ephemerals or facultative perennials; densely hairy, not succulent, not halophytic.

  3. Corollas yellow-orange; inflorescence tips curled. ____ **Amsinckia**

  3′ Corollas white or bluish, inflorescences various.

   4. Nutlets 4, at least 2 of them fringed with spines or teeth.

    5. Nutlets erect, the marginal teeth or spines barbed; corollas pale blue or sometimes white; stems erect with a well-developed main axis. ____ **Lappula**

    5′ Nutlets spreading, the marginal teeth not barbed; corollas white; stems mostly spreading, lacking a strong main axis. ____ **Pectocarya**

   4′ Nutlets 1–4, the margins not fringed with spines or teeth (with blunt finger-like fringes in *Cryptantha holoptera*).

    6. Corollas less than 1 cm wide; nutlets 1–4, variously deltoid to ovoid. ____ **Cryptantha**

    6′ Corollas 1 cm or more in width; nutlets 2, globose. ____ **Heliotropium convolvulaceum**

AMSINCKIA FIDDLENECK

Annuals, usually with 1 to several main axes and relatively large, bristly, often pustulate-based hairs. Inflorescences conspicuously helicoid. Flowers self-fertile, the corollas bright yellow or orange. Nutlets 4, all alike, intricately ornamented. Cotyledons deeply divided, the seedling thus appearing to have 4 cotyledons.

Western North America with greatest diversity in California, and 1 or more species in South America; 15 species. Some are weedy and established in distant parts of the world. There is great plasticity in sculpturing of the nutlets, which led Wilhelm Suksdorf (1931) to recognize about 235 species, almost all of them buried in synonymy. *Amsinckia* seems to be closely related to the yellow-flowered members of subgenus *Oreocarya* in *Cryptantha,* namely members of the *C. flava-confertiflora* complex. The 2 *Amsinckia* species ranging into the Sonoran Desert are regarded as evolutionarily derived in the genus. No other Sonoran Desert borages have yellow or orange corollas. Reference: Ray & Chisaki 1957.

1. Calyx lobes 5, about equal; back of nutlets with ragged-edged ornamentations, arched with a high keel. ______ **A. intermedia**

1′ Calyx lobes 3–5, unequal, often 2- or 3-toothed at tip due to fusion of lobes; back of the nutlets with smooth-edged bumps (tesselated), not appearing arched, the keel not raised or only slightly so. ______ **A. tessellata**

Amsinckia intermedia C.A. Fischer & Meyer var. **echinata** (A. Gray) Wiggins [*A. echinata* A. Gray] DEVIL'S LETTUCE, FIDDLENECK; *CEDKAM.* Usually 30–50 cm, sometimes 1 m or more in wet places. Calyx lobes all alike. Corolla tube about 10-nerved below insertion of the stamens. Nutlets 2.4–2.6 mm, relatively deep and arched, the dorsal side (back) with a high ridge crest, sculptured with sharp, ragged ornamentations, the roughenings not crowded.

Mostly along large washes and floodplains; Sonoyta region, the northeastern part of the Pinacate region, and southeast to Quitovac. Also farther west in Arizona near the international border and expected at higher elevations in the adjacent granitic ranges in Sonora such as the Sierra del Viejo.

Two loosely segregated geographic races seem worthy of consideration. Var. *intermedia,* in western North America, for the most part barely enters the Sonoran Desert at its western margin. It is generally replaced by var. *echinata* in the desert in Arizona and northwestern Sonora; but some desert specimens from adjacent California key to var. *intermedia.* Var. *echinata* has a thin, ragged, irregularly toothed crest on the dorsal keel and margins of the nutlets. In var. *intermedia* the nutlets are tuberculate but not ragged-toothed. Suksdorf "recognized over 100 species that fall within" *A. intermedia* (Higgins 1979:312).

El Papalote, *F 86-103.* 10 km SW of Sonoyta, *F 88-187.* Cerro Colorado, *F 93-237.* Tinajas Altas, *Van Devender 5 Mar 1983.*

Amsinckia tessellata A. Gray. CHECKER FIDDLENECK; *CEDKAM.* Resembling *A. intermedia,* the plants often larger and more robust. Calyx lobes often unequal, often with 2 of the lobes united. Corolla tube with about 20 veins below insertion of the stamens. Nutlets 2.8–3.0 mm, the back tesselated (checkered)—with a mosaic of crowded, broad, irregular, and smooth-edged bumps like cobblestone wrinkled by an earthquake, and sometimes with smooth, transverse ridges.

Sonoyta region and higher elevations in the Pinacate volcanic field, scattered as low as 200 m during favorable years. Rocky to gravelly, sandy, or cinder soils; slopes, pediments, flats, and arroyo beds. Eastern Washington, Idaho, and Utah to Baja California and northern Sonora.

Carnegie Peak, *F 19940.* Near Elegante Crater, *Burgess 5593.* Cerro Colorado, *F 93-238.* Arroyo Tule, *F 19207.* Sonoyta, *F 88-17A.* 10 km SW of Sonoyta, *F 88-145.*

CRYPTANTHA

Ephemerals, annuals, or herbaceous perennials sometimes slightly woody at the base, with coarse, stiff, often pustulate-based hairs or bristles. First leaves opposite, although often in a basal rosette, the later leaves alternate. (Among Sonoran Desert species the basal rosette leaves invariably wither by fruiting time and usually die soon after commencement of flowering.) Inflorescences often helicoid. Corollas mostly white (ours). Calyx glassy-spined, tightly enclosing the ovary and nutlets like a miniature bur, falling as a unit among the Sonoran Desert species. Nutlets erect, 4, or 1–3 by abortion (rarely 2 ovules from the beginning), attached to an elongated gynobase (elongation of the receptacle) for most or all of their length, the attachment scar usually seen as a ventral groove, the groove open or closed.

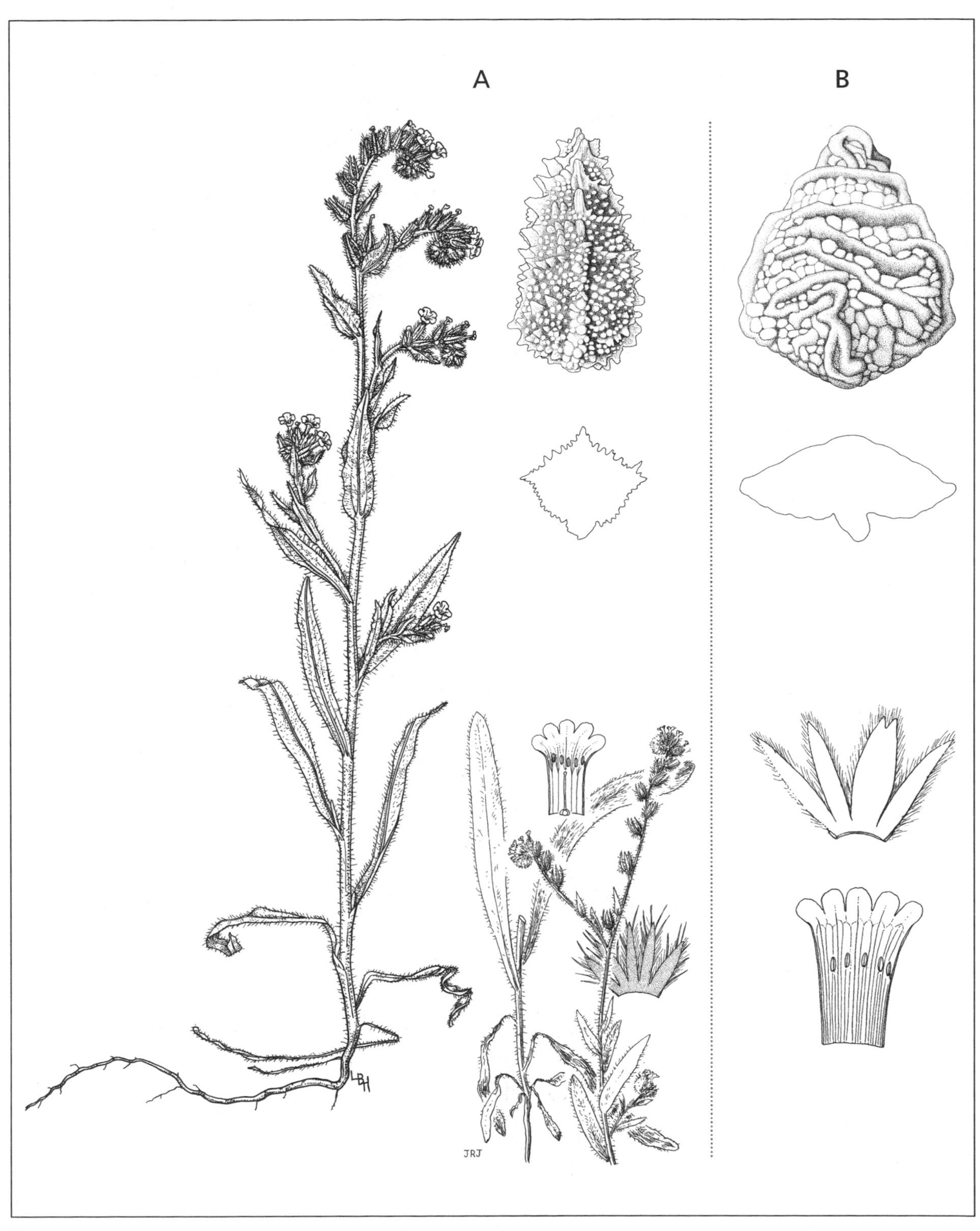

Amsinckia*: (A) *A. intermedia var. ***echinata*; (B) *A. tessellata*.** Note expanded calyx and corolla, and nutlet dorsal view and cross-section for both species. LBH (Parker 1958), JRJ (Hitchcock et al. 1959), and MBJ.

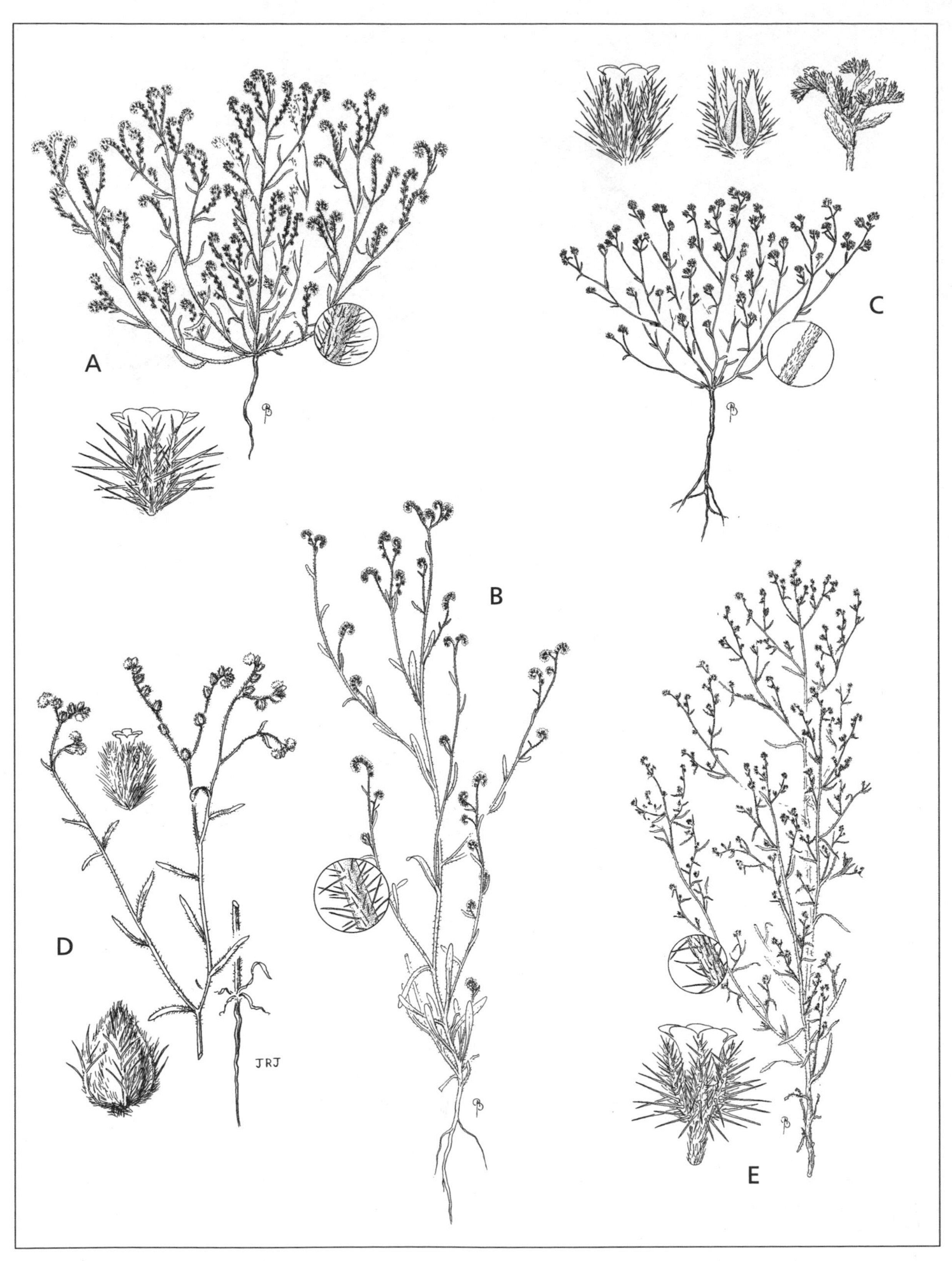

Cryptantha:* *(A) C. angustifolia; (B) C. barbigera; (C) C. micrantha; (D) C. pterocarya; (E) C. racemosa. A, B, C, E, BA (Cronquist et al. 1984); ***D,*** JRJ (Hitchcock et al. 1959).

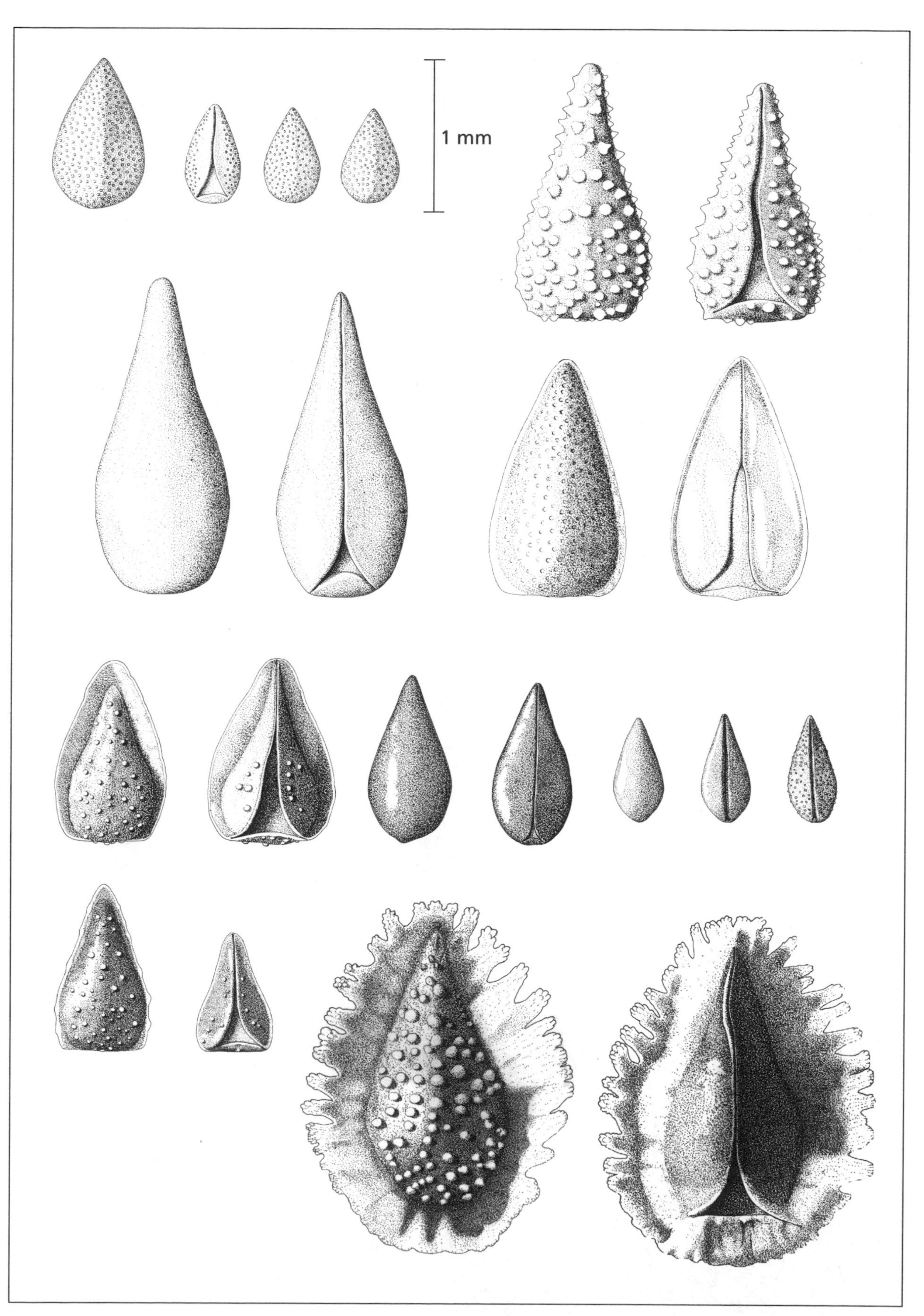

Cryptantha nutlets. Left to right: top row: ***C. angustifolia, C. barbigera; second row: C. ganderi, C. costata;*** third row: ***C. holoptera, C. maritima, C. micrantha*** subsp. ***micrantha;*** bottom row: ***C. racemosa, C. pterocarya*** var. ***cycloptera.*** MBJ.

Heliotropium curassavicum.
FR (Felger & Moser 1985).

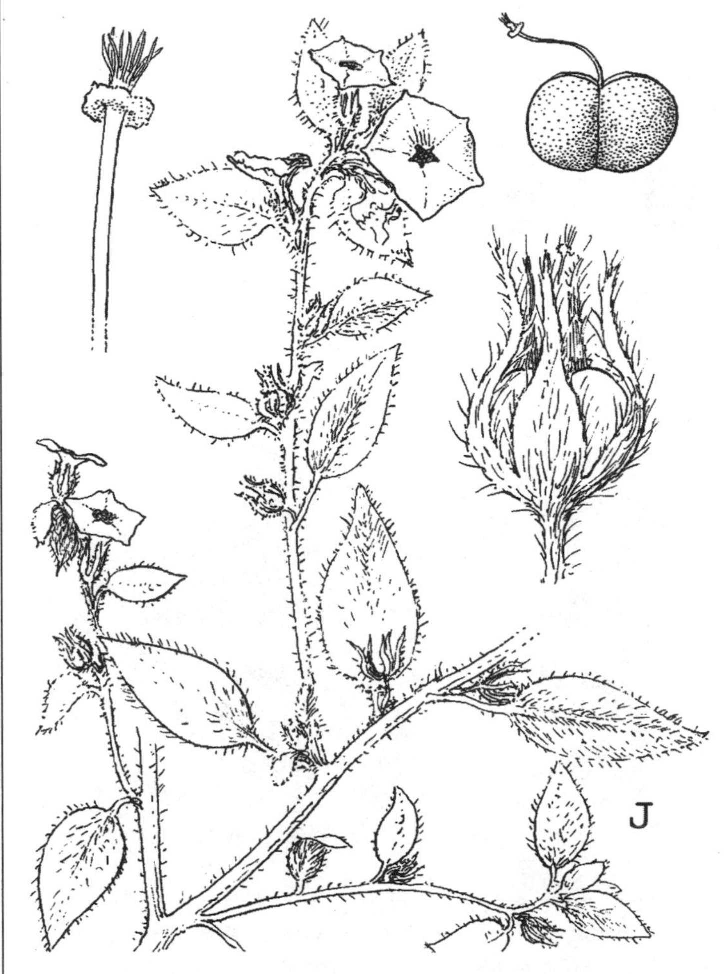

Heliotropium convolvulaceum.
JRJ (Abrams 1951).

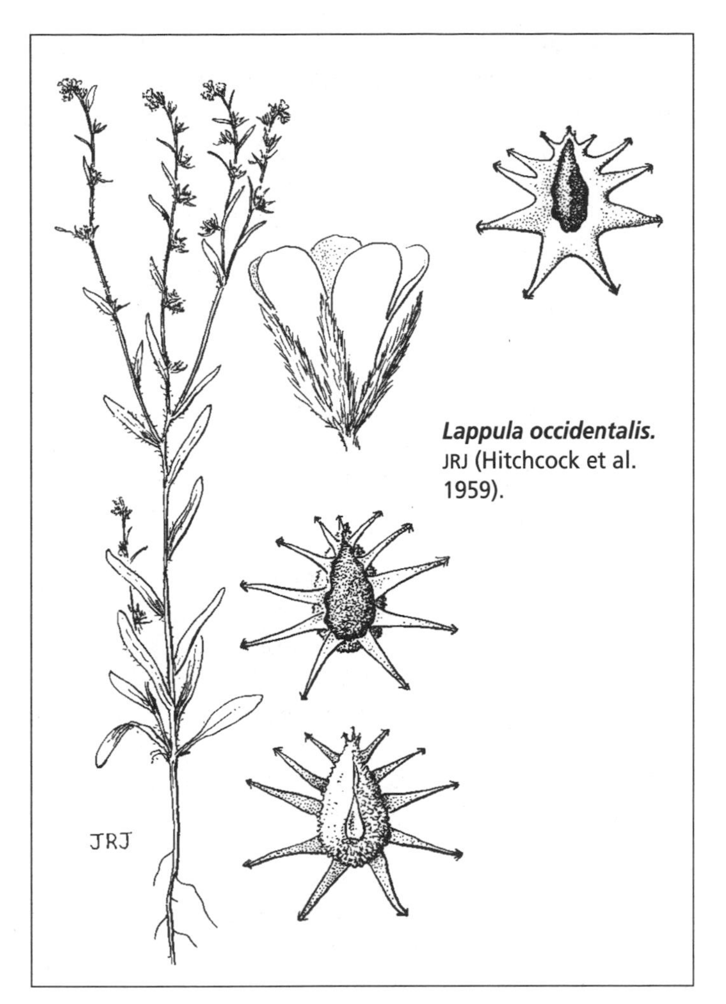

Lappula occidentalis. JRJ (Hitchcock et al. 1959).

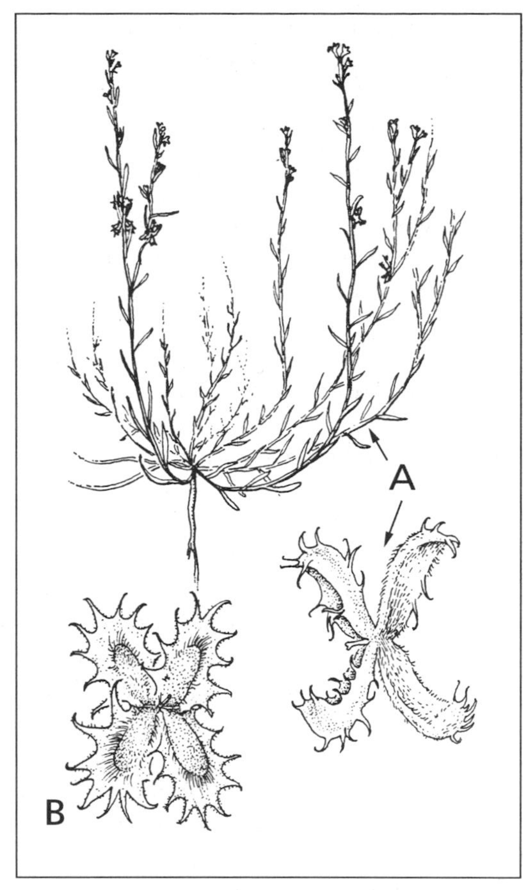

Pectocarya: (A) P. heterocarpa; (B) P. platycarpa. BA. © James Henrickson 1999.

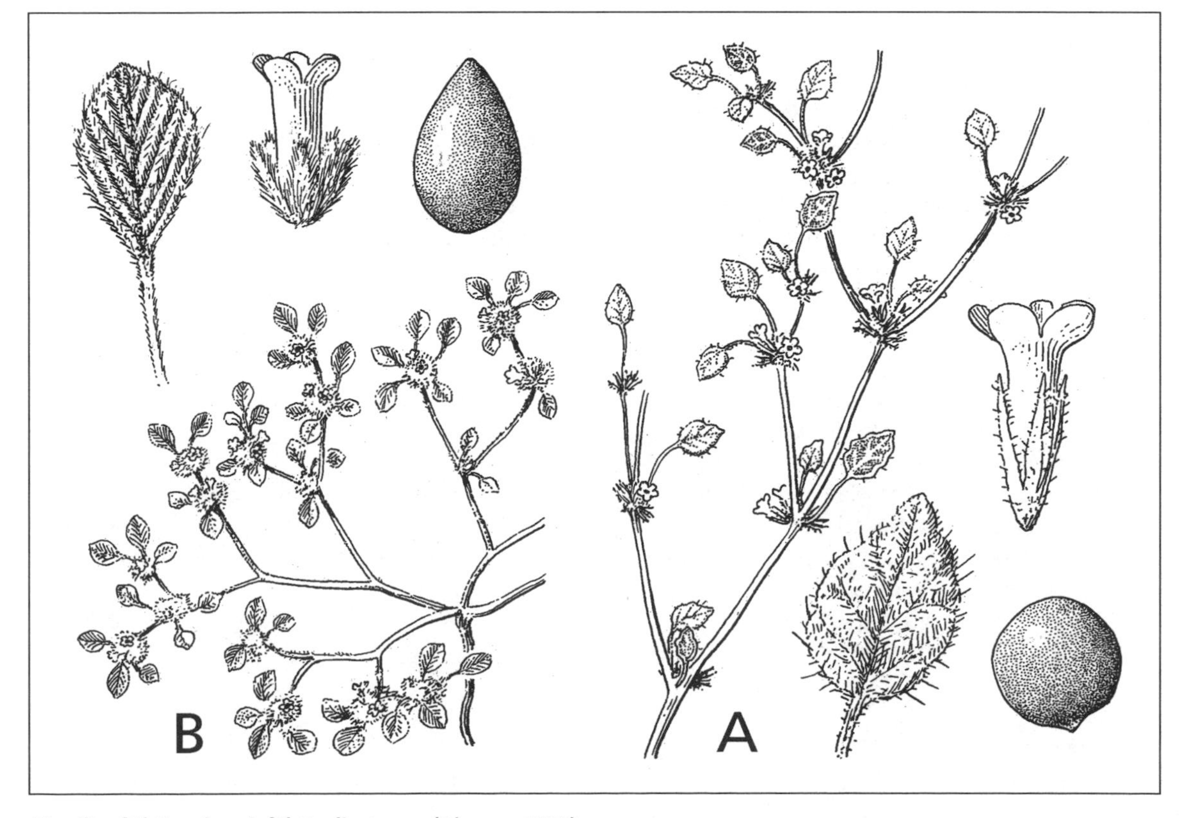

Tiquilia: (A) T. palmeri; (B) T. plicata. JRJ (Abrams 1951).

Mostly in dry habitats, many in deserts; 150 species, mostly in western North America, the others in South America. *Cryptantha* seems to have evolved in southwestern North America and then spread to deserts in Chile and Argentina. Wiggins (1964) includes 20 species in the *Flora of the Sonoran Desert,* 4 of which probably do not extend into the desert proper; 9 of the remaining 16 species occur in our the flora area. Diversity of *Cryptantha* increases northward and westward from the flora area, but southward the number of species drops off sharply. Only 1 species, *C. grayi,* occurs in Sonora south of the Guaymas region (see *C. angustifolia*). These distributions seem to be influenced by winter-spring rainfall. The Sonoran Desert species grow only during the cooler seasons of the year. *C. holoptera* and *C. racemosa* are facultative perennials or ephemerals, whereas the others in the flora area are strictly ephemerals.

The genus has 5 sections, 2 of which occur in the Sonoran Desert: section *Eremocarya* includes *C. micrantha,* and section *Krynitzkia* includes the remaining Sonoran Desert species. *Krynitzkia* includes the 53 species of North American *Cryptantha* monographed by Johnston in 1925. The species in the flora area are distinctive and well marked. Examination of the nutlets provides easy and positive identification. Once plants are matched with their nutlets one usually can distinguish the species by the appearance of the plants alone.

The shapes and ornamentations, or lack thereof, of the nutlets may be adaptations for dispersal and/or defense against seed predators such as *Pogonomyrmex* ants. Smooth nutlets, such as those of *C. ganderi,* might be more difficult to grasp and carry off. Among the *Krynitzkia,* some have homomorphic nutlets (those within each fruit are all alike), and others have heteromorphic nutlets (nutlets within each fruit are not all alike), providing more than one strategy for dispersal and survival. The odd nutlet (the one that differs from the others) remains attached to the gynobase enclosed in the spiny calyx, an adaptation that might provide added protection from a seed predator. The lightweight calyx with its enclosed nutlet(s) may be wind-dispersed like a miniature tumbleweed. Apparently the odd nutlet seed germinates inside its calyx like the seeds in a bur of *Ambrosia* or *Cenchrus.* A larger-sized seed, which should have more energy reserves to survive longer (perhaps until the next year or for several years), would probably benefit from the protection of the calyx. Perhaps smaller nutlets are better suited for quick germination and do not need as much protection. In some, such as *C. angustifolia,* the odd nutlet is larger but otherwise similar in shape and ornamentation to the others. *C. micrantha* may have both rough and smooth nutlets in the same fruit. Most *Cryptantha* species in the flora area show at least some degree of nutlet heteromorphy or reduced numbers of nutlets (fewer than 4 per fruit). Reference: Johnston 1925.

1. Nutlet margins sharp-edged, knife-like, or winged.
   2. Nutlet margins knife-edged but not winged; inner face of nutlet flat; back of nutlet smooth and markedly convex. ______ **C. costata**
   2′ Nutlet margins winged (narrowly so in *C. racemosa*); face not flat; back studded with bumps.
      3. Calyces broad (lobes ovate to lance-ovate); wing of nutlet usually as wide as body and fringed with finger-like projections; ephemerals. ______ **C. pterocarya**
      3′ Calyces narrow (lobes lanceolate); wing of nutlet entire; ephemerals to short-lived perennials.
         4. Nutlet margins with a ribbon-like wing; bark not exfoliating or moderately exfoliating with age. ______ **C. holoptera**
         4′ Nutlet margins with a very narrow beadlike wing; bark conspicuously exfoliating. ______ **C. racemosa**

1′ Nutlet margins rounded or angled but not knife-edged or winged.
   5. Inflorescence tips obviously curled; flowers without bracts between them.
      6. Plants often about as wide as tall, branching from near base; flowers crowded in inflorescence; nutlets 4 per fruit, the odd nutlet larger. ______ **C. angustifolia**
      6′ Plants often taller than wide; flowers often not crowded in inflorescence; nutlets 1 or 2, or if 3 or 4, then all of the same size.
         7. Nutlets rough (tuberculate), (1–3) 4 per fruit. ______ **C. barbigera**
         7′ Nutlets smooth, 1 (2) per fruit. ______ **C. ganderi**

5′ Inflorescences not prominently curled; bracts subtending at least some flowers.
 8. Stems thicker, not wirelike; dye in roots and stems not conspicuous; bracts few; nutlets 1. ______ **C. maritima**
 8′ Stems very slender, wirelike; roots and stems with purple dye; each flower subtended by a leaf-like bract; nutlets 4. ______ **C. micrantha**

Cryptantha angustifolia (Torrey) Greene [*Eritrichium angustifolium* Torrey] NARROW-LEAF CRYPTANTHA, DESERT CRYPTANTHA. Highly variable in size depending on soil moisture, several to ca. 35 cm, the larger plants globose and bushy; densely covered with sharp, stiff, glassy hairs. Leaves slender and relatively inconspicuous. Inflorescences helicoid. Calyx lobes relatively slender, much longer than nutlets. Corollas ca. 4 mm wide. Nutlets 4 per fruit, all alike except the odd nutlet almost always larger (the 4 nutlets rarely of the same size, e.g., *Kamb 1992*), more or less triangular, the surfaces light brown, tuberculate (evenly dotted with tiny whitish bumps), the margins round-angled (rarely sharp-edged but not prominently knife-edged), the odd nutlet 1.0–1.4 mm, the others (0.65) 0.8 (0.9) mm.

Common across many habitats nearly throughout the flora area: rocky, sandy, and gravelly soils; volcanic and granite ranges to peak elevations, hills, plains, washes and arroyos, playas, and notably abundant and vigorous on shifting dunes. Often one of the first wildflowers to appear after a rain, and sometimes persisting into June, especially near the coast.

Deserts and dry regions of southwestern United States, Baja California, and Sonora south to the vicinity of Guaymas. This is one of those itchy plants that gets in your socks and sleeping bag. It is a "well-marked species having its closest relation in *C. grayi*" (Johnston 1925:37). In Sonora *C. grayi* (Vasey & Rose) Macbride occurs mostly in subtropical scrub but extends into the Sonoran Desert along the coast as far north as the vicinity of Tastiota (28°20′ N). Both species often grow together from Tastiota southward to the Guaymas region. *C. grayi* has homomorphic and smaller nutlets.

Bahía de la Cholla, *Gould 4147*. Dunes N of Gustavo Sotelo, *Ezcurra 14 Apr 1981* (MEXU). Campo Rojo, *F 19754*. Cerro Colorado, *Kamb 1992*. Tinajas de los Pápagos, *Turner 59-39*. Dunes NE of Sierra del Rosario, *F 20420*. E of San Luis, *Asplund Apr 1965*.

Cryptantha barbigera (A. Gray) Greene [*Eritrichium barbigerum* A. Gray] BEARDED CRYPTANTHA. Often 15–55 cm, the inflorescences open and loose with several long branches, or sometimes 3–6+ cm when water stressed. Larger leaves often 3–17 × 0.4–1.4 cm, linear to lanceolate, the apex obtuse; often not forming a basal rosette. Inflorescences helicoid. Corollas often 4 mm wide. Nutlets (1.7) 1.8–2.2 mm, (1–3) 4 per fruit, homomorphic, the surfaces brown to gray with prominently raised bumps (tuberculations) throughout, the margins rounded, the inner face not flat, the ventral groove and nutlet itself slightly crooked.

Common and widespread to peak elevations, in many habitats but not on moving dunes. Southward along the coast of Sonora to the vicinity of Las Cuevitas (29°40′ N). Baja California and southeastern California to southwestern Utah, Arizona, southern New Mexico, and northern Sonora.

Cryptantha barbigera may be conspecific at least with *C. intermedia* (A. Gray) Greene (*C. barbigera* var. *fergusonae* Macbride; Higgins 1979). Johnston (1925) pointed out intergradation between *C. barbigera, C. intermedia,* and *C. nevadensis* A. Nelson & Kennedy, and suggested that the latter two might be treated as varieties of *C. barbigera.* He thought that *C. barbigera* was merely a small-flowered form of *C. intermedia. C. barbigera* has larger, blunter leaves, and more densely hairy inflorescences than does *C. intermedia.* Higgins (1979) basically verifies Johnston's observations. Also see *C. ganderi.*

Moon Crater, *F 19262*. Arroyo Tule, *F 19219*. Pinacate Peak, *F 19426b*. Elegante Crater, *F 19686*. Tinajas de Emilia, *F 19747*. Sonoyta, *F 86-84*.

Cryptantha costata Brandegee. RIBBED CRYPTANTHA Stems erect, the branches somewhat stiff and straight (not as open and loose as in *C. barbigera* or robust plants of *C. angustifolia*); taproot well developed. Leaves linear to narrowly lanceolate, the larger ones 2–4.5 cm × 1–4 mm. Inflorescences moderately helicoid. Nutlets (1.2) 1.6–1.9 mm, 4 per fruit, homomorphic or 1 sometimes larger, the surfaces shiny and smooth, the margins knife-edged, the ventral face flat, the dorsal surface high and convex.

Near Puerto Peñasco northward and westward across the Gran Desierto to the Río Colorado. Mostly in areas of moving sand—sandy plains, dunes, and interdune troughs. Not known elsewhere in Sonora. Also southeastern California, southwestern Arizona, and northeastern Baja California.

20 mi E of San Luis, *Bezy 458.* El Golfo, *F 75-89.* N of Sierra del Rosario, *F 89-66.* Dunes NE of Puerto Peñasco, *F 88-229.*

Cryptantha ganderi I.M. Johnston. DUNE CRYPTANTHA. Often 10–30 cm, vegetatively somewhat similar to *C. barbigera* but usually with somewhat smaller, narrower leaves and more rigid stems. Larger leaves 4.5–8 cm × 2.2–5 mm, not in a basal rosette. Inflorescences helicoid. Corollas pure white, 2.8–3.5 mm wide. Nutlets 2.6–2.9 mm, smooth, 1 (2) per fruit (ovules or immature nutlets rarely 3).

Widely scattered across the flora area, and barely extending into adjacent Cabeza Prieta Refuge and slightly farther northward on the Mohawk Dunes; sand flats and dunes, often especially common at their interface. Also in similar habitats in southeastern California and Baja California.

The nutlets are similar in shape to those of *C. barbigera* except larger, slightly more slender, the ventral groove narrower and straight (rather than crooked), and the sinus at the base of the groove smaller. *C. ganderi* usually has only 1 nutlet whereas *C. barbigera* has 1 to 4. Perhaps *C. ganderi* is a dune-adapted sister taxon of *C. barbigera.* Johnston (1939:387) believed *C. ganderi* to be "a desert relative of the characteristically coastal *C. clevelandii.*" However, the nutlets of those 2 species are quite different.

13.5 km W of Los Vidrios, *Burgess 5605.* S of SW base of Hornaday Mts, *Burgess 6795.* Rd to Pozo Nuevo, *F 88-88.* 24 mi SW of Sonoyta, *Wiggins 8351.* NE of Puerto Peñasco, *F 88-230.*

Cryptantha holoptera (A. Gray) J.F. Macbride [*Eritrichium holopterum* A. Gray] WINGED CRYPTANTHA. Plants extremely variable in size; facultative ephemerals to short-lived perennials becoming woody at the base, the perennials more frequent at higher elevations. Herbage hirsute with conspicuously coarse hairs, the leaves gray-green. Inflorescences helicoid. Flowers often very fragrant, the corollas ca. 2 mm wide. Nutlets 1.2–2.0 mm, light brown with a sharp-edged ribbonlike wing, 4 per fruit, nearly similar or the odd nutlet as much as 15–25% longer than the others and with smaller and usually slightly fewer tubercles.

Rocky, gravelly soils; hills, slopes, arroyos, and canyons, generally more common at higher elevations, especially on north- and east-facing slopes. Pinacate volcanic field and granitic mountains; not found in the Sierra del Rosario, sand flats, dunes, and lower-elevation plains. Not known elsewhere in Sonora. Also deserts in northeastern Baja California, southeastern California, and western Arizona.

Cryptantha holoptera, C. racemosa, and *C. fastigiata* I.M. Johnston form a series of closely related but distinguishable species. These are the only Sonoran Desert *Cryptantha* surviving beyond a single winter-spring season as long-lived annuals or short-lived perennials. California populations of *C. holoptera* lack woody bases (Andy Sanders, personal communication 1990). *C. holoptera* and *C. racemosa* occur intermixed at least in the Hornaday Mountains and at Cerro Pinto. *C. fastigiata* occurs south of the flora area on Gulf of California islands and adjacent Sonora and Baja California.

Tinajas de Emilia, *F 19738.* Cerro Colorado, *F 10780.* Pinacate Peak, *F 19429.* Sykes Crater, *F 19505.* Hornaday Mts, *Burgess 6832.* Cerro Pinto, *Burgess 5615.*

Cryptantha maritima (Greene) Greene [*Krynitzkia maritima* Greene. *K. ramosissima* Greene] WHITE-HAIRED CRYPTANTHA. Highly variable in size, often with an erect main axis and branching mostly above the middle. Stems dark brown to red-brown, foliage often dark green, the dead, dark brown leaves persisting on stems. Inflorescence branches short, reaching 2 (3.5) cm, usually not helicoid or only slightly so. Fruiting sepals slender, very spiny, and much longer than the enclosed nutlet. Nutlets 1.2–1.5 mm, smooth, shiny, dark red-brown, slender, 1 per fruit.

Many habitats including rocky slopes, rocky and sandy flats, washes, and occasional on lower stabilized dunes; near sea level to peak elevations, desert plains and the volcanic and granitic ranges. Next to *C. angustifolia* this is the most widespread and common cryptantha in the region. Sonora south to the Guaymas region, western and southern Arizona, southern Nevada, southern California, and both Baja California states.

Both varieties occur in the flora area: var. *pilosa* I.M. Johnston, with dense white tufts of hairs on the calyx, and var. *maritima* distinguished by a relative lack of such hairs. Although there are intermediates and both sometimes occur freely mixed, there seems to be a weak geographic pattern to their distributions. Var. *pilosa* is more widespread in the Pinacate volcanic field than is var. *maritima*. In contrast, var. *maritima* seems to be more common along the coast and in the granitic mountains west of the Pinacate volcanic field. Both range southward along the coast of Sonora to the vicinity of Guaymas.

Specimens from the flora area have 1 nutlet per fruit. Elsewhere a second nutlet may be present; when present this nutlet is rough with a minutely tuberculate pattern, gray-brown, and apparently falls out of the calyx more readily than the smooth one (Abrams 1951).

Var. **pilosa:** Tinajas de Emilia, *F 19742*. Pinacate Peak, *F 19428*. Moon Crater, *F 19249*. Hornaday Mts, *Burgess 6517*. Cerro Pinto, *Burgess 5614*. Sierra del Rosario, *F 20666*. Var. **maritima:** 5 km E Puerto Peñasco, *Burgess 4743*. 28 mi W of Los Vidrios, *Bezy 9 Apr 1966*. Sierra del Rosario, *F 20697*. Sonoyta, *F 86-82*.

Cryptantha micrantha (Torrey) I.M. Johnston subsp. **micrantha** [*Eritrichium micranthum* Torrey. *Eremocarya micrantha* (Torrey) Greene] DWARF CRYPTANTHA, PURPLE-ROOT CRYPTANTHA. Plants 3–10 cm, branched mostly above, stems very slender, the bark peeling on the lower stems of larger, older plants; hairs small, mostly appressed. Roots and stems staining bright purple when pressed. Leaves 3–8 mm, relatively few and scattered, those of the first 1 or 2 nodes opposite and not in a basal rosette. Inflorescence branches not strongly helicoid, reaching 5 (15) mm. Flowers minute, each subtended by a leafy bract; corollas white with a yellow center, the lobes broadly obovate-spatulate, notched at the apex. Nutlets 0.9 mm, 4 per fruit, slender, the margins rounded; homomorphic and either all smooth or all rough (tuberculate), or heteromorphic with 1 rough and 3 smooth.

Sandy plains and dunes across the Gran Desierto and lowlands. Subsp. *micrantha* ranges from Oregon to Utah to Baja California and western Sonora south to the vicinity of Bahía Kino, and eastward to western Texas. This is our smallest cryptantha. Subsp. *lepida* I.M. Johnston, with larger corolla lobes, is confined to the western margin of the desert in southern California and northern Baja California.

Rd to Rancho Guadalupe Victoria (Grijalva), *F 18716A*. Base of Hornaday Mts, *Burgess 6850*. S of Moon Crater, *F 19026*. High dunes N of Sierra del Rosario, *F 20793*. NNE of El Golfo, *F 92-163A*.

Cryptantha pterocarya (Torrey) Greene var. **cycloptera** (Greene) J.F. Macbride [*Krynitzkia cycloptera* Greene] WINGED-NUT CRYPTANTHA. Unbranched, erect or with mostly few, ascending branches. Inflorescences not helicoid or only slightly so on larger plants. Calyx appearing angled due to thickened sepal midribs, the sepals broadly ovate, obtuse, and bright green fading to light yellow-tan, enlarging as the fruit matures to 4.5–6 mm. Nutlets 2.5–3.0 mm, 4 per fruit, usually homomorphic, intricately sculptured, the body studded with blunt tubercles and edged by a broad, light-colored wing edged with blunt, finger-like projections, the wings often bent as they grow crammed in the calyx, the wings and body the color of light milk chocolate speckled with darker blotches. (In occasional water-stressed plants (e.g., *F 19882*) the wing-margin of the odd nutlet is much narrowed but not wingless as in var. *pterocarya*).

Rocky and gravelly soils, volcanic and granitic soils, often common during wetter years, especially toward higher elevations in the Sierra Pinacate. Common in the Tinajas Altas Mountains so probably also in the adjacent granitic mountains in the flora area.

Widespread in Sonora, with southern records in the Sierra Seri, Hermosillo, and east of Alamos. Also both Baja California states to Washington and Utah, and eastward to New Mexico and western Texas. Readily recognized by the relatively large fruits and broad sepals. Var. *pterocarya* occupies the northern part of the geographic range, and var. *cycloptera* replaces it to the south. Var. *pterocarya*, distinguished by heteromorphic nutlets (the odd nutlet is wingless), is not known for Sonora.

Pinacate Peak, *F 19882*. Sykes Crater, *F 19500*. Arroyo Tule, *F 19542*. Tinaja Huarache, *F 19115*. Upper N slope of Hornaday Mts, *Burgess 6839*.

Cryptantha racemosa (S. Watson ex A. Gray) Greene [*Eritrichium racemosum* S. Watson ex A. Gray] Coarse perennials to 0.5+ m, often woody at the base with several upright main axes, shredding bark, and short branches above; also flowering in the first season but these mostly do not survive the summer drought. Flowering branches not helicoid. Nutlets 4 per fruit, light brown with narrow, cordlike margins, all of similar shape and ornamentation but the odd nutlet 2.0 mm, the others 1.1–1.4 mm.

Western granitic mountains including Sierra del Rosario, Cerro Pinto, and Hornaday Mountains; mostly north- and east-facing slopes, and more common toward higher elevations. A desert species in the Mojave Desert in California and Nevada, and southward in the Sonoran Desert to northern Baja California and the flora area in northwestern Sonora (not known elsewhere in Sonora). Differing from the related *C. holoptera* by having nutlets of dissimilar size with narrowly winged margins.

Hornaday Mts, *Burgess 6828*. Sierra del Rosario, *F 75-34*. Cerro Pinto, *F 89-58*.

HELIOTROPIUM HELIOTROPE

Annuals or perennials, herbaceous to shrubby; usually hairy, rarely glabrous. Inflorescences usually helicoid. Fruits dry at maturity, usually breaking into four 1-seeded or two 1- or 2-seeded segments. Warmer parts of the world, especially in arid regions; 250 species.

1. Plants densely hairy. ____ **H. convolvulaceum**
1′ Plants glabrous. ____ **H. curassavicum**

Heliotropium convolvulaceum (Nuttall) A. Gray [*Euploca convolvulacea* Nuttall. *E. convolvulacea* subsp. *californica* (Greene) Abrams. *E. aurea* Rose & Standley, *Contr. U.S. Natl. Herb.* 16:16, pl. 11, 1912. *Heliotropium convolvulaceum* var. *californicum* (Greene) I.M. Johnston] MORNING-GLORY HELIOTROPE. Ephemerals or annuals, primarily growing with hot-weather rains. Erect to spreading, loosely branched, reaching 60 cm tall and ca. 100 cm across, or sometimes reproducing at only 5 cm; densely hairy including calyx and outer corolla surfaces, the hairs spinescent, whitish to yellowish, spreading or appressed. Leaves mostly ovate to lanceolate-elongate, highly variable depending upon soil moisture. Flowers solitary in leaf axils, and especially in larger plants in terminal "leafy-bracted" racemes; showy, resembling morning-glory flowers (hence the specific name), sweet scented, opening in the afternoon. Corollas 1–1.5 cm across, circular in outline, very pale yellow with darker yellow in center and along the interplicae. Style slender, longer than the ovary, with a tuft of stiff hairs at the apex. Nutlets 2, rounded.

Seasonally common on sand flats and dunes; San Luis to El Golfo and eastward along dunes near Bahía Adair and south of Sierra Blanca (northeast of Puerto Peñasco). There are few collections from Sonora, apparently because the populations are widely disjunct. Var. *californicum* recorded from the Mojave and Sonoran Deserts in southeastern California, southern Nevada, western Arizona, and Sonora south to the vicinity of Guaymas.

The much more widespread var. *convolvulaceum* occurs at higher elevations in Arizona and Utah to Nebraska, Texas, and north-central Mexico. Var. *californicum* is distinguished by more rigid and spreading hairs. However, I am unable to consistently distinguish the varieties.

Sand hills near Adair Bay, 20 Nov 1907, *Sykes 61* (holotype of *E. aurea,* US). N of El Golfo, *F 75-71*. NW of La Salina, *F 93-1013*. NE of Puerto Peñasco, *F 92-435*.

Heliotropium curassavicum Linnaeus. *HIERBA DEL SAPO;* ALKALI HELIOTROPE; *BA:BAD 'I:WAGǏ. KAKAICU 'I:WAGǏ.* Perennial herbs or sometimes annuals; glabrous, semisucculent to succulent, bluish glaucous. Leaves nearly sessile, mostly 2.5–7.5 cm, lanceolate to oblanceolate or obovate. Flowers in several spikelike, terminal helicoid branches. Corollas ca. 2–2.5 mm wide, white with a yellow center fading purplish. Stigma sessile. Nutlets 4, ovoid. Growing and flowering with warm to hot weather any time of year.

Wetlands with alkaline or saline soils. Margins of salt marshes (esteros), playas, waterholes, alkali riverbeds and arroyo bottoms, and a weed in irrigation ditches and fields. Especially common along the Río Sonoyta drainage, margins of the Río Colorado and the Ciénega de Santa Clara, the Bahía Adair pozos, and in agricultural fields in the San Luis Valley. When the plant is in full flower in the late afternoon during a hot, humid day there is a constant whine of small hovering flies. The flowers are also visited by large numbers of other flies, and the queen butterfly (*Danaus gilippus strigosus*).

Through much of the warm regions of the Western Hemisphere and adventive in the Old World. Wiggins (1964) lists 2 varieties for the Sonoran Desert—var. *oculatum* (A. Heller) I.M. Johnston ex Tidestrom and var. *curassavicum*—although Higgins (1979:302) indicates that only var. *oculatum* occurs in southern Nevada and western Arizona. Var. *oculatum* is widespread in western North America. I find it difficult to distinguish the varieties, perhaps because the varietal characters are not apparent on herbarium specimens or because they are poorly differentiated. Wiggins (1964:1209) gives the following key to the varieties:

1. Corollas 3–7 mm in diameter, turning deep purple in throat; calyx lobes longer than fruit, erect. ____ var. **oculatum**

1′ Corollas 1.2–5 mm in diameter, white or faintly purplish; calyx lobes shorter than fruit, appressed. ____ var. **curassavicum**

3 km E of Puerto Peñasco, *F 85-799*. Cauce de Río Sonoyta, 25 km SW of Sonoyta, *Ezcurra 19 Apr 1981* (MEXU). Pinacate Junction, *F 19611*. Quitobaquito, *F 86-175*. La Soda, *F 86-523*. Ciénega de Santa Clara, *F 92-5172*.

LAPPULA Stickseed

Small annuals or rarely biennial or perennial herbs. Inflorescences leafy or with bracts. Calyx enlarging as fruit matures. Corollas usually small, white or blue. Nutlets 4, heteromorphic or homomorphic, the margins edged with spines. Mostly Eurasia, also Australia and New World; 50 species. Many weedy and adventive, their spread undoubtedly aided by grappling-hook spines edging the nutlets. **Reference:** Johnston 1924b.

Lappula occidentalis (S. Watson) Greene [*Echinospermum redowskii* (Hornemann) Greene var. *occidentale* S. Watson] Winter-spring ephemerals 25–45 cm, with a well-developed taproot. Leaves in a basal rosette, withering as the stem develops. Main axis 1 to several, erect, slender, branching above the middle; flowering branches helicoid when developing, raceme-like at maturity. Corollas white to pale blue. Nutlets burlike, the body 2.0–2.3 mm, ovate-triangular, the margins fringed with stout, barb-tipped spines, the longer spines nearly 85–90% as long as the nutlet body.

Seasonally common in Organ Pipe Monument and the adjacent Sonoyta region. Sandy-silty and gravelly soils in natural and disturbed habitats, often in washes and brushy places near floodplains. Not known farther south in western Sonora; probably common in northern Sonora east of the flora area. Western North America from Alaska and Canada to northern Mexico, also in Argentina; often weedy, widely introduced or adventive.

Most specimens from the flora area and Organ Pipe Monument more or less fit into the concept of var. *cupulata* (A. Gray) L.C. Higgins. In this variety the nutlets have marginal spines joined at the base into an inflated wing, thicker and hollow spines, and a tendency toward nutlet heteromorphy. Var. *occidentalis* has marginal spines joined directly to the body, there is no wing, and the nutlet are all alike. Most specimens from the flora area have homomorphic nutlets but some have both kinds of nutlets. It seems that the larger, winged nutlets, characteristic of var. *cupulata,* occur earlier in the season under more favorable conditions, and that even on the same plant increasing drought favors the formation of wingless nutlets like those of var. *occidentalis*. The nomenclature of this plant is complicated (see Cronquist et al. 1984:284–286; Diggs, Lipscomb, & O'Kennon 1999; and Higgins 1979). The North America plants have been called *L. redowskii* (Horneman) Greene, an Old World species not known to occur in North America (Larry C. Higgins, personal communication 1999).

Floodplain of Río Sonoyta, S of El Papalote, *F 86-169*. Quitobaquito, *F 86-178*.

PECTOCARYA COMB-BUR

Annuals. Leaves narrow, often small, lower ones opposite, appearing as a basal rosette, the others alternate. Inflorescences, at least ours, not helicoid or only moderately so. Corollas very small, white. Nutlets 4, in 2 pairs spreading apart, the margins at least partially bristly.

Western North America and Andean-Patagonian South America; 15 species. Ours are small winter-spring ephemerals with well-developed taproots and slender, ascending to spreading stems. The fruits, divided into 4 spreading nutlets with comblike fringes, are unique. **References:** Johnston 1939; Veno 1979.

1. Nutlets not all alike—the size and margins of nutlet pairs dissimilar. ____ **P. heterocarpa**

1′ Nutlets all alike, or essentially so.

  2. Nutlets conspicuously recurved, the margins absent or very narrow and about the same color as the body, the teeth distinct. ____ **P. recurvata**

  2′ Nutlets straight or nearly so, the margins conspicuous, the margins and teeth lighter colored (often yellow) than the body, the teeth joined at their bases. ____ **P. platycarpa**

Pectocarya heterocarpa (I.M. Johnston) I.M. Johnston [*P. penicillata* (Hooker & Arnott) A. de Candolle var. *heterocarpa* I.M. Johnston] MIXED-NUT COMB-BUR. Lower leaves narrowly linear, gradually and slightly broader near apex, often 2–8 cm × 0.7–1.9 mm wide near apex, the stem leaves reduced upwards; larger leaves soon deciduous. Flowers fragrant; corollas white with a yellow throat, the lobes broadly obovate and rounded at apex; stigma green. Nutlets mostly heteromorphic, often curved or slightly curled inward (not outward), one pair larger, winged and with marginal teeth, the other pair not winged and lacking marginal teeth, each nutlet within a pair slightly different. Lowermost flowers cleistogamous, (usually?) producing wingless, homomorphic nutlets, often only 2 of the nutlets developing. Reported to be diploid.

Widespread, in the lowlands to at least ca. 400 m; Sonoyta region, Pinacate volcanic field, at least some granitic mountains, and desert plains; sandy, gravelly, and rocky soils, washes, flats, and slopes. Southward along the Sonora coast at least to El Desemboque San Ignacio; northern Sonora and southwestern Arizona to southern Nevada and southwestern Utah, and southeastern California to northern Baja California Sur.

This is the most common and widely distributed comb-bur in the flora area. The plants and fruits are generally smaller than those of the other 2 species. Plants at the southern limits, in Sonora, are unusual in having all 4 nutlets winged, although one pair has narrower wings.

18 km SW of Sonoyta, *F 88-224.* Near Elegante Crater, *Burgess 5570.* Rd to Rancho Guadalupe Victoria, *F 20585.* Moon Crater, *F 19255.* Sykes Crater, *F 18947.* Hornaday Mts, *Burgess 6814.*

Pectocarya platycarpa (Munz & I.M. Johnston) Munz & I.M. Johnston [*P. gracilis* I.M. Johnston var. *platycarpa* Munz & I.M. Johnston] BROAD-WING COMB-BUR. Nutlets all alike, straight or nearly so; margins and teeth conspicuous, often yellow, the teeth joined with margin at bases. Distinguished by the generally larger-sized plants, larger and wider nutlets, and fewer and larger marginal teeth.

Widespread, especially common in the Sonoyta region. Often growing intermixed with *P. heterocarpa.* Southward in western Sonora to the vicinity of Bahía Kino; northern Sonora, Baja California, Arizona, southeastern California, southern Nevada, and southwestern Utah.

Reported to be a tetraploid and probably derived from a common ancestor of *P. penicillata* and *P. recurvata.* On the basis of morphology, *P. platycarpa* and several other North American taxa "might well be included as geographical phases of the hexaploid South American *P. linearis* (Ruiz & Pavón) DC." (Cronquist et al. 1984:286).

18 km SW of Sonoyta, *F 88-205.* 3 km S of El Papalote, *F 86-160.* 10 mi W of Sonoyta, *F 20597.* SSW of Tezontle cinder mine, *F 88-41.* Sykes Crater, *F 18948.*

Pectocarya recurvata I.M. Johnston. ARCHED COMB-BUR. Nutlets all alike (or essentially so), relatively small, narrow, conspicuously recurved (arched or curled back), margins absent or very narrow and about the same color as body of nutlet, the teeth distinct. Reported to be diploid.

Northeastern corner of the flora area near Sonoyta and disjunct at higher elevation on the north side of Pinacate Peak. Also northern Sonora east of the flora area and southward to the vicinity of Hermosillo, and Arizona, southeastern California, southern Nevada, and Baja California.

N slope Pinacate Peak, *F 19467.* 10 km SW of Sonoyta, *F 88-186.* Quitobaquito, *F 88-126.* 4 mi S of Sonoyta on Mex 2, *F 20546.*

TIQUILIA CRINKLEMAT

Perennials, sometimes flowering in first season; mostly woody subshrubs, with a forked branching pattern (pseudo-dichotomous). Herbage densely and variously hairy. Flowers essentially sessile, axillary, single or clustered. Corollas white to lavender, blue or purple. Style once-cleft, variously attached to nutlets from top to base. Stigma capitate. Fruits dry, of 1–4 nutlets.

North and South America, dry regions, mostly deserts. They "usually do not grow in close proximity to other plants, with the exception of other species of the same genus. It is not uncommon to find two, three, or even four species growing together, especially in North America. In spite of this sympatry, no hybrids have been detected" (Richardson 1977:489). Ours are in section *Tiquiliopsis,* which includes 2 additional species.

Coldenia and *Tiquilia* form a natural taxon within the subfamily Ehretoideae. They differ strikingly from other members of that subfamily in their smaller stature, smaller and xeromorphic leaves, inflorescences crowded or reduced to 1 or few flowers rather than open, distinct, scorpioid cymes, and most notably dry rather than fleshy fruits. These characters are consistent with their relatively xeric and weedy-pioneer habitats. Richardson (1977) restricts the genus *Coldenia* to a single, widespread weedy annual species of the Old World tropics.

1. Leaves with 2 or 3 (4) pairs of shallowly impressed veins; nutlets rounded, minutely papillate to dull-surfaced, not smooth and shiny. ____ **T. palmeri**

1′ Leaves with 5 or 6 (7) pairs of deeply impressed veins, appearing as if pleated; nutlets ovoid on back, smooth and shiny. ____ **T. plicata**

Tiquilia palmeri (A. Gray) A.T. Richardson [*Coldenia palmeri* A. Gray] Semiprostrate perennial herbs from very deep, thickened, blackish roots. Herbage densely pubescent with white hairs and scattered bristles with swollen, hard, white bases on leaf blades. Leaves relatively thick, the blades 3–8 mm, broadly elliptic to ovate, rhombic, or nearly orbicular, with shallowly impressed veins on the upper surface; petioles prominent, often as long as or longer than the blades. Corollas lavender with a pale yellow throat, readily falling. Nutlets 0.8–0.9 mm in diameter, 1–4 per fruit, rounded (spheroid), light brown, the surface densely and minutely papillate (best seen at 30–40× magnification) or sometimes partially smooth (the papillae dense but in patches), the attachment scar rounded and large. Style attached at base of nutlet. Flowering with spring and summer-fall rains.

Widespread on desert plains, sand flats, interdune troughs, and occasionally on lower stabilized dunes; often with galleta grass (*Pleuraphis rigida*) and creosote bush (*Larrea divaricata*), and sometimes with *T. plicata* at dune margins. Southeastern California, southern Nevada, western Arizona, Baja California, and Sonora southward on coastal dunes to Bahía Kino.

Puerto Peñasco, *F 85-762.* 24 mi SW of Sonoyta, *Shreve 7587.* MacDougal Crater, *F 10763.* Moon Crater, *F 19275.* E of El Puerto [Paso del Aguila], *F 13227.* N of Sierra del Rosario, *F 20361.* N of El Golfo, *F 75-75.*

Tiquilia plicata (Torrey) A.T. Richardson [*Coldenia plicata* Torrey] Prostrate, matlike perennial herbs, sometimes with a buried woody base, from very deep, thickened, blackish roots; new shoots arising from long, slender roots. Herbage with dense, white hairs, and scattered bristles with mineralized expanded white bases on leaf blades. Sand often adhering to stems and petioles of dune plants. Leaves thick, the blades 3–8 mm, broadly obovate to elliptic, ovate, obovate, or orbicular, with deeply impressed veins on upper surface; petioles prominent, often as long as or longer than the blades. Corollas lavender, readily falling. Nutlets 0.8–0.95 mm, 1–4 per fruit, shiny, smooth, steel gray-brown to nearly black, the back ovoid, the common (commensurate) sides flattened, the apex pointed, the attachment scar small. Style attached near base of nutlet. Flowering with spring and summer-fall rains.

Sand flats and moving dunes; replacing *T. palmeri* on higher, shifting dunes; often with *Dicoria, Ephedra trifurca, Helianthus, Larrea, Psorothamnus emoryi,* and *Prosopis glandulosa*. Puerto Peñasco to El Golfo and San Luis. Southern California, southern Nevada, western Arizona, northern Baja California, and Sonora southward to shifting beach dunes at Bahía Tepoca (30°15′ N).

Similar in habit to *T. palmeri* but the plants more delicate, the leaf blades often proportionally broader, thinner, blunter at the tip, and with a greater number and more deeply impressed veins, giving the leaf a pleated appearance.

Puerto Peñasco, *F 20957*. 8 mi SE of Tinaja de los Chivos, *Simmons 11 Apr 1965*. Near Moon Crater, *F 19006*. NW of Sierra del Rosario, *F 20421*. E of San Luis, *F 16681*. El Golfo, *F 75-85*.

BRASSICACEAE (CRUCIFERAE) MUSTARD FAMILY

Primarily annual, biennial, or perennial herbs; containing mustard oils. Leaves usually alternate, the lower leaves often in a basal rosette; without stipules. Sepals, petals, and stamens in a crosslike pattern—hence the name Cruciferae. Calyx mostly cylindrical with 4 separate, mostly deciduous sepals, the inner pair often saclike basally. Petals 4 (rarely absent), separate, mostly clawed (conspicuously narrowed basally). Stamens usually 6 with 2 outer, shorter stamens, and 4 inner ones of 2 opposite pairs. Ovary superior. Fruits (called siliques) highly variable, dry, mostly 2-chambered capsules, dehiscent by 2 valves opening from base to apex, or sometimes 1-chambered and indehiscent, or separating into 1-seeded segments. Seeds often mucilaginous when wet.

Worldwide, mostly temperate, many in dry regions; 365 genera, 3350 species. Greatest diversity in the Mediterranean region and Central Asia; other centers in western North America, temperate South America, southern Africa, and Australia. The largest genus is *Draba,* followed by *Lepidium,* and *Lesquerella*. All Sonoran Desert crucifers are winter-spring ephemerals except *Lyrocarpa* (a few herbaceous perennials approach the margins of the Sonoran Desert). The most economically important crucifers are in the genus *Brassica,* others include crambe, horseradish, radish, watercress, and ornamental garden flowers such as stock and wallflower. The family is easily recognized but the genera are often distinguished on technical characters of the fruit that can be difficult for the non-specialist to differentiate. References: Rollins 1981, 1993.

1. Fruits of 2 disk-shaped halves (spectacle-shaped) joined along less than 25% of their margins; flowers white and fragrant.
    2. Leaves with slender lobes; petals (fresh) 5–7 mm; fruiting pedicels 7–20 mm. ______ **Dimorphocarpa**
    2′ Leaves with broad, coarse teeth or lobes; petals (fresh) ca. 10–12 mm; fruiting pedicels 1.5–2.5 mm. ______ **Dithyrea**

1′ Fruits not of disk-shaped halves; flowers various colors, fragrant or not.
    3. Perennials (sometimes flowering in first season); petals twisted, 12–20` mm; fruits usually widest well above the middle, maximum width 8–14 mm. ______ **Lyrocarpa**
    3′ Winter-spring ephemerals; petals not twisted, 10 mm or less; fruits widest at about the middle, less than 7.5 mm wide.
        4. Fruits less than twice as long as wide.
            5. Fruits globose, not at all compressed, several-seeded; flowers bright yellow; plants (including fruits) with stellate hairs. ______ **Lesquerella**
            5′ Fruits compressed (flattened), 1- or 2-seeded; flowers inconspicuous, greenish or whitish; plants glabrous or with simple hairs.

6. Pedicels thickish, straight, flattened; fruits 3–4 mm long, 2-seeded, inconspicuously winged at apex. ____ **Lepidium**
6′ Pedicels slender, recurved, not flattened; fruits 5.5–7.3 mm long, 1-seeded, the margin conspicuously winged. ____ **Thysanocarpus**

4′ Fruits at least 3 times longer than wide.
7. Fruits turned downward; fruiting pedicels 0.5–3 mm; fruits terete and slender, 2.5–6.5 cm; stems leafy, the leaves pinnatifid, more or less lanceolate.
8. Flowering and fruiting pedicels 0.5–1.0 mm; herbage green, leaves very thin. ____ **Caulanthus**
8′ Fruiting pedicels 1.2–3.0 mm; herbage glaucous and often semisucculent. ____ **Streptanthella**

7′ Fruits spreading or erect (not turned downward); fruiting pedicels 4 mm or more, or if less than 3 mm then leaves all basal and broadly obovate and merely toothed.
9. Herbage with branched (dendritic) or stellate hairs; fruits 3.5–12 mm.
10. Herbage with candelabra-shaped hairs (stalked and branched above); leaves finely divided, the stem leaves well developed; fruits nearly terete. ____ **Descurainia**
10′ Herbage with forked and stellate hairs; leaves merely toothed (not divided), all basal; fruits flattened. ____ **Draba**

9′ Herbage glabrous or with simple hairs; fruits 18–70 mm.
11. Herbage glabrous or with sparse, soft hairs; petals 3–4 mm; fruits 0.8–1.3 mm wide. ____ **Sisymbrium**
11′ Herbage usually with at least some coarse hairs; petals at least 6 mm; fruits at least 2 mm wide.
12. Petals of fresh flowers uniformly pale or bright yellow (veins not conspicuous until dried); seeds in a single row in each chamber (the fruit has 2 rows of seeds). ____ **Brassica**
12′ Petals pale yellow or white, the veins dark and conspicuous even on fresh flowers; seeds in 2 rows in each chamber (the fruit has 4 rows of seeds). ____ **Eruca**

BRASSICA MUSTARD

Annual, biennial, or perennial herbs. Glabrous or pubescent with coarse, simple hairs. Lower leaves usually forming a basal rosette. Flowers usually yellow. Petals clawed. Fruits usually linear, the body mostly terete, dehiscent, the valves with a prominent (ours) or obscure midrib; beak prominent, indehiscent, often seedless, or with 1–3 seeds. Seeds rounded, many, in a single row in each chamber; in some species slightly mucilaginous when wet.

Native to the Old World; 30 species. This is the most economically important genus in the family; includes broccoli, Brussels sprouts, cabbage, cauliflower, kale, kohlrabi, mustard (greens and seed), rape seed (yielding canola oil), turnip, and other common vegetables. In the Sonoran Desert these crops grow only during the cooler times of the year.

1. Flowers bright yellow, the petals 15 mm or more; fruits less than 2.5 cm, the beak seedless. ____ **B. nigra**
1′ Flowers pale yellow, the petals 6–8 mm; fruits 3.5 cm or more, the beak bearing 1 or 2 seeds. ____ **B. tournefortii**

***Brassica nigra** (Linnaeus) W.D.J. Koch [*Sinapsis nigra* Linnaeus] BLACK MUSTARD. Plants somewhat similar to those of *B. tournefortii* but sparsely pubescent with soft hairs or nearly glabrous; reaching 50–100 cm tall, with few, widely spreading, flowering branches. Basal rosette leaves 8–30 (50) cm, the stem leaves reduced upwards; leaves highly variable, similar to those of *B. tournefortii* but usually greener and with broader lobes. Flowers bright yellow and showy. Sepals 10 mm, purple-red, erect. Petals long-clawed, 15–24 × 7 mm or more, the veins inconspicuous on fresh flowers, dark and conspicuous when dried. Fruiting racemes often 20–40 cm. Fruits erect and appressed, 20–22 mm, the beak seedless and about 3/4 as long as the body; fruiting pedicels 3.5–4.5 mm. Seeds globose, probably ca. 2 mm diameter.

Rare to infrequent and widely scattered along the roadside of Mex 2, often growing with *B. tournefortii,* and a common agricultural weed south of San Luis. Widespread but seldom common in the lowlands of Sonora, at least as far south as the Río Mayo region. Present in Arizona at least since 1884. Weedy worldwide in temperate regions and highlands in the tropics; native to Europe.

W of Sonoyta, *F 92-144B, F 92-390.*

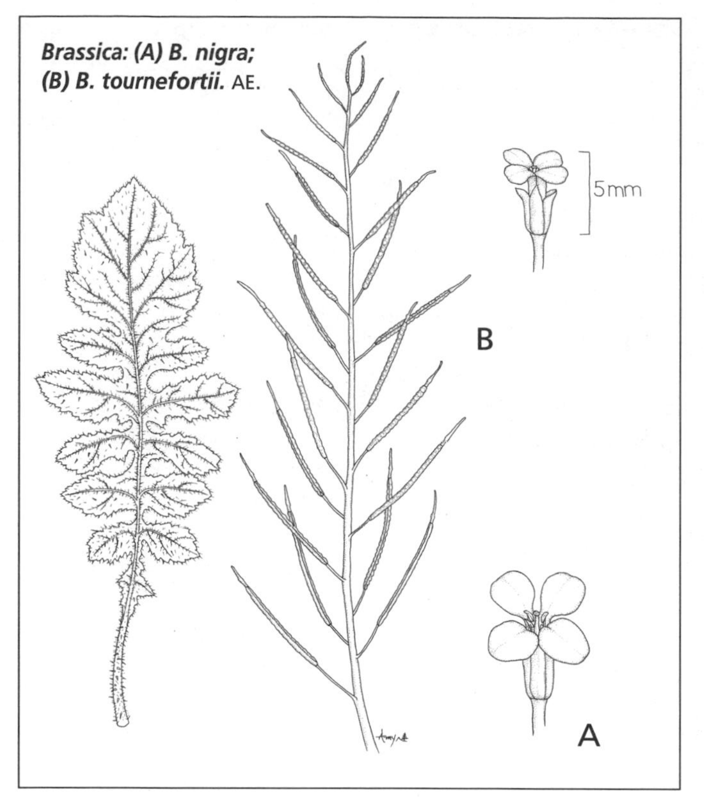

***Brassica*: (A) *B. nigra*; (B) *B. tournefortii*.** AE.

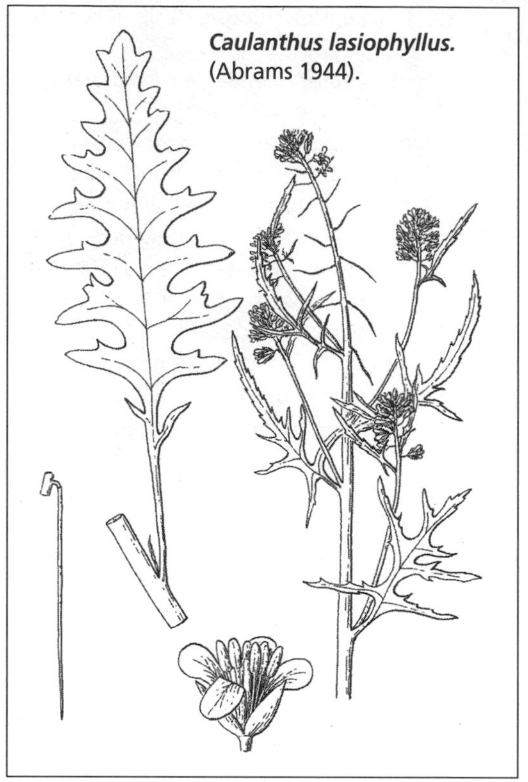
***Caulanthus lasiophyllus*.** (Abrams 1944).

***Descurainia pinnata*.** LBH (Parker 1958).

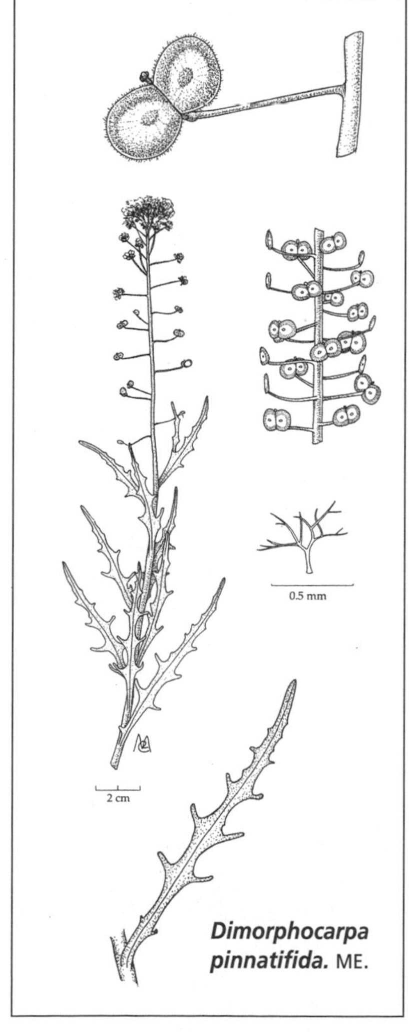

***Dimorphocarpa pinnatifida*.** ME.

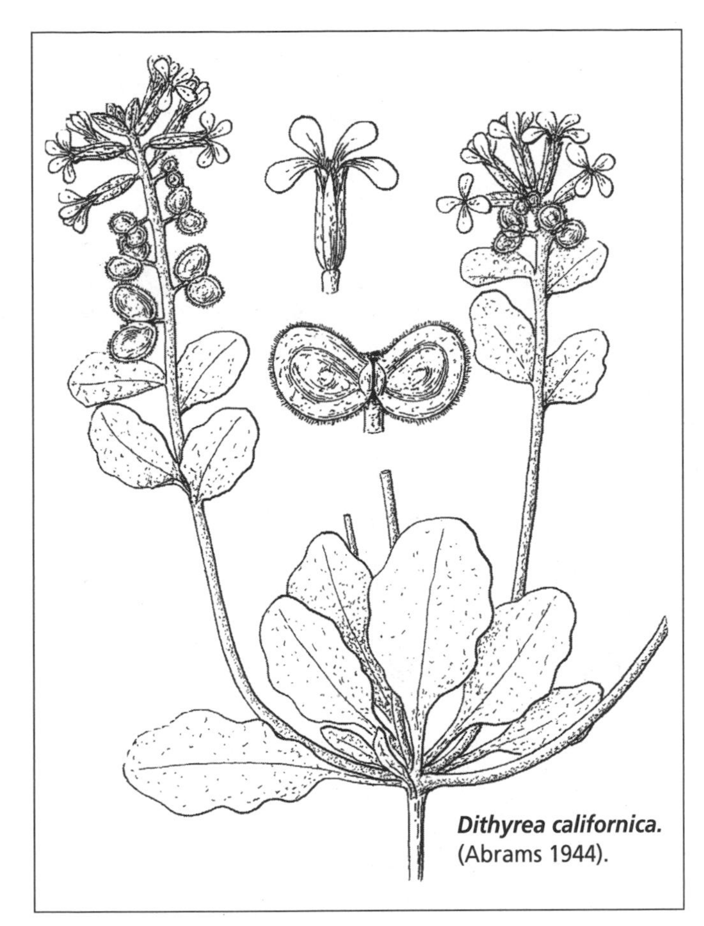
Dithyrea californica. (Abrams 1944).

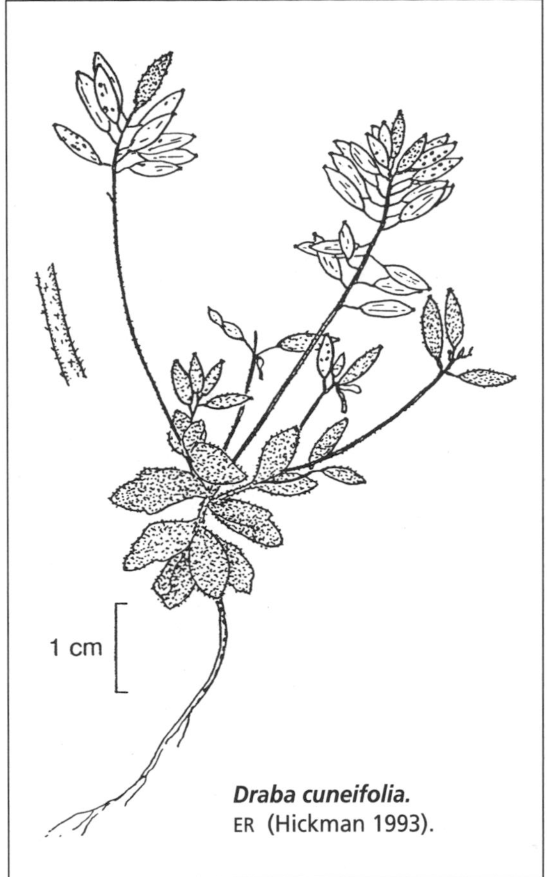

Draba cuneifolia. ER (Hickman 1993).

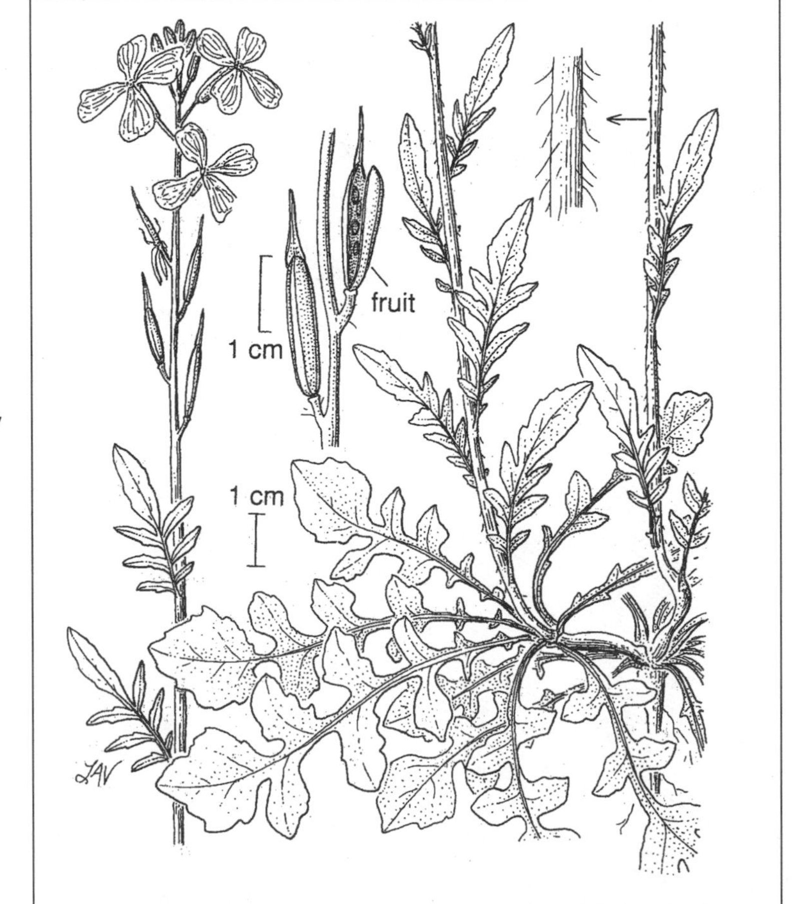

Eruca vesicaria subsp. *sativa.* LAV (Hickman 1993).

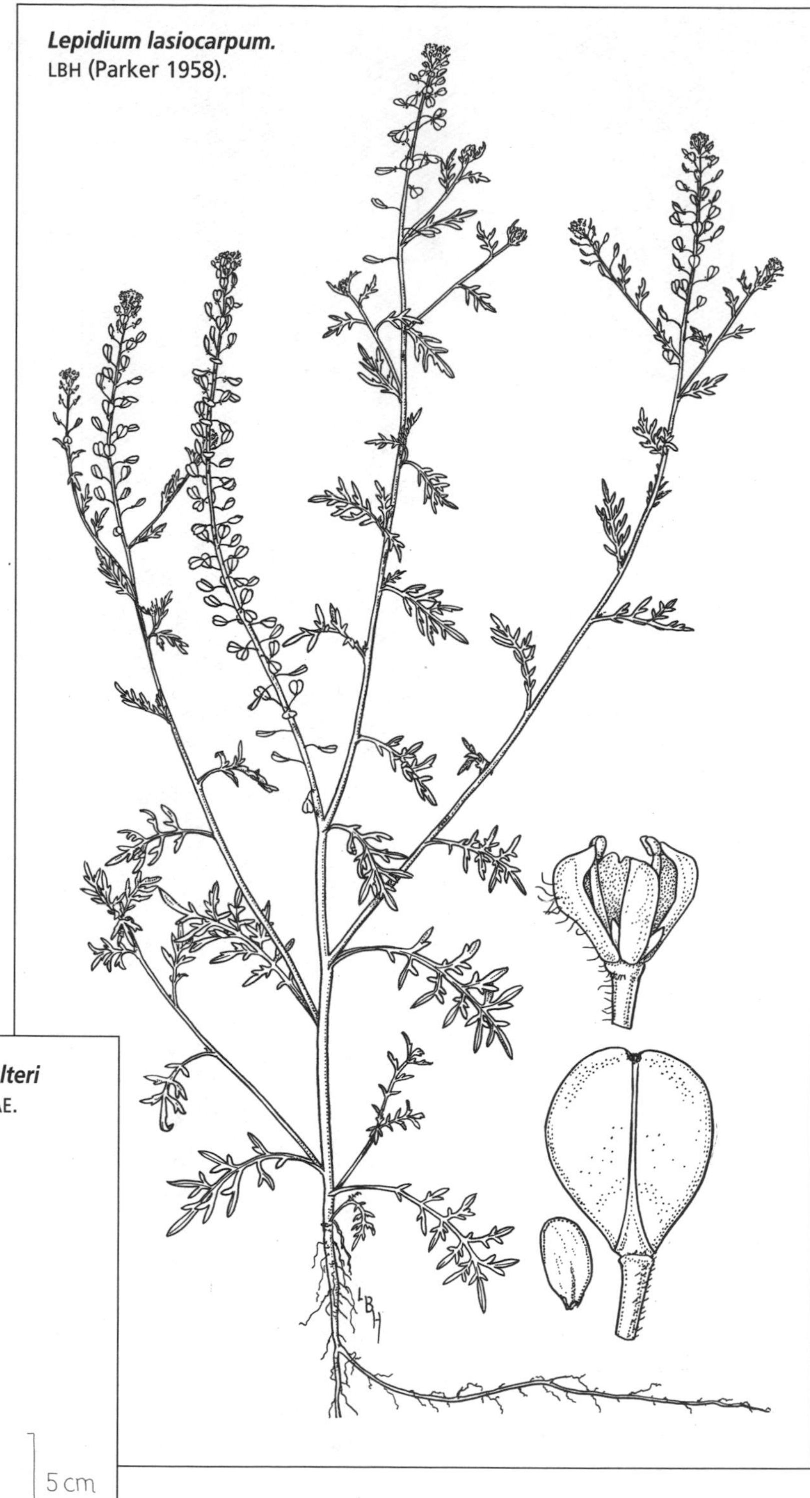

Lepidium lasiocarpum.
LBH (Parker 1958).

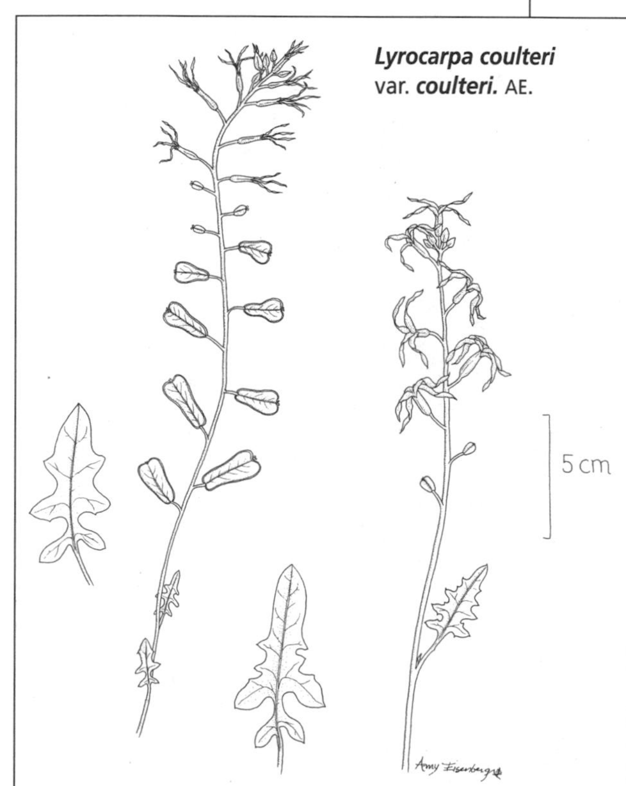

Lyrocarpa coulteri
var. ***coulteri.*** AE.

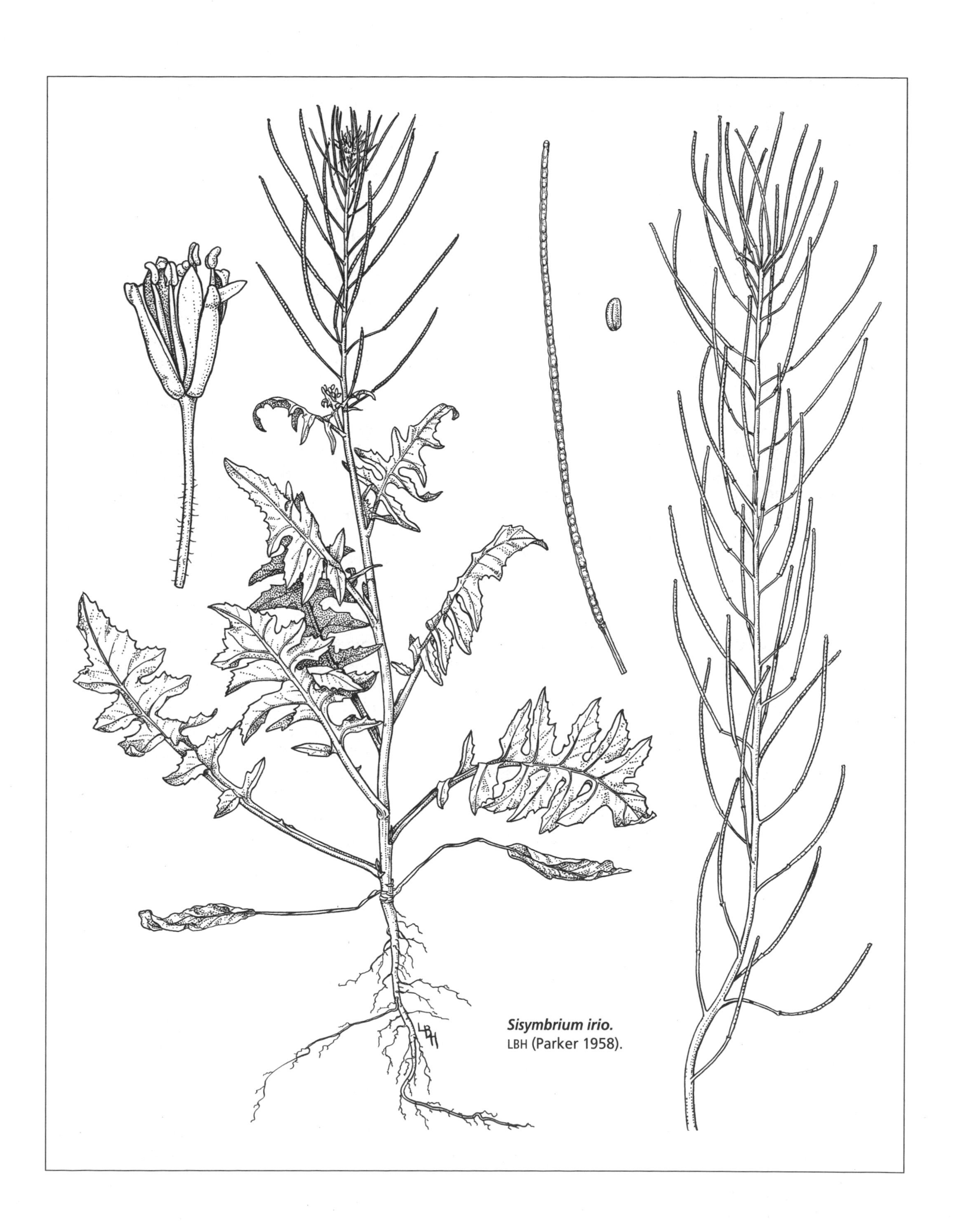

Sisymbrium irio.
LBH (Parker 1958).

***Brassica tournefortii** Gouan. *MOSTAZA;* SAHARA MUSTARD, WILD TURNIP; *MOSTA:S.* Coarse plants (15) 30–120 cm, with a well-developed taproot, the stems simple to many-branched above, the flowering branches spreading. Lower part of plant hirsute with coarse, rough white hairs, especially the lower leaf surfaces along the midrib, veins, and margins. Leaves of basal rosette (8) 15–30 (80) cm, petioled, pinnatifid with the terminal lobe usually largest (lyrate-pinnatifid), or leaves of stunted plants often obovate and merely toothed; stem leaves usually greatly reduced upwards. Sepals 3.5–4 mm, pale, almost translucent, drab purple-brown, slightly swollen basally. Petals, stamens, and stigma pale yellow. Petals 6–8 mm; corolla bilaterally symmetrical. (Unlike most other mustards, the petals are not at right angles to one another—2 of the opposite angles are less than 90° and 2 are much more than 90°.) Stamens at the same height as and touching the stigma. Fruiting pedicels (6) 12–16 (22) mm, spreading. Fruits straight, terete, 2.1–2.4 mm wide, 3.7–6 cm long including a well-developed beak 11–14 mm; beak 1- (2)-seeded. Seeds 1.3 mm in diameter, rounded, brown, mucilaginous when wet. February-May.

Well established in lowlands across the flora region: interdune troughs and dunes, especially lower dunes, crater floors, sandy flats, sandy-gravelly washes, especially abundant along roadsides, and sometimes on rockier soils and slopes. Native to the Old World, probably North Africa and Central Asia, now in warm, arid regions worldwide. It has spread almost explosively across the Sonoran Desert, especially in sandy-soil habitats. By 1938 it was well established in agricultural areas in southeastern California (Robbins, Bellue, & Ball 1951). The earliest Arizona record is 1957 from Yuma (Mason 1960), and Sonora records date from 1966. Since at least the 1970s it has been widespread in the flora area as well as in Baja California, southeastern California, southern Nevada, southwestern Arizona, and western Sonora southward to the Guaymas region. It has also reached southward to the lowlands of the Río Mayo region, and eastward to western Texas (Johnston 1990).

In open, sandy places in California "when wild turnip matures its heavy crop of seeds and becomes dry, it is easily snapped off . . . and moves with the wind as a tumbleweed" (Robbins, Bellue, & Ball 1951:216). In the flora area only larger, much-branched plants sometimes become tumbleweeds. Drought-stressed plants may produce leaves as small as 8 cm and be reproductive, whereas plants on sandy soils with sufficient soil moisture may grow to more than 1 meter across, making this species the largest herbaceous rosette plant in the region. The small, pale-colored flowers, the anomalous pattern of the petals, and stamens that touch the stigma point to probable self-fertilization. These features, together with the plasticity in plant size, undoubtedly contribute to its success in the Sonoran Desert. Zohary (1966:309) observed that in the Middle East the "desert specimens are more densely hispid and have smaller flowers; they are probably to be classed as a separate subspecies."

Bahía de la Cholla, *Vreeland 5 Mar 1966.* Moon Crater, 1970, *F 18826.* MacDougal Crater, 1986, *Turner 86-21.* Pinacate Junction, 1970, *F 19614.* Quitobaquito, 1978, *Bowers 1030.*

CAULANTHUS JEWEL FLOWER

Ephemerals to short-lived perennials; glabrous or sparsely pubescent, often glaucous, the stems leafy, hollow, sometimes swollen. Early leaves in a basal rosette. Flowers in elongated racemes. Fruits slender, much longer than wide, dehiscent. Seeds several to many, not mucilaginous. Western North America, mostly California; 14 species.

Caulanthus lasiophyllus (Hooker & Arnott) Payson var. **lasiophyllus** [*Turritis lasiophyllus* Hooker & Arnott. *Caulanthus lasiophyllus* var. *utahensis* (Rydberg) Payson. *Thelypodium lasiophyllum* (Hooker & Arnott) Greene] Erect, (8) 10–60 cm, mostly branched from above. Herbage, inflorescence branches, and buds with spreading, relatively soft, white hairs or less often glabrous. Leaves petioled, pinnatifid, the segments highly variable in shape, the larger leaves (2) 4–12 cm, the blades thin, the stem leaves well developed; plants often becoming leafless as fruits mature. Sepals greenish to white, pinkish, or purplish, often with sparse, spreading white hairs, especially near tips. Petals (2.3) 2.5–4.5 mm, narrowly oblong, white to pinkish. Filaments white to pinkish purple. Fruits (1.8) 2.5–3.5 (4.2) cm × 0.8–0.9 mm, straight, terete, spreading when young, later turning down, nearly sessile (pedicels 0.5–1 mm). Seeds ellipsoid to nearly quadrangular, 1.1 mm, the surface minutely patterned. Flowering mostly January-February; fruits mature mostly in March.

Widespread, including arroyos and washes, bajadas, and slopes. Often beneath larger shrubs or small trees such as ironwood, mesquite, and palo verde. Across much of northwestern Sonora including the Sonoyta region, granitic ranges, and Pinacate volcanic region to the peaks; not on dunes. Western Sonora southward to El Desemboque San Ignacio, and northward through Arizona to Utah, and Baja California to western Washington. One other variety in California and Baja California.

Sierra del Rosario, *F 20749*. Moon Crater, *F 88-81*. MacDougal Crater, *Turner 86-8*. Pinacate Peak, *F 19424*. Tinajas de Emilia, *F 19749*. Quitobaquito, *F 7673*.

DESCURAINIA TANSY MUSTARD

Annuals or biennials, sometimes herbaceous perennials; glabrous or with stellate, branched (dendritic), glandular, or simple hairs. Leaves pinnatifid or 1–3 times pinnate into small segments, the lower leaves petioled, the upper ones nearly sessile. Inflorescences often elongating as the fruits develop. Flowers small, white or yellow. Pedicels slender, the fruits terete, linear or club-shaped. Seeds mucilaginous when wet.

Mostly temperate North and South America, some in cold regions and few in deserts; 45 species. References: Detling 1939; Hitchcock & Cronquist 1964.

Descurainia pinnata (Walter) Britton [*Erysimum pinnatum* Walter] TANSY MUSTARD; *ṢU'UWAḌ*. Erect, 8–60 cm, mostly unbranched to few-branched above. Leaves and lower stems canescent with branched and candelabra-shaped, soft, white hairs. Stems leafy; early leaves usually in a basal rosette, these and larger stem leaves (1) 3–7 × (0.4) 2–3 cm, 1–3 times pinnatifid-divided, gradually reduced upwards, usually withering as the fruits develop. Inflorescences glabrous or glandular; racemes (3) 5–32 (45) cm. Flowers pale yellow, ca. 1.5 mm wide, petals and sepals (1) 1.5 mm. Pedicels spreading, 4–10 mm (lowermost occasionally to 12 mm). Fruits (3.5) 4–7.2 × 1.2–1.6 mm, narrowly club-shaped. Seeds 0.8 mm, ovoid, orange-brown, the surfaces minutely papillate in longitudinal lines.

Widespread; gravelly washes and arroyos, crater floors, sand flats, bajadas, and rocky slopes. Sonoyta region, granitic ranges and Pinacate volcanic field to the peaks, and desert flats; not on dunes.

Specimens from the flora area are mostly identifiable as subsp. *halictorum* (Cockerell) Detling characterized in part by leaves moderately to densely pubescent with soft white hairs, narrow leaf segments, stems and inflorescences often glandular, petals relatively small and pale yellow to whitish, and spreading fruiting pedicels. This subspecies is widespread in western North America to Arkansas and southward to northern Mexico. A number of other infraspecific taxa have been described but appear to be of varying significance. This species is widespread in North America except the Atlantic coast. Hitchcock & Cronquist (1964) identify a northern, mostly glabrous subspecies, which is replaced southward by a pubescent (canescent) subspecies; however, most floristic works do not address this simple but seemingly realistic scheme.

Sierra del Rosario, *F 75-5*. Arroyo Tule, *F 18768*. Sykes Crater, *F 18703*. S of Pinacate Peak, 975 m, *F 19935B*. Tinajas de Emilia, *F 19737*. 18 km SW of Sonoyta, *F 88-203*.

DIMORPHOCARPA SPECTACLE-POD

Annuals, biennials, or herbaceous perennials, densely pubescent with candelabra-shaped hairs. Leaves petioled below, sessile above. Pedicels slender and elongated. Petals white to pale lavender, obovate, abruptly narrowed to a claw. Fruits more or less spectacle-shaped, of 2 halves (valves), each half single-seeded, indehiscent, broadly oblong to orbicular, strongly flattened. Chromosome number based on $x = 9$. Southwestern United States and northern Mexico; 4 species. In segregating *Dimorphocarpa* from *Dithyrea*, Rollins (1979) listed more than 10 major characters distinguishing them.

Dimorphocarpa pinnatifida Rollins, *Publ. Bussey Institution*, Harvard Univ., 28, 1979. Stems (15) 30–60 (75) cm, at first upright, then sprawling. Herbage, inflorescence axis, pedicels, and sepals gray-green with dense pubescence of white hairs. Stems leafy, the upper and lower leaves similar, the lowermost ones usually withered by fruiting time. Leaves linear to narrowly lanceolate, pinnatifid with widely spaced slender lobes, larger leaves (3.5) 4–18 × 0.7–2.6 cm. Racemes 9–38 cm. Flowers white, very fragrant. Sepals spreading or moderately reflexed. Petals mostly 5–7 × 3.5–4.5 mm, broadly obovate, clawed. Fruiting pedicels (7) 10–20 mm. Fruit segments (one of the pair sometimes falling at maturity) each 5.5–8.2 mm wide and joined laterally for about 15% of their margins (the whole fruit notched above and below), the surfaces glabrous or sparsely to densely hairy with forked, branched, and stellate hairs, the margins only moderately thickened and no more densely hairy than the faces.

Sand flats to high dunes; sometimes abundant, often with *Dithyrea californica.* Puerto Peñasco to dunes near Bahía Adair and northward to ca. 20 km south of Sonoyta and the Gran Desierto dunes to nearby southwestern Arizona in the vicinity of Yuma, the Pinta Sands, and Mohawk Dunes. The flowers are smaller and the plants often much larger than those of *D. californica.*

Dunes NE of Sierra del Rosario, *F 20429* (holotype, GH). Microwave tower, 2 km N, 13 km W of Los Vidrios, *Burgess 5598.* Moon Crater, *F 19096.* MacDougal Crater, *F 9923.* 21 mi SW of Sonoyta, *Webster 22394.* E of Estación López Collada, *F 84-13.*

DITHYREA SPECTACLE-POD

Two species; *D. maritima,* a herbaceous perennial, occurs in coastal southern California and Baja California. Distinguished in part from *Dimorphocarpa* (the other spectacle pod) by having a chromosome number based on $x = 10$, coarser leaf lobes, stouter and shorter pedicels, and larger flowers.

Dithyrea californica Harvey [*D. californica* var. *clinata* (Macbride & Payson) Wiggins. *D. clinata* Macbride & Payson] Winter-spring ephemerals. Stems 15–50 cm, erect to ascending, mostly branching from the base. Herbage with slender-rayed, candelabra-shaped white hairs. Leaves semisucculent when young, with large, broad, coarse teeth or lobes, the lower leaves 4.5–15.5 cm, petioled, more or less oblanceolate to oblong, the upper leaves sessile, ovate to broadly oblong. Flowers highly fragrant, 12–14 mm wide, the petals ca. 10–12 mm, cream-white, strap-shaped, partially recurved, twisted, and crenulate (wavy). Fruiting pedicels 1.5–2.5 mm. Fruits more or less spectacle-shaped, of 2 halves (valves), each half 4.5–6.5 mm wide, indehiscent, orbicular to slightly wider than long, strongly flattened, and single-seeded, with simple, thickened (saclike or vesicular) hairs especially dense on the cordlike margin.

Sand flats and dunes across much of the flora area, but not in the Sonoyta region; often beneath creosote bushes on small hummocks. Coastal Sonora south to El Desemboque San Ignacio, western Arizona, southeastern California, southern Nevada, and both Baja California states; Sonoran and Mojave Deserts.

Sierra del Rosario, *F 20780.* Moon Crater, *F 19036.* Hornaday Mts, *Burgess 6820.* 29 km SW of Sonoyta, *Burgess 4765.* Rd to Rancho Guadalupe Victoria [Grijalva], *F 20582.* N of El Golfo, *F 75-74.*

DRABA

Annuals and biennials to long-lived herbaceous perennials. Pubescence of simple, forked, stellate, or branched hairs. Stems leafy or not. Leaves entire to toothed. Annual species often with dimorphic flowers. Fruits linear to ovate, strongly compressed to inflated. Seeds usually numerous.

Mostly North America and Eurasia, temperate and arctic regions; 300 species. References: Hartman, Bacon, & Bohnsteadt 1975; Rollins 1984.

Draba cuneifolia Nuttall ex Torrey & A. Gray. Delicate, small plants with forked and stellate hairs. Leaves 5.5–45 × 2–18 mm, in a basal rosette, sessile, thin, broadly obovate with (1) 2 or 3 pairs of shallow, coarse teeth. Flowering stems leafless; inflorescences (1.8) 3–15 cm. Flowers of two extremes, sometimes even on same plant, the smaller flowers with petals 1.2–1.5 mm, the larger flowers with petals to 3 mm; sepals and petals very quickly deciduous, the petals white. Fruits (4.5) 6–12 × (1.6) 2–2.7 mm, laterally flattened, with 2- or 3-rayed stellate to mostly short-stalked hairs. Fruiting pedicels 1.5–4.5 (6) mm. Seeds 0.6–0.7 mm, red-brown, somewhat compressed, mucilaginous when wet.

Widespread, especially in slightly protected or shaded microhabitats beneath desert legume trees or on the shady side of trees, rocks, or slopes; washes, sandy plains, bajadas, cinder flats and slopes, and sometimes on rocky slopes. Sonoyta region, Pinacate volcanic field, at least some granitic ranges, and desert flats.

Much of the United States and Mexico to Zacatecas, and Sonora south to Alamos. Rollins (1993) recognized 3 varieties, 2 in Sonora, but these not always well marked in the Sonoran Desert region and are perhaps not worthy of recognition. Many specimens from the flora area have the larger fruits characteristic of var. *integrifolia* S. Watson but with the fruits extending nearly to the base of the inflorescence axis as in var. *sonorae* (Greene) Parish. This species is distinguished from other Sonoran Desert crucifers by its broad and shallowly toothed basal leaves, small and leafless flowering stems, and relatively short, broad, and laterally flattened fruits.

Sykes Crater, *F 18944*. Arroyo Tule, *F 18752*. Pinacate Peak, *F 19346*. Elegante Crater, *F 19700*. Sierra Cipriano, *F 88-201*. Quitobaquito, *Nichol 10 Mar 1939*.

ERUCA

Annual or perennial herbs. Flowers in racemes. Sepals erect, the petals clawed with oblanceolate to broadly obovate blades and prominent veins. Fruiting pedicels erect to appressed to the raceme; fruits rather stout, longer than wide, erect, with a prominent flattened, seedless beak. Seeds in 2 rows in each chamber, mucilaginous when wet. Eurasia; 3 species.

***Eruca vesicaria** (Linnaeus) Cavanilles subsp. **sativa** (Miller) Thellung [*E. sativa* Miller] GARDEN ROCKET, SALAD ROCKET, ARRUGALA. Plants (15) 50–70+ cm, usually hispid with simple white hairs. Early leaves in a basal rosette, 5.5–20+ × 3–10+ cm. Petals 17 mm, obovate, pale yellow often fading white, with purple veins. Fruiting pedicels 4.5–6.7 mm; fruit body 17–20 mm, the beak 6–7 mm.

Sonoyta region and at San Luis; scattered to sometimes locally common near roadsides and as urban and agricultural weeds. Sometimes abundant and extensive along highways and adjacent natural habitats in southwestern Arizona. Native to the Mediterranean region, now widespread and weedy from southern Mexico to Canada. Probably a recent arrival in northwestern Sonora, although it was present in Arizona and southern Sonora in the early twentieth century. Used as a spicy salad green, especially in Italy.

10 km W of Sonoyta, *F 92-384*. S of Sonoyta, *F 92-399*. 2 km W of San Luis, *F 93-207*. Ontogata, Yaqui Valley, introduced weed, *Mackie in 1912* (UC).

LEPIDIUM PEPPERGRASS

Annual or perennial herbs or subshrubs, glabrous or with simple hairs. Flowers small, in racemes or panicles. Pedicels spreading, terete to flattened. Petals white, yellow, pink, greenish, or reduced or absent. Stamens 2, 4, or 6. Fruits orbicular to oblong, laterally compressed, dehiscent or not, often winged at tip and notched between wing lobes; seeds 1 (2) per chamber, highly mucilaginous when wet. Nearly worldwide, mostly warm-temperate regions; 175 species. References: Al-Shehbaz 1986; Hitchcock 1936, 1945.

Lepidium lasiocarpum Nuttall ex Torrey & A. Gray. SAND PEPPERGRASS; *KA:KOWANI*. Plants 5–20 (40) cm, larger plants much-branched. Herbage and inflorescences with simple, spreading, white, rather thickish hairs less than 0.4 mm; fresh herbage with a chili-like flavor. Early leaves in a basal rosette, 2.5–6+ cm, oblanceolate, bipinnatifid, quickly withering as stems develop; stem leaves smaller, oblanceolate, highly variable, usually withering as the plants mature.

Racemes mostly 2–10+ cm, numerous and often crowded on larger plants. Pedicels 1.6–3 × 0.5 mm, conspicuously flattened (even on buds), glabrous or pubescent. Flowers inconspicuous, autogamous (selfing); sepals 1.0–1.5 mm, greenish or whitish with pinkish lavender midstripes; petals white, quickly deciduous, smaller and thinner than the sepals, rudimentary, or sometimes absent; stamens 2

(4 or 6; abortive stamens often present on flowers with fewer than 6 stamens). Fruits nearly orbicular, slightly longer than wide, strongly compressed (flattened), 3–4 mm, ciliolate along margins to sparsely hairy on sides, the stigma sessile, at base of apical notch. Seeds flattened, orange-brown, obovoid, 1.5–1.7 mm, with a narrow white translucent wing or margin. Sometimes flowering as early as mid-October.

Playas, washes and arroyos, crater floors, desert plains, cinder flats and slopes, and less often on other rocky slopes; widespread, at all elevations; not on dunes. Also a common weed and in disturbed habitats. Southwestern United States and northern Mexico. Rollins (1993) recognizes 4 varieties, but for the most part they appear only weakly differentiated. Ours more or less fit into the widespread var. *latifolium* C.L. Hitchcock [var. *georginum* (Rydberg) C.L. Hitchcock].

22 mi W of Sonoyta, *F 7641.* Tinaja del Tule, *F 18749.* 8 mi W of Los Vidrios, *F 18707.* Sykes Crater, *F 18916.* Sierra Pinacate, 875 m, *F 19323.* Quitobaquito, *F 7679.*

LESQUERELLA BLADDERPOD

Annual, biennial, or perennial herbs. Usually densely pubescent, mostly with branched or stellate hairs, the plants sometimes silvery from a covering of lepidote-stellate hairs that are spectacular under high magnification. Leaves in basal rosettes and along stems. Flowers mostly yellow. Fruits usually globose, several– to many-seeded. Seeds nearly orbicular, flattened to plump, mostly lacking a margin, mucilaginous or not when wet.

North America with some in southern South America and few in arctic Asia; 105 species (including *Physaria*). Mostly temperate to arctic and often prominent in deserts. References: O'Kane, Al-Shehbaz, & Turland 1999; Rollins & Banerjee 1975; Rollins & Shaw 1973.

Lesquerella tenella A. Nelson [*L. gordonii* (A. Gray) S. Watson var. *sessilis* S. Watson. *L. palmeri* of authors. Not *L. palmeri* S. Watson] Winter-spring ephemerals. Densely stellate pubescent and also with some simple hairs. Stems 15–60 cm, erect to semi-decumbent or clambering through small shrubs (such as *Ambrosia dumosa*), occasionally reaching 1 m across. Basal leaves and lower stem leaves often 3–9 cm, petioled, the blades narrowly elliptic to obovate, entire to wavy, or shallowly or sometimes coarsely toothed; other stem leaves mostly elliptic to linear, usually entire and sessile above. Racemes (3) 9–20 cm, the flowers rather widely spaced, bright yellow, showy, ca. 9–10 mm wide; petals 8–10+ mm. Fruiting pedicels S-shaped, often (10) 15–18 mm. Fruits globose, 3.5–4.8 mm wide. Seeds red-brown, flattened, 1.8–2.4 mm in diameter, the margin a thickened, cordlike rim or narrow wing; mucilaginous when wet, especially from the rim margin. February-March, fruiting in March.

Sand and cinder flats on the northeast side of the Pinacate field to the Sonoyta region. Southwestern Utah, southern Nevada, southeastern California, western and southern Arizona, New Mexico, northern Sonora, and Baja California.

Lesquerella tenella is distinguished from its close relative *L. gordonii* by the stellate hairs of the ovary and fruits, and by the margined seeds; otherwise they are remarkably similar. The character of short-stiped (stalked) fruits in *L. gordonii* versus sessile fruits in *L. tenella* is not a practical key character. Although the karyotype of plants from northwestern Sonora is unknown, elsewhere *L. tenella* has $2n = 10$, whereas in *L. gordonii* $2n = 12$. Rollins & Banerjee (1975) show that the stellate hairs of the 2 species are distinct but more similar to each other than to other species. In general *L. tenella* replaces *L. gordonii* in western, especially southwestern, Arizona; *L. gordonii* does not range into California. Although their geographic ranges broadly overlap, they seem largely to occupy different habitats.

S of Tezontle cinder mine, *F 88-55.* 10 km SW of Sonoyta, *F 88-144.*

LYROCARPA LYREPOD

Annual or perennial herbs with dendritic hairs. Petals relatively large. Fruits green, compressed, relatively large, usually widest near apex. Northwestern Mexico, southwestern Arizona, and southeastern California; 3 species.

Lyrocarpa coulteri Hooker & Harvey ex Harvey var. **coulteri.** LYREPOD; *BAN CEÑṢAÑĬG.* Perennial herbs, sometimes flowering in the first season. Stems slender and brittle, ca. 30–75 cm. Herbage, sepals, and fruits densely covered with white, stellate, or candelabra-shaped hairs. Leaves simple, petioled, variable depending on shading, soil moisture, and position on stem, the blades thinner during wet periods; leaves mostly 2.5–7 (11) × 1.2–5 cm, pinnately lobed to deeply divided, mostly with (2) 3 (4) pairs of major segments, these sometimes again toothed or lobed; lower leaves usually with shallower lobes, the upper leaves, especially during drier conditions, becoming deeply divided.

Racemes (4) 7–45 cm, flowers highly fragrant at night. Calyx 9–11 mm. Petals often twisted, linear-attenuate, extending 12–20 mm beyond the calyx, 1.7–2 (2.5) mm wide, yellow- to purple-brown, often chartreuse at first or purple-brown at tips or along margins, becoming darker with age. Flowering and fruiting pedicels 3–14 mm. Fruits lyre-shaped, green, compressed but relatively thick, widest at top, 10–28 × 8–14 mm. Seeds 2.4–2.8 (3) mm, brown, somewhat irregularly flattened, oval to quadrangular, not mucilaginous. Growing and flowering at various seasons with sufficient moisture.

Washes, arroyos, canyons, rocky slopes, bajadas, flats, and plains; sandy, gravelly, and cinder soils, often beneath legume trees and shrubs. Sonoyta region, at least some granitic ranges, Pinacate volcanic field to near peak elevation, and desert plains. Var. *coulteri* from near Guaymas in western Sonora to southern Arizona and east-central Baja California. Two other rather weakly differentiated varieties in southeastern California and both Baja California states. The generic name derives from the lyre-shaped fruits.

S of Mex 2 [rd to Rancho Guadalupe Victoria], *Mason 1816.* Sierra Extraña, *F 19082.* Moon Crater, *F 88-85.* Sierra Pinacate, 1240 m, *F 18669.* Tinajas de Emilia, *F 19744.* 11 mi SW of Sonoyta, *F 9804.*

SISYMBRIUM

Annual, biennial, or perennial herbs; glabrous or with simple hairs. Flowers usually yellow. Fruits narrowly linear, terete or slightly flattened. Temperate regions of the world; 80 species.

***Sisymbrium irio** Linnaeus. *PAMITA;* LONDON ROCKET; *BAN CEÑṢAÑĬG.* Erect, (7) 30–60 cm, glabrous or sparsely pubescent on part of the herbage and pedicels. Early leaves in a rosette, the stem leaves well developed but reduced upwards. Leaves petioled, pinnatifid, the larger ones (3) 7–20 cm, the blades thin. Flowering stems usually branched. Sepals green; petals, filaments, and anthers yellow. Petals 3–4 mm. Pedicels very slender, spreading, 5–14 mm. Fruits (1.8) 3–5 cm × 0.8–1.3 mm, spreading. Seeds 0.7–0.9 mm, ellipsoid, yellow-orange, not mucilaginous.

Low-lying, disturbed habitats, especially as urban and agricultural weeds, and less often along roadsides. Rarely in natural desert and then usually near habitation or as a result of human activity. Recorded from Organ Pipe Monument in 1933 and in Tucson in 1908 (Felger 1990). Native to the Old World, now widespread.

Puerto Peñasco, *Dennis 14–16 Feb 1965.* Río Sonoyta S of El Papalote, *F 86-158.* Pinacate Junction, *F 19612.*

STREPTANTHELLA

The genus has a single species.

Streptanthella longirostris (S. Watson) Rydberg [*Arabis longirostris* S. Watson. *Streptanthella longirostris* var. *derelicta* J.T. Howell (see Rollins 1993)] LONG-BEAKED TWIST FLOWER. Plants upright, glabrous, 18–60 cm, simple to much-branched, the branches ascending. Stems leafy; fresh herbage bluish glaucous, the leaves often thick. Lower leaves largest, narrowly lanceolate, 2.5–10.5 × 0.1–2.2 cm, mostly pinnately parted with a few widely spaced slender lobes; upper leaves linear and entire. Racemes 4–20+ cm. Flower buds with a few appressed 2-armed (malpighian) hairs. Sepals ca. 3.2–3.5 mm, green to purple-green with white margins. Petals ca. 4–5 mm, linear, pale yellow with maroon veins. Filaments white to colorless with purple tinges. Fruits slender, terete, 2.4–6.5 cm × 1.4–1.6 mm, at maturity usually turned down on recurved pedicels 1.2–3 mm. Seeds not mucilaginous. February.

Sandy soils of washes, flats, and low stabilized dunes; northeastern side of Pinacate volcanic field, and more common to the northwest and west of the volcanic field. Eastern Washington to Wyoming, Utah, Baja California, and northern Sonora.

Readily recognized by the glaucous herbage, often thickish leaves, small purple-green and white flowers, and slender, down-turned, short-pedicelled fruits. Rollins (1993) indicates that var. *derelicta* is probably not a distinct taxon. Plants from the flora area show characters of both varieties.

Sierra del Rosario, *F 20749*. MacDougal Crater, *F 9700*. Rd to Pozo Nuevo, *F 88-67*. SSW of Tezontle Cinder Mine, *F 88-57*.

THYSANOCARPUS LACEPOD, FRINGEPOD

Annuals. Stems leafy; leaves petioled below, sessile above. Flowers minute, white to purplish in slender racemes elongating as the fruits develop; pedicels curved. Fruits 1-seeded, flattened, indehiscent, rounded in outline, with conspicuous, entire, crenately notched, or perforated wings; seeds not mucilaginous. Western North America; 4 species.

Thysanocarpus curvipes Hooker [*T. curvipes* var. *elegans* (Fischer & C.A. Meyer) Robinson] Stems mostly erect, often 15–35 cm, single to few-branched. Glabrous or sparsely pubescent with somewhat coarse, white hairs on leaves. Leaves linear-oblong or oblanceolate with few marginal teeth, the lower leaves petioled and usually in a rosette, the stem leaves sessile and clasping. Pedicels slender. Fruits nearly orbicular, slightly longer than wide, often (5.5) 6.5–7.3 mm, flat and thin, the wing usually perforated with small, evenly spaced holes.

In the northeastern corner of the flora area on the granitic bajada of the Sierra Cipriano (*F 88-196*) and in Organ Pipe Monument. British Columbia to Baja California, Arizona, New Mexico, and northern Sonora.

BURSERACEAE FRANKINCENSE OR TORCHWOOD FAMILY

Trees and shrubs with prominent resin ducts containing aromatic terpenes and essential oils. Leaves usually alternate and pinnate to tripinnate. Flowers small, radial, often unisexual.

Tropical and subtropical regions of the Americas, Asia, and Africa; 20 genera, 600 species. *Bursera* is the largest genus. Frankincense is derived from the resin of *Boswellia carteri* and related species, and myrrh from *Commiphora abyssinica* and related species. All members of the family are apparently frost sensitive. The family is most closely related to the Anacardiaceae.

BURSERA

Trees and shrubs with soft wood. Leaves without stipules. Mostly dioecious or with some bisexual flowers (polygamodioecious). Flowers small; stamens twice as many as petals (with rare exception). Fruits small, drupelike but the exocarp at maturity separating into 2 or 3 segments or valves. Seeds partially to fully enveloped in a thin, pleasant-tasting, aril-like mesocarp (pseudoaril). *Bursera* is distinguished from other members of the family by having a calyx that opens in the bud and a valved, 1-seeded, dehiscent fruit.

Americas, mostly semiarid subtropical scrub; 100 species, mostly in Mexico, with maximum diversity along the Pacific slope, where more than 80 species occur and 70 are endemic. Nine species in southern Sonora, 4 in the Guaymas region, and 1 in the desert in northwestern Sonora and adjacent Arizona and southeastern California.

The aromatic foliage and gum, or copal, have a long history of religious and medicinal uses. References: Becerra & Venable 1999; Johnson 1992; Kohlmann & Sánchez-Colón 1984; McVaugh & Rzedowski 1965; Rzedowski & Kruse 1979.

Bursera microphylla A. Gray. *TOROTE;* ELEPHANT TREE; *'UṢABAKAM.* Glabrous, short-trunked, bonsai shrubs or trees (1) 2–4 m; trunk and lower limbs fat and semisucculent, often contorted into bizarre shapes. Bark on twigs and smaller branches red-brown, becoming papery on larger limbs and trunks and peeling in large flakes or sheets, especially during spring dry season. Long shoots with relatively long internodes. Leaves mostly crowded at ends of short shoots, drought deciduous, bright green, pinnate, mostly (1.5) 2–4 cm, the rachis very narrowly winged; leaflets more or less linear-oblong to oval. Probably functionally dioecious. Flowers solitary or 2-several on slender stalks, cream-white or pale greenish yellow, ca. 2 mm wide; sepals and petals usually 3 each on female flowers, and (4) 5 each on male flowers. Fruits 7–9 mm, dull purple-brown and glaucous when fresh, the 3 leathery carpels splitting apart when fully ripe to reveal a red pseudoaril enveloping the seed. Seeds 6 mm, 3-angled. Flowering with the emerging leaves in the first summer rains.

Rocky slopes, hills, and mountains to peaks in granitic ranges and mostly below ca. 700 m in the Sierra Pinacate. Various exposures including south- and west-facing slopes, but sometimes restricted to north-facing slopes (as on crater walls). Sonoran Desert: western margin of the Colorado Desert in California, southern Arizona, and southward through Baja California Sur and western Sonora to the vicinity of Guaymas.

Locally limited by freezing weather; extensive freeze damage evident on most of the local burseras. Southward in Sonora, and especially on the Baja California Peninsula, there is a clinal increase in the height of the trunk (height of first branching). In the Cape Region of Baja California Sur *B. microphylla* is a tree often 8–10 m with a tall, well-developed trunk. There are several geographic races probably worthy of taxonomic status (e.g., Mooney & Emboden 1968). This species extends into drier habitats than does any other member of the genus.

Foliage may appear several times per year at any season following even meager rainfall, except during the coldest weather. Deeply cut branches or roots ooze blood-red sap. The sap is especially rich in terpenes and crushed leaves are highly aromatic. When a leaf is broken off a turgid stem, a tiny jet of aromatic sap may squirt from the wound. This species has been an important medicinal and ritual plant (e.g., Bean & Saubel 1972; Felger & Moser 1985).

There are several similar-appearing and related but geographically disjunct species, e.g., *B. morelensis* Ramírez of Morelos and Puebla. The Baja California elephant tree, *Pachycormus discolor* in the Anacardiaceae, shows convergent features.

Sierra del Viejo, *F 5733.* Sierra del Rosario, *F 75-6.* MacDougal Crater, *Kamb 1997.* Tinajas de Emilia, *F 19723.* S of Pinacate Peak, 875 m, *F 19318.*

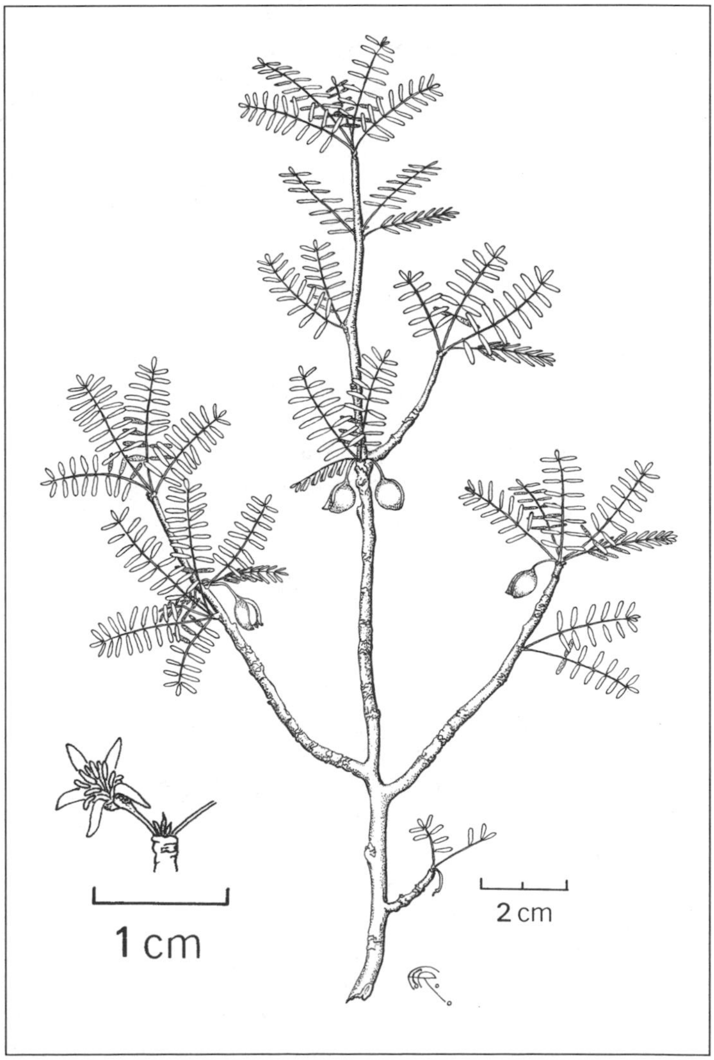

Bursera microphylla.
FR (Felger & Moser 1985).

CACTACEAE CACTUS FAMILY

By Richard S. Felger & Allan D. Zimmerman

Perennials, from several centimeters to tree heights. Stems generally very succulent; usually spiny. Spines, flowers, and branches arising from modified short shoots called areoles. Spines often differentiated into central and radial spines. Leaves (in ours) absent, or reduced and deciduous, or (elsewhere) a few genera with "normal" leaves. Tepals (perianth segments) usually numerous and intergrading from the outer tepals (sepal-like segments) to the inner tepals (petal-like segments). Stamens numerous. Ovary almost always inferior (ours) and deeply embedded in the receptacle, which often bears areoles and/or scalelike bracts. (The proximal part of the receptacle, covering the ovary, is called the pericarpel; the distal part of the receptacle is fused to the hypanthium and the stamens on its inner surface and is the floral tube.) Fruits mostly succulent, sometimes dry at maturity; seeds numerous.

Essentially New World; 122 genera, 1150+ species (Gibson & Nobel 1986). Conspicuous in hot deserts, but greatest development toward desert margins and in semiarid subtropical regions; some in the humid tropics and a minority in dry temperate regions.

Most cacti have succulent, edible fruits, although some taste better than others. These sweet, succulent tissues entice animals to consume all or part of the fruit and disperse the seeds. Sonoran Desert people used the sweet fruits and seeds of certain cacti as major food resources and prepared wine from fruits of columnar cacti such as saguaro and organpipe (Castetter & Bell 1937; Crosswhite 1980; Felger & Moser 1985).

Relatively few cacti have dry fruits. Among these species the fruits dry quickly at about the time of seed maturation, whereas fruits of most cacti remain fleshy at time of seed dispersal. Evolution of dry fruits has occurred independently among diverse cactus genera, mostly in arid and temperate regions. Six cacti in the flora area have dry fruits: *Echinocactus polycephalus, Echinomastus erectocentrus, Opuntia acanthocarpa, O. basilaris, O. echinocarpa,* and *O. ramosissima.*

Geographic affinities of our cactus flora seem to be primarily subtropical, with proto-Sonoran Desert affinities, and a minor element has inland "northern" affinities, e.g., Mojave and Chihuahuan Deserts. *Carnegiea, Lophocereus, Peniocereus striatus,* and *Stenocereus* have southern, subtropical affinities; the major factors limiting their distributions within the flora area seem to be the quantity and reliability of summer rainfall and, secondarily, freezing temperatures. In contrast, a number of cacti reach their southern limits in the flora area. These are species or species groups with inland desert affinities: *Echinocactus polycephalus, Echinomastus erectocentrus, Mammillaria tetrancistra, Opuntia basilaris,* and *O. ramosissima.*

The cactus flora, in both the flora area, with 26 species, and in the Sonoran Desert as a whole, with about 140 species, accounts for about 6% of the native flora. Although many are endemic to the Sonoran Desert, none is endemic to the flora area. Among our 26 cactus species, 18 to 20 occur in the Sonoyta region in Arizona Upland (Appendix A). The number of species declines westward and southward in the Lower Colorado Valley toward lower elevations with lower and less predictable summer rainfall. Only 5 cactus species extend to the Sierra del Rosario, and 3 of them are locally rare (Felger 1980). Even *Opuntia bigelovii,* present on all other ranges in the flora area, is absent from the extremely arid Sierra del Rosario. The Gran Desierto dune system is devoid of cacti except occasional, stunted plants of *Opuntia echinocarpa* on low semistabilized dunes.

The transition between Shreve's (1951) Arizona Upland and Lower Colorado Valley phytogeographic divisions seems to be a real biogeographic boundary for cacti as well as other plant families (Appendix A). Within the flora area 6 cacti are confined to Arizona Upland (Sonoyta region): *Echinocereus engelmannii* var. *acicularis, E. nicholii, Echinomastus erectocentrus, Ferocactus emoryi, Opuntia engelmannii,* and *Peniocereus striatus.* Three others, *Ferocactus wislizeni, Opuntia acanthocarpa,* and *O. arbuscula,* occur primarily in Arizona Upland. *Echinocereus engelmannii* var. *acicularis* intergrades with var. *chrysocentrus* along the boundary between Arizona Upland and Lower Colorado Valley. *Opuntia acanthocarpa* of the Arizona Upland in the Sonoyta region is abruptly replaced by *O. echinocarpa* near the Lower Colorado Valley boundary (although *O. acanthocarpa* also occurs in the granitic mountains in the northwestern part of the flora area, those populations are of var. *coloradensis*).

Lophocereus schottii and *Stenocereus thurberi* grow on the eastern side of the Pinacate volcanic field but are absent from the drier, western side. *Opuntia chlorotica* is restricted to higher elevations in the Pinacate volcanic field, and several others thrive in the lowlands as well as at higher elevations, e.g., *Carnegiea, Echinocereus engelmannii, Ferocactus cylindraceus, Mammillaria grahamii,* and *Opuntia bigelovii.*

The greatest number of cactus species recorded at a single site in southwestern North America is from Big Bend National Park in Texas, where approximately 18 species may grow together on limestone at about 470 m (Zimmerman, field notes). In Central Arizona (e.g, near Phoenix), one can find 10 species on a rocky hill, but it may take considerable searching to confirm the presence of the rarest ones. In the flora area 8 to 10 is the maximum number of species growing intermingled; e.g., on rocky, granitic slopes at Quitobaquito: *Carnegiea gigantea, Echinocereus engelmannii, E. nicholii, Ferocactus cylindraceus, Mammillaria grahamii, Opuntia acanthocarpa, O. bigelovii, O. fulgida, O. leptocaulis,* and *Stenocereus thurberi.* Five others grow within 1 km: *Ferocactus emoryi, F. wislizeni, Mammillaria thornberi, Opuntia engelmannii,* and *O. kunzei.* The Quitobaquito region, an area of approximately 3.5 ha, supports 18 cactus species (Felger et al. 1992). Nine species occur together in low, granitic hills to the north of Sierra Cipriano: *Carnegiea gigantea, Echinocereus engelmannii, Ferocactus cylindraceus, Mammillaria grahamii, Opuntia acanthocarpa, O. bigelovii, O. fulgida, O. leptocaulis,* and *Stenocereus thurberi.*

Seven species grow together on the *Larrea*-cactus flats north of Playa Díaz: *Carnegiea gigantea, Echinocereus engelmannii, Ferocactus wislizeni, Opuntia arbuscula, O. fulgida, O. kunzei,* and *O. ramosissima.* However, these cacti probably comprise less than 10% of the total vegetation cover. There, *O. ramosissima* accounts for an estimated 70% of the cactus cover, *O. kunzei* for about 20%, and the other 5 species for roughly 10%. Elsewhere on this open, sandy plain (which we refer to as "cactus flat") to the northeast of Cerro Colorado and Playa Díaz, about 45 km west from Sonoyta, are *Opuntia leptocaulis,* one known plant of *Peniocereus greggii,* and noteworthy and extensive populations of *Carnegiea, Echinocereus engelmannii* var. *chrysocentrus,* and *Opuntia kunzei.*

Many cactus populations in northwestern Sonora, especially in the Lower Colorado Valley region, appear to be relicts at the edges of their eco-geographic distributions. The most abundant and widespread cactus in the flora area is *Opuntia bigelovii.* Other cacti with extensive populations in the flora area are *Echinocactus polycephalus, Ferocactus cylindraceus, Mammillaria grahamii, M. tetrancistra, Opuntia echinocarpa, O. fulgida,* and *O. kunzei.* However, even these are often present in patchy distributions.

Unless otherwise stated, the descriptions and keys are based on mature plants. References: Benson 1969a, 1969b, 1982; Bravo-Hollis 1978; Bravo-Hollis & Sánchez Mejorada 1989, 1991a, 1991b; Britton & Rose 1919–1923; Castetter & Bell 1937; Gibson & Horak 1978; Gibson & Nobel 1986; Gibson et al. 1986; Grant & Grant 1979a; IOS Working Party 1986; Nobel 1988.

1. Stems usually at least 2 m, not constricted into joints or pads.
   2. Usually more than 5 m; trunk well developed, the branches none or few, usually branching more than 2 m above the ground; upper (fertile) growth with 19–25 ribs. ____ **Carnegiea**
   2′ Usually less than 4–5 (8) m; essentially trunkless, much-branched from near base; fertile growth with 5–17 ribs.
      3. Stem ribs 5–7 (8); sterile stem parts with short, thick spines, the fertile parts with long, slender, twisted, and often flattened spines. ____ **Lophocereus**
      3′ Stem ribs (14) 15–18; spines essentially uniform throughout, usually straight and terete. ____ **Stenocereus**

1′ Stems much less than 2 m, or if 2–5 m (some chollas) then constricted into many joints.
   4. Barrel cacti, usually unbranched (rarely 1- or 2-branched from base); stem more than 20 cm diameter; larger spines stout and glabrous or with very minute non-overlapping hairs; flowers and fruits hairless. ____ **Ferocactus**
   4′ Growth forms various, but the plants not large, unbranched barrel-cacti; individual stems less than 15 cm in diameter; spines not very thick and rigid, or if so (*Echinocactus*) then microscopically tomentose and flowers and fruits conspicuously woolly and prickly.
      5. Spines very stout and rigid, the surfaces covered with minute, overlapping white hairs (use lens); exterior of flowers and fruits copiously white woolly; plants forming dense mounds of multiple stems, each 9–12 cm in diameter. ____ **Echinocactus**
      5′ Spines usually not especially stout, the surfaces not covered with hairs; flowers and fruits not woolly; plants not forming dense mounds, the stems mostly less than 8 cm in diameter.

6. Chollas and prickly pears; stems constricted into joints or pads; areoles with glochids (small spines deciduous at a touch) in addition to the larger, persistent spines (or larger spines absent in *O. basilaris*). ________ **Opuntia**
6′ Not chollas or prickly pears; stems not constricted into joints or pads; glochids lacking (if spines small, then not readily deciduous).
   7. Stems less than 2 cm in diameter, more than 20 times longer than wide; spines inconspicuous, 1–8 mm, not hooked; roots tuberous. ________ **Peniocereus**
   7′ Stems 3–9 cm in diameter (if less than 3 cm, then the central spines hooked), less than 10 times as long as wide; spines conspicuous, mostly exceeding 8 mm; roots not tuberous.
      8. Most areoles with at least 1 hooked spine. ________ **Mammillaria**
      8′ Spines straight or curved, none hooked.
         9. Plants with multiple stems more than twice as long as thick; flowers and fruits spiny; rather widespread. ________ **Echinocereus**
         9′ Plants usually unbranched, usually less than twice as tall as thick, ovoid to nearly globose; flowers and fruits spineless; localized, in the flora area known only near Sonoyta. ________ **Echinomastus**

CARNEGIEA

This genus is usually interpreted to have one species, although it is probably most closely related to *Neobuxbaumia* of southern Mexico. References: Crosswhite 1980; Fontana 1980; McAuliffe 1984; McGregor, Alcorn, & Olin 1962; Nobel 1980a; Steenbergh & Lowe 1976, 1977, 1983; R. Turner 1990.

Carnegiea gigantea (Engelmann ex Emory) Britton & Rose [*Cereus giganteus* Engelmann ex Emory] *SAGUARO, SAHUARO;* SAGUARO; *HA:ṢAÑ, BAHIDAJ* (FRUIT). Giant columnar cactus often 5.5–14 m, unbranched to several mostly erect branches; stem tissue turning black when cut or injured. Growth heteromorphic, the transition from sterile to fertile growth and first branches often (2) 3–5.5 m above ground, the branches fertile (producing flowers). Sterile (lower) stem portion thick, with 11–15 ribs, these areoles not bearing flowers (except sometimes near the zone of transition to mature growth), the longer central spines (3.7) 5.5–11.0 cm long × (0.8) 1.1–2.3 mm in diameter, stout, rigid. Fertile (upper) portion of stems relatively more slender, mostly with 19–25 stem ribs, the ribs lower ("shallower"), with closely set to nearly confluent, flower-bearing areoles, the spines 2.2–3.0 cm long × 0.3–0.4 mm in diameter, bristlelike, flexible.

Flowers 1 per areole, often many per rib and mostly crowded, nocturnal, usually remaining open until about midday or later depending on temperature. Flowers 8.5–13.0 × 6.5–8.0 cm; receptacle narrowly funnelform, with an unusually large region of solid tissue between the ovary and nectar chamber. Outer tepals and scales on the receptacle (pericarpel and exterior of floral tube) green, often with reddish tips; each scale with a small tuft of wool in its axil. Larger (inner) tepals 23.0–35.5 × 11.8–18.2 mm; inner tepals, filaments, style, and stigmas white; pericarpel and floral tube green at anthesis, spineless or sometimes with a few bristlelike spines. Flowering mostly late April-May, first buds and flowers emerging on east side of stems.

Fruits just before dehiscence 6.0–9.5 × 3.5–4.0 cm, ellipsoid to obovoid, mostly green, sometimes red especially toward apex, spineless, or sometimes with few, weak, flexible, persistent spines in some or rarely most areoles. (Stunted fruits or those developing under stress seem prone to develop spines.) Areoles 24–54 per fruit (number may be correlated with fruit size), the receptacular bracts less than 4 mm, triangular, tightly appressed against the pericarpel, subtending (and partly hiding) small wooly tufts. Dried floral remnant often persistent (the margin of its dilated base can be used like a knife for slicing open the fruit). Fruit rind (pericarpel) bright red inside, inedible, splitting open at maturity into 3 or 4 thick, spreading to recurved lobes. Fruit pulp (funiculi) juicy, bright red, sweet, and edible. Seeds 1.8–2.1 mm in diameter, dark red-brown to blackish, shiny and nearly smooth, edible (digestible when the seed coat is broken, e.g., ground on a grinding stone or *metate*). Fruits mostly ripe June-early July, rarely a few in August.

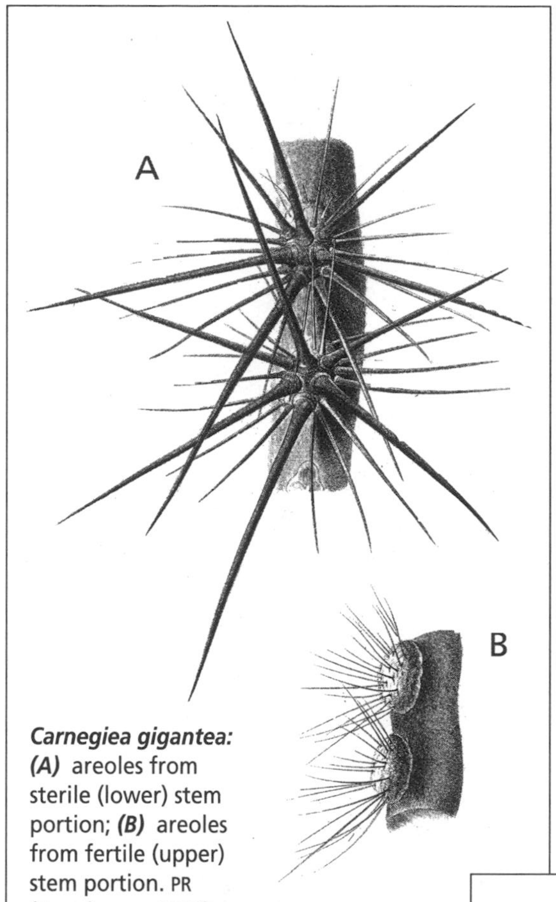

Carnegiea gigantea:* *(A) areoles from sterile (lower) stem portion; ***(B)*** areoles from fertile (upper) stem portion. PR (Engelmann 1859).

Echinocactus polycephalus var. polycephalus, Pinacate volcanic field. DTM, 1907 (courtesy Raymond Turner).

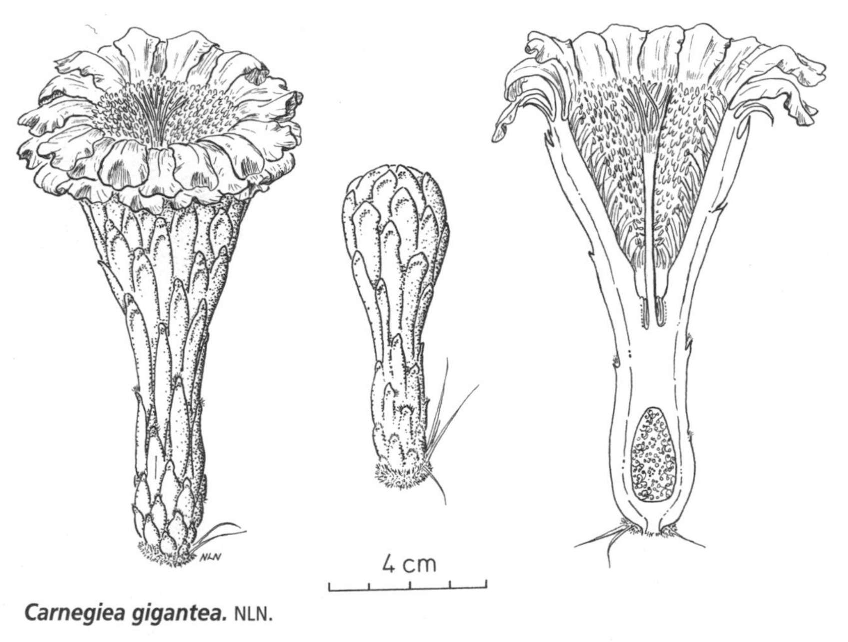

Carnegiea gigantea. NLN.

Carnegiea gigantea. Young plant among galleta grass (*Pleuraphis rigida*) as a nurse plant; "cactus flat" north of Cerro Colorado. BT.

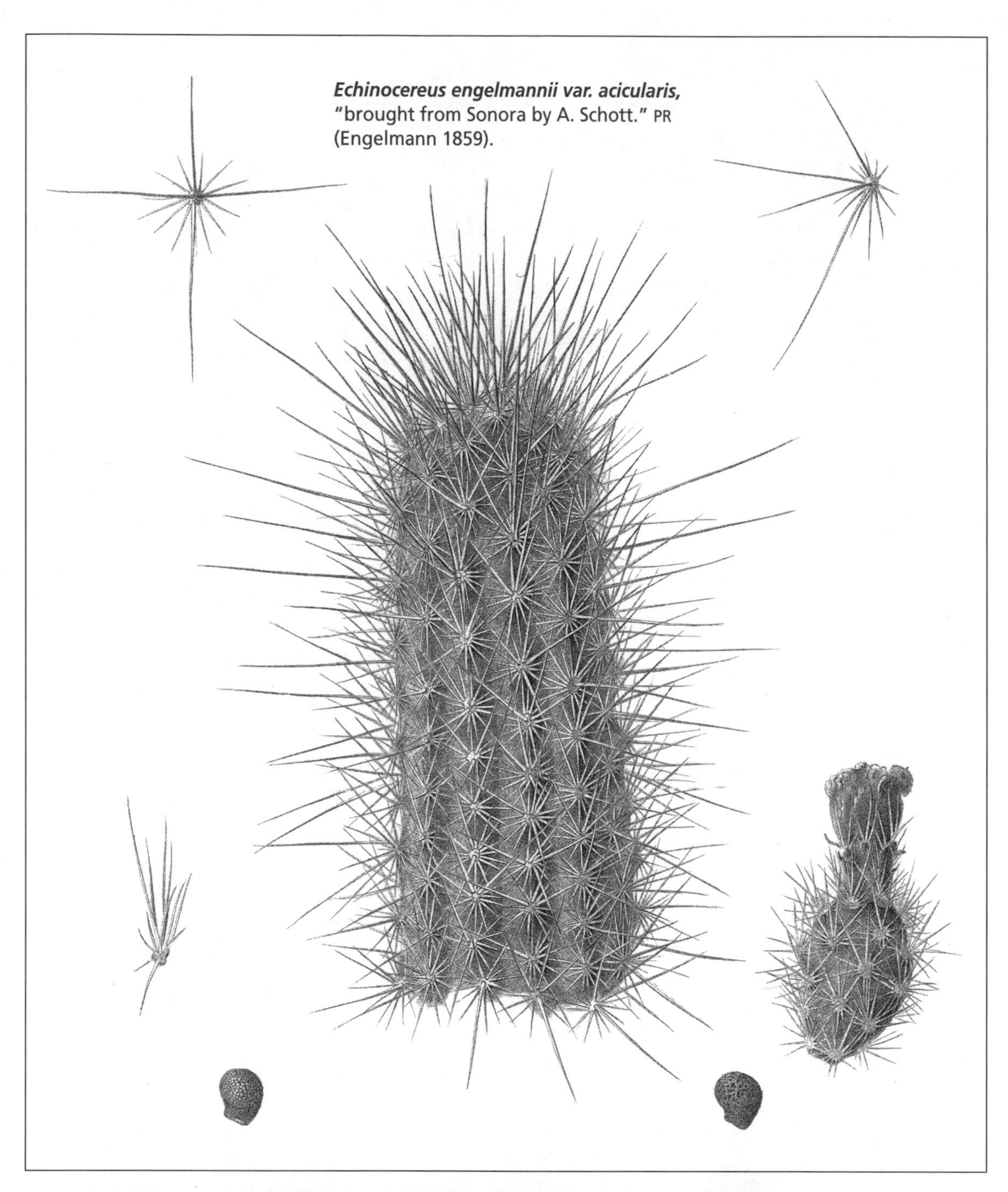

Echinocereus engelmannii var. acicularis,
"brought from Sonora by A. Schott." PR
(Engelmann 1859).

Echinomastus erectocentrus var. acunensis. Young plants in shelter of a large dead plant; near Sonoyta.BT.

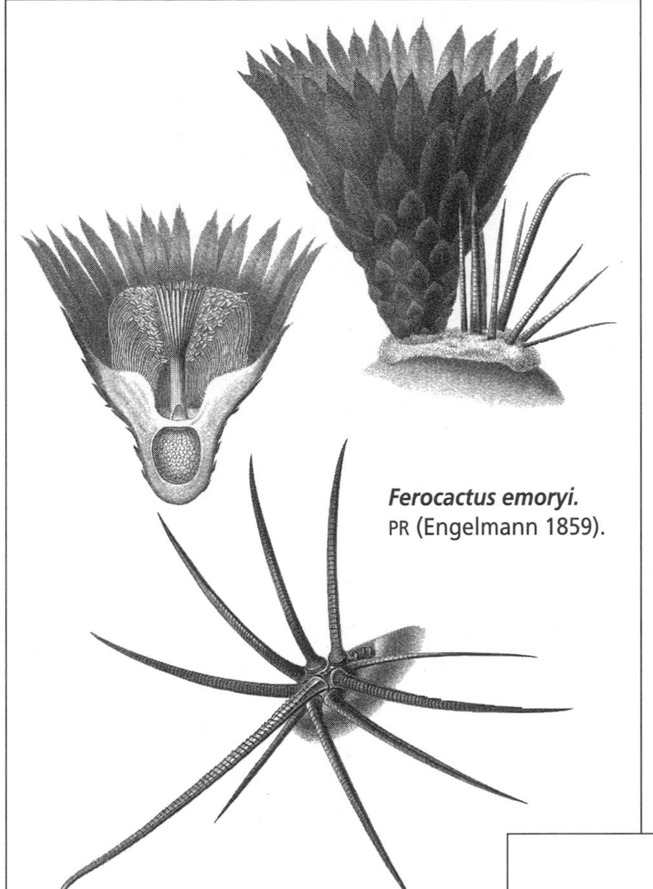

Ferocactus emoryi.
PR (Engelmann 1859).

Lophocereus schottii var. ***schottii,*** Campo Rojo. RSF.

Lophocereus schottii var. schottii. BT.

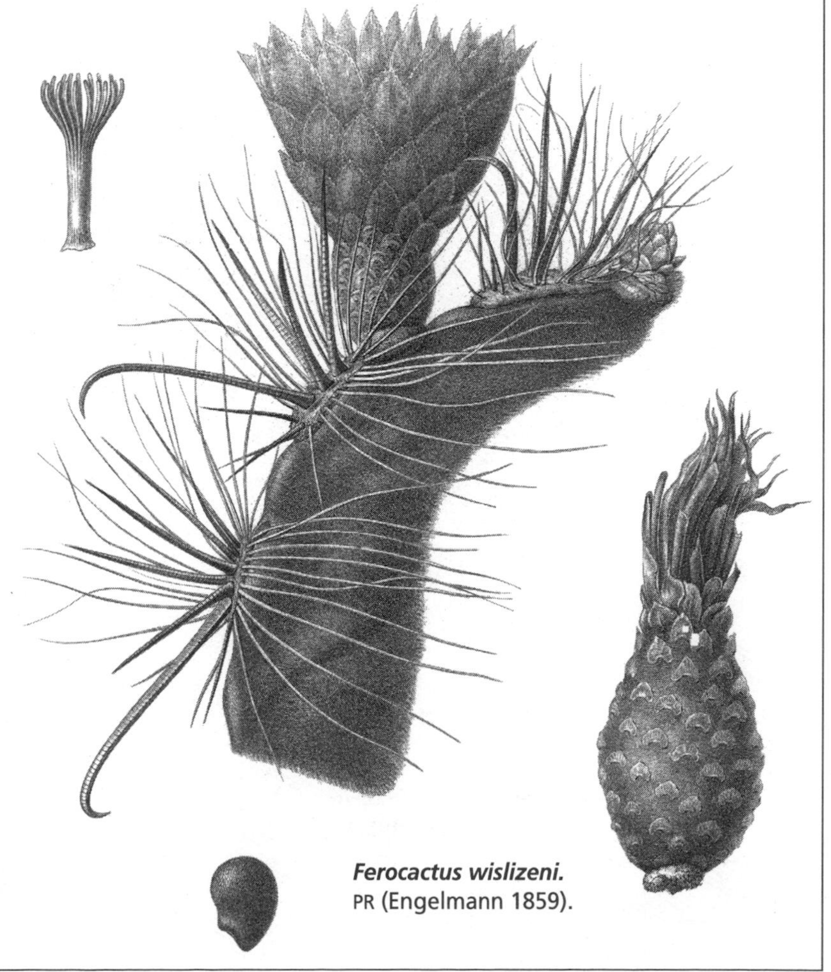

Ferocactus wislizeni.
PR (Engelmann 1859).

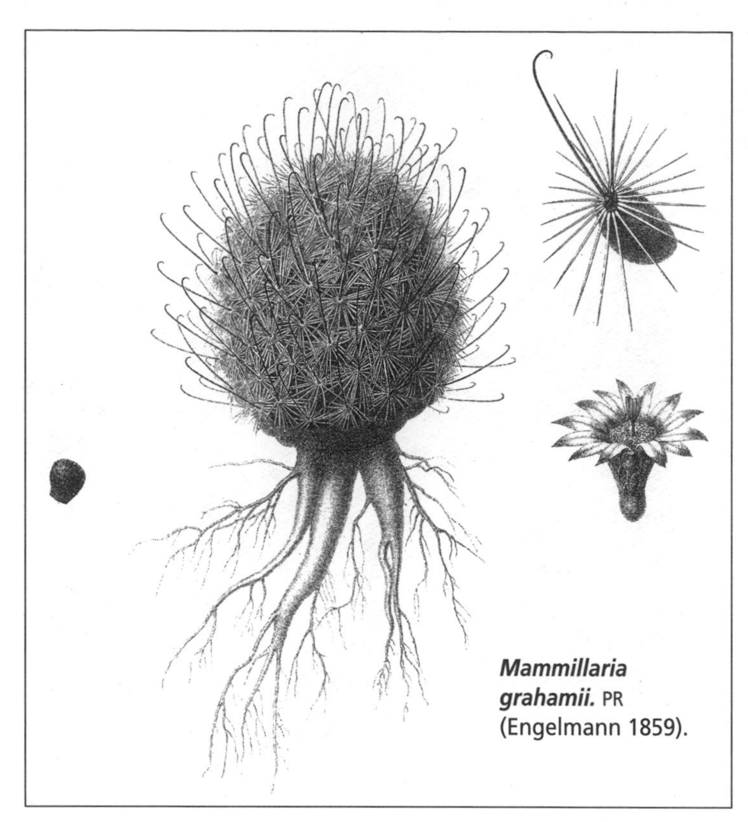

Mammillaria grahamii. PR (Engelmann 1859).

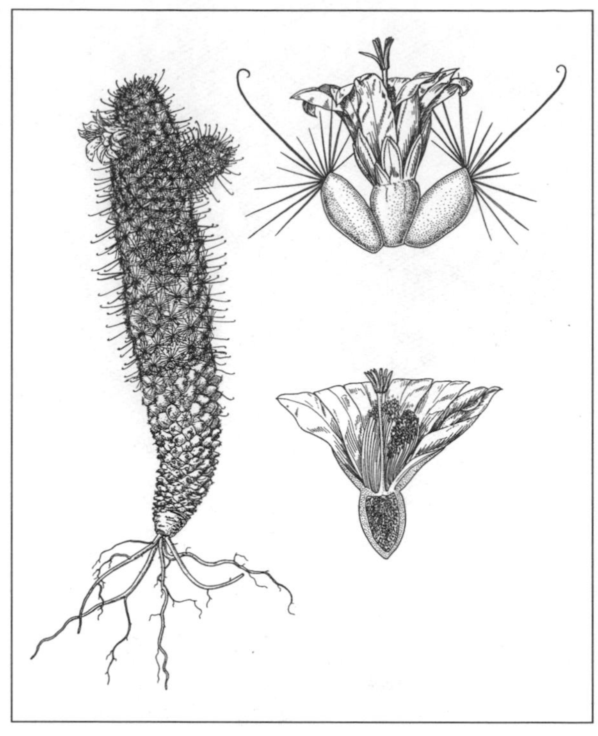

Mammillaria thornberi. LBH (Benson 1969a).

Mammillaria tetrancistra. PR (Engelmann 1859).

Opuntia acanthocarpa. RHP (Benson 1969a).

Opuntia bigelovii var. bigelovii, Sierra Pinacate. MAD.

Opuntia chlorotica, Sierra Pinacate. PKL.

Opuntia echinocarpa. RHP (Benson 1969a).

Opuntia engelmannii var. ***engelmannii.*** LBH (Benson 1969a).

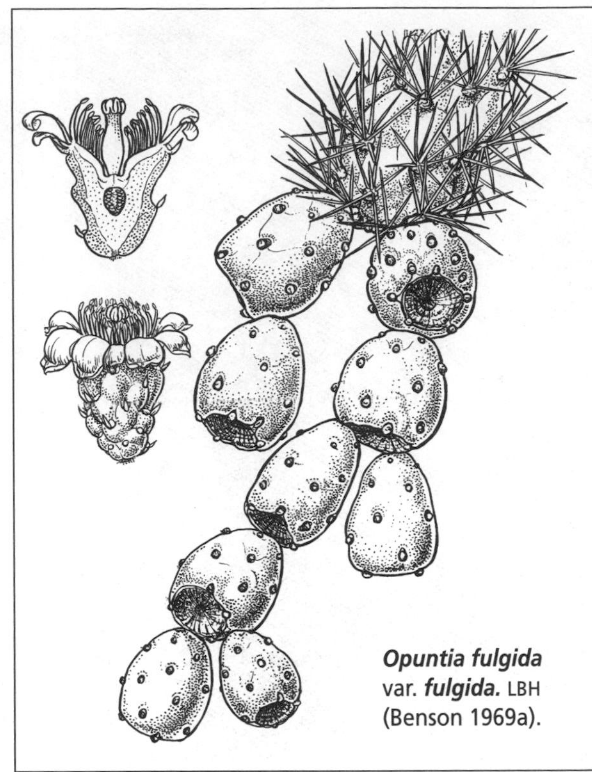

Opuntia fulgida var. ***fulgida.*** LBH (Benson 1969a).

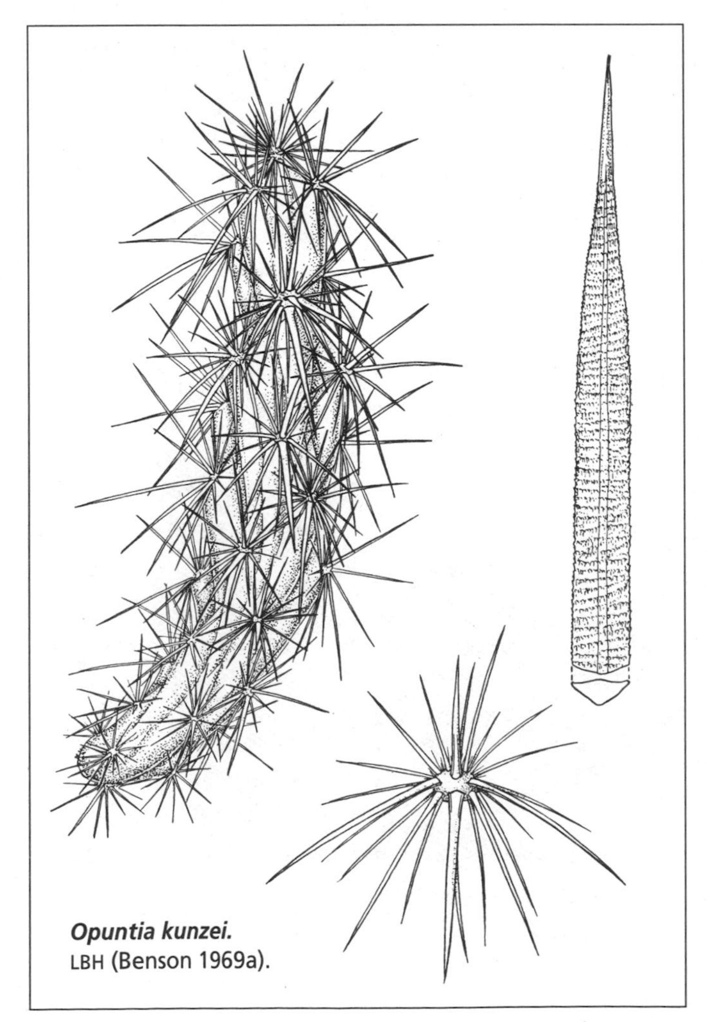

Opuntia kunzei. LBH (Benson 1969a).

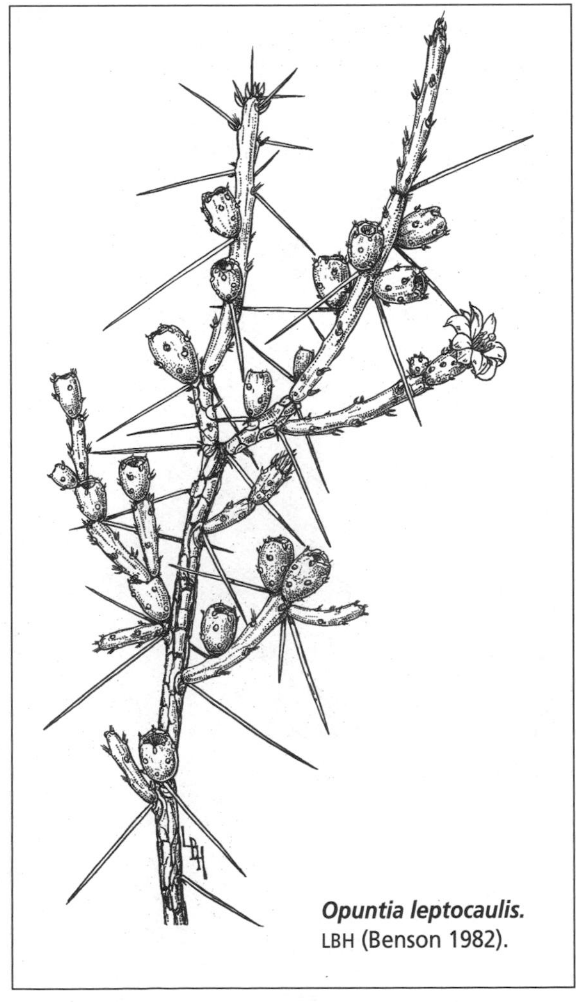

Opuntia leptocaulis. LBH (Benson 1982).

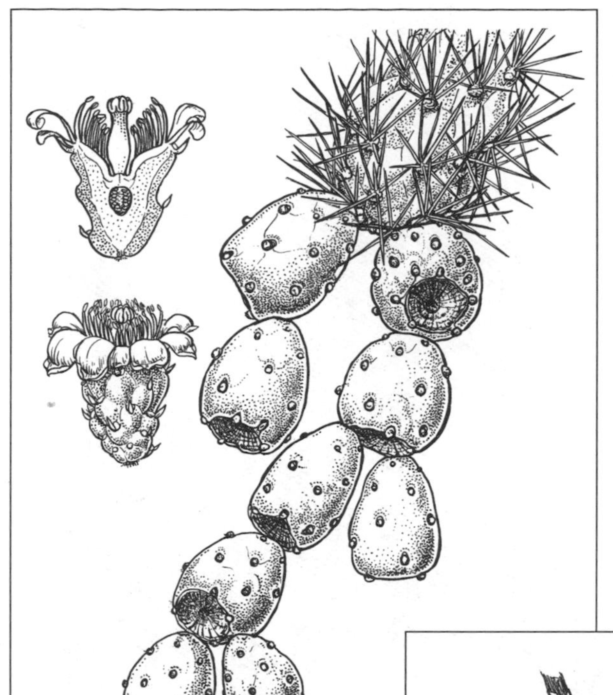

Opuntia ramosissima.
LBH (Benson 1982).

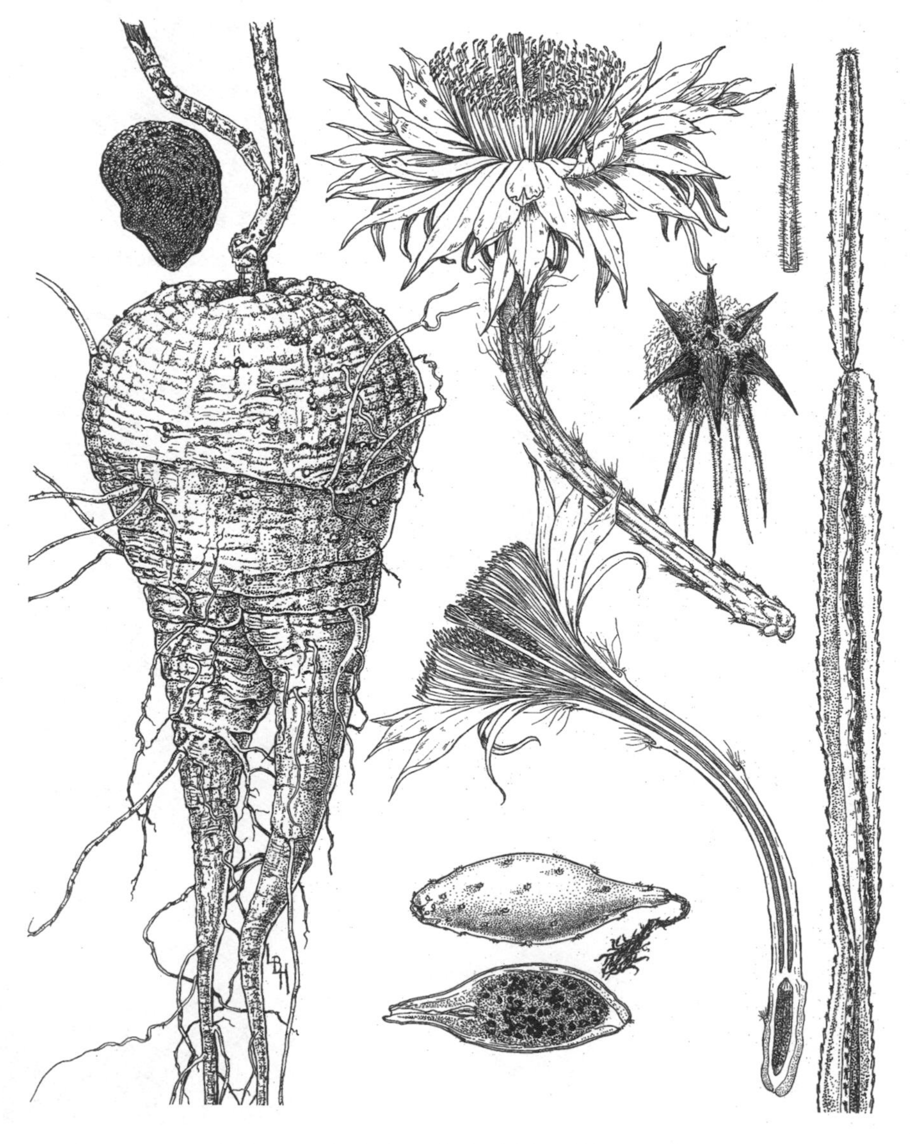

Peniocereus greggii var. transmontanus.
LBH (Benson 1982).

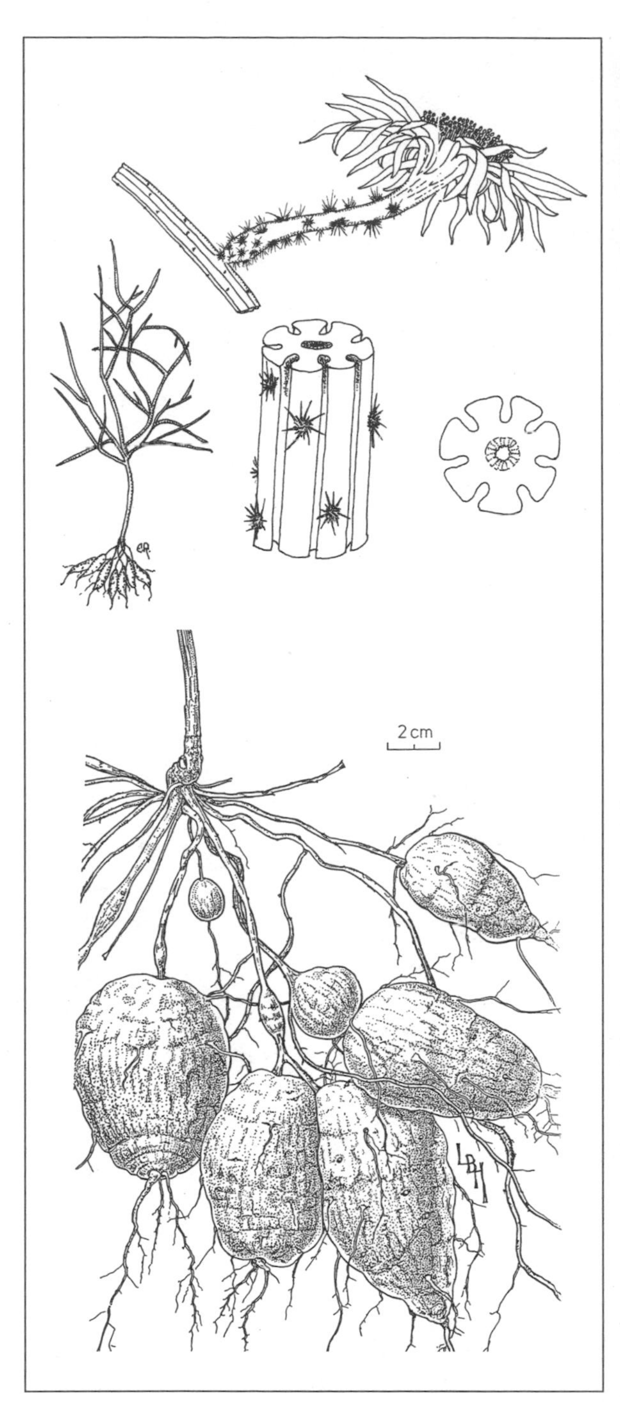

Peniocereus striatus. FR (Felger & Henrickson 1997) and LBH (below).

Stenocereus thurberi, near Sonoyta. RHP (Kearney & Peebles 1960).

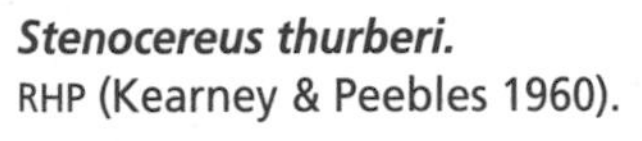

Stenocereus thurberi.
RHP (Kearney & Peebles 1960).

Common from the Sonoyta region to the northern and eastern sides of the Pinacate volcanic field and especially dense at ca. 600–700 m on the western side of the Sierra Pinacate; less common at lower elevations on the western and southern sides of the Pinacate volcanic field. Saguaro density drops off substantially in the arid granitic mountains northwest of the Pinacate volcanic field. Six saguaros were found in the Sierra del Rosario, all but 1 among rocks at the highest peak; they evidently were not reproducing. Sonoran Desert in Arizona and Sonora, and scarcely extending into California near the Colorado River.

As with other large cacti, saguaros begin life in safe sites beneath desert shrubs and trees, often legumes such as ironwood (*Olneya*) and palo verde (*Parkinsonia* spp.), which serve as nurse plants, or sometimes in the shelter of rock crevices. However, sometimes the usual nurse plants are absent, such as at MacDougal Pass and in the cactus flat north of Playa Díaz—a sand flat devoid of larger desert shrubs or trees. These sandy-soil habitats are favorable habitats for cacti, with 6 other cactus species present. In such places *Pleuraphis rigida,* a large perennial grass, is the most frequent nurse plant. Common saguaro nurse plants in the flora area other than legume trees and shrubs include *Ambrosia deltoidea, A. dumosa, Bebbia juncea, Krameria grayi, Larrea divaricata, Opuntia fulgida, O. kunzei,* and *Peucephyllum schottii.*

The most important pollinator is probably the white-winged dove (*Zenaida asiatica*), although the flowers also are visited by insects (Fleming, Tuttle, & Horner 1996). The fruit is delicious and was harvested for food and wine making by Sonoran Desert people.

Sonoyta, *Mearns 2720* (US). Observations: Sierra Blanca; Sierra Cipriano; Cactus Flat, 45 km W of Sonoyta; Elegante Crater; Tinaja del Tule; Sierra del Viejo.

ECHINOCACTUS

Stems small to large, globose or cylindrical, solitary to densely branched. Flowers and fruits at stem apex. Outer tepals and receptacular bracts bristle- or spine-tipped, the bracts subtending copious cotton-like tomentum. Fruits succulent to drying soon after maturity. Seed coat smooth to tuberculate. California to Texas south to Puebla; 6 species, 2 in the Sonoran Desert. Reference: Chamberland 1991, 1997.

Echinocactus polycephalus Engelmann & J.M. Bigelow var. **polycephalus.** MANY-HEADED BARREL CACTUS, COTTONTOP CACTUS. Forms clusters or mounds 0.5–1.0+ m wide, often containing 6–50 (100+) stems or "heads." (An exceptional plant on the east side of the Pinacate volcanic field is 1.5 m across, 0.9 m high, and has 205 stems.) Individual stems 9–12 (19) cm in diameter; stem ribs 12–18. Spines nearly obscuring stem surface, very stout, rigid, straight to slightly curved near tip and often twisted but not hooked, ridged below to flattened, the central spine especially stout, often 5.3–8.5+ cm. Spine surfaces red but obscured by felt-like covering of overlapping, short, white hairs imparting a dull pink-gray color. (Vestiture of the spines suddenly becomes translucent when wet, revealing the bright red spine surfaces. The spines turn red in the rain but soon dry and revert to their usual dull color.) Young areoles densely white woolly, older areoles less so.

Flowers often 5.5–5.8 cm, confined between the dense spines and thus not opening fully, the inner tepals 2.4–2.6 cm, the interiors of flowers bright, clear yellow (inner tepals, stamens, stigma, and style), the outer bristly tepals yellow with a reddish midstripe. Fruits densely and persistently white woolly; fruits plus the persistent, aristate outer tepals 3.5–4.0 cm. Seeds blackish (dark red-brown under magnification and bright light), (3.0) 3.5–4.0 mm maximum diameter, with a low ridge or crest on "back," the seed surface papillate and often angled due to compression at times of development. Late May-June.

Arid lowlands through much of the flora area; not in the Sonoyta region. Rocky flats, especially desert pavements and sometimes on sand flats. Lower elevations of the Pinacate volcanic field, sometimes to 300 m, and especially common on the west side of the lava shield; also common near the granitic mountains west of the Pinacate volcanic field and at Puerto Peñasco. Not known elsewhere in Mexico.

Two well-marked varieties: var. *polycephalus* in northwestern Sonora, southeastern California, southwestern Arizona, and southern Nevada, is replaced by var. *xeranthemoides* J.M. Coulter to the north in Nevada and the Grand Canyon region; intermediates in the Lake Mead area. A third member of this closely related group of dry-fruited species, *E. parryi* Engelmann, occurs in northern Chihuahua.

Individual flowers last 1 or 2 days. Seeds mature about one month after flowering. The fruits are slightly fleshy after ripening and slowly dry over a period of months. Fruit set apparently varies from year to year. The plants generally flower abundantly but in at least some years set few fruits. The dry, seed-bearing fruits are tightly held among the closely set spines for 1 year or more; the seeds fall as the fruits disintegrate. Large and small plants alike seldom re-establish after transplanting; attempts to cultivate it in Tucson fail, apparently due to too much rainfall.

Punta Peñasco, *Keck 4216* (CAS). Bahía de la Cholla, *Zimmerman 21 Jun 1971*. SW of Los Vidrios, *Pinkava 15510* (ASU). N of Moon Crater, *F 88-91*. 36 mi E of San Luis, *F 5759B*.

ECHINOCEREUS HEDGEHOG CACTUS

Stems mostly less than 50 cm (rarely to 2+ m), solitary to many, usually erect, sometimes prostrate to pendent. Flowers adapted for daytime pollination, usually medium to large and showy, the stigma usually green. Receptacle spiny at anthesis, the spines not enlarging during fruit maturation. Fruits succulent, spiny, but spine clusters usually readily fall away from ripe fruits, the dried perianth persistent. Ripe fruits of most species sweet and delicious. Seeds blackish, tuberculate, mostly less than 2 mm. Oaxaca to California and South Dakota; 50 species. References: Buxbaum 1975a; Grant & Grant 1979b; Taylor 1985.

1. Spines dull yellow, brownish, grayish, often bicolored, becoming (or remaining) gray (rarely blackish) with age; inner tepals deep magenta, very showy; seeds 1.3–1.5 mm. ____ **E. engelmannii**

1′ Spines uniformly yellow, often blackish with age; inner tepals pale pink; seeds 1.1–1.3 mm. ____ **E. nicholii**

Echinocereus engelmannii (Parry ex Engelmann) Lemaire [*Cereus engelmannii* Parry ex Engelmann] STRAWBERRY HEDGEHOG CACTUS; *'I:SWIGǏ*. Forms clusters of several to many stems, mostly cespitose. Stems 15–45 cm, often 7 cm in diameter, the areoles often 6–9 (15) mm apart. Spines moderately dense; dull yellow to brownish, whitish, or grayish, fading (or remaining) grayish with age. Central spines 2–6 (9), the longer (lower) central spine 3–10.5 cm, often fading whitish, twisted or straight, flattened or terete, the other spines terete. Flowers very showy, 7.5–9.0 × 5–9 cm; inner tepals bright purple-magenta (often varying from paler to darker in same population), 37–75 × 14–20 mm; tube typically 17 × 18 mm, its nectar chamber 4 × 3 mm (widest basally). Fruits ovoid, 4.0–4.5 × 2.5–2.7 cm, spine clusters mostly readily deciduous, the rind red at maturity, the pulp lavender-pink (whitish until fruits mature, becoming infused with pink or red from the fruit skin). Seeds 1.3–1.5 mm. March-early April; fruits ripening late May-early June.

Upon ripening the fruit pulp and the seeds are quickly consumed by a variety of animals, especially birds which poke a hole in the side of the fruit. Ants quickly finish off what the birds leave, and hollowed-out fruits are common.

Sporadically, Sonoyta westward to the Sierra del Viejo (along Mex 2), Pinacate volcanic field and granitic mountains at all elevations, and Bahía de la Cholla. Usually on rocky substrates but also in deep, sandy-silty soils; mountains, hills, bajadas, and sometimes desert flats. Southern California, Nevada, southwestern Utah, western Arizona, Baja California and northeastern Baja California Sur, and Sonora southward to the vicinity of Guaymas.

The taxonomy of this variable species is controversial. The distribution of the varieties as mapped by Benson (1982) does not seem to reflect the situation in nature. The boundaries with certain other species to the east are poorly known. Two varieties occur in the flora area. Although their extremes are well marked, they appear to intergrade.

Echinocereus engelmannii var. **acicularis** L.D. Benson. Often more robust and appearing less spiny than var. *chrysocentrus*. Central spines 2–4; lower central spine in each areole terete or nearly so.

In the flora area more or less restricted to the Sonoyta region (Arizona Upland). Central Arizona and in northern Sonora at least as far south and east as Tajitos (between Sonoyta and Caborca); southern and eastern limits in Sonora unknown. In its extreme form this variety has only 2 central spines. Plants intermediate with var. *chrysocentrus* occur near the western and southern margins of the Sonoyta region, at the approximate boundary of the Arizona Upland and Lower Colorado Valley. Benson's (1982) southwesternmost records of *E. fasciculatus* (Engelmann ex B.D. Jackson) L.D. Benson, in Organ Pipe Monument, are actually *E. engelmannii* var. *acicularis*.

18 km SW of Sonoyta, *F 88-194*. 1.5 km W of Sonoyta, *F 85-953*. Quitobaquito, *F 89-245*. 2.6 mi NW of Tajitos, *Wiggins 8291* (DS).

Echinocereus engelmannii var. **chrysocentrus** (Engelmann & J.M. Bigelow) Rümpler [*Cereus engelmannii* var. *chrysocentrus* Engelmann & J.M. Bigelow] Central spines 4–6 (9), the largest one flattened or sharply angled and often paler than the others.

Widespread through much of the flora area but not in the Sonoyta region; the Pinacate volcanic field to at least ca. 850 m, desert plains and flats to the north and east, and the larger western, granitic mountains but absent from Cerro Pinto and the Sierra del Rosario. Southwestern quarter of Arizona, adjacent southeastern California, and documented in Sonora only from the flora area; expected in northeastern Baja California.

Cactus flat, 2 km E of Pinacate Junction, *F 90-50*. N side Sierra del Viejo (16 km E of El Sahuaro), *F 85-728*. SE of Carnegie Peak, 850 m, *F 19929*. Tinajas Altas Mts, *Blackwell & Akens 15 Apr 1932*. Observations: Represo Cipriano; Campo Rojo; NE of Elegante Crater.

Echinocereus nicholii (L.D. Benson) B.D. Parfitt [*E. engelmannii* (Parry ex Engelmann) Lemaire var. *nicholii* L.D. Benson] GOLDEN HEDGEHOG CACTUS. Branching near ground level, forming clumps often reaching 45–80 × 50–100 cm or more with 16–24 or more stems; individual stems 6–9 cm in diameter, erect to ascending. Areoles (8) 10–25 mm apart. Spines conspicuously yellow, often golden, relatively monochromatic, the older spines darker, often blackish; central spines 4–6, the longer ones 4.0–7.2 cm, often twisted and somewhat flattened. Inner tepals pink (much paler than those of *E. engelmannii*). Fruits remaining spiny after ripening, ovoid, 2.3–3.4 × 1.7–2.3 cm, the skin green, becoming somewhat bronze when fully ripe where exposed to sun, the pulp remaining whitish. Seeds 1.1–1.3 × 0.8–1.2 mm. Flowering March-April, the fruits ripening in June.

Exposed slopes and upper bajadas of granitic hills and mountains in the Sonoyta region in the vicinity of Sonoyta, Sierra Cipriano, Sierra de los Tanques, and in adjacent Organ Pipe Monument and Cabeza Prieta Refuge. The largest and densest populations occur in the Cabeza Prieta Refuge.

Often most numerous on north-facing slopes and growing with the superficially similar *E. engelmannii* as well as *Carnegiea gigantea, Ferocactus cylindraceus, F. emoryi, Mammillaria grahamii, Opuntia acanthocarpa, Stenocereus thurberi,* and sometimes *Echinomastus erectocentrus*. Southern Arizona and western Sonora southward at least to the Sierra Seri (opposite Isla Tiburón).

The plants are usually conspicuously more robust than *E. engelmannii;* they are easily recognized by their yellow spines and can be readily separated from *E. engelmannii* by a greater distance between the areoles on each rib, flowers somewhat smaller, the inner tepals pale pink instead of magenta, the basal portion of the floral cup green instead of colored, and smaller seeds with large, distinct papillae instead of low, coalescent papillae. *E. nicholii* is diploid ($2n = 22$) whereas *E. engelmannii* is tetraploid ($2n = 44$). This difference in ploidy level represents a reproductive barrier. Hybridization would result in a sterile triploid ($2n = 3x = 33$), effectively blocking the flow of genes between the parent taxa (Parfitt 1987).

1 mi SW of Sonoyta, *Wiggins 8334* (DS). Quitobaquito, *F 90-39*. Observations: Sierra Cipriano; Sierra de los Tanques.

ECHINOMASTUS

Dwarf barrel or pincushion cacti, usually unbranched. Tubercles usually more or less confluent, forming (8) 13–21 poorly to well-defined longitudinal ribs. Spines stout, entirely to partially hiding the stem surface, those of adult plants clearly differentiated into radials and centrals, the centrals never hooked. Flowers diurnal, in early spring; nearly apical, campanulate, or funnelform. Pericarpel and floral tube robust, green, bearing 1–20 broad, scalelike, appressed, membranous-margined bracts, their axils without hairs or spines. Inner tepals variously colored. Stigma lobes red, brown, or green, rarely white. Fruits green, brown, or dull pink, scarcely succulent, usually indehiscent or dehiscent by longitudinal splits, the pulp colorless, drying quickly after ripening; floral remnant persistent. Seeds 1.8–2.2 mm in diameter; round to kidney-shaped, blackish, and papillate, the hilum large and deep. Southwestern United States and northern Mexico; 6 species.

Although Lyman Benson (1969a) united *Echinomastus* and *Neolloydia,* the former has been shown to be related to *Pediocactus* and *Sclerocactus,* whereas *Neolloydia* sensu stricto is part of the *Ariocarpus-Thelocactus* complex. References: Anderson 1986; Glass & Foster 1975; Kladiwa & Fittkau 1971; Zimmerman 1985.

Echinomastus erectocentrus (J.M. Coulter) Britton & Rose var. **acunensis** (W.T. Marshall) Bravo [*E. acunensis* W.T. Marshall. *Neolloydia erectocentra* (J.M. Coulter) L.D. Benson var. *acunensis* (W.T. Marshall) L.D. Benson] ACUÑA CACTUS. Dwarf barrel cactus, unbranched, nearly ovoid to ovoid-cylindroid, 6.5–13.5 (27) × 6.5–8.5 cm, with 21 prominently undulate ribs. Spines obscuring stem surface; central spines (1) 2–3 (4) per areole, ascending and slightly curved, 2.5–4.4 cm; radials 13–15 per areole, straight, spreading close to the stem. Areolar gland 1 (2) per areole, pale yellow-brown, to ca. 1.5 mm in diameter, much wider than tall, closer to the spine cluster than to the flower, attended by ants.

Flowers showy, 35–60 × 39–91 mm. Inner tepals 13–21 in number, (21) 29–43 × 7.5–14.5 mm, bright rose-pink with pale margins (varying somewhat in hue from plant to plant), basally blotched with maroon or chestnut extending 30–45% of the length of the tepal. Filaments chartreuse, contrasting with the rest of the flower; anthers pale yellow. Stigma lobes 10–13, flesh pink to dull red, covered with unicellular papillae of bright green, brownish green, or pale golden green. Fruits 1.5–2.0 cm, ellipsoid-cylindroid, greenish and slightly glaucous, slightly fleshy, dehiscent through vertical splits (often indehiscent on stressed plants), quickly drying; receptacular bracts scalelike, triangular-reniform, mostly 2 × 3–4 mm, white, membranous, the margins erose. Seeds crescentic-ellipsoid, dark red-brown (nearly black), 2.0–2.3 × 1.6–1.9 mm, surfaces regularly and minutely papillate except smooth at hilum; seeds with or without a very low dorsal crest. Mid-March–early April; fruits ripening in April.

Variety *acunensis* is documented from Sonora only in the Sonoyta region, at the southwestern geographic limit for the genus. The report of it "south at least to Puerto Peñasco" (Wiggins 1964:1012) seems erroneous. Granitic soils on hillsides and rugged alluvial fans of gravelly soil and small rocks. Usually in colonies of several plants to one dozen or more in close proximity to one another in open, exposed situations. Sometimes beneath small shrubs or even larger *Echinomastus,* which serve as nurse plants; generally more robust in partial shade of nurse plants than in open, exposed situations. Growing with *Carnegiea, Echinocereus engelmannii, E. nicholii, Ferocactus emoryi, Mammillaria grahamii, M. tetrancistra, Opuntia acanthocarpa, O. fulgida, O. leptocaulis,* and *Stenocereus thurberi.* Also in scattered localities in western Pima County, including nearby Organ Pipe Monument, as well as in Maricopa and Pinal Counties of Arizona.

Variety *erectocentrus,* found in less xeric habitats in southeastern Arizona, is distinguished by its smaller tubercles, shorter and more tightly appressed spines, narrower radials, maximum of 2 centrals per areole, absence of central spines on immature plants, consistently red stigma lobes, and generally paler pink or white inner tepals. Benson (1982:792) implies that var. *acunensis* always has 3 or 4 centrals, yet plants from the flora area (close to the type locality of var. *acunensis* and matching *acunensis* in all other respects) frequently have only 1 or 2 centrals. In addition, Benson (1982:792) states and illustrates that var. *acunensis* has "petaloids coral or orange-pink," which is contrary to our observations. *E. erectocentrus* is replaced by *E. johnsonii* (Engelmann) Baxter to the north in northwestern Arizona, eastern California, southern Nevada, and southwestern Utah; these 2 taxa are only weakly differentiated.

Sonoyta, *Keck 4148* (DS). 2–11 mi SW of Sonoyta, *F 88-146, F 90-150, Shreve 7599, Wiggins 8335A* (DS).

FEROCACTUS *BIZNAGA,* BARREL CACTUS

Plants unbranched to cespitose; stems baseball-sized to 1–2 (4) m, often very stout. Areolar hairs short, forming a felt-like indumentum at stem apex; areoles near the top of the stem have smooth, conical, brown nectaries tended by ants. Outer tepals not aristate or spine-tipped (except in *F. flavovirens,* which might not be congeneric). Anthers yellow. Fruits fleshy, with yellow to reddish rinds; dried perianth persistent on the fruits. Seeds blackish, smooth or pitted.

California to Texas and southward to Oaxaca; 30 species. Sonoran barrel cacti have solitary stems or rarely produce 1 to several basal branches.

Ours have thick stems, stout spines, and yellow fruits with white pulp (funiculi). The fruits ripen after a number of months and are semipersistent; the fruits plus the dried perianth are ca. 6.0–7.5 cm. Their seeds are ca. 2.5 mm, very dark red-brown under magnification and strong light (otherwise appearing black), relatively shiny, and without a crest or ridge; the surfaces have reticulate patterns, either pitted or nearly smooth. Although the fruits do not dehisce spontaneously, a fruit removed from the plant is like a salt shaker, scattering seeds from a hole at the base (where the fruit was attached).

Although *F. cylindraceus* is common in the Pinacate region, the other 2 species only enter the northeastern part of the flora area. We have not found any really large barrel cactus plants in the flora area, although much larger, taller plants of all 3 species are rather common nearby in regions of higher rainfall.

The flower buds, flowers, fruits, and seeds are edible (Felger & Moser 1985). Cooked buds and flowers taste somewhat like brussels sprouts. The fruits of the Sonoran Desert species, eaten fresh, have a tart lemonlike flavor. The seeds, rich in oil and protein, can be ground into an edible paste. Mashed stem tissue of certain barrel cacti was used extensively by the Seri Indians as a source of emergency liquid, but liquid from *F. emoryi* was regarded as toxic (Felger & Moser 1985). Desert bighorn sheep survive the hottest, driest seasons without free water by consuming barrel cacti (Warrick & Krausman 1989).

Seed-grown plants, readily available from growers in southern Arizona and elsewhere, can be successfully transplanted. When large plants are transplanted they often do not re-establish and eventually die. References: Grant & Grant 1979b; Lindsay 1996; Nobel 1977; Taylor 1979, 1984; Taylor & Clark 1983; Unger 1992.

1. Areoles with spines all robust, rigid, and of approximately uniform thickness. ______ **F. emoryi**

1′ Areoles with slender, bristlelike spines in addition to the thick, rigid spines.

  2. Stem generally smaller in diameter and often taller than *F. wislizeni,* seldom leaning to the south; largest spine hooked or not, often markedly twisted; seeds pitted; rocky substrates—hills and mountains. ______ **F. cylindraceus**

  2′ Stem broader, usually conspicuously leaning southward; largest spine in each areole usually hooked, seldom twisted; seeds nearly smooth; flats and open desert. ______ **F. wislizeni**

Ferocactus cylindraceus (Engelmann) Orcutt [*Echinocactus viridescens* Nuttall var. *cylindraceus* Engelmann. *E. cylindraceus* (Engelmann) Engelmann. *Ferocactus acanthodes* (Lemaire) Britton & Rose. Not *Echinocactus acanthodes* Lemaire (see Taylor 1979)] *JIAWUL.* Stems cylindrical, straight, erect, the larger plants taller than wide; flowering plants (20) 45–125 cm (rarely to 150 cm in the hills above Quitobaquito), (20) 25–38 cm in diameter. Young plants (15 cm tall) often with 14 stem ribs, adult plants with 19–20 ribs. Central spines whitish, gray, pink, or dull red (blood red when wet) with yellow-white tips, sometimes straw yellow with broad red or pink bands. Central spines moderately curved, sometimes twisted, usually not strongly hooked (except on immature plants); longer central spines 4–8.5 cm. Areoles with 4 major thick, rigid central spines, these surrounded by smaller spines (subcentrals or large radials, depending on interpretation), these rigid and intergrading with the centrals; radials (outermost spines) mostly bristlelike.

Flowers shorter than the spines, often ca. 4.5 × 4.2 cm (smaller than those of the other two *Ferocactus* in the region); perianth yellow to red, the inner segments pure yellow to more often yellow with a salmon to maroon midstripe, orange, or sometimes red and with a darker red midstripe. Filaments red or yellow (see discussion following). Style and stigma lobes yellow to pale red. Seeds pitted. Usually February-April, and less vigorously late summer–early fall.

Widespread and often fairly common on hard, rocky substrates in the Pinacate volcanic field and Sierra Blanca, Sierra Extraña, and Hornaday Mountains, where it is the only *Ferocactus* present; especially common in the Sierra Pinacate from moderate elevation to the base of Carnegie and Pinacate Peaks. Also granitic hills and mountains in the Sonoyta region, sometimes with *F. emoryi*. Absent from lowland flats and northwestern granitic mountains.

Southeastern California, southern Nevada, southwestern Utah, western Arizona, Baja California, and northwestern Sonora. Four varieties have been recognized. Populations on the volcanic regions most closely fit into Benson's (1982) concept of var. *acanthodes* (var. *cylindraceus*), characterized by elongated, twisted, stout, and red to pink or gray spines. However, individuals resembling var. *eastwoodiae* (L.D. Benson) N.P. Taylor, with dull yellowish spines, or var. *lecontei* (L.D. Benson) Bravo, with shorter and more slender (red to pink or gray) spines, occur intermixed in the flora area. The varieties are not clearly differentiated and probably are not worthy of taxonomic recognition; Benson's map (1982:687) shows only partial geographic segregation.

Plants in granitic hills and mountains have dull-colored spines, mostly straw yellow, consistently yellow inner tepals, and yellow or red filaments. In dark lava rock in the Pinacate volcanic field the plants tend to have attractive pinkish to reddish spines. On the east side of the Pinacate field, such as the vicinity of Campo Rojo, some have bright yellow inner tepals, most are orange or yellow with a red midstripe, and a few have bright red-orange to red inner and outer tepals. Plants with yellow inner tepals were recorded in the lava areas along the southwest and south side of the Pinacate volcanic field, e.g., in the vicinity of Tinaja del Tule. Plants in the Pinacate volcanic field have red filaments. Over most of its geographic range, including California and Arizona, the flowers have clear yellow inner tepals and yellow filaments. In early spring the flowers have been observed to begin opening at 8 A.M. and to be fully open in about an hour.

40 mi W of Sonoyta, *Lehto 15487* (ASU). Quitobaquito Hills, *F 90-41*. Observations: Sierra Cipriano; Campo Rojo; Moon Crater.

Ferocactus emoryi (Engelmann) Orcutt [*Echinocactus emoryi* Engelmann in Emory. *Ferocactus covillei* Britton & Rose. *Echinocactus covillei* (Britton & Rose) Berger] *JIAWUL*. Stem seldom exceeding 1 m in height (rarely to 1.7 m at international border near Sonoyta). Stem 29–32+ cm in diameter, with 21 ribs. Spines all rigid and stout (bristlelike spines not present); central spines 5.5–9.5 cm (not including curve), hooked on smaller, immature plants and curved but not hooked on mature growth. Flowers (petaloids, filaments, style, and stigmas) red, often 7.5 cm wide; stigma 16-lobed. Seeds conspicuously pitted. Budding from early March; flowers summer–mid-September.

Infrequent to fairly common on rocky granitic slopes and adjacent upper bajadas in the Sonoyta region, especially just south of Organ Pipe Monument and southwest of Sonoyta in the Sierra Cipriano; occasional on hills of volcanic basalt along the Arizona border (north of Mex 2) westward to Represo Cipriano. Southwestern Arizona and western Sonora southward to the Yaqui lands southeast of Guaymas.

Plants in Arizona and Sonora south to the vicinity of Bahía Kino and Hermosillo have red flowers. Farther south (e.g., Sierra Libre, between Hermosillo and Guaymas) the plants have yellow to orange or orange-red flowers. From the Guaymas region southward the flowers are bright yellow, as in the closely related *F. pottsii* (Salm-Dyck) Backeberg from southeastern Sonora. The northern red-flowered populations and the southern yellow-flowered populations seem worthy of infraspecific taxonomic recognition.

We disagree with Taylor (1984) that *F. rectispinus* (Engelmann) Britton & Rose should be treated as a variety of *F. emoryi*. *F. rectispinus,* endemic to Baja California Sur, is morphologically and geographically clearly isolated from *F. emoryi*. Indeed, the morphological gap is at least as wide as the one separating these taxa from *F. pottsii,* which Taylor considers specifically distinct.

1 km W of Sonoyta, *F 85-711*. 10 km SW of Sonoyta, *F 88-147*. Aguajita, *F 88-443*.

Ferocactus wislizeni (Engelmann) Britton & Rose. *JIAWUL*. Stem usually less than 1 m in the flora area, very thick, often 36–65 cm in diameter, the larger plants conspicuously leaning (growing) southward. (Larger plants often topple over as the substrate erodes from beneath them.) Stem ribs 21–23. Spines dull gray, the larger spines strongly hooked, longest central spines (from curve of hook to base) 6.1–8.5 cm (reaching 12 cm on young plants 15–17 cm tall); outer radial spines slender, bristlelike. Flowers often 7 × 6.5 cm, exceeding the spines, usually red-orange overall, the inner tepals with an orange-red to red midstripe grading to yellow-orange or yellow on margins, or sometimes the inner tepals bright yellow. Filaments red. Stigma lobes 16, cream-yellow. The flowers are reportedly self-incompatible (Grant & Grant 1979b). Seeds essentially smooth with a slightly raised reticulate pattern. Late summer–fall. Black ants often carry away seeds from hollowed but still fleshy, yellow fruits.

Small playas with sandy-silty soil and creosote bush flats from the Arizona border in vicinity of Quitobaquito westward at least to Represo Cipriano and southward to the northern margin of the main mass of the Pinacate volcanic field; not on hillsides. Plants of all age classes are moderately common in the vicinity of Represo Cipriano, but only a few kilometers southward, at slightly lower elevations, such as at cactus flat toward Playa Díaz and elsewhere south of Mexico 2, this cactus becomes increasingly scarce. For example, at the cactus-flat locality the plants seem to be mostly of 1 or possibly 2 age classes (26–73 cm) with a very few smaller plants (9–12 cm), indicating that establishment is infrequent. The plants are more common in adjacent areas at slightly higher elevations to the northeast and east of the flora area. Southwestern and central Arizona to extreme western Texas and adjacent Chihuahua and northern Sonora.

In the flora area *F. cylindraceus* and *F. wislizeni* are clearly separated ecologically, and for the most part, geographically. *F. wislizeni* is locally distinguished from *F. cylindraceus* by its relatively dull gray, strongly hooked spines, much thicker stems, and tendency to lean southward. *F. wislizeni* occasionally grows with *F. emoryi,* although they generally occupy different habitats. Both species have similar-appearing flowers and may flower at the same time, although in the flora area *F. wislizeni* mostly has orange to orange-red flowers and *F. emoryi* has deeper red flowers.

Contrary to a statement by Taylor (1984:34), we find no evidence that *F. wislizeni* "in the southern part of its range . . . intergrades with" the related *F. herrerae* J.G. Ortega. There is an eco-geographic gap of perhaps more than 200 km between the southernmost population of *F. wislizeni* in northern Sonora and the northernmost *F. herrerae* in the vicinity of Guaymas. Because there is a definite morphological, ecological, and geographical gap, we feel that they should be maintained as distinct species. Although Taylor & Clark (1983) found little difference in their seeds, the several vegetative differences are conspicuous and consistent. *F. herrerae* occurs primarily in coastal thornscrub of southwestern Sonora and Sinaloa. In comparison to *F. wislizeni, F. herrerae* grows spectacularly fast in cultivation. We also consider *F. tiburonensis* (G.E. Lindsay) Backeberg on Isla Tiburón and the Sonora coast to be a distinct species.

NW margin of Playa Díaz, *F 88-396*. Quitobaquito, *F 88-444*.

LOPHOCEREUS

Columnar cacti with many erect to arching branches. Stem tissue turning black on cut surfaces; stems strongly heteromorphic: Sterile (lower) portion relatively thick, not producing flowers, the stem ribs relatively fewer, the areoles widely spaced and the spines short, stout, not twisted, and with enlarged bases. Fertile (upper) portion relatively slender, with a higher number of stem ribs; a shaggy mane of long, slender, and twisted spines; the areoles close together or confluent and producing flowers. Fertile areoles multiple-flowered (one of the few genera of cacti producing more than 1 flower per areole). Fruits relatively small, the rind red when ripe, spineless or nearly so, irregularly splitting from the top, the pulp bright red; seeds small, blackish. The pulp and seeds avidly consumed by birds and ants; hollowed-out fruits, still clinging to the plants, are commonplace.

Northwestern Sinaloa to southwestern Arizona and both Baja California states; 2 closely related species. References: Cody 1984; Felger & Lowe 1967; Lindsay 1963.

Lophocereus schottii (Engelmann) Britton & Rose var. **schottii** [*Cereus schottii* Engelmann, *Proc. Amer. Acad.* 3:288, 1856. *Pachycereus schottii* (Engelmann) D. R. Hunt] *SINITA, SINA, MÚSARO;* SENITA; *CE:MĬ.* Essentially trunkless; stems mostly erect, (1.5) 2–7 m, arising mostly at or near the base of the plant. Lower (sterile) portion of the stem 5-ribbed, the larger stems often (8.5) 14–16 cm in diameter, the spines often 5–7 mm; upper (fertile) portion 5 or 6- (7)-ribbed, 5.5–10 cm in diameter, the spines often 4–10 cm.

Flowers nocturnal, 2.5–4 × 3.8–4.2+ cm; weakly bilaterally symmetrical (the tube slightly upcurved, the style lying near the lower side of the tube); receptacular tube often 10–14 mm, interior diameter at throat ca. 8 mm; nectar chamber 1.1–1.8 mm deep, 2.8–3.0 mm wide. Inner tepals numbering ca. 25, 11–16 × 2.9–5.0 mm, whitish to pinkish, spreading open early in the evening (e.g., 8 P.M. on a hot, mid-August night), recurved by morning. Filaments, style, and stigmas white, the anthers cream colored. Lower filaments ca. 10 mm, not modified for enclosing the nectar chamber, the upper filaments ca. 6 mm. Receptacle tissue quickly blackening when injured.

Fruits ovoid to globose, 2.3–4.0 cm; receptacular bracts 7–14 (the higher numbers include a ring of additional minute bracts at base of fruit), broadly triangular, appressed, their bases conspicuously decurrent, their axils usually naked or bearing only short hairs, or occasionally a few spines. Seeds and pulp edible, the pulp sweet but not as delicious as that of organpipe or saguaro. Seeds blackish (dark red-brown under magnification and strong light), (2.2) 2.4–2.8 × 1.8–2.1 mm, shiny and smooth, with a low ridge. Flowers recorded March-December, but mostly April-September; fruits April-December, peak fruiting often in June. At peak fruiting the upper stems are loaded with bright red fruits. Even so, it is difficult to find ripe fruits in the morning before the birds and ants have hollowed them out.

Generally on fine-textured soils of level or nearly level terrain, occasionally on rocky slopes. The tallest and most massive plants of this species are on the northeast flank of the Sierra Pinacate on cinder- and ash-derived soils (about 6 km NE of Elegante Crater to the vicinity of Campo Rojo, ca. 220–420 m). Also common on gravelly soils of upper bajadas near the international border from Quitobaquito eastward to the Sierra Cubabi and scattered along the adjacent southern margins of Organ Pipe Monument, rare in the southeastern corner of Cabeza Prieta Refuge. (The only significant stand of senita in the United States occurs in Senita Basin, in Organ Pipe Monument.) A few small populations occur in sandy soils 15–25 km W of Los Vidrios and southward from the east side of the Pinacate volcanic field to coastal areas near Bahía de la Cholla, but it is rare near the coast.

Variety *schottii* is characterized by relatively thick stems, fewer stem ribs (5–8), and an essentially trunkless or basitonic branching habit. This is the northern variety, restricted to the desert in southwestern Arizona and western Sonora, Baja California, and northern Baja California Sur. It ranges southward in Sonora nearly to the Guaymas region on the coast and to Hermosillo farther inland.

Variety *schottii* is replaced to the south by var. *australis* (K. Brandegee) Borg (var. *tenuis* G.E. Lindsay), which ranges southward into northwestern Sinaloa. A similar scenario occurs in southern Baja California Sur. Var. *australis,* primarily in thornscrub, usually forms a short trunk, a relatively open branching system, and the stems are more slender with 6–10 ribs. The shift in architecture and morphology, from stems with a higher surface-to-volume ratio in the south to stems with a lower surface-to-volume ratio in the north, is interpreted as an adaptation to increased aridity northward. The northern distribution in Arizona and northern Sonora is probably limited by freezing weather. Presumably the western distribution in northwestern Sonora and western Arizona is limited by drought.

The flowers open shortly after sunset and generally close around sunrise, but may stay open longer on cooler spring days. Flowers sometimes open when a flowering stem is taken from midday summer heat to an air-conditioned room. Much like the well-known mutualism between yuccas and yucca moths, there is an obligate relationship between sinita and a pyralid moth, *Upiga virescens* (Fleming & Holland 1998). The larvae feed on sinita ovules. The female moths collect pollen on specialized abdominal scales and actively deposit pollen on the flower stigmas. As much as 90% of the fruit set may be due to these moths. Patrolling ants attracted to extra-floral nectaries in the areoles may provide protection from herbivorous insects (Turner, Bowers, & Burgess 1995). Birds and ants consume the pulp and seeds, leaving only the hollowed-out fruit case.

The 5-ribbed (sterile or juvenile) stems are boiled in water and the liquid drunk as a remedy for diabetes, cancer, and other ailments. Considerable quantities are harvested in northern Sonora and sold in Tijuana and elsewhere in Mexico.

Sierra de Sonoyita, *Schott July 1855* (holotype, MO, not seen). 14 mi SW of Sonoyta, *F 2817*. Campo Rojo, *Sanders 5706* (UCR). Observations: NW of Pinacate Peak, 750 m; Tecolote lava flow, with *Jatropha cuneata* as a nurse plant; 2.6 mi N of El Oasis, then 0.7 mi W of Mex 8 on "Dunas" rd.; E side of Cerros Batamote; Small dune, 5 km inland and ca. 10 km W of Puerto Peñasco.

MAMMILLARIA

Mostly small cacti, the stems mostly globose to short-cylindroid, simple or forming clusters, covered with non-grooved, prominent tubercles, with white "milky" latex in certain species (not in ours). Tubercles each with a terminal spine-bearing areole and subtending a very reduced, axillary, flower-bearing areole that sometimes also bears a few bristlelike spines. Central spines straight or hooked. Flowers diurnal, often relatively small for the family. Fruits succulent, globose to elongate, spineless, mostly red or pink and edible. Seeds usually pitted.

Western North America to northern South America, the vast majority in semiarid subtropical regions of Mexico; 150 species. The taxonomy of the genus is far from stable—there are many more names in current use than there are biological species. Six subgenera are currently recognized; ours belong to subgenus *Mammillaria. M. grahamii* and *M. thornberi* are in series *Ancistracanthae* but are otherwise not especially closely related. Although *M. tetrancistra* seems best placed near series *Ancistracanthae,* it is taxonomically isolated with no known close relatives. References: Craig 1945; Hunt 1971, 1978, 1983–87.

1. Seeds 1.6–2.4 mm, the lower 1/3 covered with a corky cuplike base, the seed surface reticulate and wrinkled but not pitted. (Note: Among the 3 species dried fruits can often be found below the spine clusters in the tubercle axils.) ____ **M. tetrancistra**

1′ Seeds slightly less than 1 mm, without a corky base, the seed surface pitted.
   2. Stems commonly 3.5–6.0 cm thick; hook of central spine 2–3 mm across; stigma green; widespread in the flora area, rocky places and plains. ____ **M. grahamii**
   2′ Stems commonly (2) 3.0–3.5 cm thick; hook of central spine 1.5 (2) mm across; stigma magenta; plains near Organ Pipe Monument. ____ **M. thornberi**

Mammillaria grahamii Engelmann [among the many synonyms are *M. grahamii* var. *arizonica* Quehl. *M. grahamii* var. *oliviae* (Orcutt) L.D. Benson. *M. microcarpa* "Engelmann" of various authors. *M. milleri* (Britton & Rose) Bödeker. *M. sheldonii* (Britton & Rose) Bödeker] *CABEZA DE VIEJO;* FISHHOOK CACTUS; *BA:BAN HA-'I:SWIGĬ, BAN CEKIDA* Stems relatively firm, solitary or with 1 to several branches, globose to several times taller than wide, commonly 5–15 (30) × 3.5–6.8 cm; underground portion of stem shorter than in *M. tetrancistra,* easily distinguished from the root system. Central spines 1–3 per areole, at least 1 of them strongly hooked; subcentrals 1–3, located only above the central(s), sometimes transitional to the central(s) but usually straight and barely distinguishable from the radials.

Inner tepals showy, pink to rose purple; outermost tepals with minutely fringed margins. Stigma green. Fruits ultimately red and several times longer than wide, often 25 × 8 mm, but sometimes remaining small and green (albeit with mature, viable seeds) for long periods prior to ripening. Seeds 0.75–0.95 mm, rounded-obovoid, blackish, pitted. In southern Arizona several flushes of flowers during summer, each about 5 days after soaking rains (Mark Dimmitt, personal communication 1990), also often in April.

Widespread; hills and mountains to near peak elevations in both the volcanic and granitic ranges, upper bajadas, valley plains, and sometimes upper floodplains; not on dunes or sandy plains, and not in the Sierra del Rosario. Especially common on the east and north sides of Sierra Pinacate, more numerous toward higher elevations. Plants in granitic mountains tend to have pale spines whereas those in volcanic areas tend to have dark spines. Southeastern California to western Texas, northern and western Chihuahua and Sonora nearly to the Sinaloa border.

Growing with *M. thornberi* from Quitobaquito to the nearby Río Sonoyta, and elsewhere with *M. tetrancistra*. Seeds of *M. grahamii* and *M. thornberi* are indistinguishable.

Dark-colored spines: 4.7 mi S of Mex 2 at 32 mi W of Sonoyta, *Pinkava 15502* (ASU). Sykes Crater, *F 20006*. Elegante Crater, *F 19701*. 2.5 km NW of Campo Rojo, *F 86-413*. Sierra Pinacate, 1000 m, *F 18666A*. **Light-colored spines:** Sonoyta *F 13340*. Sierra del Viejo, *F 88-99*. Quitobaquito, *F 86-173*.

Mammillaria tetrancistra Engelmann. CORKSEED FISHHOOK CACTUS. Stem single or with 1 to several branches, rather flabby, commonly 5–10 × 3.5–6.0 cm; underground portion of stem usually thickened, flabby, rootlike, often 5–8 cm; upper part of roots also often thickened. Spines dark or light colored depending largely on substrate color; central spines hooked, 1–3 (4) per areole; subcentral spines (barely distinguishable from the radials) several to 12 or more per areole, radiating in all directions and often resembling a supplementary ring of radial spines, giving the areoles the superficial appearance of having more radial spines than are really present.

Inner tepals showy, pink to rose purple; outermost tepals with margins more broadly fringed than those of *M. grahamii*. Stigma green. Seeds (not including strophiole) 1.6–2.4 mm at widest point, the surface rugose-roughened over a reticulate cellular pattern and not pitted; basal ca. 1/3 of seed covered with a light-colored, corky aril-like cup (the *strophiole,* a specialized outgrowth or proliferation of hilum tissue), the strophiole pale colored and spongy when fresh, drying brown and corky and remaining firmly attached to the seed.

Widespread, volcanic and granitic regions on hills and mountains to peak or near peak elevations. Not known elsewhere in Sonora. Also northeastern Baja California, western Arizona, southeastern California, southern Nevada, and southwestern Utah.

Often growing with *M. grahamii,* such as on the east side of the Pinacate volcanic field above Campo Rojo and in many of the granitic ranges. *M. tetrancistra* extends into more arid habitats and regions, including the Sierra del Rosario, than does *M. grahamii; M. tetrancistra* ranges into more extreme desert than any other species of *Mammillaria.*

With practice, locating subcentral spines radiating from the lower (abaxial) 2/3 of a spine cluster is sufficient to allow positive identification of *M. tetrancistra.* However, the absence of the diagnostic "extra" subcentral spines does not necessarily indicate *M. grahamii* because some plants of *M. tetrancistra,* especially when young, have their subcentral spines confined to the upper parts of the areoles as does *M. grahamii.* In that case distinguishing *M. tetrancistra* specimens from the relatively unrelated *M. grahamii* depends on flowers, fruits, seeds, and subtleties: In *M. tetrancistra* (1) the stem base is relatively smooth, soft, extensive, and deeply buried (often wedged between rocks); (2) the stems and tubercles are surprisingly soft-skinned and flaccid, not firm like those of *M. grahamii;* (3) the pith and cortex are mucilaginous (tangible even on the smallest cut surfaces, such as tubercle sections); (4) the radial and subcentral spines do not radiate as neatly and regularly as those of *M. grahamii,* giving the plant a relatively "shaggy" appearance and/or a "salt-and-pepper" color pattern rather than the "well-groomed" appearance of *M. grahamii;* (5) the tepals are narrower and more numerous than those of *M. grahamii;* (6) the outer tepals have relatively large fringes; (7) the floral remnant is deciduous, leaving a relatively large and conspicuous abscission scar on the fruits, unlike other *Mammillaria;* (8) the seeds are unlike those of any other cactus. The deceptively similar vegetative features of these distantly related mammillarias exemplify the difficulties sometimes inherent in cactus identification.

Sierra del Rosario, *F 20395*. 2.5 km NW of Campo Rojo, *F 86-412*. Sierra Pinacate, 1000 m, *F 18666*. Observation: 8 km SW of Sonoyta.

Mammillaria thornberi Orcutt [*M. fasciculata* of authors. Not *M. fasciculata* Engelmann, a name originally applied to *Echinocereus fasciculatus*] *CABEZA DE VIEJO;* THORNBER FISHHOOK CACTUS; *BA:BAN HA-MAUPPA.* Stems often (2) 3.0–3.5 cm in diameter, more than twice as long as wide, usually somewhat flaccid, commonly branching from the base and along the stem to eventually form small, many-stemmed colonies, these apparently sometimes breaking apart, the pieces forming adventitious roots, or occasionally remaining solitary or with only a few branches. Hooked central spines 1 per areole. Inner tepals white with bright rose pink midlines (broader than those of the 2 other mammillarias in the region). Stigma magenta. Seeds 0.9 mm, similar to those of *M. grahamii.* August.

Fairly common along the floodplain of the Río Sonoyta in former mesquite bosque from west of Sonoyta to south of El Papalote and northward to Quitobaquito. Often beneath small desert shrubs such as *Ambrosia deltoidea* and *Atriplex polycarpa.* Although at least 4 species of cacti are common in the adjacent sandy desert plain, other cacti are absent or rare in this former bosque. Southwestern Arizona and northwestern Sonora southward to sandy plains southwest of Caborca. Closely related to but obviously distinct from *M. yaquensis* R.T. Craig of southwestern Sonora and northwestern Sinaloa. They are disjunct by several hundred kilometers.

Old floodplain above Río Sonoyta, S of Quitobaquito, *F 89-13.* Quitobaquito, *F 86-174.* Quitovac, *Nabhan 241.* 20 km SW of Caborca, *Burgess 7000.*

OPUNTIA *CHOLLA, NOPAL;* PRICKLY PEAR

Trees, shrubs, or low clump- and mat-forming cacti. Stems with determinate, rhythmic (seasonal or annual), constricted growth increments resulting in individual stem segments: joints if cylindroid, pads (cladodes or phylloclades) if flattened. Glochids (small, readily detachable spines with tiny retrorse barbs) in at least the most recently formed areoles. Leaves on young growing stem segments, these 1 per areole, succulent, usually subulate or conical, soon deciduous as the shoots mature. Flowers mostly diurnal (except *O. fulgida, O. leptocaulis,* and perhaps others), with sensitive stamens (filaments closing inward when touched). Fruits mostly fleshy, but some with dry fruits. "Seeds" relatively large, covered with a white to brown (light tan in ours) bony aril-like structure, the hilum region often notched.

The bony aril-like covering of *Opuntia* seeds forms as flaps along the margins of the curled distal part of the funiculus (Flores & Engleman 1976). The seed proper has a nearly smooth, dark brown testa and a shape much like that of any other large cactus seed, but it is fused to the inner surface of the much larger and variously shaped aril. Descriptions of *Opuntia* "seeds" pertain to the aril and should not be directly compared with seed descriptions for other genera.

South America to Canada; tropical to temperate and desert regions; 160+ species. There is no modern treatment for all the species. Three well-marked major groups occur in the Sonoran Desert: *Opuntia, Cylindropuntia,* and *Corynopuntia.* These are variously treated as subgenera or separate genera. Prickly pears, *Opuntia* (*Platyopuntia*), have restricted distributions in northwestern Sonora, either at higher elevations (*O. chlorotica*) or barely entering the flora area from the northeast (*O. engelmannii*) or near the border in the northwest (*O. basilaris*). Some of the larger prickly pears, such as *O. ficus-indica* in south-central Mexico, are used extensively for food: the young pads (*nopales*) as a green vegetable, and the fruits (*tunas*) for their sweet pulp. *O. ficus-indica* thrives under cultivation in the flora region if given sufficient supplemental water.

The chollas, *Cylindropuntia,* include several species that are common and widespread in the flora area. This subgenus is characterized by the presence of spine sheaths. Its center of diversity is in the Sonoran Desert. Elsewhere in the Sonoran Desert nearly all cholla species from the flora area probably hybridize with one another when phenology and distributions overlap (Marc Baker, personal communication 1995). However, we have not found hybrids within the confines of the flora area.

The club chollas, *Corynopuntia* (section *Clavatae* of subgenus *Cylindropuntia,* sensu Benson 1982) are represented in the flora area only by *O. kunzei.* There is a spine sheath but it is only rudimentary. References: Flores & Engleman 1976; Grant & Hurd 1979; Parfitt & Baker 1993; Parfitt & Pickett 1980; Rebman 1995.

1. Prickly pears, the stem segments flattened or compressed (cladodes or pads); surfaces relatively flat, not tuberculate. ____ Subgenus **Opuntia:**
   2. Pads 9–12.5 cm, bluish to purplish, minutely hairy (use a lens); larger spines lacking (with glochids only); inner tepals rose pink. ____ **O. basilaris**

   2′ Pads usually 15 cm or more, green and glabrous; larger spines and glochids both present; inner tepals yellow to yellow-orange.
      3. Plants erect with a short but well-developed trunk; pads more or less orbicular; spines yellowish, becoming blackish with age, mostly pointing downward; higher elevations of Sierra Pinacate. ____ **O. chlorotica**

3′ Plants often sprawling-spreading, trunk not well developed; pads obovate to elliptic; spines mostly white with brownish bases, or if yellow, then relatively robust and the largest ones per areole porrect and/or projecting in various directions; Sonoyta region. ________________ **O. engelmannii**

1′ Chollas, the stem segments (joints) cylindroid, often tuberculate.

4. Plants usually forming spreading colonies much wider than tall, with 2–6 joints per stem; larger spines flattened, scabrous with tiny transverse ridges; spine sheaths vestigial and ephemeral, confined to young spine tips. ________________ Subgenus **Cornyopuntia: O. kunzei**

4′ Plants usually taller than wide and mostly not forming spreading colonies (except sometimes in *O. ramosissima*), with more than 6 joints per stem; spines terete, the surfaces barbed; sheaths covering most of each spine. ________________ Subgenus **Cylindropuntia:**

5. Fruits proliferating as perennial pendent chains, remaining green and fleshy when ripe, present all year. ________________ **O. fulgida**

5′ Fruits single, not proliferating as pendent chains (occasionally with 2 fruits together), dry or fleshy when ripe, generally not present all year.

6. Joints (3) 4–6 cm in diameter (excluding spines); main axis or trunk well developed, straight, and erect. ________________ **O. bigelovii**

6′ Joints 3 cm or less in diameter; main axis or trunk not well developed, or if so, then not straight and erect.

7. Joints 1.5 cm or more in diameter, their surfaces partially to fully obscured by spines, usually evenly spiny from tip to base of each joint; fruits dry at maturity.

8. Branching relatively lax, the stem tubercles mostly 15–30 mm; filaments red; inner tepals orange-brown to dull golden yellow; basal tubercles of fruits much longer than upper tubercles; Sonoyta region and the northwestern granitic mountains. ______ **O. acanthocarpa**

8′ Branching relatively dense (compact), the stem tubercles 11–16 mm; filaments green to yellow; inner tepals greenish yellow; basal tubercles of fruits nearly equal to the upper ones; widespread but absent from range of *O. acanthocarpa* in the Sonoyta region. ________________ **O. echinocarpa**

7′ Joints mostly l cm or less in diameter, their surfaces readily visible through sparse spines, the lower areoles of each joint likely to bear only glochids; fruits dry or fleshy.

9. Tubercles sharply defined with rhomboid (diamond-shaped) outlines; each areole in a groove; fruits dry at maturity. ________________ **O. ramosissima**

9′ Tubercles relatively obsolete, the stem surfaces smooth; areoles not in grooves; fruits fleshy at maturity.

10. Ripe fruits green or mostly green (sometimes tinged with pink or pale red); terminal joints mostly 7–12 mm in diameter; usually with a short but thick and woody trunk; northeastern quarter of the flora area. ________________ **O. arbuscula**

10′ Ripe fruits bright red-orange throughout; joints mostly 4.5–6.5 mm in diameter; without a well-defined trunk; widespread. ________________ **O. leptocaulis**

Opuntia acanthocarpa Engelmann & J.M. Bigelow. *CHOLLA;* BUCKHORN CHOLLA; *CIOLIM*. Shrubby chollas, often 1.0–1.5 m; trunk poorly defined, not straight; branching at various heights. Joints mostly 5–16 × (2) 2.5–2.8 (3) cm, highly variable in size, progressively shorter above. Stems moderately reddish purple during cooler, drier seasons. Midjoint stem tubercles 15–30 mm, more than twice as long as broad. Spines moderately to quite dense, often partially obscuring stem surfaces, usually dull brown to yellow beneath the sheaths. Flowers ca. 5–6 cm in diameter. Inner tepals uniformly orange-brown to dull golden yellow (in the floral area). Filaments dull red with green only at very base, the anthers dull golden yellow. Fruits spiny, drying at maturity; basal tubercles of fruits much longer than upper tubercles. Seeds cream colored to light tan, darkening with age, irregularly angled and moderately compressed, often thickly disk-shaped, highly variable in size within a given fruit, often (4.6) 5.0–5.5 (6.8) × (3.7) 4–5 × ca. 2 mm. April-May; fruits drying May-June.

Common and widespread on rocky hillsides, bajadas, and plains in the Sonoyta region with a westward extension along the Arizona border to the vicinity of the Represo Cipriano. Also on rock slopes and bajadas in the northwestern granitic mountains from the Sierra del Viejo northward into adjacent Arizona (e.g., the Tinajas Altas Mountains). From the vicinity of Sonoyta it extends southeastward to about the vicinity of Caborca. Southwestern Utah, western and central Arizona, northwestern Sonora, southern Nevada and southeastern California. Populations in northeastern Baja California have been segregated as *O. sanfelipensis* J. Rebman. "All four presently recognized varieties of *O. acanthocarpa* are diploid, but *O. sanfelipensis* is a hexaploid" (Rebman 1995:128; 1999).

The plants in the Sonoyta region are intermediate in spination between var. *coloradensis* L.D. Benson and var. *major* (Engelmann & J.M. Bigelow) L.D. Benson; however, these varieties are not well marked. Var. *major* is characterized by shorter, more densely spiny, less robust, and more spreading branches; the spines vary from yellowish to commonly dark with dark sheaths. Var. *major* generally replaces var. *coloradensis* to the east. Var. *coloradensis* has yellow spines and sheaths, these more inflated, and forms more robust plants with a stricter (upright) growth habit. Populations in the Tinajas Altas Mountains and the adjacent Sierra del Viejo can be ascribed to var. *coloradensis*. There seems to be clinal intergradation between a robust, densely spiny lowland race and an Arizona Upland race that is sparsely armed with brown spines. The neotype locality of var. *major* is so close to the Sonoyta region that the identification of those populations as var. *major* seems obligatory. However, this neotype locality is confusingly situated in or near the zone of intergradation; thus the epithet *major* is used for the eastern race more or less by convention.

Opuntia acanthocarpa might be confused with *O. echinocarpa*. The fruit umbilicus (depression of flower scar on fruits) is usually shallow in *O. acanthocarpa*. The seeds, at least among specimens from the flora area, are remarkably similar. In both species the developing seeds form facets or angles where they press against one another tightly packed in the nearly dry ovary of the developing fruit. (The fruits lack a succulent matrix that might keep the developing seeds apart.) The seeds of *O. echinocarpa* are often somewhat smoother, with few angles or facets, more nearly disk-shaped, and perhaps with fewer seeds per fruit than in *O. acanthocarpa*. However, the differences do not seem reliable enough to use as key characters. Stem tubercle length is an "easy" character that can be seen at any time of year, but flower color (in the flora area), especially the filaments, is probably the most diagnostic feature.

These species grow intermixed at Cerro Pinto and Sierra del Viejo along Mex 2 and northward into southern Yuma County. Across most of the rest of the flora area they have mutually exclusive but nearly contiguous distributions. *O. acanthocarpa* var. *major* occurs in Arizona Upland (Sonoyta region), whereas *O. echinocarpa* occurs at lower elevations westward and southward in the Lower Colorado Valley.

Var. **major:** 2 km W of Sonoyta, *F 85-713*. 11 mi SW of Sonoyta (Sierra Cipriano), *F 14440*. Quitovac, *Nabhan 298*. Observation: 27 mi W of Sonoyta. Var. **coloradensis:** Observations: Sierra del Viejo; Cerro Pinto.

Opuntia arbuscula Engelmann. *SIVIRI;* PENCIL CHOLLA; *WIPINOI.* Shrubby chollas, often 1–2.5 m, as wide or wider than tall, intricately branched with a dense crown. Trunk usually short but often well developed, sometimes 15 cm to the lowest branch and reaching 16 cm in diameter. Bark smooth and dark bronze on older branches and young trunks, becoming dark gray, scaly, and flaking on older trunks. Much-branched above, the joints (3.5) 4–15 cm × (7) 8.5–12.3 mm, green to yellow-green all year, becoming shorter upwards. Spines 2.9–5.0 cm, relatively sparse, red-brown at base and pale yellow-brown at apex, the sheaths translucent, pale brown; lower areoles on each joint spineless or with sporadic spines. Inner tepals greenish yellow to yellow-brown. Fruits fleshy, often 2.3–2.8 cm × 10.7–12 mm, green even when ripe, or sometimes with slight reddish or yellowish tinge, often persistent for ca. 1 year, bearing a few readily deciduous slender spines 17.8–26.7 mm, 0–6 seeded, the umbilicus deep. Seeds 4.9–5.3 mm in diameter. April–early May.

Sonoyta region and sporadically westward along the floodplain of the Río Sonoyta to the bajada plain northwest of Playa Díaz. Scattered, seldom common; small washes in low hills of decomposed granitic soil, and shallow drainageways in sandy-silty *Larrea* flats. Central Arizona to the coastal plain southeast of Guaymas.

Especially large, thick-trunked plants occur along the broad, shallow drainageway at the bottom of the expansive *Larrea* bajada-plain northwest of Playa Díaz. Like the few other *O. arbuscula* of known chromosome numbers, these are hexaploid. They are unusual because mitosis follows meiosis without the usual formation of cell walls and is followed by numerous additional mitotic divisions, indicating sexual reproduction in this population is unlikely (D. Pinkava, personal communication 1988). As with many chollas, during years of drought the flower buds often dry without opening.

2.8 mi SW of Sonoyta, *F 87-305.* 5 km SE of Pinacate Junction, *F 88-261* (*n* = ca. 33, D. Pinkava, 1988). Observation: 1.5 km SSW of Quitobaquito.

Opuntia basilaris Engelmann & J.M. Bigelow var. **basilaris.** BEAVER-TAIL PRICKLY PEAR. Small prickly pears, often 25–55 cm across, the individual stems consisting of 1 or rarely 2 pads (cladodes). Often forming a very thick, short caudex giving rise directly to thick, stubby pads. Pads 8–14 (mean = 11.6) × 4–9.8 (mean = 7.9) cm (12 plants, 24 pads), reaching 18 cm long in cultivation. Stem surfaces minutely and densely puberulent, pale bluish (the color does not rub off), somewhat pale purplish during cooler, drier months. Areoles bearing dense tufts of glochids, otherwise spineless. Flowers 7.0–7.5 cm in diameter, showy, opening for 1 or 2 days; inner tepals bright rose pink, obovate, 4.0–4.5 cm, the larger ones 2.5–3.2 cm wide, the apices apiculate and erose; outer tepals thick and fleshy, greenish brown. Filaments red; anthers pale yellow. Style pale rose pink, darker on second day; stigmas white, opening on second day. Pericarpel 3.5–4.8 cm, broadly cylindrical, greenish. Fruits dry at maturity. Seeds 7–9 mm wide, puffy, looking like compressed garbanzo beans. Early April; fruits ripening in late May.

Fairly common on open exposed dissected bajadas and pediments, lower slopes of the Sierra del Viejo, to about halfway up the mountain on Cerro Pinto, and similar habitats in adjacent Arizona. Not known elsewhere in Sonora; also western Arizona, southern California, southern Nevada, southwestern Utah, and northern Baja California. Var. *basilaris,* the most wide-ranging of 5 varieties recognized by Benson, nearly encompasses the distribution of the entire species.

The Sonoran plants have a thick caudex that seems unusually large in relation to the plant size. In drought the pads shrivel and curl inward, and some or even most of the pads drop off in severe drought, almost like deciduous leaves. In cultivation the pads can become substantially larger than those found in the wild. The showy red stamens are noteworthy. *O. basilaris* is taxonomically isolated with no known close relatives (Donald Pinkava, personal communication 1991).

Sierra del Viejo, *F 16918.* Observations: Cerro Pinto; Tinajas Altas Mts.

Opuntia bigelovii Engelmann var. **bigelovii.** *CHOLLA GÜERA;* TEDDYBEAR CHOLLA; *HAḌṢAḌKAM.* Chollas often 0.5–1.8 (2.2) m; trunk erect, stout, straight, and beset with dead, persistent branches (joints) with blackened spines. Upper joints breaking off at the slightest touch, relatively short and thick, 7.5–16.0 × 3.5–5.0 cm, green all year; tubercles as broad as long and very low. Spines sheathed, extremely dense and obscuring the stem surface, dull yellow, seemingly brighter among the inland populations and duller in the coastal populations. Leaves 2.2–3 mm, slender, subulate, red-tinted, very quickly deciduous. Extra-floral nectaries (areolar glands) green when fresh, seen only on new growth.

Flowers yellow-green, 5–6 cm × 36–40 (55) mm, attached more firmly than the ultimate joints. Pericarpel at anthesis 20–32 × 20–23 mm, green, obovoid–barrel-shaped, strongly tuberculate. Areoles of the pericarpel 48–56 in number, lacking larger spines but bearing (1) readily deciduous spine-glochid transition forms in lower and central parts of areoles, these 9.2–14 mm, sometimes apically sheathed, but the sheaths extremely ephemeral and disintegrating at a touch; and (2) dense, 2-parted (bilaterally symmetrical) tufts of ordinary glochids ca. 2–3 mm. Receptacular bracts ca. 3 mm, deltoid, inconspicuous, strongly ascending, pale yellow-green tinted with red. Tepals ca. 34, the inner tepals 8–10, their insertions widely separated and not completely ringing the receptacle, not hiding the short outer tepals. Margins of outermost tepals minutely fringed to entire except near tips where erose-serrate. Inner tepals 24–29 × 10.0–12.5 mm, oblanceolate to obcuneate, pale yellow-green to fairly bright yellow-green at bases, the margins essentially colorless and erose. Stamens sensitive, the filaments brilliant grass green, the anthers bright yellow, sometimes with a slight orange tinge; anthers at least sometimes (Pinacate region) stunted, lacking pollen. Style thick, ca. 3 cm. Stigma green with about 7 lobes. Fruits 27.5–35.3 × 20.5–25.5 mm, only moderately fleshy, yellow, solitary, spineless or with slender deciduous spines to 15 mm, the umbilicus 6.4–11.4 mm deep. Seeds variously present or absent. March-April.

This is the most common and widespread cactus in northwestern Sonora, occurring nearly throughout the flora area except dunes, most of the *Larrea* flats and plains, and the Sierra del Rosario. Common near sea level to peak elevations on Pinacate Peak, the granitic ranges, and the Sonoyta region; desert pavements, sandy plains and flats, bajadas, hills, and steep slopes. Notably dense stands occur on cinder soils on the steep slopes of the main peaks of the Sierra Pinacate and their eastern and southern flanks. Bahía de la Cholla is named for the extensive, dense population on the sand flats around the bay; these plants, often 50–80 cm, are shorter than those of inland populations. Southeastern California, southern Nevada, southern and western Arizona, eastern Baja California, northeastern Baja California Sur, and Sonora south to coastal mountains south of Tastiota (vicinity of 28°08′ N). Var. *ciribe* (Engelmann) W.T. Marshall (*O. ciribe* Engelmann) replaces var. *bigelovii* southward in the Sierra de la Giganta in Baja California Sur (Rebman 1995).

Opuntia bigelovii is a clonal species that is very successful in terms of abundance and geographic distribution. The upper or younger joints fall at a touch, and the spines are difficult and painful to pull out of flesh. The plants propagate prolifically from readily rooting fallen joints. All plants sampled at Organ Pipe Monument are triploid (Marc Baker, personal communication 1990); consequently there is very little chance of the seeds being viable. Pollen production is highly variable in this species, at least the triploid plants produce relatively few pollen grains. In the Tucson region pollen production may be higher in seasons of high rainfall and lower during drier periods (Mary Kay O'Rourke, personal communication 1991). This species and *O. fulgida* share the character of having fewer inner tepals than most other opuntias. This reduction in the perianth seems to be correlated with reliance on asexual reproduction.

Cactus-*Larrea* flat, 2.5 mi S of Mex 2 at 32 mi W of Sonoyta, *Lehto 15507* (ASU). Sierra Blanca, *F 88-226*. Bahía de la Cholla, *Shelton 3 Apr 1958* (POM). Observations: Hills W of Sonoyta; Elegante Crater; Sierra Extraña; Sierra del Viejo; 12 km W of Cerro Pinto on Mex 2.

Opuntia chlorotica Engelmann & J.M. Bigelow. PANCAKE CACTUS. Prickly pears often 1–1.6 m, as wide or wider than tall, with a short, thick trunk. Pads erect, oval, obovate, or nearly round in outline, often wider than tall, (10) 13–23 × (10) 13–25 cm, slightly glaucous, not turning reddish or purplish. Areoles all bearing spines, or a few lowermost areoles spineless. Spines straw colored, mostly pointing downward; persisting on older pads and trunk and turning grayish or blackish, those of lower areoles often long and shaggy. Flowers large, inner tepals yellow with a faint red midstripe even when freshly open; stigmas green. Fruits fleshy, glaucous, rose pink. April.

Sierra Pinacate above ca. 750 m, mostly on the northern side of the mountain mass. Elsewhere in northern Sonora at similar elevations, including the Sierra del Viejo near Caborca and more extensively in northeastern Sonora. Also Baja California, northwestern Chihuahua, southeastern California, southern Nevada, Arizona, New Mexico, and southwestern Utah. In the Sonoran Desert region characteristic of desert–oak woodland ecotone and often in oak woodland.

Ferguson (1988) treated *O. santa-rita* Griffiths & Hare and *O. gosseliniana* F.A.C. Weber as varieties of *O. chlorotica*. *O. gosseliniana* and *O. santa-rita* have betacyanin pigments imparting a purplish color to the plants during winter and dry seasons; *O. chlorotica* lacks this color. Perhaps the observed intermediate forms represent hybridization rather than incomplete evolutionary divergence.

Opuntia echinocarpa Engelmann & J.M. Bigelow [*O. wigginsii* L.D. Benson] SILVER CHOLLA. Chollas mostly 0.2–0.8 m and wider than tall in the Sierra del Rosario region and the west side of the Pinacate volcanic field, and often reaching 0.8–1.0 m in the eastern and central portion of the volcanic region and the granitic mountains. Primary stem(s) of larger plants mostly erect, or sometimes stiffly horizontal just above the soil surface. Joints (1.5) 5–13 × 1.8–3.2 cm, sometimes with growth continuous over two seasons; stem tubercles 11.5–16.0 mm (dried specimens), usually less than twice as long as broad. All areoles densely spiny, the spine sheaths yellowish or whitish (silvery). Stem leaves thick, green, often 7 mm, very quickly deciduous. Flowers 1–4 from a single joint, often 5.3–6.5 cm wide; inner tepals glossy, shiny pale greenish yellow; outer tepals pale greenish rose colored; filaments green to greenish yellow (not brilliant green as in *O. bigelovii*). Fruits spiny, dry at maturity, often 15–20 × 17–22 mm. Seeds cream colored to very light tan, irregularly angled and compressed, often thickly disk-shaped, highly variable in size and shape even in the same fruit, often (4.8) 5.0–5.8 mm maximum diameter and ca. 2 mm or more in thickness. March-April; fruits ripening in June.

Across most of the lowlands of the flora area but not on moving dunes and not in the Sonoyta region (see *O. acanthocarpa*). Sandy and rocky soils near mountain bases and sandy soils of open desert plains; less common on slopes—occasional on south side of Sierra Pinacate to 875 m; rare on low dunes. Especially common at the base of the granitic mountains and the western and southern flanks of the Pinacate volcanic region. Not known elsewhere in Sonora; otherwise northeastern Baja California, southeastern California, southern Nevada, southwestern Utah, and western Arizona.

Stunted and/or immature plants of *O. echinocarpa* probably account for most of the specimens assigned to *O. wigginsii,* such as the small plants in the Sierra del Rosario region.

Sierra del Rosario, *F 75-28A*. Dunes NE of Sierra del Rosario, *F 20431*. Tinaja del Tule, *F 18777*. 18 mi N of Puerto Peñasco, *F 13182*. 23 mi SW of Sonoyta, *Wiggins 8356* (DS).

Opuntia engelmannii Salm-Dyck ex Engelmann. *NOPAL;* DESERT PRICKLY PEAR; *NAW, 'I:BHAI* (FRUIT). Robust prickly pears, trunkless, usually spreading or sprawling, often 1–1.5 × 1–3 m. Stems compressed; pads (cladodes) ca. 1 cm thick, obovate to elliptic, (18) 20–33 × 16–23 cm, green all year. Larger spines 3.0–6.5 (7.7) cm, robust, straight or curved, yellowish to red-brown basally, whitish or yellowish distally, commonly bleaching all white with age (spines on partially shaded parts of pads often not bleached). Flowers often 7.5–9.5 cm wide; inner tepals uniformly bright yellow, changing to yellow-orange (apricot color) as the flowers age; filaments and anthers pale yellow, the style white, the stigma green; flowering in spring, mostly April. Fruits 5.5–7.5 × 3.5–4.0 cm, purple-red including pulp, the pulp juicy and sweet, ripening in August.

Northern Mexico and southwestern United States. Parfitt & Pinkava (1988) recognize 6 varieties, with var. *engelmannii* being the most widespread. They treat *O. phaeacantha* Engelmann as a separate species.

Opuntia engelmannii var. **engelmannii** [*O. discata* Griffiths. *O. phaeacantha* var. *discata* (Griffiths) Benson & Walkington] Spines mostly straight, reddish brown, often lighter colored (white-coated) near tip, often 3–5 per areole, the larger ones 3.0–6.2 cm.

Extending into the Sonoyta region from adjacent Arizona and farther east in northern Sonora; infrequent, mostly in small, widely scattered, and often clonal colonies, although seedlings are sometimes encountered. Often on sandy-gravelly soils of upper bajadas or their shallow washes, and sometimes along the upper floodplain of the Río Sonoyta, less often on rocky slopes.

Texas to southwestern Utah and southern California and southward to the adjacent border states in Mexico. This is the common desert prickly pear of much of Arizona and Sonora. Although widespread in northern Sonora, it has been found in only a few localities on the Sonoran coast. In western Sonora it occurs southward to Isla Alcatraz in Bahía Kino and mountains near El Desemboque San Ignacio (Felger & Moser 1985). The plants in the flora area have relatively small pads, long spines, and diverse spine colors, perhaps reflecting ancient introgression from *O. phaeacantha.*

2 km W of Sonoyta, *F 85-712*. Quitobaquito, *F 90-430*. Observation: 1.6 km SSW of Quitobaquito.

Opuntia engelmannii var. **flavispina** (L.D. Benson) B.D. Parfitt & Pinkava [*O. phaeacantha* var. *flavispina* L.D. Benson] YELLOW-SPINED DESERT PRICKLY PEAR. Differing from var. *engelmannii* by the relatively fewer and bright yellow (not white coated) spines (often only 1 per areole and porrect, seldom more than 3 large spines per areole), and the plants somewhat more robust. Plants to 1.5 m, the branches held well above the ground. Larger spines 4.5–7.7 cm, often curved. Flowers and fruits similar in size and color to those of var. *engelmannii.*

This taxon is known for certain only from southern Arizona but occurs within 100 m of the international border at Quitobaquito, where it grows with var. *engelmannii;* juvenile plants of both varieties are present. There is also a small population in the Agua Dulce Mountains in the Cabeza Prieta Refuge, close to the Sonora border. The type collection is from the Ajo Mountains in Organ Pipe Monument.

Although the population at Quitobaquito is small, with about one dozen adult plants or colonies, the plants are vigorous and seem to be increasing due to discontinuation of cattle grazing (Felger et al. 1992). The Quitobaquito, Agua Dulce Mountains, and Ajo Mountains populations are at former O'odham village sites, and it is tempting to speculate that these prickly pears represent selections and former plantings by the local people.

Quitobaquito, *Baker 7625*. N slope of Agua Dulce Mts near Agua Dulce Pass, *F 92-737*.

Opuntia fulgida Engelmann var. **fulgida.** JUMPING CHOLLA, CHAIN-FRUIT CHOLLA; *HANAM.* Shrubby to arborescent chollas, often 1–3 (4) m; trunk well developed but not straight, sometimes reaching 20–22 cm in diameter, and often with several major branches. (Some are tall enough that you can stand in the shade of the spreading jumping cholla, being careful of spiny joints overhead and fallen joints littering the ground.) Smaller branches and joints, mostly at the crown or top of the plants, green all year, falling at the slightest touch; tubercles longer than wide and prominent. Spines usually obscuring the stem surface; sheaths large, golden yellow to straw colored.

Flowers 2.5–5.0 cm wide. Inner tepals bright purplish pink, spreading widely and sometimes curling back, relatively few and not totally ringing the flower, the larger tepals often 14–20 × 6–7 mm. Filaments bright purplish pink, or the inner filaments sometimes white; anthers white. Stigma white. Flowering during warmer times of year, often following summer rains, and also mid-April–early May. (During summer the flowers open in the late afternoon, sometimes not until 6 P.M., remain open during the night, and wither sometime before dawn. The tepals, stamens, and style usually fall as a unit. The flowers are sometimes also open in the early morning. The flowers are visited by medium-sized moths at night and by medium-sized bees during daylight.) Fruits often 3–6 × 2–4.8 cm, mostly obovoid, remaining green and fleshy, proliferating in hanging chains of (3) 6–15 fruits (the longer chains swaying with the wind), often persisting for several years.

Propagation vegetative, from readily rooting fallen joints and fruits; reproduction by seed not observed in the flora area. *O. fulgida* is mostly diploid; a small percentage is triploid and presumably sterile or nearly so. The diploids are commonly of low fertility and production of viable seed is a rare event (Baker & Pinkava 1987).

Widespread in the flora area but the distribution patchy, mostly in lowlands, on rocky slopes, sandy and clayish-soil flats, low stabilized dunes, alluvial fans and playas, plains, and less often on low hills: Sonoyta region, Pinacate volcanic field, desert plains (especially north of the Pinacate volcanic field), and bases of some granitic ranges. Extensive jumping cholla "forests" occur on certain low sandy hills or plains (e.g., sand slope ca. 9 km W of Los Vidrios). These great stands consist mostly of plants of the same age class and are subject to massive die-offs. Similar die-offs are recorded for southern Arizona, and the plants are estimated to have a life span of about 40 to 80 years. For a summary discussion of the biology of this species see Turner, Bowers, & Burgess (1995).

This species ranges from southern Arizona to northwestern Sinaloa and is common in many desert and thornscrub regions of Sonora. It is very rare in southwestern New Mexico, where extremes of winter freezing preclude long-term establishment. In the Baja California Peninsula it is replaced by the closely related and similar-appearing *O. cholla* F.A.C. Weber.

The plants in the flora area are var. *fulgida,* characterized by inflated, papery-sheathed, large spines densely covering the stems. Var. *mamillata* (Schott) J.M. Coulter, not known from the flora area, has fewer and shorter spines not obscuring the stem surface and tight-fitting sheaths. Elsewhere the two varieties often grow intermixed and intermediates are often present. Apparently var. *fulgida* does not extend south of the Guaymas region. Var. *mamillata* ranges to Sinaloa but does not reach the large sizes common for var. *fulgida* in the flora area.

28 km S[W] of Sonoyta, *Lott 20 Nov 1965* (ASU). 25 mi S[W] of Sonoyta, *Shelton 3 Apr 1958* (RSA). Cactus-*Larrea* flat, 32 mi W of Sonoyta, *Lehto 15508* (ASU). Aguajita, *F 87-264.* Observations: Bahía de la Cholla; Elegante Crater; MacDougal Pass.

Opuntia kunzei Rose [*O. stanlyi* Engelmann var. *kunzei* (Rose) L.D. Benson. *O. stanlyi* var. *wrightiana* (Baxter) L.D. Benson. *O. wrightiana* Baxter. *O. stanlyi* var. *peeblesiana* L.D. Benson, in part] DESERT CLUB CHOLLA. Cholla-like plants of low and spreading colonies often 1.0–3.5 m wide. Branches erect, stout, and tough, often 30–57 cm with 2–6 joints per stem or chain, the joints separable only by tearing the woody axis; branching mostly from near base of plant, but new branches held aloft and often branching from ca. 6–10 cm above ground level. Joints green all year, the larger ones often 8.5–16.0 × 3.3–4.8 cm, more or less cylindroid with elongated, prominent tubercles. Spines stout and very sharp and rigid, ca. 20–30 per areole, most of the larger ones deflexed downward, sheathed only at the tips (a key character of the *Corynopuntia* or club chollas); new spines purple-red at base. Larger spines 40–48 × 1.5–2.0 mm maximum width, flattened or diamond-shaped in cross section; smaller spines terete;

glochids few, in upper parts of areoles. Leaves quickly deciduous, subulate, strongly tinted with red, ca. 10 mm.

Flowers 44–55 mm wide. Inner tepals pale yellow or cream colored, the larger ones ca. 24–35 mm; outer tepals thickish with thickened reddish tips, grading into the terete, reddish, and reduced pericarpel leaves. Filaments sensitive, green basally, cream-white to yellow distally; anthers pale yellow. Stigma cream colored, 12-lobed. Fertile fruits 4.5–8.0 × 2.4–3.0 cm, separating relatively easily from the stems, fleshy, obovoid, the pulp yellowish, the skin dull yellow (green where not exposed to sunlight), the areoles with spine clusters instead of glochid tufts; floral remnants persistent. Sterile fruits with floral remnants but lacking seeds, usually cylindroid (more narrowly cylindroid than the fertile fruits), remaining green, and more firmly attached to the stems than the seed-bearing fruits. Seeds 3.3–4.9 mm and ca. 3/4 to about as wide, 1.5–2.0 mm thick, light tan, smooth. May-June; ripe fruits observed late August–May.

As the fleshy, yellow fruits ripen in spring almost all of them (at least during dry years) are chewed open on one side, presumably by small mammals, and the seeds missing. Differences in fruit ripening and persistence may be related to soil moisture and temperature or season.

Distribution patchy, mostly on sandy soils. Across the northern part of the flora area from near Sonoyta to the west side of the northwestern granitic mountains, with extensive stands on the sandy cactus-dominated flats north and northeast of Cerro Colorado; also common on low stabilized dunes with *Larrea* on the west side of the Pinacate lava shield, from near Sierra Extraña northward to Mex 2. Also fairly common on broad, high lava mesas at ca. 800 m along the south side of Pinacate Peak as well as occasional, small colonies on cinder soils and sandy pockets elsewhere at higher elevations in the Sierra Pinacate.

Northwestern Sonora near the coast to Bahía San Jorge (vicinity of 31°10′ N) and eastward to Tubutama (vicinity of 30°53′ N, 111°27′ W), northeastern Baja California north to San Felipe, and southwestern Arizona in Yuma and La Paz counties and Organ Pipe Monument in western Pima County.

There are 3 closely related, allopatric, and non-intergrading species among the club-cholla cacti included in *O. "stanlyi"* (*O. emoryi* Engelmann) by Benson (1969a, 1982). *O. kunzei* differs from the others in having (1) strongly ascending stems (usually 3–5 joints per chain held aloft); (2) numerous large spines on the fruits, instead of accrescent glochid tufts and/or rare and inconspicuous spines; and (3) tubercles more or less coalescent into ribs. Moreover, the joints of *O. parishii* Orcutt are only half as large, and the plants as a whole are less than 1/4 the size of those of *O. kunzei. O. emoryi* (known in the literature as *O. stanlyi* var. *stanlyi;* Pinkava & Parfitt 1988) has only half as many spines per areole as either of the other species.

Benson (1969a, 1982) indicated sympatry between *O. kunzei* and another club-cholla taxon, *O. stanlyi* var. *peeblesiana* L.D. Benson; he asserted that var. *peeblesiana* "intergrades completely with var. *kunzei* in some areas" (Benson 1982:361). It now appears that the collections of "var. *peeblesiana*" in apparent sympatry with *O. kunzei,* including specimens from Quitobaquito, were merely the small extremes within normal populations of *O. kunzei.* This explains the "complete intergradation" reported by Benson. However, the remainder of Benson's var. *peeblesiana* was based on populations of *O. parishii,* which is ecologically segregated from *O. kunzei,* not known to intergrade with it, and has a different ploidy level (Pinkava et al. 1992).

Populations of *O. parishii* occur in the northern part of Organ Pipe Monument and the O'odham reservation in western Pima County, Arizona. *O. parishii* does not occur in the lower elevations of the adjacent Pinacate region or anywhere else within the range of *O. kunzei.* These "upland" club chollas (*O. parishii*) are otherwise known only from disjunct parts of the Mojave Desert. Therefore, *O. parishii* appears to be a relict species disjunct across the Colorado Desert, whereas *O. kunzei* is a species restricted to the intervening lowlands.

Presumably the closest relative to *O. kunzei* is *O. emoryi,* which has a patchy distribution on deep soils from southeastern Arizona to Trans-Pecos Texas. *O. kunzei* is a tetraploid species, with $2n = 44$.

33 mi S[W] of Sonoyta, *Keck 4191* (DS, UC). Cactus-*Larrea* flat, 32 mi W of Sonoyta, *Lehto 15504* (ASU). Aguajita, $n = 22$, *Baker 7613* (ASU). Quitobaquito, *Mearns 2735* (DS). Observations: MacDougal Pass; NW side of Sierra Extraña; SE of Moon Crater; Tinaja del Tule; 12 km W of Cerro Pinto.

Opuntia leptocaulis de Candolle. *TASAJILLO;* DESERT CHRISTMAS CACTUS; *'AJI WIPINOI, CE'ECEM WIPINOI.* Slender-stem chollas, 0.5–1.0 m (reaching 1.5+ m when growing through shrubs), larger branches mostly basal. Stems cylindrical, 4.8–6.6 mm in diameter (sometimes shrivelling to 3 mm in drought), green all year, the surfaces relatively smooth with low, poorly defined tubercles. Areoles spineless or producing a single spine often 1.5–8.3 cm with a deciduous yellowish sheath. Small lateral branches less spiny than the main axes, often bearing only glochids. Young plants with a single, somewhat tuberous root.

Flowers 1.5–2.0 cm wide, usually opening late in the afternoon and remaining open in the early evening (in late May observed opening at 4 P.M. and closing at 7:30 P.M.). Inner tepals pale yellow-green to cream colored; outer tepals red-purple. Stamens very sensitive, rapidly converging when touched. Filaments, anthers, style, and stigma cream colored. Fruits (10) 14–15 (22.5) × 8 mm, broadly ellipsoid to obovoid, fleshy, the skin red, the pulp pale yellow-orange, with widely separated tufts of small glochids; often also with larger fruits 25–30 × 8.4–10.5 mm, obovoid to clavate; both classes of fruits several-seeded or sometimes seedless, and sometimes producing a short green joint or a second fruit from 1 or more of the upper areoles. Seeds 4.3–4.5 mm in diameter. April-May; fruits mostly ripening November-December, often persisting at least until May.

Widespread and sometimes common; lowland flats, margins and floodplains of washes, bajadas, coastal plains, hills, and mountains across the northern part of the flora area and the Pinacate volcanic field. In many habitats but often concentrated along drainageways; usually absent from the hottest, driest habitats. Oklahoma and Texas to Arizona and southward in Mexico to Puebla; widespread in the deserts and lowlands of Sonora to northwestern Sinaloa.

This species is the most slender-stemmed of the chollas. Despite geographical and cytological variation, we do not recognize infraspecific taxa, although Bravo-Hollis (1978) lists 8 varieties. Reports of *O. leptocaulis* from the Baja California Peninsula are actually the closely related *O. lindsayi* J. Rebman (Rebman 1995).

Cactus-*Larrea* flat, 32 mi W of Sonoyta, *Lehto 15506* (ASU). 4.5 mi N of Cerro Colorado, *F 10417.* NW side Sierra Pinacate, 650 m, *F 18662.* San Luis, *Breedlove 1369* (DS). Quitobaquito, *F 90-38.* Observations: W of Sonoyta; granitic hill at Pinacate Junction; Represo Cipriano.

Opuntia ramosissima Engelmann. DIAMOND CHOLLA. Chollas, often 0.3–0.6 (1) m, usually broader than tall, sometimes with short, thick, woody trunks, the principal branches mostly basal. Joints cylindroid, 6.5–9.1 mm in diameter, dull greenish to gray-green. Spines sometimes none, or 1 per areole in upper portion of each joint, (1.5) 2.0–8.7 cm, with deciduous straw-colored sheaths. Areoles linear, recessed, each forming a conspicuous longitudinal groove in the upper end of its subtending tubercle, the glochids mostly buried in the groove; young stem and floral areoles with small, dense tufts of white hairs. Tubercles very low but sharply defined, their outlines diamond-shaped.

Flowers 2.5–3.0 cm wide. Inner tepals pale yellow-brown, some suffused medially near apex with red-purple. Filaments bright green, the anthers bright yellow. Style and stigma cream colored. Developing fruits with accrescent spines; mature fruits characteristically dry, 17.5–42 × (6.5) 8.3–10.7 mm, mostly persistent, densely spiny, the longer spines 14–22 mm. May; flowers opening in mid to late afternoon, often about 3 P.M. and closing 1–2 hours after sunset. Fruits apparently ripening early summer. (During the exceptionally wet El Niño winter of 1991–92, ripe, fleshy fruits persisted until late January and early February.)

Widely scattered, disjunct populations or colonies, mostly on sandy flats and bajadas, occasionally on small granitic hills. Lower elevations at the periphery of the Pinacate volcanic field, and granitic soils near bases of some of the granitic mountains; especially common in the creosote bush and cactus flats between Pinacate Junction and the Tezontle cinder mine, and the desert plains bordering the northwestern granitic mountains such as Cerro Pinto. A single colony of several plants known from the Sierra del Rosario. Also adjacent to the flora area south of Sonoyta but not known elsewhere in Sonora. Also southwestern Arizona, southern Nevada, southeastern California, and northeastern Baja California.

10 mi S of Sonoyta on Mex 2, *F 20554*. Cactus-*Larrea* flat, 32 mi W of Sonoyta, *Lehto 15505* (ASU). Pinacate Junction, *F 20606*. N of Cerro Colorado, *F 10417*. Sierra del Rosario, *F 75-28*. Organ Pipe Monument, 1 mi S of Papago Well Rd, *Steenbergh 1-1062* (ORPI).

PENIOCEREUS

Stems slender, erect to scandent or climbing, with few ribs, or cylindrical without ribs. Roots tuberous, single, and sometimes large, or clustered and dahlia-like, or many and potato-like on a "normal" branched root system. Flowers relatively large with a long tube, nocturnal or nearly so, white (except *P. viperinus*) and fragrant as far as known. Fruits relatively large in relation to stem diameter, fleshy, the pericarp (rind) red when ripe (at least in ours), the pulp sweet, juicy, and edible. Seeds blackish.

Through much of the arid and seasonally dry lowlands of Mexico to western Texas, southern New Mexico, and southern Arizona; 16 species, 4 in the Sonoran Desert. The genus is strangely absent from the Atlantic coast, including the Yucatán Peninsula. References: Buxbaum 1975b; Pinkava 1995; Sánchez-Mejorada R. 1973, 1974.

1. Stems ca. 20 mm in diameter, 4- or 5-angled. ______ **P. greggii**
1′ Stems 4.5–8 mm in diameter, with 6–9 flat-topped ribs separated by narrow furrows or grooves. ______ **P. striatus**

Peniocereus greggii (Engelmann) Britton & Rose var. **transmontanus** (Engelmann) Backeberg [*Cereus greggii* Engelmann var. *transmontanus* Engelmann] *SARAMATRACA, REINA DE LA NOCHE, HUEVO DE VENADO;* DESERT NIGHT-BLOOMING CEREUS; *HO'OK WA:'O*. One to several stems arising from a single very large, tuberous root. Stems sticklike, single to few-branched, seeming to mimic dead branches of shrubs or trees, 30–50 (150) cm, ca. 2 cm in diameter, the surface purplish brown and minutely but very densely pubescent with short, whitish hairs; stem ribs 4–7. Spines straight, the radials appressed, the 1 or 2 lower ones in each areole often light tan, 3.0–4.5 mm, the upper ones stouter, thickened at base, often blackish, the uppermost often 1.0–1.2 mm; central spine erect, bulbous-based, blackish, ca. 1 mm.

Flowers large and white with a long, slender, purplish brown tube; spectacularly fragrant, opening shortly after sunset and wilting quickly after sunrise. Fruits ca. 10 × 4 cm, ovoid, indehiscent, the skin bright red, the pulp juicy, sweet, and brilliant red; dry perianth and tube (i.e., floral remnant) persistent on fruit. Seeds 3.0–3.5 mm, blackish, obscurely keeled or ridged, rugose at least dorsally due to wrinkling or folding of the whole testa, but individual cells varying from weakly convex to flat with sharply recessed interstices. Flowering synchronously, late May-July, the fruits ripening August-September, rarely persisting until March.

Widely scattered at lower elevations on the north and east sides of the Pinacate volcanic region, rarely at higher elevations on the north side of the Sierra Pinacate, and in the Sonoyta region. Rocky lava soil, sandy flats, and fine-textured clayish soils; growing through desert shrubs or sometimes completely exposed. In addition to being cryptic and usually overlooked, it is genuinely rare in the flora area. Six plants are known from the vicinity of Aguajita Wash, but otherwise only about 12 plants have been located in the flora area during a span of about 2 decades, each at a different site. Throughout its range it characteristically grows at low density, but the flora area is at the eco-geographic limit for the species. It is more numerous in Organ Pipe Monument. Sphinx moths are presumably the pollinators.

Variety *transmontanus* is largely a Sonoran Desert taxon in southern Arizona and the northern third of Sonora, and extends into the desert grassland bordering the Chihuahuan Desert in southeastern Arizona. The allopatric var. *greggii* is a Chihuahuan Desert taxon, occurring from southeastern Arizona eastward through southern New Mexico, western Texas, Chihuahua, northeastern Durango, Coahuila, and Zacatecas. *P. greggii* has not been found in western, coastal regions of Sonora. The southern limit in Sonora is in the vicinity of Hermosillo.

The root was "in much repute as a remedy for syphilis throughout [the] northwest portion of Mexico" (Mearns 1892–93). The root is used as a remedy for rheumatism, cut into slices and applied as a poultice to the palm of the hand or sole of the foot. The tuberous root has also been used medicinally for diabetes and respiratory ailments, and the fruits are eaten fresh (Felger et al. 1992).

Sonoyta, *MacDougal 15* (US). 32 mi W of Sonoyta, *Lehto 15503* (ASU). 2 mi S of Tinajas de los Pápagos, *F 20140*. Sierra Pinacate, NW of main peak, 650 m, *F 18661*. Aguajita, *F 90-574*. Tinajas Altas, *Mearns 2811* (US). Observations: S of Pinacate Junction; Pinacate Peak, below summit, *Jonas Lüthy;* MacDougal Pass.

Peniocereus striatus (Brandegee) Buxbaum [*Cereus striatus* Brandegee. *Wilcoxia striata* (Brandegee) Britton & Rose. *Neoevansia striata* (Brandegee) Sánchez-Mejorada. *Cereus diguetii* F.A.C. Weber. *Wilcoxia diguetii* (F.A.C. Weber) Diguet & Guillaumin] *SACAMATRACA, SARAMATRACA; 'I:KUL*. About 12 or more potato-like tuberous roots strung on clusters of slender connecting roots. Stems ca. 0.8–1 m, or to ca. 2 m beneath desert legume trees, mostly with several to many branches. Stems 4.5–8.0 mm in diameter (1–2 years old), at first greenish, with minute hairs, soon becoming brownish or grayish and glabrous, with 6–9 flat ribs separated by deep furrows or grooves, the stomata restricted to the grooves. Areoles circular, ca. 1 mm in diameter; spines 5–15 per areole, innocuous and bristlelike, mostly soon deciduous, the 2 centrals shorter than the radials, the radials closely appressed, 5–8 mm, white or brown-tipped.

Flowers nocturnal, ca. 8 × 8 cm, salverform; pericarpel areoles bearing short white-woolly hairs and spines 3–6 mm; floral tube areoles bearing felt-like white hairs and bristle-like white spines (often twisted and flattened) 6–16 mm. Fruits ca. 5 cm, ovoid, the rind red when ripe, the floral remnant or its base persistent, the ripe pulp red, juicy, sweet, and edible (Felger & Moser 1985). Fruiting areoles deciduous, short woolly, and spiny. Seeds 1.9–2.3 mm, dark red-brown to blackish, dorsally ridged, finely impressed-reticulate, sometimes partly wrinkled (resulting in localized bulging cells or cell groups). Summer; fruits ripening late summer–early fall.

This cactus should be sought in the Sonoyta region on rocky slopes, probably southeast-facing, or the base of hills perhaps in the Sierra Cipriano or the Sierra de los Tanques. It has been recorded from the vicinity of Sonoyta, but the habitat has probably been urbanized or otherwise seriously altered. Near the Sonora border in Arizona from nearby Organ Pipe Monument and the O'odham reservation, western Sonora and northwestern Sinaloa, both Baja California states, and Gulf of California islands.

The flowers are remarkably large for such skinny stems. This cactus often grows through shrubs such as *Lycium,* and its stems seem to mimic stems of those shrubs. Although seldom seen, it is fairly common in western Sonora, occupying a variety of habitats including coastal plains, desert plains, bajadas, and rocky slopes. The plants are frost sensitive, and the Arizona populations are often freeze damaged, especially those in more open, exposed, or "colder" microhabitats.

The unusual stem morphology of *P. striatus* is remarkably similar to that of *Euphorbia cryptospinosa* Bally, a spurge known only from the vicinity of Voi, Kenya (Felger & Henrickson 1997). These plants share an unusual water-conserving adaptation: the stomata-bearing grooves close off as the stem shrivels during extended drought, resulting in a cylinder-like non-stomata-bearing surface and effectively reducing water loss and the surface-to-volume ratio of the stems.

"Vicinity of Sonoita, Sonora, a few miles south of the international boundary," *Evans in 1939:* In 1988, in Phoenix, Arizona, J. Whitman Evans told Allan Zimmerman that he made the 1939 collection "on the road to the old Indian shrine" near Sonoyta.

STENOCEREUS

Stems columnar, arching, or creeping, sparsely to densely branched. Flowers mostly medium-sized to large, radial, the limb or inner tepals mostly white to pink, opening widely, primarily nocturnal or flowers sometimes diurnal, radial to weakly bilateral, and bright pink to red. Fruits succulent with usually deciduous spine clusters; pulp (funiculi) usually sweet.

Strikingly diverse in habit and pollination syndromes, but united by their anatomy and stem chemistry (Gibson 1991). Southern Arizona to Honduras, West Indies, and northern South America; 24 species. References: Alcorn, McGregor, & Olin 1962; Cody 1984; Gibson 1990, 1991; McDonough 1964; Nobel 1980b, 1982; Parker 1987a, b; Wiggins 1937.

Stenocereus thurberi (Engelmann) Buxbaum [*Cereus thurberi* Engelmann. *Lemaireocereus thurberi* (Engelmann) Britton & Rose] *PITAHAYA DULCE;* ORGANPIPE CACTUS; *CUCUWIS.* Freely branched columnar cacti often 3–4.5 m, the branches mostly basal or nearly so (in the flora area); also often branching from the upper stems, apparently as a result of freeze killed stem tips. (The largest organpipe known to us in the region is on the Ives Flow, and measured 6.9 m tall, with 110 stems.) Stems 9.0–15.6 cm in diameter at midstem height; stem ribs of low relief, (14) 15–17 (19) (juvenile plants often with 12–14 ribs). Spines slender, grayish to nearly black, spreading in all directions, longer spines 1–4 (5) cm, straight (or occasionally a few spines twisted); stem surfaces readily visible; young areoles with dark red-brown glandular hairs, these turning blackish within 1 or 2 years.

Flowers 7–9 × 6–7 cm, nocturnal, closing in early morning; inner tepals white or pale pinkish lavender with white margins and bases; scales on tube purplish red; nectar chamber relatively large and producing copious nectar. Stamens, style, and stigma creamy white. Fruits globose, 4.0–6.4 cm in diameter, very succulent, indehiscent or splitting irregularly; pericarpel (skin and cortex) thin, red when ripe; pulp of mature fruits delicious, sweet, juicy, and bright red; floral remnant persistent. Unripe fruits green and spiny, the spines accrescent before ripening; fruiting areoles readily deciduous from mature fruits, each with 18–32 spines emerging from a dense tuft of short, ribbonlike white hairs, the larger spines often 17–22 mm, stout (0.4–0.65 mm in diameter), usually slightly twisted, stiff, and grayish. (Occasional plants consistently produce yellow-spined fruits.) Pulp, seeds, and rind edible. Seeds 2.0–2.5 × 1.4–1.7 mm, blackish, with a weakly developed ridge along "back." March-December, mostly May-June; fruit production peaking July-early August, sometimes minor fruiting peak again in September.

Common in the Sonoyta region on the decomposed granite soils of low hills and bajadas near the international border between Sonoyta and Quitobaquito, and in the Sierra de los Tanques and Sierra Cipriano. Extensive populations occur in the granite mountains to the east of the flora area, e.g., the Sierras San Francisco, Pozo, and Cubabi. Also small, isolated populations at the southeastern and eastern sides of the Pinacate volcanic field including the basaltic Cerros Batamote. Northern Sinaloa and southwestern Chihuahua and most of Sonora to southwestern Arizona, Gulf of California islands, and Baja California Sur to southern Baja California.

Pollinated during the night, probably primarily by the now-endangered lesser long-nosed bat (*Leptonycteris curasoae*), and if open during daylight hours then primarily by the white-winged dove (Fleming, Tuttle, & Horner 1996). The long-nosed bat also feeds on the fruits and is probably a major seed disperser (Tom Bethard and William D. Peachey, personal communication 1994). The fruits are also consumed by birds, especially the white-winged dove, ants, and desert bighorn.

Conspicuous constrictions (scars), common on the stems, are the result of freeze damage. The stem tip, where the surface-to-volume ratio is highest, is the part most vulnerable to frost. Most of the plants in Organ Pipe Monument and along the adjacent Sonora border have stems that show repeated freeze damage. The northern distribution is obviously limited by freezing temperatures. The western limits, as in the Pinacate range, however, seem to be determined by aridity and more specifically by deficient summer rains.

In the thornscrub and tropical deciduous forest south of the Sonoran Desert the organpipes have well-developed trunks, branching occurs well above the ground, and the plants reach much greater heights than those farther north. Proceeding northward into the desert, the plants are progressively shorter until in the northernmost populations these cacti are trunkless and the branches arise at or near the base. Trunkless plants are undoubtedly better protected against freezing weather. Where vegetation is dense, as in the southern portion of the range in tropical deciduous forest, there may be an advantage in having the flowering stems above the forest canopy where they are more accessible to bats.

Organpipe fruits have been highly esteemed by Sonoran Desert people. The fruit was eaten fresh, made into wine, or preserved by drying. Unlike other columnar cacti in the Sonoran Desert region, the rind is relatively thin and edible, and the small seeds are consumed along with the fruit pulp (Castetter & Bell 1937; Felger & Moser 1985; Felger et al. 1992).

Sonoyta, *Mearns 2722* (US). Observations: Ives Flow, Pinacate volcanic field; Cerros Batamote, east-facing slopes; Granitic hill at Punta Pelícano, Bahía de la Cholla, 1 rather stunted plant ca. 2 m tall.

CAMPANULACEAE BELLFLOWER FAMILY

Mostly herbs. Leaves alternate, simple, without stipules. Flowers usually 5-merous; corollas sympetalous; stamens free, or filaments and often also the anthers united in a tube; ovary inferior. Fruit usually a capsule; seeds numerous and minute.

Worldwide, few in deserts; 70 genera, 2000 species. Some are garden plants, e.g., lobelia (*Lobelia*) and bellflower (*Campanula*).

NEMACLADUS

Small annuals. Leaves mostly in a compact basal rosette, the stem leaves bractlike or absent. Pedicels slender, longer than the flower. Corollas bilateral, 2-lipped, the upper lip 3-lobed, the lower 2-lobed. Filaments united in a column, the anthers spreading. Western United States and Mexico, mostly southwestern United States; 12 species. Reference: McVaugh 1939.

Nemacladus glanduliferus Jepson var. **orientalis** McVaugh. THREADSTEM. Winter-spring ephemerals, 3.5–18 cm, glabrous or sparsely to moderately pubescent with short white hairs at base. Stems threadlike, often much-branched, upright or sometimes spreading. Herbage usually dark olive-green to purple-brown. Basal leaves 3–10 mm, oblanceolate with toothed margins, soon drying, the stem leaves bractlike. Raceme stems moderately zigzag, the pedicels ascending to spreading. Calyx segments green, 0.8–1.5 mm; corollas about twice as long as the calyx, the lobes pointed, white with maroon-purple tips. Filament tube ca. 1 mm; anthers whitish, becoming purplish.

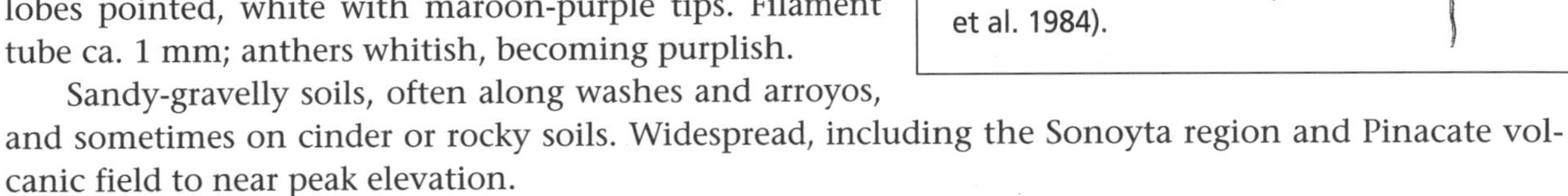

***Nemacladus glanduliferus* var. *orientalis*.** BA (Cronquist et al. 1984).

Sandy-gravelly soils, often along washes and arroyos, and sometimes on cinder or rocky soils. Widespread, including the Sonoyta region and Pinacate volcanic field to near peak elevation.

Variety *orientalis,* the most widespread of the 3 varieties, occurs in inland southern California, Baja California, southwestern Utah, southern and western Arizona, and northern Sonora southward along the coast to the vicinity of Bahía San Agustín. This variety is distinguished by its smaller calyx segments, and stiffly spreading-ascending and straight or scarcely curved pedicels; it is replaced by the other 2 varieties westward in southern California and Baja California.

Moon Crater, *F 19280*. Arroyo Tule, *F 19203*. Sykes Crater, *18977*. S of Pinacate Peak, 875 m, *F 19359*. NW of Cerro Colorado, *F 19644*. El Papalote, *F 86-119*. Observation: Bahía San Agustín, vicinity of 28°15′ N, *Phil Jenkins,* 1992.

CAPPARACEAE CAPER FAMILY

Annual or perennial herbs, shrubs, or sometimes trees; often strong- or foul-smelling with watery sap containing mustard-oil glucosides. Leaves usually alternate (ours), simple or palmately compound with 3 or more leaflets; stipules absent or small, or sometimes modified into spines. Flowers mostly in racemes with bracts, sometimes solitary and axillary, mostly bisexual, radial to moderately bilateral. Sepals and petals each 2–6, mostly 4, mostly separate, sometimes united below, or petals rarely absent. Stamens 4 to many, usually longer than the petals. Ovary superior, the ovary and fruit usually narrowed into a prominent stipe (stalk or column). Fruits fleshy or dry, 1 to many-seeded.

Tropical and subtropical regions, few in temperate, mostly arid climates; 45 genera, 675 species. Two-thirds of the species are in *Capparis* and *Cleome*. The family includes 2 subfamilies. The Capparoideae, usually woody with indehiscent fruits, includes *Capparis*. The Cleomoideae, mostly herbaceous with dehiscent fruits, includes *Cleome* and *Wislizenia,* and seems most closely related to the crucifers (Brassicaceae). Reference: Gómez 1953.

1. Woody shrubs; leaves simple, thickish and leathery, dark green above, lighter colored below, with crowded silvery to brownish scales. **Capparis**

1′ Herbs or shrubs; leaves with 3 leaflets or reduced to 1 leaflet, not thick and not leathery, the same color on both surfaces, glabrous or pubescent with glandular or non-glandular hairs but not with scales.
   2. Fruits of capsules, the body more than 1.5 cm long and longer than wide. **Cleome**
   2′ Fruits of 2 indehiscent nutlets, the body less than 0.7 cm long and wider than long. **Wislizenia**

CAPPARIS CAPER BUSH

Shrubs or trees, glabrous or pubescent, often with stellate hairs or peltate scales. Leaves simple, petioled, the blades usually thick and leathery; often with spinose stipules. Flowers generally white to yellow and large. Sepals 4, the petals usually 4, the stamens usually many, often long. Fruits berrylike or podlike. Tropics and subtropics worldwide; 250 species. Some have edible fruits; the flower buds of *C. spinosa,* native to the Mediterranean region, are the capers used as condiments. References: Burkart, Troncoso, & Bacigalupo 1987; Cozzo 1946; Hunziker 1984.

Capparis atamisquea Kuntze [*Atamisquea emarginata* Miers] *PALO HEDIONDO.* Much-branched, woody shrubs often 2–2.5 m, sometimes to 4–5 m with several trunks, each 12–15 cm in diameter. Twigs straight, stout, rigid, brittle, and woody, often branching at right angles, often bluntly thorn-tipped, otherwise unarmed. Young stems, lower leaf surfaces, outer surfaces of sepals, and pistils and fruits densely covered with transparent-winged silvery to yellowish peltate scales. Evergreen or eventually leafless in extreme drought. Leaves (0.8) 1–3.0 cm on short shoots, or often 2.5–5.5 cm on long shoots; petioles stout, 1–2 mm; blades leathery, linear-oblong, dark green above, dull silvery gray below, the margins entire except the apex sometimes moderately notched, the midrib prominent; stipules none.

Flowers solitary to few in leaf axils, moderately bilateral. Sepals essentially separate, green, the outer pair cupped and much larger and thicker than the inner pair. Petals and stamens white; petals 4 (6), often 6.5 mm. Fertile stamens 6 (7), the filaments often 5.5 mm, plus 0–3 staminodes. Ovary green. Body of fruit 8.0–11.3 × 6.0–7.7 mm, oval with a small, pointed tip, olive-green below the dense, silvery scales; exocarp separating from the aril when ripe, giving the ripe fruits a spongy texture. Exocarp and aril with a strong, spicy odor like creosote bush, and a sharp horseradish taste; ripe exocarp splitting to reveal a bright red, fleshy aril containing 1 (2 or rarely 3) seed(s). Seeds firmly embedded in the aril, 3.8–5.0 × 3.2–4.3 mm, nearly orbicular to broadly kidney-shaped, blackish when cleaned. May-June; fruiting simultaneously in August.

Scattered along the former mesquite bosque on the old floodplain several meters above the Río Sonoyta in the Sonoyta Valley, at least in areas south of Quitobaquito. Also locally common nearby on sandy desert flats, dense brush in washes near Aguajita Spring on both sides of the international fence, and sandy flats bordering the east side of Aguajita Wash; rare at Quitobaquito. The largest shrubs (4–5 m) are along the Río Sonoyta and in mesquite thickets at the international border at Aguajita. The larger ones along the river probably grew through mesquites. Seedlings and young plants are infrequent but not rare in the shelter of dense brush. The Río Sonoyta–Aguajita population is the northernmost extension of the species, which seems to be limited by freezing weather. Southward through western Sonora to northwestern Sinaloa, southern Baja California and Baja California Sur, also Bolivia, Chile and Argentina.

The flowers attract honeybees, native bees, large orange-winged tarantula hawk wasps (*Pepsis* and *Hemipepsis*), and many other insects. The red arils with the enclosed seeds are eaten by the house finch and verdin, the probable seed dispersers. The leaves are the only known food for larvae of the pierid butterfly, *Ascia howarthi* (Bailowitz 1988).

Floodplain above Río Sonoyta, SSW of Quitobaquito, *F 89-12.* Aguajita Wash at international fence, *F 88-26.* Quitobaquito, *Harbison 27 Nov 1939.*

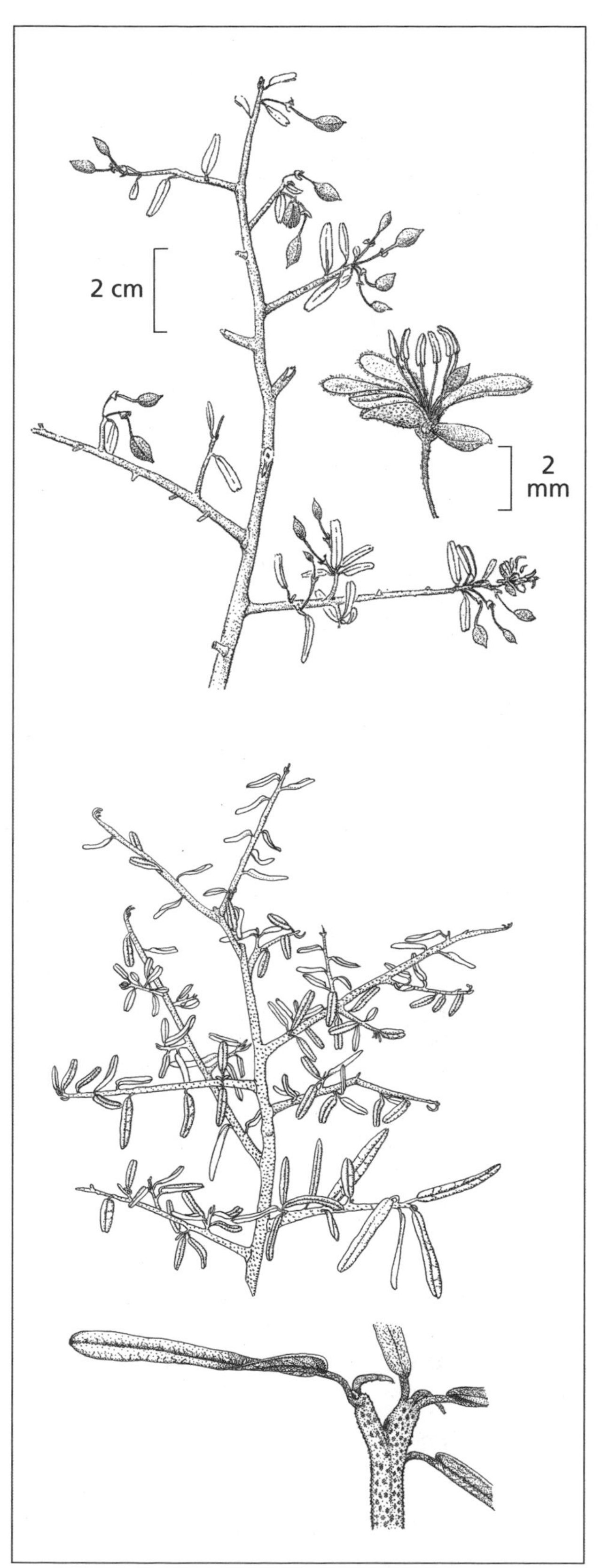

Capparis atamisquea. NE (below) and ILW (Wiggins 1980).

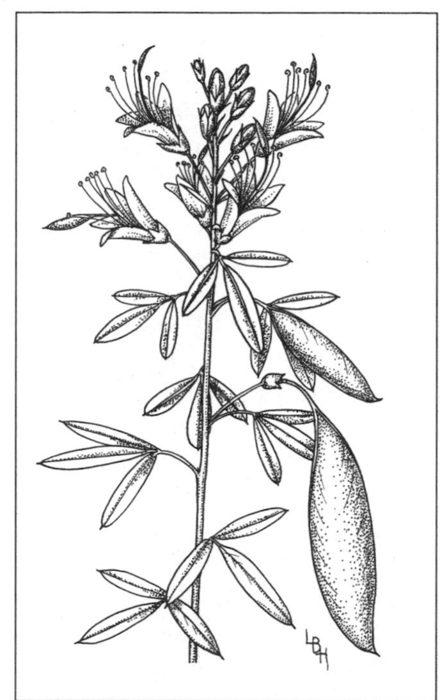

Cleome isomeris. LBH (Benson & Darrow 1945).

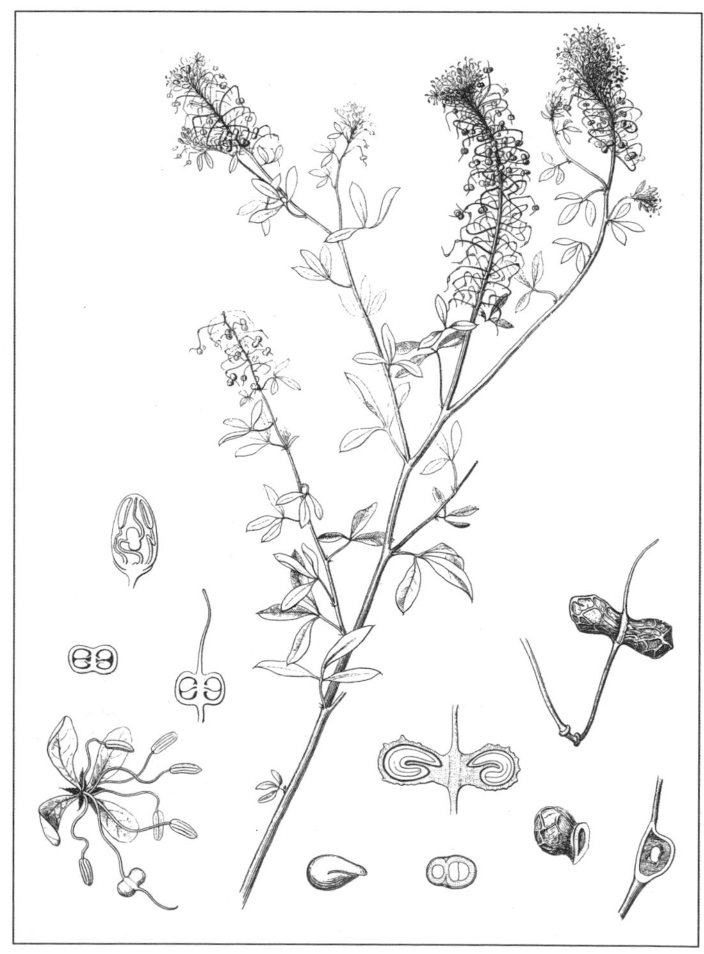

Wislizenia refracta subsp. ***pp90palmeri.*** IS (Gray 1852; HUNT/US).

CLEOME

Annual or perennial herbs, sometimes shrubs or small trees. Leaves palmately compound or simple (unifoliolate). Flowers in racemes, the pedicels with small leaflike bracts. Sepals and petals each 4. Stamens 6 (rarely 4 or many). Fruits of long-stiped capsules; seeds few to many. Both hemispheres, mostly tropical and subtropical; 150 species. References: Iltis 1957, 1960.

1. Woody shrubs, not glandular, the herbage glaucous.________________________________**C. isomeris**
1′ Annuals, the herbage conspicuously glandular-viscid (sticky), not glaucous. ________________**C. viscosa**

Cleome isomeris (Nuttall) Greene [*Isomeris arborea* Nuttall. *I. arborea* var. *angustata* Parish] *EJOTILLO;* BLADDERPOD BUSH. Shrubs often 1.2–1.8 (2) m. Herbage glaucous, minutely puberulent, especially when young, having a strong odor of mustard oil or green beans when crushed. Leaves evergreen to tardily deciduous; petioles prominent; leaflets 3, ovate to lanceolate or elliptic, 1.5–3.5 cm, with age becoming somewhat thickened and glabrous, the margins entire.

Flowers moderately bilateral, showy, crowded into terminal racemes, each flower subtended by a leaflike bract. Pedicels slightly shorter than the petals. Sepals united more or less to the middle, green to yellow, and persistent. Petals 1.3–1.8 cm, bright yellow, the lower pair slightly larger and more widely spreading than the upper pair. Stamens 6, yellow, longer than the petals, arising from a green, nectar-producing disk. Ovary at first yellow and nearly sessile, becoming green and the stipe greatly elongating after the anthers open. Capsules with a slender stipe 1.3–1.8 cm, the capsule body (1.5) 2.3–3.7+ cm, bladdery-inflated, pendulous, tardily dehiscent; seeds few. Flowers visited by large orange-and-black spider wasps (*Pepsis*), which stick their entire head into the center of the flower, and also by honeybees, which gather the pollen. Flowering with new growth February-March and fall.

In Sonora known only from higher elevations in the Sierra Pinacate, mostly above 750 m. Especially common on cinder flats and slopes, and arroyos at the bases and lower slopes of the main peaks at 900–1100 m and extending nearly to the summit. Also northern Baja California and California in the Mojave and Colorado Deserts but absent from the lowest, driest parts of the desert; also in chaparral and oak woodland along the Pacific coast. There are 3 weakly differentiated varieties based on fruit variation. Ours are var. *isomeris,* with elongated fruits (see Benson & Darrow 1981).

Carnegie Peak and vicinity: *F 86-460; Simmons 26 Jan 1965.*

***Cleome viscosa** Linnaeus. STICKY BEE PLANT. Hot-weather ephemerals or annuals, taprooted, erect, and conspicuously sticky glandular pubescent. “The plants . . . coarse, viscid and strong smelling (somewhat with the odor of burning *Cannabis*) herbs characterized by their palmately compound leaves with five leaflets when mature” (Holmes 1981:187). Petioles prominent; leaflets mostly 2.5–4.0 cm, broadly obovate, entire, glandular-ciliate. Flowers yellow, solitary in axils of upper leaves, the upper stems appearing as terminal racemes; pedicels longer than the flowers. Sepals essentially separate. Petals 9–10 mm, longer than the sepals. Stamens, including staminodes, ca. 9–13, shorter than the petals. Ovary linear, densely studded with stalked glands. Capsules 4.5–8.5 cm, erect, linear-cylindrical. Seeds 1.5 mm wide, brown, wrinkled, resembling a curled-up beetle larva.

Rare in the flora area; roadsides (W of Sonoyta, *F 86-338*), probably from seeds falling from passing trucks and not established as reproducing populations. However, it seems to be spreading as a roadside and agricultural weed through much of Sonora and Sinaloa, and in the Guaymas region and northwestern Sinaloa it has become established in natural habitats. It was not recorded for the Sonoran Desert by Wiggins (1964).

WISLIZENIA JACKASS CLOVER

The genus has 1 species. *Wislizenia* seems closely related to *Cleomella,* e.g., *C. angustifolia* Torrey and *C. longipes* Torrey of north-central Mexico and Texas. *Wislizenia* has testiculate (didymous) fruits of 2 indehiscent, single-seeded nutlets whereas *Cleomella* has several-seeded, dehiscent capsules. Reference: Keller 1979.

Wislizenia refracta Engelmann. Foul-smelling annuals or perennials; glabrate. Petioles prominent; leaves with 1–3 leaflets, the margins entire; stipules small, bristlelike, deciduous. Inflorescences of densely flowered racemes, mostly without bracts or sometimes a few leafy bracts below. Flowers bright yellow, showy. Sepals 4, united at base, the petals 4, clawed at base, longer than the sepals; corollas bilateral. Stamens 6, much longer than the petals. Ovary with a slender, elongated, pedicel-like stipe much longer than the petals; style elongated and persistent. Fruits of 2 divergent halves, each closely enclosing 1 (2 or 3) seed(s) and falling as a nutlet, the halves often unequal in size, cylindrical to usually broadest at tip and usually with a few short, blunt but prominent terminal horns or sometimes smooth. Seeds ovoid, dull brown.

Western Texas to southern Nevada and southern California and northern Chihuahua to both Baja California states, mostly in desert or semidesert. One species with 3 geographically segregated subspecies intergrading at their zones of contact.

1. Often shrubs; leaflets 1, or sometimes 3 among lower leaves, mostly linear to narrowly elliptic or lanceolate; anthers 1.8–2.1 mm; nutlets each 3.7–5.5 mm at longest axis; seeds 2.5–2.7 mm. ____ **W. refracta** subsp. **palmeri**

1′ Mostly annuals; leaflets 3, broadly obovate; anthers 1.2–1.8 mm; nutlets each 1.8–2.8 mm; seeds 1.5–1.6 mm. ____ **W. refracta** subsp. **refracta**

Wislizenia refracta subsp. **palmeri** (A. Gray) Keller [*W. palmeri* A. Gray, *Proc. Amer. Acad.* 8:622, 1873. *W. refracta* var. *palmeri* (A. Gray) I.M. Johnston. *W. mammillata* Rose ex Greene] Sometimes facultative annuals, mostly perennials and tending towards gigantism and developing into shrubs 1.5–1.8 (2+) m, often with a well-developed woody trunk ca. 5–10 cm in diameter. Leaflets mostly 1 by reduction of lateral leaflets, sometimes 3 among lower branches; blades linear to narrowly oblong to narrowly elliptic or lanceolate, the larger ones often 2.5–6 cm × 4–8 mm, those of older leaves on lower branches occasionally becoming succulent. Raceme axis sometimes reaching 30 cm. Sepals 1.0–1.5 mm, ovate; petals 4.5–6.0 mm. Anthers 1.8–2.1 mm. Nutlets each 3.7–5.5 mm at longest axis; seeds 2.5–2.7 mm. Flowering any season.

Coastal dunes and sand flats at Bahía Adair, often on alkaline or partially saline soils, beach dunes in the vicinity of Estero Morúa, and dunes at Laguna Prieta. Subsp. *palmeri,* reported from southeastern California, both Baja California states, and coastal Sonora southward to the Guaymas region, is often herbaceous. Our narrow-leaved, perennial "giants," especially those near the shore at Bahía Adair, are markedly larger than plants of subsp. *palmeri* in other regions. These "giants" include the type collection of *W. palmeri.*

Lower Colorado [River], Mexico, *Palmer in 1869* (isotypes of *W. palmeri,* US *6722, 6723;* probably near the mouth of the river at Puerto Isabel, Sonora, see McVaugh 1956:180–181). Laguna Prieta, *Ezcurra 20 Apr 1985* (MEXU). 5 km N of El Golfo, *F 85-1062A.* La Salina, *F 86-552.* Strand at Estero Morúa, 6 km SE of estero mouth, *F 91-45.*

Wislizenia refracta subsp. **refracta.** Non-seasonal annuals, sometimes short-lived perennials, often reaching 0.5–1 m, the main axis often well developed. Leaflets mostly 3, broadly obovate, 1–4 × 0.5–1.5 cm. Sepals 0.8–1.5 mm, ovate-lanceolate; petals 3.5–5.5 mm. Anthers 1.2–1.5 (1.8) mm. Nutlets each 1.8–2.8 mm; seeds 1.5–1.6 mm. Flowering nearly any season, especially late spring and early summer and late fall, often long after most other herbaceous plants have dried and withered.

Sandy and often alkaline or partially saline soils. Locally common in broad gravelly washes and floodplains, low stabilized or semistabilized dunes and sand flats, and sometimes in disturbed habitats such as abandoned agricultural fields. Especially common at Quitobaquito, Aguajita Wash, and the floodplains and riverbed of the Río Sonoyta from Sonoyta nearly to its terminus near Estero Morúa. Widely scattered in the lowlands through the flora area except along Bahía Adair and near the Río Colorado, where it is replaced by the giant form of subsp. *palmeri.* Nevada and southeastern California to New Mexico and western Texas, and south to northern Sonora and northern Chihuahua.

Some specimens are unusual in having some fruits with 3 or 4 instead of the usual 2 carpels (nutlets). These were collected during winter and spring and may be the result of freeze damage. The 2 subspecies in the flora area are narrowly separated geographically and occupy slightly different habitats. I have not seen plants in the region showing intermediate characters.

Río Sonoyta riverbed, 18 mi northward from Puerto Peñasco, *F 91-45*. Elegante Crater, *Soule 31 Jan 1983*. 1 km E of MacDougal Crater, *F 10063*. 10 mi NW of Sonoyta, *Henrickson 2352* (RSA). Quitobaquito, *Peebles 14558*.

CARYOPHYLLACEAE PINK FAMILY

Annual or perennial herbs, the stem often swollen at the nodes. Leaves almost always opposite (see *Drymaria*), simple, entire, usually narrow, the opposite petiole bases often connate (joined) by a transverse line across the node; stipules membranous or absent. Flowers usually bisexual and radial; sepals 4 or 5, separate or united; petals 4 or 5, separate, the apex often notched (bilobed) or deeply cut (fringed), or sometimes petals reduced or absent. Stamens usually as many as or twice as many as calyx lobes, but sometimes fewer or more, often fused with perianth bases. Ovary superior. Fruit a capsule, the seeds few to numerous or sometimes a 1-seeded utricle.

Worldwide except the more tropical regions, largely temperate Northern Hemisphere, maximum diversity in the Mediterranean and Near East; 89 genera, 2000 species. Some are popular garden flowers such as carnations and pinks.

1. Glabrous or with glandular patches or short hairs on stems, but the hairs not glandular; calyx glabrous.
   2. Plants glabrous; stems spreading to prostrate; stipules conspicuous, papery and white; petals absent. ____ **Achyronychia**
   2′ Plants with short hairs, especially below; stems erect to ascending; stipules absent; petals red-purple. ____ **Silene**

1′ Plants glandular hairy, or if herbage tending to be glabrous then at least the sepals conspicuously glandular hairy.
   3. Compact, dwarf, matlike plants, usually less than 6 cm wide and less than 2.5 cm high; stems at least partially obscured by stiff leaves. ____ **Loeflingia**
   3′ Plants spreading-prostrate, usually more than 6 cm wide and more than 3 cm tall; stems clearly visible, the leaves not stiff.
      4. Sand sticking to stems and leaves; at least the lower leaves appearing whorled, the stipules inconspicuous or absent; dry soils: sand flats and dunes. ____ **Drymaria**
      4′ Sand not sticking to stems and leaves; leaves opposite, the stipules conspicuous; moist soils. ____ **Spergularia**

ACHYRONYCHIA

Mexico and southwestern United States; 2 species. *A. parryi* is a prostrate-spreading perennial of dry regions in north-central Mexico. Reference: Hartman 1993.

Achyronychia cooperi. (Abrams & Ferris 1960).

Achyronychia cooperi Torrey & A. Gray. FROST MAT, SAND MAT. Winter-spring ephemerals with a well-developed taproot; glabrous, (3) 8–30 cm across, at first spreading, becoming prostrate and matlike. Stems radiating from center, often few- to many-branched, leafy and bearing flowers nearly throughout. Leaves 3–18 mm, opposite, green, spatulate, those of each pair unequal in size; stipules white, 2.5–3.0 mm, conspicuous. Flowers in dense axillary clusters, the calyx 2.3–3.5 mm, the sepals united, thick, and green below, the lobes with conspicuous papery, white wings resembling the stipules; petals absent. Stamens 10 to ca. 18, threadlike, the fertile stamens 1 or 2. Fruits indehiscent, with 1 seed (0.9) 1.0–1.1 mm, ovoid, smooth, whitish and pale brown. Plants breaking apart as the seeds mature.

Widespread through the lowland desert of the flora area; sandy, gravelly soils of washes, flats, and dunes; sometimes on cinder flats and occasionally on rocky soils. Western Sonora southward to the Guaymas region, both Baja California states, southern California, western Arizona, and southern Nevada.

The thin, papery, white stipules and small, white flowers crowded along the length of the stem give the plant a "frosty" appearance. The single-seeded fruit is unusual in the family. The plants resembling some herbaceous *Euphorbia* (*Chamaesyce*) in growth form.

Puerto Peñasco, *Burgess 4737.* 24 mi SW of Sonoyta, *Shreve 7586.* Hornaday Mts, *Burgess 6865.* Moon Crater *F 19247.* N of Sierra del Rosario, *F 20789.* N of El Golfo, *F 75-76.*

DRYMARIA

Annual or perennial herbs. Leaves opposite or appearing whorled, usually with persistent or deciduous small stipules. Sepals 5, separate; petals (0) 3–5, white, usually 2-cleft. Seeds 1 to many. Mostly western North and South America; 48 species. Unlike most members of the family, *Drymaria* is largely subtropical, although none occur in California. Duke (1961) arranged the genus into 17 series; *D. viscosa* is the only species in series *Viscosae.* Reference: Duke 1961.

Drymaria viscosa S. Watson. Winter-spring ephemerals with a well-developed taproot. Stems 3–17 cm, delicate, spreading, mostly 2 to several appearing to arise from each node. Stems, leaves, pedicels, and sepals with stalked glandular hairs, with sand sticking to the glandular-sticky surfaces. Leaves appearing whorled at least from all but the uppermost nodes, semisucculent, linear-spatulate, tapering basally to the clasping petiole, mostly 5–10 (22) × less than 1–1.5 mm, the first leaves in a basal rosette and larger than the stem leaves; stipules 0.5–1.5 mm, transparent-membranous, slender, and soon deciduous. (The microscopic stipules are perhaps sometimes absent rather than deciduous.) Flowers minute, white, closing with midday heat; sepals ovate, green in the middle with broad white margins; petals delicate, pure white, Y-shaped with a long, slender claw and an expanded bifid blade. Seeds 0.5–0.6 mm including the projecting radicle (seen as a strange little projection of the embryo), often 8–12 per capsule. The seeds, resembling grains of sand, are striking compared with seeds of other *Drymaria* species because they are smooth and pale cream or buff instead of sculptured and dark brown.

Widespread in the flora area, sand flats to higher dunes, and in adjacent Arizona only near the Pinacate lava flow. Coastal Sonora southward to the vicinity of Tastiota (28°20′ N) and both Baja California states.

E of Estación López Collada, *F 84-15.* Dunes 29 km SW of Sonoyta, *Burgess 4761.* S of Mex 2 [rd to Rancho Guadalupe Victoria], *Mason 1832.* Moon Crater, *F 19034.* NE Sierra del Rosario, *F 20430.* Cabeza Prieta Refuge, Pinacate lava flow, *Hodgson 2080* (DES).

LOEFLINGIA

Annuals; low, spreading, and glandular hairy. Flowers small and cleistogamous (at least in North America). Branching of 2 types: dichotomous at lower nodes, with a sessile flower in the axis, and "monochasial, as one branch of the dichotomy becomes reduced or obsolete at several successive nodes" (Barneby & Twisselmann 1970:400). Seeds numerous. Several species in Eurasia and north Africa, and 1 in western North America. Reference: Barneby & Twisselmann 1970.

Loeflingia squarrosa Nuttall subsp. **cactorum** Barneby & Twisselmann. Dwarf spring ephemerals, quickly forming a relatively deep taproot and compact, low, cushionlike mats mostly 1–5 cm wide. Leaves often 3–15 mm, reduced upwards, narrowly linear, mostly stiff and thickened (subulate), with a short bristle tip; first leaves often in a loose basal rosette, thinner, and soon withering; stipules white, bristlelike. Flowers essentially sessile. Sepals 5, separate, resembling the leaves, mostly recurved, the outer 3 sepals mostly 4–5 mm with a small stipule-like bristle on each margin, the inner sepals slightly smaller; petals reduced to small scales (absent among other populations). Seeds 0.4–0.45 mm, light brown, ovoid.

In the northeast corner of our area in the Sonoyta region; seasonally common in open, partially barren, sandy-gravelly decomposed granite soils on bajadas of the Sierra Cipriano and nearby at the base of the Sierra Cubabi. Subsp. *cactorum* ranges from desert to grassland in southern Arizona and southward in Sonora to Hermosillo. The Sierra Cipriano population is the westernmost record in Sonora; in southern Arizona it ranges west to the vicinity of Tinajas Altas. There are 4 geographically segregated subspecies in western North America. Barneby & Twisselmann (1970) point out that this species occurs primarily on otherwise barren ground, which is certainly the case in Arizona and Sonora.

S side of Sierra Cipriano, *F 88-214.* 4 mi S of Sonoyta on Mex 2, *F 20550.* Tinajas Altas, *F 92-608.*

SILENE CATCHFLY, CAMPION

Annual to perennial herbs, usually glandular. Leaves opposite or sometimes appearing whorled, entire, united basally; stipules none. Flowers (1) few to many in cymes, radial to moderately bilateral. Sepals 5-lobed, often inflated in fruit. Petals white to red or purple, narrowed to a claw at base, usually abruptly flared above into an entire to several-lobed blade, often with appendages at junction of claw and blade. Stamens 10, fused with petals at base of ovary. Capsules opening by 6 (8 or 10) teeth, seeds numerous. Mostly north temperate regions; 500 species. Reference: Hitchcock & Maguire 1947.

Silene antirrhina Linnaeus. SLEEPY CATCHFLY. Spring ephemerals to 60 cm, often much smaller. Stems erect with ascending branches; smaller, drought-stressed plants usually unbranched and with 1 or more flowers. Herbage sparsely to moderately pubescent with short, often recurved, white hairs, the leaf margins often ciliate near the base, the upper part of the plants tending to be glabrous. Stems with red-purple glandular-sticky bands below the nodes, the bands trapping insects and seeds that have fallen from the capsules. Early leaves in a basal rosette, stem leaves narrowly oblanceolate, the upper leaves reduced; larger leaves 2–5 cm × 4–8 mm. Calyx 7–10.5 mm with short lobes. Petals dark red-purple, the blades ca. 2 mm. Seeds kidney-shaped, 0.6–0.8 mm maximum width, dark reddish.

Barely entering the flora area north of Sierra Cipriano; localized along small, sandy-gravelly washes dissecting low granitic hills and bajadas. Not known elsewhere in lowland western Sonora, and absent from the drier parts of the Sonoran Desert. Abundant and often weedy across much of North America to nearly 3000 m.

8.5–12 km SW of Sonoyta, *F 92-411, F 92-420.*

SPERGULARIA SAND SPURRY

Low annual or perennial herbs. Leaves linear with membranous stipules. Seeds numerous. Worldwide, mostly in temperate regions, the majority from North and South America; 40 species. Often on saline or alkaline soils. Reference: Rossbach 1940.

Spergularia salina J. Presl & C. Presl, 1819 [*Arenaria rubra* Linnaeus var. *marina* Linnaeus, 1753. *Spergularia marina* (Linnaeus) Bessler, 1822. *S. marina* (Linnaeus) Grisebach, 1843] SALT-MARSH SAND SPURRY. Spring annuals; at first upright then spreading-sprawling to 30–60+ cm in width. Stems slender; leaves linear, 2–4 (5) cm × 0.7–1.5 mm; stipules 2.5–3.5 mm, triangular, conspicuous. Herbage sparsely to densely glandular hairy, the pedicels and calyx densely glandular hairy. Flowers ca. 8–10 mm wide. Sepals, petals, and stamens 5. Sepals green with broad, nearly transparent, white margins. Petals lavender-pink. Seeds 0.7 mm, red-brown, glandular-papillate, the embryo visible curved around edge of seed; seeds partially to fully crested with a white, erose wing to 0.35 mm wide, or the crest absent, varying even within the same plant or capsule. March to mid-May.

Wetland habitats in sandy-gravelly soils: the Río Sonoyta riverbed, often near *Baccharis salicifolia, Salix gooddingii, Scirpus americanus,* and *Tamarix ramosissima;* the lower Río Colorado; as a weed in agricultural fields and irrigation ditches south of San Luis; and the Quitovac pond; not known elsewhere in Sonora. Widespread along North American coasts and sporadically inland; also South America and Eurasia. Possibly introduced in the Americas (e.g., Hitchcock & Cronquist 1964:300).

S of San Luis, *F 91-68.* Wellton-Mohawk drainage canal, *F 90-206.* Río Sonoyta, *F 86-80.* Quitovac, *Nabhan 162.*

CHENOPODIACEAE GOOSEFOOT FAMILY

Annual to perennial herbs, shrubs, or rarely small trees, often succulent. Leaves alternate (mostly) or opposite; sometimes scalelike and appearing leafless; stipules none. Flowers usually small and inconspicuous, bisexual or unisexual and the plants monoecious, dioecious, or polygamous; mostly wind-pollinated. Sepals mostly 5, often fleshy and green or white, or rarely absent in female flowers; petals none. Stamens usually as many as sepals. Stigmas 2–5. Fruits 1-seeded.

Worldwide, diverse in desert and semiarid regions, especially Eurasia and Australia; 100 genera, 1300 species. Many are halophytes and/or weeds. *Atriplex, Chenopodium,* and *Suaeda* are among the larger genera. Economic members include beets, chards, quinoa, and spinach. References: Henrickson 1977a; Shishkin 1970; Standley 1916.

1. Plants appearing leafless; stems jointed and very succulent.
   2. Branching alternate. ______ **Allenrolfea**
   2′ Branching opposite. ______ **Salicornia**

1′ Plants leafy; stems not jointed, succulent or not.
   3. Shrubs, obviously woody (borderline cases will key both ways).
      4. Leaves rounded in cross section or somewhat flattened, about as thick as wide; twigs semisucculent, not woody and not spinescent; flowers bisexual and with unisexual female flowers, the flowers and fruits not winged. ______ **Suaeda moquinii**
      4′ Leaves flat, or at least broader than thick; twigs not succulent, often woody, sometimes spinescent; flowers unisexual, the female flowers and/or fruits winged.
         5. Twigs usually not spinescent; leaves scurfy gray-green; female flowers and fruits enclosed by 2 sepal-like bracts, these not like a collar. ______ **Atriplex**
         5′ Twigs spinescent; leaves green; female flowers and fruits with a single broad, collarlike wing. ______ **Sarcobatus**
   3′ Herbaceous or slightly woody at base.
      6. Leaves opposite.
         7. Plants prostrate-spreading with deep and thickened rhizomes; sepals ca. 4 mm, not widest at apex; alkaline-saline inland habitats. ______ **Nitrophila**
         7′ Plants erect with a single, slender taproot; sepals less than 2 mm, conspicuously widest near apex; tidal zones. ______ **Suaeda puertopenascoa**
      6′ Leaves alternate (may be opposite at first few nodes).
         8. Flowers with 1 sepal and 1 stamen; leaves hastate lanceolate-linear (at least the lower, larger leaves); winter-spring ephemerals. ______ **Monolepis**
         8′ Sepals and stamens more than 1; leaves not hastate lanceolate-linear.
            9. Female flowers and fruits enclosed in 2 sepal-like bracts. ______ **Atriplex**
            9′ Female flowers and fruits not enclosed in sepal-like bracts.
               10. Leaves and younger stems with soft, white hairs; fruiting calyx lobes hooked at tips. ______ **Bassia**
               10′ Plants glabrous, glaucous, or mealy but lacking slender white hairs; calyx lobes not hooked.
                  11. Leaves with an obvious blade and petiole, blades expanded and much wider than petioles. ______ **Chenopodium**
                  11′ Leaves sessile or subsessile, narrow, linear or scalelike.
                     12. Adult leaves and bracts sharp-tipped (spinose), not fleshy; fruiting sepals broadly winged and not fleshy. ______ **Salsola**
                     12′ Leaves and leafy bracts fleshy, not spine-tipped; fruiting sepals fleshy and not winged. ______ **Suaeda**

ALLENROLFEA

Leafless-appearing shrubs; 1 species in western North America, 2 in Argentina; arid and semiarid saline and alkaline habitats. **References:** James & Kyhos 1961; Jansen & Parfitt 1977.

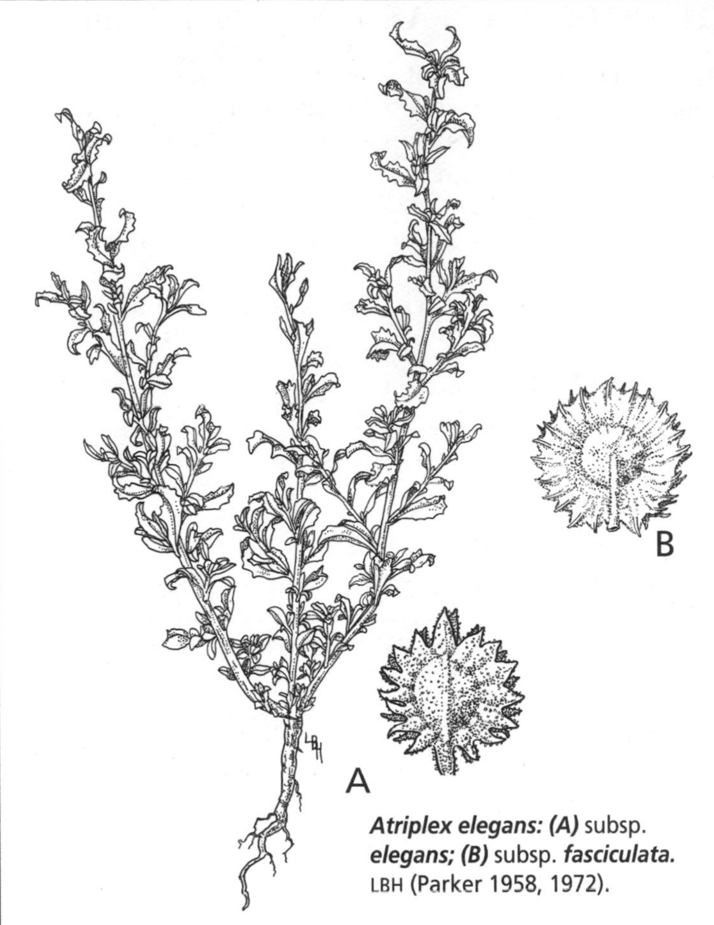

Atriplex elegans: (A) subsp. ***elegans; (B)*** subsp. ***fasciculata.*** LBH (Parker 1958, 1972).

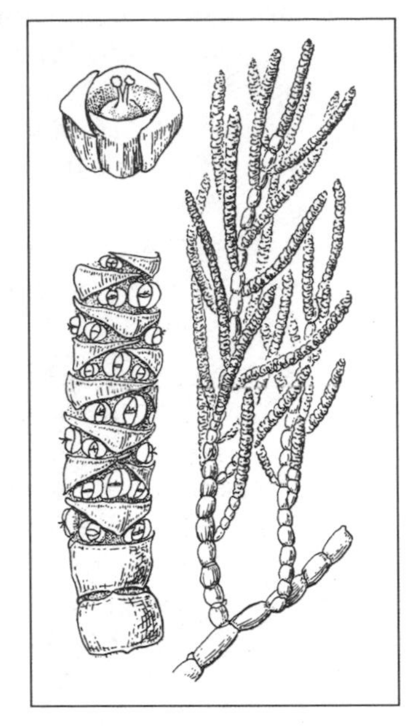

Allenrolfea occidentalis. BA. © James Henrickson 1999.

Atriplex canescens. FR (Felger 1980).

Atriplex barclayana. NLN (Felger & Moser 1985).

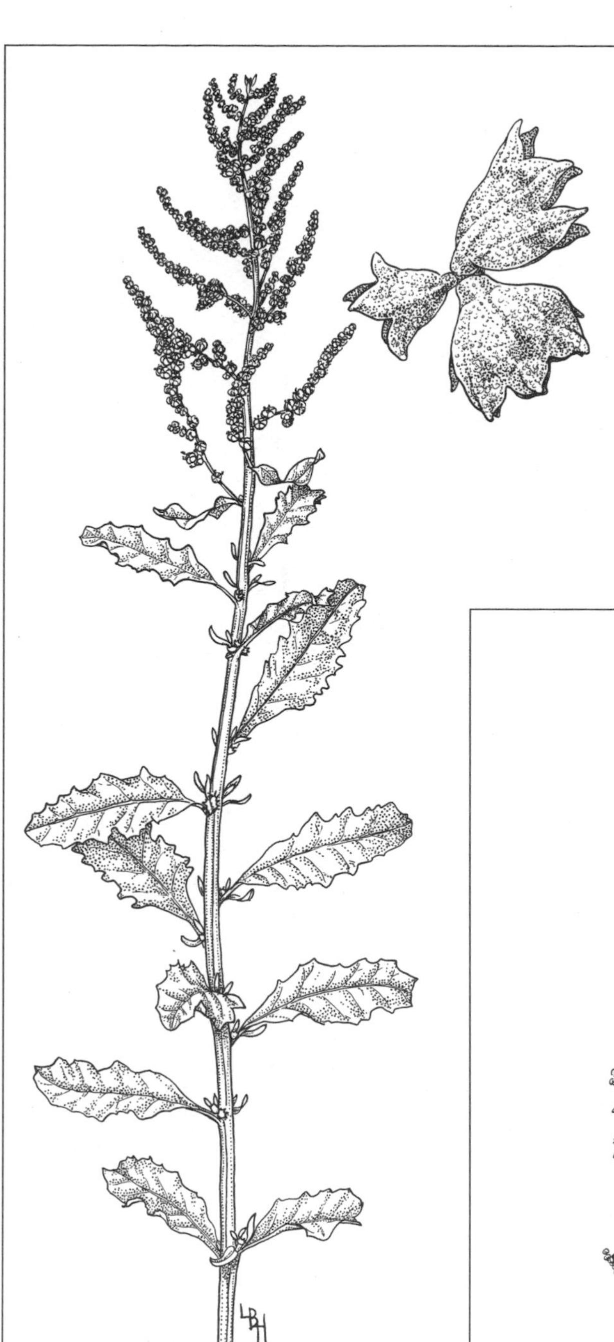

Atriplex wrightii with a cluster of three fruits. LBH (Parker 1958).

Atriplex polycarpa. LBH (Benson & Darrow 1981; Humphrey 1960).

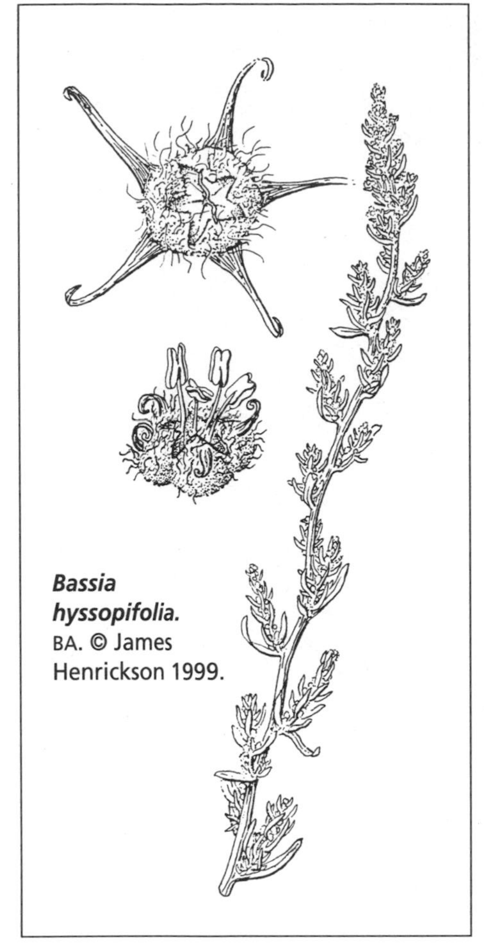

Bassia hyssopifolia. BA. © James Henrickson 1999.

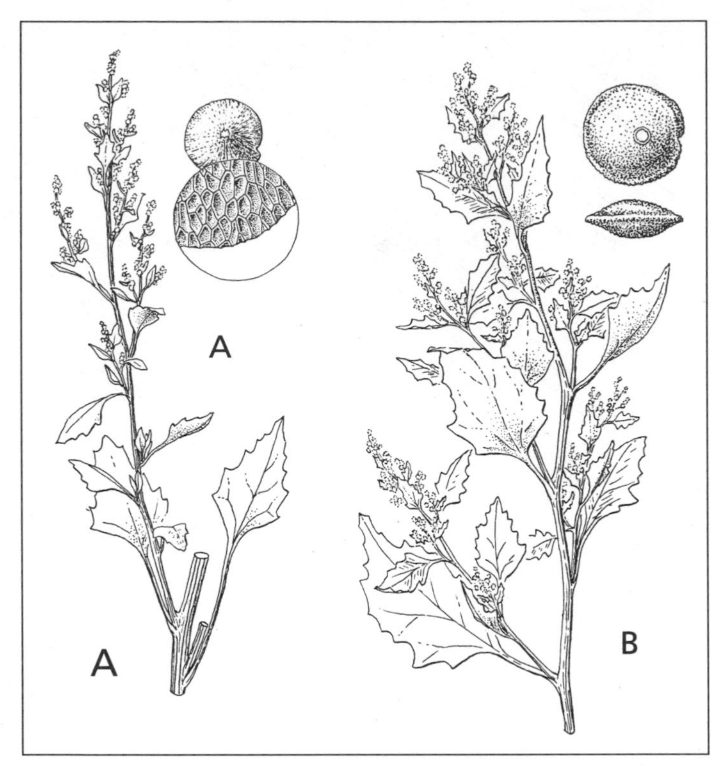

Chenopodium: (A) C. berlandieri var. ***sinuatum; (B) C. murale.*** BA. © James Henrickson 1999.

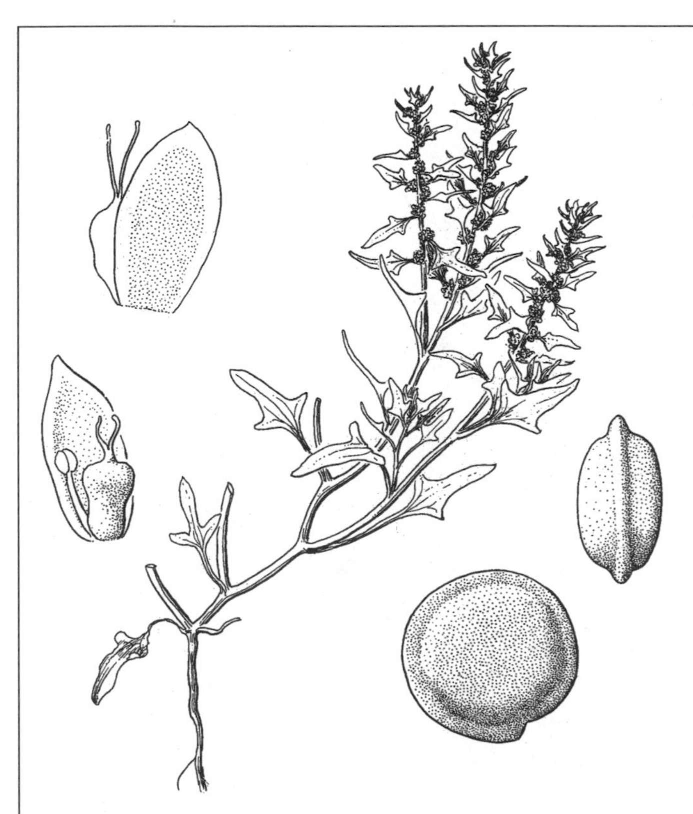

Monolepis nuttalliana. JRJ (Hitchcock & Cronquist 1964).

Nitrophila occidentalis. (Abrams 1944).

Sarcobatus vermiculatus.
LBH (Ezcurra et al. 1988).

0.5 cm

1.0 cm

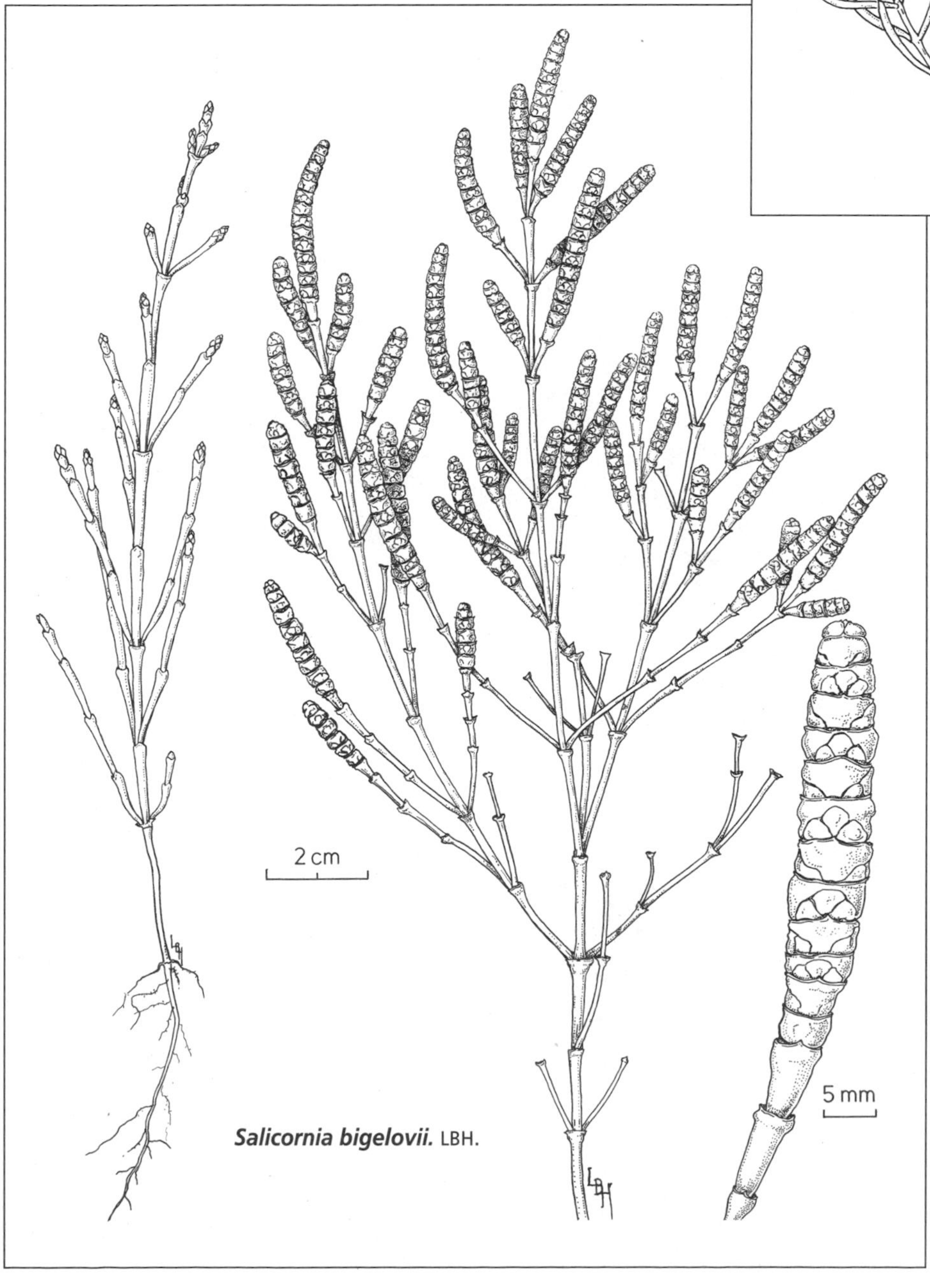

2 cm

5 mm

Salicornia bigelovii. LBH.

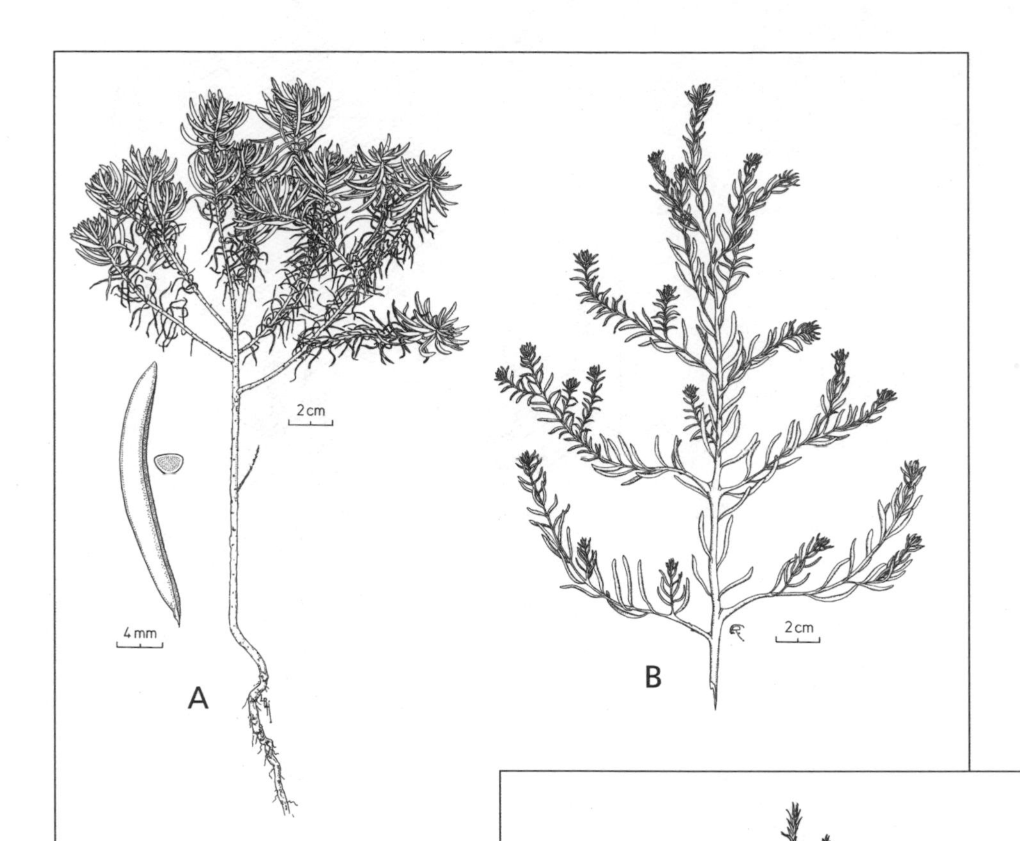

Suaeda: (A) S. esteroa,
LBH; ***(B) S. moquinii,***
FR. (Felger & Moser 1985).

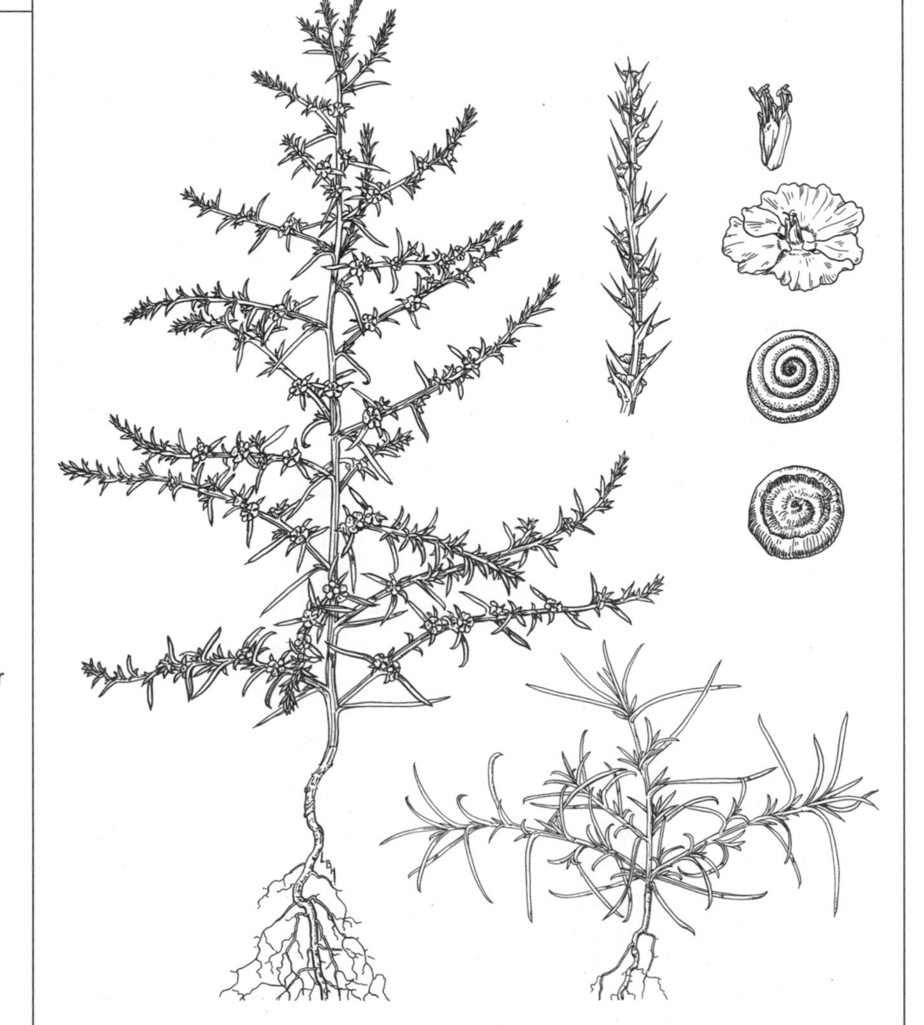

Salsola tragus. Note male flower above female flower, seed, and fruit. LBH (Parker 1958, 1972).

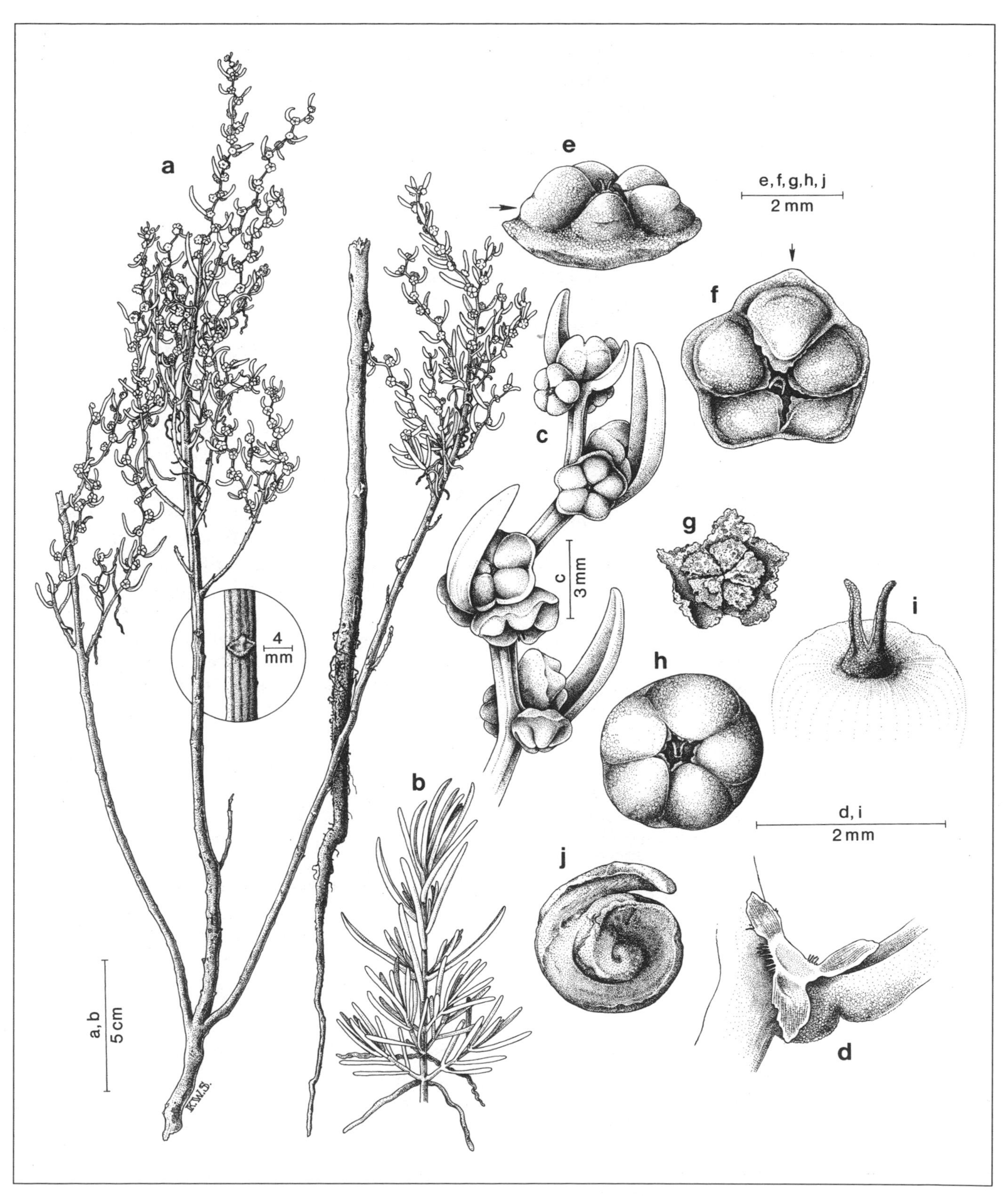

Suaeda puertopenascoa* (Ferren 2807): *(A) habit; ***(B)*** vegetative stem; ***(C)*** portion of inflorescence; ***(D)*** bractlets and base of bract; ***(E)*** flower, fresh material, oblique view; ***(F)*** flower, fresh material, top view; ***(G)*** dried flower, top view; ***(H)*** fresh flower; ***(I)*** stigmas and crest of ovary; ***(J)*** immature seed. KS (Watson & Ferren 1991).

Allenrolfea occidentalis (S. Watson) Kuntze [*Halostachys occidentalis* S. Watson. *Allenrolfea mexicana* Lundell] *CHAMIZO;* IODINE BUSH. Dense, much-branched shrubs 0.5–1.5+ m, often broader than tall, with alternate branching; older branches woody. Younger stems succulent, divided into green to red-orange beadlike segments; appearing leafless. Flowers bisexual, crowded on short axillary and terminal spikes. Calyx lobes 4 (5), succulent, spongy when dry, the inner pair smaller; stamens 1 or 2, the stigma lobes 2 (or 3?). Seeds shiny brown, ovoid, 0.9–1.0 × 0.6–0.7 mm, enclosed in a thin, membranous, semipersistent pericarp. Peak flowering in fall; seeds ripening simultaneously in midwinter.

Low-lying, saline, wet sandy-muddy soils: mud flats between the shore and desert, alkaline soils near the Río Colorado, Laguna Prieta, and bordering esteros, saline playas, and pozos. Along the shore *Allenrolfea* occurs just inland from, or overlapping, the zone of *Salicornia subterminalis*. Widespread in coastal Sonora. Coastal deserts and inland alkali sinks; Oregon to Texas and southward to Baja California Sur, northwestern Sinaloa, and the Chihuahuan Desert in Mexico.

Allenrolfea superficially resembles perennial species of *Salicornia* but has alternate branching, often becomes more shrubby, and grows on higher ground. Fleshy tissue on *Allenrolfea* and *Salicornia* stems is derived from reduced, modified leaf bases, 1 per node for *Allenrolfea* and 2 for *Salicornia*.

Banks of Río Colorado, 20 mi N of El Golfo, *F 75-61*. Laguna Prieta, *Roth 4*. Bahía Adair, 31°35′ N, 113°52′ W, *López-Portillo 9 Jul 1984*. Puerto Peñasco, *F 20966*.

ATRIPLEX SALTBUSH

Annuals to woody shrubs; glabrous or scurfy gray or whitish due to inflated hairs that collapse upon drying. Leaves mostly alternate, the lowermost ones sometimes opposite, a few (not ours) with opposite leaves throughout. Monoecious or dioecious. Female flowers and fruits enclosed in a pair of sepal-like accrescent bracts (as the fruits develop the bracts enlarge, thicken, and often grow various ornamentations such as wings and crests).

Nearly worldwide, mostly temperate and subtropical, many in arid and semiarid regions; 150 species. Many are halophytes, and most thrive on alkaline and/or saline soils. Some are important forage plants. Atriplexes are diverse and often abundant in the southwestern United States and northern Mexico. References: Hall & Clements 1923; Tiedemann 1984.

1. Woody shrubs, usually dioecious.
   2. Suffrutescent, usually less than 1 m and as broad as or broader than tall; bark at base becoming fissured and corky; wood usually not hard. ____ **A. barclayana**
   2′ Woody shrubs, plants often more than 1 m, or if 1 m or less then usually taller than or as tall as wide; bark at base smooth; wood hard.
      3. Fruiting bracts 4-winged.
         4. Large shrubs, commonly 1.5 m or more in height; larger leaves often more than 3 cm × 4–6 (8) mm; mature fruits with bracts often 1–2 cm. ____ **A. canescens**
         4′ Small shrubs, commonly 1 m or less in height; larger leaves often 1–3 (4+) cm × 2–4 mm; mature fruits with bracts often 0.4–0.5 (0.8) cm. ____ **A. linearis**
      3′ Fruiting bracts not 4-winged, more or less discoid, marginally toothed or not.
         5. Leaves mostly more than 3 cm, the petioles well developed; bracts not toothed, or teeth shallow and much broader than long. ____ **A. lentiformis**
         5′ Leaves mostly less than 2 cm, sessile to subsessile; bracts conspicuously toothed, teeth mostly longer than wide. ____ **A. polycarpa**

1′ Herbaceous or scarcely woody at base, dioecious or monoecious.
   6. Perennials with a woody base or sometimes flowering in first year; usually dioecious but sometimes monoecious. ____ **A. barclayana**
   6′ Annuals; monoecious.
      7. Fruiting bracts orbicular, evenly toothed all around the margin. ____ **A. elegans**
      7′ Fruiting bracts widest above middle, more or less truncate with coarse teeth arising above middle.
         8. Stems upright to prostrate, less than 30 cm; leaves to 2 cm, uniformly scurfy grayish on both surfaces; fruiting bracts 1.5 mm or less in width. ____ **A. pacifica**
         8′ Stems upright, usually more than 40 cm; larger leaves more than 2.5 cm, darker and green above, grayish below; fruiting bracts more than 2.5 mm wide. ____ **A. wrightii**

Atriplex barclayana (Bentham) D. Dietrich [*Obione barclayana* Bentham] *SALADILLO;* COAST SALTBUSH. Suffrutescent perennials, probably not long lived and sometimes flowering the first year; commonly 0.5–1+ m across and usually broader than tall. Lower stems woody, the wood soft, the bark fissured. Herbage semisucculent, often silvery gray-green. Leaves mostly ovate to obovate, (1.5) 2.0–3.5 (4.0) cm, mostly entire, sometimes with a few broad shallow teeth or shallowly lobed. Usually dioecious but occasionally monoecious throughout the range of the species. Fruiting bracts highly variable in size, shape, and ornamentation; mealy, often thick and somewhat spongy, the body 2.5–4.5 mm and more or less as wide, mostly obovoid, with 3 to ca. one dozen apical teeth, and 0 or 2 to several teeth on each face. Flowering at various seasons.

Abundant along the shore, including sandy beaches, strand and dune habitats, and upper margins of estero vegetation, and less common in the tidal exchange zone. Also on rocky slopes bordering the shore and sometimes extending to several kilometers inland such as at La Salina. Shores of islands and both coasts of the Gulf of California to Sinaloa, as well as the Pacific coast of the Baja California Peninsula.

Hall & Clements (1923) recognized 6 subspecies, distinguished by morphology of the fruiting bracts and leaf shape. Ours tend to key to subsp. *sonorae* (Standley) H.M. Hall & Clements. However, I am unable to make sense of the classification among Sonoran populations, and characters of more than one subspecies are sometimes present on a single plant. This species is a promising fodder crop for seawater irrigation (e.g., Aronson, Pasternak, & Danon 1988).

Estero Peñasco, *Norris 3 Dec 1972.* Bahía de la Cholla, *Pickens 1 Jul 1965.* 8.5 km N of Gustavo Sotelo, *Ezcurra 14 Apr 1981* (MEXU). El Golfo, *Bezy 363.* La Salina, *F 86-557.*

Atriplex canescens (Pursh) Nuttall [*Calligonum canescens* Pursh] *CHAMIZO CENIZO;* FOUR-WING SALTBUSH; *'ONK 'I:WAGĬ.* Shrubs, frequently 1.5–2 m, moundlike, much-branched and tardily drought deciduous. Leaves scurfy gray-green, entire, narrowly spatulate to narrowly oblong, larger leaves reaching 3.0–4.5 (5) cm × (2) 4–8 mm. Dioecious. Fruiting bracts conspicuously 4-winged, often 1–2 cm and nearly as wide. Mostly spring and summer.

Dunes and sandy soils, inland and along the shore. Often more than 2 × 3 m on shifting dunes with *Ephedra trifurca, Pleuraphis rigida,* and *Prosopis glandulosa.* On beach dunes often with *Frankenia palmeri.* Also the Río Colorado delta, including the northern and eastern margins of the Ciénega de Santa Clara.

A polymorphic species, mostly arid and semiarid regions, generally on alkaline and sandy soils; western North America and northern Mexico to Zacatecas, and western Sonora to Sinaloa. A number of subspecies have been described, and there is considerable variation in ploidy levels and size and shape of leaves and fruiting bracts. Some of the variation may be in response to environmental factors, but much of it is genetic (e.g., Dunford 1984; Stutz & Sanderson 1979). The dune plants from the flora area are notable in their gigantism, as are other gigas populations from shifting dunes, such as in certain areas of the Chihuahuan Desert (James Henrickson, personal communication 1999) and in central Utah (e.g., Stutz, Melby, & Livingston 1975). Plants of *A. canescens* and other species of *Atriplex* are able to change from one sex to another, notably from female to male, in response to stress such as drought (Freeman, McArthur, & Harper 1984).

Caucc a 31°44′ N, 113°03′ W, *Ezcurra 20 Nov 1980* (MEXU). S of Moon Crater, *F 19102.* Puerto Peñasco, *F 20959.* Bahía de la Cholla, *F 16829.*

Atriplex elegans (Moquin) D. Dietrich [*Obione elegans* Moquin] *CHAMIZO CENIZO;* WHEELSCALE ORACH; *'ONK 'I:WAGĬ.* Annuals, maturing May-October. Leaves 0.8–3 (4.7) cm, mostly oblong to oblanceolate. Monoecious. Fruiting bracts wheel-shaped, 2.4–5.0 mm wide, compressed, orbicular, the margins crenulate to toothed throughout, the teeth small to large and coarse.

Gravelly, sandy soils of dry riverbeds and large washes; margins of larger playas; and disturbed, weedy habitats including roadsides. Agricultural areas south of San Luis, near Sonoyta, along Mex 2 south of Organ Pipe Monument, and Playa Díaz. Perhaps native along the Río Sonoyta and at Playa Díaz, but otherwise weedy in the flora area.

Two subspecies; plants with characteristics of both occur in the flora area: Subsp. *elegans* with thinner leaves, the lower leaves with wavy and often toothed margins, the upper leaves often entire, and the fruiting bracts coarsely toothed. Subsp. *fasciculata* (S. Watson) H.M. Hall & Clements (*A. fasciculata* S. Watson) having generally thicker leaves with entire margins, and crenulate to shallowly small-toothed fruiting bracts. Younger and more robust plants tend to resemble subsp. *fasciculata*, whereas older and perhaps more water-stressed plants tend to key to subsp. *elegans*. The variation seems to be continuous, apparently sometimes changing between early- and late-season growth. The subspecies show geographic segregation but with a very broad region of overlap. Subsp. *elegans* reported for Durango, Chihuahua, and western Texas to Arizona, southeastern California, and northern Sonora southward to the vicinity of Guaymas. Subsp. *fasciculata* known from southern California, northern Baja California, northwestern Sonora, and southern Arizona.

Subsp. **elegans:** Sonoyta, *F 85-703*. Playa Díaz, *F 88-398A*. 10 km S of San Luis, *F 92-533*. Quitobaquito, *Nichol 28 Apr 1939*. Subsp. **fasciculata:** Sonoyta, *F 85-950*. Quitobaquito, *F 86-207*.

Atriplex hymenelytra (Torrey) S. Watson [*Obione hymenelytra* Torrey] DESERT HOLLY. Mistaken reports of this saltbush in northwestern Sonora (Munz 1974; Wiggins 1964) seem to be based on a Thurber collection (GH) labeled "Gila, Sonora" (Hall & Clements 1923:334). Thurber collected along the Gila River in early summer 1852. The Gadsden Purchase, approved in 1854, included land south of the Gila River that had been part of Sonora, Mexico, at the time of Thurber's collection. Desert holly occurs in arid regions of the southwestern United States and nearby northeastern Baja California but not Sonora (see Turner, Bowers, & Burgess 1995).

Atriplex lentiformis (Torrey) S. Watson subsp. **lentiformis** [*Obione lentiformis* Torrey. *A. lentiformis* subsp. *breweri* (S. Watson) H.M. Hall & Clements] *CHAMIZO;* QUAIL BUSH, LENS-SCALE SALTBUSH. Dense, mound-shaped shrubs reaching 2–3 × 2–4 m. Leaves 3–5 × (1.5) 2–4 cm; petioles well developed, the blades triangular, triangular-hastate, rhombic, ovate, or oblong, gray-blue, and relatively thin. Dioecious or sometimes monoecious. Fruiting bracts (2) 3–3.5 mm wide, often lens-shaped, sometimes longer than wide, flattened with shallow, blunt teeth, and not winged. Flowering and fruiting in spring and summer-fall.

Portions of the Río Sonoyta drainage, and especially common in sandy-silty soils of the Río Colorado delta region and plains adjacent to the river; not known elsewhere in Sonora. Often in agricultural areas, especially south of San Luis. Also northern Baja California, southern California, western Arizona, southern Nevada, and southwestern Utah. Distinguished from other Sonoran Desert saltbushes by its robust habit, the large, broad, relatively thin leaf blades and well-developed petioles. Another subspecies occurs from the Mojave Desert of California to Utah.

25 km SW of Sonoyta, *Ezcurra 19 Apr 1981* (MEXU). Sonoyta, *F 85-709*. 18 km W of Sonoyta, *F 86-198*. E side of Pinacate region, 31°45′15″ N, 113°20′ W, *F 87-30*. 18 km S of San Luis, *F 85-1037*.

Atriplex linearis S. Watson [*A. canescens* subsp. *linearis* (S. Watson) H.M. Hall & Clements] NARROW-LEAF SALTBUSH. Shrubs 0.5–1.3 m. Leaves linear to narrowly oblong, the larger leaves often 1–3 (4+) cm × 2–4 mm. Dioecious. Fruiting bracts 4-winged, often 5 (8) mm and about as wide. Similar to *A. canescens* and distinguished by its overall smaller size, smaller and narrower leaves, smaller fruits, and often more pronounced (more deeply laciniate) teeth on the bracts. Mostly spring and summer.

Mostly on semi-saline, siltlike and/or sandy soils, occasionally in rocky habitats with sandy soil pockets; lowland flats, especially near the coast including inland margins of esteros, the delta region of the Río Colorado, near Sonoyta, and at Quitobaquito; not on dunes. Southeastern California, both Baja California states, southwestern Arizona, and coastal Sonora south to the Río Mayo delta and perhaps northwestern Sinaloa.

Often treated as a subspecies of *A. canescens* but I regard them as distinct species. They are geographically sympatric but occupy different habitats in many places in the Sonoran Desert, from at least near Phoenix, Arizona, to the Guaymas region. I have not seen plants of intermediate character.

Sonoyta, *F 85-937.* Moon Crater, *F 10589.* NE margin of Ciénega de Santa Clara, *F 92-988.*

Atriplex pacifica A. Nelson. PACIFIC ORACH. Winter-spring to early-summer ephemerals; often semisucculent, densely short-branched, forming tangled masses, as small as 5 cm and upright to spreading and 40 cm across. Leaves mostly elliptic, obovate, to oblanceolate, uniformly scurfy gray on both surfaces, larger leaves 7–20 mm, the margins mostly entire; lower leaves sometimes petioled. Monoecious. Fruiting bracts (1.0) 1.2–1.5 mm wide, mostly obovoid, truncate at tip, cuneate at base, with 3 or 5 apical teeth, the margins otherwise entire; faces of bracts mealy, each usually with a median keel or ridge and often a prominent tubercle or spine on each side of the ridge.

Roadsides, natural desert pavements, bajadas, and arroyos or washes. Distribution patchy in Sonora and Arizona, often of small, localized populations indicating colonization by individual plants. The earliest collection from Arizona or Sonora is 1966. Perhaps an invader from the Pacific coast. Arizona in Organ Pipe Monument and Cabeza Prieta Refuge; Sonora in the flora area and farther south near the coast between Puerto Libertad and El Desemboque San Ignacio. Widespread and common in coastal areas of the Pacific side of the Baja California Peninsula and southern California, these plants often robust and 30–100 cm in diameter.

The 3 herbaceous atriplexes in the flora area are somewhat similar. *A. pacifica* is basically an early- to late-spring fruiting plant, *A. elegans* fruits from spring to fall, and *A. wrightii* fruits primarily in fall. *A. elegans* is easily distinguished by its wheel-like fruits but is vegetatively similar to *A. pacifica. A. wrightii* is often a much larger plant; the fruits are somewhat similar to those of *A. pacifica* but larger and may have more teeth.

8 km W of Sonoyta, *F 92-143.* 38 km W of Sonoyta, *F 85-715.* 1 mi S of Los Vidrios, *F 84-41.* Moon Crater, *F 19229.* Tinaja del Tule, *F 18744C.* Sierra Blanca, *F 87-18.* 20 mi N of Puerto Peñasco, *F 91-145.*

Atriplex polycarpa (Torrey) S. Watson [*Obione polycarpa* Torrey] *CHAMIZO CENIZO;* DESERT SALTBUSH; *'ONK 'I:WAGĬ.* Shrubs with slender twigs and gray-green to whitish leaves, becoming leafless or nearly so during extended drought. Leaves highly variable, elliptic to spatulate, the larger ones 5–20 mm; dry-season and short-shoot leaves seldom more than 10 mm. Dioecious. Fruiting bracts usually 4–6 mm wide, somewhat orbicular to obdeltoid, often with 7–17 fingerlike blunt teeth, varying in size and shape and often obscured by dense scurfy white hairs. Often with characteristic pink galls in the upper branches and inflorescences. Flowering in various seasons, especially with hot weather following rains.

One of the more common and widespread desert perennials in the region. In many habitats: sandy to rocky soils of flats, washes, and slopes; desert plains; and to peak elevations in the Pinacate volcanic field and granitic ranges; not on dunes. Southern Nevada, southern California, Baja California Peninsula, southern and western Arizona, and western Sonora south to the Guaymas region.

W of Sonoyta, *F 85-963.* N of Puerto Peñasco, *F 13172.* Moon Crater, *F 10586.* Quitobaquito, *F 5725.*

***Atriplex wrightii** S. Watson. Bushy annuals, often germinating in winter-spring and maturing early summer to fall, commonly 0.4–1.0 m. Leaves often 2.5–6 (7) cm, ovate, elliptic, elliptic-spatulate or narrowly oblanceolate, greenish above, grayish below, the margins irregularly toothed, the veins prominent. Lower leaves often petioled, upper leaves progressively shorter and narrower. Monoecious. Fruiting bracts 2.5–2.9 mm wide, more or less obdeltoid, with 5 or 7 terminal teeth.

In the vicinity of Sonoyta and probably not native to this region, as it occurs only in disturbed habitats. Western Texas to central and southern Arizona and northern Sonora. One of the few annuals that commonly continues growing through the really hot, dry early-summer season.

Sonoyta, *F 85-928, F 86-308.* Quitovac, *Nabhan 43.*

Bassia

Annuals to suffrutescent perennials. Leaves alternate, sessile, and entire. Flowers bisexual or female. Calyx lobes 5, becoming enlarged in fruit and developing a spine on the back. Seeds horizontal in the fruit. Native to the Old World; 20 species. Reference: Collins & Blackwell 1979.

***Bassia hyssopifolia** (Pallas) Kuntze [*Salsola hyssopifolia* Pallas] SMOTHER WEED. Annuals, usually germinating in spring and maturing in fall. Leafy with a well-developed main stem, with conspicuous white hairs, the stems glabrate with age. Leaves 1.0–2.5 cm, linear-oblanceolate, flat, the larger leaves usually deciduous by flowering time. Calyx spines hooked even on young flowers. Fruits 2.0–3.5 mm wide; seeds 1 mm wide, discoid.

In northwestern Sonora primarily an agricultural weed, less common in urban areas and other severely disturbed lowland habitats; not in natural vegetation. Native to Eurasia; widespread weed in western North America since its reported introduction in Nevada around 1915.

Sonoyta, *F 85-699*. S of San Luis, *F 85-1039*. Ciénega de Santa Clara, *F 92-518*.

Chenopodium Goosefoot

Usually annual or perennial herbs. Herbage usually mealy with inflated whitish hairs, glandular, or glabrous. Leaves alternate, usually petioled; leaf blades flat, thin or thick. Flowers bisexual. Seeds lens-shaped, usually dark brown or blackish (light colored in domesticated quinoa), contained in a translucent ovary wall, either separable or not.

The name *goosefoot* derives from the leaf shape of certain species. Worldwide; 150 species. This is the largest genus in the family. Many are weedy. Some are edible; young plants of several species are much used in Mexico as *quelites,* and some are utilized for their edible seeds. *Epazote* (*C. ambrosioides*), is a common condiment in Mexico, and quinoa (*C. quinoa*) is an important Andean "grain." Many have been used for medicine and some may be toxic. References: Aellen & Just 1943; Reynolds & Crawford 1980; Wahl 1954; Wilson 1980.

1. Leaves dark green to red-green and shiny on both surfaces, or sometimes grayish below; seed margins acute with a distinct "rim"; seed surface dull even after removal of pericarp. ________ **C. murale**

1′ Leaves pale and mealy below or on both surfaces; seed margins obtuse; seed surface shiny blackish after removal of pericarp.

  2. Leaves bicolored, green above, pale below; seed surfaces without honeycomb depressions. ________ **C. album**

  2′ Leaves not bicolored, pale on both surfaces; seed surfaces covered with a coarse but minute cellular pattern. ________ **C. berlandieri**

***Chenopodium album** Linnaeus. LAMB'S QUARTERS. Winter-spring annuals, often reaching 1–1.5 m. Stems often pink-and-green striped. Leaves often 4–10 cm long, the blades green above, grayish below, lanceolate to ovate with coarsely toothed margins. Seeds 1.0–1.4 mm wide, shiny below the persistent pericarp, the margins obtuse to moderately acute-keeled.

Weed in irrigated wheat fields in the San Luis region (6 km S of San Luis, *F 91-67*). Native to Europe and now widely naturalized in North America, especially in agricultural fields. Often confused with *C. berlandieri; C. album* lacks the regular minute honeycomb depressions on the pericarp.

***Chenopodium berlandieri** Moquin var. **sinuatum** (Murray) Wahl [*C. petiolare* var. *sinuatum* Murray] PIT-SEED GOOSEFOOT. Winter-spring ephemerals, often 0.5–1 m, coarse and erect, the stems often pink- or red-striped. Leaves often 3–10 cm, the blades mealy and pale on both surfaces, mostly lanceolate, ovate, or rhombic, the petioles well developed. Sepals keeled, mostly enclosing the fruit. Seeds 1.2–1.4 mm wide, surfaces covered with a coarse but minute cellular pattern (like minute honeycomb depressions), shiny blackish after removal of pericarp, the margins obtuse (without a distinct rim).

Gravelly washes in the Río Sonoyta valley near Sonoyta, often in disturbed habitats, and expected in the San Luis region as an agricultural weed. Native to North America, but probably not native to the flora area.

The pericarp usually sticks to the seed; the discoid area on top of the seed, especially around the style, is usually yellow with a honeycomb or alveolate pattern of pitting. The pitting is caused by collapse of the very large cells of the upper portion of the pericarp, leaving only the lateral walls; this subtle character is best seen on fully mature fruits using a lens.

El Papalote, *F 86-109*. Sonoyta, *F 85-707*.

***Chenopodium murale** Linnaeus. *CHUAL, CHOAL;* NET-LEAF GOOSEFOOT; *'ONK 'I:WAGĬ, KAWPDAM.* Winter-spring ephemerals, sometimes persisting into summer in shaded habitats or at the coast. Mostly less than 50 cm, or to 1 m in wheat fields near San Luis. Herbage green or often red-green, sometimes semisucculent, the stems sometimes red striped. Leaves glabrous or sometimes gray scurfy below, often 3–10 cm, the blades mostly ovate to rhombic, irregularly toothed, the petioles well developed. Sepals often keeled, mostly partially spreading at maturity to reveal part of the seed. Seeds 1.2–1.4 mm wide, blackish, mostly devoid of dried pericarp, mostly with minute papillae but not alveolate, dull even after removal of pericarp, the margins acute with a thin rim.

Widespread urban and agricultural weed in disturbed habitats including areas of cattle grazing, also in natural habitats including gravelly washes or arroyos, playas, and waterholes. Reportedly native to the Old World and adventive from Canada to Guatemala. Long established and widespread in the Sonoran Desert (see Felger & Moser 1985).

Ejido Luis Encinas Johnson, *Ortiz 9 Apr 1993*. 6 km S of San Luis, *F 91-66*. Tinajas de los Pápagos, *Turner 59-21*. Arroyo Tule, *F 19205*. El Papalote, *F 86-132*. Puerto Peñasco, *F 85-797*.

MONOLEPIS

Annuals. Leaves mostly alternate. Flowers minute, bisexual or unisexual; floral parts greatly reduced, the calyx usually of 1 persistent green sepal, the stamens 1 (2) or none. Mostly temperate western North America and introduced elsewhere in the world; 3 species.

Monolepis nuttalliana (Schultes) Greene [*Blitum nuttallianum* Schultes] *PATATA;* POVERTY WEED; *'OPON*. Winter-spring annuals, semisucculent, spreading to semiprostrate with age and size, the stems often 5–25 cm. Larger leaves mostly 1–5 cm, the petioles expanding into lanceolate to linear blades with hastate "ears," the smaller leaves often entire; leaves reduced upwards to leafy bracts. Flowers in dense, sessile, axillary clusters. Seeds 0.9–1.2 mm wide, flattened, lens-shaped, dull beneath a thin, persistent pericarp with a honeycomb-alveolate pattern.

Vicinity of Sonoyta in the Río Sonoyta riverbed and an urban-agricultural weed. Alaska and Canada to northern Mexico, and Argentina; adventive in the Old World. Desert populations in western North America have much narrower and smaller leaves than those from more humid, temperate regions. *Monolepis* might be confused with *Cistanthe monandra* (Portulacaceae), which has entire leaves and multiple-seeded fruits.

Río Sonoyta W of Sonoyta, *F 86-157*. Sonoyta, *F 87-15*.

NITROPHILA

Herbaceous succulents; flowers bisexual. Calyx lobes 5; stamens 5; stigmas 2. Two species in western North America, 6 in Chile and Argentina. Among the chenopods in the flora area, only *Nitrophila* and *Salicornia* have opposite leaves, bracts, and branches, although *Suaeda puertopenascoa* has opposite vegetative leaves.

Nitrophila occidentalis (Moquin) S. Watson [*Banalia occidentalis* Moquin] ALKALI WEED. Succulent perennials from thick, succulent rhizomes to 2.5 cm in diameter and 15+ cm deep. Stems much-branched. Leaves sessile, linear, semiterete, mostly 1.0–2.5 cm, the lower leaves clasping the stems. Flowers pink and green, 1 or 3 in leaf axils, subtended by 2 leaflike bracts of unequal size. Calyx 2.5–4.0 mm, the sepals all similar or slightly dissimilar in length, fleshy, green outside, pink within, persistent when dry with a midrib keel. Seeds 1.4–1.5 × 1.2 mm, D-shaped, smooth and shining, dark red-brown to blackish. At least March-April.

Dense colonies in alkaline or saline wetland habitats, the roots often in thick clay-like wet mud. Margins of the Río Colorado and Ciénega de Santa Clara, pozos and seeps of Bahía Adair salt flats, and abundant at Laguna Prieta with *Distichlis spicata*. Also Quitobaquito and localized along the nearby Río Sonoyta, and at Quitovac. Not known elsewhere in Sonora. Also saline or alkaline wetlands in Oregon to Nevada to northeastern Baja California and Arizona.

The aerial portions are often freeze killed in winter, but regrowth is rapid in spring. *Nitrophila* is the first plant to colonize newly formed Bahía Adair pozos, often growing where the underground aquifer breaks through the overlying clay and reaches the surface. Coyotes dig holes in the greener, more luxuriant colonies to get fresh water, which also may lead to the formation of a new pozo. Although the leafy stems may sprawl across salt-encrusted soil, the roots are deeply seated in non-saline mud. *Nitrophila* thus can be an indicator of fresh and potable water only 20–50 cm below the surface (Ezcurra et al. 1988).

Banks of Río Colorado, 20 mi N of El Golfo, *F 75-60.* Laguna Prieta, *Roth 1.* 6 km NW Gustavo Sotelo, *López-Portillo 7 Jul 1984* (MEXU). Salada of Río Sonoyta, 22 mi W of Sonoyta, *Shreve 7600.* Río Sonoyta S of El Huérfano, *F 92-980.* La Salina, *F 86-539.* Quitobaquito, *Nichol 28 Apr 1939.*

SALICORNIA PICKLEWEED, GLASSWORT, SAMPHIRE

Annual or perennial halophytes, herbaceous to shrubby. Stems succulent, jointed, green or turning red- to orange-green, with opposite branching; fleshy part of stems actually reduced and modified leaves with 2 scale leaves per node. Flowers bisexual or unisexual, minute, in clusters of 3 (ours) to 7 on opposite sides of each node and sunken in fleshy tissue; flowering joints forming cylindrical terminal or subterminal spikes. Calyx fleshy, nearly surrounding the flower and fruit, obpyramidal, the fruiting calyx dry and spongy but readily soaking up water.

Worldwide, seacoasts and inland saline depressions in temperate, tropical, and arid regions; 65 species. Three species in the Gulf of California. Many of the species are poorly distinguished. Some authors distinguish *Arthrocnemum,* with 15 species. References: Ball & Tutin 1959; Moss 1954.

1. Annuals (without rhizomes), usually with a single main axis and branching above the lower 1/3 of plant, the branches widely spaced and diffuse; middle flower conspicuously taller and inserted higher than the lateral ones. ______ **S. bigelovii**

1′ Perennials with conspicuously woody stems and/or rhizomes, few-branched from base or densely branched throughout; the 3 flowers of each cluster about equal height and inserted at about the same level.

  2. Stems much-branched, woody at least below, forming dense, low, and spreading shrubs, rhizomes lacking; seeds 1.3–1.4 mm, lens-shaped, shiny and glabrous; tidally wet soil to adjacent lowermost desert edge. ______ **S. subterminalis**

  2′ Unbranched to relatively few-branched clusters of stems along slender rhizomes, not shrubby; seeds 1.1–1.3 mm, ellipsoid, the surface dull and usually with short, curved hairs (best seen at 20× or higher magnification); tidally wet soil. ______ **S. virginica**

Salicornia bigelovii Torrey. PICKLEWEED. Erect annuals often 20–50+ cm, almost always with a single prominent main axis, usually taller than wide, sparsely to moderately branched, the branches spreading more or less at right angles. Stems (actually modified leaf tissue) bright yellow-green, very succulent and glassy looking; almost every branch ending in a spike. Middle flower of each triad cluster inserted higher and at least half again taller then the 2 lateral flowers. Seeds 1.8–2.0 mm, ellipsoid (about half as wide as long), light tan, opaque due to lack of perisperm, nearly filled by the "folded" embryo, and with short, thickish, transparent, and often curled hairs; seeds usually ripening as the plants mature and die.

Saline wet soils of quiet bays and esteros, often partially to fully submerged at high tides, usually in large stands. Commonly germinating with highest tides, November-January and rapidly growing to full size; 2 or more size classes may be present. Often flowering May-June, the seeds mostly mature September-October. Flowering and seed maturation often variable from low to high tidal zones—often about 1 month later in the higher tidal zones. Seed size tends to decrease southward, such as at Bahía Kino and farther south in more tropical areas.

Salicornia bigelovii appears to be the only annual member of the genus in the Gulf of California, ranging southward to Sinaloa, and northward on the Pacific coast to southern California. There are a handful of apparently closely related and often difficult to distinguish annual salicornias worldwide. Selections of *S. bigelovii* from the Gulf of California show promise as a seawater irrigated oilseed crop (Glenn et al. 1991).

Estero Morúa, *Yensen 19 Oct 1974.* Puerto Peñasco, *F 20969.* Bahía de la Cholla, *López-Portillo 9 Jul 1984.* Bahía Adair, *López-Portillo 9 Jul 1984.*

Salicornia subterminalis Parish [*Arthrocnemum subterminale* (Parish) Standley] Densely branched, shrubby perennials; often low and spreading or large, spreading moundlike clumps, to ca. 0.5–1+ m. Upper stems often glaucous, the older succulent stems sometimes pale reddish. Seeds 1.3–1.4 mm, glabrous, lens-shaped, laterally compressed, filled with perisperm, shiny and glabrous, translucent, rich amber colored (much darker than the seeds of *S. bigelovii* and *S. virginica*), the curved embryo readily visible within the seed.

Upper tidal zone of saltscrub among estero vegetation onto the lowermost edge of the saltscrub-desertscrub ecotone; vicinity of Puerto Peñasco to the Río Colorado delta. Widespread along the shores of the Gulf of California; coastal salt marshes, California to Sinaloa, and inland areas in California.

Estero Morúa, *Yensen 19 Oct 1974.* Puerto Peñasco, *F 20971.* 6 mi by rd NE of La Salina, *F 86-533.*

Salicornia virginica Linnaeus [*S. ambigua* Michaux. *S. pacifica* Standley] Perennials, often 30–50 cm, usually with a small cluster of stems, or sometimes a single stem, arising at widely separated nodes along a slender subsurface rhizome; knotty-woody at the base but otherwise not woody and not shrubby. Stems erect to ultimately decumbent, often sparsely branched, succulent and green above, or often red-orange with age, the lower stems devoid of the succulent, modified leaves; stems apparently rooting where touching the substrate. Seeds 1.1–1.3 mm, ellipsoid, brown and dull, opaque because of lack of perisperm, nearly filled by the conduplicate (folded) embryo, with short, hooked hairs best seen with 20× or higher magnification; these hairs often appressed when dry and spreading when wet.

Estero margins, often with the other 2 salicornias in tidally wet mud, and the Río Colorado delta extending into brackish water with *Distichlis palmeri.* Esteros through most of the Gulf of California but not as common as the other 2 species. Alaska to the Gulf of California, Atlantic and gulf coasts of United States, western Europe, and the Mediterranean.

The seeds are sometimes tightly held by the dry, spongy calyx and can be removed after moistening. Seeds of *S. bigelovii* and *S. virginica* are similar, although those of the latter are slightly smaller; the plants, however, are substantially different in habit. Distinguishing *S. virginica* from *S. subterminalis* may be difficult without fully developed seeds. Mature seeds of both species readily fall away and are often not present on specimens. Seeds of *S. virginica* are rough, not smooth like those of *S. subterminalis,* and sometimes hairs are present only at the base of the seed of *S. virginica,* or the hairs may be reduced to papillae. To further complicate matters, immature seeds of *S. subterminalis* may have hairlike wrinkles that might be mistaken for the appressed hairs of *S. virginica.* For reliable identification it is best to use the growth habit and seed characters.

Near mouth of Santa Clara Slough, *F 90-188*. 5 mi S of El Doctor, *Watson 9 Mar 1987*. Shore, 6 mi NE of La Salina, *F 86-532*. Estero Morúa, *F 91-141*.

SALSOLA

Annuals to shrubs or small trees. Flowers bisexual. Calyx usually 5-merous, often developing a horizontal collarlike wing persistent in the fruit. Stamens 5 or fewer. Native to the Old World; 130–150 species. Reference: Mosyakin 1996.

***Salsola tragus** Linnaeus [*S. australis* R. Brown. *S. iberica* Sennen & Pau. *S. kali* of authors. Not *S. kali* Linnaeus. *S. kali* var. *tenuifolia* Tausch. *S. pestifer* A. Nelson] *CHAMIZO VOLADOR;* RUSSIAN THISTLE, TUMBLEWEED; *HEJEL 'E'EṢADAM, WOPO'ODAM ṢA'I.* Warm-weather annuals, globose at maturity, breaking off near ground level to become tumbleweeds. Sparsely pubescent to glabrous. Lower or seedling leaves often 2–5 cm, linear to threadlike, semisucculent; upper leaves 1.0–2.5 cm, firm, spinescent-tipped. Flowers in upper leaf axils, subtended by ovate, spinescent bracts 5–8 mm. Sepals 5, essentially separate, erect and membranous in flower; fruiting calyx pink to white, with prominently veined wings often 5.0–6.5 mm wide, with 2 wings smaller than the other 3. (The wing formation, unique among our chenopods, is from a horizontal keel on the back of each sepal.) Seeds horizontal, falling with the calyx.

Disturbed habitats; settlements, farms, and roadsides; not in natural habitats. Germinating with winter-spring and summer rains, maturing at the end of summer. Widespread in the Sonoran Desert. This C_4 plant is native to Eurasia and has become a troublesome worldwide weed, especially on disturbed soils in hot, dry regions.

Sonoyta, *F 85-927*.

SARCOBATUS GREASEWOOD

This genus has a single species.

Sarcobatus vermiculatus (Hooker) Torrey [*Batis* (?) *vermiculata* Hooker. *S. baileyi* Coville. *S. vermiculatus* var. *baileyi* (Coville) Jepson] Much-branched shrubs 1.2–2.5 m, with rigid, woody, thorn-tipped twigs. Leaves alternate, succulent, mostly 1–2.5 cm, linear to linear-spatulate. Monoecious. Male flowers in slender spikelike racemes terminal on short shoots, the calyx none, stamens probably 2 or 3, beneath a peltate scale 1.5–2.0 mm wide. Female flowers crowded in axils and often bending the smaller branches with their weight; stigmas 2, the calyx distinctively winged, developing into a flared collar 1.3–1.6 cm wide around middle of fruit, the wings dry and papery at maturity, persistent and tightly enclosing the seed and falling with it. Seeds 3+ mm in diameter, orbicular, vertical (erect), and compressed (flattened); the tightly coiled embryo clearly visible.

Low dunes and sandy flats, often adjacent to salt flats. In a relatively narrow zone across approximately 100 km in the coastal region between Puerto Peñasco and the Río Colorado delta. Not known elsewhere in Mexico. The closest known population is around the confluence of the Salt and Gila Rivers near Phoenix and Sacaton in Maricopa and Pinal Counties, Arizona. Also California to New Mexico and Washington, Alberta, and North Dakota. Greasewood is a characteristic element of the Great Basin Desert and the southern part of the Mojave Desert and has a limited distribution in the Sonoran Desert (see Benson & Darrow 1981).

Variety *baileyi* is distinguished by branched white hairs on the leaves and fruits, and the leaves and staminate spikes are reported to be shorter, but this character does not correlate with pubescence. Populations with characteristics of var. *baileyi* occur in central Arizona, Nevada, and southern California. The Sonoran specimens are glabrous.

Gustavo Sotelo, *Ezcurra 15 Sep 1980* (MEXU). 11 mi by rd W of Estación Gustavo Sotelo, *F 84-10*.

SUAEDA SEA-BLITE, SEEPWEED

Annual or perennial herbs or shrubs. Leaves and stem tips succulent; leaves usually alternate or opposite below. Flowers small, usually bisexual, sometimes intermixed with unisexual female flowers, in clusters of 3 or more subtended by small membranous bracts in axils of leaflike bracts. Sepals 5, green or greenish, usually succulent. Stamens 5. Stigmas 2 or 3. Embryo flat, spirally coiled. Seeds shiny blackish and/or dull brown, horizontal or vertical.

Worldwide, tropical to arctic, many in aridlands, mostly along shores and inland saline or alkaline soils and depressions; 100 species. A taxonomically difficult genus in need of revision. North American suaedas fall into 2 or perhaps 3 readily distinguishable major groups, or sections. Some difficulty in classifying suaedas is due to variation in characters influenced by environmental conditions such as salinity, temperature, and season. References: Bassett & Crompton 1978; Hopkins & Blackwell 1977; Iljin 1936.

1. Globose perennial shrubs, usually 1–1.5` m in height and width, branched throughout with many spreading, interlacing branches; herbage and calyces minutely pubescent to glabrous; all seeds shiny blackish and convex on both faces; inland and coastal. ______ **S. moquinii**

1′ Annual or short-lived perennial herbs or subshrubs, usually taller than wide, mostly less than 1.2 m, branched from above, the branches relatively few and often ascending; glabrous; seeds dull brown and flattish and sometimes intermixed with some shiny blackish seeds; coastal.

  2. Herbage often becoming reddish; seeds 1.5–2.0 mm wide, dull brown and sometimes also with some shiny blackish seeds. ______ **S. esteroa**

  2′ Herbage often becoming yellowish to orange but not red; seeds 2.4–3.0 mm wide, all dull brown. ______ **S. puertopenascoa**

Suaeda esteroa Ferren & Whitmore. Succulent annuals, perhaps sometimes weakly perennial, (15) 20–45 cm, glabrous, the main axis often well developed, the main axis and branches often ascending to decumbent. Branches and leaves alternate. Leaves mostly 2.0–2.5 (4.0) cm, sessile, green, often glaucous, turning yellowish to red to red-purple, linear, semi-cylindrical with the upper surfaces flattened to slightly concave.

Most branches terminating in dense, leafy-bracted inflorescences, these bracts often 1.5–2.0 cm. Flowers 3–8 in dense, axillary, sessile clusters, the individual flowers subtended by minute membranaceous bracts, bisexual or occasionally unisexual, ca. 2 mm wide when fresh. Sepals very succulent, hooded with membranaceous marginal flanges, often with 4 smaller sepals and 1 larger sepal (flowers thus bilaterally symmetrical). Styles 2 or 3, short and thick. Seeds horizontal in ovary, with 2 kinds on the same plant: (1) mostly dull brown, 1.5–2.0 mm wide, disk-shaped, the faces flattened and slightly concave, the coiled embryo prominent beneath a very thin seed coat adherent to the pericarp; and (2) less often shiny blackish, 1.7 mm wide, the faces conspicuously convex, the embryo not visible, the seed coat well developed and readily separable from the pericarp. Mostly July-November, also February-April; seeds mostly ripening late fall–March.

Saltscrub of estero margins, often inundated by the highest tides; Bahía Adair southward along the Sonora coast. Saltscrub and mangrove esteros throughout the Gulf of California and apparently southward to Altata, Sinaloa; also the Pacific coast of the Baja California Peninsula and southern California. Mostly annuals in the Gulf of California and perennials on the Pacific coast. However, the Gulf of California plants sometimes continue growing vegetatively after flowering.

Bahía Adair, 16 km WNW of Gustavo Sotelo, *López-Portillo 9 Jul 1984* (MEXU). Estero Peñasco, *F 20965*. Bahía de la Cholla, *Hill 8 Jul 1966*. Estero Morúa, *Cliffton 3 Apr 1977*. Estero Cerro Prieto, *F 86-515*.

Suaeda moquinii (Torrey) Greene [*Chenopodium moquinii* Torrey. *Suaeda torreyana* S. Watson. *S. ramosissima* (Standley) I.M. Johnston. *S. torreyana* var. *ramosissima* (Standley) Munz] *QUELITE SALADO;*

DESERT SEEPWEED; *S-CUK 'ONK*. Shrubby perennials, often (0.8) 1.2–2 m, semi-hemispherical, much-branched throughout, the branches often spreading and interlacing. Stems slender and brittle. Herbage and calyces minutely and densely pubescent to sometimes glabrous, succulent, green to glaucous blue-green and often reddish purple. Leaves alternate, thick and succulent, the long-shoot leaves moderately flattened with rounded margins, often 1–3 cm, the internodes often 2+ cm, the short-shoot leaves terete, usually crowded, often 3–8 mm and beadlike; leaves, especially the larger ones, narrowed at base to a short petiole or subsessile.

Flowering branches slender, paniculate. Flowers 1 to ca. 10 per cluster, often functionally unisexual. Sepals succulent with membranous margins, hooded and unequal in size in female flowers, the fruiting calyx bilateral, often 1.3–1.6 mm across; sepals of male flowers spreading. Stigmas usually 3, thickish and linear, papillose to sometimes pubescent on densely pubescent plants. Filaments of male flowers longer than anthers. Seeds 1.0–1.5 mm, erect, shiny, blackish. Flowering mostly during warmer months, especially August-September.

Inland margins of esteros and bays, upper beaches, beach dunes, washes or arroyos, and sometimes rocky slopes near the shore; above the high tide zone. Also inland on alkaline and semi-saline soils, including the Sonoyta region and mesas near the Río Colorado and the river floodplains; especially common in disturbed habitats as an urban and farmland weed.

In the flora area mostly pubescent, the distinguishing character of *S. torreyana* var. *ramosissima* (e.g, Wiggins 1964). Plants near the coast, especially in the Puerto Peñasco region, have exceptionally dense and somewhat glandular pubescence; similar plants occur elsewhere, e.g., southeastern California. Var. *ramosissima* reported from southeastern California and southwestern Arizona to Baja California Sur and western Sonora to Sinaloa. Glabrous plants occur inland near the international border at Quitobaquito, the floodplain of the Río Sonoyta, and south of Yuma. However, pubescent plants have also been collected from the same places. The lack of hairs may be seasonal and attributable to luxuriant growth resulting from lower salinity, higher soil moisture, and perhaps lower temperatures. The glabrous form ranges from Alberta through the western United States to northern and central Mexico. *S. moquinii* is in the *S. frutescens* complex, which includes Old World species; these species belong to the section *Limbogermen*.

Sonoyta, *F 85-947*. 15 km W of Sonoyta, *F 89-10*. Puerto Peñasco, *F 20960*. Quitobaquito, *Darrow 2926*.

Suaeda puertopenascoa M.C. Watson & Ferren, *Madroño* 38:30–31, 1991. Perennial succulents, probably short-lived, 40–110 cm, with a taproot, a well-developed, erect main axis, relatively few, mostly ascending branches, and mostly branched above midheight; glabrous, often glaucous, greenish and often becoming yellowish to yellow-orange with age. Vegetative branches and leaves mostly opposite; vegetative leaves 3–5 (7) cm, sessile, linear, terete to flattened or concave above (adaxial surface).

Inflorescences compound, on a separate branch or branches from the vegetative growth, the flowering branches and leafy bracts alternate. Flowers in sessile clusters of 3 in axils of leaflike bracts, bisexual, slightly to moderately bilateral, ca. 2–3 mm wide when fresh, subtended by 3 minute, unequal membranaceous bracts. Sepals very succulent, hooded (bulbous toward apex and nearly covering the rest of the flower), usually not opening (not spreading apart to reveal stamens and ovary). Sepals fused below to form a basal disk having horizontal thickenings or wings. Styles 2, linear, short and thick (longer and more slender when cultivated in a greenhouse). Fruiting calyces 3–4 (5) mm wide. Seeds 2.4–3.0 mm wide, horizontal in ovary, dull brown, disk-shaped, the faces flattened and slightly concave, the coiled embryo clearly visible beneath a thin, membranous seed coat adherent to the pericarp. Mostly June-July, the fruits ripening mostly October-November, sometimes flowering and fruiting at other seasons.

Known only from esteros in northwestern Sonora; Estero la Pinta, ca. 25 km SE of Puerto Peñasco to Estero las Lisas, ca. 40 km NW of Puerto Peñasco. Carolyn Watson and Wayne Ferren searched other esteros in the Gulf of California but found no other populations. *S. puertopenascoa* joins the ranks of the very few species endemic to the flora area, with a range even narrower than that of *Distichlis palmeri,* the only other halophyte endemic to the Gulf of California. Watson & Ferren (1991) report:

> *Suaeda puertopenascoa* grows as linear groupings or scattered individuals in low marsh habitats along margins of tidal lagoons and banks of tidal channels represented by the lower littoral zones. It colonizes open sand to silt substrates and usually occurs upslope from barren tidal flats and channel bottoms. Plants established in the low zones of the tidal gradient are generally taller (110 versus 40 cm) and more erect in growth form than those found in the uppermost limits. It often stands higher than the frequently associated species *Batis maritima* L., *Distichlis palmeri* Fassett, *Salicornia virginica* L. and *S. bigelovii* Torr.

Estero Cerro Prieto, *Ferren & Watson 2807* (isotype). Estero las Lisas, *Watson 891010-1* (paratype). Estero la Cholla, *Ferren & Roberts 2833.* Estero la Pinta, *Watson & Ferren 891009-1.*

Suaeda puertopenascoa and *S. esteroa* occur in many of the same estuaries but can be distinguished by morphological, phenological, and ecological characteristics. Both are in section *Heterosperma.* Characters of this section include annual and herbaceous perennial habit, bilaterally symmetrical flowers, 2 or 3 stigmas arising directly from the top of the ovary, sepals often with appendages (hooded in ours), seeds generally dimorphic, and stems usually not branched from the base. *S. esteroa* and *S. puertopenascoa* have these characters except that in the latter there is only 1 seed type. However, the seed is dull brown and flat rather than shiny blackish and biconvex like that of many monomorphic-seeded suaedas. Elsewhere seed type has been correlated with season: most Old World species, especially across Eurasia, produce shiny blackish seeds in summer and early fall, and dull brown seeds with thinner seed coats in late fall (Shishkin 1970). Both *S. esteroa* and *S. puertopenascoa,* grown from seed in adjacent field plots at Puerto Peñasco, maintained their distinctive characteristics (Watson & Ferren 1991). There are no known intermediates.

At least among the North American members of section *Heterosperma,* most of the presently recognized species form allopatric series of poorly differentiated taxa, their distinguishing features mostly subjective. It is unusual to have well-marked sympatric species. Differences between *S. puertopenascoa* and *S. esteroa* are as follows:

S. puertopenascoa	*S. esteroa*
Perennials, flowering 2nd year.	Mostly annuals.
Plants larger.	Plants smaller.
Branches mostly erect to ascending.	Stems mostly decumbent to ascending.
Vegetative branches and leaves opposite.	Branches and leaves alternate.
Leaves terete to flattened above, 3–5 (7) cm.	Leaves flattened above, 2–3 (4) cm.
Herbage and calyces turning yellowish to yellowish orange.	Herbage and calyces mostly turning reddish, sometimes yellowish to orange.
Flowers in clusters of 3.	Flowers in clusters of 3–8.
Flowering calyces 2–3 mm wide.	Flowering calyces ca. 2 mm wide.
Fruiting calyces 3–5 mm wide.	Fruiting calyces 2–3 mm wide.
Stigmas 2.	Stigmas 2 or 3.
Fused sepal bases forming a basal disk.	Fused sepal bases not forming a basal disk.
Dull brown seeds 2.4–3.0 mm wide.	Dull brown seeds 1.6–2.0 mm wide.
Shiny blackish seeds not present.	Shiny blackish seeds present.
Mostly flowering late.	Mostly flowering early.
Low tidal zone.	High tidal zone.

CONVOLVULACEAE MORNING GLORY FAMILY

Annual or perennial herbs, mostly vines, in seasonally dry tropics some shrubs and trees; obligate parasites with little or no chlorophyll and vegetative parts much reduced in *Cuscuta*. Leaves alternate, simple, entire to deeply parted, lacking stipules. Flowers usually bisexual, radial, usually 5-merous, often showy. Calyx persistent, the sepals separate or united. Corollas sympetalous, usually funnelform; among ours (except *Cuscuta*) the parts folded within the bud (plicae) are glabrous, whereas those parts exposed in the bud (interplicae) form a star-shaped pattern in the expanded corolla and are often pubescent. Ovary superior; pistil 1. Fruits mostly dry capsules; seeds 1–4 (6).

Worldwide, mostly tropical; 56 genera, 1600 species. Well represented in the Sonoran Desert, but only entering the margin of the flora area except for *Cuscuta*. References: Austin 1992, 1998; Austin & Pedraza 1983.

1. Parasitic, the stems uniformly yellow or orange; leaves reduced to scales or absent. ______ **Cuscuta**
1′ Not parasitic, the stems green or brown; leaves well developed.
  2. Stems conspicuously vining; corollas pink or blue. ______ **Ipomoea**
  2′ Stems not vining; corollas white or blue.
    3. Flowers solitary in leaf axils, the pedicels appearing continuous with the calyx, shorter than flowers or capsules; corollas white; halophytes on saline or alkaline soils, mostly low-lying coastal wetlands and agricultural areas. ______ **Cressa**
    3′ Flowers 1 or 2 on slender, filiform peduncles much longer than flowers, the pedicels longer than capsules or flowers; corollas blue; dry desert soils. ______ **Evolvulus**

CRESSA

Annual or perennial herbs. Stems not twining. Herbage with appressed hairs. Leaves sessile or nearly so, entire, small or reduced to scales. Flowers solitary, axillary in bracts of inflorescences; pedicels with 2 small bracteoles. Sepals united only at base. Corollas white. Styles 2, separate, the stigmas knoblike. Capsules usually 1-seeded by abortion of other ovules (occasionally 2–4-seeded). Tropical and warm-temperate regions worldwide; 3 or 4 species.

Cressa truxillensis Kunth ALKALI WEED. Perennials, dying back during adverse times to thickened, underground stems and/or rootstocks, often 8–15+ cm below the surface; lower stems often semi-woody. Herbage silvery gray-green with dense silky hairs. Stems erect to spreading, 8–25 cm in open, sunny habitats, mostly branching from near the base; or, when growing through and branching over the tops of other halophytes, with weak, slender stems to 75 cm. Leaves mostly ovate to elliptic, sometimes lanceolate, (3.5) 6–10 mm, usually reduced upwards; older leaves sometimes succulent. Flowers in upper axils, 6–8 mm including the short pedicel, with 2 small bracts just below the calyx. Calyx leafy, persistent, the dry brown shriveled corollas also persistent. Corolla lobes reflexed; corolla lobes, style, and stigma pure white; anthers reddish, turning purplish lavender. Capsules ca. 5 mm, shiny brown, glabrous or glabrate except for a small tuft of white hairs at apex.

Often common in tidally wet saline mud and less commonly in sandy soils at edges of esteros, bays, and low-lying coastal soils in the flora area and elsewhere on the shores of the Gulf of California. Also common in the lower Río Colorado and delta including margins of the Ciénega de Santa Clara, banks of the Río Hardy, sometimes in small inland playas but near the coast, and an agricultural weed south of San Luis, especially in fine-textured silty-clay alkaline soils. In contrast to the situation in northwestern Sonora, this species has apparently become relatively rare in nearby southwestern Arizona (Austin 1992). Western North America and South America.

In the early 1990s *Cressa* was a common weed in sandy soil of seawater-irrigated experimental plots at the Environmental Research Laboratory adjacent to Estero Morúa. *Cressa* had been evaluated earlier as a potential halophytic seed crop in these plots. Substantial seed crops were obtained, but no economic value was found and further evaluation was terminated. This is probably the first report of a

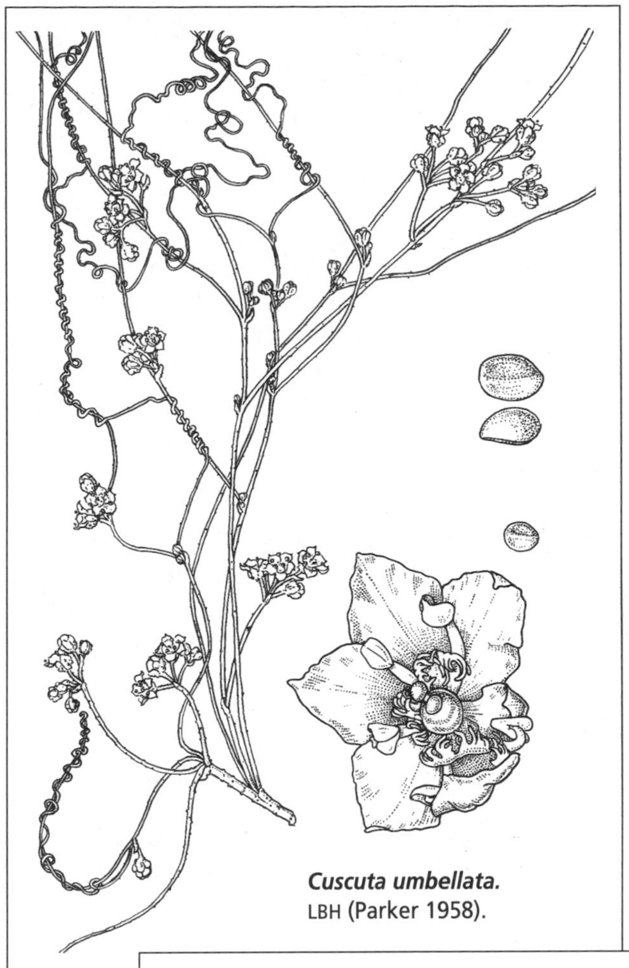

Cuscuta umbellata. LBH (Parker 1958).

Cressa truxillensis. Plant, FR; flowers, BA (Cronquist et al. 1984).

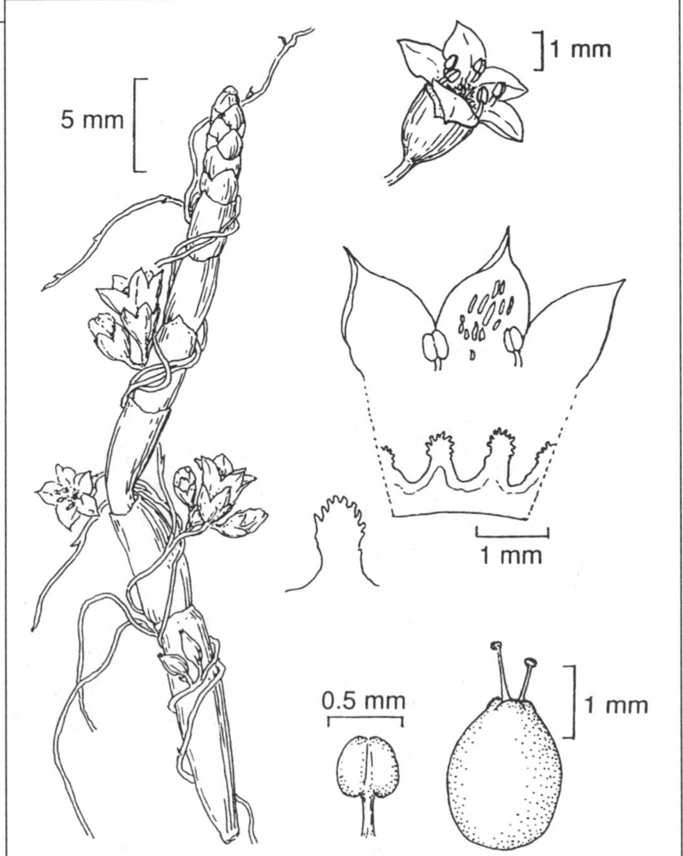

Cuscuta salina on Salicornia. LAV (Hickman 1993).

weed in seawater-irrigated agriculture. *C. cretica* Linnaeus of the Old World and *C. truxillensis* are only subtly different.

Estero Morúa, *F 91-40.* Estero Peñasco, *F 20963.* Estero Cerro Prieto, *F 90-163.* 5 km S of Riito, *F 85-1043.* Ciénega de Santa Clara, *F 92-989.* Colorado River, Sonora, opposite mouth of Hardy River, 29 May 1894, *Mearns 2842* (DS).

CUSCUTA DODDER

Annuals or sometimes perennials. Obligate parasites, the chlorophyll absent or much reduced, glabrous or some with papillae, the vegetative morphology highly reduced. Stems slender and twining, yellow to orange or white, forming numerous small, suckerlike haustoria. Leaves reduced to scales. Flowers small, mostly 5-merous, mostly white to yellowish, in few– to many-flowered cymose clusters. Sepals united or sometimes separate; corolla tube usually with small scales (corolla appendages) in a ring below the stamens and alternate with the corolla lobes. Filaments attached to the corolla tube and alternate with corolla lobes. Styles 2, separate or sometimes united, or the stigma sometimes sessile. Capsules usually thin walled and rounded, opening near the base (circumscissile), irregularly dehiscent, or indehiscent, with 1–4 seeds (when only 1 seed develops it tends to be rounded, but when 2 or more develop the common sides tend to be flattened and the seeds thus angled). Cotyledons absent or extremely reduced.

Worldwide, mostly tropical and subtropical, some in temperate regions, the majority in the New World; 145 species.

Dodders begin life as non-parasitic plants, largely by digesting the food reserves in the seed (some seem to have a little bit of chlorophyll). As the tiny, slender seedling stem grows it swings around in a circle, and if it touches a host plant, it attaches by means of a haustorium and invades the host tissue, after which the root and lower portion of the stem rapidly die. The rest of the plant's life is spent as a parasite with no connection to the soil. However, among some dodders the seeds germinate *in situ* among the tangled mass of stems of the previous year or season.

Because there is so little variation in the vegetative morphology, the taxonomic characters are based almost entirely on flowers, fruits, and inflorescences. Partly because the flowers and fruits are small these characters can be difficult to distinguish. For purposes of identification, considerable emphasis has been placed on the corolla (or staminal) appendages or scales. These scales are perhaps analogous to the variations in glandular trichomes and other structures in diverse Convolvulaceae and even different families such as Hydrophyllaceae (Dan Austin, personal communication 1998). The function of these floral scales is not known. In *C. tuberculata* and *C. umbellata* the scales surround the ovary and fill the spaces at the summit of the corolla tube. Perhaps the scales serve to protect the developing ovary and restrict entry of certain insects into the chamber containing the ovary. The ratio of the length of the styles to the ovary is constant for each species.

The taxonomy of the cuscutas is based largely on the life work of Truman G. Yuncker. His monographs are remarkable achievements, although the genus seems over-classified. Hadac & Chrteck (1970, 1973) proposed splitting the genus into 4 genera by elevating Yuncker's subgenera and sections. Beliz (1986) concluded that embryological work does not support such a division. Beliz strongly supported Yuncker's classification of subgenera and sections but found that many of his subsections do not reflect evolutionary relationships. Ours are in the subgenus *Grammica,* characterized by 2 separate styles each with a rounded, knoblike (capitate or peltate) stigma. *Grammica,* the largest of the 3 subgenera, is on all continents but is most diverse in the New World. *C. salina* is in section *Cleistogrammica,* distinguished by indehiscent capsules, and the others are in section *Grammica* with dehiscent (circumscissile) capsules.

Some Sonoran Desert cuscutas show considerable size variation, which appears to correlate with temperature and water availability through the host plants. The 3 species in the flora area have extremely slender yellow to orange stems, 5-merous flowers in dense clusters, obvious pedicles that vary from shorter to longer than the flower, white to yellow-white perianths, calyces and corollas united below, and seeds that tend to be bright yellow-brown when young and darker red-brown with age.

Ours are annuals or ephemerals, responding to warm weather. They are frost sensitive and I have not seen them alive during the cooler months of winter.

Some dodders are serious crop weeds, especially on legumes and flax. Non-native, weedy cuscutas can be expected to show up in agricultural areas. The seeds are often disseminated as contaminants with crop seeds. References: Beliz 1986; Kuijt 1969; Yuncker 1932, 1965.

1. Calyx lobes keeled, conspicuously shorter than the corolla tube; perianth, especially the corolla tube, obviously papillate. ______ **C. tuberculata**

1′ Calyx lobes not keeled, about as long to longer than the corolla tube; perianth smooth, not papillate.

  2. Flowers, especially the calyx, mostly longer than wide; capsules indehiscent, 1-seeded; localized; growing on *Suaeda moquinii*. ______ **C. salina**

  2′ Flowers, especially the calyx, as wide to mostly slightly wider than long; capsules dehiscent, with 4 or fewer seeds; widespread; growing on various desert herbs. ______ **C. umbellata**

Cuscuta salina Engelmann var. **salina** [*C. salina* var. *squamigera* (Engelmann) Yuncker] *WEPEGI: WAṢAI.* Warm-weather annuals. Flowers 2.7–3.2 mm. Calyx 1.8–2.8 mm, the lobes 0.9–1.3 mm, about as long as the corolla tube, the tips acute. Corolla lobes 1.5–1.8 mm, spreading, with age becoming reflexed, narrowly triangular, the tips acuminate; scales united (bridged) at their bases, oblong, relatively small and narrow, fringed with short, knoblike papillae nearly to the base. Perianth, filaments, and styles pure white, the anthers and stigma bright yellow. Capsules indehiscent, with a single seed 1.1–1.2 mm. June-December.

Seen only on *Suaeda moquinii* at Sonoyta and Quitobaquito. I have seen it during different years at the same, highly localized area northwest of the pond at Quitobaquito. Apparently annuals but the dry, dead, tangled masses may persist on the perennial host plants (the parasite apparently perishes in winter following the first frost or cold weather) and perhaps the seeds germinate epiphytically in the manner reported by Wiggins (1964) for the apparently closely related *C. veatchii* from Baja California.

Although Wiggins (1964) reported *C. salina* from Sonora, I do not know of any other Sonoran records. Elsewhere it often parasitizes other chenopods such as *Allenrolfea, Atriplex canescens,* and *Salicornia.* Baja California to British Columbia and Utah, and southern Arizona, where the populations apparently are few and widely disjunct. Yuncker (1965) recognized 3 varieties: 2 along the Pacific coast and the inland and widespread var. *salina.*

Beliz (1986) reports that the perianth and ovary of fresh flowers produce conspicuous white latex from laticifers. On dry flowers these laticifers appear as brown transparent lines or dots. This character is not seen on other species in the flora area. Additional diagnostic characters include the ovary thickened at the insertion of styles, and styles shorter than the ovary.

Sonoyta, *F 86-394.* Quitobaquito, *Harbison 29 Nov 1939.*

Cuscuta tuberculata Brandegee. *WEPEGI: WAṢAI.* Hot-weather ephemerals. Flowers 3.0–3.3 mm, in dense clusters, the pedicels about as long as to shorter than the flowers. Perianth segments acute. Calyx 1.7–1.8 mm, the lobes 0.9–1.2 mm, conspicuously shorter than the corolla tube, keeled (the calyx thus appearing angular), the keel usually with a tubercle drying the same color as the flower (sometimes difficult to see on dried specimens, but at least some flowers should have keeled sepals). Corolla tube cylindrical, minutely but obviously papillate (especially the part surrounded by the calyx), the lobes narrowly triangular with acute tips to sometimes oblong with obtuse tips, occasionally reflexed; scales spatulate, about half as long as the corolla tube, fringed to the base. Capsules circumscissile, with 4 or fewer seeds each (0.5) 0.8–0.9 mm.

Probably locally common but not as widespread as *C. umbellata.* On summer-fall ephemerals, mostly *Boerhavia.* Sonoyta region and the northeast side of the Pinacate volcanic field. New Mexico, southwest and south-central Arizona, northern Mexico including Baja California Sur, and western Sonora at least as far south as the vicinity of Hermosillo.

Yuncker (1965) placed *C. tuberculata* in the subsection *Leptanthae,* together with *C. leptantha* Engelmann and *C. polyanthemos* Schaffner (from Sinaloa). *C. leptantha,* characterized by an absence of calyx tubercles, seems very closely related to *C. tuberculata* and replaces it in central and southern

Sonora. Near the apparent zone of contact, such as at Hermosillo, some specimens appear intermediate, with the sepals bearing tubercles that form an interrupted, irregular partial keel.

W of Sonoyta, *F 86-315.* Sykes Crater, *F 19998.* Organ Pipe Monument, 16 mi N of Lukeville, *F 10532.*

Cuscuta umbellata Kunth. DESERT DODDER; *WEPEGI: WAṢAI.* Hot-weather ephemerals. Flowers 2.5–5.0 mm, in dense clusters (the relatively large variation in size appears to correlate with vigor and size of the host plants). Perianth smooth (without papillae or tubercles), the lobes usually acute but sometimes obtuse. Calyx cuplike, 2.5–3.5 mm, the lobes 1.0–2.7 mm, about as long to mostly longer than the corolla tube. Corolla tube shallow, the lobes 2.2–2.6 mm, ovate-lanceolate, soon reflexed; scales spatulate, about half as long as the corolla tube, fringed on upper half. Capsules circumscissile, with 4 or fewer seeds each 1.0–1.3 mm.

On many summer-fall ephemerals and several perennial herbs. Sonoyta to the Pinacate volcanic field and adjacent areas. Arizona, New Mexico, and Texas to northern South America and the West Indies. This is the most common and widespread *Cuscuta* in northwestern Sonora. There are several varieties. Ours are var. *reflexa* Yuncker, characterized by having generally larger flowers and styles conspicuously exserted from the corolla tube and longer than the ovary.

24 km W of Los Vidrios, *F 85-1007.* NE of Tezontle cinder mine, *F 86-381.* MacDougal Crater, *F 10479.* Sykes Crater, *F 20035.* Suvuk, *Nabhan & Burgess 373.* Sonoyta, *F 85-940.*

EVOLVULUS

Annual to perennial herbs or small shrubs. Stems prostrate to erect, not twining. Leaves simple. Flowers pedicelled or sessile in axils, or on slender, 1- to few-flowered peduncles. Styles 2, bifid, the 4 stigmas mostly long and threadlike. Capsules rounded, 1–4-seeded.

Tropical to temperate regions of the Americas; 100 species. Two species, ours and *E. nummularius,* introduced and widespread in the Old World. References: Austin 1990; Ooststroom 1934.

Evolvulus alsinoides Linnaeus var. **angustifolia** Torrey [*E. alsinoides* var. *acapulcensis* (Willdenow) van Ooststroom] *OREJA DE RATÓN.* Small tufted perennials, the stems very slender and wiry, usually less than 25 cm, dying back to near ground level in drought. Herbage densely pubescent with appressed and long spreading hairs. Leaves 7–23 mm, lance-linear to elliptic, smaller above, sessile or very short petioled. Peduncles 14–31 mm, the pedicels 3–6 mm. Flowers opening an hour or so after dawn, fading with daytime heat. Sepals 2.0–2.5 mm. Corollas 8–10 mm across, rotate, pale blue with a white center, glabrate except the prominent starlike interplicae. Ovary, style, and anthers pure white, the filaments pale blue. Capsules globose, 2.5–3.0 mm in diameter. Flowering in warmer months with sufficient moisture.

Occasional colonies along small washes or arroyos, low hills, and sandy-gravelly bajadas, mostly in the Sonoyta region, also occasionally along small washes at the north side of the Pinacate volcanic field. Var. *angustifolia* from Texas to southern Arizona and northern Mexico including Sonora and Chihuahua to Guanajuato.

Now pantropical, *E. alsinoides* is native to the New World and is one of the most widespread and weedy members of the genus in tropical America. This species is extremely polymorphic. At least a few populations in Asia and Australia appear to have differentiated since post-Columbian times into new infraspecific taxa, in a manner akin to the English sparrow in North America—e.g., *E. alsinoides* var. *linifolius* (Linnaeus) Baker.

Corolla and sepal size can be used to separate certain closely related taxa. For example, var. *angustifolia* is characterized in part by its especially short sepals. It could be confused with *E. arizonicus,* which has larger corollas and sepals 3.0–3.5 mm. Corolla size of fresh (not wilted) flowers is also diagnostic. Variation within *E. alsinoides* is not well defined, although var. *angustifolia* is interpreted as the narrow-leaf variety that includes the Arizona and Sonora populations.

16 mi S of Sonoyta on Mex 2, *F 7537.* Observations: Bajada of Sierra Cipriano; 1 km NE of Tezontle Cinder Mine.

IPOMOEA MORNING GLORY

Twining annual or perennial herbs to woody vines, or sometimes shrubs or trees. Corollas entire to somewhat 5-lobed, folded fanlike in bud. Fruit a several-seeded capsule. Tropical and warm-temperate regions of the world; 500 species. The sweet potato is *I. batatas.*

1. Sepals and peduncles with spreading, ascending, or reflexed hairs; corollas blue. ________ **I. hederacea**
1′ Sepals and/or pedicels and peduncles glabrous; corollas pinkish; agricultural weed. ________ **I. ×leucantha**

Ipomoea hederacea Jacquin. *TROMPILLO MORADO;* MORNING GLORY; *BI:BHIAG.* Summer annuals; vines twining 2–3+ m in mesquites and other shrubs or trees; densely to sparsely hairy throughout. Leaves petioled, the blades (2) 3.5–13 cm and about as wide, 3-lobed to 3- or 5-parted. Sepals 15–25 mm, lanceolate, narrowed above the base, the expanded, basal portion hirsute with long spreading hairs. Corollas often 3.0–3.5 cm wide, light blue. Capsules 9 mm. Seeds 4.5–4.8 mm, dark brown to blackish, wedge-shaped, with minute hairs.

Densely vegetated, brushy arroyo bottoms and playas along the northern part of the Pinacate volcanic field. North and South America; introduced in the Old World.

6 km W of Los Vidrios, *F 92-965.*

***Ipomoea ×leucantha** Jacquin. Warm-weather annuals, tenaciously twining; glabrous or sparsely hairy. Leaves petioled, the blades 3.5–6.0 cm, mostly cordate-ovate, often 3-lobed. Sepals 9–10 mm, the outer pair lanceolate and often falcate at tip. Corollas 1.5–2.0 cm wide, pinkish. Flowers opening in early morning, fading with midmorning heat. Common weeds in irrigation ditches and agricultural fields south of San Luis (*F 85-1032*).

Widespread weed in disturbed habitats from southern United States, including Arizona, to Argentina and Peru. This plant is a stable hybrid between *I. trichocarpa* Elliott (*I. cordato-triloba* Dennstaedt) and *I. lacunosa* Linnaeus (Abel & Austin 1981; Austin 1978). *I. lacunosa* is usually totally autogamous (selfing) whereas *I. cordato-triloba,* with larger flowers, is usually totally allogamous (outcrossing). The honeybee, introduced from the Old World, carries pollen from one species to the other and is the only insect known to move between the two. However, certain other pollinators, when present, will visit *I. lacunosa* and the hybrid (Dan Austin, personal communication 1990.)

CRASSULACEAE STONECROP FAMILY

Succulent herbs, mostly perennials, some shrubs; often glabrous. Leaves alternate, opposite, or sometimes whorled. Flowers mostly bisexual, radial, mostly 4- or 5-merous. Sepals and petals often separate, or sometimes united at base or to above middle. Ovary usually superior with 4 or 5 carpels separate or united only at base. Seeds very small.

Greatest diversity in southern Africa, with spectacular variation as leaf succulents; 25 genera, 900 species. Secondary centers of diversity in the Mediterranean region and mountains of Mexico and Asia. Many are cultivated, especially in mild semiarid Mediterranean climates. These are CAM plants but many or most tolerate neither the extreme summer heat nor the winter freezing in northern Sonora and southern Arizona.

1. Small, delicate winter-spring ephemerals; leaves opposite on stems, less than 5 mm. ________ **Crassula**
1′ Thick-stemmed perennials; leaves alternate in a rosette, more than 3 cm. ________ **Dudleya**

CRASSULA

Leaf succulents and many with succulent stems; mostly small perennials and subshrubs, some shrubs and a few annuals. Leaves opposite, usually united at base. Mostly in semiarid subtropical regions in southern Africa; 300 species.

New World crassulas represented by 12 native and 1 adventive species placed either in the subgenus *Disporocarpa* of *Crassula* or in the genus *Tillaea.* The segregation of *Tillaea* from *Crassula* hangs on little more than geographic distribution and the generally smaller size of the plants of *Tillaea. Tillaea,* with 30 species, occurs in both the New and Old Worlds, whereas *Crassula* sensu stricto is restricted to the Old World. Some New World species, following the treatment of Bywater & Wickens (1984), are difficult to distinguish, and the authors suggest using seed characters best seen with a scanning electron microscope. As annoying as this may seem, these are tiny plants and there may be little to work with once the succulent or aquatic plants are dried and plastered onto a herbarium sheet. To further complicate matters, there are developmental differences in some key characters, such as elongation of pedicels between early and late stages in the life cycle. References: Bywater & Wickens 1984; Moran 1992.

Crassula connata (Ruiz & Pavón) A. Berger [*Tillaea connata* Ruiz & Pavón. *T. erecta* Hooker & Arnott. *T. erecta* var. *eremica* Jepson. *Crassula connata* var. *eremica* (Jepson) Bywater & Wickens. *C. erecta* (Hooker & Arnott) A. Berger] Diminutive winter-spring ephemerals (0.5) 1.0–5.5+ cm, the individual plants green, yellow-green, or reddish (different colored plants often intermixed). Leaves very fleshy, beadlike, united at base, the larger ones mostly 2.0–3.5 mm. Flowers 1 mm in diameter; sepals (1.2) 1.4–1.6 mm and longer than the carpels; petals white to nearly transparent, lanceolate, shorter than the sepals. Most or all flowers on a plant sessile or subsessile, some with a minority of flowers on pedicels to 3 mm (pedicel length seems to be influenced by soil moisture). Seeds 0.32–0.38 mm, brown, ellipsoid, shiny with faint longitudinal lines (longitudinally ridged when viewed with a scanning electron microscope).

Widespread during favorable years, including the Sonoyta region, Pinacate volcanic field, granitic ranges, desert plains, and lower dunes. Mostly in sandy, gravelly, or cinder soils, usually on desert flats and bajadas, sometimes lower dunes, arroyo beds, and soil pockets in rocky areas. This is one of the smaller terrestrial plants in the Sonoran Desert.

Oregon to Argentina in many habitats, sea level–3900 m. Bywater & Wickens (1984) recognize 5 varieties, although Moran (1992:229) found that "the varieties seem to have little geographic meaning and seem too poorly defined to be useful."

10 mi SW of Sonoyta, *Zimmerman 1475.* Arroyo Tule, *F 19226.* Elegante Crater, *F 19672.* Lava flow 8 km N, 6 km E of Pinacate Peak, *Burgess 5589.* 10 km W of Estación Gustavo Sotelo, *F 84-2.*

Crassula connata. JRJ (Abrams 1944).

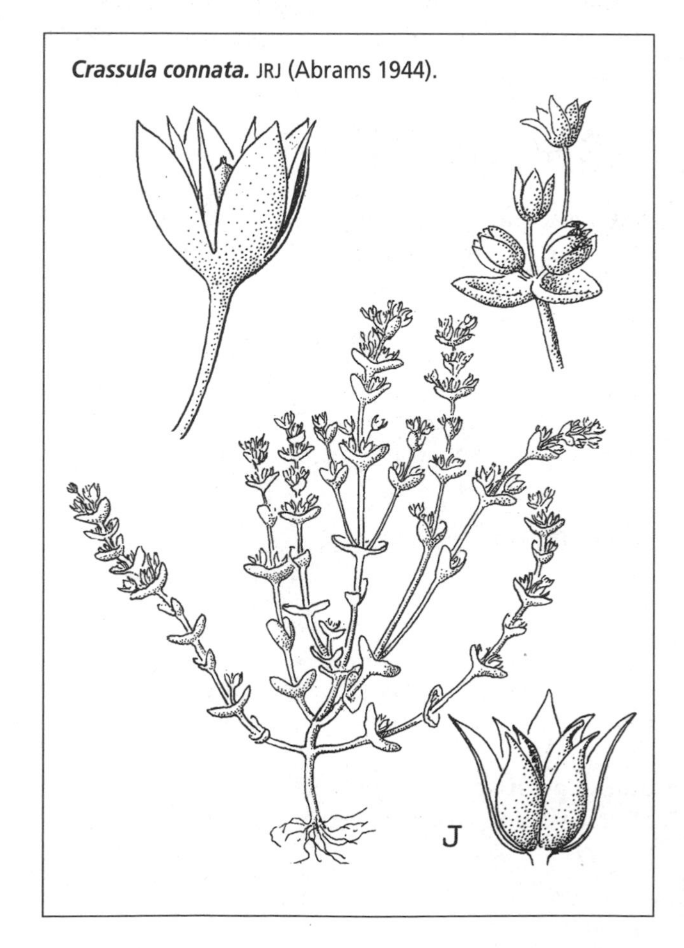

Dudleya arizonica. MBJ.

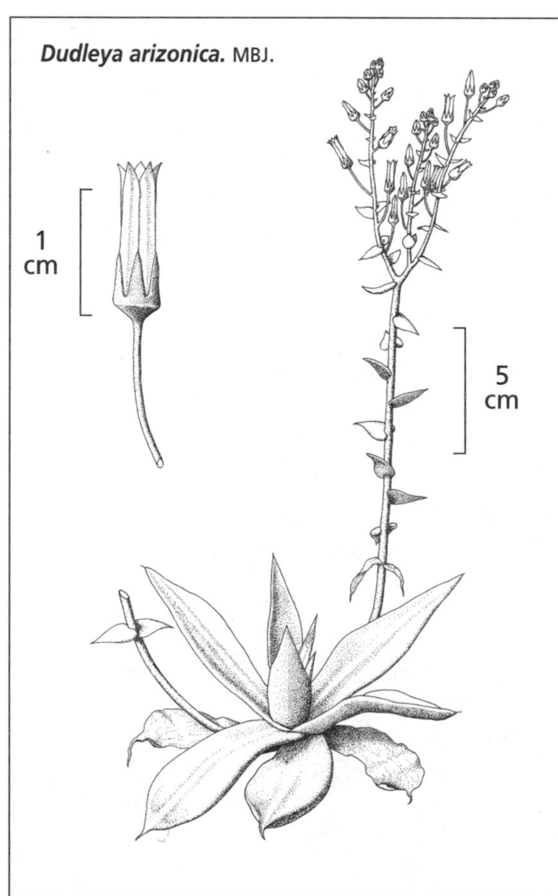

Dudleya

Herbaceous perennials with solitary or clustering rosettes on short, thick stems, or stems elongated in some maritime species. Leaves thick, succulent, clasping, the old dry leaves persistent. Flowering stems axillary with reduced, fleshy leaves. Flowers 5-merous. Sepals equal in size, mostly appressed to the corolla, the sepal disk wider than the corolla base. Corollas mostly cylindrical, white to yellow or red, the petals fused at base, the lobes convolute. Stamens 10. Seeds many.

Mostly on the Pacific side of California and both Baja California states, with a few in adjacent states; 40 species. Dudleyas are summer-dormant plants attuned to a Mediterranean climate. Ours is the only *Dudleya* in mainland Mexico.

Dudleya has a counterpart in *Echeveria,* a diverse genus of succulents mostly of intermediate elevations in montane subtropical Mexico to Argentina. *Dudleya* is distinguished by its persistent leaves, equal-sized sepals, calyx disk wider than the corolla base, and convolute petals (in bud one petal edge is outside the next and the other inside in an overlapping pattern). Echeverias are not summer dormant, and in cultivation they mostly do not thrive in the hot Sonoran Desert summer.

Dudleya arizonica Rose [*D. pulverulenta* (Nuttall) Britton & Rose subsp. *arizonica* (Rose) Moran. *Echeveria arizonica* (Rose) Kearney & Peebles] Rosettes 1 to several from a short, thick stem and thickened root. Leaves several per rosette, 4.0–10.5 cm, lanceolate to broadly ovate, very thick and succulent, with a white-waxy bloom, whiter during dry periods, greener during winter-spring wet periods. Flowering stems usually 1–3 per rosette, 9–30 (50) cm, mostly erect, with leaves partially clasping, thick, succulent, triangular-ovate, usually reddish, gradually reduced upwards, the larger ones 2–3 (4) cm, shriveling as the flowers develop.

Buds and developing inflorescences conspicuously glaucous-whitish; flowering stem and its leaves, bracts, and pedicels usually coral pink at anthesis. Panicles (flower-bearing portions) 3.5–18 cm. Pedicels erect, (8) 9–16 mm. Flowers 13–16 mm, cylindrical-campanulate, erect. Calyx mostly 5–6 mm, very fleshy, greenish glaucous or sometimes coral pink above. Corollas often 13–14 mm, greenish yellow to yellow at base, yellow-orange above, to yellow-orange or red-orange toward tips, nearly tubular, united to about the middle, the lobes scarcely spreading at tips. Anthers red prior to pollen shedding. Capsules 7.5–12 mm including the 1.5–2.0 mm persistent styles. Seeds 0.7 mm, light brown. March–early April; seeds ripe in May.

The plants grow wedged into crevices in granitic rock faces, often on steep north-facing cliffs and slopes. As with *Agave, Dudleya* occurs in granitic mountains on either side of the Pinacate volcanic field (Sierra del Viejo to the northwest, Sierra Blanca and Bahía de la Cholla to the southeast) but is absent from the intervening volcanic region. Northwestern Sonora southward in coastal mountains to Cerro Tepopa (vicinity 29°22′ N; Felger & Moser 1985), western Arizona, southern Nevada, southwestern Utah, southeastern California, and northern Baja California.

The plants respond to cool-season rains. New leaves are relatively elongated and greener during the winter-spring wet season. During drought or through the summer the leaf tips dry and the leaves become short and stubby. In extreme drought the plants are almost without living leaves, and the old dead leaves form a protective cover over the stem and growth bud.

Dudleya arizonica is an inland, allopatric desert segregate of the much larger and spectacular white-leaved *D. pulverulenta* of the Pacific coast cliffs and mountains of Baja California and southern California. *D. arizonica* differs in being a much smaller plant with fewer leaves, a root caudex thick and fleshy like a taproot, leaves and inflorescences not nearly as white-waxy, erect rather than pendent pedicels and flowers, smaller flowers, and corollas yellow to red-orange rather than deep red. *D. arizonica,* at least in Sonora, seems to have a much shorter flower-to-seed cycle than does *D. pulverulenta.* On the Pacific coast *D. pulverulenta* begins producing its inflorescences in late winter, flowers at the end of April, and the seeds are not ripe until around July (Mark Dimmitt, personal communication 1992). The two taxa apparently intergrade near the western edge of the desert in California, where *D. arizonica* tends to have reddish flowers. *D. arizonica* obviously has its evolutionary ties with *Dudleya* of the Pacific margin.

Granitic promontory at S side of Cholla Bay, $n = 17$, *Moran 1951* (DS). Sierra Blanca, *F 87-27.* Sierra del Viejo, *F 85-726.*

CROSSOSOMATACEAE

Xerophytic glabrous shrubs. Leaves simple, alternate or opposite. Flowers solitary or in small clusters, bisexual or some of them unisexual, radial, perigynous, the floral tube short or the hypanthium forming a thickened nectary disk. Sepals and petals each 3–6, the sepals persistent, the petals white and deciduous. Stamens few to many, attached to the nectary disk. Pistils 1–5 (9), separate, stalked (stipitate), the stigmas capitate. Seeds with a conspicuous fringed aril.

Arid and semiarid western United States and northwestern Mexico; 3 genera, 8 species. A taxonomically isolated family readily identified by the separate pistils. Reference: Mason 1992.

CROSSOSOMA

Leaves alternate, entire, the stamens numerous. Two species; *C. californicum* on islands off California and Baja California.

Crossosoma bigelovii S. Watson. RAGGED ROCK-FLOWER. Much-branched shrubs mostly 1–1.5 m, with long and short shoots, the short shoots sometimes changing to long shoots. Leaves mostly 7–15 (21) mm, subsessile with petioles less than 1 mm, alternate on long shoots, clustered on short shoots, tardily drought deciduous, thickish, grayish or glaucous green, elliptic to oblong or obovate, mostly acute. Flowers fragrant; sepals 5, broadly ovate, cupped and turning down at anthesis; petals 5, 12.5–13.5 mm, obovate, rounded at apex; stamens ca. 15+, the filaments white, the anthers yellow. Pistils 2–5; fruits 7.0–11.5 mm, often purplish. January-February; fruits mostly ripe in April.

Granitic mountains at the northeastern corner of the flora area on Sierra Cipriano and farther west on the Sierra del Viejo. Mostly north-facing slopes in canyon bottoms and steep slopes of side canyons, often at bases of north-facing cliffs; about 1/3 of the way up these mountains to peak elevations. Deserts and desert-woodland ecotone across much of northern Sonora southward to the Sierra El Aguaje near Guaymas, northwestern Chihuahua, western and southern Arizona, western Nevada, inland southern California, and Baja California.

Sierra Cipriano, *F 91-21*. Sierra del Viejo, *F 85-722*.

CUCURBITACEAE GOURD FAMILY

Annuals or herbaceous perennials, some with a softwood caudex or thick stem, the aerial parts frost sensitive. Glabrous or variously pubescent, often scabrous. Stems mostly climbing or trailing, usually with tendrils. Leaves alternate, petioled, usually palmately veined and lobed to sometimes palmately compound; stipules none. Monoecious or dioecious. Flowers radial, the sepals and petals each usually 5, separate or united. Stamens usually 5, often appearing as 3 with 2 pairs of united stamens, the filaments separate or united in a column, the anthers often appearing united. Corollas mostly yellow, white, or green. Ovary inferior. Fruits highly variable; seeds 1 to many.

Mostly tropical and subtropical; 120 genera, 760 species. Included are cucumbers, gourds, loofas, melons, pumpkins, squashes, watermelons, and a number of minor vegetables. References: Jeffrey 1980; Whitaker & Davis 1962.

1. Mostly growing during cooler times of the year; flowers less than 0.5 cm wide; fruits obovoid-oblique (bilaterally symmetric in cross section), ca. 1 cm; seeds 1 (2). ____ **Brandegea**

1′ Mostly growing with warmer weather; flowers more than 2 cm wide; fruits oblong or round (radially symmetric in cross section), 4 cm or more; seeds many.

  2. Fruits oblong, 4–5 cm, the rind soft; weed in agricultural fields. ____ **Cucumis**

  2′ Fruits round, 8–9 cm, the rind hard; widespread. ____ **Cucurbita**

BRANDEGEA

Annual or short-lived perennial vines with slender stems and unbranched tendrils. Monoecious. Flowers small, white. Stamens 3, the filaments fused into a short column, the anthers horseshoe-shaped, separate but adjacent to each other (often appearing fused on dried specimens). Fruits small, 1- (2)-seeded, obovoid-oblique, prominently beaked. Southwestern United States and northwestern Mexico; 3 species.

Brandegea bigelovii (S. Watson) Cogniaux [*Elaterium bigelovii* S. Watson] DESERT STARVINE. Annual vines growing with cool-season rains (possibly also with late-summer rains), with a single, elongated, fleshy, white taproot often reaching 2.0–2.5+ cm in diameter. Leaves simple, the blades 2.5–7.0 cm and about as wide, thin, uniformly green, prominently short-scabrous, shallowly to deeply 3- or 5-lobed, the lobes broader and shallower on stems of juvenile growth, becoming narrower (to broadly linear) and deeper on upper stems; petioles reaching 2–4 cm. Flowers white, delicately fragrant, the male and female flowers in the same axil; male flowers 3–4 mm across, in several-flowered racemes; female flowers slightly larger, solitary on slender pedicels, 1 (2 or 3) per axil. Fruits green, drying light tan, the body (6) 7.0–7.5 mm, thin-walled, laterally compressed, with a few to ca. 20 soft, forward-projecting prickles, the beak (4) 5–7 mm. Seed 5 mm, obdeltoid, compressed, nearly filling the body of the fruit, brown mottled and sculptured. September-April depending on soil moisture.

Often in smaller playas and sandy-gravelly arroyos and washes, seasonally festooning shrubs and trees in green curtains. Sandy soils in lowlands nearly throughout the region, and to ca. 420 m in the Sierra Pinacate; not on dunes except interdune troughs at the western and southern edges of the Pinacate volcanic shield. Not on rock slopes and not at higher elevations. Mojave and Sonoran Deserts; southeastern California to Baja California Sur, southwestern Arizona, and western Sonora southward to the vicinity of El Desemboque San Ignacio.

This is the most common and widespread vining plant in northwestern Sonora. I have not found it growing with hot-weather rains in northwestern Sonora, although in southern California deserts new growth and even flowering may occur in late August. In cultivation in Tucson the plants actively grow and flower from February through summer and the following seasons, although during the extreme heat of summer, flowering ceases and most leaves die even on well-watered, shaded plants. Although the roots are relatively large and thick, the plants are not perennial.

El Papalote, *F 86-332*. 45 mi W of Sonoyta, *Turner 59-6*. Suvuk, *Ezcurra 22 Sep 1982* (MEXU). Cerro Colorado, *Webster 22316*. Moon Crater, *F 19265*. Sierra Extraña, *F 19081*.

CUCUMIS

Annual or perennial vines, the stems trailing or climbing with unbranched tendrils. Usually monoecious. Fruits fleshy. Native to Africa and southern Asia; 30 species. The genus includes cucumbers and many kinds of melons, some of which are grown in the San Luis region. Seeds many.

†**Cucumis melo** Linnaeus var. **dudaim** (Linnaeus) Naudin [*C. dudaim* Linnaeus] *MELÓN DE COYOTE, MELONCILLO DE COYOTE;* DUDAIM MELON, STINK MELON. Annual vines. Herbage scabrous. Leaves often ca. 10–15 cm, the blades about as wide as long, broadly ovate to 3-lobed, cordate at base. Flowers yellow. Fruits 4–5+ cm, fleshy, oblong, yellow marbled with brown, with a strong cantaloupe fragrance, the rind soft and smooth; not edible.

Agricultural fields near San Luis, Arizona, and the Imperial Valley in California, and expected in fields south of San Luis; a potentially serious agricultural weed, especially in cotton and asparagus fields. Probably native to southern Asia. This species includes numerous melon cultivars including cantaloupe, casaba, honeydew, muskmelon, and Persian melon.

CUCURBITA GOURDS, SQUASH

Annual or perennial vines. Stems spreading, often rooting at nodes or climbing, the tendrils branched or unbranched. Leaves large, simple. Monoecious. Flowers large, yellow, solitary (male flowers sometimes on short branches and appearing clustered). Stamens 3. Fruits gourdlike with a hard rind in non-cultivated species; seeds many, ovate, compressed, smooth.

New World; 20 species. Includes squashes, pumpkins, and various gourds. The large green-striped cushaw squash, *C. argyrosperma* Huber, has been cultivated in the Sonoran Desert since ancient times.

It has been grown along the lower Río Colorado, at Quitovac, and undoubtedly at Quitobaquito and at settlements along the Río Sonoyta; it is distinguished by the enormous corky peduncle and large, rather irregular gourds. References: Bemis & Whitaker 1969; Whitaker & Bemis 1964.

Cucurbita digitata A. Gray. *CHICHI COYOTA;* COYOTE GOURD; *'ADAWĬ, 'AD.* Perennials from a deep, thickened, turniplike root. Stems relatively coarse, sprawling across the ground to several meters, occasionally climbing shrubs and trees. Herbage scabrous. Tendrils 2–5-branched from base. Leaves digitately cleft, the lobes (6) 8–13 cm, the upper surfaces whitish near the midrib due to dense hairs and greenish near margins, the lower surfaces more or less uniform, the petioles 4–9 cm; leaves of juvenile plants and first growth of season with much shorter and broader lobes (like those of *C. palmata* S. Watson). Corollas bright yellow, 7–10 cm across. Gourds 8–9.4 cm in diameter, rounded, smooth, green with whitish stripes and mottling, yellow at maturity with a thin but hard-shelled rind. Seeds 9–11 × 5.5–6.5 mm, smooth, compressed, whitish. Growing and flowering with warm weather, often beginning in spring even in dry years, dying back in severe drought.

Widespread but seldom very common. Sonoyta region, Pinacate volcanic field, near the bases of the western granitic mountains, and desert plains; not on dunes. Large gravelly washes or arroyos, sand flats, and disturbed sites including roadsides. Sometimes common on nearly barren cinder soils to near peak elevation in the Sierra Pinacate.

Northern Sonora, northern Baja California, southern Arizona, and southern New Mexico. Intergrading with *C. palmata* near the Río Colorado in southeastern California, southwestern Arizona, and northeastern Baja California. These and *C. cordata* S. Watson and *C. cylindrata* L.H. Bailey, both of Baja California, form a closely knit group of allopatric, mostly intergrading xerophytic cucurbits (Bemis & Whitaker 1965).

El Papalote, *F 86-329.* 10 mi S of Tinajas de los Pápagos, *F 18914.* NE of Tezontle cinder mine, *F 85-987.* S slope of Pinacate Peak, 1200 m, *F 19405.* N of Cerro Pinto, *F 85-735.* 18 mi E of San Luis, *F 77-13.*

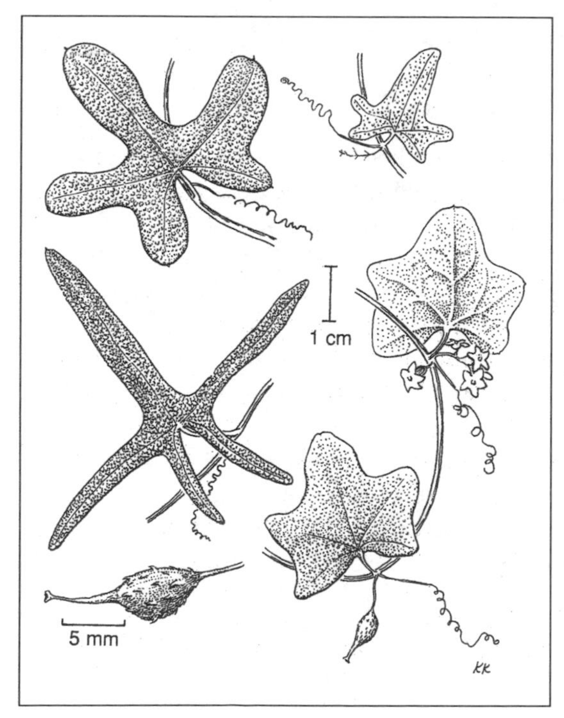

Brandegea bigelovii. KK (Hickman 1993).

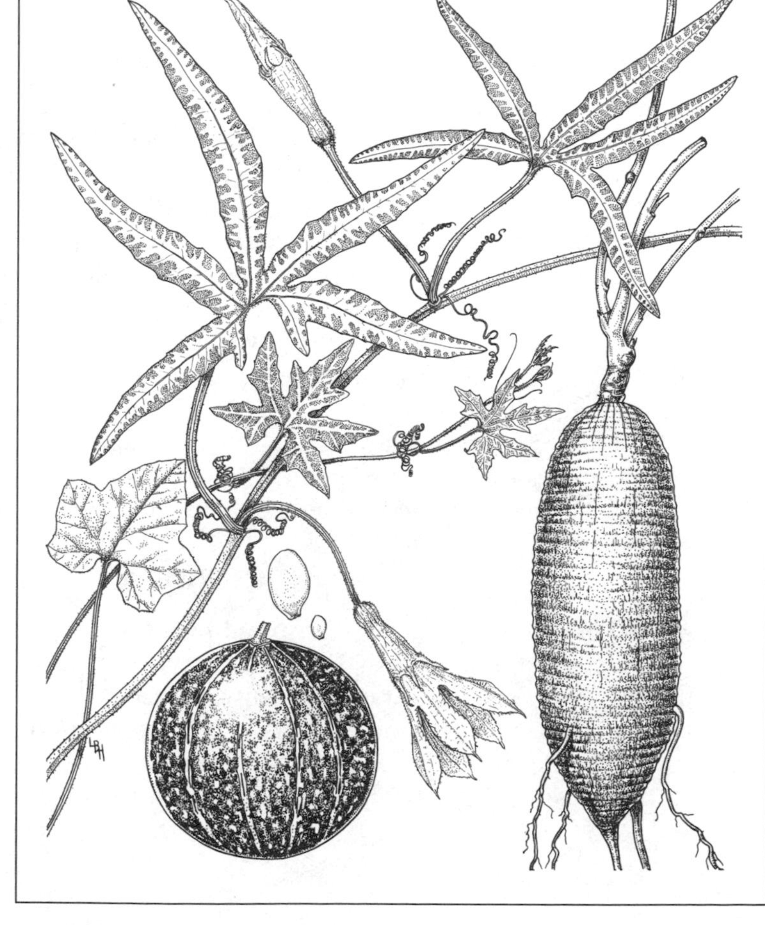

Cucurbita digitata. Note single leaf and portion of juvenile stem at lower and middle left. LBH (Parker 1958).

EUPHORBIACEAE SPURGE FAMILY

Plants of extremely diverse habit, herbs to large trees, many with diverse succulent growth forms, especially in dry regions of Africa. Many with milky sap. Leaves alternate, or less commonly opposite or whorled, simple (ours) or compound, usually with stipules, the stipules often small or quickly deciduous. Inflorescences basically cymose but sometimes greatly reduced with a cuplike cyathium (involucre). Monoecious or dioecious. Different flower parts often reduced, sometimes greatly so; perianth often inconspicuous, the tepals separate or united below, or the petals or entire perianth absent. Ovary superior, commonly 3-chambered; styles usually 3, simple or branched, the stigmas often 3. Fruits usually of capsules, the segments (mericarps) usually 3 (1 or 2 by abortion), separating elastically from a persistent column, with 1 (ours) or 2 seeds per chamber or segment. Seeds often with a knoblike basal appendage (caruncle).

Worldwide, mostly tropical and subtropical; 317 genera, 7500 species. *Acalypha, Euphorbia,* and *Jatropha* are among the larger genera. Includes Brazilian rubber tree (*Hevea brasiliensis*) and manihot (*Manihot esculenta*). Many members of this family are poisonous. References: Steinmann & Felger 1997; Webster 1975, 1994.

1. Shrubby, or semishrubby, often 1 m or more.
  2. Sap milky.
    3. Leaves opposite, the blades about as wide as long; stems thick and knobby-gnarled. ____**Euphorbia misera**
    3′ Leaves alternate, the blades more than twice as long as wide; stems straight, not unusually thick, and not knobby-gnarled.
      4. Sap thick and conspicuously milky; woody shrubs often 2–2.5 m; leaves 6–15 mm wide. ______**Sebastiania**
      4′ Sap thin and watery-milky; bushy perennials, scarcely woody at base, usually less than 1 m; leaves less than 3 mm wide.____________**Stillingia linearifolia**
  2′ Sap not milky.
    5. Leaf margins evenly serrated; female flowers subtended by a leafy, toothed bract; styles multiple-branched with many threadlike segments. ____________**Acalypha**
    5′ Leaf margins entire, or with a few shallow lobes or minute teeth near apex; female flowers not enclosed by such a bract; styles entire or 2-cleft (bifid).
      6. Plants glabrous; leaves linear, mostly 1–2 (2.8) mm wide. ____________**Stillingia linearifolia**
      6′ Plants glabrous or hairy; leaves linear-lanceolate or spatulate or broadly kidney-shaped, usually more than 2 mm wide (if less than 2 mm wide then densely hairy).
        7. Herbage, flowers, and fruits glabrous or with short, unbranched hairs; leaves cuneate or kidney-shaped to orbicular; sap copious, watery or bloodlike.____________**Jatropha**
        7′ Herbage, flowers, and fruits with 2-armed or stellate hairs; leaves lanceolate to elliptic; sap not copious, not watery or bloodlike.
          8. Stems tough, not brittle; herbage and fruits densely pubescent with stellate hairs; dioecious. ____________**Croton**
          8′ Stems brittle; herbage sparsely hairy, young herbage and fruits with 2-armed hairs; monoecious. ____________**Ditaxis brandegeei**

1′ Herbaceous, mostly less than 1 m.
  9. Leaves alternate, opposite, or whorled; sap milky; flowers in cyathia (compact, cuplike inflorescences about as wide as tall).____________**Euphorbia**
  9′ Leaves alternate; sap not milky (or slightly milky in *Stillingia*); flowers not in cyathia.
    10. Plants with stinging, stellate hairs; stems slender, vining. ____________**Tragia**
    10′ Plants glabrous or the hairs 2-armed and not stinging; stems not vining.
      11. Plants conspicuously hairy with 2-armed hairs. ____________**Ditaxis** (in part)
      11′ Plants glabrous.
        12. Petals 5, white, the styles 3, each 2-cleft (bifid); rare, coastal. ______**Ditaxis serrata** var. **californica**
        12′ Petals absent, the styles 3, entire; widespread in lowlands.____________**Stillingia**

Croton wigginsii. AE.

Ditaxis lanceolata. JRJ (Abrams 1951).

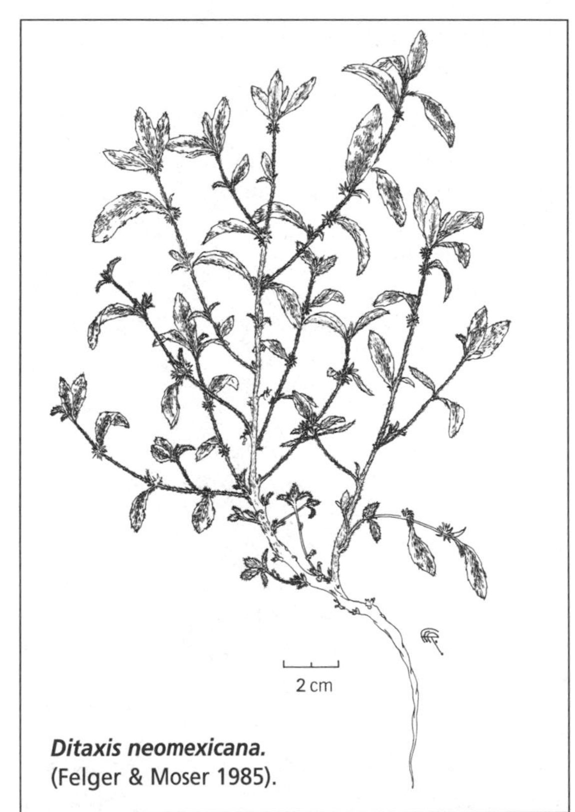

Ditaxis neomexicana. (Felger & Moser 1985).

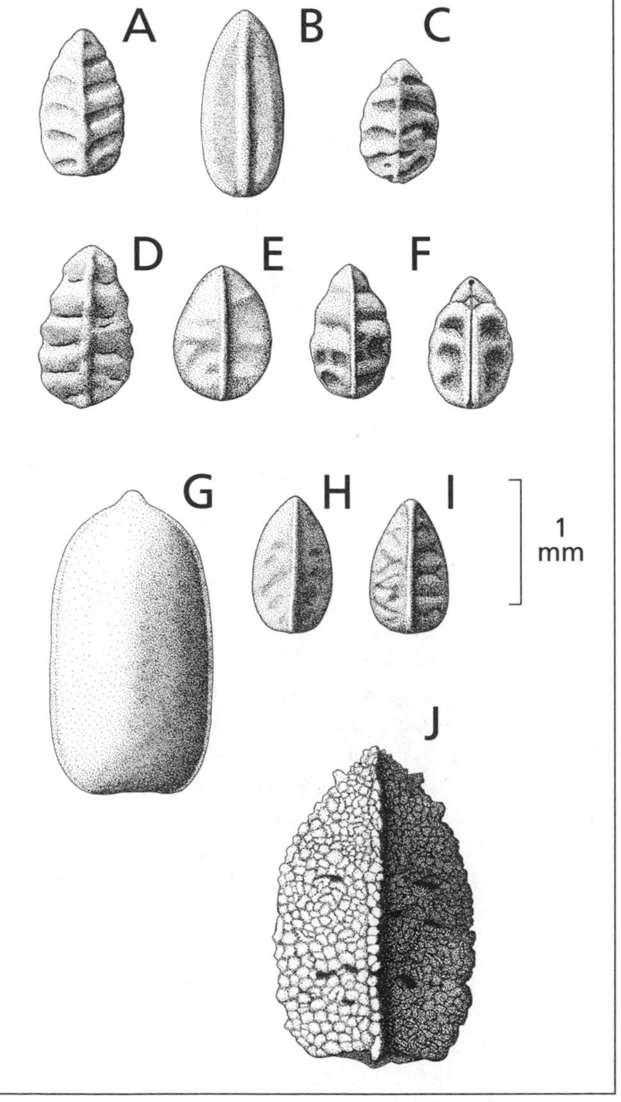

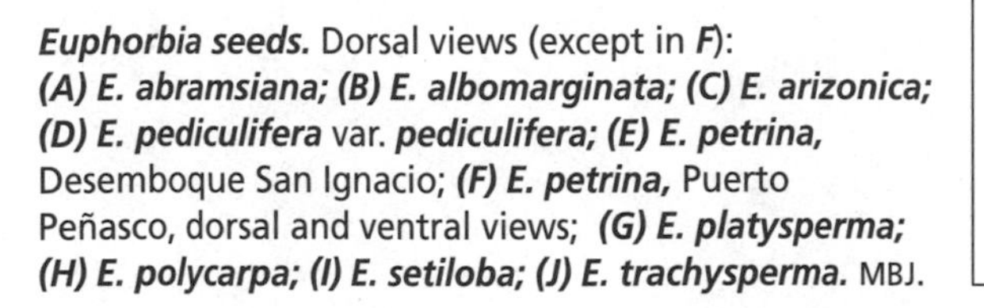

Euphorbia seeds. Dorsal views (except in ***F***): ***(A) E. abramsiana; (B) E. albomarginata; (C) E. arizonica; (D) E. pediculifera*** var. ***pediculifera; (E) E. petrina,*** Desemboque San Ignacio; ***(F) E. petrina,*** Puerto Peñasco, dorsal and ventral views; ***(G) E. platysperma; (H) E. polycarpa; (I) E. setiloba; (J) E. trachysperma.*** MBJ.

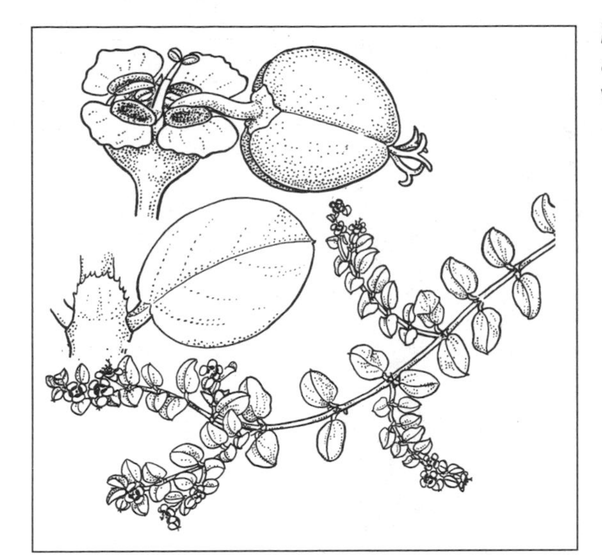

Euphorbia albomarginata. Note cyathium with capsule and involucral glands with large appendages, and leaf with large, ragged-edged stipule. LBH (Parker 1958).

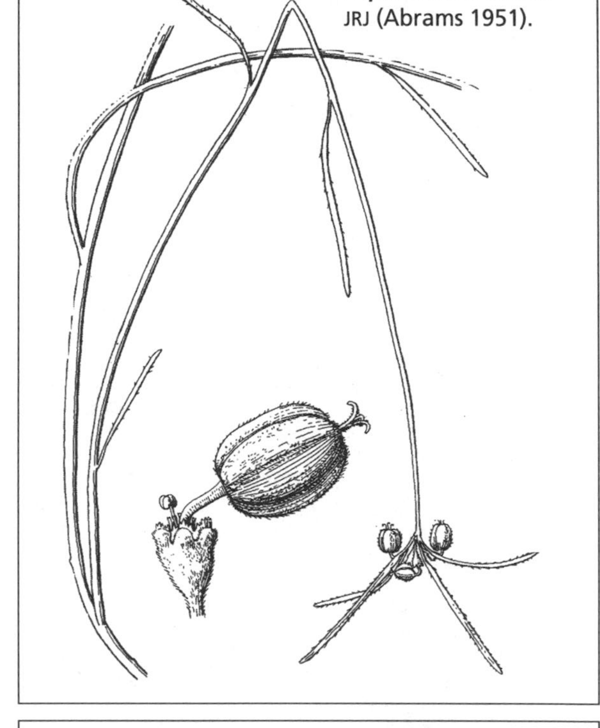

Euphorbia eriantha. JRJ (Abrams 1951).

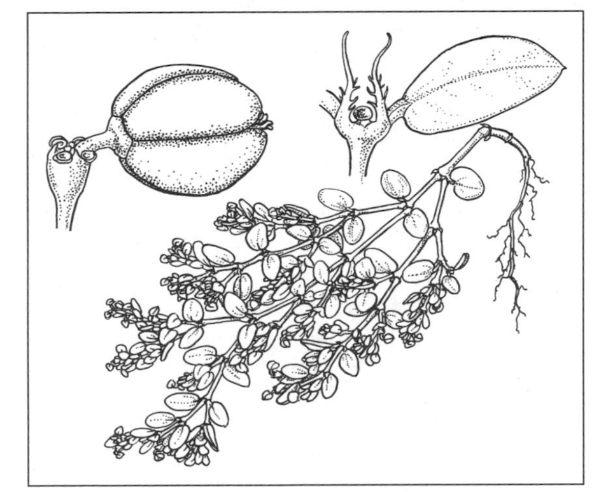

Euphorbia micromera. Note cyathium with capsule and small involucral glands lacking appendages, and leaf with pair of narrowly triangular stipules. LBH (Parker 1958).

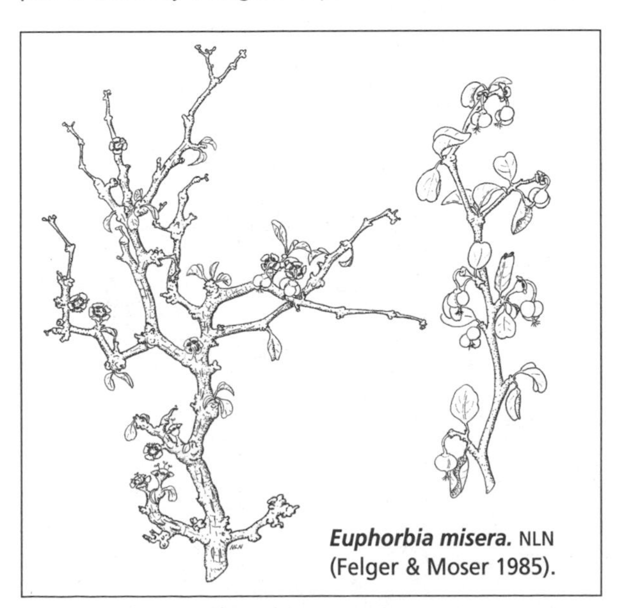

Euphorbia misera. NLN (Felger & Moser 1985).

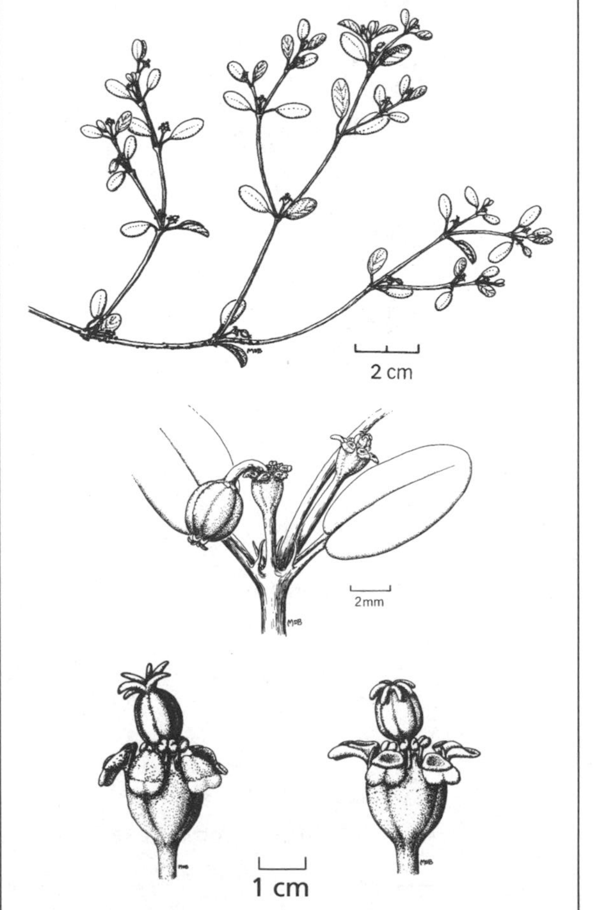

Euphorbia platysperma. Note appressed stigma lobes and spreading involucral appendages of the slightly older of the two cyathia. MB.

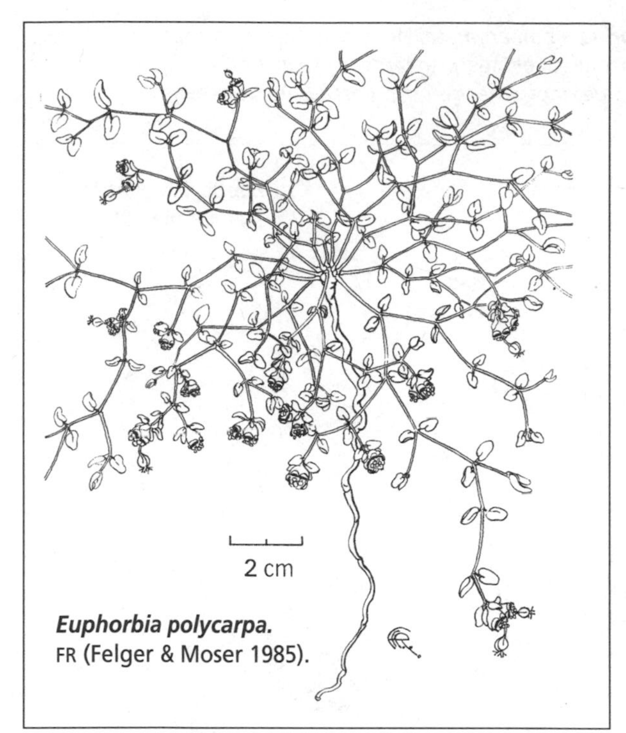

Euphorbia polycarpa.
FR (Felger & Moser 1985).

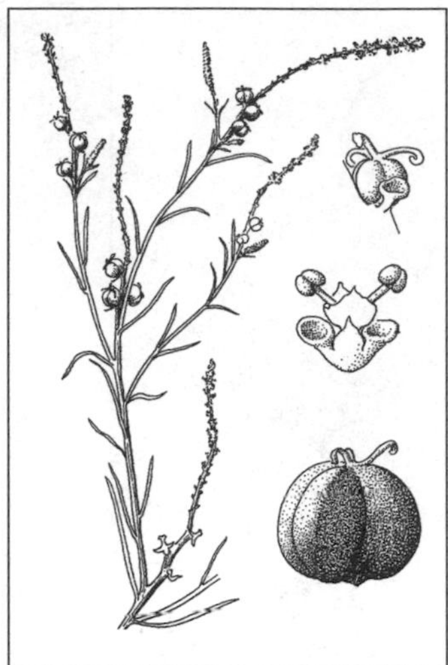

Stillingia linearifolia.
JRJ (Abrams 1951).

Stillingia spinulosa.
JRJ (Abrams 1951).

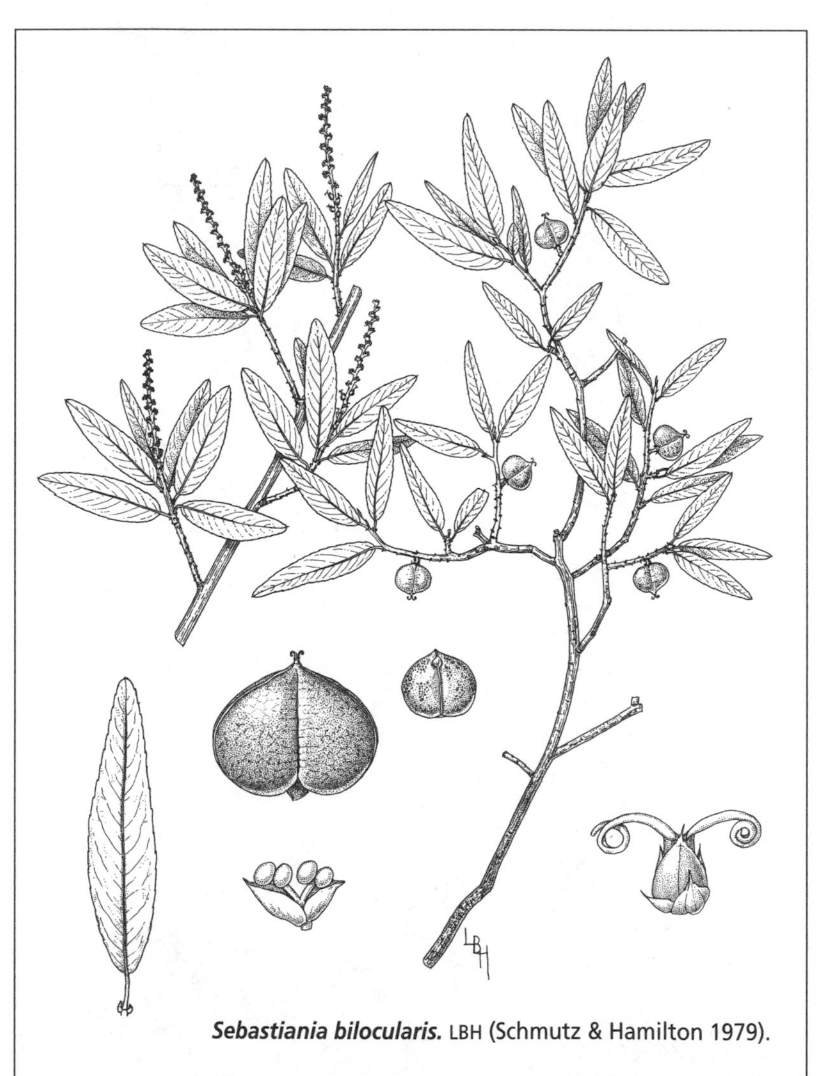

Sebastiania bilocularis. LBH (Schmutz & Hamilton 1979).

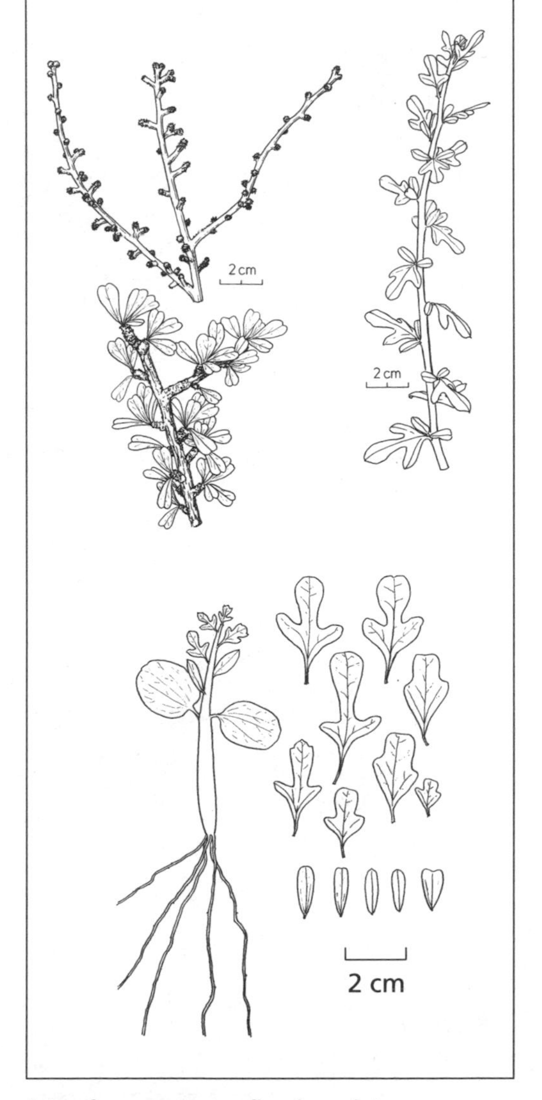

Jatropha cuneata. Leafless branch in dry season, branch with leafy short shoots, and a long shoot with summer-fall rains. LBH (Felger & Moser 1985).

Acalypha

Annuals to small trees; sap not milky. Leaves alternate, petioled, the blades usually toothed. Mostly monoecious. Inflorescences often of axillary or terminal spikes or racemes. Flowers usually subtended by bracts, those of female flowers enlarging with age. Calyx usually small; corollas none. Styles usually 3, with few to many threadlike branches. Fruit a capsule, usually 3-seeded. Seeds usually with a caruncle.

Worldwide, tropical to temperate, relatively few in deserts; 430 species. Copperleaf (*A. amentacea* var. *wilkesiana*), a shrub widely cultivated in the tropics for its large, spectacular, reddish mottled leaves, is grown farther south in Sonora. It does not tolerate the occasional winter freezing in northwestern Sonora, and the leaves scorch in the sun. Reference: Levin 1994.

Acalypha californica Bentham [*A. pringlei* S. Watson] *HIERBA DEL CÁNCER;* COPPERLEAF. Shrubs often 1–1.5 m, with slender, woody stems. Leaves and pubescence variable with soil moisture and temperature. Young stems and leaves densely pubescent with both glandular and non-glandular unbranched hairs in highly varying ratios; herbage viscid-sticky and tawny brown during drought. Leaves larger, thinner, greener, and less glandular during periods of warm weather and high soil moisture; eventually leafless in drought. Petioles 0.5–3 (3.5) cm; leaf blades 1.5–5.5 cm, ovate to triangular-ovate, the base cordate to rounded or truncate, the tip blunt and obtuse to acute, the margins crenulate-toothed.

Flower spikes axillary; flowering stalks (peduncles and spikes) in lower leaf axils 1.7–4.0 cm with a solitary female flower at or near base of peduncle, the peduncle about 1/3 as long as the densely flowered staminate spike; scales (bracts) subtending male flowers usually reddish. Upper spikes (in leaf axils at branch tips) shorter and of pistillate flowers only; bracts below female flowers broad and evenly serrate- to crenate-toothed, densely glandular to glandular only on margins; stigmas conspicuous, 5–8 (9) mm, pale lavender when fresh, drying reddish, laciniate into threadlike segments. Male flowers red-green, the stamens white. Seeds 2 mm, obovoid. Plants often freeze damaged, sometimes freezing to ground level. Flowering various seasons, most luxuriantly with summer-fall rains.

Arizona Upland in the vicinity of Sonoyta and Sierra Cipriano, seldom common; granitic mountains, low hills, and upper bajadas, or more often along small drainageways. Northwestern Sinaloa through western Sonora to southwestern Arizona, Gulf of California islands, and Baja California Sur to southern California.

Sierra Cipriano, *F 91-25*. 2.6 mi W of Sonoyta, *F 86-403*.

Croton

Herbs, shrubs, and trees; usually with stellate hairs or lepidote scales; often aromatic. Leaves usually alternate, simple; stipules usually present, sometimes very reduced. Monoecious or dioecious. Flowers mostly in terminal spikes or spikelike racemes. Petals mostly present in male flowers (absent in the *C. californicus* group), and usually absent or greatly reduced in female flowers. Fruit a capsule, usually 3-seeded. Seeds with a caruncle.

Warm regions worldwide, especially in the tropics, with centers of diversity in Mexico and Brazil; 800 species.

Many have been used medicinally. Croton oil is obtained from *C. tiglium*. The cultivated tropical "croton" belongs to the genus *Codiaeum*. *Croton californicus* and *C. wigginsii* are sister taxa and are not closely related to *C. sonorae*. These 3 species are somewhat unusual in the genus in lacking stipules, which seems correlated with the relatively smaller, xeromorphic leaves.

1. Monoecious; small shrubs with rigid, woody branches; leaves sparsely to moderately pubescent with stellate hairs, the blades visibly green because of sparseness of hairs; flowers with petals; near Sonoyta, bajadas and slopes. ____ **C. sonorae**

1′ Dioecious; herbaceous to bushy, the branches flexible, not rigidly woody; leaves densely covered with silvery stellate-lepidote scales; flowers without petals; dunes and sand flats.

  2. Stems divergent, the branches spreading, the plants often as wide as or wider than tall; leaves linear; seeds 4.0–4.5 mm; sand flats and beach dunes. ____ **C. californicus**

  2′ Stems mostly erect, the branching pattern strict, the plants taller than wide; leaves linear-lanceolate to linear (largest leaves narrowly lanceolate); seeds 4.4–7.8 mm; inland dunes. ____ **C. wigginsii**

Croton californicus Müller Argoviensis. *HIERBA DEL PESCADO;* SAND CROTON. Bushy perennials, also flowering in first season or year, herbaceous to suffruticose, 0.5–1.0 m, mostly branched from below, often densely branched with mostly erect to ascending stems. Densely pubescent with silvery-gray, lepidote-stellate hairs. Leaves often 2.5–11 cm × 2–8 mm, petioled, the blades linear to linear-lanceolate; stipules apparently absent. Dioecious. Flowers of both sexes without petals. Seeds 4.0–4.5 mm, gray-brown with blackish mottling. Various seasons.

Low stabilized dunes, sand flats, and the riverbed and floodplain of the Río Sonoyta to the east and southeast of the Pinacate lava shield. Southward on coastal dunes of western Sonora to Sinaloa. Also western and southern Arizona, southwestern Utah, southern Nevada, and California southward to Baja California Sur.

This species belongs to the *C. dioicus* Cavanilles complex, characterized by the absence of petals in the male as well as female flowers. Relationships among the species and infraspecific taxa in this complex remain to be worked out (see *C. wigginsii*). Several varieties are described for *C. californicus;* specimens from northwestern Sonora possess narrow leaves and may be referable to var. *tenuis* (S. Watson) A.M. Ferguson.

The lepidote trichomes cover almost every surface of the plants. These trichomes have radiating, hollow, conical, elongated hairs (as seen in a scanning electron microscope image) that break open when crushed. The plants have a strong odor, and were used as a fish poison by the Seri (Felger & Moser 1985). Presumably the material responsible for the strong odor and biological activity is contained in the hollow hairs.

8.8 mi N of Batamote Crossing, *F 90-168.* Río Sonoyta, 15 mi northward from Puerto Peñasco, *F 91-50A & B.* 23 mi SW of Sonoyta, *Wiggins 8353* (DS).

Croton sonorae Torrey. Shrubs mostly 1.0–1.5 m, the wood hard and the bark dark gray to nearly black. Leaves drought deciduous, 2.5–4.0 cm, the blades lanceolate to ovate, densely stellate-haired when young, becoming sparsely haired or glabrate, especially the upper surfaces; stipules absent. Monoecious. Flowers of both sexes with petals. Seeds 4.7–5.5 mm, broadly ovoid, shiny, mottled at maturity (nearly ripe seeds uniformly brown to gray). Flowering mostly with summer rains.

Arizona Upland in the Sierra Cipriano; not common, mostly in small arroyos in the upper bajadas and lower rocky slopes. Southern Arizona to Sinaloa, Baja California Sur, and disjunct in Guerrero, Puebla, and Oaxaca. The plants in the flora area are relatively small and stunted, probably as a result of the arid conditions and repeated freeze damage.

11–16 mi S of Sonoyta, *F 7573, F 16728.*

Croton wigginsii L.C. Wheeler [*C. arenicola* Rose & Standley, *Contr. U.S. Natl. Herb.* 16:12, 1912. Not *C. arenicola* Small, 1905] DUNE CROTON. Shrubs or subshrubs 1.0–2.5 m, slender and usually taller than wide, often woody at base; erect with sharply ascending, straight branches. Densely pubescent with silvery-gray lepidote-stellate scales. Leaves gradually drought deciduous, highly variable in size depending upon soil moisture; larger leaves lanceolate to narrowly lanceolate, 4.5–8.5 × 0.8–1.6 cm; dry-season leaves linear-lanceolate to linear, 1.5–6.5 cm × (0.5) 1.5–1.8 (3) mm; minute, deciduous cylindrical glands in the position of stipules. Dioecious. Flowers of both sexes without petals. Stamens 8–15 per flower. Seeds 4.5–7.8 mm, oblong, mottled. Various seasons.

Low stabilized to high shifting dunes across most of the Gran Desierto, sometimes on sand flats. Otherwise only near the international border in adjacent Yuma County, and the Algodones Dunes in southeastern California.

The large range in seed size is notable. The larger seeds, probably more than twice as heavy as the smaller ones, are from large-leaved, robust plants, whereas the smaller seeds are from relatively stunted, drought-stressed plants. *C. wigginsii* is an easily recognized allopatric segregate of *C. californicus,* but relationships between these and other members of the *C. dioicus* complex are unresolved. I suspect there are plants of intermediate morphology at the zones of contact between *C. californicus* and *C. wigginsii.*

Adair Bay, sandhills, 20 Nov 1908, *Sykes 62* (holotype of *C. arenicola* Rose & Standley, US). S of Moon Crater, *F 19013*. N of Sierra del Rosario, *F 75-30*. 20 mi E of San Luis, *Bezy 457*. Beside microwave tower, 2 km N, 13.5 km W of Los Vidrios, *Burgess 5602*. Punta Borrascoso, *F 84-39*.

DITAXIS

Annual or perennial herbs to shrubs or rarely small trees (ephemerals to subshrubs in the Sonoran Desert), sometimes with unusual reddish and/or bluish green coloring, with 2-armed hairs, often very dense on the calyx, corolla, ovary, and fruits; sometimes also with unbranched hairs, or sometimes glabrous (3 taxa). Leaves alternate. Mostly monoecious. Flowers in axillary, usually short racemes subtended by bracts, the lower flowers female, the upper ones male. Sepals 5, petals 5, usually entire, sometimes reduced in female flowers; glands 5, opposite sepals and interior to the petals. Male flowers with the stamens usually in 2 whorls of 5 each (ours), the filaments united into a column, the upper part of each filament free. Styles 3, each 2-cleft (bifid). Fruit a capsule splitting at maturity into three 1-seeded segments. Seeds rounded to ovoid or ovoid-pyramidal, without a caruncle.

Warmer regions in the Americas; 50 species. Its origin seems to have been in South America, and it is largely a genus of summer-rainfall regions.

The malpighian hairs are unique among Sonoran Desert Euphorbiaceae; these hairs are 2-armed and often glassy, with the opposite arms lying close to the surface (appressed). The Sonoran Desert species and many others have female flowers with petals shorter than to as long as the sepals, and male flowers with petals longer than the sepals. The female flowers are conspicuously larger than the male flowers. Most Sonoran Desert *Ditaxis* are facultative ephemerals to short-lived herbaceous perennials dying back severely during times of drought. They are non-seasonal, capable of responding quickly to rainfall during warm weather at any time of the year. *Ditaxis* has been treated as a subgenus of *Argythamnia*. References: Ingram 1970, 1980.

1. Glabrous, or new growth and ovaries hairy and rest of plant glabrous or very sparsely hairy.
   2. Shrubs, more than 40 cm, more than 3 times as tall as wide; stems erect, the first-season stems 4.0–7.0 mm in diameter, semisucculent. ______ **D. brandegeei**
   2′ Annuals or short-lived perennials, less than 30 cm, about as wide as or wider than tall; stems spreading, larger first-season stems less than 1.7–3.0 mm in diameter, tough and wiry, not at all succulent. ______ **D. serrata** var. **californica**

1′ Densely and coarsely hairy throughout.
   3. Subshrubs; stems mostly erect and straight; petals of male flowers united to the staminal column at base, appearing to arise above the glands. ______ **D. lanceolata**
   3′ Herbaceous; stems mostly ascending to spreading, or sometimes the main axis at first erect but the branches spreading and seldom straight; petals of male flowers free from the staminal column, appearing to arise between and alternate with the glands.
      4. Leaves mostly ovate-elliptic, the tips mostly pointed, not truncate; seeds usually with a reticulate pattern of shallow craters with fine radiating lines, hairs (if present) not saclike and papillate; widespread but mostly not on sand flats and dunes. ______ **D. neomexicana**
      4′ Leaves mostly obovate to spatulate, the tips mostly more or less truncate; seeds smooth to sometimes faintly patterned, the narrower end often with saclike or minutely papillate white hairs; mostly sand flats and dunes. ______ **D. serrata** var. **serrata**

Ditaxis brandegeei (Millspaugh) Rose & Standley var. **intonsa** I.M. Johnston [*Argythamnia brandegeei* Millspaugh var. *intonsa* (I.M. Johnston) J.W. Ingram] Slender, erect, sparsely branched shrubs 1.2–2.5 m, sometimes with a woody trunk to 3 cm in diameter. Stems few-branched, 4–7 mm in diameter in first season, herbaceous and pithy, brittle and tough, semisucculent, with long internodes. Young herbage sparsely hairy, the capsules densely or occasionally sparsely covered with coarse hairs, the plants otherwise glabrate; herbage of a strange blue-green color often becoming accentuated on drying, or sometimes reddish. Leaves sparse, on upper nodes only, quickly drought deciduous, mostly (3) 5.0–9.5 × (0.7) 1.0–2.8 cm, petioled, the blades lanceolate to sometimes elliptic, the margins with small blunt gland-tipped teeth; stipules glandular, hornlike, 0.4–0.5 mm, quickly deciduous or perhaps sometimes absent. Flowers yellow-green, the male flowers with laciniate-fringed petals. Seeds 3.5–4.0 mm, thick, brown, ovoid-pyramidal with a flat base and a pattern of shallow craterlike depressions toward the base. At least in spring.

Widely scattered, seldom common; northwestern side of the Pinacate volcanic complex and the northern and western granitic ranges. Usually on rocky soils or rock slopes, cliffs, and crater walls, often in rock crevices. Not known elsewhere in Sonora. Also southwestern Arizona, and the gulf coast and adjacent islands of both Baja California states.

The plants are frost sensitive and freezing weather seems to be a major limiting factor. The Arizona and Sonora plants match var. *intonsa,* distinguished from var. *brandegeei* "only in possessing a few scattered, very coarse, appressed, malpighian hairs on the leaves and young stems, and in the densely setose-hispid capsules" (Wiggins 1964:786). The term *intonsa,* 'unshaven', refers to the hairy ovaries and capsules. However, an occasional Sonoran specimen (e.g., *Turner 59-308*) is sparsely hairy or the ovaries glabrous or glabrate. Var. *brandegeei* is characterized as glabrous. Both occur in the Baja California Peninsula and do not seem to be geographically segregated, although var. *brandegeei* seems to be more widespread. The laciniate-margined petals are unusual in the genus.

Hornaday Mts, *Burgess 6520.* Sykes Crater, *F 18999.* MacDougal Crater, *Turner 59-308.* Granite hill, 65 mi W of Sonoyta, 19 Mar 1936, *Shreve 7606.* Sierra Nina, *F 89-49.*

Ditaxis lanceolata (Bentham) Pax & K. Hoffmann [*Serophyton lanceolatum* Bentham. *Argythamnia lanceolata* (Bentham) Müller Argoviensis. *Ditaxis palmeri* S. Watson] Perennials, also flowering in the first season, often forming densely branched clumps to 1 m, or sparser and taller when scrambling through brush. Stems slender and brittle, the herbage silvery hairy during dry seasons, the new growth greener, more sparsely pubescent, and the leaves larger and thinner following hot, wet weather. Leaves 5–42 × 2–9 mm, linear-lanceolate to broadly lanceolate. Seeds 2.3–2.5 mm, reticulate with shallow craters having radiating lines, the edges slightly angled, the base flat (the seed will sit on end; seeds "chunkier" than those of *D. neomexicana*). Various seasons.

Widespread and common, including the volcanic and granite ranges and the Sonoyta region; absent from dunes. Slopes, flats, arroyos, and washes; rocky soils, usually in pockets of fine-textured soil, or in sandy-gravelly soils; often among and beneath shrubs and trees. Heavily browsed by native animals, including chuckwallas and rabbits, and often reduced to dense, knobby stubs. The new growth is occasionally freeze damaged. Nearly throughout the Sonoran Desert: southeastern California and western Arizona through the Baja California Peninsula and Sonora south at least to the Guaymas region. Wiggins (1964:786) mentions that the plants are "apparently sometimes dioecious," but I have not seen such plants.

Sierra del Rosario, *F 20720.* Hornaday Mts, *Burgess 6860.* Tinajas de los Pápagos, *Mason 1835.* Tinajas de Emilia, *F 19733.* Sykes Crater, *F 18930.*

Ditaxis neomexicana (Müller Argoviensis) A. Heller [*Argythamnia neomexicana* Müller Argoviensis] Non-seasonal ephemerals or small, short-lived herbaceous perennials. Much-branched and densely pubescent with stiff, coarse hairs. Leaves mostly 13–34 mm, elliptic to oblanceolate, the apex acute to obtuse (pointed, not truncate), the margins entire or with a few small teeth. Leaves longer (to 5 cm), broader, greener, not as thick, and less hairy during warm, wet conditions. Male flowers: sepals green, the petals longer than the sepals, obovate, white with red-purple veins, the glands 0.5 mm, transparent-membranous, with age becoming yellow-brown and thickened, the staminal column 1.5 mm, 2 whorls of stamens near apex of column. Female flowers: sepals 3.2–4.0 mm, the petals 1.5–3.0 mm. Seeds 2.0 × 1.5–1.6 mm, ovoid with a pointed tip, brown with low hairs forming either fine radiating lines from minute craterlike pits or a reticulate pattern (the radiating lines often not formed on immature seeds).

Widespread and often common, including the granitic ranges, volcanic complex, and Sonoyta region; rocky slopes including south exposures, arroyos and gravelly washes, bajadas, and desert flats. Sometimes on sand flats but not on dunes. Lowlands of southwestern North America. It is closely allied to *D. serrata.*

11 mi SW of Sonoyta, *F 9802.* Cerro Colorado, *F 10793.* S of Pinacate Peak, 875 m, *F 19345.* Chivos Butte, *F 19626.* W of MacDougal Crater, *F 9927.* Sierra del Rosario, *F 75-4.*

Ditaxis serrata (Torrey) A. Heller var. **serrata** [*Aphora serrata* Torrey. *Argythamnia serrata* (Torrey) Müller Argoviensis. *Ditaxis odontophylla* Rose & Standley, *Contr. U.S. Natl. Herb.* 16:12, 1912] Robust non-seasonal ephemerals to short-lived herbaceous perennials, herbage strigose with very dense silvery hairs, and with a well-developed taproot. Stems with age spreading to decumbent. Leaves 7–50 mm, often broadly cuneate-spatulate, sometimes obovate to ovate, the apex commonly somewhat truncate and toothed, the margins with few small teeth. Male flowers: sepals 1.8–3.0 mm, the petals 2.5–2.8 mm, the glands 0.5 mm, lanceolate, thin and membranous. Female flowers: calyx sometimes with callosities below the sinus between sepals, the sepals often 2.2 mm, the petals with dense, stiff hairs near apex on lower surface or glabrate. Seeds 2.1–2.3 × 1.7–1.8 mm, ovoid with a pointed tip, brown with a few hairs, the hairs minute, white, blunt, and often saclike or minutely papillate, often dense around the hilum.

Sand flats and lower dunes; generally replacing *D. neomexicana* on sandy soils. Also crater floors, mostly in the playas and occasionally on rocky slopes adjacent to sandy habitats. Southward in Sonora on sandy soils to the vicinity of El Desemboque Río de la Concepción; also southwestern Arizona, southeastern California, Baja California, and perhaps Baja California Sur and Gulf of California islands. (The Baja California Sur and certain island populations have pointed leaves and may be a different taxon.)

Ditaxis serrata is distinguishable from *D. neomexicana* by its more robust habit of growth, stouter and deeper taproot, lighter-colored foliage, usually broader and blunter (commonly truncate) leaves, and differences in the seed surfaces. In addition there may be minor floral differences. The local distributions are narrowly separated by differences in habitat, and in places of contact (such as the alluvium of the floor of MacDougal Crater) there are plants that look like *D. serrata* but have seeds somewhat intermediate in character. In these plants the seeds have shallow, craterlike depressions with very faint radiating lines as seen in *D. neomexicana,* but the saclike, short hairs show a closer affinity to *D. serrata.* Leaves of the first dozen or so internodes of the seedlings or young plants of *D. serrata* are often elliptic with acute or obtuse (pointed) tips like those of *D. neomexicana* rather than truncate (blunt). Ingram (1970:942) states that *D. neomexicana* is doubtfully distinct from *D. serrata* of "Arizona, California, and northwestern Mexico; apparently forming an intergrading complex that needs further intensive study." The type locality for *D. serrata* is "sandy plains near Fort Yuma, California."

10 mi N of El Golfo, *F 75-78.* 25 mi E of San Luis, *F 16704.* Sierra del Rosario, *F 75-5.* MacDougal Crater, *F 10461.* Moon Crater, *F 20168.* Punta Peñasco, *Shreve 7596.* Papago Tanks, 14 Nov 1908 [*sic.* 1907], *MacDougal 36* (holotype of *D. odontophylla,* US).

Ditaxis serrata var. **californica** (Brandegee) V.W. Steinmann & Felger [*Argythamnia californica* Brandegee. *Ditaxis californica* (Brandegee) Pax & K. Hoffmann] Annuals or possibly short-lived perennials with a well-developed taproot; glabrous, the herbage drying the same strange blue-green color as that of *D. brandegeei.* Larger leaves 2.5–3.5 cm. Male flowers: petals 2.5–3.5 mm, mostly longer than the sepals, the glands 0.6–0.7 mm, lanceolate, thin and membranous, the staminal column 1.5 mm. Female flowers: calyx often with a callus (thickened, warty tissue) below the sinus between sepals, the flowering sepals 3.0–3.7 mm, thick, with whitish margins, the petals 2.5 mm, white, thin. Seeds 2.3 × 1.8 mm, brown, hairs on some seeds bright blue (this does not seem to be an artifact) near the hilum and along the ventral suture. Flowering at least in March.

The single collection from Sonora consists of 2 plants from a coastal dune 3 km east of Puerto Peñasco (*Yatskievych 80-43*). Otherwise southeastern California and the eastern flanks of the Sierra Juárez in northeastern Baja California.

Var. *californica* is essentially a glabrous form of the more widespread var. *serrata* (Steinmann & Felger 1995). An analogous situation occurs in Texas, where *D. humilis* var. *laevis* is a glabrous expression of the pubescent, more common and widespread var. *humilis.*

EUPHORBIA SPURGE

Great variation in growth form, ephemerals to perennial herbs, shrubs, trees, and especially in drier regions of Africa, a fantasy of succulent growth forms often convergent with New World cacti; with milky sap throughout. Leaves alternate, opposite, or whorled, the blades symmetrical or not.

Several to many flowers aggregated into a headlike inflorescence or cyathium. Each cyathium with fused bracts forming the involucre surrounding the flowers and bearing at its apex 4, 5, or rarely fewer glands, these commonly with a petaloid (petal-like) appendage on the outside margin. (The cyathia, or flower heads, with their petaloid appendages, have the gross appearance of small flowers.) Individual flowers unisexual, the floral parts greatly reduced. Each cyathium with a few to many minute male flowers and a single, much larger female flower, or the plants sometimes dioecious. Each male flower reduced to a single stamen fused to a short pedicel. The female flower reduced to a pedicel and pistil (ovary, styles, and stigmas; a vestigial calyx sometimes present). Pistil usually extending well beyond the edge of the cyathium and often turning down and away from its male nest mates. Styles 3, each divided into 2 branches. Fruit a 3-seeded capsule; seeds with or without a caruncle.

Worldwide; one of the world's largest genera with 2000 species. Five fairly well-marked subgenera, 2 in the flora area: *Agaloma* with 2 species and *Chamaesyce* with 13 species. This is the most diverse genus in the flora area.

Controversy over the lumping or splitting of *Chamaesyce* and *Euphorbia* has spanned a century. All *Chamaesyce* have sympodial growth: The apical meristem of the main stem of the seedling aborts after the first true leaves are formed. The aerial portion of the plant develops "from lateral branches originating in the region of the cotyledonary nodes. Sympodial growth continues throughout the life of the plant: each terminal bud of a branch aborts and is alternately replaced by a bud from either side of the stem apex" (Koutnik 1985:188). No other species in the Euphorbia family have this growth form. This branching pattern contributes to the prostrate or flat, spreading habit of growth so common in the *Chamaesyce*. Other characters include the C_4 photosynthetic pathway and associated Kranz anatomy, often asymmetric leaf bases, presence of stipules (rarely caducous or absent, as in *E. setiloba*), and seeds without a caruncle.

For the most part the *Chamaesyce* are pollination generalists; the major pollinators are probably small bees, wasps, and flies, but ants and butterflies probably also do their share (Krombein 1961; Grady Webster, personal communication 1996). However, at least some are self-pollinating, especially the weedy ones, and presumably the majority of all the species are self-compatible (e.g., Ehrenfeld 1976, 1979). It is an attractive hypothesis that there is a trend for species with reduced appendages to be selfing and those with larger appendages to be outcrossing.

Chamaesyce are worldwide, with 250 species. The hot bed of their evolution is in southwestern North America in subtropical semiarid to arid regions. *Chamaesyce* are especially numerous in the Sonoran and Chihuahuan Deserts, with 33 species in the Sonoran Desert. For the most part the different species of *Chamaesyce* can be readily identified by their distinctive seeds. As in the case of *Cryptantha* (Boraginaceae), rather technical characters are used in the key, but with a little practice one can distinguish the species by the gross appearance of the plants. Our *Chamaesyce* are distinct from one another, with no indication of intermediates or hybrids, although several species often occur together. Seven species are similar in their overall leaf shape and size of the plants, seeds, and leaves: *E. abramsiana, E. arizonica, E. micromera, E. petrina, E. polycarpa, E. prostrata,* and *E. setiloba*. *E. arizonica* and *E. setiloba* appear to be related, being linked by features such as cyathia narrowed at the apex, stipules absent or very reduced, and glandular hairs. The members of 2 pairs of sister taxa (species pairs)—*E. albomarginata* and *E. serpens,* and *E. micromera* and *E. polycarpa*—are differentiated by rather subtle characters.

The seed coats of 13 of the 15 species of *Euphorbia* in the flora region produce varying amounts of mucilage when wet. Upon drying this mucilage tenaciously adheres the seed to the substrate. The 2 species with non-mucilaginous seeds are *E. misera* and *E. platysperma,* both with relatively large seeds. (The latter has both the largest seeds and smallest geographic range among our *Chamaesyce*.) Presumably the mucilaginous seeds play a role in dispersal, including possible long-distance dispersal. References: Ehrenfeld 1976, 1979; Huft 1984; Koutnik 1984, 1985; Krombein 1961; Webster, Brown, & Smith 1975; Wheeler 1936, 1941.

1. Shrubs or erect annuals, the growth habit not sympodial; leaves alternate (often crowded on short shoots) or alternate below and whorled above, symmetrical. ______ Subgenus **Agaloma:**
  2. Annuals, less than 1 m; leaves alternate below, whorled above; seeds with a caruncle. ______ **E. eriantha**
  2′ Shrubs, often 1 m or more; leaves alternate (crowded on short shoots); seeds without a caruncle. ______ **E. misera**

1′ Annual or perennial herbs, prostrate or not, with sympodial growth habit; leaves opposite, asymmetric toward base; seeds lacking a caruncle. ______ Subgenus **Chamaesyce:**
  3. Seeds more than 2 mm.
    4. Stems orange, many and sprawling; seeds smooth, flattened, about twice as long as wide; dunes. ______ **E. platysperma**
    4′ Stems green, usually with a single strong, erect or upright main axis; seeds sculptured, chunky, almost as wide as long; playas and arroyos, not on dunes. ______ **E. trachysperma**
  3′ Seeds less than 2 mm.
    5. Stems prostrate, rooting at nodes; leaves all or mostly orbicular or nearly so; stipules conspicuous, united into white scales.
      6. Perennials; cyathia more than 1 mm wide, with 10 or fewer stamens; widespread but localized. ______ **E. albomarginata**
      6′ Annuals; cyathia ca. 0.6–1.0 mm wide, with 15–30 stamens; probably rare in the flora area. ______ **E. serpens**
    5′ Stems prostrate or not, not rooting at nodes; leaves not orbicular, often ovate to obovate or oblong; stipules inconspicuous, slender.
      7. Petaloid appendages absent; involucral glands 0.3 mm wide or less.
        8. Seeds smooth (see species account for *E. micromera*).
          9. Involucral glands usually round, dotlike, 0.1–0.2 (0.25) mm wide. ______ **E. micromera**
          9′ Involucral glands usually oval, 0.3 mm or more wide. ______ **E. polycarpa**
        8′ Seeds with transverse ridges.
          10. Capsules glabrous. ______ **E. abramsiana**
          10′ Capsules hairy.
            11. Capsules densely hairy throughout. ______ **E. petrina**
            11′ Capsules hairy only or mostly on ridges, especially toward capsule base. ______ **E. prostrata**
      7′ Petaloid appendages present (sometimes not developed in young cyathia or drought-stressed plants; these plant key out in both choices); involucral glands various.
        12. Cyathia narrowed or constricted at apex, urn-shaped or narrowly turbinate (shaped like a top).
          13. Ephemeral to perennial; petaloid appendages entire to slightly lobed (rounded); glandular hairs slightly enlarged at tip (club-shaped). ______ **E. arizonica**
          13′ Strictly ephemeral; petaloid appendages with triangular, pointed segments, giving the cyathia a star-shaped appearance; glandular hairs not enlarged at tip. ______ **E. setiloba**
        12′ Cyathia not narrowed or constricted at apex.
          14. Seeds smooth. ______ **E. polycarpa**
          14′ Seeds with conspicuous transverse ridges and grooves.
            15. Herbage, cyathia, and capsules hairy; cyathia 1.2–1.5 mm wide, the glands 0.6–0.9 mm wide, the appendages usually 0.5 mm or more in width (rarely absent or reduced on immature cyathia). ______ **E. pediculifera**
            15′ Plants glabrous or hairy; cyathia 0.4–0.9 mm wide, the glands 0.1–0.4 mm wide, the appendages none to 0.4 mm wide (measured perpendicular to gland).
              16. Capsules hairy on ridges. ______ **E. prostrata**
              16′ Capsules glabrous.
                17. Plants mostly spreading to prostrate; leaf margins entire to few toothed mostly near apex; involucral glands 0.1–0.15 mm wide. ______ **E. abramsiana**
                17′ Plants erect to ascending; leaf margins evenly serrated all around; involucral glands 0.2–0.4 mm wide. ______ **E. hyssopifolia**

Euphorbia abramsiana L.C. Wheeler [*E. pediculifera* Engelmann var. *abramsiana* (Wheeler) Ewan. *Chamaesyce abramsiana* (L.C. Wheeler) Koutnik] Small, warm-weather ephemerals (also in spring in southern Sonora) often forming prostrate mats with red-brown herbage in dry, open habitats, but upright to spreading and green among dense vegetation and in shaded places. Herbage glabrous or hairy. Leaves 2.5–12 (14) mm, ovate-elliptic to oblong, entire to minutely toothed mostly toward the leaf apex, with a reddish blotch near the center. Cyathia inconspicuous, 0.4–0.5 mm wide, the involucral glands dotlike, rounded or nearly so, 0.1 (0.15) mm wide, the appendages absent to 0.2 mm wide, white to pink. Capsules glabrous, bright green with red margins (angles) and furrows, the angles rather sharp. Seeds 1.0–1.2 mm, ashy grayish-white to tan, resembling a mealy bug, with a sharply angled crest and transversely ridged; mucilaginous when moistened.

Seasonally abundant in the larger playas, such as Playa Díaz and adjacent alluvial flats, large washes, and an urban and agricultural weed. Essentially a Sonoran Desert–Sinaloan thornscrub species, extending into open or weedy woodland vegetation in southern Arizona. Southeastern California through much of the Baja California Peninsula, and southern Arizona to Sinaloa. The minute cyathia, reduced petaloid appendages, and glabrous capsules are distinctive features (see *E. prostrata*).

Sonoyta, *F 85-948.* El Papalote, *F 86-322.* Pinacate Junction, *F 86-344.* Playa Díaz, *F 86-364.*

Euphorbia albomarginata Torrey & A. Gray [*Chamaesyce albomarginata* (Torrey & A. Gray) Small] *GOLONDRINA;* RATTLESNAKE WEED. Perennials and also flowering in first season; deeply set, thickened roots, the stems dying back in drought; new growth appearing any time of year, often locally carpeting the ground with prostrate stems rooting at the nodes. Herbage green, glabrous. Leaves short petioled, the blades 2–8 mm, more or less broadly ovate to orbicular, often with a red blotch in the middle, the margins entire. Stipules relatively large, united into a triangular white scale with a fringed margin. Cyathia solitary at nodes, the appendages white and showy. Seeds 0.9–1.4 mm, the upper surface with a rounded ridge crest, moderately flattened on either side of the ridge, and excavated on both sides of the septum on the ventral surface; mucilaginous when moistened. Various seasons.

Usually in poorly drained, fine-textured soils, especially in playas including those of crater floors, clayish soils in arroyo bottoms, and sometimes in dried pools in roadside ditches; also an agricultural weed. Widespread in western North America. Closely related to *E. serpens.*

MacDougal Crater, *Turner F 86-33.* Playa los Vidrios, *Ezcurra 19 Ap 1981* (MEXU). Sykes Crater, *F 20020.* Pinacate Junction, *F 86-342.* Playa Díaz, *F 86-372.* Suvuk, *Nabhan 369.*

Euphorbia arizonica Engelmann [*Chamaesyce arizonica* (Engelmann) Arthur] *GOLONDRINA*. Perennials flowering in first season. Herbage reddish; stems, leaves, cyathia, and capsules pubescent, the hairs glandular and slightly enlarged at the tip with several septa between red beadlike segments. Stems usually ascending-spreading. Leaves short petioled, the blades 1.2–6+ mm, mostly broadly ovate or nearly orbicular or oblong, the margins entire; stipules reduced to minute "spurs" or absent. Cyathia urn-shaped (narrowly turbinate to urceolate, taller than wide, slightly narrowed at apex just below the glands), ca. 1 mm high, pink to reddish, the appendages showy, petal-like, white fading to dark pink, the margins entire or sometimes shallowly lobed. Seeds 0.9–1.1 mm, transversely ridged, not conspicuously mucilaginous when wet but adhering tenaciously after drying.

Rocky places in the Pinacate volcanic region, especially along better-vegetated arroyos or canyons and gravelly-sandy arroyo beds; distribution apparently patchy in the flora area. Southern California to western Texas, south to Baja California Sur, southern Sonora, and Durango.

The cyathia are similar to those of *E. setiloba* except larger and the petaloid appendages entire or shallowly lobed rather than laciniate-margined. The reduced or seemingly absent stipules, reddish herbage, dense pubescence, and bright white to pink appendages serve to distinguish it. The absence or reduction of stipules is an unusual character among the *Chamaesyce* and one shared with *E. setiloba.* The seeds are similar to those of *E. abramsiana* but differ in being slightly smaller and having deeper and more irregular furrows.

Tinaja Huarache, *F 19117*. Tinaja del Tule, *F 18746*. Hourglass Canyon, *F 19153*. Sierra Pinacate, 1150 m, *F 18676*.

Euphorbia eriantha Bentham. BEETLE SPURGE. Non-seasonal ephemerals, often taller than wide; often (15) 25–70 cm, the main axis usually erect, well formed, and stout, with few to many branches above. Leaves 2.5–6.0 cm, narrowly linear, the lower ones alternate and very quickly deciduous, the upper ones in a whorl beneath the flower heads; stipules minute. Young growing tips and inflorescences, including cyathia and ovaries, densely pubescent with short, white, appressed hairs. Involucral glands 3 (5), often inconspicuous and obscured by hairs and also often by herbaceous flaps. Anthers and stigmas maroon red, the red stigma protruding from the white-pubescent ovary before the female pedicel has elongated and before the male flowers (stamens) appear. Seeds mottled white, gray, and blackish, with low bumps and a minute cellular pattern, 3.5–4.0 mm including a conspicuous whitish caruncle; mucilaginous when wet.

Widespread and often common nearly throughout the lowland areas except dunes, and to at least 875 m in the Sierra Pinacate; sand flats, washes, desert pavements, hills, and cinder slopes. Baja California Sur to southeastern California and eastward to southwestern Texas, Coahuila, Durango, and northwestern Sinaloa and lowlands of Sonora.

11.8 mi SW of Sonoyta, *F 9826*. Cerro Prieto, Bahía de la Cholla, *Parker 13 Mar 1984*. Sierra Pinacate, 875 m, *F 19353*. 3 mi E of Sierra Extraña, *Simmons 515*. Cerro Pinto, *F 85-1023*. Sierra del Rosario, *F 20719*.

***Euphorbia hyssopifolia** Linnaeus [*Chamaesyce hyssopifolia* (Linnaeus) Small] Warm-weather ephemerals with an erect to ascending main axis to several-branched, to ca. 60 cm; glabrous. Leaves 4–20 (30) mm, lanceolate to oblong, sometimes with a red blotch at center, the margins evenly serrated. Cyathia 0.4–0.9 mm wide, the involucral glands 0.2–0.4 mm wide, oval, pink to maroon, the appendages 0.3–0.6 mm wide, broader than long, white to pink, darkening with age. Seeds 1.3–1.4 × 1.0 mm and nearly as thick with few shallow transverse depressions; mucilaginous when wet.

Disturbed habitats, including roadside depressions near Sonoyta, and a garden weed; not native to the flora area. Widespread in Sonora and southern Arizona from desert to mountain habitats. Arizona to southeastern United States and South America; also adventive in the Old World.

15.7 mi SW of Sonoyta, *F 91-138*. Puerto Peñasco, *F 91-143*.

Euphorbia micromera Boissier [*E. polycarpa* var. *micromera* (Boissier) Millspaugh ex Orcutt. *Chamaesyce micromera* (Boissier) Wooton & Standley. *Euphorbia parishii* Greene] Small herbaceous perennials or mostly non-seasonal ephemerals resembling *E. polycarpa*. Prostrate to procumbent or spreading, often hairy, especially the capsules and new growth. Leaves 1.2–5.0 mm, the larger ones few and on larger stems near the center (base) of the plant, the blades mostly broadly ovate to oblong, the margins entire. Involucral glands 0.12–0.25 mm wide, dotlike, rounded or sometimes oval, maroon, without appendages. Seeds 0.9–1.0 mm, resembling those of *E. polycarpa*, mucilaginous when wet.

Widespread and common in sandy, gravelly, and rocky soils; Sonoyta region, Pinacate volcanic complex, granitic ranges, and desert plains; not on dunes. Southeastern California to Utah and western Texas, Baja California, northern Sonora, and the Chihuahuan Desert. Wheeler's (1941) report of it from Peru is inaccurate (Brako & Zarucchi 1993).

Puerto Peñasco, *F 85-769*. MacDougal Crater, *F 10739*. 4.5 mi N of Cerro Colorado, *F 10416B*. 19 mi E of San Luis, *F 16695*. 24 km W of Los Vidrios, *F 85-1014*.

Euphorbia micromera and *E. polycarpa* probably diverged from a common ancestor in the not-too-distant past, perhaps during the Pleistocene when they may have been geographically isolated. *E. micromera* ranges farther north than does *E. polycarpa*, and *E. polycarpa* ranges farther south than does *E. micromera*. Especially in arid situations, *E. polycarpa* may produce plants lacking petaloid appendages, and the glands may be smaller than usual. Although most *Chamaesyce* have distinctive seeds, those of

E. micromera and *E. polycarpa* appear identical. *E. micromera* tends to be more numerous in relatively arid habitats, whereas *E. polycarpa* tends to be more numerous in somewhat less arid habitats. *E. micromera* has been reported to be annual (ephemeral), but both species occur as annuals or perennials. *E. parishii* of California, supposedly distinguished from *E. micromera* primarily by its greater number of staminate flowers, does not seem to be a distinct species. Differences between *E. polycarpa* and *E. micromera* are summarized as follows:

E. polycarpa	*E. micromera*
May have larger and smaller leaves on same plant.	Mostly smaller leaves.
Plants glabrous or hairy.	Plants usually hairy.
Stipules on ventral side of node wholly fused, triangular with a broadened base.	Ventral stipules fused basally, subulate.
Cyathia relatively larger and broader (more campanulate).	Cyathia smaller, not as broad.
Involucral glands oval, usually larger.	Involucral glands rounded, smaller.
Appendages on glands usually present, as broad or broader than glands (sometimes absent or reduced).	Appendages absent or vestigial.
Styles longer, the branches separate (best seen on mature capsules).	Styles shorter, like a nubbin at top of capsule, difficult to see separate branches.

Euphorbia misera Bentham. CLIFF SPURGE. Shrubs with copious milky sap, often 0.7–1.4 m, the stems thickish, semisucculent, and flexible, appearing gnarled due to the knobby short shoots; multiple stemmed from the base, resembling *Jatropha cuneata* in architecture. Herbage with short hairs. Leaves alternate but crowded on short shoots, drought deciduous, the blades ovate-orbicular, about as wide as long, 7–13 mm, the petioles 1/4 to 1/2 as long as blades; stipules to 1 mm on long shoots, scale-like, obscured by white-wooly hairs, and inconspicuous on short shoots. Cyathia predominantly yellow-green, 2.0–3.8 mm wide, the glands 5 in number, 2.3–3 mm wide, yellow-green, the appendages conspicuous, white to yellow-white, reaching 2.3 mm long × 3.2 mm wide (green flies observed eating nectar from the cyathia disk). Seeds 3.0–3.5 mm, ovoid-globose, whitish to brown, somewhat pitted, not mucilaginous. Growing and flowering after rains at any season.

Rocky volcanic and granitic hills and small mountains on the coast, and sometimes on coastal dunes adjacent to granitic slopes; abundant at Punta Peñasco and Bahía de la Cholla. South of the flora area it also grows on coastal dunes. It seems to be frost sensitive, which may explain its absence farther inland. Southern California to mid–Baja California, and Sonora southward along the coast to Cerro Tepopa and Isla Tiburón. Farther south in Baja California and in the Hermosillo-Guaymas region in Sonora it is replaced by the closely related *E. californica* Bentham.

Punta Peñasco, *Shreve 7597*. Bahía de la Cholla: *Fishbein 14; Lehto 3714* (ASU).

Euphorbia pediculifera Engelmann var. **pediculifera** [*Chamaesyce pediculifera* (Engelmann) Rose & Standley] *GOLONDRINA;* LOUSE SPURGE. Non-seasonal ephemerals to short-lived perennials. Young herbage, capsules, and cyathia densely pubescent, the hairs thickish, short, white, and appressed to spreading. Stems usually erect when young, then spreading to prostrate. Herbage often red-brown to gray-brown. Leaves at least twice as long as wide, petioled, the blades 5.5–15.5 mm, ovate to obovate or oblong, the margins entire or sometimes with a few small, irregular teeth. Cyathia 1.2–1.5 mm wide, the glands maroon, oval, 0.6–0.9 mm wide, the appendages rather showy, white, fading pink, usually considerably wider and longer than the glands, sometimes as large as 2.0–2.5 × 1.0–1.2 mm. Seeds 1.0–1.2 mm, chunky, conspicuously cross-ridged, the dorsal ridge rounded and somewhat flattened; mucilaginous when wet.

Widespread and often common on a variety of substrates nearly throughout the flora area, including the volcanic and granitic ranges to peak elevations; absent from dunes. Sonoran Desert in southeastern California and Arizona, and southward to Sinaloa and the Cape Region of Baja California Sur. Var. *linearifolia* S. Watson, from the Guaymas region, differs in having narrower and often longer leaves.

Sierra Extraña, *F 19087.* Tinaja del Tule, *F 19196.* Sykes Crater, *F 18991.* S side of Carnegie Peak, 900 m, *F 19906.* Elegante Crater, *19691.* Suvuk, *Burgess 6338.* 21 mi W of Sonoyta, *Lehto 19237.*

Euphorbia petrina S. Watson [*Chamaesyce petrina* (S. Watson) Millspaugh] Non-seasonal ephemerals; usually prostrate or nearly so, densely pubescent with relatively long, slender, spreading hairs on herbage, cyathia, and capsules. Herbage red-brown. Leaves petioled, the blades mostly 2.2–6.0 mm, broadly ovate to oblong or elliptic. Cyathia crowded, difficult to distinguish individually, 0.6 mm wide, the glands 0.3 mm wide, oval, often obscured by hairs, without appendages. Seeds 0.9–1.0 mm, with low transverse ridges, the ventral side constricted near apex; mucilaginous when wet.

Coastal dunes and sand flats, in natural and disturbed habitats; Puerto Peñasco to the Río Colorado delta. Coastal Sonora to Sinaloa, islands in the Gulf of California, and the Baja California Peninsula. Seeds from plants farther south tend to be broader and slightly smoother. This species superficially resembles *E. polycarpa* but is distinguished by the unusual red-brown color, minute cyathia, absence of petaloid appendages, dense and relatively long spreading white hairs, and distinctive seeds.

Puerto Peñasco, *F 85-770.* Bahía de la Cholla, *F 16842.* 2 km westward from El Golfo, *F 93-05.* 1 km N of El Doctor, *F 93-253.*

Euphorbia platysperma Engelmann ex S. Watson [*E. eremica* Jepson. *Chamaesyce platysperma* (Engelmann ex S. Watson) Shinners] DUNE SPURGE. Ephemerals or annuals, possibly short-lived perennials, with deeply buried roots and often with dune-buried stems; forming loose, spreading mounds 0.5–1.0+ m across; glabrous. Sand adheres to the glandular-sticky buried portion of the stems forming a sand jacket. Stems slender, flexible, pale orange, arching or ascending, becoming semiprostrate with age. Leaves petioled, the blades 5–16 × 2.2–7.2 mm, elliptic to oblong or obovate, relatively thin, the midrib prominent. Peduncles prominent; cyathia 1.2–1.8 mm wide, yellow-green, solitary in the leaf axils, the glands and appendages yellowish, the glands darker, ca. 0.5–0.9 mm wide, the appendages usually larger than the glands. Seeds 2.4 × 1.4 mm, gray-white, smooth, the inner face flat, with a prominent medial ridge; not mucilaginous. October-May.

Abundant on shifting dunes of low to medium height, and sometimes on higher dunes; beach dunes at El Golfo nearly to high tide level. Endemic to dunes of the Gran Desierto and adjacent areas: vicinity of El Golfo northward to the U.S. border region about 30 km east of San Luis and southward to the northwest side of the Sierra del Rosario, then eastward to the southern and western margins of the Pinacate volcanic shield. Very rarely encountered as waifs in dunes in southeastern California, and a single collection from northeastern Baja California.

The stems are often freeze killed, and freezing weather may account for the failure of this species to become established in southeastern California. The seeds are relatively large in comparison to those of other *Chamaesyce* species in the flora area.

S of Moon Crater, *F 19011.* Dunes, vicinity of Sierra del Rosario, *F 75-33.* 25 mi E of San Luis, *F 16707.* Punta Borrascosa, *F 84-40.* El Golfo, *F 75-84.* Baja California: Sierra de las Pintas, *Gentry 8733, 8733a.*

Euphorbia polycarpa Bentham [*Chamaesyce polycarpa* (Bentham) Millspaugh. *Euphorbia polycarpa* var. *hirtella* Boissier. *E. intermixta* S. Watson. *E. polycarpa* var. *intermixta* (S. Watson) L.C. Wheeler] *GOLONDRINA; VI'IBGAM.* Non-seasonal ephemerals to small perennial herbs, the taproot well developed. Herbage and capsules glabrous or hairy. Stems prostrate to ascending, the larger plants much-branched. Leaves 1.5–6.8 mm, petioled, the blades broadly ovate or orbicular to oblong, the margins entire. Involucral glands 0.3–0.6 mm wide, dark maroon (almost black, or rarely yellow in summer), oval (wider than long), with conspicuous white appendages, or the appendages sometimes reduced or absent on drought-stressed plants. Seeds 0.8–1.0 mm, fairly smooth but dull rather than shiny, mucilaginous when wet. Flowering any time of year.

Nearly throughout the flora area, in many habitats from near sea level to peak elevations in the Pinacate volcanic region and the granitic ranges; absent from higher dunes. One of the most abundant and widespread plants in the region; sandy to rocky soils including cinder soils. Especially common on gentle slopes, but also on rock walls and cliffs.

Southern California to Baja California Sur, southern Nevada and Arizona, and western Sonora south to the Río Mayo delta region. Sometimes these small plants grow from crevices, such as west-facing crater walls, spreading close to dark lava rock too hot to touch with bare hands in the summer. Seeds placed on damp paper germinated within 6 hours.

The pubescent plants have often been called var. *hirtella,* but this variation shows no geographic segregation, ranges along a continuum from glabrous to densely pubescent, and glabrous or sparsely to densely pubescent plants sometimes occur freely intermixed. The type collection of *E. intermixta* appears to be nothing more than first-season specimens. Wheeler (1936) and Wiggins (1964) recognized several additional varieties, all from the Baja California Peninsula and its adjacent islands, and some of these may be worthy of continued recognition (Steinmann & Felger 1997).

Sierra del Viejo, *F 16916.* Sierra del Rosario, *F 20381.* Moon Crater, *F 1924.* Molina Crater, *Kamb 2012.* Tinaja del Tule, *F 18905.* Campo Rojo, *Webster 24273* (DAV). Pinacate Peak, *F 19406.*

***Euphorbia prostrata** Aiton [*Chamaesyce prostrata* (Aiton) Small] Warm-weather ephemerals (short-lived perennials elsewhere); stems upright to spreading among dense ephemeral vegetation (elsewhere prostrate). Stems, capsules, and young leaves moderately hairy with short, crinkled, white hairs. Leaves 4.5–8.5 mm, obovate-elliptic to oblong, entire (to serrated elsewhere), petioled. Cyathia minute, 0.4 mm wide, the glands pink, transversely elliptic, 0.15 mm wide (long axis), without appendages (elsewhere with short appendages). Capsules green with white hairs on the angles, especially near the capsule base. Seeds 0.7–0.8 mm, ashy gray-white to tan, with a sharply angled crest and relatively short transverse ridges; not conspicuously mucilaginous when wet but adhering tenaciously after drying.

Native to southeastern United States, the West Indies, and South America; also a weed and naturalized in warm regions of the world. In the flora area known from a single collection. Also in mostly disturbed habitats elsewhere in Sonora, Arizona, Baja California, and California. The plants might be confused with *E. abramsiana,* from which they can be distinguished by the usually larger, broader leaves, pubescent capsule margins, and smaller seeds with sharper ridges. *E. prostrata* has been incorrectly called *E. chamaesyce* Linnaeus, which is an Old World plant.

Pinacate Junction, *F 86-343B.*

***Euphorbia serpens** Kunth [*Chamaesyce serpens* (Kunth) Small] Resembling *E. albomarginata* but differing by the strictly annual habit, somewhat thicker and more glaucous leaves, slightly smaller cyathia (less than 1 mm wide), narrower appendages, smaller capsules, fewer stamens per cyathium (fewer than 10 verses 15–30), and slightly smaller and thicker seeds; mucilaginous when wet (specimens from southern Sonora).

A cosmopolitan weed. In the New World from Illinois to Colorado and southward to South America. Almost certainly not native to the flora area. The only record for northwestern Sonora is a specimen collected by Edward Palmer (*954 in 1889,* F) at Colonia Lerdo.

Euphorbia setiloba Engelmann ex Torrey [*Chamaesyce setiloba* (Engelmann ex Torrey) Millspaugh ex Parish] *GOLONDRINA;* FRINGED SPURGE. Non-seasonal ephemerals; densely glandular pubescent throughout including capsules. Herbage reddish in drier or cooler seasons, green to yellow-green in hot seasons. Stems slender and wirelike, with age often nearly black and usually prostrate. Leaves petioled, the blades 1.1–6.5 mm, broadly ovate to elliptic or oblong; stipules minute conical "spurs" to

0.2 mm or often absent. Cyathia 0.8–1.0 × 0.5–0.8 mm, urn-shaped (see *E. arizonica*), usually reddish; glands 0.2–0.3 mm wide, the appendages white, fading pink, the margins deeply divided into elongated, pointed segments, the cyathium looking like a tiny star-shaped flower. Seeds 0.8–0.9 mm, with a few low furrows or sometimes smooth, the dorsal side high; not conspicuously mucilaginous when wet but adhering tenaciously upon drying.

Growing with both spring and summer rains, but the plants killed or damaged by even the slightest frost. Widespread to peak elevations; Sonoyta region, Pinacate volcanic complex, granitic ranges, and plains, absent from dunes. Sometimes on rocky slopes, but most often in sandy, gravelly, silty, or rocky soils of washes and arroyos, and on cinder flats and slopes. Southeastern California, southern Nevada, and southwestern Utah to western Texas and southward to Baja California Sur and Sinaloa.

Sierra del Rosario, *F 75-6*. Cerro Colorado, *F 10786*. Sykes Crater, *F 19997*. Arroyo Tule, *F 18757*. Moon Crater, *F 18631*. El Papalote, *F 86-126*.

Euphorbia trachysperma Engelmann [*Chamaesyce trachysperma* (Engelmann) Millspaugh] Summer ephemerals, glabrous, 30–50+ cm tall, usually taller than wide, the stems relatively stout, the branches mostly ascending and straight. Longer internodes 3.5–6 cm. Leaves petioled; larger leaves with blades 3–5 cm, narrowly lanceolate, the margins minutely toothed (serrulate) at least near the apex. Cyathia 1.5 mm wide, the glands green, round, 0.3–0.4 mm in diameter, the appendages pinkish, larger than the glands. Seeds 2.1–2.6 mm, chunky, gray to brown, with a white waxy surface when fresh, strongly 4-angled (quadrangular in cross section) with flattish faces and sharply angled ridges, the ridges slightly irregularly ragged; copiously mucilaginous when wet.

Abundant during the hot, humid, summer rainy season in playas and low-lying flats subject to temporary flooding, such as Playa Díaz and adjacent alluvial flats. Also in coastal, low-lying savanna-like habitats along the coast of the southern half of western Sonora, from the vicinity of Punta Baja at least as far south as the Río Yaqui region. Southern Arizona to southern Sonora, and probably northwestern Sinaloa.

Pinacate Junction, *F 86-340*. 2 mi S of Tinajas de los Pápagos, *F 86-491*.

JATROPHA

Shrubs or trees, sometimes herbaceous perennials with tuberous roots; sap clear or yellow to red but not milky. Stems usually flexible. Leaves alternate, the stipules often well developed, sometimes modified as glands. Monoecious or sometimes dioecious. Inflorescences cymose, or the female flowers solitary. Sepals 5, petals 5. Stamens in 2 whorls, the filaments united below into a column. Fruit a capsule, usually 3-seeded, the seeds with a caruncle. Ours are monoecious, the stipules are greatly reduced or absent, and the capsules are 1- or 2-seeded by abortion of 1 or 2 carpels or ovules.

Mostly tropical and subtropical, many in semiarid regions and some in deserts, worldwide, mostly tropical Americas and many in Africa; 175 species. References: Dehgan & Grady 1979; Dehgan & Webster 1978; McVaugh 1945.

1. Petioles about as long as the blades; blades kidney-shaped, as broad or broader than long. ____ **J. cinerea**

1′ Short-shoot leaves sessile or subsessile, long-shoot leaves with petioles less than half as long as blades; blades spatulate, about twice as long as wide. ____ **J. cuneata**

Jatropha cinerea (Ortega) Müller Argoviensis [*Mozinna cinerea* Ortega. *M. canescens* Bentham. *Jatropha canescens* (Bentham) Müller Argoviensis] *SANGRENGADO;* ASHY LIMBERBUSH; *KOMAGĬ WA:S*. Multiple-stemmed shrubs 1.0–1.8 m; stems rather thick and semisucculent. Young herbage, inflorescence branches, calyces, and corolla buds moderately to densely pubescent with short white-woolly hairs. Leaves quickly drought deciduous, the petioles 2.5–8.0 cm, the blades 3–9 × 4.0–11.5 cm, more or less kidney-shaped, the upper surface greener and with sparser hairs than the often densely pubescent, gray-green lower surface, the margins with few, small glands; long-shoot leaves on widely spaced nodes, usually larger than the short-shoot leaves, often shallowly 3-lobed, the lobes often with a few shallow teeth. Stipules absent (I have not found stipules on any Sonoran specimen, although other authors state they are present).

Leaves produced after rains during warm times of the year; long shoots produced only with summer-fall rains. Male flowers in compact, many-flowered clusters on slender branches, the corollas pinkish white or rose colored, fading to white. Female flowers solitary, the sepals 4–6 mm, linear to oblong, densely pubescent outside, glandular inside, the corollas dark pink inside. Fruits often 2-lobed, 2-seeded (presumably 1 carpel aborts); seeds rounded, ca. 1 cm in diameter. Flowering in summer–early fall following rains, the fruit ripening in the same season.

Low hills, bajadas, and sandy plains in the Sonoyta region, and barely entering Arizona in Organ Pipe Monument; often along the margins of small washes. Also higher elevations in the Sierra Pinacate, with several hundred shrubs on the steep, unstable slope of the Pinacate Peak cinder cone, and occasional rare and widely scattered plants on other high-elevation cinder slopes. These high-elevation plants are mostly 0.6–1.5 m and show signs of repeatedly freezing to the ground. Southward and eastward in Sonora to Sinaloa, and both Baja California states.

The *J. cinerea* complex of northwestern Mexico includes several allopatric taxa in subsection *Canescentes* of section *Loureira* in the subgenus *Curcas*. McVaugh (1945) treated *J. cinerea* and *J. canescens* as conspecific, whereas Dehgan & Webster (1978) regarded them as closely related but distinct species. Following their system, ours would be *J. canescens* "a disjunct relictual species . . . occurring as isolated populations in northern parts of the Sonora Desert and in Magdalena Island" (Dehgan & Webster 1978:32). However, these plants are neither "disjunct" nor "relictual" but continuous in distribution from the flora area to Sinaloa. The leaves of *J. canescens* are densely pubescent on both surfaces whereas leaves of *J. cinerea* are pubescent only on the lower surface (see Dehgan & Webster [1978] for additional key characters). Further investigation is needed to distinguish Dehgan & Webster's (1978:32) "introgressive hybridization between the two taxa" from clinal variation. Southward, in Baja California Sur as well as in Sonora, beyond the desert and into higher elevations, the shrubby *J. cinerea* becomes larger and more arborescent and grades into several other taxa (Steinmann & Felger 1997).

11.8 mi SW of Sonoyta, *F 9806*. 5 km W of Sonoyta, *F 85-956*. NE of Pinacate Peak, 925 m, *F 86-422*. Quitobaquito, *Schott Aug 1855* (cited by Torrey 1858:198; a specimen at F, cited by McVaugh 1945:289, not seen).

Jatropha cuneata Wiggins & Rollins. *SANGRENGADO, MATACORA;* LIMBERBUSH; *WA:S.* Multiple-stemmed shrubs often 1.5–2.5 m. Roots thick, almost tuberous. Stems and roots oozing copious blood-like sap when cut (hence the local common name, a corruption of *sangre de drago* 'dragon's blood'). Stems thick and semisucculent, beset with knobby short shoots. Glabrous or glabrate, or young stems and leaves with short white hairs. Short-shoot leaves 4–21 (41) × 2.3–9.0 (11) mm, sessile or short petioled, spatulate-cuneate, the margins entire except the apex often shallowly notched; short-shoot leaves appearing after rains at almost any time of year except during the coldest periods, and quickly drought deciduous. Seedlings and long shoots growing with hot-weather rains; long-shoot leaves often 20–56 × 14–26 mm, usually 3- (5)-lobed or cleft, petioled, and quickly drought deciduous. Stipules absent (or difficult to find) or minute, reddish, spur-shaped glands.

Corollas white to pink. Male flowers several, in subsessile to short inflorescences usually shorter than the leaves, the corollas white. Female flowers solitary or in pairs, the fruiting sepals 3–6 mm. Fruits rounded, 1-seeded, the seed rounded, 9–10 mm with a minute white-waxy caruncle. Flowering with rains, July-August.

Common on rocky slopes and upper bajadas nearly throughout the flora area; Sonoyta region, Pinacate volcanic complex, and granitic ranges. Mostly at lower elevations, extending to ca. 800 m on some south-facing slopes in the Sierra Pinacate. The plants are frost sensitive and often show freeze damage.

Endemic to the Sonoran Desert region surrounding the Gulf of California: southwestern Arizona to northwestern Sinaloa, most of the Baja California Peninsula, and islands in the Gulf of California.

Wiggins and Rollins (1943) segregated *J. cuneata* from the seemingly closely related *J. dioica* Cervantes (*J. spathulata* (Ortega) Müller Argoviensis) of the Chihuahuan Desert. *J. dioica* ranges from central Texas to Oaxaca. As with so many other vicariad species of similar distributions (west of the continental divide in northern Mexico versus east of the divide), *J. cuneata* is highly sensitive to freezing temperatures whereas *J. dioica* is relatively much more frost hardy. When cultivated in Tucson, *J. cuneata* perishes during usual winter freezing weather, whereas *J. dioica* is generally winter-hardy. There is little doubt that freezing weather is the major factor limiting the northern distribution of *J. cuneata*. At the higher elevations in the Pinacate mountains the shrubs are relatively dwarfed and show signs of repeatedly freezing to the ground with subsequent recovery. Dehgan & Webster (1978) placed these 2 species, plus *J. cardiophylla* of Arizona and Sonora, in section *Mozinna* of the subgenus *Curcas*. Splints prepared from the stems of *J. cuneata* were used extensively for all-important utilitarian baskets (Felger & Moser 1985; Nabhan et al. 1982).

Sykes Crater, *M. Burgess 14 Feb 1982.* S of Pinacate Peak, 875 m, *F 19352.* 3 mi S from boundary monument 180, *Kamb 1995.* MacDougal Crater, *Kamb 1999.* 11 mi SW of Sonoyta, *F 9816.* 45 mi W of Sonoyta, *Turner 59-4.* Sierra del Viejo, *F 16912.*

SEBASTIANIA

Mostly trees and shrubs, the sap milky. Leaves usually alternate, simple, petioled, the margins usually glandular-toothed, the stipules small. Mostly monoecious, rarely dioecious. Inflorescences spikelike, with female flowers at the base. Sepals present; petals absent. Male flowers with 2 stamens. Female flowers with a 2- or 3-chambered ovary, the fruit a 2- or 3-seeded capsule. Seeds usually with a minute caruncle.

Warmer parts of the world; 100 species. Seeds of the Chinese tallow or vegetable-tallow tree, *S. sebiferum,* are used for oil. Several South American species yield rubber.

Sebastiania bilocularis S. Watson [*Sapium biloculare* (S. Watson) Pax] *HIERBA DE LA FLECHA; 'INA HITÁ*. Multiple-stemmed shrubs, often 2.0–2.5 (3) m. Sap thick and apparently caustic. Leaves (1.8) 3–6 (11) cm × 6–15 (25) mm, subsessile to short petioled, mostly lanceolate to elliptic, sometimes narrowly so, shiny green and glabrous, often turning reddish, the margins with minute gland-tipped teeth and sometimes also with a few much larger, scattered, reddish brown glands; often with a pair of (or 1) rounded yellowish to reddish glands, 0.2–0.5 (0.9) mm wide, at or near the base of the blade; stipules scalelike, red-brown, fringed, ca. 1.5 mm. Inflorescences bright yellow to reddish, 2–4.5 cm. Seeds rounded, ca. 6.5 mm. Various seasons.

The plants are frost sensitive; sometimes the young growth and occasionally larger stems are freeze killed. The local distribution seems to be influenced primarily by freezing weather and soil moisture. Fairly common in much of the Sonoyta region and rare and localized at the northeastern edge of the Pinacate volcanic complex. Upper bajadas and floodplains, often along margins of washes, especially on gravelly soil. Southwestern Arizona, Sonora southward to the Guaymas region and the Río Yaqui drainage southeast of Hermosillo, and both Baja California states.

El Papalote, *F 86-333.* 35 mi W of Sonoyta, *Webster 19700.* 13.9 mi SW of Sonoyta, *Reichenbacher 904.* Observation: E of Crater Salvatierra, Oct 1984, *Dan Lynch* (photo).

STILLINGIA

Mostly herbaceous perennials, rarely annuals, sometimes shrubs or small trees. Sap milky but thin and watery (ours) or clear. Glabrous; leaves alternate or sometimes opposite or whorled above, mostly with glandular stipules. Monoecious. Flowers in terminal or axillary spikelike racemes, with female flowers at the base, the flowers subtended by bracts. Petals absent. Stamens 2, united at base. Styles 3, simple (entire). Fruit a 2- or 3-seeded capsule. Seeds with the caruncle present (ours) or absent.

Mostly tropical and dry regions of the Americas, several in Oceania; 30 species. Ours lack the paired leaf glands characteristic of most of the genus. Reference: Rogers 1951.

1. Perennials flowering in the first season; leaves alternate, linear, the margins entire or minutely toothed near tip. ______ **S. linearifolia**

1′ Spring ephemerals; leaves alternate below, opposite to whorled above, ovate, margins spinose-toothed. ______ **S. spinulosa**

Stillingia linearifolia S. Watson. Short-lived perennials or less often facultative ephemerals or annuals. Usually with many broomlike mostly erect, slender stems arising from near the base, forming dense bushes usually with sparse foliage, (15) 30–75+ cm. Leaves 1.5–5 (7.0) cm × (0.6) 0.9–2.8 mm, alternate, linear, entire to minutely toothed at apex; stipules 0.4 mm, represented by stalked glands, the stalk purple-red, the gland yellow-white. Spikes slender, 2–7 cm; female flowers red to green, subtended by fleshy, maroon, cuplike glands, the fruit red to green; portions of spikes with male flowers often reddish, the anthers maroon, the pollen bright yellow. Capsules 3-seeded, the seeds 2.6 mm, rounded-ovoid, light tan, smooth. Various seasons, most spring.

Sand flats, low dunes, and less commonly in interdune troughs and washes, occasionally on high dunes. Widespread and common at low elevations in sandy soils through most of the flora area; not in the Sonoyta region. Eastward to the vicinity of Quitovac and southward along the coast of Sonora to the vicinity of Puerto Libertad, southeastern California, Baja California, northern Baja California Sur, western Arizona, and southern Nevada.

30 mi SW of Sonoyta, *Shreve 7591.* Estero Morúa, *Webster 22381.* Gustavo Sotelo, *Ezcurra 14 Ap 1981* (MEXU). Sierra del Rosario, *F 20698.* Moon Crater, *F 10580.*

Stillingia spinulosa Torrey. Spring ephemerals, usually with stout branches 2.5–18 cm, arising from near the base; taproot well developed, often stout. Herbage sticky-viscid, sand often adhering to the lower stems and leaves. Leaves alternate below and opposite or whorled above, relatively thick and firm, the blades 2–3 cm × 2.7–11.0 mm, ovate to obovate, green with conspicuous veins, the apex acute to acuminate, the margins spinose-toothed; stipules absent (reported as glandular but not seen in ours). Seeds 3.1–3.6 mm, oblong, mottled becoming whitish with age, minutely wrinkled and ridged.

Sandy soils, often on rolling sand plains with *Larrea.* Common in the open sandy desert west of the Pinacate volcanic complex. Not known elsewhere in Sonora. Also southwestern Arizona, southeastern California, and northeastern Baja California.

30 mi E of San Luis, *F 5764.* 10 mi N of El Golfo, *F 75-73.* NE of Sierra del Rosario, *F 20357.*

TRAGIA

Perennial herbs, shrubs, or vines often with hairs stinging like nettles. Leaves alternate. Monoecious. Petals none. Seeds without a caruncle. Worldwide, mostly tropical and subtropical; 150 species. References: Miller & Webster 1967; Urtecho 1996.

Tragia nepetifolia Cavanilles var. **dissecta** Müller Argoviensis. *ORTIGUILLA;* NOSEBURN. Small perennial vines with slender, twining stems. Leaves gradually drought deciduous, petioled, the blades often 1.0–3.5+ cm, thin, mostly triangular-ovate or triangular-lanceolate to broadly lanceolate or ovate, the margins coarsely toothed. Inflorescences axillary, 1.5–4.5 cm including the peduncle. Seeds 2.5 mm, nearly globose with 2 small, flat “faces” near base, shining coppery yellow with a minute reticulate pattern.

Entering the flora region in the Sierra Cipriano (*F 91-30*), mostly in canyon bottoms, often shaded beneath large boulders and shrubs; common in the Ajo Mountains. *Tragia* is the only plant in northwestern Sonora with nettle-like stinging hairs. The irritation is short lasting.

According to Urtecho (1996), var. *dissecta* differs from the other 3 varieties of *T. nepetifolia* in its deeply toothed leaves and red male flowers. Var. *dissecta* is known from southeastern Arizona, eastern Sonora, and Chihuahua to Zacatecas and San Luis Potosí. The plants in northwestern Sonora and southwestern Arizona may belong to a different taxon.

FABACEAE (LEGUMINOSAE) LEGUME FAMILY

Annual or perennial herbs, shrubs, trees, or vines, many with nitrogen-fixing bacteria in root nodules. Leaves and branches usually alternate. Leaves mostly once or twice pinnately compound, often with pulvini at base of petioles, pinnae, and individual leaflets. (Pulvini are swollen areas, often dark or differently colored, that swell or contract to open or close leaf parts, often in response to water stress and night-day cycles; pulvini may sometimes be present but not functional.) Stipules usually present, often well developed and persistent. Flowers usually bisexual. Sepals usually 5 and united. Petals mostly 5, often separate. Stamens 5 or 10 to numerous. Pistil single, superior, simple (1 carpel), the ovules attached along one side. Fruit usually a pod, very diverse, dry and dehiscent (opening) along one or both edges, or indehiscent and often with mealy or fleshy pulp. Seeds 1 to many, often very long-lived, the seed coat often hard and impervious to water and air (scarification, e.g., nicking or abrading the seed coat, often enhances germination).

Worldwide with 3 subfamilies; 650 genera, 16,500 species. Nearly 1/3 of the species fall into 6 genera: *Acacia, Astragalus, Crotalaria, Indigofera, Mimosa,* and *Senna.* All 6 occur in the Sonoran Desert and 3 (*Acacia, Astragalus,* and *Senna*) are in the flora area. The 34 species in the flora are in 17 genera listed below by subfamily. Three species are non-native. The number of species reported for the flora area (for genera with more than one species) is in parentheses.

Subfamily **Caesalpinioideae.** Trees and shrubs, sometimes vines or herbs; seldom nitrogen-fixing. Leaves usually once or twice pinnate. Flowers bilaterally symmetrical but often not conspicuously so, often large and showy. Stamens usually 10 or fewer, mostly separate. Largely tropical and subtropical: *Hoffmannseggia* (2), *Parkinsonia* (3), *Senna.*

Subfamily **Mimosoideae.** Largely trees and shrubs, mostly nitrogen-fixing. Cyanogenic glycosides present in many taxa. Leaves usually twice pinnate; leafstalks often with crateriform or cup-shaped nectary glands, the center often raised and nectar filled when fresh, later becoming sunken. Flowers radial except the pistil, mostly small, sessile or nearly so, and close together in spikes, heads, and racemes. Stamens usually 10 or numerous, usually much longer than the petals and forming the showy part of the flower, the anthers small and often with a deciduous gland at the tip of the tissue connecting the 2 anther cells. Seeds often with a horseshoe-shaped groove on each side or face. Mostly tropical and subtropical: *Acacia* (2), *Calliandra, Prosopis* (3).

Subfamily **Papilionoideae.** Annual to perennial herbs, vines, shrubs, or trees, almost all nitrogen-fixing. Leaves pinnately or digitately once-compound or simple. Flowers bilaterally symmetrical. Upper petal (banner or standard) flanked by 2 lateral petals (wings), the 2 lower petals coalesced into a boat-shaped structure (keel) enclosing the stamens and style. Stamens mostly 10, monodelphous, or diadelphous (9 filaments united into an open sheath surrounding the ovary with the 10th filament partially or wholly separate). Tropical to temperate: *Astragalus* (6), *Dalea* (2), *Lotus* (3), *Lupinus* (2), *Marina, *Melilotus, Olneya, *Pediomelum, Phaseolus, Psorothamnus* (3), *Sesbania.*

Legumes form important components of the local desert vegetation. Many are nitrogen-fixing and provide major nitrogen (protein) sources for the ecosystem. The most abundant and important legumes in the vegetation of northwestern Sonora are *Parkinsonia* spp., *Dalea mollis, Lupinus arizonicus, Olneya tesota, Prosopis glandulosa,* and *Psorothamnus* spp. The larger Sonoran Desert legumes, such as the common trio of *Olneya, Parkinsonia* and *Prosopis,* are important nurse plants for an array of herbs and larger perennials including columnar cacti. Destruction of the large desert legumes is seriously damaging local desert ecosystems.

The Sonoran Desert flora includes 281 legume species, or about 11% of the total flora. Legume species make up only 7% of the flora of northwestern Sonora. The flora of Organ Pipe Monument contains 36 legume species (Felger et al. 1997), representing 8% of the total flora. However, approximately 37% or 13 of the legume species in Organ Pipe do not occur in the flora area, and a similar percentage of ours do not occur in Organ Pipe. Most of the Organ Pipe legumes not found in northwestern Sonora are locally restricted to the Ajo Mountains, an area of higher and more predictable annual rainfall. Most of the legumes of northwest Sonora that do not occur in Organ Pipe either are "Californian" elements characteristic of dune or sand habitats or occur near the Río Colorado.

Legumes rival the grass family in economic importance, providing food, dye, fiber, insecticides, medicine, and timber. Major legume food and fodder crops include alfalfa, beans, clover, peanuts, peas, and soybeans, all of which are papilionoids. Five legumes in the flora area featured importantly in the economies of local Sonoran Desert peoples: desert bean (*Phaseolus filiformis*), ironwood (*Olneya*), mesquite (*Prosopis* spp.), palo verde (*Parkinsonia microphylla*), and screwbean (*Prosopis pubescens*) (e.g., Castetter & Bell 1942, 1951; Felger & Moser 1985; Felger et al. 1992; Rea 1997).

Several legume trees cultivated in northwestern Sonora are worthy of mention. *Guaje—Leucaena leucocephala* (Lamarck) de Wit—a mimosoid shrub or small tree, has twice-pinnate green leaves, capitate white to yellow flower heads, and clusters of green to red-brown, flattened, dehiscent pods 15–20 cm long. The green pods are much used in southern Mexico as a condiment. It is cultivated in San Luis, and appears to be spreading along the Colorado River near Yuma.

Guamúchil—Pithecellobium dulce (Roxburgh) Bentham—is widely cultivated in arid to semiarid tropical and subtropical regions around the world. It is a large mimosoid tree with paired spines persisting to form horizontal ringlike ridges on smooth, pale-gray bark. The leaves are formed of a single pair of pinnae, each with 2 large leaflets. It is grown as a shade tree and for the highly prized pods, which produce a pink or white, spongy, sweet edible aril (pulp). The pods ripen in May and June. *Guamúchil* is one of the most commonly cultivated large trees in the lowlands and foothills of Sonora, and is a common street and garden tree in Puerto Peñasco, San Luis, and Sonoyta. It is frost sensitive and for this reason usually not grown in Arizona. The pods are sold in quantity in Sonoran markets. A long bamboo fruit-gathering pole can often be seen leaning against a *guamúchil* tree at harvest time. The pole has a transverse stick tied to it, much like that of a columnar cactus fruit-picking pole. Other commonly cultivated legume trees include *Bauhinia variegata, Delonix regia, Parkinsonia aculeata,* and South American *Prosopis* in section *Algarobia* (Appendix C).

References: Barneby 1989; Burkart 1952; Isely 1998; McVaugh 1987; Polhill & Raven 1981.

1. Trees or woody shrubs, mostly more than 1 m tall (ambiguous cases key out in both places).
    2. Small shrubs to 1.5 m in height; unarmed (no spines or thorns).
        3. Leaves once pinnate, tomentose, densely hairy, grayish, gland-dotted and staining yellow-orange when rubbed or crushed; flowers purplish, papilionoid. ____ **Psorothamnus emoryi**
        3′ Leaves twice pinnate, green and not densely hairy, not gland-dotted, not staining yellow; flowers pinkish or yellow, mimosoid or caesalpinioid.
            4. About as wide as tall, usually less than 0.8 m; much-branched throughout, branches often short and stubby; flowers whitish to pinkish, mimosoid; pods with a prominent cordlike rim, not gland-dotted. ____ **Calliandra**
            4′ Usually taller than wide, commonly 1–1.5 m; larger stems arising from base, slender and rushlike; flowers yellow, caesalpinioid; pods gland-dotted, the rim not cordlike. ____ **Hoffmannseggia microphylla**
    2′ Trees or large shrubs, usually more than 2 m; armed with spines or thorns or the twigs thorn-tipped (individual branches may be unarmed); (*Psorothamnus arborescens* not armed).
        5. Bark green, at least on upper limbs; flowers caesalpinioid. ____ **Parkinsonia**
        5′ Bark not green (except sometimes on first or second year's growth); flowers mimosoid or papilionoid.
            6. Leaves once pinnate, with 1 leaflet, or plants leafless; flowers papilionoid.
                7. Trunk often massive, the bark shredding; branches leafy, the leaves pinnate, with prominent spines at base of at least some leaves. ____ **Olneya**
                7′ Trunk not massive, the bark smooth; leafless or with a few, scattered leaves with 1 leaflet; branchlets sharp-pointed but otherwise unarmed (or unarmed and pinnate in *P. arborescens*, a shrub doubtfully present in the flora area). ____ **Psorothamnus**

6′ Leaves twice pinnate; flowers mimosoid.
8. Spines recurved, usually laterally compressed, between nodes, single (not paired); pods conspicuously flattened (more than 5 times wider than thick). ______ **Acacia greggii**
8′ Spines straight, round in cross section, at nodes, usually paired; pods terete or nearly so (if somewhat flattened, then less than twice as wide as thick).
9. Leaflets 1.5–3 mm; flowers in globose heads, the stamens more than 30; pods dehiscent, straight, 4–6 mm maximum width, constricted between each seed. ______ **Acacia constricta**
9′ Leaflets at least 4.5 mm; flowers in spikes, the stamens 10; pods indehiscent, tightly coiled or straight to moderately curved, 7.5–10 mm maximum width, not constricted between each seed. ______ **Prosopis**

1′ Herbaceous annuals and perennials, not woody.
10. Perennials.
11. Leaves twice pinnate; flowers caesalpinioid. ______ **Hoffmannseggia glauca**
11′ Leaves once pinnate.
12. Pods papery, bladdery-inflated; corollas lavender-pink; flowers papilionoid; (probably annual but may appear perennial). ______ **Astragalus magdalenae**
12′ Pods not papery, not bladdery-inflated; corollas of various colors.
13. Leaves even pinnate with (4) 6 (8 or 10) leaflets; flowers caesalpinioid; anthers large, with terminal pores. ______ **Senna**
13′ Leaves odd pinnate with 3, 5, or 9–23 leaflets; flowers papilionoid; anthers small, opening longitudinally, without pores.
14. Leaflets 9–23; herbage gland-dotted; flowers dark blue; pods 1.8–2.5 mm, indehiscent, 1-seeded. ______ **Marina**
14′ Leaflets 3 or 5; herbage not gland-dotted; flowers yellow or orange; pods more than 8 mm, dehiscent, the valves rolled up after dehiscence, with more than 2 seeds.
15. Stems upright; leaflet apex blunt (truncate to notched); flowers 16–18+ mm, yellow; pods 2.5–4.5 cm. ______ **Lotus rigidus**
15′ Stems spreading to procumbent; leaflet apex pointed (acute); flowers 6–8 mm, salmon orange; pods ca. 0.9–1.2 cm. ______ **Pediomelum**

10′ Ephemerals (annuals).
16. Leaves with 3 leaflets.
17. Stems twining; pubescence of minute, hooked hairs; keel (of corolla) and style bent and twisted. ______ **Phaseolus**
17′ Stems not twining; glabrous or pubescence of straight hairs; keel and style straight.
18. Glabrous; leaflets serrated at least toward apex; flowers many in elongate racemes. ______ **Melilotus**
18′ Pubescent; leaflets entire; flowers several in sub-capitate spikes. ______ **Pediomelum**
16′ Leaves with 5 or more leaflets.
19. Leaves digitately compound. ______ **Lupinus**
19′ Leaves pinnately compound.
20. Herbage conspicuously gland-dotted; pods less than 3 mm, 1-seeded.
21. Flowers subtended by teardrop-shaped, dark reddish brown, persistent glands; corolla about as long to scarcely longer than the calyx; midrib of calyx lobes extending in awnlike plumose bristles. ______ **Dalea**
21′ Flowers not subtended by dark glands; corolla clearly longer than the calyx; calyx lobes not bristle-tipped. ______ **Marina**
20′ Herbage not gland-dotted; pods at least 10 mm, with more than 2 seeds.
22. Leaves 7–23 mm; pods elastically dehiscent, the valves twisting after opening; stipules represented by small but conspicuous dark glands. ______ **Lotus**
22′ Leaves at least (20) 30 mm; pods indehiscent or dehiscent, the valves not twisting after opening; stipules not glandlike.
23. Leaflet pairs 2–11; pods less than 3.5 cm. ______ **Astragalus**
23′ Leaflet pairs 30 or more; pods at least 15 cm. ______ **Sesbania**

Astragalus magdalenae:
(A) var. ***magdalenae;***
(B) var. ***peirsonii.*** AE.

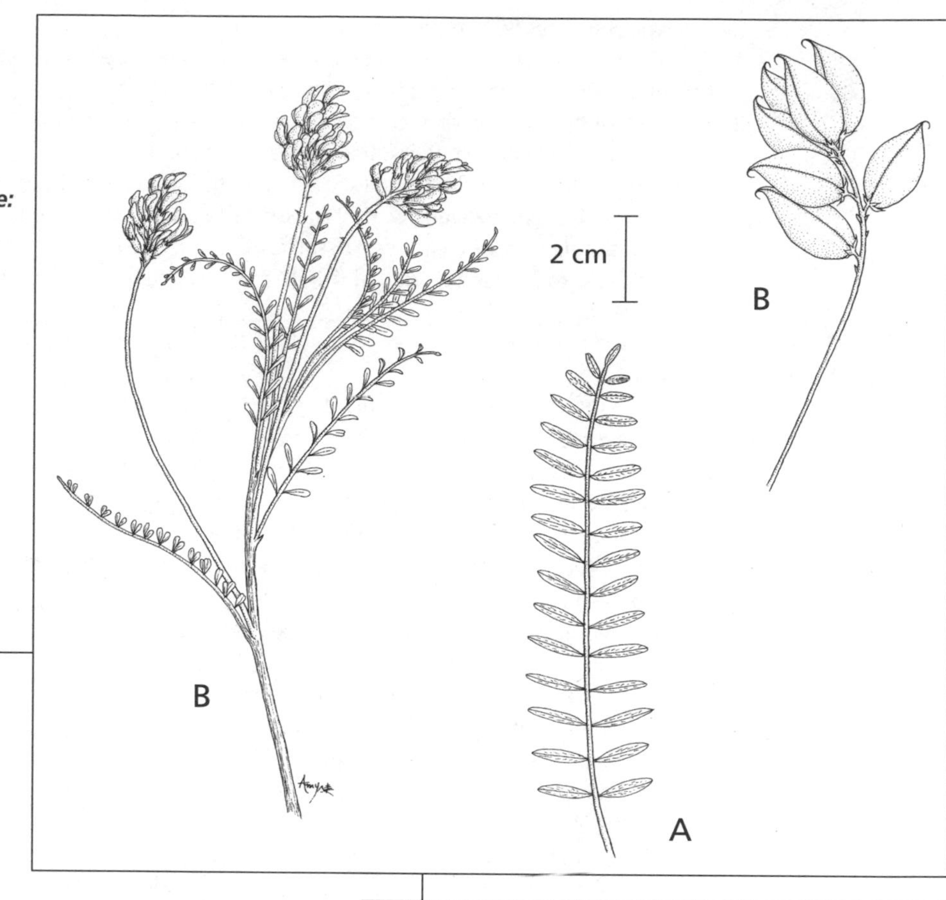

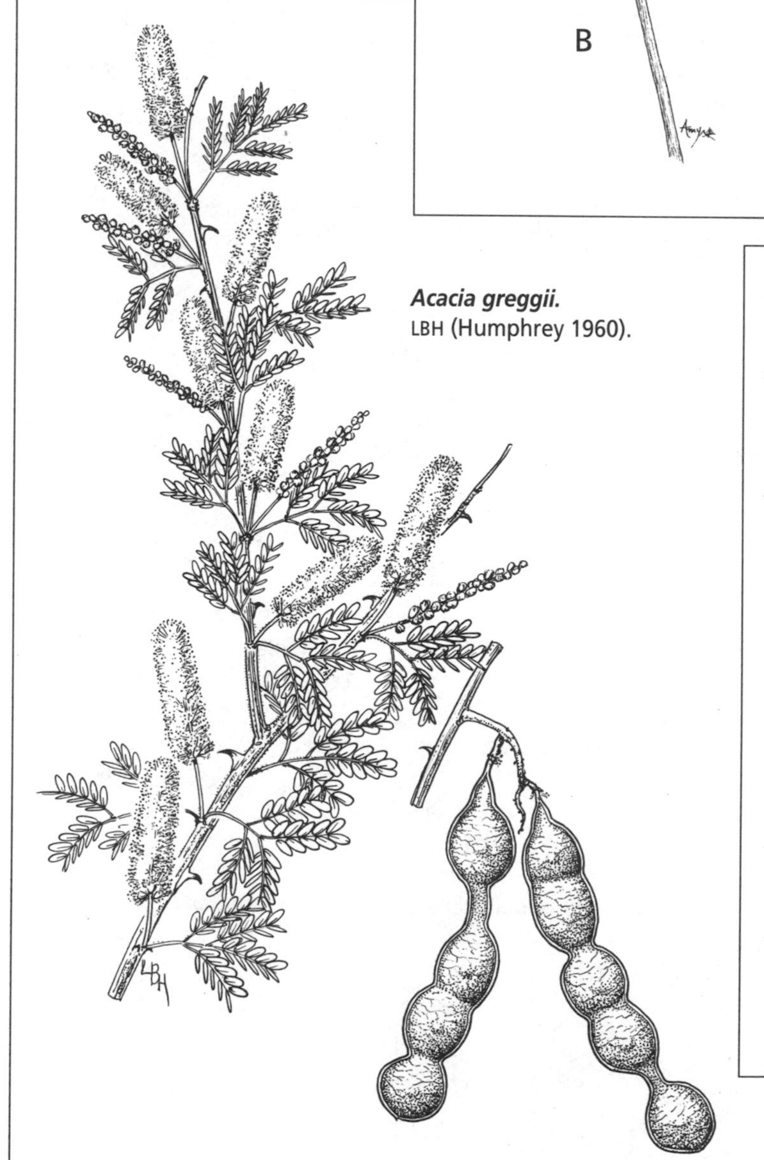

Acacia greggii.
LBH (Humphrey 1960).

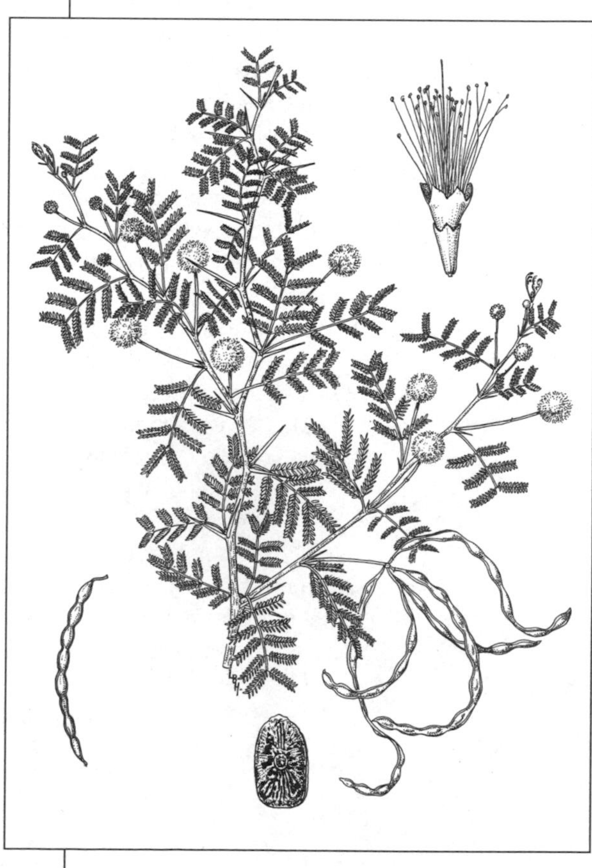

Acacia constricta. LBH (Benson & Darrow 1945; Parker 1958, 1972).

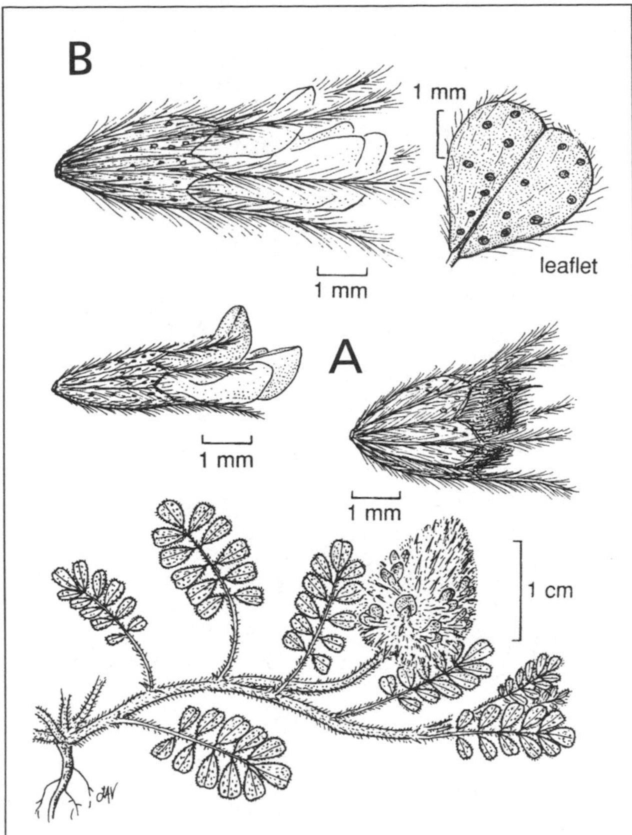

Dalea: (A) D. mollis; (B) D. mollissima.
LAV (Hickman 1993).

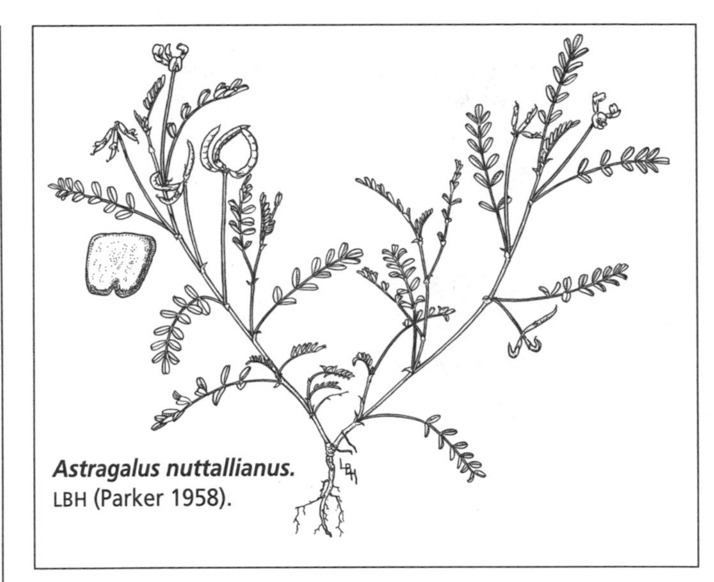

Astragalus nuttallianus.
LBH (Parker 1958).

Calliandra eriophylla var. ***eriophylla.*** LBH (Benson & Darrow 1945).

Lotus:* (A) *L. rigidus;* (B) *L. salsuginosus var. ***brevivexillus;* (C) *L. strigosus*** MBJ.

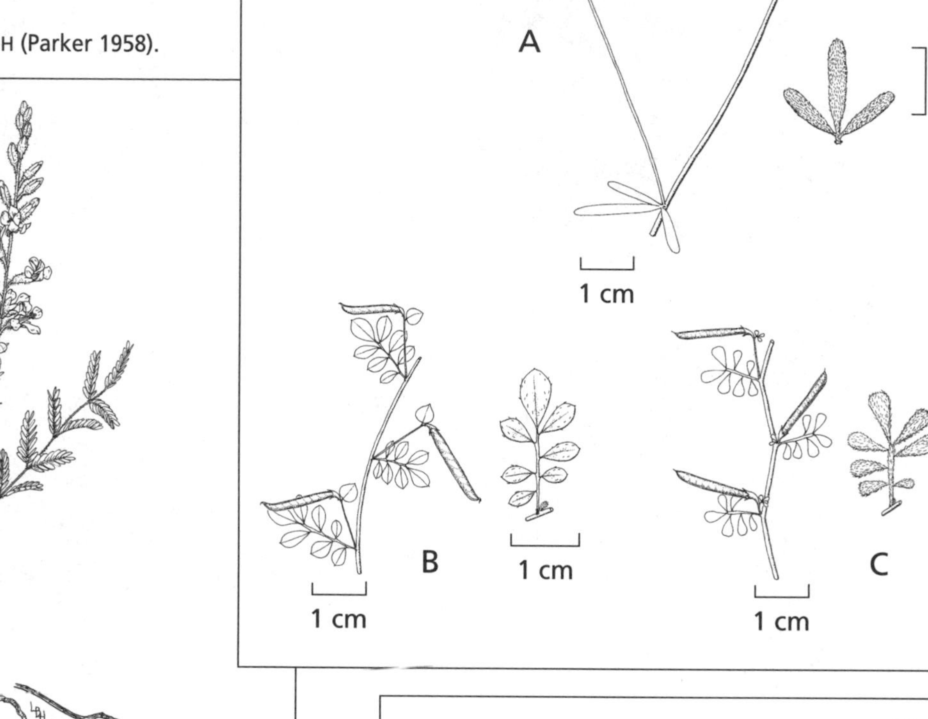

Hoffmannseggia glauca. LBH (Parker 1958).

Olneya tesota. MBJ, cmm, and NLN.

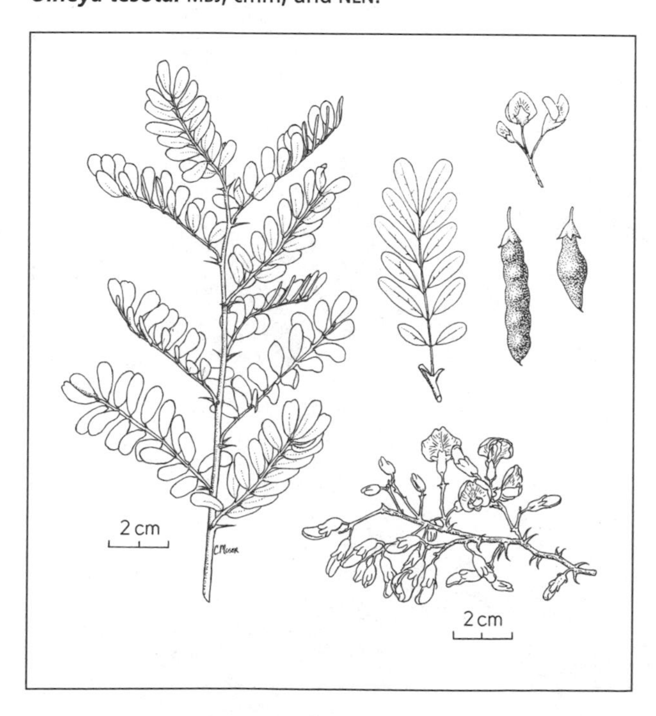

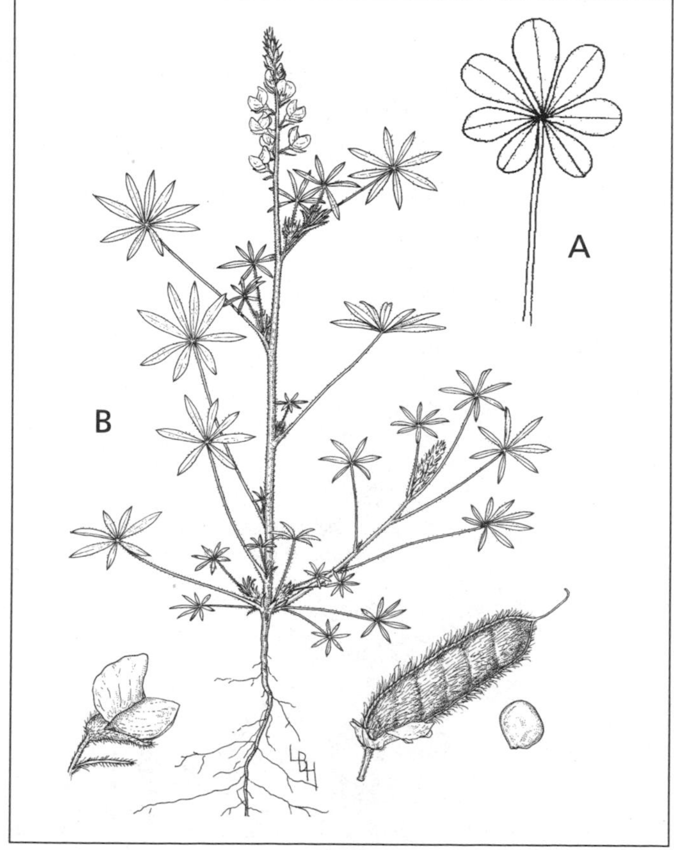

Lupinus:* (A) *L. arizonicus, AE; ***(B) L. sparsiflorus,*** LBH (Schmutz & Hamilton 1979).

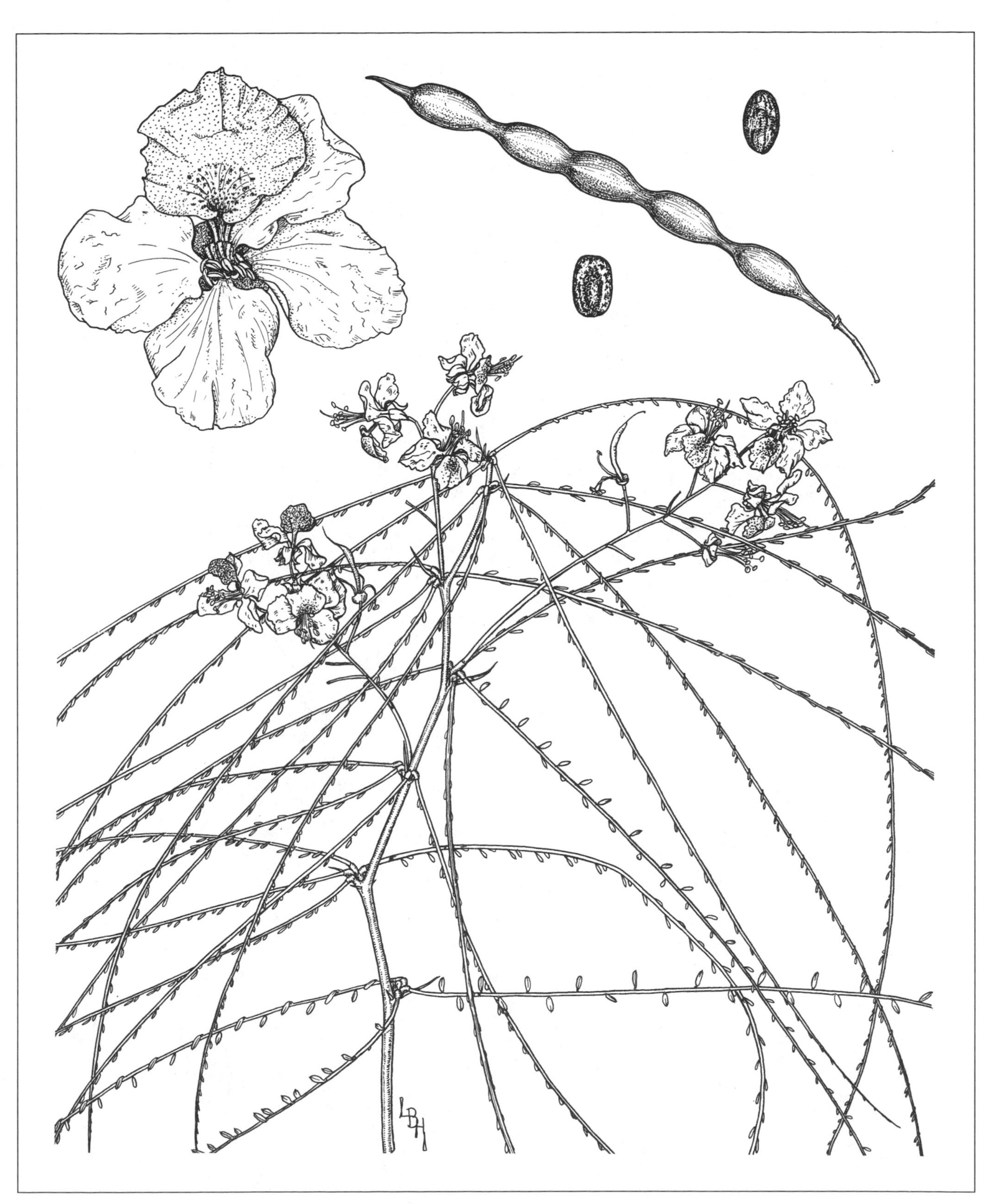

Parkinsonia aculeata. LBH (Benson & Darrow 1945).

Parkinsonia florida. LBH (Benson & Darrow 1945).

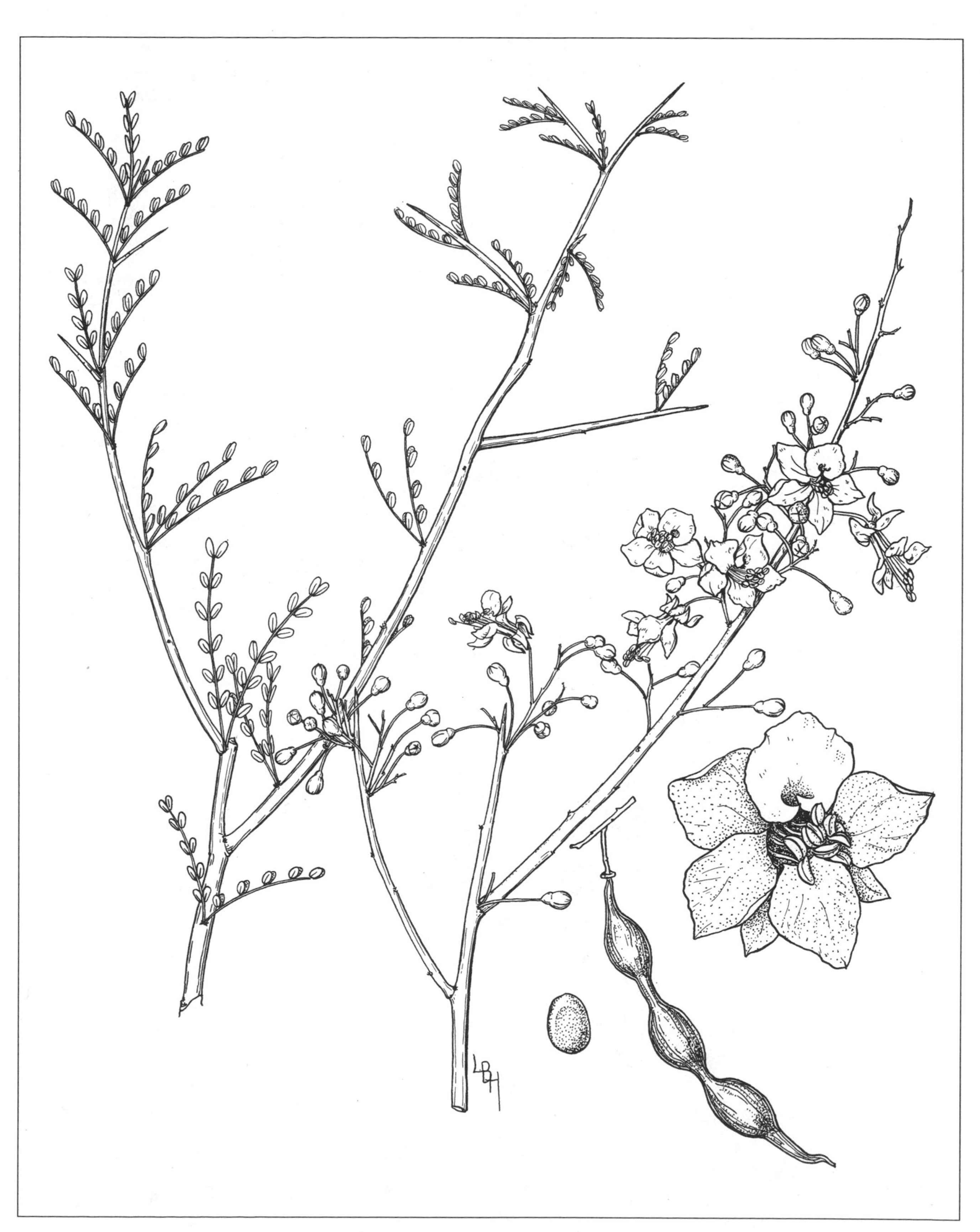

Parkinsonia microphylla. LBH (Benson & Darrow 1945).

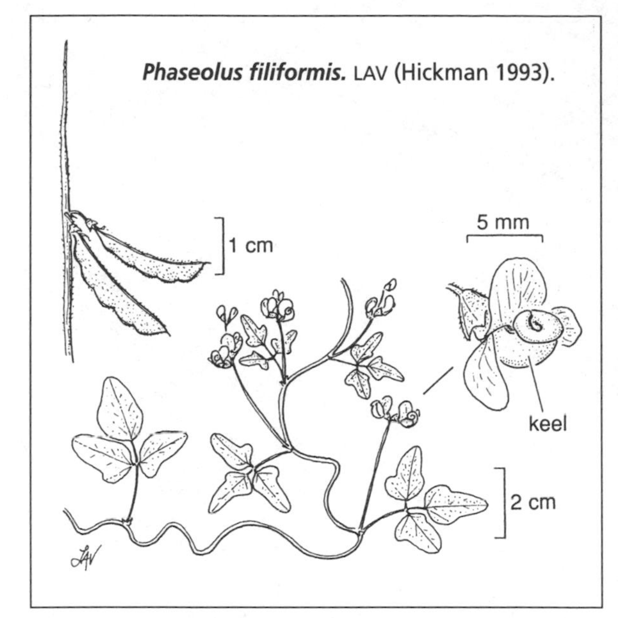

Phaseolus filiformis. LAV (Hickman 1993).

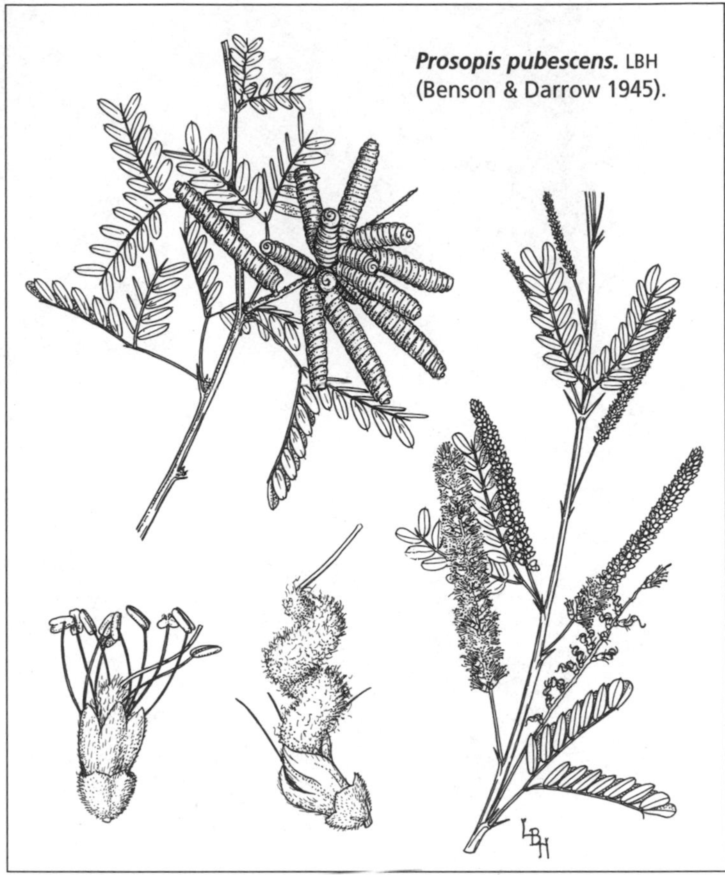

Prosopis pubescens. LBH (Benson & Darrow 1945).

Prosopis: (A) P. glandulosa var. ***torreyana; (B) P. velutina.*** mbj.

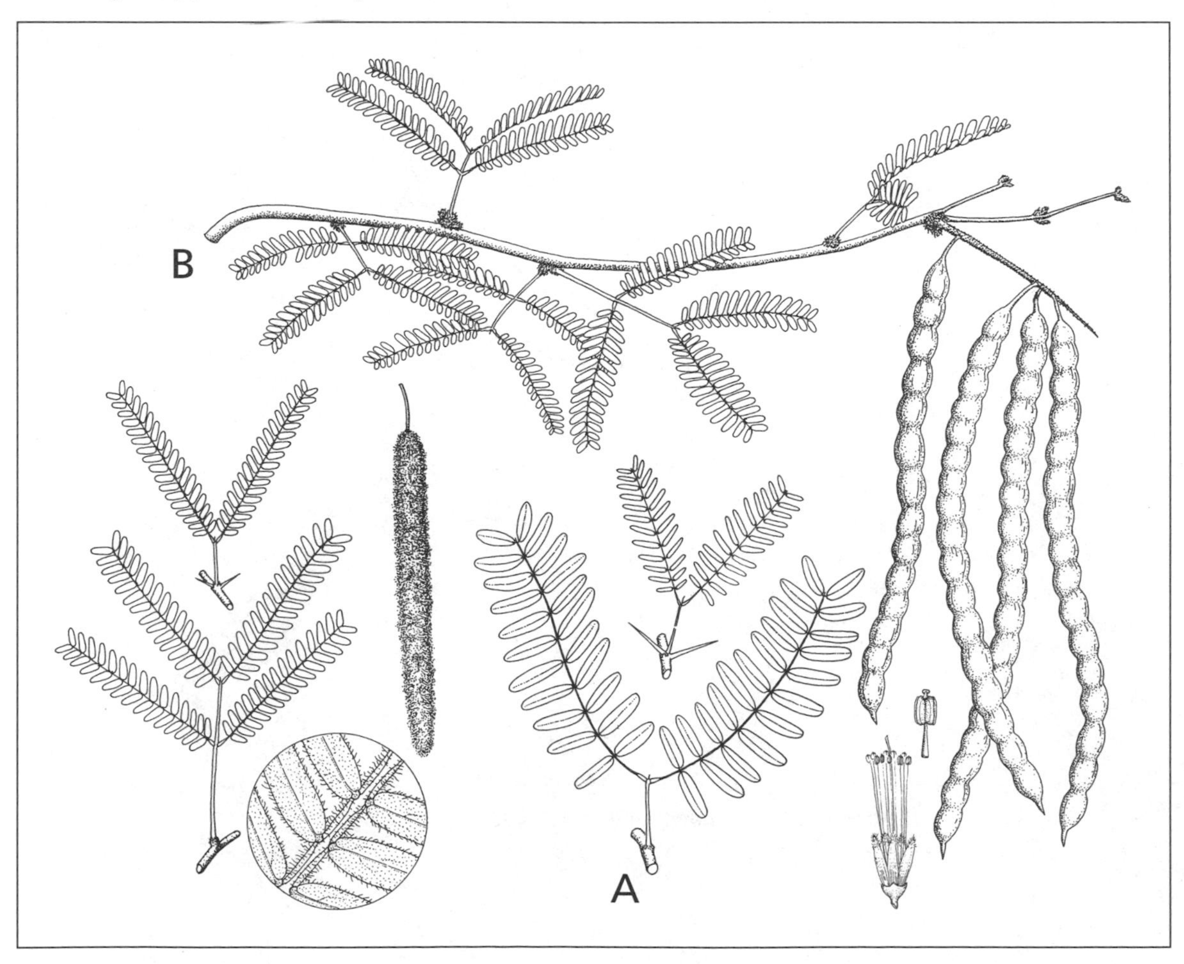

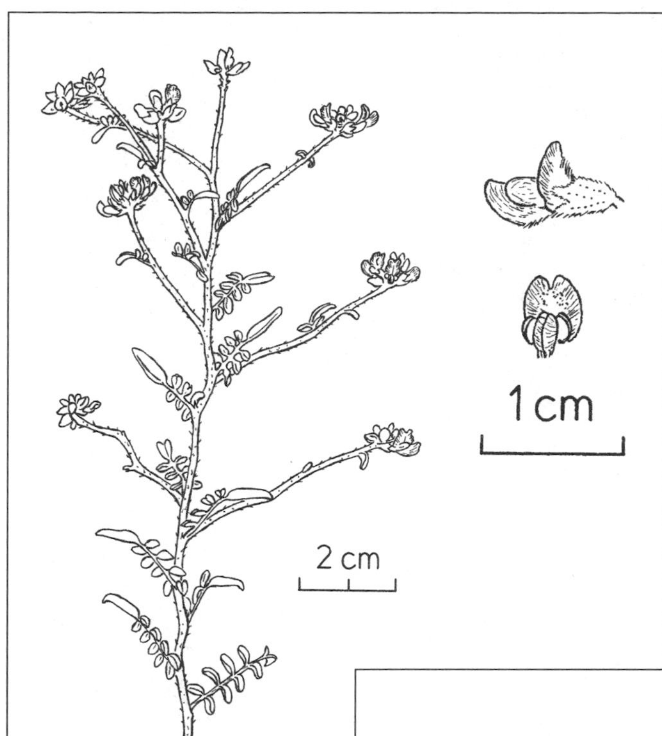

Psorothamnus emoryi
var. ***emoryi.*** NLN.

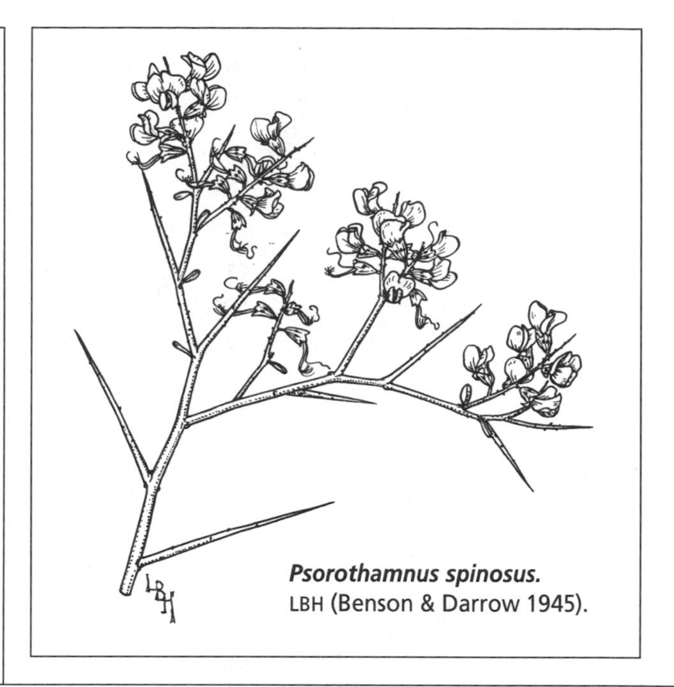

Psorothamnus spinosus.
LBH (Benson & Darrow 1945).

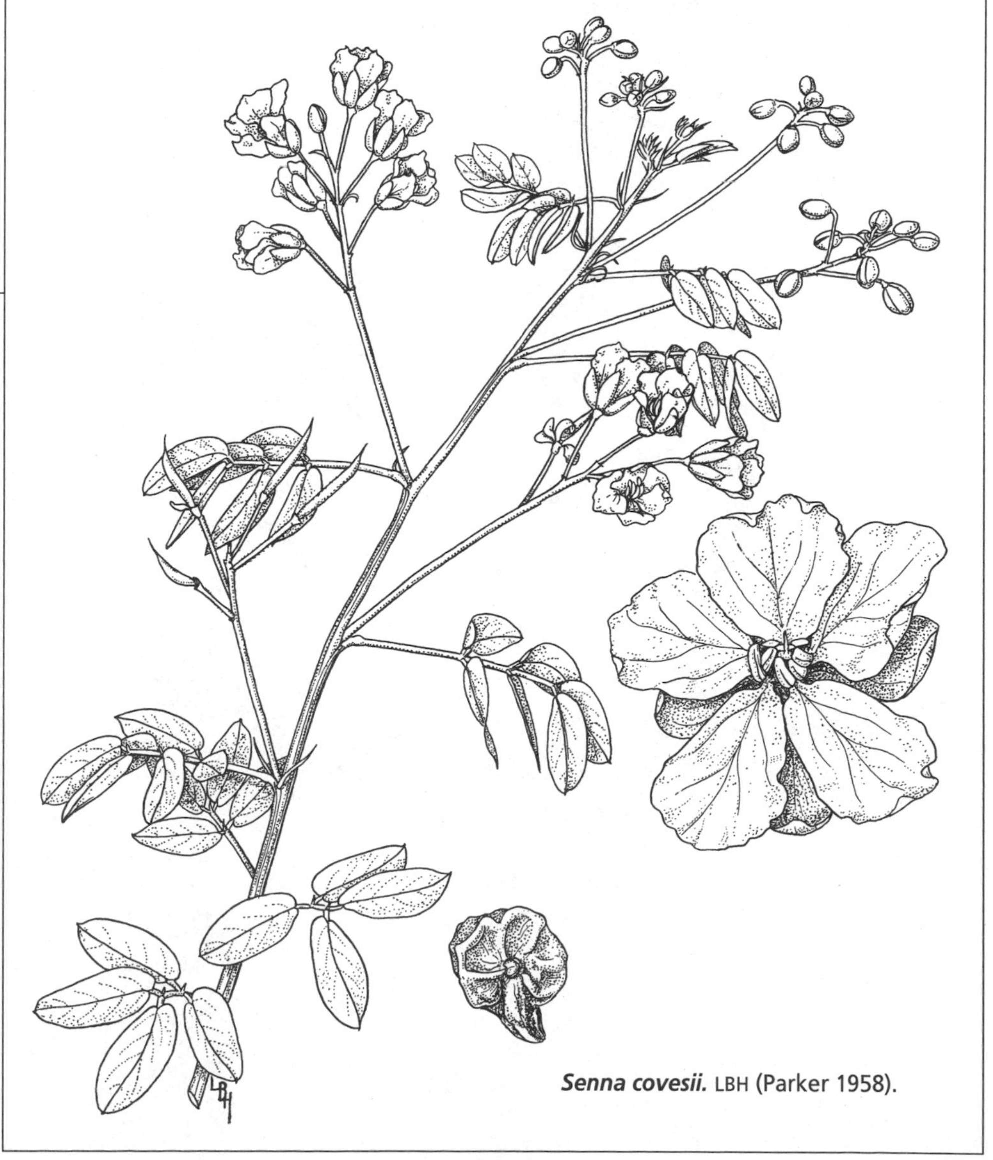

Senna covesii. LBH (Parker 1958).

Sesbania herbacea. LBH (Parker 1972).

Acacia

Mostly trees and shrubs. Often armed with spines or prickles. Leaves twice pinnate (ours) or variously modified; leafstalks usually with nectar-producing glands; leaflets often small and numerous; stipules often spinescent. Flowers mimosoid, usually yellow or white; calyx usually shallowly lobed, the stamens usually numerous and separate, the anthers very small, sometimes with a terminal gland. Pods highly variable, dehiscent or not.

Largely subtropical to tropical, many in arid and semiarid regions; 1200 species, mostly in warmer parts of Australia, Africa, and the Americas. After *Astragalus, Acacia* is the largest legume genus in the world. However, there is a trend to raise the 3 subgenera to generic status. Subgenus *Heterophyllum,* with about 900–1000 species, is almost entirely Australian; many have highly modified and diverse leaves. Subgenus *Acacia,* which includes *A. constricta,* has about 75 species mostly in Africa and the Americas. Subgenus *Aculeiferum,* which includes *A. greggii,* has about 175 species, mostly tropical in Africa and the Americas. References: Isely 1969; Vassal 1981.

1. Spines stipular, paired, straight, and round in cross section; flowers bright yellow, in globose heads; pods not strongly flattened, 4–6 mm wide. _______ **A. constricta**

1′ Spines (prickles) internodal, single, recurved and laterally compressed; flowers dull, pale yellow, in elongated spikelike racemes; pods conspicuously flattened, 10–20 cm wide. _______ **A. greggii**

Acacia constricta Bentham. WHITE-THORN ACACIA; *GIDAG* Shrubs to ca. 3 m, symmetrical with mostly straight branches. Bark smooth, red-brown. Spines stipular, the larger ones 1–3 cm, paired, straight, and terete, usually white. Herbage glandular and viscid, at first sparsely to moderately pubescent with small white hairs, glabrate with age. Leaves winter deciduous and very tardily drought deciduous, 3.4–4.0 cm, the pinnae 3–6 pairs, the leaflets many, 1.5–3 mm; petioles with a prominent nectary gland. Flowers bright yellow, in rounded heads ca. 1 cm in diameter. Pods 4.5–13.5 cm × 4–6 mm maximum width, straight to slightly curved, constricted between each seed, moderately compressed, the fresh pods reddish with glistening viscid glands, gradually dehiscent. Seeds 4.5–5.2 mm, mottled brown, gray, and black, moderately compressed, somewhat oblong. Flowering in spring, the pods ripening in June; also flowering in late summer depending on rainfall and producing pods in fall.

Entering the flora area in Arizona Upland at Sonoyta (SE side of town, *F 89-2*). Low hills and mostly along small washes. Widespread and abundant in Arizona and Sonora, mostly in desert regions. Arizona to Texas and through much of arid and semiarid Mexico but not south of Sonora on the west coast, and disjunct in northern Baja California and southern Baja California Sur. The glandular pods are unusual among the acacias. The flowers of this species are fragrant but produce little nectar and attract few insects (Simpson 1977a).

Acacia greggii A. Gray. *UÑA DE GATO;* CATCLAW; *'U:PAḌ.* Large, irregularly branched shrubs with crooked branches, often 2–5 (6) m; wood strong and hard, the heartwood reddish. At least some branches armed with small, sharp, recurved, usually laterally compressed, solitary prickles scattered along the stems between the nodes. Young twigs, leaves, and inflorescences moderately to densely pubescent with short hairs. Leaves tardily winter deciduous, often (1.5) 2.5–3.5 cm, the pinnae 2 or 3 pairs, the leaflets 3.5–6 (8) mm, 4–6 (7) pairs per pinna; petioles with a prominent nectary gland. Flowers in dense, cylindrical, spikelike racemes, often 1.8–3 cm, the peduncle often 1–2.5+ cm; flowers pale yellowish, at first sessile or nearly so, becoming very short-pedicelled. Pods (3.5) 6–15.5 × 1–1.6 (2+) cm, flattened, ribbonlike, usually curved or curled, sometimes moderately constricted between some seeds, indehiscent or tardily semidehiscent. Seeds oval to orbicular, conspicuously flattened, dark brown, 10.3–12.7 × 3.4–3.9 mm thick. Mostly flowering in spring, the pods ripening in June.

Larger washes and arroyos, less often along smaller drainageways, and occasionally on north-facing mountain slopes. Sonoyta through most of the Pinacate volcanic field nearly to the peak, also the larger granitic ranges including the Sierra del Viejo. Deserts across northern Mexico including Sonora southward to the Guaymas region, and western Texas to southwestern Utah and southeastern California.

Isely's (1969) 3 varieties of *A. greggii* are not entirely satisfactory, and the keys and distributional patterns are confusing. Ours are var. *arizonica* Isely, characterized by densely hairy leaves, but because "the leaflets of part of Gregg's original gathering are puberulent, a taxonomic distinction is difficult to sustain" (Barneby 1989:7).

Tinajas de Emilia, *F 19729*. Cerro Colorado, *Lincoln 17 Jun 1964*.

ASTRAGALUS MILKVETCH, LOCOWEED

Annual to perennial herbs with a well-developed taproot (ours are winter-spring ephemerals); also some shrubs in the Old World. Leaves usually odd pinnate, without pulvini. Inflorescences of axillary racemes or spikes. Flowers papilionoid. Calyx 5-toothed. Stamens diadelphous. Pods highly variable, inflated or not, sessile or the base variously elongated into a stipe or stipelike structure. Seeds 2 to many.

A taxonomically difficult genus of perhaps 2500 species (Sanderson & Wojchiechowski 1996), one of the larger genera in the world and the largest in the family; temperate and cool-temperate parts of the world except Australia and southern Africa (where replaced by closely related genera). Centers of diversity include the former Soviet Union–Iranian region, western North America, and western South America. The genus is largely xerophytic and well represented in North American deserts, especially the Great Basin Desert, but is poorly represented in the southern part of the Sonoran Desert. Only 1 species, *A. nuttallianus,* reaches the southern limits of the Sonoran Desert in Sonora. Barneby (1964:20) pointed out that since about the year 1800 there has been no attempt to prepare a comprehensive phylogenetic treatment of the world's *Astragali* and that it might be difficult to gain the necessary field information because of political dissension. "Seeds of some desert *Astragali* retain their viability for thirty or forty years, perhaps longer, germinating readily if thoroughly soaked" (Barneby 1964:19). The ability of the seeds to remain dormant for many years probably contributes to their success in arid regions.

Many species of *Astragalus* are toxic to livestock and cause erratic behavior, and for this reason are called locoweed. Our milkvetches apparently are not toxic. Several species, especially *A. gummifera,* in the arid or semiarid regions of western Asia yield gum tragacanth. A few are cultivated in temperate regions as garden plants.

The following key should be used together with distributional information. It is difficult to construct a key to cover drought-stunted plants. References: Barneby 1964, 1989.

1. Plants diminutive, the stems very slender (filiform); pods strongly curved, not inflated; racemes 1–3 flowered, the flowers 5–6 mm or less. ____ **A. nuttallianus**

1′ Plants mostly not diminutive, the stems stouter; pods inflated or not, but not strongly curved; racemes usually more than 2- or 3-flowered, usually elongated; flowers mostly more than 5 mm.

  2. Pods incompletely 2-chambered with a well-formed septum; near the Río Colorado in the western margin of the flora area. ____ **A. lentiginosus**

  2′ Pods 1-chambered, without a septum.

    3. Racemes with more than 10 flowers; plants often large and robust. ____ **A. magdalenae**

    3′ Racemes with (2) 3–9 (10) flowers; plants usually smaller than in 3.

      4. Flowers ca. 5 mm, the corollas whitish tinged with lavender-pink, drying straw colored; pods turned upward (ascending); near the Río Colorado in the extreme northwestern part of the flora area. ____ **A. aridus**

      4′ Flowers 6–8 mm, the corollas mostly purplish both fresh and dried; pods spreading to turned down (declined).

        5. Leaves and pods with stiff, thickish, appressed hairs mostly less than 0.6 mm; pods obviously swollen (bladdery), papery, thin and not firm, translucent, purple flecked; widespread in the flora region. ____ **A. insularis**

        5′ Leaves and pods with soft, slender, spreading hairs mostly more than 0.7 mm; pods scarcely swollen, papery but thickish and firm, not translucent, not purple flecked; extreme western part of the flora region. ____ **A. sabulonum**

Astragalus aridus A. Gray PARCHED MILKVETCH. Plants with silver-satiny hairs often 0.7–1.0 mm. Leaves (2) 3–9 cm; leaflets (7) 9–17, mostly oblanceolate to elliptic, 7–13 mm. Racemes (3) 4–9-flowered. Flowers 5+ mm; corollas whitish tinged with lavender-pink, drying straw colored. Pods 10–17 mm, ascending, turgid and scarcely inflated, with (3) 4–6 (7) seeds.

Dunes and open sandy plains. Mainly in the Colorado Desert in southeastern California and adjacent Arizona and Baja California. Barneby (1964:888) maps 2 collections in the northwestern corner of the flora area along the Río Colorado. *A. aridus* is closely related to *A. sabulonum* and they are not always readily distinguishable—the flowers of *A. aridus* are generally slightly larger.

Astragalus insularis Kellogg var. **harwoodii** Munz & McBurney ex Munz. SAND LOCOWEED; *KOPONDAKUḌ*. Quickly developing a deep taproot and often surviving longer into the pre-summer drought than most other spring ephemerals. Herbage green to silvery (strigulose), the pubescence mostly of hairs much less than 0.5–0.6 mm, firm, white, and mostly appressed. Leaves (3) 4–11 cm; leaflets 13–19 in number, 6–12 (20) mm. Racemes elongated, with (5) 8–9 (10) flowers or pods. Flowers ca. 7–8 mm, the banner petal with dark spots and a white eye between lavender veins, the corolla otherwise mostly rose-violet (lavender) and drying violet. Pods (16) 20–24 mm, papery-membranaceous, greatly swollen (inflated), purple speckled, with minute rough, coarse hairs, and more than 8 seeds.

Sand plains and shifting dunes across much of the flora area, sometimes at the edge of local playas, and occasionally on rocky soils near the shore; vicinity of Sonoyta to Puerto Peñasco and westward across the Gran Desierto dunes and sand flats to the Río Colorado. Var. *harwoodii* occurs in southeastern California, southwestern Arizona, northeastern Baja California, Islas Angel de la Guarda and San Lorenzo, and northwestern Sonora southward to Desemboque Río Concepción and the Altar Valley.

This species, with 3 varieties, occurs in northern Baja California Sur, Baja California, southeastern California, southwestern Arizona, and northwestern Sonora. A few seeds collected by Ivan Johnston in 1921 germinated in a greenhouse in 1956 (Barneby 1964:893). Distinguished from *A. magdalenae,* the only other widespread bladdery-pod locoweed in the flora area, by its smaller plants, smaller and purple-flecked pods, and fewer-flowered racemes.

25 mi E of San Luis, *Sanders 5639* (UCR). N of Sierra del Rosario, *F 20777.* Hornaday Mts, *Burgess 6825.* 40 mi S[W] of Sonoyta, *Keck 4222.* Punta Peñasco, *Wiggins 8383* (UC).

Astragalus lentiginosus Douglas ex Hooker var. **borreganus** M.E. Jones. FRECKLED MILKVETCH; *KOPONDAKUḌ*. Robust unless drought stressed, erect, unbranched, ca. 15–30 cm, or much-branched and somewhat bushy to 60+ cm when well watered; stems relatively stout. Herbage and pods with dense, rather coarse white hairs. Leaves 9–14 cm; leaflets (7) 9–19 (23) in number, mostly obovate, 9–17 mm. Racemes relatively open, often (6) 10–21 cm, with (3–5) 6–30 (or more?) flowers. Flowers 16 mm, the corollas showy, deep pink-violet or purple, fading bluish, the banner with a white spot at the base (white between pink-purple veins). Pods 16–19 × 5–6 mm, nearly rounded in cross section, not inflated (or scarcely so), slightly curved upward and held upright, with coarse white somewhat silky hairs; incompletely 2-chambered, the septum readily visible, dividing 3/4 of the pod. Seeds probably 10–20.

Sands of expansive flats and lower, usually stabilized dunes in the western margin of the flora area within 20+ km of the Río Colorado; also the coastal badlands fringing the Mesa de Sonora between El Golfo and La Salina. Not known elsewhere in Sonora. Also in the southern part of the Colorado Desert in southeastern California and the Yuma region of southwestern Arizona.

Diagnostic features of this sand-adapted plant include the long, open racemes of white-silky, narrow, upturned and slightly upcurved, 2-chambered pods. The septum is unique among our *Astragali.* This is the most beautiful locoweed in the region, with bright, showy flowers contrasting against the silvery foliage. It would be a handsome winter-spring annual for desert gardens. This excessively polymorphic milkvetch, "hardly a species in any conventional sense of the term" (Barneby 1964:912), ranges across the Columbia Basin and Great Basin, Colorado Plateau, and deserts of southeastern California, Arizona, and extreme northern Mexico. Barneby distinguishes 36 varieties.

12 km ESE of El Golfo, *F 92-188.* 9 km NNE of El Golfo, *F 92-181.* 7 mi E of San Luis, *F 91-59.* Near San Luis, *Kearney 8434.*

Astragalus magdalenae Greene. KOPONDAKUḌ. Large, robust ephemerals with dense silky or satiny hairs. Racemes often with 13–22 flowers, the flowers ca. 15 mm. Corollas magenta to lilac with white markings on banner and keel. Pods yellowish white, papery, and conspicuously inflated; seeds probably ca. 10–20.

This is the largest and most robust locoweed species in the flora area, distinguished by the relatively large-sized and large number of flowers, and very inflated and ascending pods. The mature pods readily fall away, so that the largest pods are often missing from specimens. Average seed size apparently increases in a cline from south to north, the range being 1.6–5.5 mm (Barneby 1964). This species occurs along both coasts of the Baja California Peninsula, coastal northwestern Sonora, and inland in the flora area to the Salton Sea trough of California.

Barneby recognized 3 varieties: the southernmost, var. *magdalenae,* is replaced in the northern Gulf of California region by var. *niveus,* which in turn is replaced to the north and inland by var. *peirsonii.*

1. Leaflets widest at middle, 10–21 mm; coastal dunes. ________ var. **magdalenae**
1′ Leaflets widest above middle, 3–9.5 (14) mm; inland dunes. ________ var. **peirsonii**

Astragalus magdalenae var. **magdalenae** [*A. magdalenae* var. *niveus* (Rydberg) Barneby. *A. niveus* (Rydberg) Barneby] SATINY MILKVETCH. Stems often curved upright and somewhat spreading. Larger leaves 6–15 cm; leaflets 17–23 (25) in number, mostly elliptic (broadest at about the middle), 10–21 mm. Racemes often with 13–22 flowers. Petals deep violet (lighter in color than those of *A. lentiginosus* var. *borreganus*), except the wing petals pale violet with white on the upper half and a white spot or "flag" on the banner. Pods 18–30 mm. Seeds often 2.5–3.5 mm. Common on dunes and sand flats along the coast.

Barneby (1964:862) admitted that the difference between var. *niveus* and var. *magdalenae* is "insecure." Var. *niveus* has slightly larger seeds, slightly larger corollas with minor and difficult to document color differences, and clumped and erect rather than diffuse and spreading stems. These characters seem to broadly intergrade and apparently do not show a geographic segregation. I favor combining the two taxa.

Variety *magdalenae* sensu stricto occurs along both coasts of Baja California Sur to the southeastern coast of Baja California and the opposite Sonora coast at least in the vicinity of El Desemboque San Ignacio. Var. *niveus* is reported from the region surrounding and near the head of the gulf in northeastern Baja California, northwestern Sonora, and extreme southwestern Arizona.

Estero Morúa, *Webster 22383*. Puerto Peñasco, *Burgess 4753*. Bahía de la Cholla (Cholla Bay), *Bell 17543* (RSA). [Punta] Borrascoso, *F 92-209*. El Golfo, *Johnson 5116* (DAV).

Astragalus magdalenae var. **peirsonii** (Munz & McBurney) Barneby [*A peirsonii* Munz & McBurney] DUNE MILKVETCH. Stems, leaves, and racemes tending to be straight and sharply ascending or erect. Leaves often 6–16 cm, the rachis often flattened with more surface area than the leaflets. Leaflets 11–15 (17) or fewer, widely spaced, mostly obovate to cuneate (widest above middle), often greatly reduced, reaching 3–9.5 (14) × 1.2–2.5 mm, the terminal leaflet greatly reduced and often decurrent and scarcely differentiated from the rachis, or sometimes larger than the adjacent leaflet pair. Racemes with at least 8–14 flowers. Pods 23–32 × 13–15 mm. Seeds probably 4.5–5.5 mm (reported by Barneby 1964 for California plants).

Dunes of the Gran Desierto south of Sierra Blanca westward at least to the vicinity of the Sierra del Rosario. Otherwise known only from dunes in southeastern California (Algodones Dunes in Imperial County, and Borrego Valley in San Diego County) and near Yuma, Arizona.

This dune endemic differs from var. *magdalenae* in its strict (erect) branching habit with straight stems, usually reduced and widely spaced leaflets widest near the apex, flattened rachises, decurrent terminal leaflet, stouter peduncles, larger pods, and larger seeds. It apparently has the largest seeds of any American *Astragalus*. The robust habit of the plant together with the large seeds is yet another case of gigantism among the Gran Desierto dune flora.

13.8 mi NE of Puerto Peñasco, then 7.6 mi NW, *Reichenbacher 1654.* 7.5 mi SE of Tinaja de los Chivos, *Simmons 11 Ap 1965.* S of Moon Crater, *F 19010.*

Astragalus nuttallianus de Candolle var. **imperfectus** (Rydberg) Barneby [*Hamosa imperfecta* Rydberg] SMALL-FLOWERED MILKVETCH. Plants usually diminutive, the stems very slender; hairs appressed, white and firm. Leaves often 2–5.5 (7) cm; leaflets 7–11 (13), mostly all alike, elliptic to linear-elliptic, 3.7–17.6 mm, the apex pointed (acute to obtuse). Racemes with 1–3 flowers, the peduncles very slender, 2.5–5 (8) cm. Flowers 5–6 mm, semi-cleistogamous; calyx tube 1–1.5 mm, the teeth 0.7–1.5 (2.0) mm, the hairs appressed and short (0.2–0.3 mm); petals lavender-pink. Pods often 12–20 mm, strongly curved, not inflated, tardily dehiscent or semidehiscent, semipersistent. Seeds mostly 12–16 per pod.

Puerto Peñasco to the Sonoyta region and throughout the Pinacate volcanic field to peak elevation. Often on sandy, gravelly, or cinder soils; plains or flats, washes, slopes, roadsides; weedy in agricultural areas.

Readily separated from all other Sonoran Desert *Astragali* by its small size, slender stems, delicate leaves, long slender peduncles, tiny flowers, and indehiscent semipersistent pods. *A. nuttallianus* occupies the southwestern United States and the northern half of Mexico. Barneby (1964) recognizes 10 varieties. Specimens from the flora area seem to best fit the description of var. *imperfectus* because of the homomorphic elliptic leaflets, short and appressed calyx hairs, and short calyx teeth; however, the short calyx tube is more consistent with var. *cedrosensis* M.E. Jones. Similar plants occur in western Sonora southward to the Guaymas region. Var. *imperfectus* occurs in southeastern California, Nevada, the western half of Arizona, and northeastern Baja California. Var. *cedrosensis* is known from the western margin of the Colorado Desert in southeastern California, both Baja California states, west-central Sonora, and western Pima County in southwestern Arizona. Characters used to distinguish these two varieties (length of calyx tube and dimorphic leaflets versus homomorphic leaflets) seem weakly differentiated.

Arroyo Tule, *F 19223.* Campo Rojo, *F 92-481B.* Pinacate Peak, summit, *F 19447.* 7 mi NE of mouth of Estero Morúa, *F 91-49.*

Astragalus sabulonum A. Gray [*Phaca lerdoensis* Rydberg in *N. Amer. Fl.* 24:356, 1929] GRAVEL MILKVETCH. Plants with silky hairs 0.7–1.0 mm. Leaves 3–7 cm; leaflets 9–15, elliptic to obovate, 6.8–12.2 mm. Racemes relatively open with (2) 3–5 (7?) flowers. Flowers 6–8 mm, the corollas lavender-purple with a white eye at the base of the banner (white between lavender-purple veins). Pods curved downward, papery and firm, small, hollow or swollen but not really inflated, densely pubescent with short soft hairs, and with juicy inwardly filamentous valves drying to a leathery or stiffly papery texture. Seeds usually more than 8.

Sand flats and low stabilized dunes on the Mesa de Sonora northeast of El Golfo and in similar habitat on the Mesa de Andrade in the Río Colorado delta. Also nearby southeastern California, and disjunct in northern Arizona, Nevada, Utah, and northwestern New Mexico. An unusual feature of *A. sabulonum* is its enormous elevational range, from near sea level to 2000 m.

Colonia Lerdo, *MacDougal Feb 1904* (NY, type of *Phaca lerdoensis,* cited by Barneby 1964:886, not seen). Mesa de Andrade, *F 93-212.* 13 km NNE of El Golfo, *F 92-185.* 12 km ESE of El Golfo, *F 92-188.*

CALLIANDRA

Trees, shrubs, and subshrubs, and some herbaceous perennials; usually unarmed. Leaves bipinnate; without leafstalk glands. Flowers mimosoid, in heads, compact racemes, or umbels. Stamens often numerous, united below into a tube, often long and showy; pollen shed in 8-grained polyads. Pods flattened with thick cordlike margins, the valves separating elastically and curling back but not twisted. Americas, mostly tropical or subtropical, few in deserts; 135 species. Reference: Barneby 1998.

Calliandra eriophylla Bentham var. **eriophylla.** *HUAJILLO;* FAIRY DUSTER. Much-branched dwarf woody shrubs ca. 0.5–0.8 m, with firm but flexible stems and grayish bark. Young twigs, peduncles, and leafstalks densely to moderately pubescent with short white hairs, the leaflets sparsely haired. Leaves gradually drought deciduous and frost sensitive, the pinnae (1 or 2) 3 pairs; leaflets 2.5–6 mm, 6–15 pairs per pinna. (Long-shoot leaves may be much larger than short-shoot leaves.) Flowers sessile, in few-flowered, pedunculate heads. Stamens ca. 1.5–2.0 cm, showy, whitish to pinkish, opening at night, drooping with daytime heat; anthers small. Pods often 4.0–5.6 cm. Flowering after rains, mostly in spring.

Mostly on slopes and along arroyos, often with northern exposures. Scattered colonies, mostly at higher elevations in the Sierra Pinacate; more common in the Sonoyta region and in Organ Pipe Monument. Through much of the Sonoran Desert, mostly in less xeric habitats and at elevations just above the desert.

Southeastern California to southwestern New Mexico and southward including northern Baja California to Chiapas. A second, weakly differentiated variety in southern Texas. The 8-grained pollen packets adhere to butterfly wings (Turner, Bowers, & Burgess 1995).

Cerro basáltico [Pinacate region], 32°09′ N, 113°52′ W, *Ezcurra 29 Ap 1981* (MEXU). 11 mi SW of Sonoyta, *F 90-156*.

DALEA

Annual or perennial herbs, or shrubs. Herbage and usually the flowers dotted with blister-like glands (secretory vesicles), with spirally twisted and usually reddish brown hairs, aromatic when bruised. Leaves usually odd pinnate (leaflets lacking the small wavy lines seen in *Marina*). Flowers papilionoid, in spikes or sometimes racemes; flowers subtended by a bract and usually 2 bractlets. Calyx falling with the fruit, the tube 10-ribbed, with 5 ribs extending into teeth, the veins of the ribs joining (anastomosing) toward the tip of the calyx teeth. Stamens usually 10, fertile, monodelphous (all united into a tube). Ovules 2, only 1 developing into a seed; pods indehiscent.

North and South America, largely semiarid climates, with greatest diversity in Mexico, especially pine-oak regions; 165 species. Ours belong to section *Theodora* with 5 species, centered in the Sonoran and Chihuahuan Deserts and semiarid northern Sierra Madre Occidental. Reference: Barneby 1977.

1. Leaflet margins entire and flat; floral bracts 0.7–1.0 mm wide; calyx tube 1.8–2.4 mm long, the calyx teeth (bristles) 1.1–3.1 mm long; wing petals usually notched and with a gland. ____ **D. mollis**

1′ Leaflet margins undulate (wavy); bracts 0.8–1.6 mm wide; calyx tube 2.4–2.9 mm long, the calyx teeth 2.6–4.3 mm long; wing petals entire, glandless. ____ **D. mollissima**

Dalea mollis Bentham [*D. mollis* subsp. *pilosa* (Rydberg) Wiggins] SILKY DALEA. Non-seasonal ephemerals, with a deep, orange-colored taproot and nodulated secondary roots. Stems 10–30+ cm, erect to prostrate with age. Herbage with slender, spreading, white hairs and dotted with swollen, maroon glands. Leaves 1.5–2.7 (3) cm; leaflets (7 or 9) 11 or 13, thickish, often glaucous, mostly cuneate to obovate or obcordate, 4.0–7.0 mm, the apex often truncate to notched, the lower surface gland-dotted, the upper surface glandless, the margins entire and flat.

Flowering at a very early age and continuing as long as soil moisture is available. Racemes (1) 1.5–3.5 cm, densely flowered, appearing furry due to very dense, long, whitish to tawny, tightly spiralled, silky hairs on calyx tube and teeth, the hairs often obscuring the flowers and fruits; inflorescences and pedicels with conspicuous teardrop-shaped dark maroon-brown glands remaining after the flowers or fruits have fallen. Floral bracts 0.7–1.0 mm wide, often extending into an awnlike bristle. Pedicels very short, stout and persistent. Flowers (3.5) 4.5–6.5 mm, the corolla scarcely longer than the calyx-teeth bristles. Calyx tube 1.8–2.4 mm (measured from pedicel to sinus between calyx teeth), with rows of small iridescent orange glands, the margins and midribs dark maroon, the teeth triangular, the midribs extending into stout, densely long-haired bristles (awns) 1.1–3.1 mm. Wing petals notched with a gland, or the gland and notch sometimes absent. Upper, or exposed, portion of petals violet, drying red-purple. Pods 2.5 mm, obovoid. Variously flowering at least September-April.

Widespread, to peak elevations in the Sonoyta region, Pinacate volcanic field, granitic ranges, desert flats, and low to high dunes; fine textured, sandy, gravelly, and rocky soils. Sonoran Desert: southwestern Arizona, southeastern California, Baja California, Baja California Sur, and Sonora south to the vicinity of Guaymas.

In the northern part of its range *D. mollis* sometimes occurs with *D. mollissima*. In general *D. mollis* is replaced by *D. mollissima* to the north, mostly in the southwestern United States, although there is a considerable area of sympatric contact. To the east, from Arizona to Texas and southward into the Chihuahuan Desert region in Mexico, *D. mollis* is replaced by *D. neomexicana* A. Gray; they are apparently allopatric. In Arizona *D. neomexicana* occurs at elevations above the Sonoran Desert. *D. mollis* seems to be more closely related to *D. neomexicana* than to *D. mollissima*.

Bahía de la Cholla, *Gould 4146*. 21 mi SW of Sonoyta, *Bowers 2594*. W of Suvuk, *Ezcurra 6 Ap 1981* (MEXU). Pinacate Peak, *F 19409*. MacDougal Pass, *Turner 59-11*. Moon Crater, *F 19278*. Sierra del Rosario, *F 20733*.

Dalea mollissima (Rydberg) Munz [*Paroselea mollissima* Rydberg] Non-seasonal ephemerals; resembling *D. mollis* but generally more robust and larger in all dimensions, the leaflets more glaucous and their margins undulate (wavy). Leaflets 11–15 per leaf, 5–8+ mm. Floral bracts greenish, lanceolate, 0.8–1.6 mm wide. Calyx tube 2.4–2.9 mm, the bristles (awns) 2.6–4.3 mm. Keel petals about as long as the calyx, thus hardly showing; wing petals entire, glandless.

Sandy plains with *Larrea* and *Ambrosia dumosa* in the extreme northwestern corner of the Gran Desierto near San Luis; not known elsewhere in Sonora. Western Arizona, southern Nevada, Baja California, and southeastern California.

19.5 mi E of San Luis, *F 16700*.

HOFFMANNSEGGIA

Unarmed perennial herbs or shrubs, generally with stalked glands. Leaves twice pinnate, the pinnae odd-pinnate; stipules small. Flowers caesalpinioid, in terminal racemes or panicles. Pedicels not jointed. Flowers yellow, the petals often red-flecked. Stamens 10, separate. Pods flattened, tardily dehiscent or indehiscent.

North and South America, mostly subtropics and warm temperate regions; 23 species. References: Barneby 1989; McVaugh 1987; Simpson & Miao 1997; Ulibarri 1979.

1. Herbaceous perennials from a tuberous root; stems less than 30 cm; leaves 6 cm or more, with 7 or more pinnae. **H. glauca**

1′ Small shrubs without tuberous roots; stems mostly more than 1 m; leaves less than 5 cm, with 3 pinnae. **H. microphylla**

Hoffmannseggia glauca (Ortega) Eifert [*Larrea glauca* Ortega. *Caesalpinia glauca* (Ortega) Kuntze. *H. densiflora* Bentham. *H. falcaria* Cavanilles] *CAMOTE DE RATÓN;* HOG POTATO; *'I:KOWĬ.* Herbaceous from deeply buried rhizomes and edible, potato-like tuberous roots; with stalked reddish glands and usually sparse, short white hairs. Stems renewed annually or seasonally, usually several, ca. 15–20 cm including the erect terminal inflorescence. Leaves several near the stem base, 6–12 cm, often held upright, long petioled, with (7) 9 or 11 pinnae, the terminal pinna longer and with as many as 11 leaflet pairs, the other pinnae with 4–9 leaflet pairs; leaflets oblong, 2.8–5.5 (7) mm. Flowers in racemes, the petals predominantly bright yellow, turning red or red-flecked with age; stamens often red. Pods 2.5–3 cm, compressed, readily deciduous, indehiscent. Flowering during warm weather.

Several widely scattered, localized populations in sandy soils and playa flats, and occasionally in weedy urban-agricultural habitats. Southeastern California to Kansas and southward to northern Mexico including Baja California, northern Baja California Sur, and northern Sonora to San Luis Potosí; also Peru, Chile, and Argentina.

Sierra Blanca, lado oeste, pastizal efímeras, *Equihua 24 Aug 1982*. Sonoyta, *F 85-934*. 10 km NE of Sánchez Islas, abandoned ranchito, *F 92-214*.

Hoffmannseggia microphylla Torrey [*Caesalpinia virgata* E.M. Fischer] Small shrubs 1–1.5 m, the stems mostly arising from the base, numerous, rushlike, rigid, erect, and with green bark. New shoots, leafstalks, inflorescences, and buds with short white hairs but not stipitate glandular, the leaflets sparsely hairy or glabrate. Leaves few and widely spaced, quickly drought deciduous, 1.5–4.5 cm, with 3 pinnae, the terminal one 10–32 mm with 6–11 leaflet pairs, the lateral pinnae 3.5–18 mm with 3–9 leaflet pairs; leaflets 2–3 (5) mm. Inflorescences reaching 10–25 cm, racemes or panicles with 1 to few lateral branches. Flowers 13–15 mm wide, yellow, the banner with a yellow-brown arc, the sepals turning orange with age. Pods 1.5–2.0 cm, crescent-shaped, densely dotted with thick-stalked yellowish glands, thin-walled, dehiscent. Flowering various seasons including spring.

Sierra del Rosario among rocks at the base of the western slopes (*F 75-20*), and the south side of Cerro Pinto (*F 75-44*). Southward in Sonora to the Sierra Bacha south of Puerto Libertad, Yuma County in southwestern Arizona, southeastern California, and eastern Baja California. The green stems and sparse and highly ephemeral foliage give it the aspect of a miniature palo verde (*Parkinsonia*). *H. microphylla* is most closely related to *H. intricata* Brandegee and *H. peninsularis* (Britton) Wiggins of the Baja California Peninsula and islands of the Gulf of California.

LOTUS DEERVETCH, TREFOIL

Annual or perennial herbs to subshrubs. Leaves odd pinnate, the leaflets often 3–5; without pulvini; stipules membranous or leafy, or (all Sonoran Desert species) represented by small but conspicuous, dark glands. Flowers papilionoid; mostly in small axillary umbels, the peduncle often with a leaflike bract, or flowers solitary in axils. Stamens 10, diadelphous. Pods elastically dehiscent and the valves coiling (ours) or indehiscent; seeds 1 to many. Worldwide, mostly north temperate; 100 species.

1. Perennials; stems upright, the internodes notably longer than the leaves; northwestern granitic mountains. **L. rigidus**

1′ Winter-spring ephemerals; stems ascending to prostrate, the internodes not notably longer than the leaves; widespread.

  2. Leaflets broadly elliptic or nearly so, widest at or slightly above the middle, the tip rounded to pointed (acute). **L. salsuginosus**

  2′ Leaflets obovate, broadest at or near apex and gradually tapered to base, the tip blunt (truncate) and sometimes notched. **L. strigosus**

Lotus rigidus (Bentham) Greene [*Hosackia rigida* Bentham] DESERT ROCK-PEA. Small upright, summer-dormant perennials, 20–70 cm; stems firm, brittle with age, sparsely leafy, the internodes long. Herbage, inflorescences, and calyces strigose with appressed white hairs. Leaves and ultimately many twigs drought deciduous. Leaves sessile or the petioles very short (1–2 mm), the leaflets 3 or 4 (basically odd pinnate even though 4 leaflets may be present), thickish, linear-oblong, 11–15 mm. Peduncles (3) 5–10 cm, resembling the twigs, 1-flowered, with a 1–3-foliolate bract below the flower. Flowers 1.6–1.8+ cm, the corolla at first uniformly bright yellow, soon developing a red-orange nectar guide, with age the corolla turning red-orange or pink tinged with orange or red-orange. Pods 1.8–4 cm, strigose with appressed, moderately dense, white hairs 0.2–0.4 mm; seeds 6–12 per pod, 1.5–2.0 mm, rounded or quadrangular, and papillate. Early spring.

North sides of Sierra del Viejo and Cerro Pinto, among rocks on steep slopes and cliffs, especially at higher elevations. Also in the adjacent granitic mountains in Yuma County and the Ajo Mountains. Southeastern California, both Baja California states, northern Sonora, Arizona, southern Nevada, and southwestern Utah; desert to oak chaparral.

Cerro Pinto, *F 89-55*. Sierra del Viejo, *F 85-729*. 67 mi W of Sonoyta on Mex 2, *Prigge 7254*.

Lotus salsuginosus Greene var. **brevivexillus** Ottley [*L. humilis* Greene. *Hosackia humilis* (Greene) Abrams] Winter-spring ephemerals. Stems often 6–33 cm. Herbage often semisucculent, glabrous or sparsely hairy except for short, white hairs on the youngest stem tips. Leaves 14–28 mm, the leafstalk moderately flattened, the petioles 2.0–2.5 mm, the leaflets 3–7 (or lower "pair" represented by a single leaflet), thickish, broadly elliptic, 6–12.5 mm, the apex broadly rounded to acute with a small, blunt point. Peduncles slender with a leaflet-like bract on the joint (node) just below the flowers, with 2–4 flowers. Flowers 5.5–7.0 mm, the corolla bright yellow, or the banner sometimes turning pale orange. Pods 10.6–16.7 mm. Seeds 5–8 per pod, 1.3–1.4 mm, rounded, mottled, and smooth.

Sandy-gravelly washes, cinder flats and slopes, bajadas, and sometimes on rocky slopes. Sonoyta region and the northern and eastern sides of the Pinacate volcanic field to higher elevations. Baja California to southern California, Arizona, and northwestern Sonora to Isla Tiburón and the Sierra Seri.

El Papalote, *F 86-113*. Tinajas de Emilia, *F 19752*.

Lotus strigosus (Nuttall) Greene [*Hosackia strigosa* Nuttall. *Lotus tomentellus* Greene. *Hosackia tomentella* (Greene) Abrams. *Lotus strigosus* var. *tomentellus* (Greene) Isely] DESERT LOTUS. Winter-spring ephemerals; roots conspicuously nodulated. Stems spreading, becoming prostrate, 4–30+ cm. Herbage sparsely to densely hairy (cinereous to strigose) depending on soil moisture. Leaves 7–23 mm, the leaf-stalk flattened, the petioles 4.0–7.5 mm, the leaflets (3) 5–7, relatively thick, obovate, 4.5–13 mm, the apex truncate (blunt) and sometimes notched. Flowers solitary and sessile in axils, or 1 or 2 on slender peduncles with a leaflet-like or sometimes 3-foliolate bract on the joint (node) just below the flower(s). Flowers 7.5–9 mm, the corolla bright yellow with a red nectar guideline on the banner. Pods 12–27 mm. Seeds 8–11 per pod, 1.5–2.2 mm in diameter and dimorphic: (1) nearly round, yellow-green, and minutely granulate, or (2) somewhat kidney-shaped or thickly U-shaped, mottled brown on buff, and smooth. Seed morphology seems constant for an individual plant and possibly among local populations.

Widespread, sometimes carpeting the desert; Sonoyta region, Pinacate volcanic field, granitic ranges, desert plains, and dunes; near sea level to the summit of Sierra Pinacate. Rocky slopes, cinder soils, sandy to rocky flats, lower dunes, and interdune troughs. Sonora south to Bahía Kino and Isla Tiburón, and Baja California to southeastern California and Arizona.

Sierra del Rosario, *F 20339*. MacDougal Crater, *Turner 86-1*. Pinacate Peak, *F 19446*. Elegante Crater, *F 19687*. Sierra Extraña, *F 19066*. Bahía de la Cholla, *Gould 4145*.

LUPINUS LUPINE

Annual or perennial herbs, or small shrubs. Leaves digitately compound, the leaflets usually 5–12 (17), the margins entire or notched at tip; stipules fused to petiole base. Flowers papilionoid, usually showy in terminal racemes, with 1 or 2 floral bracts fused to the calyx base. Banner petal erect. Stamens 10, the filaments fused (monodelphous), the anthers dimorphic (every other one longer and basifixed or shorter and versatile). Stigma encircled by a brush of short hairs or papillae. Pods laterally compressed, obliquely depressed between the 2–12 seeds, elastically dehiscent, the valves coiling to eject the seeds. Seeds resembling pebbles.

Mostly North and South America, a few in the Mediterranean region and tropical African highlands; 150+ species. Some are domesticated as ornamental garden plants, for forage and fodder, food for humans, and fiber. The several Sonoran Desert species are winter-spring ephemerals with a stout taproot and conspicuous root nodules. References: Christian & Dunn 1970; Dunn, Christian, & Dziekanowski 1966.

1. Leaflets obovate to elliptic, the tips mostly blunt to broadly rounded, often notched or apiculate; flowers pale lavender-pink. ______ **L. arizonicus**

1′ Leaflets linear to narrowly obovate or narrowly elliptic, the tips acute; flowers bright blue. ______ **L. sparsiflorus**

Lupinus arizonicus (S. Watson) S. Watson [*L. concinnus* var. *arizonicus* S. Watson. *L. sparsiflorus* Bentham var. *arizonicus* (S. Watson) C.P. Smith. *L. arizonicus* subsp. *sonorensis* J.A. Christian & D.B. Dunn, *Trans. Missouri Acad. Sci.* 4:87, fig. 1, 1971] *LUPINO;* ARIZONA LUPINE; *TAȘ MA:HAG*. Plants 12–60+ cm, the herbage often semisucculent, with relatively coarse hairs and the leaflets moderately hairy or glabrate. Stems commonly erect to ascending, solitary to much-branched. Leaflets (5) 7–9 per leaf, (13) 15–37 × 4–12 mm, obovate, the tip mostly blunt to broadly rounded and sometimes notched to apiculate (with a soft, blunt projection at the tip). Corollas 7–9 mm, pale lavender-pink. Pods (9) 15–24 mm. Seeds 4–6 per pod, 2.8–3.1 mm, smooth and marbled.

Seasonally common throughout the flora area; to peak elevation on Sierra Pinacate. Often abundant on sand flats, dunes, and larger gravelly washes; also arroyos, bajadas, cinder flats and slopes, and less common on rocky slopes. Southwestern Arizona and southeastern California to the Cape Region of Baja California Sur and coastal Sonora southward to the vicinity of Tastiota.

Exceptional winter-spring rainfall often brings spectacular displays of this lupine across dunes and sand flats, especially on the west side of the Pinacate region, where it may form 100% ground cover. Such plants may reach more than 1 meter in height with many branches, but the flowers, pods, and young herbage are often ravaged by enormous populations of sphinx moth caterpillars and other lepidopterans. During one of these El Niño springs the lupines on the sand flats southwest of Sierra Blanca were stripped of flowers, fruits, and leaves by hoards of caterpillars of the painted lady butterfly (*Vanessa cardui*), the most widely distributed species of butterfly in the world. These caterpillars were preyed upon by great numbers of *Ammophila* wasps, which were busily dragging paralyzed caterpillars across the desert floor into already prepared underground burrows.

Lupinus arizonicus seems to be closely related to *L. sparsiflorus* and *L. concinnus* Agardh of the southwestern United States and northwestern Mexico.

Cerro Pinto, *Burgess 5616.* Sierra del Rosario, *F 20440.* Hornaday Mts, *Burgess 6808.* Pinacate Peak, *F 19445.* Elegante Crater, *F 9699.* N of Rocky Point, *Clark 12683* (UNM, type of subsp. *sonorensis,* not seen). Aguajita Wash, *F 88-284.*

Lupinus sparsiflorus Bentham [*L. sparsiflorus* subsp. *mohavensis* Dziekanowski & D.B. Dunn. *L. sparsiflorus* var. *mohavensis* Welsh] MOJAVE LUPINE. Plants 12–45 cm, with coarse white to golden yellow hairs. Stems erect to ascending, solitary to much-branched. Leaflets 6–10 per leaf, 11–40 × 2–6 mm, linear to narrowly obovate or narrowly elliptic, the apex mostly acute. Corollas 8–11 mm, bright blue with a white blotch (eye) at the base of the banner. Pods 10–23 mm. Seeds 3–8 per pod, 2.4–2.8 mm, smooth (but less so than in *L. arizonicus*) and marbled.

Common along roadsides and washes in the Sonoyta region, but generally only during El Niño years. More common eastward in Sonora and to the north in Arizona. Northern Sonora, Arizona, Nevada, southwestern Utah, and southern California to Baja California Sur.

Although the geographic ranges of the two lupines overlap, I have not found them growing together in the flora area. Young, vigorous plants of *L. sparsiflorus* are often somewhat semisucculent, although not as succulent as *L. arizonicus.*

6 km W of Sonoyta, *F 92-235.* 8 km SW of Sonoyta, *F 92-409.* 10 mi S of Sonoyta on Mex 2, *F 20562.*

MARINA

Annual to perennial herbs or shrubs; gland-dotted and with relatively stiff, short, straight or curly hairs that remain unchanged by drying (not spirally twisted and reddish brown as in *Dalea*). Leaves odd pinnate, the terminal leaflet stalked; leaflets marked on upper surface with tiny wavy lines representing veins (these sometimes difficult to see on small leaflets of drought-stressed plants). Flowers papilionoid, small, brightly colored, pedicelled; subtending bracts quickly deciduous. Calyx 5-toothed and 10-ribbed, with prominent glands between the calyx ribs; veins of ribs not joining (anastomosing) toward the tip of calyx teeth (these veins often difficult to see on *M. parryi* because of the dense hairs). Stamens mostly 10. Pods small, 1-seeded.

Subtropical and arid tropical, desert, and dry-winter climates, from Guatemala to southwestern United States, all native and mostly endemic to Mexico; 38 species.

Series *Chrysorrhizae,* with 17 species including the few western Sonora species, is characterized by short, firm pedicels, a pubescent calyx, and a gland-sprinkled banner. *M. parryi* belongs to a group of 10 closely related species dispersed around the Gulf of California, especially in the southern half of Baja California Sur and the foothills of the Sierra Madre Occidental in Sinaloa. Reference: Barneby 1977.

Marina parryi (Torrey & A. Gray ex A. Gray) Barneby [*Dalea parryi* Torrey & A. Gray ex A. Gray] Non-seasonal ephemerals, mostly with winter-spring rains, occasionally short-lived perennials, with a yellow taproot. Stems erect to spreading, commonly 30–50 cm, the plants sometimes reproducing at only 8–10 cm, rarely reaching 1.5 m across; dotted with maroon glands. New shoots usually densely pubescent with short, firm, appressed hairs; but during summer rainy seasons sometimes soon glabrate with only the stem tips pubescent. Leaves 2.5–3.5 cm, green to grayish depending on density of hairs, the leaflets 9–23 per leaf, 2.0–6.5 mm. Flowers 2.8–5.9 mm. Calyx 2.5–3.5 mm, densely hairy but with a wide range in density and length of hairs. Corollas dark blue and white. Pods 1.8–2.5 mm.

Sonoyta region, Pinacate volcanic field, granitic ranges, and desert plains. Widespread; open rocky hills and slopes to peak elevations, bajadas, especially along arroyos and washes, sandy plains, and lower dunes. The geographic range nearly coincides with the boundaries of the Sonoran Desert as defined by Shreve (1951), and over most of the Sonoran Desert it is the only species of *Marina* present.

Hornaday Mts, *Burgess 6824.* Tinaja del Tule, *F 18903.* S of Carnegie Peak, 900 m, *F 19804.* Papago Tanks, *Turner 59-34.* W of Sonoyta, *F 85-954.*

MELILOTUS SWEETCLOVER

Annual or biennial herbs; dry herbage fragrant. Leaves pinnate, the leaflets 3, with veins usually ending in marginal teeth. Flowers papilionoid, small, yellow or white, often fragrant. Pods rounded to ovoid, indehiscent or tardily dehiscent, 1 to few-seeded. Native to the Old World; 20 species. Some are widespread weeds and some are grown for fodder, forage, and soil-enrichment with a number of cultivars.

***Melilotus indica** Allioni. *TRÉBOL AGRIO;* YELLOW SWEETCLOVER, SOURCLOVER; *PU:WL.* Cool-weather ephemerals; glabrous, the stems mostly erect to ascending. Leaflets 1–3 cm, ovate to obovate, the margins serrated especially toward apex, or occasionally some nearly entire. Flowers yellow, pealike, 3–4.5 mm, turning downward after anthesis, in spikelike racemes. Pods reflexed, ovoid, wrinkled, 2.5 mm, 1-seeded.

Mostly disturbed wetland habitats; Sonoyta (*F 85-706*), sandy riverbanks and floodplain of the Río Sonoyta, and at Quitobaquito (*F 7659*). Wetland habitats through much of the Sonoran Desert. Native to the Mediterranean, now worldwide especially in temperate regions.

OLNEYA

The genus has 1 species; flowers papilionoid. Its closest relatives are *Coursetia, Genistidium, Peteria,* and *Robinia.* References: Lavin 1988; Lavin & Sousa 1995.

Olneya tesota A. Gray. *PALO FIERRO;* IRONWOOD; *HO'IDKAM.* Trees, often 7–8.5 m, very long-lived and slow growing, with 1 to several massive trunks, shredding gray bark, and dense gray-green foliage. Herbage with dense, short, relatively coarse whitish hairs, the leaflets sometimes sparsely hairy and appearing glabrous except under magnification. Spines (3) 4–11 mm, seemingly stipular in origin, single or paired, slender to thickened at base, sharp, straight to slightly curved, or spines absent from some twigs. Leaves odd or even pinnate, tardily drought deciduous in extreme drought or deciduous prior to flowering, 2.8–6+ cm; leaflets opposite, subopposite, or sometimes alternate, 6–16 per leaf, (7) 10–20 mm, oblong to mostly oblong-spatulate, often slightly asymmetric (especially toward base), the apex entire to notched or sometimes with a minute apiculate projection.

Inflorescence branches, pedicels, sepals, and portion of banner petal exposed in bud densely pubescent with short white hairs. Flowers mostly pinkish to purplish, ca. 1.5 cm, pedicelled, in short dense racemes or panicles. Calyx purple-brown, the lobes about as long as the tube, spreading at anthesis. Petals pinkish to purplish on exposed portions of the blades, white below; banner petal with a thickened succulent white callus at junction of blade and claw and an invaginated chartreuse nectar guide on the banner above the callus, a whitish band surrounding and highlighting the nectar guide and margined with red-purple apically, the flat blade of the banner folding back later in the day "behind" the "protruding" nectar guide; wing and keel petals red-purple (darker than the other petals). Stamens 10, the upper one free, the filaments including the tube white; anthers yellow-orange. Style pale green, the distal portion with spreading hairs forming a pollen brush, the stigma yellow-orange like the anthers. Ovary white, densely studded with white thick-stalked globular glands except at its distal end.

Pods short and thick, often 2–4 cm, nearly terete, often moderately constricted between the seeds, dark reddish-brown, with stout glandular hairs and shorter non-glandular hairs; mostly tardily dehiscent, the valves spreading from apex but not twisting. Seeds 1 or 2 (3 or 4) per pod, ca. 7–8 mm, ellipsoid, mostly about 50% longer than wide, dark reddish brown, darkly mottled, often appearing nearly black except in bright light. Masses of pale, pink to purple-pink flowers, attracting myriads of bees and other insects, usually late April–May; fruits ripening June-July.

Widespread, to near peak elevations but mostly at lower elevations. The largest ironwoods and greatest concentrations are in xeroriparian arroyos and larger washes; they are less common on rocky slopes and rare in interdune troughs and lower dunes. Ironweed is one of the few trees in the region that offers worthwhile shade; a host of animals and smaller plants find shelter and protection beneath ironwoods (Búrquez & Quintana 1994). It often grows with mesquite (*Prosopis velutina* and *P. glandulosa*) and blue palo verde (*Parkinsonia florida*) to form spinescent legume gallery forests.

Southeastern California to the Cape Region of Baja California Sur and southwestern Arizona to southwestern Sonora. Essentially endemic to the Sonoran Desert, extending beyond the desert in the lowlands of the Cape Region and the lower Río Mayo region.

Trees and branches that will produce mass flowering become leafless or nearly so in late spring and early summer shortly before the buds appear. At the same time new leaves appear on adjacent non-flowering ironwoods. Leafout on fruiting trees occurs with the summer rains after the pods fall. Many of the great ironwood stands in Sonora have been decimated by woodcutters (Nabhan & Carr 1995). The wood is in demand for sculpting by both the Seri Indians and their imitators, and for firewood and charcoal. The wood is extremely hard, does not float, and burns with a hot flame (Record & Hess 1943).

Sierra del Rosario, *F 20742.* 63 mi E of San Luis, *F 5735.* Near Batamote crossing, *Sanders 5707.* 17 mi SW of Sonoyta, *F 90-170.*

PARKINSONIA PALO VERDE

Large shrubs to trees with smooth green bark and relatively soft wood. Twigs with paired stipular spines, or stipular spines absent and the twigs spinescent. Short shoots very reduced. Leaves twice pinnate with 1–3 pairs of pinnae, petioled or the pinnae sessile; individual leaflets, pinnae, or leaves drought deciduous. Flowers caesalpinioid, in reduced to open axillary racemes, the pedicels appearing jointed (articulated) due to the elongated pedicel-like hypanthium base. Banner petal yellow or whitish, and often with orange spots or flecks, the entire banner often fading orange or orange-flecked, the other petals yellow. Stamens 10, distinct; anthers orange. Pods indehiscent or tardily partially dehiscent; seeds 1 to several.

Warm to tropical desert and semiarid regions; 10 species. The greatest number of species occur in the Sonoran Desert. For more than a century *Cercidium* and *Parkinsonia* have been variously treated as one or two genera. The work of Julie Hawkins and her colleagues strongly supports a single, monophyletic genus. *Parkinsonia* sensu stricto includes 3 species in southern and eastern Africa and the pantropical *P. aculeata*. *Cercidium* sensu stricto ranges from the southwestern United States to Argentina. A magnificent horticultural selection of a hybrid found in the Tucson region, called *Cercidium* cv. "Desert Museum," involves hybridization of *P. aculeata, P. florida,* and *P. microphylla* (Dimmitt 1987). References: Carter 1974; Dimmitt 1987; Hawkins 1996; Hawkins et al. 1999.

1. Leaves more than (10) 20 cm. **P. aculeata**
1′ Leaves less than 7 cm.
   2. Twigs not spinescent at tip; paired stipular spines often present; leaves petioled, the leaflets mostly 5–9 mm. **P. florida**
   2′ Twigs spinescent at tip; stipular spines absent; petiole absent, the pinnae sessile, the leaflets mostly 1–3.3 mm. **P. microphylla**

***Parkinsonia aculeata** Linnaeus. *BAGOTE, GUACAPORO, HUACAPORI, RETAMA;* MEXICAN PALO VERDE. Trees with a well-developed trunk and smooth, green bark on branches and trunk. Growth buds and youngest herbage with small white hairs, soon becoming glabrous or glabrate. Leaves of long shoots often bearing a pair of sharp stiff stipular spines, these often breaking off or falling away or sometimes not developing, the actual leafstalk (petiole and primary leaf rachis) fused and modified into a very stout, stiff, sharp spine often 1–3 cm, bearing near its base 1 or 2 (3) pairs of long slender pinnae (each pinna resembling an individual leaf). Secondary rachises (10) 20–45 cm, flattened, stringy and strap-like, the pulvini conspicuously swollen and flexible; each secondary rachis bearing numerous small leaflets. Leaflets quickly drought deciduous, the phyllode-like secondary rachises tardily drought or winter deciduous but the leafstalk spine persistent. Short-shoot leaves mostly spineless.

Inflorescences of racemes (5) 10–16 cm, relatively few-flowered; pedicels jointed near base of flower. Flowers showy, 27–35 mm wide; sepals and petals bright yellow, the banner petal at first with basal red-orange spots or flecks, with age most or all of the banner and the sepals turning red-orange. Filaments (at least at first) and the ovary green, the anthers pale orange to somewhat rose colored. Pods few-seeded, more or less indehiscent or tardily semidehiscent. Massive flowering in April; sporadically few-flowered in summer and fall.

Extensively cultivated and sometimes reproducing at ranches and in towns, but apparently not reaching maturity in the flora area unless given supplemental water. Rare and widely scattered along highways and roads as non-reproducing young trees to ca. 3–4 m. In the mid-nineteenth century Arthur Schott found it growing in the hills along the Río Colorado near Fort Yuma, and described it as "valued by the Mexican Indians as a febrifuge and sudorific, and also as a remedy for epilepsy" (Torrey 1858:59).

Cultivated and weedy across much of the rest of Sonora, and locally naturalized especially in subtropical nondesert areas. Occurring in many arid subtropical and tropical regions of the world following introduction as an ornamental, hedging, and shade tree, and subsequent invasion and naturalization. Probably native to Central America and southern Mexico (Julie Hawkins & Colin Hughes, personal communication 1999).

Rancho los Vidrios, *Ezcurra 20 Sep 1980* (MEXU). 4 km N of Puerto Peñasco, *F 85-777.* 3 mi S of Pinacate Junction, *F 86-375.* Observation: 18 mi E of San Luis.

Parkinsonia florida (Bentham ex A. Gray) S. Watson [*Cercidium floridum* Bentham ex A. Gray] *PALO VERDE AZUL;* BLUE PALO VERDE; *KO'OKOMADK, KALISP.* Large shrubs to small trees reaching 7–10 m, with a well-developed trunk. Bark of twigs and branches bluish green. Leaves short petioled, usually with 1 pair of pinnae, each 5–12 mm and usually with 3 or 4 pairs of leaflets, the leaflets 5–9 mm; small, often stalked or cylindrical glands at the base of pinnae and between the leaflets (on upper surfaces), darkening when dried. Flowers (18) 22–28.7 mm wide. Calyx green to yellow-green, the lobes reflexed. Petals bright yellow, the banner with a few small orange-red spots basally, or the spots absent. Filaments and petals with long white hairs at their bases; filaments bright yellow. Ovary sparsely pubescent at base, otherwise glabrate. Pods 3–7 (10) × 1.1–1.4 cm in width, moderately flattened, mostly indehiscent, generally not constricted between the seeds; seeds 1–6.

Blazing yellow masses of flowers in spring, the peak usually late March–early April (about two weeks in advance of flowering of *P. microphylla*); also sometimes flowering sparsely in fall. Prodigious crops of pods usually ripening in May and June.

One of the few common desert trees in the flora area, mostly along watercourses: best developed along major washes, dry streamways, riverine floodplains, and other deep-soil xeroriparian habitats; also sandy soils of dune-lava interfaces and interdune troughs; occasionally on dunes, rocky slopes, and cinder flats or slopes. Widespread across the flora area but apparently no farther west than the vicinities of Sierra del Rosario and northwestern granitic mountains.

Northwestern Sinaloa to southern Arizona and southeastern California, and Baja California Sur. The blue palo verde of Baja California Sur and the south side of Isla Tiburón often has denser foliage and larger leaflets, and seems to differ in characters of the inflorescences and flowers. It is known as *Cercidium floridum* subsp. *peninsulare* (Rose) Carter. Hybrids between *P. florida* and *P. microphylla* have been found in the bottom of Molina and MacDougal Craters, as well as elsewhere in the Sonoran Desert (Jones et al. 1998; Hawkins et al. 1999).

Molina Crater, *Kamb 2014*. Campo Rojo, *F 19757*. Moon Crater, *Burgess 6329*. Sierra del Rosario, *F 20739*. Papago Tanks, *Turner 59-30*. Molina Crater, *Kamb 2014* (DS, UC, putative hybrid between *P. florida* and *P. microphylla*).

Parkinsonia microphylla Torrey [*Cercidium microphyllum* (Torrey) Rose & I.M. Johnston] *PALO VERDE;* FOOTHILL PALO VERDE; *KE:K CEHEDAGĬ.* Small desert trees or large shrubs to 5 m, with thick limbs branching from near the base of a very short, thick trunk; bark yellowish green. Leaves lacking a petiole, with 1 pair of pinnae, each 1–5.3 cm and with 4–8 (10) pairs of leaflets, the leaflets 1–3.5 (6) mm, broadly elliptic to broadly oblong or sometimes orbicular; pulvini conspicuous at the base of leaflets, less conspicuous at the base of pinnae; individual leaflets and ultimately the leafstalks drought deciduous. (Because the leaves lack petioles, each sessile pinna might be confused for a single, once-pinnate leaf.) Youngest, emerging herbage protected by short, stiff, white hairs and glands, the twigs glabrate with age; glands clustered at leaflet bases, 0.1 mm in diameter, globular, and translucent on youngest twigs, dark brown with age.

Peduncles and pedicels with short white hairs similar to those on the young herbage. Flowers (10) 12–18 mm wide. Calyx tube green to yellow-green, the lobes pale yellow, spreading to reflexed (not as strongly reflexed as in *P. florida*). Banner (upper) petal white, with age turning pale yellow and then orange-flecked or not, the other 4 petals pale yellow. Filaments pale yellow, with long white hairs below. Ovary and young green fruit densely white pubescent. Style and stigma pale yellow. Pods sparsely pubescent, 3–9.5 × 0.9–1.1 cm, indehiscent to tardily and partially dehiscent, constricted between the seeds, the apex often extending into a slender, sterile snout. Seeds 1–5 per pod, hard, ovoid, 8.3–9.4 × 6.8–7.1 mm. Masses of pale yellow flowers, mostly April–early May, the pods ripening late May–June.

This is the most numerous small desert tree in the region; widespread including hills and mountains, canyons, arroyos, and upper bajadas, from the Sonoyta region through most of the Pinacate volcanic field to the peak, and the granitic mountains. Sonoran Desert: southwestern Arizona to the Guaymas region in Sonora, both Baja California states, and California at a few stations along the Colorado River.

Sierra Cipriano, *F 91-29*. Pinacate Junction, *F 11182*. Sierra del Viejo, *F 5734*. Campo Rojo, *Sanders 5643* (UCR).

PEDIOMELUM INDIAN BREADROOT

Herbaceous perennials with swollen, tuberous, and often edible roots. Leaves usually palmately compound with 3–7 leaflets. Flowers papilionoid. Pods 1-seeded, dehiscent by transverse rupturing of the pod. Western North America, Mexico to Canada, especially the Great Plains; 21 species. Reference: Grimes 1990.

***Pediomelum rhombifolium** (Torrey & A. Gray) Rydberg [*Psoralea rhombifolia* Torrey & A. Gray] Herbaceous perennials, or probably facultative annuals in the flora area. Roots slender and fibrous; this species known to have swollen, deeply buried tuberous roots ca. 1 × 2 cm. Pubescence of appressed to spreading white hairs. Stems striate, mostly procumbent with age, at least 23–35 cm. Leaflets 3, glandular punctate above, mostly broadly lanceolate, the larger ones 2.0–2.5 cm; petioles 1.5–4.5 cm. Several-flowered headlike spikes on peduncles 2–5 cm. Calyx 4–5 mm, the petals orange, 6.5–8 mm, clawed at base. Body of pod rounded, 4–5 mm, with a beak 5–8 mm.

Suvuk and near Puerto Peñasco; probably not established. Louisiana and Texas to Oaxaca and Veracruz, and Baja California Sur. In Sonora otherwise south of the desert near Alamos and Navojoa. Mostly a species of juniper or sparse oak woodland. It tends to be a pioneer species in disturbed habitats. The flowers are selfing but still produce lots of nectar—at least in greenhouses.

Suvuk, sandy soil, *Nabhan 25 Sep 1982* (DES). 4 km N of Puerto Peñasco, crack in asphalt at edge of highway, *F 85-773*.

PHASEOLUS *FRIJOLES,* BEANS

Mostly twining annual or perennial herbs. Pubescence of both small straight hairs and small hooked hairs (uncinate; visible with 20× magnification). Leaves usually pinnate with 3 leaflets. Flowers papillionoid. Keel of corolla bent laterally across the flower, spirally twisted above the bend, and coiled at tip; style coiled within the keel. Stamens 10, diadelphous. Pods usually explosively dehiscent, the valves twisting; the cultivars mostly indehiscent.

Tropical and subtropical America, mostly in Mexico and very few in deserts; 50 species. The tiny hooked hairs and twisted keel are diagnostic. Includes the common bean or *frijol,* and lima, runner, scarlet, and tepary beans.

Domesticated teparies (*P. acutifolius* A. Gray var. *latifolius* Freeman) were grown along the Río Colorado, at the Suvuk floodwater field, at Quitobaquito, and undoubtedly also at Sonoyta. Wild teparies occur in the Ajo Mountains. References: Delgado-Salinas et al. 1999; Lakey 1984; Maréchal, Mascherpa, & Stainier 1978; Nabhan 1985; Nabhan & Felger 1978.

Phaseolus filiformis Bentham [*P. wrightii* A. Gray] DESERT BEAN; *BAN BAVĬ, CEPULIÑ BAVĬ.* Non-seasonal ephemerals with a well-developed taproot. Herbage and inflorescences moderately to sparsely pubescent. Stems twining or sprawling. Leaflets 3 per leaf, (1) 2–5 cm, thin, highly variable, entire to deeply 3-lobed, often smaller and narrower on leaves closer to the stem tips. Flowers ca. 6–9 mm, the petals pink. Pods 1.5–3 cm, slightly curved, laterally flattened, explosively dehiscent, the valves twisting. Seeds 3–3.9 × 1.6–3.4 mm, flattened, mottled brown and black.

Common and widespread, near sea level to peak elevations in the volcanic and granitic mountains. Often well developed along washes, sometimes twining into overhanging branches and even on saguaros. Also rocky slopes, bajadas, sand flats, and playas; often very common on cinder slopes, especially at higher elevations. Southward in Sonora to the Guaymas region, both Baja California states, and Arizona to Texas. The Pinacate people used the seeds for food.

S of Papago Tanks, *F 18734.* Campo Rojo, *Starr 726.* Elegante Crater, *F 19679.* Pinacate Peak, *19390.* Suvuk, *Nabhan 367.* Cholla Bay, *Gould 4137.*

PROSOPIS *MEZQUITE,* MESQUITE

Mostly trees or shrubs; usually armed with spines, thorns, or prickles. Leaves twice pinnate (rarely leafless or nearly so in South America), with 1 to several pairs of opposite pinnae, the leaflets usually small; petiole with a circular nectar gland at the base of the lower pinnae; pulvini well developed at base of the pinnae and leaflets. Flowers mimosoid, small, usually white to yellow, usually in densely flowered globose heads to elongate spikes or racemes. Calyx shallowly toothed. Stamens 10 in 2 series and separate, the anthers with a small terminal gland. Flowers usually protogynous (the slender style protrudes from the bud before the stamens and petals emerge). Ovary hairy or sometimes glabrous. Pods indehiscent, straight to curved or coiled, usually with carbohydrate-rich mesocarp; multiple seeded.

Hot, arid and semiarid regions; 41 in the Americas and 4 from Africa to India. The mesquites (including *P. glandulosa* and *P. velutina*) belong to section *Algarobia,* with about two dozen species in South America and 6 in North America. The pods are straight to curved and each seed is enclosed in a leathery to bony endocarp. One species, *P. juliflora* (Swartz) de Candolle sensu stricto, bridges the gap from southern Mexico to northern South America. This tropical species extends northward on the Pacific coast of Mexico to coastal mid-Sinaloa. The *Algarobia* seem to be in an active state of evolution, and many of the taxa broadly intergrade at their geographic boundaries. Screwbean (*P. pubescens*) is in section *Strombocarpa,* characterized by tightly coiled pods, with 9 species in North and South America.

The 3 species in northwestern Sonora are winter deciduous, although sometimes tardily so. Brilliant green new foliage appears with warm weather, usually in March and April, and the new shoots can be very fast growing.

Stored mesquite seeds can remain viable for half a century if dry and still enclosed in their endocarp (Martin 1948). The sugary mesocarp and tough endocarp of the pod are adapted for animal dispersal. Germination of the endocarp-enclosed seeds is greatly enhanced after passing through the digestive tract of animals. Although the major Pleistocene herbivores with which the mesquites probably co-evolved are gone, coyotes and other animals consume large quantities of the fruit. Today cattle fill the role of the extinct Pleistocene eaters of mesquite pods (e.g., Mooney, Simpson, & Solbrig 1977). During the late 1970s and early 1980s relatively dense stands of same-sized and presumably same-aged mesquite shrubs became common in playas, alluvial plains, and sand flats in areas of intense cattle grazing. In some of the harsher habitats, such as the playas in the bottom of MacDougal Crater, local mesquite populations wax and wane with cycles of wetter or drier decades. These fluctuations occur in a place free from human disturbance and cattle grazing (R. M. Turner 1990).

The mesquites produce high-quality hardwood that has been the preferred cooking fuel in desert regions of southwestern North America. The mesocarp (pulp) of the pods provided a major carbohydrate (calorie-rich) component in native diets. Although the seeds were apparently not widely used, the people living in the Pinacate region probably included mesquite seeds in their diet. The gyratory crusher, a specialized grinding stone developed in ancient times, is well suited for processing substantial quantities of mesquite pods and seeds. Gyratory crushers are especially common among the artifacts left by the Pinacate people (Felger 1977; Hayden 1969). In addition, mesquite was used for shelter, weapons, tools, fiber, dye, cosmetics, medicine, and other practical as well as aesthetic purposes in both North and South America. References: Bell & Castetter 1937; Benson 1941; Burkart 1976; D'Antoni & Solbrig 1977; Felger 1977; Felger & Moser 1985; Johnston 1962a; Rzedowski & Calderón de Rzedowski 1988; Simpson 1977b.

1. Flowers sessile, bright yellow, the petals united below; pods tightly coiled. ______ **P. pubescens**
1′ Pedicels ca. 0.5 mm, flowers cream to dull yellow, the petals separate; pods straight or nearly so, not coiled;
   2. Leaves jugate (with a single pair of pinnae); herbage glabrous or hairy. ______ **P. glandulosa**
   2′ At least some leaves bijugate (with 2 pairs of pinnae); herbage hairy. ______ **P. velutina**

Prosopis glandulosa Torrey var. **torreyana** (L.D. Benson) M.C. Johnston [*P. juliflora* (Swartz) de Candolle var. *torreyana* L.D. Benson] *MEZQUITE;* WESTERN HONEY MESQUITE; *KUI.* Woody shrubs or trees. Bark dark brown to blackish, irregularly checkered to shredding, often with oily black oozing bacterial slime flux from wounds beneath the bark; younger stems sometimes exuding light-colored gum. Paired or sometimes single thorns (apparently modified branches) above leaves in axillary buds (short shoots), or sometimes absent or reduced from some or most stems, and especially prominent among juvenile and salt-stressed plants; these thorns often 0.8–4 cm, straight, rigid, and sharp, at first dark colored, fading whitish. Twigs and leaves glabrous or hairy (see following). Leaves with 1 pair of pinnae (jugate); leaflets about 5 times longer than wide, 4.5–31 × 1.3–6 (8) mm (the smaller measurements are from drought-stressed plants), (6) 11–19 pairs per pinna.

Inflorescence a dense, spikelike cylinder often (4) 6–9 cm, often several in a cluster. Flowers numerous, crowded, pale yellowish, 4.5–5 mm, the pedicels ca. 0.5 mm. Stamens longer than the petals; ovary and inner surface of petals densely white hairy. Pods straight to moderately curved, often (7) 10–20 cm × 7.5–10 mm maximum width, slightly compressed, slightly to conspicuously constricted between seeds; mesocarp sweet and well developed. Seeds encased in a tough, leathery endocarp drying hard and bony; cotyledons bright green even within the endocarp in the fruit. Flowering April-May, and sporadically through early fall; pods ripening in early summer and sporadically through fall.

Lowlands nearly throughout the flora area except toward Arizona Upland near Sonoyta and Organ Pipe Monument. Most numerous along drainageways, on medium-height dunes, and in some playas and areas of cattle grazing. Trees along the Río Colorado and in major washes, otherwise mostly large shrubs. Forming spreading shrubby mounds, often 4–8 m wide, capping small dunes.

Populations from the vicinity of the Río Colorado and the western part of the dunes are glabrous or glabrate, and have proportionally longer and more widely spaced leaflets, thus falling into the usually accepted concept of *P. glandulosa* var. *torreyana.* Farther eastward, approaching the domain of *P. velutina* and encompassing most of the flora area including the Pinacate volcanic field, the mesquites have the jugate leaves and elongated and often widely spaced leaflets of *P. glandulosa.* These plants also have pubescent leaves and twigs characteristic of *P. velutina,* indicating intermediate populations. There seems to be a tendency for these intermediate populations to have proportionally broader (and shorter) leaflets to the east, where they approach *P. velutina*—so that except for the jugate leaves one could place them with *P. velutina.* Even during favorable times, however, these mesquites produce only jugate leaves characteristic of the *glandulosa* taxon. Although there is a continuum in leaflet characters, the bijugate versus jugate condition is closely correlated with the Arizona Upland and the Lower Colorado Valley phytogeographic zones, respectively. *P. glandulosa,* especially the western var. *torreyana,* and *P. velutina* are very closely related; perhaps they should be treated as subspecies or varieties.

There are 2 geographic races of *P. glandulosa.* Var. *torreyana* primarily occurs from Baja California Sur to southern California, western Arizona, and southwestern Utah; in Sonora it occurs from the Río Colorado southward along the coastal fringe of the state to Guaymas and then inland, beyond the desert, through the thornscrub to northern Sinaloa, and then eastward into New Mexico, western Texas, and northern Mexico. East of the continental divide it is largely replaced by var. *glandulosa* from Texas to Kansas, northeastern Mexico, and the Chihuahuan Desert region; the area of overlap and intergradation is large. Plants of var. *glandulosa,* at least from Texas, are much more cold hardy than are those of var. *torreyana.*

Laguna Prieta, *F 85-747.* Herbage pubescent, intermediate with *P. velutina:* 13.9 mi W of Los Vidrios on Mex 2, *F 89-45;* MacDougal Crater, *Burgess 6325;* Molina Crater, *Kamb 2015;* Pinacate Junction, *F 19613;* Tinajas de Emilia, *F 19731.*

Prosopis pubescens Bentham. *TORNILLO;* SCREWBEAN; *KUWIDCULIS.* Large woody shrubs or small trees to 6 m. Spines stipular, paired, straight, white, often stout, reaching 1–2.5 cm. Leaves jugate, occasionally bijugate; leaflets 4.5–13 mm, 4–9 per pinna. Inflorescences of cylindrical spikes 3–4.5 cm. Flowers sessile, the petals, sepals, filaments, anthers, and anther glands bright yellow; stamens well exserted, the filaments with some reddish flecks, the style and ovary hidden by dense white hairs. Pods in clusters of (3) 6–20+; individual pods tightly coiled, cylindrical, (2) 3–4 cm × 4–5.5 mm. Flowering in May and sporadically through summer and fall; pods ripening mostly during summer months.

Riparian or semiriparian habitats; generally in alkaline or partially saline soils. At the La Salina pozos and a few other pozos near Bahía Adair, Laguna Prieta, sparsely scattered along the shore between La Salina and El Golfo—particularly at canyon mouths, along the Río Sonoyta, Quitobaquito, and the Río Colorado, where it was once abundant. In Sonora otherwise at Quitovac. Also northeastern Baja California and southern California to southwestern Utah, Arizona, Texas, and northern Chihuahua.

The mesocarp of the pods is sweet and was a major carbohydrate source for earlier people. The wood served for fuel and house construction, and the roots and bark as medicine (Bell & Castetter 1937). Mearns (1892–93) gives the following account:

> 'Tornilla.' Screw Bean. Jan. 28, 1894. This tree grows abundantly along the Rio Grande at El Paso and thence down to Fort Hancock, Texas. Going west, on the boundary line, we did not see any more 'Screw-Bean' until we reached the Santa Cruz River. It was quite common at Tucson. West of the Santa Cruz it was not again seen until we reached Quitovaquito, on the Sonoyta River. In the rich 'bottom' of the Colorado River, at Fort Mohave, I saw forests of screw beans. Some of the trees would probably have measured from one to two metres in circumference at the base. The Mojave Indians had gathered immense piles of them for food. They were stacked up in their villages, and covered with a thatch to protect them from weather. I was told that they ground them up into a meal, and mixed [the flour] with other materials for use as food. At Tucson and Quitobaquito the trees seldom exceeded 1/3 of a metre in diameter, or 8 metres in height, while the average size was about one-half as much.

Screwbean has been virtually extirpated from the Tucson region since at least the 1950s. *P. pubescens* is morphologically and geographically isolated from other members of its section, the nearest of which is *P. palmeri* S. Watson in Baja California Sur.

La Salina, *F 84-22*. Laguna Prieta, *F 85-749*. Río Sonoyta, 25 km SW of Sonoyta, *Ezcurra 19 Ap 1981* (MEXU). 20 mi N of El Golfo, *F 75-55*. El Papalote, *F 86-327*.

Prosopis velutina Wooton [*P. juliflora* var. *velutina* (Wooton) Sargent] *MEZQUITE;* VELVET MESQUITE; *KUI.* Large woody shrubs or trees. Similar to *P. glandulosa* var. *torreyana* but (1) the herbage and inflorescences pubescent, (2) at least some and usually most leaves bijugate (new growth, spring, and drought leaves tend to be jugate), and (3) the leaflets proportionally broader, not attaining as large a size, usually more closely spaced, and tending to average more leaflets per pinnae; economic uses likewise similar. Leaflets about 3 times longer than wide, often 6–13 × 2.0–3.5 mm, 14–24 pairs per pinna. Flowering April-May, and sporadically through fall. Pods mostly ripening in early summer and sporadically through summer and fall.

Common in the Sonoyta region and spreading along roadsides. Largest and densest along drainageways, also on bajadas and sometimes rocky slopes. Large mesquite trees were once common along the floodplain of the Río Sonoyta. Widespread in Arizona, northern Sonora southward to the vicinity of Guaymas, and extreme southwestern New Mexico and northwestern Chihuahua.

4.5 mi S of Sonoyta, *F 14425*. 17 mi SW of Sonoyta, *F 90-171*. Aguajita, *F 87-270*.

PSOROTHAMNUS

Shrubs, subshrubs, 1 tree, and 1 herbaceous perennial; usually densely hairy and with lenslike glands especially on the stems. Leaves odd pinnate or reduced or largely leafless in *P. spinosus*. Flowers papilionoid, the petals pink to vivid blue or purple. Ovules 2 to several; pods indehiscent, 1- or 2-seeded. Highly modified xerophytes in all 4 North American deserts; 9 species. Reference: Barneby 1977.

1. Large shrubs or small trees; twigs ending in sharp thorns; mostly leafless or leaves few and reduced to 1 leaflet. ____ **P. spinosus**

1′ Shrubs, scarcely woody at base; twigs spinescent or not; branches leafy, the leaves pinnate.

  2. Spinescent; doubtfully present in the flora area. ____ **P. arborescens**

  2′ Unarmed; common and widespread. ____ **P. emoryi**

Psorothamnus arborescens (Torrey ex A. Gray) Barneby var. **arborescens** [*Dalea arborescens* Torrey ex A. Gray] Stiffly and irregularly branching, often thorny shrubs 0.3–1 m. Young branchlets usually purplish. Leaves 1.0–3.5 cm, with (3) 4–7 leaflets. Flowers relatively large with indigo-blue petals. Known from Sonora from a single, out-of-place collection: Puerto Peñasco, *H.L. Cook in 1934* (SD, see Barneby 1977:33). Var. *arborescens* occurs in the Mojave Desert of California. There are 3 other varieties; none occur near Sonora.

Psorothamnus emoryi (A. Gray) Rydberg var. **emoryi** [*Dalea emoryi* A. Gray] EMORY INDIGO BUSH. Small, densely branched shrubs, mostly 1–1.2 m. Herbage tomentose, grayish white with very densely tangled hairs obscuring the surfaces, and sparsely dotted with minute red-orange glands; crushed herbage and flowers sweet smelling and imparting a fugitive yellow-orange stain. Leaves tardily drought deciduous, 1.8–4.5 cm; leaflets (3) 5–11 or sometimes reduced by 1 leaflet and even pinnate on upper branches, the margins entire or with rounded, shallow teeth, the paired leaflets 2–9 mm, mostly obovate to spatulate, blunt-tipped, the terminal leaflet (7) 12–32 mm, linear to narrowly oblong; stipules 0.7–1.0 mm, deciduous, inconspicuous and obscured by hairs. Flowers in dense capitate clusters. Calyx 4.5–6.1 mm, with oval, red-orange glands between the ribs. Corollas dark purple, longer than the calyx. Pods ca. 2.5 mm, the seeds ca. 1.5 mm, smooth, brown. Flowering after rains, spring and summer.

Widespread and abundant on sand flats and shifting dunes. Var. *emoryi* occurs in southeastern California to northern Baja California Sur, extreme southwestern Arizona, and coastal Sonora southward to Bahía Colorado (28°17′ N). Southward in Baja California Sur it intergrades with var. *arenarius* (Brandegee) Barneby. This species is characteristically on coastal dunes in the Gulf of California region. It is sometimes parasitized by *Pilostyles thurberi* (Rafflesiaceae).

16 mi N of El Doctor, *F 92-510.* 13 km NNE of Gustavo Sotelo, *Ezcurra 15 Ap 1981* (MEXU). Puerto Peñasco, *F 85-760.* N of Sierra del Rosario, *F 75-37.*

Psorothamnus spinosus (A. Gray) Barneby [*Dalea spinosa* A. Gray] SMOKE TREE. Small hardwood trees 3–6 m, the trunks or major limbs often several, erect to ascending, often moderately irregular and twisted, sometimes reaching 22–35 cm in diameter; bark smooth and gray, with age the trunk often somewhat fissured and rough. New growth appearing during hot weather. Upper and smaller stems much-branched, the twigs silvery to bluish gray, thorn-tipped and painful to handle (perhaps due to mild toxins), densely hairy with short appressed white hairs, when crushed with an odor akin to turpentine and sweeter than pine; pedicels and calyces also with short white hairs. Essentially leafless, the leaves very sparse, present on fresh young growth, quickly deciduous, reduced to 1 leaflet often 5–25 mm, oblanceolate, bluish gray, thick and almost succulent, dotted with orange glands. (This leaflet is analogous with the enlarged terminal leaflet of other *Psorothamnus* species.) Seedlings and young plants usually leafier with leaves reaching 20–65 mm. Inflorescences of few-flowered, spinescent, twig-like racemes. Flowers 12 mm, showy; calyx 4.5–5.0 mm, the tube prominently ribbed with several large orange-brown blister glands around the middle, the lobes reflexed with age; corollas brilliant indigo-blue. Pods often 5.0–5.5 mm with several orange-brown blister glands around the middle, and a tail-like persistent style, 1(2?)-seeded, the seeds 4.0–4.3 mm, oblong-ovoid, smooth, brown, and mottled. Spectacular, mass flowering in June-July.

Sandy and gravelly soils; often along larger washes, floodplains, sand flats, and interdune troughs. Especially common between the dunes and the southern and western margins of the Pinacate lava shield, along the floodplain of the Río Sonoyta near the Arizona border nearly to its terminus east of Puerto Peñasco, and the highway and washes in the valleys between granitic mountains northwest of the volcanic field. Seed germination is assisted by scarification, which could certainly be accomplished during flash floods. In some places smoke trees have spread along highway shoulders, the seeds apparently scarified by highway traffic. Not known elsewhere in Sonora. Also southwestern Arizona, southeastern California, southern Nevada, and Baja California.

In Sonoyta, Mearns (1892–93) recorded "*mangle*" as the local name. However, "*mangle*" in Mexico is usually applied to mangroves such as *Avicennia* and *Rhizophora.*

Puerto Peñasco, *F 85-779.* 21 mi N of Puerto Peñasco, *F 13208.* 17 mi SW of Sonoyta, *F 90-173.*

SENNA

Herbs, shrubs, or trees. Leaves even pinnate, often foul smelling, the leafstalk often with nectary glands. Pedicels without bracts. Inflorescences of axillary racemes or sometimes a terminal panicle. Flowers caesalpinioid, usually yellow, often with 6 or 7 fertile stamens and 3 sterile, smaller stamens (staminodes), the fertile anthers opening by terminal pores, the filaments shorter than to about as long as the anthers. Fruits varied, dehiscent or not.

Mostly in the Americas; 250 species (most were included in *Cassia*). *Senna* and *Solanum* are the only genera in our flora with buzz-pollinated flowers. Pollination is accomplished by "buzzing," or drumming, pollen-collecting bees, the vibrations causing the pollen to be ejected from the anther pores. Reference: Irwin & Barneby 1982.

Senna covesii (A. Gray) H.S. Irwin & Barneby [*Cassia covesii* A. Gray] *HOJASEN, DAISILLO; KO'OWĬ TA:TAMĬ.* Perennials from a semiwoody base, probably not long-lived; dormant during colder and drier months. Stems mostly erect, semiwoody, dying back in drought; herbage densely velvety hairy. Leaves drought deciduous, the leafstalk with a long-stalked orange nectary gland between the leaflets of each pair; leaflets (2) 3 (4) pairs, elliptic to elliptic-oblong, the larger ones 1.5–2.8 (3.5) cm. Flowers 1.5–2 cm wide, pale yellow. Pods 2.2–3.4 cm, splitting along both sutures, with dense short appressed to spreading hairs. Flowering response nonseasonal during warmer weather.

Mostly on rocky soils of slopes, bajadas, and roadsides; sometimes along gravelly washes; Sonoyta region and occasional in the Pinacate volcanic field, seldom common. Sinaloa to Arizona, Nevada, southeastern California, and Baja California.

4.5 km W of Sonoyta, *F 85-150.* Observation: 1 mi S of Tinaja de los Pápagos.

SESBANIA

Annual or perennial herbs, shrubs, or trees with soft often pithy wood. Leaves long and slender, even pinnate with many leaflets. Flowers papilionoid, on axillary racemes. Pods woody and winged to often slender and elongated, dehiscent to indehiscent with partitions between the seeds. Warm regions of the world; 50 species.

These legumes are noted for very fast growth under hot, humid conditions. Various sesbanias are used for food, medicine, fiber, firewood, forage, green manure, ornamental plants, reforestation, dye, gum, tannin, pulp, and paper; some are noxious weeds. References: Isely 1990; Rydberg 1924.

Sesbania herbacea (Miller) McVaugh [*Emerus herbacea* Miller. ?*Sesban exaltata* (Rafinesque) Rydberg. ?*Darwinia exaltata* Rafinesque. ?*Sesbania macrocarpa* Muhlenberg. *Sesban sesban* (Linnaeus) Britton. *Sesban sonorae* Rydberg] *BEQUILLA, BAIQUILLO;* COLORADO RIVER HEMP, COFFEE WEED. Giant feathery-leaved, wandlike hot-weather annuals, reaching 3–4 m with sufficient moisture, but often 0.5–1 m. Stem strong and stout, unbranched or with slender, flowering branches above. Larger leaves 15–35 cm, with ca. 30–41 leaflet pairs; leaflets 10–35 mm, linear-oblong. Racemes 4–10 cm, slender, several-flowered. Corollas dull yellow, the banner 16–17 mm, flecked with purple-brown. Pods often in clusters of several or sometimes single, often 15–21 cm, slender with a short beak, dehiscent, the valves not twisting. Seeds many, 3.2–4.5 mm, oblong, smooth, brown and mottled.

This giant legume herb grew in great thickets along the lower Río Colorado. The stems yield a smooth, lustrous, and very strong fiber known as wild hemp or Colorado River hemp. Colonia Lerdo "was started in June 1873 by San Francisco capitalist Thomas H. Blythe and Mexican Gen. Guillermo Andrade, who dreamed of building an agricultural empire on the lower Colorado, harvesting the wild hemp . . . that abounded in the bottomlands" (Lingenfelter 1978:60). In November 1922, Aldo Leopold found that "one can gather quite a lot of the beans [seeds] by simply holding the hands cupped and letting them rain in from overhead" (Meine 1988:208). Today *Sesbania* is a common weed of roadsides and irrigation ditches south of San Luis.

This species, as broadly interpreted by McVaugh (1987), is widespread from southern United States to northern South America. It is one of the largest annual plants in the Sonoran Desert and seems to be the only species of *Sesbania* in Sonora. According to Isely (1990:70) this species "is probably an annual derivative of the tropical, somewhat shrubby *S. emerus* and perhaps should be considered a regional variant of that species."

5 km S of San Luis, *F 85-1033.* Colorado River, Sonora below mouth of Hardy River, 27 Mar 1894, *Mearns 2837* (DS). Yuma, Colorado River, *Brandegee Ap 1884* (DS). California: Paloverde, Colorado Desert, plentiful—called "wild hemp" and used for chicken feed, 31 Oct 1904, *Shellenger 22* (UC).

FOUQUIERIACEAE OCOTILLO FAMILY

The family has 1 genus. **Reference:** Henrickson 1972, 1977b.

FOUQUIERIA

Spiny shrubs or trees. Long shoots with alternate, widely spaced nodes; long-shoot leaves with well-developed petioles, each forming a rigid spine from the petiole and leaf-blade midrib. Short shoots in axils of the spines, their internodes reduced and not apparent, the leaves clustered and sessile to short petioled. Flowers subtended by 3 deciduous bracts; sepals 5, separate and overlapping; petals fused below into a tube, with 5 lobes. Stamens 10 or more, uneven in length. Ovary superior. Fruit a capsule, with 6–15 flat, papery-winged seeds. Dry regions of Mexico, 11 species, 1 extending into the United States. Includes the *cirio* or boojum tree, *F. columnaris*.

The system of branching and spine formation in *Fouquieriaceae* is unusual but apparently not unique. Convergence occurs among *Sesamothamnus* in the Pedaliaceae, native to eastern and southern Africa, which has similar leaf-spine formation, and some of the species approach certain *Fouquieria* species in gross morphology.

Fouquieria splendens Engelmann subsp. **splendens.** OCOTILLO; *MELHOG*. Unique long-lived desert "shrubs" often 3–5 m, with slender few-branched wandlike spiny branches arising from a very reduced trunk. Short-shoot leaves appearing at almost any time of year following a ground-soaking rain, the leaves turning yellow and falling quickly as the soil dries or with freezing or cold weather. Long shoots produced during times of hot weather and high soil moisture, such as in the wake of a fall hurricane-fringe storm. Inflorescences of dense panicles at branch tip, often (6) 19–24 cm, with conspicuous leafy bracts falling at about anthesis. Flowers bright red-orange, the corolla lobes reflexed. Mostly late February–March, attracting hummingbirds; fruits ripening in late spring.

One of the most common long-lived perennials in the region. In a wide range of habitats and soils from near sea level to peak elevations; granitic and volcanic slopes, cinder cones, crater walls, bajadas, rocky plains and pavements, and sand flats overlying rocky substrate. Absent from dunes and seldom on deep sand soils, but sometimes abundant on sand flats such as southeast of Sierra Blanca and at MacDougal Pass. Notable stands of very large ocotillos grows on cinder slopes below the south and southeast sides of the Sierra Pinacate peaks, and between MacDougal and Sykes Craters.

The long, straight stems are made into living fences and animal corrals—the stems readily form roots and sprout new stems. The usual harvesting practice is to cut all of the stems near the base of the plant. When cut close to the base the plants usually die, but when harvested ca. 1 m or more above ground level they usually survive to resprout. The larger plants may be more than a century old (see Felger & Moser 1985:303). Although the cortex below the waxy bark may contain chlorophyll, photosynthesis ceases in the drought-induced leafless condition (Mooney & Strain 1964).

There are 3 subspecies. The most wide ranging is subsp. *splendens,* encompassing nearly the entire range of the species. It extends though most of the Sonoran and Chihuahuan Deserts and ranges into elevations well above the deserts; in Sonora southward to the Guaymas region. The other 2 subspecies are in the southwestern and southern reaches of the Chihuahuan Desert.

Km 83 on Mex 8 [ca. 10 km N of Puerto Peñasco], *Lockwood 94*. Bahía de la Cholla, *Yatskievych 80-29*. Sierra Extraña, *F 88-69A*. MacDougal Crater, *F 10764*.

FRANKENIACEAE FRANKENIA FAMILY

Herbs and shrubs; mostly halophytes in warm regions of the world, the distribution interrupted, mostly in Mediterranean climates and temperate-tropical deserts; 3 genera, 85 species. Related to the tamarisks (Tamaricaceae).

FRANKENIA

Annual or perennial herbs or subshrubs with salt-excreting glands. Leaves opposite, simple, sessile or petioled, and entire, the bases membranous and often united, the margins often inrolled (revolute). Flowers small, mostly bisexual (ours), radial, sessile (the American species). Sepals united into a tube with 4 or 5 short lobes. Petals 4 or 5, separate, each with an appendage near the base. Stamens 4–7. Ovary superior. Fruit a capsule enclosed by the persistent calyx. Worldwide, mostly halophytes in temperate and subtropical salt marshes and salt deserts; 80 species. Reference: Whalen 1987.

1. Woody dwarf shrubs without rhizomes; stems stiff; leaves mostly 2.5–3.5 mm, nearly terete. ____________ **F. palmeri**

1′ Herbaceous perennials with rhizomes; stems flexible; leaves mostly 6–12 mm, flattened but strongly revolute. ____________ **F. salina**

Frankenia palmeri S. Watson. Dwarf woody shrubs, 30–50 cm; shallow-rooted. Leaves mostly 2.5–3.5 mm, essentially evergreen, broadly oblong, nearly terete and beadlike with strongly revolute margins, and grayish with thick, minute white hairs and encrusted with salt; short-petioled. Flowers small and starlike. Calyx 3–5 mm, the lobes very short. Petals 5, white, 5 mm, the lobes irregularly and minutely toothed at tips. Anthers red. Flowering at least January-May, mostly April.

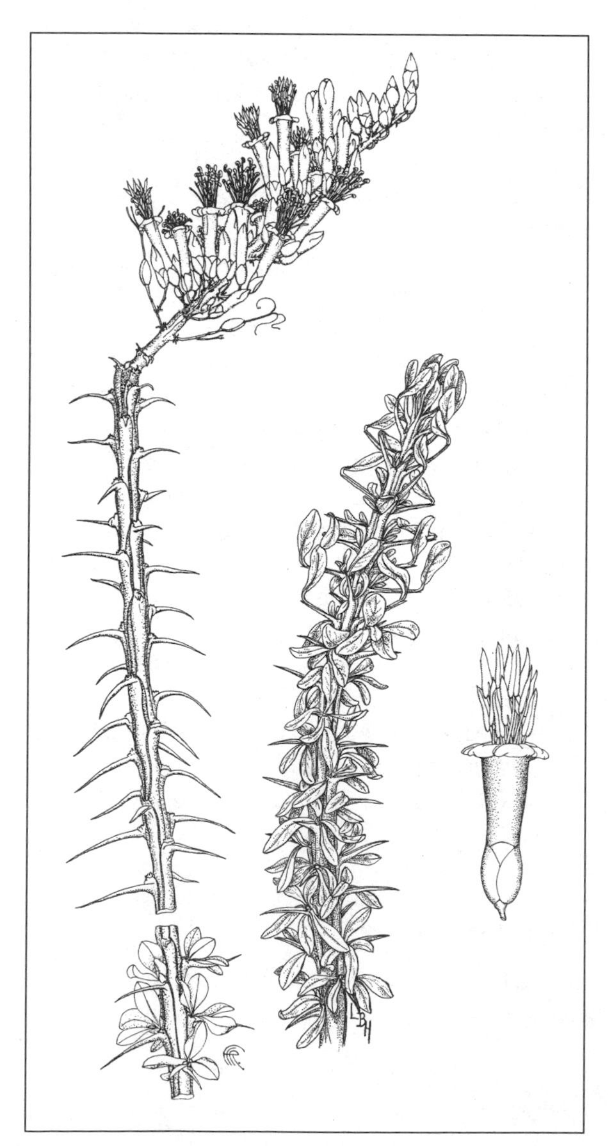

Fouquieria splendens **subsp. *splendens.*** Note spine formation from petiole midrib (center), and leaves emerging from axillary short shoots (lower left). LBH (Humphrey 1960) and FR (Henrickson 1972).

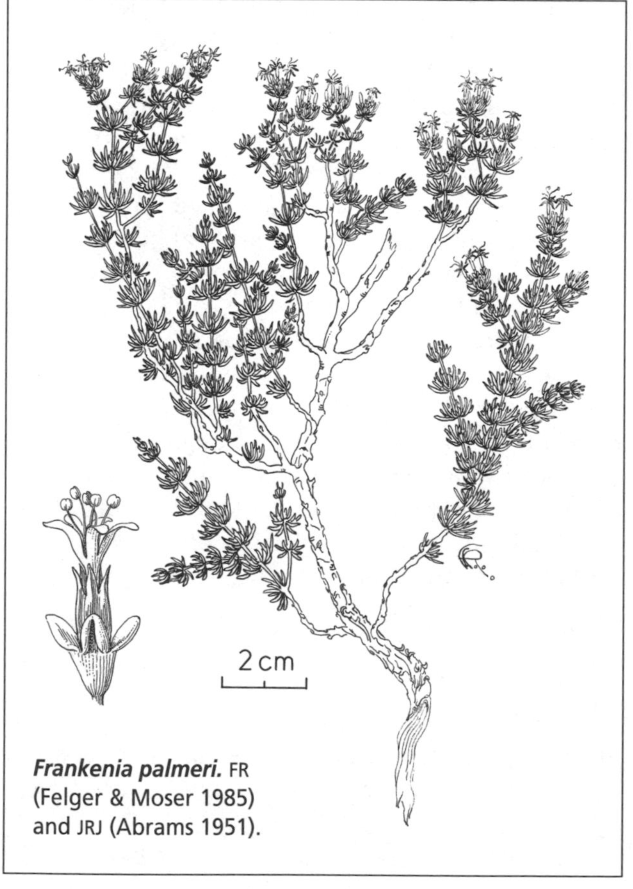

Frankenia palmeri. FR (Felger & Moser 1985) and JRJ (Abrams 1951).

Forming a zone of low, gray, and monotonous coastal vegetation on sand and rocky granitic soils; also on coastal volcanic hills; between the coastal halophytic saltscrub and the lower margin of mixed desertscrub or creosote bush scrub. Coastal margins of southern California, both coasts of the Baja California Peninsula, and southward in Sonora to Punta Baja (between Bahía Kino and Tastiota).

The flowers are visited by flies and small wasps including ichneumonids (see *Lycium californicum,* Solanaceae). The leaves excrete salt and on nights with dew, water condenses on the leaves and drips down the stems onto the ground. When dry the leaves are gray and the stems brownish, but when wet the leaves turn green and the stems become blackish. The occurrence of coastal maritime dew seems to limit its distribution. In most places *F. palmeri* ranges no more than 1–3 km inland from the shore. However, in central Baja California it occurs where the Vizcaino maritime fog extends far inland.

Bahía de la Cholla, *Gould 4150.* Puerto Peñasco, *Shreve 7595.* Observation: La Salina.

Frankenia salina (Molina) I.M. Johnston [*Ocymum salinum* Molina. *F. grandifolia* Chamisso & Schlechtendal] ALKALI HEATH. Herbaceous perennials with thickened, rhizomatous rootstocks. Stems slender, wiry, and nearly black, the leafy stems seasonal. Leaves 6–12 mm, mostly linear-oblong to narrowly elliptic or oblanceolate, the lower margins strongly revolute; short-petioled. Calyx 5.0–7.5 mm, the lobes very short. Petals 5, pink, exserted 2–2.6 mm beyond the calyx tube, the tips entire to minutely toothed. Filaments, style, and stigmas white, the anthers lavender-purple. Seeds mucilaginous when wet. Flowering at least July and September.

Common in tidally inundated soils of esteros and bays among other maritime halophytes. Southward along the coast of Sonora to Estero Sargento at 29°20′ N. Also coastal salt marshes and inland saline soils in California, Baja California, and northwestern Baja California Sur, and disjunct at the southwestern edge of the Chihuahuan Desert in Coahuila and Zacatecas, and coastal Chile.

Puerto Peñasco, *F 20962.* Bahía de la Cholla, *López-Portillo 9 Jul 1984* (MEXU). Estero Cerro Prieto, *F 90-162.* SW of Gustavo Sotelo, *López-Portillo 9 Jul 1984* (MEXU).

GENTIANACEAE GENTIAN FAMILY

Annual to perennial herbs, rarely shrubs, mostly glabrous (ours). Leaves mostly opposite, simple, usually entire, often clasping the stem; stipules none. Flowers usually bisexual and radial, often showy, 4- or 5- (7)-merous except the pistil, the calyx persistent, united below, the corolla united below. Ovary superior. Fruit a capsule; seeds minute, often numerous.

Mostly in moist, cool, often montane, arctic or alpine habitats; few in deserts and hot climates. Worldwide; 75 genera, 1200 species. Apart from 2 other *Centaurium* species at the desert edge, ours are the only 2 members of this family in the Sonoran Desert. They are worthy of cultivation.

1. Leaves green, not glaucous; petals bright pink; anthers twisting after dehiscence. ________ **Centaurium**
1′ Leaves glaucous blue-green; petals white; anthers not twisting. ________ **Eustoma**

CENTAURIUM CENTAURY

Annual or biennial herbs. Leaves sessile, often clasping. Inflorescences of terminal cymes. Flowers pedicelled to subsessile, white to pink or rose, (4)- 5-merous except the pistil. Anthers twisting spirally after dehiscence. Worldwide in moist, often montane habitats; 50 species.

Centaurium calycosum (Buckley) Fernald [*Erythraea calycosa* Buckley. *C. calycosum* var. *arizonicum* (A. Heller) Tidestrom. *C. arizonicum* A. Heller] Annuals, 12–45 cm, the stems erect, single to much-branched; herbage green. Leaves 1–4 cm, lanceolate or oblanceolate to ovate or obovate, entire. Flowers showy; pedicels often about as long as the calyx tube. Calyx lobes linear, separate nearly to the base. Corollas 5-lobed, the lobes 7–12 mm, bright pink, spreading, the throat yellow-green, or corollas rarely white. Filaments and style white, the anthers and stigma yellow. Capsules 1 cm, oblong, the walls relatively thin and smooth; seeds 0.35–0.45 mm, ovoid, minutely reticulate and pitted, gray, iridescent. March-November.

Locally abundant in wet alkaline soil at seeps and springs from Quitobaquito to Williams Spring in Organ Pipe Monument. Also recorded from Yuma and elsewhere along the lower Colorado River, and probably once along the lower Río Colorado, although not known for certain from the state of Sonora. In many life zones in widely separated alkaline wetlands in southwestern United States and northern Mexico.

Quitobaquito, *Bowers 1308*. Williams Spring, *F 86-272*. Yuma, *Brown 29 Apr 1906*.

EUSTOMA CATCHFLY GENTIAN

Herbaceous perennials with glaucous herbage and leafy stems. Flowers showy, long pedicelled, 5- or 6-merous except the pistil. Capsules ellipsoid, many-seeded, the seeds pitted. Southern United States to northern South America; 3 species. Reference: Shinners 1957.

Eustoma exaltatum (Linnaeus) G. Don [*Gentiana exaltata* Linnaeus] CATCHFLY GENTIAN; *HAWAÑ TA:TAḌ*. Stems erect, 30–75 cm; herbage glaucous blue-green, winter deciduous. Leaves 2–6 cm, the lowest leaves short petioled, the stem leaves sessile, clasping, and often fused basally, obovate to broadly elliptic or oblong, thickish, entire. Flowers 2.5–4 cm. Calyx deeply cleft, the lobes slender, keeled. Corollas cream-white, with 5 (6) large erect to partially spreading lobes. Anthers yellow. Capsules 1.0–1.5 cm, oblong, thickened, tough and rough-surfaced; seeds 0.5 mm, rounded, deeply pitted, gray, iridescent. June-September.

Locally abundant in wet alkaline soil at seeps and springs from Quitobaquito to Williams Spring; not known elsewhere in Organ Pipe Monument or northwestern Sonora. Recorded on the U.S. side of the lower Colorado River and probably formerly present on the Mexican side; also in a few widely separated alkaline wetland habitats from western Sinaloa to Arizona. Southern California to Florida, Mexico to Venezuela, and the West Indies. Across much of its range the corollas are blue or lavender; those with white petals are known as forma *albiflorum* Benke.

Quitobaquito: *F 86-213; Mearns 7 Feb 1894* (US, not seen). Yuma, *Pritchard 20 Jul 1961*.

GERANIACEAE GERANIUM FAMILY

Herbs or shrubs. Leaves usually opposite; often with stipules (ours). Inflorescences umbellate or cymose. Flowers usually bisexual, mostly radial, the sepals 5, separate and persistent, the petals 5, separate, or rarely absent. Stamens usually in 1–3 whorls of 5 each, one or more whorls often without anthers. Ovary superior. Fruit a capsule, typically beaked or lobed and distinctive by its usually elastic dehiscence and separation of the "mericarps."

Both hemispheres, largely tropical and temperate, especially Africa; 14 genera, 730 species. The garden and greenhouse geraniums belong to the genus *Pelargonium*. Ornamental geraniums are sometimes cultivated in Sonoran Desert gardens but they suffer from the extreme heat of summer and are frost sensitive.

ERODIUM FILAREE, STORKSBILL

Annual or perennial herbs. Leaves usually opposite, toothed and lobed or pinnately dissected, the annuals with early leaves in basal rosettes. (Opposite leaves in the pair, especially among *E. texanum*, tend to be unequal in size.) Flowers in axillary umbels, mostly radial or nearly so (ours), the pedicels recurved. Upper 2 petals slightly smaller than the other 3. Stamens 10: 5 fertile and 5 sterile. Carpels at maturity long-beaked, separating elastically from the base and twisting spirally. The long, corkscrew-like beak on each fruit segment uncoils when moistened and screws the sharp-pointed and heavier, seed-bearing end into the ground.

Mostly Old World; 75 species. Several are important forage plants, and some Old World perennials are cultivated as ornamentals in temperate regions, especially in rock gardens. Some annuals have become troublesome weeds. Reference: Guittonneau 1972.

1. Leaf blades pinnately dissected, much more than twice as long as wide, the petiole shorter than the blade. **E. cicutarium**

1′ Leaf blades ovate to 3-lobed or 3-parted, less than twice as long as wide, the petiole longer than the blade. **E. texanum**

***Erodium cicutarium** (Linnaeus) L'Héritier ex Aiton [*Geranium cicutarium* Linnaeus] *ALFILERILLO;* FILAREE, STORKSBILL; *HOHOI 'IPAD*. Winter-spring ephemerals. Stems sometimes to 30+ cm, usually much shorter or essentially stemless. Herbage with glandular and non-glandular white hairs, the glands minute. Leaves pinnatifid, mostly twice divided, much longer than wide, the blades 2–15+ × 0.8–4 cm, the petioles 1.8–7.5 cm. Umbels (1) 3–7-flowered. Flowers 11–14 mm across, the petals lavender-pink, readily falling, slightly longer than the sepals; fruiting sepals 5.0–6.5 mm. Beak of fruits 2.7–3.8 cm. During highly favorable conditions the plants occasionally reach 2 m across with lower leaves to 27 cm, peduncles 8–14 cm, and fruit beaks 3.5–4.5 cm.

Widespread to peak elevations; often especially common in disturbed habitats including roadsides and as an urban and agricultural weed, also in many natural habitats including sandy-gravelly washes and sand flats; apparent maximum local development in the Sonoyta region and on cinder soils toward higher elevations in the Sierra Pinacate. Western Sonora south at least to the vicinity of El Desemboque San Ignacio. Native to the Mediterranean region and introduced in early Spanish colonial times as a forage plant; widely naturalized in the New World, mostly in non-tropical regions.

N of Puerto Peñasco, *F 87-22*. W of Sonoyta, *F 20599*. Summit of Pinacate Peak, *F 19441*. SE of Carnegie Peak, 975 m, *F 19944*.

Erodium texanum A. Gray. FALSE FILAREE, DESERT STORKSBILL. Winter-spring ephemerals. Stems sometimes reaching 25+ cm, but usually much shorter or essentially stemless. Herbage with small, coarse white hairs, not glandular. Leaf blades 9–21 mm, ovate to heart-shaped or rounded in outline, usually 3-lobed or parted, the margins toothed; petioles 10–42 mm. Umbels 2–3-flowered. Petals pink-purple, readily falling, longer than the sepals; fruiting sepals 5.5–9 mm. Beak of fruit (2.6) 3.2–5.0 cm.

Erodium texanum. FB.
© James Henrickson 1999.

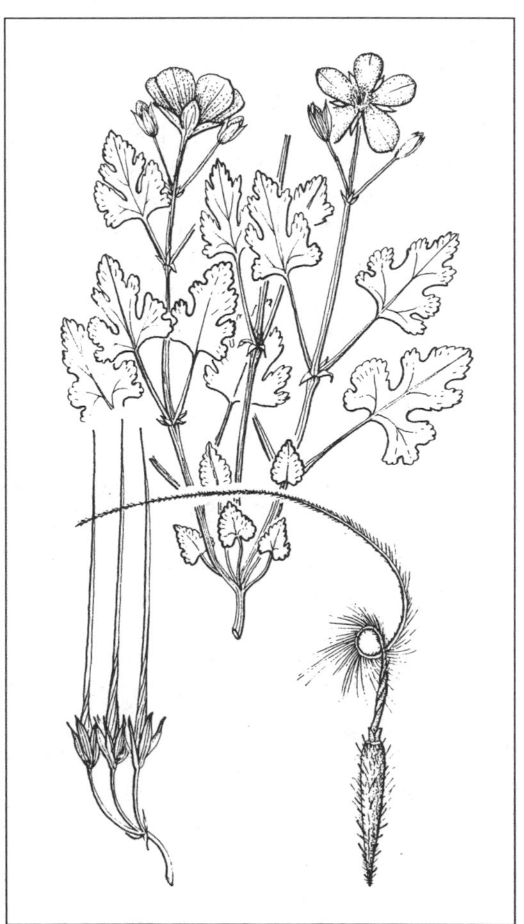

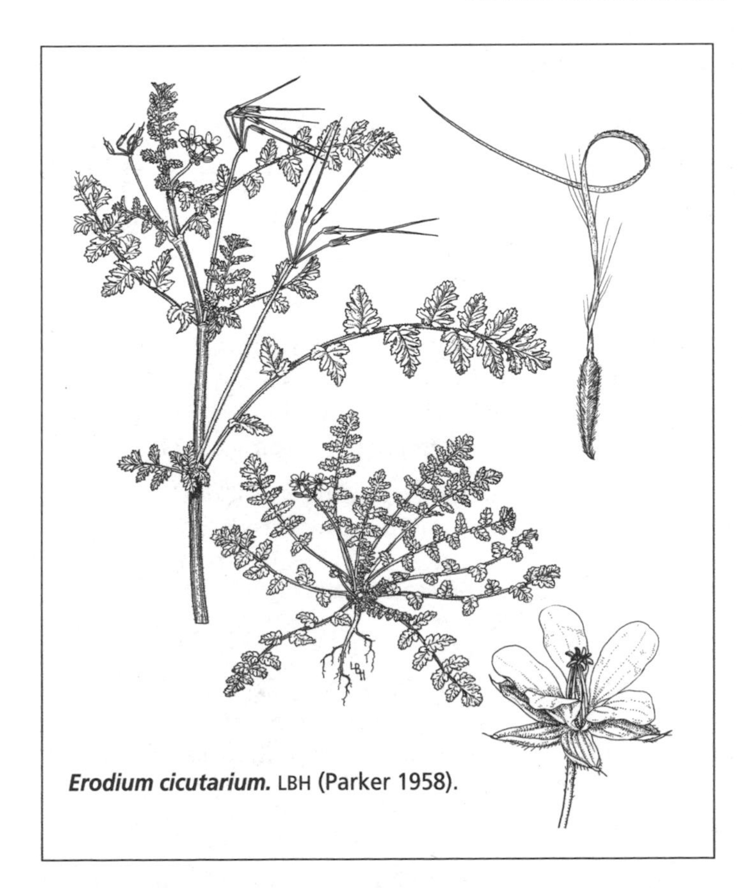

Erodium cicutarium. LBH (Parker 1958).

Widespread, mostly at lower elevations and usually on sandy or fine-textured soils, sometimes among rocks and on desert pavements, less common on dunes. *E. texanum* is generally more common at low elevations in open and dry habitats, whereas *E. cicutarium* seems more common toward higher elevations in the Sierra Pinacate. However, their local distributions broadly overlap.

Southeastern California to southwestern Utah and Texas to northern Mexico including Baja California and western Sonora south to the vicinity of El Desemboque San Ignacio. Plants in western Sonora have relatively small flowers; those from Texas tend to have much larger flowers.

MacDougal Crater, *F 9898.* S of Pinacate Junction, *F 19665.* Moon Crater, *F 19237.* Sierra del Rosario, *F 20751.*

HYDROPHYLLACEAE WATERLEAF FAMILY

Annual or perennial herbs and a few shrubs and trees; often rough or glandular hairy and smelly. Leaves mostly alternate, or partly or wholly opposite, simple to pinnately compound; stipules none. Inflorescences of variously modified, often helicoid cymes, or less often the flowers solitary and axillary. Flowers radial, the sepals, petals, and stamens usually 5-merous. Sepals united and often cleft (lobed). Petals united, the stamens attached to the corolla tube, the tube often with a scalelike appendage at the base of each filament. Ovary superior (except *Nama stenocarpum*), the carpels 2; style terminal, usually single and bifid, or in some *Nama* species 2-cleft nearly to the base (or the style sometimes 3-branched in *N. stenocarpum*). Fruit a capsule. Seeds few to many.

Widely distributed, concentrated in dry regions of western United States; 18 genera, 275 species. *Phacelia* is the largest genus. A few hydrophylls are grown as garden plants in temperate climates. The Sonoran Desert species grow during the cooler seasons and most are ephemerals. Ours are winter-spring ephemerals with blue, lavender-purple, or white flowers. Most of the *Phacelia* species, but not the other genera, have helicoid inflorescences.

1. Stems square in cross section, weak and scrambling, with small and often larger hooked hairs on stem angles; calyx with conspicuous auricles (sepal-like appendages between calyx lobes). ________ **Pholistoma**

1′ Stems rounded in cross section, usually not weak and not scrambling, the hairs not hooked; auricles not present.

  2. Leaves sessile or the blade tapering into the petiole, the margins entire (sometimes inrolled). ________ **Nama**

  2′ At least the lower leaves petioled, the petiole and blade clearly differentiated, the blades pinnately lobed, pinnatifid or dissected, or toothed to wavy.

    3. Leaves opposite below; ovary 1-chambered; calyx lobes cleft to about the middle or 3/4 distance to base, the tube evident; stamens not exserted from corolla tube. ________ **Eucrypta**

    3′ Leaves all alternate; ovary 2-chambered; calyx lobes divided nearly to base, the tube very short or absent; stamens exserted or not. ________ **Phacelia**

EUCRYPTA

Delicate spring annuals; 2 species. Herbage aromatic and viscid-glandular, the leaves pinnatifid, opposite below, alternate and reduced upwards with petiole bases clasping the stem. Cymes often few-flowered. Calyx divided about 1/2 to 3/4 of the way to the base, the lobes enlarging slightly in fruit. Stamens not protruding from corolla. Style bifid near apex. Seeds several to 16. Reference: Constance 1938.

1. Leaf lobes shallowly pinnatifid, the tips obtuse-angled rather than rounded; calyx opening wide (like a dish) at maturity to expose the capsule, the lobes broadly spreading, without stalked glands; seeds dimorphic. ________ **E. chrysanthemifolia**

1′ Leaf lobes entire or sometimes few-toothed, the tips rounded; calyx lobes erect and enclosing the capsule (only tip of capsule visible), usually with stalked glands; seeds all alike. ________ **E. micrantha**

Eucrypta chrysanthemifolia (Bentham) Greene var. **bipinnatifida** (Torrey) Constance [*Phacelia micrantha* var. ? *bipinnatifida* Torrey] Highly variable in size depending on soil moisture and shading. Stems slender, erect to spreading or decumbent with age, 4–15 (25) cm. Leaves pinnatifid to partially bipinnatifid, mostly 1.5–7 × 0.5–3 (4.3) cm. Calyx divided about 2/3 of the way to the base, at maturity (4.5) 5.5–6.5 mm across, rotate, the lobes spreading to clearly reveal the capsule; calyx pubescent but without stalked glandular hairs. Corollas pale lavender. Capsules at maturity separating into 2 conic-hemispherical halves with stout white hairs. Seeds dimorphic within the same capsule: several seeds chunky, blunt-ended, and irregularly lumpy-wrinkled, dark brown with age, 1.0–1.1 mm, and 2 seeds light brown, 1.2 mm, flattened and concave on one side, and sharp-edged and strongly convex (almost conic) on the other side, essentially smooth on both sides (the convex side minutely reticulate).

Arroyo beds, north-facing rocky slopes, sand or cinder flats, generally in protected and often shaded habitats such as beneath shrubs; widespread in the Sonoyta region, granitic ranges, and Pinacate volcanic field to peak elevation. Var. *bipinnatifida* in deserts in southern Californica, Nevada, Baja California, Arizona, and to southern Sonora in thornscrub. Replaced by var. *chrysanthemifolia* on the Pacific coast of California and Baja California.

In addition to characters given in the key, *E. chrysanthemifolia* in the flora area differs from *E. micrantha* in that the plants tend to be larger, usually with larger and more "fernlike" or more finely dissected leaves, differently shaped capsule valves, and fewer and differently shaped seeds. Although they often grow intermixed, *E. chrysanthemifolia* seems to favor slightly more mesic and cooler microhabitats and be the more common of the two at higher elevations on Sierra Pinacate.

Tinaja de los Chivos, *F 18809*. Arroyo Tule, *F 19210*. Tinajas de Emilia, *F 19741*. S of Carnegie Peak, 900 m, *F 19891*. Quitobaquito, *F 88-134*.

Eucrypta micrantha (Torrey) A. Heller [*Phacelia micrantha* Torrey] Glandular-viscid, with stalked glandular hairs intermixed with non-glandular hairs. Stems often 5–23 (28) cm, slender, erect to ascending, sometimes spreading on large plants or when shaded. Leaves pinnatifid, 1.5–5 × 0.5–2.6 cm. Calyx usually divided about halfway or more to the base, beset with stalked glandular hairs as well as non-glandular hairs; fruiting calyx 3.0–4.5 mm (to 6 mm in the Ajo Mountains in nearby Arizona), not spreading open at maturity (revealing only the tip of the capsule). Corolla lobes white, pale violet, or lavender, the throat yellow with yellow nectaries and often nectar filled in the morning. Capsules splitting but the 2 carpels (halves) not falling free, the halves obovoid, obtusely pointed at tip. Seeds up to 16 per capsule, all alike, brownish black, 0.8–1.0 mm, incurved and cylindrical, with sharply sculptured transverse ridges.

Sonoyta region, across the Pinacate volcanic field, and at least some granitic ranges; rocky soils and gravelly washes, often in shady places, wet soil near waterholes, on north-facing slopes, and sometimes on sandy or cinder flats but then usually in shade of trees such as palo verdes. Southeastern California to Nevada, Utah, and Texas, and southward to Baja California and Sonora at least to Isla Tiburón and the vicinity of Hermosillo.

Sierra del Rosario, *F 20750*. Arroyo Tule, *F 18745*. Sykes Crater, *F 19499*. Pinacate Peak, summit, *F 19432*. Tinajas de Emilia, *F 87-37*. SW of Tezontle cinder mine, *F 88-44*.

NAMA PURPLE MAT

Delicate annuals to robust woody perennials. Leaves alternate, entire (ours) to toothed. Flowers solitary or in short cymes. Calyx divided nearly or entirely to the base (except in *N. stenocarpum*), the lobes linear, lanceolate, or spatulate. Corollas readily deciduous, the lobes rounded and usually spreading, purple to white. Stamens not exserted. Styles 2, or single and very shallowly to deeply 2-cleft (often 3-cleft in *N. stenocarpum*). Seeds many.

Mostly in dry regions of southwestern North America, a few in South America, 1 in Hawaii; 45 species. Six species in the Sonoran Desert, all winter-spring ephemerals. References: Bacon 1984; Hitchcock 1933.

1. Ovary partially inferior, fused to the lower portion of calyx; calyx lobes firm, the tips recurved in fruit; style one, 2- or 3-branched toward apex; wetland habitats. ________ **N. stenocarpum**

1′ Ovary wholly superior, free from calyx; calyx lobes soft, the tips erect; styles 2, distinct to the base; widespread.

2. Plants prostrate, matted; longer stem hairs 0.4 mm; corollas bright lavender-pink; seeds dark brown, about as wide as long. ________ **N. demissum**
2′ Plants erect with ascending branches, sometimes spreading-prostrate with age, but not matted; longer stem hairs 0.8–1.2 mm; corollas lavender; seeds yellowish, about twice as long as wide. ____**N. hispidum**

Nama demissum A. Gray [*N. demissum* var. *deserti* Brand] *MORADA;* PURPLE MATSEED. First flowering as rosette plants, often developing slender stems 2–10 cm, prostrate to prostrate-ascending at tips. Stem hairs (0.3) 0.4 (0.5) mm, often not straight, relatively soft. Leaves 1.5–4 cm, very narrowly spatulate, gradually narrowed to a winged petiole, the upper leaves smaller, sessile, and mostly in compact terminal clusters. Corollas bright lavender-pink, 10–12 mm. Basal portion of filaments (the part fused to corolla tube) narrowly winged. Styles 2, distinct to the base. Seeds 0.55–0.6 × 0.5 mm, chunky and ovoid, dark brown at maturity, the surface lumpy with transverse grooves or pits.

Widespread, often on cinder soils, sand flats, crater floors, and larger washes; extending to higher elevations on cinder slopes in the Sierra Pinacate. Most abundant and often forming spectacular displays on otherwise barren cinder slopes and flats. Not known elsewhere in Sonora; also Baja California and southeastern California to Utah and Arizona.

Cerro Colorado, *Webster 22322.* Elegante Crater, *F 19669.* MacDougal Crater, *F 9725B.* Sykes Crater, *F 18995.* Moon Crater, *F 19270.*

Nama hispidum A. Gray [*N. hispidum* var. *spathulatum* (Torrey) C.L. Hitchcock. *N. coulteri* A. Gray. *N. hispidum* var. *coulteri* (A. Gray) Brand] *MORADA;* BRISTLY NAMA First flowering as rosettes, often developing stems 5–30 cm, erect to ascending or spreading with age. Larger stem hairs (0.8) 1.0–1.2 mm, dense, bristly, straight. Leaves 1.5–4.6 cm, narrowly spatulate, gradually narrowed to a winged petiole, the upper leaves smaller, sessile. Corollas lavender, (10) 13–15 mm. Styles 2, distinct to the base. Seeds 0.5–0.6 mm, ellipsoid-ovoid, about twice as long as wide, yellowish, the surface minutely reticulate.

Common and widespread on rocky, gravelly, and sandy soils of low hills, plains, washes, interdune troughs, and lower dunes; Sonoyta region, Pinacate volcanic region, granitic ranges, and desert plains. Southeastern California to Baja California Sur, Nevada and Utah to western Texas and northern Mexico including Sonora and Sinaloa. Our plants mostly key to var. *spathulatum* (Torrey) C.L. Hitchcock, distinguished by stiff, harsh hairs and linear-lanceolate calyx lobes, but the varieties are of doubtful taxonomic significance.

In addition to characters in the key, *N. hispidum* can be distinguished from *N. demissum* by its larger and more robust habit, usually thicker stems, stouter and stiffer hairs, especially on the stems, leafier stems, often larger flowers, and wingless filament bases. *N. coulteri,* from Baja California Sur and southern Sonora (Hermosillo and Guaymas southward), seems to be a synonym of *N. hispidum. N. coulteri* is reportedly distinguished by the adnate (fused) lower portion of the filaments being broader than the free (upper) portions—this character is difficult to see or not present.

Sierra del Rosario, *F 20799.* Moon Crater, *F 19269.* Tinajas de los Pápagos, *Turner 59-38.* 30 mi W of Sonoyta, *F 7606.* NNE of Puerto Peñasco, *Burgess 4755.* Quitobaquito, *F 7682.*

Nama stenocarpum A. Gray. Winter-spring ephemerals. Stems erect, becoming ascending-decumbent, leafy, to 10+ cm. Herbage and calyces with soft, spreading, relatively long, slender white hairs. Leaves mostly narrowly oblanceolate, 2.5–6 cm × 5–10 mm, gradually narrowed at base to a winged petiole, the margins entire but often wavy. Calyx lobes 3.5–6 mm, persistent and becoming tough, the tips spreading, the lower, or tubular, portion of the calyx fused to the ovary, the ovary thus partially inferior. Corollas relatively inconspicuous, pale bluish white to pale bluish lavender, 5.5–6 mm. Style single, the distal one-third portion 2- or 3-branched. Seeds numerous, 0.3–0.4 mm, light brown, chunky, the sides irregularly flattened to elliptic-flattened, the surface minutely pitted (alveolate).

Small populations in wet sandy-muddy soil at Tinaja del Tule (*F 19194, F 20139*) and nearby at Tinaja de los Chivos (*F 18789*). Widely scattered in wetlands in the Sonoran Desert, often in large but local populations. Southern Texas to southern California and northern Mexico including Sonora,

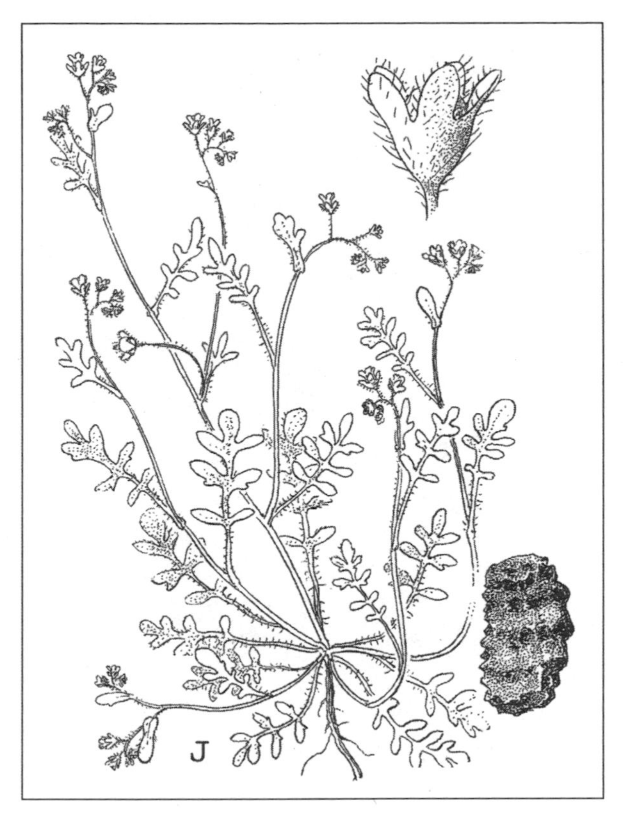

Eucrypta micrantha. JRJ (Abrams 1951).

Nama hispidum. LBH.

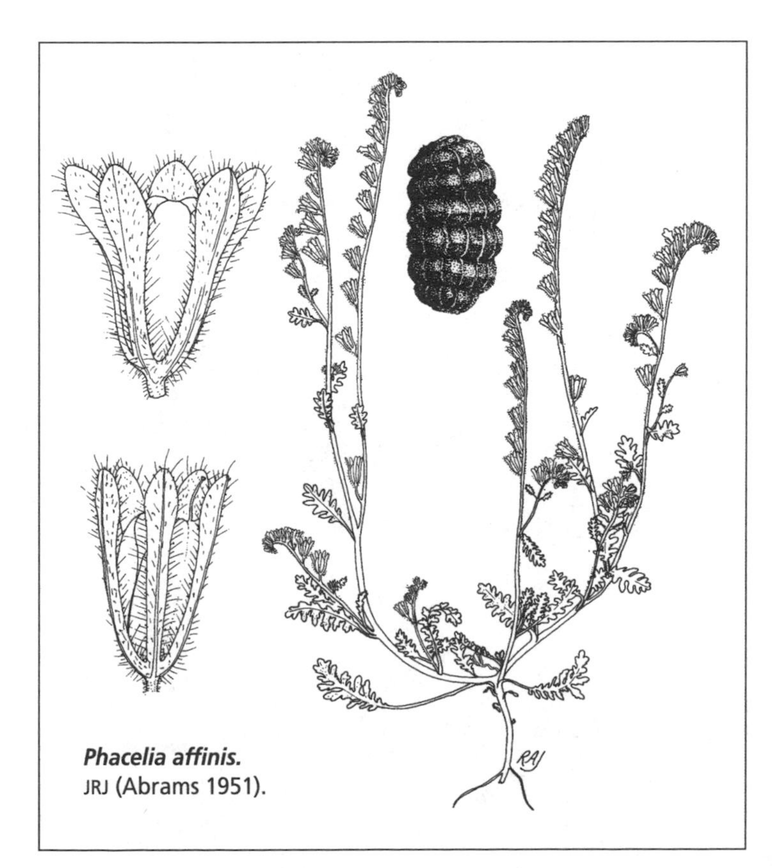

Phacelia affinis.
JRJ (Abrams 1951).

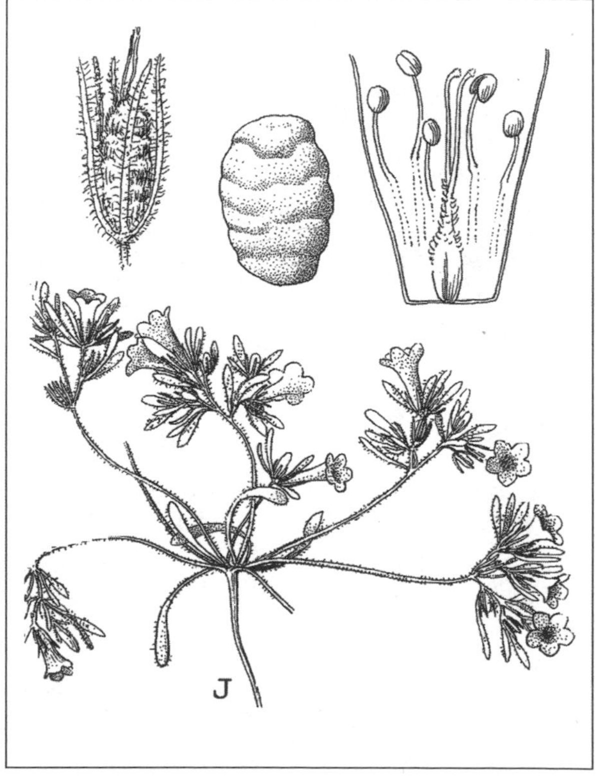

Nama demissum. JRJ (Abrams 1951).

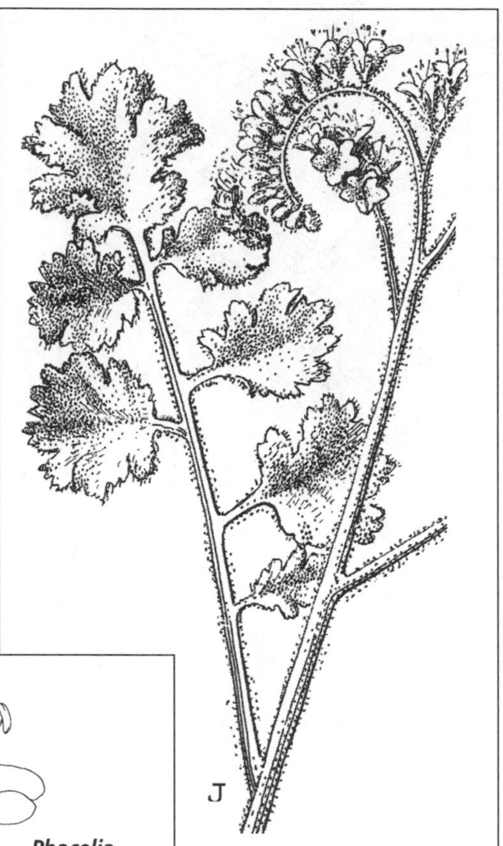

Phacelia pedicellata. JRJ (Abrams 1951).

Phacelia distans. JRJ (Abrams 1951).

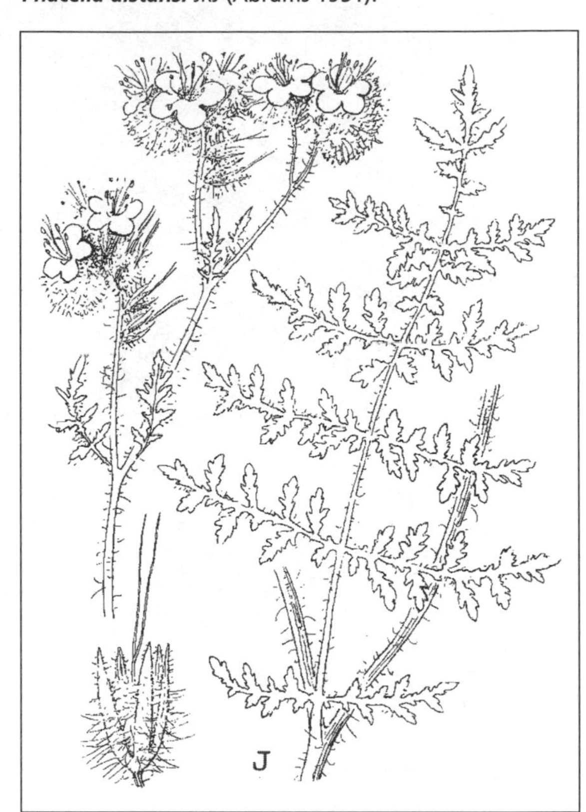

Phacelia ambigua. RAJ (Cronquist et al. 1984).

Pholistoma auritum* var. *arizonicum. ER (Hickman 1993).

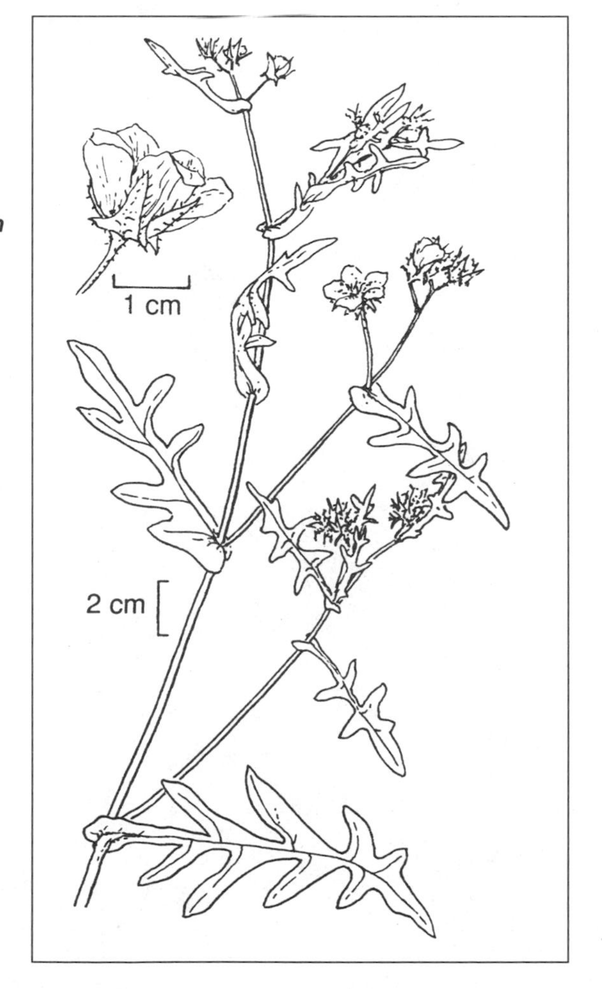

Sinaloa, and both Baja California states; in Arizona known from only one collection at Yuma in 1881 (Kearney & Peebles 1960:705).

This species is unusual among the hydrophyl family in having the lower, or tubular, portion of the calyx united to the ovary so that the ovary is partially inferior, and in sometimes having 3 style branches. It is taxonomically isolated—Hitchcock (1933) placed it in a monotypic section.

PHACELIA

Annual to perennial herbs, usually hairy and often glandular. Leaves mostly alternate (ours), entire to bipinnate (ours petioled, at least the lower, larger leaves). Inflorescence branches generally helicoid (curled at the tip like a scorpion tail—in the shape of a flattened spiral, straightening as they grow, the flowers on one side), the flowers few to numerous. Calyx lobed almost to the base. Corollas blue to white or yellow. Stamens protruding or not. Styles very shallowly to very deeply bifid. Seeds 1 to many.

New World with greatest diversity in western North America; 175 species. Ours are all winter-spring ephemerals with lavender or white petals and mostly helicoid inflorescences. Species with glandular-hairy herbage can cause dermatitis. References: Atwood 1975; Howell 1946.

1. Stamens not protruding; seeds solid, more or less terete in cross section, similar on both sides.
   2. Seeds many per capsule.
      3. Stems not thick and succulent; leaf blades pinnately lobed to pinnatifid, more than twice as long as wide; flowers few to many in moderately helicoid inflorescences. ____ **P. affinis**
      3′ Stems thick and succulent; leaf blades with wavy margins to shallowly toothed, about as long as wide; flowers few, not on helicoid branches. ____ **P. neglecta**
   2′ Seeds 4 or fewer per capsule.
      4. Herbage with both longer white hairs and shorter gland-tipped hairs; fruiting calyx lobes (7.5) 9.5–10.5 mm; higher elevations of the Sierra Pinacate. ____ **P. cryptantha**
      4′ Herbage with longer white hairs, these sometimes basally expanded, the glands sessile; fruiting calyx lobes to ca. 6 mm; Sonoyta region. ____ **P. distans**

1′ Stamens usually protruding, readily visible; seeds boat-shaped, concave and sculptured on the ventral side (excavated on either side of central ridge), convex on other side.
   5. Fruiting pedicels mostly 1.0–1.5 mm, shorter than the capsules; fruiting sepals 3–5 mm, about as long as to 1/4 longer than capsules. ____ **P. ambigua**
   5′ Fruiting pedicels 4–7 mm, longer than the capsules; fruiting sepals 5–6.5 mm, about twice as long as the capsules. ____ **P. pedicellata**

Phacelia affinis A. Gray. Mostly several-branched from near the base, (2) 4–20 cm. Herbage, inflorescences, and calyx with dense stiff white hairs and sessile glands. Leaves mostly basal and on the lower stem, (1.5) 3–6 cm, pinnately lobed to pinnatifid, mostly narrowly oblong; upper leaves reduced. Cymes moderately helicoid, especially on larger plants; flowers few to numerous. Calyx lobes oblanceolate to spoon-shaped, glandular. Corollas white (occasionally pale purple) with a pale yellow-green throat. Filaments whitish, the anthers included (not protruding) and cream colored. Seeds many, nearly 1 mm, brown, reticulate, transversely corrugated.

Widespread in the Pinacate volcanic field, mostly on cinder flats and slopes to peak elevation, arroyos, and sometimes on desert pavement. Northwestern Sonora to the Sierra del Viejo near Caborca and on Isla Tiburón; Arizona, southwestern New Mexico, southwestern Utah, southern Nevada, and southeastern California to Baja California Sur.

Arroyo Tule, *F 18776A*. Campo Rojo, *F 19761*. Pinacate Peak, *F 19430*. Elegante Crater, *F 19673*. SW of Tezontle cinder mine, *F 88-38*.

Phacelia ambigua M.E. Jones [*P. crenulata* Torrey ex S. Watson var. *ambigua* (M.E. Jones) J.F. Macbride. *P. minutiflora* J.W. Voss. *P. ambigua* var. *minutiflora* (J.W. Voss) N.D. Atwood] DESERT HELIOTROPE. Herbage stinky and irritating, sticky-viscid with spreading firm white hairs, and minutely glandular above. Stems often 10–30 (45+) cm, mostly erect, simple to several-branched. Larger leaves basal and often also on the lower stem, often 3–15 cm, pinnatifid or the lower segments cleft to the midrib or into separate leaflets, the lobes variously toothed; leaves reduced upwards. Inflorescences helicoid, simple to several-branched. Flowering and fruiting pedicels often 1.0–1.5 mm. Flowers moderately fragrant; corollas lavender with a white tube, 4–8 mm. Style 4.5–11 mm, bearded at base. Fruiting sepals 3–5 mm, about as long as to 1/4 longer than the capsules. Seeds 4 per capsule, 2.2–3.5 mm, boat-shaped, red-brown, alveolate (minutely pitted and reticulate), the outer surface convex, the inner surface excavated on each side of a prominent ridge, the ridge corrugated on one side, the margin not differentiated (in *P. crenulata* the margins are incurved).

One of the most widespread and abundant winter-spring wildflowers in the region; rocky, gravelly, sandy, and cinder soils; granitic and volcanic slopes, desert pavements, flats, washes, and lower dunes. Sonora south to the vicinity of Bahía Kino and Hermosillo; also southern California, Baja California, Arizona, and southern Nevada.

Very closely related to the more northern *P. crenulata*. The varieties of *P. ambigua,* here treated as synonyms, are also sometimes placed in *P. crenulata*. Ours mostly fall into var. *minutiflora,* generally characterized by having smaller flowers, although robust plants growing during times of high rainfall may resemble *P. ambigua* sensu stricto. Var. *minutiflora* is reported from northwestern Sonora, southwestern Arizona, southeastern California, and Baja California.

22 mi W of Sonoyta, *F 7639*. Pinacate Peak, *F 19402*. Sykes Crater, *F 18941*. Moon Crater, *F 19256*. Sierra del Rosario, *20693*. Cerro Pinto, *F 89-60*.

Phacelia cryptantha Greene. Mostly erect, often 8–30+ cm; leafy and profusely flowering. Herbage sticky, with both longer non-glandular white hairs and shorter gland-tipped hairs. Leaves well distributed along the stem, 2–6+ cm, pinnate with toothed to pinnatifid divisions, the lower segments often of separate leaflets with toothed or cleft margins. Inflorescence branches slender, helicoid, with dense, long white hairs. Flowers at first nearly sessile, soon developing slender pedicels. Calyx lobes enlarging in fruit to (7.5) 9–10.5 mm. Corolla lobes pale lavender to whitish. Stamens slightly protruding from corolla or not. Seeds 4 per capsule, 1.9 mm, red-brown, broadly ovate, pitted and reticulate, the back convex, the ventral surface angled and not excavated.

Sierra Pinacate mostly above 750 m, mostly on cinder soils including steep and often rugged slopes and small arroyos in cinder fields (Pinacate Peak and vicinity, *F 19431, F 19480*). The nearest Arizona populations are apparently in oak scrub in the Kofa Mountains. Also California, southern Nevada, southwestern Utah, west-central and northwest Arizona, and western Sonora southward to the Sierra Seri (29°20′ N); primarily a Mojave Desert species. Resembling and probably closely related to *P. distans;* in the flora area their distributions are widely separated.

Phacelia distans Bentham. CATERPILLAR PHACELIA. Erect and simple to much-branched, and spreading to procumbent. Herbage moderately sticky and often scabrous with conspicuous white hairs, sometimes with swollen white bases (pustulate), and also sessile glands, golden when fresh. Stems leafy, often semisucculent and relatively stout on robust young plants, otherwise often relatively delicate, often 15–45+ cm including inflorescences. Leaves usually relatively thin and fernlike, 6–17 cm, 1 or 2 times pinnatifid, the segments pinnately lobed or toothed to pinnatifid. Cymes helicoid. Calyx lobes enlarging moderately in fruit, reaching 6+ mm. Corollas (7.0) 8–9.5 mm, pale violet to blue, the lobes spreading. Stamens usually not or scarcely exserted. Seeds 4 or fewer, ca. 2 mm, red-brown, narrowly ovoid, pitted, the back convex, the ventral side angled and not excavated, or one side slightly concave.

Arizona Upland in the Sonoyta region; often common along sandy-gravelly washes and bajadas, less often on rocky slopes. In many life zones in California, Baja California, southern Nevada, western and southern Arizona, and western and northern Sonora southward to Hermosillo and the Sierra El Aguaje north of Guaymas. Apparently closely related to and replaced by *P. gentryi* Constance in southeastern Sonora.

W of Sonoyta, *F 89-7*. Río Sonoyta, 21 mi SW of Sonoyta, *Webster 22398*. El Papalote, *F 86-102*. 10 km SW of Sonoyta, *F 88-155*. Sierra Cipriano, *F 88-208*.

Phacelia neglecta M.E. Jones. Plants not stinky. Stems 2–6 cm, solitary to few-branched, thick and succulent; upper part of root and lower stem especially thick. Herbage and calyces with both non-glandular hairs and hairs tipped with red glands. Leaves semisucculent, the petioles mostly 8–24 mm, the blades mostly 5–25 mm wide, simple, broadly ovate to orbicular, the margins wavy to shallowly toothed. Cymes mostly not helicoid with the flowers few and on one side, or longer cymes of larger plants scarcely helicoid. Corollas white, the stamens barely exserted. Seeds numerous, 0.7–0.9 mm, red-brown, more or less terete in cross section, blunt-ended, with bumpy transverse ridges, resembling tiny beetle pupae.

Cinder flats and desert pavements in the Pinacate volcanic field; not known elsewhere in Sonora. Also southeastern California, southern Nevada, and western Arizona. Closely related to *P. pachyphylla* A. Gray, a generally larger plant with longer inflorescences of southeastern California and northeastern Baja California.

13.8 mi NE of Puerto Peñasco, then 6 km W, *F 92-426.* S of Pinacate Junction, *F 20614.* Cerro Colorado, *Quirk 3 Mar 1985* (DES). Tinaja de los Chivos, *F 18801.*

Phacelia pedicellata A. Gray. Often robust, the stems often 10–45+ cm, relatively thick and semisucculent, and often leafy. Herbage viscid glandular hairy and stinky. Leaves often 5–16+ cm, pinnatifid to pinnately compound, and semisucculent. Inflorescences helicoid. Flowers on slender pedicels with dense, spreading hairs, the fruiting pedicels 4–7 mm. Corollas pale blue-lavender. Stamens and styles protruding. Fruiting sepals 5.0–6.5 mm, about twice as long as capsules. Seeds 4 per capsule, 2.5–3.1 mm, boat-shaped, red-brown, pitted, the outer surface convex, the inner surface excavated on both sides of a prominent ridge, corrugated on margins and on one side of the ridge, the margins lighter colored.

Mostly on north- and east-facing slopes and gravelly washes; Pinacate volcanic field and granitic ranges to peak elevations; common at higher elevations on Sierra Pinacate, often in shaded places. Also southeastern California, eastern Baja California, and Arizona.

This is the largest and most robust phacelia in the region. The calyx lobes are somewhat elongated, approaching those of the closely related *P. scariosa* Brandegee in length but narrower than that species. *P. scariosa* occurs from the Guaymas region to northwestern Sinaloa and in Baja California Sur; those from Isla Tiburón and the opposite mainland appear to be intermediate in character.

2.8 km NW of Cerro Colorado, *F 19631.* Tinaja de los Chivos, *F 18793.* Sykes Crater, *F 18963.* Pinacate Peak, *Starr 729.* Sierra del Rosario, *F 20676.* Observation: Cerro Pinto.

PHOLISTOMA

Annuals; stems square in cross section, semifleshy, brittle, weak and prostrate or scrambling, with or without recurved hairs. Leaves pinnate, the lower leaves opposite, the upper ones alternate; petioles often winged and clasping. Flowers solitary or in few-flowered cymes. Calyx lobes divided nearly to the base, the sinus between sepals with or without an auricle (leaflike appendage). Capsules globose; seeds mostly 1–6, globose. California, Baja California, Arizona, and northwestern Sonora; 3 species. Reference: Constance 1939.

Pholistoma auritum (Lindley) Lilja var. **arizonicum** (M.E. Jones) Constance [*Nemophila arizonica* M.E. Jones] Herbage not aromatic; pubescence not glandular, of stout spreading white hairs of varying sizes, enlarged at the base, the smaller and larger hairs often curved or recurved, the larger pedicel hairs mostly recurved. Stems 10–40 (75) cm. Leaves thin, often 3.0–5.5 (8) cm, pinnate, the lobes with few coarse, shallow teeth, the petiole bases broadly winged or expanded (especially on the upper leaves) and clasping the stems. Pedicels 8–27 (ca. 50) mm. Calyx enlarging in fruit, the fruiting calyx ca. 1.5 cm wide, the lobes often 5.5–8 mm, the auricles conspicuous, often 4.0–6.5 mm. Corollas bright blue, 12–15 mm wide. Style bifid near tip, bearded below with stout spreading white hairs. Seeds 2.5 mm in diameter.

Sierra Pinacate mostly above 800 m on cinder slopes and among rocks and lava rubble, especially in protected and often shaded habitats; often on north-facing slopes and growing in the protection of shrubs (Pinacate Peak and vicinity, *F 19439, F 19454, F 19479*). Not known elsewhere in Sonora; also western and central Arizona. Var. *auritum,* distinguished in part by its larger stature, stouter stems, larger flowers, and more prickly herbage, occurs west of the deserts in California and Baja California.

KOEBERLINIACEAE

The family has a single species.

Koeberlinia spinosa Zuccarini. CORONA DE CRISTO; CRUCIFIXION THORN, ALLTHORN. In northwestern Sonora known from a single shrub 2 × 3 m. Branches and twigs stiff, interlacing, and woody, the twigs yellow-green, slender, and thorn-tipped. Twigs and inflorescences minutely pubescent. Leaves alternate, scalelike, very quickly drought deciduous, the plants essentially leafless. Inflorescences of small semi-persistent racemes. Flowers pale yellow on slender pedicels; sepals 4, deciduous; petals 4, ca. 4.5 mm; stamens 8. Fruits globose berries on a short stipe, drying capsule-like, 3.0–3.5 mm in diameter (the berries probably 5–6 mm in diameter), 2-chambered, each 1-seeded (or perhaps 2–4-seeded), the style persistent as a slender beak. Flowering April; fruits in May.

Fairly common in many other desert areas of Sonora, and southward in thornscrub near the coast nearly to the Sinaloa border. Also Baja California Sur to southern California, eastward to Texas and southward to central Mexico, and the Chaco of Argentina, Bolivia, and Paraguay. The wood is exceedingly hard and produces copious oily black smoke when burned (Felger & Moser 1985). Two varieties are described, and Correll & Johnston (1970:1074) allude to further diversity, but the distinctions are not clear-cut.

North side of Playa Díaz, fine textured clay-alluvium, sparse vegetation with *Larrea, F 88-398.*

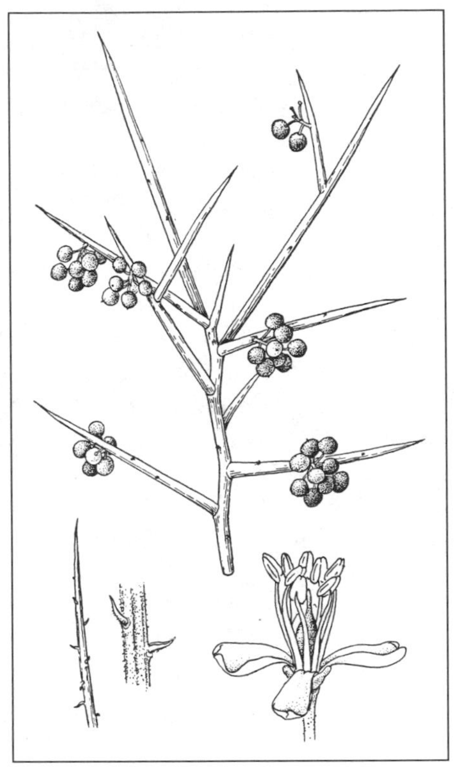

Koeberlinia spinosa. FB. © James Henrickson 1999.

KRAMERIACEAE RHATANY FAMILY

The family has a single genus. The pollinators are primarily *Centris* bees, which collect saturated fatty acids from floral glands. These bees also collect oil from flowers of the closely allied Polygalaceae and the Malpighiaceae. All these families are in the order Polygalales. References: Buchmann 1987; Simpson 1989.

KRAMERIA RHATANY

Shrubs to perennial herbs with high tannin contents, reported as root parasites on other shrubs. Leaves alternate, simple with one exception; stipules none. Flowers bilateral, the sepals and petals separate, the sepals usually 5 and showy, petal-like and larger than the petals; petals 5, the 2 lower ones highly modified as fleshy oil-secreting glandular structures (elaiophores). Stamens usually 4, the anthers opening by 1 or 2 terminal pores (ours) or short slits. Fruits dry, 1-seeded, usually burlike with slender spines like a miniature sea urchin. Mostly in dry regions, southwestern United States to Argentina and Chile; 17 species.

The anthers have chambers that dehisce terminally into a common tubular chamber, described by Simpson (1989:13) as follows:

As an anther ages, the basal portion of the tube constricts, forcing the pollen out in a cylindrical mass. Irritation of the anthers can also trigger extrusion of the pollen when an anther is mature. This feature has the effect of packing *Krameria* pollen into the proboscidial fossae of female centridine bees when they straddle the flowers. Fully dehisced anthers are terminated by a scarious collar, the vestige of the tube through which the pollen was shed.

1. Branches tough and knotty with many very short spur branches; claws of the 3 upper petals fused basally; spines of fruit with barbs along the upper part of shaft. ____ **K. erecta**

1′ Branches mostly straight and without knotty spur branches; the 3 upper petals separate; spines of fruit with barbs in a terminal cluster. ____ **K. grayi**

Krameria erecta Schultes [*K. parvifolia* Bentham] RANGE RHATANY. Dwarf shrubs, often 0.3–0.5 m, usually less than 1 m across, with many short, crowded, spreading branches. Stems tough and woody with gray bark; upper branches knotty due to many very short spur branches. Herbage densely pubescent and grayish with short white hairs. Leaves 3–9 × 0.8–1.3 mm, drought deciduous, sessile, and linear. Flowers showy, 1.5 cm wide, solitary or in short racemes with leafy bracts. Sepals bright magenta-purple inside, white hairy outside. Upper petals expanded above into dull lavender blades about as wide as long, narrowed into thick bright green claws, the claws united below; oil-gland (lower) petals thick and very dark purple. Filaments whitish, the anthers dull cream colored. Styles magenta-purple. Fruits more or less globose and moderately compressed, ca. 6 mm wide, the spines ca. 3.5 mm with small barbs more or less evenly distributed along the upper part of the shaft. Flowering at various seasons.

Plains with sparse desertscrub, upper bajadas, gravelly sandy plains adjacent to mountains, rocky hills, and mountains. Widespread but often in small, aggregated colonies. Often growing with *K. grayi.* Sonoyta region, Pinacate volcanic field, and granitic ranges. Southeastern California, Nevada, southwestern Utah, and Texas, and southward to Baja California Sur, Sinaloa, Zacatecas, San Luis Potosí, and Durango.

Sierra del Rosario, *Burgess 6874.* Hornaday Mts, *Burgess 6843.* Tinaja Huarache, *F 19560.* 2 km SE Cerro Colorado, *Ezcurra 17 Ap 1981* (MEXU). Quitobaquito, *F 88-468.*

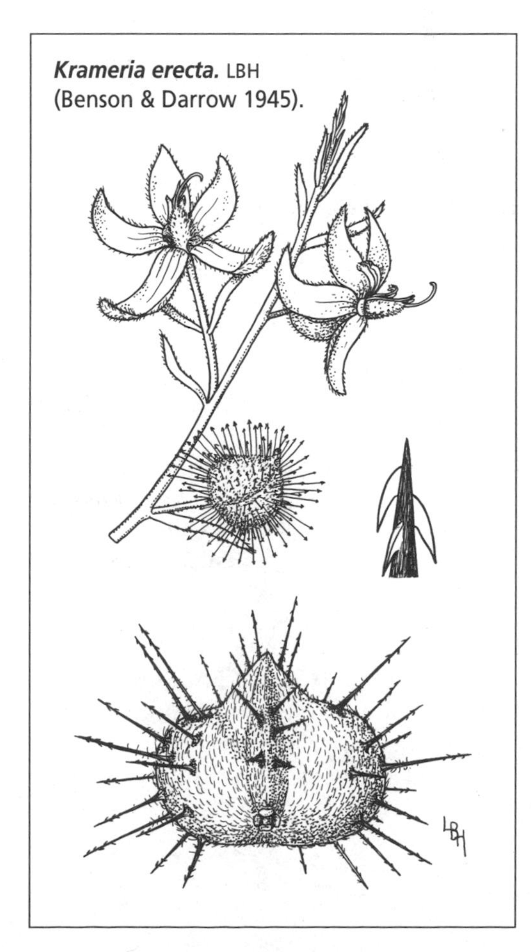

Krameria erecta. LBH (Benson & Darrow 1945).

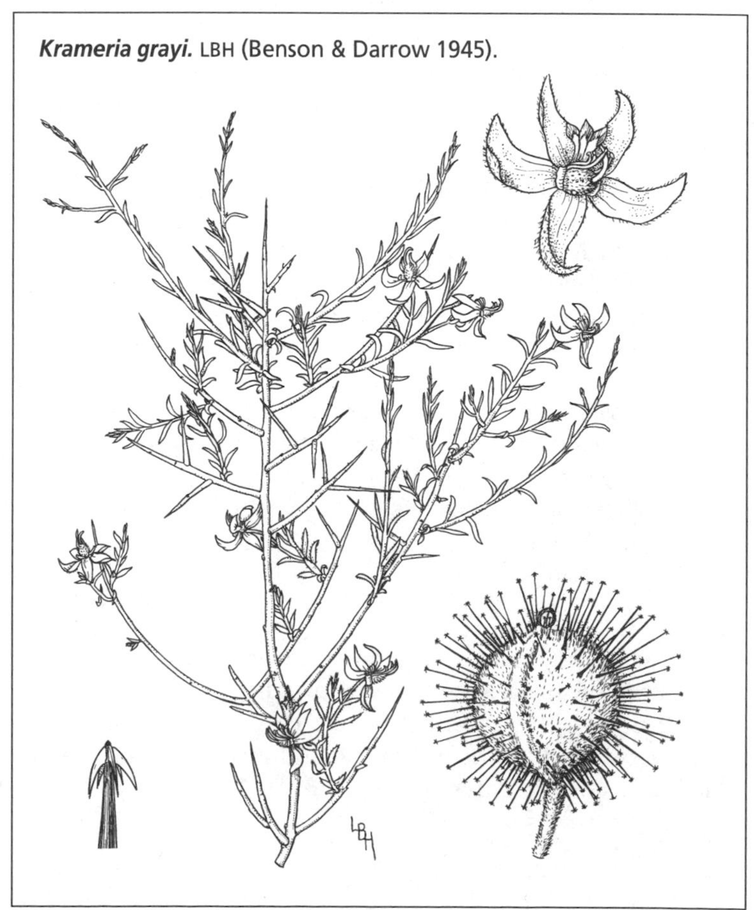

Krameria grayi. LBH (Benson & Darrow 1945).

Krameria grayi Rose & Painter. *CÓSAHUI;* WHITE RHATANY; *'ẸDHO, HE:Ḍ.* Shrubs often 0.4–1 m tall and more than 1 m across, with thick, radiating lateral roots. Herbage densely pubescent with short, gray hairs, the stems glabrate with age. Stems intricately branched at mostly right angles, straight, slender, and sparsely leaved or often leafless, the tips spinescent. Leaves 5–10 (14) × 1.1–2.4 mm, soon drought deciduous, sessile, linear to oblong, often grayish. Flowers showy, 1.5 cm wide, solitary or in short racemes with leafy bracts. Sepals bright magenta-purple inside, white hairy outside. Upper petals spear-shaped (narrowly spatulate), distinct, bright chartreuse with purplish tips; oil-gland (lower) petals extremely thick and slablike, very dark purple. Filaments and style thickened and purple. Anthers dull in color. Fruits more or less globose, ca. 1 cm wide, the spines with a cluster of small terminal barbs. Flowering various seasons.

Widespread and common through most of the flora area except dunes; generally most common at lower elevations; rocky slopes to near peak elevations, pavements, bajadas, and sand flats. Sonoyta region, Pinacate volcanic field, and granitic ranges. Sonora southward to Bahía Kino and Hermosillo, where it is abruptly replaced southward by the closely related but distinct *K. sonorae* Britton. Also southeastern California to western Texas, Nevada, southwestern Utah, Baja California Sur, Chihuahua, and Coahuila. The roots are a source of a reddish brown dye used in basketry (Felger & Moser 1985).

Sierra del Rosario, *F 75-23.* 63 mi E of San Luis, *F 5739.* Crater de la Luna, *Ezcurra 17 Ap 1981* (MEXU). Hornaday Mts, *Burgess 6854.* Pinacate Peak, *F 19380.* MacDougal Crater, *F 10457.*

LAMIACEAE (LABIATAE) MINT FAMILY

Annual or perennial herbs, shrubs, or rarely trees; usually aromatic. Stems typically square in cross section. Leaves opposite, sometimes whorled, usually simple; stipules none. Sepals united into a tube, mostly 2-lipped with 5 lobes or teeth. Corollas usually bilateral, 5-lobed, mostly 2-lipped, the lower one usually larger. Stamens attached to the corolla tube, 4 and paired, or 2 and sometimes with an additional pair of staminodes. Style often cleft at tip into 2 equal or unequal lobes, or 1 lobe vestigial. Fruits usually of 4 small 1-seeded nutlets.

Worldwide, especially in seasonally dry temperate and tropical regions, abundant from the Mediterranean region to central Asia; 250 genera, 6700 species. More than half the species belong to 8 genera of which *Salvia* and *Hyptis* are the largest. Economic labiates include horehound, lavender, mint, rosemary, thyme, and a number of ornamental garden plants.

1. Herbaceous perennials or annuals.
   2. Leaves mostly basal, the stem leaves greatly reduced; inflorescences of widely spaced (interrupted) headlike flower clusters; stamens 2. ____ **Salvia columbariae**
   2′ Stem leaves well developed; inflorescences racemose, the flowers more or less evenly spaced and not densely clustered; stamens 4. ____ **Teucrium**

1′ Shrubs, at least the lower branches woody.
   3. Leaves often whitish; stem tips, leaves, and especially pedicels and calyx tubes densely pubescent with branched white hairs. ____ **Hyptis**
   3′ Leaves greenish; plants glabrous or with very short simple hairs.
      4. Plants glabrous, not gland dotted; inflorescence racemose; fruiting calyx inflated and hollow like a bag. ____ **Salazaria**
      4′ Herbage with minute hairs and dotted with tiny golden glands; inflorescence headlike; calyx not inflated. ____ **Salvia mohavensis**

HYPTIS

Annual to perennial herbs, small to large shrubs, or rarely trees (*H. arborea,* a tree to 18 m, is the largest member of the family). Flowers usually in dense axillary clusters. Calyx lobes 5, equal in length. Corollas 2-lipped, the lower lip like a sac or pouch. Stamens 4 and fertile, the upper pair shorter.

New World, mostly tropical and mostly in Brazil, a few naturalized as weeds in the Old World tropics; 350 species. Reference: Epling 1949.

Hyptis emoryi Torrey [*H. emoryi* var. *amplifolia* I.M. Johnston. *H. emoryi* var. *palmeri* (S. Watson) I.M. Johnston] *SALVIA;* DESERT LAVENDER. Shrubs often 1.5–2.5 (3) m, with many straight, slender woody stems arising from the base. Branched white hairs densely covering leaves, twigs, pedicels, calyces (especially the tube); some hairs on parts of the corollas and filaments. Leaves ovate, grayish or whitish to olive-green depending on moisture conditions, gradually and tardily drought deciduous; larger leaf blades mostly 1.5–3 cm, the largest leaves on new primary-growth stems sometimes with blades to 5 cm and petioles to 3 cm. Leaf margins evenly crenulate to broadly toothed. Lower leaf surfaces with prominent veins. Flowers wonderfully fragrant, in dense axillary and terminal clusters. Pedicels as long as to longer than the calyx. Calyx 4.6–6 mm, the lobes slender and purplish. Corollas longer than calyx, bright lavender-blue and showy against the white foliage and calyces. Flowering profusely and for lengthy time periods at any season with even scant moisture.

Widely distributed in the Sonoyta region, Pinacate volcanic field, and granitic ranges. Best developed along major gravelly washes and arroyos; also rocky slopes, mostly along drainageways, as well as cliffs, lava flows, crater walls including south-facing cliffs, and to 860+ m in the Sierra Pinacate. Baja California Sur and the Guaymas–lower Río Yaqui region in Sonora northward to southeastern California, southern Nevada, and southwestern and central Arizona.

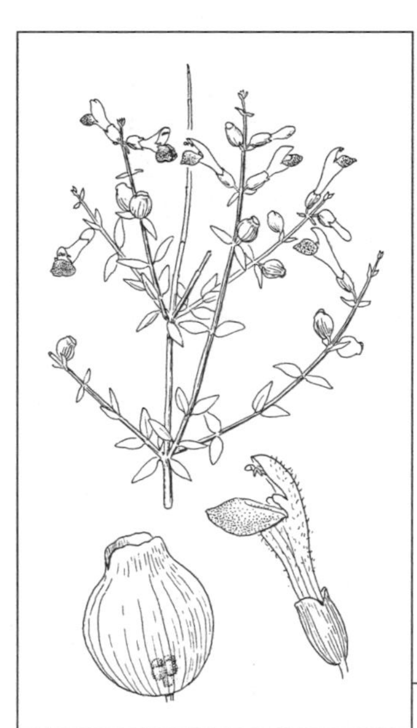

Salazaria mexicana. BA.
© James Henrickson 1999.

Salvia columbariae.
RAJ (Cronquist et al. 1984).

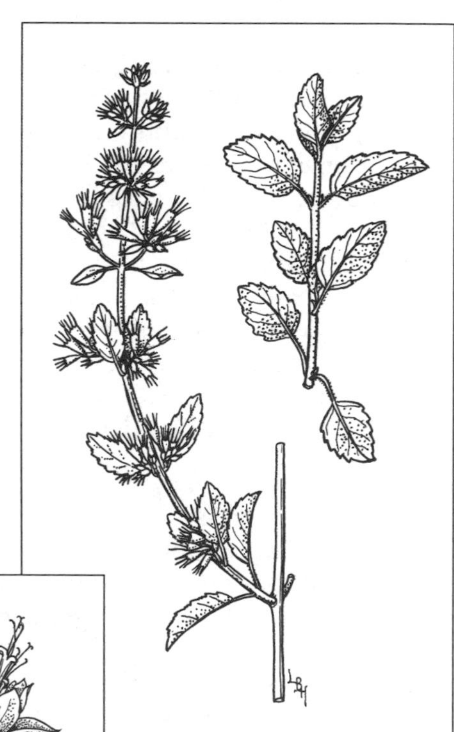

Hyptis emoryi.
LBH (Benson & Darrow 1945).

Salvia mohavensis.
LBH (Benson & Darrow 1945).

Leaves produced during times of favorable temperature and soil moisture tend to be greener and thinner than dry-season leaves. The foliage and stem tips are often freeze damaged. (It is surprisingly frost sensitive in cultivation.) The flowers are frequented by Costa's and black-chinned hummingbirds and myriad insects. Flowering seems limited only by freezing weather and drought.

This species is the northernmost member of this predominately tropical genus in western North America. It is a characteristic xeroriparian shrub of the Sonoran Desert and one of the major nectar-producing plants of the region. Its distributional limits closely approximate that of the Sonoran Desert. Populations from beyond the desert in southeastern Sonora pass into *H. albida* Kunth, a sometimes scarcely distinguishable shrub that replaces *H. emoryi* to the south and east in Mexico. *H. albida* is distinguished by its thinner leaves and softer or more "cottony" branched hairs. *H. emoryi* is sometimes treated as a synonym of *H. albida*. The varieties of *H. emoryi* are based on minor vegetative characters. Epling (1949) neither recognized nor discussed the varieties, although Wiggins (1964) did recognize them.

Sierra del Viejo, *F 16894*. Sierra del Rosario, *F 20683*. Papago Tanks, *Turner 59-18*. Sykes Crater, *F 18957*. Tinajas de Emilia, *F 19727*. 11 mi S of Sonoyta, *F 9805*.

Salazaria

The genus has one species.

Salazaria mexicana Torrey. BLADDER SAGE. Moundlike shrubs 1–2 m; densely and divaricately branched, the twigs slender and brittle, often spinescent and with rather sparse foliage; pubescent with short, rather thick, and mostly appressed white hairs. Leaves entire, quickly drought deciduous, the larger ones 1–1.8+ cm. Flowers 16–22 mm long; flowering calyx often dark rose, about half as long as the flower, the fruiting calyx enlarging to become globose and hollow like a paper bag, ca. 15–20 mm, fading to white or pink. Corollas, 2-lipped, lobes of upper lip dark blue, the lower lip white with purple-blue nectar guides. Stamens 4. Massive flowering at various seasons following rains.

Steep rocky slopes, small arroyos, and cinder flats and slopes at higher elevations in the Sierra Pinacate and on the north side and mostly higher elevations in the Sierra del Viejo. Deserts in southeastern California to Utah and southward to Baja California Sur and northwestern Sonora only in the flora area, and disjunct in Texas and Coahuila.

W of Tinajas de Emilia, *Simmons 26 Jan 1965*. Pinacate Peak, *F 19886*. Sierra del Viejo, *F 16876*.

Salvia

Mostly herbaceous perennials and shrubs, about 7% are annuals; usually aromatic. Flowers often in whorls, inflorescences of spikes, often interrupted (flower clusters separated by bare stem), panicles, racemes, or terminal headlike clusters. Corollas 2-lipped. Stamens 2, arching beneath the upper lip, often with an elongated connective (portion of filament connecting the anther sacs of a stamen), one of the anther sacs sometimes reduced.

Worldwide; 900 species; 500+ in the Americas (including 275+ in Mexico), mostly in the subgenus *Calosphace* and these mostly in tropical-subtropical montane regions, very few in deserts. The 2 species in the flora area are not closely related: *S. columbariae* is in section *Pycnosphace,* and *S. mohavensis* is in subsection *Eplingia* of section *Audibertia;* both are in subgenus *Calosphace*. *Salvia* is the largest genus in the family. Many are cultivated, mostly in humid regions. Reference: Epling 1938, 1939; Fernald 1900; Starr 1985.

1. Winter-spring ephemerals, herbaceous; leaves mostly in a basal rosette. ____ **S. columbariae**
1′ Woody shrubs; stems leafy. ____ **S. mohavensis**

Salvia columbariae Bentham. *CHÍA;* CHIA; *DAPK.* Winter-spring ephemerals, 5–35+ cm with unevenly distributed, mostly short, white hairs, and sessile golden to red-orange, globular glands, especially dense on the lower leaf surfaces and calyces. Leaves mostly basal, 4–9 (12) cm, once or twice pinnatifid, the petioles prominent and narrowly and often indistinctly winged, the stem leaves substantially reduced upwards. Flowers in dense, headlike clusters (1.5) 2–4 cm across, terminal and widely separated on leafless stems—the clusters appearing skewered on the flowering stems. Floral bracts nearly orbicular with spinescent-awned tips. Calyx 9–12 mm, firm, the upper lip with a 2-pronged spinescent tip. Corollas dark blue.

Sonoyta region, Pinacate volcanic field, granitic ranges, and desert plains. Washes, plains, and bajadas, and generally most common on cinder soils at higher elevations. Western Sonora southward to the vicinity of El Desemboque San Ignacio, and Baja California, California, Arizona, Nevada, and Utah: desert to chaparral, coastal sagescrub, and woodland communities.

18 mi S of Sonoyta [Caborca road], *Shreve 7579*. Campo Rojo, *Ezcurra 16 Ap 1981* (MEXU). E of Carnegie Peak, 750 m, *F 19790*. Pinacate Peak, *F 19465*. Aguajita, *F 88-281*.

Salvia mohavensis Greene. MOJAVE SAGE. Compact, much-branched small shrubs, often 50–80 cm, mostly broader than tall, the branches thick and woody, often with a thick trunk formed of several coalesced stems. Herbage moderately aromatic. Stem tips, leaves, and outer surfaces of flower bracts and calyces speckled with small round glands and minute white hairs; glands glistening translucent to whitish opaque when fresh, and becoming jewel-like golden and glistening when dry; glands absent or essentially so from inner surfaces of bracts and calyces. Petioles narrowly winged; leaf blades mostly lanceolate to ovate, 1.2–1.8 cm, green, thickish, and wrinkled, the veins of the lower surfaces prominently raised, the margins minutely crenulate.

Inflorescences of compact, rounded, headlike spikes terminal on slender stems, with crowded bracts and flowers. Floral bracts green to whitish or occasionally pink, 1–2.5 cm; bracts of outer flowers ovate, the inner ones much narrower. Calyx 12–17 mm, the corollas pale to dark violet-blue, 20–23+ mm; stamens longer than the corolla. Spring and fall.

Pinacate Peak and adjacent cinder hills above 1000 m; among rocks and on cinder slopes except south and southwest exposures (*F 19415, F 86-455. Starr 728*). Not known elsewhere in Sonora; also in the Mojave and Sonoran Deserts in southeastern California, southern Nevada, and western Arizona.

The lack of glands on the inner surfaces of the bracts and calyces may play a role in pollination. In contrast, specimens from California and Arizona do have glands on the inner surfaces of the bracts and calyces. These glands seem to be characteristic of the shrubby mints (subsection *Eplingia*) of southwestern North America. In addition to the glands, the leaves of the *Eplingia* species have a characteristic microscopic fantasy landscape of glandular hairs.

TEUCRIUM

Herbs, mostly perennial, to small shrubs. Stems square to rounded. Corollas seemingly with only a lower lip, the middle lobe enlarged and often elongated, the pair of smaller lateral lobes actually derived from part of the upper lip. Stamens 4.

Worldwide, temperate and tropical; 100 species, 8 native in the New World. Includes germander, medicinal and dye plants, and temperate garden ornamentals. Reference: McClintock & Epling 1946.

1. Annuals, pubescent at least on upper part of plant; corollas 11 mm or less, blue to nearly white. ________ **T. cubense**

1′ Perennials, glabrate except pedicels and few hairs on newest growth and corolla; corollas more than 15 mm, white or whitish, often with lavender nectar guides. ________ **T. glandulosum**

Teucrium cubense Jacquin subsp. **depressum** (Small) McClintock & Epling [*T. depressum* Small. *T. cubense* var. *densum* Jepson] Non-seasonal ephemerals, the stems 8–ca. 45 cm, square, soon spreading to ascending-decumbent. Herbage and inflorescences glandular, the inflorescences and sometimes the new herbage with coarse, white, curly to crinkled hairs. Leaves 3–8 cm, variously lobed or incised; lower leaves broader and not as deeply cut as upper leaves, the upper leaves with fewer but larger lobes. Flowers in leafy-bracted terminal racemes. Pedicels (0.8) 2–8 mm. Calyx 4.5–6.5 (7.0) mm, deeply parted into prominent teeth. Corollas pale blue to nearly white, 8–11 mm. Nutlets 2.2–2.5 mm, somewhat obovoid, glandular, with short bristly hairs on reticulate ridges.

Pinacate region, generally on fine-textured soils in low, seasonally wet and often poorly drained habitats including roadside depressions, playas including those in craters, riparian arroyos, and waterholes. Highly localized but often densely crowded. Subsp. *depressum* occurs in north-central and northwestern Mexico including northeastern Baja California, and southeastern California to western Texas including southern Arizona. This species extends to southern South America; there are 5 subspecies.

Ciénega 25 mi W of Sonoyta, *Shreve 7603*. Pinacate Junction, *F 86-345*. 2 km N, 3.6 km E of Los Vidrios, *Burgess 6794*. Sykes Crater, *F 18918*.

Teucrium glandulosum Kellogg. Perennial herbs, sometimes dying back severely in drought, commonly 50–75 cm including inflorescences. Stems slender and brittle, square, mostly erect from a much-branched caudex. Outer calyx surfaces and leaves, especially lower surfaces, glandular-punctate with glistening golden glands. Pedicels pubescent, the youngest herbage, buds, and corollas with few scattered hairs, otherwise glabrous. Leaves tardily drought deciduous, (1.6) 2.3–6 cm, thin, sessile, mostly lanceolate; lower leaves sometimes with 1–3 prominent teeth on each side, the larger leaves with a pair of large lobes; upper leaves smaller and usually entire. Flowers in racemes (13) 18–50 cm, crowded above. Pedicels (7) 9–27 mm, longest in the lower part of the inflorescences, glandular and also with short white hairs. Calyx 5–8 mm, deeply parted into prominent teeth. Corollas 18–26 mm, white with pale purple nectar guides, or sometimes all white. Nutlets 3.0–3.4 mm, oval, glandular, hairy at apex. At least spring and fall.

Localized in isolated rocky arroyos, canyon bottoms, and north-facing cliffs in the Sierra Pinacate, mostly above 850 m (*F 19313, F 86-441*). Southward in Sonora in scattered, mostly wetland habitats to the vicinity of El Desemboque San Ignacio, both Baja California states, one locality in California; Arizona in the Castle Dome Mountains and Cabeza Prieta Refuge. This is one of the few species in this large genus that is confined to a desert.

LENNOACEAE SANDFOOD OR LENNOA FAMILY

Herbaceous perennials and 1 annual, obligate parasites on roots of various annuals and perennials, very fleshy, lacking chlorophyll. Roots mostly of two kinds: pilot roots (often relatively long and extending to a host root), and haustorial roots (short outgrowths from pilot roots forming a parasitic union); root hairs none. Leaves sessile, usually reduced to scales with stomata. Inflorescences densely flowered, pyramidal to caplike or discoid. Flowers radial or nearly so, the sepals mostly separate, the petals united. Pistil with 5–16 carpels, each with 2 chambers. Fruit a capsule, irregularly opening around the middle; seeds 1 per carpel chamber, minute, mostly 0.5–1.0 mm.

Two genera; *Lennoa* is annual and has 1 species. Southwestern United States through Mexico, and "a small number of populations in Guatemala, Colombia, and Venezuela, which probably represent a relatively recent, secondary spread" (Yatskievych 1985:74). Seasonally dry to desert habitats. This distinctive family is distantly related to the Boraginaceae and Verbenaceae. References: Kuijt 1969; Yatskievych 1982, 1985; Yatskievych & Mason 1986.

PHOLISMA

Southern California to northern Baja California, southwestern Arizona, Sonora, and Sinaloa; 3 species. References: Felger 1980; Kearney & Peebles 1960; Nabhan 1985; Torrey 1855.

Pholisma sonorae (Torrey ex A. Gray) Yatskievych [*Ammobroma sonorae* Torrey ex A. Gray, *Pl. Thurb.* 327, 1854] *CAMOTE DE LOS MÉDANOS;* SANDFOOD; *HIA TATK* (derived from *hia* 'dunes', and *tatk* 'root'). Stems subterranean, succulent, ca. 0.5–1.5 m, often 1–2.5 cm in diameter, erect or angled from point of attachment, each stem terminating at or near the soil surface in a single caplike inflorescence; lower stems often forming seemingly undifferentiated, slender, and sometimes branched coralloid roots. Stems and leaves purple-brown near the soil surface and white below. Leaves mostly scalelike below, mostly linear to linear-spatulate above, often 8–55 mm; uppermost leaves densely pubescent with elongated glandular hairs.

Inflorescence forming a concave, discoid, mushroom-shaped receptacle 3–15 (20) cm in diameter, sand colored, partially buried to slightly above the sand surface, or the larger receptacles folding and convoluting inward from the margin to sometimes cristate (like a crested cactus). Perianth 7–9 mm; calyx segments slender, with very dense, elongated, thick, tangled and spreading white hairs, mostly

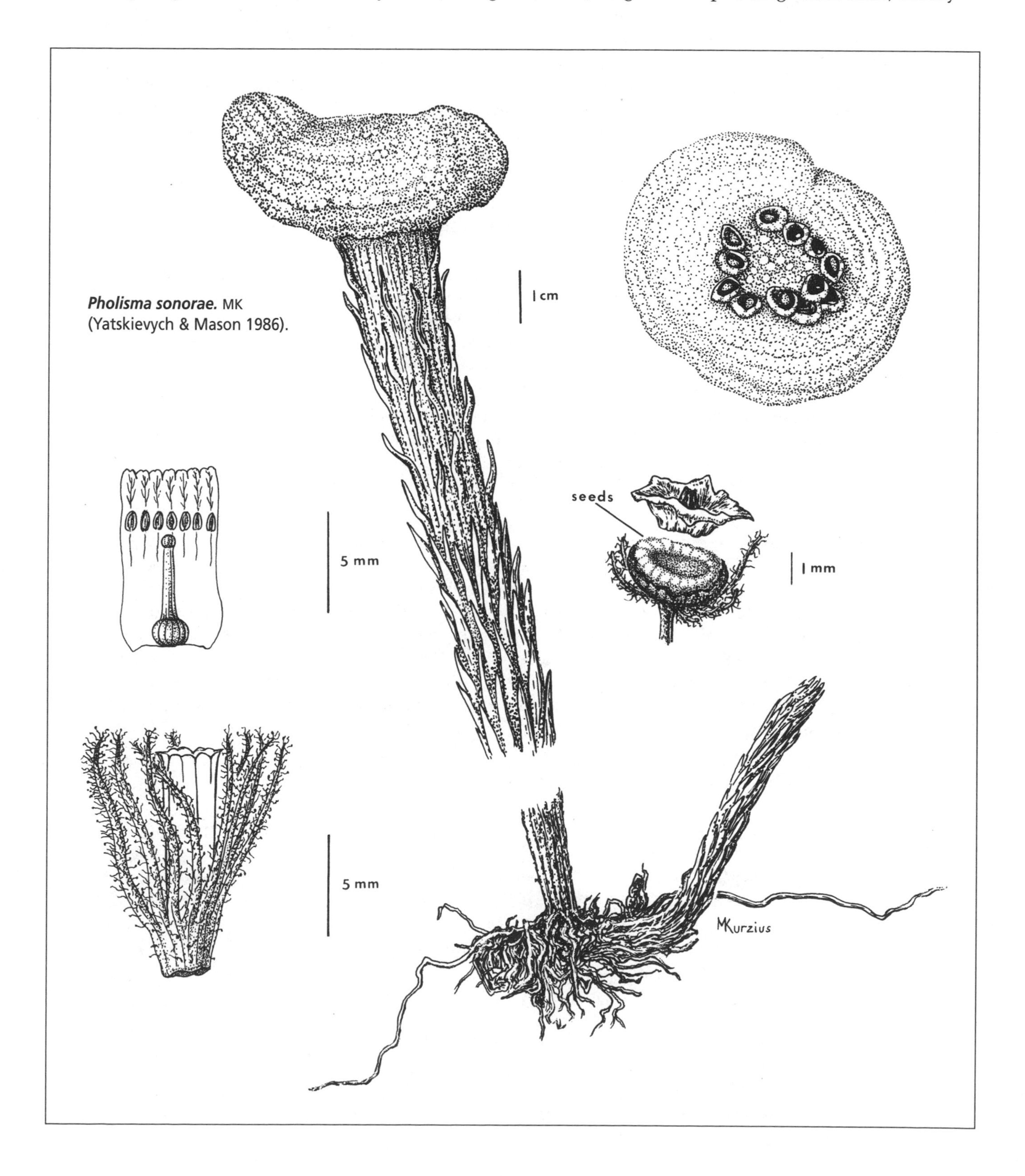

Pholisma sonorae. MK (Yatskievych & Mason 1986).

ca. 1 mm; corollas somewhat tubular, 3.3–4.0 mm wide, purplish with white margined lobes. Stamens as long as or shorter than the corolla tube; filaments white, the anthers yellow. Ovary reported to have 6–16 carpels. Seeds 1.1–1.2 mm, compressed, nearly discoid. Reproductive March-May, or failing to appear during drought.

In the flora area parasitic on *Ambrosia dumosa, Eriogonum deserticola, Pluchea sericea, Tiquilia palmeri,* and *T. plicata.* Flowers visited by a small syrphid fly. Across most of the Gran Desierto on sand flats, stabilized lower dunes, and sometimes on moving dunes. Often abundant east of San Luis along Mex 2, sometimes growing through the pavement at the edge of the road. Not known in Sonora beyond the flora area. Also extreme southwestern Arizona, southeastern California, and extreme northeastern Baja California. Sometimes common in disturbed habitats, such as banks of irrigation canals in the Imperial Valley of California and citrus orchards in southwestern Arizona (Yatskievych 1985).

Some time after the seeds mature, in late spring or early summer, the plants "die back to a small, relatively undifferentiated mass of tissue in and at the infection site" (Yatskievych 1985:75). The site of attachment may be more than 1 meter below the surface. Blowing sand and wind may bury or move the seed-bearing inflorescences, which often break off near the surface, and animals undoubtedly also are agents of dispersal. Yatskievych (1985:76) reports that harvester ants frequently visit "mature inflorescences of all species of Lennoaceae" and "dismember portions and carry them away to their nests." He also reports occasional excavation by mammals, presumably rodents and perhaps coyotes, and that rodents take the seeds into their underground burrows. Experimental germination has not been successful.

Following times of sufficient soil moisture, the long, thick, succulent stems rapidly elongate to 1 m or more. Cothrun (1969) reported a sandfood plant that outweighed its host (*Tiquilia plicata*) by a factor of more than 40 times (fresh weight). Fresh stems are more than 80% water (Yatskievych 1985). The amazing thing is that host plants, for all members of the Lennoaceae, have been reported to be strikingly healthy and sometimes even more robust than non-parasitized plants in the same population. Yatskievych (1985:75–76) reported that

> Cothrun's extreme example . . . involved the host transmitting about 32 times its own fresh weight in water to the parasite, discounting the large quantity of water lost by the parasite to the environment as transpiration . . . [and] it seems probable that the abundant stomates present on the scale-leaves of the developing parasite stem function to absorb water from the surrounding matrix. . . . The relative robustness of many host[s] of Lennoaceae raises further speculation that, at least in times of drought stress, the host is able to siphon water from the parasite through the haustorial connection. If the parasite is able to absorb water from the surrounding sand and some of this water is available to the host, then the relationship is not parasitic in the strictest sense.

If the host is actually more robust than non-parasitized plants, there might be an advantage to this strange relationship. It may be that the host is water limited, and with an additional water reservoir it can produce excess photosynthate.

Sandfood is one of the culinary oddities of the world. It was highly esteemed by the O'odham and Cocopa (Castetter & Bell 1942, 1951). The westernmost O'odham (the "Sand People") often camped in the dunes in spring in order to gather sandfood (Lumholtz 1912:331). The stem is somewhat sweet and juicy, providing food as well as liquid for thirsty travelers. Yatskievych reports that *P. sonorae* accumulates greater amounts of sugars than do the other species in the family. Sandfood was eaten boiled, baked in hot ashes, or dried for later use. Uncooked or undercooked immature stems are very bitter. Another member of the family, *Lennoa madreporoides,* ranging from southwestern Chihuahua to south-central Mexico, is sold in local marketplaces and cooked as a vegetable (Dressler & Kuijt 1968; Yatskievych 1985).

Hills around Adair Bay, *A. B. Gray 17–19 May 1854* (holotype, NY, not seen; photos, ARIZ). Punta Borrascosa, *F 84-37.* The Great Camote, sand dunes of Sierra del Rosario, 1910, *Lumholtz 4* (G). 20 mi W of La Joyita, *Myhrman 22.* Lerdo, *24–26 Apr 1889, Palmer 932* (G). Arizona boundary, 7.5 mi E of San Luis, *Peebles 8433.* NW of El Doctor, *Reichenbacher 1661.* 9 mi SE of Chivos Tank, *Simmons 11 Ap 1965.*

LOASACEAE STICK-LEAF OR LOASA FAMILY

Herbaceous perennials and some annuals. Hairs silicified or calcified, barbed (these usually glochidiate, with the barbs in whorls), needlelike, or gland-tipped and sometimes stinging. Leaves alternate (ours) or opposite, simple; stipules none. Flowers radial. Sepals persistent, often 5, the petals separate or united, often 5 (or seemingly 10 including the staminodes). Stamens 5 to many, the filaments narrow or petal-like in the outermost whorls. Ovary inferior, the style 1. Fruits dry, dehiscent, 1 to many-seeded, or indehiscent and 1-seeded.

Temperate, tropical, and arid regions, New World except 1 small genus in Africa; 15 genera, 260 species. *Mentzelia* is one of the larger genera; 3 genera in the Sonoran Desert. References: Christy 1998; Ernst & Thompson 1963.

1. Perennials; flowers white; stamens 5. ______ **Petalonyx**

1′ Ephemerals or annuals (perennial in *M. puberula* or perhaps *M. multiflora*); flowers green, yellow, or orange; stamens many.

  2. Stems and petioles thickened and succulent or semisucculent; leaf blades as wide as or wider than long, the upper surface glistening; petals united below, the corolla yellow with green lobes. ______ **Eucnide**

  2′ Stems and petioles not succulent; leaf blades longer than wide, dull; petals separate, whitish, yellow, or orange. ______ **Mentzelia**

EUCNIDE ROCK NETTLE

Annual or perennial herbs to small shrubs; hairs smooth and needlelike, barbed, or sometimes stinging. Petals united basally to a short staminal tube, or fused below into a tube. Stamens few to many. Fruits opening terminally by 5 valves. Seeds minute, numerous.

Tropical to subtropical or deserts, often on cliffs. Guatemala to the southwestern United States; 11 species in 3 sections. Reference: Thompson & Ernst 1967.

Eucnide rupestris (Baillon) H.J. Thompson & W.R. Ernst [*Loasella rupestris* Baillon. *Sympetaleia rupestris* (Baillon) S. Watson] VELCRO PLANT. Winter-spring ephemerals, and perhaps also with summer rains. Roots unusually small relative to plant size. Stems and petioles glassy, succulent, and brittle. Herbage, inflorescences, ovaries, and calyx bases with stout, glasslike hairs, the larger hairs smooth and needlelike, the shorter hairs with a terminal whorl of barbs like miniature grappling hooks causing pieces of the plant to stick like Velcro. Leaf blades 3–7 cm wide, nearly rounded or cordate at the base, broadly toothed to shallowly lobed, relatively thin and bright yellow-green, the upper surfaces glistening. Flowers 2 cm, short-tubular; corollas dull yellow below, the lobes short and bright green.

Pinacate volcanic field; often seasonally and locally common among rocks and crevices on rock faces and cliffs including crater walls, especially with north- and east-facing exposures, ca. 200–650+ m. Sonoran Desert and thornscrub; gulf side of Baja California Sur to Imperial County, California, islands in the Gulf of California, and northwestern Sinaloa and western Sonora to Cabeza Prieta Refuge and one record in Organ Pipe Monument in southern Arizona. *E. rupestris* is in section *Sympetaleia,* which includes 2 other species, both restricted to Baja California Sur.

Papago Tanks, *Spaulding 76-30.* Arroyo Tule, *F 18823.* Sykes Crater, *F 20010.* Elegante Crater, *F 19692.* Campo Rojo, *Joseph 24 Feb 1984.* Pinacate Mt, 21 Nov 1907, *MacDougal 74* (US).

MENTZELIA *PEGA PEGA,* BLAZING STAR

BY RICHARD S. FELGER AND BARRY PRIGGE

Annuals or perennials, usually herbaceous or somewhat woody at the base, rarely small trees. Stems often with peeling, whitish epidermis. Hairs barbed (glochidiate) but not stinging, the plants adhesive, especially the leaves and capsules (fresh leaf pieces can stick on clothing like Velcro). Flowers solitary and axillary or in terminal cymes usually with bracts. Petals 5 or 10, yellow or orange to whitish. Stamens few to many, the filaments slender to petal-like staminodes, the staminodes present or not. Fruit a capsule, the seeds few to many, variously sculptured, angled, flattened, or winged.

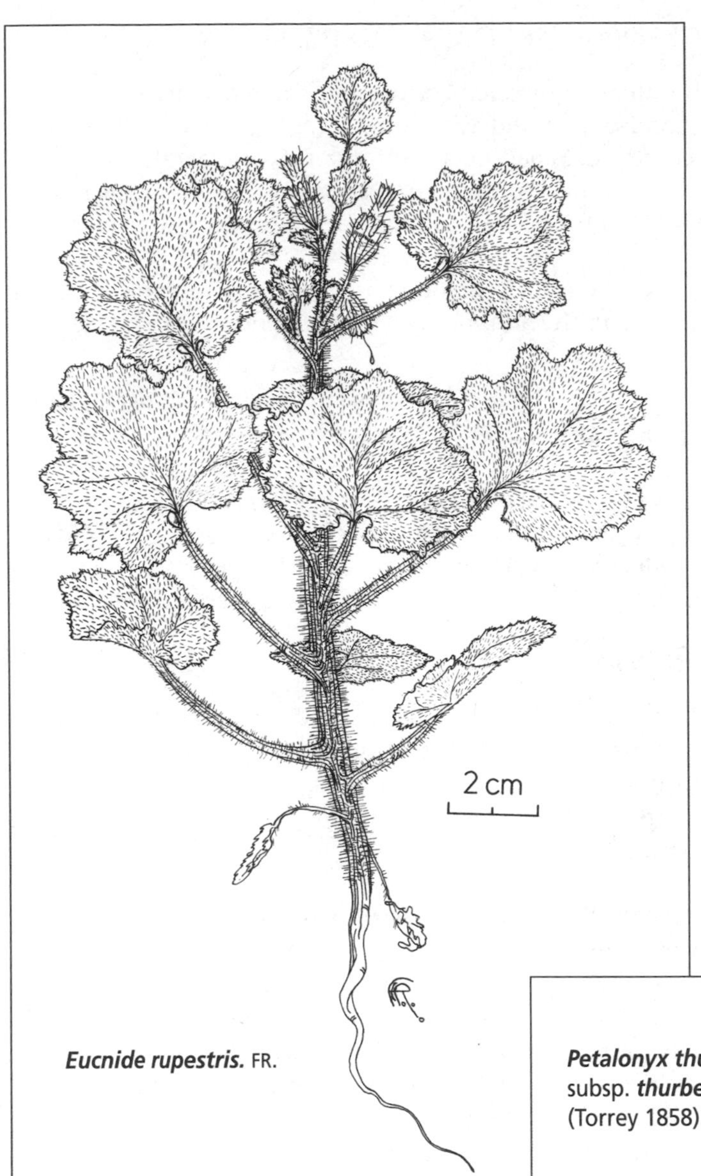

Eucnide rupestris. FR.

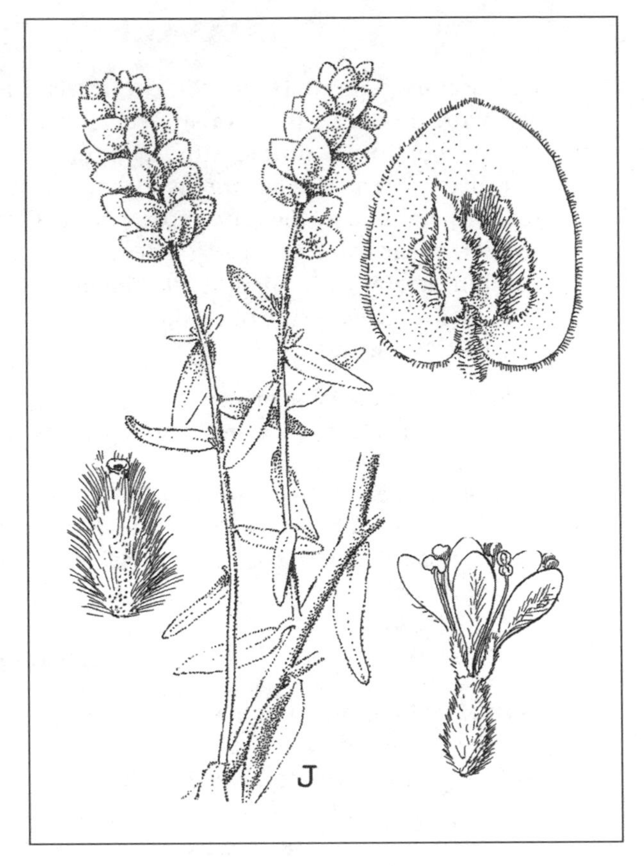

Petalonyx linearis. JRJ (Abrams 1951).

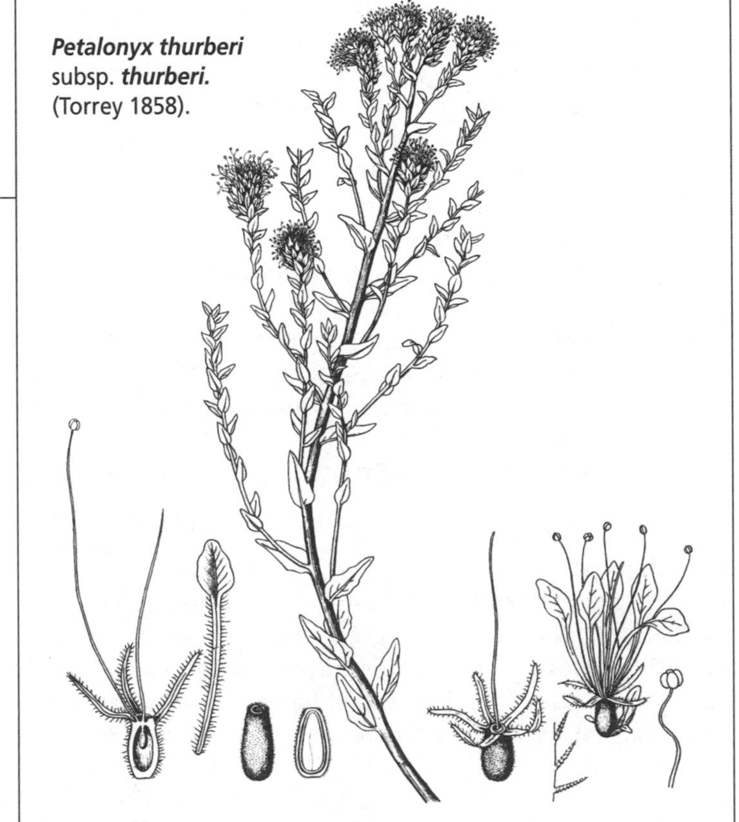

Petalonyx thurberi
subsp. ***thurberi.***
(Torrey 1858).

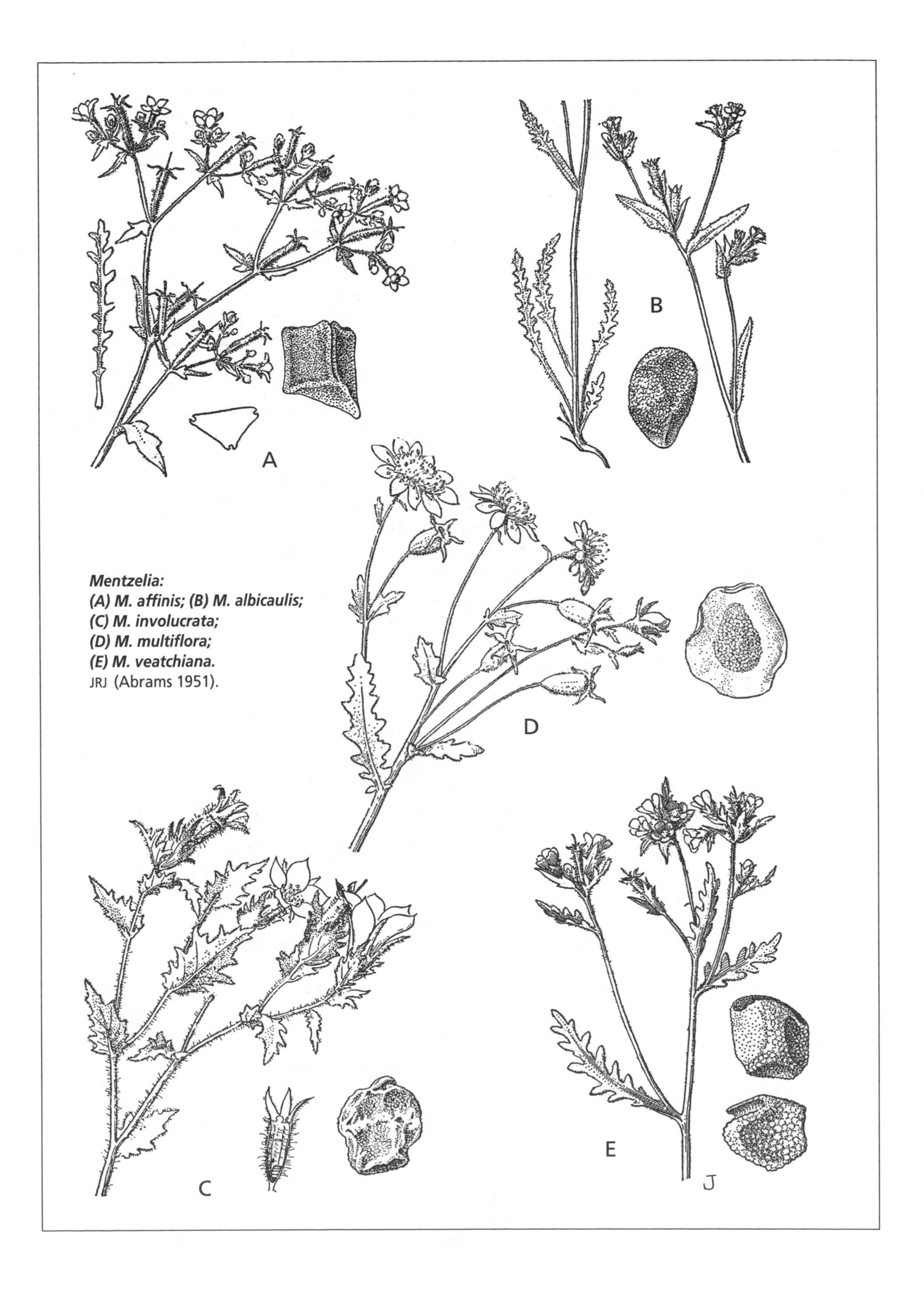

Mentzelia:
(A) M. affinis; (B) M. albicaulis;
(C) M. involucrata;
(D) M. multiflora;
(E) M. veatchiana.
JRJ (Abrams 1951).

Temperate and tropical America, largely in drier regions and mostly in the southwestern United States and Mexico; 60 species. Sonoran Desert mentzelias are placed into 4 easily recognized sections and several species groups (see key to the species). For many or most mentzelias there is a transition from basal-rosette leaves to the uppermost bracts: basal-rosette leaves and sometimes a few of the lower stem leaves are petioled, whereas upward there is a gradation to smaller, sessile, and even clasping leaves or bracts.

The flowers are usually closed during midday. Among the members of sections *Mentzelia* and *Trachyphytum* the flowers open in the morning, in section *Bartonia* the flowers open in the late afternoon, and in section *Bicuspidaria* the flowers are open both in the morning and afternoon.

Distinguishing some of the seemingly narrowly defined taxa within section *Trachyphytum* is sometimes difficult, especially among the polyploid complexes. Members of this section have narrow capsules, slender stems, 5 petals and no petaloid staminodes, yellow or orange flowers, and among ours the flowers are moderately to very small. The taxa are often sympatric. Thompson & Roberts (1971: 550) warn their readers that in the deserts of southern California some of these "species form a polyploid complex which can only be distinguished with difficulty; up to 5 species may grow together." They report different ploidy numbers and sterility barriers linked to distinctive seed morphology, providing strong evidence for placing these taxa at the species rather than the infraspecific level (Thompson & Roberts 1971; Zavortink 1966).

Identification relies heavily on characters of the mature seeds and capsules. Among the *Trachyphytum,* seeds at the base of the capsule tend to be triangular prisms with a longitudinal groove along each angle. The capsule is so narrow at its base that seed shape is largely the result of the developing seed(s) being confined between the walls of the triangular capsules. The grooves along the seed angles are formed by the placenta pressing against the seeds while they are still plastic and developing. In *M. affinis,* with its very slender capsules and seeds in a single row (like a stack of coins), all the seeds are triangular prisms. In most of the other *Trachyphytum* only the lowermost seeds are triangular prisms (or at least a somewhat different shape than the upper seeds). For this reason it is important to look at seeds toward the apex of the capsule, which are often diagnostic; unfortunately these seeds tend to fall out of mature capsules.

Differences in seed surface textures are also important key characters. Immature seeds of all the *Trachyphytum* are smooth and become dark, almost black, upon drying. Immature seeds are often more or less the same shape as mature seeds, but the surface sculpturing is not yet developed. In some taxa the seeds are tesselated—with a few (often only 1 or 2) darker, almost purple papillae, or cells, among the light-colored papillae. Seeds of some taxa have a lip or noselike flap. When soaked in water the seeds of the various taxa swell to become spherical.

Some of the *Trachyphytum* have straight capsules, whereas others have bent or recurved capsules. Although capsule curvature can be a helpful character, it should be used with caution. Immature capsules are usually straight, and sometimes mature ones at the stem tips are also straight whereas only those in the axils of the branches are curved to 90° or more.

Among the *Trachyphytum,* the petals tend to elongate with age, so that petal shape on the first day may be ovate but become lanceolate with age. The petals are yellow to orange, and some have a darker orange to red spot at the base. Petal size and the bicolored pattern may be of taxonomic significance, but these characters are usually not discernable on herbarium specimens. References: Darlington 1934; Glad 1976; Hill 1976; Thompson & Roberts 1971; Zavortink 1966.

1. Stems slender (often very slender), minutely pubescent (may appear glabrous without magnification); flowers and capsules sessile; capsules narrowly cylindrical or gradually and slightly expanded above; petals usually small (less than 8 mm). ______ Section **Trachyphytum:**
   2. Seeds in a single row in capsules, all precisely prismatic (3-sided prisms) and sharply angled, conspicuously and regularly grooved along the 3 longitudinal edges; capsules not narrowed at the base. ______ **M. affinis**

   2′ Seeds in 2 or more rows toward apex (1 row at base), angular or not, at least those toward apex of fruit not precisely prismatic, not grooved on all 3 sides or if grooved on all 3 sides then the 3 "corners" not all the same shape (caution: seed(s) at base of capsule sometimes prismatic as in *M. affinis*); capsules slightly narrowed toward the base.

3. Bracts uniformly green; mature capsules in lower axils curved; seeds not tessellate (or occasionally 1 or 2 dark-colored papillae); widespread including higher elevations. ______________ **M. albicaulis complex**

3′ Bracts, especially the larger ones, with a white area at base; capsules straight; seeds tessellate (with some dark-colored papillae against a light-colored background); higher elevations of Sierra Pinacate. ______________ **M. veatchiana**

1′ Stems slender or not, conspicuously pubescent (scabrous hispid); flowers and capsules pedicelled; capsules not narrowly cylindrical; petals usually medium to large (usually 1 cm or more).

4. Bracts 1.5–3 cm, with a white basal area, the margins with slender lobes; corolla satiny, silvery yellowish or white. ______________ Section **Bicuspidaria: M. involucrata**

4′ Bracts less than 1.3 cm, all green and entire, or bracts absent; corolla yellow or orange.

5. Petals orange, 5, the filaments all slender (but outer 5 stamens larger than the inner ones); floral bracts absent; seeds not winged. ______________ Section **Mentzelia: M. adhaerens**

5′ Petals and petaloid filaments bright yellow, the petals 5 or seemingly 10 (5 petals + 5 petaloid staminodes); bracts linear to linear-triangular; seeds broadly winged. ______________ Section **Bartonia:**

6. Caudex often thickened but lower stems not woody; larger leaves in a basal rosette (5.5) 9–15 cm; petals seemingly 10; seeds 3.3–3.5 mm; widespread, sandy soils to rocky slopes. ______________ **M. multiflora**

6′ Lower stems often woody like a dwarf shrub; larger leaves well distributed on stems, (1.8) 2.5–6.5 cm, the basal rosette very quickly withering in first season; petals 5 (the larger staminodes clearly narrower than the petals); seeds 2.4–3.1 mm; granitic slopes and cliffs in the northwestern part of the flora area. ______________ **M. oreophylla**

Mentzelia adhaerens Bentham. Non-seasonal ephemerals, sometimes reaching 1+ m across. Stems relatively slender, conspicuously hairy (scabrous-hispid), the epidermis peeling from older stems. Leaves dark green, often 2–4.5 × 0.8–2.6 cm, petioled, the blades thin, more or less ovate to lanceolate and coarsely toothed; apparently not forming a basal rosette. Flowers pedicelled, mostly solitary in axils of dichotomous (2-forked) branches; floral bracts absent. Calyx lobes 6–8 mm. Petals 5 in number, 11–15 mm. Petals, style, and filaments orange to yellow-orange, the anthers whitish, the style green. Stamens all anther bearing; filaments all slender. Capsules 8–14 mm, obovoid (obconic). Seeds ca. 2.7 mm, wingless, irregularly ovoid, lumpy, and granulated.

Widely scattered in the Pinacate volcanic field; arroyo beds at lower elevations of the mountains, and cinder slopes at middle or especially higher elevations; also at Puerto Peñasco. Southward in western Sonora to the Guaymas region, and Baja California from ca. 50 km north of San Felipe southward through Baja California Sur; not known from the United States. On the slopes of Sierra Pinacate, honeybees and syrphid flies were gathering pollen only.

WSW of Campo Rojo, 550 m, *F 92-482*. Tinaja del Tule, *Burgess 6289*. Emilia Camp [Campo Rojo], *Simmons 27 Jan 1965*. Elegante Crater, *Webster 22362* (DAV). Suvuk, *Dimmitt 3 Ap 1989*. Puerto Peñasco, *Dennis 23*.

Mentzelia affinis Greene. TRIANGLE-SEED BLAZING STAR. Spring ephemerals. Stems very slender, minutely pubescent (visible at 10×), glabrate and shiny white with age. Leaves highly variable, the lower leaves more or less narrowly elliptic to lanceolate, deeply and irregularly lobed, the upper leaves few-toothed to entire and sometimes broadly ovate. Bracts green throughout, variable, ovate to lanceolate, entire to lobed, shorter than the capsules. Petals 5, yellow, small. Filaments all slender. Capsules sessile, cylindrically triangular, not narrowed at the base, straight or curved, 11–15 × 1.5 mm. Seeds 0.9–1.3 mm, about as wide as long, in a single row in the capsule, angular, prismatic, conspicuously 3-angled and grooved along each angle, the ends abruptly truncate (flat and smooth as if sliced in cross sections); mottled and smooth in appearance with flat-topped papillae. Flowers opening in the early morning, closing later the same morning, and apparently opening again the next day.

Sandy-gravelly soil along washes, in arroyo beds and their floodplains, and probably more widespread. Sonoyta region, the Pinacate volcanic field, and likely westward near the Arizona border south of Cabeza Prieta Refuge; widespread in Organ Pipe Monument and probably widespread in the flora area. Deserts, grasslands, and woodlands; southern California, Baja California, Nevada, Arizona, and northern Sonora.

The prismatic, groove-edged seeds in a single row are unique. Specimens from the flora area seem somewhat small for plants of this species. This member of section *Trachyphytum* has not been studied in depth, and it may be polymorphic.

Río Sonoyta, *F 86-85.* El Papalote, *F 86-108.* Aguajita Wash, *F 88-287.* Arroyo Tule, *F 19540.* Cabeza Prieta Refuge, NE of Tule Well, *McLaughlin 2976.*

Mentzelia albicaulis Douglas ex Hooker. WHITE-STEM BLAZING STAR. Spring ephemerals. Stems whitish, very slender, minutely pubescent, with age whiter and glabrate. Leaves extremely variable, entire to variously lobed, toothed, or pinnatisect. Bracts lanceolate to broadly ovate to narrowly or broadly triangular, uniformly green, entire to few-toothed. Petals often 3–7 mm, bright yellow, sometimes with a dark orange or reddish spot at the base. Capsules mostly 15–30 × 1.3–2.0 mm, sessile, cylindrical-triangular, gradually and slightly wider at the apex, curved downward 90° or more at maturity, especially those in stem axils. Seeds 0.8–1.2 mm in width and length, in 2 or more rows toward the capsule apex, with several or more surfaces or faces variously concave or convex, the edges sharply angled or rounded, the surfaces not tessellate or some seeds with a few dark papillae. Flowers opening in the early morning, closing later in the same morning, and apparently opening again the next day.

Widespread; deserts and rolling sand plains, dunes of low to medium height, washes, and cinder flats and slopes to peak elevations. British Columbia to South Dakota, Baja California, northern Sonora, Texas, and probably Chihuahua.

The *M. albicaulis* species complex represents a fascinating but taxonomically difficult assemblage. Within the *M. albicaulis* complex, 3 of the most difficult taxa to distinguish are *M. albicaulis* sensu stricto, *M. desertorum,* and *M. obscura.* These taxa are distinguishable with certainty only by characters of the mature seeds. Members of the *M. albicaulis* complex tend to be most diverse in desert regions. Although widespread in various life zones in western North America, *M. albicaulis* sensu stricto has not been found in the flora area, although *M. desertorum* and *M. obscura* have.

Mentzelia desertorum (Davidson) H.J. Thompson & J. Roberts [*Acrolasia desertorum* Davidson. *M. arenaria* Zavortink, nomen nudum (the name not published)] Often compact and much-branched from the base. Leaves linear-lanceolate, the upper leaves ovate-lanceolate, few lobed to entire. Floral bracts entire, ovate to ovate-lanceolate. Petals bright yellow, probably with an orange spot at the base. Seeds sharply angled, most or all faces concave with smooth surfaces (the papillae or polygonal cells flat on top); some seeds with a small "nose" or flap, but this not as prominent as in *M. obscura.* $2n = 18$.

Widespread in sandy flats, often with *Larrea,* and in the Pinacate volcanic region, including higher elevations; not known elsewhere in Sonora. Also Baja California, southeastern California, and western Arizona.

31 mi E of San Luis, *Raven 11646* (LA). 13 mi N of Puerto Peñasco, *Raven 14819* (LA). NE of Sierra del Rosario, *F 20358.* S of Mex 2 (rd to Rancho Guadalupe Victoria), *Mason 1823.* S of Pinacate Peak, 875 m, *F 19355.*

Mentzelia obscura H.J. Thompson & J. Roberts [*M. deserticola* Zavortink, nomen nudum (the name not published)]. Very similar to *M. desertorum.* Petals reaching 7 mm, mostly pure yellow, probably sometimes with a faint orange spot at base. Upper seeds with a "nose" or ledgelike lip on 1 edge, the faces or surfaces slightly convex with small rounded papillae (polygons), the edges irregular and rounded (the angles more rounded and seeds plumper than in *M. desertorum*). $2n = 36$.

Widespread and to peak elevation in the Pinacate volcanic field; not known elsewhere in Sonora. Otherwise southeastern California and Baja California to Utah and western Arizona.

Pinacate Peak, *F 19452.* 7.5 km NE of Elegante Crater, *F 88-251.* Cerro Colorado, *F 93-229.* Arroyo Tule, *F 19215.*

Mentzelia involucrata S. Watson. SILVER BLAZING STAR. Winter-spring ephemerals. Stems stout, branching from the base, shiny white, hairy (scabrous-hispid), not peeling. Leaves narrowly lanceolate to elliptic, coarsely toothed to pinnatifid, the lower ones (3) 5–18 cm. Bracts 1.5–3.0 cm, with a broad whitish membranous center, green margins, and several slender pectinate lobes. Flowers pedicelled. Calyx lobes 8–16 mm. Petals (1.7) 2–3.5 cm, satiny silvery white to pale yellow. Style stout. Outer filaments expanded toward the apex into 3 teeth, the anther on the middle tooth. Capsules 1.5–2.2 cm. Seeds thick but flattened, ovoid, not winged, whitish and rough surfaced, usually constricted like a figure 8, often 3.0–3.5 mm. (Drought-stressed plants can be quite small with proportionally smaller parts.)

Widespread, Pinacate volcanic field to peak elevation and the granitic ranges. Usually in localized populations along sandy-gravelly washes, on cinder slopes and flats, and sometimes in sandy soils of crater floors, sand flats, and rocky slopes. Southward in western Sonora to Cerro Tepopa (29°21′ N), Baja California, southeastern California, and southwestern Arizona.

Variety *megalantha* I.M. Johnston is characterized by having larger petals (3.5–4.5 cm). The largest-petaled plants are from the western part of the Colorado Desert in California and adjacent northern Baja California. Although there is a general trend for increase in petal size westward, there is considerable local variation in petal size, and the larger petals are on well-watered robust plants.

28 mi W of Los Vidrios, *Bezy 446.* Sierra del Rosario, *F 20741.* Pinacate Junction, *F 20605B.* Campo Rojo, *Ezcurra 16 Ap 1981.* (MEXU). MacDougal Crater, *Turner 86-31.* Pinacate Peak, *F 19384.* SE of Moon Crater, *F 18861.* Elegante Crater, *F 19677.*

Mentzelia multiflora (Nuttall) A. Gray subsp. **longiloba** (J. Darlington) Felger [*M. longiloba* J. Darlington. *M. pumila* (Nuttall) Torrey & A. Gray, in part] Robust winter-spring ephemerals or perhaps sometimes short-lived perennials (elsewhere usually perennials, e.g., in vicinity of Caborca); stems few to many, often 40–75+ (100) cm, white with barbed short hairs. Larger leaves in basal rosettes, (5.5) 9–15 cm, narrowly lanceolate to oblanceolate and pinnately lobed to parted, the stems leafy with leaves often about half as long as the rosette leaves. Flowers pedicelled, in terminal corymbs, several per cluster. Bracts linear to linear-triangular, 8–12 mm. Calyx lobes 8–13 mm. Petals bright yellow, often with orange stripes, seemingly 10 (the 5 inner ones staminodes with anthers at apex), larger petals 12–26 cm. Stamens bright yellow, numerous; inner filaments slender, the outer filaments broadly winged and grading into the petals. Capsules 1–2 cm × 8–10 mm. Seeds numerous, 3.3–3.5 mm, the body flat, light brown, 1.2–1.4 mm, surrounded by whitish wings 0.8–1.0 mm wide, the wings transparent when wet.

Widespread; rolling sand plains with *Larrea* and on dunes of low to medium heights, roadsides, rocky soils and slopes, and cinder flats and slopes to higher elevations in the Sierra Pinacate. Subsp. *longiloba* occurs in northwest Sonora, northeast Baja California, southeast California, and southwest Arizona. This polytypic species ranges from California to Wyoming and south to Texas and adjacent Mexico. *M. pumila* sensu stricto occurs in northeast Arizona (see Felger 1980).

E of San Luis, *Raven 11636* (LA). 10 mi SE of Santa Clara near mouth of Colorado River, *Day 12 Ap 1966* (UCR). NE of Sierra del Rosario, *F 20415.* 30 mi W of Sonoyta, *F 7614.* Moon Crater, *F 19251A.* S of Carnegie Peak, 925 m, *F 19821.* Campo Rojo, *F 87-56.* 13 mi N of Puerto Peñasco, *n = 9, Raven 14818* (LAM).

Mentzelia oreophila J. Darlington [*M. puberula* J. Darlington] Annuals or perennials often 30–60 cm, with woody stems branching throughout, sometimes like a dwarf shrub with thickened woody trunklike lower stems. Stems white, leafy. Leaves (1.8) 2.5–6.5 cm, mostly dull grayish-silvery glaucous, more or less ovate-lanceolate to elliptic or obovate, coarsely toothed to shallowly lobed, the uppermost leaves sessile, the lower ones gradually narrowed basally, the base not clasping the stem. Bracts linear. Flowers opening in late afternoon. Petals and stamens bright yellow on interior surfaces, pale on backs, the petals 5 in number, 10–13 mm, rounded at apex. Inner filaments slender, the outer filaments broadly winged and grading into the petals but clearly narrower than the 5 petals. Capsules 5–9 mm, rounded at the base. Seeds many, horizontal in the capsules, 2.4–3.1 mm, conspicuously winged, papery and white when dry, when wet the wings transparent and the body yellow-brown.

Crevices on steep rock slopes and walls on the north side of Cerro Pinto, ca. 300–500 m (*Burgess 5610. F 89-57*); not known elsewhere in Sonora but expected in adjacent granitic mountains. Also western Arizona including the nearby Butler and Tinajas Altas Mountains, and deserts in California, Nevada, Utah, and north-central Baja California.

Mentzelia veatchiana Kellogg [*M. gracilenta* Torrey & A. Gray var. *veatchiana* (Kellogg) Jepson] Stems often straight, moderately slender, erect to ascending, several to many from the base. Leaves (among ours) pectinately pinnatifid, the larger leaves often 8–10+ cm. Bracts toothed, rather broad, with pronounced to minute often modest but obvious whitish area surrounding the midrib at the base, the uppermost bract often adnate (fused) to capsule. Corollas 21–31 mm wide, bright yellow-orange with a red-orange center. Filaments green, the anthers yellow. Capsules straight, 18.5–20+ mm, slightly expanded to 3.5–5.0 mm wide toward the apex. Seeds in 2 or more rows toward the capsule apex, ca. 1.7–1.8 × 1.0–1.3 mm, irregularly angled, the surfaces tessellate, the papillae rounded to pointed.

Higher elevations in the Sierra Pinacate, often common on open cinder slopes (*F 19408, F 19939. Sherbrooke 9 Apr 1983*); not known elsewhere in Sonora. Baja California including Pacific coastal areas northward and inland to southern Oregon, Nevada, and Arizona.

Members of the *M. gracilenta* complex are characterized by a white spot or area without chlorophyll at the base of the floral bracts; toothed bracts; congested (crowded) inflorescences; and straight, erect, and somewhat fat capsules. The white area at the bract base is best observed in the field; it is seen in the laminar area on both sides of the midrib and is not to be confused with the midrib itself, which is white (like the stems) on all members of *Trachyphytum*. The size of the bracts and the basal white area varies with moisture conditions. Well-watered plants are often more robust and with stouter stems than those of other *Trachyphytum* species in the flora area. The diploid *Trachyphytum* are mainly Californian (California, Baja California, extreme western Arizona, and parts of Nevada), whereas the polyploids have more extensive ranges and occur farther eastward and northward in western North America. *M. veatchiana* is tetraploid.

PETALONYX SANDPAPER PLANT

Shrubs or subshrubs, perennial but some may flower in the first year; conspicuously scabrous (rough like sandpaper) and hispid with short and long hairs. Leaves petioled or sessile. Inflorescence racemose or spicate; each flower subtended by an involucre of 3 bracts, the outermost largest and innermost smallest, the bracts enveloping the buds. Petals whitish, 5, clawed, the claw partially joined to form a tube or the petals separate. Stamens 5. (Some members of the genus are unusual in that the stamens appear to be outside the petals: near the base of the petals the stamens pass through a slit between the petals.) Fruits small, cylindrical, constricted near the apex then flared above, indehiscent, 1-seeded. Southwestern United States and Mexico; 5 species. Reference: Davis & Thompson 1967.

1. Leaves short petioled to sessile, entire, linear-oblong to lanceolate; petals separate. __________ **P. linearis**

1′ Leaves sessile, toothed, or if entire then triangular; upper part of claws of petals forming a tube. ______ **P. thurberi**

Petalonyx linearis Greene. NARROW-LEAF SANDPAPER PLANT. Perennials and often flowering during the first season, forming a dense, woody-based, rounded bush 0.3–0.5 m, with many stiff, erect-ascending brittle and leafy twigs; tardily drought deciduous. Leaves 1.2–3.4 cm, linear-oblong to lanceolate, more than twice as long as wide, the margins entire. Outer floral bract ovate, cordate (heart-shaped) at the base, the margins entire or sometimes with a few small teeth. Petals, filaments, style, and stigma white; anthers cream colored. Calyx lobes ca. 1.5 mm. Petals ca. 3 mm, separate, narrowed to a slender claw. Stamens about as long as the petals. Seeds ovoid, 2.1–2.4 (2.8) mm, brown, tightly enclosed in the dried floral tube. Flowering at least spring and fall.

Pinacate volcanic field, often localized on exposed rocky slopes, tuff breccia rims of major craters, and cinder flats and slopes to peak elevation. Also lava rock at Puerto Peñasco and granitic slopes in the northwestern part of the flora area. Not known elsewhere in Sonora except on Islas Tiburón and San Esteban; otherwise southeastern California to northern Baja California Sur and southwestern Arizona.

Cerro Pinto, *F 75-46*. MacDougal Crater, *Kamb 2009*. Chivos Butte, *F 19622*. Carnegie Peak, *F 19823*. NE of Crater Elegante, *Ezcurra 13 Ap 1981* (MEXU). 29 mi W of Sonoyta, *F 10540*. Punta Peñasco, *Keck 4211* (RSA).

Petalonyx thurberi A. Gray subsp. **thurberi.** SANDPAPER PLANT; *HAḌṢADKAM*. Bushy perennials to woody-based shrubs (0.5) 1.2–1.8 m with several to many few-branched stems, or when on dunes to 2.5 m with fewer stems. Larger, lower leaves of new shoots 2–5 cm, elliptic to lanceolate, sessile, coarsely toothed, soon deciduous; leaves gradually reduced upwards, often 0.6–2 cm, sessile, broadly lanceolate to ovate, few-toothed to entire, the uppermost ones triangular. Outer floral bract more or less ovate, coarsely few-toothed. Petals, filaments, style, and stigma white; anthers cream colored. Calyx lobes ca. 2 mm. Petals 6.0–6.5 mm, the claws free basally and connivent (coming together but not organically united) above to form a slender tube. Stamens long exserted. Flowering at least late spring and fall.

Gravelly soils of larger washes, shifting sands of larger dunes, less frequent on semistabilized smaller dunes. Especially common on dunes between the south flanks of the Pinacate volcanic shield and Bahía Adair and Puerto Peñasco; not seen in the western dunes such as near the Sierra del Rosario. Occasionally on the northeast side of the Pinacate volcanic field and more common but in widely scattered small populations in Arizona Upland near the international border between Quitobaquito and Sonoyta. Otherwise known in Sonora only from Quitovac. Also Baja California, southern California, southern Nevada, and western and southern Arizona.

Plants from the Gran Desierto dunes are unique in their substantially woodier, few-stemmed, taller, and more slender growth habit, commonly reaching 2.5 + m in height. Subsp. *gilmannii* (Munz) W.S. Davis & H.J. Thompson is known from Inyo County, California.

13.8 mi NE of Puerto Peñasco, then 7.6 mi W, *F 88-233*. 34 mi N of Puerto Peñasco, *F 13211*. Dunas, 31°39′ N, 113°38′ W, *Ezcurra 15 Apr 1981*. El Papalote, *F 86-331*. S of Moon Crater, *F 19015*. Observation: 15 km NNE of El Golfo.

LYTHRACEAE LOOSESTRIFE FAMILY

Tannin-bearing herbs, seldom shrubs or trees. Leaves generally opposite, sometimes whorled, rarely alternate, simple and often entire; stipules absent. Flowers radial (ours) or sometimes bilateral. Sepals united in a hypanthium enclosing the ovary but free from it, with 4–7 lobes usually alternating with toothlike external appendages. Petals often crumpled in the bud, separate or absent. Fruit usually a capsule; seeds usually numerous.

Tropical regions of the world and a few in temperate regions; 27 genera, 600 species. Very few in deserts and these in wetland habitats. Henna is made from the leaves of *Lawsonia inermis* of the Old World. The crape myrtle, *Lagerstroemia indica,* widely cultivated as a flowering tree, thrives in the Sonoran Desert if given sufficient water. Reference: Graham 1964.

1. Annuals, probably less than 0.5 m tall; leaves opposite. ____ **Ammannia**
1′ Perennials, often subshrubby, more than 1 m tall; leaves alternate (opposite below). ____ **Lythrum**

AMMANNIA

Annuals with opposite leaves and small flowers, the calyx with small horn-shaped appendages between the lobes. Warm tropical and temperate regions; 30 species. Reference: Graham 1985.

†**Ammannia robusta** Heer & Regel. Warm-weather wetland herbs; glabrous. Leaves broadly linear, sessile, entire, probably less than 10 cm long. Flowers small, 2 each in upper leaf axil; petals bright purple-pink; stamens 4. Capsules globose, probably 5 mm in diameter or smaller. Recorded from the vicinity of the Colorado River near Yuma (California: Bard, in wet valley lands, *Thornber 22 Sep 1912*). It was undoubtedly also in the river wetlands on the Mexican side of the border, and must have been plentiful as the seeds were utilized for food by the Mohave and Yuma people (Castetter & Bell 1951). Widespread in the Americas.

LYTHRUM

Slender annual or perennial herbs, or semishrubby. Stems brittle, 4-angled or winged. Leaves entire, mostly opposite or alternate above. Flowers generally showy, purple to white in axillary clusters or leafy terminal inflorescences. Calyx forming a cylindrical, ribbed floral tube or hypanthium. Petals 4–8, attached to the top of the hypanthium, or rarely absent. Seeds minute, many.

North America and Old World; 36 species. Some are grown for their showy flowers in temperate-region gardens.

Lythrum californicum Torrey & A. Gray. CALIFORNIA LOOSESTRIFE. Glabrous, winter-dormant perennials, becoming semiwoody at the base, with erect slender brittle stems reaching 1.2–2 m, and peeling, tan-colored bark. Leaves sessile or short petioled, mostly 1.8–7 cm, linear-lanceolate, and alternate except the lower ones opposite and often larger and broader. Racemes often ca. 40–50+ cm, occasionally with 1 or few ascending branches. Calyx lobes inconspicuous and minute, broader than long, the calyx appendages toothlike, ca. 1 mm or less in length, longer than wide, and larger and more prominent than the sepal lobes. Petals (5) 6 or 7 in number, ca. 4–5 mm, purple, and showy. Stamens often 6 or 7, the filaments red. Capsules 6.5–8.0 mm, ribbed, narrowly club-shaped, short pedicelled. Seeds ca. 1 mm. Flowering with new growth in late spring.

At several of the La Salina pozos, where it forms dense colonies in permanently wet soil at the edge of freshwater pools. This population may be a relict of more extensive populations that probably existed along the lower Río Colorado. It occurs in wetland habitats along the Colorado River in nearby Arizona and southeastern California. Southwestern North America, mostly non-desert habitats.

La Salina, *F 84-33, F 86-556.* California: Colorado River just E of Ft. Yuma, *McLaughlin 3043.*

MALPIGHIACEAE MALPIGHIA FAMILY

Woody vines, shrubs, or trees, often with double hairs (malpighian hairs: T-shaped, attached in the middle with 2 opposite-pointing arms). Leaves usually opposite, simple, entire; stipuled. Flowers bilateral, often showy, 5-merous, on jointed pedicels with 2 bractlets. Sepals 5, often with paired, fleshy oil glands. Petals 5, often clawed with fringed or toothed margins. Ovary superior, of (2) 3 (5) carpels, each with a single ovule. Fruits dry with winged to nutlike mericarps or sometimes fleshy.

Mostly tropical and subtropical, largely in the Americas; 68 genera, 1100 species. New World species have sepals with oil glands; their principal pollinators are oil-gathering anthophorid bees such as *Centris* (see *Krameriaceae*). At least 14 species, with both dry and fleshy fruits, occur in southern Sonora; 6 species, all with dry fruits, are found in the Sonoran Desert; 1 species extends into northwestern Sonora and Arizona; none reach California. References: Buchmann 1987; Robertson 1972.

JANUSIA

Slender semiwoody to woody vines. Flowers yellow, weakly bilateral, in axillary clusters or solitary. Fruits of 2 or 3 one-winged nutlets or samaras (indehiscent, 1-seeded, winged fruits). Southern United States to Argentina; 20 species, 3 in the Sonoran Desert. Some in South America produce cleistogamous flowers, the North American species do not. Reference: Anderson 1982.

Janusia gracilis A. Gray. *FERMINA*. Perennial vines, dying back severely in drought, climbing into shrubs, or non-vining in open habitats and as short as 30 cm; densely to moderately hairy. Stems few to many, often intertwined, slender, tough. Leaves gradually drought deciduous, 1.5–4.5 cm, short petioled, the blades linear to narrowly lanceolate, with 2 (6+) toothlike red-brown glands near the base, 0.1–0.5 mm; stipules minute or absent. Flowers 11.5–13.6 mm in diameter. Sepals green, 4 of them with a pair of fleshy, oval glands 0.5–0.9 mm (in drought 2 sepals often with only 1 gland), the fifth sepal glandless. Petals and stamens bright yellow, the petals 5–6 mm, subequal. Samaras 2 or 3, each 1.0–1.4 cm, the wings papery, green to red-green. (No other plant in the flora area has samaras.) Various seasons.

Lythrum californicum. JRJ (Abrams 1951).

Janusia gracilis. BA. © James Henrickson 1999.

Widely scattered; rocky slopes (often with north exposures), canyons, and washes. Sonoyta region, Pinacate volcanic field to ca. 800 m, and on the north side of the higher granitic mountains. Baja California Sur to northeastern Baja California, Sonora from the Guaymas region to southern Arizona and eastward and southward to western Texas, Chihuahua, Coahuila, Durango, and Zacatecas.

Sierra Cipriano, *F 91-9*. Between Tinajas de Emilia and Carnegie Peak, 750 m, *F 19791*. Hourglass Canyon, *F 19167*. Sierra del Viejo, *F 16875*. Quitobaquito, *F 88-125*.

MALVACEAE MALLOW FAMILY

Herbs to trees with soft wood, mostly with stellate hairs, also with forked or simple hairs. Leaves alternate, simple but sometimes deeply cut, usually stipuled. Ours lack nectaries, but elsewhere nectaries sometimes on leaves, pedicels, or calyces. Flowers often subtended by an involucel of (2) 3 or more bracts or bractlets (the floral bracts or bractlets) immediately below the flower at the apex or top of the pedicel. Flowers usually radial. Sepals united below, 5-toothed to parted; petals 5, separate except where fused to stamen tube at or near base. Stamens many, the filaments united into a column surrounding and hiding the ovary and style, the upper portions of filaments free from the column. Ovary superior, with 3 to many carpels (each apparent carpel sometimes actually of 2 adjacent carpel halves fused together). Fruits usually capsules (opening through longitudinal splits between the carpels or along the inner seam of the carpels into the middle or "top" of the capsule) or schizocarps (separating into single-carpelled segments or mericarps). Seeds kidney- or top-shaped.

Worldwide, temperate and mostly tropical, some in deserts; 115 genera, 1500 species. Readily recognized by the filament tube, plus many have stellate hairs. Includes cotton (*Gossypium*), extensively cultivated in the Sonoran Desert including the flora area, edible species of hibiscus, marsh mallow, okra, and garden plants such as hibiscus and hollyhock.

Malvaceae are prominent in Sonora. The family is usually divided into 5 tribes, 2 of which occur in the flora area: the Hibisceae and Malveae. Some Malveae have two strategies for seed dispersal: The seed in the lower and usually thicker-walled chamber of the mericarp is often tightly held and dispersed with the mericarp. In contrast, the seed(s) in the upper, dehiscent chamber readily fall as the mericarp wings dry and split apart (see *Horsfordia* and *Sphaeralcea*). References: Bates 1968; Fryxell 1988; Kearney 1951.

1. Perennials; fruit a capsule of 5 persistent carpels spreading at maturity, the carpels splitting open but remaining attached, the lower and upper portion of each carpel essentially similar. ______Tribe **Hibisceae: Hibiscus**

1′ Perennials or ephemerals; fruit a schizocarp, the segments (mericarps) separating at maturity. ______Tribe **Malveae:**

 2. Winter-spring ephemerals.

  3. Stems very slender and usually not erect; flowering and fruiting pedicels very slender and bent at a conspicuous joint; fruits resembling a miniature paper lantern. ______**Herissantia**

  3′ Stems not unusually slender and usually erect or upright; pedicels not threadlike and not jointed or bent; fruits not like a paper lantern.

   4. Leaf blades longer than broad, broadest near the base (short to long ovate), often more or less 3-lobed to 3-cleft.

    5. Petals white to pale lavender; mericarps indehiscent, the upper and lower halves identical. ______**Eremalche exilis**

    5′ Petals orange; mericarps dehiscent, the lower and upper portions very different (lower portion reticulate and not opening, upper part membranous-papery, smooth and opening). ______**Sphaeralcea**

   4′ Leaf blades as broad as long, widest at about the middle (orbicular to kidney-shaped).

    6. Plants densely hairy, especially lower leaf surfaces, the hairs stiff, coarse, yellowish, mostly 1–2 mm (especially on stems); petals 1.5–2 cm, rose-purple with a large dark reddish spot near the base of each petal. ______**Eremalche rotundifolia**

    6′ Plants glabrate or sparsely hairy with soft white hairs mostly less than 1 mm; petals ca. 0.5 cm, white to pale lavender, not spotted at base. ______**Malva**

 2′ Perennials, herbaceous to shrubby (sometimes flowering in first season during summer-fall rains).

  7. Stems very slender and weak; lower leaves petioled, upper leaves of flowering branches sessile; flowering and fruiting pedicels very slender, bent at a conspicuous joint; fruits resembling a miniature paper lantern. ______**Herissantia**

  7′ Stems thicker than in 7, usually not weak; pedicels usually not unusually slender and often not bent at joint; fruits not as in 7 (plants lacking one or more of preceding characters).

   8. Herbaceous root perennials often propagating by rhizomes; leaf blades conspicuously asymmetric at base; mericarps indehiscent, lacking dorsal cusps. ______**Malvella**

   8′ Occasionally herbaceous, mostly bushy or shrubby, without rhizomes; leaf blades symmetric (or nearly so); mericarps dehiscent, with dorsal cusps or wings.

    9. Lower and upper halves of mericarps similar, not reticulate on sides. ______**Abutilon**

    9′ Lower and upper portions of each mericarp markedly dissimilar, the lower chamber reticulate, the upper part with flared membranous wings and smooth.

     10. Floral bracts absent. ______**Horsfordia**

     10′ Floral bractlets 3 per flower. ______**Sphaeralcea**

ABUTILON INDIAN MALLOW

Herbaceous perennials to shrubs (ours) or small trees, the herbage usually densely hairy. Pedicels jointed (in ours the joints usually best seen on fruiting pedicels, but sometimes not readily evident; the fruits usually break off or bend down at the joint, perhaps to more readily release the seeds). Floral bracts absent. Corollas often yellow or orange. Fruit a schizocarp (sometimes appearing capsulelike), the mericarps usually twice as long as the column (place of attachment), not inflated, usually with 3–6 seeds each.

Mostly in tropical and subtropical regions of the Americas, Africa, Asia, and Australia; 200 species. *A. californicum* and *A. palmeri* are in section *Mexabutilon* but otherwise do not seem closely related, whereas *A. incanum* and *A. malacum,* in section *Oligocarpa,* appear to be closely related.

Flowering and fruiting responses occur with sufficient soil moisture during warm weather. They are frost sensitive and dormancy or dieback may follow freezing weather. References: Fryxell 1983; Strong 1977.

1. Mericarps 5, the tips blunt and rounded; calyx lobes not overlapping; fruiting calyx shorter than the mericarps.
   2. Leaf tips pointed; petals with a maroon spot at base; fruiting calyx 2.8–4 mm, ca. 1/4 as long as the mericarps. ______ **A. incanum**
   2′ Leaf tips blunt; petals uniformly yellow-orange; fruiting calyx 4–6 mm, ca. 1/2 to 3/4 as long as the mericarps. ______ **A. malacum**

1′ Mericarps 7 or more, the tips slender and pointed; calyx lobes overlapping basally; fruiting calyx about as long as mericarps.
   3. Leaf blades not especially velvety; stems and mericarps with hairs 0.1–0.4 mm. ______ **A. californicum**
   3′ Leaf blades velvety; stems and mericarps with hairs 0.8–2.5 mm. ______ **A. palmeri**

Abutilon californicum Bentham. Shrubs often 1.5–2.5 (3) m, often few-branched, spindly, and in the shade of ironwood trees (*Olneya*). Densely stellate pubescent, the hairs relatively sparser on robust leaves of well-watered plants. Leaf blades broadly ovate to nearly orbicular, 2–7.5 cm, greener above, the apex acute-acuminate to obtuse, the margins toothed. Pedicel lengths highly variable, the joint often inconspicuous at flowering time, the pedicel thickened above the joint, the joint usually conspicuous in fruit, 1.5–5 mm below calyx base. Petals yellow to orange. Fruiting calyx about as long as mericarps. Mericarps with uniformly minute stellate hairs plus longer simple hairs on the margins, and with slender acuminate to awnlike tips. Fruits 11.5–13.5 × ca. 15 mm; mericarps 8–10.

Arroyos and large, brushy washes with desert trees; Pinacate volcanic field, occasional on the south side of Sierra Pinacate to 875 m, and at Sierra Blanca. Gulf coast of Sonora from the vicinity of Guaymas to the Pinacate region, Isla Tiburón and several other gulf islands on the Baja California side, through most of the Baja California Peninsula, and Isla Socorro of the Islas Revillagigedos.

Abutilon californicum is a Gulf of California segregate of the more widespread, inland, southern, and tropical *A. abutiloides* (Jacquin) Garcke ex Britton & Wilson. Inland and southward in Sonora *A. californicum* is replaced by *A. abutiloides,* which ranges from Oaxaca to southern and central Arizona, southern Texas, and the Caribbean. Apparently *A. californicum* does not occur in Arizona; *A. abutiloides* is in the Ajo Mountains. The distinctions between these taxa are subtle (see Fryxell 1988); they are geographically segregated but it remains to be seen whether or not they intergrade at the zones of contact in Sonora.

N of Moon Crater, *F 10574.* S of Pinacate Peak, 875 m, *F 19316.* Tinaja de los Chivos, *F 18622.* Campo Rojo, *Webster 24267* (DAV). Sierra Blanca, *F 20228.*

Abutilon incanum (Link) Sweet [*Sida incana* Link. *A. pringlei* Hochreutiner] Shrubs often 1–1.6 m, usually taller than wide with slender stems. Herbage densely and velvety pubescent with minute stellate hairs. Leaf blades often 3–10 cm, ovate, cordate at the base, acute to more often acuminate at the tip, greener above, gray-green below, the margins toothed; petioles prominent. Flowers solitary in leaf axils and in large, open, and sparse terminal inflorescences. Pedicel often jointed 2.5–3.0 mm below the calyx base, the joint occasionally obscure, the pedicel usually thickened above joint even in bud. Corollas pale orange with large maroon spots in center, the petals reflexed with age; filaments red near bases, yellow near tips; stigmas red. Fruits 6.5–8 mm × ca. 5–7 mm; mericarps 5, with blunt tips; fruiting calyx 2.8–4.0 mm.

Sonoyta region; common on north-facing slopes of Sierra de los Tanques, mostly at higher elevations. Western Arizona to Texas, northern Mexico, and Hawaii.

Sierra de los Tanques, 460 m, *F 89-16.* Cabeza Prieta Refuge, Agua Dulce Mts, *F 92-728.*

Abutilon malacum S. Watson. Small shrubs, 0.5–1 m. Herbage densely stellate pubescent. Leaf blades often 1.5–6 cm, broadly ovate to orbicular, cordate at base, obtuse at tip, dull green above, the lower surface somewhat yellow-green with prominent raised veins, the margins toothed; petioles prominent. Flowers solitary in leaf axils or in a short, compact, few-flowered terminal inflorescence. Pedicels 12–30 mm, the joint often 2–9 mm below calyx base, the joint occasionally obscure, the pedi-

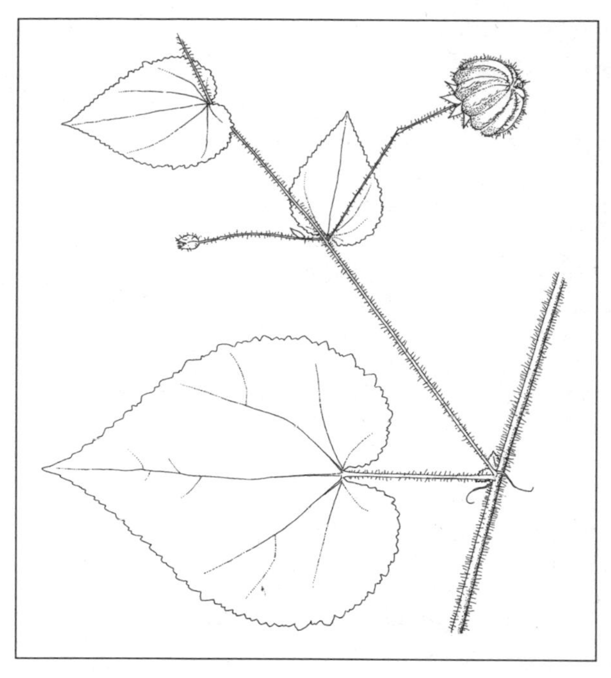

Herissantia crispa. Note jointed pedicel and sessile branchlet leaves. MBJ.

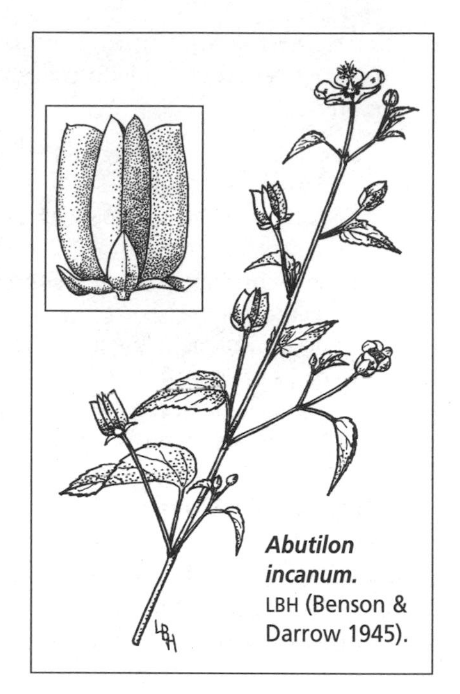

Abutilon incanum. LBH (Benson & Darrow 1945).

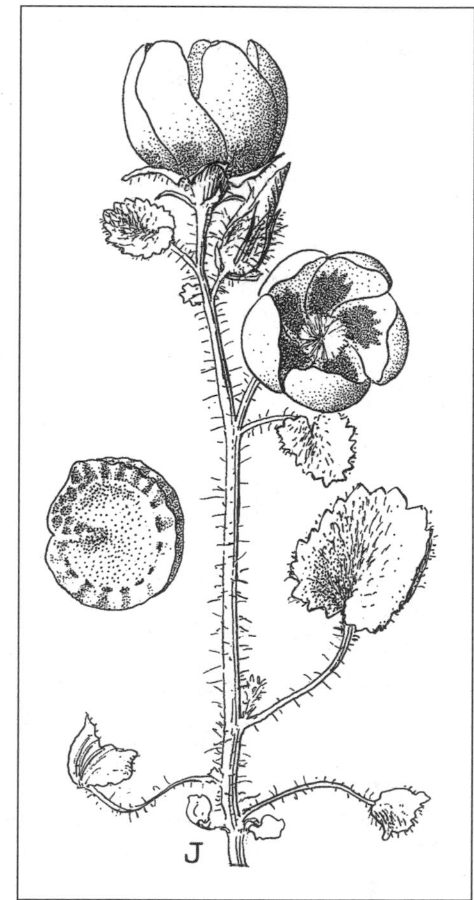

Eremalche rotundifolia. JRJ (Abrams 1951).

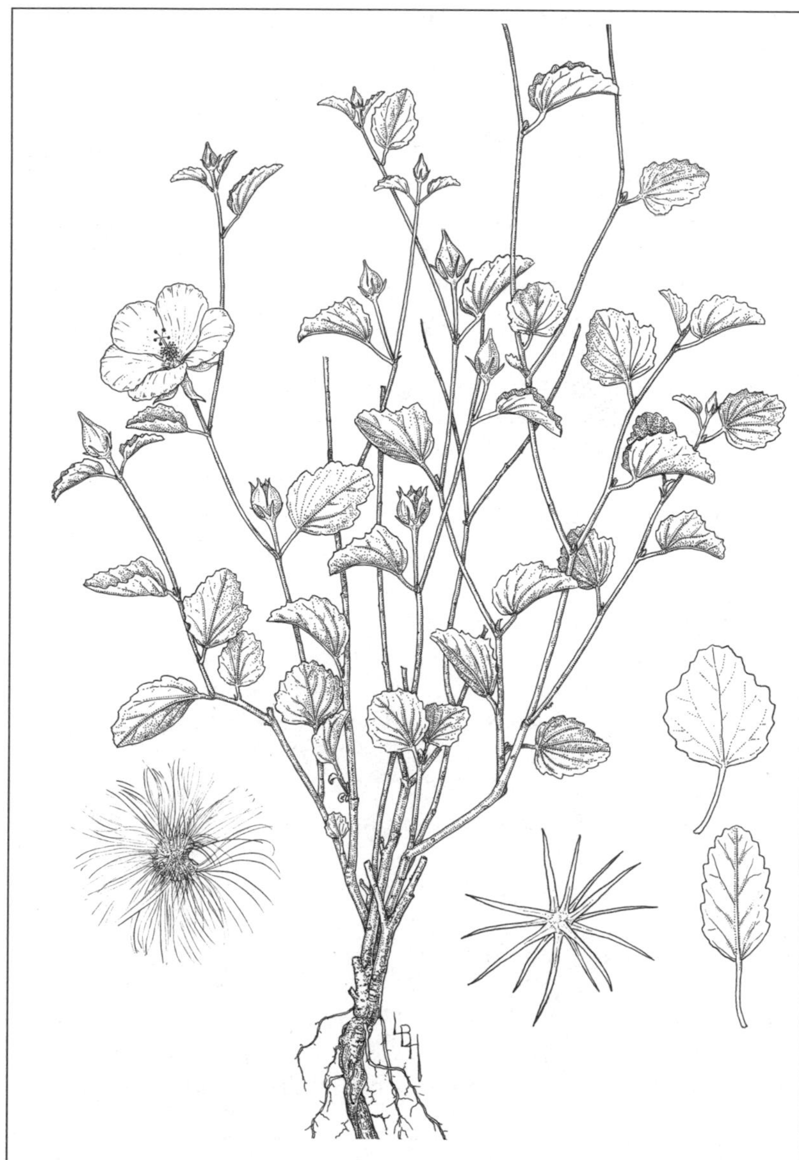

Hibiscus denudatus var. ***denudatus.*** LBH (Benson & Darrow 1981).

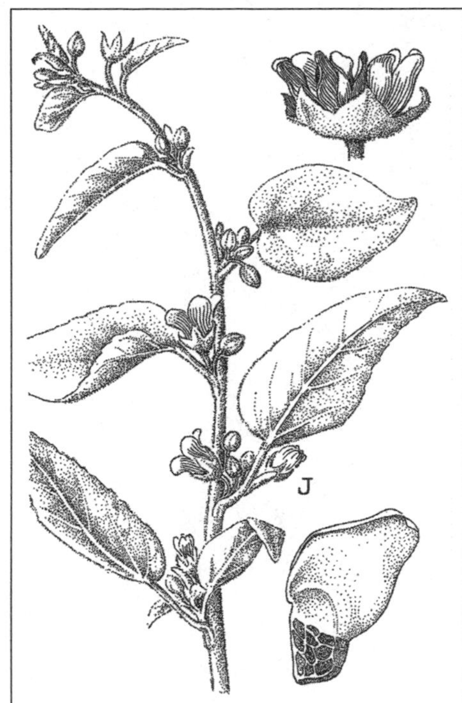

Horsfordia newberryi.
JRJ (Abrams 1951).

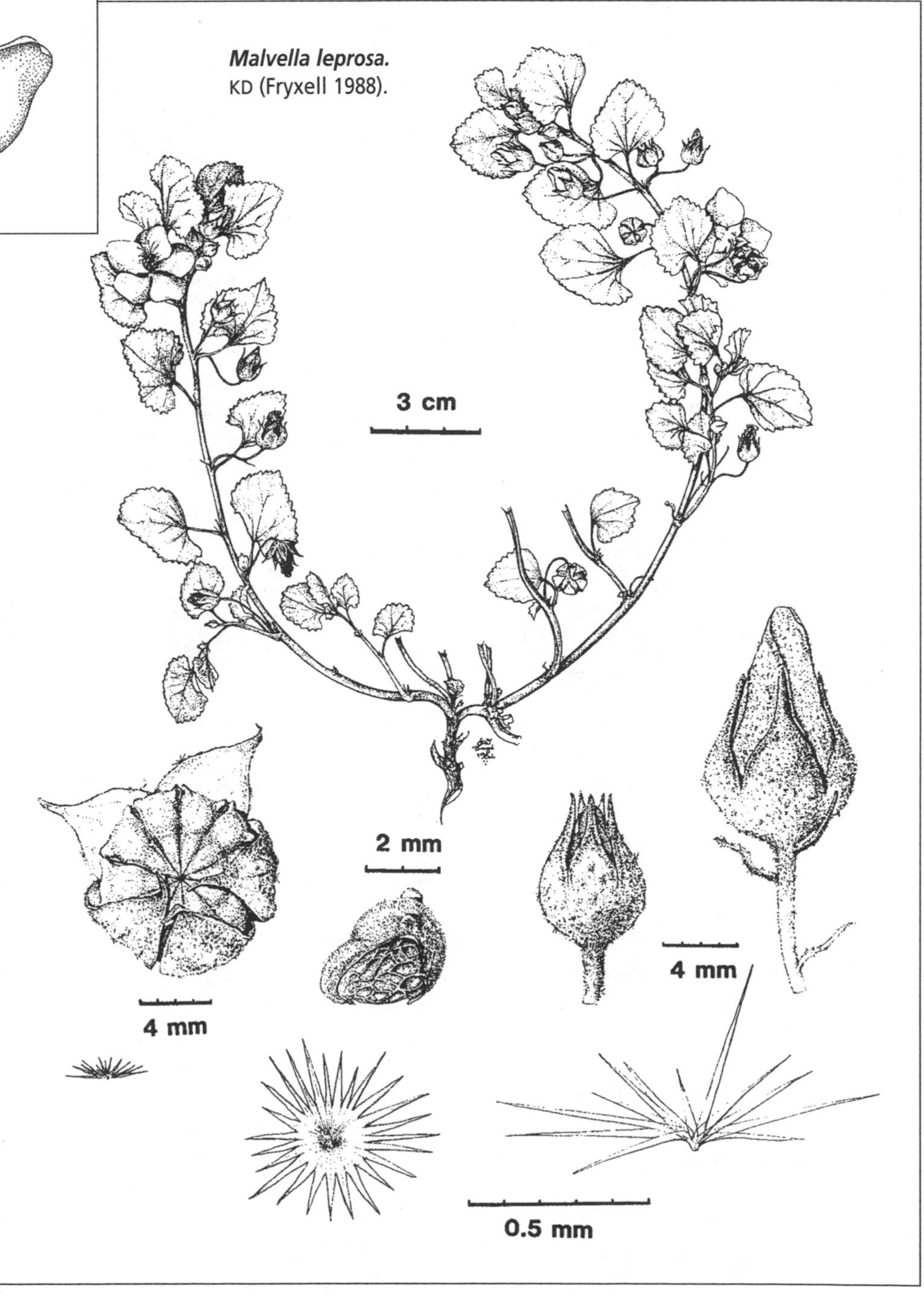

Malvella leprosa.
KD (Fryxell 1988).

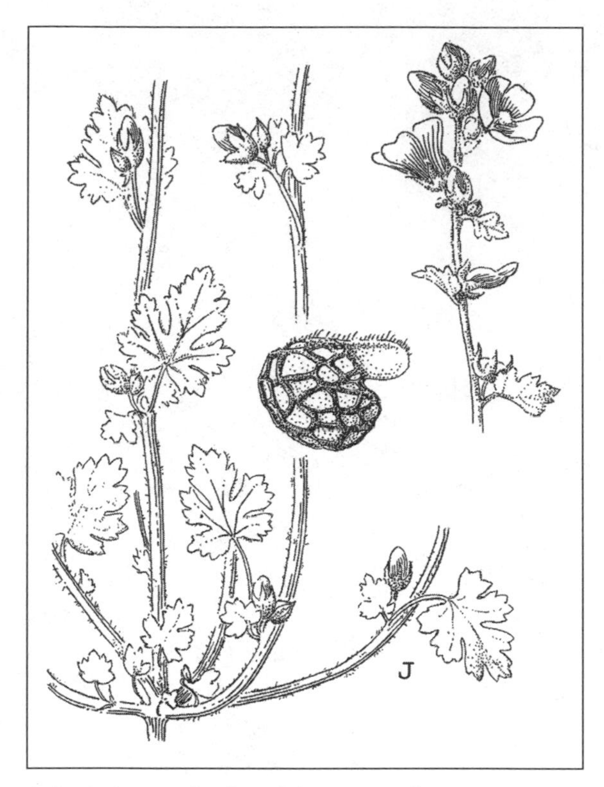

Sphaeralcea coulteri. JRJ (Abrams 1951).

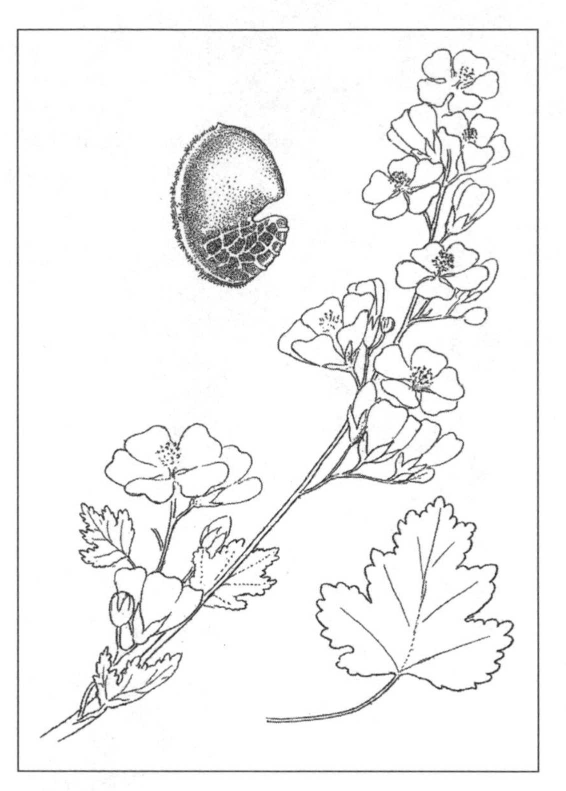

Sphaeralcea ambigua subsp. ***ambigua.*** JRJ (Abrams 1951).

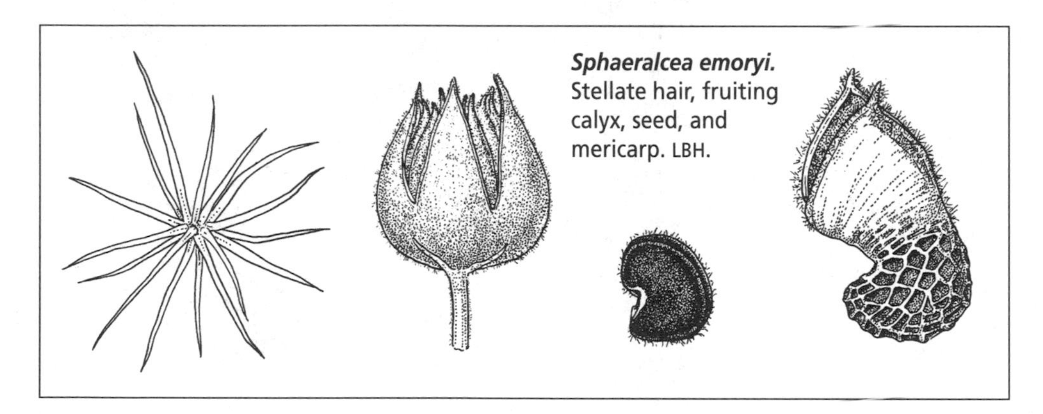

Sphaeralcea emoryi. Stellate hair, fruiting calyx, seed, and mericarp. LBH.

cel usually thickened above the joint. Corollas and stamens uniformly yellow-orange, the corollas spreading, ca. 1.5 cm in diameter. Fruiting calyx 4–5 mm. Fruits 6–8 mm; mericarps 5, the tips blunt.

Low, rolling hills near Sonoyta, localized along brushy wash margins and occasionally along roadsides; also in the Agua Dulce Mountains in adjacent Arizona. In Arizona and northern Sonora often on limestone or granitic slopes. Primarily a plant of the Chihuahuan Desert; southern Texas and New Mexico, limited areas of southern Arizona, and southward to San Luis Potosí, Durango, and northern Sonora.

Although *A. malacum* might be confused with *A. incanum,* they are distinct and generally occupy different habitats. *A. malacum* has a more compact growth habit, stouter and thicker stems, coarser and larger stellate hairs, yellow-green rather than gray-green herbage, relatively broader and usually thicker leaf blades with blunter tips and prominently raised veins on the lower surface, much shorter inflorescences, usually thicker peduncles and pedicels, slightly larger flowers, longer calyces, corollas of a single color, petals spreading rather than reflexed, and generally slightly smaller fruits.

1–5 mi W of Sonoyta, *F 86-319, F 88-12.* 9 km SW of Sonoyta, *F 92-81.* Cabeza Prieta Refuge, Agua Dulce Mts, *F 92-729.*

Abutilon palmeri A. Gray [*A. aurantiacum* S. Watson. Not *A. aurantiacum* Linden ex Turczaninow. *A. macdougalii* Rose & Standley, *Contr. U.S. Nat. Herb.* 16: 13, pl. 4, 1912] Sparsely branched, open and scarcely woody subshrubs to 1.5 (2) m. Densely stellate pubescent, the hairs on the stems and mericarps long and slender. Petioles of larger leaves (2.5) 4–13 cm; leaf blades broadly ovate to nearly orbicular, the larger ones 4.5–15 cm, minutely to coarsely toothed, deeply cordate at base, acuminate-acute to obtuse at tip, conspicuously velvety, usually light green and somewhat lighter colored below. Flowers showy, in large, slender-stemmed, open and sparsely branched terminal panicles rising well above the foliage, and also solitary in leaf axils. Pedicel length highly variable, the joint often inconspicuous at flowering time; pedicel often thickened above joint and the joint usually conspicuous at fruiting time. Petals pale orange, ca. 2–2.5 cm. Fruiting sepals broadly ovate, about as long as the mericarps; fruits ca. 1.3–1.5 cm wide, the mericarps about 10.

Sandy-gravelly arroyos and washes, and rock crevices of arroyos and canyons, often at waterholes, and on the volcanic flanks of the Sierra Pinacate to 800 m in canyons and arroyos on the southern side of the mountain mass; often with *Acacia greggii*. Southwestern Arizona to Sinaloa, and southeastern California to the Cape Region of Baja California Sur; disjunct in Tamaulipas.

Tinajas de los Pápagos, *Turner 59-26*. Tinaja del Bote, *F 18645*. Tinaja de los Chivos, *Simmons 539*. Tinajas de Emilia, *F 19728*. Pinacate Mts, 22 Nov 1907, *MacDougal 47* (holotype of *A. macdougalii*, US).

EREMALCHE

Herbaceous annuals with stellate and often simple hairs, the stems hollow. Leaves petioled. Involucel of 3 persistent, separate bractlets. Flowers mostly solitary in axils. Corollas white to pink (drying lavender). Fruits globose-depressed; mericarps 1-seeded, disk-shaped, indehiscent or eventually the thin lateral walls rupturing, blackish; seed filling the central cavity of the mericarp, the seed margins sculptured. California, southwestern Arizona, northwestern Sonora, and Baja California; 3 species. Reference: Kearney 1956.

1. Pubescence of mostly stellate hairs mostly less than 0.5 mm long; petals less than 0.5 cm, of one color and lacking spots; leaf blades cleft to about the middle. ____ **E. exilis**
1′ Stems with simple hairs 1.5–3.0 mm long; petals 1.5–2 cm, with a large spot or blotch at the base; leaf blades shallowly crenulate to toothed. ____ **E. rotundifolia**

Eremalche exilis (A. Gray) Greene [*Malvastrum exile* A. Gray] Spring ephemerals with mostly stellate hairs mostly less than 0.5 mm. Stems to 60 cm, several to many, mostly branching from near the base, spreading to decumbent, slender, often few-branched. Leaf blades palmately 3- or 5-lobed, the lobes divided to toothed or entire; petioles longer than the blades. Floral bractlets linear, about 2/3 as long as sepals (bractlets resembling those of *Malva parviflora*). Pedicels about as long as leaves. Petals 3–5 mm, white to pale lavender and without basal spots. Fruits 4.0–4.5 mm in diameter; mericarps semicircular, 1.5–1.8 mm in diameter.

Common in localized habitats on sandy plains along the west side of the Pinacate volcanic field, from the vicinity of Moon Crater to the Pinta Sands in Cabeza Prieta Refuge; not known elsewhere in Sonora. Also western and southern Arizona, Baja California, southern California, Utah, and Nevada.

S of Moon Crater, *F 19089*. MacDougal Pass, *F 92-314*. 1 km SW of MacDougal Crater, *F 92-342*. 0.5 mi S of Mex 2 [rd to Rancho Guadalupe Victoria], *Mason 1828*.

Eremalche rotundifolia (A. Gray) Greene [*Malvastrum rotundifolium* A. Gray] DESERT FIVE-SPOT. Spring ephemerals, highly variable in size, 4–50 cm. Stems erect, simple or few-branched, with coarse, mostly simple hairs often 1.5–3.0 mm. Leaf blades orbicular, shallowly crenulate to coarsely toothed, with coarse stellate hairs. Floral bractlets linear, ca. 2/3 as long as sepals. Pedicels longer than the leaves. Flowers showy, the corollas pale pink-lavender with dark spots in the center, the petals ca. 1.5–2 cm. Fruits 8.0–11.5 mm in diameter; mericarps semicircular, 2.6–3.0 mm in diameter.

Desert pavement mesas, occasionally on sand or cinder flats. Locally common on the southwest side of the Pinacate volcanic field, for example in the vicinity of Tinaja del Tule, Tinaja de los Chivos, and Moon Crater (*F 19232*). Not known elsewhere in Sonora; Mojave Desert and the northwestern part of the Sonoran Desert in southeastern California, western Arizona, and Baja California.

HERISSANTIA

Erect to decumbent or trailing perennial herbs to shrubs, sometimes facultative annuals. Hairs soft and stellate, sometimes also with long straight hairs. Pedicels conspicuously jointed. Floral bracts absent. Fruits globose, inflated, and papery, the mericarps readily separating at maturity.

Southern United States and the West Indies to Argentina and Bolivia; 5 species. Four species have relatively restricted distributions and one is a widespread generalist and the only one in the North American deserts. Reference: Brizicky 1968.

Herissantia crispa (Linnaeus) Brizicky [*Sida crispa* Linnaeus. *Abutilon crispum* (Linnaeus) Medicus. *Bogenhardia crispa* (Linnaeus) Kearney] BLADDER MALLOW. Mostly short-lived perennials, also facultative annuals. Stems very slender, upright to trailing, arching or semivining and often growing in shrubs, to 1 m from a scarcely woody base. Stems, petioles, and pedicels with velvety long simple hairs and shorter stellate hairs. Leaves long petioled on vegetative, non-flowering stems, and sessile and usually much smaller on secondary, flowering stems; blades broadly ovate, relatively thin, 3–7 cm, cordate at base, acute to acuminate-acute at tip, the margins crenate-toothed; stipules (2) 3–7 (10) mm, stringlike, green, relatively persistent. Flowers solitary in leaf axils, the flowering and fruiting pedicels very slender and conspicuously jointed, with a prominent elbowlike bend at the joint. Flowers ca. 1.5 cm wide, the petals pale yellow-orange (elsewhere the petals are white, Fryxell 1988). Fruits mostly 1.3–1.5 cm in diameter, resembling a miniature paper lantern, at maturity quickly falling apart into separate mericarps. Flowering and fruiting mostly during warm, moist weather.

Widespread but seldom common; Sonoyta region, the Pinacate volcanic field, and granitic ranges. Arroyos and washes, upper bajadas, and rocky slopes usually in soil pockets; often beneath shrubs and desert trees. Absent from dunes and sand flats, and usually not on open desert flats. Arizona to Texas and Florida, and the West Indies to Argentina and Bolivia; perhaps adventive over much of its distribution in the New World, also adventive in the Old World including tropical Asia; often in disturbed, weedy habitats.

MacDougal Crater, *F 10654*. SSW of Tinajas de los Pápagos, *F 10617*. Tinaja del Bote, *F 18646*. Arroyo Tule, *F 18816*. Moon Crater, *F 10573*.

HIBISCUS

Annual to perennial herbs, shrubs, or trees; usually with stellate hairs. Stipules awl-shaped to leaflike, falling early. Floral bracts usually relatively large and conspicuous. Fruit a capsule of 5 persistent carpels, at maturity spreading open, star-shaped. Seeds glabrous or woolly (ours), kidney-shaped to nearly round.

Warm regions of the world, a few in temperate areas; 200 species. The 3 species in the Sonoran Desert (ours plus *H. biseptus* S. Watson) are in section *Bombicella,* which includes 21 species from the southern United States to northwestern South America, the West Indies, plus 1 in the Old World. The *Bombicella* are subshrubs of relatively arid habitats, mostly with showy flowers and silky-haired seeds.

The genus includes common garden ornamentals. *H. rosa-sinensis* "is perhaps the best known ornamental shrub of the tropics and subtropics throughout the world. It is cultivated widely in Mexico in all frost-free areas and exists in a profusion of color forms, both single and double" (Fryxell 1988). It is cultivated in northwestern Sonora but requires substantial irrigation and is often damaged by freezing weather. Other hibiscus species are grown as fiber crops. *H. sabdariffa,* known as *jamaica* or *flor de jamaica,* cultivated in humid areas of Mexico, serves as a vegetable and condiment; the calyces are used to make *té de jamaica*. Reference: Fryxell 1980.

1. Leaves deeply 3-lobed to parted (except on juvenile growth); floral bracts conspicuous, more than 1 cm, longer than the carpels; corollas yellow. ________ **H. coulteri**

1′ Leaves broadly ovate to obovate, not deeply lobed or parted; floral bractlets inconspicuous, less than 0.5 cm, shorter than the carpels; corollas white to pink. ________ **H. denudatus**

Hibiscus coulteri Harvey ex A. Gray. DESERT ROSE-MALLOW. Perennials, also flowering in the first season, to ca. 1 m and usually few-branched. Stems slender and woody; pubescence dense, whitish, more or less evenly distributed around the stem, of stellate hairs with 4 conspicuously appressed arms. Leaves deeply 3-parted, the segments more than twice as long as wide; juvenile and new-growth leaves not as deeply parted. Floral bracts slender, stiff, longer than the carpels. Calyx 1.6–2.0 cm, deeply divided into lanceolate lobes. Corollas often 5 cm wide, bright yellow with a maroon center, showy, the petals mostly 2.5–3+ cm; flowers closing at night, opening after sunrise. Warmer, moist times of year.

Entering the flora area in the rocky hills and mountains southwest of Sonoyta, e.g., Sierra Cipriano (*F 16740, F 91-12*). Widespread in Organ Pipe Monument and the vicinity of Sierra Cubabi eastward in northern Sonora and southward to the Sierra del Viejo near Caborca. Sonora to Zacatecas, southern Texas, southern Arizona, and Baja California.

Hibiscus denudatus Bentham var. **denudatus.** ROCK HIBISCUS. Multiple-stemmed subshrubs, branched from the base, 0.3–1 m, dying back to a thickened woody base in extreme drought. Herbage pale yellow-green, densely pubescent with stellate hairs; rainy-season leaves often greener. Leaves petioled, (2) 2.5–4.5 cm, broadly ovate to oblong-ovate, rounded to obtuse at tip, the margins toothed; tardily drought deciduous. Floral bractlets usually appressed and inconspicuous, the longer ones 1–3 mm, the tips often breaking off early, sometimes difficult to distinguish in color and texture from base of calyx. Calyx 10–16 mm, deeply divided into lanceolate lobes. Corollas white to pinkish with maroon spots in the center, the petals often 2–2.7 cm. Seeds covered with long, dense silky hairs 3–4 mm. Various seasons, especially spring.

Rocky slopes to peak elevations throughout the flora area; slopes including south exposures, bajadas, and sometimes along gravelly washes and desert flats. One of the most drought-tolerant perennials in the region. Var. *denudatus* occurs from southern Arizona to southeastern California and southward to the Cape Region of Baja California Sur and in Sonora to the Guaymas region. Var. *denudatus* is distinguished from var. *involucellatus* A. Gray by its much shorter and more fragile (deciduous) floral bractlets and generally larger-sized plants. Var. *involucellatus* extends from the vicinity of Tucson and northeastern Sonora eastward into the Chihuahuan Desert; it often has persistent, spreading bractlets 2.5–4.5 mm, which are often darker than the calyx.

Bahía de la Cholla, *F 16831.* Tinajas de Emilia, *F 19748.* Sykes Crater, *F 18922.* MacDougal Crater, *F 9735.* 61 mi E of San Luis, Sierra del Viejo, *F 16891.* Sierra del Rosario, *F 20376.*

HORSFORDIA

Tall spindly shrubs or subshrubs, densely stellate pubescent. Floral bracts none. Pedicels jointed but sometimes obscurely so, the joint best developed at fruiting time. Fruit a schizocarp, the mericarps separating at maturity, 1–3-seeded, the lower chamber indehiscent with firm, wrinkled walls and a single tightly held seed, the upper chamber dehiscent, usually (ours) splitting into expanded membranous wings.

Centered in the Sonoran Desert: the Baja California Peninsula, Sonora, perhaps northwestern Sinaloa, southeastern California, and southern Arizona; 4 species. Reference: Fryxell 1985.

1. Larger leaves mostly broadly ovate and often cordate at base; petals pink or sometimes nearly white, drying bluish to pale lavender; mericarps 1-seeded. ______ **H. alata**

1′ Larger leaves mostly lanceolate and not cordate at base; petals orange when fresh; mericarps 2–3-seeded. ______ **H. newberryi**

Horsfordia alata (S. Watson) A. Gray [*Sida alata* S. Watson] PINK VELVET MALLOW. Sparsely branched slender shrubs 1.5–3.5 m; often grayish pubescent. Leaves velvety, larger ones 6–10+ cm; petioles prominent, the blades usually broadly ovate, cordate at the base, often thickish, but new-growth summer-fall leaves often thin and scarcely or not at all cordate (uppermost leaves of flowering branches often lanceolate). Petals white to pale pink or pale lavender, drying bluish to pale lavender, 14–18+ mm. Filaments, anthers, and style white, or free portion of filaments pale lavender; stigma often dark purple-red. Fruits 10–15 mm wide. Mericarps 1-seeded, the upper chamber forming ovules but these aborting. Various seasons.

Gravelly washes, shallow arroyos in rocky bajadas, and canyons; near Sonoyta through much of the lowlands of the Pinacate volcanic field and granitic mountains to at least 600 m. Sonora south to the vicinity of Isla Tiburón, southeastern California, southwestern Arizona, and both Baja California states.

The plants are frost tender and the new growth is often killed by freezing weather. Both *Horsfordia* species may flower and fruit and even become woody in the first year. Their habit—tall, slender-stemmed, sparsely branched, spindly shrubs—is unique for the flora of the region. They often grow intermixed.

Sierra del Viejo, *F 16877*. Sierra del Rosario, *F 75-25*. Tinajas de los Pápagos, *Turner 9772*. N of Moon Crater, *F 10572*. 12 mi SW of Sonoyta, *F 9821*. Quitobaquito, *Bowers 1526*.

Horsfordia newberryi (S. Watson) A. Gray [*Abutilon newberryi* S. Watson] ORANGE VELVET MALLOW. Slender shrubs, usually several times taller than wide and few-branched, to 3 m; often yellowish pubescent. Leaves velvety, larger ones 5.5–10 (15) cm, the petioles prominent, the blades often rather thick, lanceolate, not cordate at base or scarcely so in some larger leaves. Petals bright yellow-orange, 7–8 (10?) mm. Fruits 8–15 mm wide. Lower mericarp chamber 1-seeded, the upper chamber 1- or 2-seeded. Various seasons.

Rocky slopes, usually canyons and arroyos; Sonoyta region, Pinacate volcanic field, and granitic ranges. Sonora south to the Guaymas region, both Baja California states, southeastern California, and southwestern Arizona.

There are two strategies for seed dispersal: One seed is tightly held in the lower, indehiscent part of the mericarp (as in *H. alata*) and dispersed with the mericarp. The upper seed or seeds are loose in the upper, dehiscent part of the mericarp and readily fall as the mericarp wings dry and split apart. The flowers are smaller, and the leaves are usually thicker and narrower than those of *H. alata* and not cordate at the base.

Hornaday Mts, *Burgess 6831*. 24 km W of Los Vidrios, *F 85-1017*. Sierra Blanca, *F 20225*.

MALVA MALLOW

Annual, biennial, or perennial herbs with a well-developed taproot. Flowers subtended by 2 or 3 bractlets. Petals notched at tip. Fruit a schizocarp, depressed-globose and disk-shaped; mericarps 1-seeded, indehiscent. Warm regions of the Old World, a few adventive and weedy in the New World; 100+ species.

***Malva parviflora** Linnaeus. *MALVA, QUESITO;* CHEESEWEED; *TASH MA:HAG, HA:HAGAM CU'IGAM*. Winter-spring ephemerals (elsewhere often annuals to biennials); glabrate to moderately pubescent with soft, mostly stellate white hairs. Petioles longer than the blades, the blades nearly orbicular to kidney-shaped, usually palmately 5- or 7-lobed and veined. Flowers 1 to several in leaf axils. Floral bractlets slender. Corollas 20–30% longer than the flowering calyx, the petals ca. 5 mm, white with pale pink-lavender tinge, with a glabrous claw, the blade obovate, deeply notched with broad lobes. Calyx enlarging in fruit. Fruits 7–8 mm wide.

Disturbed habitats including roadsides and especially weedy in the vicinity of Sonoyta, Puerto Peñasco, ranches and agricultural areas, relatively rare in natural vegetation. Native to Eurasia and naturalized around the world. Elsewhere in Mexico the young leaves and fruits are eaten as greens. Spira & Wagner (1983) germinated seeds 180–200 years old.

Puerto Peñasco, *F 20955*. W of Sonoyta, *F 88-17*. Pinacate Junction, *F 88-61B*. Quitobaquito, *F 7663*.

MALVELLA ALKALI MALLOW

Perennial, decumbent herbs. Herbage gray or silvery with appressed stellate hairs or more often with specialized peltate and often ciliate (fringed) scales. (In these elegant lepidote scales, the basal portions of the radii are webbed like an umbrella with protruding ribs.) Leaves petioled; base of leaf blades obviously asymmetric. Floral bractlets 1–3, quickly deciduous or absent. Petals with stellate hairs on parts exposed in bud. Fruit a schizocarp, globose-depressed, flattened at each end, the mericarps indehiscent, 1-seeded. Warm regions of the New World and 1 species in southern Europe and Middle East; 4 species. Reference: Fryxell 1974.

1. Leaf blades nearly as wide as to wider than long; plants with mostly stellate hairs and some peltate scales. ________ **M. leprosa**
1′ Leaf blades narrowly triangular, longer than wide; herbage and calyces with silvery peltate scales throughout. ________ **M. sagittifolia**

Malvella leprosa (Ortega) Krapovickas [*Malva leprosa* Ortega. *Sida leprosa* (Ortega) K. Schumann. *S. leprosa* var. *hederacea* (Douglas ex Hooker) K. Schumann. *S. hederacea* (Douglas ex Hooker) Torrey ex A. Gray] Scurfy perennials from relatively deep roots, often propagating from branching rhizomes. Stems, leaves, and calyces densely covered with stellate hairs; calyces, upper surfaces of leaf blades, and often the stems with peltate scales. Stems 6–30 cm, erect, or becoming spreading and decumbent; larger stems 1.3–2 mm in diameter. Leaf blades broadly ovate, about 2/3 as wide as long to orbicular and broader than long (nearly kidney-shaped) or asymmetrically triangular-ovate, 1.6–4 × 2–5 cm, the margins toothed. Bractlets linear, shorter than the sepals, sometimes seen on buds or flowers, usually difficult to see and very quickly deciduous, or absent. Calyx 8–11 mm, the lobes broadly ovate, acute at tip. Petals cream colored, 1.5–1.8 cm. Fruits red-brown, 6–7 mm in diameter. Warmer months.

Hard-packed clay soils of playas, playa sediments in crater floors, and an agricultural weed south of San Luis. On Playa Díaz north of Cerro Colorado it is abundant on the flats among *Euphorbia albomarginata.* Central Mexico to Washington and Idaho, also in South America.

The stems and leaves are thicker than those of *M. sagittifolia,* and its distribution in our area is much more restricted. Within the flora area the 2 species have not been found growing together.

Playa Díaz, *F 10807, F 86-366.* 18 km S of San Luis, *F 85-1040.*

Malvella sagittifolia (A. Gray) Fryxell [*Sida lepidota* A. Gray var. *sagittifolia* A. Gray] Perennials from a relatively deep, somewhat thickened taproot; often propagating from branching rhizomes. Stems, leaves, and calyx dotted with elegant silvery peltate scales, their margins erose-dentate (fringe-toothed). Stems slender, spreading and becoming nearly prostrate, 4–30 cm. Leaf blades triangular, much longer than broad, 1.5–4.5 × 0.5–1.5 cm near base with sagittate ("eared") lobes, the apex acute, the margin otherwise entire to few- or minutely toothed. Floral bractlets absent, or perhaps sometimes vestigial and quickly deciduous below young buds. Petals mostly white, 1–1.3 cm, lower (outer) surface of each petal with a pink pie-shaped wedge dotted with stellate hairs where the corolla was exposed in bud. Stamens and styles white, the stigma rounded, dotlike, and bright red. Fruits probably 5–6 mm in diameter. Warmer months.

Locally abundant in silty-clay soils in playas and other poorly draining fine-textured soils such as playa sediments in crater floors. Southern Colorado to Arizona and Sonora, eastward to western Texas, and southeast to Coahuila and Durango.

NW of Sierra Blanca, *Equihua 24 Aug 1982.* Pinacate Junction, *F 86-352.* S of Tinajas de los Pápagos, *F 18735.*

SPHAERALCEA GLOBEMALLOW

Mostly perennial herbs or subshrubs and some annuals, often densely pubescent with stellate hairs. Floral bractlets 3, separate. Petals of various colors. Fruit a schizocarp, the mericarps 1–3-seeded, with a ventral notch; each mericarp usually differentiated into (1) an upper, dehiscent section (the carpel walls smooth and spreading apart at maturity, the seed(s), if present, falling away early), and (2) a lower, indehiscent section (the body of the mericarp, reticulate and retaining the seed); thus 2 strategies for seed dispersal. Seeds kidney-shaped.

Western United States and northern Mexico, also South America, especially Argentina; 40 species. References: Kearney 1935; Kearney & Peebles 1939.

In 1935 Kearney recognized and described a number of subspecies, and in 1939 Kearney & Peebles transferred those subspecies to varieties. As with most mallows, leaf measurements and descriptions are for the larger leaves rather than the smaller ones in the upper stems or near the inflorescences. The perennials flower at various seasons depending on moisture, especially in spring.

1. Ephemerals to annuals (may appear perennial), often taller than wide; mericarps 1-seeded, the dehiscent section short and stubby, less than half as large as the body.
   2. Spring ephemerals, several cm to 1+ m tall, stems usually slender; leaf blades about as broad as long, more or less ovate to orbicular but often shallowly 3-lobed and variously toothed, usually thin and green. **S. coulteri**
   2′ Robust ephemerals (to short-lived perennials?), stems relatively stout, often 1–2.5 m tall; leaf blades longer than wide, usually hastate (2 large, "eared" lobes at base), thin and green when vigorously growing, becoming thickish, wrinkled, and yellowish in drought. **S. orcuttii**

1′ Perennials; mericarps 2- or 3-seeded, the dehiscent section more than half as long as the body.
   3. Plants often as tall to much taller than broad; leaf blades usually conspicuously longer than wide; petals pale reddish or reddish orange; mericarps 2.7–4.3 mm; mostly in disturbed habitats including the Sonoyta region and on clayish soils. **S. emoryi**
   3′ Mostly globose bushes, often at least as broad as tall; leaf blades usually as wide as long, more or less rounded to ovate; petals orange or lavender; mericarps 4.0–5.5 mm; widespread. **S. ambigua**
      4. Herbage often with yellowish or straw-colored hairs; petals orange, the anthers whitish; widespread except higher elevations of Sierra Pinacate. **S. ambigua** subsp. **ambigua**
      4′ Herbage with whitish hairs; petals lavender, the anthers purple; higher elevations of Sierra Pinacate. **S. ambigua** subsp. **rosacea**

Sphaeralcea ambigua A. Gray subsp. **ambigua** [*S. macdougalii* Rose & Standley, *Contr. U.S. Nat. Herb.* 16:13, plate 5, 1912] DESERT GLOBEMALLOW. Bushy perennials, woody at base, often globose and reaching 1 (1.3) m. Stems and upper leaf-blade surfaces often with a yellow cast, hairs of lower surfaces usually whiter. Leaf blades mostly 1.8–6 cm, more or less ovate, about as wide as long, often more or less 3-lobed, thinner and larger during more favorable conditions and thicker, smaller, and often wrinkled during drier conditions, the margins variously toothed to wavy. Petals orange, 2–3 cm; anthers white. Mericarps 2-seeded (3 seeds reported for the species elsewhere), 4.0–5.5 mm, the upper, dehiscent section larger than the body.

Rocky slopes, small to large arroyos and canyons, bajadas and flats near hills and mountains, and sometimes on desert flats such as playa margins; Sonoyta region, Pinacate volcanic field, granitic ranges, and desert flats.

This highly variable species is the most widespread perennial globemallow in the Sonoran Desert. By default ours are subsp. *ambigua,* but the orange rather than red petal color is at odds with Kearney's descriptions. Subsp. *ambigua* is a desert plant, usually well below 1000 m, and is the most widespread, or "generalist," of the several infraspecific taxa; southwestern Utah and southern Nevada to southeastern California, Baja California, and northwestern Sonora southward at least to Bahía Kino.

Along the coast in the vicinity of Puerto Peñasco and Bahía de la Cholla, many plants appear intermediate with subsp. *versicolor* Kearney, and more strongly so during drought. Subsp. *versicolor,* known from Islas San Esteban and Angel de la Guarda as well as portions of the Sonoran coast, is characterized by thick, wrinkled, yellow pubescent and bicolored leaves.

Sierra del Rosario, *F 20667.* Arroyo Tule, *Burgess 6297.* Hornaday Mts, *Burgess 6837.* Tinajas de Emilia, *F 87-32.* Papago Tanks: *F 9770;* 16 Nov 1907, *MacDougal 45* (holotype of *S. macdougalii,* US 574253). Playa Díaz, *F 88-397.* Bahía de la Cholla, *F 16825.*

Sphaeralcea ambigua subsp. **rosacea** (Munz & I.M. Johnston) Kearney [*S. rosacea* Munz & Johnston] Similar to subsp. *ambigua,* differing as follows: Plants often more robust. Stems usually more densely white woolly, the herbage, inflorescence branches, and calyces with white stellate hairs. Leaf blades markedly wrinkled (rugose) but flat when growing during moist conditions, ovate to somewhat triangular, mostly deeply 3-lobed, (2) 3–4.5 cm, and about as wide as to slightly narrower than long, the margins often coarsely toothed (appearing ragged). Bractlets and calyx base densely hairy to glabrate. Sepals often more attenuate, 11–17 mm. Petals pink-lavender to lavender, drying blue-lavender, 1.8–3 cm. Anthers dark purple. Mericarps 4.2–6.2 mm, 2-seeded.

Sierra Pinacate above 600 m, sometimes descending to 220 m, usually at elevations above the range of subsp. *ambigua;* mostly among rocks and cinder surfaces. Not known elsewhere in Sonora; otherwise southern Arizona, southeastern California, and northern Baja California.

Pinacate Peak, 1100 m, *F 19379*. Hourglass Canyon, 600 m, *F 19142*. S of Carnegie Peak, 900 m, *F 19879*. S of Pinacate Peak, 875 m, *F 19303*. Arroyo Tule, *F 18822*. Tinajas de los Pápagos, *F 9771*.

Sphaeralcea coulteri (S. Watson) A. Gray [*Malvastrum coulteri* S. Watson, 1876] *MAL DE OJO;* ANNUAL GLOBEMALLOW; *HAḌAM TATK*. Winter-spring ephemerals, sometimes germinating as early as September; with a well-developed taproot. Highly variable, stems 5–60 cm, occasionally 1.8–2 m in well-watered situations such as crater-floor playas. Leaf blades usually greenish, ovate to orbicular, about as wide as long, usually relatively thin and not wrinkled, variously lobed but not hastate (upper leaves may have proportionally longer blades). Flowers showy, opening in early morning, sometimes around 8 A.M., the petals orange. Mericarps 1-seeded, about as long as wide, 2.0–2.7 mm, the dehiscent section much smaller than the body.

One of the most common spring wildflowers, often carpeting the desert in fields of orange; sandy or fine-textured and poorly drained soils, playas (especially in craters), arroyo margins, sand flats, cinder flats and slopes, and often in disturbed, weedy habitats. Lowlands nearly throughout the flora area, but not on moving dunes. Southeastern California to Baja California Sur, and southern Arizona to Mazatlán, Sinaloa. The mericarps of *S. coulteri* and *S. orcuttii* are essentially identical.

Moon Crater, *F 19090*. MacDougal Crater, *Turner 86-24*. Sykes Crater, *F 20034*. Tinajas de los Pápagos, *Turner 59-35*. 30 mi W of Sonoyta, *F 7605*. 21 mi SW of Sonoyta, *Webster 22395*. Quitobaquito, *F 7649*.

***Sphaeralcea emoryi** Torrey. *MAL DE OJO*. Annuals to perennials, herbaceous to subshrubby 0.5–1 (1.5) m. Stems erect to floppy and curving, especially in shaded habitats and with age. Leaves highly variable, the blades conspicuously longer than wide and broadest near the base, 3-lobed and relatively thin, or ovate and thicker with drier conditions. Petals pale red-orange. Mericarps 2- or 3-seeded, 2.7–4.3 mm, longer than wide, the dehiscent section about as large as the body.

Clay soils in roadside playa-like ditches at Pinacate Junction, disturbed habitats in the Sonoyta region, and especially well established at Quitobaquito. Although this distribution seems like that of a non-native plant, it is seasonally abundant in wholly natural vegetation along the large braided washes of Cabeza Prieta Refuge near the Sonora border. Southern Nevada, southeastern California, Baja California, and Arizona to Sinaloa.

Pinacate Junction, *F 19596*. NW side of Sonoyta, *F 85-941*. Quitobaquito, *F 86-206*.

Sphaeralcea orcuttii Rose, 1893. Winter-spring annuals, often 1–2.5 m, commonly appearing perennial because of their large size, or perhaps short-lived perennials. Quickly developing a deep, stout taproot. Herbage often yellowish due to dense, yellowish pubescence. Leaf blades longer than broad, often subhastate. Flowers mostly crowded in long, slender, many-flowered terminal inflorescences, the petals orange. Mericarps 1-seeded, about as long as wide, 2.5–3.0 mm, the dehiscent section much smaller than the body. Mostly spring, often to May or June.

Sandy plains with *Larrea* and low stabilized or shifting dunes of moderate height; mostly to the south and west of the Pinacate volcanic shield. Southward in western Sonora to the vicinity of Punta Baja (between Bahía Kino and Tastiota); also southwestern Arizona, southeastern California, and Baja California.

It is closely related to *S. coulteri* and the only characteristic that really seems to separate them is size. The mericarps are nearly identical, but those of *S. orcuttii* tend to be slightly larger. In *S. orcuttii* the leaf blades may be relatively thin and not wrinkled on well-watered robust plants, but usually the blades are relatively thick and wrinkled with a yellow cast due to the hairs. The thinner leaves of *S. coulteri* may be related to their shorter life span. *S. orcuttii* continues to thrive and produce flowers much later into the late-spring drought than most spring ephemerals, and it sometimes survives into early summer, probably in part because of the large, deep taproots. Perhaps *S. orcuttii* should be treated as a synonym of *S. coulteri*.

30 mi E of San Luis, *Sanders 5642*. N of Sierra del Rosario, *F 20436*. W of MacDougal Crater, *F 9932*. Hornaday Mts, *Burgess 6803*. Bahía de la Cholla, *F 16825*. Puerto Peñasco, *F 85-778*.

MARTYNIACEAE UNICORN-PLANT FAMILY

Annual or perennial herbs with sticky hairs. Leaves without stipules. Inflorescences of terminal racemes. Flowers usually showy, bilateral, the corollas sympetalous and 2-lipped. Ovary superior with a basal nectar-producing disk. Fruit a capsule, often with terminal or lateral horns, hooks, or prickles for animal dispersal. Seeds sculptured. Southern United States to South America; 3 genera, 12 species.

PROBOSCIDEA DEVIL'S CLAW, UNICORN PLANT

Herbage viscid-sticky with mucilage-filled hairs, the surfaces slimy. Leaves opposite below, mostly alternate above, large and long petioled. Calyx unequally 5-lobed, loosely subtending the corollas. Corollas relatively large, in open, often short racemes (terminal but often appearing axillary because of continuing growth of the stem). Fertile stamens 4 with 1 staminode. Style single, the stigmas 2, sensitive and snapping shut on contact. Fruits drying as woody capsules with 2 long claws. Seeds tenaciously held in the capsule body. Southern United States, Mexico, and Peru; 9 species.

The green fruits develop a long beak, hooked at the tip. As the fruits mature and dry, the glandular-sticky green pericarp peels away from the woody capsule and the beak splits into 2 claws. These unique capsules can be found at any time of year. The large claws may hook onto an animal's leg; possible dispersal agents include bighorn sheep, pronghorn, coyotes, and jackrabbits. References: Bretting 1982, 1985; Bretting & Nilsson 1988; Nabhan et al. 1981.

1. Perennials with a large tuberous root; corollas yellow. ______ **P. altheaefolia**
1′ Ephemerals, roots rather small and not tuberous; corollas pale lavender. ______ **P. parviflora**

Proboscidea altheaefolia (Bentham) Decaisne [*Martynia altheaefolia* Bentham. *P. arenaria* (Engelmann) Decaisne] *GATO, UÑA DE GATO, TORITO;* DESERT UNICORN PLANT, DEVIL'S CLAW; *BAN 'IHUGGA.* Perennials from a large thick deeply set tuberous root, the shoots emerging about when the summer rains begin and withering with post-summer drought. Stems and petioles semisucculent. Leaves often with petioles 4–11 cm, the blades 2–6 cm, broadly ovate to orbicular or kidney-shaped and shallowly lobed. Flowers 4 cm, showy, the corollas bright yellow inside the tube and on lobes, with brown-purple speckles and dark yellow-orange nectar guides, the tube often bronze colored outside. Capsule body 4.0–6.6 cm, the claws 9–14 cm. Seeds 6.0–9.2 mm, obovoid, blackish, and warty.

Sandy-gravelly soils of arroyo beds, larger washes, crater floors, cinder flats and slopes, and sandy flats; widely scattered, often localized. Sonoyta region, desert plains westward at least to the area south of Cabeza Prieta Refuge, and the Pinacate volcanic field to near peak elevation. Southeastern California to Baja California Sur, eastward to western Texas, Chihuahua, Coahuila, and through much of Sonora, also Peru.

The Seri Indians peeled the fleshy root and ate the cortex fresh (Felger & Moser 1985). The capsules are usually smaller and more delicate than those of *P. parviflora,* but under favorable conditions they can be about the same size.

Suvuk, *Equihua 22 Sep 1982* (MEXU). Pinacate Peak, *F 86-463.* MacDougal Crater, *F 10498.* Cerro Colorado, *F 10800.* E of Pinacate Junction, *F 86-337.* 32 mi W of Sonoyta, cactus flat, *Lehto 15509* (ASU). Quitobaquito, *Bowers 1387.*

Proboscidea parviflora (Wooton) Wooton & Standley subsp. **parviflora** [*Martynia parviflora* Wooton] *UÑA DE GATO, CUERNITOS, TORITO, AGUARO;* DEVIL'S CLAW; *'IHUG.* Summer ephemerals, highly variable in size, reaching more than 1 m across with favorable conditions; roots relatively small and poorly developed. Stems and petioles thick and semisucculent. Larger leaves 12–30 cm; petioles about as long as the blades, the blades broadly ovate, shallowly lobed. Corollas 3 cm, pale lavender with purple blotches and white-and-yellow nectar guides. Capsule body 5–6 cm, the claws 10–15+ cm. Seeds 8 mm, obovoid, blackish, warty.

Sonoyta region to the northeastern side of the Pinacate volcanic field. Mostly in brushy areas and sandy-gravelly or clay soils along larger arroyos and washes, apparently not common. This subspecies

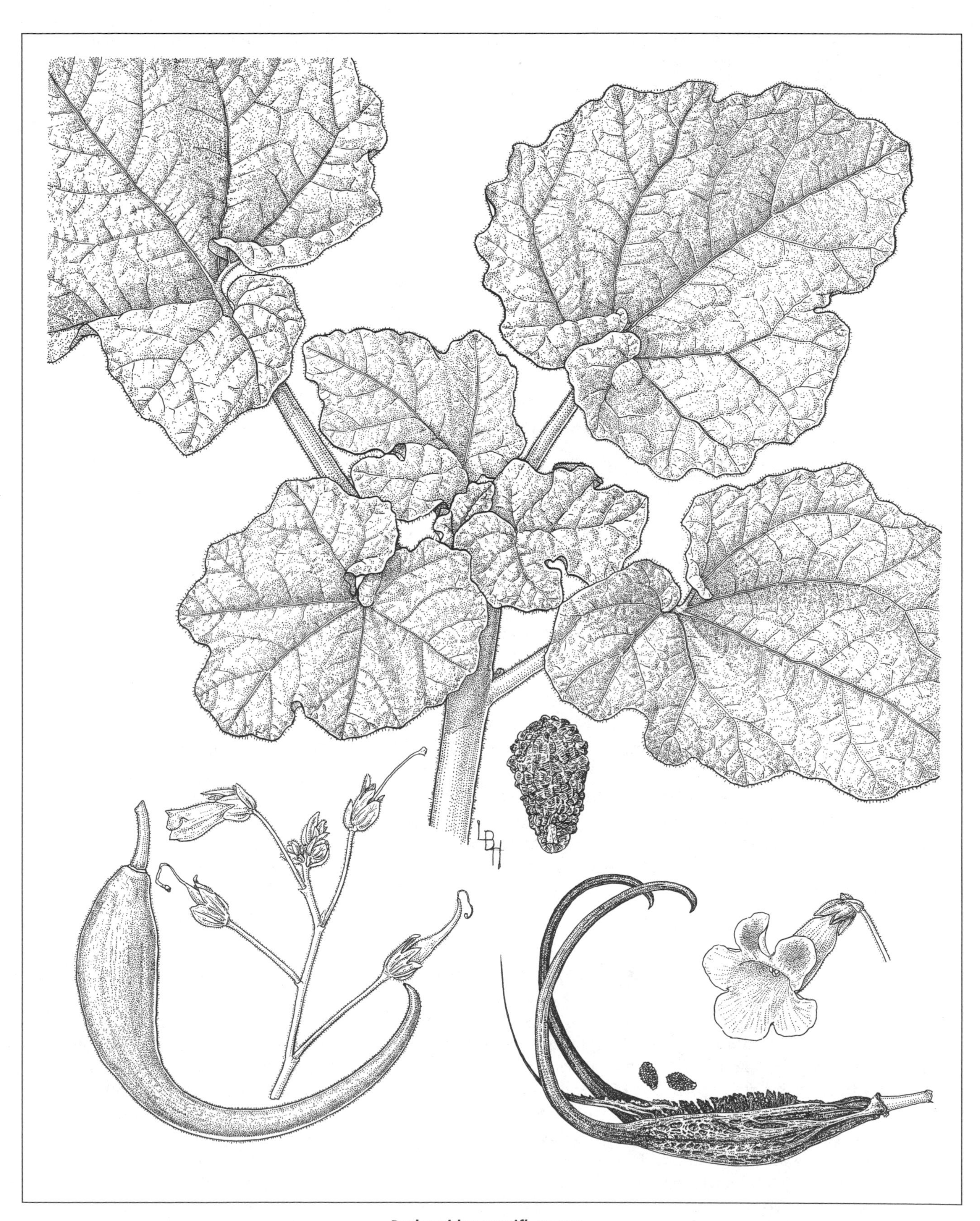

Proboscidea parviflora. LBH.

occurs from southern Utah to central California, eastward to Trans-Pecos Texas and Chihuahua, and southward to northern and western Sonora.

Suvuk, *Equihua 22 Sep 1982* (MEXU). Organ Pipe Monument, Dos Lomitas, *Warren 17 Nov 1974.*

Southward in coastal Sonora and Sinaloa subsp. *parviflora* is replaced by subsp. *sinaloensis* (Van Esseltine) Bretting, and another subspecies occurs in Baja California Sur. The cultivar, var. *hohokamiana* Bretting, is grown by Native Americans in the southwestern United States; it is not known in the wild and has much-elongated claws, white seeds, and reduced seed dormancy. Each claw yields a blackish strip used in basketry. It has been grown in O'odham gardens, perhaps at Quitobaquito and Suvuk.

MOLLUGINACEAE CARPETWEED FAMILY

Herbs or rarely shrubs. Leaves simple, entire. Flowers mostly small and inconspicuous, usually bisexual and radial. Ovary superior. Sepals usually 5; petals mostly absent. Fruits dry, often capsules.

Worldwide, mostly tropical and subtropical, greatest diversity in Africa; many are weedy; 10 genera, 100 species. Reference: Bogle 1970.

MOLLUGO CARPETWEED, INDIAN CHICKWEED

Annual or perennial herbs with slender, much-branched stems. Leaves whorled to alternate; stipules none. Flowers axillary on threadlike pedicels, without petals. Sepals 5, distinct; petals none. Stamens 3–5. Capsules multiple-seeded.

Mostly tropical and subtropical; often dry, open, disturbed habitats; 15 species. Reported to be self- and insect-pollinated with wind-dispersed seeds. Reference: Bakshi & Kapil 1954.

***Mollugo cerviana** (Linnaeus) Séringe [*Pharnaceum cerviana* Linnaeus] THREAD-STEM CARPETWEED, INDIAN CHICKWEED. Delicate summer-fall ephemerals; glabrous. Stems 4–13.5 cm, ascending to spreading, threadlike, often orange. Leaves somewhat glaucous, the first leaves 3.5–11 × 1–2 mm, linear-spatulate to spatulate, in a basal rosette, the stem leaves 6–14.5 × 0.3–1.1 mm, linear to linear-spatulate, whorled, often opposite above. Sepals persistent, 1.2–1.7 mm, thin, green especially along veins, with white membranous margins. Stamens 5. Capsules globose, 1.2–1.5 mm wide. Seeds 0.33–0.42 mm wide, minutely reticulate with faint striae (low, thin lines) on the dorsal side, uniformly red-brown.

Sandy, gravely, or rocky soils in the lowlands of the Sonoyta region, Pinacate volcanic field, granitic ranges, sand flats, interdune troughs, and shifting dunes to moderate height. Widespread in the Sonoran Desert; southward in Sonora to the Guaymas region. Widespread in the New World; reported as introduced from the Old World. It is unusual among hot-weather ephemerals in having a basal rosette of leaves.

Tinajas de los Pápagos, *Ezcurra 5 Nov 82* (MEXU). MacDougal Crater, *F 10495.* Aguajita Wash, *F 88-419.*

NYCTAGINACEAE FOUR O'CLOCK FAMILY

Herbs, shrubs, trees, and vines, the nodes often swollen. Leaves simple, usually opposite, sometimes alternate (opposite in ours or sometimes subopposite in *Boerhavia*); stipules none. Involucral bracts often showy and sepal- or petal-like, united or separate. Flowers radial or bilateral, usually bisexual. Sepals united, often constricted above the ovary, the upper part corolla-like, the lower part closely enclosing the ovary and easily mistaken for the ovary itself; petals none. Stamens 1 to many. Ovary superior but hidden by the enclosing calyx tube (or its base), thus appearing inferior. Fruits 1-seeded and indehiscent, resembling an achene or nut, often enclosed in the persistent fleshy to hard or leathery base of the calyx tube, the collective structure called an anthocarp (referred to here as the "fruit").

Mostly New World, warm and tropical regions; 34 genera, 350 species. *Bougainvillea spectabilis* is one of the most widely cultivated ornamental plants in the Sonoran Desert. References: Standley 1918; Toursarkissian 1975; Willson & Spellenberg 1977.

1. Floral tube at least 10 cm. ________ **Acleisanthes**
1′ Floral tube less than 2 cm.
   2. Involucral bracts united into a tube with 5 teeth, persistent; fruits ovoid or oval, nearly smooth. ________ **Mirabilis**
   2′ Involucral bracts separate, 1–5, sometimes reduced and/or soon deciduous; fruits not rounded and smooth, variously winged, angled, grooved, or ornamented.
      3. Fruits 5-winged, each wing as wide as or wider than the fruit body; stigmas longer than wide (fusiform). ________ **Abronia**
      3′ Fruits not winged, or the wings clearly narrower than the fruit body, or the wings 2 and inrolled; stigmas as wide as long (capitate or peltate).
         4. Stems slender, weak, prostrate or trailing; flowers in clusters of 3, the cluster resembling a single flower; fruits ellipsoid and strongly convex (sausage-shaped, with a single deep cavity or groove formed by a pair of inrolled wings). ________ **Allionia**
         4′ Stems erect to spreading, sometimes decumbent but not trailing; flowers often clustered but each flower clearly separate; fruits more or less club-shaped, not grooved or the grooves 4 or 5.
            5. Ephemerals or herbaceous perennials; stems and leaves glandular-sticky; perianth pink or white to red-purple; fruits not glandular or with glandular hairs; widespread. ________ **Boerhavia**
            5′ Shrubby; glabrous and not sticky; perianth yellow-green; fruits with large, peglike sticky glands; Sonoyta region, rare. ________ **Commicarpus**

ABRONIA SAND VERBENA

Annual or perennial herbs. Leaves petioled, the opposite leaves at a node unequal in size. Flowers in umbellate clusters subtended by 5 (4 or 6) separate bracts, fragrant, showy, pink to purple (ours), white, or yellow, the perianth lobes 5, deeply notched at apex, the stigmas fusiform, the stamens not exceeding the perianth tube. Fruits winged to ridged, often beaked. Western North America, generally on sandy soils; 25 species. References: Galloway 1975; Tillet 1967; Wilson 1972.

1. Perennials; perianth 4–5 mm wide, the lobes ca. 1.5 mm; upper beaches and beach dunes. ________ **A. maritima**
1′ Spring ephemerals; perianth 8–10 mm wide, the lobes 4–5 mm; coastal and inland. ________ **A. villosa**

Abronia maritima Nuttall ex S. Watson subsp. **maritima.** COASTAL SAND VERBENA. Perennials forming dense matlike colonies of trailing stems. Stems and leaves very thick and succulent; stems often buried and with sand sticking to glandular hairs. Leaves (2) 2.5–10 cm including petioles, the petioles mostly shorter than to sometimes equaling the blades, the blades reaching 4–5 cm wide, broadly elliptic-oblong with wavy margins. Flowers showy, in dense, rounded heads, the perianth 7–8 mm wide, bright purple-magenta, the lobes ca. 1.5+ mm. Anthocarps sculptured and winged, in ball-shaped clusters. Warm seasons.

Beach strand above the tide line and seaward-facing beach dunes, the colonies often capping sand hummocks or small dunes. Strand at Estero Morúa, especially the southeast side, occasional between Punta Borrascoso and El Golfo, and perhaps on remote beaches of Bahía Adair. Development in the Puerto Peñasco region and recreational use of off-road vehicles has probably destroyed many beach-side populations. Sandy seashores, southern California to Nayarit, the Islas Tres Marías, and the Gulf of California.

Johnson (1978) distinguished 2 "forms" of subsp. *maritima*. The "Gulf" form occurs along the Gulf of California and southward; it has a shorter, bright purple-pink to purple-magenta perianth with non-reflexed lobes, and smaller, crenately lobed leaves. The "Pacific" form occurs along the Pacific coast of Baja California Peninsula and southern California. Another subspecies is found in the Cape Region of Baja California Sur.

5 km E of Puerto Peñasco, *Boyer & Turk Apr 1990.* Playa Encanto, Estero Morúa, *F 91-44.* 5 km NW of El Tornillal, *F 92-193.*

Abronia villosa S. Watson var. **villosa.** SAND VERBENA. Spring ephemerals. Herbage viscid pubescent, the blades often becoming glabrate. Highly variable in size depending upon soil moisture, the stems 15–100 cm, the smaller plants erect to ascending, the larger plants with long trailing-decumbent

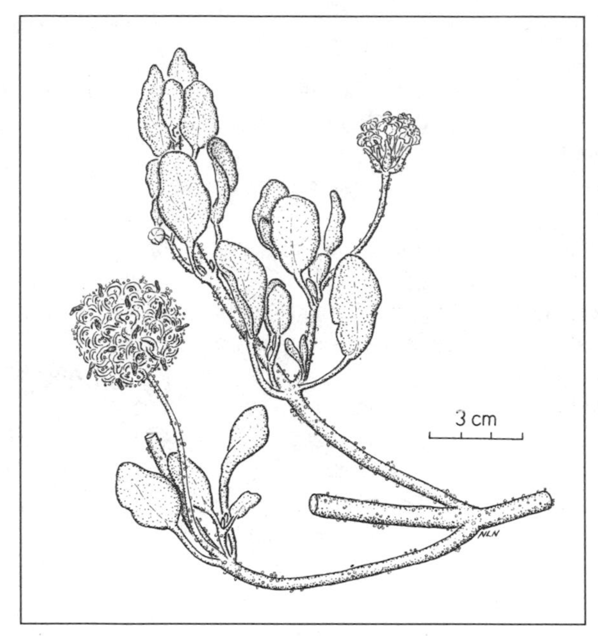

Abronia maritima subsp. ***maritima.*** NLN (Felger & Moser 1985).

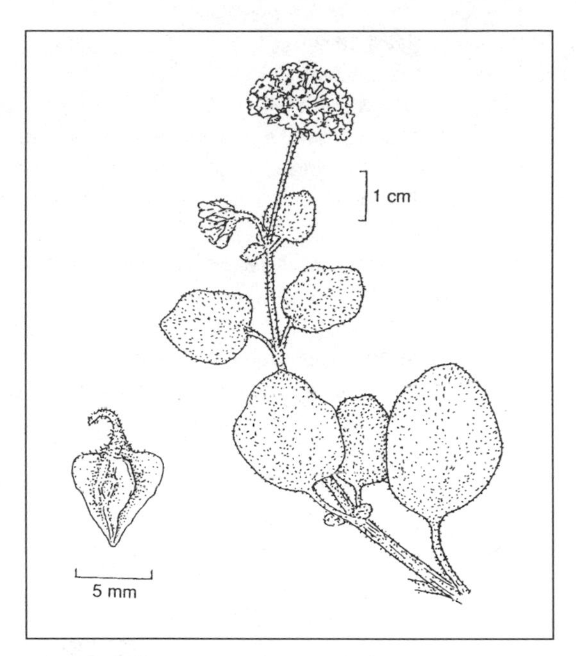

Abronia villosa. ER (Hickman 1993).

stems. Stems and leaves semisucculent. Leaves 2.5–7.0 cm, the petiole often as long as or longer than blade; blades more or less ovate, the margins mostly crenulate to shallowly sinuate but sometimes entire. Flowers showy, very fragrant, in relatively open clusters 4–5 cm across, the perianth 10–15 mm wide, pinkish purple to pale magenta or pale purple, the lobes 4.0–5.5 mm.

Widespread on sandy soils including high dunes, beach dunes, rolling sand hills, and sand flats. Flowers frequently visited by the painted lady butterfly (*Vanessa cardui*). Mojave and Sonoran Deserts in southern California, Baja California, southern Nevada, southwestern Utah, western Arizona, and northwestern Sonora south to El Desemboque San Ignacio.

Some Sonoran specimens or populations approach *A. gracilis* Bentham, known from the Pacific maritime-influenced regions of the Baja California Peninsula. *A. villosa* is distinguished by its hard, sculptured (rugose-reticulate), and beaked fruits, whereas *A. gracilis* has softer, smoother, and essentially beakless fruits. These characters show up on fully mature, dry fruits, but at this stage they readily fall away and are missing from many specimens. At least from the vicinity of Puerto Libertad to El Desemboque San Ignacio, the populations are unusual in having large pink floral bracts enclosing the flower buds and young fruit.

Sierra del Rosario, *F 20428.* 8 km N of Gustavo Sotelo, *Ezcurra 14 Apr 1981* (MEXU). 5 km E of Puerto Peñasco, *Burgess 4749.* Moon Crater, *F 19031.* Hornaday Mts, *Burgess 6806.* MacDougal Crater, *F 9906.* 19 mi E of San Luis, *Galloway 1225.*

ACLEISANTHES

Perennial herbs or low shrubs, the stems branching in a forked pattern. Leaves thick, the opposite leaves at a node nearly equal to unequal; petioled. Flowers nocturnal, the perianth white with a long slender tube and a relatively small 5-lobed limb, or the flowers sometimes cleistogamous and relatively short. Anthocarps narrowly ellipsoid, shallowly 5-ribbed or 5-angled. Southwestern United States and Mexico; 7 species. Reference: Smith 1976.

Acleisanthes longiflora A. Gray. ANGEL TRUMPETS. Perennial herbs, woody at base with thick, knotty roots. Stems slender and brittle, erect-ascending to sometimes prostrate. New herbage often with moderately dense, short white hairs, the leaves soon glabrate; stems and leaves sometimes shiny white waxy. Leaves often 15–35 mm, opposite leaves at a node nearly equal, the blades triangular to lanceolate, moderately thickened, the margins entire to wavy. Perianth tube 10.5–13.5 cm, the limb 1.8 cm wide. Anthocarps cylindrical-ellipsoid, 5-ribbed. Warm seasons, especially with summer rains.

Rocky, gravelly soils, roadsides, and bajadas; entering the northeastern margin of the flora area near the base of Sierra Cipriano southwest of Sonoyta. Also in granitic mountains southeast of Sonoyta, at Sierra del Viejo near Caborca, and expected eastward in northern Sonora; also near the Sonora border in the Agua Dulce Mountains in Cabeza Prieta Refuge. Sonoran and Chihuahuan Deserts and Tamaulipas thornscrub; Tamaulipas to Texas, Arizona, Sonora, and Durango. *A. longiflora* is aptly named—the long, slender floral tube is unique among the flora area. This species also produces minute, cleistogamous flowers (Spellenberg & Delson 1977).

11 mi SW of Sonoyta, *F 90-151.* 8 mi SE of Sonoyta [on Mex 2], *Spellenberg 3612* (LL). Cerro del Viejo, SW of Caborca, *Gentry 14464.*

ALLIONIA

Annual or mostly perennial herbs. Stems slender and prostrate-trailing with forked branching. Flowers in 3s, the easily separated flowers appearing like a single flower, each forming a wedge-shaped third of the cluster, the trios axillary on slender peduncles. Each flower in the trio subtended by a sepal-like bract. Fruits ellipsoid and sausage-shaped, one side convex and relatively smooth, the other side with a pair of broad, inrolled wings forming a deep cavity, the wings toothed or not. Southwestern North America to Argentina and Chile; 2 or 3 closely related species. Reference: Turner 1994.

Allionia incarnata Linnaeus [*A. incarnata* var. *nudata* (Standley) Munz. *A. incarnata* var. *villosa* (Standley) B.L. Turner] WINDMILLS, TRAILING FOUR-O'CLOCK. Ephemerals, annuals, or short-lived perennial herbs with a stout taproot, dying back to the roots during drought. Glandular hairy and sticky viscid throughout except the flowers, often with sand sticking to the herbage. Stems sometimes reaching more than 1 m. Leaves petioled, variable, the larger ones (1.7) 2.5–7.2 cm, the blades usually ovate; opposite leaves of a pair unequal in size.

Flowers opening in the early morning and fading by late morning on hot days, the perianth showy, bright lavender-pink, bilateral (the "inner" part of the perianth limb much shorter than the "outer" portion); the trio of flowers often 1–2 cm wide. (Flower size and peduncle length highly variable depending on soil moisture and temperature.) Stamens exserted, the filaments and style lavender-pink, the anthers yellow. Fruits 3.0–4.8 mm, light tan and firm, obovoid-oblong, the opposite wings each with 3 (4) broad teeth folded over a deep cavity, or the teeth sometimes not developing (immature?), the cavity with a double row of conspicuous, short-stalked glands; all 3 fruits in the cluster often developing. (The fruits are unique among Sonoran Desert nyctages in being bilaterally rather than radially symmetrical. With magnification the fruit structure is spectacular; the glands inside the cavity are often relatively large, whitish, and translucent.) Warmer months, dormant during extreme drought and colder months.

Widespread, mostly at lower elevations; Sonoyta region, Pinacate volcanic field, granitic ranges including the Sierra del Rosario, and desert flats. Rocky, cinder, and sandy soils; washes and arroyos, plains, crater floors, and slopes; absent from dunes. Southwestern North America to Argentina and Chile; with larger- and smaller-flowered forms and much variation in fruit ornamentation.

Suvuk, *Nabhan 376.* S of Pinacate Junction, *F 11175.* Sykes Crater, *F 18974A.* N of Moon Crater, *F 19282.* Sierra del Rosario, *F 75-17.* E of La Joyita, *Prigge 7249* (UCR).

BOERHAVIA SPIDERLING, *JAUNILIPIN*

Annual or perennial herbs, usually branched from the base, usually pubescent with glandular-sticky areas at least on the stems. Leaves opposite or subopposite, usually petioled, the opposite leaves of a pair often unequal in size. Inflorescences with small bracts. Perianth white to pink, purple, or red. Stamens 1–5, collapsing onto the stigma as the flower fades (apparently self-fertilizing if the flower has not been cross-pollinated). Fruits obovoid or obpyramidal, usually 5-angled and/or grooved (furrows or sulci), exuding mucilage (at least ours) when wet.

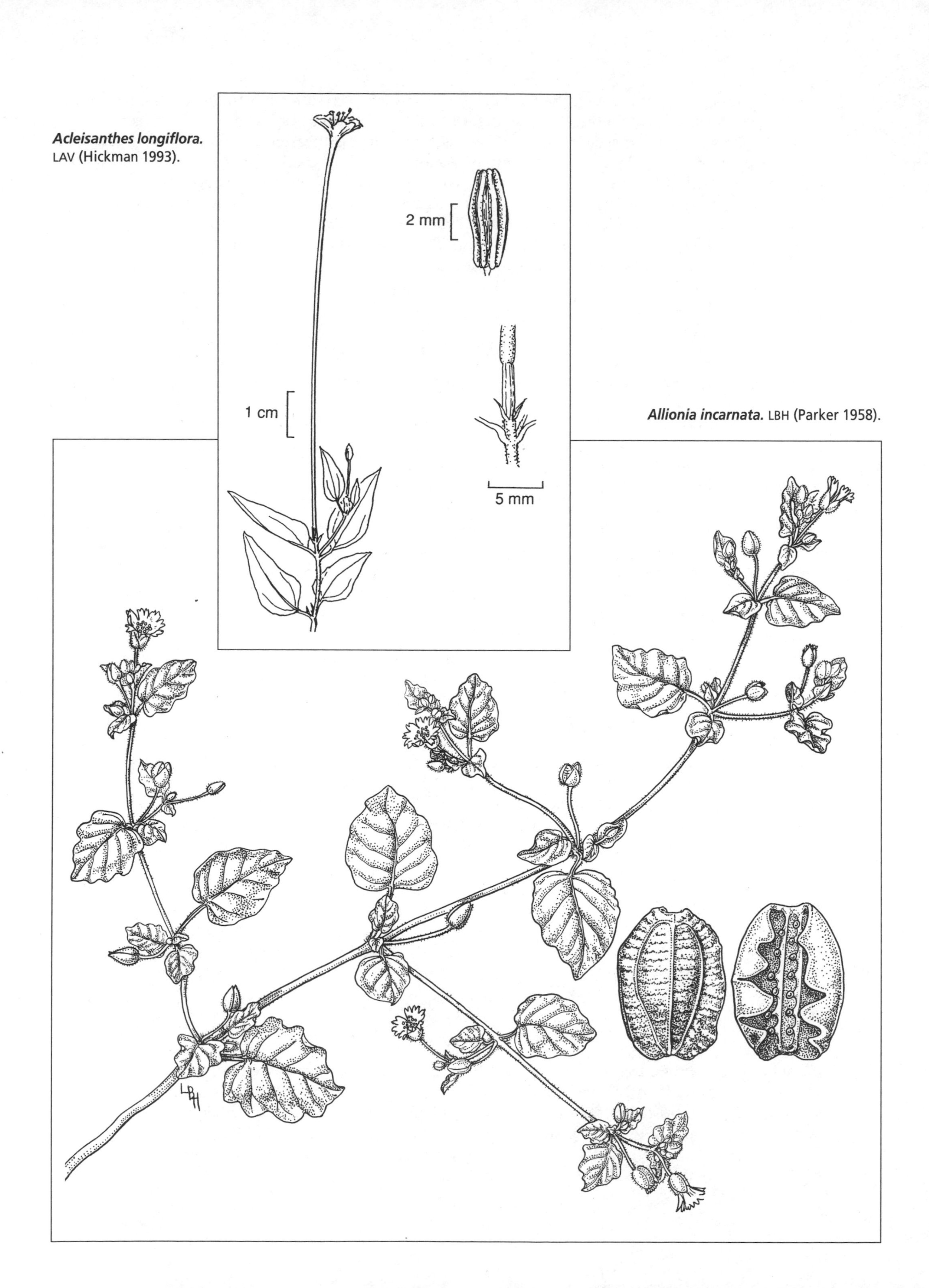

Acleisanthes longiflora. LAV (Hickman 1993).

Allionia incarnata. LBH (Parker 1958).

Warm regions of the world, mostly New World tropics and subtropics, greatest diversity in southwestern North America; some weedy species now widespread; 40 species.

Species definitions are often difficult. Like so many other members of the nyctage family, there are relatively few morphological gaps within the genus. At least those with larger flowers produce nectar and are insect pollinated. The boerhavias as a whole, especially those with smaller flowers, are probably capable of selfing and the smaller-flowered ones may be entirely selfing (Richard Spellenberg, personal communication 1995).

Sonoran Desert boerhavias grow and flower with the summer rains. Ours have well-developed taproots and flowers that open in the early morning and fade with midmorning heat. At least some of our boerhavias may develop enlarged, inflated gall fruits. Adams (1972:261) reported that gall fruits of *B. diffusa* (*B. coccinea*) and other species "may become infected by a gall midge, *Asphondylia* sp." Reference: Fosberg 1978.

1. Annuals or perennials; perianth dark red-purple; fruits sticky with glandular hairs. ______ **B. diffusa**
1′ Hot-weather ephemerals; perianth white to pale pink; fruits glabrous, not sticky.
  2. Flowers in umbellate or subumbellate clusters. ______ **B. erecta**
    3. Fruits (2.9) 3.4–4.7 mm, often substipitate (fruit base narrowed into a "stalk" or stipe), often bright green to reddish green when fresh. ______ **B. erecta** var. **erecta**
    3′ Fruits 2.0–2.9 mm, lacking a stipelike base. ______ **B. erecta** var. **intermedia**
  2′ Flowers on elongated racemose branches.
    4. Fruits narrowly obovoid (slender), mostly 5-angled, the angles or ridges (area between furrows) broadly rounded to obtuse, the furrows narrow to nearly closed. ______ **B. spicata**
    4′ Fruits broadly obovoid (chunky), mostly 4-angled, the angles sharp-edged, the furrows broad, open, and roughened inside. ______ **B. wrightii**

***Boerhavia diffusa** Linnaeus [*B. caribaea* Jacquin. *B. coccinea* Miller] SCARLET SPIDERLING. Warm-weather annuals or perennials (probably short-lived in the flora area) with thickened roots, sprawling, openly branched, often 1–2 m across. Some with densely hairy stems, others glabrous. Inflorescences diffuse and much-branched, the branches very slender, glabrous or glandular hairy; flowers in subumbellate clusters at ends of threadlike branchlets. Perianth bright red-purple. Fruits 2.8–3.5 mm, very sticky with exudate from short-stalked glandular hairs; immature fruits rounded, the mature ones narrowly obovoid, prominently ribbed (the ribs raised and smooth), the tip rounded.

Weeds in disturbed habitats in the Sonoyta region and elsewhere in agricultural areas; not seen among natural vegetation and probably not native to northwestern Sonora. Widespread and mostly weedy in warm regions of much of the world. Although often a facultative annual in the flora area, or only a short-lived perennial, elsewhere in the Sonoran Desert and worldwide it is strongly perennial once established. Different flower-color phases occur in various parts of the world; e.g., white-flowered or ocher-flowered races in New Mexico, and deep pink in eastern Sonora and the Old World (Richard Spellenberg, personal communication 1998).

Woodson, Schery, & Kidd (1961:55) found "no tangible differences between *B. caribaea* and *B. coccinea* of the New World and *B. diffusa* of the Old World. The species appears plainly to be an effusive pantropic weed, the dispersal of which has been greatly facilitated by the viscid anthocarps. It is probably, but not certainly, of American origin."

Vicinity of Sonoyta, *F 85-957, F 88-391, F 90-152.*

Boerhavia erecta Linnaeus. Hot-weather ephemerals. Stems often 20–100 cm, erect to spreading or decumbent. Leaf blades ovate to narrowly lanceolate. Stems, petioles, and sometimes leaf blades with moderately to densely glandular patches. Flowers white to pale pink, 2 to several in umbellate or subumbellate clusters. Fruits 5-angled, the tip blunt (truncated). Warm regions of Latin America and southern United States, and weedy in parts of the Old World.

Boerhavia erecta var. **erecta.** Fruits (2.9) 3.4–4.7 mm, 5-angled, often bright green to red-green when fresh, often substipitate (base of fruits weakly differentiated into a narrowed and nearly terete stalk).

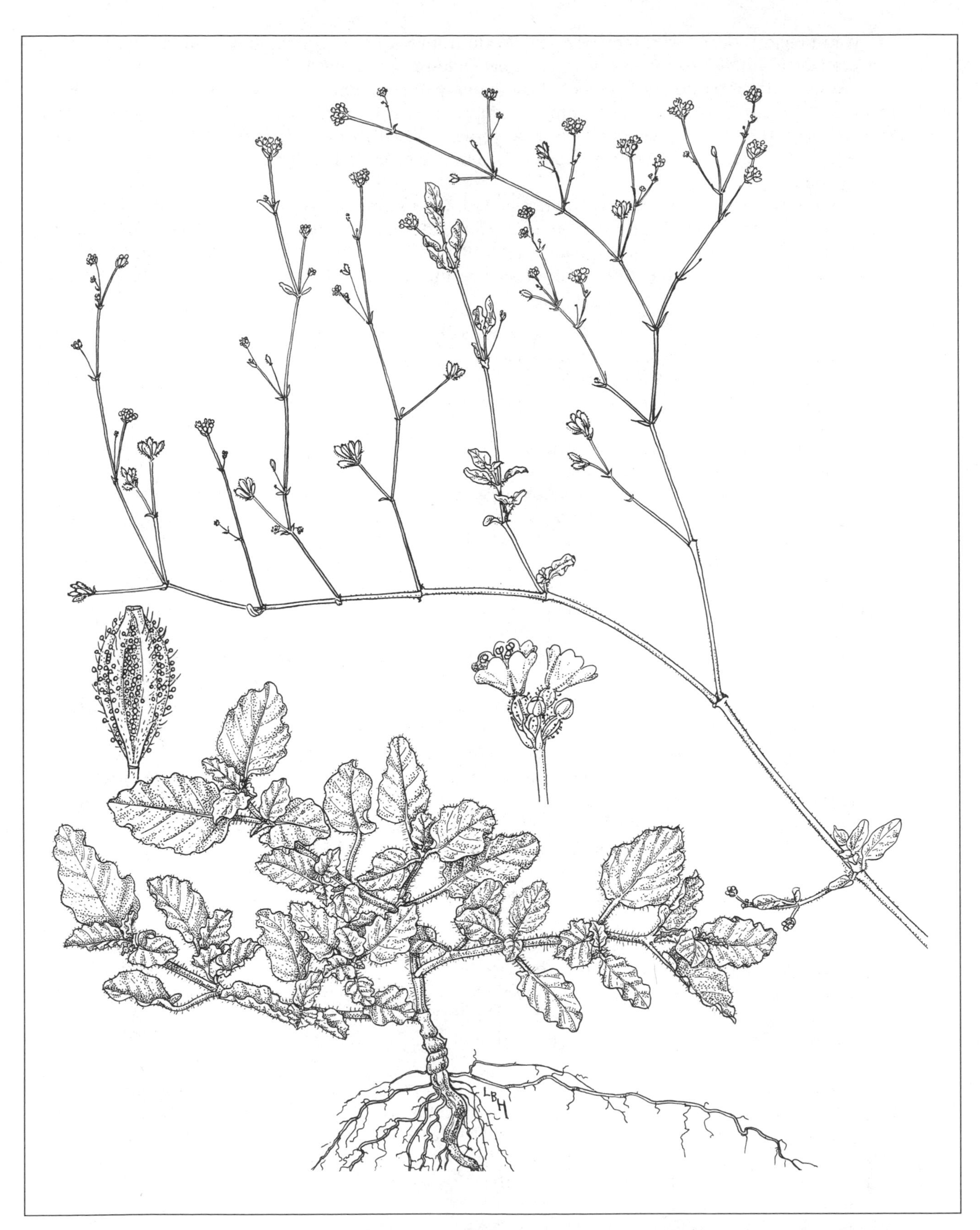

Boerhavia diffusa. LBH (Parker 1958).

Washes, sandy flats, crater floors, bajadas, sometimes on rocky slopes and in soil pockets among rocks, and as an agricultural and urban weed. Widespread in the Sonoyta region and the Pinacate volcanic field. Arizona to southeastern United States and southward through much of South America.

Suvuk, *Nabhan 384.* NE side of Sonoyta, *F 85-931.* MacDougal Crater, *F 10766.* 6 km W of Los Vidrios, *F 92-969.* SSW of Tinajas de los Pápagos, *F 10626.* Aguajita Wash, *F 88-424.*

Boerhavia erecta var. **intermedia** (M.E. Jones) Kearney & Peebles [*B. intermedia* M.E. Jones. *B. triqueta* S. Watson] *MAKKUMI HA-JEWEḌ.* Distinguished from var. *erecta* by the smaller fruits (2.0–2.9 mm) without a stipitate base, the fruits generally duller, not as dark green when fresh, and browner at maturity, and the plants usually more delicate.

Fairly widespread on sand flats and in arroyos and washes near Sonoyta and the Pinacate volcanic field, and as an agricultural weed. Southeastern California to Texas and southward through much of Mexico.

The two varieties are often sympatric. Perhaps var. *intermedia* represents smaller, drought-stressed plants or a xerophytic ecotype. Some specimens from the flora area seem intermediate. In drought-stressed plants and/or late in the season the number of flowers per cluster may be reduced, often to a single flower but with some 2- or 3-flowered umbellate or subumbellate clusters present. These plants key out to *B. triqueta.* In some cases a plant may produce several-flowered clusters early in the season and then produce mostly solitary flowers as drought sets in.

W of Sonoyta, *F 88-393.* Suvuk, *Burgess 6341.* Cerro Colorado, *F 10787.* MacDougal Crater, *F 10732.* Aguajita, *F 88-413.*

Boerhavia spicata Choisy [*B. spicata* var. *palmeri* S. Watson. *Senkenbergia coulteri* Hooker f. *B. coulteri* (Hooker f.) S. Watson. *B. watsoni* Standley] *MAKKUMI HA-JEWEḌ.* Summer-fall ephemerals, often 30–50+ cm. Leaves often 3–6+ cm, the blades ovate to ovate-deltate. Herbage moderately glandular-sticky. Inflorescences of racemose branches. Bracts beneath flowers deciduous. Flowers white to pale pink. Fruits 2.1–2.5 mm, narrowly obovoid, the furrows nearly closed to open and roughened (rugulose) inside, the ridges 5, broad, rounded or broadly obtuse (more angled when immature), the tip rounded.

Widespread; Sonoyta region, Pinacate volcanic field, desert plains, also agricultural and probably urban areas. Mostly sandy to silty soils of washes, arroyos, crater floors, and disturbed habitats.

Boerhavia spicata (sensu stricto or var. *spicata*) ranges from southern Arizona to Sinaloa, Chihuahua, and the Gulf of California side of the Baja California Peninsula. *B. coulteri* is reported from southeastern California through much of southern Arizona to southern Sonora and the gulf side of the Baja California Peninsula.

Boerhavia spicata var. *palmeri,* better known in the literature as *B. coulteri,* is continuous with "typical" var. *spicata.* Although extremes in the variation can be distinguished, there are many specimens that cannot consistently be placed in one or the other taxon. Usual differences are as follows:

var. *spicata*	var. *palmeri*
Angles of fruit acute.	Angles of fruit obtuse.
Furrows on fruits relatively broad and rugulose.	Furrows nearly closed.
Branches tending to be stout.	Branches, especially flowering branches, tending to be slender and delicate.

MacDougal Crater, *F 10721.* Molina Crater, *F 10648.* Suvuk, *Nabhan 360.* Aguajita Wash, *F 88-415.*

Boerhavia wrightii A. Gray. *MAKKUMI HA-JEWEḌ.* Summer-fall ephemerals, sometimes persisting until freezing weather; highly variable in size, flowering as small as 10 cm high, sometimes reaching 1 m in height and 1.5+ m in width. Herbage extremely glandular-sticky throughout. Inflorescences of racemose branches, glandular-sticky. Larger (lower) leaves often 2.5–7.5 cm, the blades ovate to oblong or lanceolate, darker green above; margins crenulate. Bracts pinkish, semipersistent or sometimes soon deciduous, often 2/3 as long as the fruits. Flowers pale pinkish white to pink. Fruits broadly ovoid, rel-

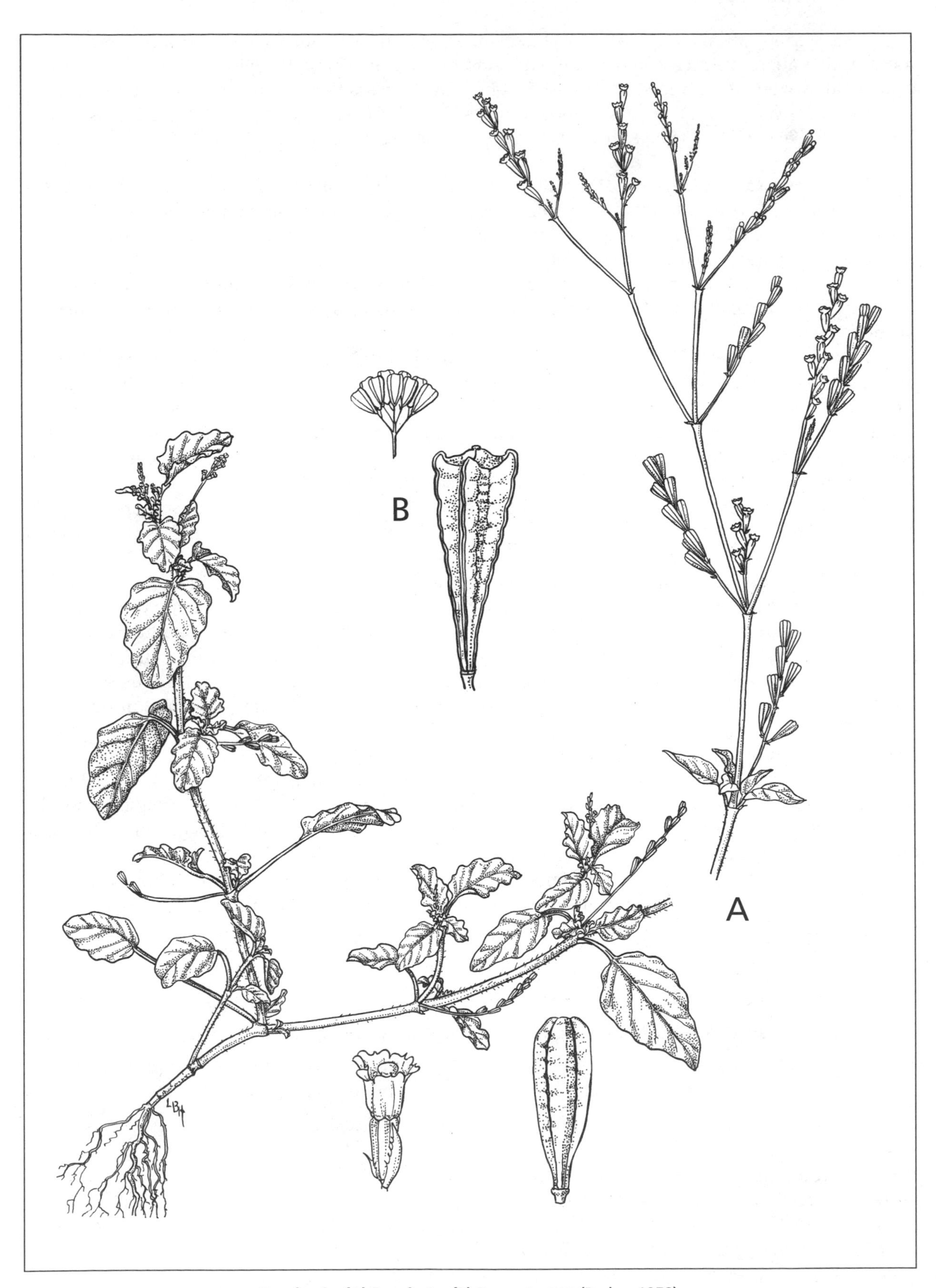

Boerhavia: (A) B. spicata; (B) B. erecta. LBH (Parker 1958).

atively short, squat, and chunky, 2.0–2.9 mm, 4(5)-angled, the angles sharp-edged, the furrows open, broad, and often roughened.

Seasonally abundant throughout most of the flora area, natural desert and disturbed habitats; absent from higher dunes. Mostly in the lowlands; rocky, gravelly, and sandy soils; desert plains, drainageways, roadsides, and rocky slopes. Northwestern Sonora south at least to the vicinity of El Desemboque San Ignacio; southeastern California and southern Nevada to western Texas and Chihuahua, probably also northeastern Baja California.

This is the most common boerhavia in the region, and the only one known from the Sierra del Rosario. The fruit is about 50% wider than that of *B. spicata.*

Suvuk, *Nabhan 377.* Cerro Colorado, *F 10802.* MacDougal Crater, *F 10761.* Tinaja de los Chivos, *F 18619.* 24 km W of Los Vidrios, *F 85-1008.* Sierra del Rosario, *F 75-26.* Aguajita Wash, *F 88-412.*

COMMICARPUS

Perennial herbs or shrubs. Flowers green to pale yellow or white to pink. Fruits club-shaped, 10-ribbed, terete or nearly so, with viscid, knobby glands causing the fruit to stick to fur, feathers, and clothing. Warm regions worldwide, largely African; 25 species. Reference: Harriman 1999.

Commicarpus scandens (Linnaeus) Standley [*Boerhavia scandens* Linnaeus] Perennials, usually growing through other shrubs, often reaching 1.5–1.8 m; glabrate. Stems slender, straight, and brittle, the internodes usually long. Leaves often 3.0–7.5 cm, the petioles prominent, the blades nearly triangular to ovate, semisucculent. Flowers in umbellate clusters, the perianth 6–8 mm wide, nearly white in early morning, soon becoming pale yellow-green, collapsing with daytime heat. Stamens 2, long-exserted. Fruits 8–10 mm, producing relatively small amounts of mucilage when wet. Warmer seasons, especially with summer rains.

Dense, brushy vegetation bordering major drainageways along the international border from Lukeville to Quitobaquito and just east of Sonoyta. Widespread in Sonora. Both Baja California states, Arizona to Texas, and southward to northern South America and West Indies.

1.6 km W of Lukeville, *F 91-136.* Quitobaquito, *Harbison 27 Sep 1939* (SD). Aguajita Wash, rare, *F 88-401.*

MIRABILIS FOUR-O'CLOCK

Annual or perennial herbs, often woody at base, the roots often tuberous, the stems forking. Leaves entire, the lower ones petioled, the upper ones sessile. Involucral bracts united, calyx-like, 5-lobed; 1 to many flowers per involucre; involucres often clustered at stem tips. Calyx petal-like, longer than the involucre, white, pink, red, yellow, or purple. Stigmas capitate. Fruits rounded to elongated, smooth to slightly angled or ridged.

Warmer parts of the Americas and 1 in the Himalayas; 50 species (including *Oxybaphus*). The garden four-o'clock is *M. jalapa.* Ours have single-flowered involucres, apparently representing an evolutionarily advanced condition of highly reduced inflorescences.

1. Involucres 6–9 mm, the lobes 2–3 (4) mm. ____ **M. bigelovii**
1′ Involucres 10–15 mm, the lobes 5–9 mm. ____ **M. tenuiloba**

Mirabilis bigelovii A. Gray var. **bigelovii.** Herbaceous perennials. Stems erect to spreading, slender, often dying back to rootstock during drought. Leaves semisucculent, tardily drought deciduous, larger ones 2–4+ cm including petioles, the blades deltate to ovate. Involucres green, enlarging moderately as fruits develop, reaching 6–9 mm, the lobes 2–3 (4) mm. Perianth white to pale pink, collapsing with the heat of the day; anthers yellow, the filaments and style white, the stigma rose colored. Fruits seedlike, 3.8–4.4 × 2.5–3.0 mm, ovoid, faintly reticulate to mottled dark brown and gray, sometimes faintly glaucous. Flowers often visited by honeybees; various seasons except during cold weather.

Generally among rocks on slopes and along washes in upper bajadas; to peak elevations in the Pinacate volcanic field, granitic ranges, Sonoyta region, and Organ Pipe Monument; not in the Sierra del Rosario. Var. *bigelovii,* the most wide-ranging of the 3 varieties, extends south in Sonora to the Sierra Seri. This species ranges from northern Baja California Sur to southern California, Nevada, southwestern Utah, much of Arizona, and northern Sonora. The stems are more slender and the leaves thinner and smaller than those of *M. tenuiloba.*

Pinacate Peak, *F 19410.* Sierra del Viejo, *F 85-731.* Quitobaquito, *F 86-180.*

Mirabilis tenuiloba S. Watson. LONG-LOBED FOUR O'CLOCK. Herbaceous perennials, sometimes flowering in the first year, often dying back to rootstock during drought; roots stout, somewhat thickened. Herbage, especially the new growth, and inflorescences densely pubescent with sticky, glandular hairs. Leaves often 3–6+ cm (including petioles), tardily drought deciduous, the blades thick and semisucculent, mostly deltate. Involucres green, enlarging moderately to 10–15 mm as fruits develop, the lobes 5–9 mm. Perianth white, collapsing with the heat of the day. Fruits 4.4–5.4 × 3.0–3.6 mm, oval, brown, nearly smooth when fresh, to moderately reticulated or with low ridges when dry. Various seasons except during cold weather.

In Sonora known only from the Sierra del Rosario (*F 20652, F 20737, F 75-1*); rocky canyons and slopes, often with north and east exposures, occasionally in gravelly washes adjacent to rocky slopes. Also the western part of the Colorado Desert in southern California, Baja California, and some islands in the Gulf of California; in Arizona known only from the Tinajas Altas Mountains, where it grows intermixed with *M. bigelovii* (Felger 1993). Plants of *M. tenuiloba* are larger, more robust, and generally with larger, thicker, and more yellowish leaves and larger "fruits" than *M. bigelovii.*

OLEACEAE OLIVE FAMILY

Trees, shrubs, woody vines, rarely perennial herbs. Leaves usually opposite, simple to pinnately compound; stipules usually absent. Flowers radial, bisexual or sometimes unisexual. Calyx 4(–15)-parted or sometimes none. Corollas sympetalous, 4–6 lobed, or sometimes none. Stamens 2 (4), borne on the petals. Ovary superior. Fruits 2-chambered, highly variable, the seeds 1 (2–4) per chamber.

Both hemispheres, temperate and tropical; 29 genera, 600 species. The olive (*Olea europaea*), originally from the Mediterranean region, is widely grown in the Sonoran Desert, as is the Arizona velvet ash (*Fraxinus velutina*).

MENODORA

Perennial herbs to small shrubs, usually glabrous. Leaves simple, alternate or often opposite below, sessile or subsessile. Flowers showy, solitary or inflorescences corymbose or paniculate. Calyx deeply divided into linear lobes. Corollas 5- or 6-lobed. Stamens 2. Fruits inflated, membranous-papery, testiculate capsules of 2 hemispheres each 2- (4)-seeded, indehiscent or circumscissile.

Arid and semiarid, usually mountainous regions, 20 species; southwestern United States and Mexico (14 species), southern South America, and southern Africa. References: Meyer 1957; Steyermark 1932; Turner 1991.

Menodora scabra Engelmann ex A. Gray [*M. scoparia* Engelmann ex A. Gray] TWINBERRY, BULL BALLS. Subshrubs, mostly with slender, erect branches, 0.5–1.0 m. Stems leafy, dying back in drought. Herbage and calyces scabrous-puberulent (short, rough sandpaper-like hairs). Leaves gradually drought deciduous, reduced upwards, mostly 1.2–4.5 cm × 2–5 mm, sessile or subsessile, narrowly elliptic to narrowly oblong, the apex acute; margins entire. Flowers 1.2–1.5 cm wide; calyx with 7–10 (13?) slender lobes; corollas bright yellow. Capsule hemispheres opening around the middle. Seeds 4 per chamber, 4–6 mm, rounded on the back, somewhat flattened, with a narrow (water-absorbing?) wing. Warm weather.

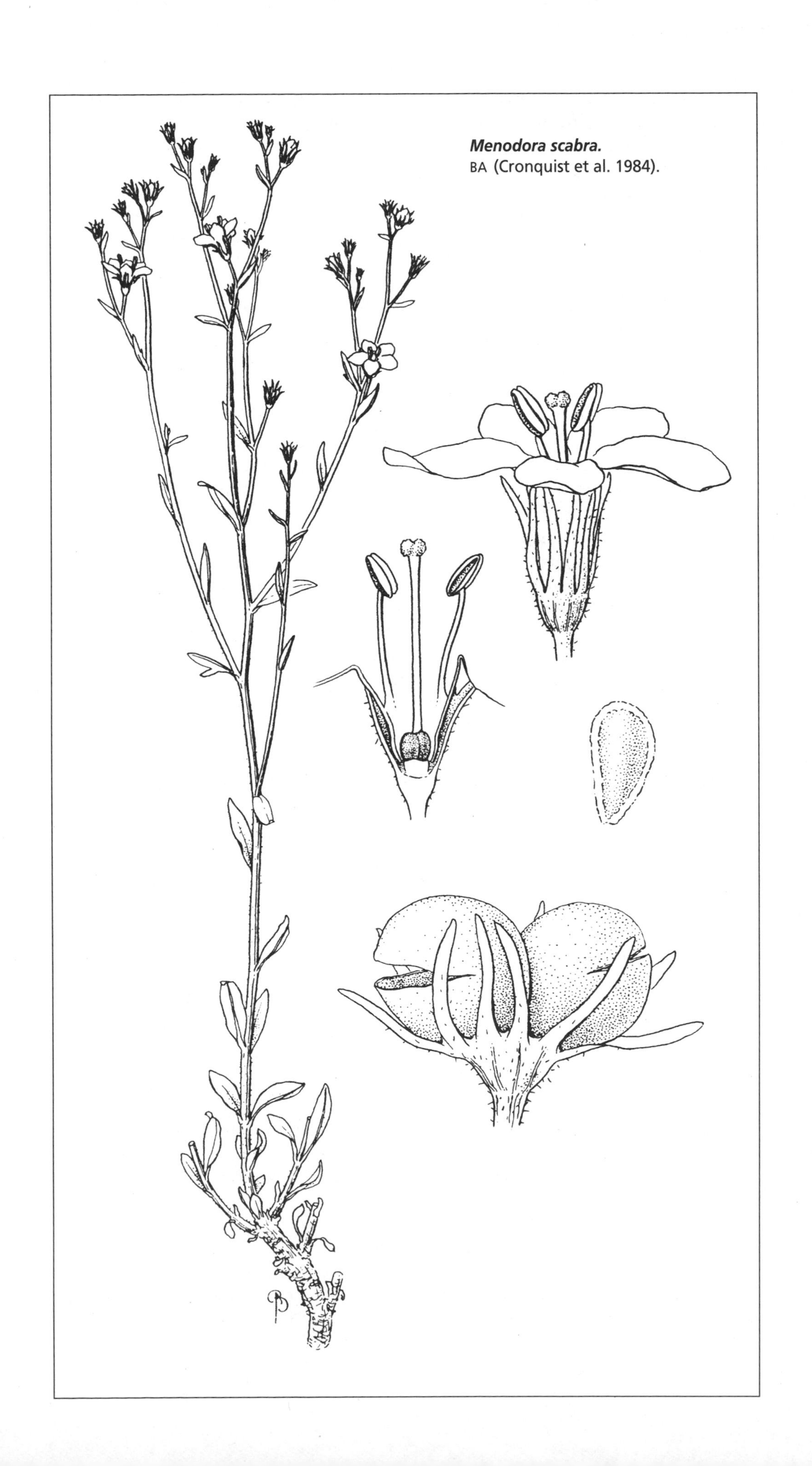

Menodora scabra.
BA (Cronquist et al. 1984).

Granitic mountains in the Sonoyta region, mostly on steep north-facing slopes and along small washes in upper bajadas, and farther west on steep north-facing canyons on the north side of Sierra del Viejo. Southward in western Sonora in granitic mountains nearly to Puerto Libertad and the Sierra del Viejo (southwest of Caborca), and eastward across the northern part of the state. Also Colorado and Utah to Baja California Sur, Arizona to southwestern Texas, Durango, and Nuevo Leon. Within the genus *M. scabra* is a generalist; its geographic range is the largest of any of the North American species.

Sierra del Viejo (along Mex 2), *F 85-730*. Sierra de los Tanques, *F 89-21*. Sierra Cipriano, *F 88-221*.

ONAGRACEAE EVENING PRIMROSE FAMILY

Mostly annual, biennial, or perennial herbs, or rarely shrubs or small trees. Leaves alternate, opposite, or in basal whorls. Flowers usually bisexual, mostly radial, the parts mostly in 2s or 4s, usually with a more or less elongated floral tube (hypanthium) above the ovary, the tube producing nectar near its base; floral parts above the ovary usually deciduous after flowering. Sepals and petals each mostly 4, separate above the hypanthium (summit of floral tube; the terms *sepal* and *petal* refer to the lobes or free portions of the calyx and corolla). Ovary inferior. Pollen grains usually among cobwebby threads. Style 1, stigma 2- or 4-lobed or discoid, capitate or elongated. Fruit mostly a capsule, sometimes a berry or nutlike.

Mainly temperate and subtropical; worldwide with greatest diversity in the New World, especially western United States and Mexico; 18 genera, 675 species. About half the species are self-pollinated, the others variously pollinated by birds, bees, moths, flies, or other insects. Ours are in tribe Onagreae, characterized in part by the absence of stipules, similarities in chromosome behavior, and a basic chromosome number of $x = 7$.

Many of the Sonoran Desert species are cool-weather ephemerals. The few summer-active species tend to occur in wetland habitats (see *Gaura*). References: Lindsay, MacSwain, & Raven 1963; Munz 1965; Raven 1964.

1. Petals (2.0) 2.5–6.0 cm, monochrome yellow or white without spots or other markings. ______________ **Oenothera**

1′ Petals 0.2–1.7 cm, colors various, often with spots or other markings.

  2. Plants with a solitary axis or branched from below the middle; fruit a dehiscent capsule, (12) 15 mm or more, widest at the base or about same thickness throughout, many-seeded; stigma hemispherical or capitate. ______________ **Camissonia**

  2′ Plants with a solitary axis or branched from above the middle; fruits indehiscent, nutlike and hard, 8–10 mm, widest above the middle, 1- or 2-seeded; stigma 4-lobed. ______________ **Gaura**

CAMISSONIA SUN CUP

Annual or perennial herbs with a taproot, or rarely subshrubs. Leaves basal and/or along the stems, alternate, usually reduced upwards. Flowers opening near dawn or near sunset, fading with midmorning or midday heat (remaining open longer on cooler days); stigmas receptive and anthers shedding pollen simultaneously and immediately. Flowers usually 4-merous. Floral tube prolonged beyond the ovary, deciduous after flowering. Petals yellow, purplish, pinkish, or white. Stamens and style yellowish, the stigma almost always green. Stamens 8. Stigma capitate or hemispheric. Capsules straight or contorted, dehiscent, many-seeded.

New World centering in the western United States and adjacent Canada and Mexico; 62 species in 10 sections. The 6 species in the flora area are winter-spring ephemerals, or if surviving as short-lived perennials, then active growth and flowering ceases during the hotter months.

The smaller-flowered species, such as *C. chamaenerioides,* are selfing (autogamous). The long, slender, nectar-filled floral tube (hypanthium) and evening-opening (vespertine) flowers of *C. arenaria* suggest it is pollinated primarily by hawk moths, as in other evening primroses (e.g., our *Oenothera* species) with similar flower structure and biology. Other vespertine species, both white- and yellow-flowered, are undoubted visited by smaller moths. The outcrossing (allogamous) species open an hour or more before sunset and/or after sunrise, at which time they are pollinated by certain solitary, ground-nesting bees. These bees tend to nest in sandy soils, and the lowland distribution of certain of the *Camissonia* may be largely controlled by the habitat needs of their pollinators. The female bees gather pollen to stock the larval cells in their underground nests. References: Raven 1962, 1969.

1. Capsules conspicuously pedicelled.
   2. Largest leaves in a basal rosette, the stem leaves reduced; leaf blades more than 3 times longer than wide, pinnately lobed to divided. ____ **C. claviformis**
   2′ Largest leaves on stems well above the base, the leaves not in a basal rosette; leaf blades about as wide as long, rounded to ovate or somewhat triangular, toothed or serrated.
      3. Flowers including pedicels 6.5–8.0 cm; hypanthium often 25–35 mm. ____ **C. arenaria**
      3′ Flowers including pedicels 3.8–4.7 cm; hypanthium 4–8 mm. ____ **C. cardiophylla**

1′ Capsules sessile.
   4. Leaf blades usually conspicuously red spotted or reddish; flowers minute, petals 2.0–2.5 mm, white. ____ **C. chamaenerioides**
   4′ Leaf blades sometimes moderately red spotted; flowers small but not minute, petals 5.0–7.5 mm, yellow or whitish.
      5. Stems thickish, the epidermis shiny whitish and peeling; inflorescences usually densely flowered, the flowers crowded, the petals white, the hypanthium 4.5–7.0 mm; capsules woody, 12–20 × 2–3 mm wide at base, abruptly tapering. ____ **C. boothii**
      5′ Stems slender, the epidermis green, not peeling; inflorescences few-flowered, the flowers not crowded, the petals bright yellow, the hypanthium 0.7–0.9 mm; capsules not woody, (32) 40–85 × 1.0–1.4 mm wide throughout. ____ **C. californica**

Camissonia arenaria (A. Nelson) P.H. Raven [*Chylismia arenaria* A. Nelson. *Oenothera arenaria* (A. Nelson) P.H. Raven] Robust ephemerals to short-lived, subshrubby perennials, 15–60 cm, with soft, spreading, long white hairs. Stems leafy. Petioles (2.5) 3.5–6.0 cm; leaf blades 2.0–5.5 cm, orbicular to deltate, the bases cordate, the margins toothed. Flowers vespertine, 6.5–8.0 cm including the pedicel; hypanthium often 25–35 mm; petals ca. 10.0–16.5 mm, bright yellow, drying pinkish. Capsules (2.0) 3.0–4.6 cm, straight or nearly so; fruiting pedicels 2–20 mm.

Granitic ranges in the northwestern part of the flora area. Rocky slopes, mostly at higher elevations, occasionally in gravelly washes and upper bajadas; usually localized. Ephemerals in the more arid habitats such as the Sierra del Rosario. Not known elsewhere in Sonora; also southeastern California and western Arizona.

Similar to *C. cardiophylla* but more robust, the flowers larger, and the hypanthium much longer. There is a clear morphological gap with no known intermediates. Although allopatric in the flora area, *C. arenaria* and *C. cardiophylla* are sympatric in Arizona and California. These 2 species comprise the section *Lignothera*.

Sierra del Rosario, *F 20663*. 53 mi SE of San Luis, Tinajas Altas Range, $n = 7$, *Raven 11650*. Cerro Pinto, *F 89-56*.

Camissonia boothii (Douglas) P.H. Raven subsp. **condensata** (Munz) P.H. Raven [*Oenothera decorticans* (Hooker & Arnott) Greene var. *condensata* Munz] WOODY BOTTLE-WASHER. Mostly (6) 8–23 cm. Stems thickish, the epidermis shiny silvery whitish, peeling away. Young herbage, lower leaf surfaces, and inflorescences densely to sparsely hairy with small glandular as well as larger non-glandular short white hairs, becoming glabrous or glabrate with age. Leaves larger and petioled below, 3–14 cm, becoming smaller and nearly sessile upwards; leaf blades mostly narrowly oblanceolate to narrowly elliptic, the margins minutely toothed or sometimes entire. Inflorescences densely flowered, buds on larger plants nodding. Flowers 16–29 mm, presumably vespertine, noticeably fragrant; hypanthium 4.5–7.0 mm; sepals fading pinkish; petals 5.6–7.5 × 3–4.2 mm, white, narrowed at base. Capsules 12–20 mm, 2–3 mm wide at base, rigid and woody, obscurely 4-angled, bent downward and often twisted, abruptly tapering, tardily dehiscent. Skeletons of stems and their woody fruits long lasting.

Mostly on gravelly and sandy soils of desert flats, washes, and lower dunes, sometimes on cinder or volcanic soils; seldom very common. Open desert from the vicinity of Puerto Peñasco to the south

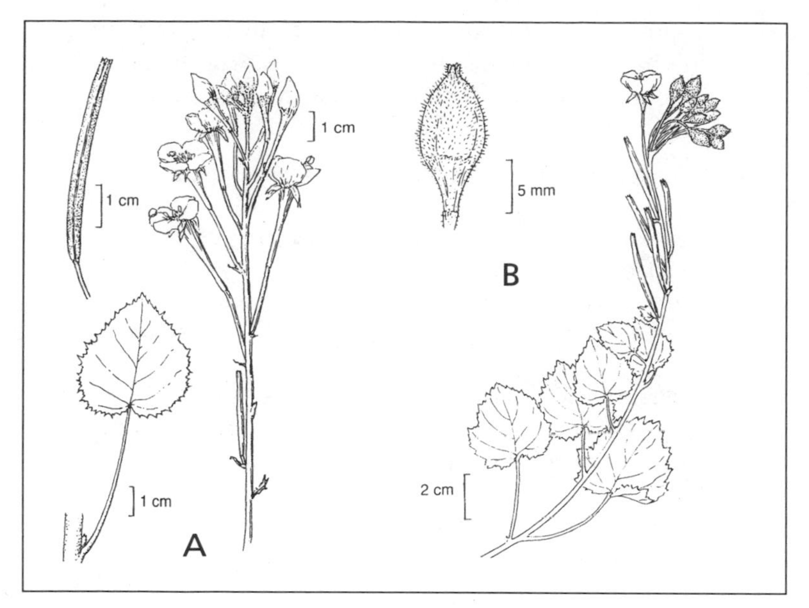

Camissonia: (A) C. arenaria; (B) C. cardiophylla subsp. cardiophylla. ER (Hickman 1993).

Camissonia boothii subsp. ***condensata.*** BA (Cronquist, Holmgren, & Holmgren 1997).

Camissonia chamaenerioides. RAJ (Cronquist, Holmgren, & Holmgren 1997).

Camissonia californica. ER (Hickman 1993; left) and MBJ.

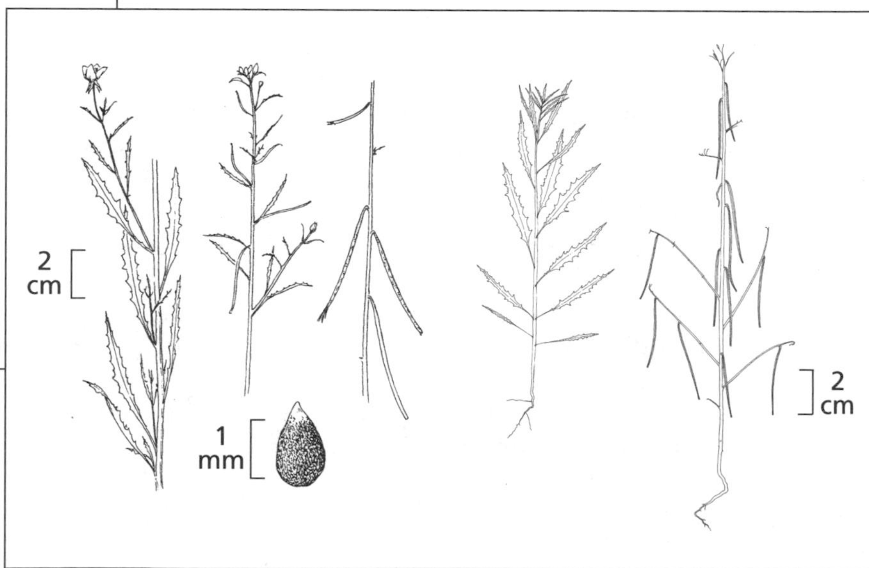

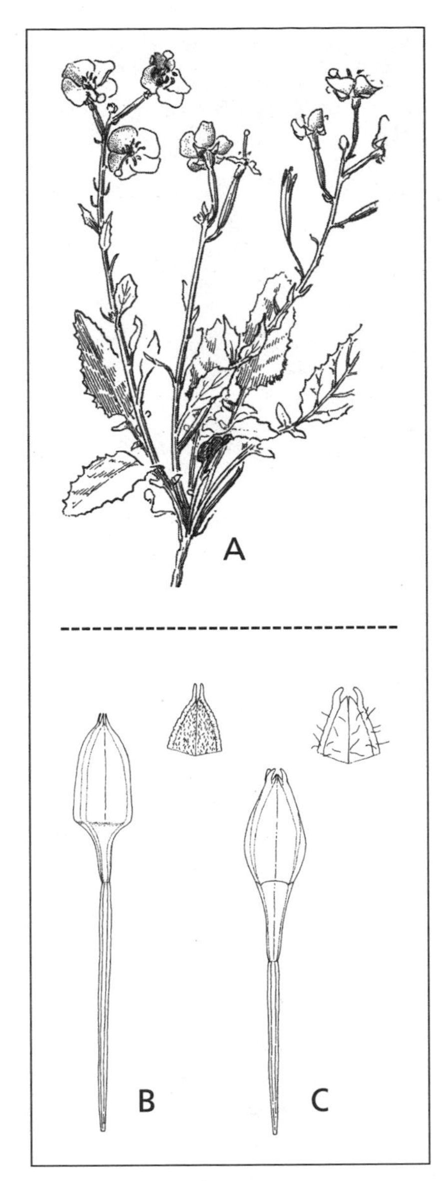

Camissonia claviformis:* *(A) plant, JRJ (Abrams 1951); ***(B)*** buds of subsp. ***peeblesii*** with short appressed hairs and terminal caudate sepal tips; ***(C)*** subsp. ***rubescens*** with long spreading hairs and caudate tips arising below apex of sepals. MBJ.

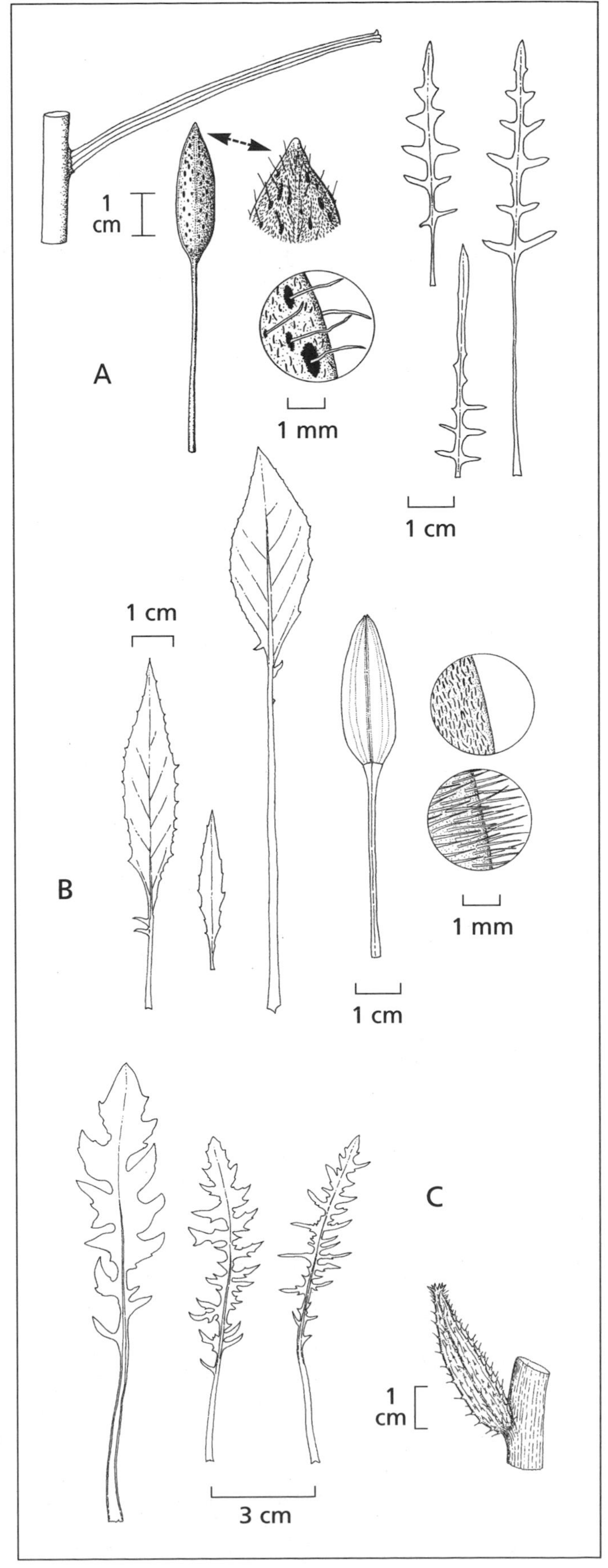

Oenothera:* *(A) O. arizonica; (B) O. deltoides subsp. ***deltoides,*** sepal with appressed hairs 0.1–0.2 mm long (upper circle); sepal surface from a plant with spreading, ribbonlike hairs 1.2–1.5 mm long (lower circle); ***(C) O. primiveris.*** MBJ.

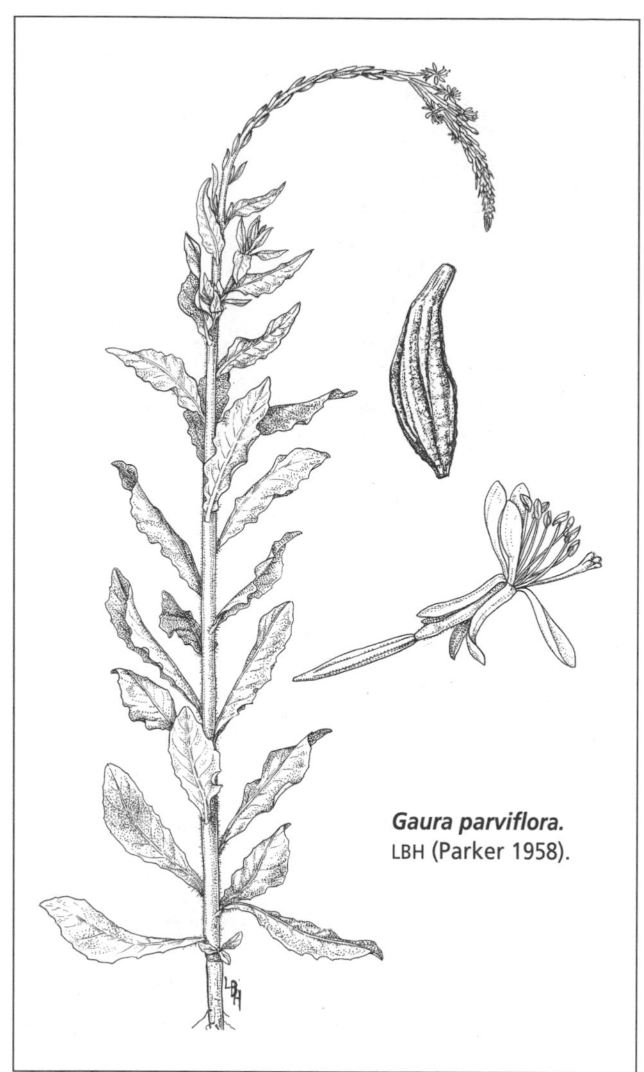

Gaura parviflora.
LBH (Parker 1958).

Oenothera deltoides subsp. ***deltoides. Palafoxia arida*** on right and juvenile ***Dicoria canescens*** left. PLK.

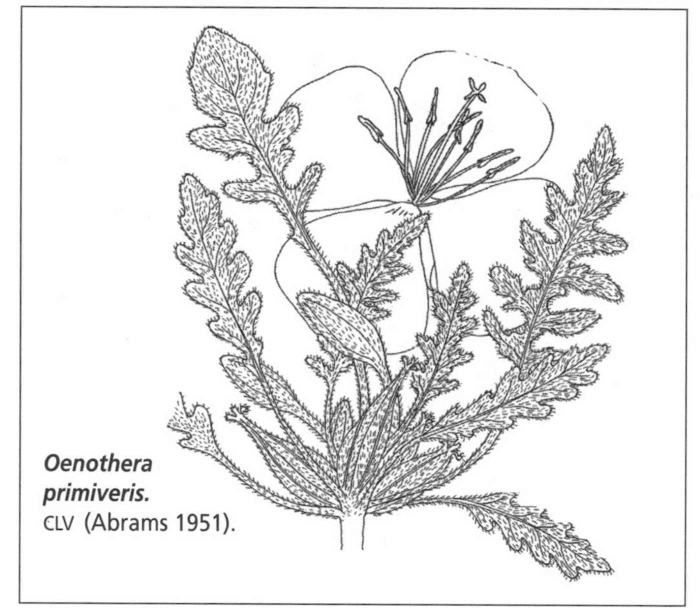

Oenothera primiveris.
CLV (Abrams 1951).

Argemone cf. gracilenta.
LBH (Parker 1958).

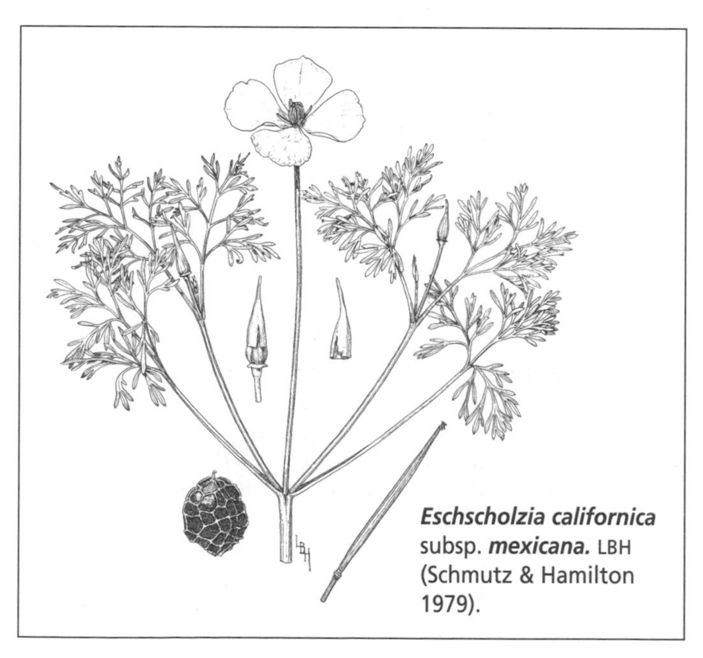

Eschscholzia californica subsp. ***mexicana.*** LBH (Schmutz & Hamilton 1979).

and west sides of the Pinacate volcanic field and adjacent desert flats. Not known elsewhere in Sonora; also northeastern Baja California, southeastern California, southern Nevada, southern and western Arizona, and southern Utah. Five other subspecies in the western United States. Subsp. *condensata* seems to be the most highly modified subspecies; it has the thickest, woodiest, and most tardily dehiscent capsules. Section *Eremothera* is made up of 5 species including *C. boothii* and *C. chamaenerioides,* but these two species are not closely related.

MacDougal Crater, *Mason 1834.* S of Papago Tanks, *F 18911.* Sierra Extraña, *F 19083.* Moon Crater, *F 19253.* 41 mi SW of Sonoyta, *Shreve 7593A.*

Camissonia californica (Nuttall ex Torrey & A. Gray) P.H. Raven [*Eulobus californicus* Nuttall ex Torrey & A. Gray. *Oenothera leptocarpa* Greene] Slender and often much taller than wide, 15–70+ cm, with an erect main axis; solitary, or sparsely branched, with ascending straight branches, or sometimes much-branched. First leaves in a basal rosette, or often not forming a basal rosette. Young herbage, buds, and immature fruits glabrate or sparsely pubescent with short white hairs as well as small glandular hairs and becoming glabrous with age. Larger (lower) leaves often 3–24 × 0.4–3.7 cm, petioled, the blades linear to narrowly elliptic, the margins pinnately and coarsely lobed and toothed; stem leaves reduced above. Plants leafy when young, leafless or nearly so at flowering time (warmer, drier weather). Flowers reportedly vespertine, observed to be closed until at least 10 P.M. and open at 5 A.M., often 15–18 mm wide. Petals 5–7 mm, bright yellow, sometimes with red flecks, fading orange, drying pink. Style, stigma, anthers, and filaments bright yellow. Ovary slender, 15–30 mm, the hypanthium 0.7–0.9 mm. Capsules (3.2) 4–8 cm × 1.0–1.4 mm, slender, straight to slightly curved, turning downward.

Seasonally abundant and widespread across most of the flora area including higher elevations of the Sierra Pinacate; not on moving dunes. Rocky, gravelly, sandy, and cinder soils of hills, mountains, bajadas, plains, lower dunes, washes, and arroyos. Southern California, Baja California, Arizona, and Sonora south to the vicinity of Isla Tiburón and Hermosillo.

This species is one of 4 in section *Eulobus,* the others are endemic to Baja California. *C. californica* is "mostly self-pollinating, and apparently always self-compatible" and "rarely visited by insects" (Raven 1969:198). As the flower matures the anthers collapse on top of the stigma. The 4 species in this section share some striking similarities, such as general size, the leafless condition at flowering time, and similar architecture, with certain crucifers (e.g., *Brassica, Caulanthus, Sisymbrium,* and *Streptanthella*).

Tinajas Altas Mts (Sonora), *Bezy 450.* Sierra del Rosario, *F 20716.* Hornaday Mts, *Burgess 6807.* S of Pinacate Peak, 875 m, *F 19329.* Moon Crater, *F 88-79.* Cerro Colorado, *Kamb 1993.* Tinajas de Emilia, *F 19732.* 29 km SW of Sonoyta, *Burgess 4760.*

Camissonia cardiophylla (Torrey) P.H. Raven subsp. **cardiophylla** [*Oenothera cardiophylla* Torrey] Ephemerals to short-lived perennials, (5) 12–30+ cm, with short, glistening, yellow glandular hairs among long, spreading, white hairs (villous). Stems leafy. Petioles (0.8) 1.5–5.5 cm; leaf blades (0.5) 1.2–4.0 cm, broadly ovate to nearly orbicular, about as wide as long, the bases cordate, the margins coarsely toothed. Flowers vespertine, pungently fragrant, 3.8–4.7 cm including the pedicel; hypanthium 4–8 mm; petals 6–10+ mm, bright yellow, often drying pink. Capsules 3–4 cm, slender, cylindrical; fruiting pedicels 3–16 mm.

Widespread in the Pinacate volcanic field, especially at middle to higher elevations, best developed on north-facing slopes. Gravelly-sandy arroyo beds and sometimes washes, cinder slopes, rugged lava slopes, and crater walls. Northern Baja California Sur, Baja California, southeastern California, southwestern Arizona, and northwestern Sonora nearly to Puerto Libertad.

There are two other more or less allopatric subspecies. Southward in Sonora, in the vicinity of the Sierra Bacha south of Puerto Libertad to the Sierra Seri and on Isla Tiburón, subsp. *cardiophylla* passes into subsp. *cedrosensis* (Greene) P.H. Raven, otherwise known from Baja California. Members of this species are reported as self-compatible but are often outcrossing. *C. cardiophylla* is most closely related to *C. arenaria.*

Cerro Colorado, *F 19636.* Campo Rojo, *F 19758.* Elegante Crater, *F 19680.* Pinacate Peak, *F 19420.* Cerro López, *F 18876.* Sykes Crater, *F 18986.* Arroyo Tule, *F 18817.*

Camissonia chamaenerioides (A. Gray) P.H. Raven [*Oenothera chamaenerioides* A. Gray] WILLOW-HERB PRIMROSE. Plants (5) 12–50 cm, with glandular hairs and small coarse non-glandular hairs near the inflorescence. Herbage often reddish; stems very slender. Leaves 1.5–7+ cm × 1–15 mm, green to reddish with dark red spots, the blades more or less elliptic, entire to sparsely and shallowly toothed or crenulate. Flowers opening near sunset. Sepals reflexed, cream-white inside, pink outside. Petals often 2.0–2.5 mm, whitish, often with a broad pink midstripe or markings, or turning pink with age. Filaments white. Stigma at first white, becoming yellowish to pinkish by morning, surrounded by the anthers at anthesis (both at the same height, the anthers often touching the stigma). Capsules 2.8–6.0 cm × 0.7–1.0 mm, sessile, slender, terete, straight.

Widespread at all elevations; not on moving dunes. Southeastern California to western Utah and western Texas, Baja California and Sonora southward to the Guaymas region.

This is the smallest-flowered evening primrose in the Sonoran Desert; the floral structure and modifications are characteristic of self-pollinated flowers. The pink rather than green stigma is unusual for the genus (also see *C. californica*). It is the only member of the genus in Texas. This species, in section *Eremothera,* is closely related to *C. refracta* (S. Watson) P.H. Raven, which has larger flowers, is outcrossing, and differs in characters of leaf shape, pubescence, and seed size. *C. refracta* occurs in Arizona southward to the vicinity of Yuma.

Sierra Cipriano, *F 88-220.* NE of Elegante Crater, *F 92-145A.* S of Pinacate Peak, 875 m, *F 19309.* MacDougal Crater, *F 9721.* Sykes Crater, *F 18982.* SE of Moon Crater, *F 18870.* Sierra del Rosario, *F 20747.*

Camissonia claviformis (Torrey & Frémont) P.H. Raven [*Oenothera claviformis* Torrey & Frémont] Stems branching mostly from the base, (10) 15–46+ cm. Leaves in a basal rosette, reduced above; rosette and lower leaves (3.5) 6–30 cm, the blades more or less elliptic, pinnately dissected, variously lobed or toothed, the lower segments of larger leaves often of separate leaflets, the smaller and/or upper leaves often entire; basal leaves often withering by time of flowering and fruiting. Veins on lower leaf surfaces conspicuously brown due to crowded oil cells. Inflorescences ascending, nodding at tips. Flowers vespertine, pedicelled, attractive, often (12) 15–18 mm wide, the petals white or yellow, the stamens and style of the same color as petals; center (hypanthium) orange-brown. Fruiting pedicels and capsules spreading to erect-ascending; fruiting pedicels 4–20 mm, the capsules 13.5–27.0 mm, moderately club-shaped (slightly enlarged at apex), often 1.8–2.0 mm wide near apex, straight or slightly curved. (This description covers the species as it occurs in the flora area.) Chromosome number $n = 7$ for all subspecies.

Widespread in the lowlands throughout the flora area. Most numerous on sandy and gravelly soils of arroyos, bajadas, flats, and dunes. Oregon and Idaho to Baja California and Sonora south at least to the vicinity of El Desemboque San Ignacio.

The seven "yellow-flowered subspecies occupy two relatively small areas peripheral to the distribution of the species as a whole, whereas the bulk of its area is occupied by the five relatively similar white-flowered, evening-opening subspecies" (Raven 1962:98). Subsp. *peeblesii* is white-flowered whereas the others in the flora area are yellow-flowered. Raven (1962:98) found that the 12 subspecies "represent modally distinct natural entities, [although] it is difficult to express these differences in the form of a key because the variation is often overlapping." The plants are reported to be self-incompatible. Three subspecies occur in the flora area. *C. claviformis* is the most diverse and widespread species in the genus. It is in section *Chylismia,* which is the largest section in the genus with 18 species.

Typical of many plants in arid regions, *C. claviformis* shows increased photosynthetic capacity and a decrease in leaf longevity. Experiments on plants in Death Valley, California, showed that they fix CO_2 by the C_3 photosynthetic mechanism at an exceptionally high rate at midday in spring (Mooney, Ehleringer, & Berry 1976).

1. Plants glandular hairy, especially on leaves and inflorescence; petals white, fading to pink, often drying purplish. ______ subsp. **peeblesii**

1′ Plants glandular hairy or not; petals yellow, remaining yellow or becoming reddish with age.

2. Plant villous (hairs long, slender, and spreading); petals bright yellow; sepals with caudate tips arising below the apex. ______ subsp. **rubescens**

2′ Plants strigose (hairs small, short, appressed, and coarse) and glandular pubescent or sometimes glabrous; petals pale yellow; sepals with caudate tips arising at the very apex or the tips absent. ______ subsp. **yumae**

Camissonia claviformis subsp. **peeblesii** (Munz) P.H. Raven [*Oenothera claviformis* var. *peeblesii* Munz] Pubescence of translucent glandular hairs ca. 0.1 mm as well as sparse to moderately dense, appressed, relatively thick, white hairs 1–3 mm (strigose) on new growth and inflorescence tips including buds. Leaves often somewhat thick and drying dark green or dark bluish green. Sepals with caudate or apiculate tips projecting from the very end of the sepal (terminal rather than below the apex), or the tips absent. Petals white, becoming pink with age, drying pale purplish. Filaments and anthers white, the anthers hairy.

Washes and low open desert, especially in sandy soils, across much of the lowlands of the flora area; Puerto Peñasco nearly to Sonoyta and westward to the western granitic mountains including Sierra del Rosario. Also eastward in northwestern Sonora, southern Arizona, and northeastern Baja California. Sometimes growing intermixed with subsp. *rubescens* and subsp. *yumae.*

Puerto Peñasco, *Dennis 26.* 13 mi S of Sonoyta on Mex 2, *Raven 14821* (RSA). MacDougal Crater, *Turner 86-2.* SW of Sierra Extraña, *F 88-70.* Sierra del Rosario, *F 20743.* 25 mi E of San Luis, *Asplund 15 Apr 1965.* 58 mi NW of Sonoyta on Mex 2, *Raven 11655* (RSA).

Camissonia claviformis subsp. **rubescens** (P.H. Raven) P.H. Raven [*Oenothera claviformis* subsp. *rubescens* P.H. Raven, *Univ. Calif. Publ. Bot.* 34:103, 1962] Villous throughout (young, vigorous plants sometimes sparsely pubescent), the hairs (0.2) 0.4–1.5 (2.0) mm, slender, spreading, and white, and sometimes also with shorter glandular hairs above. Basal lateral leaf segments ("leaflets") sometimes well developed. Sepals with conspicuous free caudate tips arising just below apices. Flowers open at dawn, often (12) 15–18 mm wide. Petals, filaments, and anthers bright yellow, the petals turning brick red after pollination (or perhaps sometimes not turning reddish). Style and stigma at first green, becoming yellow.

Widespread; sand flats, lower shifting dunes, crater floors, washes, open desert flats, and low hills; about 35 km north of Puerto Peñasco to the southern and western flanks of the Pinacate volcanic region, the adjacent east side of the Gran Desierto dunes, and northward to southwestern Arizona.

27 mi NE of Puerto Peñasco, *Raven 11686* (RSA). 24 mi S[W] of Sonoyta, *Keck 4178* (holotype of subsp. *rubescens,* DS). Sykes Crater, *F 18959.* Hornaday Mts, *Burgess 6822.* Moon Crater, *F 88-78.* 52 mi W of Sonoyta, *Raven 11657* (RSA).

Camissonia claviformis subsp. **yumae** (P.H. Raven) P.H. Raven [*Oenothera claviformis* subsp. *yumae* P.H. Raven] Strigose and glandular pubescent, or only glandular pubescent, or sometimes glabrate or glabrous: glandular hairs translucent, 0.1–0.2 mm, the non-glandular hairs sparse to dense, at least on new growth and usually also on inflorescence tips including buds, these hairs 0.15–0.3 mm, appressed, relatively thick, and white. Sepals with caudate (apiculate) tips projecting from the very end of the sepals (terminal rather than below the apex), or the tips absent. Petals pale yellow, becoming reddish, or often not changing color after pollination.

Sand flats, dunes including coastal and inland moving dunes, and washes; at least the southern and western part of the flora area, from Puerto Peñasco to the west side of the Pinacate volcanic field, El Golfo, and San Luis. Plants on higher dunes tend to be glabrous or glabrate and more robust. Also southeastern Imperial County, California, southern Yuma County, Arizona, and northeastern Baja California.

Sometimes growing with subsp. *peeblesii* and subsp. *rubescens*. The glandular and strigose pubescence and terminal caudate sepal tips, or lack thereof, indicate relationships with subsp. *peeblesii*. They are distinguished by petal color.

23 mi SE of San Luis on Mex 2, *Raven 11639*. 10 mi N of El Golfo, *F 75-77*. Sierra del Rosario, *F 20353*. SW of Sierra Extraña, *F 88-71A*. Bahía de la Cholla, *Gould 4148*.

GAURA

Annual, biennial, or perennial herbs. Stem leaves alternate. Inflorescence a spikelike raceme, not leafy. Flowers mostly opening near sunset, fading within one day, mostly bilateral, the petals held in the upper half of flower, stamens and style confined to lower half, or flowers rarely radial; usually 4-merous. Petals usually white, flushed with red after pollination, usually clawed (narrowed) at base. Each filament usually with a flap of sterile tissue near its base. Stigma 4-lobed. Fruits hard, woody, indehiscent, nutlike, often 1- or 4-seeded. Temperate and subtropical North America; 21 species. References: Carr, Crisci, & Hoch 1990; Raven & Gregory 1972.

Gaura parviflora Douglas ex Lehmann [*G. parviflora* var. *lacnocarpa* Weatherby] LIZARD TAIL, VELVET WEED. Non-seasonal ephemerals with a stout taproot, slender, erect, often reaching 1–2 m, often unbranched or with a few branches from above the middle, with glandular and long spreading non-glandular hairs. Leaves soft and velvety, elliptic to oblanceolate or obovate; lower leaves 4–20 cm and often deciduous by flowering time, narrowed to a winged petiole, the margins shallowly toothed to entire; first leaves, at least in winter-spring plants, in a basal rosette. Inflorescences erect, slender, densely and many-flowered, 19–33+ cm. Flowers small, the tube 2–3.5 mm, the sepals 3.4–4.5 mm, reflexed (at least in daytime), green with pink tinges, the petals 2.0–3.6 mm, pale to bright pinkish red in the daytime and at least sometimes when first open at dusk (perhaps at first white); the filaments and style bright pink, the stigma white. Fruits 8.4–10.2 mm, spindle-shaped, widest above the middle, 1- or 2-seeded.

Riparian or semiriparian habitats and low-lying poorly drained temporarily wet clayish or silty soils of riverbanks, crater-floor playas, and roadside depressions; often locally and seasonally abundant, especially in playas of larger craters. Most of the United States and northern Mexico, adventive in many parts of the world.

"Presumably the scales, which serve in most species of *Gaura* to exclude all but long-tongued insects from the nectar in the floral tube, have been greatly reduced in *G. parviflora* owing to its strict autogamy, very small flowers, and absence of nectar" (Raven & Gregory 1972:23). In *G. parviflora* there are minute, often difficult-to-see papillae where the scales in other species would be located. Another apparent adaptation to the self-pollinating habit is the relatively small and radial rather than bilateral flower. It "was probably originally native of the shortgrass prairie in the interior of the United States, and spread widely from there as a weed of cultivated and waste areas. Owing to its strict autogamy, it easily becomes established from a single fruit" (Raven & Gregory 1972:26).

The technically correct name for this species is *G. mollis* James. However, there is a proposal to retain (conserve) the well-known name *G. parviflora* as the correct one (Warren Wagner, personal communication 1999).

Sykes Crater, *F 18917*. MacDougal Crater, *Turner 86-36*. Ciénega, 25 mi W of Sonoyta, 17 Mar 1936, *Shreve 7602*. Sonoyta, Río Sonoyta, *F 85-696*. Río Sonoyta S of El Papalote, *F 86-156*. Quitobaquito, *F 86-174A*. Observation: Pinacate Junction.

OENOTHERA EVENING PRIMROSE

Ephemeral to perennial herbs, sometimes woody at base (winter-spring ephemerals in the Sonora portion of the Sonoran Desert), with a well-developed taproot. Leaves alternate or in a basal rosette. Flowers mostly vespertine, pollinated by hawk moths, and fading within a day. Flowers axillary. Floral tube prolonged beyond the ovary, deciduous after flowering. Sepals 4, reflexed in anthesis. Petals 4. Stamens 8. Stigma of 4 linear lobes. Capsules mostly 4-valved, dehiscent or sometimes short and only partially dehiscent, usually many-seeded.

New World, mostly temperate; those in deserts growing during cooler seasons; 120 species. References: Gregory 1963; Klein 1970; Raven 1970; Wagner, Stockhouse, & Klein 1985.

1. Petals yellow; capsules (1.8) 2.8–4.6 cm × 6.4–7.5 mm at base, thick, upright, straight, tapering to a conspicuously narrowed tip; ovaries and fruits with large papilla-based hairs. ____________ **O. primiveris**

1′ Petals white; capsules 4.0–7.5 cm × 2.0–3.5 mm at base, slender, often curved and usually moderately bent downward, tapering only slightly; hairs of ovaries and fruits not papillate-based.

  2. Buds purple-spotted; petals 2–3.6 cm; Cabeza Prieta Refuge and probably adjacent Sonora. ____ **O. arizonica**

  2′ Buds not purple-spotted, of one color; petals 2–5 (6) cm; widespread. ____________ **O. deltoides**

†Oenothera arizonica (Munz) W. L. Wagner [*O. deltoides* Torrey & Frémont var. *arizonica* Munz. *O. californica* S. Watson subsp. *arizonica* (Munz) W. Klein. *O. avita* (W. Klein) W. Klein subsp. *arizonica* (Munz) W. Klein] ARIZONA EVENING PRIMROSE. Plants with basal rosettes and developing leafy stems; stems spreading to decumbent, often reaching 30–60 cm and 2.5–4 mm in diameter, with shiny silvery whitish, smooth, peeling epidermis, the younger herbage and inflorescences often sparsely hairy. Leaves pinnatifid, somewhat regularly cleft to parted more than halfway to the midrib, the stem leaves (3.5) 5.5–15+ × 0.5–3.8 cm, the basal leaves often reaching 15–30 cm with very long, slender petioles. Leaves and bud sepals pilose with both smaller and larger white hairs, the larger hairs 0.9–1.5 mm, flat, and ribbonlike. Larger plants with flowers in upper stems. Buds with conspicuous purple spots usually surrounding the larger hairs. Petals 2.5–3.6 cm, white, turning pale pink with age. Capsules 5.0–7.5 cm × 2.0–2.3 mm at base, narrowly subcylindrical, nearly straight to moderately curved, spreading to downwardly bent, woody at maturity.

Sandy soils with *Larrea* and *Pleuraphis* in Cabeza Prieta Refuge along the international border and undoubtedly along adjacent margin of the flora area in Sonora. Sandy soils in southern Arizona from the vicinity of Yuma to Pima and Cochise Counties and southward in western Sonora (but eastward and inland from the flora area, or perhaps disjunct) to coastal dunes of Sonora from the vicinity of El Desemboque San Ignacio southward to Tastiota (28°20′ N).

Oenothera arizonica somewhat resembles *O. deltoides* but differs in its more slender, delicate, and often longer stems; often smaller, narrower, more regularly and deeply cleft to divided leaves mostly with narrower segments; often smaller flowers; and purple-spotted, longer-haired buds (sepals). *O. arizonica* is distinguished from the perennial taxa of the *Anogra* complex, currently treated as 3 subspecies of *O. californica,* by its annual habit with a taproot, stems thickened near the base and tapering toward the apex, buds with conspicuous purple spots, each spot at the base of a long hair, longer attenuate capsules, and pinnatifid leaves (Wagner 1998).

Cabeza Prieta Refuge: Pinta Sands, *F 92-36.* N of Las Playas, *Darrow 15 Apr 1941.* W side of Pinacate lava flow [at international border], *McLaughlin 2982.*

Oenothera deltoides Torrey & Frémont subsp. **deltoides.** DUNE PRIMROSE, WHITE DESERT PRIMROSE, DEVIL'S LANTERN. Plants with a basal rosette, and usually developing leafy stems, the stems of well-watered plants reaching 50+ cm, 8–15 mm in diameter, stout, semisucculent, developing silvery whitish, smooth, peeling epidermis. (During the exceptional El Niño spring of 1992, plants in the low dunes north of Sierra del Rosario measured 45–92 cm tall × 65–172 cm across.) Young herbage and sepals (buds) with dense white hairs, these on different plants either small and appressed or large and spreading, the older leaves glabrous or sparsely to moderately hairy. Leaves 3.5–22 cm, often reaching 2–3 cm wide, the blades lanceolate or elliptic to ovate (narrowly so in drought-stunted plants), often coarsely toothed or lobed, sometimes nearly entire, or the lower segments sometimes distinct and the leaves lyrate-pinnatifid. Petioles of lower leaves 3.0–10.5 cm, often longer than the blades; petioles of upper leaves shorter than the blades. Leaves often reduced upward, but the stem leaves of large, robust plants scarcely or not at all reduced.

Corollas white, often turning pale pink, with pale yellow around the tube forming an eye, the petals 2.0–5.5 (6) cm, reaching 6.5 cm wide, broadly obovate, the apex notched. Anthers white to pale yellow, the filaments pale yellow toward the base and mostly white above; style white, the stigma branches pale yellow. Capsules woody and stiff at maturity, nearly straight to curved, spreading to usually moderately bent downward, 4.0–6.5 cm × 2.0–3.5 mm at the base, narrowly cylindrical (tapering only slightly). The dry, basketlike skeletons may persist for several years, hence the name devil's lantern.

Seasonally abundant and widespread on moving dunes and rolling sand plains and flats; southward in coastal Sonora to the vicinity of 30°40′ N. This subspecies also occurs in northeastern Baja California, southeastern California, western Arizona, and southern Nevada. It is the southernmost of the several subspecies. The flowers are generally cross-pollinated.

This is one of the most conspicuous spring wildflowers in northwestern Sonora. During favorable years it forms a major portion of the biomass on the dunes. The large white flowers open shortly before dusk and remain open until the late-morning heat of the following day, or nearly all day in cool weather. Toward the end of the season the plants are ravaged by hoards of sphinx moth caterpillars (see *Lupinus*). Great masses of evening primroses carpeted the dunes during the 1992 El Niño, but the caterpillars destroyed almost the entire fruit (seed) crop.

El Golfo, *Bezy 364.* E of San Luis, *Gloyd 27 Mar 1960.* Sierra del Rosario, *Lumholtz 13* (GH). Moon Crater, *F 18828.* 30 km N of Puerto Peñasco, *Equihua 10 Apr 1981* (MEXU). Bahía de la Cholla, *Gould 4140.*

Oenothera primiveris A. Gray. YELLOW DESERT PRIMROSE. Leaves in basal rosettes, nearly stemless or often developing stout, erect, leafy stems sometimes 10–20 (30) cm. Taproot thick and well developed. Pubescence dense, of bristly spreading papillate-based white hairs. Leaves (3.3) 5–27 cm, the larger ones 3.5–7.0 cm wide, mostly pinnatifid into toothed or rounded lobes, narrowed into a long, winged petiole expanded at the very base. Petals and stigma yellow, the petals (1.8) 3.5–5.5 cm, notched at apex (flower, leaf, and plant size correlated with soil moisture). Ovaries and fruits densely hairy with large spreading white hairs, each from a large conical fleshy translucent red to pink or colorless papilla (gland). Capsules (1.8) 2.8–4.6 cm × 6.4–7.5 mm wide at base, thick and woody, upright, straight, 4-angled, tapering to a conspicuously narrowed tip. The dry skeletons, consisting of part of the taproot and stem and a cluster of woody capsules, may persist for several years. Flowers opening at dusk, closing the following morning.

Sand flats, playas, gravelly-sandy washes, and sometimes on cinder soils, often common but seldom abundant; Sonoyta region, Pinacate volcanic field including higher elevations, granitic ranges, and desert flats. Generally absent from dunes, where it is replaced by the more abundant *O. deltoides.*

Southwestern Utah to southeastern California, Baja California, Sonora south to the coastal plain west of Hermosillo, and southern Arizona to western Texas. Ours are generally identifiable as subsp. *bufonis* (M.E. Jones) Munz, distinguished by larger and generally cross-pollinated flowers. However, some specimens of drought-stressed plants from the flora area (e.g., *F 19252* and *F 20589*) have smaller flowers characteristic of subsp. *primiveris.* Subsp. *primiveris* (subsp. *caulescens* (Munz) Munz) has smaller, self-pollinated flowers with petals mostly 1.5–2.5 cm long.

S of Pinacate Peak, 875 m, *F 19351.* MacDougal Crater, *F 9716.* Sierra del Rosario, *F 20801.* Moon Crater, *F 19252.* 1 km S of 10 km by rd W of Los Vidrios, *F 20589.*

OROBANCHACEAE BROOMRAPE FAMILY

Annual or perennial succulent, herbaceous root parasites lacking chlorophyll; roots absent or coralloid and without root hairs. Leaves reduced to scales, alternate but tending to form a series of tight spirals especially in species with thick succulent stems. Flowers bilateral, calyx (1) 4–5-lobed; corollas 5-lobed and 2-lipped. Stamens 4, sometimes with a rudiment of a fifth stamen. Ovary superior. Fruit a capsule; seeds usually numerous and minute.

Mostly subtropical to temperate in the Northern Hemisphere; best developed in the Old World; 17 genera, 230 species. Related to the Scrophulariaceae and differing mostly in its obligate parasitic habit, lack of chlorophyll, and parietal rather than axile placentation. Reference: Thieret 1971.

OROBANCHE BROOMRAPE

Annuals; the above-ground portions often thick; roots often poorly developed. Calyx 4–5-lobed (5 in the New World). Worldwide, mainly subtropical and warm temperate, greatest diversity in Eurasia; 140 species.

Orobanche cooperi (A. Gray) Heller [*Aphyllon cooperi* A. Gray. *O. ludoviciana* Nuttall var. *cooperi* (A. Gray) Beck. *O. ludoviciana* var. *latiloba* Munz. *O. multicaulis* Brandegee var. *multicaulis*] *FLOR DE TIERRA;* DESERT BROOMRAPE; *MO'OTAḌK.* Stems very thick and succulent, the basal portion below ground and often ca. 3 cm wide, the above-ground stem and inflorescence 10–37 cm, suffused with purple-brown, unbranched or with several branches usually from near the base. Plants, especially the flowers, glandular hairy. Flowers 2–3 cm, the lower ones pedicelled, the upper ones sessile. Corollas purple and white, the throat marked with yellow. Stigmas shallowly cup-shaped, often bilobed. Shoots appearing in spring, flowering and withering during April or earlier.

Mostly in sandy, gravelly, and cinder soils, sometimes in pockets of sandy soil in rocky areas; sand flats, arroyo beds and washes, cinder flats and slopes, lower dunes, crater floors, and playas. Common nearly throughout the flora area except on higher moving dunes and very rocky habitats; near sea level–900+ m in the Sierra Pinacate. Western Sonora southward to El Desemboque San Ignacio. Deserts and semiarid regions from southern California to Texas and northern Mexico including northern Sonora, Baja California, and Chihuahua.

In northwestern Sonora it is commonly parasitic on two bursages, *Ambrosia deltoidea* and *A. dumosa,* and is sometimes on *A. ilicifolia.* Elsewhere it has been recorded as a parasite on *Hymenoclea, Larrea,* tomatoes, etc. (Wiggins 1964), but mostly on *Ambrosia* and *Hymenoclea.* In Texas it "flowers at almost any time of the year when moisture conditions are favorable" (Correll & Johnston 1970:1452).

Bahía de la Cholla, *Gould 4134.* Elegante Crater, *Sherbrooke 10 Apr 1983.* S of Pinacate Peak, 875 m, *F 19327.* Moon Crater, *F 10583.* Hornaday Mts, *Burgess 6859.* Quitobaquito, *F 7656.* Sierra del Rosario, *F 92-225.* SE of El Golfo, *Burgess 701.* 20 mi E of San Luis, *Norris & Bucher 31 Mar 1973.*

PAPAVERACEAE POPPY FAMILY

Mostly annual and perennial herbs, with diverse alkaloids in milky (latex) colored or colorless sap. Leaves mostly alternate; stipules none. Flowers often relatively large, radial. Sepals usually 2 or 3, enclosing the bud before it opens, usually falling as the flower opens. (The unopened flower might be mistaken for a fruit because the calyx may be crowned by style-like horns.) Petals separate, often twice as many as sepals, overlapping and often crumpled in bud, or rarely absent. Stamens often many. Ovary superior. Fruit usually a capsule. Seeds usually numerous.

Mostly temperate and tropical in the Northern Hemisphere; 25+ genera, 210 species. Includes temperate-climate garden plants such as ornamental poppies in *Papaver* and the opium poppy, *P. somniferum,* a winter-spring cash crop in the Sierra Madre Occidental of northwestern Mexico. Reference: Kiger 1997.

1. Plants coarse and spiny. ____ **Argemone**
1′ Plants delicate and spineless. ____ **Eschscholzia**

ARGEMONE PRICKLY POPPY

Robust annual or perennial herbs (1 shrub species) from taproots; glaucous, often prickly-spiny throughout and thistlelike with toxic yellow or orange latex. First leaves usually in a basal rosette;

leaves sessile, the upper ones clasping. Flowers usually large. Sepals 2–6, mostly 3, each with a subterminal spinescent horn; petals usually 6. Stamens 20–250+. Capsule valves opening terminally; seeds numerous.

United States to South America and 1 in Hawaii; 32 species. Often in disturbed habitats and difficult to determine the original, natural distribution. The plants are avoided by livestock because of chemical deterrents and the prickly spines. The seeds can float for several days. References: Ownbey 1958, 1997.

1. Petals white; stamens 150 or more; latex pale lemon-yellow. ______ **A. gracilenta**
1′ Petals yellow; stamens 40–75; latex bright yellow. ______ **A. ochroleuca**

Argemone gracilenta Greene. *CARDO;* WHITE-FLOWERED PRICKLY POPPY, COYBOW'S FRIED EGG. Robust perennials often to 1+ m, also flowering in first season. Herbage, sepals, and fruits glaucous and densely prickly-spiny. Sap lemon-yellow, drying black. Leaves 8–20 (40) cm, pinnately lobed, thistlelike. Petals 3–5 cm, white; stamens bright yellow, ca. 150 or more. Capsules 3.0–4.5 cm.

Widely scattered and often seasonally abundant in low-lying, poorly drained soils including playas, washes, and roadside ditches, often in disturbed habitats, and also cinder flats and slopes of Pinacate and Carnegie Peaks. Often flowering in late spring or early summer after most other wildflowers have dried up, continuing sporadically through the summer. Sonora from the Guaymas region northward to western and southern Arizona, and both Baja California states.

Tinajas de los Pápagos, *Burgess 6288.* MacDougal Crater, *F 9905.* S slope of Pinacate Peak, 1100 m, *F 19381.* NE of Sierra del Rosario, *F 20346.*

***Argemone ochroleuca** Sweet subsp. **ochroleuca.** MEXICAN PRICKLY POPPY. Facultative annuals or perennials. Herbage, sepals, and fruits glaucous and conspicuously prickly-spiny. Sap bright yellow. Leaves 8–15+ cm, pinnately lobed, thistlelike. Petals ca. 2.5 cm, yellow; stamens yellow, 40–75. Capsules ca. 2.5–3.5 cm.

Weed in disturbed urban and agricultural habitats, not widespread and not native in the flora area. Weedy in southern Arizona and Sonora, and southward primarily in western and central Mexico; 1 other subspecies. This species now weedy and adventive worldwide, the original distribution difficult to determine. The plants and flowers are smaller than those of the white-flowered prickly poppy.

Río Sonoyta at Sonoyta, *F 86-93.*

ESCHSCHOLZIA GOLD POPPY

Annual or perennial herbs with a well-developed taproot; spring ephemerals in the Sonoran Desert. Mostly glabrous, the sap colorless. Leaves 3-times dissected. Flowers generally showy, opening during the day, closing in cloudy weather, mostly yellow to orange, perigynous (the perianth and stamens on a ring surrounding ovary). Sepals 2, united into a cap pushed off by the opening flower. (The calyx breaks away along a horizontal rift at the base when forced by pressure of the expanding corolla, which begins crumpled in the bud.) Petals usually 4. Capsules slender, ribbed, 2-valved, dehiscent, often explosively upon drying. Seeds elaborately sculptured and diverse. Western North America; 12 species. Reference: Clark 1997.

1. Receptacle with a ringlike winged rim; petals 15–40 mm (smaller when drought-stunted); stamens 20 or more; capsules 4.7–8 cm; flowers mostly 1 per stalk (scape), otherwise stemless; Sonoyta region. ______ **E. californica**
1′ Receptacle not winged; petals 3–8 mm; stamens mostly fewer than 12; capsules 2–5 (6) cm; stems of larger plants branched, leafy, and with several or more flowers; widespread. ______ **E. minutiflora**

Eschscholzia californica Chamisso subsp. **mexicana** (Greene) C. Clark [*E. mexicana* Greene] MEXICAN POPPY; *HO:HI 'E'ES.* Herbage and capsules bluish glaucous. Leaves mostly basal, 6–11 cm. Flowering stalks mostly 7–20 cm, mostly leafless, 1-flowered, or with several 1-flowered branches or pedicels; receptacles forming a cup with a spreading winged rim. Flowers showy, the petals, stamens, style, and stigma bright yellow-orange, the petals 1.5–4.0 cm (as small as 0.7 cm when drought-stunted at end of season), the stamens ca. 20–24. Capsules 4.7–8.0 cm. Seeds 1.1–1.3 mm, similar to those of *E. minutiflora.*

Northeastern corner of the flora area near Sonoyta, mostly during springs with exceptional rainfall; especially along roadsides and granitic bajadas. More common eastward and northward at slightly higher elevations. Inland southwestern North America: southeastern California, southern Nevada, and southwestern Utah to northern Sonora, northwestern Chihuahua, and western Texas.

Disjunct and separated by the intervening severe desert from the more western subsp. *californica,* which ranges from Washington to northwestern Baja California. *E. californica* probably extended its range into the Sonoran Desert region during glacial times and has been isolated by increasing aridity. The California or Mexican poppy is sometimes cultivated and often included in wildflower seed packages.

10 mi SW of Sonoyta, *F 91-52.* Lukeville, *F 88-07.* Quitobaquito, *Nichol 10 Mar 1939.*

Eschscholzia minutiflora S. Watson. LITTLE GOLD POPPY. Herbage bluish glaucous, highly variable, the larger plants with multiple-flowered leafy-branched stems 15–45 cm. Leaves 3–11 cm, reduced upwards. Petals 3.2–8.0 mm, yellow-orange. Stamens often 8–10. Capsules mostly 2.3–5.0 (6.0) cm. Seeds globose, 1.0–1.2 mm, dark brown with a reticulate pattern of grayish white saclike hairs, swelling when wet.

Gravelly washes, cinder flats and slopes, sand flats, and sandy-gravelly bajadas; Pinacate volcanic field to the summit, Sonoyta region, and intervening desert flats. Southeastern California to southern Nevada and southwestern Utah south to Baja California and northwestern Sonora.

Drought-stressed plants may reproduce when only 2–3 cm tall. South of the flora area *E. minutiflora* seems to intergrade with populations identifiable as *E. parishii* Greene, distinguished primarily by having larger flowers. These plants extend southward to the vicinity of El Desemboque San Ignacio and Isla Tiburón.

10 mi SW of Sonoyta, *F 91-53.* Sierra Extraña, *F 19080.* SE of Moon Crater, *F 18858.* Sykes Crater, *F 18979.* Pinacate Peak, *F 19435.* Elegante Crater, *F 19688.*

PEDALIACEAE SESAME FAMILY

Annual or perennial herbs to small trees, usually with mucilage-secreting stalked glands. Stipules none. Flowers bilateral, the calyx synsepalous, the corollas sympetalous often with 2-lipped lobes. Ovary superior. Fruit a capsule. Mostly tropical, especially dry tropical regions of the Old World; 13 genera, 60 species. Reference: Manning 1991.

SESAMUM SESAME

Herbaceous; leaves opposite below, often alternate above. Flowers axillary, usually solitary. Calyx 5-parted, the lobes slightly unequal, the corolla 2-lipped, the stamens 4 plus 1 staminode. Native to Africa and Asia; 21 species. Reference: Ihlenfeldt & Grabow-Seidensticker 1979.

***Sesamum orientale** Linnaeus [*S. indicum* Linnaeus] *AJONJOLÍ;* SESAME. Warm-weather ephemerals, erect, often 30–45+ cm, with long to relatively short glandular hairs, and leaf blades also minutely scabrous. Lower leaves 8–12 (15) cm, ovate, coarsely toothed to lobed, the upper leaves smaller and narrower; petioles prominent. Calyx persistent. Corollas white and sometimes tinged with pink inside, ca. 2.5 cm. Capsules 2.0–3.2 × 0.7 cm, mostly in pairs (1 in each leaf axil), erect, hard-walled, more or less cylindrical, opening at apex. Seeds 2.9–3.5 mm, white to cream or slightly pinkish.

Infrequent along highways, growing from seeds falling from passing trucks, and not reproducing. Sesame is an important cultivated crop in Sonora and elsewhere in Mexico. Naturalized in parts of the New World. Seegeler (1989) showed that *S. orientale* is the correct name for sesame.

W of Sonoyta on Mex 2, *F 85-958, F 86-312.*

PHYTOLACCACEAE POKEWEED FAMILY

Perennial herbs to trees, often glabrous. Leaves alternate, simple, and entire; stipules absent (ours) or reduced. Flowers mostly in racemes, usually small, radial; sepals 4 or 5, all or some united below; petals usually absent; stamens 4 to many. Ovary superior. Fruits fleshy or dry.

Mostly tropical and subtropical, best developed in the Americas and thinly distributed in the Old World; broadly interpreted the family includes 26 genera and 92 species.

STEGNOSPERMA

Glabrous shrubs with semi-fleshy leaves. Sepals 5, united at the base, the petals 5, quickly deciduous. Stamens 10, united basally into a ring. Ovary of 5 carpels, each 1-seeded or some ovules not developing. Fruits red, fleshy, drying to a leathery capsule. Seeds blackish with a fleshy white or red aril. Dry tropical to subtropical and mostly coastal regions of Mexico, Central America, and the West Indies; 3 species.

Stegnosperma halimifolium Bentham [*S. watsonii* D.J. Rogers] *CHAPACOLOR*. Shrubs mostly forming large much-branched dense mounds or colonies 2+ m tall, 3–5+ m across. Evergreen to nearly leafless in extreme drought; leaves 1.5–4.5+ cm, petioled, the blades broadly elliptic to obovate, the apex rounded to notched; wet-season leaves relatively thin and green, the leaves of drier seasons semi-succulent, glaucous, and smaller, the petioles and veins often red. Flowers ca. 1 cm wide, fragrant, white, in terminal racemes to 12 cm (mostly in wet seasons), or in short axillary clusters (mostly in drier seasons). Flowering sepals greenish, often red tinged, the margins membranous white, the petals, stamens, and stigma white. Capsules 6–7 mm wide, ovoid, red. Seeds 3.5–3.8 mm, shiny blackish, more or less ovoid and slightly compressed laterally, the aril at first white, becoming bright red. Flowering response nonseasonal depending on soil moisture and temperature, at least in March, April, October, and November.

Several widely scattered small but well-established populations in the Pinacate volcanic field. Usually along the sides of arroyos in lava rock, mostly in protected niches with north-facing exposures. Dense embankments of these shrubs hanging from lava ledges near Tinaja Huarache produce masses of fragrant, star-shaped white flowers in spring, attracting great numbers of honeybees. Also among rocks at 2 sites, each with 2 to several shrubs, at higher elevation in the Sierra Pinacate. This is the northernmost location for this frost-sensitive shrub; the young shoots are often freeze damaged.

Common along the shores and coastal desert on both sides of the Gulf of California except in its northernmost reaches, and sometimes extending 100–150 km inland; also in mountains in northeastern Baja California.

Stegnosperma watsonii, distinguished from *S. halimifolium* by axillary rather than terminal inflorescences, seems to be based upon specimens collected during the dry season.

Tinaja Huarache, *Simmons 10 Nov 1963.* Trail between Chivos and Tule tanks, *F 18825.* Near SW side of Carnegie Peak, 1100 m, *F 19824.*

PLANTAGINACEAE PLANTAIN FAMILY

Mostly annual or perennial herbs. *Plantago* plus 2 small genera. Reference: Cronquist et al. 1984.

PLANTAGO PLANTAIN, INDIAN WHEAT

Annual or perennial herbs. Leaves usually in close spirals resembling a basal rosette, simple, parallel veined (apparently representing an expanded petiole). Flowers in spikes, usually wind-pollinated, the individual flowers small and inconspicuous, subtended by bracts, radial or occasionally bilateral, and 4-merous. Sepals often green or papery, the corollas sympetalous, papery, and persistent. Ovary superior. Fruit a circumscissile capsule. Seeds 1 to many, mucilaginous when wet.

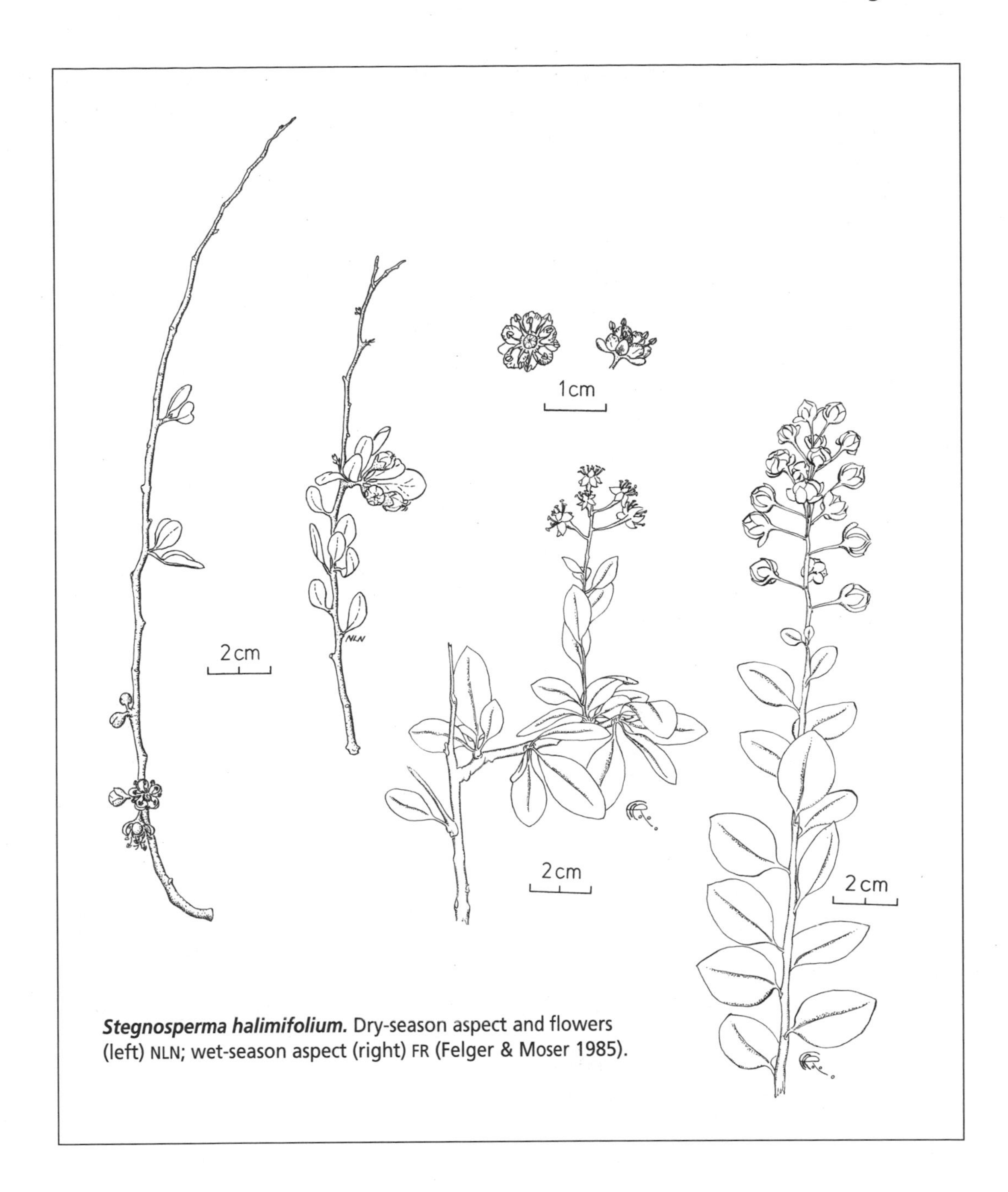

Stegnosperma halimifolium. Dry-season aspect and flowers (left) NLN; wet-season aspect (right) FR (Felger & Moser 1985).

Worldwide, largely temperate regions and mountains in the tropics; many in wetlands and few in deserts; 250 species. Eight species are listed for the Sonoran Desert but only *P. ovata* and *P. patagonica* are truly common in desert habitats, the former being the more widespread and common. Psyllium seed of commerce is produced from *P. afra* and *P. ovata* of the Old World. References: Rahn 1979a, b.

1. Bracts broadly ovate to obovate, with broad papery-membranous margins, none longer than the sepals; seeds shiny; stamens exserted. ____ **P. ovata**

1′ Bracts linear to narrowly oblong, all green (or sometimes with minute membranous wings at base), bracts of lower flowers usually much longer than the sepals; seeds dull; stamens not exserted. ____ **P. patagonica**

Plantago ovata Forsskal, 1775 [*P. insularis* Eastwood, 1898. Not *P. insularis* (Grenier & Godron) Nyman, 1881. *P. fastigiata* E. Morris, 1900. *P. insularis* var. *fastigiata* (E. Morris) Jepson, 1925] *PASTORA;* WOOLLY PLANTAIN, INDIAN WHEAT; *MUMṢA*. Winter-spring ephemerals, highly variable in size depending on soil moisture, usually with a well-developed slender taproot. Herbage, flowering stems, and inflorescences moderately to densely pubescent with loosely woolly and silky silvery-white hairs. Plants generally appearing stemless, but sometimes late in the season forming a short, leafy stem rarely reaching 7.5 cm tall. Leaves (4) 5–17 cm × 1–10 mm, erect to ascending or spreading, linear to linear-lanceolate, gradually narrowed below to a winged petiole; margins entire.

Flowering stems (peduncle or scape plus spike) (3) 4–20 (30) cm, leafless, mostly many-flowered, usually several or sometimes many per plant, erect to ascending. Floral bracts 2.5–3.0 mm, the bracts and sepals similar, broadly ovate to slightly obovate, with a thickened green midstripe and broad papery-membranous white margins. Corollas membranous-papery and brown, the lobes broad and spreading. Fresh petal lobes transparent white to pale pink, the style, stigma, and its pollen brush lavender-red, the flowers protogynous (the stigma protruding from the unopened corolla before the stamens expand). Filaments lavender, the anthers cream-white, vibrating in the slightest breeze. Seeds (1.8) 2.1–2.5 (2.7) mm, 2 per capsule, elliptic to ovoid, shiny yellowish to orange-brown, the outer face convex, the inner face flat and excavated.

Growth may begin in October, and some plants may germinate with late spring rains and straggle into June, especially near the coast; peak flowering and fruiting is in late winter and early spring.

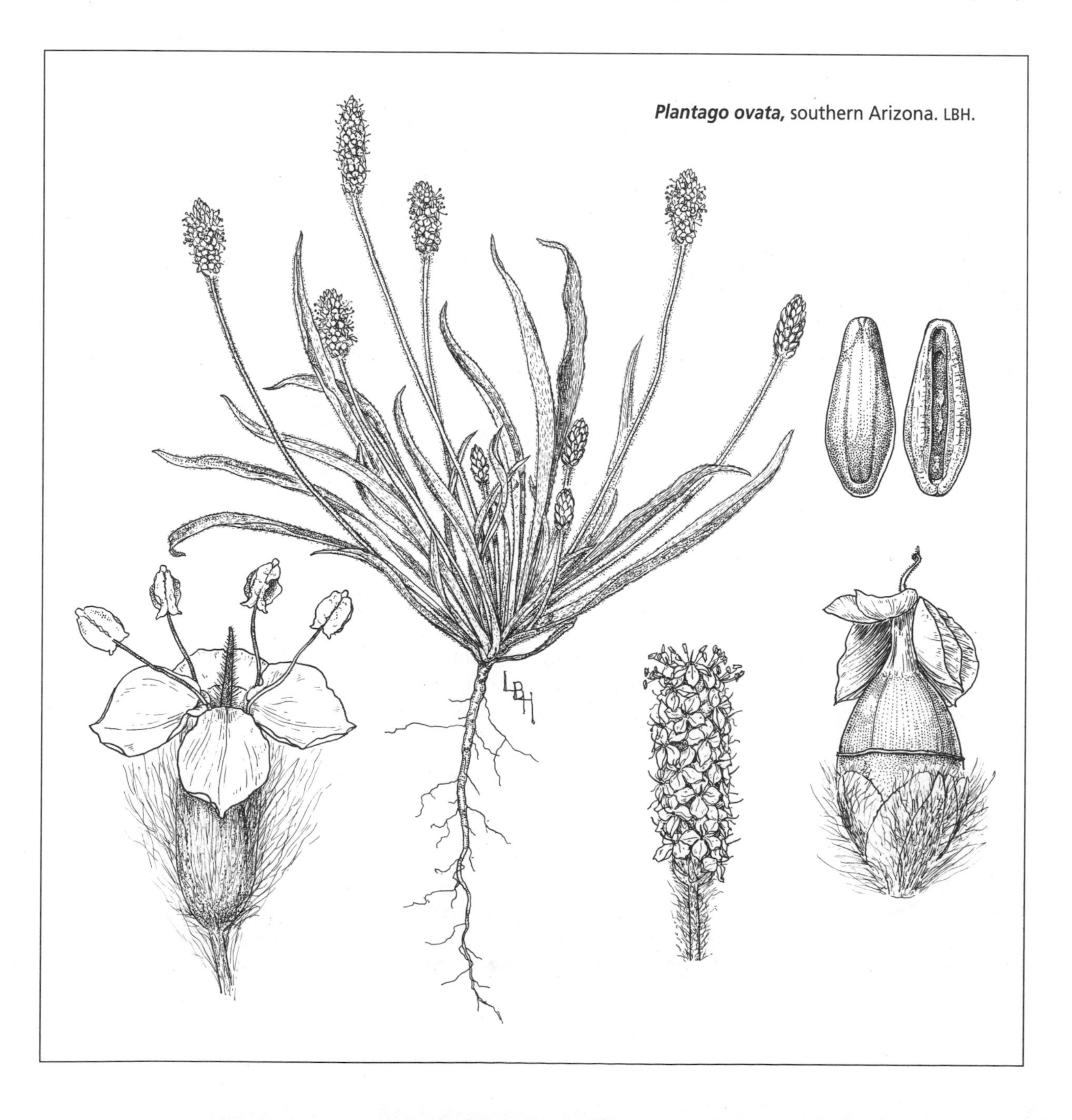

Plantago ovata, southern Arizona. LBH.

Widespread and often abundant, near sea level to 860+ m in the Sierra Pinacate; often carpeting the desert. Dunes, desert plains, washes, slopes, and roadsides.

One of the most ubiquitous winter-spring ephemerals in the Sonoran and Mojave Deserts. Southwestern North America from California to Baja California Sur, southern Nevada, southwestern Utah, western and southern Arizona to western Texas, and Sonora south to the Guaymas region. In the Old World from Pakistan to North Africa, southern Spain, and the Canary and Madeira Islands.

As with other plantagos, when water contacts the seed coat it immediately forms a jacket of slime (mucilage) that on drying tenaciously glues the seed to any available substrate. This species shows great variation in plant size, leaf width, and pubescence. Plants from the Gran Desierto dunes have especially narrow leaves, often only 1.0–1.5 mm wide.

The New World populations, long known as *P. insularis* Eastwood, are closely related to the Old World *P. ovata*. Although there are differences, Rahn (1979b) concluded that the two populations are best treated as a single species without infraspecific taxa. In the event that the New World populations are treated as a distinct species, then the correct name is *P. fastigiata* E. Morris. Bassett & Baum (1969) claimed that the North American population resulted from Old World introduction by California settlers in the late eighteenth and early nineteenth centuries. However, seeds of this species have been recovered from Sonoran Desert fossil packrat middens, including some from the Hornaday Mountains radiocarbon dated as old as 10,000 years B.P. (Van Devender et al. 1990).

Plantago ovata has some unique features. It is the only known species within the Plantaginaceae having the basic chromosome number 4, other species have 6 or 5. Of the species belonging to *Plantago* section *Albicans, P. ovata* is the only one also found outside the Old World (Rahn 1979b).

Cultivated varieties or forms of *P. ovata,* of Old World origin, produce larger plants and seeds with more mucilage (which is the commercial product) than do wild populations.

Sierra Cipriano, *F 88-199*. Puerto Peñasco, *F 20956*. 24 mi SW of Sonoyta, *Shreve 7583*. Sierra Pinacate, 875 m, *F 19358*. Moon Crater, *F 19276*. MacDougal Crater, *Turner 86-14*. N of Sierra del Rosario, *F 20785*. 10 mi N of El Golfo, *F 75-66*.

Plantago patagonica Jacquin [*P. purshii* Roemer & Schultes. *P. purshii* var. *oblongata* (E. Morris) Shinners] *PASTORA*. Winter-spring ephemerals, resembling *P. ovata*. Herbage with tawny brown hairs. Flowering stems leafless, erect to ascending, often (6) 9–20 (27) cm. Bracts all green or with membranous wings, linear to nearly oblong and slightly wider at base, the bracts subtending the lower flowers often more than twice as long as the sepals, becoming shorter upwards; midspike bracts 3.5–5.0 (6.0) mm. Flowers cleistogamous (self-pollinating in the bud, the anthers not exserted beyond the corolla mouth). Corollas moderately bilateral. Seeds 2.0–2.2 mm, 2 per capsule, dull dark brown.

Sandy-gravelly granitic bajadas. Entering the flora area in the Sonoyta region in the Sierra Cipriano (*F 88-200*) growing with *P. ovata*. Also eastward in northern Sonora probably to northern Chihuahua and in Baja California. Widespread north of Mexico to southern Canada, and in Argentina. In Sonora and Arizona generally at higher elevations than *P. ovata*.

POLEMONIACEAE PHLOX FAMILY

Annual or perennial herbs, rarely shrubs, vines, or small trees; often foul-smelling from gland-tipped hairs. Leaves opposite or alternate, sometimes in basal rosettes, or opposite below and alternate above, simple and entire to dissected or compound; stipules none. Flowers in variously modified cymes or solitary. Calyx, corolla, and stamens usually 5-merous. Sepals usually fused below into a tube. Corollas sympetalous, mostly radial, the filaments attached to the tube. Ovary usually 3-carpelled with 3 chambers. Style 1; stigma lobes 1 per carpel (stigmas 3-lobed in ours). Fruit usually a capsule; seeds 1 to many, often mucilaginous.

Mostly North America, centered in the temperate western part of the continent, greatest diversity in California; some in Central America and western South America, few in Eurasia; 21 genera, 300 species.

Ours are all winter-spring ephemerals with taproots, and show biogeographic affinity with regions of winter rainfall to the west and north, especially California and the Great Basin. The seeds are small, not winged, and except those of *Aliciella* are highly hygroscopic and upon contact with water immediately produce an envelope of slime, or mucilage, which as it dries tenaciously cements the seed to anything it touches. These species tend to have relatively thick, wrinkled seed coats that expand when wet. All of the herbaceous South American species have mucilaginous seeds. References: Cronquist et al. 1984; Grant 1959; Porter 1997.

1. Leaves opposite; flowers nocturnal. ______ **Linanthus**
1′ Leaves alternate; flowers diurnal.
   2. Inflorescences woolly; corolla lobes blue. ______ **Eriastrum**
   2′ Inflorescences not woolly; corolla lobes not blue.
      3. Leaves and calyx lobes not bristle-tipped; inflorescence leaves few and reduced, the inflorescences open and greatly overtopping leaves.
         4. Leaves more or less ovate, toothed but not dissected, to about 2 times longer than wide; corollas bright pink-red. ______ **Aliciella**
         4′ Leaves pinnately dissected, more than 3 times longer than wide; corolla lobes white to violet. ______ **Gilia**
      3′ Leaves and calyx lobes bristle-tipped; flowers sessile in leaf axils or in dense, compact leafy-bracted heads.
         5. Leaves broadest near the tip; most leaves with some 2- or 3-forked bristles; corollas radial or nearly so. ______ **Langloisia**
         5′ Leaves linear, not expanded above; leaf bristles all single; corollas bilateral, 2-lipped. ______ **Loeseliastrum**

ALICIELLA

Annual to perennial herbs with taproots; mostly glandular pubescent. Leaves generally alternate, at first in a basal rosette, reduced above, entire to once-pinnatifid. Calyx membranous between the green midribs. Corollas radial or occasionally slightly bilateral. Seeds small or minute, not (or only slightly) mucilaginous when wet.

Rocky Mountains, intermountain and desert southwestern United States, and barely extending into adjacent Mexico; 21 species in 2 subgenera. Subgenus *Gilmania,* with 2 species, is characterized in part by hollylike leaves with aristate teeth, filaments papillose below the anthers (at least the longest anthers), and yellow pollen. It includes *A. ripleyi* (Barneby) J.M. Porter, in California and Nevada, and *A. latifolia.* Reference: Porter 1998.

Aliciella latifolia (S. Watson) J.M. Porter subsp. **latifolia** [*Gilia latifolia* S. Watson] Plants (2.5) 5.0–25+ cm, branched or unbranched when drought stunted, glandular pubescent. Smaller plants with mostly basal leaves, larger plants with moderately leafy stems. Leaves alternate (lower leaves occasionally subopposite), (0.7) 1.5–6.5+ × 1.2–3.2 cm; lower leaves petioled, the petioles often winged, the upper leaves often sessile; leaf blades broadly ovate, coarsely toothed to laciniate, the teeth 1–2 mm. Inflorescences of few- to many-flowered panicles; flowers mostly pedicelled. Calyx 4–6 mm, the segments narrow and sharp-pointed. Corollas bright red-pink (magenta internally, pale pink externally). Filaments papillose toward apex. Dried anthers white. Seeds 0.6–0.7 mm, yellow-tan, glistening, plump, and more or less ovoid.

Widely scattered in the Pinacate volcanic field, sometimes locally common; not known elsewhere in Sonora. Otherwise northeastern Baja California, southeastern California, western Arizona, southern Nevada, and southwestern Utah. A second subspecies in Utah.

NW of Cerro Colorado, *F 19643.* S of Pinacate Junction, *F 19662.* Chivos Butte, *F 19618.* Sykes Crater, *F 18966.*

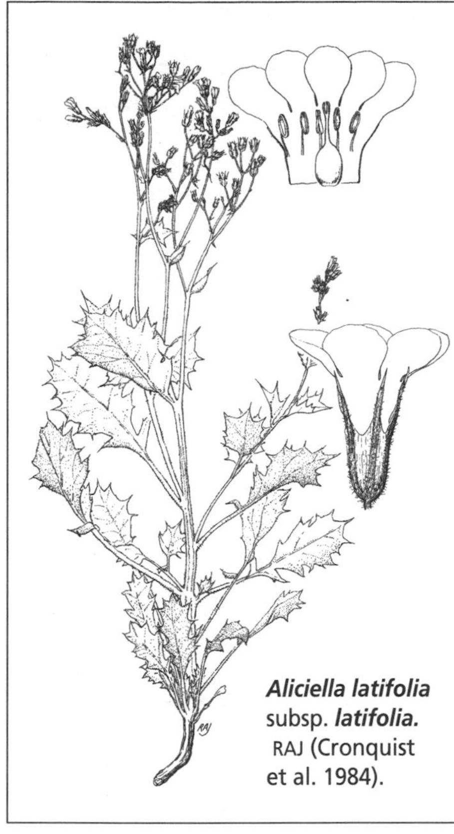
Aliciella latifolia subsp. ***latifolia.*** RAJ (Cronquist et al. 1984).

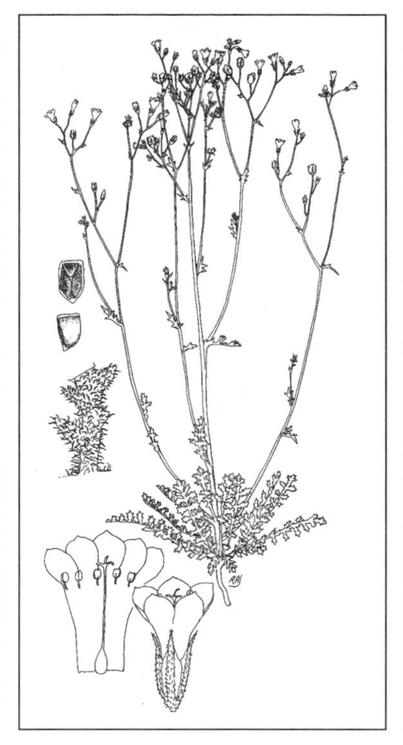
Gilia stellata. RAJ (Cronquist et al. 1984).

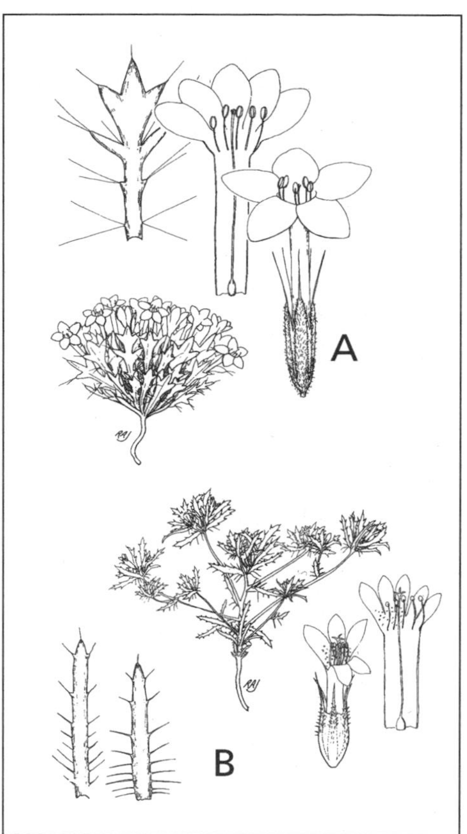

Polemoniaceae: (A) Langloisia setosissima; (B) Loeseliastrum schottii. RAJ (Cronquist et al. 1984).

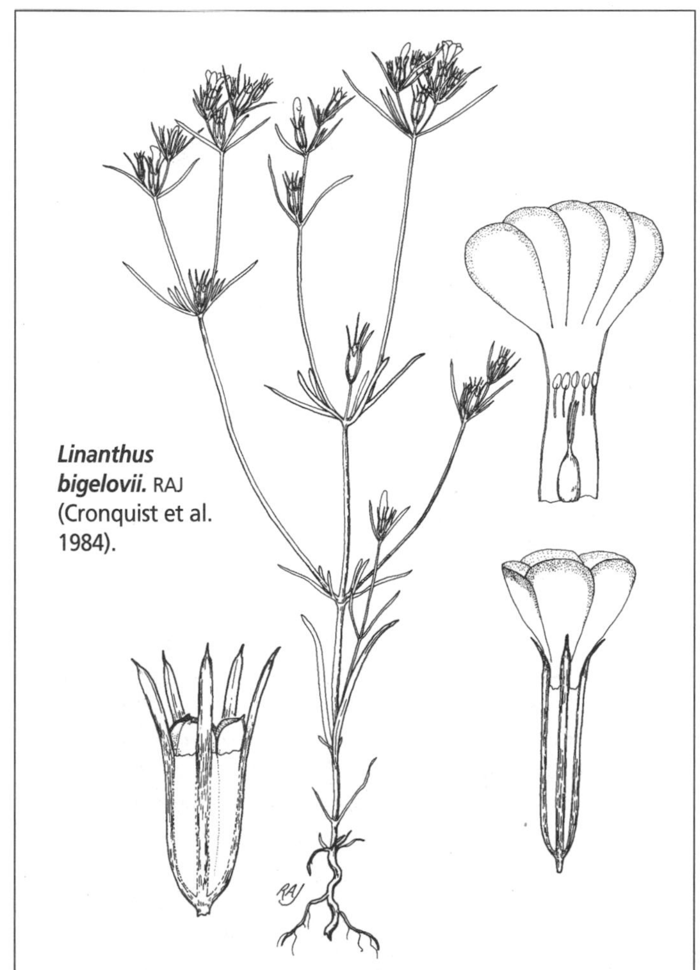
Linanthus bigelovii. RAJ (Cronquist et al. 1984).

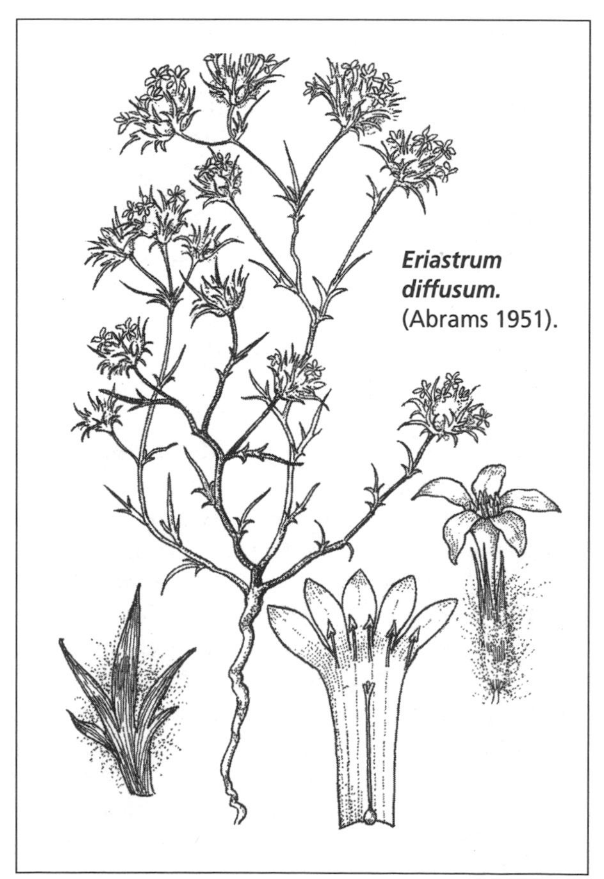
Eriastrum diffusum. (Abrams 1951).

ERIASTRUM

Annual or perennial herbs, some with woody bases. Leaves alternate, linear, and entire to pinnately toothed to dissected, the segments spinulose-tipped. Flowers usually sessile in compact, woolly heads subtended by leafy bracts. Calyx lobes often spinulose. Corollas radial or 2-lipped. Seeds mucilaginous when wet.

Western North America, greatest diversity in California; 14 species. References: Harrison 1972; Mason 1945; Patterson 1993.

Eriastrum diffusum (A. Gray) H. Mason [*Gilia filifolia* Nuttall var. *diffusa* A. Gray] Plants (3) 5–20 cm, unbranched when stunted, otherwise diffusely branched above, the main axis short, the branches spreading. Stems sparingly white woolly, glabrate with age. Leaves simple, narrowly linear, or pinnatifid with 3 or 5 narrowly linear segments; lower leaves largest, 1.5–4.0 cm, usually deciduous at or soon after flowering. Flower heads white woolly, the bracts similar to the leaves, 8–17 mm. Corollas slightly bilateral, 8–10 mm, the tube yellowish, the lobes blue, 3.0–3.5 mm. Stamens 1–2 mm, the filaments attached just below the corolla lobes, the anthers 0.5–0.6 (0.8) mm; anthers and pollen pale yellow.

Widespread; sandy flats, gravelly washes and arroyo beds, roadsides, lava rubble, alluvium of crater floors, and cinder slopes to the summit of Pinacate Peak. Not on moving dunes. Southeastern California to western Colorado, southward to northern Baja California, northern Sonora, northern Chihuahua, and Texas. Southward in western Sonora to the vicinity of Puerto Libertad.

Mason (1945) segregated 3 subspecies but they are only weakly correlated with geographic distribution. Ours mostly fall into subsp. *diffusum,* having relatively short corollas and anthers. Some plants, especially toward the northwestern part of the flora area, approach subsp. *jonesii* H. Mason, with larger corollas and anthers.

MacDougal Crater, *Turner 86-11.* Arroyo Tule, *F 18753.* Pinacate Peak, *F 19472.* 30 mi W of Sonoyta, *F 7619.* 41 mi SW of Sonoyta, *Shreve 7592.* Sierra Cipriano, *F 88-219.*

GILIA

Annual to perennial herbs with taproots. Leaves usually alternate, mostly in basal rosettes, reduced above, entire to highly dissected. Calyx membranous between the green midribs. Corollas radial or occasionally slightly bilateral. Seeds small or minute, ovoid to angular, brownish, generally mucilaginous when wet. Western North America with 35 species, plus 4 species in temperate South America. References: Day 1993; Grant & Grant 1956.

1. Lower part of plant with some cobweb-like hairs; calyx (2.6) 3.0–5.5 mm; corollas and anthers white. ____ **G. minor**

1′ Hairs bent or crinkled, not cobweb-like; calyx 3.5–7.0 mm; corollas white tinged with violet, the anthers blue after dehiscence. ____ **G. stellata**

Gilia minor A.D. Grant & V.E. Grant. Generally resembling *G. stellata.* Lower part of plants with at least some cobwebby hairs. Herbage, at least when drought stressed, conspicuously infused with reddish purple; upper stems, leaves, and calyces with stalked glandular hairs. Inflorescences of few- to many-flowered panicles; pedicels short to mostly long even on the same branchlet. Calyx (2.6) 3.0–5.5 mm. Corollas 5.6–6.2 mm wide, pure white with faint red tinge in the inner, lower part of the tube; filaments, anthers, style, and stigmas white.

Cinder soils at higher elevations in the Sierra Pinacate (*F 19461b, F 19914, F 92-443*). Northwestern Sonora southward to the vicinity of Caborca, western Arizona, and southern California.

Gilia minor represents a desert species of the many self-pollinating and often difficult-to-distinguish taxa belonging to the "*G. inconspicua*" complex. This complex is widespread in inland western United States and northern Mexico in northern Baja California, Sonora, and Chihuahua, and is also in southern South America. The Grants recognized about 30 taxa within this complex; Cronquist and colleagues (1984) submerged many of them into *G. inconspicua* (J.E. Smith) Sweet, a name of dubious standing (Mark Porter, personal communication 1998).

Gilia minor differs from *G. stellata* in having cobwebby hairs, usually darker red-purple herbage, leaves apparently not as finely divided, almost pure white and smaller flowers, white anthers (pollen), and the sepals may be smaller. They grow intermixed in the Sierra Pinacate.

Gilia stellata A. Heller. STAR GILIA. Plants (4) 6–35 cm. Herbage green to sometimes red-green. Lower leaves (1.0) 1.5–6.0 × 0.8–2.5+ cm, in a dense basal rosette, 2- or 3-times dissected, the segments sharp-pointed with sharp white tips 0.3–0.4 mm; stem leaves relatively few and much reduced. Lower leaves and lower stems with crinkled or bent (geniculate) white hairs (not cobwebby); upper stems, leaves, and calyces with stalked glandular hairs. Inflorescences of few to many-flowered panicles; pedicels range from short to mostly long even on the same branchlet. Calyx 3.5–7.0 mm, the segments narrowly triangular and pointed. Corollas 7.8–9.2 mm wide, white tinged with violet, the lobes 2.5–3.5 mm, the tube (including the moderately expanded throat) 3.5–5.0 mm, pale yellow with 5 lavender to red spots below. Stamens attached just below corolla sinuses (not in the throat), the filaments white, glabrous; anthers white before dehiscence, blue after dehiscence due to the blue pollen. Style white. Seeds (1.1) 1.3–1.5 mm, orange, lumpy, and angular like a miniature orange gumdrop that has been in a backpack for several days, extremely mucilaginous when wet.

Common during favorable years from the Sonoyta region to the western granitic ranges, and through most of the Pinacate volcanic field to peak elevation. Southeastern California, southern Nevada, and southwestern Utah to Baja California, western and central Arizona, and northwestern Sonora to the vicinity of Caborca. *G. stellata* is diploid ($n = 9$). As with many other gilias, the flowers are protandrous (stamens dehisce before stigma is receptive), indicating outcrossing rather than selfing as in *G. minor.*

Sierra Cipriano, *F 88-222.* Tinajas de Emilia, *F 19712.* Pinacate Peak, *F 92-444.* Hourglass Canyon, *F 19139.* Arroyo Tule, *F 19217.* Sierra de Tinajas Altas, E of San Luis, *Breedlove 15968* (DS).

LANGLOISIA

A genus of one species closely related to *Loeseliastrum* (Porter 1997). The plants of *Langloisia* are generally more compact than those of *Loeseliastrum schottii.* Reference: Timbrook 1986.

Langloisia setosissima (Torrey & A. Gray ex Torrey) Greene subsp. **setosissima** [*Navarretia setosissima* Torrey & A. Gray ex Torrey] Plants compact, 2–5 cm, often much-branched, the main axis very short or lacking. Leaves alternate, simple, narrowly obovate, the larger ones 1–5 cm, pinnately toothed and lobed, the segments triangular and prominently bristle-tipped; bristles white, (3) 5–9 mm, 2- or 3-forked on lower lobes of all except first leaves. Calyx segments bristle-tipped. Corollas 10–16 mm, radial or nearly so, pink-lavender and showy. Anthers and pollen white to blue. Capsules 5–7 mm, the valves separating completely at maturity. Seeds 1.1–2.0 mm, ellipsoid to short and squat depending on position in capsule, pale orange-yellow. Flowering March, dying by mid-April.

Mostly on cinder soils in the northeastern part of the Pinacate volcanic region; not known elsewhere in Sonora. Also southeastern California, southern Nevada, southwestern Utah, and western Arizona. Subsp. *setosissima* is replaced by subsp. *punctata* (Coville) Timbrook north and northwestward in southern California, Nevada, Idaho, and eastern Oregon.

1 mi S of Pinacate Junction, *F 19653.* Elegante Crater, *F 19949.*

LINANTHUS

Annual or perennial herbs, mostly slender and small. Leaves opposite or the upper ones sometimes alternate, mostly palmately parted to trifid, rarely simple and linear. Flowers diurnal or nocturnal. Pollen orange, yellow, or white. Seeds 1 to many per chamber, often mucilaginous when wet. Western North America and Chile, greatest diversity in California; 35 species. Reference: Wherry 1961.

Linanthus bigelovii (A. Gray) Greene [*Gilia bigelovii* A. Gray] Stems very slender, erect, simple to few-branched or sometimes much-branched with sufficient soil moisture, 4–20 cm. Herbage and calyces glabrous or with glandular hairs. Leaves opposite, simple, narrowly linear, (1) 1.5–5.0 cm ×

0.4–0.8 mm, the margins entire. (Elsewhere this species may have leaves with 3 linear segments cleft nearly to the base.) Flowers solitary or in few-flowered cymes, mostly sessile. Calyx (3.3) 5.0–6.5 mm, membranous with green midribs and lobes, the lobes slightly elongating in fruit. Corollas white, the lobes 4–7+ mm. Anthers deep inside the corolla tube, the portion of tube surrounding anthers red-purple; pollen and anthers yellow-orange. Capsules (3.5) 4.5–6.0 mm. Seeds more or less kidney-shaped, lumpy and wrinkled, orange-brown, 0.7–1.3 mm.

Dunes, granitic ranges, Pinacate volcanic field, and Sonoyta region: lower dunes, interdune troughs, sand flats, bajadas, desert pavements, and scattered on rocky slopes at various elevations, sometimes common on cinder soils at higher elevations in the Sierra Pinacate. Southeastern California to southwestern Utah and western Texas, and southward to Baja California and Sonora to the vicinity of El Desemboque San Ignacio and Hermosillo.

The flowers open shortly after sunset, after the oenotheras, and close before dawn. They produce a strong, rather overpowering perfume, sometimes sweet and sometimes smelling like stale urine.

Sierra del Rosario, *F 20710*. Sykes Crater, *F 18994*. Moon Crater, *F 19245*. Hornaday Mts, *Burgess 6799*. S of Pinacate Peak, 875 m, *F 19339*. 8 km NNE of Elegante Crater, *F 88-260*.

LOESELIASTRUM

Compact annuals, often much-branched. Leaves alternate, simple, pinnately bristle-toothed. Flowers in dense leafy-bracted clusters. Calyx segments bristle-tipped. Corollas bilateral. Anthers and pollen yellow. Valves of capsules separating completely at maturity. Seeds mucilaginous when wet. Desert and semidesert in southwestern United States, Baja California, and northwestern Sonora; 2 species.

Loeseliastrum schottii (Torrey) Timbrook [*Navarretia schottii* Torrey. *Langloisia schottii* (Torrey) Greene] Plants 2.5–8+ cm, often with a short main axis and compact to spreading branches. Leaves broadly linear, 1–2 cm; marginal bristles 1.5–3.5 mm, single. Corollas 8–9 mm, moderately 2-lipped, white to pink and showy. Seeds 1 mm.

Sand and gravelly soils; broad washes and floodplains, flats, low dunes, and interdune troughs in the Gran Desierto west of the Pinacate volcanic region and eastward to the Río Sonoyta drainage east of Puerto Peñasco; not known elsewhere in Sonora. Also deserts in southern California, southern Nevada, southwestern Utah, western Arizona, and Baja California.

N of Sierra del Rosario, *F 20769*. 10 mi N of El Golfo, *F 76-65*. Rd to Microondas La Lava, *F 92-299*. 6 km W of MacDougal Crater, *F 92-347*. Río Sonoyta, 18 mi NE of Puerto Peñasco, *F 91-51C*.

POLYGONACEAE BUCKWHEAT FAMILY

Annual or perennial herbs, shrubs, lianas, and large tropical trees. Stems often swollen at nodes with fused stipules enclosing (sheathing) the stem or the stipules reduced or absent. Leaves simple, alternate, or occasionally opposite or whorled, usually entire. Flowers mostly small, radial, bisexual, or sometimes unisexual. Calyx in 1 or 2 series, often petal-like; petals none. Pistil solitary, superior. Fruits 1-seeded, an achene (ours) or sometimes fleshy.

Mostly north temperate, also in the tropics, well developed in many desert or semiarid temperate regions; 50 genera, 1150 species. The largest genera are *Eriogonum* (240 species), *Polygonum* (200), *Rumex* (150), and *Coccoloba* (125). *Coccoloba* barely enters the edge of the Sonoran Desert near Guaymas; the others are widely distributed in the Sonoran Desert. The family includes buckwheat (*Fagopyrum esculentum*), rhubarb (*Rheum*), and timber trees (e.g., *Triplaris*). Queen's wreath or San Miguelito (*Antigonon leptopus*), native in the southern part of the Sonoran Desert, is locally cultivated as an ornamental vine for its fast growth and showy pink bracts and flowers. References: Reveal 1978, 1989a, b.

1. Stipules evident and sheathing the stems.
   2. Leaves sessile or nearly so; calyx 1.5–3 mm, the margins entire and not winged. ____ **Polygonum**
   2′ Petioles well developed, 1.8–23 cm; calyx 4+ mm, the margins winged or with several prominent spines. ____ **Rumex**

1′ Stipules absent.
  3. Teeth of involucres spine-tipped; involucres single-flowered. ____ **Chorizanthe**
  3′ Teeth or bracts of involucres blunt, not spine-tipped.
    4. Flowers not in woolly headlike clusters, subtended by involucres of bracts united to form a tube; stamens 9; ephemerals to woody perennials. ____ **Eriogonum**
    4′ Flowers in woolly headlike clusters subtended by separate bracts; stamens 3; spring ephemerals. ____ **Nemacaulis**

CHORIZANTHE SPINEFLOWER

Unusual ephemeral to perennial herbs. Leaves entire, mostly basal, some also alternate along stem; stipules none. Involucres usually with 3 or 6 spine-tipped teeth, these often recurved; usually dimorphic. (The few involucres in the lower axils are solitary, larger, and less modified than those in the upper axils, and are often not used for taxonomic diagnosis.) Flowers 1 per involucre, the perianth barely or not protruding from the involucre. Stamens often 9, sometimes 6 or 3. Achenes 3-angled.

Arid and semiarid western North America and Chile, mostly in California; 60 species. The North American species are annuals with a taproot and leaves in a basal rosette, whereas the South American species are mostly perennials. The 3 northwestern Sonora species are in section *Acanthogonum* of subgenus *Amphietes;* this section includes 39 species, mostly in California.

Ours are small spring ephemerals with minute flowers. The larger leaves are soon deciduous, usually withering by early to mid-March. Flowering mostly February-March, the plants mostly mature in April. By the time they are in full flower with onset of fruiting, the weather is warm and the plants are leafless or nearly so. The 3 species in the flora area often occur together. References: Goodman 1934; Reveal & Hardham 1989.

1. Stems stout, obscured by spines, not breaking apart, the dried plants (skeletons) tough and persistent; involucral tube as wide as long. ____ **C. rigida**
1′ Stems slender, not spiny, at maturity fragile and completely breaking apart; involucral tube longer than wide.
  2. Involucre 6-toothed, the teeth 1 mm; leaves narrowly oblanceolate. ____ **C. brevicornu**
  2′ Involucre 3-parted, the teeth (segments) 2.0–7.5 mm; leaves broadly ovate to orbicular. ____ **C. corrugata**

Chorizanthe brevicornu Torrey subsp. **brevicornu.** BRITTLE SPINEFLOWER. Plants (3) 5–25 cm tall, often becoming as wide or wider than tall. Stems ascending, much-branched above (except when stunted); lower stems often reddish green; upper stems, inflorescence branches, and involucres yellowish green. Internodes of fresh, young growth slightly swollen just below nodes (reminiscent of a miniature inflated stem of *Eriogonum inflatum* or *E. trichopes*). Dry mature stems brittle, breaking apart at nodes. Herbage with appressed white hairs. Basal rosette leaves (1.0) 1.8–8.0 cm × (1.0) 1.3–5.0 mm, narrowly oblanceolate without a distinct petiole, semisucculent on young robust plants, very quickly withering; stem leaves reduced above.

Involucres solitary at nodes, each subtended by a pair of recurved (hooked) spinose bracts 2–3 mm. Involucral tube straight or slightly curved (even on the same plant), 3.5–4.0 mm, somewhat cylindrical but 3-angled with 6 prominent ribs, each rib extending into a 1 mm recurved spinose tooth. Perianth white, barely protruding from the involucre. Stamens 3. Achene cylindrical, tightly enclosed by and shorter than the involucre.

Widespread across the flora area and to peak elevation in the Sierra Pinacate; not on dunes. Sandy, gravelly, or rocky soils, especially common on desert pavements and cinder flats and slopes. Mojave, Sonoran, and Great Basin Deserts from Mono County, California, to southwestern Utah, and southward to Baja California and northwestern Sonora at least to the vicinity of El Desemboque San Ignacio. Subsp. *spathulata* (Small ex Rydberg) Munz has broader leaves; it replaces subsp. *brevicornu* at elevations and latitudes above the desert from east-central California to Idaho and southeastern Oregon. The geographic range is the largest of any species in the genus.

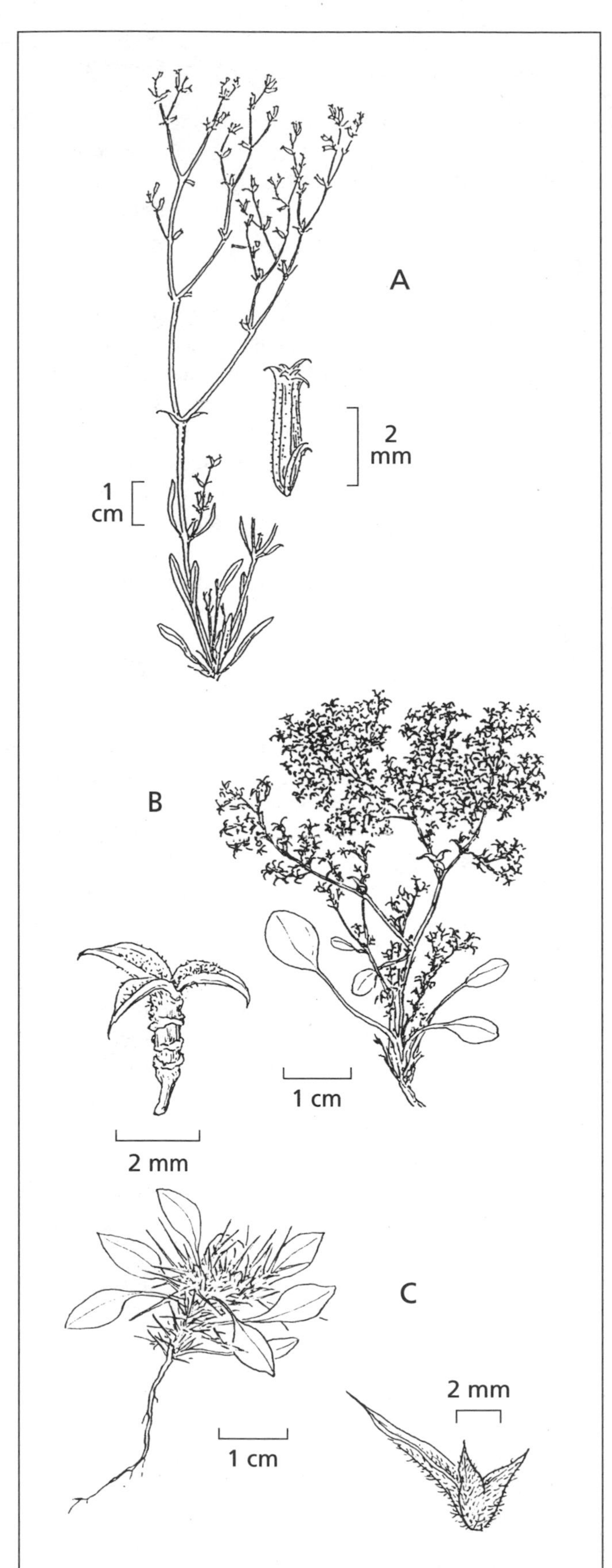

Chorizanthe: ***(A) C. brevicornu*** subsp. ***brevicornu;*** ***(B) C. corrugata; (C) C. rigida.*** ER (Hickman 1993).

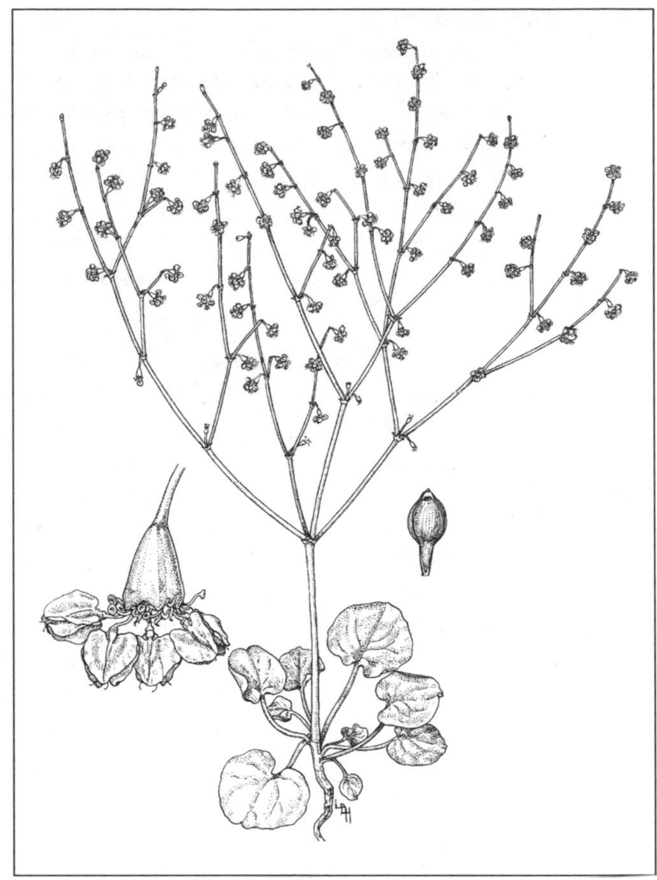

Eriogonum deflexum. LBH (Parker 1958).

Eriogonum wrightii. BA. © James Henrickson 1999.

Eriogonum deserticola. RSF.

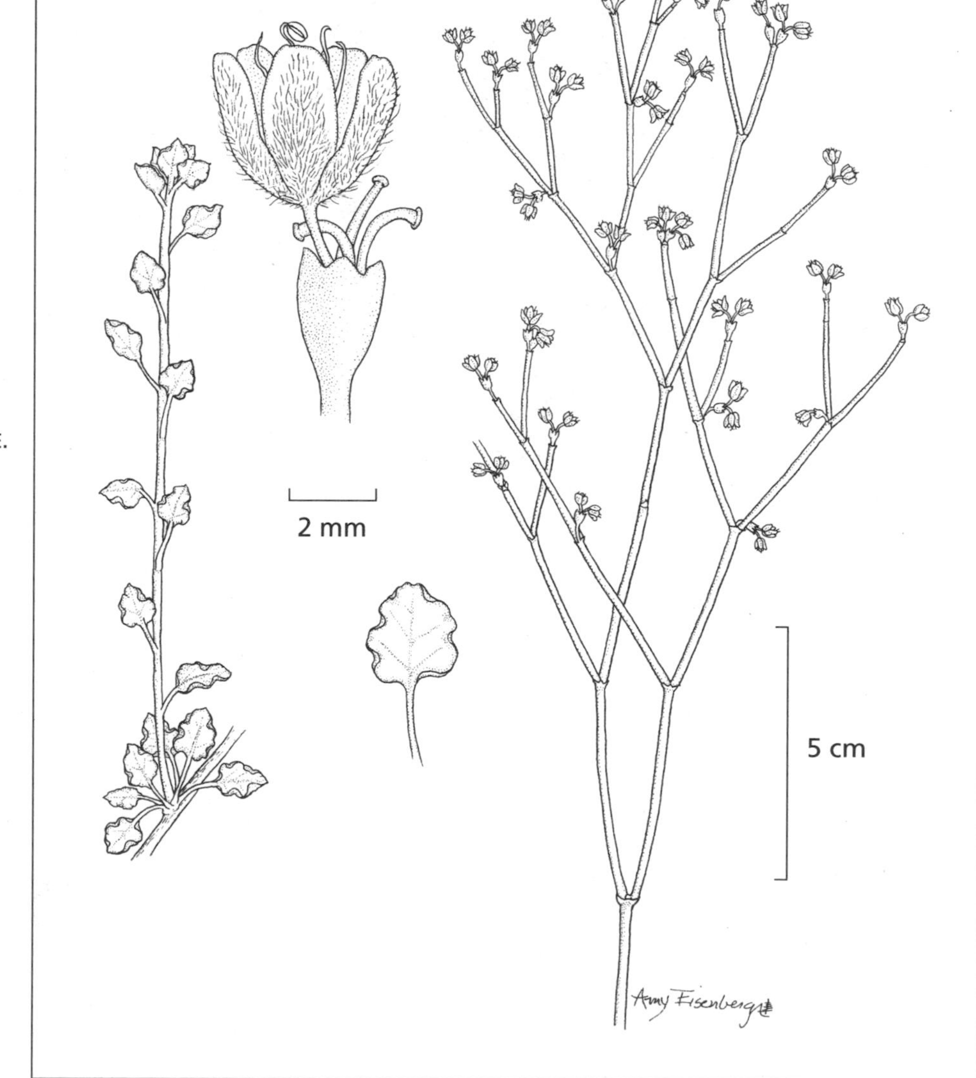

Eriogonum deserticola. AE.

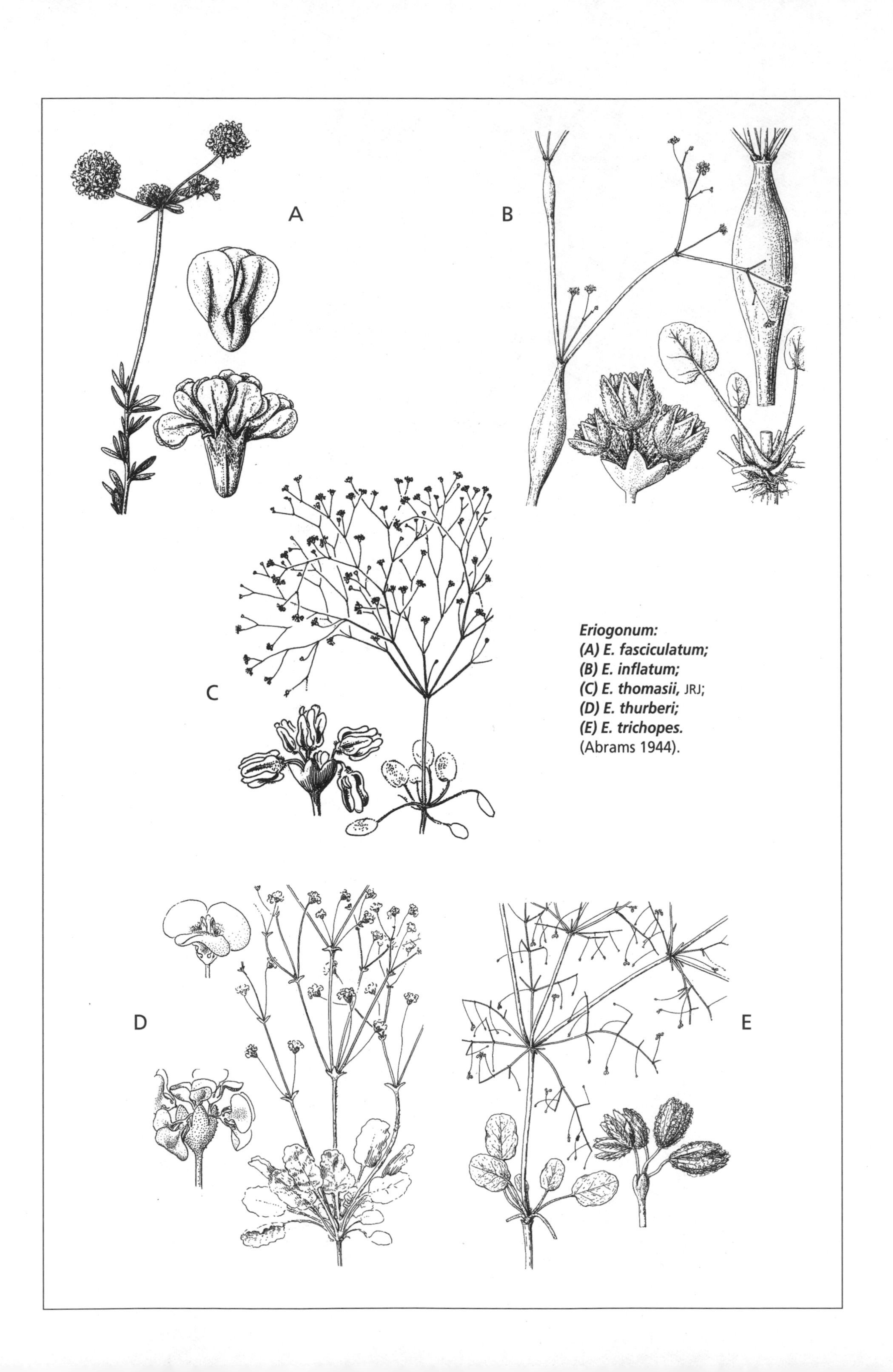

Eriogonum:
(A) E. fasciculatum;
(B) E. inflatum;
(C) E. thomasii, JRJ;
(D) E. thurberi;
(E) E. trichopes.
(Abrams 1944).

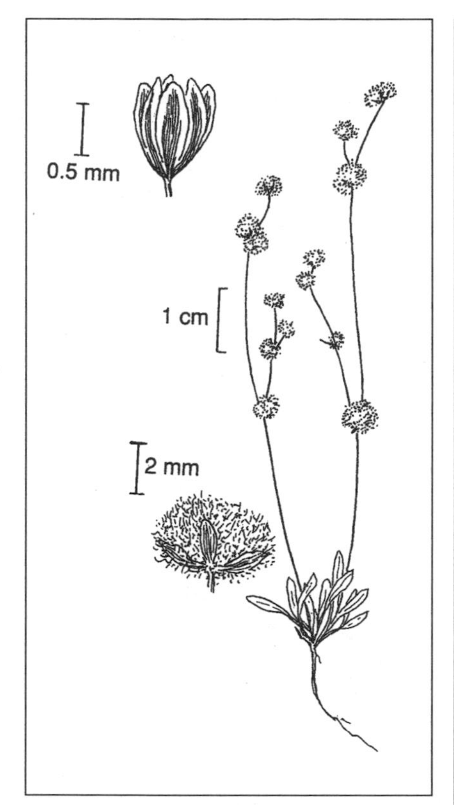

Nemacaulis denudata var. ***gracilis.***
ER (Hickman 1993).

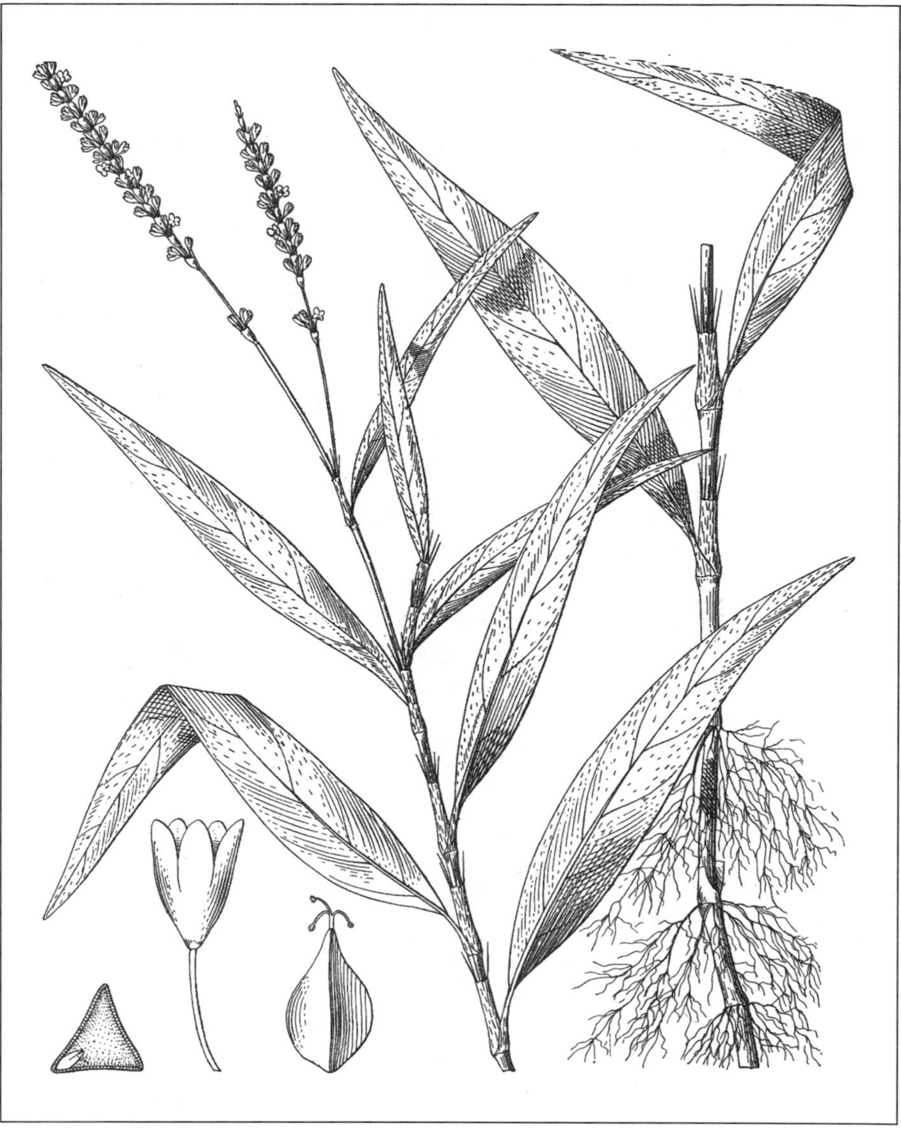

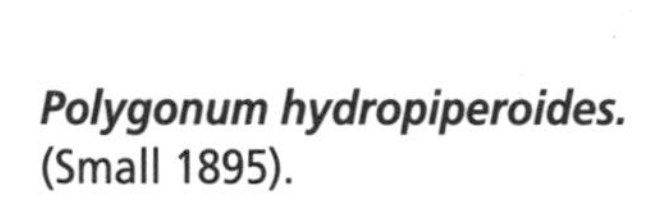

Polygonum hydropiperoides.
(Small 1895).

The dry mature plants completely break apart and "disappear" during April or May. The propagules consist of stem pieces with 1–3 (or more) internodes and an involucre with its achene. The rigid, sharp recurved teeth at both the base and tip of the involucres act like miniature grappling hooks, readily hooking onto skin, clothing, fur, or feathers. Apparently the seed germinates while enclosed in the involucre. There seems to be weak dimorphism in size and weight of disseminules. The lower stems, which are stouter and have longer internodes than the upper stems, produce stouter involucres in their axils than do the upper-stem axils.

Sierra del Rosario, *F 20686.* Cerro Pinto, *F 89-54.* S of Pinacate Junction, *F 19666.* Sykes Crater, *F 18936.* SE of Moon Crater, *F 18842.* SE of Carnegie Peak, 850 m, *F 19924.* Elegante Crater, *F 19671.*

Chorizanthe corrugata (Torrey) Torrey & A. Gray [*Acanthogonum corrugatum* Torrey] WRINKLED SPINEFLOWER. Plants (1.5) 3–15 cm, compact, intricately much-branched above, as wide or wider than tall. Stems densely white hairy below, often glabrate above and with age. Leaves of basal rosette densely white woolly, conspicuously petioled, the blades (5) 8–30 mm, broadly ovate to orbicular; stem leaves reduced above. Involucres at flowering time densely crowded, yellow-green, hat-shaped, drying red-brown, firm, and very quickly disarticulating. Involucral tube 2.0–3.7 mm, corrugated with transverse wrinkles, cylindrical, with 3 lobes (teeth) 2.0–7.5 mm, the tips recurved and spinose; lobes of lower involucres markedly larger and broadly elliptic to ovate, those of upper involucres smaller and narrower. Perianth white, barely protruding from the involucral tube. Stamens 6. Achene ca. 2.5 × 0.4 mm, cylindrical, smooth but not shiny, tightly enclosed in the hard involucral tube, the tip 3-angled and minutely tuberculate; seed disseminated with the involucre and apparently germinating while enclosed in it.

Desert pavements and gravelly soils, usually less common on rocky slopes and sand flats. Lower elevations in the Pinacate volcanic region and granitic ranges. Generally aggregated in small, isolated colonies. Many or perhaps most of the plants germinate in close proximity to their parent plant. Southeastern California, western Arizona, northeastern Baja California, and northwestern Sonora southward to the vicinity of Caborca and Sierra Bacha (29°43′ N, south of Libertad).

NW of Cerro Colorado, *F 19638*. S of Pinacate Junction, *F 20619*. MacDougal Crater, *Turner 86-28*. Sierra Extraña, *F 19050*. Sierra del Rosario, *F 20752*.

Chorizanthe rigida (Torrey) Torrey & A. Gray [*Acanthogonum rigidum* Torrey] RIGID SPINEFLOWER, SPINE-HERB. Plants (1) 2–10 cm with stout stems soon becoming rigid and obscured by crowded, stiff, straight spines. Larger leaves in a basal rosette as well as along the stem, the leaves soon withering; leaves petioled, the blades 6–32 mm, ovate to orbicular, densely white woolly below, usually less densely to sparsely white hairy above. Bracts spinelike, (4) 7–26 mm. Involucres 3-angled and 3-toothed or segmented, the tube 2 mm, about as wide as long and reticulated, the segments triangular to leaflike, unequal, the longest one 4–20 mm. Perianth yellow, scarcely protruding. Stamens 9. Achenes 1.4–1.5 mm, shiny red-brown, strongly 3-angled or lobed and beaked. The dead, dry skeletons may persist for several years, resembling miniature ocotillos.

Nearly throughout the region including the Sonoyta region, Pinacate volcanic field, granitic ranges, and desert flats; absent from dunes. Rocky, sandy, and gravelly soils, and especially common on desert pavements. South in Sonora to Isla Tiburón and the opposite mainland; also Baja California, Arizona, southern Nevada, southwestern Utah, and southeastern California.

Several plants 5 cm tall had taproots more than 25 cm deep. The size, length and width, weight, and even the shape of the involucral segments are highly variable, resulting in different dispersal characteristics. The seed is held tightly in the involucre, even on dry dead plants more than a year old. Seedlings often germinate at the base of old dry plants that seem to serve as nurse plants.

S of Pinacate Junction, *F 20616*. S of Pinacate Peak, 875 m, *F 19350*. S of Moon Crater, *F 19106B*. Hornaday Mts, *Burgess 6818*. Sierra del Rosario, *F 20778*. 19 mi E of San Luis, *F 16691*.

ERIOGONUM WILD BUCKWHEAT

Annual or perennial herbs and shrubs. Leaves basal or along stems, alternate to whorled, the margins entire or sometimes wavy or scalloped; without stipules. Flowers in involucres, these with 4–10 teeth or lobes (ours 4- or 5-toothed or -lobed). Sepals (tepals) petal-like, in 2 series of 3 segments each. Stamens 9; styles 3. Achenes mostly 3-winged or 3-angled.

A diverse North American genus mainly in western North America; 240 species. References: Reveal 1968, 1969, 1970, 1976.

1. Shrubs, at least the lower stems definitely woody; branches leafy (often drought deciduous), the leaves not in basal rosettes.
    2. Sparse, spindly shrubs often more than 1 m; leaf blades mostly about as wide as long; involucres 4-lobed, the flowers yellow; sand soils and dunes. ______ **E. deserticola**
    2′ Much-branched shrubs usually 1 m or less; leaf blades at least twice as long as wide; involucres 5-lobed or 5-toothed, the flowers white to pink; rocky habitats.
        3. Involucres (flowers) in dense, headlike clusters on well-developed peduncles; involucres 2.5 mm. ______ **E. fasciculatum**
        3′ Involucres solitary and sessile at nodes (flowering branch may appear racemose); involucres 1.0–2.0 mm. ______ **E. wrightii**

1′ Ephemerals or perennial herbs, the stems not leafy, the leaves in basal rosettes.
  4. Annuals or perennials; leaves hairy but not woolly; involucres 4- or 5-lobed.
    5. Perennials and flowering in first year; involucres 5-lobed; tepals (1.5) 2.0–2.7 (3.5) mm. ____ **E. inflatum**
    5′ Annuals; involucres 4-lobed; tepals 1.2–1.8 mm. ____ **E. trichopes**
  4′ Annuals; leaves woolly, at least below; involucres 5-lobed.
    6. Outer tepals 1.8–2.3 mm, glabrous on outside. ____ **E. deflexum**
    6′ Outer tepals 1.2–1.9 mm, with hairs on outside, some of them glandular.
      7. Involucres 0.8–2.3 mm; outer tepals longer than wide, broadest and swollen (inflated) at base. ____ **E. thomasii**
      7′ Involucres 2.0–3.2 mm; outer tepals as wide as or wider than long, broadest above middle, narrowed at base to a claw. ____ **E. thurberi**

Eriogonum deflexum Torrey var. **deflexum.** SKELETON WEED. Ephemerals to annuals, often 5–50+ cm, with a stout well-developed deep taproot. Leaves in a basal rosette, 1.5–5.0 cm, soon deciduous, the petioles prominent and usually longer than the blades, the blades orbicular, often cordate, especially the larger leaves; lower leaf-blade surfaces densely grayish-white woolly, the upper surfaces densely to sparsely woolly and often darker green. Flowering stems much-branched, especially above, forming more or less flat-topped inflorescences, the stems leafless, glaucous, the larger ones relatively thick but not swollen. Involucres, pedicels, and outer tepal lobes glabrous. Involucres 5-lobed, 1.5–2.2 mm, the pedicels 0.5–10 mm. Perianth white to pink, not swollen, the outer segments 1.8–2.3 mm, orbicular to broadly elliptic, cordate, blunt at tip. Often flowering late spring and sometimes into early summer.

Sonoyta region through most of the Pinacate volcanic region to peak elevation. Not found in the southernmost portions of the Pinacate field nor in the western granitic ranges.

The plants usually dry up by early summer, but some may survive the summer and flower through fall. It is often the only annual flowering or surviving after really hot weather sets in by mid-May. The deep taproot and leafless condition probably contribute to its ability to withstand summer heat and drought.

This polytypic species occurs from eastern California to Utah and south through Arizona to extreme western New Mexico, Baja California, and northern Sonora. The half-dozen or so varieties are more or less geographically segregated. Var. *deflexum* occupies much of the range of the species, in the Mojave and Sonoran Deserts in southeastern California, Nevada, northern Baja California, northwestern Sonora, southern Nevada, southern Utah, and Arizona. Eastward to southwestern New Mexico and in northern Sonora it is replaced by the weakly defined var. *turbinatum* (Small) Reveal, which generally has long peduncles bearing numerous flowers (Reveal 1989b:363). Some specimens from the Sonoyta region approach var. *turbinatum.*

W of Sonoyta, *F 85-952.* NW of Cerro Colorado, *F 19633.* Pinacate Junction, *F 19608.* Tinajas de Emilia, *F 19725.* Pinacate Peak, *F 19444.* Sykes Crater, *F 20000.*

Eriogonum deserticola S. Watson. DESERT BUCKWHEAT, GIANT DUNE BUCKWHEAT. Sparsely branched, spindly shrubs (1.0) 1.5–2.2 m, often with a slender "trunk." Herbage densely white woolly, the stems glabrate with age. Leaves (1.0) 1.5–3.6 cm, the blades mostly nearly orbicular, somewhat truncate to broadly cuneate at the base, thickish, the margins crenulate to crenulate-lobed; petioles mostly longer than the blades, winged, the wings gradually narrowed basally. Inflorescences open and relatively sparsely flowered. Involucres minutely 4-lobed, 1.4–3.0 mm, glabrous or sparsely to densely white hairy, sessile to pedicelled, (5) 6–11-flowered. Perianth segments oblong, the tips rounded, the midstripe broad, green to red-brown, and densely white hairy, the margins broad, relatively thin, bright yellow, mostly glabrous; outer segments 3.6–4.4 mm. Filaments similarly bright yellow, the anthers cream colored. Achenes smooth, red-brown, 4.3 mm, the body more or less lens-shaped, the beak prominent, 1.0–1.2 mm. Various seasons including April and September-December.

Widespread and often abundant on low to high shifting dunes and sand flats with small hillocks across the Gran Desierto; not on coastal dunes. The spindly trunks take on grotesque shapes as sand blows out from beneath the crown. The exposed lateral roots sometimes extend 6+ m. Essentially endemic to the Gran Desierto dunes in Sonora, extreme southern Yuma County, Arizona, and the Algodones Dunes in southeastern California.

This species seems to be taxonomically isolated with no obvious close relatives. It shares interesting features with some members of the *E. corymbosum* Bentham complex. Following Reveal's work, Munz (1968) placed *E. deserticola* next to *E. plumatella* Durand & Hilgard, a Mojave Desert species of northwestern Arizona and adjacent California.

13.8 mi NE of Puerto Peñasco on Mex 8, then 7.6 mi by rd W to shifting dunes, *F 88-228*. S of Moon Crater, *F 19016*. 11 mi SE of López Collada, *F 86-527*. N of Sierra del Rosario, *F 75-45*.

Eriogonum fasciculatum Bentham. MOJAVE BUCKWHEAT. Low spreading woody shrubs 0.5–0.8 (1.0) m, often compact and much-branched with leafy stems and shredding bark, the branches woody. Leaves fascicled, sessile or nearly so, linear to narrowly oblanceolate, white woolly below, glabrous or canescent above, the margins mostly revolute, the margins and midrib on lower surface thick and prominent. Involucres 5-toothed, 2.5 mm. Flowers crowded into compact, often clustered heads on long, leafless stems (scapes).

Utah to California, Arizona, northwestern Sonora, and Baja California. Several varieties (or subspecies) mostly in California. Two in the flora area:

Eriogonum fasciculatum var. **fasciculatum.** Similar to var. *polifolium* but differing primarily in having glabrous or less hairy parts: First-season or first-year stems, inflorescence axes (scapes), and peduncles sparsely hairy or sometimes glabrous. Leaves 4–20 mm, relatively dark green, drying red-brown, the upper surfaces glabrous or occasionally sparsely hairy on some leaves of a plant. Involucres 2–4 mm, glabrous or sparsely hairy. Flowers 2.5–3.4 mm, pink, the outer perianth segments glabrous or sparsely hairy, somewhat white hairy at base only. At least in spring.

Coastal granitic hills at Puerto Peñasco and Bahía de la Cholla, mostly on north-facing slopes. This variety, not known elsewhere in Sonora, has its main area of distribution along the Pacific side of Baja California and southern California. Ours are slightly more pubescent in some of their parts than are plants of the California populations but otherwise fit well into var. *fasciculatum*. It may be that the Sonoran population is more closely related to nearby *polifolium* populations than to the coastal California variety. The glabrous aspect seems to be an adaptation to a coastal habitat. Otherwise, it is an interesting disjunction or possibly a Pleistocene relict.

Puerto Peñasco, *Dennis 38*. Bahía de la Cholla: *Bittman 420, F 13153, F 16821*.

Eriogonum fasciculatum var. **polifolium** (Bentham) Torrey & A. Gray [*E. polifolium* Bentham. *E. fasciculatum* subsp. *polifolium* (Bentham) S. Stokes] First-year stems, inflorescence branches, and peduncles conspicuously pubescent (canescent). Leaves 3–15 (18) mm, white hairy on both surfaces, canescent above, densely woolly below. Involucres 5-toothed, 2.5 mm, densely to moderately white hairy. Flowers white to pink, 3.0 (3.5) mm, the outer perianth segments densely white hairy mostly toward the base and along the broad midrib. Achenes probably 2.0–2.5 mm (as reported for Arizona by Reveal 1976). Spring and fall.

Granitic mountains and Pinacate volcanic field; mostly on north- and east-facing slopes, most common toward higher elevations and in steep canyons, absent from the lowlands and smaller mountains. East-central California to southwestern Utah and south to western and southern Arizona, Baja California, and northern Sonora. This is the most common shrubby *Eriogonum* in northern Sonora. The only other shrubby *Eriogonum* with pubescent flowers in the Sonoran Desert and Arizona is *E. deserticola*.

Sierra del Viejo, *F 16900*. Sierra del Rosario, *F 20383*. E side Sierra Pinacate, 860 m, *Ezcurra 16 Apr 1981* (MEXU).

Eriogonum inflatum Torrey & Frémont. DESERT TRUMPET Herbaceous perennials, also flowering in first season. Rootstock and caudex usually well developed and woody, the stems often many. Stems glaucous to greenish; the first internode usually erect or nearly so, often (7) 9–31 cm, the upper part inflated (swollen) or not with no apparent pattern. First node commonly producing 2 or 3 (4–6) branches. Leaves basal or nearly so, green to reddish green, soon withering, the dry leaves semipersistent, the blades oblong to orbicular or kidney-shaped with wavy margins; petioles prominent. Involucres 5-toothed, 1.2–2.0 mm. Flowers yellow with red to green bases; tepals mostly 2.0–2.7 mm (as short as 1.5 at early anthesis to 3.5 mm as fruit develops), covered with whitish-translucent fleshy curved to curled hairs, the margins glabrous and bright yellow. At least spring and summer.

Widespread; upper bajadas, hills and mountains, mostly on arid rocky slopes and canyons; Sonoyta region, Pinacate volcanic field, and granitic ranges. One of the latest-flowering herbaceous plants during late-spring drought. Deserts; Baja California Sur to eastern California and eastward to Colorado and much of Arizona, and northwestern Sonora southward to the vicinity of El Desemboque San Ignacio.

Plants with inflated stems are known as var. *inflatum,* whereas var. *deflatum* I.M. Johnston has been applied to those without inflated stems. This species somewhat resembles *E. trichopes* but is readily distinguished by its perennial habit, larger flowers, 5-lobed involucres, red-tinged flowers, mostly erect or ascending stems, plants often taller than wide (rather than broad and spreading), and habitat differences. *E. inflatum* usually occurs on rocky, gravelly, or cinder soils, whereas *E. trichopes* generally occurs on sandy soils. They do not occur together.

Sierra del Rosario, *F 20697.* Sierra del Viejo, *F 16906.* Desert pavement, 32°05′ N, 113°41′ W, 340 m, *Ezcurra 29 Apr 1981* (MEXU). Campo Rojo, *Ezcurra 16 Apr 1981* (MEXU). Elegante Crater, *F 19693.* MacDougal Crater, *F 10514.*

Eriogonum thomasii Torrey. Delicate and often diminutive winter-spring ephemerals, mostly (4) 8–15 cm (robust, well-watered plants to 25+ cm). Leaves in a basal rosette, densely white woolly, the blades about as wide as long, circular to kidney-shaped, upper surfaces not as hairy and often greener than lower surfaces, the petioles prominent. Flowering stems 1, or sometimes several to many, much-branched above, the branches very slender and glabrous. Involucres 5-lobed, glabrous, 0.8–1.2 mm. Flowers at first yellow, becoming pink below and white above, 1.5–1.9 mm, with age the outer tepals becoming swollen at base, longer than wide, with thickish, short, and stout minute glandular hairs below, otherwise glabrous.

Widespread, often locally common; Sonoyta region, desert flats, Pinacate volcanic field to peak elevation, and granitic ranges including Sierra del Rosario. Rocky slopes, desert pavements, gravelly washes, sandy flats, cinder flats and slopes; not on dunes. Southward in Sonora to the vicinity of Caborca and El Desemboque San Ignacio. Also western Arizona, Baja California, southeastern California, southern Nevada, and southwestern Utah.

20 mi E of San Luis, *Bezy 476.* Sierra del Rosario, *F 20725.* Moon Crater, *F 19246.* Sykes Crater, *F 18929.* Pinacate Peak, 1100 m, *F 19396.* Elegante Crater, *F 19684.* NW of Cerro Colorado, *F 19637.*

Eriogonum thurberi Torrey. Diminutive winter-spring ephemerals about the same size as and similar in general appearance to *E. thomasii.* Leaf blades ovate, broadly elliptic, or nearly orbicular, white woolly, sometimes moderately woolly and greenish, the margins nearly entire to crenate-lobed, the petioles prominent. Peduncles and involucres densely glandular hairy, the hairs (stalks) whitish, the glands pinkish. Involucres 5-lobed, 2.0–3.2 mm. Flowers 1.2–1.3 mm, about as wide as long, the outer tepals narrowed at the base to a claw, white to pink.

Sandy-gravelly bajadas, washes, and roadsides in the Sonoyta region. Sonora southward to Puerto Libertad, southern and western Arizona, southern California, and Baja California.

Sierra Cipriano, *F 88-216.* 18 mi S[W] of Sonoyta, *Shreve 7580.* Puerto Libertad, *Bowers 1622.*

Eriogonum trichopes Torrey var. **trichopes.** YELLOW TRUMPET. Winter-spring or spring–early summer ephemerals, rarely surviving through the summer and flowering in fall. Plants usually broader than tall, 1 to several main stems, much-branched above into very slender stems, extremely variable in size depending on soil moisture, (ca. 10) 30–50 (100+) cm across. Stems yellow-green, glabrous, sometimes glaucous, the upper part of the first internode or larger branches sometimes inflated (especially on vigorously growing plants), the first internode usually erect or nearly so, (2) 5–10 cm, the lateral branches whorled at most nodes, spreading nearly at right angles, the lower 1 to several nodes commonly producing 3 to ca. 17 branches, and even the upper nodes usually whorled. Leaves coarsely hairy, the larger leaves in a basal rosette; blades broadly oblong to circular, (0.5) 1.5–5.5 cm, the margins crenate to nearly entire; petioles (1) 4–13 cm.

Involucres glabrous, 4-lobed, 1.1 mm, on slender-elongate spreading pedicels. Flowers yellow to greenish yellow, occasionally red-tinged; tepals 1.2–1.8 mm, the outer tepals covered with whitish-translucent fleshy curled hairs. Usually March–late April.

Seasonally abundant on sandy plains of open desert, gravelly washes, and on crater floors including alluvium and playa sediments; Sonoyta region, lower elevations of the Pinacate volcanic field, bajadas of the granitic ranges, desert flats, and less common on low dunes. Southward in Sonora to the Infiernillo coast and the vicinity of Hermosillo; also Baja California and southeastern California to southern Nevada, southwestern Utah, Arizona, and southwestern New Mexico. Var. *hooveri* Reveal, also an annual, occurs along the California coast.

Dry plants, especially larger ones, often become tumbleweeds. The thick taproot snaps off just below the soil surface. Although many seeds will already have been shed, substantial numbers remain, providing two seed-dispersal strategies. In early September 1992, I watched dust devils lift these tumbleweeds into the sky until they were no longer visible.

As with *E. inflatum,* the basal leaves are quickly drought deciduous, and flowering plants are essentially leafless. *E. trichopes* and *E. inflatum* are closely related but clearly distinct, with well-marked morphological gaps and different habitats; *E. trichopes* stems are more slender and yellowish rather than glaucous green, the plants are ephemerals rather than perennials, and they are much more densely branched with spreading rather than erect-ascending branches.

Sierra del Rosario, *F 20773.* MacDougal Crater, *Turner 86-17.* Hornaday Mts, *Burgess 6796.* Moon Crater, *F 19243.* 11.5 mi S of Tinajas de los Pápagos, *F 18912.* N of Puerto Peñasco, *F 13180.*

Eriogonum wrightii Torrey ex Bentham var. **pringlei** (J.M. Coulter & E.M. Fisher) Reveal [*E. pringlei* J.M. Coulter & E.M. Fisher] Messy-looking much-branched small shrubs to ca. 1 m, the bark shredding, the twigs slender and mostly straight. Herbage densely white woolly. Leaves readily drought deciduous, the plants essentially leafless or with relatively few and reduced leaves in dry seasons; larger leaves on primary shoots 8–25 mm; leaf blades narrowly to broadly lanceolate or elliptic, the upper surfaces often more sparsely woolly than lower surfaces, gradually narrowed at the base into a sometimes prominent petiole, the smaller leaves sessile. Involucres 5-ribbed and 5-toothed, solitary in axils (the floriferous stems appear racemose), more or less cylindrical, 1.0–1.7 mm (to 2.0 mm at higher elevations in the Sierra Pinacate), the perianth white to pink. Achenes short-beaked and ovoid, about as long as the involucre. Various seasons but perhaps not during summer.

Pinacate volcanic region and granitic ranges, often locally abundant; canyons and rocky slopes, mostly with northern exposures and especially common at higher elevations, but sometimes (e.g., Sierra del Rosario) even in west-facing canyons. A Sonoran Desert endemic; northwestern Sonora including the Sierra Cubabi and south-central and southwestern Arizona.

This species has 11 varieties, from northern California to the central Baja California Peninsula, and from southern Nevada and Arizona to west Texas and central Mexico. Var. *pringlei* is distinguished by its smaller involucres, flowers, and achenes, and usually fewer and smaller leaves.

Cerro Pinto, *F 13214.* Sierra del Rosario, *F 75-33.* Campo Rojo, *Ezcurra 16 Apr 1981* (MEXU). Carnegie Peak, 850 m, *F 19928.* N of Pinacate Peak, 950 m, *F 86-434.* S end of Sierra Cubabi, *F 16725.*

NEMACAULIS

The genus has 1 species. It differs from *Eriogonum* in having 3 rather than 9 stamens, and the involucral bracts separate rather than united into a tube. Reference: Reveal & Ertter 1980.

Nemacaulis denudata Nuttall var. **gracilis** Goodman & L.D. Benson. WOOLLY HEADS. Delicate spring ephemerals; erect to ascending and spreading, often as wide or wider than tall, with a well-developed taproot. Stems mostly (7) 12–25 cm, slender, and moderately white hairy below, threadlike and glabrous above. Leaves, nodes, bracts, and involucres densely white woolly; hairs of involucres often turning brown with age. Leaves mostly in a basal rosette, (1.5) 2.6–5.0 (6.5+) cm × 2.6–5.5 (10) mm, narrowly oblanceolate, often reddish beneath the dense white hairs, turning green or greenish red when wet (the hairs instantly absorb water to become appressed and transparent), the midrib prominent; margins mostly undulate to sometimes entire. Stem leaves few and reduced or absent; flowering stems usually much-branched and diffuse, the lower internodes relatively long.

Inflorescence nodes each with 3 bracts subtending a single glomerule consisting of involucral bracts in whorls of 5. Involucral bracts separate, each subtending a single flower; bracts often yellow-green when young, otherwise red with conspicuous yellow-membranous margins, the larger, outer bracts 1.9–3.0 mm. Flowers yellow, buried in wool of glomerule, 0.7–1.3 mm, on slender pedicels; perianth segments 6, united at base. Seeds 0.75–0.9 mm, smooth and shiny, dark brown to blackish, plump, shaped like a fat teardrop, the tip 3-angled. Flowering mostly March, and near the coast sometimes through May.

Sand flats and shifting dunes nearly throughout the flora area but not in the Sonoyta region. Occasionally in gravelly-sandy arroyo beds in the Pinacate volcanic field including higher elevations. Not known elsewhere in Sonora; otherwise southwestern Arizona, southern California (mostly inland deserts), Baja California, and northern Baja California Sur. Var. *denudata* occurs along the Pacific coast of southern California and Baja California, where intermediates and plants characteristic of both varieties often occur intermixed.

Puerto Peñasco, *Van Devender 31 May 1983.* Dunes 24 mi SW of Sonoyta, *Shreve 7584.* Hornaday Mts, *Burgess 6812.* S of Pinacate Peak, 875 m, *F 19359.* Moon Crater, *F 19244.* N of Sierra del Rosario, *F 20795.*

POLYGONUM KNOTWEED, SMARTWEED

Annual or perennial herbs, sometimes herbaceous vines or shrubs. Leaves alternate, simple, and entire; stipules fused, forming sheaths around the stem above the usually swollen nodes. Sepals petal-like, usually white or pink. Achenes lens-shaped or 3-angled.

Mostly in the Northern Hemisphere; 200 species. The genus is taxonomically difficult. Often weedy and/or in aquatic or wetland habitats. A few are garden plants in temperate regions. Reference: Small 1895.

1. Annuals from a single taproot; leaves mostly less than 1 cm long, the largest leaves 2.0–6.5 cm, thickish and glaucous; mostly an urban and agricultural weed, not restricted to riparian places. ________ **P. argyrocoleon**

1′ Perennials, rooting along lower nodes; leaves usually 10–16 cm, thin and green; emergent from shallow water or in wet riverine soil.

   2. Perianth greenish to light pink; spikes 2–6 mm wide. ________ **P. hydropiperoides**

   2′ Perianth deep pink; spikes 7–20 mm wide. ________ **P. persicaria**

***Polygonum argyrocoleon** Steudel ex Kunze. SILVERSHEATH KNOTWEED, PERSIAN WIREWEED. Nonseasonal annuals with a stout taproot. Stems slender and erect to spreading, with 1 to several or many branches. Stems and leaves glaucous, highly variable in size and number. Leaves sessile or nearly so, narrowly elliptic to narrowly lanceolate, mostly less than 1 cm, sometimes 2.0–6.5 cm, soon deciduous; upper leaves usually reduced to bracts, the plants leafless or nearly so in drought or dry habitats. Stipular sheaths membranous and lacerate. Flowers in small axillary clusters on terminal spikelike stems. Calyx pink, 2–3 mm. Achenes 3-angled, ca. 2 mm, dark brown, smooth, and shiny.

Río Sonoyta and major gravelly washes in the Sonoyta region, occasional in roadside depressions, margins of Río Colorado wetlands, and an agricultural weed in the Sonoyta and San Luis Valleys. Native to the Near and Middle East, now widespread and weedy in southwestern United States and northwestern Mexico.

El Papalote, *F 86-165*. Sonoyta, *F 85-709B*. Pinacate Junction, *F 19605*. Ciénega de Santa Clara, *F 92-512*. S of San Luis, *F 85-1038*.

Polygonum hydropiperoides Michaux. WATER PEPPER. Facultative annuals or herbaceous perennials, to 1+ m across, glabrate or sparsely hairy. Stems arching to reclining and leafy, often rooting at lower nodes, the nodes becoming swollen in late summer. Leaves petioled, thin, green, tapering at both ends, the larger leaves 10–16 × 1.3–2.2 cm. Stipular sheaths with upper margins ciliate or entire, even on the same plant. Flowers 2.0–2.5 mm, in terminal racemes. Sepals white to pale pink, the veins inconspicuous even at fruiting time. Styles 2-branched. Achenes 2.5–3.0 mm, somewhat lens-shaped to moderately 3-angled, shiny brown to blackish. Warmer months.

Fairly well established but not common along the Río Sonoyta in the vicinity of Sonoyta; emergent from shallow water or in wet soil at water's edge. Widespread in North America; rare in western Sonora.

Sonoyta: Río Sonoyta, *F 86-83*, Presa Derivadora, *F 86-299*.

*†**Polygonum persicaria** Linnaeus [*P. fusiforme* Greene] Similar to *P. hydropiperoides*, and part of the highly variable species complex that includes *P. lapathifolium* Linnaeus, with many named and often difficult to identify variants. Native to Euro-Siberian and Mediterranean regions and probably portions of North America, now adventive in many parts of the world. Many are selfing and with cleistogamous flowers. *P. fusiforme*, known from southern Arizona and the Colorado River in California and Arizona, is distinguished by its swollen nodes, but it does not seem to be distinct from *P. persicaria*, even at the infraspecific level. Once common along the Colorado River at least as far south as Yuma and presumably along the Mexican portion of the river.

RUMEX *CAÑAIGRE,* DOCK, SORREL

Annual or perennial herbs. Perianth with an outer whorl of 3 smaller sepals and an inner whorl with 3 larger lobes, these inner ones, termed valves, enlarging in the fruit, often variously winged or toothed, and generally with a conspicuous tubercle-like grain or warty bump called a callosity.

Worldwide, tropical to alpine tundra, mostly temperate; 200 species. Many are weeds and wetland plants in dry western North America. Some have edible leaves, e.g., French sorrel (*R. scutatus*), and others are medicinal plants and sources of tannin. Reference: Rechinger 1937.

*†**Rumex crispus** Linnaeus. *CAÑAIGRE;* CURLY DOCK. Annuals or perennials differing from *R. inconspicuus* in their usually larger basal leaves, taller stems, and valves with entire margins. Kelly (1977) reports it was one of the Río Colorado "delta food plants" harvested by the Cocopa. Also known from the Quitovac oasis (*Nabhan 258*). Reported as native to Eurasia and widely naturalized elsewhere in the world.

Rumex inconspicuus Rechinger f., *Candolla* 11:231, 1948. *CAÑAIGRE, HIERBA COLORADO;* DOCK; *WAKONDAM.* Winter-spring annuals in western Sonora with a very stout, deep taproot. Flowering stems 13.5–55+ cm, reaching 1.5 cm in diameter, erect, usually red-green and leafy. Leaf blades 6–30 × 1.5–10 cm, more or less elliptic, the margins entire to toothed or irregularly and shallowly lobed; petioles 1.8–23 cm. Inflorescences of spikelike panicles of densely clustered flowers, or under better-watered conditions, inflorescences more open and branched. Mature, dry fruits tawny red-brown, semipersistent. Valves hard, relatively thick and tough, 4.0–4.3 × 1.9–2.3 mm (not including the teeth), with prominent reticulate veins, the callosity 2.3–3.0 × 0.8–1.2 mm, usually covering 1/2 of valve surface; valve margins with stout, rigid teeth 1.6–2.2 mm.

Locally along the Río Sonoyta from Sonoyta downstream at least 18 km, and sporadically nearly to its terminus east of Puerto Peñasco; also wetlands in the Río Colorado delta. Sandy or muddy wet soil, often emergent from shallow water. Southwestern Arizona, Baja California, Sonora, and Sinaloa.

This species was described from a Baja California Sur collection made in 1939 by Howard Scott Gentry. *R. inconspicuus* appears to be native, although apparently closely allied to the Eurasian *R. den-*

tatus Linnaeus (Sanders 1998), with which it has been confused. Gentry collected *R. inconspicuus* in Sinaloa in 1944 (*Gentry 7006*), and Pinkava et al. (1975) reported a 1957 collection from Arizona (identified as *R. dentatus*).

Río Sonoyta: Sonoyta, *F 86-95;* S of El Papalote, *F 86-161.* Ciénega de Santa Clara, *F 92-522.*

PORTULACACEAE PURSLANE FAMILY

Usually annual (ours) or perennial herbs, mostly succulent. Leaves simple, usually entire (ours), alternate or opposite or in basal rosettes; stipules none. Flowers usually radial. Sepals almost always 2, distinct or basally fused, usually persistent. (The sepals have been interpreted as sepal-like bracts.) Petals mostly 5 (2–15) or rarely more. (The petals have been interpreted as being derived from sepals or staminodes.) Stamens few to many. Pistil 1; ovary superior or partially inferior (in *Portulaca*); stigmas 2–9. Fruit usually a capsule, circumscissile (opening more or less around the middle, the top coming off like a lid) or 2 or 3 valves opening longitudinally from the tip. Seeds 1 to many, variously shaped—e.g., rounded to kidney-shaped, lens-shaped, snail-shaped (narrowed at one end like a thick comma)—or variously sculptured or ornamented to smooth and shiny, and sometimes with a strophiole (appendage).

Worldwide, best developed in western North America and the Andes; 20 genera, 400 species. More than half the species belong to *Cistanthe, Portulaca,* and *Talinum,* all of which occur in the Sonoran Desert. References: Bogle 1969; Carolin 1987; Hershkovitz 1993.

1. Winter-spring ephemerals; larger leaves in a basal rosette, mostly more than 2 cm, often 2.5–7 cm, the stem leaves reduced; capsules opening longitudinally. ____ **Cistanthe**

1′ Summer-fall ephemerals; leaves not in a basal rosette, the larger leaves mostly 1.5 (2.5) cm or less, the stems leafy; capsules circumscissile. ____ **Portulaca**

CISTANTHE

Annual and perennial herbs (winter-spring ephemerals in the flora area), mostly succulent (ours), the sap somewhat to quite mucilaginous (not as mucilaginous as in *Portulaca*). Leaves in basal rosettes and usually also alternately scattered on flowering stems, especially toward the base, sessile or petioles winged and clasping the stem. Inflorescence a simple to 2-branched helicoid cyme. Bracts usually 2 and markedly unequal. Sepals 2, persistent, membranous to succulent and green, with or without membranous margins. Petals (2) 5 (12). Stamens 1 to many. Style 1, the stigmas 2–4, sometimes barely distinct (see *C. ambigua*). Capsule valves as many as stigmas, mostly opening longitudinally from the tip (ours). Seeds 1 to many, usually glabrous (ours); strophiole absent (ours) or present.

Temperate and desert regions; western North America and South America; 45–50 species. Hershkovitz (1990) transferred the genus *Calyptridium* to *Cistanthe,* section *Calyptridium. Calyptridium,* with 8 species in western North America, is largely a nondesert taxon, although *C. monandra* and *C. parryi* extend into the Sonoran Desert. *C. ambigua* is in section *Amarantoideae* of *Cistanthe,* which includes 4 species in South America and 1 in North America. References: Hershkovitz 1990, 1991a, b; Howell 1956.

1. Leaves very thick, kidney-shaped in cross section or appearing terete (may appear flat when dry); flowers in umbellate clusters shorter than the leaves; stamens 5; capsules 3-valved. ____ **C. ambigua**

1′ Leaves thickened but definitely flattened; flowers in raceme-like cymes longer than the leaves; stamens 1; capsules 2-valved.

   2. Capsules more than twice as long as sepals; sepals (1.5) 2.0–3.3 mm. ____ **C. monandra**

   2′ Capsules less than twice as long as sepals; sepals (2.5) 3.0–6.0 mm. ____ **C. parryi**

Cistanthe ambigua (S. Watson) Carolin ex Hershkovitz [*Claytonia ambigua* S. Watson. *Calandrinia ambigua* (S. Watson) J.T. Howell. *C. sesuvioides* A. Gray] Stems and leaves extremely succulent, light green, the stems leafy and thick, usually several, branched from the base, reaching 12–20 (30) cm (as

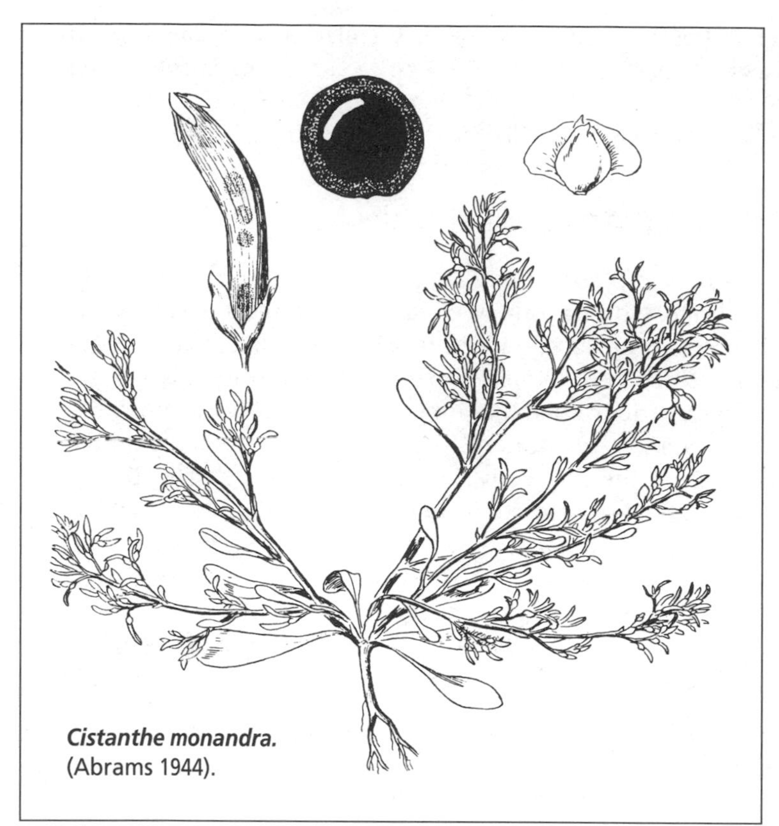
Cistanthe monandra. (Abrams 1944).

Portulaca oleracea var. *oleracea.* LBH (Parker 1958).

short as 3 cm when drought stunted). Leaves linear-spatulate, thick, broadly channeled below, nearly reniform in cross section, the first leaves in a basal rosette, the larger leaves 3–7 × 0.5–1.0 cm.

Inflorescences compact, umbel-like, shorter than the longer leaves. Fresh flowers 9 mm wide. Sepals green and succulent, sometimes slightly pink at tips, the margins membranous and white; petals 3–5 in number, 4 mm, white and semifleshy when fresh, transparent and membranous when dry. Stamens 5, the interior base of each with a nectary producing a glistening nectar drop; anthers of fresh flowers touching the stigma. Filaments, style, and stigma white, the anthers yellow. Stigmas cryptically 3-lobed (often appearing capitate). Capsules and the persistent floral remnant 5–6 mm, the capsule body 4–5 mm, with 3 membranous valves slightly shorter than the calyx; ovules ca. 20–23, with ca. 8 to all developing into seeds. Seeds nearly lens-shaped, 0.8–0.9 × 0.6–0.7 mm, shiny blackish with a low-relief reticulate pattern near the margin, notched at hilum.

The only record for Sonora is a well-established population on the north side of the main mountain mass in the Sierra del Rosario (*F 20718, F 92-383*). During years of favorable winter-spring rains it is locally abundant, with many thousands of individuals on otherwise nearly barren, white, granitic desert pavement. Otherwise Baja California, southeastern California, and Yuma County in southwestern Arizona, where it grows in sandy as well as rocky soils.

Cistanthe monandra (Nuttall) Hershkovitz [*Calyptridium monandrum* Nuttall] Herbage and capsules often reddish. Stems mostly ascending to prostrate, reaching 5–13 cm. First leaves in a basal rosette, the stems leafy with the leaves reduced upwards; leaves succulent, flattened, linear-spatulate, the larger ones 2–7 cm. Sepals (1.5) 2.0–3.3 mm. Petals 2 or 3 (4), thin and white, drying as a transparent, membranous cap carried to the top of the elongating capsule. Stamens 1 per flower, the filament white, the anther yellow. Capsules 4.5–7 mm, the valves 2, linear. Seeds 3–5, lens-shaped, 0.6 mm wide, shiny blackish with a low-relief reticulate pattern near the margin, notched at the hilum.

Sandy soil flats at the northwestern margin of the Pinacate volcanic region. Not known elsewhere in Sonora but expected eastward in the northern part of the state. Also Arizona, California, Nevada, and Baja California; largely nondesert habitats.

Seeds of *C. monandra* are flattened but thicker and about 25% smaller in diameter than those of *C. parryi* var. *arizonica;* in both species the body of the seed is smooth and shiny, but in *C. monandra* the wings are slightly roughened with a reticulate pattern.

MacDougal Pass *F 92-301.* Cráter Trébol [Molina], *F 92-331.* 1 km W of MacDougal Crater, *F 92-339.*

Cistanthe parryi (A. Gray) Hershkovitz var. **arizonica** (J.T. Howell) Kartesz & Gandhi [*Calyptridium parryi* A. Gray var. *arizonicum* J.T. Howell] Stems usually several from the base, reaching 10–20 cm, often spreading to prostrate. Leaves spatulate, the basal rosette leaves 2.5–4.5 cm, the stem leaves alternate and reduced. Inflorescences longer than the leaves, sometimes curved at tip. Flowers nearly sessile. Sepals (2.5) 3.0–6.0 mm, green to red-green, succulent including the white to pink margins (membranous when dry). Petals 4, pink. Stamens 1 per flower, the filament pink like the petals, the anther bright yellow. Stigma white, sessile, 2-parted; stamen and stigma at the same height and often touching. Capsules 2-valved, about 1.5–1.8 times as long as the sepals. Seeds about one dozen per capsule, shiny blackish, 0.8 mm wide, disk-shaped with a narrowly to broadly winged margin.

Sandy, gravelly, or cinder soils; arroyo beds, crater floors, and cinder flats and slopes in the Pinacate volcanic field to peak elevation; often common at higher elevations but mostly occasional and widely scattered. This is the only record for this species in Sonora.

Cistanthe parryi ranges from northern California to Baja California, southern Nevada, Utah, southern Arizona, and northern Sonora. There are 4 geographically segregated varieties. Documented collections of var. *arizonica* are from widely separated places (i.e., Santa Cruz and eastern Pima Counties in southern Arizona, the Pinacate Region, and Baja California). Disjunct distributions are not unusual among the varieties of this species (Howell 1956). Var. *arizonica* is distinguished from the other varieties by its completely smooth, shiny seeds and non-articulated (not jointed) pedicels.

Cerro Colorado, *Webster 22341.* Arroyo Tule, *F 19227.* SE of Pinacate Peak, 200 m, 31°55′ N, 113°25′ W, *Webster 22300* (DAV). S of Pinacate Peak, 875 m, *F 19377.* Campo Rojo, *F 92-478.*

PORTULACA PURSLANE

Succulent annual or perennial herbs, glabrous or hairy, the sap mucilaginous. Leaves mostly alternate, narrowed at the base into a short petiole, articulated at the point of attachment, not clasping. Sepals 2, fused to lower part of ovary. Flowers brightly colored, opening in sunlight, the petals often lasting only a few hours and then deliquescent (dissolving or melting away). Petals usually (4) 5 (6). Stamens 6 to numerous. Ovary partially inferior; styles deeply 3–9-parted. Capsules circumscissile. Seeds usually many, usually kidney-shaped to snail-shaped and somewhat laterally compressed; strophiole present.

Worldwide, mostly tropical and subtropical with greatest diversity in the Americas; 60 species. The cultivated moss rose (*P. grandiflora*) is sometimes grown in local gardens. Ours are summer ephemerals. References: Danin, Baker, & Baker 1978; Kelley 1990; Legrand 1962; Matthews & Levins 1985, 1986.

1. Leaves terete (appearing flat when dry); leaf axils and flower clusters densely white hairy. ________ **P. halimoides**

1′ Leaves thick but flattened; plants glabrous or glabrate.

  2. Capsule rim surrounded by a collarlike wing 1–2 mm wide, the capsule opening above the middle, the lid flattish, saucerlike. ________ **P. umbraticola**

  2′ Capsule rim without a collar-like wing, the capsule opening at about the middle, the lid conical.

    3. Seeds 0.6–0.8 mm in diameter, reddish brown to blackish, with low star-shaped tubercles, their rays touching or interlocking. ________ **P. oleracea**

    3′ Seeds 0.8–1.1 mm in diameter, blackish, studded with peglike projections slightly separated from one another. ________ **P. retusa**

Portulaca halimoides Linnaeus [*P. parvula* A. Gray] DWARF PURSLANE. Plants 2–12 cm, often broader than tall, usually much-branched. Stems relatively slender, scarcely succulent when mature, often bright pinkish red. Leaves succulent, green, nearly terete but slightly flattened when fresh, 7–23 mm. Long, silky white hairs in leaf axils and surrounding flowers. Sepals red-pink; petals, anthers, and stigma golden yellow, the petals ca. 4 mm, about as long as to slightly longer than the sepals; flowers opening about midmorning (ca. 10 A.M.). Capsules 1.3–1.6 mm wide, the lid separating slightly below the middle. Seeds 0.5–0.6 mm at maximum width, reddish brown to iridescent (metallic) blackish blue, cochleate (snail-shaped), studded with star-shaped tubercles, their radii (arms) short; strophiole white.

Sonoyta region to peak elevation in the Pinacate volcanic field; rocky, gravelly, sandy, or cinder soils; seasonally common in widely scattered and apparently localized habitats including floodplains, washes, bajadas, and slopes. Arid and tropical America from Brazil and Peru to Florida, and southwestern North America: Arizona and Sonora southward to the vicinity of Hermosillo, the Baja California Sur Peninsula, southeastern California, western New Mexico, southern Utah, and western Colorado. The southwestern North American populations, as well as those from other arid regions such as deserts in Argentina, apparently consist of small, relatively dwarfed plants.

Legrand (1962) treated *P. parvula,* a western North American taxon, as a synonym of the more wide-ranging *P. halimoides. P. parvula* of North American authors, with small yellow petals, has often been confused with *P. pilosa* Linnaeus (*P. mundula* I.M. Johnston is a synonym). *P. pilosa* occurs in eastern Colorado, New Mexico, and Texas to Missouri; also northeastern Mexico including Nuevo León, San Luis Potosí, and Coahuila; it is distinguished by red petals longer than the sepals, larger flowers, and generally larger plant size.

Mex 2 at 1 km E of rd to Cerro Colorado, *F 20122.* Tinaja de los Chivos, *F 18616.* 2 mi S of Tinajas de los Pápagos, *F 86-487.* Pinacate Peak, *F 86-453.* Aguajita Wash, *F 88-433.*

***Portulaca oleracea** Linnaeus var. **oleracea.** *VERDOLAGA;* PURSLANE; *KU'UKPALK.* Erect to spreading or prostrate with age and size; glabrous except for a few inconspicuous axillar hairs (visible with magnification). Stems few to many, usually much-branched, relatively thick and succulent except in stunted plants. Herbage green, or reddish late in the season. Leaves mostly 1.0–1.5 cm, alternate or subopposite, flattened, spatulate to obovate, the larger leaves quickly deciduous, the apex truncate to moderately notched. Sepals green, winged toward apex, the wings persistent on the capsule lid. (The capsules tend to form narrower wings during drought and also late in the season.) Petals, stamens, and stigma yellow, the petals 5.0–5.5 mm, broadly oblong to obovate, deeply notched at apex. Capsules separating at about the middle, the capsule lid conical and somewhat constricted at about the middle, with at least 1 seed usually remaining with the lid, the others falling quickly. Seeds 0.6–0.8 mm wide, numerous, dull blackish, granulate, with a thin white strophiole.

Mostly disturbed habitats as an urban and agricultural weed, and wet clayish mud in roadside ditches (e.g., Pinacate Junction) and sometimes along larger washes. Widespread in the Sonoran Desert, mostly in disturbed habitats. The plants have a propensity for self-pollination and long-term seed viability, and lack specialized seed dispersal mechanisms. Matthews, Ketron, & Zane (1993:174) concluded that it is "a polymorphic species and is not divisible into subspecies based on seed surface as the primary morphological trait."

Worldwide in tropical to warm-temperate climates, mostly weedy. It is often difficult to determine which populations might be native and which are introduced or invasive. Native to the Old World and present in the New World in pre-Columbian times. *P. oleracea* is one of the world's most serious and ubiquitous weeds. The plants, used as a potherb, are sold in markets in Sonora and southern Arizona. Fosberg & Renvoize (1977) differentiated several varieties in the Old World. In addition, there are horticultural varieties selected for use as vegetables.

Pinacate Junction, *F 85-993, F 86-351.* Aguajita, *F 88-432.*

Portulaca retusa Engelmann. Resembling *P. oleracea* but the plants usually more slender and erect, the seeds larger (0.8–1.1 mm in diameter) with longer and more slender peglike papillae or projections (these slightly separate from each other), and the capsule lid not retaining seeds. Several widely separated collections from northern Sonora; perhaps more widespread in the state but seldom collected because the plants disappear quickly after the summer rains cease. Known in the flora area from 2 collections.

Legrand (1962) indicates that *P. retusa,* widespread in the Americas, is probably a synonym of *P. oleracea,* and Matthews, Ketron & Zane (1993) conclude that it is part of the *P. oleracea* complex. Perhaps our *retusa*-like plants belong to a native population whereas the *oleracea*-like plants are non-native. Wiggins (1964) and other authors indicate that *P. retusa* can be distinguished by persistent sepal wings that fall with the capsule lid, but this character is also seen among ordinary "weedy" *P. oleracea.*

Rancho los Vidrios, *Equihua 26 Sep 1982.* Mex 2 at 1 km E of rd to Cerro Colorado, *F 20123.*

†Portulaca umbraticola Kunth [*P. coronata* Small. *P. lanceolata* Engelmann. *P. umbraticola* subsp. *lanceolata* (Engelm.) J.F. Matthews & Ketron] This distinctive portulaca should be sought in the more favorable habitats in the Sonoyta region or middle elevations in the Sierra Pinacate. It is distinguished by flat leaves, a conspicuous collarlike wing 1–2 mm wide surrounding the capsule rim, capsules opening above the middle to shed a flattish, shallow saucerlike lid, and dull gray seeds. Widespread in the Americas. Matthews & Ketron (1991) recognized three subspecies.

RAFFLESIACEAE RAFFLESIA FAMILY

Obligate parasites devoid of chlorophyll. Stems often filamentous like a fungal mycelium in the host's tissues; only the flowers or short flowering stems break through the host plant's stem or root to become visible. Mostly dioecious. Flowers unisexual, radial, subtended by bracts, the calyx corolla-like; corolla none. Stamens numerous. Seeds tiny, numerous, often thousands per fruit.

Worldwide, mostly Old World, tropical and subtropical, few in deserts and temperate regions; 8 genera, 50 species. This strange family includes the world's largest single flower, *Rafflesia arnoldii* of Malaysia, which reaches 1 meter in width. At the opposite end of the scale is *Pilostyles thurberi,* one of the smallest plants in the Sonoran Desert. Like many other carrion-imitating fly-pollinated flowers, those of the Rafflesiaceae are red-brown and foul smelling. Many of the species are taxonomically poorly known. Reference: Kuijt 1969.

PILOSTYLES

Leaves reduced to floral bracts. Flowers small, with a thick central column expanded at apex into a convex disk; female flowers and fruits larger than male flowers. Anthers in male flowers many, below the margin of the column disk. Ovary inferior.

Mostly New World tropics, also Africa, Iran, and Australia, tropical and subtropical, some in deserts; 18 species. Parasitic on legumes. Many are poorly known and difficult to distinguish. Reference: Rutherford 1970.

Pilostyles thurberi A. Gray. Presumably perennial; in the Sonoran Desert parasitic on *Psorothamnus emoryi.* Buds, flowers, and fruits purple-brown and somewhat fleshy. Dried flowers and fruits semipersistent on host stems, leaving a raised craterlike scar. Male and female flowers similar but ovary not developing in male flowers. Flowers with 3 nearly identical whorls, the outer whorl of bracts, the inner 2 of sepals, plus a ringlike nectary in both male and female flowers. Flowers and fruits 2.5–4.3 × 1.5–4.0 mm. Fruit a globose capsule, breaking apart irregularly. Seeds 0.3 mm, yellow to light brown, ovoid, obtusely pointed at tip, at least 50 per capsule. Flowering in April.

The largest population in the region is along the lower Río Colorado valley within 1 kilometer or so of the river channel south of Riito (other potential habitat has been converted to farmland). Additional localities include dunes near Punta Borrascosa at the west end of Bahía Adair, and at least one interdune valley or corridor between the southwestern edge of the Pinacate lava shield (at Moon Crater) and Bahía Adair. The only other locality in Sonora is on the north side of Cerro Tepopa near El Desemboque San

Ignacio. At these places fog or dew settles at night during the cooler times of the year. In the Sonoran Desert otherwise in the vicinity of Yuma and Wellton in southwestern Arizona, southeastern California, northeastern Baja California. Also parasitic on *Dalea formosa* and *D. frutescens* in the southern Great Plains and the Chihuahuan Desert in western Texas and eastern New Mexico to central Mexico.

The purple-brown buds, flowers, and fruits dot the lower stems of the host plants and sometimes extend into the smaller branches but usually not into the leafy portions. The wartlike visible parts of the parasite resemble an infestation of scale insects.

Punta Borrascosa, *F 84-38*. 8 mi S of Riito next to the Río Colorado, *F 75-51*.

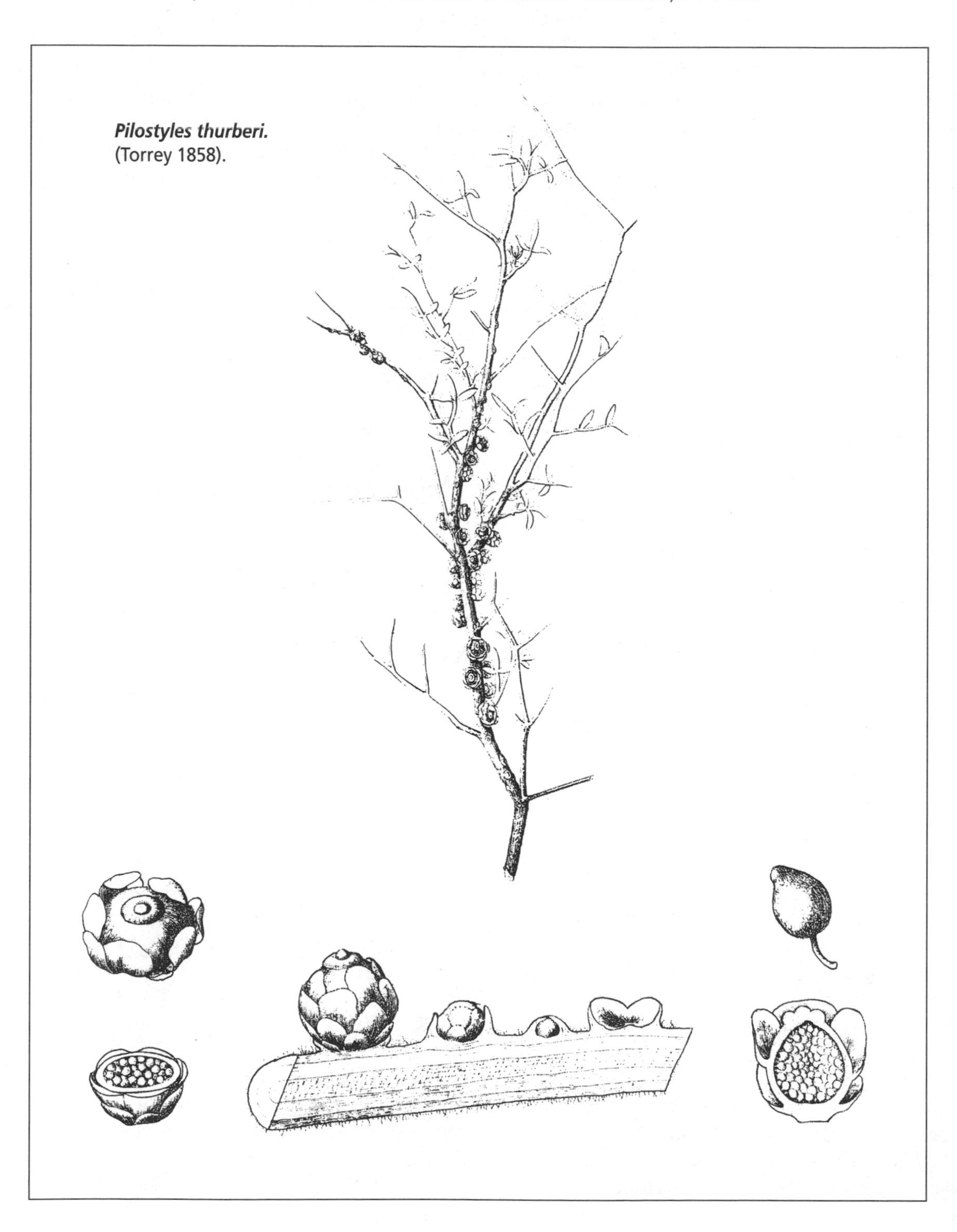

Pilostyles thurberi.
(Torrey 1858).

RANUNCULACEAE BUTTERCUP FAMILY

Mostly annual or perennial herbs, sometimes vines or small shrubs. Leaves alternate, sometimes in basal rosettes. Flowers highly variable, radial or bilateral, usually bisexual, the floral parts usually separate. Stamens usually many. Pistils 1 to many; ovary superior. Worldwide, mostly north temperate and arctic; 60 genera, 2500 species. Reference: Whittemore & Parfitt 1997.

1. Perennials; leaf blades about as broad as long, deeply cleft; dry desert habitats. __________ **Delphinium**
1′ Spring annuals; leaf blades linear, entire; wetland habitats; extirpated from the flora area. __________ **Myosurus**

DELPHINIUM LARKSPUR

Annual or mostly perennial herbs with tough fibrous to fleshy roots. Stems mostly erect. Leaves basal to alternate, usually palmately parted. Inflorescences of racemes or panicles. Flowers bilateral, the sepals 5, the upper one forming a spur, the petals deciduous, of 2 pairs, the upper pair forming a spur enclosed by the calyx spur. Stamens numerous. Pistils mostly 3, the fruits many-seeded.

North America and Old World, mostly temperate and arctic; 300 species. Includes delphinium and larkspur, sometimes grown as spring flowers in Sonoran Desert gardens. All are apparently toxic. Reference: Warnock 1997.

Delphinium scaposum Greene. BARE-STEM LARKSPUR. Perennials from thickened, fibrous-fleshy roots, with 1 to several erect stems 35–60 (85) cm. Herbage glabrous or soft pubescent. Leaves basal, relatively fleshy, the petioles 2.0–13.5 cm, the blades 2.5–6.0 cm wide and about as long, with 3–5 deeply cleft primary divisions, the lobes broad, each with a nipple-like tooth. Racemes usually 5–12-flowered. Flowers showy, 2.5–3.0 cm wide. Sepals deep azure blue, upper pair of petals white with blue markings, lower petals dark blue like the sepals. Filaments dark blue, with thin whitish wings below. Flowering March-April; plants winter-spring active, summer dormant.

Decomposed granitic soil of upper bajadas in the Sonoyta region; usually along small arroyos and often on north-facing banks. Colorado, Utah, New Mexico, Arizona, Nevada, and northern Sonora.

10–12 km SW of Sonoyta, *F 88-169, F 92-416.*

MYOSURUS MOUSE-TAIL

Dwarf annuals. Leaves basal, simple, narrow, entire. Flowers single on slender scapes, inconspicuous, the fruiting receptacle elongated (cylindrical) and spikelike (like a mouse tail), the pistils numerous. (The single flower somewhat resembles the many-flowered spikes of *Plantago.*) Sepals petal-like, usually 5, spurred; petals 5 or fewer, inconspicuous. Stamens 5 to many.

Temperate wetlands worldwide; 15 species. Reference: Whittemore 1997b.

Myosurus minimus Linnaeus. Delicate spring annuals, tufted, emergent from shallow water. Leaves erect, linear, 3–8+ cm × 0.4–0.9+ mm. Scapes erect, 3.0–6.3+ cm. Spikelike receptacle 1.5+ cm × 2.6+ mm. Sepals 1.5–1.6 mm, the spur ca. 0.5 mm. Stamens usually 5. Achenes with a minute, scarcely apparent beak.

Recorded from Quitobaquito when the oasis was owned and managed by the Orozco family (with *Poa annua* in marshy area bordering alkaline pool, 18 Mar 1945, *Gould 2986*). It grew with other small and likewise locally extirpated herbaceous plants requiring open wetland habitat (Felger et al. 1992). Vernal pools and other wetland habitats; Alaska and Canada, across the United States to Baja California, Eurasia, and Africa.

RESEDACEAE MIGNONETTE FAMILY

Annual, biennial, or perennial herbs, or small shrubs. Leaves alternate or appearing fascicled in axils, simple, the blade decurrent on petiole, with a pair of minute stipules and often also small stipule-like basal teeth. Inflorescences of terminal bracteate racemes or spikes. Flowers bilateral. Ovary superior, usually open at the top, the carpels 2–6, distinct or fused at margins, each with its own stigma.

Especially diverse in the Mediterranean region and eastward, also Africa except humid tropics, and 1 species in the New World; 6 genera, 75 species. *Reseda odorata,* mignonette, is a garden ornamental grown for its fragrance. The family, with floral parts often in 2s and 4s, is related to the crucifers and capers in the Capparales. References: Abdallah & De Wit 1967–68; Leister 1970.

OLIGOMERIS CAMBESS

Annual or perennial herbs or dwarf shrubs. Flowers minute, bisexual or plants monoecious. Two species, 1 in southern Africa.

Oligomeris linifolia (Vahl ex Hornemann) J.F. Macbride [*Reseda linifolia* Vahl ex Hornemann. *R. subulata* Delile, nomen nudum. *Oligomeris subulata* (Delile) Boissier] DESERT CAMBESS. Ephemerals, recorded in the flora area October-May (annual or biennial in Middle Eastern deserts). Glabrous, semiglaucous, semisucculent, (5) 10–40 cm, mostly erect, slender and few-branched from the base or much-branched when well watered, with a well-developed taproot. Herbage sometimes turning orange; stems leafy. Leaves often appearing fascicled because of dense clusters of axillary short shoots, which may develop into longer branches. Leaves narrowly linear, often thickened, larger ones mostly 1.5–3.5 (4.3) cm × 6–12 mm; stipules a pair of inconspicuous whitish bristles and larger leaves also with a pair of stipule-like bristles.

Spikes slender, densely flowered, simple or often with few short appressed branches. Flowers sessile, moderately bilateral (often not readily apparent on dried specimens). Bracts and sepals fleshy,

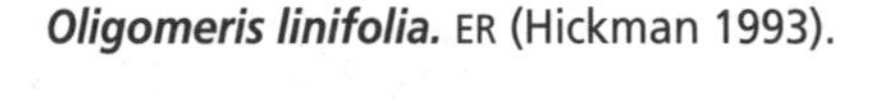
Oligomeris linifolia. ER (Hickman 1993).

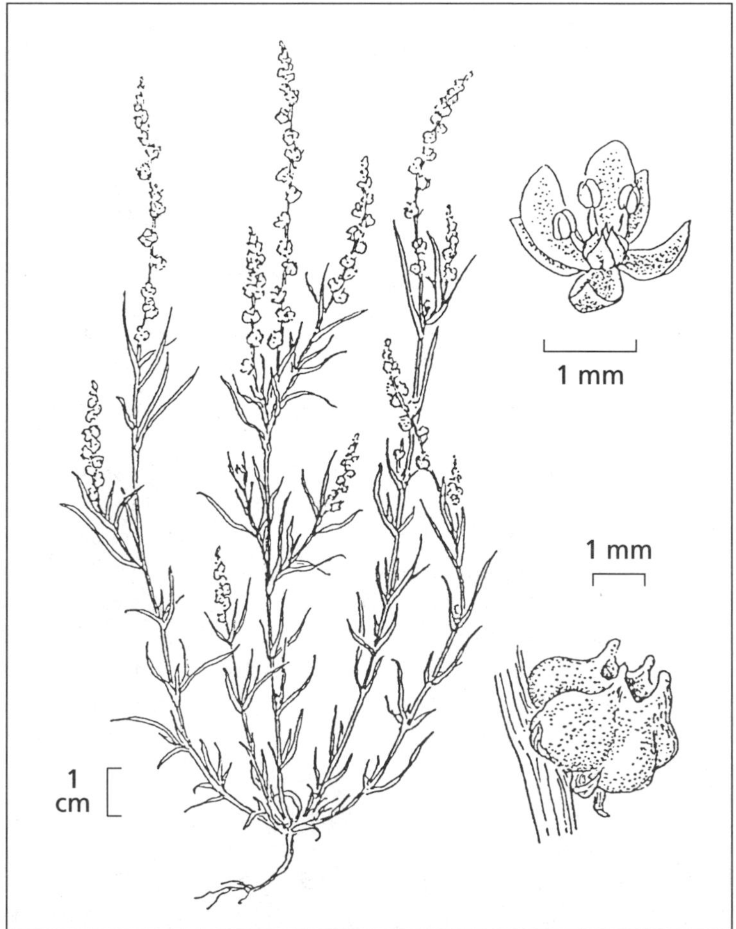

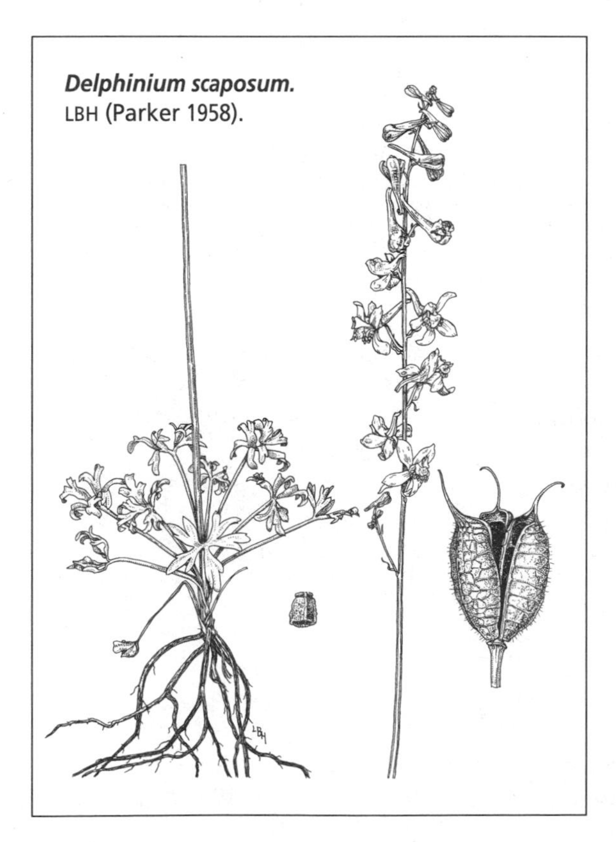
Delphinium scaposum. LBH (Parker 1958).

green with white margins, 1–2 mm; sepals 4, the petals 2, ca. 1 mm, white. Stamens 3. Fruits (capsules) (2.3) 2.5 (3.0) mm wide, globose, open at top, green becoming light tan when dry, moderately persistent, of 4 carpels each conspicuously swollen and saclike at the base and terminating in a short, broad tooth surmounted by a stigma, the teeth unequal in size. Seeds 0.5–0.6 × 0.45 mm, numerous, shiny blackish when fully ripe, smooth, obovoid, folded at hilum.

Widespread in the volcanic and granitic regions and across the open desert; mostly lower elevations such as lower slopes, desert plains, pavements, gravelly soils of arroyo beds, sand flats, and often abundant in playas including those of crater floors. Sometimes on lower dunes but not on higher dunes. Mostly deserts, also Mediterranean climates; northwestern Mexico and southwestern United States, and north Africa and the Middle East to India, also southern Africa.

It has been suggested that this species is not native to North America because of the great geographic disjunction. The idea that it is non-native was questioned by Parish (1890), who pointed out it was well established in natural desert habitats and did not demonstrate "weedy" attributes (also see Nelson & Kennedy 1908). Torrey (1858) reported it along the Gila River in southern Arizona and the Rio Grande in Texas. The Seri Indians gathered the seeds in substantial quantities during favorable years; these were toasted and ground, mixed with water, and consumed as *atole* (Felger & Moser 1985).

Bahía de la Cholla, *Lockwood 86*. Quitobaquito, *F 86-185*. Moon Crater, *F 19236*. MacDougal Crater, *Turner F 86-25*. E of Los Vidrios, *F 19598*. Sierra del Rosario, *F 20692*. Laguna Prieta, *F 85-748*.

RHAMNACEAE BUCKTHORN FAMILY

Trees, shrubs, and sometimes woody vines, rarely herbaceous perennials. Leaves usually alternate (ours), simple, pinnately veined or several main veins from the base; stipules usually present, small or sometimes modified into spines. Flowers mostly in axillary clusters, usually bisexual, radial, small; green, yellow, or white, or sometimes pink to blue (e.g, *Ceanothus* in California); usually with a short hypanthium resembling a calyx. Sepals 4 or 5, united at the base, the lobes usually triangular and quickly deciduous. Petals 4 or 5, or sometimes absent. Stamens alternate with sepals and equal in number, usually fitting into a cavity or depression of the subtending petal. Fruits often drupes or drupelike or sometimes capsules.

Worldwide, mostly tropical and subtropical, some in deserts and temperate regions; 55 genera, 900 species. The Sonoran Desert genera are mostly of tropical origin. Reference: Brizicky 1964.

1. Fruits dry capsules, 3-seeded. ______ **Colubrina**

1′ Fruits fleshy drupes, 1-seeded.

  2. Leaves spatulate, widest above middle; fruits 3.0–4.5 mm, the fleshy part bitter. ______ **Condalia**

  2′ Leaves ovate to narrowly elliptic, widest below to middle of leaf; fruits 8–10 mm, the fleshy part edible but not very sweet. ______ **Ziziphus**

COLUBRINA SNAKEWOOD

Unarmed shrubs or small trees (the short shoots of aridland species sometimes spinescent). Leaves petioled with 3 main veins or the veins pinnate; stipules usually deciduous. Fruits dry, dehiscent, separating into three 1-seeded segments. *Colubrina* "probably includes the least specialized members of the Rhamnaceae. The flowers are of the usual rhamnaceous nature, being small and borne in cymes or thyrses, with a floral cup at the rim of which are borne the 5 sepals, 5 petals, and 5 stamens, with no remarkable specialization save for a thick disk which nearly fills the cup [with nectar] at anthesis and hides the 3-celled ovary. The ovary has been described as inferior or partly so, although at their maturity the cup and disk, which are somewhat accrescent, surround and adhere only to the basal fifth to half of the fruit" (Johnston 1971:2).

About 30 species, 21 in the New World, 1 in Hawaii, the others in Madagascar and southeast Asia. Reference: Johnston 1971.

Colubrina californica I.M. Johnston [*C. texensis* (Torrey & A. Gray) A. Gray var. *californica* (I.M. Johnston) L.D. Benson] CALIFORNIA SNAKEWOOD. Divaricately branched shrubs 2–3 m; tardily drought deciduous. Twigs and petioles densely white hairy, the upper leaf blade sparsely hairy to sometimes glabrate, the lower surface, especially the veins, densely hairy. Leaves alternate or fascicled in short shoots, the petioles 0.8–2.7 mm, the blades 8–31 × 5–21 mm, green to gray-green, broadly elliptic or oval to obovate, the veins pinnate, prominent on lower surfaces, the margins entire to serrate, the teeth tipped with a small gland; stipules 0.8–1.5 mm, scalelike, brown, soon deciduous. Flowers in small axillary clusters. Capsules 7.0–8.7 mm, globose, the style bases persistent. Flowering spring and probably following summer rains.

A small population southwest of Sonoyta along arroyo margins among low desert hills (8 km SW of Sonoyta, *F 92-80*). Widely scattered in western Sonora southward to Cañón del Nacapule near San Carlos (Felger 1999). Also Arizona, southeastern California, southern Baja California and northern Baja California Sur.

CONDALIA

Much-branched hardwood shrubs, rarely small trees. Long shoots of primary branches bearing secondary thorn-tipped twigs, both with alternate leaves and relatively widely spaced nodes; short shoots with fascicled leaves in leaf axils of the secondary twigs. Leaves usually with pinnate veins, the margins usually entire; stipules small, not spinose. Flowers small, produced from axils of short shoots. Calyx tube broad and very short, with a flat disk and 5 lobes. Petals usually none. Fruit a small drupe.

Eighteen species, 5 in South America and West Indies, 13 in Mexico and southwestern United States. Reference: Johnston 1962b.

Condalia globosa I.M. Johnston var. **pubescens** I.M. Johnston. *CRUCILLO; KAWK KOAWUL*. Shrubs to small trees reaching 2.5–4.5+ m, more or less symmetrical and much-branched, often with short, thick trunks. Branches and twigs rigid, the twigs thorn-tipped. Leaves spatulate, 3.3–22.0 × 1.5–4.0 mm, crowded on very short spur-branches (short shoots), the margins entire. Partly evergreen to gradually deciduous in severe drought; larger leaves petioled, quickly drought deciduous, the smaller leaves subsessile. Lower leaf surfaces with convex and conspicuous veins, the upper surfaces with the midrib concave. Lower surface of fresh, young but mature leaves often yellowish with minute glistening golden glands. Flowers in small clusters, yellowish green, ca. 3 mm wide, the disk at anthesis awash with sticky glistening nectar; petals none. Pedicels slender, 2.7–6.0 mm. Fruits globose, dark brown or blackish, very bitter, 3.0–4.5 mm plus the persistent style 0.5 mm; disk at base of fruit ca. 2 mm wide at maturity. Flowering during warmer months, especially with rains; fragrant and attracting hoards of insects.

Occasional in the Sonoyta region along arroyos, bottomlands, and the Río Sonoyta, and uncommon on the east side of the Pinacate volcanic field.

Variety *pubescens* apparently differs from var. *globosa* only in having hairy leaves, although var. *pubescens* sometimes may develop into small trees whereas var. *globosa* does not seem to become arborescent. Larger arborescent individuals of var. *pubescens* occur in remote places where there has been little or no woodcutting. Many of the larger condalias in the Sonoyta region have regrown from large axe-cut stumps. The hard wood is highly desirable for firewood. The plants are very slow growing.

This species ranges from southeastern California and southwestern Arizona to northwestern Sinaloa and through the Baja California Peninsula. Apparently only var. *pubescens* occurs in the northernmost part of the range and only var. *globosa* occurs in the southernmost part (e.g., south of Guaymas). Both varieties, however, occur intermixed over the majority of the distribution of the species. *C. globosa* is a member of the *C. spathulata* A. Gray complex that includes a series of 4 closely related allopatric species from central Mexico to southwestern United States. Ours is the most western and xerophytic member of the complex.

6 km NE of Elegante Crater, *F 10062*. 2.6 mi W of Sonoyta, *F 86-404*. Aguajita Spring, *Bowers 1046* (ORPI). 20 mi SE of Ajo, "... common in the vicinity of Sonoita, Sonora, and there attains a height of about 20 feet," 6 May 1939, *Peebles 14329*.

Ziziphus

Evergreen to deciduous shrubs and trees. Mostly with stipular spines. Leaves alternate, opposite, or fascicled in short shoots, mostly with 3 prominent veins branched from the leaf base, otherwise the veins pinnate. Flowers in axillary clusters. Fruit a fleshy drupe, usually edible.

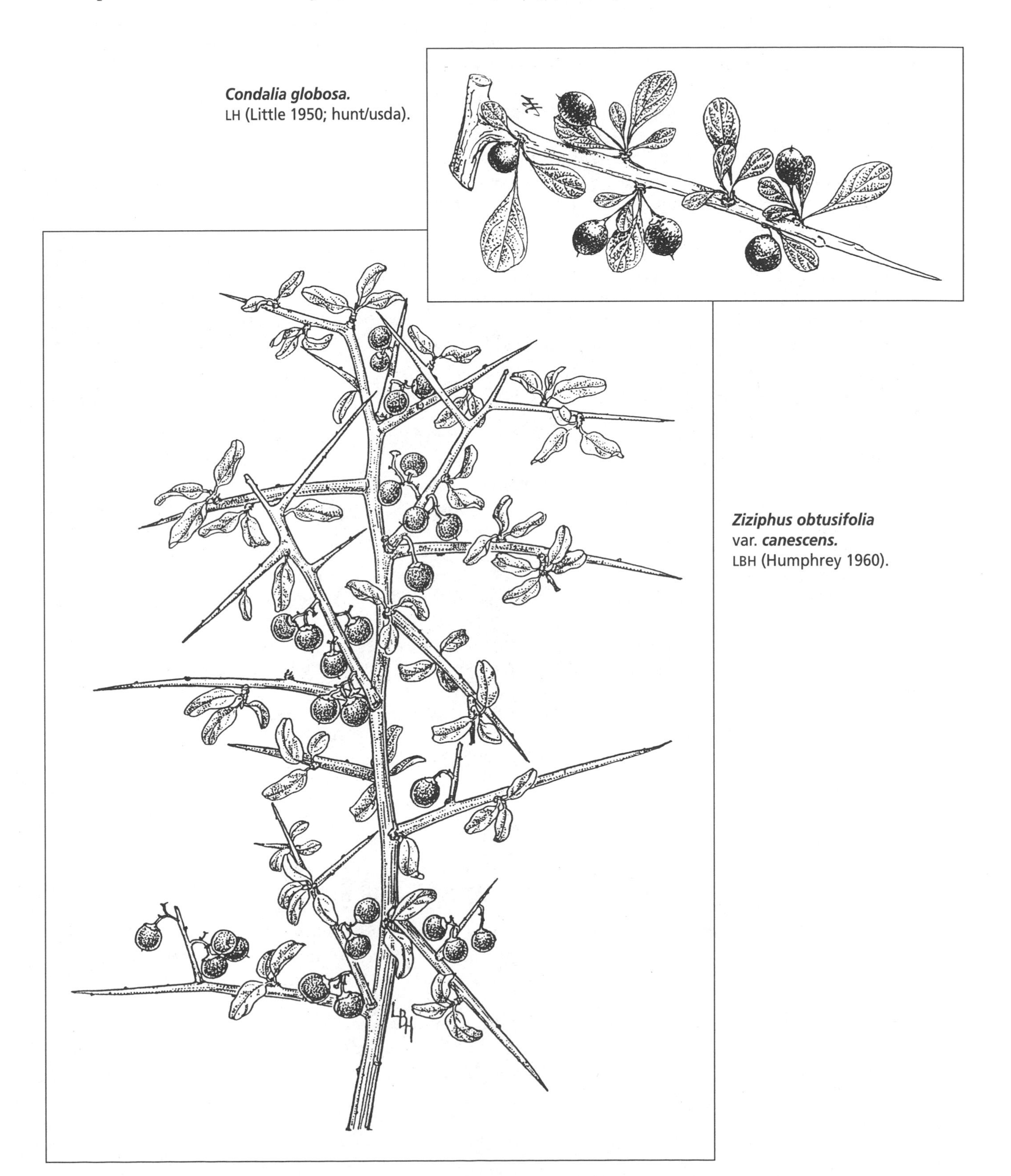

Condalia globosa.
LH (Little 1950; hunt/usda).

Ziziphus obtusifolia
var. ***canescens.***
LBH (Humphrey 1960).

Mostly tropical to subtropical and some in deserts, rarely in dry-temperate regions. The New and Old World groups are distinctive; 90 species, 7 in North America.

Most Old World species have paired stipular spines, whereas many New World species have paired thorn-tipped stipule-like lateral branches. In a small group of New World species, including *Z. obtusifolia,* the thorn-tipped branchlets are solitary at the nodes. Jujube (*Z. jujuba*) in China, and *ber* (*Z. mauritiana*) in arid zones of India, are economically significant fruit trees. Jujube is sometimes grown in Sonora and Arizona. Many *ber* varieties would be worthwhile introductions to Mexico, especially Sonora. Jujube is extremely frost hardy but *ber* is frost sensitive. Reference: Johnston 1963.

Ziziphus obtusifolia (Hooker ex Torrey & A. Gray) A. Gray var. **canescens** (A. Gray) M.C. Johnston [*Z. lycioides* A. Gray var. *canescens* A. Gray. *Condalia lycioides* (A. Gray) Weberbauer var. *canescens* (A. Gray) Trelease. *Condaliopsis lycioides* (A. Gray) Suessenguth var. *canescens* (A. Gray) Suessenguth] *ABROJO;* GRAYTHORN; *'U:S JEWEDBAḌ.* Sprawling, messy-looking shrubs with stiff branches forming dense thorny tangles. Stems gray-green, the twigs thorn-tipped and commonly spreading at right angles. Young twigs or branches, leaves, inflorescences, pedicels, and exposed portions of buds densely pubescent with short spreading white hairs; twigs and branches glabrate with age. Leaves sparse and quickly drought deciduous, the shrubs often nearly leafless; alternate on long shoots and in several-leaved fascicles on short shoots, the blades triangular-ovate to narrowly elliptic, mostly 8–22 mm, the midrib prominent, the lateral veins pinnate, the margins entire or with a few small teeth, especially on the long-shoot leaves. (Leaf edges are often chewed by insects and at first glance might not appear entire.)

Inflorescences in short-shoot axils, of short subumbellate clusters usually less than 1 cm, the old inflorescences with the basal disk of the fruit semipersistent. Flowers 2.5–3.0 mm wide. Sepal lobes broadly triangular, surrounding a flat disk. Petals white, clawed, sticking out laterally from not quite open flowers, later loosely enfolding the stamens and quickly falling after the anthers mature (petals often gone by the time the stigma expands). Fruits 8–10 mm, glaucous, dark blackish blue to purple-brown, the pericarp thin, fleshy, and edible but not very sweet. Flowering at least May-September, the flowers visited by honeybees, native bees, large orange-winged spider wasps (*Pepsis* or *Hemipepsis*), and other insects.

Sonoyta region, mostly locally, along washes, floodplains, and sometimes in roadside ditches; also higher elevations in the Sierra Pinacate and occasional elsewhere, mostly roadsides and arroyo margins; not in the western part of the flora area.

Variety *canescens* differs from var. *obtusifolia* in its thicker leaves, greater average number of flowers per cluster, and thicker and longer peduncles. Var. *canescens,* for the most part a Sonoran Desert element, ranges from southeastern California, southern Nevada, and northern Arizona to southern Sonora and the Cape Region of Baja California Sur. It is nearly ubiquitous in Sonora except at higher elevations. Var. *obtusifolia,* largely a Chihuahuan Desert element, replaces our variety in extreme southeastern Arizona, northeastern Sonora, New Mexico, Texas, and north-central and northeastern Mexico.

Bahía de la Cholla, *F 13162.* 4 mi W of Los Vidrios, *F 86-384.* Sonoyta, *F 86-399.* Quitobaquito, *F 86-186A.*

RUBIACEAE MADDER FAMILY

Trees, shrubs, or vines, and some herbs. Leaves opposite and decussate (often appearing whorled because of resemblance of stipules to leaf blades), simple, mostly entire, and stipuled. Flowers radial, usually bisexual. Calyx and corolla mostly 4- or 5-lobed; stamens 4 or 5 or more, borne on the corolla. Ovary nearly always inferior; fruits fleshy or not.

Worldwide, mostly tropical and subtropical; 500 genera, 6500 species, 11 genera in the Sonoran Desert region. Includes *Cinchona* (quinine), *Coffea* (coffee), and many ornamentals such as *Gardenia. Rubia tinctorum* (madder) was the source of the red dye alizarin, now made artificially. Reference: Dempster 1979.

GALIUM BEDSTRAW

Mostly annual or perennial herbs, rarely shrubs. Stems slender, square at least when young. Leaves appearing whorled (2 leaves and 2 leaflike stipular appendages). Flowers small, usually 4-merous, bisexual or unisexual and the plants dioecious or monoecious. Calyx minute or absent; corollas usually rotate, almost always deeply 4-parted. Fruits 2-lobed, 2-seeded, dry or fleshy, indehiscent, sometimes bristly or hairy.

Worldwide, mostly temperate regions; 400 species; 8 species in the Sonoran Desert region. Some used as sources of dye, others medicinal. Reference: Dempster & Ehrendorfer 1965.

Galium stellatum Kellogg var. **eremicum** Hilend & J.T. Howell. DESERT BEDSTRAW. Dense to open, untidy small shrubs or subshrubs, irregularly spreading and sprawling, reaching 50–80 cm. Stems slender, square with white-margined corners. Leaves and young stems scabrous with short stiff white hairs. Vegetative or long shoots woody with long internodes and brittle perennial stems forming the framework of the plants, with peeling stringy bark and relatively large leaves, the old dry leaves whitish and often persisting. Fertile or short shoots bunched at nodes of the woody long shoots, with very short internodes, dying back or deciduous during drought seasons, and with slender and relatively flexible stems and smaller leaves. Leaves sessile, ovate to lanceolate, narrowed to a subspinose tip, mostly 2–8 (10+) × 0.8–3.0 mm (includes short- and long-shoot leaves); lower leaf surfaces with a prominent white midrib and moderately inrolled margins. Dioecious. Flowers in small leafy panicles among the short shoots. Corollas 2.5–3.0 mm wide, whitish to pale yellow with purplish netlike veins. Ovaries and fruits densely covered with spreading straight silky white hairs, the hairs longer than the fruit; fruits dry, 3–5 mm wide including the hairs. Flowering in spring.

Usually north- and east-facing slopes among rocks and along rocky drainageways; most common toward higher elevations in the granitic ranges and the Pinacate volcanic field. Mostly in desert mountains; southwestern Utah, southern Nevada, western and southern Arizona, southern New Mexico, southeastern California to northern Baja California Sur, and northern Sonora southward in coastal mountains to Cerro Tepopa (29°21′ N).

Galium stellatum var. *eremicum* is the shrubbiest of all the New World galiums, and *G. stellatum* is the most southern and xeromorphic member of the 18 North American species of section *Lophogalium*. Within this group *G. stellatum* var. *eremicum* seems to be a generalist; it has the largest geographic range and is mostly diploid ($2n = 22$). Its apparent nearest relatives are tetraploids. Var. *stellatum* is endemic to Isla Cedros off the Pacific coast of Baja California.

Sierra del Viejo, *F 13215*. Sierra del Rosario, *F 20380*. Pinacate Peak, 900+ m, *F 19805*. Elegante Crater, *F 19697*. Hornaday Mts, *Burgess 6838*. Puerto Peñasco, *Dennis 31*.

RUTACEAE RUE OR CITRUS FAMILY

Mostly shrubs or trees. Leaves and other organs gland-dotted, producing essential oils. Leaves alternate or opposite, simple or compound. Flowers mostly bisexual, radial, and with a disk. Sepals and petals each often 3–5, or petals rarely absent. Stamens often 3–20. Ovary superior. Fruits various including capsules, leathery-skinned berries (*Citrus*), drupes, etc.

Worldwide, mostly tropical and subtropical, especially diverse in South Africa and Australia; 160 genera, 1700 species. Five genera in the Sonoran Desert region but only *Thamnosma* occurs well within the desert. Various kinds of citrus are cultivated in the region.

THAMNOSMA

Small shrubs, conspicuously gland-dotted. Leaves alternate, simple, deeply divided or entire, and linear to reduced. Flowers pedicelled. Sepals and petals each 4; stamens in 2 series, 4 long and 4 short. Ovary 2-chambered, 2-carpelled. Style slender and elongate. Fruits of 2 inflated chambers.

Six species with an astonishing distribution: southwestern North America, Arabia, Socotra, Somalia, and southern Africa. *T. montana* is the most xerophytic member of the family in North America.

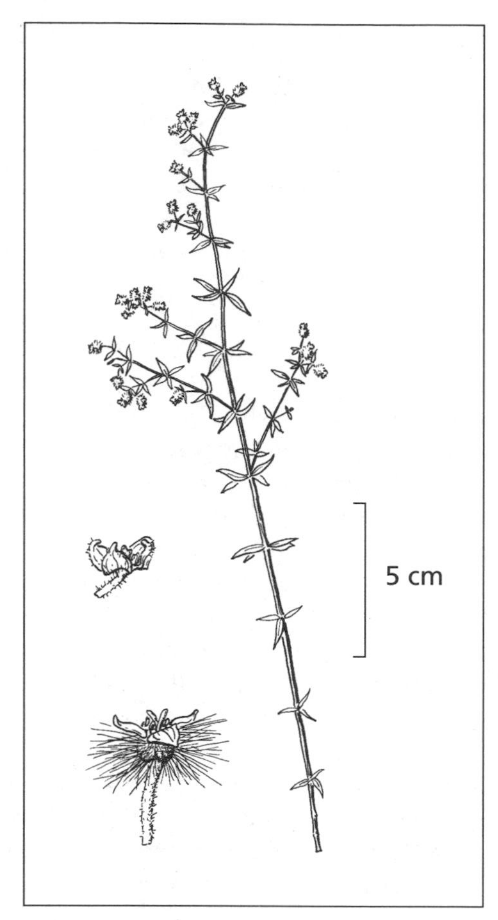

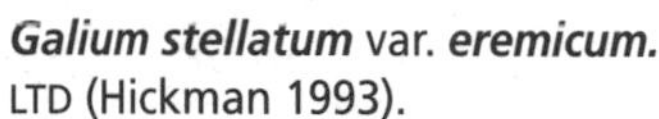

Galium stellatum var. *eremicum.*
LTD (Hickman 1993).

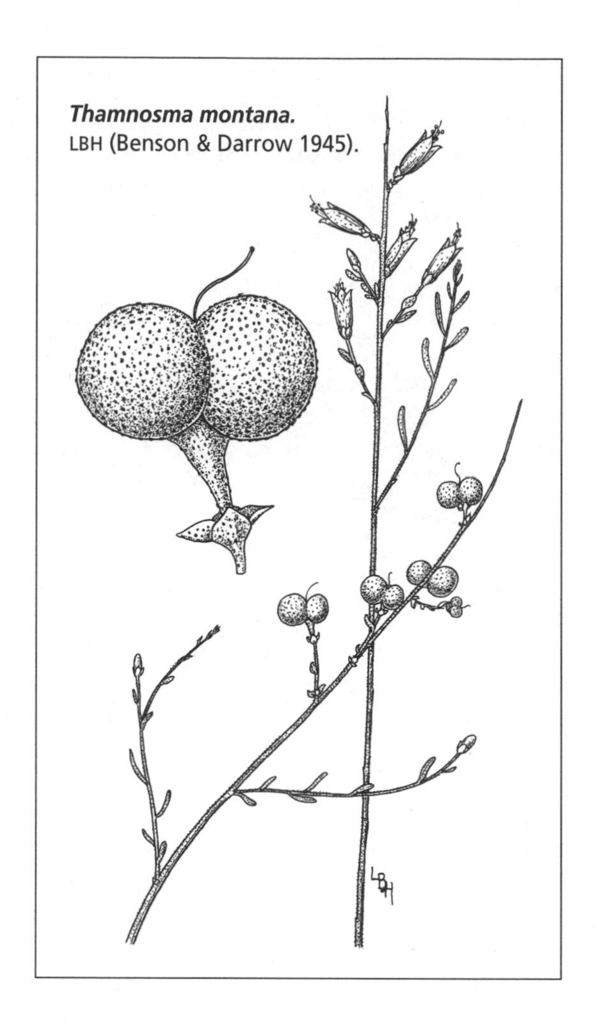

Thamnosma montana.
LBH (Benson & Darrow 1945).

Thamnosma montana Torrey & Frémont. TURPENTINE BROOM. Erect to globose much-branched subshrubs, often 50–80 cm. Stems a strange yellowish green color throughout; herbage and flowers stinky, or new growth often lemon-scented when crushed, sometimes moderately glaucous between densely and evenly dotted glands. Glands of herbage 0.3–0.4 mm wide, raised and at first globose with a clear glistening viscid center, the center collapsing with age or when dried, the glands then doughnut-shaped; similar or flatter glands on inflorescences, bracts, sepals, petals, ovaries, and fruits. Leaves sparse, linear or very narrowly spatulate, 6–23 × 0.8–1.2 mm, reduced above to scales, very quickly shed, the plants usually nearly leafless.

Inflorescences of few-flowered raceme-like branches near stem tips, or flowers solitary in axils. Sepals, petals, and stamens glistening dark indigo blue-purple, contrasting spectacularly with the stems; pollen yellow. Calyx 1.7–2.5 mm, thickish and persistent, the lobes broadly ovate-rounded with yellowish tips, the corollas likewise thick, 8.0–11.5 mm. Style 7–10 mm, pale violet, single, slender, twisted, and fragile but persistent on the fruit; stigma capitate, glistening chartreuse, glandular with protruding papillae. Fruits tough and leathery, with a pedicel-like stipe and 2 gland-dotted globose chambers, each 5–7 mm wide, opening tardily along a slit at the top of each sphere. Seeds ca. 5 mm, 1–3 in each chamber. Flowering January-April, and fruiting in the same season. Flowers protogynous—the style and stigma protrude from the unopened flower, then the corolla opens, the filaments elongate, and the anthers dehisce. Flowers stinking at midday, visited by large syrphid flies and honeybees (the bees vigorously sticking their head all the way into the corolla, spreading apart the petals to get inside).

Western granitic ranges, in localized pockets on north-facing mountainsides, from the bases to peaks, with various slope exposures. Not known elsewhere in Sonora. Also Arizona including the Tinajas Altas Mountains, southeastern California, southern Nevada, southwestern Utah, and Baja California; rocky habitats in desert, grassland, and oak-juniper woodland.

Sierra Nina, *F 89-48.* El Puerto, "Sierra Tuseral," *F 13216.* Desert pavement, 32°05′ N, 113°41′ W, *Ezcurra 29 Apr 1981* (MEXU). Sierra del Viejo, *F 85-730.* Observation: Cerro Pinto.

SALICACEAE WILLOW FAMILY

Trees and shrubs. Leaves winter deciduous, alternate, simple, usually petioled, and with stipules (often soon deciduous). Dioecious. Flowers in catkins, each flower subtended by a scalelike bract, and wind-pollinated. Calyx vestigial; petals none. Fruit a small capsule. Seeds minute, each with a tuft of long, silky hairs adapted for wind dispersal.

Mostly temperate regions of the Northern Hemisphere; 2 genera. Even the Sonoran Desert and subtropical Mexican species are winter deciduous, pointing to their north-temperate affinity.

Great floodwater gallery forests of willow and cottonwood once stretched along the Río Colorado and its tributaries, from the myriad delta channels northward to elevations well above the desert. With all that water and heat the growth of these trees must have been incredibly fast and their biomass enormous—fueling the teeming life of the delta and Gulf of California. One imagines biological richness like that of the Nile Delta before it too met its twentieth-century fate. The leaves turn yellow in winter, the cottonwood more brilliantly than the willow, and new leaves burst forth early in spring. The blaze of yellow and flush of green along the rich waterways must have been spectacular.

Not until the Río Yaqui and the Río Mayo in southern Sonora, south of Guaymas and south of the desert, did one again encounter comparable willow and cottonwood forests. Remnants of the Yaqui and Mayo riverine forests still existed in the 1990s. In these southern rivers Frémont cottonwood is replaced by the distinctive Yaqui cottonwood, *Populus mexicana* subsp. *dimorpha,* but the willow trees are the same species. References: Leopold 1949; Rea 1983; Sykes 1937.

1. Leaf blades more or less deltate, about as long as wide; catkins drooping. ____________ **Populus**
1′ Leaf blades linear to lanceolate, more than twice as long as wide; catkins erect. ____________ **Salix**

POPULUS *ÁLAMO;* COTTONWOOD, POPLAR

Trees, mostly large, winter deciduous. Early-season leaves often different from late-season leaves; petioles usually long, terete or often flattened lengthwise. Flowers in drooping, stalked catkins, appearing in spring before the leaves.

Northern Hemisphere; 40 species; includes aspen, cottonwood, and poplar. Female trees produce copious quantities of seeds surrounded by white hairs, resulting in wind-borne cottony drifts. Reference: Eckenwalder 1977, 1992.

Populus fremontii S. Watson subsp. **fremontii** [*P. arizonica* Sargent. *P. arizonica* var. *jonesii* Sargent. *P. fremontii* var. *pubescens* Sargent. *P. macdougalii* Rose] *ÁLAMO;* FRÉMONT COTTONWOOD; *'AUPPA.* Large, fast-growing trees with soft wood. New shoots and leaves hairy, glabrate with age. Petioles 3–9 cm, laterally compressed, especially near the blade; leaf blades 4.5–17.0 cm long and about as wide, shiny green, more or less triangular in outline, the base truncate to cordate, the apex short acuminate to acute. Early-season leaves often with larger marginal teeth than the late-season leaves. Leaves turning golden yellow in early winter. Usually flowering in February, the leaves enlarging as flowers fade, the fruits ripening late March–early April. Also propagating by root sprouts.

This is the largest native tree in the Sonoran Desert. Formerly abundant along the Río Colorado and its delta. Historic photographs indicate these trees reached at least 20 m. These forests vanished during the first half of the twentieth century as the river was dammed and diverted. It is presently a common tree in the agricultural areas south of San Luis, and south of the agricultural area in the lower part of the delta are some thousands of young trees to ca. 10 m tall where the river water has a salinity of 1000–2000 ppm (see discussion of the delta in the introduction). In the late 1980s there was a single 12 m tree at Laguna Prieta, and well-established colonies occur at Quitobaquito, here and there

Salix gooddingii. LBH (Benson & Darrow 1945).

Salix exigua subsp. ***exigua.*** LBH (Benson & Darrow 1945).

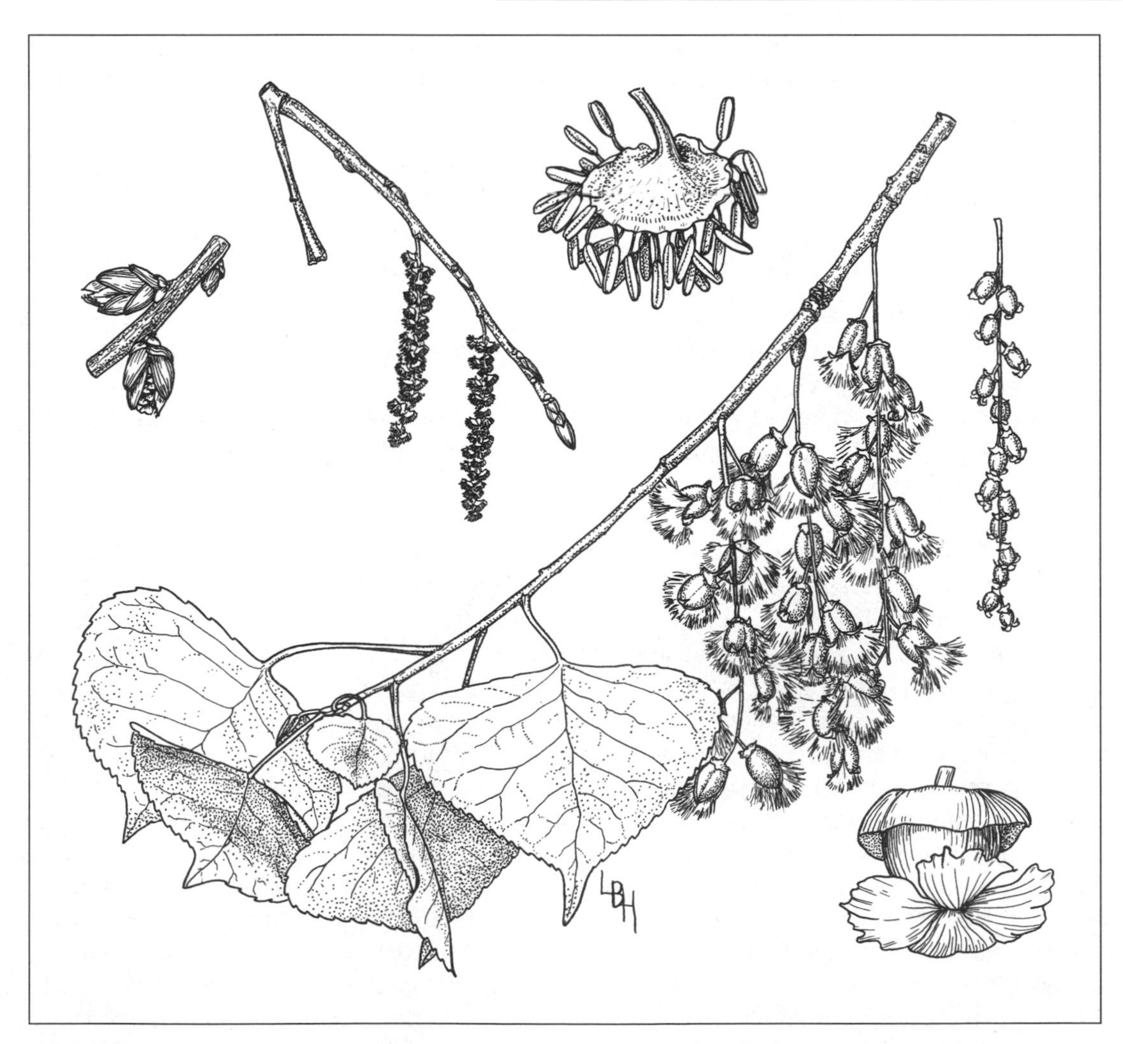

Populus fremontii subsp. ***fremontii.*** Note male flower above, and female flower at lower right. LBH (Benson & Darrow 1945).

along the Río Sonoyta near Sonoyta, and at Quitovac. I know of no other native or long-established cottonwoods southward near the coast of Sonora until the Río Yaqui.

The Quitobaquito and Quitovac cottonwoods, and some or all of those at Sonoyta were planted from cuttings. The Quitobaquito trees are all pistillate. In his plant catalog for January 21, 1894, at Sonoyta, Edgar Mearns wrote, "It is extensively planted along acequias here, and is said to be the cottonwood of the Gila River near Gila Bend" (Mearns 1892–93:#187, US). Approaching Sonoyta in late February 1910, Carl Lumholtz (1912:287) wrote, "Already at a distance the attractive light green color of the cotton-wood trees (Spanish, álamo) and willows were evident, the new leaves being half-grown." In April 1986 I was told by local inhabitants that the cottonwoods at Presa Derivadora were brought from the Río Colorado.

In Eckenwalder's "broad view," *P. fremontii* subsp. *fremontii* occurs west of the continental divide, primarily in California and Utah southward to Baja California and northern Sonora. Subsp. *mesetae* Eckenwalder replaces subsp. *fremontii* in southern New Mexico and southwestern Texas, and ranges southward to the Valley of Mexico.

Laguna Prieta, *F 85-745*. Sonoyta: *Anderson 13 Nov 1985; Mearns 2738* (#187 in his plant catalog notebook, US, specimen not located). Colonia Lerdo, *MacDougal Feb 1904* (US). Colorado River south of International Boundary, Sonora shore, *MacDougal 27 Mar 1905* (US). Quitobaquito, *Peebles 14563.*

SALIX *SAUCE, SÁUZ,* WILLOW

Trees or shrubs, dwarfed in arctic-alpine regions. Leaves often lanceolate. Catkins appearing before or with the leaves. Northern Hemisphere, mostly colder and temperate regions, a few in the Southern Hemisphere; 400 species.

Three willows occur in the Sonoran Desert: *S. bonplandiana* Kunth, *S. exigua,* and *S. gooddingii*. The wide-ranging *S. bonplandiana* does not extend into lowland western Sonora. References: Argus 1986, 1995; Dorn 1976, 1977.

1. Shrubs 1.8–3 m lacking a well-defined trunk; ultimate stems erect or ascending; leaves 0.8–2.8 mm wide, often grayish; stamens 2; capsules sessile or subsessile with "pedicels" to 0.3 (0.4) mm. ________ **S. exigua**

1′ Shrubs or trees to 8+ m with a well-developed trunk; ultimate stems often drooping; leaves 8–17 mm wide, green; stamens 4–9; capsules with "pedicels" 1.0–1.3 mm. ________ **S. gooddingii**

Salix exigua Nuttall subsp. **exigua.** SANDBAR WILLOW, COYOTE WILLOW. Shrubs 1.8–3 m, stems erect or ascending. Young shoots including leaves densely pubescent with short white appressed hairs; stems becoming glabrate with age, the leaves often silvery to grayish green, becoming glabrate above. Stems often reddish brown, or yellow-brown especially when young. Leaves short petioled, linear, mostly 3.5–12.0 cm × 0.8–2.8 mm, the midrib prominent on the lower surface (more so than in *S. gooddingii*); margins entire to small-toothed. Male flowers with 2 stamens. Ovary glabrous or glabrate, the stigma sessile. Fruits sessile to substipitate with the pedicel-like stipe to 0.3 (0.4) mm. Flowering spring and sporadically until fall.

Dense colonies of slender-stemmed, spindly shrubs in wet soil at the larger La Salina pozos and occurring as agricultural weeds south of San Luis; also at Yuma and probably once along the lower Río Colorado and the delta. Elsewhere in Sonora at one waterhole on Isla Tiburón and in the northeastern part of the state as far south as the Sierra Aconchi; also northwestern Sinaloa in the Río Fuerte delta. Alberta, western United States and northwestern Mexico. This species, with 1 other subspecies, ranges from Alaska across much of Canada and the United States to northern Mexico.

La Salina ["Pozo Borrascosa"], *F 86-547.* 18 km S of San Luis, *F 86-1036.* Isla Tiburón, Sauzal waterhole, *F 10076* (incorrectly reported earlier as *S. gooddingii* by Felger & Lowe [1976] and Felger & Moser [1985]).

Salix gooddingii C.R. Ball [*S. gooddingii* var. *variabilis* C.R. Ball. *S. nigra* Marshall var. *vallicola* Dudley] *SAUCE, SÁUZ;* GOODDING WILLOW; *CE'UL.* Large shrubs to trees. Female flowers and fruits with pedicel-like stipes (0.8) 1.0–1.3 mm. Young herbage hairy, soon glabrate. Leaves short petioled, narrowly lanceolate, 5.5–11.0 cm × 8–17 mm. Male flowers with 4–9 stamens. Peak flowering in February-March.

Quitobaquito and Río Sonoyta near Sonoyta; formerly along the waterways of the Río Colorado and presently scattered in irrigated agricultural areas in the lower Río Colorado valley; also Quitovac. Southwestern United States and northern Mexico including both Baja California states, Sonora, and Sinaloa.

Early photos show willows along the Río Colorado and its delta reaching at least 20 m. In the early 1990s there were still 3–8 m trees and shrubs of a remnant willow forest at Sonoyta among urban refuse and weedy tamarisk shrubs along flowing water in the deeply cut Río Sonoyta channel. The largest willow at Sonoyta, at the small lake behind Presa Derivadora, is 15 m with a trunk 60 cm in diameter. There are few unequivocally native willow populations near the Sonora coast south of the flora area and north of the Río Yaqui (see *S. exigua*). *S. nigra* Marshall, the black willow, replaces *S. gooddingii* in the central and eastern United States.

Colonia Díaz, *Mearns 2840* (UC). Río Sonoyta at Sonoyta, *F 85-691*. Río Sonoyta, 21 km W of Sonoyta, *F 85-977*. Quitobaquito, *Darrow 2396*.

SAPINDACEAE SOAPBERRY FAMILY

Trees, shrubs, or vines, or rarely herbaceous. Leaves almost always alternate, usually compound, seldom entire; stipules usually none. Flowers often unisexual. Fruits highly variable, dry to fleshy. Worldwide, greatest diversity in Asia and America, tropical and subtropical, few in deserts; 140 genera, 1325 species.

DODONAEA

Shrubs or small trees. Herbage usually resinous-glutinous. Leaves simple or pinnate; stipules none. Flowers inconspicuous; petals none. Fruits capsular. Mostly Australian, several from Madagascar, Java, and Hawaii, 1 in the Americas; 50 species. References: Lippold 1978; West 1984.

Dodonaea viscosa Jacquin [*D. angustifolia* Linnaeus f. *D. viscosa* var. *angustifolia* (Linnaeus f.) Bentham] *JARILLA;* HOP BUSH. Leafy shrubs ca. 1.5–2 m with shredding-peeling bark, the branches mostly erect-ascending. Herbage, especially new shoots, ovaries, and developing fruits, densely resinous-glutinous. Leaves simple, 3–9 cm × 3–9 mm, linear to linear-oblanceolate, narrowed at base and essentially sessile, the midrib prominent, the margins entire. Mostly dioecious but with some bisexual flowers. Flowers pedicelled, in small clusters among the leaves, yellow-green; sepals 2 mm, often 4 in number. Capsules 1–2 cm wide, papery, 3-winged, often pinkish red, drying straw colored; producing massive fruit crops in spring. Seeds 3.0–3.5 mm across, dull blackish, thickly lens-shaped with a low crest.

Sierra Cipriano in the northeastern corner of the flora area; north-facing slopes to peak elevation, mostly canyons and cliff bases. In the Sonoran Desert region often in the ecotone between desert and oak woodland. Worldwide in tropics and subtropics. There is much variation in leaf size. The narrow-leaved plants from Arizona and northwestern Mexico have been called var. *angustifolia;* the broader-leaved var. *viscosa* occurs within the same region. However, *D. angustifolia* is sometimes considered a species distinct from *D. viscosa*.

Sierra Cipriano, *F 91-26*. 10 mi SW of Sonoyta, *Van Devender 1 Jun 1983*.

SAURURACEAE LIZARD'S-TAIL FAMILY

Perennial herbs with rhizomes; aromatic with ethereal oil cells and tannins. Leaves alternate, simple, the stipules fused to the petiole. Flowers small, in terminal, bract-bearing spikes or racemes; perianth none. Eastern Asia and North America, temperate and subtropical wetlands; 5 genera, 7 species. References: Buddell & Thieret 1997; Wood 1971.

Anemopsis

This genus of 1 species is the only representative of the family in western North America.

Anemopsis californica (Nuttall) Hooker & Arnott [*Anemia californica* Nuttall. *Anemopsis californica* var. *subglabrata* Kelso] *HIERBA DEL MANSO; VA:VIS.* Herbaceous perennials with thick creeping aromatic rootstocks and long above-ground stolons. Leaves mostly basal; petioles 3–80 cm, the blade ovate, 5–15 (30) × 3.0–7.5 (14) cm (much smaller when drought stressed), the margins entire. Inflorescence a spike simulating a single flower (*Anemopsis* is Greek for 'anemone-like'). Spikes many-flowered, compact, thick and conical, mostly 3.0–3.8 cm, subtended by petal-like bracts, these (2.0) 2.5–4.0 cm, white, aging to green; flowers fragrant. Spikes 1 to several on mostly erect stems 12–110 cm with a few reduced leaves. Ovaries sunken in rachis of spike. Mass flowering in late spring and summer. Winter dormant, the leaves freeze killed.

Wetlands in muddy soil and emergent from shallow, slow-moving or stagnant fresh water (often alkaline or brackish). Extensive populations in the Ciénega de Santa Clara and marshy places at the edge of the Río Colorado below Mesa de Sonora (e.g., vicinity of El Doctor), Laguna Prieta, freshwater pozos at La Salina, Río Sonoyta below Cerro el Huérfano, and at Quitobaquito. Often with *Distichlis spicata* and beneath *Prosopis pubescens*. At Laguna Prieta it covers several hectares of wet mud at the edge of the nearly dry laguna. Semi-saline and alkaline soils in southwestern and midwestern United States and northwestern Mexico.

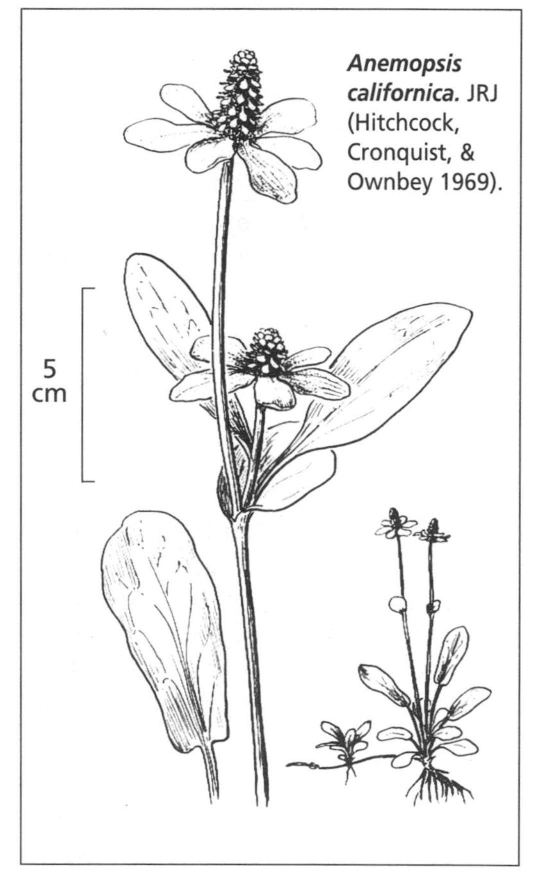

Anemopsis californica. JRJ (Hitchcock, Cronquist, & Ownbey 1969).

Hierba del manso is one of the most esteemed medicinal plants in western North America (e.g., Felger & Moser 1985; Felger et al. 1992; Ford 1975). It has long been cultivated in Sonora and is grown in cool, shaded gardens in Sonoyta as well as kitchen gardens in ejidos and farms in the Río Colorado valley. The term *manso* may be translated as 'gentle', 'lamblike', or 'tame'. The plants also have been used as a source of tannin.

Mearns (1892–93) reported that "the Mexicans at the village of Sonoyta . . . call it '*Yerba del Manzo,*' and assert that it imparts properties to the water which render it deleterious to the teeth. It grows on marshy ground, and emits a strong odor when trodden upon. Mexicans use it medicinally. Found on the boundary line from Lake Palomas (Mimbres Valley) westward to Quitovaquito, in all marshy grounds." Tom Childs recounted that the "Sand Papago" went to La Salina (Salina Grande) for "yerba mansa" (Van Valkenburgh 1945). Lumholtz (1912:264–265) gives the following account of his experience with *Anemopsis* at La Salina in February 1910:

> A plant . . . called by the Mexicans *herba del manso* was a singular growth in these pozos. Its large root, which has a strong medicinal scent, like that which characterizes an apothecary shop, is perhaps the most popular of the many favorite remedies of northern Mexico. It is used internally to cure colds, coughs, or indigestion, as well as externally for wounds or swellings, and is employed in a similar way by the Indians. Of the latter, those who lived in the dune country are said to have been in the habit of chewing bits of this root, as elsewhere tobacco is chewed. These plants grew here in great numbers and to enormous proportions; some of their roots were as much as three feet long and very heavy. The root finds a ready sale everywhere and my Mexicans were not long in gathering as many of the plants as they could carry on their animals. One of the men, whose horse was well-nigh exhausted, walked himself in order to put a load of fifty pounds on his horse.

La Salina, *F 86-540*. Laguna Prieta, *Roth 8*. 20 mi N of El Golfo, *F 75-54*. Río Sonoyta, S of Cerro el Huérfano, *F 92-978*. Sonoyta, cultivated, *F 86-142*. Quitobaquito, *Mearns 2786* (US).

SCROPHULARIACEAE FIGWORT FAMILY

Mostly annual or perennial herbs, some vines, shrubs, and trees. Leaves opposite or alternate, sometimes whorled or in basal rosettes, simple, entire to dissected, without stipules. Sepals usually 4 or 5, separate or fused; corollas sympetalous, usually 4- or 5-lobed, usually bilateral, often 2-lipped. Fertile stamens (2) 4, sometimes with a fifth sterile stamen. Ovary superior. Fruit often a capsule (ours) with numerous small smooth or sculptured seeds.

Worldwide, mostly temperate and tropical montane regions; 220 genera, 4000 species. Includes garden flowers such as snapdragon (*Antirrhinum majus*) and foxglove (*Digitalis purpurea*). Reference: Holmgren 1984.

1. Woody shrubs on Pinacate Peak. __________ **Keckiella**
1′ Ephemeral or perennial herbs.
  2. Stems and leaves glandular hairy and sticky.
    3. Corollas purplish blue, ca. 1 cm; body of seed tuberculate with irregular ridges. __________ **Antirrhinum cyathiferum**
    3′ Corollas yellow, 2.5–3.5 cm; body of seed smooth. __________ **Mohavea**
  2′ Stems and leaves glabrous or moderately glandular or white hairy when young but not sticky glandular.
    4. Plants vining, the pedicels 4–6+ cm and coiling like tendrils; corollas yellow. __________ **Antirrhinum filipes**
    4′ Plants not vining, the pedicels not coiling; corollas pinkish, white, or sometimes yellow.
      5. Plants 40–250 cm; larger leaves more than 4 cm; corollas 2 cm or more. __________ **Penstemon**
      5′ Plants less than 35 cm; leaves less than 2 cm; corollas 1.2 cm or less.
        6. Pedicels as long as or longer than calyx, especially fruiting calyx; sepals united and enclosing the fruits; corollas pink or yellow and showy. __________ **Mimulus**
        6′ Pedicels inconspicuous, much shorter than calyx; sepals separate (or essentially so), not enclosing the fruits; corollas white, inconspicuous. __________ **Veronica**

ANTIRRHINUM SNAPDRAGON

Annual or perennial herbs, erect or sometimes vining. Leaves mostly opposite below and alternate above (ours). Sepals united only at base. Corollas 2-lipped, the tube swollen below, the throat with a prominent palate (the projecting part of the lower lip that closes the throat). Stamens 4. Seeds intricately sculptured with irregular corky and tuberculate ridges.

Fifteen species in western North America and 21 in the western Mediterranean. The garden snapdragon, *A. majus,* is grown as a winter-spring annual in the Sonoran Desert. Reference: Thompson 1988.

1. Stems erect to ascending and self-supporting, not threadlike and not vining; pedicels straight, shorter than the flowers or fruits; corollas purplish blue. __________ **A. cyathiferum**
1′ Stems threadlike and vining; pedicels twining and many times longer than the flowers or fruits; corollas yellow. __________ **A. filipes**

Antirrhinum cyathiferum Bentham. DESERT SNAPDRAGON. Non-seasonal ephemerals, 3–25 cm, viscid glandular hairy and foul smelling; stems moderately thick, the branches ascending. Leaves opposite at lower 1 or 2 nodes, alternate and usually gradually smaller upwards, the blades ovate to broadly lanceolate, often with apical glands (dark purplish areas), the petioles conspicuous; larger leaves 1.5–4 (5.5) × 0.8–1.6 cm. Pedicels shorter than the flowers or fruit, turning downward, inverting the fruit. Corollas 1 cm, purple-blue with darker veins, the lip with 2 yellow spots and hairy at the entrance to the throat. Seeds whitish, becoming dark brown with age, 1.9–2.5 mm, with a wide cup-shaped wing surmounted by a linear body, the body 1.5–1.7 mm, tuberculate on irregular ridges.

Sonoyta region, Pinacate volcanic field, granitic ranges, and desert plains; rocky, cinder, gravelly, and sandy soils; desert pavements, plains, arroyos, and slopes. Often abundant on cinder soils at higher elevations in the Sierra Pinacate. Endemic to the Sonoran Desert: western Sonora south to the Guaymas region, both Baja California states, southwestern Arizona, and rare in southeastern California.

Sierra del Rosario, *F 20715*. MacDougal Crater, *F 10730*. Arroyo Tule, *F 19224*. Near Crater Elegante, *Starr 732*. Cholla Bay, *Bittman 419*. Quitobaquito, *Harbison 27 Nov 1939* (SD).

Antirrhinum filipes A. Gray [*Asarina filipes* (A. Gray) Pennell. *Neogaerrhinum filipes* (A. Gray) Rothmahler] CLIMBING SNAPDRAGON. Winter-spring ephemerals. Seedlings and young plants or lower stems sparsely to densely villous with white hairs, the stems and leaves otherwise glabrate and odorless. Stems vining and threadlike, with long internodes. Leaves opposite and broadly lanceolate to ovate at the several lower nodes, narrower upwards and linear-lanceolate to linear; longer leaves (1.5) 3–5 cm, the petioles conspicuous. Pedicels very slender, often more than 4–6 cm, twining like a tendril when contacting other stems—even those of the same plant, cactus spines, etc. Flowers with glandular hairs and sparse non-glandular white hairs; corollas bright yellow and showy, 1.5–1.6 cm. Seeds whitish, becoming dark brown with age, 1.1–1.5 mm, with several thickened, parallel ridgelike wings on one side, the body irregularly tuberculate on the other side.

Mountains in the Sonoyta region and adjacent Organ Pipe Monument, and the Pinacate volcanic field including higher elevations; creosote bush plains, crater floors, and cinder flats and slopes. Seldom common, apparently growing mostly during wetter years. Not known elsewhere in Sonora; also southeastern California, southern Nevada, southwestern Utah, and western Arizona.

Thompson (1988:1180) reports, "Plants are frequently found with clusters of filiform shoots bearing exclusively cleistogamous flowers, the shoots originating from the hypocotyl; chasmogamous flowers are simultaneously produced by the main stem axes above." However, I have not seen cleistogamous flowers on Sonoran plants.

N face of Pinacate Peak, 875 m, *F 19369*. Arroyo Tule, *F 19225*. Cerro Colorado, *Webster 22305*. 11 mi SW of Sonoyta, *F 9800*.

KECKIELLA

Shrubs or perennials with woody bases. Leaves opposite or sometimes alternate above, entire to toothed, somewhat tough. Nectary a hypogynous disk. Sepals 5, essentially separate (united only at base). Corollas strongly 2-lipped, relatively large, showy, and yellow, whitish, or red. Fertile stamens 4, the filaments hairy at their bases, the fifth stamen a sterile staminode, glabrous or densely hairy. Capsules firm, ovoid.

Keckiella, named in honor of David D. Keck, is a genus of 7 species "native to California and immediately adjacent areas, where they are shrubby members of the chaparral and similar xerophytic or sub-xerophytic communities of the Mediterranean climatic area" (Straw 1966:87). *Keckiella,* also treated as a subgenus of *Penstemon,* is distinguished by its shrubby habit, nectar-producing disk at the base of the flower, lack of glandular hairs on the corollas, and filaments all prominently hairy at their bases. References: Keck 1936; Straw 1966, 1967.

Keckiella antirrhinoides (Bentham) Straw subsp. **microphylla** (A. Gray) Straw [*Penstemon microphyllus* A. Gray. *P. antirrhinoides* Bentham subsp. *microphyllus* (A. Gray) D.D. Keck. *Keckia antirrhinoides* (Bentham) Straw subsp. *microphylla* (A. Gray) Straw] BUSH PENSTEMON. Woody shrubs ca. 2 m, much-branched, the stems slender and brittle, often branching at right angles. Bark light tan. Leaves ca. 1.5–3.0 cm, more or less ovate, thickish, canescent with sparse to moderately dense short hairs, the margins entire; facultatively drought deciduous. Sepals acute. Corollas bright yellow, 2+ cm across. Sterile fifth stamen conspicuously bearded. Dry, dead capsules persistent. Flowering April and October.

Well established but restricted to the northeast slope of Pinacate Peak, among chaparral-like scrub of the major north-facing cliff and among rock rubble at the cliff base (1050–1100 m, *F 86-442, F 92-500*). Not known elsewhere in Sonora. The nearest population is in the Ajo Mountains in southern Arizona.

Keckiella antirrhinoides subsp. ***micro-phylla.*** LBH (Benson & Darrow 1945).

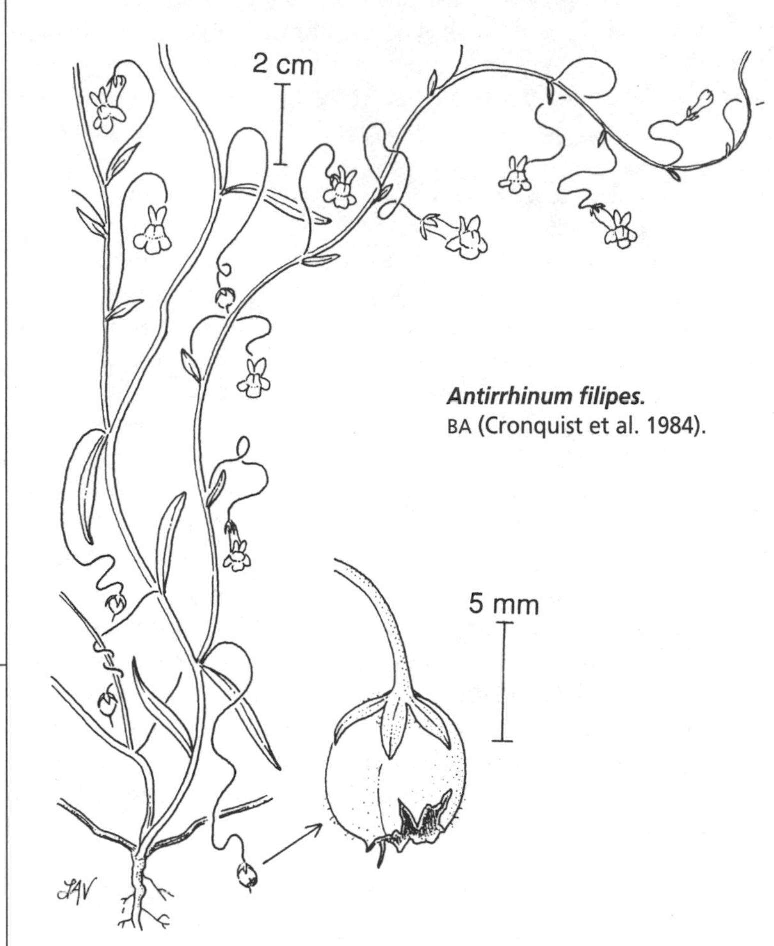

Antirrhinum filipes. BA (Cronquist et al. 1984).

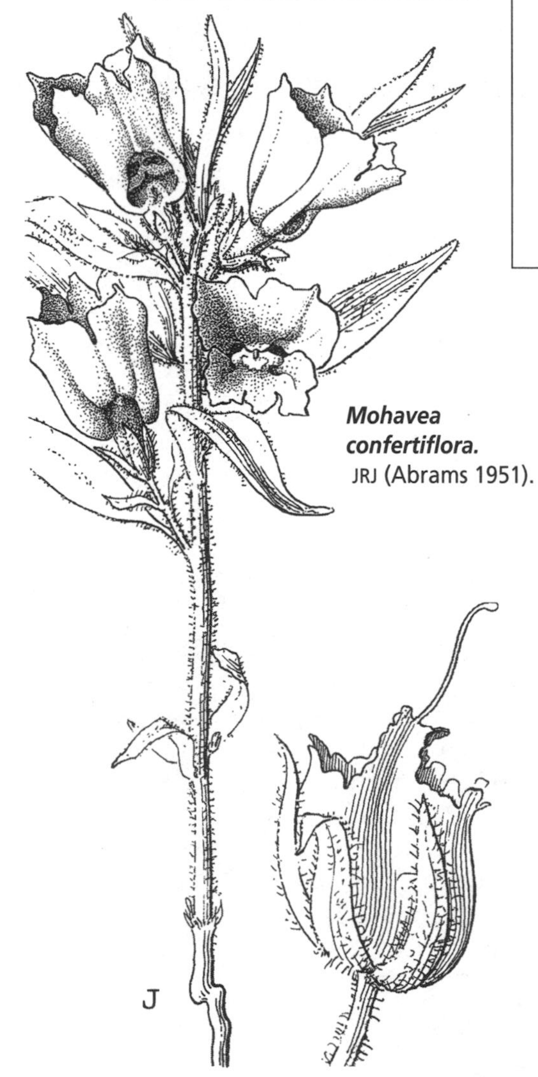

Mohavea confertiflora. JRJ (Abrams 1951).

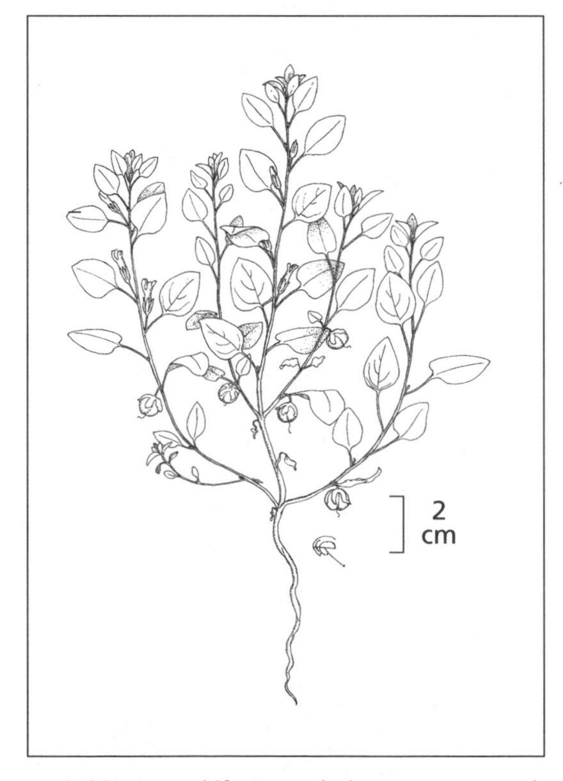

Antirrhinum cyathiferum. FR (Felger & Moser 1985).

This subspecies is the only member of *Keckiella* that occurs inland, east of the western part of California or Baja California. Subsp. *microphylla* otherwise in southeastern California, north-central Baja California, and west and central Arizona. It intergrades with subsp. *antirrhinoides* at the area of contact in southwestern California.

MIMULUS MONKEY FLOWER

Annual or perennial herbs. Leaves opposite. Inflorescences of flowers in pairs from axils of opposite leaves, or sometimes of bracteate racemes. Calyx often inflated, the sepals united into a strongly 5-angled or pleated tube. Corolla slightly to strongly 2-lipped. Stamens 4, fertile. Seeds numerous, small.

Mostly western North America, 70 species in California, especially in nondesert regions, a few in South America and widely disparate in the Old World; 150 species. Reference: Grant 1924.

Mimulus rubellus A. Gray. Diminutive winter-spring ephemerals, 1.5–8.0 cm. Herbage reddish, sparsely short-haired. Leaves 5–10 mm, variable, the lower ones often broadly obovate, short petioled, the other leaves sessile and narrower, mostly narrowly obovate to oblanceolate; margins entire to broadly and bluntly shallow-toothed. Calyx cylindrical, slightly enlarging in fruits, 4–6 mm, the lobes short and sparsely ciliate. Corollas protruding 4.5–6.0 mm from the calyx, some populations with pink corollas, but mostly bright yellow with a nectar guide of maroon-red spots along each side of the mid-ventral channel in the throat, or the throat white with pink and the palate (the projecting part of the lower lip that closes the throat) yellow.

Cinder flats and slopes, mostly higher elevations in the Sierra Pinacate and extending to lower elevations between Elegante Crater and Campo Rojo; often locally abundant in favorable years. Not known elsewhere in western Sonora. The nearest population is in the Ajo Mountains. Southeastern California to Baja California Sur, Nevada, Utah, Wyoming, Colorado, Arizona, New Mexico, perhaps western Texas, and adjacent northern Mexico; often at elevations above desert.

Sierra Pinacate, higher elevation, *F 92-455.* 6 km NE of Elegante Crater, *F 92-147.*

MOHAVEA

Winter-spring desert ephemerals. Herbage and sepals glandular hairy and sticky. Leaves alternate, entire. Flowers solitary in upper leaf axils. Stamens 2. Seeds disk-shaped, surrounded by an incurving wing. Mojave and western part of Sonoran Deserts; 2 species. *Mohavea breviflora* Coville has smaller flowers. Seed morphology points to a close relationship with *Antirrhinum.*

Mohavea confertiflora (Bentham) A. Heller [*Antirrhinum confertiflorum* Bentham] GHOST FLOWER. Winter-spring ephemerals, upright, simple to moderately branched, to ca. 30 cm but mostly much smaller. Stems thick and semisucculent. Leaves sessile, ovate to lanceolate, mostly narrowly so, the lower ones (1.0) 2.5–3.5 cm; upper leaves subtending flowers, linear or nearly so, (2.0) 4.5–6+ cm. Flowers showy, the corollas 2.5–3.5 cm, pale yellow with a red spot below a bright yellow callus or nectar guide, and maroon striations and speckling on the pale cream-yellow background color. Capsules thin-walled, ovoid, 1.2–1.4 cm. Seeds 1.9–2.2 mm, resembling a miniature cockroach egg case, with a folded, cuplike wing, the body smooth and about 3/4 as long as the wing.

Pinacate volcanic field to the summit but the distribution not continuous, best developed on loose, gravelly cinder soils. Southeastern California, Baja California including Isla Angel de la Guarda, western Arizona, and northwestern Sonora southward to the vicinity of Puerto Libertad.

The flowers seem large for the size of the plant. In California desert regions it grows within the range of *Mentzelia involucrata.* Both are pollinated by male bees in the genus *Xeralictus. Mentzelia* provides nectar but *Mohavea* flowers, which are thought to mimic those of *Mentzelia,* offer no nectar (Brown & Gibson 1983).

Campo Rojo, *Soule 19 Mar 1983.* Pinacate Peak, *F 19385.* Chivos Butte, *F 19620.* Sykes Crater, *F 18985.*

PENSTEMON BEARDTONGUE

Perennial herbs (rarely functionally annual), sometimes woody at base. Leaves opposite, rarely whorled, at first sometimes in a basal rosette, the upper leaves rarely alternate; leaf margins toothed to entire; lower leaves often petioled, the upper ones sessile. Inflorescences of oppositely branched panicles or racemes, the flowers subtended by usually conspicuous bracts. Sepals 5, nearly separate. Corollas showy, moderately to strongly 2-lipped. Fertile stamens 4, the fifth a sterile staminode, usually bearded (the "beardtongue"); nectaries of glandular hairs on the filaments. Capsules firm. Seeds numerous, irregularly angled.

Alaska to Guatemala, mostly western United States with greatest diversity in Utah; 250 species. This is the largest genus of flowering plants endemic to North America. Apart from the 2 species listed here, no other penstemons occur in western Sonora. Reference: Holmgren 1984.

1. Leaf margins entire, the leaves all separate. ____ **P. parryi**
1′ Leaf margins toothed, the upper leaves joined at their bases. ____ **P. pseudospectabilis**

Penstemon parryi (A. Gray) A. Gray [*P. puniceus* A. Gray var. *parryi* A. Gray] DESERT PENSTEMON, PARRY'S PENSTEMON; *HEWEL 'E'ES.* Winter-spring ephemerals in the flora area, 45–60+ cm. Herbage glaucous and glabrous. First leaves in a basal rosette. Leaves entire, mostly lanceolate to elliptic, petioled below, the larger leaves 5–12+ cm, sessile and narrowed upwards. Flowers on erect, slender, raceme-like panicles. Corollas 2 cm, bright rose-pink. Capsules 7–9 mm. Seeds 1.1–1.8 mm, dark red-brown, minutely tuberculate beaded, irregularly angled due to the developing seeds pressing on one another.

Arizona Upland along the international border from the vicinity of Sonoyta to Quitobaquito; mostly in floodplains of larger washes; not common. Southern Arizona and western Sonora southward to the vicinity of El Desemboque San Ignacio and eastern Sonora southward to the Río Mayo region.

El Papalote, *F 86-15*. Quitobaquito, *Harbison 30 Nov 1939.*

Penstemon pseudospectabilis M.E. Jones var. **pseudospectabilis.** MOJAVE BEARDTONGUE. Robust herbaceous perennials, flowering in the first year; summer dormant. Herbage glaucous and glabrous. Stems erect, usually stout. Leaves 7.0–16.5+ × 3.3–5.0 cm, the margins coarsely toothed; lower leaves petioled and separate, the middle ones sessile, and the upper leaves connate (leaves of the opposite pair joined at their bases). Pedicels, calyces, and corollas sparsely to moderately glandular hairy. Corollas bright rose-purple, 2.5–3.1 cm. Staminode glabrous except for very sparse glandular hairs basally. Capsules 10–14 mm. Flowering during cooler months, from mid-October but mostly February-April.

Abundant on cinder soils in the Sierra Pinacate near Carnegie and Pinacate Peaks, often forming fields of spectacularly large, robust plants reaching 1.8–2.5 m (*F 86-459; Sherbrooke 9 Apr 1983*). Also granitic slopes of Sierra del Viejo, especially higher elevations, here mostly ca. 1.0 m (*F 85-721*). Western Arizona, southeastern California, and northwestern Sonora.

The gigantism in the Sierra Pinacate seems to be environmentally influenced. When cultivated in Tucson these plants thrive but reach only about half the height of their Pinacate parents. Subsp. *pseudospectabilis,* distinguished by glandular hairs on the pedicels and calyces, is replaced eastward in Arizona, northern Sonora, and New Mexico by the usually less robust subsp. *connatifolius* (A. Nelson) D.D. Keck.

VERONICA SPEEDWELL

Annual to perennial herbs. Stems leafy, the leaves opposite. Inflorescence a terminal raceme or lateral from upper leaf axils, the floral bracts alternate or subopposite, leaflike or reduced. Sepals 4, separate or sometimes united into a short tube. Corolla tube short, irregularly 4-lobed. Stamens 2. Mostly temperate, especially diverse in the Old World; 200 species. Reference: Pennell 1921.

Veronica peregrina Linnaeus subsp. **xalapensis** (Kunth) Pennell [*V. xalapensis* Kunth] NECKLACE WEED, PURSLANE SPEEDWELL. Delicate winter-spring ephemerals, (6) 10–30 cm. Leaves and lower stems glabrate, the upper stems with slender stalked glands. Leaves 6–24 mm, usually deciduous by fruiting time, oblong to elliptic or oblanceolate, the margins entire to shallowly toothed. Racemes elongated, mostly terminal, reaching 15–25 cm on larger plants, the floral bracts alternate and gradually reduced upwards. Pedicels short. Corollas and stamens white, the corollas smaller than the calyx. Fruiting sepals 2.5–4.0 mm. Capsules 2.5–4.0 mm, moderately flattened, wider than long, obcordate. Seeds yellow-orange, many, (0.6) 0.7 (0.8) mm, smooth, flattened on one side, ridged on the other side.

Often locally forming 100% coverage in sandy or clayish muddy wet soils at the margins of waterholes, arroyo beds, and temporarily wet depressions such as playas and roadside ditches; localized in the Sonoyta region and the Pinacate volcanic field. This species occurs in temperate North and South America and Eurasia; subsp. *xalapensis* ranges from Alaska and Canada to Central America.

Tinajas de Emilia, *F 87-39.* 9 km SW of MacDougal Crater, *Ezcurra 4 Oct 1983.* Pinacate Junction, *F 88-59.* Quitobaquito, *Gould 2987.*

SIMAROUBACEAE QUASSIA FAMILY

Trees and shrubs. Leaves usually alternate, simple or pinnate. Flowers bisexual or unisexual, usually small and radial. Fruit a drupe, berry, capsule, or samara, the seeds usually solitary. Warmer regions worldwide, few in deserts; 22 genera, 120 species.

CASTELA

Shrubs or small trees, mostly with spinescent twigs. Leaves alternate, simple, or essentially leafless. Dioecious. Inflorescences of panicles. Flowers 4–8-merous, sepals 4–8, pistils mostly as many as sepals, each with 1 ovule, the pistils separate below, united by the styles, the stigmas separate. Fruits of separate drupes, deciduous or persisting capsule-like for several years.

Dry tropics to deserts in the Americas; 15 species, including 3 in the Sonoran Desert region. Reference: Moran & Felger 1968.

Castela emoryi (A. Gray) Moran & Felger [*Holacantha emoryi* A. Gray] *CORONA DE CRISTO;* CRUCIFIXION THORN. Woody shrubs 1.8–2.3 m, with well-developed trunks, the wood very hard and twisted. Twigs rigid, thick, thorn-tipped, moderately to densely pubescent with short white hairs, glabrous with age. Seedlings and young plants with slender, moderately leafy stems, the leaves to 1 cm and quickly drought deciduous; mature plants essentially leafless, the new growth with few, quickly deciduous scale leaves. Flowers and fruits crowded on paniculate branches ca. 1.5–15 cm, the flowering branches pubescent and pinkish. Flowers 8–9 mm wide, the petals often 7, cupped, mostly cream-yellow to greenish or rose-pink; stigmas chartreuse, the ovaries green, turning red as the petals fall; flowers tended by black ants. Fruits in dense, short clusters, persisting for several years or more in the outer branches, with woody carpels in a starlike pattern. Flowering late spring.

At the southeastern margin of the flora area in sandy creosote bush flats. In Sonora rare and known from only 4 localities in the northwestern part of the state (Turner, Bowers, & Burgess 1995). Also southwestern Arizona, southeastern California, and northernmost Baja California.

31°29′ N, 113°20′ W, 7.2 km E of Mex 8 at ca. 20 km N of Puerto Peñasco: *Ezcurra 15 Mar 1982* (MEXU); *F 91-44.*

Castela emoryi. LBH (Benson & Darrow 1945).

SOLANACEAE NIGHTSHADE OR POTATO FAMILY

Herbs, shrubs, trees, and vines. Leaves usually alternate (often fascicled on short shoots), simple (ours) or pinnate; stipules none. Flowers bisexual or occasionally functionally unisexual, usually 5-merous; often showy. Calyx usually 5-lobed, and persistent. Corollas usually 5-lobed, usually plicate (plaited or folded) in bud, often rotate (the tube essentially absent, the limb flat and circular) or tubular, usually radial or nearly so. Stamens usually 5, the filament bases fused to the corolla. Ovary superior, the style 1, the stigma 2-lobed. Fruit usually a berry or capsule; seeds usually many.

Worldwide, mostly tropical and temperate, greatest diversity in Central and South America; 94 genera, 2200 species. Includes chilies (red peppers), eggplant, potato, tobacco, tomato, and garden plants such as petunia, as well as many medicinal and poisonous plants. References: D'Arcy 1973, 1986; Hawkes, Lester, & Skelding 1979; McGregor, Gentry & Brooks 1986; Nee 1993.

1. Herbage densely covered with stellate hairs and the plants often with at least some straight slender spines or prickles, the twigs not thorn-tipped. ____ **Solanum**

1′ Glabrous or if hairs present then not stellate; leaves and twigs without spines or prickles, but some (*Lycium*) with thorn-tipped short shoots or twigs.
- 2. Woody shrubs.
  - 3. Leaves less than 5 cm, sessile or nearly so, or gradually narrowed to a petiole-like base less than 0.5 cm. ____ **Lycium**
  - 3′ Leaves more than 7 cm, petioles well developed, clearly differentiated from the blade, more than 3 cm. ____ **Nicotiana glauca**
- 2′ Plants herbaceous.
  - 4. Plants prostrate, rooting at nodes; leaves 0.5–1.5 (2.0) cm, short petioled to nearly sessile (petiole much shorter than blade). ____ **Calibrachoa**
  - 4′ Plants not prostrate, not rooting at nodes; leaves 2.5–15 cm, or if less than 2.5 cm then the petiole about as long or longer than blade.
    - 5. Corollas 12–16 cm; fruits spiny. ____ **Datura**
    - 5′ Corollas less than 4 cm; fruits not spiny.
      - 6. Corollas tubular; fruit a capsule (dry). ____ **Nicotiana**
      - 6′ Corollas as broad as or broader than long or deep (not tubular); fruit a berry (fleshy).
        - 7. Fruiting calyx not growing around fruit; corollas ca. 0.5 (1) cm wide; anthers opening by terminal pores, the filaments less than 1/3 as long as anthers. ____ **Solanum americanum**
        - 7′ Fruiting calyx partially to fully surrounding fruit; corollas ca. 1–2 cm wide; anthers opening by full-length slits, the filaments as long as or longer than the anthers.
          - 8. Fruiting calyx partially growing around and tightly enclosing the berry; corollas greenish yellow with a woolly pad at center. ____ **Chamaesaracha**
          - 8′ Fruiting calyx completely and loosely growing around the berry like an inflated paper lantern; corollas various colors, without a woolly pad (except *P. lobata* which has purplish corollas). ____ **Physalis**

CALIBRACHOA

Herbs or subshrubs, erect, sprawling or procumbent, glabrous or pubescent with simple glandular hairs. Leaves mostly linear or spatulate. Flowers small, the calyx lobed halfway or more; corollas salverform to cup-shaped, partly violet; stamens 5. Fruit a capsule opening apically; seeds numerous, minute. Tropical to warm-temperate regions of the Americas; 37 species. Reference: D'Arcy 1989.

***Calibrachoa parviflora** (Jussieu) D'Arcy [*Petunia parviflora* Jussieu] Annuals, probably non-seasonal (elsewhere also perennial); stems spreading-prostrate, rooting at nodes, reaching 50 cm. Leaves alternate, 5–20 mm, mostly linear to spatulate or oblanceolate, often semisucculent; margins entire or nearly so. Calyx lobes much longer than the tube. Corollas ca. 8 mm wide, showy, the tube yellowish, the lobes purplish. Seeds 0.5 mm, chunky, brown, conspicuously reticulate with minute iridescent "windows" or facets.

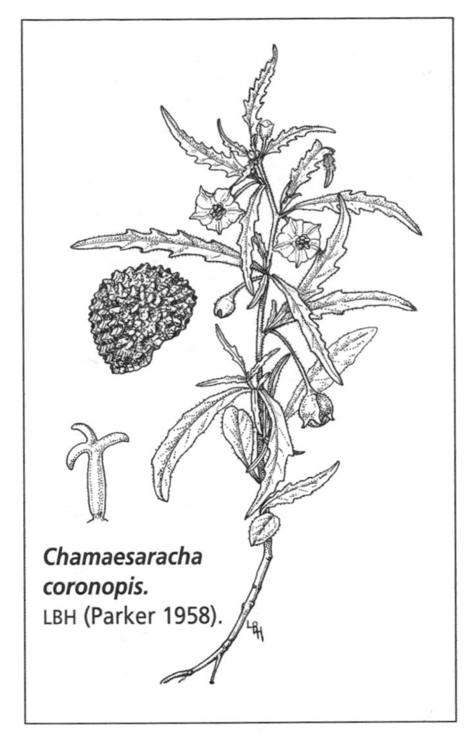

Chamaesaracha coronopis.
LBH (Parker 1958).

Datura discolor.
LBH (Parker 1958).

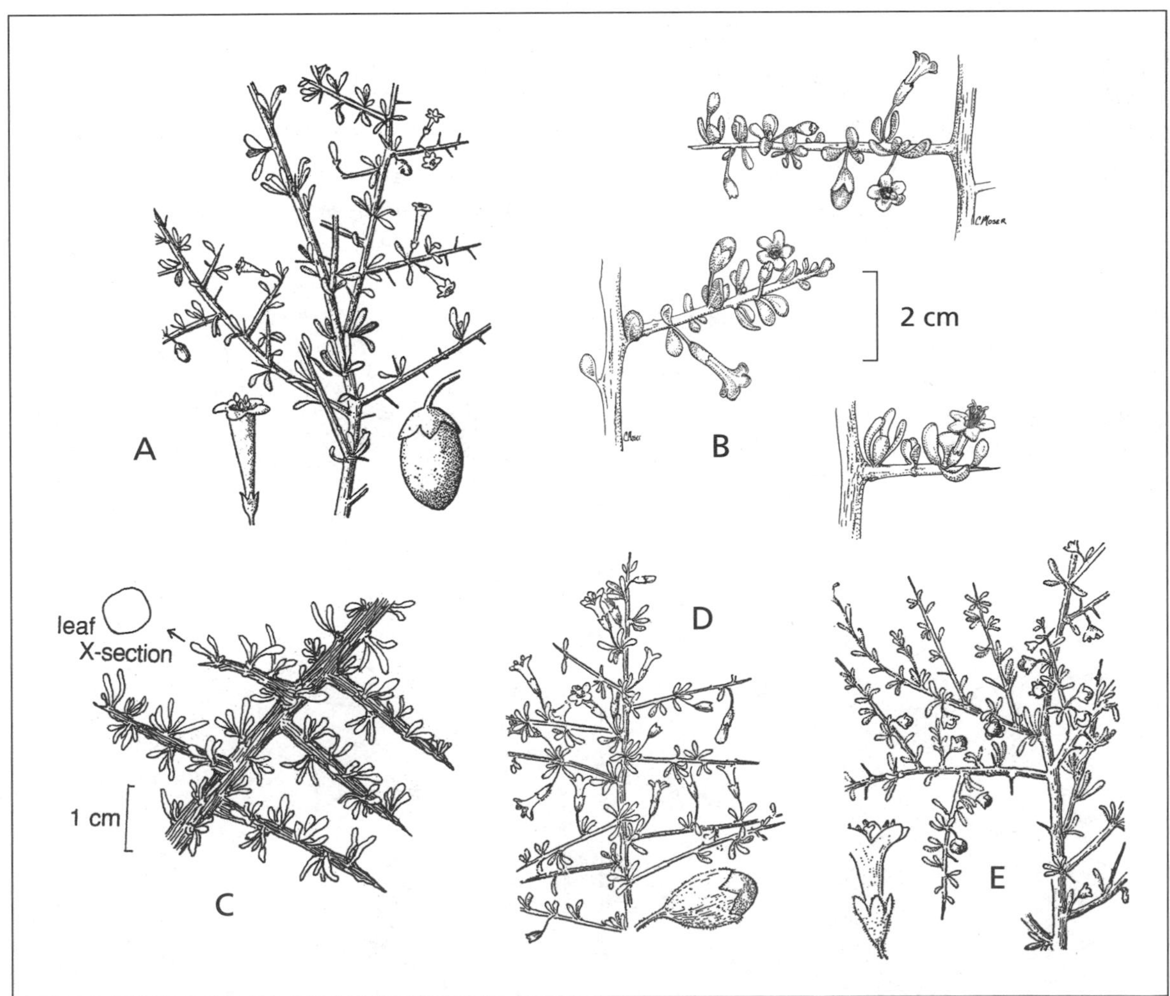

Lycium: (A) L. andersonii var. ***andersonii; (B) L. brevipes*** var. ***brevipes; (C) L. californicum*** var. ***californicum; (D) L. fremontii*** var. ***fremontii; (E) L. parishii*** var. ***parishii. A, D, E,*** JRJ (Abrams 1951); ***C,*** ER (Hickman 1993); ***B,*** CMM (Felger & Moser 1985).

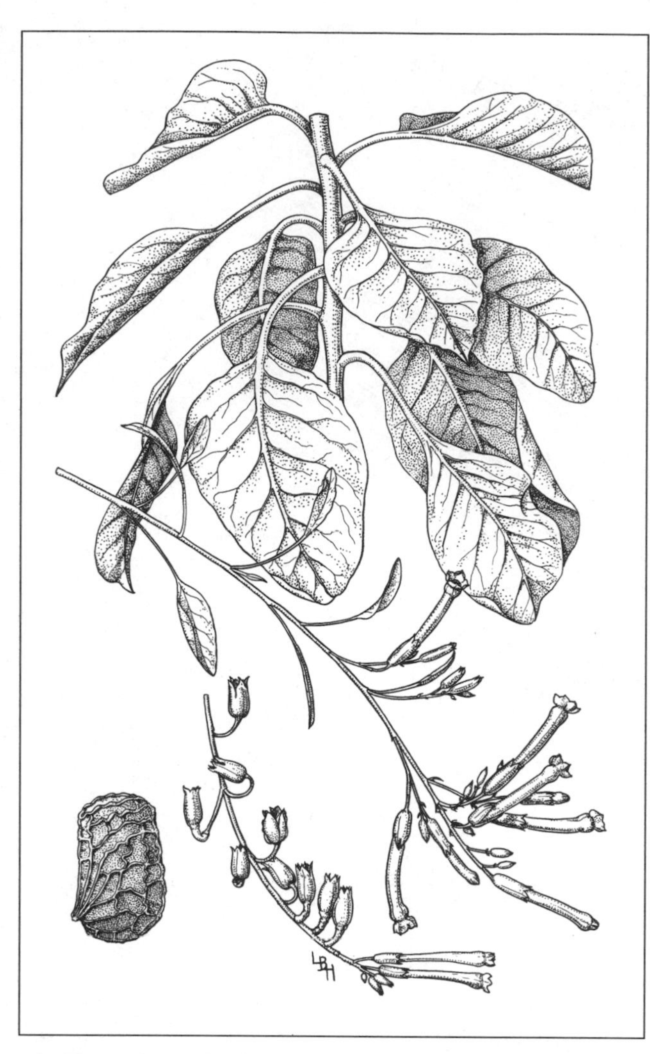

Nicotiana glauca. LBH (Benson & Darrow 1945; Parker 1972).

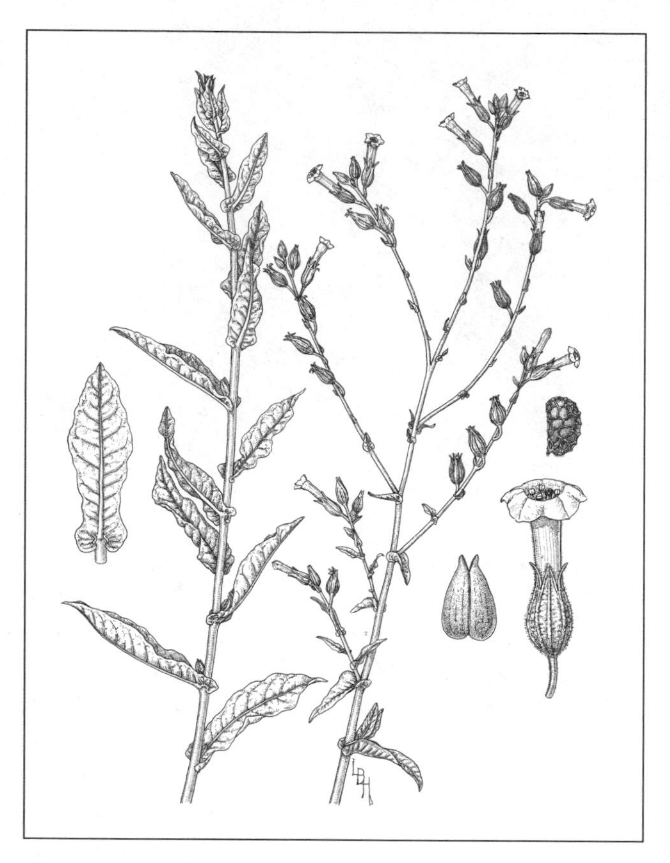

Nicotiana obtusifolia. LBH (Schmutz & Hamilton 1979, in part).

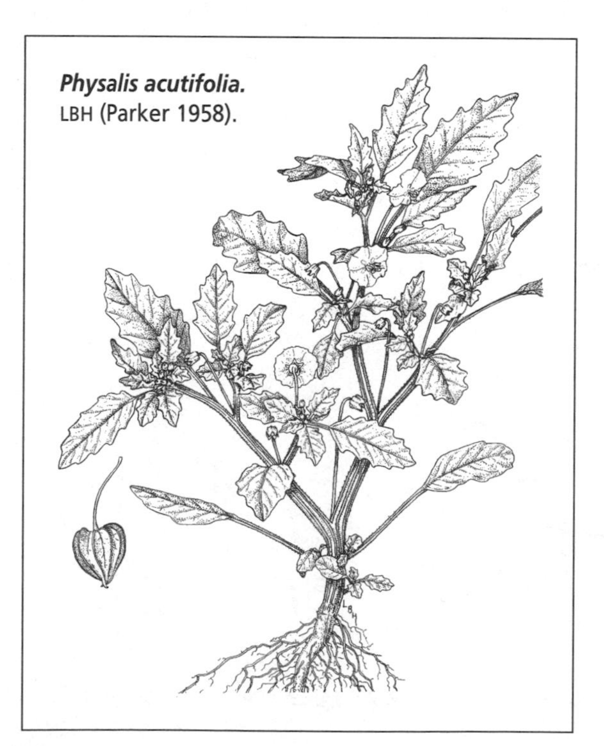

Physalis acutifolia.
LBH (Parker 1958).

Physalis crassifolia var. ***versicolor.***
JRJ (Abrams 1951).

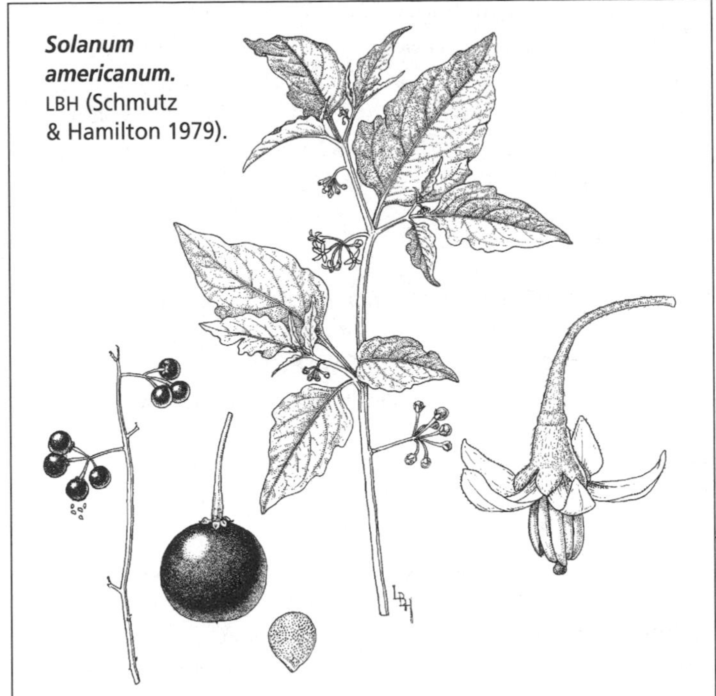

Solanum americanum. LBH (Schmutz & Hamilton 1979).

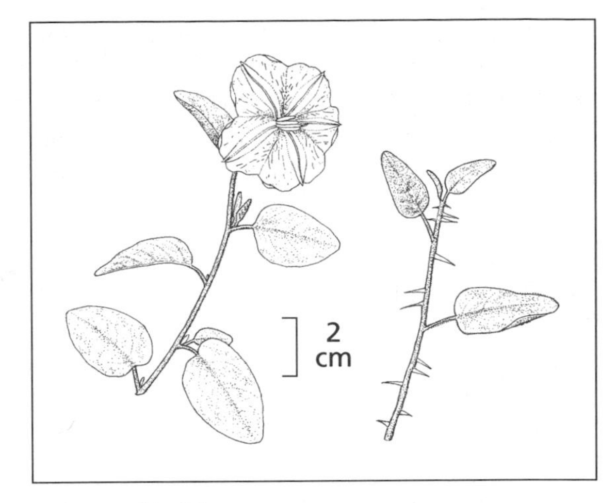

Solanum hindsianum. MBJ.

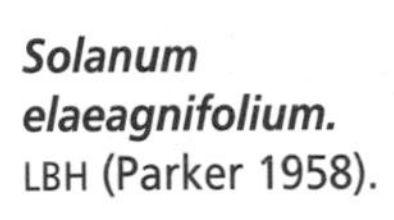

Solanum elaeagnifolium. LBH (Parker 1958).

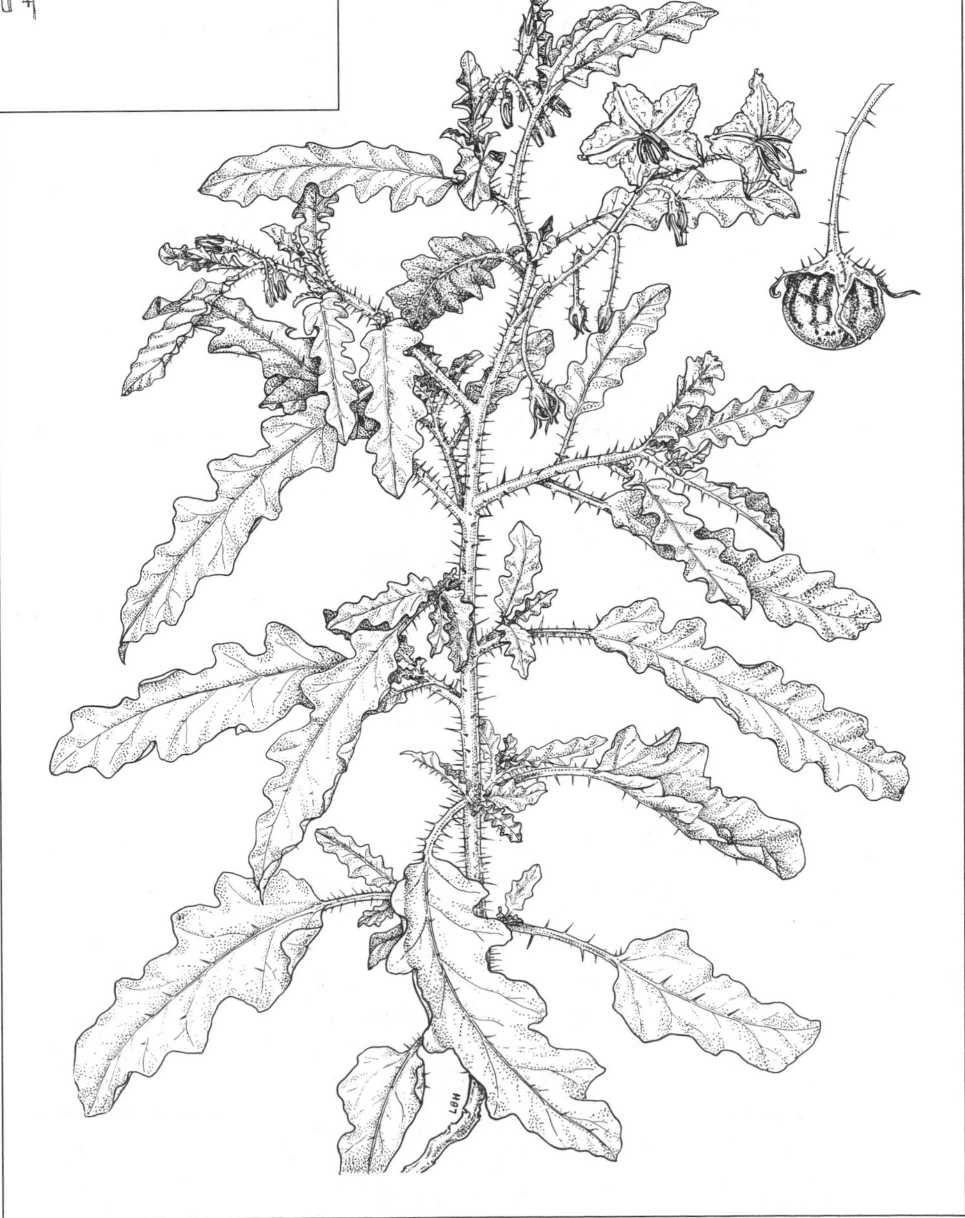

About half a dozen plants found in 1988 on moist soil of alkaline flats at Quitobaquito (*F 88-317*); these plants were not present the following year. *Calibrachoa* might also occur in the nearby riverbed or floodplain of the Río Sonoyta. Recorded along the banks of the Colorado River at Yuma (25 Apr 1916, *Swingle 260*) and perhaps formerly along the lower Río Colorado in Sonora. California to Florida and southward to South America. "This species . . . a low growing paludal weed bearing little resemblance to the garden petunia . . . is probably a native of South America" (D'Arcy 1989:465).

CHAMAESARACHA

Herbaceous perennials with simple, branched, or stellate and sometimes glandular hairs. Petioles winged. Calyx 5-lobed, moderately enlarged and partially or mostly covering the berry (not inflated and bladdery as in *Physalis*). Anthers yellow. Fruits small, usually whitish rather dry berries, few-seeded and bearing seeds only near the base. Mexico and southwestern United States; 6 species. Reference: Averett 1973.

†Chamaesaracha coronopus (Dunal) A. Gray [*Solanum coronopus* Dunal] FALSE NIGHTSHADE. Herbaceous perennials from deep roots, the stems often partially buried in sand. Herbage with widely scattered branched hairs with thick, short, stocky, white cells. Stems spreading, often 10–55 cm. Leaves 2.5–11 cm, mostly narrowly elliptic, sessile to petioled, the margins mostly pinnately lobed or sometimes wavy. Flowers 1 (2) in axils, on slender pedicels 2.0–3.5 cm. Calyx partially enclosing the fruit. Corollas 1.5–2.5 cm wide, rotate or circular in outline (without lobes), cream-yellow, with a darker, thicker star-shaped center and with woolly pads alternating with the filament bases. Berries globose. Growing and flowering during warmer times of the year depending on soil moisture.

Low-lying silty soils of playas and adjacent sand plains and low dunes. At the southern margin of Cabeza Prieta Refuge along the Mexican border and undoubtedly on the adjacent Sonora side of the border, although there are no specimens from western Sonora. Kansas and Colorado to Utah and southward to northern Mexico. Three varieties will be described by James Henrickson (personal communication 1999). Ours are var. *coniodes* (Moricand ex Dunal) J. Henrickson ined., characterized by leaves with a close and moderately dense pubescence of relatively slender stalked branched-stellate hairs to 0.3 mm long, sometimes mixed with longer branched hairs. This taxon occurs in central Texas and western Oklahoma to northern Tamaulipas with disjunct plants in the dunes of southwestern Arizona. The plants are robust and appear fast growing. *C. coronopus* is found in the Chihuahuan Desert region in northern Zacatecas, San Luis Potosi, Nuevo León, Coahuila, and Chihuahua north to Texas and Oklahoma, as well as in New Mexico, Colorado, southern Utah, Kansas, and Arizona.

Cabeza Prieta Refuge: Pinta Sands, *F 92-26*. Pinacate Flats, *Hardy* [*Hardies*] *& Goodding 3 Dec 1935*.

DATURA JIMSONWEED

Robust annuals or perennials; herbage usually foul smelling. Flowers large, solitary, erect, fragrant, usually white, and nocturnal. Capsules usually rounded and prickly. Seeds typically small, kidney-shaped, blackish, and hard, often with a white aril (strophiole).

New World, tropical to warm temperate, especially Mexico and southwestern United States, adventive in the Old World; 10 species. The plants contain dangerously narcotic and poisonous alkaloids. Reference: Hammer, Romeike, & Tittel 1983.

Datura discolor Bernhardi. *TOLOACHE;* DESERT DATURA; *KOTADOPĬ.* Non-seasonal ephemerals, frost sensitive and responding poorly to cold weather; 8–60 (100) cm. Larger leaves often 7–20 cm, the petioles prominent. Flowers 8–17 cm, the corollas white with a purple flush in the throat. Flowering calyx tubular, ribbed, 7.0–9.5 cm; most of the calyx falling away as the fruit develops, leaving the calyx base to form a bractlike skirt 1.0–1.5 cm around the base of the fruit. Fruits globose, turning down at maturity, covered with spiny projections, (3.5) 4–5 (7) cm wide including spines. Seeds with a conspicuous white aril (relatively large in comparison with most other daturas).

Widespread, Sonoyta region, desert plains, Pinacate volcanic field, granitic ranges, and dune areas; often along sandy-gravelly arroyo beds or washes; sometimes on cinder flats and slopes and lower stabilized dunes. Southeastern California and southern Arizona to Baja California Sur and southern Mexico, and the Caribbean.

This species differs from most other daturas in its smaller flowers and generally smaller-sized plants. It appears to be the only *Datura* native to the Sonoran Desert. Together with several other nocturnal bat or moth-pollinated plants such as *Ceiba acuminata* in the Bombacaceae and three cacti (*Carnegiea gigantea, Peniocereus greggii,* and *Stenocereus gummosus*), this *Datura* is one of the largest-flowered plants in the Sonoran Desert.

Sierra del Rosario, *F 20759*. Sykes Crater, *F 20026*. Suvuk, *Nabhan 361*. Aguajita, *F 86-282*.

LYCIUM *SALICIESO, WOLFBERRY, BOX THORN*

Mostly woody, thorny shrubs. Leaves usually thickish, often succulent or semisucculent, entire (ours) or sometimes minutely toothed, fascicled on short shoots and widely spaced on long shoots. Flowers borne in the leaf axils, single or groups of 2–6 per leaf fascicle (or more in South America). Flowers (4) 5 (6)-merous. Calyx cup-shaped to tubular, (2–4) 5-toothed or lobed. Filaments longer than the anthers. Fruits 2 to many-seeded, a berry (fleshy) or a drupe (hard and scarcely fleshy). Seeds usually somewhat flattened.

Worldwide, mostly deserts and warm semiarid regions, a few in more humid regions, many semihalophytic; 100 species. *L. chinensis,* the Chinese wolfberry, is cultivated in the eastern Orient as a leaf vegetable, and a few species are grown in temperate climates as ornamental shrubs. *Lycium* can be propagated from hardwood cuttings or seeds.

The genus seems to be of South American origin. The species in North America are generally placed in two sections. Section *Sclerocarpellum,* which includes *L. californicum* and *L. macrodon,* has scarcely fleshy, few-seeded fruits, with the seeds enclosed in a hardened endocarp. The remaining species in the flora area are in section *Lycium.* Among members of this section the fruits are fleshy, smooth, rounded to ovoid or ellipsoid, orange to red, mostly edible, and multiple-seeded. The seeds are light tan, almost whitish, minutely pitted, mostly ovoid-triangular, and moderately flattened. The maximum number of seeds per fruit seems to be a significant taxonomic character.

Sonoran Desert lyciums are densely branched, unpleasantly thorny shrubs mostly 1–2.5 m. They are drought deciduous and early-spring flowering, peak flowering often in February and early March. Some, especially *L. andersonii, L. brevipes,* and *L. californicum,* may flower at any time of year; others may flower sporadically in fall. References: Chiang-Cabrera 1981, 1983; Hitchcock 1932.

1. Fruits scarcely fleshy or with a thin fleshy pericarp, greenish glaucous or red-orange, 1- or 2-seeded; corollas white or white and green.
    2. Shrubs mostly 1.0–1.3 m or less; calyx lobes (2 or 3) 4, the corolla lobes 4; fruits not constricted, 2–4 mm, orange to orange-red; fascicled leaves rounded in cross section and often beadlike. ______ **L. californicum**
    2′ Shrubs often more than 1.5 m; calyx lobes 5, the corolla lobes 5; fruits constricted near or below middle, ca. 10 mm, greenish glaucous; leaves flattened. ______ **L. macrodon**

1′ Fruits fleshy throughout, orange to red-orange, several to many-seeded; corollas white to lavender.
    3. Calyx lobes about as long as or longer than the calyx tube (lobes sometimes shorter on some flowers on a shrub, but some longer-lobed calyces usually present).
        4. Calyx usually 4- or 5-keeled, the sinuses between lobes of flowering calyx rounded, the calyx lobes unequal in size; flowers often very congested on stems. ______ **L. brevipes**
        4′ Calyx usually not keeled (or not conspicuously so), sinuses between lobes of flowering calyx narrow, acute, the calyx lobes equal; flowers not congested on stems. ______ **L. parishii**
    3′ Calyx lobes shorter than the tube (sometimes as long as the tube in *L. berlandieri*).
        5. Margins of corolla lobes densely but minutely ciliate woolly; probably once along the Río Colorado, now extirpated. ______ **L. torreyi**
        5′ Corolla lobes smooth or sometimes minutely ciliate but not woolly.
            6. Leaves, pedicels, and calyces glandular hairy; pedicels 4–16 mm.
                7. Corollas mostly white, the tube pubescent inside; stamens (staminate plants) exserted, the free portion of the filaments very densely hairy below. ______ **L. exsertum**
                7′ Corollas lavender, the tube glabrous inside; stamens (staminate plants) barely or not exserted, the free portion of the filaments glabrous or moderately hairy. ______ **L. fremontii**

6′ Plants glabrous; pedicels mostly 1–7 mm.
8. Flowers slender, longer than wide, the corolla tube narrowly cylindrical or nearly cylindrical, the lobes lavender; filaments glabrous or sparsely hairy at base of free portion; common and widespread. ______ **L. andersonii**
8′ Flowers as wide as to wider than long, the corolla tube campanulate (conspicuously expanded above), the corollas (including lobes) whitish; filaments densely hairy at base of free portion (perhaps in mountains in the Sonoyta region; not treated in the text). ______ **L. berlandieri** Dunal

Lycium andersonii A. Gray var **andersonii** [*L. andersonii* var. *deserticola* (C.L. Hitchcock) Jepson. *L. andersonii* var. *wrightii* A. Gray] *SALICIESO;* DESERT WOLFBERRY; *S-TOHA KOAWUL.* Often 1.2–2.5 m; glabrous, the bark mostly light tan. Leaves mostly linear-spatulate, 5–30 × 1.5–4 (6) mm. Pedicels 1.0–7.5 mm. Flowers slender, nearly tubular; calyx tube 1.6–2.6 mm, the teeth (lobes) 0.4–1.3 mm and often with minutely ciliate margins (calyx often cleft, forming 2 larger lobes, each with 2 or 3 teeth, the calyx then appearing 2-lipped). Corolla tube funnel-shaped, mostly very narrow and nearly tubular, greenish white, 5.5–12.5 mm (on dried specimens; variation in part due to artifacts of drying), the lobes 1.4–2.5 mm, lavender, 4 or 5 (sometimes even on the same plant; the fifth lobe sometimes larger than the others). Stamens about as long as the tube to moderately but conspicuously exserted; basal part of free portion of filaments glabrous or moderately pubescent with short white hairs, the fused portion densely hairy. Fruits often to ca. 1 cm, ovoid, fleshy, bright orange, with multiple seeds each 1.3–1.5 mm.

This is the most common and widespread *Lycium* in the region: often especially common along major washes with palo verde and ironwood, also along minor washes and arroyos, scattered on bajadas and rocky slopes, and sometimes on sand flats and lower stabilized dunes. Sonoyta region, Pinacate volcanic field, granitic ranges, and sometimes on desert flats and dune areas and to higher elevations at least in the Sierra Pinacate. This is the only *Lycium* in the Sierra del Rosario, where it is abundant along the larger washes.

Variety *andersonii* occurs through most of the species range: northern Sinaloa and Baja California Sur to Utah, Nevada, and California. In the flora area this species is distinguished by its relatively narrow leaves, narrow and nearly tubular flowers, non-exserted or moderately exserted stamens, and glabrous habit. *L. andersonii* might be confused with *L. berlandieri,* which has not been found in the flora area.

Chiang-Cabrera (1981) lists 4 varieties and cites specimens of 3 varieties from the flora area, distinguishing them as follows:

1. Flowers 4-lobed, 4–8 mm, leaves broadly spatulate to obovate. ______ var. **wrightii** A. Gray
1′ Flowers 4- or 5-lobed, 8–16 mm, leaves linear-terete, linear-spatulate or narrowly spatulate.
2. Leaves 3–16 mm, linear-terete to narrowly spatulate. ______ var. **andersonii**
2′ Leaves 20–35 mm, narrowly spatulate to spatulate. ______ var. **deserticola.**

Chiang-Cabrera (1981:111) reports that var. *deserticola* "is probably not distinct from var. *andersonii,* from which it differs only in its somewhat longer leaves." These two varieties are not geographically segregated. The larger leaves, at least in the flora area, result from better-watered conditions. Key characters of var. *wrightii* and var. *andersonii* can be found on plants in the same population and even on the same plant. The fourth variety, var. *pubescens* S. Watson (distinguished by its pubescent leaves, pedicels, and calyces) occurs in the gulf coast region of southern Baja California, Baja California Sur, and on Tiburón and neighboring islands. It is the most distinctive and probably the most geographically segregated of the varieties. In my opinion only var. *andersonii* and var. *pubescens* are worthy of recognition.

Sierra del Rosario, *F 75-10.* Tinaja de los Chivos, *F 18627.* Hourglass Canyon, *F 19160.* S of Carnegie Peak, 900+ m, *F 19909.* MacDougal Crater, *Kamb 2003.* Tinajas de los Pápagos, *Turner 59-25.* 24 mi SW of Sonoyta, *Wiggins 8344* (UC).

Lycium brevipes Bentham var. **brevipes** [*L. richii* A. Gray] Leaves, pedicels, and calyces with short glandular hairs. Leaves broadly to narrowly spatulate to oblanceolate, 6–40 × 4–8 mm. Flowers and fruits often very crowded, especially near the branch tips. Pedicels 3–5 mm. Calyx, corolla, and stamens 4- or 5-merous, even on the same branch. Calyx tube often 4- or 5-keeled, especially when dried, the tube 2.5–3.5 mm, the lobes 1.5–4.5 mm, the sinuses (gaps) rounded between the lobes. Corolla tube (2.2) 4.5–5.5 mm, the lobes lavender, spreading, (2.5) 3–4 mm. Stamens well exserted, the anthers conspicuous, the free base of the filaments hairy. Fruits fleshy, orange to red, the seeds many, 1.5–2.2 mm. Flowering warmer months except during drought.

Inland margins of saltscrub or semi-halophytic communities, beach dunes, arroyos, sandy flats and rocky slopes, banks of the lower Río Colorado, along the banks and floodplain of the Río Sonoyta, and widely scattered on desert plains and near the bases of the granitic and volcanic ranges. Abundant and widespread in coastal habitats of the Gulf of California and inland in the Sonoran and Mojave Deserts, usually in xeroriparian or alkaline semiriparian habitats. Sinaloa, Sonora, and Baja California Sur to southern California, Gulf of California islands, and south-central Arizona. Var. *hassei* (Greene) C.L. Hitchcock occurs on the islands and coastal areas of southern California.

Distinctive characters of *L. brevipes* include calyx and corolla lobes usually long relative to the tube, spreading corolla lobes, keeled calyx tube and lobes, glandular hairs, and often congested or crowded flowers. It is closely related to *L. parishii; L. brevipes* usually has larger flowers, larger and greener leaves, and stouter twigs and branches.

20 mi N of El Golfo, *F 75-58.* 5 km S of El Doctor, *F 85-1056.* Río Sonoyta S of El Papalote, *F 86-197.* Bahía de la Cholla (Pelican Point), *Bittman 424.* Puerto Peñasco, *F 85-767.*

Lycium californicum Nuttall ex A. Gray var. **californicum** [*L. californicum* var. *arizonicum* A. Gray] Densely branched shrubs as wide as or wider than tall, mostly 0.7–0.9 m in height along the coast, or reaching 1.3 m at the Arizona border; branching at right angles, the twigs relatively short, very rigid, stout, and thorn-tipped. Leaves very succulent; short-shoot (fascicled) leaves mostly 2.5–8.0 mm, very succulent, terete, often beadlike and pear-shaped to globose; larger long-shoot leaves often 6.5–21.5 × 1.7–3.5 mm, often linear-terete to linear-oblong or spatulate or narrowly oblanceolate.

Flowers subsessile, 6.5–7.0 mm when fresh including stamens and subsessile pedicel, or 5 mm excluding pedicels and stamens or style. Fresh calyces 2.6–4.3 mm, succulent, the angles of the tube and lobes rounded, often becoming keeled upon drying; calyx lobes 4, or sometimes 2 opposite lobes reduced, or occasionally the calyx 2-lobed. Corollas white, 4-lobed, 4.0–5.3 mm wide, the lobes spreading, sometimes with purple markings along midrib of upper (adaxial) surfaces. Stamens conspicuously exserted, the filaments white; anthers and stigma chartreuse-yellow. Style white, well exserted. Fruits 3–4 mm, orange, rounded to oval, 2-seeded, the seeds embedded in a bony endocarp; the endocarps 2.5–3.5 mm, whitish, oblong-obovoid, grooved along the lower 2/3 of their common surface, the outer surfaces convex and smooth.

Sandy soils of the coastal plain northwest of Bahía de la Cholla in a zone parallel to the shore at the inland margin of the zone of *Frankenia palmeri.* Here dew is commonplace during the winter months and there is a well-developed microphytic crust on the soil. The flowers of *L. californicum* and *F. palmeri* are similar in color, size, shape, and phenology, and both have exserted anthers, styles, and stigmas; perhaps they share pollinators. Several disjunct populations of *L. californicum* occur in similar habitats southward along the Sonora coast. Also an inland population ca. 80 km northward on La Abra Plain in Organ Pipe Monument and adjacent Sonora along the international border east of Aguajita, on sandy desert plains dissected by small washes and low, broad floodplains.

Chiang-Cabrera (1981) recognizes 4 geographically segregated varieties of *L. californicum,* one of which is disjunct in the interior of northeastern Mexico. Var. *arizonicum* A. Gray occurs in disjunct populations in central and south-central Arizona, with the southwestern records being from La Abra Plain. Var. *californicum* occurs in southwestern California, the Baja California Peninsula, and Puerto Libertad in Sonora. Farther south on the coast of Sonora, south of Bahía Kino, it is replaced by var. *carinatum* (S. Watson) F. Chiang. Var. *arizonicum* is characterized by its geographic range and smaller, thicker, and more succulent leaves. However, the leaf differences do not seem to hold up in the field and I follow Hitchcock (1932) in combining vars. *arizonicum* and *californicum.*

8 km W of Puerto Peñasco, 4.5 km inland, *F 90-55.* La Abra Valley near Santo Domingo, 13 Apr 1941, *McDougall 87.* Organ Pipe Monument, La Abra Plain 3.7 km E of Aguajita, 5–10 meters N of international fence, *F 90-42.*

Lycium exsertum A. Gray. Densely branched shrubs often 2.5 m. Leaves, pedicels, and calyces glandular hairy. Leaves 1.5–4 cm, spatulate to obovate. Pedicels 4–12 mm. Calyx tube 3.5–4.0 mm, the lobes 0.5–2.0 mm, unequal; calyx rupturing and spreading as the fruit develops. Corollas, filaments, anthers (before anthesis), and style white, or the corollas tinged with pale violet, the stigma green. Corolla tube 7.5–8.5 + mm, hairy inside, narrowed below, moderately flared above, the lobes 1.5 mm, spreading. Stamens exserted on staminate plants; free portion of filaments very densely hairy below. Fruits fleshy, orange, many-seeded. Flowering at least in February.

Southwest of Sonoyta, along small arroyos dissecting low rolling hills (8.5 km SW of the Río Sonoyta at Sonoyta, *F 92-79*). The nearest known population is in Alamo Canyon in the Ajo Mountains. Widely scattered in Sonora southward to the vicinity of Ciudad Obregón, Arizona except the eastern and northern parts of the state, and southern Baja California.

This species is functionally dioecious in a manner similar to *L. fremontii,* to which it is closely related. *L. exsertum* can be distinguished by its usually pendent flowers, shorter pedicels, more campanulate (less tubular) calyx, mostly white and more flaring corollas, long-exserted stamens (on male flowers), and densely hairy filaments. It is a species with diploid and tetraploid members, with $n = 12$ or 24, whereas *L. fremontii* is a polyploid (Chiang-Cabrera 1981). *L. exsertum* is readily distinguished from *L. brevipes* by its longer pedicels, non-keeled calyces, mostly white and more slender and elongated corollas, and non-congested flowers. *L. megacarpum* Wiggins, from the Baja California Peninsula and islands in the Gulf of California, distinguished by being glabrous, is probably different only at the subspecific level (Chiang-Cabrera 1981).

Lycium fremontii A. Gray var. **fremontii.** FRÉMONT WOLFBERRY; *KOAWUL.* Densely branched shrubs ca. 1.5–2.0 m (to 4 m growing into mesquite trees along washes such as at El Papalote just south of Aguajita Spring). Leaves, pedicels, and calyces glandular hairy, the wet plants smelling like a wet dog. Leaves often 1.0–3.5 (4.5) cm, spatulate to obovate, the larger leaves often succulent. Pedicels 6–16 mm. Calyx tube 3.4–5.5 mm, the lobes 0.5–6.0 mm; calyx rupturing and spreading as the fruit develops. Corolla tube 4–12 mm, glabrous inside, the lobes 2.3–3.0 mm, spreading, pale to dark lavender. Stamens (on male plants) not exserted or barely so; free portion of the filaments glabrous or moderately hairy. Filaments and style white, the stigma green. Fruits fleshy and edible, orange, usually longer than wide, many-seeded. Seeds 1.7–2.1 mm, light tan, compressed, the surfaces minutely pitted and tuberculate (closely resembling those of *L. parishii*). Flowering (January) February-March (May), and often also October-November; fruiting about one month after flowering.

Mostly along arroyos, larger washes, river floodplains, sandy plains near major watercourses, and other riparian or semiriparian habitats. Sonoyta region, Laguna Prieta, and along the lower Río Colorado. Southern California to Baja California Sur, the southwestern 2/3 of Arizona, and western Sonora southward to the Río Mayo region. Another taxon, var. *congestum* C.L. Hitchcock, occurs in Baja California and Baja California Sur.

Readily recognized by the glandular hairs, long pedicels, often large flowers, and non-exserted stamens. The fruits are edible but there is considerable variation in sweetness among different plants (Felger & Moser 1985).

Hitchcock (1932:294) noted the "strikingly dimorphic" flowers, and Chiang-Cabrera (1981:137) confirmed "the presence of incipient dioecy, with a tendency for size reduction in functionally female flowers." Plants with included styles and slightly exserted stamens are functionally male and do not produce fruit. In contrast, plants with smaller flowers, exserted styles, and included sterile anthers do produce fruits. Var. *fremontii* is known as an octoploid ($n = 48$) and decaploid ($n = 60$), whereas other North American lyciums are mostly diploid ($n = 12$) or sometimes tetraploid ($n = 24$).

Laguna Prieta, *F 85-737.* Sonoyta: *F 88-14; Mearns 2733* (US, cited by Hitchcock 1932, not seen). El Papalote, *F 88-23.* Sonoyta River at Santo Domingo, *Mearns 2717* (US, cited by Hitchcock 1932, not seen).

Lycium macrodon A. Gray var. **macrodon.** *S-CUK KOAWUL.* Mostly 1.5–2.3 m, usually with stout rigid branches and twigs. Leaves (especially their bases, pedicels, and calyces) minutely and sparsely to relatively densely glandular hairy, or the leaves glabrous or glabrate. (Better-watered plants tend to be glabrous or less pubescent.) Leaves linear-spatulate to obovate or oblong, gradually narrowed basally to a short petiole or subsessile. Long shoots usually with relatively large blue-glaucous succulent leaves, the larger ones 25–50 × 8–13+ mm. Short-shoot leaves mostly smaller, thinner (not as succulent), and greener, the larger ones 16–46 × 4–7 mm.

Pedicels often 3 mm. Calyx about 2/3 as long as the corolla tube, the lobes slender and much longer than the tube. Corolla tube 8.5–10.5 mm, glabrous to sometimes sparsely hairy outside, white with green veins in the throat, the lobes white, often with slight bluish tinge, triangular and reflexed. Stamens and style scarcely or not at all protruding from corolla; filaments glabrous. Stigma green. Fruits (pericarp) hard, 1 cm when ripe, conspicuously glaucous, constricted (appearing pinched off) below the middle, notched at the tip, with 2 chambers, each producing 1 or 2 seeds; lower part of fruit producing ovules but these aborting. Endocarps often 5.5–6.0 mm, bony and sculptured, not grooved and not flattened, the seeds often 3.5–4.0 mm, flattened, orbicular to oblong. Flowering mostly February-March, the fruits ripening early April–early May.

Sandy and sandy-gravelly soils; sand flats, low stabilized dunes and arroyos. Locally common in sandy soils of a zone nearly surrounding the western and northeastern sides of the Pinacate lava shield; scattered elsewhere near the periphery of the Pinacate volcanic field and in the Sonoyta region. Var. *macrodon* occurs in southern Arizona and northwestern Sonora. It is replaced southward by var. *dispermum* (Wiggins) Chiang, which extends to southwestern Sonora.

Rd to Pozo Nuevo, *F 88-64.* Rancho Pozo Nuevo, *F 88-77.* MacDougal Pass, *Turner 59-2.* El Papalote, *F 88-21.* 22 mi S[W] of Sonoyta, *Keck 4156* (UC). Quitobaquito, *F 88-311.*

Lycium parishii A. Gray var. **parishii.** *SALICIESO;* PARISH WOLFBERRY. Shrubs 1.5–2 m. Leaves, pedicels, calyces, and first-season stems glandular hairy; older twigs and stems often light grayish and striate (occasionally dark colored) and relatively slender. Leaves ovate to obovate or spatulate, narrowly to broadly so, 3–18 mm. Pedicels 2–5 mm. Calyx tube 1.5–3.5 mm, the lobes mostly more than 2/3 to nearly twice as long as the tube, blunt-tipped, usually (1.0) 1.8–6.5 mm; calyx sinuses mostly acute in flower, obtuse in fruit. Corolla tube 4.5–8.5 mm, the lobes 1.3–3.5 mm, lavender. Stamens exserted and readily visible due to spreading of corolla lobes, the free portion of filaments glabrate or sparsely hairy near base. Fruits 5–9 mm, globose-ovoid, fleshy, orange, many-seeded. Seeds 1.7–2.2 × 1.3–2.0 mm, whitish tan, compressed, variable in shape, somewhat D-shaped, the surfaces minutely tuberculate (closely resembling those of *L. fremontii*). Flowering February-March, sometimes also August-November; fruiting about one month after flowering.

Common in the Sonoyta region along washes and sandy-gravelly soils of bottomlands and bajadas; often scattered on north-facing granitic slopes. Also in the Tinajas Altas Mountains in adjacent Arizona, and the Baja California side of the Río Colorado opposite San Luis. Inland southern California, southwestern Arizona, Baja California, and northwestern Sonora south at least to the Sierra Bacha south of Puerto Libertad.

Vigorously growing plants or branches may have flowers with unusually large, leafy, and round-ended calyx lobes, but during tough times the lobes can be very short and stubby, occasionally as short as 0.5 mm. Sometimes the same branch may have some flowers with large, long-lobed calyces and others with smaller, short-lobed calyces. This species is readily recognized by the glandular hairs and relatively long calyx lobes—often larger in relation to the calyx tube than other local species. In addition the shrubs have a distinctive grayish cast and slender and light-colored twigs. It could be confused with *L. brevipes* if only fruiting material is present because the fruiting calyx of *L. parishii* sometimes can be keeled. Compared to *L. brevipes,* in *L. parishii* the sinus between the calyx lobes is usually more acute, the corolla tube is longer, and the flowers are not crowded as in *L. brevipes.* The opened (dissected) flower shares features with *L. andersonii:* both have corollas, stamens, and anthers of similar sizes and shapes, and the free portion of their filaments is only sparsely hairy. *L. parishii* var. *modestum* (I.M. Johnston) F. Chiang of the Chihuahuan Desert is "completely isolated geographically from var. *parishii*" (Chiang-Cabrera 1981:109).

W of Sonoyta, *F 88-13.* 10 km SW of Sonoyta, *F 88-192.* Sierra de los Tanques, *F 89-24.* Quitobaquito, *Hodgson 217* (DES).

†Lycium torreyi A. Gray. Large shrubs resembling *L. fremontii,* but differing in corolla-lobe margins having silvery-white edges due to woolly ciliate hairs, an often shorter calyx tube, the flowers not dimorphic, and a different chromosome number ($n = 12$ for *L. torreyi,* Chiang-Cabrera 1981). Leaves lanceolate-acute, sometimes quite large. Fruits multiple-seeded. Generally a riverine plant along the Colorado River and Rio Grande. Known from the lower Colorado River in Arizona and California, and probably formerly on the Mexican side of the river.

NICOTIANA TOBACCO

Annual or perennial herbs, rarely shrubs. Leaves mostly entire, usually viscid pubescent. Flowers usually opening and fragrant at night. Calyx 5-lobed, persistent. Corolla 5-lobed, often tubular (ours). Fruit a capsule; seeds numerous, minute. Mostly in the Americas, tropical and warm temperate; 60 species. Includes commercial tobacco, *N. tabacum,* and several ornamental garden annuals. Reference: Goodspeed 1954.

1. Shrubs; herbage glabrous; corollas yellow. ______ **N. glauca**
1′ Herbs; herbage viscid pubescent; corollas white or cream.
    2. Spring ephemerals; leaves sessile to short petioled, the stem leaves not clasping; flowers nocturnal. ______ **N. clevelandii**
    2′ Perennials, sometimes flowering in first season; leaves all sessile, the stem leaves clasping (leaf base wrapping around the stem); flowers diurnal. ______ **N. obtusifolia**

Nicotiana clevelandii A. Gray. *TABACO DEL COYOTE;* DESERT TOBACCO; *BAN WIWGA.* Winter-spring ephemerals, 10–75 cm, mostly with a single main axis, or the more robust plants often with several axes. Stems and leaves viscid (sticky) hairy. Leaves variable, quickly wilting, lanceolate to elliptic or ovate, the larger ones (4) 5–13 cm, the leaf tip acute to sometimes obtuse. Early leaves in a basal rosette; lower leaves larger and with winged petioles, the stem leaves reduced upwards and sessile. Calyx lobes slender, 1 lobe wider than the others and conspicuously longer than the capsule, the other lobes about as long as to shorter than the capsule. Corollas white, ca. 12–20 mm, nocturnal, closing in the morning depending on temperature. Capsules 5.8–8.0 mm.

Widespread in lowlands, usually on sandy or gravelly soils; desert flats, washes, low dunes, and disturbed habitats. Often in shade on the north side of large ironwoods (*Olneya*). Southern California to southeastern Arizona and southward through Baja California Sur, and Sonora southward to the Guaymas region.

Sierra del Rosario, *F 20745.* Moon Crater, *F 18832.* MacDougal Crater, *Turner 86-4.* 0.5 mi S of Mex 2 [rd to Rancho Guadalupe Victoria], *Mason 1820.* Puerto Peñasco, *Yatskievych 80-41.*

***Nicotiana glauca** Graham. *PALO LOCO, JUAN LOCO;* TREE TOBACCO. Open, sparsely branched shrubs or small trees, 2–3+ m, glabrous except the corollas. Leaf blades (4) 5.5–16 (22.5) cm, ovate, glaucous, smooth, somewhat thickish; petioles (3) 3.5–7.5 (9.5) cm. Calyx 7–16 mm, green; corollas tubular, 28–40 mm, with gland-tipped hairs outside, flared and then constricted just below the short broadly obtuse lobes, the lobes and the expanded upper tube at first (at anthesis) yellow-green, the rest of the tube yellow, the corollas fading yellow throughout. Anthers yellow-green before anthesis. Stigma bilobed, green; styles and filaments white. Capsules ca. 1 cm. Warmer months, the flowers frequented by hummingbirds.

Urban and agricultural weeds, occasionally along roadsides and other disturbed habitats. Originally from northwestern Argentina and southern Bolivia, now adventive in warm regions of the world.

Observations: Puerto Peñasco, Jun 1985; Sonoyta, Oct 1985; Vicinity of San Luis, Oct 1985.

Nicotiana obtusifolia M. Martens & Galeotti [*N. trigonophylla* Dunal. *N. palmeri* A. Gray. *N. trigonophylla* var. *palmeri* (A. Gray) M.E. Jones] *TABAQUILLO DE COYOTE;* DESERT TOBACCO; *HA-WIWGA.* Herbaceous perennials or sometimes facultative annuals or ephemerals, often 0.8–1 m. Herbage sticky glandular pubescent. Leaves sessile, mostly 6.0–12.5 cm, the bases of stem leaves clasping (wrapping around the stem), the upper leaves reduced, mostly not clasping. Flowers diurnal, open all day. Calyx and its lobes as long as to much longer than the capsules. Corollas 15–22 mm, cream-white. Capsules 5.8–8.0 mm. Flowering in winter and spring, the plants usually dying back severely during drought and often flowering again with summer rains. Apparently germinating during the winter-spring season.

Widespread but seldom common, scattered in different habitats. Often in sandy-gravelly soils especially in shaded places: along washes, among rocks along arroyos, and on rocky slopes. Sonoyta region, Pinacate volcanic field, and granitic ranges. Sometimes growing out of crevices in barren rough lava. California deserts to Nevada and Texas southward to Nayarit. *N. obtusifolia,* an earlier name for what was previously known as *N. trigonophylla,* is based on a plant from the southern end of the distribution (Michael Nee, personal communication 1993).

Sierra Extraña, *F 19046.* MacDougal Crater, *Kamb 2008* (DS). Hornaday Mts, *Burgess 6836.* Punta Peñasco, *Keck 4212* (DS).

PHYSALIS *TOMATILLO,* GROUND CHERRY, HUSK TOMATO

Annual or perennial herbs. Leaves petioled, the blades mostly toothed, usually soft in texture. Flowers pendent, mostly solitary in leaf axils, small, and pedicelled. Calyx conspicuously 10-veined, enlarging and growing over the fruit to form a bladderlike structure like a miniature paper lantern. Corollas rotate, often yellowish, sometimes blue or white, usually dark-spotted and hairy toward center. Anthers yellow or bluish to purplish, opening by longitudinal slits. Fruit a globose, many-seeded berry.

Worldwide, greatest diversity in Mexico; 80 species. Some grown for their edible fruits. Tomatillo (*P. philadelphica*), extensively cultivated in the more humid parts of Mexico, is a major ingredient in green salsa. References: Averett 1979; Waterfall 1958, 1967.

1. Ephemerals; erect, usually taller than wide with a single main axis; leaves thin, 5–18 cm; anthers bluish or greenish to purplish; well-watered, weedy/disturbed habitats.
   2. Herbage smooth, glabrous or glabrate; corollas white with a yellow center; anthers bluish. ________**P. acutifolia**
   2′ Herbage clammy or "slimy" with spreading hairs; corollas pale yellow with maroon spots in center; anthers purplish. ________**P. pubescens**

1′ Perennials and sometimes flowering in first season; often spreading, usually as wide as or wider than tall with several or more branches or major axes; leaves often slightly thickened, 1.5–10 cm; anthers yellow; desert and semiriparian habitats.
   3. Bushy perennials, lower branches usually slightly woody; herbage with short, straight, glandular hairs about same color as herbage; corollas yellow. ________**P. crassifolia**
   3′ Perennials from a thickened underground root, stems all herbaceous; herbage scurfy with spherical white "hairs," especially on young growth and young calyces; corollas purple. ________**P. lobata**

Physalis acutifolia (Miers) Sandwith [*Saracha acutifolia* Miers. *P. wrightii* A. Gray] Warm-weather ephemerals, often 25–60+ cm; erect with a well-developed main axis. Younger herbage, flowering calyces, and fruiting calyx bases with short thick white hairs, the younger plants sometimes with elongated simple hairs near the base of the stem, the plants otherwise glabrous or glabrate. Leaves relatively thin, variously lanceolate to narrowly ovate, coarsely and often deeply toothed, the larger ones often 5–18 cm. Pedicels slender and elongate, the fruiting pedicels 2–5 cm. Corollas 10–15 mm wide, white with a yellow center. Anthers blue-gray, the pollen yellow.

Generally weedy in well-watered agricultural and urban habitats, often locally common. Southeastern California to Texas and Sinaloa; weedy over most of its range. Perhaps not native to the flora area, or originally native along the rivers. The elongate fruiting pedicel is one of the best distinguishing characters.

Sonoyta: *F 85-929, F 86-302.* Rancho los Vidrios, *Equihua 26 Sep 1982.*

Physalis crassifolia Bentham var. **versicolor** (Rydberg) Waterfall [*P. versicolor* Rydberg] *TOMATILLO DEL DESIERTO;* DESERT GROUND CHERRY. Perennial herbs slightly woody at base, and sometimes flowering in the first season; plants globose, drought deciduous, dying back severely in drought, the stems slender, spreading, and much-branched. Leaves highly variable, 1.5–9 cm, the petioles as long as to much longer than the blades, the blades often broadly ovate and shallowly toothed. Corollas and anthers yellow, the corollas 15–22 mm wide. Fruiting calyx (2) 2.5–3.5 cm. Fruits edible, resembling the domesticated tomatillo (*P. philadelphica*) but much smaller and the flavor inferior. Flowering and fruiting response non-seasonal.

Sonoyta region, as well as granitic and volcanic ranges but apparently most common in the Pinacate volcanic field, where it extends nearly to the summit. Cliffs, rocky slopes, and rocky arroyos, often on north-facing exposures, tuff breccia and basalt cliffs of Pinacate craters, and less common in gravelly soils of arroyos. Var. *versicolor* occurs in northwestern Sinaloa, western Sonora, western Arizona, southern Utah, southeastern California, and northeastern Baja California.

The type locality of var. *versicolor* is Guaymas, whereas that of var. *crassifolia* is Bahía Magdalena, Baja California Sur. Specimens from the Vizcaino and other Pacific coast fog-influenced regions of the Baja California Peninsula have thicker leaves and entire or nearly entire leaf margins, which are often used as key characters for var. *crassifolia*. These plants seem distinct, and in my opinion the concept of var. *crassifolia* should be restricted to those Baja Californian populations. This scheme calls for placing the Sinaloa, Sonora, northeastern Baja California, and apparently all U.S. populations in var. *versicolor*. Another geographic taxon, var. *infundibularis* I.M. Johnston, occurs on the east-central side of the Baja California Peninsula and some of the gulf islands. It is distinctive for its much larger, funnel-shaped, and longer corollas; it does not seem conspecific with other varieties of *P. crassifolia*.

Cerro Colorado, *F 10789*. Elegante Crater, *F 19670*. Tinajas de Emilia, *F 19736*. SE of Carnegie Peak, 850 m, *F 19925*. Tinaja del Tule, *F 18748*. MacDougal Crater, *F 10704*. Cerro Pinto, *Burgess 5613*.

Physalis lobata Torrey [*Quincula lobata* (Torrey) Rafinesque] Low perennial herbs from a deep, thick root; growth apparently non-seasonal. Stems herbaceous, rather weak, often 18–26 cm, spreading to decumbent. Herbage granular scurfy with rounded white trichomes ("hairs"), especially dense on young herbage, pedicels, and calyces. Leaves relatively thick and sometimes semisucculent, commonly ovate to spatulate, 4–10 cm, the margins undulate or shallowly to deeply lobed or parted, the petioles prominent and winged. Corollas lavender, 2–3 cm wide, the center with a star-shaped pad of white hairs. Filaments lavender, the anthers yellow. Style pale white, the stigma capitate, yellow. Fruiting calyx 2 cm. Flowering spring and summer-fall.

Locally abundant in poorly drained clayish silty soils of Playa Díaz and sandy soils of the large bajada-plain leading into the playa from the west. Northern Mexico to Arizona, Colorado, and Texas.

Sometimes placed in the monotypic genus *Quincula,* distinguished from *Physalis* by the granular-scurfy herbage, capitate rather than slightly 2-lobed stigma, characters of the seed, habit, and chromosome number. It might be confused with *Chamaesaracha coronopus;* both have a hairy pad at the center of the corolla but show many differences including flower color and leaf texture.

0.5 km S of Pinacate Junction, *F 86-360*. Playa [Díaz] just N of Cerro Colorado, *Sherbrooke 10 Apr 1983*. Playa 2 mi W of Cerro Colorado, *Kamb 2016* (UC).

***†Physalis pubescens** Linnaeus [*P. pubescens* var. *integrifolia* (Dunal) Waterfall] HAIRY GROUND CHERRY. Warm-weather, water-loving annuals with clammy ("slimy") herbage, soft hairs, thin and quickly wilting leaves, pale yellow corollas with 5 dark maroon spots in the center, and green to purple anthers.

Wiggins (1964:1312) reports it "near Yuma and on the delta of the Colorado River in both Baja California and Sonora." It grows in riverine and other wetland habitats in southern Sonora. Widespread weed in the warmer regions of the world; probably native in eastern and midwestern United States to Latin America. Apparently not native in western North America. Several varieties; ours might be var. *grisea* Waterfall. The varieties, however, do not seem worthy of recognition (Cronquist et al. 1984; McGregor, Gentry, & Brooks 1986).

SOLANUM NIGHTSHADE

Herbs, shrubs, or sometimes small trees or vines; often spiny and with stellate hairs. Leaves usually alternate, simple or compound (ours alternate, simple, and petioled). Flowers often showy, the calyx usually 5-lobed, the corollas 5-lobed, mostly rotate, radial to slightly bilateral. Filaments short; anthers usually connivent (close together but not united) in a cone around the style, each anther cell usually opening by a terminal pore (ours; the flowers are "buzz-pollinated," see *Senna,* Fabaceae). Fruit a berry. Seeds many, flattened, more or less kidney-shaped.

One of the larger genera in the world, and the largest in the family. Largely tropical; 1200 species. Includes potato; eggplant; and various medicinal, ornamental, weedy, and highly toxic plants. Both *S. elaeagnifolium* and *S. hindsianum* seem to have their closest ties with the otherwise Australian *S. ellipticum* group of solanums (Whalen 1984). References: Schilling 1981, 1990b.

1. Annuals, without spines; new growth moderately pubescent, otherwise sparsely hairy or glabrous, the hairs simple; corollas white, ca. 0.5–1.0 cm wide.________ **S. americanum**
1′ Perennials, often spiny; herbage densely pubescent with stellate hairs; corollas lavender (very rarely white), ca. 2–5+ cm wide.
   2. Herbaceous perennials, less than 1 m, dying back to rootstock annually or seasonally; corollas 2–3 cm wide; weedy habitats. ________ **S. elaeagnifolium**
   2′ Shrubs, usually more than 1 m; corollas 3–5 cm wide; widespread in natural habitats. ________ **S. hindsianum**

Solanum americanum Miller [*S. nodiflorum* Jacquin] *CHICHIQUELITE;* BLACK NIGHTSHADE; *CU:WĬ WU:PUI.* Ephemerals or annuals (also short-lived perennials elsewhere), relatively open and sparsely branched, to 0.5–1.0 m; unarmed. Herbage, especially new growth, inflorescences, outer calyces, and portion of corolla exposed in bud with sparse to moderately dense short white unbranched hairs; older herbage and especially the larger leaves glabrate. Leaves often 5–20 cm, the blades variable, mostly ovate, green, often very thin, the margins usually irregularly blunt-toothed, the petioles prominent. Inflorescences usually of short umbellate clusters. Fruiting calyx covering 1/3 or less of fruit. Corollas white, 5 (10) mm wide, star-shaped with the lobes longer than the short tube. Anthers pale yellow, 1.2–1.3 mm before dehiscence, 1.5–1.6 mm after dehiscence. Berries 5–8 mm, globose, blackish or purplish.

Growing and flowering with warm weather, frost sensitive and dormant during cooler weather. Aguajita Spring and Quitobaquito in moist, often shaded locations, also an urban and agricultural weed. In the Sonoran Desert usually in wetland habitats. Probably "native to South America from which a limited number of genotypes have spread to colonize other areas of the world" (Schilling 1990b:257). "The black nightshades form a complex of weedy species also variously known as garden nightshades, the *Solanum nigrum* complex, and more formally as *Solanum* section *Solanum*" (Schilling 1990b:253). The widespread New World native populations are diploids whereas typical European *S. nigrum* is hexaploid.

Quitobaquito, *Bowers 1329.* Aguajita Spring, *Warren 10 Nov 1983.*

Solanum elaeagnifolium Cavanilles. WHITE HORSE-NETTLE, SILVER-LEAF NIGHTSHADE, BULL NETTLE. Herbaceous perennials, dying back to below-ground rootstocks in winter and severe drought; densely stellate hairy. Stems, petioles, leaf midribs, pedicels, and calyces often but not always with straight, slender spines or prickles. Leaves 3–10 cm, lanceolate to oblong, the margins entire to sinuate-lobed or cleft. Corollas lavender, 2–3 cm wide. Anthers yellow. Fruits round, yellow when ripe.

Common weed at Sonoyta (*F 86-400*) and Puerto Peñasco (*F 85-795*), and probably elsewhere in urban and agricultural areas. A widespread weed, probably native to parts of southwestern and midwestern United States and northern Mexico (Boyd, Murray, & Tyrl 1984).

Solanum hindsianum Bentham. *MALA MUJER, TOMATILLO ESPINOSO.* Sparsely branched shrubs 1.5–3 m; densely stellate pubescent. At least some stems, petioles, leaf midribs, and calyces often with straight, relatively slender prickles to stout spines. Leaves mostly 3–13 cm, mostly lanceolate to ovate or oblong, the margins entire or nearly so, or with irregular shallow teeth or lobes; leaves noticeably thinner and larger during periods of high soil moisture and warm weather. Flowers showy, the corollas 3–6 cm wide, largest during periods of high soil moisture, lavender (plants with white corollas very rare). Anthers 7.0–10.5 mm, bright yellow. Stigma green. Fruits 2 cm, round, mottled with dark and light green. Various seasons depending on soil moisture and temperature.

Widespread and common in the Pinacate volcanic field, to 875 m on the south side of Sierra Pinacate, granitic ranges within and adjacent to the Pinacate field (e.g., Sierra Blanca and Hornaday Mountains), and the Sonoyta region as far north as Cerro Cipriano (12 km S of Sonoyta). Arroyos, washes, bajadas, and rocky slopes. Western Sonora, most of the Baja California Peninsula, and 1 population in Organ Pipe Monument. The northern limit seems to be determined by freezing weather. Freeze damage is commonplace, and plants near Sonoyta are sometimes freeze killed nearly to the ground.

Arroyo Tule, *F 19222.* Tinajas de Emilia, *F 19726.* Pinacate Mts, *MacDougal 19 Nov 1907* (US). S of Pinacate Peak, 875 m, *F 19343.* 12 mi S[W] of Sonoyta, *Shreve 7581.* Puerto Peñasco, *Prigge 7276.*

STERCULIACEAE CACAO FAMILY

Mostly trees and shrubs, usually with stellate hairs or peltate scales. Leaves alternate; stipules present. Flowers usually radial. Sepals mostly 5, the petals 5. Fertile stamens 5, the filaments usually united into a tube around the superior ovary (but free from it); staminodes often present. Fruits dry capsules or fleshy and indehiscent.

Mostly tropical and subtropical regions worldwide; 67 genera, 1500 species. The family includes cacao (*Theobroma cacao*), the source of cocoa and chocolate, and cola nut (*Cola nitida*).

AYENIA

Shrubs or herbs. Flowers small, often long-stalked (pedicel and stipe). Leaves simple, the margins toothed or serrated. Flowers mostly small, intricate, and unusual. Calyx 5-lobed, rather petal-like. Corollas parachute-like, the petals of slender claws with expanded blades forming a hood over the stamens. Fertile stamens and staminodes each 5. Fruit a 5-chambered capsule separating into five 1-seeded carpels. New World, mostly tropical and subtropical; 70 species. Reference: Cristobal 1960.

Ayenia compacta Rose. Annuals to short-lived perennials, scarcely woody, erect, slender, few-branched, often 50–80 cm, or much shorter when browsed with stiff, stubby, spreading branches. Stems with firm, downward-pointing white hairs. Leaves gradually drought deciduous, petioled, the blades, especially the lower surfaces, with stellate and simple hairs; blades mostly linear to narrowly lanceolate but with sufficient moisture sometimes ovate to elliptic, the margins toothed; wet-season leaves often 14–38 mm, the dry-season leaves often 7–18 mm. Flowers in small axillary clusters, maroon, ca. 3 mm wide. Capsules rounded, 5-lobed, 4.6–5.1 mm wide, pale yellow-green with stellate hairs and studded with stout blunt green to dark maroon spinescent tubercles 0.1–0.3 mm long. Seeds dark red-brown to nearly black, 2.5–2.9 mm, resembling small beetle pupae. Growing and flowering with warmer weather, especially summer-fall rains.

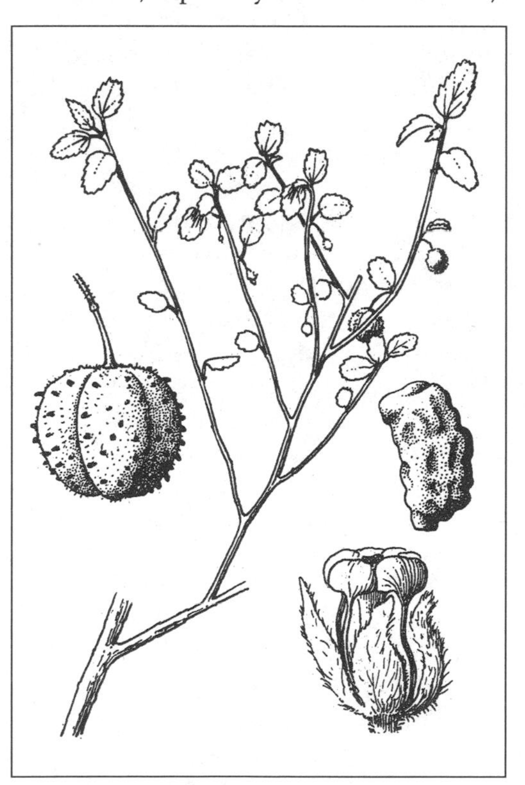

Ayenia compacta. JRJ (Abrams 1951).

Granitic slopes, mostly north-facing, often locally common and not widespread. Arizona Upland in hills and mountains south and southwest of Sonoyta, more common in the mountains in Organ Pipe Monument, and the western granitic mountains. Western Sonora from the vicinity of Bahía Kino northward, western Arizona, Baja California, islands in the Gulf of California including Islas San Esteban and Tiburón, and southeastern California.

This is the smallest, and most northern and xerophytic *Ayenia* in North America. It is replaced by the closely related *A. filiformis* S. Watson in more favorable regions to the east and south. The two species differ in the length of the capsule tubercles (0.5–1+ mm long on *A. filiformis* versus less than 0.5 mm in *A. compacta*), a character that seems to intergrade among specimens from western Sonora, as well as among drought-stressed specimens of *A. filiformis* from southern Sonora. Should they really be treated as distinct species? *A. filiformis* occurs in the Ajo Mountains, and the plants from the Sierra de los Tanques may belong to this taxon. It ranges from Sinaloa to Coahuila northward to Arizona, New Mexico, and Texas.

Sierra de los Tanques, *F 89-20.* 67 mi W of Sonoyta, *Prigge 7262* (UCLA).

TAMARICACEAE TAMARISK FAMILY

Trees or shrubs. Leaves alternate, persistent, mostly small, without stipules. Flowers small, mostly bisexual, radial, with a fleshy nectar-producing disk. Ovary superior. Fruit a capsule with many small seeds. Native to the Old World, mostly xerophytes and halophytes in arid and semiarid regions; 4 genera, 75 species. The family is generally considered to be related to the Frankeniaceae. Reference: Crins 1989.

TAMARIX SALT CEDAR, TAMARISK

Leaves scalelike, mostly with salt-excreting glands, clasping or sheathing the stem. Flowers in dense racemes or spikes, usually with bracts, usually bisexual, 4- or 5- (rarely 6)-merous. Seeds many, minute, with feathery hairs. Deserts in temperate and subtropical regions; 50 or fewer species. References: Baum 1967, 1978.

1. Trees more than 5 m, with massive limbs and trunk; leaf completely encircling stem, the blade reduced to a cusp less than 0.5 mm; flowers white, sessile. ______ **T. aphylla**

1′ Shrubs, rarely small trees, seldom more than 3–4 m; leaf not completely encircling stem, the blade evident, 0.7–3.0 mm, triangular-ovate; flowers usually pink, sometimes white aging to pink, with pedicels 0.6–0.8 mm. ______ **T. ramosissima**

***Tamarix aphylla** (Linnaeus) H. Karsten [*Thuya aphylla* Linnaeus] *PINO;* ATHEL TREE, SALT CEDAR. Trees to 7+ m, the trunk well developed and often massive. Branchlets (actually the leaf bases) dotted with salt-excreting glands. Flowers sessile, 5-merous, white except anthers rose-purple, the perianth 2.5–3 mm, the petals longer than the sepals. Capsules 3.3–4.0 mm. Flowering in fall.

Extensively planted in the Sonoran Desert as a shade tree. It grows readily from cuttings of almost any size. Once established the trees often persist, especially on dunes near the coast. Salt cedar trees, apparently several decades old and planted at widely spaced intervals along the railroad across the Gran Desierto, were growing vigorously in the late 1980s and 1990s. During the cooler months dew condenses on the salt-encrusted twigs and drips off onto the soil. I have not seen seedlings in the Sonoran Desert. Probably these trees have been grown from a single clone (see Benson & Darrow 1981). Native to North Africa and the eastern Mediterranean.

Río Sonoyta 21 km W of Sonoyta, *F 85-969*. 6 mi W of Estación Gustavo Sotelo, *F 84-8*.

***Tamarix ramosissima** Ledebour. *PINO SALADO, SALADO;* SALT CEDAR, TAMARISK; *'ONK 'U'US, VEPEGI: 'U'US*. Shrubs 2.5–5 m. Branchlets (short shoots) or ultimate twigs winter and tardily drought deciduous, their internodes shorter than, and obscured by, the overlapping scale leaves. Long shoots perennial. Leaves and young stems with regularly spaced, alkali- or salt-excreting glands (readily seen with 10× magnification); these glands glistening gold when fresh and first developing, soon becoming white with the buildup of alkali or salt. Scale leaves of ultimate branchlets (short shoots) 0.7–1.4+ mm, those of the long shoots mostly 2.3–3.0 mm. Flowers 5-merous, the perianth 1.5–2.0 mm, pinkish white to pink, the petals longer than the sepals, the filaments pink, the anthers white, the ovary often dark rose-purple, the style yellow or pink, the stigma trilobed, white; floral disk glands or nectary maroon, producing glistening nectar droplets; pedicels 0.5–0.8 mm. Seeds 0.5 mm, yellow-brown. Capsules 3.7–4.5 mm. Flowering January-October.

Abundant along the Río Sonoyta, along the Río Colorado and its delta, as an agricultural weed, in roadside depressions, in wetland habitats including Laguna Prieta and the La Salina pozos, and rarely in playas in the Pinacate volcanic field. Incredibly, it has even reached the bottom of Elegante Crater. Native to the Old World, now widespread, weedy, and invasive in many of the warm, dry parts of the world, especially in disturbed desert riparian habitats.

I am including all the shrubby, pink-flowered tamarisks in southwestern North America in *T. ramosissima*. The plumose-tufted minute seeds resemble the propagules of *Arundo* and *Phragmites,* which seem to have similar dispersal strategies.

Tamarix aphylla. LBH.

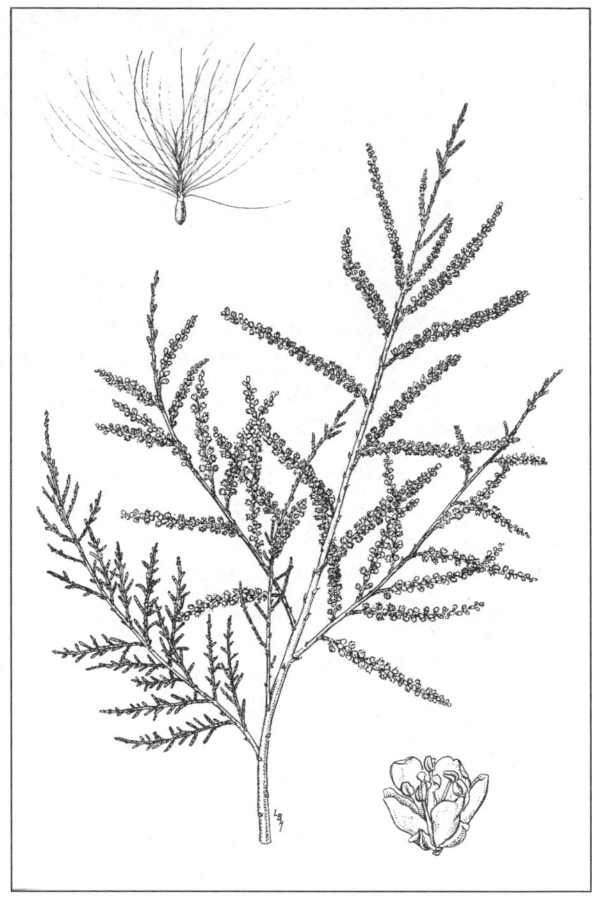

Tamarix ramosissima. LBH (Parker 1958).

Río Sonoyta, 20 km W of Sonoyta, *F 85-980.* Laguna Prieta, *Ezcurra 25 Apr 1981* (MEXU). La Salina, *Ezcurra 28 Jun 1982.* 5 km S of El Doctor, *F 85-1057.* 10 mi E of San Luis, *F 77-16.* Delta of Colorado River, California (?), *Sykes 9 Apr 1935.* Quitobaquito, *West 96.* Elegante Crater, playa sediment, *Christopher Eastoe* (photo, Mar 1985).

ULMACEAE ELM FAMILY

Trees or shrubs. Leaves almost always alternate, simple, the blade usually asymmetric at base, the stipules shed as the leaves unfold. Flowers in small axillary clusters, wind- or sometimes self-pollinated, small, the sepals united; petals none; styles 2-branched. Fruits 1-seeded. Two subfamilies: Ulmoideae, with dry fruits (includes elm trees, *Ulmus*), and Celtoideae, with fleshy fruits. Worldwide in temperate and tropical regions and extending into deserts; 18 genera, 150 species.

CELTIS HACKBERRY

Trees or shrubs, spinescent or not; often winter deciduous. Leaf blades usually with 3 main veins. Flowering with new growth; flowers unisexual or bisexual, both often on the same branch. Female flowers with a large sessile or nearly sessile 2-branched feathery stigma. Stamens held by the cupped sepals, the filaments expanding like tension-held springs as they mature, straightening rapidly and flinging dry, powdery pollen from the anthers. Fruit a drupe, the seeds primarily bird-dispersed. Tropical and temperate regions worldwide, some in deserts; 60 species. References: Romanczuk & del Pero 1978; Sherman-Broyles, Barker, & Schulz 1997.

Celtis pallida Torrey subsp. **pallida** [*C. tala* Gillies var. *pallida* (Torrey) Planchon] *CUMBRO, GARAMBULLO;* DESERT HACKBERRY. Shrubs 2.0–2.5 m. Young twigs, thorns, and inflorescences with stiff, appressed white hairs, glabrate with age. Twigs often zigzag, often with single or paired thorns, these, especially the larger ones, often with 1 (2 or 3) node(s) often bearing flowers or a small leaf. Leaves gradually drought deciduous, 2.2–4 (7) cm, the blades more or less ovate to broadly elliptic, scabrous (dotted with hard, minute glands often sprouting a stiff forward-pointing hair; rough like sandpaper), the margins with forward-pointing teeth to sometimes entire, the petioles short. Flowers green, inconspicuous, in small axillary clusters shorter than the leaves. Fruits 8 mm, bright orange, the pericarp rather thin, fleshy, not hard-walled, edible and moderately sweet and tart, the seed hard. Flowering March and with summer-fall rains.

Along the international border near Sonoyta; a small population scattered along gravelly soil of arroyos, often beneath *Olneya* and *Prosopis*. Widespread and common in adjacent Organ Pipe Monument and eastward in Sonora. Nearly throughout the rest of Sonora except higher elevations. Semiarid and arid regions; central Arizona to southern Texas and Oaxaca, both Baja California states, southern Florida, West Indies, and Paraguay and Argentina. An additional subspecies occurs in South America. Benson & Darrow (1945, 1981) aligned *C. pallida* with *C. tala* of South America, but Romanczuk & del Pero (1978) showed that *C. tala* is a different species restricted to South America.

1.5–5 km W of Sonoyta: *F 88-16, F 93-01.*

URTICACEAE NETTLE FAMILY

Annual to perennial herbs, rarely shrubs or small trees; often with stinging hairs. Leaves opposite or alternate, usually with stipules. Monoecious, dioecious, or polygamous. Flowers small, usually greenish, without petals, mostly wind-pollinated. Male flowers mostly with 4 or 5 sepals and as many stamens; filaments inflexed in bud, straightening suddenly and elastically to fling the mature pollen from the anthers. Female flowers mostly 4-merous or the perianth absent. Fruit an achene (1-seeded). Nearly worldwide, mostly tropical and subtropical; 45 genera, 800 species. Reference: Boufford 1997.

PARIETARIA PELLITORY

Annual or perennial herbs, glabrous or hairy with non-stinging hairs. Leaves alternate, simple, petioled, the blades entire and dotted with punctiform cystoliths that appear as minute blisters under magnification; stipules none. Flowers in small axillary clusters, inconspicuous, green, subtended by green bracts, with bisexual and unisexual flowers in the same cluster. Calyx 4-parted; stamens 4. Achenes hard and shiny, more or less ovoid, enclosed in the persistent perianth. Worldwide, mostly temperate and subtropical; 20 species. References: Boufford 1992; Hinton 1969.

Parietaria floridana Nuttall [*P. hespera* D.B. Hinton var. *hespera*] DESERT PELLITORY. Delicate winter-spring ephemerals, the roots relatively small, the stems erect to spreading or semiprostrate. Sparsely to densely hispid, the hairs fine to relatively stout with enlarged bases. Herbage rapidly wilting when cut. Branching largely from near base, the stems semisucculent, often 8–60 cm. Leaf blades (5) 8–30 mm, thin, soft, broadly lanceolate to mostly ovate, sometimes nearly circular to broadly and nearly triangular, the base commonly truncate, the lower, major veins arising from the base of the blade; petioles prominent. Bracts green, (1.4) 2–5 (7) mm. Seed-bearing calyces brown, variable in size even on the same stem, 11–26 mm. Achene smooth and shiny, light brown, ovoid, plump but slightly compressed lengthwise, 0.9–1.2 mm.

Widespread in the Sonoyta region and the Pinacate volcanic field to higher elevations. Often locally common, especially at higher elevations and in protected niches such as arroyo beds, shady nooks, and in the protection of spiny shrubs and large rocks. Southern United States and northern Mexico.

Key characters for the several seemingly closely related annual taxa largely involve leaf size and shape and do not consistently serve to distinguish them. *P. floridana* seems to form a continuous

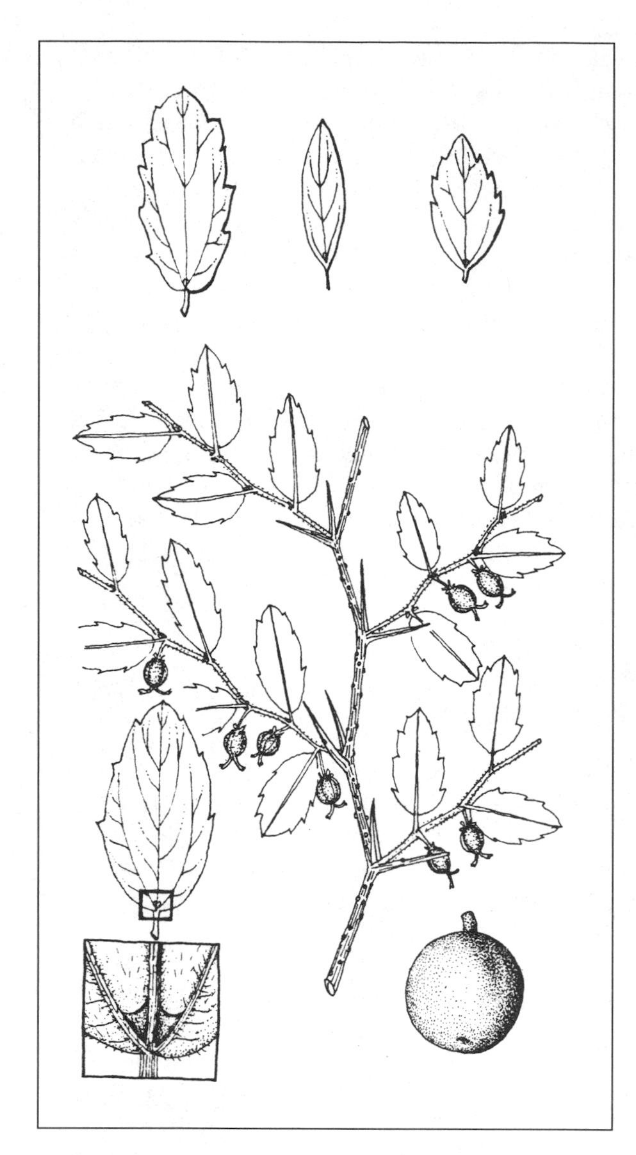

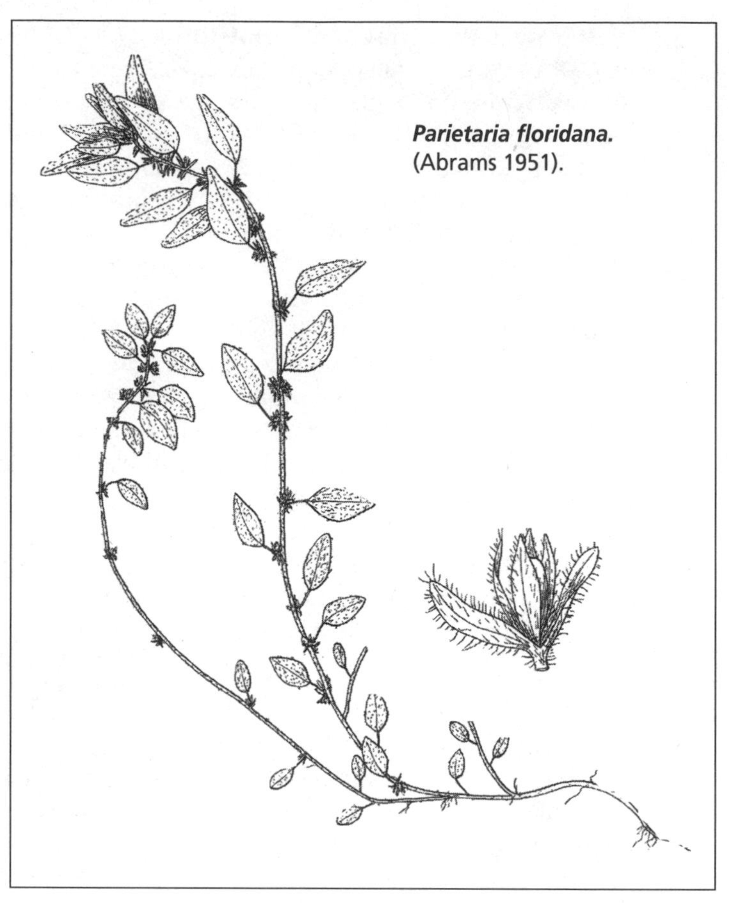

Parietaria floridana. (Abrams 1951).

Celtis pallida subsp. ***pallida.*** FB. © James Henrickson 1999.

graded series with *P. pennsylvanica* Muhlenberg ex Willdenow, which extends across much of the rest of North America, and Welsh et al. (1987) raise doubts concerning their distinctiveness.

18 km SW of Sonoyta, *F 88-217*. El Papalote, *F 86-106*. Tinaja del Tule, *Burgess 6293*. Sykes Crater, *F 19507*. Tinajas de Emilia, *F 87-35*. Sierra Pinacate, 900+ m, *F 19422*.

VERBENACEAE Verbena Family

Plants of diverse habit including herbs (mostly perennials), vines, shrubs, or trees. Branchlets mostly 4-angled (square in cross section). Leaves mostly opposite (ours), simple or compound; stipules none. Calyx mostly 4- or 5-lobed or toothed. Corollas sympetalous, mostly 4- or 5-lobed, usually bilateral. Stamens usually 4. Ovary superior; style 1, the stigma often 2-lobed. Fruits often dry, mostly enclosed in the calyx, separating into 1-seeded nutlets (ours), drupes, or capsules.

Worldwide, mostly tropical; quite a few in aridlands and relatively few in temperate regions; 90 genera, 2000 species. Among the tropical trees in the family is teak (*Tectona grandis*), one of the world's more valuable hardwoods.

1. Stems creeping, rooting at nodes; hairs 2-armed; corollas markedly bilateral; fruits separating into 2 nutlets. __________ **Phyla**
1′ Stems not creeping; hairs solitary, not 2-armed; corollas only moderately bilateral; fruits separating into 4 nutlets.
   2. Calyx more than 5 mm; corollas lavender-pink, 8–14 mm wide; style more than 5 mm; nutlets blackish, 3.3–3.5 mm, the ventral surface dark, not white papillate. __________ **Glandularia**
   2′ Calyx 3.5 mm or less; corollas pink or blue, 2.5–5 mm wide; style 1.5 mm or less; nutlets brown, 1.7–2.2 mm, the ventral surface white papillate. __________ **Verbena**

GLANDULARIA

Mostly perennial herbs, some annuals. Branches and leaves opposite. Leaves simple to bipinnatifid, usually toothed to cleft, with stiff simple and sometimes glandular hairs. Inflorescences spicate, usually not especially slender or elongated in fruit. Each flower subtended by a bract. Calyx tubular, 5-angled, irregularly 5-toothed. Corollas 5-lobed, salverform (the tube abruptly expanded into a flat limb), moderately bilateral, pink, red, purple, yellow, or white. Fruits splitting into four 1-seeded nutlets, the nutlets nearly cylindrical, mostly blackish. Base chromosome number: $x = 5$. Primarily flowering in spring and again in late summer and fall.

Twenty-two species in North America north of Guatemala, with greatest diversity in northern Mexico and southwestern United States, and 50 species in South America. *Glandularia* is often not separated from *Verbena,* from which it can be distinguished by differences in seed morphology, chromosome number, style length, and reproductive modes. *Glandularia* seems to have originated in South America. Reference: Umber 1979.

Glandularia gooddingii (Briquet) Solbrig [*Verbena gooddingii* Briquet. *V. gooddingii* var. *nepetifolia* Tidestrom] DESERT VERBENA. Short-lived herbaceous perennials or non-seasonal ephemerals, pilose with relatively long spreading white hairs plus stalked and sessile glands on the herbage and calyces. Leaves 2.5–4.5+ cm, broadly ovate, deeply toothed to laciniately parted. Inflorescences reaching 5.0–8.5 cm, single or with 3 spikes, the flowers fragrant, crowded into a headlike spicate cluster. Bracts 5.0–7.5 mm. Calyx 5.5–7.0 mm. Corollas lavender-pink, the tube longer than the calyx, with hairs inside and outside, the flat limb ("face") 8–14 mm wide, the lobes shallowly to conspicuously notched, the throat white and yellow-green, nearly closed off by a dense ring of white, moniliform hairs (resembling a string of beads) guarding the entrance; inner surface of corolla tube with non-moniliform downward-pointing white hairs, these dense just below the anthers and along a line extending inward from the sinus between the 2 larger, lower corolla lobes. Style 5.5–7+ mm. Nutlets 3.3–3.5 mm, dark brown to nearly black, the dorsal side resembling an ear of corn with an alveolate pattern and smooth ridges. Flowering spring and fall.

Cinder soils at higher elevations in the Sierra Pinacate, seldom common (above 780 m: *F 18657, F 86-436; Joseph 20 Mar 1983*). Common northward in southern Arizona and eastward in Sonora. Southeastern California, Baja California, Nevada, southwestern Utah, western Texas, New Mexico, Arizona, and Sonora southward to the Río Mayo region.

PHYLA

Prostrate or creeping perennial herbs. Herbage with malpighian (2-armed) hairs. Leaves simple. Flowers subtended by bracts in dense spikes on well-developed axillary peduncles. Corollas of 2 unequal lips. Fruits of 2 tardily separating 1-seeded nutlets. North and Central America; 15 species, some of which seem poorly defined.

***Phyla nodiflora** (Linnaeus) Greene [*Verbena nodiflora* Linnaeus. *Lippia nodiflora* (Linnaeus) Michaux] Small creeping herbs rooting at nodes, sparsely to rather densely pubescent. Larger leaves 2–4 × 0.5–1.8 cm, with 3 or 4 teeth on each side of the leaf. Peduncles 1.6–5 cm, longer than the subtending leaves. Corollas white to pink, in small conical heads.

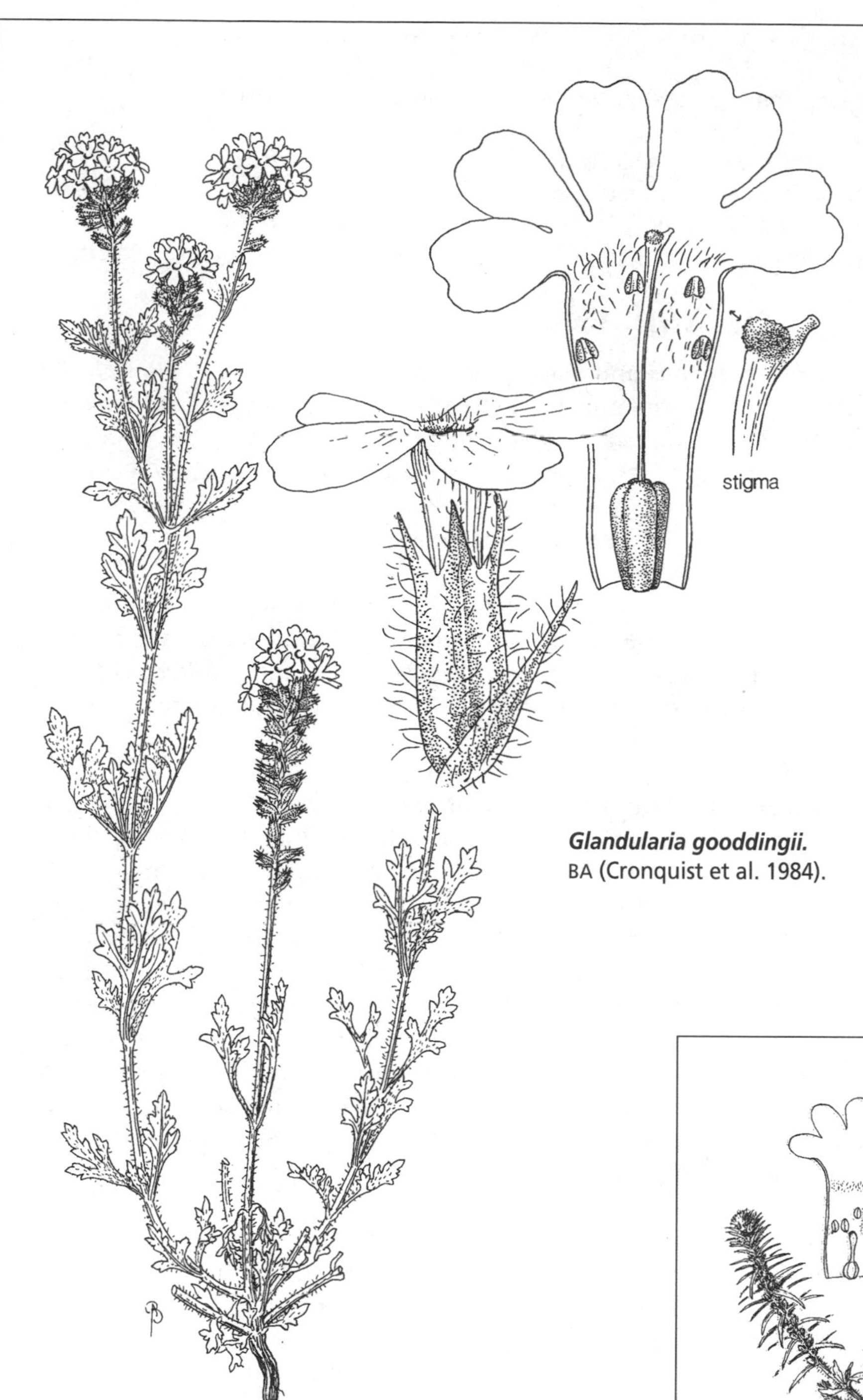

Glandularia gooddingii. BA (Cronquist et al. 1984).

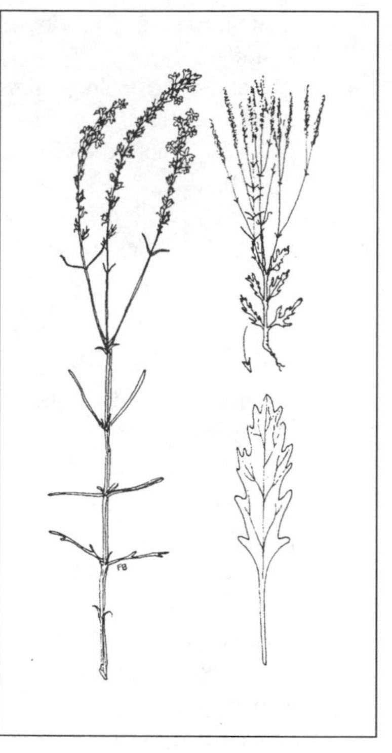

Verbena officinalis subsp. ***halei.*** FB. © James Henrickson 1999.

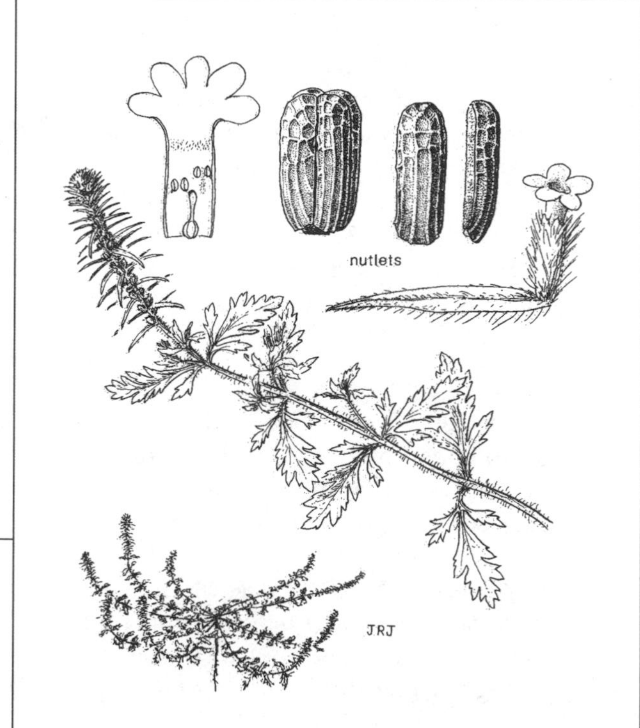

Verbena bracteata. JRJ (Hitchcock et al. 1959).

A weed in well-watered gardens, often growing with Bermuda grass (*Cynodon*) in lawns and ditches. Probably flowering most of the year. Apparently more densely pubescent in drier habitats or seasons. Widely scattered in lowland Sonora in well-watered artificial or wetland habitats. Worldwide in tropical and subtropical regions, mostly in low, wet areas, often weedy. Ours are var. *incisa* (Small) Moldenke, of southwestern United States and Mexico, distinguished by having leaves longer than wide and narrowly wedge-shaped.

Sonoyta, *F 91-1*. Puerto Peñasco, *F 91-144*.

VERBENA

Prostrate to erect herbs, usually perennials in North America. Leaves usually toothed to dissected. Flowers in terminal, usually elongated, densely flowered spikes, each flower subtended by a narrow bract. Corollas usually funnelform, sometimes nearly salverform, with a flat, 5-lobed, weakly bilateral limb. Style 3 mm or less. Fruits separating into four 1-seeded nutlets (similar to those of *Glandularia*), the nutlets usually falling as a unit, brown or occasionally blackish, small (1–2 mm), the common faces angled and reaching to the apex. Base chromosome number: $x = 7$ in North America.

Mostly in North and South America with greatest diversity in North America, where it seems to have originated (see *Glandularia*); 200 species. Reference: Barber 1982.

1. Plants usually spreading; spikes less than 12 cm, the bracts 6.0–10.5 mm; nutlets 2.1–2.2 mm. ____ **V. bracteata**

1′ Plants usually erect; spikes usually 15–30 cm, the bracts 1.7–2.7 mm; nutlets 1.7–1.8 mm. ____ **V. officinalis**

Verbena bracteata Lagasca & Rodriguez. Spring ephemerals, low and spreading to prostrate, the stems and inflorescences 25–40 cm, hirsute or strigose with coarse spreading white hairs throughout. Leaves, flowers, and fruits crowded. Leaves often 2–6 cm, more or less lanceolate, deeply toothed or cleft. Inflorescences of several-branched spikes less than 12 cm. Bracts 6.0–10.5 mm. Calyx 2.6–3.0 mm. Corollas 2.5–3.0 mm wide, pink, inconspicuous and nearly hidden by bracts. Style 0.6+ mm. Nutlets brown, 2.1–2.2 mm, resembling a miniature ear of corn, the ventral side densely and minutely white prickly papillate, the dorsal side striate below (raised longitudinal veins), alveolate above.

In the flora area known from two records from the southeastern flank of the Pinacate volcanic region: near Tanque Romero (Víboras), 200 m, 31°40′ N, 113°25′ W (*Webster 24263*), and Rancho los Vidrios (*Ezcurra 8 Apr 1981*). Southern Canada to northern Mexico, mostly in disturbed habitats, and in Arizona mostly at dirt cattle tanks (represos). The nearest record is from the Cabeza Prieta Refuge. The Sonoran Desert populations are few and widely separated, and perhaps result from dissemination by birds visiting waterholes.

Verbena officinalis Linnaeus subsp **halei** (Small) S. Barber [*V. halei* Small. *V. menthaefolia* Bentham] Ephemerals to short-lived perennials, few-branched, slender and erect, often 60–100 cm, the leaves, flowers, and fruits rather widely spaced. Leaves strigulose with sparse to moderately dense pubescence of appressed stiff white hairs, the stems glabrate or sparsely hairy. Larger, lower leaves 7–8+ cm, once- or twice-pinnatifid. Inflorescences of several to many slender, mostly erect spikes often 15–30 cm; bracts 1.7–2.7 mm. Calyx 2.4–3.5 mm, hirsute with short, appressed non-glandular white hairs. Corollas bright blue, 4–5 mm wide. Style 1.1–1.5 mm. Flowering-fruiting calyx 2.4–3.5 mm. Nutlets 1.7–1.8 mm, brown, the common ventral surfaces white papillate. Flowering various seasons including spring and summer.

Wetland habitats such as the banks of the Río Sonoyta, a few widely scattered waterholes, and sometimes in clay soils of playas. According to Barber (1982), subsp. *halei* ranges from the southeastern United States to extreme southeastern California and southward through northern Mexico, and subsp. *officinalis* is native to the Old World and introduced in the New World.

Río Sonoyta at Sonoyta, *F 86-97*. Tinajas de los Pápagos, *Turner 59-22*.

VISCACEAE MISTLETOE FAMILY

Chlorophyll-bearing hemiparasites (water parasites), perennials with brittle stems. Leaves opposite, simple, entire, sometimes reduced. Monoecious or dioecious. Flowers small, mostly green or yellow, insect-pollinated, the perianth parts reduced, especially on female flowers. Ovary inferior. Fruits fleshy and sticky, usually 1-seeded berries; seeds lacking a well-defined ovule, the cotyledons reduced and the embryo in mature fruits already green. Seeds sticky at one end, typically germinating on dry branches of the host where birds have roosted.

Worldwide, mostly tropical; 7 genera, 400 species. *Phoradendron* is the largest genus. The traditional Christmas mistletoe is *Viscum album* in Europe and *Phoradendron serotinum* in North America. Two families of mistletoe are usually recognized, the other being the generally tropical and usually large-flowered or showy mistletoes, Loranthaceae. Seven species of Loranthaceae occur farther south in Sonora, mostly in the southern part of the state. References: Kuijt 1969, 1982.

PHORADENDRON MISTLETOE

Parasites on dicot and conifer trees and shrubs. Stems jointed. Flowers usually embedded in a jointed spike, usually several in the axil of each bract. Male flowers with a 3-lobed calyx, the anthers sessile and fused to calyx lobe bases, opening by large longitudinal pores; female flowers with the calyx fused to the ovary. Fruit a berry, globose, sessile, and with mucilaginous pulp; seeds bird-dispersed. New World, mostly tropical; 230 species. Reference: Wiens 1964.

1. Leaves scalelike, 1–2.5 mm; on desert trees. ______ **P. californicum**
1′ Leaves well developed, 3–6 cm; on cottonwoods and willows. ______ **P. macrophyllum**

Phoradendron californicum Nuttall [*P. californicum* var. *distans* Trelease. *P. californicum* var. *leucocarpum* (Trelease) Jepson] *TOJI;* DESERT MISTLETOE; *HA:KWOḌ, TO:KĬ.* Branches arching to drooping, often forming much-branched masses festooning common desert trees, especially legumes including mesquites and screwbean (*Prosopis* spp.), ironwood (*Olneya*), catclaw (*Acacia greggii*), and palo verdes (*Parkinsonia* spp.); rarely on creosote bush (*Larrea*) and *Condalia*. Stems terete, at first silvery-green pubescent with minute, appressed scalelike hairs, soon glabrous and green to reddish green. Leaves 1.0–2.5 mm, closely appressed to the stem, at first green or yellow-green and quickly drying as persistent scales or remaining green only at base.

Dioecious or occasionally monoecious. Flowers fragrant. Calyx thick, fleshy, and yellow-green. Anthers short and yellow. Fruits globose, 4.5–5.5 mm when fresh, the fresh pulp viscid and translucent white, salmon colored on exposed surfaces and whitish to yellow-white where not exposed to sunlight. Seeds 4 mm, oblong, the surface white, the radicle haustorium (the seedling root that attaches to the host stem) 0.5 mm, green, and starting to protrude as the fruit ripens. Flowering and fruiting non-seasonally, with massive, simultaneous flowering often December-February and fruits ripening in late winter to early spring.

Distributed with its hosts throughout most of the flora area including dunes. Southern California to southwestern Utah and southward to the Cape Region of Baja California Sur and northwestern Sinaloa.

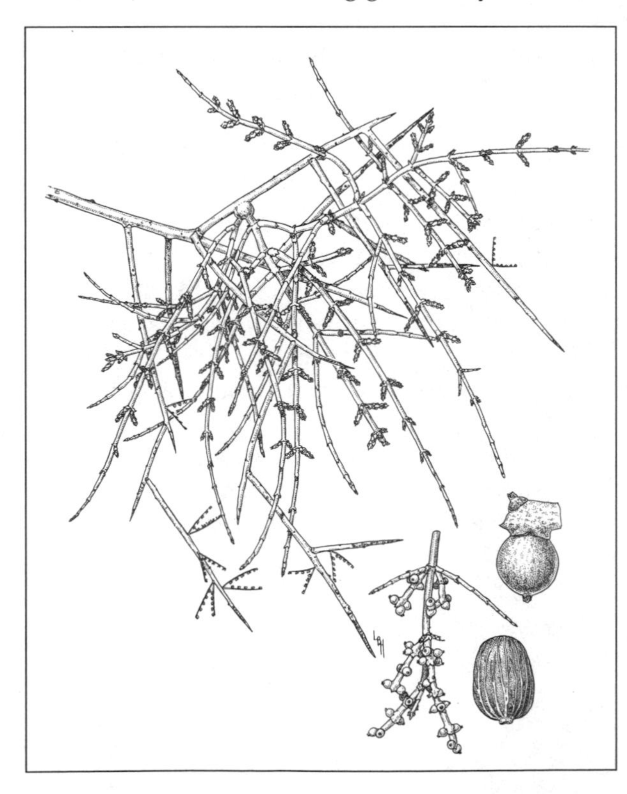

Phoradendron californicum **on** ***Parkinsonia microphylla.***
LBH (Benson & Darrow 1981).

In winter the phainopepla (*Phainopepla nitens*) feeds "largely on the sticky berries of mistletoe; the favorite perches are easily spotted by the pyramids of these seeds glued to the branches below" (Phillips, Marshall, & Monson 1964:139). Mearns (1892–93:#220) recorded "mistletoe from *Prosopis juliflora* [*P. velutina*], Quitovaquito . . . January 30, 1894. The waxwing [sic] (*Phainopepla nitens*), mockingbird, house finch, and several species of woodpeckers and other birds feed largely on this mistletoe during the winter." The leaves of this desert mistletoe are greatly reduced in comparison to other phoradendrons in more mesic habitats and regions. *P. californicum* is most closely related to *P. olae* Kuijt, of Puebla (Kuijt 1997). Their geographic ranges are disjunct by more than 1400 km. There are no other known closely related species.

Elegante Crater, *Burgess 5565*. MacDougal Crater, *F 9764*. Sierra del Rosario, *F 20723*. Laguna Prieta, *F 85-758*. Quitobaquito, *Mearns 220* (US).

†Phoradendron macrophyllum (Engelmann) Cockerell [*P. flavescens* (Pursh) Nuttall var. *macrophyllum* Engelmann. *P. tomentosum* (de Candolle) Engelmann ex A. Gray subsp. *macrophyllum* (Engelmann) Wiens] BIG LEAF MISTLETOE. Large mistletoe, the leaves 3–6 cm, thick, green, elliptic-obovate. Fruits ca. 6 mm in diameter. On cottonwoods (*Populus*) and willows (*Salix*). Coahuila and Durango to Chihuahua, Sonora, and Baja California, mostly near the U.S. border, and western Texas to Colorado and California. Recorded from Yuma (Wiens 1964) and undoubtedly once in cottonwoods and willows along the Río Colorado in Mexico. I have not seen mistletoe in any of the cottonwoods and willows in the flora area. This mistletoe is apparently extirpated along the lower Colorado River along with all but a remnant population of its cottonwood host.

ZYGOPHYLLACEAE CALTROP FAMILY

Herbs, shrubs, and some trees. Leaves opposite (ours) or less commonly alternate, simple, with 2 or 3 leaflets or mostly even pinnate; stipules usually well developed. Flowers often solitary in leaf axils (ours), usually bisexual, radial, and 5-merous. Ovary 1, superior; style 1, the stigmas 1 to several. Sepals and petals usually separate, the petals often clawed and twisted at the claw to stand perpendicular to the axis; nectary disk usually well developed. Stamens often 2 or 3 times as many as the petals. Fruits often capsules or schizocarps, mostly of 5 carpels; seeds usually 1 to several per carpel or mericarp.

Worldwide, mostly tropical semiarid and desert climates; 27 genera, 250 species. The largest genera are *Fagonia, Tribulus,* and *Zygophyllum* (Old World). Members of this family are especially prominent elements of desert vegetation on every continent. References: Porter 1963, 1972; Scholz 1964.

1. Woody shrubs; leaflets 2 or 3.
   2. Usually less than 1 m; leaflets 3, the stipules leaflet-like and spinescent; Sierra del Rosario. ______ **Fagonia densa**
   2′ Usually more than 1 m; leaflets 2 (fused leaflets appearing as 1), the stipules triangular and neither leaflet-like nor spinescent; widespread. ______ **Larrea**

1′ Herbaceous; leaflets 3 or more.
   3. Perennials sometimes flowering in first year; stipules spinescent; leaflets 3 (1 or more leaflets per leaf may be shed in drought); petals rose-pink. ______ **Fagonia**
   3′ Summer-fall ephemerals; stipules not spinescent; leaflets 6 or more; petals yellow or orange.
      4. Fruits knobby but not spiny, separating into 10 segments (rarely fewer) and leaving a persistent axis when falling. ______ **Kallstroemia**
      4′ Fruits with sharp spines, separating into 5 nutlets (rarely fewer) and leaving no central axis when falling. ______ **Tribulus**

FAGONIA

Perennial herbs or dwarf shrubs, occasionally annuals, mostly spiny, usually intricately branched. Stems of first season or year photosynthetic, slender, and striate. Leaves digitately compound, mostly with 3 leaflets (ours) or sometimes more or leafless; stipules spinescent. Petals clawed, usually pink to purple; pedicels well developed. Fruit a capsule, moderately inflated, deeply 5-lobed, obovoid, and beaked, resembling a miniature dome of a mosque. Seeds flat, glistening with a hard, nearly transparent "varnish" or envelope becoming mucilaginous (slimy) when wet and adhering tenaciously when dry.

Exclusively in deserts of North and South America, Africa, India-Pakistan, and the Mediterranean region; absent from intervening regions; 30 species. No other genus is so wide ranging yet so closely restricted to hot, arid deserts. The mucilaginous seeds are probably a major factor in the unique distribution (Bray 1898).

In the Sonoran Desert I find 3 well-marked species groups. These groups, probably equivalent sections or subgenera with representatives also in other regions are (1) *F. densa* and *F. palmeri;* (2) *F. barclayana* and *F. pachyacantha;* and (3) *F. californica* and *F. villosa*. Within each species group the taxa are allopatric whereas sympatry is commonplace among those of the different species groups. Exceptions occur among the *californica* group, as discussed further.

Fagonia densa and *F. palmeri* are dwarf shrubs. The other Sonoran Desert species are short-lived perennials, although flowering may occur during the first season. Growth and flowering may occur at any season following rains, except during times of freezing weather. *Fagonia* species occupy the hotter, drier, and more exposed habitats, often on south- and west-facing rock slopes. References: El Hadidi 1966, 1974; Schreiber 1974; Standley 1911.

1. Stipular spines straight, reaching 5–12+ mm.
   2. Plants as tall or taller than wide, erect, the stems woody; leaflets glandular pubescent, not succulent, 1 mm wide. ____ **F. densa**
   2′ Plants more than twice as wide as tall, sprawling to semiprostrate, the stems slightly woody only at the base; leaflets glabrous with age, often semisucculent, to 5–10 mm wide. ____ **F. pachyacantha**

1′ Stipular spines mostly slightly curved (at least some on each plant), 1.5–3.0 mm. ____ **F. californica**
   3. Herbage mostly glabrous or essentially so, mostly not glandular (in ours); lateral leaflets often narrower than the middle leaflet; fruiting pedicels less than 7 mm. ____ **F. californica** subsp. **californica**
   3′ Herbage glandular pubescent; lateral leaflets about as wide as the middle leaflet; fruiting pedicels 8–20 mm. ____ **F. californica** subsp. **longipes**

Fagonia californica Bentham subsp. **californica** [*F. laevis* Standley. *F. californica* subsp. *laevis* (Standley) Wiggins] Perennials often reaching 30–50 cm across, much-branched with very slender, brittle stems. Stem tips, stipules, petioles, pedicels, and sepals glabrous or sometimes with some small stalked glands, the stems becoming scabrous; fruits often minutely hairy but not glandular. Leaflets 1.0–2.5 mm wide, the middle one often slightly wider. Stipules mostly 1.5–3.0 mm, at least some slightly curved. Fruiting pedicels 1.5–6.5 mm. Seeds (2.7) 2.9–3.2 × 1.9–2.3 mm, broadly ovate, pale tan to dark brown with age.

Widespread; volcanic and granite ranges to their summits. Rocky soils on all slope exposures, mostly open, xeric habitats of hills and mountains, crater walls, upper bajadas, and desert pavements. Arizona, southeastern California, both Baja California states, and western Sonora south to the Sierra Seri.

Porter (1963) treated the Baja California members of the *californica* group as distinct species. Although his treatment works well for the peninsula, there are problems in trying to sort out the populations in Sonora and Arizona. Scattered more or less throughout the range of *F. laevis* in Sonora and Arizona are specimens with glandular pubescent young leaflets, which according to Porter's work keys to *F. californica* rather than *F. laevis*. I find neither ecological nor geographic segregation between them in Arizona and Sonora, and therefore treat them as a single taxon. The plants in the flora area are usually glabrous or glabrate, but in adjacent areas there is a perplexing array of glabrous and glandular-pubescent plants.

Sierra del Rosario, *F 20704*. Sierra del Viejo, *F 16902*. MacDougal Crater, *Burgess 6328*. Sierra Pinacate, 1100 m, *F 18653*. Cerro Colorado, *Webster 22309*. Bahía de la Cholla, *F 16843*.

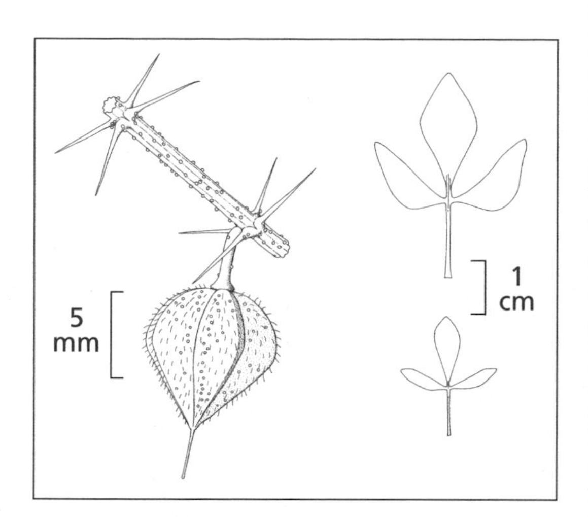

Fagonia pachyacantha. MBJ.

Fagonia californica subsp. ***californica.***
JRJ (Abrams 1951).

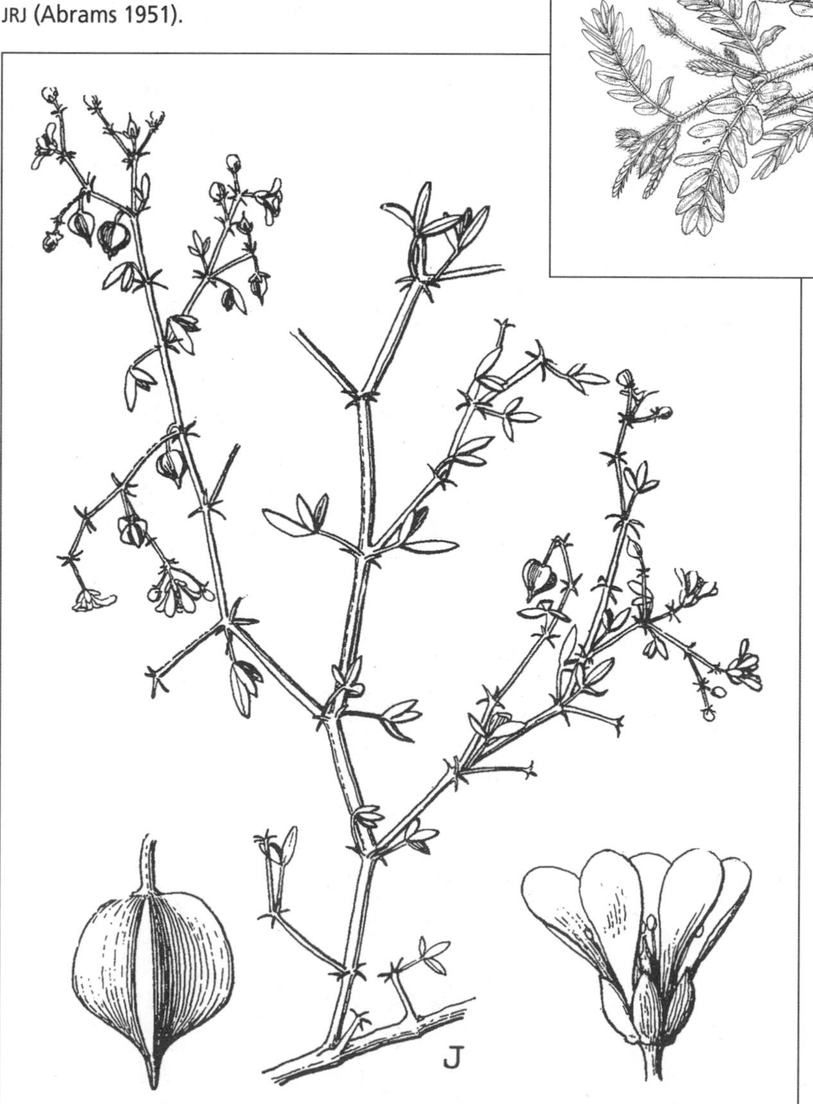

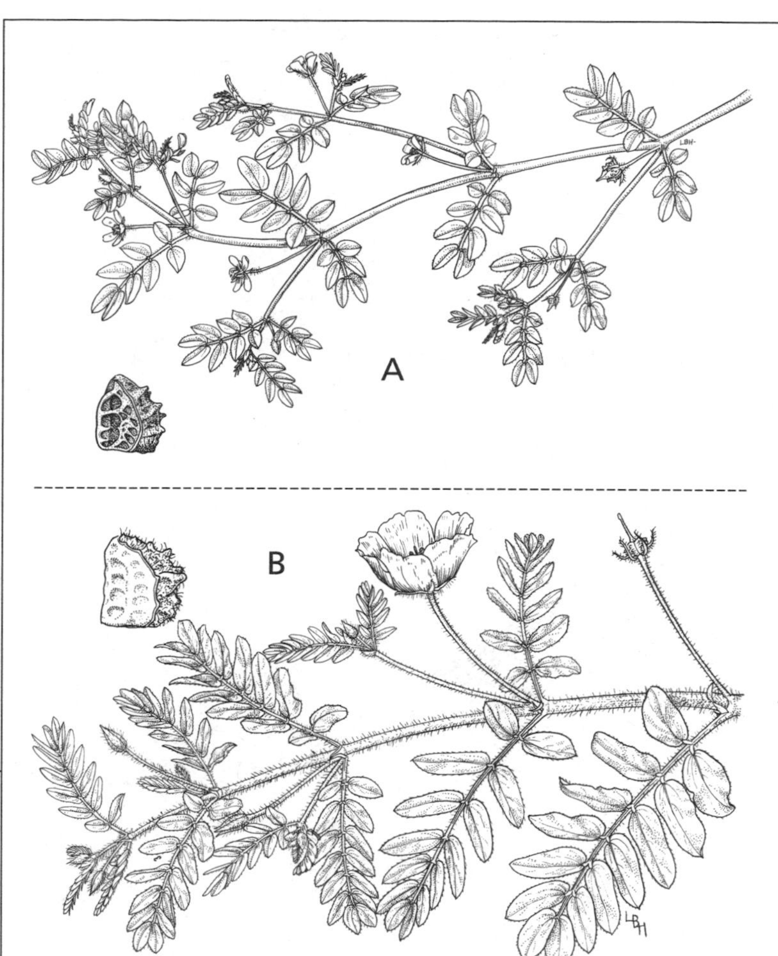

Kallstroemia: (A) K. californica;
(B) K. grandiflora. LBH (Parker 1958).

Larrea divaricata subsp. ***tridentata.*** BA. © James Henrickson 1999.

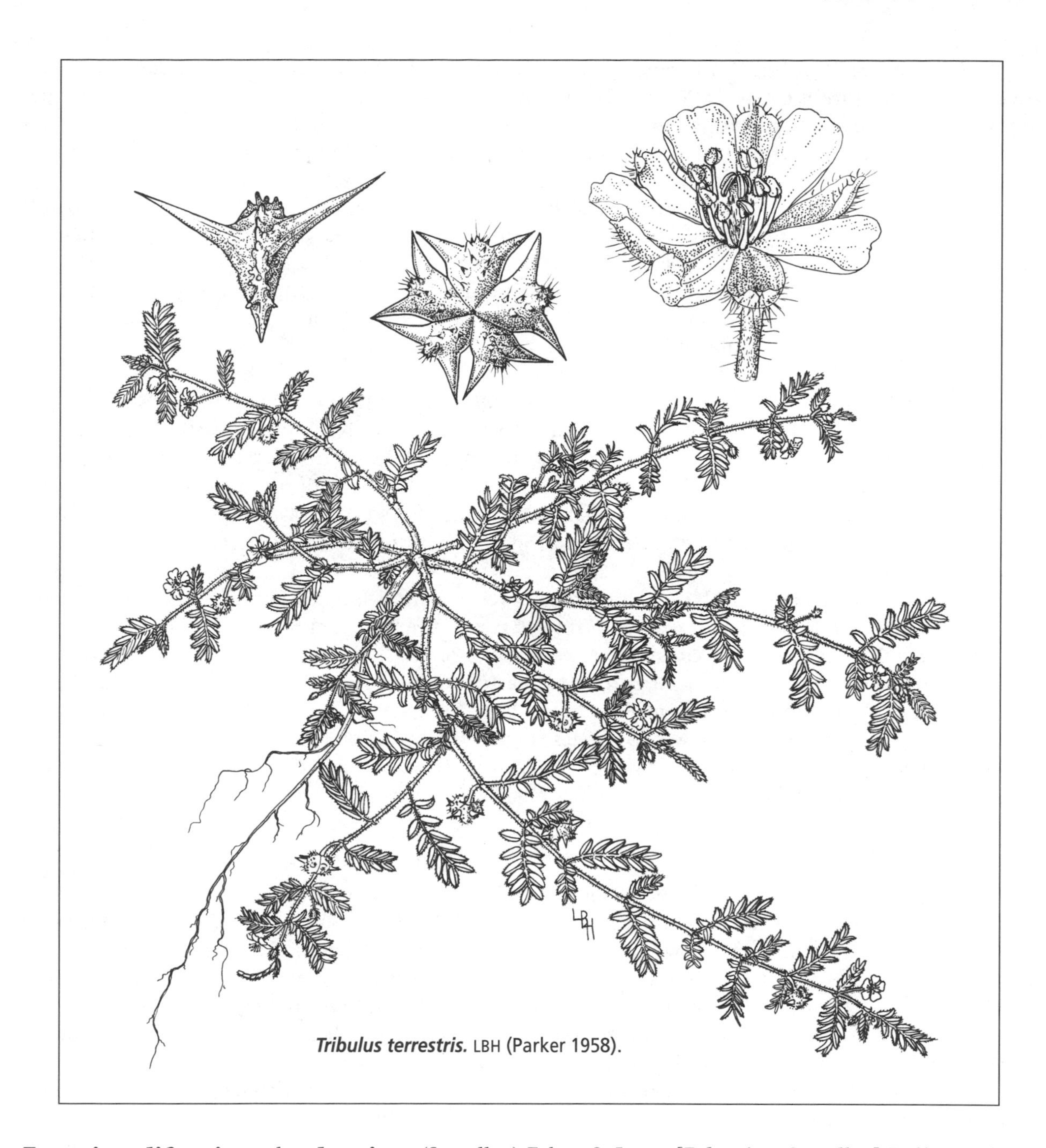

Tribulus terrestris. LBH (Parker 1958).

Fagonia californica subsp **longipes** (Standley) Felger & Lowe [*F. longipes* Standley] Differing from subsp. *californica* in its more slender habit of growth, narrower (often linear) leaflets (lateral and terminal leaflets usually about the same width), lighter-colored herbage, longer petioles, and longer pedicels which elongate to 8–20 mm at fruiting time. Also differing from subsp. *californica* in the flora area by its glandular-pubescent herbage, pedicels, and fruits.

Rocky soils of dry hillsides, bajadas, and roadsides in Arizona Upland near Sonoyta. Southward in Sonora to Isla Tiburón and the opposite Sierra Seri; also Arizona and southeastern California.

Although extreme specimens are readily identified, many show perplexing gradation in characters of the *F. californica-laevis-longipes* complex. The *californica-longipes* boundary is especially blurred. *F. villosa* D.M. Porter of Baja California Sur, characterized as villous, is perhaps the best-defined segregate of the *californica* group. Drought-stressed (e.g., late-season) plants of subsp. *longipes* may have pedicels as short as 5 mm, but are identifiable by their glandular hairs.

12 mi SW of Sonoyta, *F 9818*.

Fagonia densa I.M. Johnston. Perennials and also flowering in the first season. Densely branched dwarf hardwood shrubs often 0.5–1.0 m, with numerous closely set upright branches. Herbage yellow-green, conspicuously glandular pubescent. Stipular spines 5–11+ mm, straight and sharp, resembling the leaflets or petiole. Leaflets linear, 1 mm or less in width, soon deciduous from the more prominent petiole.

Steep granitic slopes in the Sierra del Rosario (*Burgess 6871; F 89-63*), common to the peak on all slope exposures, and particularly abundant on south and west exposures; not known elsewhere in Sonora. Also along the arid Gulf Coast of Baja California and adjacent islands.

Because the leaflets are often not present, the stem tips, petioles, and stipules are probably the major photosynthetic organs. Its closest relative, *F. palmeri* Vasey & Rose, abruptly replaces it to the south on the Baja California Peninsula and Gulf of California islands.

Fagonia pachyacantha Rydberg [*F. californica* var. *glutinosa* Pringle ex Vail] Forming spreading or semiprostrate mats to 1.0–1.5+ m across during favorable times. New growth with golden yellow glands (rarely glabrous or glabrate during times of high rainfall); leaves bright green, glabrate with age. Leaflets lanceolate, ovate, or obovate, often becoming semisucculent during favorable seasons, the middle leaflet reaching 1.5–2.5 × 0.5–1.2 cm. Stipular spines straight, at least some on each plant 5–12 mm, often stout, with age bending back toward the stem but not curved. Pedicels 1.5–5.0 mm. Seeds dark brown, 2.6–2.7 × 1.9–2.0 mm.

Exposed, open habitats in the lowland, rocky, western and southern flanks of the Pinacate volcanic field and the granitic ranges to the northwest; not in the Sierra del Rosario. Often growing with *F. californica* subsp. *californica*. Southward in western Sonora to Cerro Tepopa (vicinity 29°22′ N) and Isla Tiburón, extreme southwestern Arizona, southeastern California, and Baja California. Its apparent closest relative, *F. barclayana* (Bentham) Rydberg, abruptly replaces it to the south in Sonora and on the Baja California Peninsula. These two species are allopatric with no indication of intermediates.

Sierra del Viejo, *F 16903*. Sierra Extraña, *F 19072*. Mex 2, km marker 117, W of Sonoyta, *Yatskievych 81-187*. Tinaja del Tule, *F 18908*.

KALLSTROEMIA

Annuals or occasionally perennial herbs (summer ephemerals in the Sonoran Desert). Leaves of each pair often of different sizes or one abortive; pinnate with 2–10 leaflet pairs, those of the lowermost pair often markedly different. Each flower open only part of one day, the pollen and stigma maturing simultaneously; most (ours) capable of self-pollination. Fruits with 5 (6) carpels, each divided by a septum between the 2 ovules to form 10 (12) half-carpel segments (nutlets), the fruits thus breaking into 1-seeded nutlets; the nutlets wedge-shaped, indehiscent, knobby (tuberculate) but not spiny, falling away from the persistent axis and beak (the styliferous axis).

New World, dry tropical to arid regions; 17 species, 3 in the Sonoran Desert. This is the largest genus of Zygophyllaceae in the New World. Reference: Porter 1969.

1. Petals yellow, 4–6 mm; sepals usually deciduous; beak of fruit less than 5 mm; fruiting pedicels 1.0–2.3 cm. ________ **K. californica**

1′ Petals bright orange with a darker base, (15) 20–35 mm; sepals persistent; beak of fruit (5) 8–12 mm; fruiting pedicels (2) 3–7 cm. ________ **K. grandiflora**

Kallstroemia californica (S. Watson) Vail [*Tribulus californicus* S. Watson. *K. californica* var. *brachystylis* (Vail) Kearney & Peebles] *MAL DE OJO;* CALIFORNIA CALTROP. Stems sprawling to prostrate, often 20–50 cm, sometimes to 1 m. Leaves 2.0–4.5 (6) cm, with 3–6 (7) pairs of leaflets. Petals 4–6 mm, yellow to yellow-orange. Fruiting pedicels 1.0–2.3 cm. Body of fruit 4–5 mm; beak of fruit usually shorter than to about as long as the body.

Widespread but not known from dunes or higher elevations. Larger washes, arroyo bottoms, playas, roadsides, alluvial and sand flats, and sometimes on rocky slopes. Southwestern Texas to southern Utah and southeastern California, and southward to Nayarit and Baja California Sur.

25 mi E of San Luis, *F 16715*. 24 km W of Los Vidrios, *F 85-1013*. MacDougal Crater, *F 10745*. Tinaja de los Chivos, *F 18615*. Suvuk, *Nabhan 25 Sep 1982*. Playa Díaz, *F 86-369*.

Kallstroemia grandiflora Torrey ex A. Gray. *BAIBORÍN, MAL DE OJO;* ORANGE CALTROP. Stems often (15) 30–100 (150) cm, procumbent with age. Leaves 4.5–12 cm, with 5–9 pairs of leaflets. Sepals green in bud, becoming pale orange after the flowers open. Flowers showy, opening about an hour after dawn and fading by mid-afternoon (or remaining open all day on cloudy days), with greenish nectaries between the sepals and petals. (Ants and a variety of flying insects eagerly feed at the nectaries.) Corollas (3) 5–7 cm wide, deep dark orange with high color saturation, fading to pale yellow-orange by mid-afternoon, the corolla center and filaments dark orange-red; individual petals often 2.0–3.6 cm wide. Anthers yellow, often open by 8 A.M.; ovary and style green. Fruiting pedicels (2) 3–7 cm. Body of fruits 4–5+ mm and knobby; beaks 8–12 mm (as short as 5 mm when drought stressed).

Widespread and often common but not known from dunes or higher elevations. Larger washes, arroyo bottoms, playas, roadsides, alluvial flats, sand flats, and slopes; sometimes carpeting the desert with orange. Southwestern Texas to southeastern California and southward to Guerrero; not known from the Baja California Peninsula.

Suvuk, *Nabhan 387*. Pinacate Junction, *F 85-1001*. MacDougal Crater, *F 10432*. SW of Tinajas de los Pápagos, *F 10602*. Tinaja de los Chivos, *F 18614*.

LARREA

Shrubs and a prostrate subshrub, with very hard, brittle wood. Stems swollen at nodes, at first angled or moderately squarish in cross section with forward-pointing hairs, becoming terete and glabrous with age. Herbage resinous and highly aromatic. Leaves opposite, odd pinnate or bifoliolate (appearing simple and 2-lobed), with 2–17 leaflets. Stipules at first green, soon becoming red-brown, triangular to broadly lanceolate to broadly ovate or orbicular, persistent after the leaves fall; glands on the inner (adaxial) surface secrete resin, making the plants highly glutinous. Sepals 5, united below; petals 5, separate, yellow, narrowed below to a twisted claw. Stamens 10, the filaments each with a well-developed yellow scale or appendage, these serving to cup nectar produced at the base of the style. Fruit with 5 mericarps tardily separating or remaining associated, each 1-seeded, sparsely to densely hairy.

Four species, their distribution closely circumscribing the major hot deserts of the Americas. *Larrea* is South American in origin and the single most widespread, abundant, and characteristic woody genus across most of the New World deserts: Argentina, Chile, Bolivia, Peru, Mexico, and southwestern United States. All 4 species occur in South America and *L. divaricata* is also in North America. References: Barbour 1968, 1969; Felger & Lowe 1970; Hunziker et al. 1972; Mabry, Hunziker, & DiFeo 1977; Vasek 1980; Wells & Hunziker 1976.

Larrea divaricata Cavanilles subsp. **tridentata** (Sessé & Moçiño ex de Candolle) Felger & Lowe [*Zygophyllum tridentatum* Sessé & Moçiño ex de Candolle. *L. tridentata* (Sessé & Moçiño ex de Candolle) Coville] *GOBERNADORA, HEDIONDILLA;* CREOSOTE BUSH; *ṢEGAI, ṢEGOI*. Multiple-stemmed, trunkless shrubs often 0.8–2+ m. Stems slender, the nodes with dark rings formed of the persistent stipules and their glandular exudate. Leaves and young stems with white hairs, the hairs soon submerged in gummy, varnish-like glandular exudate, the exudate especially thick and viscid during dry seasons. (Also see *Ambrosia deltoidea* and *Baccharis sarothroides,* which show a somewhat similar pattern of hairs holding or "trapping" gummy exudate and then "drowning" in the exudate.) Stipules 1.2–2.2 mm, broadly ovate-triangular with an acute to narrowly obtuse, short acuminate tip, cordate with the basal lobes often overlapping, at first green like the leaves, then yellow and soon red-brown, glandlike and covered with resin, and tightly appressed (appearing stuck to the stem by the sticky glandular exudate) to somewhat spreading. (Stipules at the growing stem tips are green, clasping, leafy-glandular organs protecting the enclosed apical meristem, and more viscid than the subtending emerging leaves.) Leaves 5–12 mm, subsessile or with very short petioles, with 2 sessile, more or less lanceolate, slightly curved (falcate) leaflets united at their broad bases (appearing as a simple leaf).

Fresh flowers 2.4–3.0 cm wide. Petals, stamens, and filament scales bright yellow; sepals and style green. Stamens at first often spreading or hanging down between the petals, becoming erect with the anthers close together, the style at first equaling the stamens, with age projecting beyond them. Ovary and fruits covered with silky white hairs, the fruits 8–10 mm, obovoid, appearing as small fuzzy white balls. Seeds 4.5–5.0 mm, brown, not producing mucilage.

This is the single most abundant major perennial across the desert floor; it also occurs to the summits of the Sierra Pinacate and granitic ranges, and on all except the highest dunes. Subsp. *tridentata* is widespread in the three major warm deserts of North America and is the primary element in mapping and defining these deserts. Subsp. *divaricata* occupies deserts in South America.

The resinous leaves shrivel during extreme drought, and the plants eventually may become essentially leafless. During times of high soil moisture and warm temperatures there is rapid growth of lush, green foliage; growth is facultative and largely non-seasonal. Massive flowering often occurs in spring and again with hot-weather rains. After a rain the aroma of terpenes from the wet foliage imparts a magical quality to the crisp desert air. The herbage has long been appreciated by local people for medicinal purposes.

There has been considerable controversy over the taxonomic rank of the North American *Larrea*. Some botanists confine *L. divaricata* to South America and call the North American plants *L. tridentata*. Most *L. divaricata* from South America have obtuse to somewhat orbicular stipules whereas the North American populations have acuminate stipules. Hybrids between them have stipules of intermediate morphology, as does *L. divaricata* from Peru.

The *L. divaricata-tridentata* complex exists as chromosome races with apparently largely discrete geographic ranges. It has been shown to consist of diploid (South American and Chihuahuan Deserts, $2n = 26$), tetraploid (Sonoran Desert, $2n = 52$), and hexaploid (Mojave Desert, $2n = 78$) populations. Some hexaploid populations are also present in the Colorado Desert of southeastern California, a region usually included in the Sonoran Desert. There are, however, some notable exceptions, such as in Baja California where all three ploidy levels can be found; there are also significant areas with no reported chromosome counts (Hunter 1998).

Creosote bush remains have been recovered from packrat middens in southeastern California that date to 21,000 years ago. Kim Hunter (1998) documented polyploidy distribution since the last glaciation by inferring chromosome number on the basis of size of stomatal guard cells from both modern and fossil leaves. Using this method, the 21,000-year-old fossils were deduced to be tetraploids, the present-day Sonoran Desert ecotype. Creosote bush was also present 18,700 years ago in the Tinajas Altas Mountains (Van Devender 1990) and these too were tetraploids (Hunter 1998).

In 1970, Charles Lowe and I co-authored a change in the taxonomic rank of *L. tridentata* to a subspecies of *L. divaricata,* a seemingly logical course of action. Lowe probably was aware that members of the International Biological Program (IBP) were planning to publish a similar taxonomic change, but in his opinion we had come up with the idea first and he encouraged me to rush into print with the nomenclature change. Having done so much work on *Larrea,* the IBP group was obviously annoyed. Several years later, after presenting detailed analysis, various authors of the IBP group (Mabry, Hunziker, & DiFeo 1977) concluded that *L. divaricata* and *L. tridentata* are "semispecies" but resisted the subspecific status.

Based on uncontrolled putative hybrids obtained by Tien Way Yang from transplanted plants in his garden in Tucson, the IBP group concluded that there is a sterility barrier between the North American and South American plants. Dr. Yang collected 152 seeds from his garden; 130 did not germinate, and it was concluded that "in all probability a large proportion of the 130 seeds that did not germinate represent lethal genotypes" (Mabry, Hunziker, & DiFeo 1977:34). However, there are several complications: (1) no control was run (or at least no such data were published); (2) as most growers of desert plants know, *Larrea* seeds are often notoriously difficult to germinate (also see Barbour 1969); and (3) within the North American populations there are obviously additional sterility barriers between ploidy levels. Although Yang observed bees visiting his plants at the "La Creciente Experimental Plot," which was at his home, the plants were grown in partial shade and most certainly not under "natural" conditions. Nevertheless, Yang is an expert at germinating seeds of desert plants and has extensive knowledge and experience with *Larrea.*

In summary, there are several chromosome races in North America with both North and South America having diploid populations. Furthermore, the South American populations are more diverse biochemically than the North American populations. Differences in leaflet shape and pubescence for North and South American populations have been reported, but I find these characters taxonomically unusable—the full span of this variation occurs in the flora area and seems to be influenced by environmental conditions. The variation in stipule shape seems to be the "best" morphological character by which the North and South American populations can realistically be distinguished. These resin-producing organs may have played a significant role in the success of *Larrea* in aridlands with intense herbivore pressure. It seems reasonable to attach taxonomic and evolutionary importance to these structures.

Some people wonder how *Larrea* got from South America to North America, but this one case of intercontinental disjunction is no more amazing than many others (see Solbrig 1972). Although many of the North and South American intercontinental disjuncts are small, sticky-seeded annuals or facultative annuals (e.g., *Fagonia* and various Polemoniaceae), others have larger seeds comparable in size to those of *Larrea* (e.g., *Capparis atamisquea*).

El Golfo, *Bezy 371*. 35 mi E of San Luis, *F 90-179*. Sierra del Rosario, *F 20711*. Moon Crater, *F 19040*. Tinajas de los Pápagos, *Turner 59-28*. Sykes Crater, *F 18956*. Pinacate Peak, 1100 m, *F 19387*. 7 km E of Los Vidrios, *F 11183*.

Larrea divaricata subsp. **tridentata** var. **arenaria** (L.D. Benson) Felger comb. nov. [based on *Larrea tridentata* (Sessé & Moçiño ex de Candolle) Coville var. *arenaria* L.D. Benson, in L. Benson & R.A. Darrow, *Trees and Shrubs of the Southwestern Deserts*, 3rd ed., p. 124, 1981] DUNE CREOSOTE BUSH. Differing from all other taxa of *Larrea* in having an erect branching pattern—reaching 3–5 (6) m, the major stems and branches erect (vertical) and straight (not spreading or ascending), and with age often falling over or leaning. Leafy branchlets generally becoming pendulous.

On shifting sands and dunes of moderate height in portions of the Gran Desierto west of the Sierra del Rosario; *Larrea* is absent from high, unstable dunes. Otherwise known only from Imperial County, California, west of Yuma, on the eastern and western margins of the Algodones Dunes as well as farther westward. The California and Sonora populations grow in essentially identical habitats.

This dune endemic differs most strikingly from the usual (non-dune) creosote bush in its erect growth habit. Both seedlings and vegetative propagation are commonplace. Due to the shifting sands that often expose the lower stems and roots, the several-meter-tall stems sometimes topple over, become buried by drifting sand, and produce roots and shoots at nodes which develop into new plants. Exposed roots also may produce new shoots that develop into new plants. The secondary, pendulous branches characteristically have relatively small leaves but within the range of those of var. *tridentata*. The two varieties occupy different habitats and do not occur intermixed, but intergrade where the habitats meet.

The unique characteristics of var. *arenaria* appear to be genetically controlled. Small plants from several collections from the California population retained their unique growth form when transplanted to non-sandy soils in Tucson. These plants, with stems about 20–50 cm when transplanted, grew to full size within only 2 to 3 years. Likewise, plants cultivated in Tucson and Phoenix from seeds collected at the type locality as well as second-generation cultivated plants produced the slender, erect growth habit characteristic of the parent plants. Seedlings and first- and second-year plants already show the unique erect growth habit.

California: Imperial County, sand dunes south of the Freeway near the Glamis Road interchange, *Benson, 16,722* (holotype of var. *arenaria*, POM 321,664, cited by Benson in Benson & Darrow 1981:124; not seen by Felger and the specimen has not been located).

TRIBULUS CALTROP

Herbaceous annuals and perennials. Leaves pinnate. Flowers with 5 glands between the stamens at the base of the ovary. Ovary with 5 carpels. Fruits horizontally flattened, with 5 indehiscent nutlets (or fewer by abortion); nutlets several-seeded, leaving no central axis when falling, and with stout, sharp spines. Native to the Old World, mostly in Africa and southwestern Asia; 25 species, 2 species in Sonora.

***Tribulus terrestris** Linnaeus. *TORITO, TOBOSO;* PUNCTURE VINE, GOATHEAD; *JEWEḌ HO'IDAG, TO:L 'A'AG.* Hot-weather ephemerals with spreading to prostrate stems. Leaves 1–4.5 cm, with 4–7 leaflet pairs, the leaflets of the lower pair unequal in size. Flowers yellow, the petals 5 mm or less. Fruits 15–18 mm wide, intricately sculptured and spiny, at maturity breaking into 5 spiny nutlets, each with (1) 2 larger spines. After separation the vicious tacklike nutlets land with the largest spines upward.

Disturbed habitats, including roadsides and urban and agricultural areas; not established in natural habitats within the flora area. This species has spread to warm regions worldwide. The scientific name for this obnoxious weed translates as "tribulation of the earth."

7 km N of Puerto Peñasco, *F 85-786*. Sonoyta, *F 85-939*. 2.8 mi W of Pinacate Junction, *F 86-334*. 35 mi E of San Luis, *F 90-175*.

ANGIOSPERMS
FLOWERING PLANTS: MONOCOTYLEDONS

KEY TO THE FAMILIES

1. Fully submerged aquatic plants; leaves linear to threadlike.
   2. Leaves and stems with coarse teeth, the leaf blades linear but not threadlike. ____________**Hydrocharitaceae**
   2′ Leaves entire (or with microscopic teeth), the blades threadlike.
      3. Leaves alternate; fruits asymmetric or not, nearly round, ovoid (about as wide as long), not toothed.____________**Potamogetonaceae**
      3′ Leaves opposite; fruits asymmetric, flattened, oblong (longer than wide), with small teeth on crest.____________**Zannichelliaceae**

1′ Plants terrestrial or emergent from shallow water; leaves variable.
   4. Leaves large and many in rosettes, thick, tough, and succulent with a stout terminal spine, or the leaves long, linear, and flat and the plants with woody trunks; mountain habitats but absent from the Pinacate lava.
      5. Leaves thick, firm, and succulent, with a stout terminal spine.____________**Agavaceae**
      5′ Leaves flat, pliable, the tip not spinescent. ____________**Nolinaceae**
   4′ Plants not as in 4; widespread, wetland to desert habitats.
      6. Leaf blades arrow-shaped (sagittate); petals white, separate, and quickly deciduous; plants emergent from shallow water in Río Colorado delta, probably extirpated from the flora area. ____________**Alismataceae**
      6′ Leaf blades gradually narrowed from base to tip.
         7. Perennials from underground scaly bulbs; individual flowers showy, with petal-like segments, whitish, blue, or yellow; desert habitats. ____________**Liliaceae**
         7′ Annuals or perennials but not from underground scaly bulbs (sometimes with tuberous roots in wetland habitats); individual flowers inconspicuous, the perianth membranous or scalelike, or reduced to bristles, or minute or absent, not petal-like; plants of desert or wetland habitats.
            8. Cattails; leaves linear, erect, more than 1 m tall, thickened and pithy, not spinescent, not triangular in cross section; wetland habitats.____________**Typhaceae**
            8′ Grasses, sedges, and rushes; leaves less than 1 m, or if 1 m or more, then not thickened and pithy (*Scirpus* stems are leaflike and pithy but triangular in cross section); various habitats.
               9. Rushes; perianth segments 6 and clearly visible as 3 outer and 3 inner segments, brownish to membranous, sometimes with a green midstripe; fruits many-seeded. ______**Juncaceae**
               9′ Grasses and sedges; perianth often not evident, reduced to bristles, microscopic scales, or absent; fruits 1-seeded.
                  10. Sedges and bulrushes; stems triangular to terete, solid (pithy); leaf sheaths usually closed; each flower subtended by a single bract; wetland habitats or at least temporarily wet soils. ____________**Cyperaceae**
                  10′ Grasses; stems terete, hollow or solid; leaf sheaths usually open; each flower subtended by at least 2 bracts (the lemma and palea); wetland to desert habitats. ____**Poaceae**

AGAVACEAE AGAVE FAMILY

Herbaceous perennials to trees. Leaves simple, sessile, often in rosettes, often long-lived (perennial), with parallel and often tough fibrous veins. Inflorescences often large, sometimes massive. Flowers 3-merous, radial (ours) to bilateral, bisexual or unisexual and the plants dioecious. Tepals petal-like and similar, distinct or united at base. Stamens 6. Ovary inferior to superior; fruits dry and dehiscent (capsules) or fleshy and indehiscent, often many-seeded.

Warm and mostly arid or semiarid regions of the New World; perhaps 18 genera and 400 species depending on definition of family limits. *Agave* is the largest genus. In the following key and discussions, leaf widths were measured at the lower third of the leaf but above the expanded leaf base. References: Gentry 1972; Hernández 1995; McVaugh 1989.

1. Leaves 7+ cm wide, the margins spiny; ovary inferior. ____ **Agave**
1′ Leaves less than 1 cm wide, the margins entire or minutely serrated but not spiny; ovary superior. __**Hesperoyucca**

AGAVE *MAGUEY,* CENTURY PLANT

Rosette-forming plants, solitary to suckering, small to very large, usually monocarpic (dying after flowering). Leaves succulent and tough, usually ending in a stout spine; margins entire to toothed (often horny and very hard, sharp prickles). Inflorescences many-flowered, spikelike (spicate) or paniculate (branched). Flowers mostly relatively large. Tepals united at base, forming a short to long tube often filled with nectar. Ovary usually inferior. Stamens and anthers usually large, the anthers versatile (attached near the middle and swinging freely on the filament); style and stamens much longer than the tepals. Seeds flattened, blackish.

Mostly in Mexico and adjacent regions; 200 species, more than two dozen in Sonora. Like so many Sonoran Desert plants of tropical origin, the number of species drops off sharply as one proceeds into the drier regions of the desert. Northwestern Sonora is the only region of its size in Mexico that has only a single species. Most agaves, at least the larger-flowered ones, are bat-pollinated. References: Burgess 1988; Gentry 1982; Nobel 1988; Pinkava & Gentry 1985.

Agave deserti Engelmann subsp. **simplex** Gentry. *LECHUGUILLA, MEZCAL;* DESERT AGAVE; *'A'UḌ.* Rosettes solitary or with a few offsets. Leaves 30–95 in number (larger plants), thick, glaucous gray-green, linear-lanceolate, often (30) 45–60 × 7–9 cm; marginal spines often 6–14 mm and curved, dark red-brown, with age turning gray, the terminal spine stout. Flowering stalks 3–5 m, the inflorescences paniculate, the branches 8–30 cm. Fresh flowers including the stamens often 6–8 cm, the tepals and anthers yellow. Capsules 3.3–6.2 cm. Seeds 5.2–6.7 × 4–5 mm, D-shaped, the margins with a raised, irregularly fluted narrow wing. Flowering April-June; fruiting June-July.

Mostly on north-facing slopes and most numerous toward higher elevations. Higher granitic ranges along Mex 2 including the Sierra del Viejo and the Tinajas Altas range in adjacent Arizona. Smaller disjunct populations in the Hornaday Mountains and Sierra Blanca, also in the Sierra Cipriano.

This subspecies occurs in scattered granitic hills and mountains in northwestern Sonora eastward to the vicinity of Quitovac and southward to the mountains north of Puerto Libertad, southeastern California, and southwestern Arizona. Two other subspecies in southern California and Baja California. Subsp. *simplex* is distinguished by very thick and relatively short leaves, and few or no offsets. Southward along the Sonora coast it is replaced by *A. subsimplex* Trelease. Gentry places *A. deserti* in the Deserticolae species group, with 10 species endemic to the Sonoran Desert region.

Sierra del Viejo (44 mi W of Los Vidrios), *Bezy 374.* Sierra Blanca, *F 87-26.* Sierra Cipriano, *F 91-22.*

HESPEROYUCCA

Hesperoyucca, with a single species, has traditionally been treated as a distinct section of the genus *Yucca.* Morphological and molecular comparisons clearly place *Hesperoyucca* outside of *Yucca* and show that it shares a more recent ancestor with *Hesperaloë* than with *Yucca* sensu stricto. *Yucca* includes 45 species mostly in dry regions of Mexico and southwestern United States. *Hesperaloe* has 5 species, all in Mexico, 2 of them in eastern Sonora.

Key characters of *Hesperoyucca* include a monocarpic, agave-like growth habit, dehiscent capsules, slender styles, capitate stigmas with long papillae, the styles, ovaries, and unripe fruit all papillate, each filament fused below to the opposite tepal, pollen agglutinated in coherent masses, and pollina-

Agave deserti subsp. ***simplex,*** Sierra Blanca. RSF.

Hesperoyucca whipplei, Sierra del Viejo. RSF.

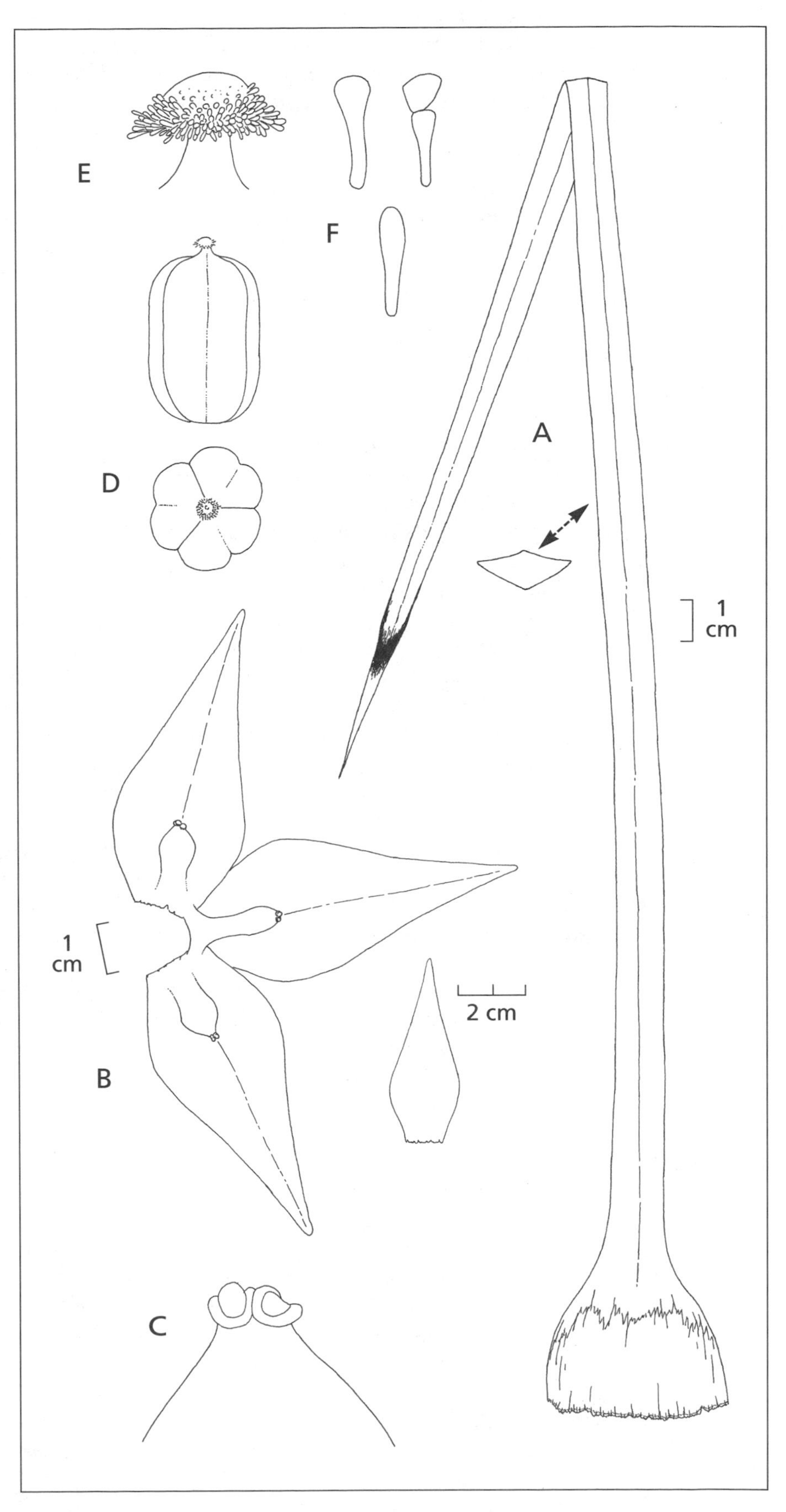

Hesperoyucca whipplei (F 88-100): ***(A)*** leaf; ***(B)*** three tepals spread open with adjacent stamens; ***(C)*** tip of filament with opened anther sacs, one with a pollinium; ***(D)*** ovary, side view (1 cm long) and cross-section; ***(E)*** stigma (4.3 mm wide) and style; ***(F)*** papillae of stigma, the longer ones 0.6 mm. MBJ.

tion by prodoxid yucca moths. Species of *Yucca* sensu stricto have a highly specialized and distinctively recessed and hidden stigma. The famous *Yucca*–yucca moth mutualism seems to have evolved independently in *Yucca* and *Hesperoyucca*. References: Baker 1986; Benson & Darrow 1981; Bolger, Neff, & Simpson 1995; Clary & Simpson 1995; De Mason 1984; McKelvey 1938, 1947; Webber 1953.

Hesperoyucca whipplei (Torrey) Baker [*Yucca whipplei* Torrey. *Y. newberryi* McKelvey] SPANISH BAYONET. Rosettes solitary, dying after flowering (monocarpic). Leaves mostly 35–60 × less than 1 cm, several hundred per plant, crowded but evenly spaced, firm, striate, semisucculent, glaucous, linear, keeled above and more conspicuously below; terminal spine stout and sharp. Leaf margins translucent yellow, edged with minute teeth or serrations, or at least some flowering plants with entire leaves or the distal 1/2 to 2/5 of the leaf entire. Flowering stalks 2–4 m, the main axis 10+ cm in diameter near the base; main axis, branches, and pedicels conspicuously glaucous; upper 1/2 of flowering stalk with crowded lateral branches 30–45 cm; main axis with leaflike bracts below, these reduced above to bracteoles subtending the flowers; bracts and bracteoles glaucous, often purple-green, drying and turning downward at about anthesis, and relatively persistent.

Flowers showy, white except the yellow pollen mass. Pedicels 1.3–4.0 cm, terete, longer at base of flowering branches. Base of flowers with a stipelike structure (giving the pedicel a jointed appearance) derived from fused perianth bases, (1.5) 2.0–3.5 mm at anthesis, enlarging in fruit to 15+ mm and thicker than the pedicel. Tepals broadly lanceolate, often 4.0–4.5 × 1–2 cm. Stamens 10–14 mm, unequally inserted on alternate tepals. Filaments pure white, attached below (near base) on their outer surfaces to the adjacent tepal, this lower, fused portion not swollen; upper, free portion conspicuously swollen, 5–9 × 2.5–3.8 mm. Anther sacs 2, each 1.2–1.5 mm, about as wide as long, thickish and white, dish- or bowl-shaped after opening, each with a bright yellow pollinium (pollen mass) sitting in the "dish." Style, pistil, and unripe fruit minutely papillate. Ovary at anthesis pure white, ca. 1 cm. Style slender, 1.4 mm. Stigma capitate, 4.2–5.0 mm wide, the center pure white and viscid, surrounded by nearly transparent club-shaped glistening crystalline-like papillae reaching 0.6 mm. Fruits green-glaucous, the dry capsules 3.8–4.3 × 2.9–3.2 cm, woody. Seeds 7.0–8.5 × 5.5–6.5 mm, D-shaped, flattened, blackish, the margins with a raised, irregularly fluted, narrow, often whitish wing. Flowering February-April but not during years of severe drought.

Limited to the north-facing granitic rock slopes in the Sierra del Viejo (east of San Luis); the plants scattered and with maximum density toward higher elevations. Bighorn sheep eagerly eat the branches, flowers, and green fruits when they can reach them. The seeds are strikingly similar to those of the agaves growing on the same slopes. The yucca seeds are distinguishable by their slightly larger size and often whitish margins.

This is the only known population of this species in mainland Mexico, and it has not been found in the nearby granitic mountains in southwestern Arizona. The Sonoran population is the smallest and most isolated of any *H. whipplei* population, which raises some interesting questions regarding the mutualistic yucca moths. However, members of this species can apparently set seed without yucca moths (Webber 1953). This species is widespread in southern California and Baja California, and there is an outlier population in northwestern Arizona. There are several allopatric subspecies distinguished primarily by geographic distribution and differences in branching habit, size, and leaf character. The Sonoran plants most closely match *Yucca whipplei* subsp. *whipplei,* which occurs in southern California and northwestern Arizona.

Relationships between the disjunct Sonoran and Arizonan populations have not been studied, although there are no apparent morphological differences. I am unable to distinguish our plants from those of subsp. *whipplei* in the southwestern part of the Colorado Desert in California (such as at Anza Borrego). However, when plants grown from seed were cultivated side by side in Tucson, the Sonoran plants formed more compact rosettes with shorter leaves than those of subsp. *whipplei* from California. During full glacial times this species probably had a continuous distribution across Death Valley (Woodcock 1986), and the Sonoran population was probably also continuous with the present-day Arizonan and Californian populations.

The "hearts," emerging flowering stalks, flowers, and seeds were widely utilized for food by Native Americans in southern California and Baja California (e.g., Bean & Saubel 1972; McKelvey 1934; Zigmond 1981).

NE slopes and upper bajada of Sierra del Viejo (Sierra de los Alacranes): *Burgess 6877; F 85-719, F 88-100.*

ALISMATACEAE WATER PLANTAIN FAMILY

Aquatic or wetland annuals or herbaceous perennials. Ovary superior; sepals 3, persistent; petals 3, deciduous. Pistils usually separate, usually 1-ovuled and developing into achenes. Worldwide; 11 genera, 75 species. Reference: Haynes & Holm-Nielsen 1994.

SAGITTARIA ARROWHEAD

Perennials or sometimes annuals in seasonally dry habitats, with milky juice and often with edible corms (bulbs). Leaves in deeper water submerged and ribbonlike, in shallower water emergent and the blades linear to arrow-shaped (sagittate) with long, spongy petioles. Flowers in whorls of 3, mostly unisexual; petals white. Stamens many. Pistils many; achenes flattened, usually winged and beaked. Mostly New World; 25 species. Reference: Bogin 1955.

Sagittaria longiloba Engelmann ex Torrey. This species is characterized by annual corms, sagittate emergent leaves with blades to ca. 25 cm, the terminal lobe acute and often about half as long as the large and relatively narrow basal lobes, and an achene with a minute, horizontally projecting lateral beak or the beak vestigial.

Although *Sagittaria* was obviously abundant in the delta of the Río Colorado, there is but a single record, and it has long been extirpated from the region. Vasey and Rose (1890) report a collection by Edward Palmer in 1889 from "near Lerdo," that they called *S. variabilis* Engelmann. Palmer reported that "the bulbs of this plant are much used by the Cocopa Indians either raw or roasted." According to Robert R. Haynes (personal communication 1992) the specimen (at US) has huge corms. Nebraska to California and southward to Nicaragua and Venezuela.

CYPERACEAE SEDGE FAMILY

Mostly perennial grasslike plants, often with rhizomes. Stems triangular or less often cylindrical (terete), usually solid. Leaves mostly 3-ranked, differentiated into a sheath and a blade, the blade usually grasslike or sometimes reduced or absent, the sheath closed around the stem, at least to begin with. Inflorescences frequently subtended by an involucre of 1 or more bracts, the bracts often leaflike or sometimes stemlike. Flowers (florets) small, sessile, in spikes or spikelets, subtended by small bracts (scales), arranged in spirals or 2-ranked, almost always wind-pollinated. Perianth none or few to many (often 6) bristles or scales. Stamens usually 3, sometimes 1 or 2. Ovary superior. Style 2-branched, the ovary 2-carpelled, and the fruit lens-shaped, or the style 3-branched, the ovary 3-carpelled, and the fruit 3-angled. Fruit a 1-seeded achene.

The 9 species in the flora have very broad geographic distributions; in the flora area these species are restricted to temporarily or permanently wet habitats. Worldwide, most abundant in temperate climates, generally in wet or moist places; 115 genera, 4350 species. The sedge genera in the flora area are treated in a broad, traditional sense. References: Bruhl 1995; Cronquist et al. 1977; Tucker 1987. (Note: measurements of spike or spikelet lengths refer to the scale-bearing or floriferous portions.)

1. Leaves reduced to basal sheaths, the blades lacking; inflorescence a single terminal spikelet. ____ **Eleocharis**

1′ Leaf blades present but sometimes relatively short and near base of stem; inflorescence of more than one spikelet.

 2. Scales of spikelets 2-ranked (distichous), not spiraled; perianth bristles absent. ____ **Cyperus**

 2′ Scales of spikelets spiraled; perianth bristles present. ____ **Scirpus**

CYPERUS SEDGE, NUT SEDGE

Mostly perennial or sometimes annual herbs. Stems triangular to terete, solid. Blades mostly grasslike. Inflorescences terminal (sometimes appearing lateral when the terminal bract resembles an extension of the stem), the spikelets usually in simple to compound umbellate clusters, or the branches reduced and the inflorescence appearing headlike; inflorescence branches (rays) often slender. Scales (bracts) of spikelets 2-ranked. Perianth absent. Stamens 3 or occasionally 1 or 2. Style 2- or 3-branched. Worldwide, mostly warm regions; 600 species.

Some have edible tubers, many are utilized for weaving mats and baskets, the aromatic oils of some are used in perfumes and for medicine, and others are serious agricultural weeds. Writing paper in ancient Egypt was made from *C. papyrus*. The several subgenera are sometimes treated as genera. References: Tucker 1983, 1994.

1. Plants with creeping rhizomes; lower leaves reduced to sheaths, the upper leaf with a short, narrowly linear blade; style 2-branched; achenes lens-shaped. ______ **C. laevigatus**

1′ Plants cespitose, tufted, or apparently solitary (from long fragile rhizomes); blades well developed including those of lower leaves; style 3-branched; achenes 3-sided.
  2. Plants mostly 3–10 cm; flowers subtended by scales with recurved awn tips. ______ **C. squarrosus**
  2′ Plants (15) 30–150 cm; scales straight, the tips awnless and not recurved.
    3. Roots without tubers; spikelets at maturity breaking into 1-flowered segments; scales conspicuously constricted around top of achene, the lower part of scale clasping the achene. ______ **C. odoratus**
    3′ Roots with small round tubers at ends of slender rhizomes; spikelets not breaking up into 1-flowered segments; scales not constricted above achene.
      4. Each rhizome producing a single tuber; spikelet scales dull brown, the keels sometimes green. ______ **C. esculentus**
      4′ Each rhizome producing a chain of several tubers; spikelet scales reddish to purplish brown with a green keel. ______ **C. rotundus**

Cyperus esculentus Linnaeus var. **esculentus.** *COQUILLO AMARILLO, CEBOLLÍN, COQUILLO;* YELLOW NUTGRASS, YELLOW NUT SEDGE; *WAṢAI ṢU:WĬ.* Summer-fall annuals in the flora area, perhaps sometimes perennials (elsewhere usually perennials), often (15) 30–60 cm. Stem base forming a single tuberous root below soil surface, with rhizomes growing out from this root, each rhizome ending in a single small, round tuber. Leaves basal, well developed. Involucral bracts leafy, of different lengths, often broader than the basal leaves. Spikelets 7–20+ × 1.0–1.8 mm, moderately flattened, the scales dull brown (especially when mature), the keels sometimes green (especially on young spikelets). Stamens 3. Style 3-branched, the achenes 3-angled.

Widely scattered in low, temporarily or permanently wet habitats in and around the Pinacate volcanic field, such as playas, sandy-silty soils at waterholes, arroyo beds and streamways, and rain pools at roadside ditches; also disturbed habitats at least around Sonoyta. Probably native in our region. Common along the west coast of Sonora; worldwide, and sometimes considered native to the Old World and adventive in the New World.

"This species and *C. rotundus* are the only New World species of *Cyperus* that produce stolons and tubers" (Tucker 1983:41). *C. esculentus* is a serious weed in many parts of the world; it is more tolerant of wet soils and extends farther into cold temperate regions in both hemispheres than does *C. rotundus*. In croplands the propagation of *C. esculentus* by rhizomes and tubers has been shown to be more imporant than reproduction by seeds (Horak & Holt 1986).

The tuberous roots have been utilized as food for people and livestock nearly worldwide. Several varieties are described but are generally not recognized (e.g., Diggs, Lipscomb, & O'Kennon 1999; Tucker 1994). However, the domesticated or cultivated form, var. *sativus* Boeckeler, known as chuffa, earth almond, rush nut, or tiger nut, does seem worthy of recognition. It is grown in Spain, Italy, and elsewhere for its nutty-flavored tubers rich in starch, sugar, and oil. Unlike the wild variety it is not weedy and is much less floriferous.

Suvuk, *Nabhan 362.* Pinacate Junction, *F 86-353.* Tinajas de los Pápagos, *F 86-480.* 2 mi S of Tinajas de los Pápagos, *F 86-490.*

Cyperus laevigatus Linnaeus. COQUITO; FLAT SEDGE. Perennials with creeping rhizomes, usually forming dense colonies (3) 10–20 (75) cm. Lower leaves reduced to reddish or brownish sheaths, the upper leaf with a green, narrowly linear blade. Inflorescences compact, appearing lateral, with a cluster of sessile, scarcely compressed spikelets 4–15 mm; involucral bracts 2, very unequal, the lower one 1.6–8.5 cm, resembling a continuation of the stem, the other bract 0.4–2.3 cm. Stamens 2. Style 2-branched, the achenes lens-shaped. Flowering and fruiting almost any time of year.

Open places at scattered wetland habitats—this species apparently cannot compete with most larger and probably faster-growing riparian plants. Common in wet soil along the Río Sonoyta and lower Río Colorado. Locally dense colonies in wet soil immediately surrounding many of the Bahía Adair pozos. Individual plants or colonies at the pozos show extreme variation in size and are often severely grazed by rabbits. Also at other oases such as Aguajita, Quitobaquito, and Quitovac. Tropical to warm-temperate regions around the world.

Sonoyta, Presa Derivadora, *F 86-309*. Quitobaquito, *F 20603*. Río Sonoyta, 20 km W of Sonoyta, *F 85-981*. La Soda, *F 86-522*. La Salina, *F 86-549*. 5 mi S of El Doctor, *F 85-1046*.

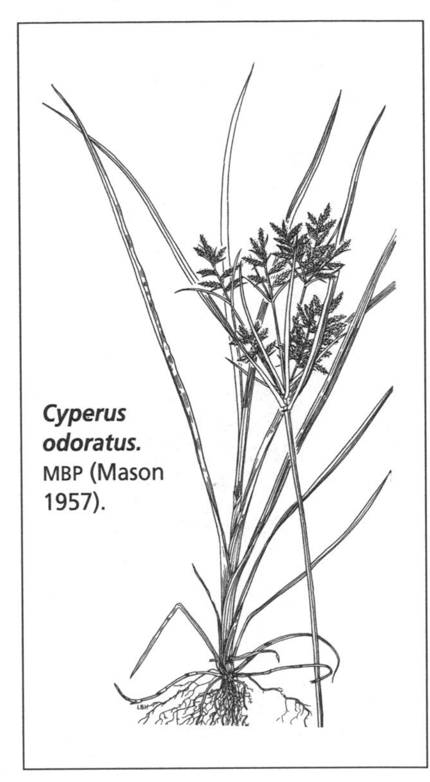

Cyperus odoratus. MBP (Mason 1957).

Cyperus odoratus Linnaeus [*C. ferax* Richard. *Torulinium odoratum* (Linnaeus) S.S. Hooper] Perennials or facultative annuals, reaching 1.5 m but often much smaller. Bracts subtending inflorescence large, longer than inflorescence branches, to 12 mm wide; rays with umbellate clusters of once-compound branches. Spikelets variable in size, reaching ca. 20–30 × ca. 1 mm, linear, nearly terete, at maturity breaking into segments of 1 floret and its scale plus the next lower internode. Scales straw colored with a broad green midstripe. Stamens 3. Style branches 3, the achenes unequally 3-angled. Flowering and fruiting at least in September.

Cyperus esculentus var. *esculentus.* LBH.

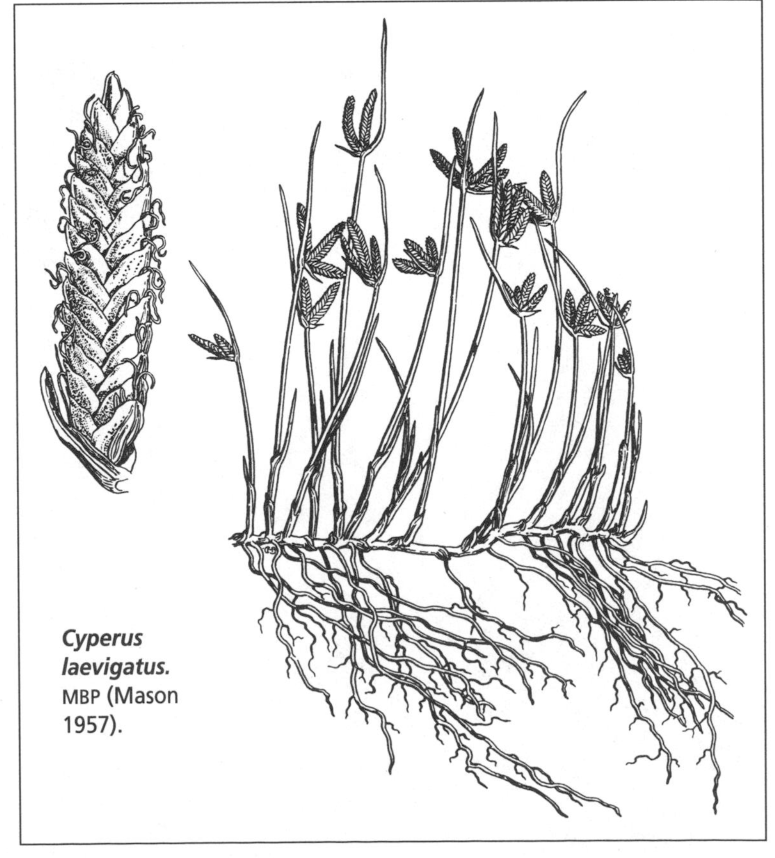

Cyperus laevigatus. MBP (Mason 1957).

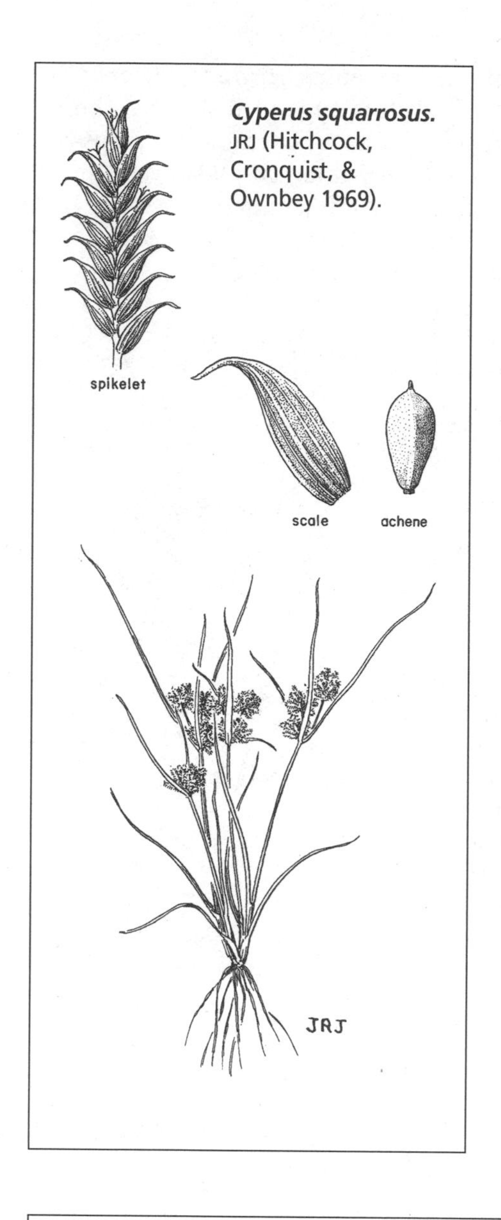

Cyperus squarrosus.
JRJ (Hitchcock, Cronquist, & Ownbey 1969).

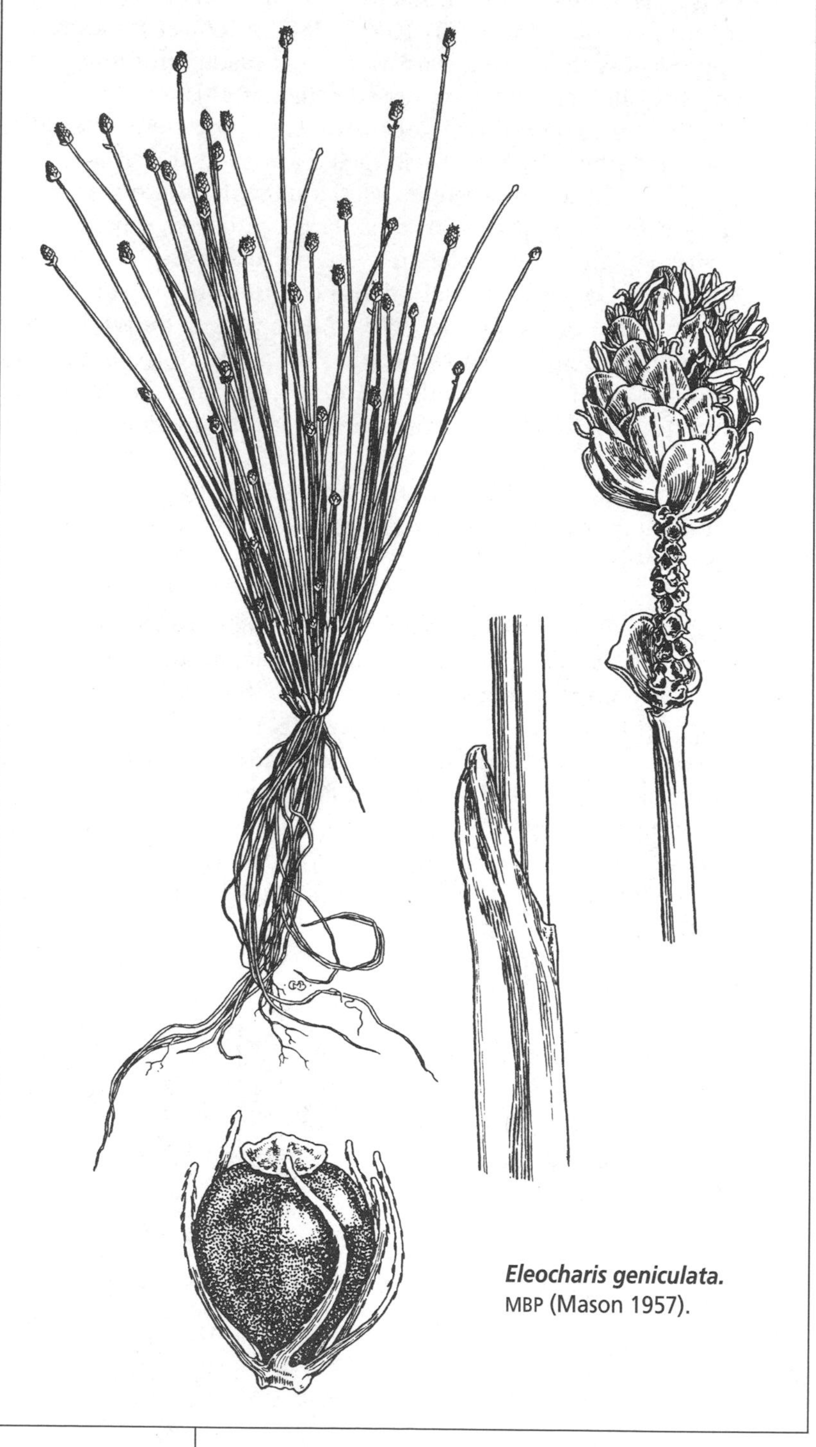

Eleocharis geniculata.
MBP (Mason 1957).

Eleocharis rostellata.
JRJ (Hitchcock, Cronquist, & Ownbey 1969).

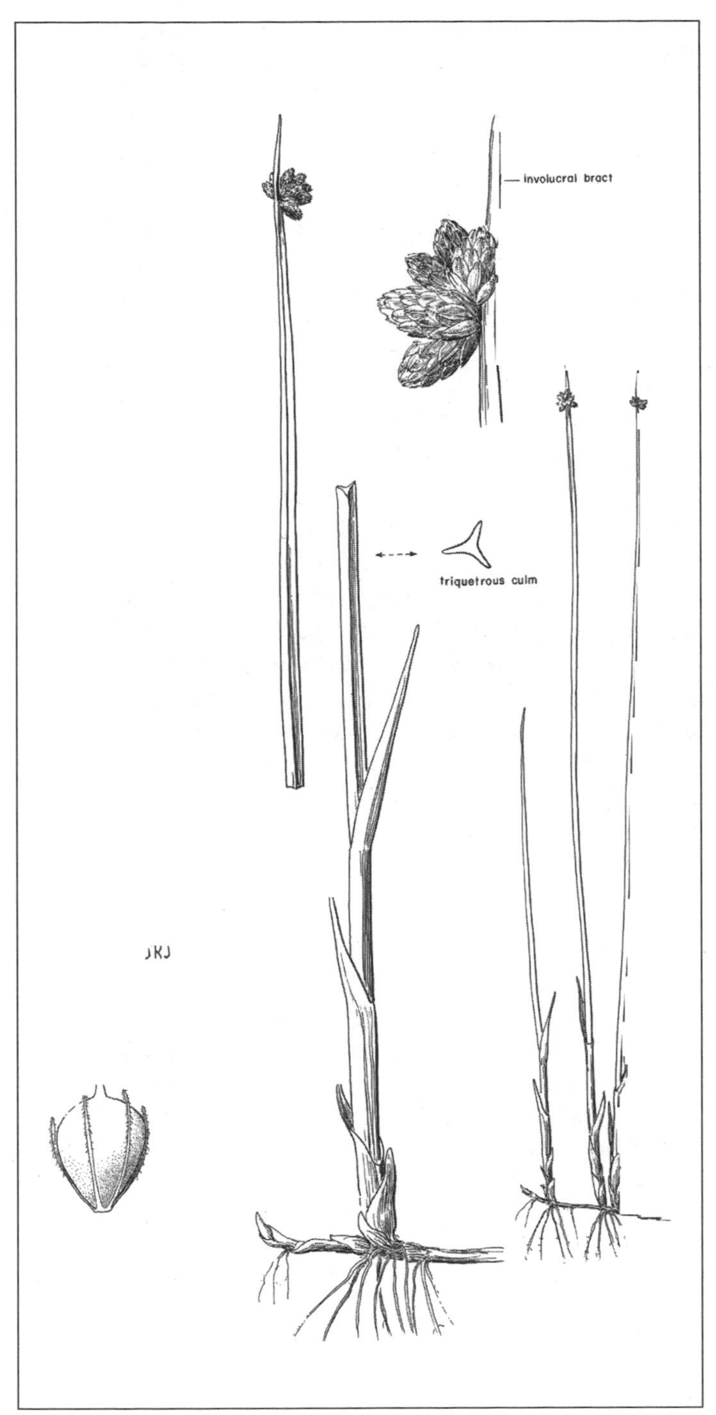

Scirpus americanus. JRJ (Hitchcock, Cronquist, & Ownbey 1969).

Scirpus maritimus. JRJ (Hitchcock, Cronquist, & Ownbey 1969).

In wet soil along irrigation ditches at Sonoyta (*F 86-298*). One of the most widespread and common sedges in the southwestern United States and the west coast of Mexico. Tropical and temperate regions nearly worldwide.

***Cyperus rotundus** Linnaeus. *COQUILLO MORADO, CEBOLLÍN;* PURPLE NUTGRASS, PURPLE NUT SEDGE. Closely related to and resembling *C. esculentus,* but each rhizome producing a chain of several tubers and the spikelet scales red- to purple-brown (especially when mature) with a green keel. Plants often 40–60 cm. Stem bases thickened, woody; stolons woody, springy when dried.

Common weed in the agricultural areas south of San Luis. This species has spread around the globe in historic times to become "the world's worst weed" (Holm et al. 1977:8); native to the Old World, and possibly also native in the New World.

5 km S of San Luis, *F 85-1035*. Yuma, *Bailey 3 Sep 1917*.

Cyperus squarrosus Linnaeus [*C. aristatus* Rottboell. *Mariscus squarrosus* (Linnaeus) C.B. Clarke] DWARF SEDGE. Non-seasonal ephemerals, diminutive and tufted, (1.5) 3–10 (18) cm. Leaves few, soft, basal or nearly so, usually less than 1 mm wide. Bracts subtending inflorescence leafy, the larger bracts longer than the inflorescence. Spikelets 4–10 (15) mm, flattened, in compact clusters, sessile or on short rays. Each spikelet scale with a prominent recurved awnlike tip, the awn tips giving an unusual "fringed" appearance to the spikelets; scales often reddish bronze to yellowish with green margins. Stamen 1, sometimes with an additional 1 or 2 stamens or staminodes. Style 3-branched, the achene 3-sided.

Temporarily or permanently wet soils; along the Río Sonoyta and at waterholes in the Pinacate region in rock crevices and sand, and during exceptionally wet times sometimes spreading onto wet soils of adjacent non-riparian habitats. Widespread in the Sonoran Desert in permanently to temporarily wet soils. Worldwide in temperate and tropical regions.

This is the smallest sedge in the Sonoran Desert. It is reported to have flowers with 1 stamen, but some specimens from our region and elsewhere in Sonora have at least some flowers with 1 or 2 additional stamens or staminodes.

Tinaja de los Chivos, *F 18788*. Tinajas de los Pápagos, *F 86-479*.

ELEOCHARIS *TULILLO,* SPIKE RUSH

Annuals or perennials, often with long slender stolons. Stems without leaves; leaves reduced to basal sheaths without blades. Inflorescence of a solitary, terminal spikelet, with a scalelike bract at the base. Flowers crowded, mostly bisexual, with a perianth of mostly 6 bristles. Stamens usually 3. Style 2- or 3-branched, the style base enlarging into a persistent tubercle capping the achene.

Worldwide, 150 species; closely related to *Scirpus*. Some produce edible tubers; the Chinese water chestnut is *E. dulcis*. Reference: Svenson 1957.

1. Annuals or seemingly perennials, the rootstock not tough; stems mostly 6–28 cm, terete; perianth bristles white; achenes dark brown to blackish, the tubercle constricted at base. ____ **E. geniculata**

1′ Perennials with tough rootstocks; stems usually more than 45 cm, oblong to flattened in cross section; perianth bristles brown; achenes pale, the tubercle continuous with the achene, not constricted. ____ **E. rostellata**

Eleocharis geniculata (Linnaeus) Roemer & Schultes [*Scirpus geniculatus* Linnaeus. *Eleocharis caribaea* (Rottboell) S.F. Blake (see Wilson 1990)] Annuals, often 6–28 cm, the stems slender and delicate, straight, mostly erect or ascending and not curved. Leaf sheaths prominent. Spikelets globose to ovoid, 3–7+ mm. Perianth bristles white, backward barbellate. Style branches 2. Achenes broadly lens-shaped, obovoid, shiny dark brown to blackish, the tubercle white, constricted at base. Flowering at least March-November.

Occasional in wet sand along the Río Sonoyta streambed, and at Quitobaquito and nearby waterholes; probably formerly more prevalent along the Río Sonoyta. Delta region of the Río Colorado along irrigation and drainage canals, and shallow water at ciénega margins. Warmer regions worldwide; this is the most widely distributed species of *Eleocharis*.

The tiny barbs on the perianth bristles hook tenaciously onto bird feathers, which seems to explain the large geographic range of this and many other cyperaceous species. I found numerous achenes attached to old bird feathers floating in small pools at Quitobaquito.

Ciénega de Santa Clara, *F 90-521.* SW of El Doctor, *F 92-999.* Quitobaquito, *Darrow 2402.*

Eleocharis rostellata (Torrey) Torrey [*Scirpus rostellata* Torrey] TRAVELING SPIKE RUSH. Perennials with tough rootstocks; forming dense and massive, grasslike mounded colonies. Stems bright, shiny green, wiry and tough, at first upright, arching over as they elongate, often 45–100+ cm, a compressed oval (oblong) in cross section when fresh, becoming flattened, thick, and ridged when dry; stem tips sometimes proliferating (forming bulbils or plantlets that may develop into new plants). Spikelets 4–12 mm. Perianth bristles unequal in length, some longer than the achene, backward barbellate, reddish brown. Style branches 3, or sometimes 2 in some florets in a spikelet. Achenes smooth to slightly roughened with a cell-like pattern apically, pale green to pale brown, vaguely 3-angled to thickly lens-shaped, the tubercle dull white, continuous with the achene body, not constricted basally. At least March-May.

Extensive colonies in wet, alkaline soils at springs and streamways at Quitobaquito, margins of some of the larger pozos at La Salina, and wetland marshes and seeps along the margin of the lower Río Colorado near El Doctor. Most of North America, Alaska to northern Mexico, West Indies, and Andean South America.

Quitobaquito, *F 87-296.* La Salina, *F 93-14.* 1 km N of El Doctor, *F 92-250C.*

SCIRPUS BULRUSH

Perennial herbs, rarely annuals; usually emergent from fresh water. Stems erect, usually solid (pithy), triangular to terete. Leaf sheaths closed, the blades well developed, reduced, or absent. Involucral bracts leafy, 1 to several, or reduced or absent. Inflorescences umbellate, or a compact cluster of many spikelets, or reduced to a single terminal spikelet. Scales of spikelets spiraled and overlapping. Perianth bristles 0–6, barbellate or ciliate. Stamens 3 or sometimes fewer. Style 2- or 3-branched. Achenes lens-shaped or 3-angled. Worldwide; 175 species. References: Koyama 1958, 1962, 1963; Schuyler 1967.

1. Leaves only at the base of stems, the blades greatly reduced—less than 15 cm; inflorescence with a single 3-angled bract appearing as a continuation of the stem, the inflorescence thus seems lateral near tip of "stem." ______ **S. americanus**

1′ Stems leafy, the stem leaves with well-developed blades—more than 20 cm; involucral bracts several, flat and leafy, the inflorescence obviously terminal. ______ **S. maritimus**

Scirpus americanus Persoon [*S. olneyi* A. Gray of western authors. *Schoenoplectus americanus* (Persoon) Volkart ex Schinz & R. Keller] *TULE;* BULRUSH; *WA:K.* Perennials with long rhizomes. Stems (0.2) 1–2+ m, triangular in cross section and pithy. Leaves few, basal or nearly so, the blades relatively reduced, 1.7–15 cm. Inflorescence a cluster of sessile or nearly sessile, ovoid spikes appearing lateral near the stem tip (actually the bract—the inflorescence subtended by a single bract appearing as a continuation of the stem). Perianth bristles mostly 4, slightly longer than the achene. Styles 2- or 3-branched. Achene body 1.7–2.3 mm, obovoid, more or less lens-shaped, the surface dull. At least April-December.

Emergent from shallow water in pozos, oases, riverbeds, and other wetland habitats. Locally common along the Río Sonoyta at scattered habitats with permanent water or wet soil, from the vicinity of Sonoyta to Cerro el Huérfano, at Quitobaquito, La Soda, La Salina, Laguna Prieta, and along the Río Colorado. The largest population occurs in the Ciénega de Santa Clara. Widespread in wetland habitats in the Americas.

Schuyler (1974) concluded that the bulrush previously known as *S. olneyi* A. Gray should properly be *S. americanus* Persoon, and that the plant western authors have called *S. americanus* should be called *S. pungens* Vahl. *S. pungens* is closely related to *S. americanus.*

Río Sonoyta at Cerro el Huérfano, *F 92-973.* El Papalote, at international fence, *F 86-104.* La Soda, *F 86-520.* La Salina [Pozo Borrascosa], *Ezcurra 28 Jun 1982.* Laguna Prieta, *Roth 6.* Banks of Río Colorado, 20 mi N of El Golfo, *F 75-56.*

Scirpus maritimus Linnaeus [*Bolboschoenus maritimus* (Linnaeus) Palla] SALT-MARSH BULRUSH. Perennials with prominent horizontal rhizomes and leafy, triangular stems arising from a thickened tuberous root. Leaves well developed, the blades flat, often 20–40 cm. Involucral bracts several, flat, leaflike, and well developed. Spikes 2–4 cm × 6.0–8.5 mm. Floral scales membranous, partially transparent, straw colored to pale reddish brown with an awn of variable length, often ca. 2/3 as long as the scale. Perianth bristles often 6, ca. 1/2 to 3/4 as long as the achene. Styles 2- (3)-branched. Achene body 2.3–2.7 mm, obovoid, more or less lens-shaped, the surface shiny. Warmer months.

Common in wetland habitats in the delta region including the Ciénega de Santa Clara and La Salina pozos, often in highly alkaline or brackish water. The only other record for Sonora is from the delta region of the Río Mayo (Martin et al. 1998). In the southwestern United States and western Sonora it grows in fresh, saline, and alkaline marshes including the Salton Basin and along the Colorado River. The tuberous roots and seeds are edible.

Members of the *S. maritimus* complex are widespread in the Northern Hemisphere and South America. Plants of western and central North America, including Sonora, "belong to the rather ill-defined" var. *paludosus* (A. Nelson) Kükenthal (Cronquist et al. 1977:72). Another member of this complex, *S. robustus* Pursh, widespread in the New World, is often confused with *S. maritimus;* they are very similar, often hybridizing, and perhaps not distinct species (Nordlindh 1972).

La Salina (Pozo Borrascosa), *F 86-548.* Wellton-Mohawk drainage canal [near Ciénega de Santa Clara], *F 90-205.* Baja California: Delta of Río Colorado, *Bauml 1502.*

HYDROCHARITACEAE WATERWEED FAMILY

Annual or perennial aquatics. Flowers unisexual or bisexual. Leaves alternate, opposite, or whorled. Worldwide; 17 genera, 130 species; some are noxious weeds.

NAJAS WATER NYMPH

Submerged herbaceous aquatics, usually annuals. Leaves subopposite, or appearing whorled due to short internodes, sessile, the margins toothed to serrulate. Monoecious or dioecious. Flowers axillary, hidden in leaves clustered at branch tips, the perianth greatly reduced. Male flowers with a single stamen, the anther sessile or subsessile; female flowers sessile with 2–4 stigma branches, the fruits achene-like, 1-seeded.

Fresh and brackish water worldwide, the majority in the Old World; 40 species. *Najas* is sometimes placed in its own family. Two subgenera are recognized; the monotypic subgenus *Najas* is represented by *N. marina.* References: Haynes 1979; Lowden 1986.

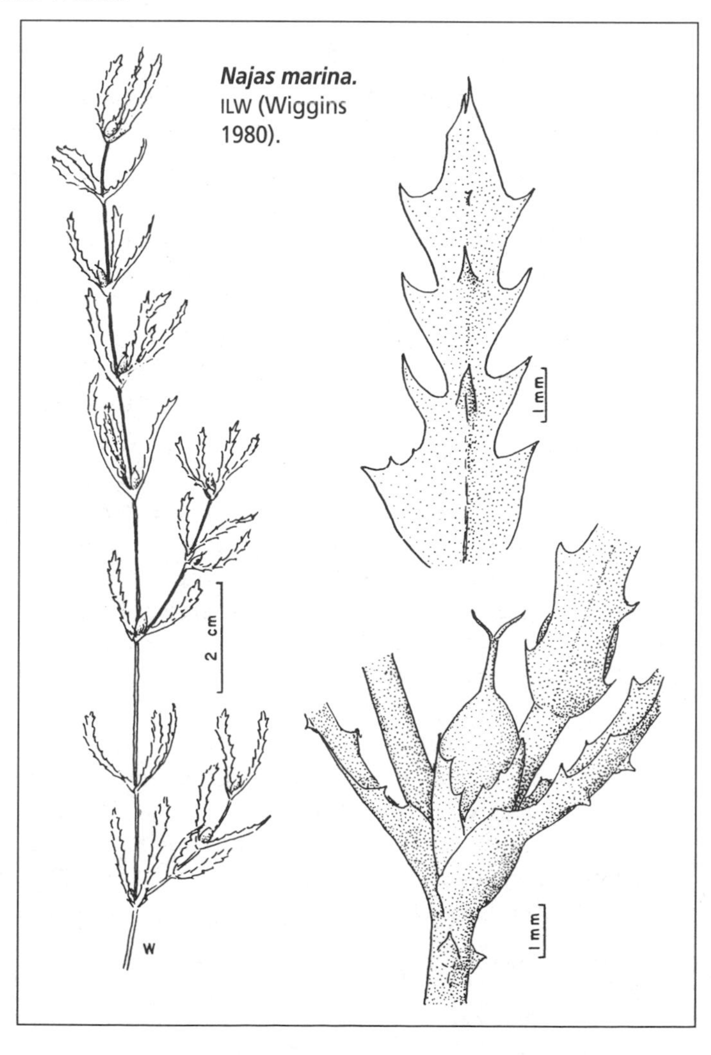

Najas marina. ILW (Wiggins 1980).

Najas marina Linnaeus. *SARGASO;* HOLLY-LEAF WATER NYMPH. Annuals (or perennials?). Stems branched, often more than 50 cm, 1.0–2.3 mm in diameter, rooting at base and from lower nodes. Stems and leaf margins and lower surfaces with coarse and relatively thick, firm spines. Leaves subopposite and mostly appearing whorled, linear, (1.2) 1.5–3.0 cm × 2.5–5.0 mm. Dioecious. Stigmas 3-branched.

Unlike any other submerged aquatic in the flora area, the stems and leaves are relatively firm and spinose. In the 1990s it was abundant in shallow, moderately brackish, open water at the north end of the Ciénega de Santa Clara. The plants were rooted in soft mud and formed locally dense, tangled masses just above the substrate during winter; in the summer they grew up to the water surface and became so dense that they fouled outboard motor propellers. Also in the Río Hardy, becoming thick in the summertime, and formerly in the pond at Quitobaquito. Warm regions worldwide; native to the Old World and perhaps introduced in the New World.

Ciénega de Santa Clara, *F 93-214*. Quitobaquito, *Pinkava 2363* (ASU). Baja California: Río Hardy, 1 mi S of El Mayor, *F 77-18*.

JUNCACEAE RUSH FAMILY

Mostly grasslike herbs, perennials or occasionally annuals. Flowers mostly bisexual and wind-pollinated, some secondarily insect-pollinated. Perianth radial, small and inconspicuous, the tepals usually 6 in 2 sets, mostly greenish to brownish, not petal-like. Stamens mostly 6, sometimes 3 or fewer. Ovary superior. Fruit a capsule.

Often in moist places, mostly temperate or cold regions, or montane in the tropics. Worldwide; 9 genera, 350 species. *Juncus* and *Luzula* are the largest genera—their greatest diversity is in the Northern Hemisphere. References: Balslev 1996; Cronquist et al. 1977.

JUNCUS RUSH

Annual and mostly perennial herbs, glabrous, and with rhizomes or sometimes cespitose. Leaves alternate, scalelike on the rhizomes, cataphyllous (with reduced blades), and/or leafy on the base of the stem with a sheath and blade, the blades often stiff and bractlike in the inflorescences. Stems often leaflike. Stamens 6 (ours), or 3 or fewer. Seeds numerous.

Mostly in moist climates in temperate and boreal regions, rare in tropical regions; in the Sonoran Desert restricted to wetlands. Worldwide; 300 species.

1. Annuals, mostly 15–20 (30) cm; stems and leaves soft and flexible, not spine-tipped. ______ Subgenus **Poiophylli: J. bufonius**
1′ Perennials, usually 40–200 cm; stems and leaves harsh, rather firm, and spine-tipped.
  2. Plants not cespitose, the stems arising from long, deeply set rhizomes; stems often compressed (oblong in cross section). ______ Subgenus **Juncus: J. articus**
  2′ Plants cespitose, forming dense pincushion-like clumps without long rhizomes; stems terete (rounded in cross section). ______ Subgenus **Thalassici:**
    3. Outer tepals mostly obtuse at apex, the margins with white membranous wings, these often erose (the edges ragged); capsules relatively broad, obovoid to nearly globose, more than 1.5 times longer than the tepals. ______ **J. acutus**
    3′ Outer tepals mostly acute at apex, the margins entire or narrowly winged; capsules relatively narrow, ovoid, about as long or only slightly longer than the tepals. ______ **J. cooperi**

Juncus acutus Linnaeus subsp. **leopoldii** (Parlatore) Snogerup [*J. leopoldii* Parlatore. *J. acutus* var. *sphaerocarpus* Engelmann] SPINY RUSH. Large cespitose tussock-forming perennial rush resembling a giant pincushion, often (0.5) 1–2.2 m. Stems firm, terete, ending in a spine-tipped involucral bract more or less continuous with the stem. (The portion below the inflorescence is stem, and the part above is bract.) Leaves arising from the base, spine-tipped, resembling and often nearly as long as the stems. Inflorescences appearing lateral near the ends of the leaflike stems. Panicles often compound, the branches unequal, the flowers in small clusters, the bracts spine-tipped. Tepals mostly obtuse, the margins with white-membranous wings. Capsules brown, stiff, and almost woody, obovoid to subglobose, ca. 1.5 times or more longer than the tepals. Growing and flowering during warmer seasons.

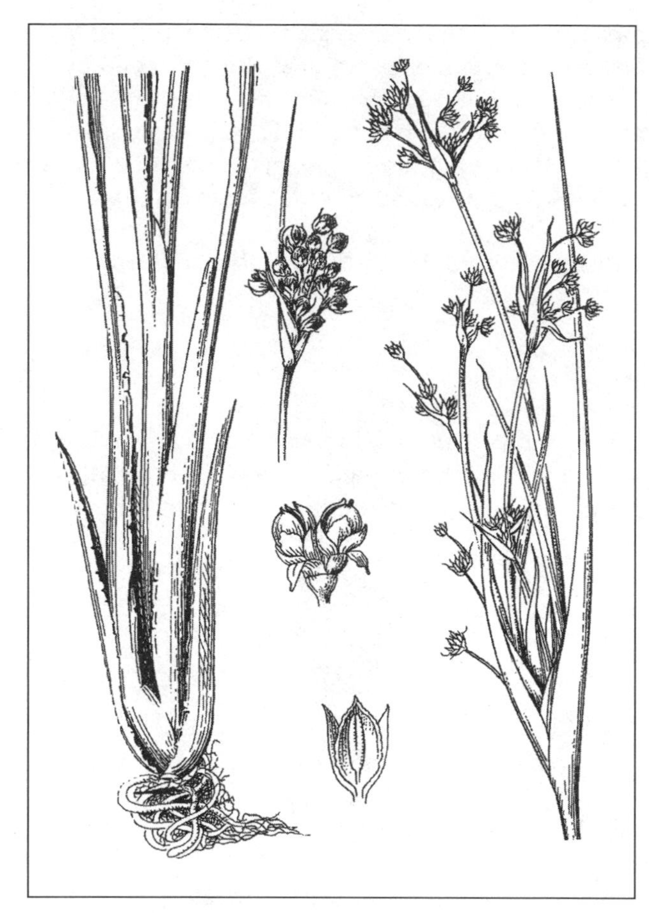

Juncus acutus subsp. ***leopoldii.*** (Abrams 1923).

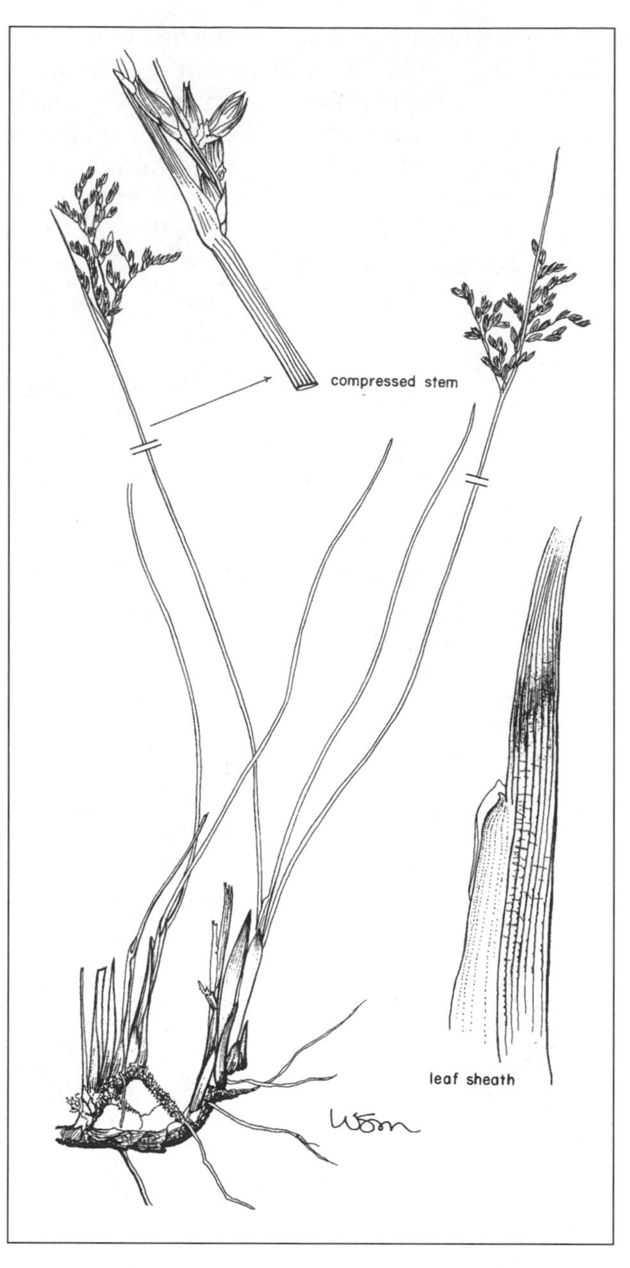

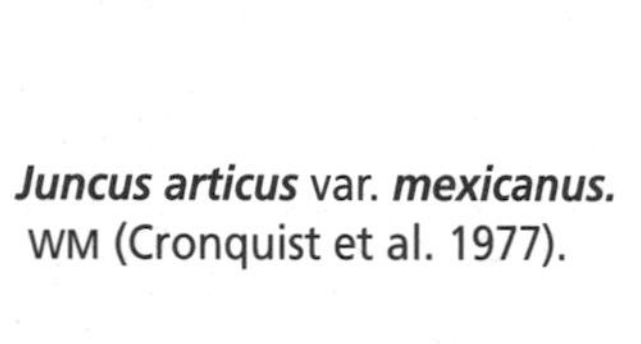

Juncus articus var. ***mexicanus.*** WM (Cronquist et al. 1977).

In northwestern Sonora found only at La Salina, where it is locally extensive in moist alkaline or saline sandy soils bordering pozos and the salt flat (*F 86-538; Equihua 14 Jul 1982*). In Sonora also in the vicinity of Bahía Kino. Subsp. *leopoldii* occurs in southern California, southern Nevada, Arizona, Sonora, and the Baja California Peninsula, and is disjunct in Puebla, Mexico, South America, and also in southern Africa. Subsp. *acutus* occurs in the Mediterranean region. This species closely resembles and is apparently closely related to *J. cooperi*. Both occur at La Salina.

Juncus articus Willdenow var. **mexicanus** (Willdenow) Balslev [*J. mexicanus* Willdenow. *J. balticus* Willdenow var. *mexicanus* (Willdenow) Kuntze] WIRE RUSH. Perennials with long deep blackish creeping rhizomes. Stems erect, 35–75 cm, firm, tough, often twisted and laterally compressed (oblong in cross section), bearing 1 or 2 basal, sheathing leaves, these brownish, the longer one often 4–8 cm. Inflorescences 3.5–8.5 cm, appearing lateral. Growing and flowering during warmer seasons.

Along the Río Sonora, apparently rare in damp soil with *Anemopsis californica*. Locally abundant in alkaline wet or damp soil at Quitobaquito springs and seeps, often with *Distichlis spicata*. Not known elsewhere in Sonora.

Juncus articus, with a number of geographic varieties, occurs in temperate-boreal regions of the Northern Hemisphere and in South America. Var. *mexicanus* extends from southwestern United States to Guatemala and in Andean South America to Patagonia.

Río Sonoyta, Cerro el Huérfano, *F 92-979.* Quitobaquito, *F 87-287.*

Juncus bufonius Linnaeus. TOAD RUSH. Annuals, cespitose and tufted, the leaves and stems soft and flexible. Larger plants with branched stems often to ca. 15–20 cm, the leaves mostly basal but some along stems. Leaves convex below, flat to channeled above. Tepals slender with a green midstripe and nearly transparent membranous margins.

Emergent from very shallow water or on wet soil at Quitobaquito (25 Mar 1944, *Clark 11501,* ORPI). Extirpated following bulldozing and deepening of the pond in 1962 and development of the dense growth of larger wetland plants following removal of cattle. Not known from western Sonora. Common, at least formerly, along the Colorado River north of Mexico. Worldwide, especially north temperate regions, also in cool tropical highlands and often weedy.

Juncus cooperi Engelmann. SPINY RUSH. Closely resembling *J. acutus* except the tepals and capsules narrower, the tepals entire or not as broadly winged, and the capsules shorter. Outer tepals acute with firm, often spinescent tips, the margins entire or with narrowly membranous wings. Capsules ovoid, about as long as or slightly longer than the tepals. Growing and flowering during warmer seasons.

Locally common in moist alkaline or saline soils at Laguna Prieta and Ciénega de Santa Clara, wetland seeps and small marshes along the lower Río Colorado and La Salina, and at Quitobaquito. The Ciénega de Santa Clara population is extensive. Otherwise known from moist saline soils in the Mojave and Colorado Deserts in southeastern California, southern Nevada, and extreme northeastern Baja California.

Laguna Prieta, *Roth 2.* 5 km S of El Doctor, *F 85-1053.* Ciénega de Santa Clara, *F 92-986.* La Salina, *F 84-26.* Quitobaquito, *F 85-1052.*

LILIACEAE LILY FAMILY

Perennials, mostly herbaceous, including many geophytes (bulbs, corms, rhizomes, etc.), sometimes woody. Flowers bisexual to unisexual, radial (ours) or bilateral, often showy, usually 3-merous; tepals usually 6 in 2 whorls, distinct or united. Ovary superior to inferior.

This "family," interpreted in a broad sense, includes about 280 genera and 4000 species worldwide. Although polyphyletic and artificial, it makes a convenient grouping for our purposes. Highest diversity of these geophytes is in fairly dry but not extremely arid, temperate to subtropical regions. *Dichelostemma* and *Triteleiopsis* are also placed in the Alliaceae, *Hesperocallis* seems to fall within the Agavaceae, and *Zephyranthes* is in the Amaryllidaceae.

These plants are poorly represented in the more arid parts of the Sonoran Desert. The "bulb" growth form seems better adapted to climates with dry summers, winter freezing, or both. There are 11 genera and 22 species of Liliaceae listed for the flora of the entire Sonoran Desert (Wiggins 1964). Most of them occur at the margins of the desert, especially along the Pacific coast in California and Baja California, and at higher elevations such as at the northern and eastern margins. Only a handful of these species extend into the arid lowlands of western Sonora: 4 in our region, 3 in the Puerto Libertad–Desemboque San Ignacio region, and none as far south as the Guaymas region. In the Libertad-Desemboque region are *Allium haematochiton, Dichelostemma pulchellum,* and *Triteleiopsis palmeri* (Felger & Moser 1985). The amaryllid spider lily, *Hymenocallis sonorensis,* is common in subtropical riparian habitats to the south and east of the Guaymas region, and sporadically farther north. It is sometimes cultivated in gardens in Sonoyta.

1. "Bulbs" (cormlets) solitary or clustered, usually 2 cm or smaller; inflorescence an umbel; flowers blue to violet.
   2. "Bulbs" solitary (1 per plant); Sonoyta region in rocky soils. ______ **Dichelostemma**
   2′ "Bulbs" (cormlets) in clusters; widespread in sandy soils, not in the Sonoyta region. ______ **Triteleiopsis**

1′ Bulb solitary, usually 3 cm or more; inflorescence racemose or flowers solitary; flowers whitish or yellow.
  3. Leaves more than 8 mm wide, the margins wavy (crisped); inflorescences multiple-flowered elongated racemes; tepals whitish; widespread in lowlands, occasional at higher elevations. _____ **Hesperocallis**
  3′ Leaves less than 2 mm wide, the margins entire; inflorescences 1(2)-flowered; tepals clear yellow; higher elevations of Sierra Pinacate. _____ **Zephyranthes**

DICHELOSTEMMA

Perennials from a deeply buried fibrous corm ("bulb"). Leaves linear. Flower stalks leafless. Flowers violet to pink or red, in small terminal umbels subtended by spathe-like bracts. Tepals 6, united into a basal tube. Stamens 3 or 6. Ovary superior. Seeds angled, blackish, in 3-angled capsules.

Western North America, especially northern California; 5 species. *Dichelostemma* is a segregate of *Brodiaea* sensu lato with 40 species in mostly semiarid to dry western North America and western South America. References: Hoover 1940; Keator 1992.

Dichelostemma capitatum (Bentham) Wood subsp. **pauciflorum** (Torrey) Keator [*Brodiaea capitata* Bentham var. *pauciflora* Torrey. *Dichelostemma pulchellum* (Salisbury) A. Heller var. *pauciflorum* (Torrey) Hoover] *COVERIA;* BLUE DICKS; *HA:D.* Bulb single, often 1.8–2 cm. Leaves ca. 5 mm wide, the green portion often 27–32 cm, with a dry, dead, curled tip often 8–14 cm. Flowers 1 cm, violet-blue, attractive, several per umbel. Stamens 6, in 2 alternating sets of 3, one with smaller anthers. Growing in winter and spring; flowering mostly in March.

Locally common on some of the better-vegetated rocky hillsides in the Sonoyta region, often with northern or northeastern exposures. Disjunct at higher elevations in the Sierra Pinacate, where it is apparently not common. Often growing among the branches of *Ambrosia deltoidea* and *A. dumosa.* Southwestern Utah, Arizona, and southwestern New Mexico to northern and central Sonora; southward in the mountains of western Sonora to the vicinity of El Desemboque San Ignacio.

Two intergrading subspecies; subsp. *capitatum* ranges from Oregon to Baja California west of the deserts. This species is the most wide-ranging member of the genus and is unique in having 6 fertile stamens. According to Keator (1992:34), *D. pulchellum* is a confused name, or *nomen confusum,* and for this reason the epithet *capitatum* is considered to be the correct one.

Sonoyta, low hills at SE side of town, *F 89-4.* Hills near 12 km SW of Sonoyta, *F 92-413.* Sierra Pinacate, higher elevation, *F 92-464.*

HESPEROCALLIS

The genus has a single species.

Hesperocallis undulata A. Gray. *AJO SILVESTRE;* AJO LILY, DESERT LILY. Bulb single, rounded, 4–6 cm. Above-ground portion of leaves 20–45 × 1–1.5 (2.0) cm, thickish, glaucous, with undulated or crisped margins. Flower stalks erect, stout, with a few reduced leaves below and bracts above, unbranched, (20) 30–60 cm, or sometimes with 1 to several branches and reaching 150–180 cm in favorable seasons in sandy soil. Flowers opening in late afternoon, closing with the heat of next morning, 6–8 cm wide when fully open. Tepals 6, united below into a tube, each 6.4–7.5 cm × 10–12 mm, white inside with a greenish purple midstripe outside. Stamens 6, about as long as the tepals, the anthers 11–13 mm, yellow. Ovary superior. Stigma, style, and filaments white; style slightly longer than tepals. Growing during winter and spring; Flowering (November) February-May.

Widespread; abundant during wet years on sandy soils, especially sand flats and dunes; also common on rocky soils such as desert pavements and low volcanic hills (the bulbs probably in pockets of sandy soil), and occasional on cinder slopes to near peak elevation in the Sierra Pinacate. Scarce or failing to appear during drier years. Southeastern California, northeastern Baja California, southwestern Arizona, and northwestern Sonora southward to the vicinity of Caborca. The bulb is slimy but edible with a good flavor.

Hesperocallis undulata. (Abrams 1923).

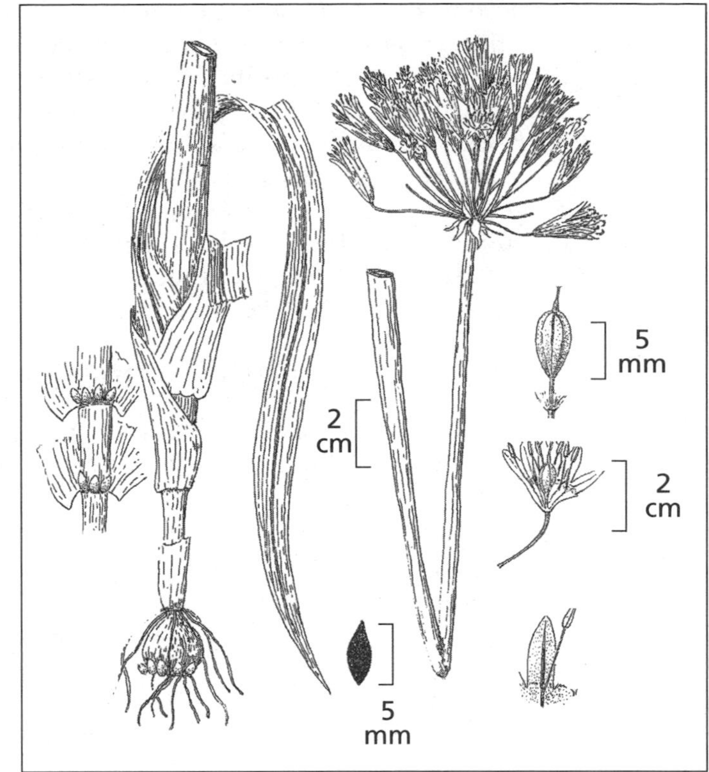

Triteleiopsis palmeri. ILW (Wiggins 1980).

25 mi E of San Luis, *Sanders 5631* (UCR). 11 mi N of El Doctor, *Bezy 359A.* NE of Sierra del Rosario, *F 20438.* S of Moon Crater, *F 19032.* Km 97 on Mex 8 [10 km N of Puerto Peñasco], *Lockwood 90.*

TRITELEIOPSIS

This monotypic genus seems to be allied to *Triteleia,* a genus sometimes included in *Brodiaea* (see *Dichelostemma*). Reference: Hoover 1941.

Triteleiopsis palmeri (S. Watson) Hoover [*Brodiaea palmeri* S. Watson] BLUE SAND LILY. Bulbs small and clustered, the tiny cormlets (bulblets) produced on top of the previous year's bulb and in axils at leaf bases; propagating (mostly?) by bulblets. Leaf margins entire, the leaves of 2 kinds: first leaves of season green, stringy (linear-filiform), limp, and prostrate, reaching 32–53 cm; later leaves appearing with flower stalks, glaucous, erect-ascending, the blades to 1 cm wide, sometimes becoming moderately fleshy, usually 15–30 cm, somewhat thickened like the leaf of a leek, wide and clasping the cormlets. Umbels many flowered on leafless stalks 35–60 cm, the upper portion often blue. Flowers 1.5–1.8 cm, deep blue, very attractive, on slender pedicels. Tepals 6, united below into a tube. Ovary superior. Pedicels, tepals, filaments, and ovary blue-purple, the anthers nearly white before dehiscence, the hypanthium at summit of the tube with a darker blue-purple ring outlined with a white ring, the hypanthium and filaments forming a somewhat scalloped collar with holes through which one can see the lighter tube that appears white in the sun and probably functions as a nectar guide. Seeds blackish, flat. Growing winter-spring; flowering February–mid-May.

Sand flats and lower dunes across most of the flora area and occasionally on gravelly granitic flats such as near the Arizona border in the northwestern part of the flora area. Abundant during favorable years. Widely distributed in both Baja California states, along coastal dunes and sand flats in Sonora southward to Cerro Tepopa (29°24′ N), and in southwestern Arizona.

A chromosome count of a specimen from dunes near Yuma, Arizona, indicates that the plants are at least sometimes triploid ($2n = 33$) and therefore sterile (Lee Lenz, personal communication 1987). Dimorphism in the leaves seems related to weather conditions. The first leaves of the season lie flat and limp on the ground, resembling green spaghetti on the warm sand. Early in the season there is a chance of freezing weather at night, although the sand surfaces are usually warm during the day. Later in the season, as the days and sands become hot, the prostrate leaves wither and the newer leaves are held aloft, well above the substrate. Often growing with *Hesperocallis*. The small *Triteleiopsis* "bulbs" are tasty, and Seri Indian children eagerly seek them (Felger & Moser 1985).

Sierra del Rosario, *F 20797*. SE of Moon Crater, *F 18856*. Hornaday Mts, *Soule 17 Mar 1983*. 21 mi S[W] of Sonoyta, *Bowers 2593*. 5 km N of Puerto Peñasco, *Burgess 4752*.

ZEPHYRANTHES RAIN LILY, ZEPHYR LILY

Bulbs rounded, with a thickish "neck." Leaves linear-filiform to narrowly strap-shaped. Flower stalks almost always 1-flowered, the flowers showy, erect, subtended by a membranous, spathe-like bract, Perianth segments 6, white to pink or yellow. Stamens 6, the anthers orange. Ovary inferior. Fruit a papery capsule, the seeds blackish, flat, D-shaped. Warmer parts of the western hemisphere; 40 species. Some are cultivated as ornamental garden plants.

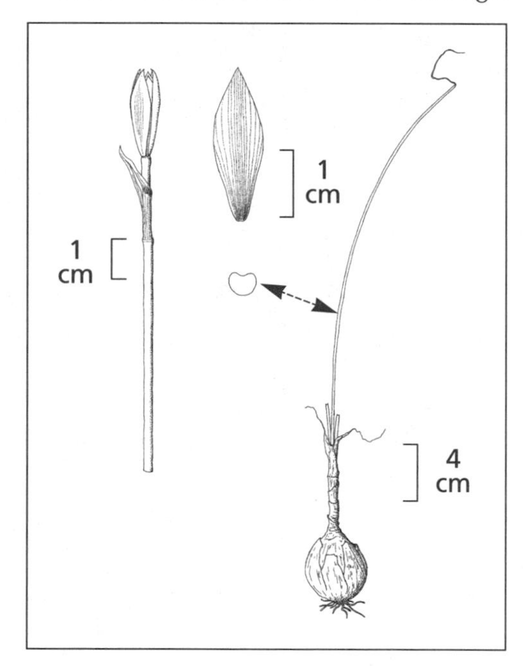

Zephyranthes longifolia, Sierra Pinacate. Note expanded tepal. MBJ.

Zephyranthes longifolia Hemsley [*Z. arenicola* Brandegee. *Z. aurea* S. Watson] Bulb single, ovoid, 3–4 cm plus a neck ca. 3.5–5+ cm extending to the soil surface. Leaves several, grasslike, 12–25 cm × 1–1.5 mm, dark green, semisucculent, nearly terete except for a shallow groove above, the tips dry, dead, persistent, and usually curled. Flower stalks 1 to several per season, 8–20 cm × ca. 2–3 mm, usually shorter than the leaves. Bract beneath flower quickly drying and splitting as the ovary develops. Flowers not opening wide; tepals, stigma, and style clear lemon-yellow, the anthers yellow-orange. Tepals all petal-like, 2.0–3.2 × ca. 1 cm. Ovary at anthesis 5–6 mm; capsules conspicuously 3-lobed, often 10 × 13 mm. Leaves produced in spring and summer; flowering with summer-fall rains, the fruits maturing 10–12 days after anthesis.

There is an extensive population at higher elevations in the Sierra Pinacate, especially on the north flanks. Not known farther south in western Sonora. The nearest population is in mountains in Organ Pipe Monument.

Central and northern Mexico including both Baja California states, and southern Arizona to western Texas. In Baja California and southern Arizona generally near or above the upper limits of the desert. Farther south, such as in Baja California Sur, it occurs at lower elevations. Cultivated plants from the Sierra Pinacate are self-compatible.

Vicinity of Pinacate Peak, 950–1150 m, *F 18694, F 86-429*.

NOLINACEAE BEAR GRASS FAMILY

Warm and mostly semiarid or arid regions; Central America to southern United States, mostly in Mexico; 4 genera, 55 species. Several species of *Dasylirion* and *Nolina* occur in Sonora; the other 2 genera restricted to Mexico. Allied to or sometimes included in the Dracaenaceae, or both placed in the Convallariaceae (includes tropical to temperate perennial herbs). Nolinaceae have dry fruits whereas Dracaenaceae have fleshy fruits and are widely distributed in tropical to subtropical regions worldwide. Reference: García-Mendoza & Galván V. 1995.

NOLINA

Perennial herbs to small trees. Leaves mostly numerous, linear with an expanded base, the margins entire to serrate or with shredding fibers. Inflorescences of compound, often large racemose panicles. Flowers small, white to green, on jointed pedicels, bisexual or unisexual and the plants dioecious; radially symmetrical, with 6 tepals in 2 whorls. Stamens short, reduced in female flowers. Ovary superior. Fruits 3-winged capsules, papery, relatively small. Seeds 1–3, rounded, brown to pale gray.

Mexico and southern United States; 24 species. The few western species beyond the desert (e.g., *N. microcarpa*) tend to have short, nearly subterranean trunks. Southward the nolinas tend to be arborescent. *Nolina* is not known elsewhere in western Sonora; 2 or 3 other species occur in north-central and eastern Sonora. References: Benson & Darrow 1981; Dice 1988; Gentry 1978; Munz 1974; Munz & Roos 1950.

Nolina bigelovii (Torrey) S. Watson [*Dasylirion bigelovii* Torrey] BIGELOW BEAR GRASS; *MOHO*. Trunk often 0.5–3 m, unbranched or sometimes few-branched, thick and fibrous, the bark thick and checkered with age. (Plants with trunks 4–4.5 m tall occur in the Tinajas Altas Mountains within 1 or 2 km of the Sonoran border, and a dead fallen plant had a trunk slightly more than 6 m long.) Leaves linear, to 1+ m × 1.5–3.0 cm, flat and easily bent, light green; margins minutely serrated, soon peeling away in persistent fibrous strips; dead leaves persistent. Dioecious. Flowering stalks 1.5–3 m, branched, usually overtopping the leaves. Pedicels slender, the tepals 2.5–3.0 mm long. Female flowers pure white with a pale green ovary; male flowers pure white with yellow anthers. Capsules 8–10 mm in diameter and about as long. Seeds 3.5–4.0 mm, oblong-ovoid, pale yellow-brown to tan. Flowering (March) May-June; fruiting June-July.

Rock slopes, cliffs, and canyons in granitic mountains in the northwestern part of the flora area including the Sierra del Rosario, Sierra Lechuguilla, Sierra del Viejo, and Cerro Pinto. The densest stands are at higher elevations on north- and east-facing slopes, generally along the sides of very steep canyons. Sometimes also on south- and west-facing slopes but much less common and concentrated in larger canyons. Not known elsewhere in Sonora. Also in similar granitic mountains in the adjacent western part of Cabeza Prieta Refuge and the Tinajas Altas Mountains, and farther north in western Arizona, southeastern California, and northeastern Baja California.

At flowering time the inflorescences are massive white plumes swarming with honeybees and large spider wasps (*Pepsis*). The bees and wasps cling and crawl all about the floriferous branches, even when blowing in the hot summer wind. Nolinas in the Sierra del Rosario continued to look healthy in April 1972, after 2 1/2 years of no measurable rainfall. Flowering ceased until after the rains resumed later in 1972 (Felger 1980).

Sierra del Rosario, *F 20653, F 75-13.* Cabeza Prieta Refuge, Heart Tank, *F 92-592A, F 92-592B.*

Nolina bigelovii, Tinajas Altas. MBJ.

POACEAE (GRAMINEAE) GRASS FAMILY

Annual and perennial herbs and some woody and arborescent plants (bamboos). Stems (culms) solid or hollow and closed at nodes. Leaves 2-ranked, consisting primarily of the sheath, ligule, and blade: sheaths partially to fully enclosing the stem; ligules usually present at junction of the sheath and blade and facing the stem, membranous throughout, or membranous at the base and topped by cilia, or entirely of hairs; blades mostly linear, flat or variously folded or inrolled.

Flowers usually small, bisexual or, when unisexual, the plants monoecious or dioecious. Flowers in spikelets, these mostly in panicles or simple to compound spikes, spikelike branches, or racemes. (Terminology for grass inflorescences treats the spikelet as the basic unit in the way the flower is treated in general terminology.) Spikelets of bracts and 1 or more flowers (florets), the bracts and florets (if more than one) 2-ranked (alternate on opposite sides of the rachilla). Bracts variously awned or not; lowermost bracts (glumes) paired or seldom reduced to 1 or absent. Floret consisting of 2 bracts, the lemma below and the palea above, these enclosing the flower itself. In spikelets of more than one floret the lowermost or uppermost floret or florets sometimes staminate or often reduced to a sterile lemma, or just a pedicel (the rudiment). Perianth segments (lodicules) greatly reduced, mostly 2 (3 in *Aristida,* most *Stipa* species, and bamboos), often swelling and spreading the floret open at anthesis. Stamens usually 3. Stigmas generally 2 and plumose. Fruit a 1-seeded grain or caryopsis, the seed fused to the fruit wall (pericarp), or rarely (e.g., *Dactyloctenium* and *Sporobolus*) the seed separates from the pericarp.

Grasses are most diverse in semiarid tropics and north-temperate regions with seasonal rainfall; 650 genera, 10,000 species. In contrast to the global situation, the majority of our grasses are annuals or ephemerals. Only 3 are not truly herbaceous: *Muhlenbergia porteri, Phragmites australis,* and *Pleuraphis rigida.* Two common grasses in the region, *Enneapogon desvauxii* and *Muhlenbergia microsperma,* invariably produce small, modified cleistogamous inflorescences at the stem bases, and another, *Leptochloa dubia,* often produces them. Various other grasses in the region produce mostly or only cleistogamous flowers, e.g., species of *Bromus* and *Sporobolus.* (Cleistogamous flowers remain closed and produce seeds without exposing the stamens and stigmas, and thus are self-pollinated.) Cleistogamous grasses tend to occur in dry areas with erratic rainfall, sparse vegetation, and lighter soils (Lord 1981).

The flora includes 38 genera and approximately 75 species of grasses, of which 32 are not native (Appendix E). No other large family in the region has such a high percentage of non-natives. The majority of these are from the Old World and tend to occur in disturbed habitats (see Felger 1990). Six non-native grasses, all of Old World origin, are established and common in undisturbed, natural habitats: *Bromus rubens, Cynodon dactylon, Eragrostis cilianensis, Polypogon monspeliensis,* and *Schismus* spp.

Aristida californica, Bouteloua aristidoides, and *Pleuraphis rigida* are among the most abundant desert grasses in the region. Other very common species, all ephemerals, are *Aristida adscensionis, Bouteloua barbata, Cenchrus palmeri, Festuca octoflora, Muhlenbergia microsperma,* and **Schismus arabicus.* Certain other ephemerals, such as **Eragrostis cilianensis, E. pectinacea, Eriochloa aristata, Leptochloa viscida, Panicum alatum,* and *P. hirticaule,* can become seasonally and locally abundant. Ephemeral grasses probably account for a major part of the biomass productivity in northwestern Sonora.

Fourteen grasses, 8 of them non-natives, enter the northeastern part of the flora area in Arizona Upland (the Sonoyta region) and do not extend into the lower elevations and drier Lower Colorado Valley: *Aristida ternipes* var. *gentilis, A. ternipes* var. *ternipes, *Avena fatua, Bromus carinatus, *B. catharticus, *B. tectorum, Chloris crinita, *Dactyloctenium aegyptium, *Eragrostis barrelieri, *Hordeum murinum, *Phalaris caroliniana, *P. minor, Poa bigelovii,* and *Sporobolus pyramidatus.* In addition, *Bouteloua trifida,* with a sizable population in the Sonoyta region, is known elsewhere in the flora area from only 1 small colony. *Aristida californica* and *Cenchrus palmeri,* although common across most of northwestern Sonora, do not enter the Sonoyta region.

Three grasses occur only at higher elevations, mostly above 800–1000 m, on Sierra Pinacate: *Bothriochloa barbinodis, Bromus berterianus,* and *Stipa speciosa. Digitaria californica* and *Muhlenbergia porteri* occur at higher elevations in the Pinacate mountains as well as mountains in the Sonoyta region but are absent from the intervening lowland desert. In recent years *Bromus rubens* has become seasonally abundant at higher elevations in the Sierra Pinacate, but is so far absent from lower elevations except in limited areas of disturbed habitats.

Following favorable rains in late summer and early fall, certain low-lying habitats—such as large playas, expansive sandy-silty alluvial fans leading into these playas, and crater floor playas—may support extensive ephemeral meadows with 100 percent ground cover made up largely of grasses. Characteristic grasses in these habitats include *Aristida adscensionis, Bouteloua aristidoides, B. barbata, Cenchrus palmeri, Chloris virgata, *Eragrostis cilianensis, E. pectinacea, Eriochloa aristata, Leptochloa panicea, L. viscida, Muhlenbergia microsperma, Panicum alatum,* and *P. hirticaule.*

No grasses are restricted to the granitic mountains or dunes. Five species are common on lower dunes and interdune troughs: *Aristida californica, Bouteloua aristidoides, B. barbata, Cenchrus palmeri, Pleuraphis rigida,* and **Schismus arabicus*. No grasses grow on the higher dunes.

Ten species of grasses, only 4 of them native, are restricted to non-maritime wetland habitats. These wetland soils are often highly alkaline. *Phragmites australis* and **Polypogon viridis* are limited to permanent-water habitats. The other eight are *Chloris crinita, *Cynodon dactylon, Distichlis spicata, *Echinochloa colonum, *E. crusgalli, Leptochloa fusca, *Poa annua,* and **Polypogon monspeliensis*. Two saltgrasses, *Distichlis palmeri* and *Monanthochloë littoralis,* grow on wet saline soils within the tidal zone.

Most of the grass species in northwestern Sonora have extensive geographic ranges (Appendix F). Two unusual wetland grasses, *Distichlis palmeri* and *Tuctoria fragilis* (*Orcuttia fragilis*), have the narrowest ecological and geographic ranges of any Sonoran Desert grasses. The *Tuctoria* is known only from a seasonally inundated dry lake in the Magdalena Plain of Baja California Sur, and *D. palmeri* only from the shores of the Gulf of California. These are the only grasses wholly endemic to the Sonoran Desert.

Aristida californica var. *californica* and *Cenchrus palmeri* are nearly but not entirely endemic to the Sonoran Desert. Both extend into northwestern Sinaloa—south of the limits of the Sonoran Desert as defined by Shreve (1951). However, it may be argued that this arid coastal fringe is an extension of the Sonoran Desert. In addition, *A. californica* var. *californica* extends into the Mojave Desert in California. No other species or infraspecific taxon of grass from the flora area has a range narrower than southwestern North America, and most are much wider ranging. *Chloris brandegeei* is endemic to Baja California Sur and the Gulf of California islands but is not confined to the desert. Furthermore, it may "only be a variant" of the widespread *C. chloridea* (Gould & Moran 1981). *Aristida parishii* is largely a Sonoran Desert species but also occurs in the Mojave Desert and extends to vegetation zones well above the desert. Thus, although grasses are successful in our extremely arid region as well as in the Sonoran Desert as a whole, local differentiation has been minimal. *Hordeum arizonicum* is nearly confined to southern Arizona but extends into elevations well above the desert in southeastern Arizona.

One higher elevation perennial grass, *Stipa speciosa,* and a few winter-spring ephemerals, *Bromus* spp. and *Festuca octoflora,* occur no farther south than the flora area, or at least no farther south in the desert in western Sonora. Four perennials, *Bouteloua trifida, Muhlenbergia porteri, Pleuraphis rigida* and *Tridens muticus,* and one ephemeral, **Schismus arabicus,* find their southern limits in the vicinity of Puerto Libertad or in the Sierra del Viejo southwest of Caborca. *Erioneuron pulchellum* seems to range no farther southward along the coast than around Bahía Kino. *Cenchrus palmeri* is not known north of the flora area.

The grass flora of northwestern Sonora contains 30 species of perennials, including 6 non-natives, and 46 species of ephemerals (annuals) including 25 non-natives. Cool-season or winter-spring ephemerals are represented by 18 species, only 5 of which are native. Twenty-seven species are summer or non-seasonal ephemerals, and 12 are not native. The summer and non-seasonal grasses tend to be C_4 plants, whereas the winter-spring grasses, such as the pooids, tend to have C_3 photosynthetic pathways (see Waller & Lewis 1979).

Six subfamilies of grasses are generally recognized, 4 of which occur in northwestern Sonora and 3 of which are well represented (Appendix E). The chloridoids are by far the most important subfamily in northwestern Sonora, with 34 species including 7 non-natives. Worldwide the chloridoids are characteristic of hot, arid climates.

Among the panicoid subfamily only *Cenchrus palmeri* and, to a lesser extent, *Panicum alatum* and *P. hirticaule* are truly common and widespread. Most panicoids in the flora area are warm-weather ephemerals, but only about half are native. Panicoids are "particularly successful in moist, humid, tropical, or subtropical habitats" (Gould & Shaw 1968:110), and are better represented farther south and east in Sonora.

Members of the tribe Andropogoneae reach greatest development in monsoon areas. *Bothriochloa barbinodis* and *Heteropogon contortus* are the only native members of this tribe in northwestern Sonora.

Sixteen of the 19 species in the pooid subfamily are winter-spring ephemerals (or annuals), 1 is a non-seasonal ephemeral, and 2 are perennial. Thirteen are non-native, and only 1, the native *Festuca octoflora,* is widespread and common in the lowland desert regions. This group has developed in cool to cold climates.

I expect that at least one *Paspalum* species was present in the delta region, and undoubtedly other unrecorded grasses have disappeared from the Colorado delta. *Setaria* and others probably have been extirpated from the Río Sonoyta region due to habitat destruction.

Grasses have played an important role in regional agriculture and food resources. Francisco de Alarcón brought wheat to the lower Río Colorado in 1540 (Hammond & Rey 1940) but that introduction was apparently lost. Eusebio Kino reintroduced wheat and other crops in the late eighteenth century (Bolton 1936). From pre-Spanish times until the early twentieth century, people along the lower Río Colorado made extensive use of Palmer grass (*Distichlis palmeri*) and cultivated maize (*Zea mays*), panic grass (*Panicum hirticaule* var. *miliaceum*), barnyard grass (*Echinochloa crusgalli*), and crowfoot grass (*Dactyloctenium aegyptium*) (Castetter & Bell 1951; Kelly 1977). Kelly (1977) mentions Cocopa use of grain from *Eriochloa aristata* and *Eragrostis mexicana* (actually probably *E. pectinacea*).

In 1701 Padre Kino and his companions found O'odham people cultivating maize (corn) along the river at Sonoyta (Burrus 1971). The missionaries introduced cattle, which thrived on grasses and other forage, but the cattle probably contributed to the demise of local stands of grasses and wetland habitat. In historic times wheat, barley, oats, maize, and sugarcane have been grown under irrigation in the Río Sonoyta and San Luis valleys. Maize and other crops have been grown at several small dryland, agricultural fields in the Pinacate volcanic region, such as the floodwater field at Suvuk and near Tinajas de los Pápagos (Bell, Anderson, & Stewart 1980; Childs 1954; Lumholtz 1912; Nabhan 1985; Julian Hayden, personal communication 1992).

Wheat (*Triticum*); maize (*Zea*); and sorghum, broomcorn, and Sudan grass (*Sorghum bicolor*) are grown in irrigated fields in northwest Sonora. Several grasses are used for lawns: Bermuda grass (*Cynodon dactylon*), Italian ryegrass (*Lolium perenne*), and St. Augustine grass (*Stenotaphrum secundatum*). Carrizo or giant cane (*Arundo donax*) has long been cultivated regionally, and fountain grass (*Pennisetum setaceum*) is common in local gardens.

References: Arnow 1987; Beetle 1983–87, 1987; Beetle & Johnson 1991; Burkart 1969; Cabrera 1970; Clayton & Renvoize 1986; Feinbrun-Dothan 1986; Gould 1951; Gould & Moran 1981; Hitchcock 1913, 1935a, 1951; Holmgren & Holmgren 1977; McVaugh 1983; Navas Bustamante 1973; Nicora 1978; Pohl 1980; Soderstrom et al. 1987; Swallen 1964; Swallen & Hernandez X. 1961; Tzvelev 1983; Van Devender, Toolin, & Burgess 1990; Yatskievych 1999; Zuloaga et al. 1994).

1. Plants bamboo-like or reedlike, 2–3 m tall; wet places.
   2. Rachillas glabrous, the lemmas with long hairs; cultivated. ______ **Arundo donax** Linnaeus (not included in species accounts; also see species account for *Phragmites*).
   2′ Rachillas with long hairs, the lemmas glabrous; wetland habitats. ______ **Phragmites**

1′ Plants not bamboo-like or reedlike.
   3. Coarse grasses; spikelets in pairs (or some in 3s), 1 sessile (bisexual or pistillate, awned or not), and 1 (2) pedicellate (smaller, staminate or sterile, not awned, deciduous or not, or reduced to a rudiment). (If you are not certain about this choice, then go to 3′; such genera key out both places.)
      4. Inflorescence a terminal panicle; awns deciduous, absent or not exceeding 1.5 cm; weeds in disturbed habitats including roadsides. ______ **Sorghum**
      4′ Inflorescence of 1 to many spikelike branches clustered at or near tip of stem; awns more or less persistent, at least 2 cm long; natural, mostly rocky habitats.
         5. Inflorescences cottony and whitish; awns ca. 2 cm long. ______ **Bothriochloa**
         5′ Inflorescences not cottony, not whitish; awns 4.5–7 cm long. ______ **Heteropogon**

   3′ Coarse or delicate grasses; spikelets not in pairs (or 3s) in combination with characters given in 3.
      6. Plants with stolons and/or rhizomes; perennials.
         7. Plants tufted, with short rhizomes or stolons but not creeping, not mat-forming; spikelets with awns 1–7 mm long; dry desert habitats.

8. Plants mostly more than 50 cm tall, with short, knotty, almost woody rhizomes (at or below ground level); at least lower portion and younger stems woolly. (Caution: older stems and leaf sheaths may become glabrous.) __________ **Pleuraphis**
8′ Plants mostly less than 40 cm tall, the rhizomes or stolons not woody; stems glabrous.
9. Plants 8–40 cm tall, the rhizomes or stolons short; flowering stems more than twice as tall as the leaves, the spikes 1-sided (comblike), 1 to several widely spaced on the flowering stem. __________ **Bouteloua trifida**
9′ Plants less than 15 cm tall, with slender, flexible stolons (above ground); flowering stems less than twice as tall as the leaves, the spikes densely crowded, not 1-sided. __________ **Erioneuron**
7′ Plants creeping and/or forming dense colonies or mats; spikelets awnless; tidal marshes, riparian, or weedy habitats, often on saline or alkaline soils.
10. Leaf blades less than 1 (1.5) cm long; inflorescences inconspicuous, hidden among leaves. __________ **Monanthochloë**
10′ Leaf blades more than 2 cm long; inflorescences conspicuous.
11. Spikelets several-flowered; glumes 4.5–8.5 mm, the lemmas 3–16 mm. __________ **Distichlis**
11′ Spikelets 1-flowered; glumes 1.4–2.5 mm, the lemmas less than 2.5 mm.
12. Inflorescences of digitately arranged slender spikes. __________ **Cynodon**
12′ Inflorescences of compact panicles, not digitate. __________ **Polypogon viridis**
6′ Plants without conspicuous stolons or rhizomes; annuals or perennials.
13. Inflorescences with spines or bristles surrounding or just below the spikelets; sometimes the spines or bristles united into a bur or fascicle enclosing the spikelets.
14. Spikelets not enclosed in burs or fascicles, but with slender bristles below many of the spikelets; spikelets breaking off above the bristles. __________ **Setaria**
14′ Spikelets enclosed in burs or fascicles, these falling as a unit with the attached bristles or spines.
15. Spikelets enclosed in burs with sharp, stiff spines; spines conspicuously united at least in lower 1/4 of the bur. __________ **Cenchrus**
15′ Spikelets enclosed in fascicles with flexible bristles; bristles separate except scarcely united at the very base. __________ **Pennisetum**
13′ Inflorescences without spines or bristles just below the spikelets; spikelets not in burs or fascicles.
16. At least some spikelets with awns.
17. Spikelets clearly 1-flowered and with 3 prominent terminal awns from a short to long central column. __________ **Aristida**
17′ Spikelets 1 to several-flowered with 1 to several awns, not both 3-awned and 1-flowered (if 3-awned, then spikelets with rudiments above).
18. Perennials, almost always obviously so.
19. Lemmas with 9 feathery awns; panicles spikelike, lead-gray or green-gray. __________ **Enneapogon**
19′ Lemmas or other spikelet parts 1- to several-awned, the awns not feathery (some spikelets in an inflorescence may lack awns).
20. Plants less than 15 cm; leaves fascicled. __________ **Erioneuron**
20′ Plants usually more than 20 cm; leaves not conspicuously fascicled.
21. Spikelets with 2 dissimilar florets each with 3 awns 1.5–2.0 cm long. __________ **Chloris crinita**
21′ Spikelets various but florets (lemmas) not 3-awned.
22. Awns at least 2 cm.
23. Awns 4.5–7 cm; old leaves semipersistent, rust or reddish brown; inflorescences solitary and spikelike. __________ **Heteropogon**
23′ Awns 2–4.5 cm; leaves drying straw colored; inflorescence of panicles or clusters of spikelike branches.
24. Awns ca. 2 cm; inflorescences cottony, of short branches clustered near top of stem. __________ **Bothriochloa**
24′ Awns 3.3–4.5 cm; inflorescence an elongated spikelike panicle, not cottony. __________ **Stipa**

22′ Awns less than 1.5 (2) cm.
25. Stems with feltlike hairs (glabrate with age); inflorescence spikelike, the axis zigzag; spikelets in clusters of 3s; widespread and abundant. ______ **Pleuraphis**
25′ Stems glabrous; inflorescence not spikelike, the axis not zigzag; spikelets solitary, or some in 2s and others in 3s on same panicle; not especially widespread.
26. Plants bushy, branching freely in upper part of plant. ______ **Muhlenbergia porteri**
26′ Plants tufted, branching mostly from base of plant.
27. Spikelets all alike, not in 2s or 3s; awns not twisted (awn column may be twisted), not deciduous. ______ **Aristida ternipes** var. **ternipes**
27′ Spikelets dissimilar, in 2s or 3s; awns twisted, readily deciduous. ______ **Sorghum**

18′ Annuals (mostly ephemerals).
28. Spikelets (excluding awns and pedicels) at least (8) 10 mm, clearly 2- to several-flowered; winter-spring ephemerals.
29. Awns more than 5 cm, stiff and scabrous. ______ **Triticum**
29′ Awns 4.5 cm or less.
30. Spikelets clustered in 3s, the 2 lateral ones staminate or reduced to lanceolate or bristlelike glumes. ______ **Hordeum**
30′ Spikelets not clustered in 3s.
31. Glumes thin and longer than the rest of the spikelet or florets (excluding awns); lemmas of firmer texture than glumes; awns 3.5–4.5 cm. ______ **Avena**
31′ Glumes shorter than the rest of the spikelet; lemmas similar in texture to the glumes; awns 2.2 cm or less.
32. Inflorescence a slender unbranched spike with sessile spikelets evenly spaced and edgewise on alternate sides of main axis; axis moderately but obviously zigzag; most spikelets of an inflorescence with only 1 glume. ______ **Lolium**
32′ Inflorescence mostly branched except in drought-stunted plants, the spikelets not edgewise and evenly spaced on opposite side of axis; axis straight; both glumes present.
33. Spikelets 12–40 mm (not including awns); lemmas 8–13 mm; awns straight, curved, or twisted. ______ **Bromus**
33′ Spikelets 8–13 mm (not including awns); lemmas 4–8 mm; awns straight. ______ **Festuca**
28′ Spikelets (excluding awns and pedicels) 2–6 (7) mm, 1 to several-flowered; winter-spring or summer ephemerals.
34. Inflorescences of 1-sided spikes or spikelike branches, these alternate (single) along main inflorescence axis or digitate at top of the main axis (occasionally 1 spike per stem).
35. Inflorescence branches (spikes) not digitate or alternate on main axis.
36. Spikelets with only 1 well-developed glume, and a conspicuous cuplike or disklike ring below. ______ **Eriochloa**
36′ Spikelets with 2 evident glumes, the lower one often smaller.
37. Awns 6 mm or less; spikelets with 1 basal fertile floret and 1 or more reduced sterile florets (the rudiment) above, the glumes and lemma of similar texture; desert habitats, widespread. ______ **Bouteloua**

37′ Awns often more than 10 mm; lower floret sterile or staminate, the lemma of similar texture as the glumes; upper floret fertile, the lemma firm and shiny, the texture strikingly different than the glumes; wetland habitats or moist soil. ________ **Echinochloa crusgalli**

35′ Spikes or spikelike branches digitately arranged at top of main axis (stunted plants occasionally with a spike).

38. Spikes (4) 5–10 (12+) and ascending (upright); rachis tip not extending beyond spikelets; spikelets (lemma of lower floret) with a tuft of long white spreading hairs, the awns 5–20 mm. ________ **Chloris virgata**

38′ Spikes (rarely 1) 2–6 and spreading (nearly at right angles to main axis); rachis tip extending beyond spikelets; spikelets glabrous, the awns 0.5–3.2 mm. ________ **Dactyloctenium**

34′ Inflorescences not obviously of 1-sided spikes.

39. Spikelets with several bisexual florets; awns ca. 1 mm. ____ **Leptochloa viscida**

39′ Spikelet with only 1 bisexual floret; awns at least 5–10 mm.

40. Panicles very densely flowered, cylindrical, spikelike, appearing "furry" like a rabbit's foot. ________ **Polypogon monspeliensis**

40′ Panicles open or interrupted with spikelike branches, not dense and spikelike, not appearing "furry."

41. Coarse grasses with dense spikelike branches at widely spaced intervals on main axis; cleistogenes none; fertile lemma hard and shiny; natural and disturbed wetland habitats. ________ **Echinochloa**

41′ Delicate ephemerals with open and filmy panicles; basal leaf axils with cleistogenes; lemma thin and dull; widespread including desert habitats. ______ **Muhlenbergia microsperma**

16′ Spikelets not awned.

42. Spikelets with 2 or more distinct, bisexual florets.

43. Glumes 6–15 mm; spikelets 2–3 cm, strongly compressed. ________ **Bromus catharticus**

43′ Glumes 5 mm or shorter; spikelets less than 1.5 cm, compressed to rounded in cross section.

44. Hot-weather annuals (ephemerals); panicle branches open or spreading (unless stunted) or ascending but not appressed.

45. Primary panicle branches 2 or more times branched, the secondary branches usually ascending to spreading; pedicels as long as or longer than the spikelets. ________ **Eragrostis**

45′ Primary panicle branches unbranched, spikelike, or with secondary branchlets closely appressed; pedicels shorter than the spikelets.

46. Inflorescence branches 2 or more, digitately clustered at top of main axis (2 or more from a single node) and sometimes with a few more branches below. ________ **Eleusine**

46′ Inflorescence branches scattered along the main axis, usually not more than 1 per node. ________ **Leptochloa**

44′ Perennials or winter-spring annuals (ephemerals); panicles contracted or compact, the branches erect and appressed (or only slightly spreading), or panicles spikelike.

47. Perennials, roots tough and well developed; spikelets (5.5) 6.5–13 mm. ____ **Tridens**

47′ Winter-spring ephemerals, roots weak and not extensive; spikelets 2–6.5 mm.

48. Glumes conspicuously shorter than the rest of the spikelet; Sonoyta region only. ____ **Poa**

48′ Glumes about as long as the rest of the spikelet; widespread. ____ **Schismus**

42′ Spikelets with 1 bisexual floret, sometimes also with 1 (2) reduced or vestigial floret(s) (the spikelet may appear 1-flowered unless dissected).

49. Perennials, the stems tough and erect with knotty bases.

50. Less than 1 m; spikelets cottony-looking with white to purplish silky hairs; hills and mountaintops. ____ **Digitaria californica**

50′ Usually 1.5–1.8 m; spikelets glabrous; common in lowland non-rocky habitats. ____ **Sporobolus airoides**

49′ Annuals, the stems not tough with knotty bases.

51. Spikelets with only 1 well-developed glume and a cuplike disk at the base. ____ **Eriochloa**

51′ Both glumes well developed, spikelets without a cuplike disk at base.

52. Glumes both longer than the lemma, conspicuously keeled and often often winged; panicles spikelike, ovoid to cylindrical and very densely flowered, the spikelets overlapping. ____ **Phalaris**

52′ One or both glumes conspicuously shorter than the lemma, the glumes not winged; panicles not spikelike.

53. Leaves generally with purple spots or bands; inflorescences racemose; spikelets densely crowded in 3–7 compact branches, these branches 1–2 (3) cm long at widely spaced intervals along the main axis. ____ **Echinochloa colonum**

53′ Leaves without purple spots or bands; inflorescence branches paniculate or digitate; spikelets more or less distributed throughout the much-branched inflorescence.

54. Inflorescences of digitate spikelike branches, the spikelets subsessile or short-pedicelled and evenly spaced in regular rows. ____ **Digitaria**

54′ Inflorescences paniculate, the spikelets pedicellate and not in regular rows.

55. Spikelets 1.5–2.5 mm, the grain readily falling; spikelets 1-flowered; lemma, palea, and grain distinct from each other. ____ **Sporobolus**

55′ Spikelets 2.3–4 mm, falling as a unit (breaking off below the glumes); spikelets 2-flowered, the lower one sterile or staminate, the lemma like the glume in texture, the upper floret hard and shiny, the fertile lemma firmly enclosing the palea and grain.

56. Plants hairy, including panicle branches, branchlets, and spikelets; prominent veins of spikelets longitudinal and transverse on upper part of spikelet forming a netlike pattern; known from adjacent Arizona. ____ **Brachiaria**

56′ Plants variously hairy but panicle branches, branchlets, and spikelets glabrous; prominent veins of spikelets longitudinal only; widespread. ____ **Panicum**

ARISTIDA *TRES BARBAS,* THREE-AWN

Annuals and perennials, often tufted, without stolons or rhizomes. Spikelets 1-flowered, readily breaking off above the glumes (often lodging in socks—people might be supplanting animals as dispersal agents). Glumes persistent, the upper and usually also the lower glume 1-veined. Lemma hard at maturity, terete, slender, often elongated into a straight or twisted awn column, the lemma apex or awn column bearing 3 awns, but sometimes the lateral awns reduced or absent.

Warm regions worldwide; 300 species, 60 in North America and 13 in Sonora. Many of the species are poorly defined, and sharp morphological boundaries are often lacking. A number of the species have extensive geographic ranges. References: Allred 1992; Henrard 1926–33, 1929–33; Hitchcock 1935b.

1. Spikelets appearing 1-awned, or lateral awns very short and stubby; perennials. ________**A. ternipes** var. **ternipes**
1′ Awns 3 and usually well developed (if lateral awns reduced, then the plants clearly are stunted ephemerals); ephemerals or perennials.
  2. Awns flattened with minutely serrated margins (seen with at least 10× magnification), 5–15 (17) mm (or awns occasionally reduced, especially the lateral ones); ephemerals. ________**A. adscensionis**
  2′ Awns terete, 12–50 mm; perennials (sometimes flowering in first season).
    3. Awn column of mature spikelets jointed (seen as a horizontal line across the column) just above the lemma body; awns 3–5 cm; sandy soils including dunes. ________**A. californica**
    3′ Awn column not jointed; awns 1.2–3 cm; rocky soils and slopes, and sometimes sand flats and washes.
      4. Panicles open and conspicuously branched, the branches mostly long and spreading at about 90° from the main axis; glumes subequal; sandy soils, not common in the flora area. ________**A. ternipes** var. **gentilis**
      4′ Panicles contracted to moderately spreading, branches mostly spreading at 45° or less, the branches short; glumes unequal or subequal, the lower glume often about 2/3 as long as upper; mostly in rocky habitats or washes.
        5. Panicles contracted to often moderately spreading; body of lemma gradually tapering, commonly without an evident awn column and not twisted (or sometimes slightly twisted), lacking a discernable neck; glumes largely equal or subequal, at least toward the apex of the inflorescence; gravelly washes at Arizona border, rare. ________**A. parishii**
        5′ Panicles contracted; upper part of awn column often twisted, the neck (narrowed part of awn column) usually evident, often 1–2 mm; glumes unequal, the upper at least 1/3 longer than lower; mostly in rocky habitats, widespread and common. ________**A. purpurea**

Aristida adscensionis Linnaeus [*A. heymannii* Regal. *A. adscensionis* subsp. *heymannii* (Regal) Tzvelev] *ZACATE TRES BARBAS, ZACATE DE SEMILLA;* SIX-WEEKS THREE-AWN. Non-seasonal ephemerals; 7–30 (60) cm, often with a purple-brown cast during cooler, drier seasons and yellow-green to green during summer-fall rainy season; glabrous except very short hairs at the ligule and minutely scabrous pedicels. Roots usually weakly developed.

Panicles contracted (narrow), with short erect-ascending branches, or larger well watered plants often with larger and somewhat spreading branches. Glumes markedly uneven, the lower one much shorter; upper glume often shorter than the lemma body, ranging from considerably shorter to rarely slightly longer than lemma. Lemmas 6.5–10.5 mm (shorter on drought-stressed plants), gradually tapered toward apex, not twisted, scabrous on keel; awn column short or virtually absent with no discernable neck. Awns usually 3, often 5–15 (17) mm, about equal to unequal in length, flattened (a useful diagnostic character and one of striking beauty under magnification), the midrib prominent, its upper (adaxial) surface bearing a row of slender hairs 0.5–0.8 mm; awn margins winged, thin, translucent, and serrulate. Drought-stressed plants or local populations sometimes with reduced or aborted lateral awns, or all awns reduced or rarely awnless.

Not on moving dunes, otherwise throughout the region from near sea level to peak elevations; flats, slopes, and gravelly washes and arroyos. One of the several most widespread and abundant ephemerals of the Sonoran Desert, and the only annual *Aristida* in the Sonoran Desert region. Warmer regions of the world, deserts to relatively humid, tropical and temperate climates. In less arid regions often short-lived perennials with stout roots.

Worldwide "a number of varieties have been proposed . . . but they intergrade one with another and their significance is doubtful" (Clayton 1972:372). Beetle (1974) recognized 7 varieties for Mexico, but I do not find them useful (also see Allred 1992). Recognizing 2 subspecies worldwide, Tzvelev (1983) restricted subsp. *adscensionis* to the New World and parts of Africa. He reported subsp. *heyman-*

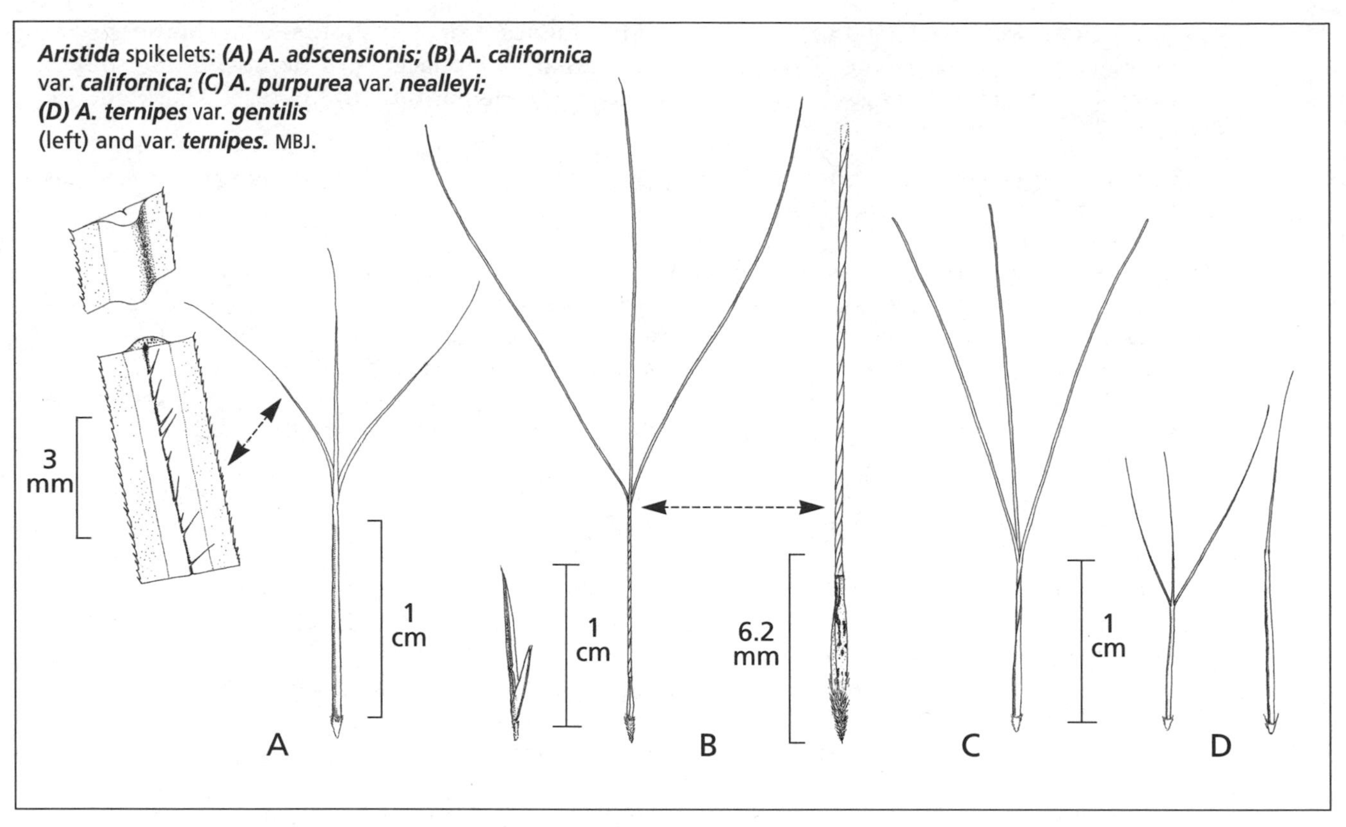

Aristida **spikelets: *(A) A. adscensionis; (B) A. californica*** var. ***californica; (C) A. purpurea*** var. ***nealleyi; (D) A. ternipes*** var. ***gentilis*** (left) and var. ***ternipes.*** MBJ.

Aristida adscensionis. Note the relatively small root system. LBH (Humphrey, Brown, & Everson 1958).

Aristida californica. Arrow points to articulation at base of lemma column. (Lamson-Scribner 1899; HUNT/H-US).

Aristida purpurea var. ***nealleyi.*** LBH (Humphrey, Brown, & Everson 1958).

Aristida ternipes var. ***gentilis.*** LBH (Gould 1951).

Avena fatua. LBH (Parker 1958).

Bothriochloa barbinodis. (Hitchcock 1951; HUNT/H-US).

Bouteloua aristidoides: plant, LBH (Gould 1951); ***(A)*** spikelet, ***(B)*** fertile lemma, ***(C)*** fertile palea, ***(D)*** sterile floret and portion of rachilla (bottom), ***(E)*** grain (caryopsis). (Griffiths 1912; HUNT/H-US2).

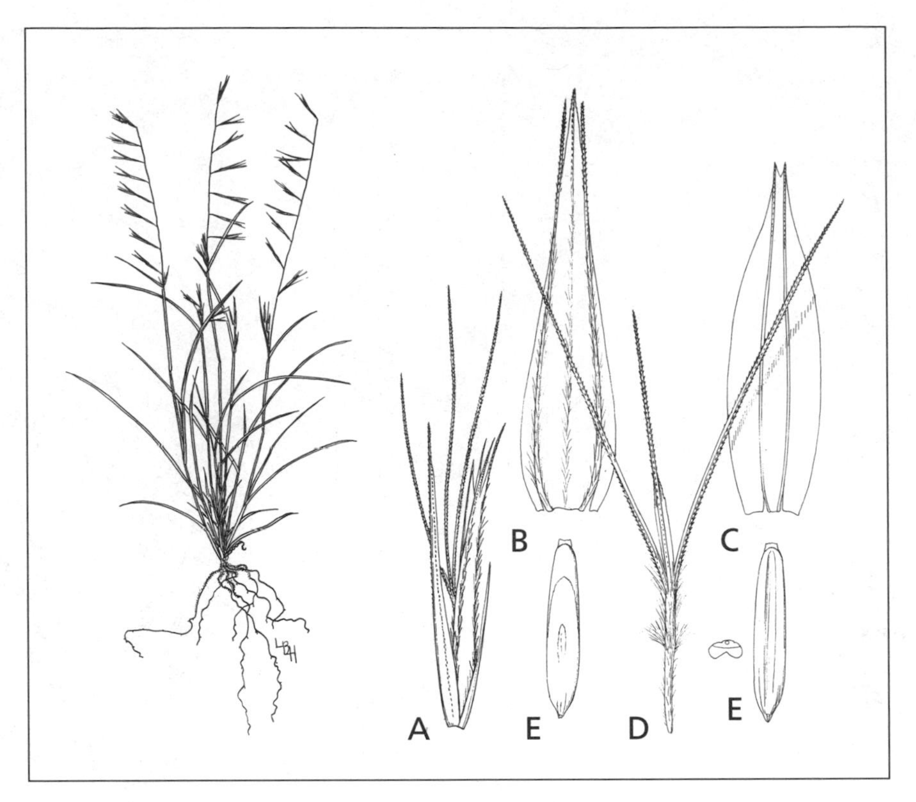

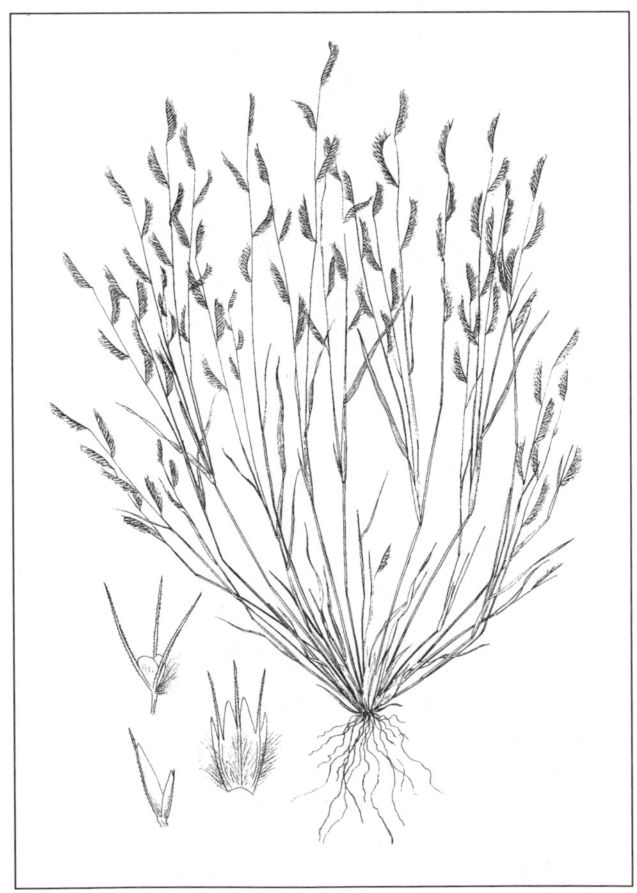

Bouteloua barbata. Note glumes, fertile lemma, and sterile floret. HUNT/H-US.

Bouteloua trifida. Note glumes, fertile and sterile florets, and grain. MDB (Lamson-Scribner 1897; HUNT/H-US).

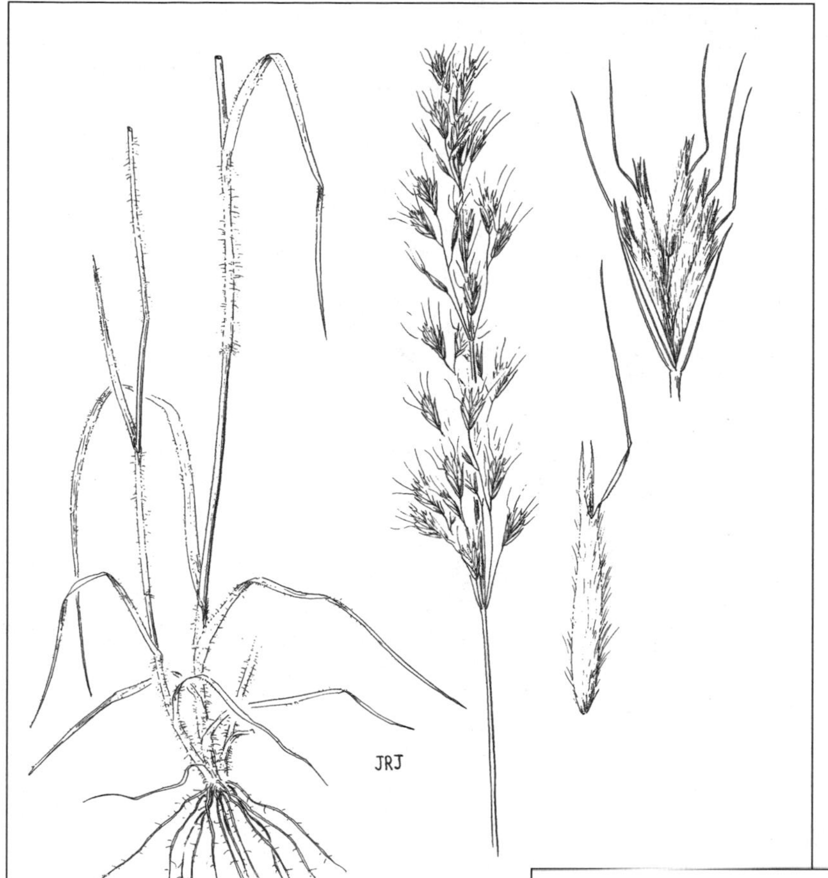

Bromus carinatus. LBH.

Bromus berterianus. JRJ (Hitchcock, Cronquist, & Ownbey 1969).

Bromus catharticus var. ***catharticus.*** LBH (Parker 1958).

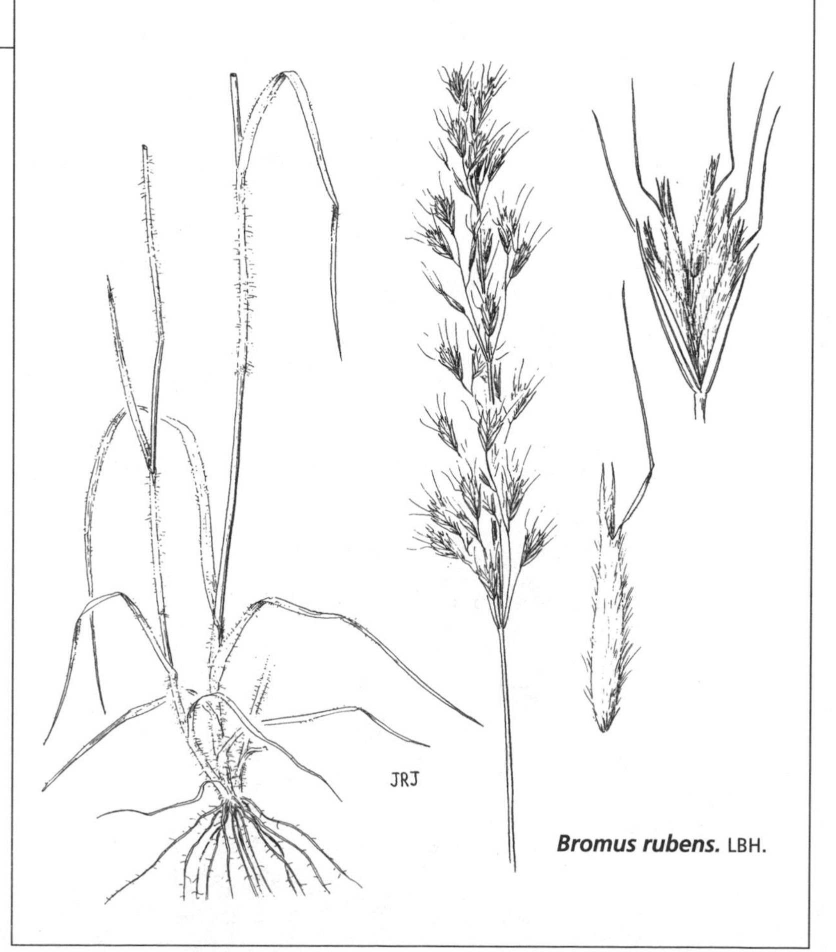

Bromus rubens. LBH.

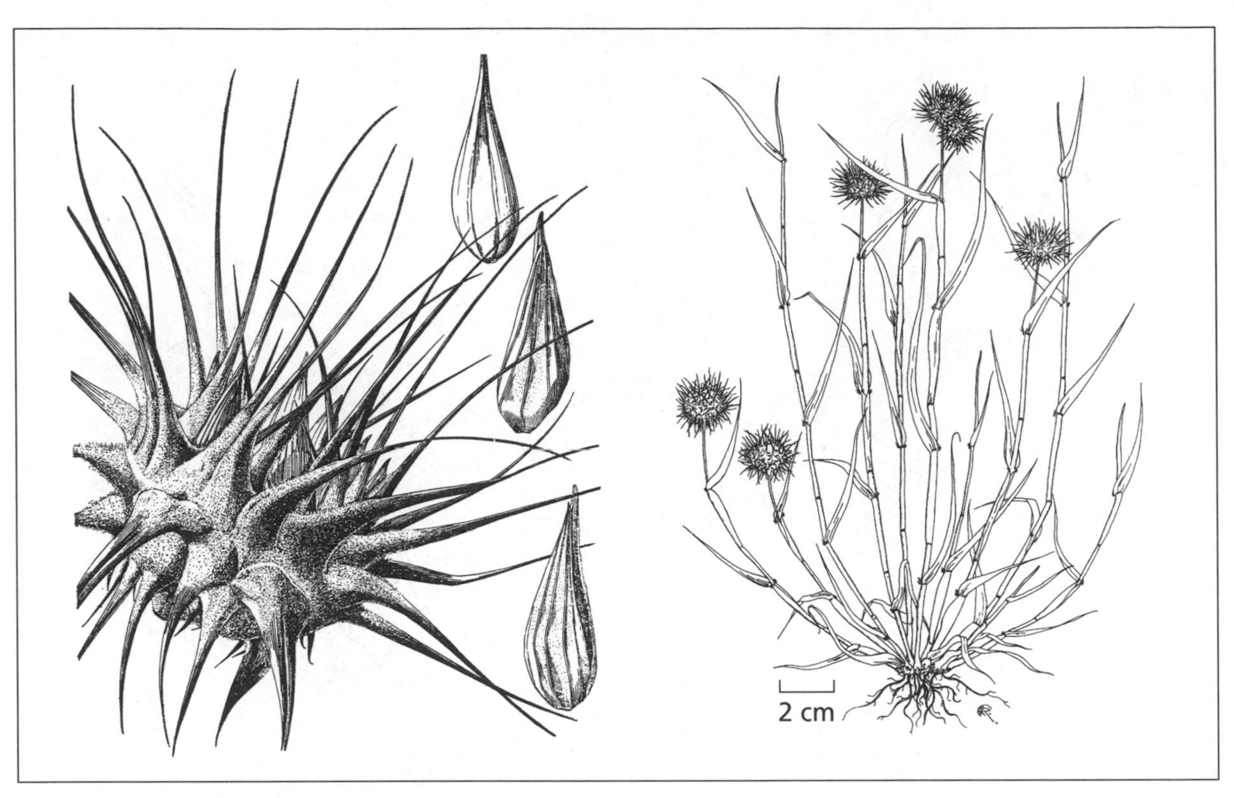

Cenchrus palmeri. Bur and two views of spikelet (below) and fertile floret. AC (Chase 1920); plant, FR (Felger 1980).

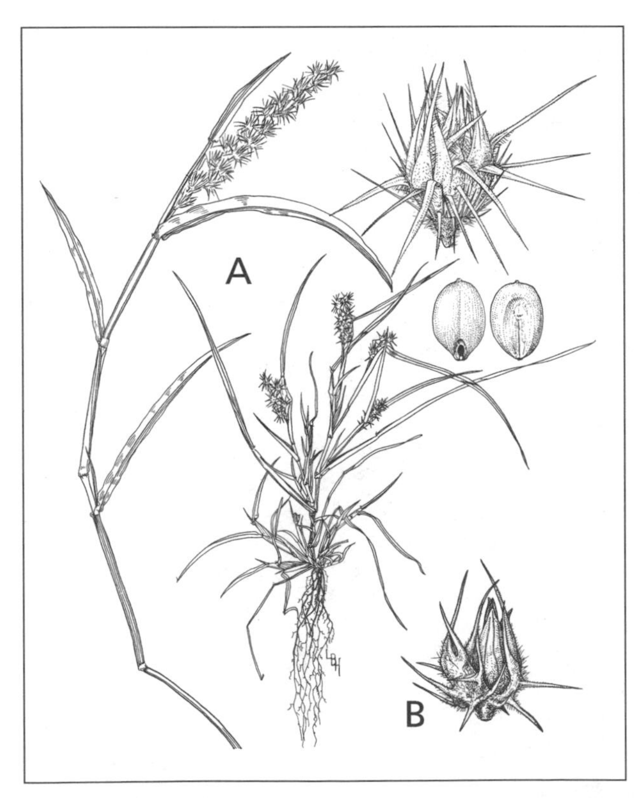

Cenchrus: (A) C. echinatus; (B) C. incertus. LBH (Parker 1958).

Chloris crinita. (Hitchcock 1951; HUNT/H-US).

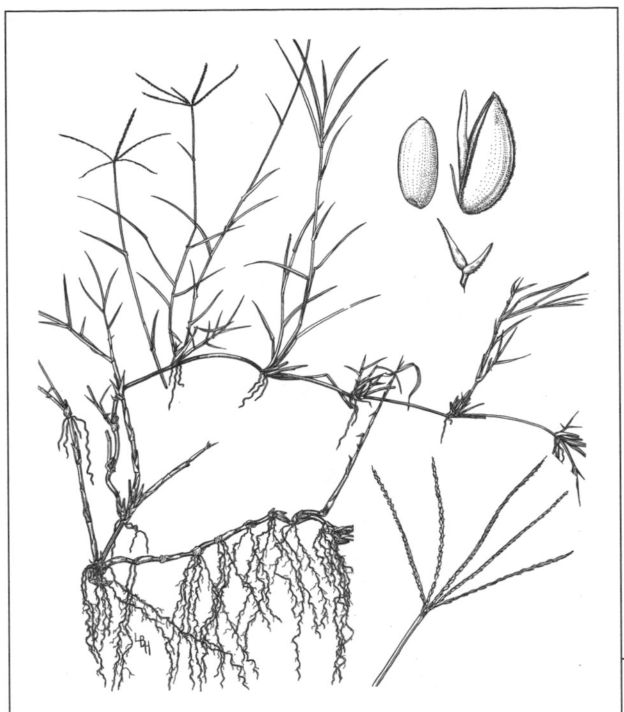

Chloris virgata. Note spikelet with detached glumes. (Hitchcock 1951; HUNT/H-US).

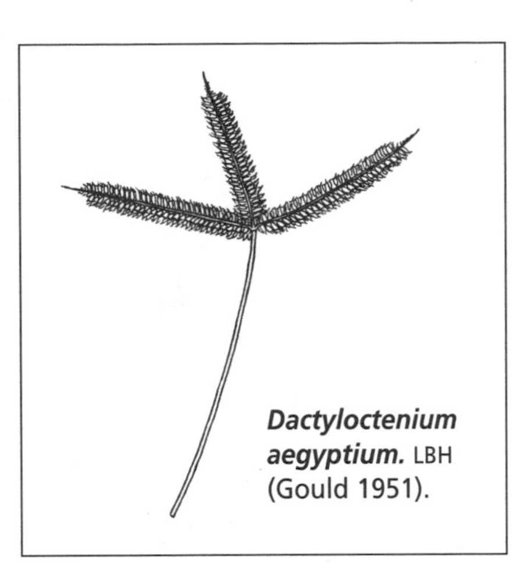

Dactyloctenium aegyptium. LBH (Gould 1951).

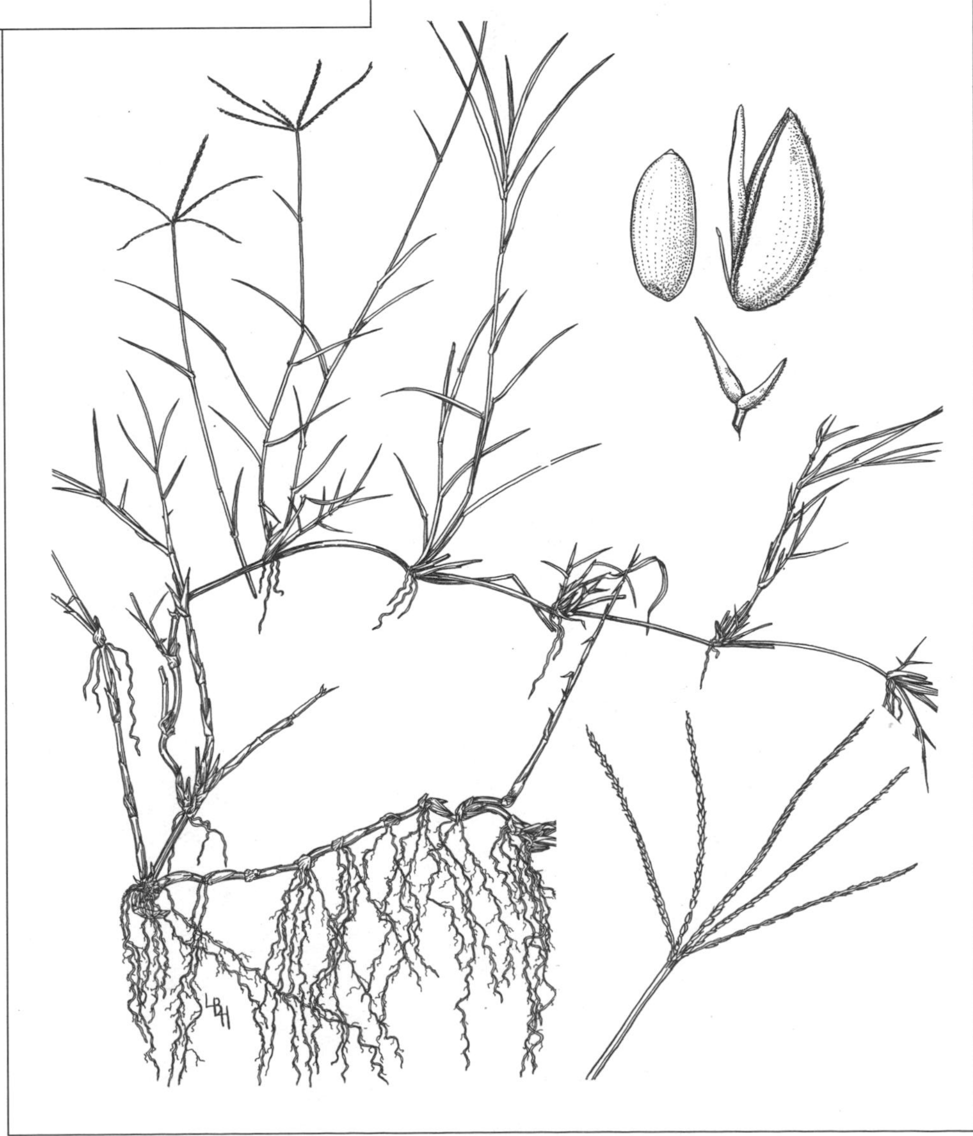

Cynodon dactylon var. ***dactylon.*** LBH (Gould 1951).

Digitaria californica var. ***californica.***
LBH (Gould 1951).

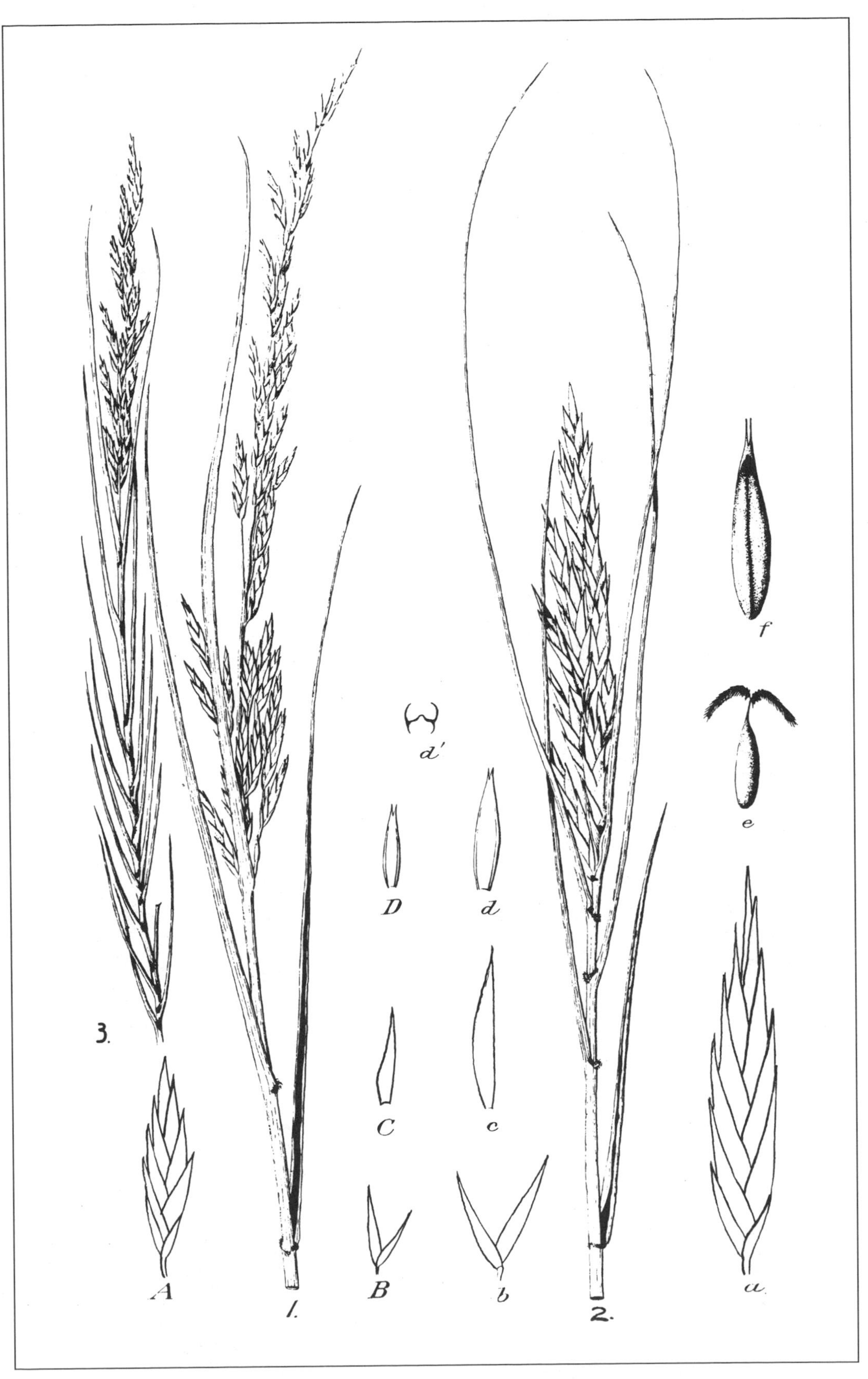

***Distichlis palmeri:* "*1.* A male plant; *2.* A female plant; *2.* A male plant with small close leaves: *A.* A spikelet of the male plant; *B.* Empty glumes of same; *C.* Flowering glumes of same; *D.* Palet [palea] of same; *a.* Female spikelet; *b.* A pair of empty glumes of same; *c.* A flowering glume of same; *d.* Palet of same; *d'.* A transverse section of the palet; *e.* Young fruit with the styles united; *f.* Older, ripe fruit, with the styles separate."** TH (Vasey 1889; HUNT/H-US).

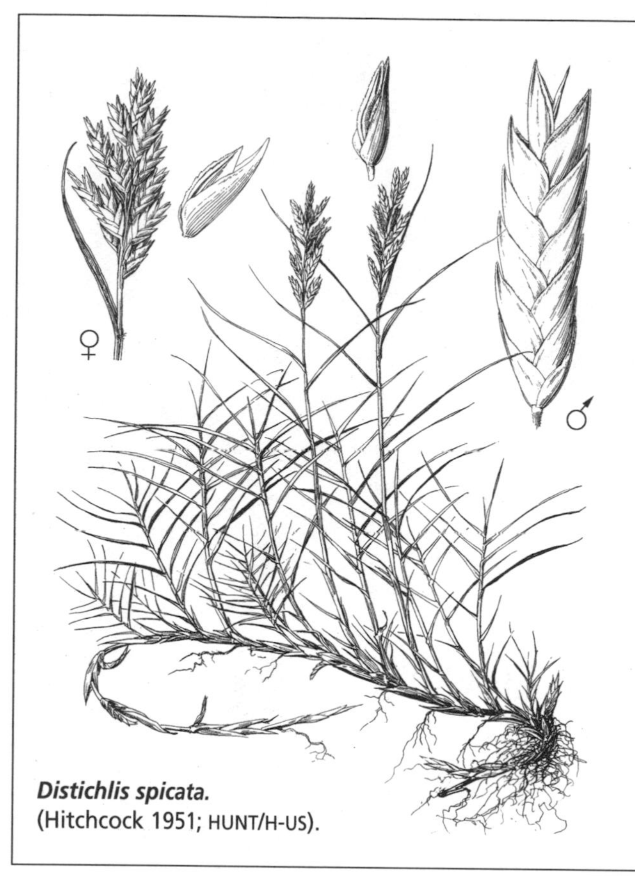

Distichlis spicata.
(Hitchcock 1951; HUNT/H-US).

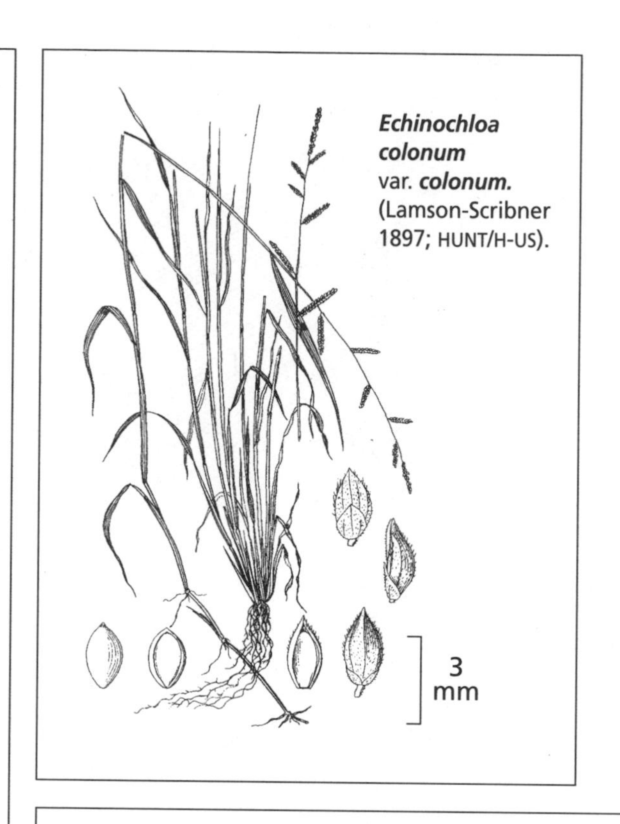

Echinochloa colonum var. ***colonum.*** (Lamson-Scribner 1897; HUNT/H-US).

Echinochloa crusgalli var. ***crusgalli.*** (Hitchcock 1951; HUNT/H-US).

Enneapogon desvauxii. Plant and cleistogen (Hitchcock 1951; HUNT/H-US) and spikelet LBH (Gould 1951).

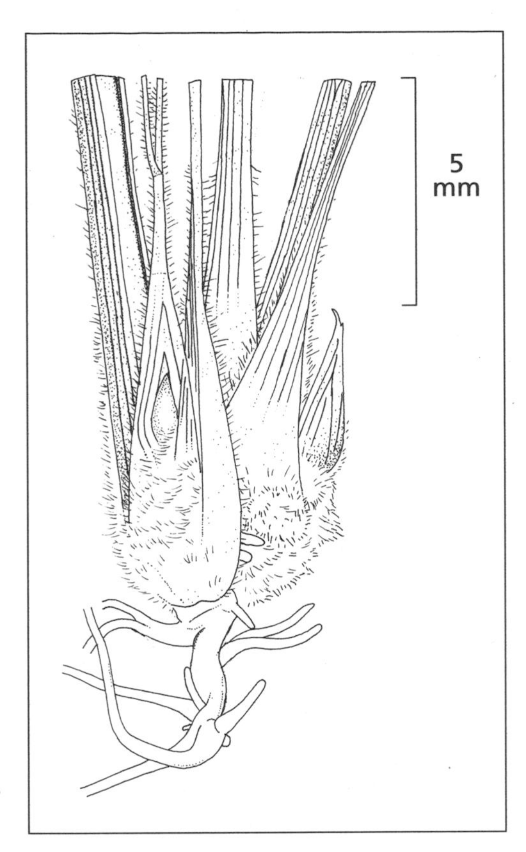

Enneapogon desvauxii. Base of plant showing two cleistogamous spikelets, one partially exposed. MBJ.

Eragrostis cilianensis. (Hitchcock 1951; HUNT/H-US).

Eragrostis pectinacea var. ***pectinacea.*** LBH (Gould 1951).

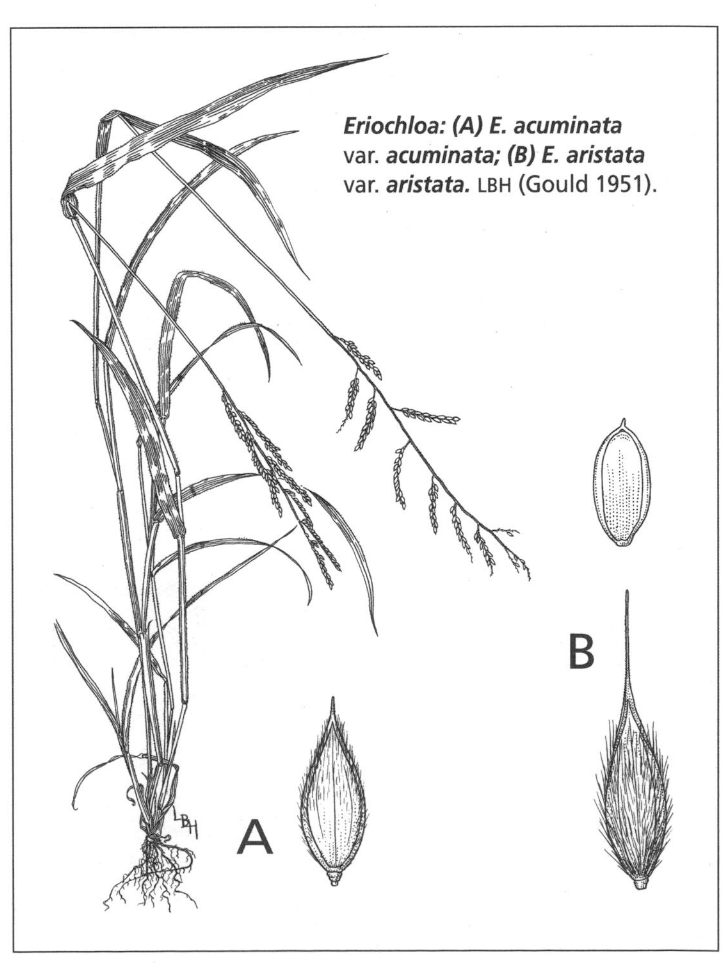

Eriochloa: (A) E. acuminata var. ***acuminata; (B) E. aristata*** var. ***aristata.*** LBH (Gould 1951).

Heteropogon contortus.
(Hitchcock 1951; HUNT/H-US).

Festuca octoflora. MWG
(Hitchcock 1951; HUNT/H-US).

Erioneuron pulchellum.
(Hitchcock 1951;
HUNT/H-US).

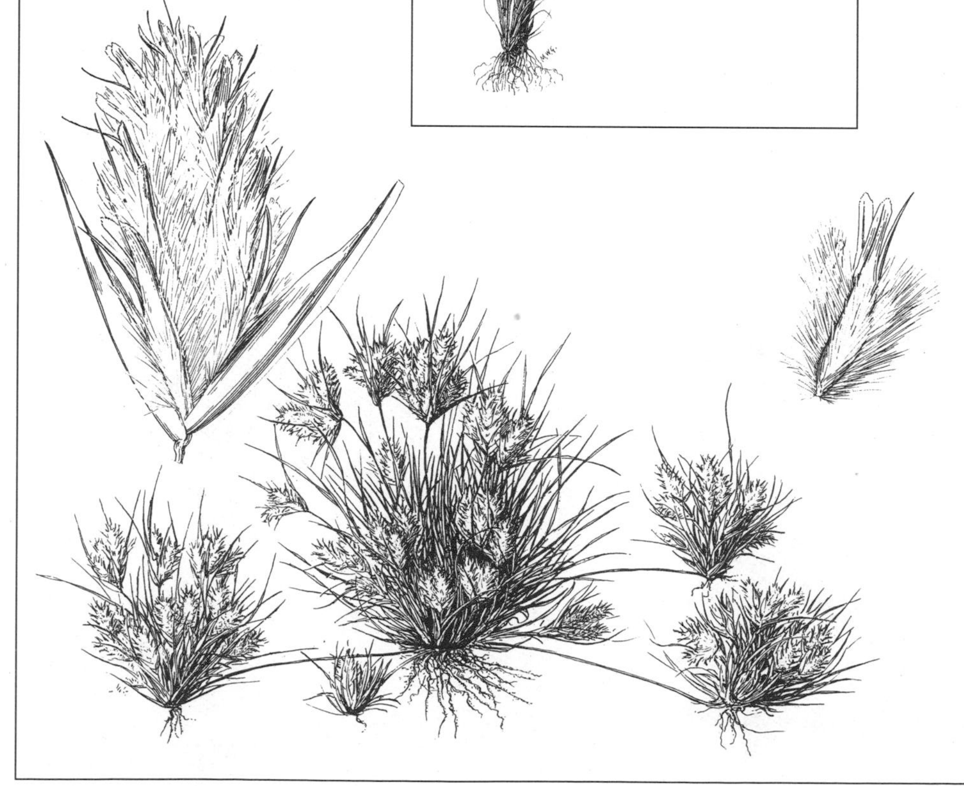

Hordeum murinum. LBH (Parker 1958).

Leptochloa fusca subsp. ***uninervia.*** LBH (Parker 1958).

Leptochloa panicea subsp. ***brachiata.*** (Hitchcock 1951; HUNT/H-US).

Leptochloa dubia. LBH (Gould 1951).

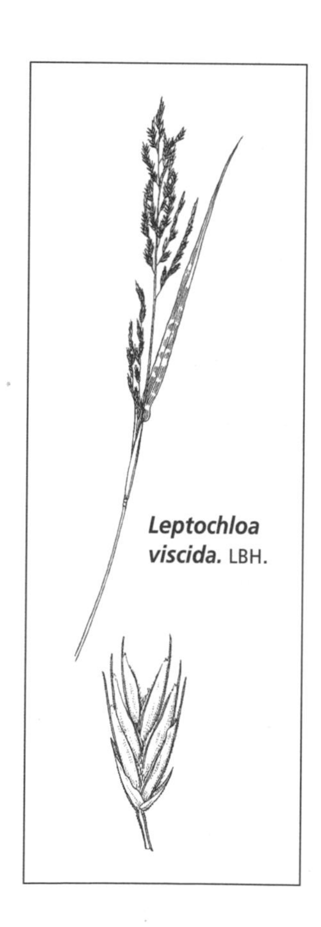

Leptochloa viscida. LBH.

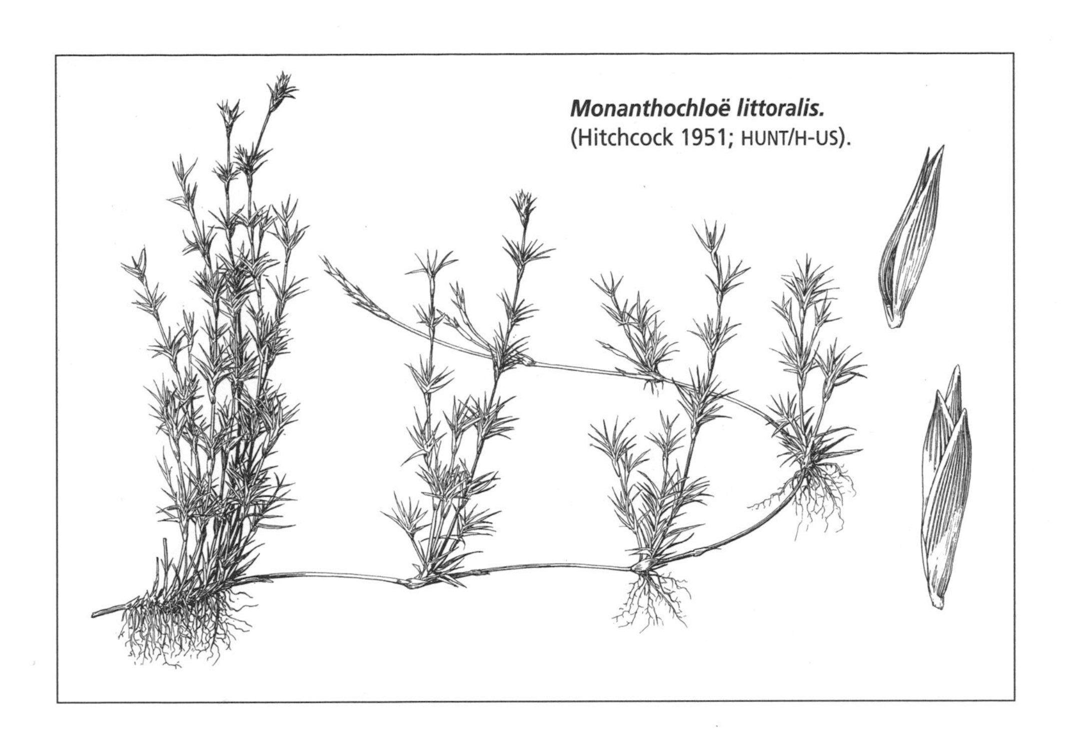

Monanthochloë littoralis. (Hitchcock 1951; HUNT/H-US).

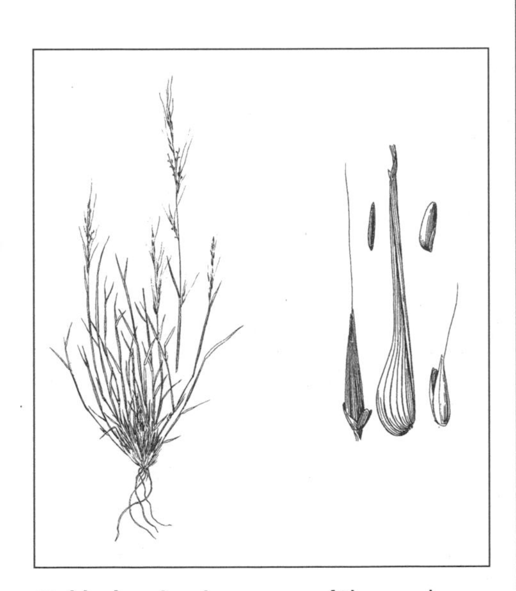

Muhlenbergia microsperma: (A) normal spikelet; ***(B)*** cleistogamous spikelets enclosed in sheath of reduced leaf; ***(C)*** cleistogamous floret. HUNT/H-US.

Muhlenbergia porteri. LBH (Gould 1951).

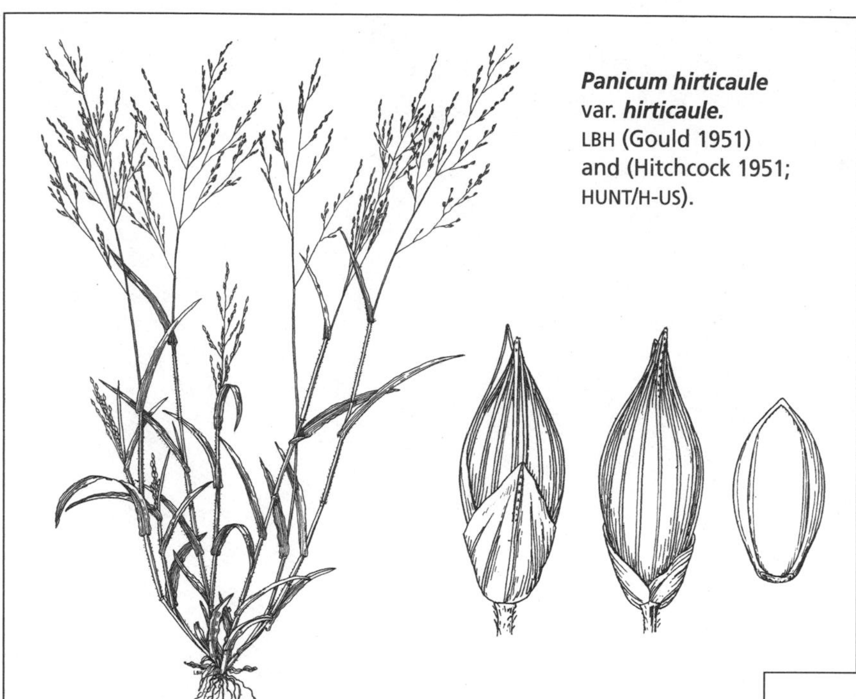

Panicum hirticaule var. ***hirticaule.*** LBH (Gould 1951) and (Hitchcock 1951; HUNT/H-US).

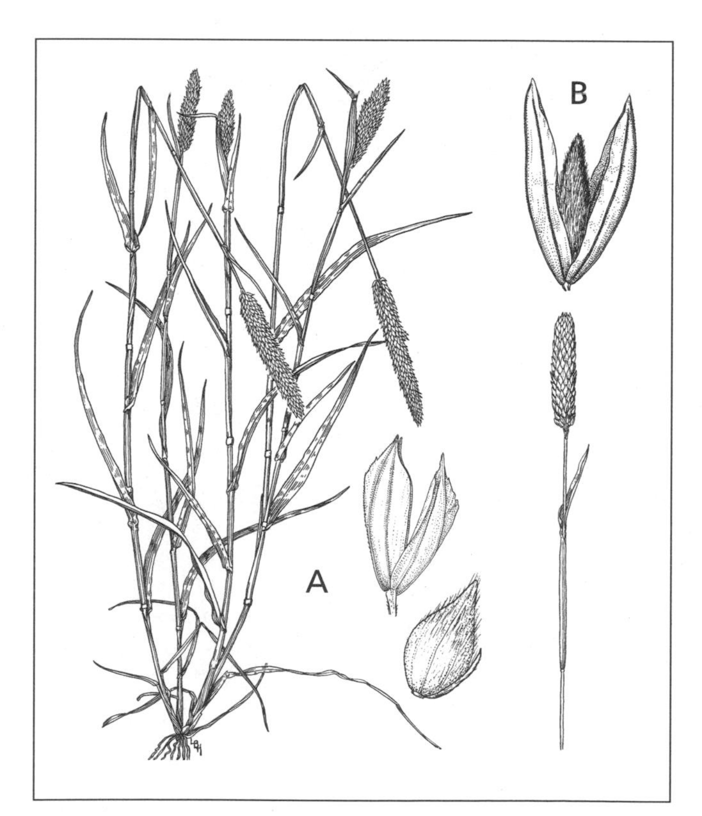

Phalaris: (A) P. minor, note glumes and detached florets, the scalelike sterile floret below the larger fertile floret; ***(B) P. caroliniana.*** LBH (Parker 1958, 1972).

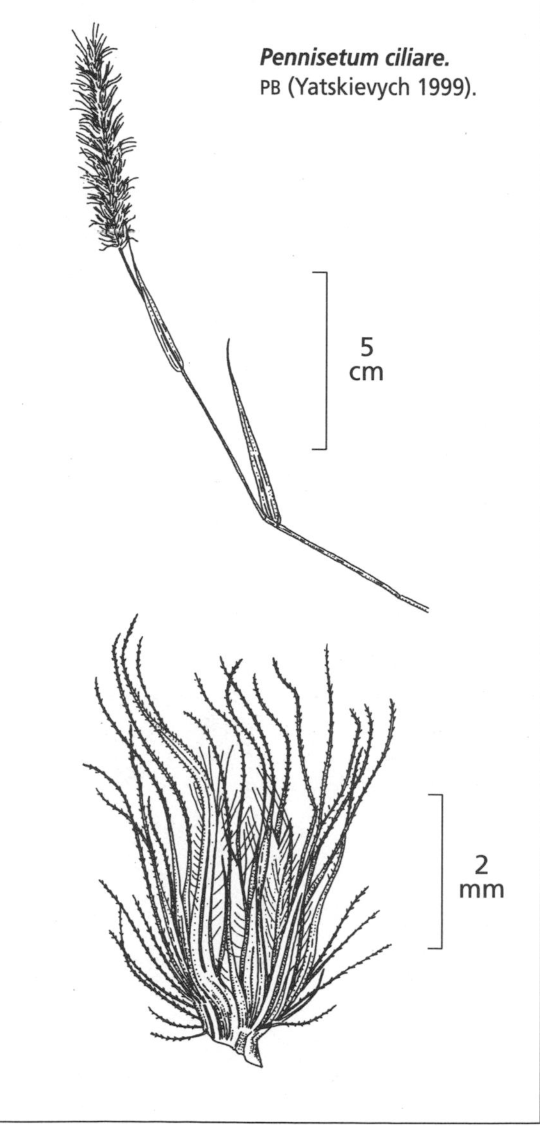

Pennisetum ciliare. PB (Yatskievych 1999).

Pleuraphis rigida.
HUNT/H-US.

Phragmites australis.
(Hitchcock 1951; HUNT/H-US).

Poa bigelovii. FLS (Lamson-Scribner 1899; HUNT/H-US).

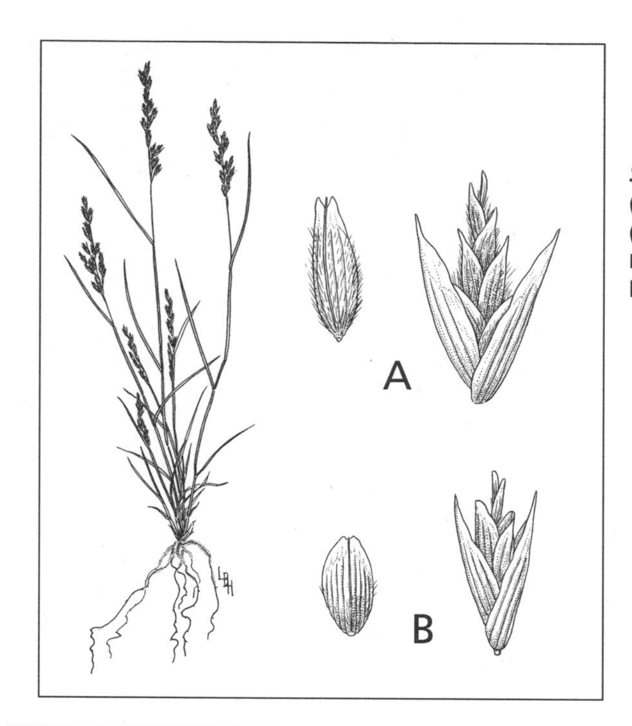

Schismus:
(A) S. arabicus;
(B) S. barbatus.
LBH (Gould 1951; Parker 1972).

Polypogon monspeliensis.
(Hitchcock 1951; HUNT/H-US).

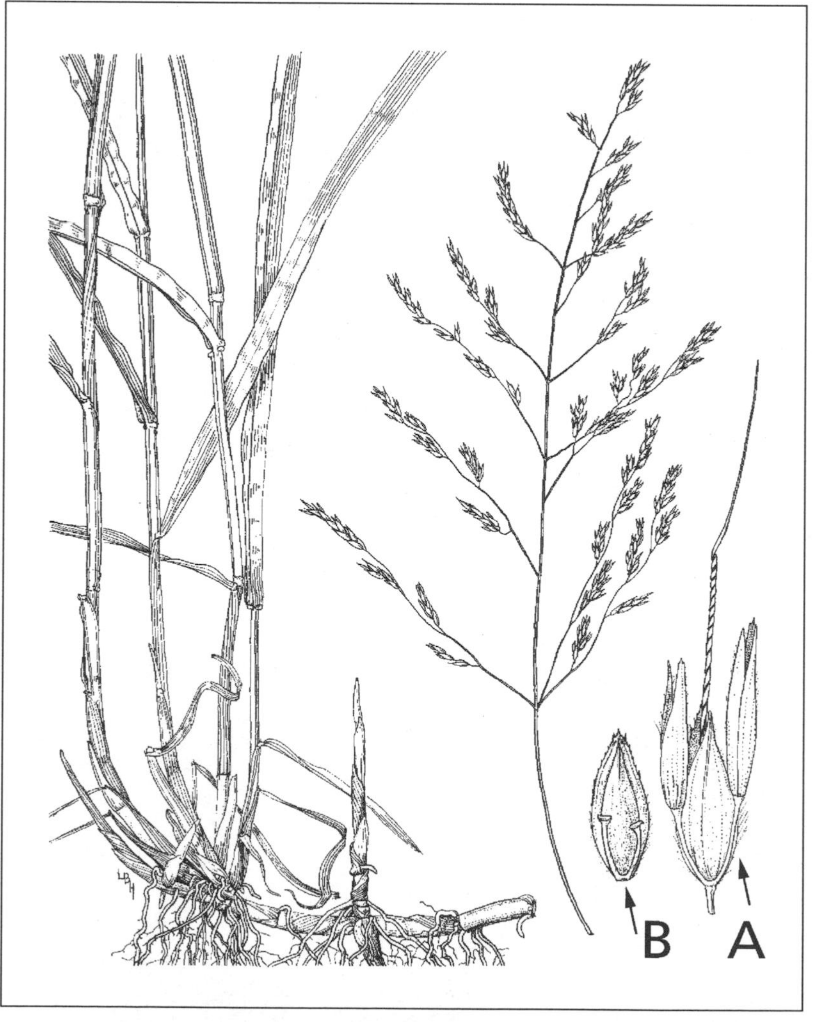

Sorghum halepense: (A) cluster of three spikelets from tip of a panicle branch, with two sterile, awnless, stalked spikelets, and one fertile, awned, sessile spikelet; ***(B)*** sessile spikelet with pair of persistent sterile spikelet pedicels. LBH (Parker 1958).

Sporobolus cryptandrus.
(Hitchcock 1951; HUNT/H-US).

Sporobolus airoides var. ***airoides.***
(Hitchcock 1951; HUNT/H-US).

Sporobolus flexuosus. JRJ (Cronquist et al. 1977).

Stipa speciosa
var. ***speciosa.***
(Lamson-Scribner
1899; HUNT/H-US).

Sporobolus pyramidatus.
JRJ (Cronquist et al. 1977).

Tridens muticus var. ***muticus.*** Plant LBH (Gould 1951) and VK (Gould & Box 1965).

nii in the Old World, distinguished by having the lemma longer than the upper glume and being scabrous only along the keel. However, these characters occur on plants from the flora area and elsewhere in the New World. Furthermore, the relative lengths of the lemma and upper glume can vary with age and growing conditions.

Sierra del Rosario, *F 20670*. Tinaja del Tule, *F 18779*. Higher elevation, Sierra Pinacate, *F 19403*. W of Sonoyta, *F 86-314*. Puerto Peñasco, *Wiggins 8378*.

Aristida californica Thurber var. **californica** [*A. peninsularis* Hitchcock] *TRES BARBAS DE CALIFORNIA;* CALIFORNIA THREE-AWN. Perennials, sometimes flowering in the first season; small tufted clumps overtopped with dense many-flowered panicles; often with a red-purple cast during the cooler, drier times of the year and green during the summer rainy season. Roots relatively thick, fleshy, well developed, encased in a jacket of sand laced together by a felt-like mat of root hairs. At least the lower internodes with fine to coarse white-woolly hairs. Upper glume 10.5–17.5 mm, the lower one shorter. Lemma jointed at base of column (the articulation, visible as a line across the column, is best seen on mature spikelets); body of lemma (portion below joint of column) (5) 5.5–6.0 (7) mm, often gray and mottled; awn column (10) 12–18 (28) mm, loosely twisted, relatively slender, often lighter colored than the lemma body; awns 3–5 cm. Various seasons depending on soil moisture.

Widespread on sand flats and lower dunes but not in the Sonoyta region; dense stands along sandy roadsides waving in the wind like a sea of feathers. Often grazed by animals. Plants growing under better-watered conditions seem to produce longer awns. Inland populations tend to have more densely white pubescent stems than those along the coast. Deserts in southeastern California, both Baja California states, southwestern Arizona, and Sonora southward along coastal dunes to Bahía Colorado (south of Tastiota) and farther south near the mouth of the Río Mayo to northwestern Sinaloa. Var. *glabrata* Vasey occurs at higher elevations and/or areas of higher precipitation and for the most part is geographically peripheral to the desert variety (Reeder & Felger 1989).

The roots of many perennial grasses inhabiting desert sands are encased in a jacket of sand laced together by mucilage and a felt-like mat of persistent root hairs. These sheaths of sand grains, known as *rhizosheaths,* create a nano-climate supporting nitrogen-fixing bacteria (Danin 1996). The rhizosheaths of *A. californica* roots are especially conspicuous and seem best developed during times of high soil moisture.

Soon after, or sometimes even before, the floret blows away from the parent plant the column breaks off at the joint, or articulation, just above the body of the lemma, leaving the seed-bearing lower part free from the long awns and thus slowing its movement. The jointed awn column is a distinctive key character of section *Arthratherum,* a group characteristic of sandy soils in widely disjunct regions around the world. There are no other species of *Arthratherum* in western North America.

E of San Luis, *F 16712*. Sierra del Rosario, *F 20755*. Moon Crater, *F 19028*. MacDougal Crater, *F 10503*. 24 mi SW of Sonoyta, *Wiggins 8355*. Bahía de la Cholla, *F 16835*.

†Aristida parishii Hitchcock [*A. purpurea* var. *parishii* (Hitchcock) Allred. *A. wrightii* Nash var. *parishii* Gould] PARISH THREE-AWN. Densely tufted perennials often 40–70 cm, with well-developed fibrous roots. Ligules with a sparse tuft of loosely tangled hairs 2–4 mm, the leaf blades narrow, firm, and usually inrolled, the upper surface with short, stiff, somewhat scabrous hairs. Young inflorescences and spikelets dark purple-brown during winter and spring. Panicles contracted to moderately spreading, overtopping the leaves. Glumes mostly (or at least some on each panicle) equal or subequal, the upper glume 11.0–13.5 mm. Lemma 9.0–13.0 mm, gradually tapering to the awn bases, sometimes narrowed into a neck and slightly twisted; awns equal or nearly so, 16–27 mm.

Aguajita Wash, rare, mostly among boulders, within 100 m of the international fence (*F 92-102*), and in similar habitats in Cabeza Prieta Refuge (Papago Well, *F 92-10*); usually solitary or in small, isolated colonies. These seem to be pioneer plants and may have become established from propagules carried downstream along major drainageways. Southern California, Baja California, southern Nevada, and Arizona. It has not been reported for Sonora.

Aristida parishii can be distinguished from *A. purpurea* var. *nealleyi* by (1) stouter and more robust plants, (2) thicker stems and inflorescences, (3) often broader, more open inflorescences with some (lower) spreading branches, (4) more densely flowered panicles, (5) darker-colored spikelets, (6) glumes on even the same plant equal to unequal, (7) the lemma apex not twisted or only weakly twisted, and lacking a well-defined neck (or beak), and (8) stouter, broader awns (see Allred 1984; Allred & Valdés-Reyna 1997). In southern Arizona *A. parishii* is often sympatric with *A. purpurea* var. *nealleyi.* Plants in the field are distinct and I have not seen any with intermediate characters.

Aristida purpurea Nuttall var. **nealleyi** (Vasey) Allred [*A. stricta* Michaux var. *nealleyi* Vasey. *A. glauca* (Nees) Walpers. *Chaetaria glauca* Nees. *Aristida purpurea* var. *glauca* (Nees) A. Holmgren & N. Holmgren. *A. wrightii* Nash] *TRES BARBAS;* PURPLE THREE-AWN. Dense, tufted perennials often 30–60 cm, with well-developed fibrous roots. Ligules and leaf blades with hairs like those of *A. parishii,* the blades narrow, firm, and inrolled. Panicles contracted (at first glance sometimes appearing unbranched), overtopping the leaves. Upper glume (8.5) 10.0–14.5 mm, usually at least 1/3 longer than the lower glume. Lemma 8.0–13.5 mm. Awns nearly equal, 2–3 cm; awn column often twisted (but not always conspicuously so), 1–2+ mm, lighter in color than the lemma body and slightly narrowed to form the neck; drought-stressed plants tend to have shorter columns.

At all elevations in the Pinacate volcanic field, granitic ranges, and the Sonoyta region. Widespread including extremely arid lowland habitats, but seldom very common; slopes, bajadas, and arroyos, mostly on rocky soils, and absent from open desert flatlands and sand flats. Often best-developed toward higher elevations in steep rocky canyons and on north-facing slopes. Southward in western Sonora to the Sierra del Viejo near Caborca, and eastward across the northern part of the state.

Variety *nealleyi,* the most wide-ranging taxon in the Purpureae complex, extends from the state of Puebla, Mexico, to southwestern United States. *A. purpurea* including its varieties ranges from Mexico to western Canada and Arkansas. Other members of the Purpureae complex occur in both North and South America. Their nomenclature presents an unbelievably tangled taxonomic trail (see Allred 1984). Some authors (e.g., Arnow 1987) do not recognize any varieties in this species.

Cerro Pinto, *F 85-1020.* Sierra del Rosario, *F 20671.* Arroyo Tule, *Burgess 6291.* Campo Rojo, *F 19756.* Pinacate Peak, *F 87-50.* Bahía de la Cholla, *F 16833.* Sierra de los Tanques, *F 89-26.*

Aristida ternipes Cavanilles. Coarse, tufted perennials 0.5–1 m, also flowering in the first season. Roots tough and wiry. Leaf blades flat to inrolled in drought, the upper surface glabrous or with short, somewhat rough hairs; ligules glabrous or with a sparse tuft of loose hairs. Panicles openly branched, the branches spreading to approximately 90°. Glumes subequal (spikelets at first often showing only one glume, the lower glume develops with a little more age). Awns 1–3 in number, 11–20 mm (when 3 awns present, one usually longer than the others). Awn column not twisted (or only slightly so), short (1 mm or less), with no discernable neck.

Southwestern United States to northern South America. This species barely enters the flora area; both varieties are plentiful at slightly higher elevations to the north and east. The varieties are distinguishable by the presence or absence of lateral awns (Allred 1994; Trent & Allred 1990). This easily discerned key character is variable. The spikelets of an individual plant, and usually those of each population, tend to be similar in pattern and size of homologous awns. In the disturbed, roadside habitat west of Sonoyta I found 3-awned plants during the relatively favorable (wet) summer-fall season of 1985, but during the very dry summer-fall season of 1988 I found only 1-awned plants in essentially the same place.

Aristida ternipes var. **gentilis** (Henrard) K.W. Allred [*A. gentilis* Henrard. *A. ternipes* var. *hamulosa* (Henrard) J.S. Trent. *A. hamulosa* Henrard] *ZACATE ARAÑA DE TRES BARBAS;* POVERTY THREE-AWN. Upper glume 12.0–14.5 mm; lemma 10–12 mm; awns 3, well developed, 12–20 mm.

Occasional in sandy soil in agricultural areas in the Sonoyta Valley (5 km W of Sonoyta, *F 85-967*). Also a small population in natural vegetation in the northwestern part of the Pinacate volcanic field, in sandy silt among mesquites and blue palo verdes in a broad wash within a lava flow (6 km W of Los Vidrios, *F 92-964*). Also nearby Cabeza Prieta Refuge but highly localized. Western Texas to southern California and southward to Honduras.

Aristida ternipes var. **ternipes.** *ZACATE ARAÑA;* SPIDER GRASS. Upper glume 10–15 mm; lemma 13–19 mm, often moderately curved, with one well-developed awn, straight or sometimes curved, 11–14 mm.

Well established among natural vegetation in the Sierra Cipriano on rocky slopes and upper bajadas (*F 91-27*), and occasional near Sonoyta in sandy roadsides (2.4 km W of Sonoyta, *F 88-390*). Widespread on rocky slopes above 600 m in Organ Pipe Monument (Bowers 1980) and the southeastern part of Cabeza Prieta Refuge. Arizona to Texas and southward to northern South America, also the West Indies.

AVENA *AVENA,* OATS

Annuals. Panicles usually with relatively large spikelets, the spikelets with 2 or 3 (6) florets, breaking apart above the glumes and between the florets. Glumes thin and usually longer than the florets. Lemma usually with a bent, twisted awn.

Old World, mostly native to temperate Eurasia and North Africa; 25 species. A number of the species are naturalized in temperate regions elsewhere in the world and are especially common along the Pacific coast of North America. Domesticated oats, *A. sativa,* are sometimes grown as a winter-spring forage crop in Sonora. Reference: Baum 1977.

***Avena fatua** Linnaeus. *AVENA LOCA;* WILD OATS; *'AGṢPI MUḌATKAM, KOKṢAM.* Winter-spring ephemerals, erect, often 50–100 cm. Panicles open and sparse. Spikelets usually on long, slender, almost threadlike and often curled pedicels, mostly 2- or 3-flowered; glumes 2–3 cm; lemmas with red-brown hairs, the awn 3.5–4.5 cm, stout, strongly twisted below, bent at about the middle.

Common agricultural weeds, especially in wheat fields in the San Luis and Sonoyta regions. Infrequent at roadsides, especially in the vicinity of Sonoyta and San Luis. Native of Eurasia; a common weed in the Pacific states of North America.

6 km S of San Luis, *F 91-70.* Sonoyta, *F 85-702.* 5 km W of Sonoyta, *F 86-152.* 7 km NE of Estero Morúa, *F 91-50A.*

BOTHRIOCHLOA

Perennials. Inflorescences of terminal panicles, the branches spikelike or sparingly rebranched. Pedicels and upper rachis joints with a central groove or membranaceous area appearing as a thin, translucent or "naked" area. Spikelets in pairs: 1 sessile and fertile, and 1 pedicelled and sterile or staminate (terminal spikelets in 3s, 1 fertile and 2 sterile); each spikelet pair falling as a unit together with its rachis segment. Glumes of both spikelets large and firm (indurate). Lemma of fertile floret bearing a stout twisted or straight awn, the body of the lemma thin, membranous, and greatly reduced; palea of fertile floret small or absent. Sterile (pedicelled) spikelet awnless, usually reduced, but sometimes about as large as the fertile spikelet.

Primarily Old World tropics with a minority of the species native to the Americas; few in deserts; 35 species. References: de Wet 1968; Gould 1953, 1957; McVaugh 1983.

Bothriochloa barbinodis (Lagasca) Herter [*Andropogon barbinodis* Lagasca. *A. perforatus* Trinius ex E. Fournier. *A. palmeri* (Hackel) Gould. *Bothriochloa barbinodis* var. *palmeri* (Hackel) de Wet. *B. barbinodis* var. *perforata* (Trinius ex E. Fournier) Gould. *B. barbinodis* var. *schlumbergeri* (E. Fournier) de Wet] *ZACATE POPOTILLO;* CANE BLUESTEM. Robust tufted perennials often 0.75–1 m, usually villous with dense tufts of long, white hairs at nodes, at ligules, and on inflorescences. Leaves drying red-brown, at least the bases semipersistent; leaf blades flat. Panicles white and cottony, 7–11 cm, with numerous branches clustered at the top of tall, nearly naked stems. Rachis joints and pedicels with hairs reaching 6–8 mm. Glumes of the sessile (fertile) spikelets about equal in size, 5.0–5.5 mm, but differently shaped: lower glume broad, green, and flat to concave on the back; upper glume markedly hump-backed or V-shaped with a blunt keel (ridge). Fertile lemma reduced with a stout, twisted, bent awn 1.4–2.2 cm. Sterile (pedicelled) spikelet reduced to a deciduous linear rudiment 3.5–4.0 (5) mm.

North-facing cliffs and steep slopes of Pinacate Peak and adjacent lava hills and rock arroyos above 800 m (*F 92-86, F 86-445*). Apparently absent from the hot lowlands of the Sonoran Desert, at least in western Sonora. However, it does occur on the relatively equable north-facing grassy slopes on the east side of Isla San Pedro Nolasco, northwest of Guaymas, and elsewhere in Sonora at elevations above the desert. Southwestern United States to southern Mexico, and in South America.

Although this wide-ranging species is variable, it does not seem realistic to recognize infraspecific taxa (McVaugh 1983). It is interesting to note that chromosome counts for this species are about as high as those recorded for any other grass ($2n$ = 180 or perhaps rarely $2n$ = 220, de Wet 1968). *Bothriochloa barbinodis* is closely related to *B. saccharoides* (Swartz) Rydberg (or *B. laguroides* (de Candolle) Herter, sensu Allred & Gould 1983) with which it is sometimes confused. The *perforatus* taxon is based on the presence of a gland, or pit, on the glume of the sessile spikelet; this character seems to have little taxonomic value. The Pinacate specimens do not have these glands.

BOUTELOUA *GRAMA,* GRAMA GRASS

Annuals and low tufted or sod-forming perennials, sometimes stoloniferous. Stems slender with 1-sided spicate branches (hereafter called spikes) arranged in a racemose pattern or 1 per stem. Ligule usually a ciliate ring of straight hairs. Spikelets relatively small, in 2 rows along margins of a flattened rachis, with 1 fertile floret below and often 1–3 or more reduced, sterile florets (rudiments) above. Glumes unequal. Lemma of fertile floret with 3 veins, awnless or the veins often extending into short awns, and the palea with 2 veins, awnless or the veins often extending into short awns. Basic chromosome number: x = 10. (Counts purporting x = 7 are erroneous.)

North, Central, and South America; 57 species, including 51 in Mexico and 16 in the Sonoran Desert. Grama grasses are especially abundant in semiarid regions and grassland plains and prairies, where they provide important forage. The center of distribution is in central Mexico.

The 1-sided spikes are easily recognized; even *B. aristidoides* with its small spikes has a flattened rachis. The first (lowermost) sterile floret, often with a short stalk (stipe), has a short, slender column and 3 terminal awns (these awns tend to be larger than those of the fertile floret); although the ovary and stamens are absent from the sterile florets, a reduced lemma and palea are often present. The rudiments seem to serve as dissemination devices; the 3-awned sterile floret somewhat resembles a miniature *Aristida* floret. Additional sterile florets, if present, are greatly reduced. In some species (e.g., *B. barbata*) the reduced lemma of the first and second rudiments look like tiny petals at the base of the awns, but careful examination (at high power of a dissecting microscope) usually reveals the minute pedicel of the second rudiment. References: Columbus 1999; Columbus et al. 1998; Gould 1979.

1. Perennials with hard, knotty bases consisting of short rhizomes with very short internodes; basal (lower) leaves densely clumped, nodes between them not readily visible; Sonoyta region and rare elsewhere. ______ **B. trifida**

1′ Ephemerals; basal leaves sparse, nodes between them readily visible; common and widespread.
  2. Spikes somewhat dart-shaped, with 1–4 slender spikelets closely appressed to the spike axis (spikelets scarcely spreading). ______ **B. aristidoides**
  2′ Spikes comb-shaped, with ca. 20–50 spikelets crowded in a double row, the spikelets at right angles to the spike axis. ______ **B. barbata**

Bouteloua aristidoides (Kunth) Grisebach [*Dinebra aristidoides* Kunth. *B. aristidoides* var. *arizonica* M.E. Jones] *ACEITILLA;* SIX-WEEKS NEEDLE GRAMA. Summer-fall ephemerals. (Sometimes a few plants overwinter, resume growth, and flower in spring, or rarely they germinate and grow in the spring but these plants are small and scrawny.) Highly variable in size, fast growing; roots weakly developed. Stems delicate, slender, erect to spreading-ascending, 10–30 (70) cm. Spikes 8–22 mm, 11–16 (24) per stem, slender and readily falling at maturity together with their several spikelets, or spikelets falling separately. (These spikes or spikelets can be bothersome, sticking in socks and shoes. People may be supplanting animals as dispersal agents.) Spikelets (1 or 2) 3 (4–10) per spike, appressed to the main axis; rachis flattened, ciliate, curving out from the attachment of the terminal spikelet, which it about equals or exceeds in length. Lowermost spikelet almost sessile, usually 1-flowered, awnless, and without a rudiment. Upper spikelets short-pedicelled, bearing a sessile, awnless, fertile floret and a sterile floret with a short stipe and 3 conspicuous *Aristida*-like awns 2.5–6.0 mm.

Sometimes seasonally abundant on a wide range of soils, but most numerous on sandy-gravelly soils. Throughout most of the flora area; hills and mountains, bajadas, desert flats and plains, washes and floodplains, low dunes, and interdune troughs. Also on cinder soils to the summit of Pinacate Peak. One of the most widespread and abundant hot-weather ephemerals in the Sonoran Desert. Southwestern United States to Oaxaca and Argentina, Bolivia, and Peru. The great success of this grass in the desert seems to be partly due to plasticity in size and number of structures and partly to rapid growth and reproduction. This unique grass cannot be confused with any other in the Sonoran Desert.

24 km W of Los Vidrios, *F 85-1011*. Sierra del Rosario, *F 20721*. Pinacate Junction, *F 86-356*. Arroyo Tule, *F 18770*. Molina Crater, *F 10651*. Suvuk, *Nabhan 386*.

Bouteloua barbata Lagasca [*B. sonorae* Griffiths]. *ZACATE LIEBRERO, NAVAJITA;* SIX-WEEKS GRAMA; *CUK MUDATKAM*. Summer-fall ephemerals (rarely growing with spring rains and then stunted). Roots weakly developed. Usually branched from near the base, the stems 5–50 (80+) cm, geniculate-spreading (abruptly bent upward from near the base of the plant) or even prostrate to erect and straight. Spikes (1) 4–12 per stem, (6) 10–25 (30) mm, comb-shaped, nearly straight to moderately arched. Spikelets often 20–54 per spike, 2.2–3.5 (4.5) mm (including awns), crowded, more or less in 2 rows and at right angles to axis of spike. Basal floret in the rudiment with 3 short awns on a column. Upper florets of the rudiment greatly reduced and consisting merely of tiny petal-like structures on top of awn bases of the lower sterile floret.

Seasonally abundant and widespread throughout the flora area except on the higher dunes and apparently not at higher elevations in the mountains. Often growing with *B. aristidoides*. Common and widespread throughout the Sonoran Desert. Southwestern United States to Oaxaca and in Argentina.

Bouteloua rothrockii Vasey differs from *B. barbata* in its perennial habit, hard knotty bases, and well-developed roots; it occurs at elevations above the desert. Gould (1979) considered them varieties of the same species, making the combination *B. barbata* var. *rothrockii* (Vasey) Gould. Gould & Moran (1981:90) reported var. *rothrockii* south of the desert in the Cape Region of Baja California Sur. However, specimens from the Cape Region (including those at SD and therefore presumably the ones seen by Gould) are not perennial. I have not found the perennial *B. rothrockii* within the Sonoran Desert, although it sometimes extends to the margins of the Sonoran Desert near Tucson.

Vigorous, erect-growing, straight-stemmed, and crowded stands of summer-fall ephemeral plants are often common during favorable times in the Sonoran Desert including the flora area. Some botanists have considered such plants as "annual forms" of *B. rothrockii,* largely because of the robust, erect stems, straight rather than curved spikes, and relatively greater number of spikes. I am unable to segregate these satisfactorily from "obvious" *B. barbata* with geniculate-spreading stems, fewer spikes per stem, slightly curved spikes, and generally smaller plants. These so-called "annual forms of *B. rothrockii*" are merely vigorous plants of *B. barbata*. The taller, more vigorous "annual forms," as well as intermediates and the smaller *barbata* plants, often occur together. Griffiths (1912:383) neatly summed up what seems to be going on: "Plants exceedingly variable in both size and general aspect, erect when growing thickly, but prostrate when scattered."

N of Sierra del Rosario, *F 20765*. 24 km W of Los Vidrios, *F 85-1010*. Moon Crater, *F 19233*. Sykes Crater, *F 20032*. Cerro Colorado, *F 10782*. Suvuk, *Nabhan 382*.

Bouteloua trifida Thurber. *NAVAJITA CHINA;* RED GRAMA. Small tufted perennials, mostly erect, (8) 15–40 cm, with hard knotty bases and sometimes with very short rhizomes. Leaves basal, with few along flowering stems; leaf blades flat, mostly 3–7 cm × 1.0–1.5 mm. Flowering stems slender; spikes often 2 or 3 per stem, often 2.0–3.5 cm, pectinate, bristly, purplish, each with (8) 16–23 (27+) spikelets. Glumes 2.9–4.7 mm, thin, keeled, and with membranous margins. Fertile floret and first rudiment about equal in size, each with 3 awns of 5–8 mm; additional sterile florets very reduced. Flowering response non-seasonal.

Extending into the flora area in Arizona Upland southwest of Sonoyta in small, scattered populations on arroyo banks in the granitic upper bajadas of Sierra Cipriano (*F 88-148*). Also isolated in Arroyo Tule in the southwestern part of the Pinacate region—about a dozen plants in crevices of water-eroded gray lava bedrock along the canyon bottom (*F 19190*). The southernmost record in western Sonora is from Sierra del Viejo near Caborca. Southern Utah to southern California and Texas, and southward to Baja California and drier regions of northern Mexico including northern Sonora, Chihuahua, Coahuila, Nuevo León, San Luis Potosí, and Tamaulipas.

This species is unusual among grama grasses in that it has adapted to winter as well as summer rains, at least in the western part of the Sonoran Desert region, where it seems to grow and flower at almost any time of year.

BRACHIARIA

Annuals and perennials; spikelets panicoid. Distinguished from *Panicum* by technical characters of the inflorescence and spikelet. Tropics and subtropics, mainly Old World; 100 species. The delimitation of this genus and relationships with related genera remain unresolved (Clayton & Renvoize 1986; Yatskievych 1999). Some authors include only a single species in *Brachiaria* and place most or all of the remaining species including *B. arizonica* in *Urochloa*.

†**Brachiaria arizonica** (Scribner & Merrill) S.T. Blake [*Panicum arizonicum* Scribner & Merrill. *Urochloa arizonica* (Scribner & Merrill) R.D. Webster] Hot-weather ephemerals, mostly slender, light green and soft, with rather weakly developed roots. Mostly hispid or pilose with sparse hairs at nodes. Spikelets ovoid, 3.4–3.7 mm, pale green or reddish with green cross veins toward the tip. Lower glume 1.3–1.9 mm.

Along the Sonora border in Cabeza Prieta Refuge and Organ Pipe Monument east of Quitobaquito, and expected on the Sonora side of the international fence; widespread elsewhere in Sonora. Texas to southern California and southward to Oaxaca.

Sonoyta Hills, sandy flats along arroyo [probably Gray's Ranch headquarters], 27 Aug 1943, *Clark 10891* (ORPI). Cabeza Prieta Refuge, Pinta Sands, *F 92-759*.

BROMUS *BROMO*, CHESS, BROME

Annuals, biennials, or perennials, the longevity often facultative; winter-spring ephemerals in the Sonoran Desert. Inflorescences of open or dense panicles. Spikelets large, several-flowered, laterally compressed or turgid and only slightly compressed, breaking apart above glumes and between florets; often cleistogamous. Glumes persistent, shorter than the lowermost lemma. Lemmas usually 1-awned from a notch between the bilobed apex or sometimes awnless. Ovary capped by a hairy-lobed appendage bearing subterminal stigmas.

Worldwide, mostly temperate; 150 species, including 16 in Mexico. In Mexico and Central America mostly in mountain habitats and seldom encountered below 650 m except on the Pacific slope of Baja California. *Bromus* is relatively scarce and not native in western Sonora south of the flora area. No *Bromus* survives the Sonoran Desert summer. In terms of number of species, *Bromus* is the largest genus of grass in the flora, but its contribution to the vegetation is insignificant except at higher elevations in the Sierra Pinacate. References: Matthei 1986; Pavlick 1995; Pillay & Hilu 1995; Pinto-Escobar 1986; Soderstrom & Beaman 1968; Stebbins 1981.

1. Spikelets conspicuously flattened; lemmas strongly keeled, V-shaped in cross section.
   2. Lemmas 7-veined, with awns 6–8 (11) mm. ______ **B. carinatus**
   2′ Lemmas 9–13 (or more)-veined (veins sometimes difficult to see on dried specimens; it helps to soak the lemma in water and backlight it), awnless or with awns less than 3 mm. ______ **B. catharticus**

1′ Spikelets moderately flattened; lemmas not strongly keeled, U-shaped in cross section.
   3. Awn of lemmas bent, the lower part strongly twisted; higher elevations on Sierra Pinacate. ______ **B. berterianus**
   3′ Awn straight or sometimes curved, not bent or twisted; Sierra Pinacate and elsewhere.
      4. Panicle branches and spikelets mostly erect; panicles very dense. ______ **B. rubens**
      4′ Panicle branches and pedicels conspicuously curving and drooping, the spikelets mostly not erect; panicles open. ______ **B. tectorum**

Bromus berterianus Colla, 1836 [*Bromus trinii* E. Desvaux, 1853. *Trisetum hirtum* Trinius, 1836. Not *Bromus hirtus* Lichtenstein, 1817. *Trisetobromus hirtus* (Trinius) Nevski, 1934] CHILEAN CHESS. Plants (8.5) 12–40 cm (elsewhere sometimes to 1 m). Stems 1 to several, erect, the nodes densely but often minutely pilose (with soft, slender white hairs). Leaves (sheaths and blades) pilose (in other regions this species varies from pilose to nearly smooth), the blades flat, 5–7 mm wide. Panicles often branched, contracted, and the branches erect and appressed, or the branches spreading, especially on larger plants, and with age often having many spikelets; inflorescence of smaller plants sometimes spikelike with as few as 5 spikelets. Spikelets 11–25 (27) mm (excluding awns), 2–5-flowered. Glumes conspicuously veined, the upper one with 3 veins, the lower with 1 vein; upper glume (7.5) 8–10 mm, longer and broader than the lower glume and about as long as the first lemma. Veins green, the glumes and lemmas otherwise translucent and lustrous (lacking chlorophyll). Lemmas 8.5–10.5 mm (not including awn), pilose and moderately hairy, the veins 3 or 5 but not as easy to see as on the glumes, the terminal bifid lobes slender; awn 6.5–12.5 mm, usually conspicuously twisted and bent outward at about the middle, the bend more pronounced with age.

Seasonally common in the Sierra Pinacate above 1000 m (*F 19449, F 87-52; Ezcurra 5 Oct 1981*). Often on partially barren, open cinder slopes or, especially in moderately dry years, growing through small spiny shrubs. It seems that *B. rubens* is displacing *B. berterianus*. Not known elsewhere in Sonora.

Southern Oregon to Baja California and Nevada, southern Utah, western Colorado, and Arizona but not in regions adjacent to northwestern Sonora. South America in Argentina, Bolivia, Chile, Ecuador, and Peru. In both North and South America it is a winter-spring ephemeral (annual) in regions with winter-spring rains; it ranges through many plant communities including deserts.

Bromus berterianus is polymorphic. Desvaux (1853) described 5 varieties of *B. trinii* from Chile but recent authors have not dealt with these varieties. Shear (1900) recognized 2 varieties in addition to the nominate variety for California: *B. trinii* var. *pallidiflorus* E. Desvaux and var. *excelsus* Shear (*B. berterianus* var. *excelsus* (Shear) Pavlick, 1995). Shear distinguished var. *pallidiflorus* as having more robust and larger plants and much-elongated panicles. Plants with larger spikelets, 7-veined lemmas, and bent but not twisted awns were described as var. *excelsus*. This variety is known from the Panamint Mountains in the desert in California and near Lake Mead in Arizona. Munz & Keck (1963), Munz (1974), and Holmgren & Holmgren (1977) reduced var. *excelsus* to synonymy but offered no reasons for doing so. Hitchcock (1951:838) recognized var. *excelsus* but reduced var. *pallidiflorus* to synonymy, again without comment. The opinion given by Hitchcock (probably Agnes Chase's) seems reasonable. The majority of the North American populations, including the Pinacate plants, presumably are the nominate variety (var. *berterianus*). Chromosome counts of $2n = 42$ have been obtained in both North and South America (Bowden & Senn 1962; Knowles 1944; Nicora 1978).

A number of authors, including Hitchcock (1951), report this species as native to Chile and introduced in North America. Gould & Moran (1981:28) introduce doubt by saying it is "sometimes considered as introduced in North America." Several aspects hint at its being native in North America. Nineteenth-century North American records for this species are surprisingly numerous and widespread; Soderstrom & Beaman (1968) cite 10 nineteenth-century collections from Baja California including one in 1875, and there are a number of early twentieth-century records from Arizona. Because var. *excelsus* is geographically isolated in remote, non-coastal areas of North America, introduction from South America seems unlikely. The Pinacate population, isolated on a mountaintop, seems more like a Pleistocene relict than a historic, non-native introduction. The other dozen or so Sierra Pinacate "sky island" plants are all native.

Bromus carinatus Hooker & Arnott [*B. carinatus* var. *arizonicus* Shear. *B. arizonicus* (Shear) Stebbins. *B. marginatus* Nees. *B. polyanthus* Scribner] CALIFORNIA BROME. Plants (20) 50–70 (100) cm. Spikelets conspicuously flattened when young, less so as the grain matures (not as flat as in *B. catharticus*). Lemmas ca. 10 mm, strongly keeled, V-shaped in cross section, 7-veined, with a stout awn often 6–8 (11) mm.

Roadsides and washes in the Sonoyta region; sometimes locally common during El Niño years; also Organ Pipe Monument and near the Mexican border in Cabeza Prieta Refuge. Not known south of the flora area in western Sonora, but occurring eastward in northern Sonora. Alaska to Central America including Baja California.

Bromus carinatus is facultatively cleistogamous and part of a highly variable polymorphic complex (Arnow 1987). It is commonly perennial in nondesert, moderately moist habitats. Related and perhaps conspecific taxa occur in South America.

6–12 km SW of Sonoyta *F 91-56, F 92-417.* Quitobaquito, *F 7676.*

***Bromus catharticus** Vahl var. **catharticus** [*B. unioloides* Kunth. *B. willdenowii* Kunth] RESCUE GRASS. Stems to 50 cm. Spikelets strongly compressed laterally. Lemmas conspicuously keeled, V-shaped in cross section, awnless or with an awn less than 3 mm. Veins of upper glume and lemmas 9–13 (or more), usually prominent on mature spikelets (sometimes difficult to see on immature, dried specimens).

Mostly in disturbed habitats, usually sandy or fine-textured soils; urban weeds and sometimes in playas under heavy cattle grazing or in washes. Much of the United States and Mexico; widespread in South America, where it is native. Often weedy and also cultivated as a forage grass. In nondesert regions it reaches 1+ m in height and can be biennial or a short-lived perennial. Another variety occurs in South America.

Sonoyta, *F 87-14.* 4 km S of Sonoyta, *F 92-392.* Playa los Vidrios, *Ezcurra 7 Oct 1981.* Puerto Peñasco, *F 85-771.*

***Bromus rubens** Linnaeus [*B. madritensis* subsp. *rubens* (Linnaeus) Husnot] *BROMO ROJO;* RED BROME, FOXTAIL BROME. Plants often (7) 10–25+ cm; hairy especially on leaf sheaths and stems just below the inflorescence, the hairs often soft and backward-pointing on leaf sheaths. Panicles densely contracted, brushlike, the branches and spikes mostly upright; spikelets more or less sessile, not flattened. Lemmas with slender, bifid terminal lobes, the awn 12–26 mm, stiff, stout, straight or slightly curved but neither bent nor twisted.

Seasonally abundant and widespread in natural habitats on cinder soils at higher elevations in the Sierra Pinacate, and at low elevations mostly in disturbed habitats, especially in urban and agricultural areas. This species is spreading in the flora area as well as elsewhere in Sonora and Arizona. I have seen dramatic increases in *B. rubens* at higher elevations in the Sierra Pinacate. I did not see it in northwestern Sonora during my fieldwork in the 1970s. By the early 1980s it had become established at Puerto Peñasco and was common around Sonoyta, and at least by 1985 it was at the summit of Pinacate Peak and elsewhere in the Pinacate region. In the higher elevations of the Sierra Pinacate it seems to be outcompeting the more delicate *B. berterianus.* Likewise, it has become very common eastward across the northern part of Sonora and in southern Arizona including Organ Pipe Monument, especially since the late 1970s (Bowers 1980; Felger 1990). It has been present in the town of Ajo since at least 1916. This weedy Mediterranean annual, adventive and common through much of the western United States, was established in California by 1848 (Burgess, Bowers, & Turner 1991; Frenkel 1977).

El Papalote, *F 88-25.* Crater Salvatierra, *Ezcurra 3 Apr 1982.* Sierra Pinacate: 960 m, *F 87-44;* summit, *F 87-53;* near summit, 24 Apr 1985, *Sanders 5671* (UCR).

***Bromus tectorum** Linnaeus. DOWNY CHESS, CHEAT GRASS. Panicles branched, the branchlets and pedicels slender, curved, and drooping under the weight of large red-brown spikelets. Lemmas hairy, 10–12 mm, with a straight awn 12–14 mm.

In the flora area known from a single record at El Papalote (*F 86-133*). Occasionally encountered elsewhere in northern Sonora and southern Arizona. Native to Europe and extensively naturalized in temperate North America.

CENCHRUS SANDBUR

Annuals and some perennials. Inflorescence a highly modified panicle, usually greatly contracted and spikelike. Spikelets enclosed in a spiny bur; larger spines on ours flattened at least basally. Spikelets 2-flowered, the lower floret staminate or sterile, the upper one bisexual.

The bur, the most striking feature of the genus, is formed of branchlets that have evolved into basally fused bristles and/or flattened spines. Although the bur is unique, the enclosed spikelets are typical of panicoid grasses. The seeds germinate within the bur. The young spines of ours have backward-facing barbs that are somewhat tubelike and contain a light purple toxic substance, apparently missing in mature spines, which intensifies the pain of puncture wounds. This toxin may serve to protect the developing bur while it is relatively palatable.

Warmer regions worldwide, mostly New World, often sandy soils in disturbed habitats; 20 species. Several species included in *Cenchrus* by DeLisle are placed in *Pennisetum* by Pohl (1980) and others. Among *Cenchrus* sensu stricto $2n = 34$ is usual. References: Chase 1920; DeLisle 1963; Sohns 1955.

1. Inflorescence with 1–3 burs; spines of bur (6) 9–15 mm; widespread including natural habitats. ____ **C. palmeri**

1′ Inflorescence with more than 10 burs; larger spines of bur 3.5–6.0 mm; mostly in disturbed habitats.

  2. Basal bristles slender, many in a ring, the larger flattened spines of different sizes and more or less in a single whorl. ____ **C. echinatus**

  2′ Basal bristles none to several, not forming a ring, the larger flattened spines mostly of the same size and at different positions on the bur. ____ **C. incertus**

***Cenchrus echinatus** Linnaeus. *ZACATE TOBOSO,* SOUTHERN SANDBUR. Warm-weather annuals. Inflorescences spikelike to 7.5 cm, with 18–40 burs. Burs often 4–5 mm in diameter (not including spines), with a basal ring of often 30–50 small, slender bristles; larger spines flattened, 3.5–6.0 mm.

Weedy habitats near Sonoyta and Puerto Peñasco, and expected elsewhere, especially in agricultural areas; mostly sandy or fine-textured soils. A troublesome agricultural weed near Yuma. Often weedy in much of the New World; adventive in the Old World.

This species is closely related to *C. brownii* Roemer & Schultes. Although recognized as distinct species by most agrostologists, these taxa are not readily distinguished by the novice. *C. brownii* is worldwide in warm coastal regions; in western North America it extends northward to the vicinity of Guaymas and Baja California Sur.

Puerto Peñasco, *F 85-765*. 8 km W of Sonoyta, *F 87-311*. Yuma, *Thornber 24 Sep 1912*.

***Cenchrus incertus** M.A. Curtis [*C. pauciflorus* Bentham] *GUACHAPORI, HUIZAPORI;* SANDBUR. Perennials or facultative annuals, with deep, well-developed roots. Inflorescences compact, spikelike, 3.5–7 cm, with 12–27 burs. Burs variable, 2.5–4.0 mm in diameter, shorter to longer than wide; spines highly variable, long and slender to short and broad, the larger spines 4.0–5.7 mm, the lower, smaller spines several and not in a basal ring or absent.

Well established in sandy washes in creosote bush flats north of Rancho Guadalupe Victoria in an area heavily grazed by cattle (*F 92-963*), and in similar but non-grazed habitat in the nearby Pinta Sands in Cabeza Prieta Refuge (6 Mar 1977, *Reeder 6836*). The burs are probably spread by domestic livestock. Southern United States to Panama, West Indies, and South America.

Cenchrus palmeri Vasey. *GUACHAPORI, HUIZAPORI;* GIANT SANDBUR. Non-seasonal ephemerals; at least August-May, depending upon soil moisture and winter freezing; highly variable in size, often (5)

8–30 cm. Roots unusually small. Burs 1–3 per inflorescence, 20–31 mm in diameter, with sharp stiff spines (6) 9–15 mm, dark purple-brown most of the year, green during hot, humid weather.

Lowlands through most of the flora area, but not in the Sonoyta region and not known from the U.S. side of the border. Mostly in sandy soils of upper beaches and beach dunes, sand flats, washes, crater floors, interdune troughs, lower dunes, and silty-clay flats and playas, also sometimes on rocky slopes. Coastal Sonora to northwestern Sinaloa, both Baja California states, and islands in the Gulf of California. This is one the few grasses endemic to the Sonoran Desert region (also see *Distichlis palmeri*).

No other *Cenchrus* species has so small a geographic range, such large burs, nor so few burs per inflorescence. According to DeLisle (1963), *C. palmeri* is one of the most advanced species in the genus. As with other *Cenchrus,* seedlings germinate within the bur, which may remain attached to the root of the mature plant.

Sierra del Rosario, *F 20802.* Hornaday Mts, *Burgess 6811.* Rancho Solito, *Ezcurra 20 Apr 1981* (MEXU). Moon Crater, *F 10585.* Puerto Peñasco, *Burgess 676.* 25 mi N of Puerto Peñasco, *F 13199.*

CHLORIS FINGERGRASS

Annuals or perennials. Inflorescence usually of 5–25 spikes or spikelike branches clustered at the top of the flowering stem, the spikes often digitate or whorled. Spikelets in 2 rows on one side of the rachis, breaking apart above the glumes, with 1 (2) bisexual floret(s) and usually 1 (2 or 3) sterile florets above. Glumes unequal. Fertile lemma usually awned from an extension of the midrib.

Worldwide, mostly tropical and subtropical, some in warm-temperate regions, few in deserts; 55 species. Reference: Anderson 1974.

1. Perennials; florets 3-awned. ____ **C. crinita**
1′ Annuals or ephemerals; florets 1-awned. ____ **C. virgata**

Chloris crinita Lagasca [*Trichloris crinita* (Lagasca) Parodi. *T. mendocina* (Philippe) Kurtz] Erect, tufted perennials, 1.0–1.2 m. Inflorescences with about one dozen erect, bristly, spicate branches each 9–15 cm. Spikelets crowded, with 1 fertile and 1 sterile floret (occasionally 2 in other regions), each with 3 slender awns often 1.5–2.0 cm; sterile floret cylindrical, similar to the fertile floret but much smaller.

A well-established population along the Arizona border west of Sonoyta includes several hundred plants on gravelly soil along a drainageway leading into a small alkaline-soil seep with *Distichlis spicata* (17 km W of Sonoyta, *F 86-407*). More common and widespread in the same habitat in adjacent Organ Pipe Monument. There are few records for this grass in western Sonora.

Arizona to western Texas and south to Durango, Coahuila, Sonora, and Baja California Sur, and disjunct in South America (Argentina, Bolivia, Chile, Paraguay, Peru, and Uruguay). This species is rather unusual in having 3 long awns on both the fertile and sterile florets.

***Chloris virgata** Swartz. *ZACATE LAGUNERO;* FEATHER FINGERGRASS. Warm-weather annuals or ephemerals, 10–80 cm. Stems 1 to many, often geniculate-spreading. Spikes 4–17 in number, digitate at stem apex, appearing feathery, densely flowered from base, upright, whitish to tawny with silky hairs. Spikelets 2.8–3.5 mm (excluding awns). Fertile and sterile florets differently shaped. Fertile lemma humpbacked on keel, bearing a conspicuous tuft of hair at apex and a single stout awn 5.0–7.5 mm. Sterile floret conspicuous, broad and truncate at apex, bearing a single stout awn.

Usually in sandy-silty soils in widely separated, mostly disturbed habitats including grassy playas and depressions with extensive cattle grazing; not at higher elevations and probably not native to the flora area. Western United States to South America; widespread, weedy, and adventive elsewhere including the Old World.

Sonoyta, *F 86-401.* Playa los Vidrios, *Equihua 5 Nov 1982.* Rancho Guadalupe Victoria, *Ezcurra 4 Nov 1982.* 2 mi S of Tinajas de los Pápagos, *F 86-488.*

Cynodon

Perennials with stolons and/or rhizomes and short leaves. Inflorescences of several slender spikes digitate or whorled at apex of stem. Spikelets 1-flowered, laterally compressed, in 2 rows on one side of a flattened and keeled (triangular) rachis. Lemmas laterally compressed and boat-shaped; rachilla sometimes prolonged behind the palea and bearing a tiny rudimentary floret. Native from Africa and Madagascar to India and northern Australia; 9 species. References: Clayton & Harlan 1970; de Wet & Harlan 1970.

***Cynodon dactylon** (Linnaeus) Persoon var. **dactylon** [*Panicum dactylon* Linnaeus] *ZACATE BERMUDA, ZACATE INGLÉS;* BERMUDA GRASS; *KI: WECO WAṢAI, 'A'AI HIHIMDAM WAṢAI.* Plants creeping with long stolons, scaly rhizomes, and obvious internodes, often forming extensive mats. Leaves mostly 2-ranked (distichous). Spikes 4–7, digitate, slender, often (2) 2.5–6.0 (8) cm, purplish to green. Spikelets (1.7) 2.0–2.5 mm, awnless, numerous, crowded. Growing and flowering during warmer months; winter dormant unless protected from freezing weather.

Mostly in disturbed habitats, an urban and agricultural weed, and in alkaline soils at widely scattered natural as well as disturbed wetland habitats; firmly established even at remote waterholes. Naturalized in warm regions nearly worldwide. Bermuda grass is extensively planted in warm climates for forage and as a lawn or turf grass. It is the most common lawn grass in the Sonoran Desert, where it is grown as a "summer" lawn; the leaves turn brown with the first frost. Probably native to Africa.

Harlan & de Wet (1969) recognize 6 varieties (also see Tzvelev 1983); var. *dactylon* is a worldwide weed. There are a number of cultivars of var. *dactylon* developed as turf grass and forage.

Quitobaquito, *F 7665.* Río Sonoyta at Sonoyta, *F 86-296.* Puerto Peñasco, *F 20952.* Tinajas de Emilia, *F 19745.* Pinacate Junction, *F 19599.* Tinaja de los Chivos, *F 18786.* Laguna Prieta, *F 85-741.*

Dactyloctenium

Annuals with flat leaf blades and short, thick, digitate spikes. Spikelets overlapping, with 3–5 densely crowded tiny florets, the naked rachis extending beyond the spikelets. Upper glume with an oblique awn just below the tip; lemmas strongly keeled, often with a short, recurved awn. The grain is unusual among the grasses in that the seed can be seen through the thin pericarp, which is soon shed. The seed is elegantly sculptured. Native to the Old World; 3 to 12 species depending on one's point of view.

***Dactyloctenium aegyptium** (Linnaeus) P. Beauvois [*Cynosurus aegypticus* Linnaeus] *ZACATE PATA DE CUERVO;* CROWFOOT GRASS. Glabrous summer-fall weeds. Stems spreading to ascending or decumbent and rooting at the nodes, often forming radiating mats. Spikes (1) 2–6, short and stubby, 1.5–4.5 cm, spreading at nearly right angles from stem apex. Spikelets 2.7–3.5 mm, 3–5-flowered. Glumes unequal in size and shape; lower glume awnless, the upper glume with an awn of 0.5–3.2 mm. Lemma of the first floret 2.2–3.2 mm, pointed or with awn usually less than 1 mm. Seed glistening red-brown with thin, evenly spaced ridges, these ridges often very dark brown or almost black and the sulci (minute valleys) lighter colored—reminiscent of the seed of *Mollugo cerviana* (see also *Eleusine*).

Agricultural weed and occasionally in riparian habitats in the Sonoyta Valley. Widely naturalized and weedy in the warmer parts of the world. Castetter & Bell (1951:167–168, 171–172) reported that the Cocopa living along the lower Río Colorado grew crowfoot grass under "semicultivated" conditions and harvested the grain for food. It was recorded in Yuma in 1881 and seed specimens were collected among the Cocopa in 1901 (Castetter & Bell 1951:171–172). There are many references to the use of the grain as food in hot, semiarid or tropical regions of the Old World (e.g., Tanaka 1976).

Ejido Morelia, 8 km W of Sonoyta, *F 87-307.* Quitobaquito, *F 87-289.*

Digitaria

Annuals or perennials. Inflorescences a pair or cluster of slender spikelike branches, these digitate or racemose on the main axis. Spikelets often in pairs or 3s, the pedicels often of different lengths, alternate in 2 rows on one side of a usually slender, flattened, 3-angled rachis. Lower glume greatly reduced or absent; upper glume much shorter than to as long as the spikelet. Spikelets with 2 florets, the lower floret sterile and reduced to a lemma. Sterile and fertile lemmas nearly equal in size, the fertile lemma usually acute and slightly indurate but with thin, flat margins (the margins not rolled inward as in *Panicum*).

Worldwide, tropical to warm temperate, with greatest diversity in the Old World; 220 species. One or two species of crabgrasses are expected in the flora area as agricultural or urban weeds. References: Henrard 1950; Webster & Hatch 1981.

1. Coarse tufted perennials; spikelets densely pubescent with silky hairs, appearing cottony; inflorescence branches racemose and contracted; natural habitats. ____ **D. californica**

1′ Warm-weather annuals (crabgrasses); spikelets not densely silky-haired, not cottony; inflorescence branches digitate or nearly so and spreading at least with age; probably in disturbed habitats.

  2. Lower lemma margins with silky hairs; upper glume about 3/4 as long as spikelet. ____ **D. ciliaris** (Retzius) Koeler (not included in species accounts)

  2′ Lower lemma margins and veins minutely scabrous; upper glume not more than 1/2 as long as spikelet. ____ **D. sanguinalis** (Linnaeus) Scopoli (not included in species accounts)

Digitaria californica (Bentham) Henrard var. **californica** [*Panicum californicum* Bentham. *Trichachne californica* (Bentham) Chase] *ZACATE PUNTA BLANCA,* CALIFORNIA COTTONTOP. Perennials with hard, knotty bases. Stems 50–90 cm, erect and with felt-like hairs. Blades of larger leaves 12–17 cm × 4.0–6.5 mm. Panicles 8–15 (17) cm, narrow, the branches alternate (racemose), appressed, and densely flowered (the spikelets usually obscuring the branches). One spikelet of the pair long-pedicelled, the other much shorter, or the second spikelet on portions of a branch sometimes not developing. Spikelets 3.0–3.7 mm (excluding the hairs), biconvex. Lower glume minute (0.5–0.8 mm), membranous, and glabrous, the upper glume slightly narrower and shorter than the sterile lemma. Upper glume and sterile lemma each with 3 major veins in addition to the marginal ones, and with dense silky white to pale purplish marginal hairs overtopping the spikelet.

Localized on the north side of Pinacate Peak in cliff crevices and rock rubble at the cliff base. Other small populations in the Sonoyta region near the summit of Sierra de los Tanques, the Quitobaquito Hills, and rarely in sandy soil plains between Quitobaquito and Sonoyta. Common on rocky slopes of the Sierra Cipriano and eastward and southward through much of Sonora. Colorado to Texas and Arizona, Mexico south to Puebla, and the Caribbean; also Argentina, Bolivia, Peru, and Uruguay. Not in California; the type collection is from Baja California Sur. An additional variety occurs in Argentina, Bolivia, and Paraguay.

This is the most widespread species in section *Trichachne,* which has approximately one dozen New World species. *D. californica* is often distinguished from closely related species by its 3-veined sterile lemma. The rather broad veins forming the margin are generally submarginal and actually often consist of 2 narrowly separated veins, so that technically the sterile lemma is 5- or obscurely 7-veined, linking *D. californica* with other closely related 5- or 7-veined species (see Correll & Johnston 1970; Henrard 1950).

N side of Pinacate Peak, *F 87-47.* Sierra de los Tanques, *F 89-31.* Sierra Cipriano, *F 91-38D.*

DISTICHLIS SALTGRASS

Perennials with creeping, scaly rhizomes and occasional stolons. Leaves conspicuously 2-ranked and usually overlapping, the blades often firm and sharp-pointed; hydathodes (salt-excreting bicellular hairs) sunken in leaf tissue. (In saline habitats the leaves excrete salt that often crystallizes on the leaf surface.) Dioecious. Inflorescences of narrow panicles (ours) or racemes, the spikelets laterally compressed, awnless, with few to many florets; spikelets breaking off above the glumes, the pistillate spikelets tardily breaking off and also breaking apart between the florets. Glumes and lemmas keeled.

Saline and alkaline habitats in the Americas and Australia; 4 or 5 species. The report of an African species (Beetle 1955) is in error (Chase 1958). Five saltgrass species in 4 genera grow along the shores of the Gulf of California. Three occur in the flora area: *Distichlis,* with 2 species, and *Monanthochloë.* The others are *Jouvea pilosa,* which extends northward in Sonora to the Infiernillo Channel opposite Isla Tiburón, and *Sporobolus virginicus,* which ranges northward to Bahía Kino. References: Beetle 1955; Chase 1958.

1. Staminate lemmas 7–9 mm; pistillate lemmas 12–16 mm; anthers 3.8–4.9 mm; coastal and Río Colorado delta. ______ **D. palmeri**

1′ Staminate and pistillate lemmas 3–6 mm; anthers 1.8–2.6 mm; coastal and interior. ______ **D. spicata**

Distichlis palmeri (Vasey) Fassett [*Uniola palmeri* Vasey, *Garden & Forest* 2:401, fig. 124, 1889] *ZACATE SALADO, ARROZ DEL MAR; NIPA* (COCOPA); PALMER GRASS. Forming dense colonies with creeping rhizomes. Stems mostly erect, (20) 40–65 cm (occasionally to 1.5 m growing through shrubs such as *Allenrolfea*). Leaf blades 2.5–9.5 cm, usually with salt crystals on the surface. Panicles (3) 5.0–13.5 cm. Lower glume 4.5–8.2 mm, the upper glume 7.5–8.5 mm. Staminate spikelets 13–25 mm, the lemmas 7–9 mm, the anthers 3.8–4.9 mm. Pistillate spikelets 19–32 mm, 6–9-flowered, the lemmas 12.1–16.0 mm with narrow membranaceous wings, the paleas 11.8–12.1 mm, enclosing the grain and adaxially winged, the stigmas purple. Grain 7.6–11.3 mm, narrowly ovoid, with a neck formed by the persistent fused style bases, the apices of the style bases tending to be free but closely appressed, the pericarp separable (with forceps) from the seed apex but tightly fused below. Seeds 5.4–7.7 × 1.5–2.0 mm, oblong (dissected from the surrounding pericarp). March–early April, the grain ripening one month later.

Tidally inundated muddy-sand soils of esteros (salt marshes) and tidal flats, and saline wet soils in the Río Colorado delta. Sonora southward to Estero de la Cruz at Bahía Kino and Baja California at least to the vicinity of Santa Rosalia. Also southward to the vicinity of La Paz, Baja California Sur, and adjacent islands; however, all known specimens from Baja California Sur are sterile. Although these probably are *D. palmeri,* there is no reliable vegetative character to distinguish them from *Sporobolus virginicus* and sometimes from *Jouvea pilosa.* Along the shores of the Gulf of California *Jouvea* and *S. virginicus* occur on seaward dune faces and upper beaches whereas *D. palmeri* occurs on muddy, usually tidally inundated soils.

Distichlis palmeri extends inland along the western side of the Ciénega de Santa Clara in shallow brackish water influenced by tidal inflow. Here it often grows intermixed with *D. spicata,* as well as *Phragmites australis, Scirpus maritimus, Tamarix ramosissima,* and *Typha domingensis. D. palmeri* also occurs at isolated brackish-water pozos (waterholes) dotting the salt flats at least 3.5 km inland from the mouth of the Santa Clara Slough. These are probably relicts of the once more expansive delta populations.

In certain areas of the delta and estero margins of Bahía Adair, the zone of *D. palmeri* ends sharply at the high-tide zone and it is abruptly replaced on higher, drier ground by *D. spicata.* There are no indications of intermediate forms. *D. palmeri* is distinguished from *D. spicata* by its coarser, thicker, and taller stems; broader and usually larger leaves; and consistently larger spikelets, lemmas, anthers, and grain. Cross sections of *D. palmeri* leaves confirmed the presence of hydathodes, as in *D. spicata* (John Reeder, personal communication 1992). *Sporobolus virginicus* has similar hydathodes, and its leaf anatomy is also similar to that of *Distichlis* (John Reeder, personal communication 1992).

The greatest stands of *D. palmeri* are on tidal mud flats in the delta of the Río Colorado, especially Islas Gore and Montague and the opposite delta shores, and in the large tidal wetlands at the western side of Bahía Adair. In these places there is nearly 100 percent plant cover of *D. palmeri* intermixed with other halophytes. Nearly all stems become reproductive in spring. Other populations are much less extensive. The delta once supported many thousands of hectares of *D. palmeri.* In his plant catalogue, Mearns (1892–93) recorded it "from the Colorado River, Sonora, below the mouth of the Hardy River. March 27, 1894. This species covers enormous flats or savannas about the mouth of the Colorado River. . . . The leaves are armed with such sharp rigid points as to make walking on these savannas very uncomfortable."

It was harvested in considerable quantities by the Cocopa prior to the construction of the large upriver dams. The last harvest was in the 1950s (Felger, field notes). In May and probably June great quantities of grain-containing spikelets washed ashore and accumulated in tidal windrows, where it was easily gathered. Stems were also harvested while the grain was still green, dried next to fires, and threshed (Castetter & Bell 1951:192-194). The grain was ground into coarse flour and usually consumed as a gruel (*atole*), or the flour was made into leavened or unleavened bread (Castetter & Bell 1951; Vasey 1889). Hardy (1829:347) reported "wild wheat, the taste of which is very sweet." The nutritional content compares favorably with that of wheat (Yensen & Weber 1986, 1987). Although the delta populations of *D. palmeri* have been greatly diminished in the latter half of the twentieth century, substantial quantities of spikelets still accumulate in tidal windrows in early summer.

This grass has the potential to become an important grain crop (e.g., Felger 1979). The plants grow vigorously when irrigated with fresh water and thrive with water salinities ranging to that of seawater (Glenn 1987). Under laboratory conditions germination rates are highest in fresh water, although occasional seeds will germinate in seawater. Salinities greater than 10–20 percent that of seawater severely reduce germination (Ed Glenn, personal communication 1996).

The present-day populations seem to be maintaining themselves mostly by vegetative propagation. Reproduction by seed seems to be attuned to the annual late spring–early summer floods of the Colorado River—floods that have been all but eliminated by the upriver dams. The once extensive delta stands appear to be losing ground.

Head of the Gulf of California, Horseshoe Bend of the Colorado River, about 35 miles south of Lerdo, *Dr. Edw. Palmer, Apr 1889* (lectotype, here designated, US 81764!, pistillate; note on packet: "seeds shelled out from the chaff, and used as food by the Cocopa Indians."). I chose this specimen as lectotype because one of the culms closely matches the illustration of the "female plant" in Vasey's 1899 paper (see *D. palmeri* illustration). Several other specimens are undoubtedly part of the original collection, and some have been labeled or catalogued as "syntypes" or isotypes. However, these include both pistillate and staminate collections, and although Vasey undoubtedly saw at least some of them, it is not possible to know which ones were used in the protologue. Horseshoe Bend, 35 mi S of Lerdo by river and 15 mi from mouth of Colorado River: *Palmer 28–30 Apr 1889* (UC 926627, pistillate); *Palmer 28–30 Apr 1889* (UC 926628, staminate); Apr 1889, *Palmer 924-929* (UC 40137, pistillate and staminate). Santa Clara Slough, *F 92-244.* Isla Montague, *F 92-259.* Ciénega de Santa Clara, *F 93-259.* Punta Peñasco, *Shreve 7598.* Bahía de la Cholla, *Bacigalupi 2873* (UC). Estero at Cerro Prieto, *López-Portillo 9 Jul 1984* (MEXU). Bahía Adair, 16 km NNW of Gustavo Sotelo, *Ezcurra 2 Apr 1981* (MEXU).

Distichlis spicata (Linnaeus) Greene [*Uniola spicata* Linnaeus. *D. dentata* Rydberg. *D. spicata* var. *stricta* (Torrey) Scribner. *D. stricta* (Torrey) Rydberg] *ZACATE SALADO;* SALTGRASS; *'ONK WAṢAI.* Forming dense, often large, spreading colonies with creeping rhizomes. Leaf blades mostly 4–8 cm. (Under conditions of high soil moisture, high temperatures, low salinity, and partial shade, the stems may become unusually tall and slender, the internodes elongated, and the leaf blades reaching 11–19 cm. Under relatively drier and more saline conditions the leaves are very short and densely crowded.) Panicles (1.5) 2.5–7.5 cm, the spikelets 5.5–14 (17) mm. Glumes and lemmas at anthesis whitish with 5–9 or more green veins, the spikelets soon becoming straw colored. Palea about as long as or slightly shorter than the lemma, pale whitish with pale green veins and adaxial membranous wings sometimes with small teeth along the margins. Staminate panicles usually overtopping the leaves; staminate spikelets sometimes longer than the pistillate spikelets, the lower glume 2.2–3.1 mm, the upper glume 2.7–4.6 mm, the lemmas 3–5 mm, the anthers 1.8–2.6 mm. Pistillate panicles more or less as tall as the leaves, the spikelets often with 4–9 florets, the lower glume reaching 5 mm, the lemmas to 6 mm. Grain 2.8–3.4 mm, dark brown. April-December.

Often forming dense stands at scattered alkaline or saline seeps along or near the Río Sonoyta, at Quitobaquito, Laguna Prieta, freshwater pozos (waterholes) in the salt flats bordering Bahía Adair, and wetlands along the banks of the lower Río Colorado. Sometimes on sandy or muddy soils only a few meters from or even adjacent to the margins of tidal mud flats near the shore, including the margins of the Río Colorado delta, but not in the tidal zones, where it is replaced by *D. palmeri.*

Distichlis spicata is especially abundant at the Bahía Adair pozos on slightly raised sand hummocks, either alone or with *Nitrophila occidentalis* (Ezcurra et al. 1988). Both *D. spicata* and *N. occidentalis* are pioneers in the formation of pozos. The leaves of *D. spicata* are salt excreting and the salt crystals are hygroscopic. Dew copiously condenses on the stems and the runoff may leach salts from the soil at the plants' bases. This localized desalination may be important to the early stages of the successional process in pozo formation. The plants also act as baffles, trapping sand and organic material, and the grass seems to be responsible for the relatively high peaty mounds surrounding older pozos.

Both Baja California states and widely scattered localities near the coast of Sonora southward to the Guaymas region. Both North American coasts from Canada to Mexico, the West Indies, and southern South America; also interior basins of North America including Mexico.

Beetle (1943) recognized 7 varieties. An older and more conservative system distinguished 2 taxa: var. *spicata* along the coast of much of North America and the Pacific coast of South America, and var. *stricta* mostly inland as a continental desert saltgrass in North America and southern South America. Var. *stricta* is distinguished by (1) looser or less dense panicles (less congested spikelets, slightly more open panicles), (2) broader keel wings on the pistillate paleas, (3) fewer florets per spike, and (4) more pilose pubescence in the vicinity of the leaf collar. These distinctions, and Beetle's, are mostly vegetative, minor, overlapping, intergrading, difficult to distinguish, and probably due largely to environmental conditions.

12 km W of Sonoyta, *F 86-172A.* Sonoyta River at Agua Dulce, 8 Feb 1894, *Mearns 2787* (US). La Salina [La Borrascosa], *F 84-20.* Laguna Prieta, *Roth 3.* 5 km S of San Luis, *F 85-1043.* Banks of Río Colorado, 5 km S of El Doctor, *F 85-1054.*

ECHINOCHLOA BARNYARD GRASS

Annuals or perennials. Leaf blades flat; ligules a ring of hairs or often absent. Panicles of few to many spikelike branches. Spikelets solitary or paired in 2 or 4 crowded rows on one side of a flattened rachis, subsessile, awned or awnless, breaking off below the glumes, with 2 florets, the lower one sterile and reduced or staminate, the upper one bisexual. Lower glume short, the upper glume similar to the sterile lemma. Fertile lemma smooth, shiny, with margins covering a flat palea of similar texture, but free at the apex. Grain more or less oval and hard.

Warm regions worldwide; 30 or fewer species. Gould, Ali, & Fairbrothers (1972) provide evidence that the New World taxa of *Echinochloa* are tetraploids, $2n = 36$, whereas the introduced Old World taxa are hexaploids, $2n = 54$.

Characteristic of rich, moist, and often disturbed soils, or some emergent from shallow water or with floating vegetative parts. Some Old World weedy taxa mimic rice and even germinate under flooded conditions. Others are grown for fodder and grain. Domesticated taxa, including those derived from *E. colonum* and *E. crusgalli,* are grown as millets in the Old World. Hybrids between wild and domesticated forms of *E. colonum* are interfertile, as are hybrids between wild and domesticated forms of *E. crusgalli,* but hybrids between domesticated *E. colonum* and domesticated *E. crusgalli* are sterile (Yabuno 1966). References: de Wet et al. 1983; Gould, Ali, & Fairbrothers 1972; Yabuno 1966.

1. Leaves often with purple blotches or bands; hairs or bristles (spines) on spikelets, panicle axes, and branches not papilla-based; sterile lemma and second glume often with a few hairs or spines but not beset with harsh spines; spikelets 2.2–2.6 (3) mm, awnless; panicle branches 1–2 (3) cm, simple. _______ **E. colonum**

1′ Leaves without purple markings; hairs and bristles on spikelets, panicle axes, and branches often papilla-based; sterile lemma and second glume beset with harsh spines; spikelets 2.8–4.0 mm (excluding awns and projections), often but not always long-awned; panicle branches usually 2.5–6.0 cm, usually with at least some short secondary branches. _______ **E. crusgalli**

***Echinochloa colonum** (Linnaeus) Link var. **colonum** [*Panicum colonum* Linnaeus] *ZACATE RAYADO, ZACATE PINTO, ZACATE TIGRE;* JUNGLEGRASS, JUNGLE RICE; *SO-O'OI WAṢAI.* Tufted ephemerals or annuals; non-seasonal but best developed during warm weather. Stems erect to semiprostrate, reaching 45–75 cm, often much smaller. Leaf blades 5–9 (12) mm wide, often with transverse purple bars or blotches (unique among Sonoran Desert grasses). Inflorescence axis and branches glabrous or with some hairs, these and hairs of spikelets not papilla-based. Spikelets paired, 2.2–2.6 (3.0) mm (to 4 mm on some southern Sonora specimens), sharp-pointed but not awned.

Lowland habitats; temporarily wet soils but sometimes extending onto drier soils; streambeds and larger washes, in disturbed as well as some natural habitats, often in areas of intensive cattle grazing and as an urban and agricultural weed. Native to the Old World; widespread and mostly weedy in tropical and subtropical regions worldwide. A number of cultivated or domesticated varieties and forms are known (de Wet et al. 1983).

S of San Luis, *F 85-1024.* Río Sonoyta, SW of Quitobaquito, *F 85-973.* Sonoyta, *F 85-944.* Williams Spring, *Van Devender 31 Aug 1978.*

***Echinochloa crusgalli** (Linnaeus) P. Beauvois var. **crusgalli** [*Panicum crusgalli* Linnaeus. *Echinochloa crusgalli* var. *mitis* (Pursh) Peterman. *E. muricata* (P. Beauvois) Fernald] *ZACATE DE AGUA, ZACATE DE CORRAL;* BARNYARD GRASS. Coarse, warm-weather annuals, 0.3–1.8 m. Stems erect to ascending from decumbent branches. Leaf blades uniformly green, the larger ones mostly 1.0–1.5 cm wide. Branches and main axis of panicles with stout, often papillose-based bristles. Spikelets 2.8–4.0 mm (excluding awns), awnless to awned, even on the same plant and within the same panicle; longer awns 2.6–5.8 cm, often purplish. Second glume and sterile lemma often with papillose-based hairs along veins.

San Luis Valley and Río Colorado delta region, primarily as an agricultural weed, and along irrigation and drainage canals; also in the vicinity of Sonoyta in seasonally wet mud along the riverbed, irrigation ditches, and at Presa Derivadora. I have not seen it in natural areas in the Sonoran Desert. Gould, Ali, & Fairbrothers (1972) report var. *crusgalli* as the only common, widespread variety of this species in the United States. Native to the Old World, now weedy and adventive in subtropical and warm-temperate places worldwide. A number of cultivated varieties and forms are well known in the Old World. The taxonomy of this genus is confusing, especially in the New World. Some authors have recognized "New World species" such as *E. muricata,* which does not seem to be separable from *E. crusgalli.*

In northwestern Sonora the two *Echinochloa* species usually can be distinguished by habitat and general appearance. Plants of *E. crusgalli* are usually larger and more robust than those of *E. colonum,* and have tougher (chartaceous rather than membranous) glumes, often with long awns, and usually grow in wetter habitats. In the absence of awns, the larger spikelets and numerous spines on the spikelets (upper glume and sterile lemma) of *E. crusgalli* are diagnostic.

Reports of the Cocopa harvesting grain from a cultivated and/or wild barnyard grass along the lower Río Colorado (Castetter & Bell 1951; Kelly 1977) may be based on a native barnyard grass such as the *E. muricata* "form." Castetter & Bell (1951:18) stated that in the delta region of the Río Colorado "belts of barnyard grass (*Echinochloa crusgalli*), which formerly fringed the more permanent sections of bank, have now well nigh disappeared." They reported it as a "native grass" (Castetter & Bell 1951:190).

Sonoyta, *F 86-397.* NE side of Ciénega de Santa Clara, *F 92-513.* Horseshoe Bend of the Colorado River, *Palmer 949 in 1899* (ARIZ, US).

ELEUSINE

Annuals, with 2 to several spikelike branches near the summit of stems and sometimes also with a few branches below. Spikelets in 2 rows on one side of a broad raceme, breaking apart above glumes and between florets, compressed, several-flowered. Seeds dark with thin ridges, loose inside the pericarp.

The seeds, similar to those of the related genus *Dactyloctenium,* are unusual among the grasses. Nine species, mostly in tropical Africa, 1 in South America, and one a worldwide weed. Fingergrass or ragi (*E. coracana*) is a widely grown cereal in Africa and the Orient. Reference: Phillips 1972.

***†Eleusine indica** (Linnaeus) Gaertner subsp. **indica** [*Cynosurus indicus* Linnaeus] *ZACATE DE GANZO;* GOOSEFOOT GRASS. Hot-weather glabrous annuals. Leaf blades often dark green and folded. Spikelets 5 mm, overlapping; glumes and lemmas strongly keeled. Weedy in Arizona and Sonora and expected as a weed at least in the San Luis region. Worldwide weed, native to Africa. Another subspecies occurs in Africa.

Yuma, *Palmer in 1887.*

ENNEAPOGON PAPPUS GRASS

Perennials or sometimes annuals; mostly erect and tufted. Glabrous, scabrous, or with 2-celled glandular hairs (elongated, bulbous-tipped epidermal microhairs large enough to be seen with a 10× lens—a character of subtribe *Cottinae*). Inflorescence a terminal panicle. Spikelets breaking off above glumes but not between florets; florets falling as a unit, the empty glumes often persistent. Glumes membranous, mostly 7–9 veined. Spikelets several-flowered, the lower 1–3 florets bisexual, the second or middle florets staminate or sterile, and the third or upper 1 or 2 florets sterile rudiments sometimes very reduced. Lemmas prominently 9-veined and crowned with 9 flat, usually feathery awns—making this genus unmistakable. (*Enneapogon* is derived from the Greek words for 'nine beards'.) Axillary cleistogamous inflorescences or spikelets sometimes present. Australia, Africa, Asia, and 1 species worldwide; 30 species. References: Chase 1946; Renvoize 1968; Tzvelev 1968.

Enneapogon desvauxii P. Beauvois [*Pappophorum wrightii* S. Watson] *ZACATE LOBERO;* SPIKE PAPPUS GRASS. Small perennials mostly 20–35 cm with hard, knotty bases. Herbage with glandular hairs. Blades wiry, less than 10 cm × 0.5 (1.0) mm, rolled up during drier times, or flat and reaching 1.5 mm wide during moist times (e.g., with summer–fall rains). Panicles spikelike, overtopping leaves, slender, feathery, and with soft hairs. Spikelets 4.5–7.0 mm including awns, lead-gray or gray-green, mostly 3-flowered, only the lowest floret bisexual. Glumes subequal, sparsely and finely pubescent, with well-marked green veins plus 1 or 2 faint ones, the lower glume 5–7-veined, the upper one 3–4-veined. Lemmas with dark-colored feathery awns, the longer awns 3.0–4.5 mm. (Each floret, especially when fresh or moistened, looks like a miniature octopus with an extra arm.) Grain golden brown, plump and ovoid, 1.2 mm including an abruptly narrowed tip of 0.1 mm.

Cleistogamous spikelets buried in the lower part of the leaf sheaths, the stems breaking off "at the lower nodes, carrying the cleistogenes with them" (Hitchcock 1951:227). Cleistogamous florets lacking glumes and awns, the grain splitting the sheath in which it develops, resembling the "normal" grain but larger, 1.8–1.9 mm, and lacking the short projection at the tip. The plants sometimes produce short stems with reduced panicles bearing spikelets with short-awned florets that are intermediate between chasmogenous florets and cleistogenes.

Rocky habitats and rocky-gravelly soils. Granitic ranges, mostly among rocks and especially in crevices on north-facing arroyo and canyon slopes, also on more exposed slopes; sometimes as low as 100 m, but mostly at higher elevations. Pinacate volcanic field, especially above 450 m, but sometimes common as low as 150 m. Western Sonora southward to the Sierra del Viejo (near Caborca) on limestone slopes and at Hermosillo in decomposed granite soil. (Hitchcock's 1913 report of this grass from Guaymas is based on *Palmer 511 in 1887.* However, the specimen is from Bahía de los Angeles, Baja California [Watson 1889]). Southwestern United States to Oaxaca, Mexico; disjunct in Argentina, Bolivia, and Chile; and widespread in the Old World, where Tzvelev (1968) recognized a second subspecies based on slightly sparser pubescence on the rachilla.

Sierra Extraña, *F 19071.* 2 mi S of Papago Tanks, *F 20141.* Suvuk, *Burgess 6340.* Campo Rojo, *Ezcurra 28 Oct 1982.* NE of Pinacate Peak, 760 m, *F 86-418.*

ERAGROSTIS LOVEGRASS

Diverse annuals or perennials. Ligule ciliate (a line or band of hairs). Spikelets laterally flattened, awnless (except a few elsewhere), few- to many-flowered, breaking apart above glumes and sometimes between florets. Glumes thin; glumes and lemmas deciduous, the lemma 3-veined, the palea often persistent. Warm regions of the world; 300 species.

The annual (ephemeral) species in Arizona and Sonora have minute and usually non-exserted anthers, a strong indication of a selfing mode of reproduction. Because the anthers commonly remain enclosed in the florets, they are relatively persistent. In contrast, the perennial (mostly non-desert) species have much larger (often more than twice as large), conspicuously exserted, and readily deciduous anthers and are likely to be outcrossing. The *Eragrostis* species in the flora area are ephemerals except for *E. lehmanniana,* which is not locally established.

1. Prominent large shining yellow, pink, or purplish glandular areas or rings at or below nodes, especially below the lower inflorescence branches. ____ **E. barrelieri**

1′ Plants without large shining glandular areas.
  2. Panicles usually dense, the spikelets close together, the pedicels usually shorter than mature spikelets; warty, glandular pits especially on leaf sheaths, leaf margins, stems, and keels of glumes and lemmas; spikelets 2–3.5 mm wide; grain broadly ovoid, 0.5–0.6 mm. ____ **E. cilianensis**
  2′ Panicles usually open, the spikelets usually widely spaced, the pedicels longer or shorter than spikelets; without glandular pits; spikelets 0.8–1.4 mm wide; grain ellipsoid, 0.6–0.8 mm.
    3. Pedicels mostly shorter than the spikelets; spikelets 0.8–1.1 mm wide; anthers ca. 1 mm; probably no longer present or if present locally rare, spring and summer-fall. ____ **E. lehmanniana**
    3′ Pedicels longer than the spikelets; spikelets 1.2–1.4 mm wide; anthers 0.3–0.4 mm; abundant and widespread, warmer months. ____ **E. pectinacea**

***Eragrostis barrelieri** Daveau. MEDITERRANEAN LOVEGRASS. Summer-fall ephemerals, often persisting through winter. Stems (10) 25–25 (50) cm, tufted, erect to decumbent. Rings or patches of shiny yellow, pink, or purplish glandular tissue on stems, and inflorescence axes and their branches. Leaf sheaths pilose with a dense tuft of white hairs 1.5–2.0 mm at the summit, otherwise glabrous or with sparse long hairs. Leaf blades (2) 3–10 (15) cm × 2–3 (5) mm. Inflorescences open but narrow, 3–15 × 2–6 (8) cm; branches ascending to spreading, often bearing spikelets nearly to the base. Spikelets 10–15 (20)-flowered, often 5–12 (15) × 1.0–1.5 mm, linear, slightly compressed, the rachilla persistent; lower glume 1.0–1.5 mm, the upper glume slightly longer; lemmas 2+ mm; paleas about equal in size to the lemmas, persistent. Anthers 0.3 mm. Grain 0.8 mm.

A weedy species native to southern Europe. Apparently established as a weed in Lukeville and Sonoyta and expected elsewhere as an urban and agricultural weed. Also in southern Sonora and probably elsewhere in the state. The prominent glandular rings or patches are unique among Arizona and Sonora grasses. The earliest Arizona record is from Nogales in 1926, and it was common in Tucson by the 1930s (Felger 1990).

Lukeville, *F 87-276.* Yuma, disturbed soil on lower bank above Colorado River, 2 Nov 1975, *Booth A-164.*

***Eragrostis cilianensis** (Allioni) Vignolo-Lutati ex Janchen [*Poa cilianensis* Allioni. *E. megastachya* (Koeler) Link] *ZACATE APESTOSO;* STINKING LOVEGRASS, STINK GRASS. Summer-fall ephemerals, size highly variable, to 45 cm but mostly much smaller. Leaf blades with wartlike glands along margins and at least a few glands elsewhere, including margins and/or keels (midribs) of glumes, lemmas, and leaf sheaths. Mouth of leaf sheaths bearded, the leaves otherwise glabrous; leaf blades 4–22 cm. Spikelets and inflorescence branches pale green when fresh, turning straw colored at maturity, the spikelets 3.5–21 (26) × 2.0–3.0 (3.6) mm. Lemmas 2.0–2.5 mm. Anthers 0.35–0.4 mm. Grain 0.5–0.6 mm, red-brown, broadly ovoid to subglobose and somewhat compressed laterally.

Urban and agricultural weed in the Sonoyta and San Luis Valleys, and in natural, temporarily wet, low-lying habitats in the Pinacate region including waterholes and playas, especially Playa Díaz and the surrounding alluvial flats. Also scattered along the Río Sonoyta watercourse and in disturbed sandy-soil lowland habitats within the Pinacate region. Native to the Old World, now a worldwide weed. Well established in the Sonoran Desert, often in disturbed habitats.

The plants in the flora area tend to be relatively small. They are easily recognized by the pale spikelets and inflorescence branches and the relatively large, compressed spikelets. The glands, making the plants sticky and stinky, are often used as a key character, but the glands may be scarce on Sonoran Desert plants.

19 mi SW of Sonoyta, *F 86-570*. Suvuk, *Nabhan & Burgess 25 Sep 1982*. Pinacate Junction, *F 86-365*. Tinajas de los Pápagos, *F 86-485*. Arroyo Tule, *F 20151*.

***Eragrostis lehmanniana** Nees. LEHMANN LOVEGRASS. Usually perennials, also flowering in the first year or season (ours were first-season plants), with tufted stems 50–60+ cm (elsewhere larger and often forming stolons). Stems bent at the lower nodes. Leaf blades 4.0–13.5 cm. Panicles often 15–21 cm, openly branched, the branches and pedicels ascending to spreading; pedicels less than 1/3 as long as spikelets. Spikelets slightly compressed, often dark gray-green to straw colored, often 4.0–5.8 × 0.8–1.1 mm, several to ca. 12-flowered. Lower glume 1.2–1.4 mm; upper glume 1.7–1.9 mm, usually broader than the lower one. Lemma 1.7–1.8 mm. Anthers 1.0–1.1 mm, conspicuously exserted and readily deciduous. Grain 0.6–0.8 mm, ellipsoid with one side flattened; the embryo dark brown to almost black.

Native to South Africa, introduced into Arizona by the U.S. Soil Conservation Service in the 1930s and now well established in southern Arizona and much of northern Sonora. About half a dozen small plants were found in April 1992, at Campo Rojo among old horse droppings (*F 92-476A*)—I destroyed all of them. If this grass establishes itself at higher elevations in the Sierra Pinacate, it might displace native species. The nearest well-established population is at Why, Arizona, north of Organ Pipe Monument (Felger 1990).

Eragrostis pectinacea (Michaux) Nees var. **pectinacea** [*Poa pectinacea* Michaux. *Eragrostis arida* Hitchcock. *E. diffusa* Buckley] Summer-fall ephemerals often (5) 12–37 cm. Leaf blades 4–15+ cm. Inflorescence branches and pedicels slender; panicles 7–20+ cm. Spikelets 2.6–5.5 × 1.2–1.4 mm. Lower glume half or more as long as the first lemma. Lemmas 1.4–1.8 mm. Disarticulation begins with the glumes and then the lowermost florets (except paleas) in sequence. Paleas translucent and tardily deciduous to persistent after lemmas and glumes have fallen. Anthers 0.35 mm, not exserted or barely so. Grain 0.8 × 0.3–0.4 mm, dark golden brown, ellipsoid, not grooved (immature grain sometimes folding inward upon drying).

Arroyos, playas, alluvial flats, and probably elsewhere. Sometimes abundant during late summer in temporarily wet habitats, such as Playa Díaz and other places of low-lying, poorly drained, and fine-textured soils. One of the most widespread hot-weather ephemeral grasses in Sonora. Much of the United States to South America and the West Indies.

Reeder (1986) reduced *E. tephrosanthos* Schultes to a variety of *E. pectinacea,* and due to technicalities of the International Code of the Rules of Botanical Nomenclature, it becomes var. *miserrima* (E. Fournier) J. Reeder. Var. *miserrima* has spreading pedicels whereas those of var. *pectinacea* tend to be appressed or diverging at acute angles. Both forms grow intermixed—for example, at Playa Díaz and other local playas and sand flats in the vicinity. Plants of var. *pectinacea* are much more common. One additional variety occurs in Florida.

Var. **pectinacea:** Pinacate Junction, *F 86-349A*. S of Tinajas de los Pápagos, *F 18735B*. Campo Rojo, *F 92-476B*. Var. **miserrima:** Pinacate Junction, *F 86-349B*.

ERIOCHLOA *ZACATE TAZA,* CUPGRASS

Annuals and perennials with flat leaf blades and terminal panicles. Spikelets usually in pairs (or one spikelet of the pair often aborted in dry regions) in 2 rows on one side of a slender, flattened rachis, very short-pedicelled, readily breaking off below the glumes. Lower glume reduced and fused to a rounded callus derived from a segment of the rachilla to form a unique cup-shaped disk at the spikelet base. Spikelets with 2 florets, the upper one bisexual, the lower one sterile and reduced to a lemma and sometimes with stamens. Upper glume and sterile lemma similar in size and texture. Fertile lemma firm with margins inrolled over the sides of the tightly gripped palea.

Tropical to warm-temperate regions. Worldwide but mostly New World; 20 species. The 3 species in the Sonoran Desert are ephemerals growing only during hot, wet summer weather as weeds in well-watered places, in seasonally wet low places, and in riparian or semiriparian habitats. They are among the first summer-fall grasses to shed their spikelets. The plants wither very quickly with drying conditions, and only a few weeks after the grain matures and falls there may be no sign of the plants. In the 3 species in the flora region the sterile floret is represented only by a lemma, and the fertile lemma is abruptly blunt-ended with a tiny, stubby, mucronate tip or tiny bristle. Reference: Shaw & Webster 1987.

1. Spikelets 6.5–8 mm from base of cup to awn tip. ____ **E. aristata**
1′ Spikelets 4–5 mm from base of cup to awn tip.
   2. Leaf blades glabrous or sparsely hairy on lower surface; panicles not cylindrical, the branches often spreading; fertile lemma awnless or with a stubby mucronate tip 0.1 (0.3) mm. ____ **E. acuminata**
   2′ Leaf blades hairy on lower surface; panicles subcylindric, the branches appressed; fertile lemma with a slender awn 0.7–0.8 mm. ____ **E. contracta**

***Eriochloa acuminata** (J. Presl in C. Presl) Kunth var. **acuminata** [*Piptatherum acuminatum* J. Presl in C. Presl. *Eriochloa gracilis* (E. Fournier) Hitchcock. *E. lemmonii* Vasey & Scribner var. *gracilis* (E. Fournier) Gould] SOUTHWESTERN CUPGRASS. Often 30–50 cm or more, the stems often weak. Leaf blades thin, flat, bright green, glabrous to sparsely hairy. Spikelets usually densely hairy, 4–5 mm (including cup and mucro or awn). Upper glume awnless or extending into a short mucronate tip. Fertile lemma awnless or with a mucronate tip to 0.1 (0.3) mm.

Urban and agricultural weeds in temporarily moist soil, probably not native to the flora area. Both Baja California states, mainland Mexico and southwestern United States; deserts to pine-oak woodland. Also adventive in southeastern United States and reported for Argentina (Zuloaga et al. 1994).

Sonoyta, *F 85-935.* 15 mi SW of Sonoyta, *F 91-137.* Puerto Peñasco, *F 91-142.* S of San Luis, *F 85-1029.*

Eriochloa aristata Vasey var. **aristata.** BEARDED CUPGRASS. Similar to *E. acuminata* but generally more robust. Leaf blades glabrous to sparsely hairy. Spikelets (6.0) 6.5–8.0 mm including cup and awn. Upper glume tapering to an awn 1.5–3.0 mm. Fertile lemma with a stubby mucronate tip or awn point 0.15–0.4 mm.

Temporarily wet soils in natural habitats in the Pinacate region, such as large playas and shaded semiriparian arroyos or washes, often beneath *Olneya* and *Prosopis.* Sometimes forming extensive stands in Playa Díaz and surrounding alluvial fans. Kelly (1977) reports that the Cocopa collected the grain for food in the region between the Colorado River and Somerton, Arizona. Elsewhere it is often a weed of moist swales, roadsides, and irrigated fields. Central America to southwestern United States; widespread in Sonora.

NW of Tezontle cinder mine, *F 86-376.* W of Los Vidrios, *F 92-968.* Tinajas de los Pápagos, *F 86-478.*

*†**Eriochloa contracta** Hitchcock. PRAIRIE CUPGRASS. Panicles tightly contracted, almost cylindrical. Leaf blades conspicuously hairy. Spikelets 4.3–4.5 (5) mm. Upper glume tapering to an awn 0.5–0.8 mm. Fertile lemma with an awn 0.7–0.8 mm beset with short hairs.

Irrigated lands in southwestern Arizona and northeastern Baja California, and expected in fields south of San Luis. United States and northern Mexico, and introduced elsewhere.

Yuma, *Wooton 6 Oct 1950.* W of Mexicali, *Wiggins 451.*

ERIONEURON

Perennials, low, tufted, or spreading and mat-forming, the leaf margins firm and thickened. Inflorescence a reduced panicle or raceme, compact and often shorter than the leaves. Spikelets nearly sessile, mostly several-flowered, breaking apart above glumes and between florets. Glumes 1-veined, subequal, acute to awn-tipped. Florets gradually reduced above to rudiments. Lemmas 3-veined, short-awned.

Dry regions, southwestern United States, Mexico, Bolivia, and Argentina; 4 species. References: Reeder & Crawford 1970; Tateoka 1961; Valdés-Reyna & Hatch 1997.

Erioneuron pulchellum (Kunth) Tateoka [*Triodia pulchella* Kunth. *Dasyochloa pulchella* (Kunth) Willdenow ex Rydberg. *Tridens pulchellus* (Kunth) Hitchcock] *ZACATE BORREGUERO;* FLUFF GRASS. Dwarf, tufted perennials, 5–14 cm, sometimes flowering in the first season. Stems with 1 long internode, the longer stems 7–11 cm, bearing at the top a tightly fascicled cluster of short leaves and inflorescences, arching and bending to the ground but seemingly seldom taking root and only rarely producing a second stolon. Leaf blades 1.8–7.0 cm × less than 0.5 mm, the margins firm and often white. During the driest times the leaves may perish; the dry, drought-killed leaves, stems, and inflorescences persist, acting to shelter emerging shoots, or perhaps seedlings, of subsequent seasons. Panicles compact and dense, on peduncles shorter than the longer leaves. Spikelets 6–10 mm, with (4) 6–8 florets, tardily breaking off above the glumes. Glumes and lemmas papery, sometimes purple-tinged. Each glume with a green midvein. Lemmas 3–5 mm, densely pilose with long hairs on each of the 3 green veins, the tip deeply 2-lobed with a stout awn 1–2 mm long from between the lobes.

Sonoyta region, Pinacate volcanic region, granitic ranges, and sometimes on desert flats. Widespread on rocky soils, including pavements and slopes to peak elevations, and various exposures including south- and west-facing slopes; also bajadas and occasionally on sand flats. No other perennial grass in the region extends into harsher, drier habitats. This little grass responds to even very scant rainfall at any season, although growth and flowering are most luxuriant during warm or hot weather.

Arid and semiarid southwestern United States and northwestern Mexico to Coahuila, Zacatecas, and Aguascalientes. Southward in Sonora to Isla Tiburón and the opposite Sierra Seri and vicinity of Hermosillo. Hitchcock's (1913) report of this grass from Guaymas is based on *Palmer 500 in 1887;* however, the specimen is from Bahía de los Angeles, Baja California (Watson 1889).

Valdés-Reyna & Hatch (1997) place this species in the monotypic genus *Dasyochloa* but admit that the evidence for doing so is weak. Reeder & Crawford (1970) indicate a closer affinity with *Munroa.* The morphology of *E. pulchellum* is strikingly similar to that of *Munroa,* and both are hosts to the same woolly aphids.

Sierra del Viejo, *F 16883.* Sierra del Rosario, *F 20366.* MacDougal Crater, *F 10509.* Tinajas de Emilia, *F 19735.* Pinacate Junction, *F 86-358.* Agua Dulce [Río Sonoyta], *MacDougal 11 Nov 1907* (US). Puerto Peñasco, *Dennis 37.*

FESTUCA

Mostly perennials with some annuals in dry regions. Inflorescences panicles or racemes to sometimes spikelike. Spikelets (1) 2- to many-flowered, breaking apart above glumes and between florets. Glumes unequal, the lower one often small, both of them normally shorter than the first lemma. Lemmas awned or awnless.

Worldwide, mostly temperate regions, also mountains in the tropics; 450 species. About 26 species in temperate and subtropical and often arid regions almost worldwide are generally annuals that are often segregated as *Vulpia.* These species tend to be small plants with mostly cleistogamous flowers, 1 (3) stamens, short filaments, and small stigmas. The perennials generally have chasmogamous flowers, 3 stamens, and shorter grain. Arnow (1987) and others make a convincing case for recognizing only *Festuca* but Tucker (1996) indicates *Vulpia* may be more closely related to *Lolium*.

Only 2 species, both annuals, extend into the Sonoran Desert: *F. microstachys* (*Vulpia microstachys*) at the periphery and higher elevations, and the more widespread *F. octoflora*. References: Henrard 1937; Lonard & Gould 1974.

Festuca octoflora Walter [*Vulpia octoflora* (Walter) Rydberg. *F. octoflora* var. *hirtella* Piper] SIX-WEEKS FESCUE, EIGHT-FLOWERED FESCUE. Delicate winter-spring ephemerals, 4–15 (29) cm, or when drought-stressed as small as 2.5 cm with a single spikelet. Stems mostly single or few, usually erect, with few small leaves above. Panicles interrupted, often spikelike, the branches mostly short and appressed or unbranched. Spikelets laterally compressed, 8–13 mm (excluding awns). Florets crowded, 6–13 per spikelet. Lower glume shorter than the second. Lemmas lanceolate, 4–8 mm, tapering to a slender awn 1–8 mm; awn length generally locally uniform.

Sonoyta region, Pinacate volcanic field, and granitic ranges. Common on cinder slopes and gravelly soil of arroyo beds, and especially abundant at higher elevations in the Sierra Pinacate. Across the northern part of the Sonoran Desert and western Sonora south to the vicinity of El Desemboque San Ignacio. Southern Canada to northern Mexico, and weedy in many parts of the world.

There are 3 varieties but the distinctions are "not entirely satisfactory" (Lonard & Gould 1974:221; also see Holmgren & Holmgren 1977). The two common Sonoran Desert varieties are var. *octoflora*, with the lemmas glabrous or slightly scabrous on the back and often scabrous on the margins, and var. *hirtella*, with the lemmas prominently long-scabrous to densely hairy on the back, at least near the apex. Specimens from the flora area show variation from glabrate to short-scabrous to densely hairy lemmas with no apparent pattern.

Tinajas de Emilia, *F 19713*. Elegante Crater, *F 19683*. Sierra Pinacate, 780 m, *F 19287*. Tinaja del Tule, *F 19206*. NW of Cerro Colorado, *F 19649*. Hornaday Mts, *Burgess 6798*. Observation: Sierra del Viejo.

HETEROPOGON

Annuals or tufted perennials. Leaf blades flat. Inflorescence a solitary spikelike 1-sided raceme (note that the awns are all on one side). Spikelets in pairs, one very different from the other, one larger and pedicelled and one sessile, except a few spikelet pairs at lower nodes of inflorescence staminate and sessile. Pedicelled spikelet awnless and staminate or sterile, with a slender pedicel-like callus, the actual pedicel reduced to a little stump. Sessile spikelet bisexual or pistillate with a large, stout, hairy, twisted awn, and a long, sharp callus. Each spikelet of the pair 1-flowered or appearing 1-flowered, the lower floret greatly reduced or absent. Six species in the warmer regions of the Old World, 2 of these also widespread in the New World.

Heteropogon contortus (Linnaeus) P. Beauvois ex Roemer & Schultes [*Andropogon contortus* Linnaeus] *ZACATE COLORADO;* TANGLEHEAD. Robust perennial tufted grass; stems erect, crowded, often 60–70 cm. Leaf blades flat, the dry leaves rust colored and persistent. Spikes usually strongly 1-sided, often long-pedunculate, 2.5–6.0 cm not including awns. Pedicellate spikelet staminate, its glumes bright green when fresh and drying light brown, with numerous veins and broad membranous margins; outer (lower) glume larger than the inner one, sparsely hispid to papillose-hispid especially toward tip, and tightly enclosing upper glume and floret. Sessile spikelet pistillate, brown, 2.5–6.0 mm (excluding awn), its outer glume larger than the inner glume and tightly enclosing it; lemma with a reduced body, an awn 4.5–7.0+ cm, twice bent when mature, tawny brown, with forward-pointing hairs toward base, and "a needle-sharp sessile spikelet callus which readily penetrates clothing and is a ferociously efficient dispersal mechanism" (Clayton & Renvoize 1986:359). Spikelet bases surrounded by coppery brown hairs.

Widely scattered but localized in the Pinacate volcanic complex, the granitic ranges including the Sierra del Rosario and at Bahía de la Cholla, and the Sonoyta region. Often in crevices among rocks on steep slopes and canyons, and gravelly arroyos in rocky areas. Widespread in the Sonoran Desert. Through Latin America to southwestern United States and warm regions of the Old World.

Some authors claim it is "adventive in America since the time of Columbus" (Correll & Johnston 1970:201), but it has been in southwestern North America for at least 7900 years (Van Devender, Toolin, & Burgess 1990). This species is apomictic (Emory & Brown 1958).

Tinajas de los Pápagos, 17 Nov 1907, *MacDougal 52* (US). Sierra del Rosario, *F 20387*. Bahía de la Cholla, *F 13159*. Quitobaquito, *Mearns 2752* in 1894 (US).

HORDEUM *CEBADA,* BARLEY

Annuals or perennials (winter-spring annuals or ephemerals in the Sonoran Desert). Inflorescences of spikes or spikelike racemes, bristly and terminal. Spikelets 1-flowered, relatively large, mostly in tight clusters of 3 at each node; glumes narrowed into firm bristles or awns in front of and often partially obscuring the fertile spikelet. Triplet clusters alternate in cavities of rachis in 2 longitudinal rows in "wild" species, and appearing to be in 4 or 6 rows in some cultivated forms. In wild and weedy barleys the triplet clusters fall as a unit with the stipelike rachis joint; in cultivated barleys the rachis is continuous and non-shattering. Central spikelet usually sessile and fertile, the lemma ovate, sharp-pointed, and nearly awnless to awned with long, stout bristles, the rachilla extending into a bristle behind the palea. Lateral 2 spikelets usually short-stalked, staminate or sterile (or fertile in some cultivated barleys), often represented only by glumes sometimes reduced to bristles or spines.

Worldwide, mostly temperate regions; 40 species, many of them seem to be narrowly defined. The common cultivated barley is *H. vulgare*. Both native and adventive weedy species are widespread in North America. Karen Adams (1987) found evidence of prehistoric domestication of *H. pusillum* in Arizona.

Weedy barleys occur in disturbed habitats across northern Sonora except in the most arid regions. Native hordeums seem to be rare in Sonora, or at least poorly represented in herbaria, and mostly occur in the northern part of the state or at higher elevations. Reference: Bothmer et al. 1991.

1. Margins of glumes scabrous (not ciliate); central spikelet and its florets sessile. ________ **H. arizonicum**
1′ Margins of glumes conspicuously ciliate; florests of central spikelet raised above the glumes on an elongated rachilla. ________ **H. murinum**

†Hordeum arizonicum Covas [*H. adscendens* sensu Hitchcock (1935). Not *H. adscendens* Kunth] Often 20–60 (100) cm. Spikes 5–12 cm, linear-oblong. Lateral spikelets pedicellate; central spikelet sessile, 26–32 mm including the awns, ca. 0.3 mm broad basally. Glumes all linear-subulate and scabrous (not ciliate), the awns more than 1 cm. Stamens evident, the anthers 1.6–2.0 mm.

Hordeum arizonicum is characteristic of agricultural habitats in areas of former Sonoran Desert, grassland, and oak woodland vegetation. It should be sought in riparian habitats and irrigated fields along the Río Sonoyta and the Mexico portion of the Río Colorado valley. Known for certain from Mexico only at the Quitovac oasis, and recorded from the vicinity of Yuma. Otherwise central and southern Arizona, southeastern California, and perhaps Baja California (Gould & Moran 1981). A report of *H. arizonicum* from Quitobaquito is based on a misidentified collection of *H. murinum* subsp. *glaucum* (Felger et al. 1992).

Hordeum arizonicum is hexaploid ($2n = 42$) and apparently of hybrid origin with genomes from *H. jubatum* ($2n = 28$) and *H. pusillum* ($2n = 14$), or possibly *H. intercedens*.

Quitovac, 14 May 1982, *Nabhan 283*. California: Bard [near Yuma], 1944, C. Reeder 21 (US, not seen).

***Hordeum murinum** Linnaeus subsp. **glaucum** (Steudel) Tzvelev [*H. glaucum* Steudel. *H. stebbinsii* Covas] *CEBADILLA SILVESTRE;* WILD BARLEY. Often 20–60+ cm. Spikes linear-oblong, 5.5–7 cm (excluding awns). Central spikelets 16–36 mm including awns, technically sessile (the glumes sessile, but the rachilla joint elongated, raising the floret conspicuously above the glumes, thus the 3 spikelets appearing pedicellate); central spikelet glumes with long, spreading cilia on their margins. Stamens included within the florets and persistent, the anthers of the central spikelet 0.5 mm or less, those of lateral spikelets often 1.0 mm.

Common urban and agricultural weed at least in the vicinity of Sonoyta. A few plants were found at Campo Rojo in April 1992, on the east side of the Pinacate lava shield. These were among old horse droppings, the obvious seed source. If this grass should become established at higher elevations in the Sierra Pinacate, it might displace native species (see *Eragrostis lehmanniana*). Common weed in northern Sonora, northwestern Baja California, and southern Arizona. Native to the Mediterranean and Middle East, now widespread and weedy in temperate regions of the world including western North America. *H. murinum* is a weedy annual species with 3 subspecies native to Eurasia.

Sonoyta, *F 87-16*. Quitobaquito, *Nichol 28 Apr 1939*. Campo Rojo, *F 92-478*.

LEPTOCHLOA SPRANGLETOP

Annuals or perennials, usually tufted. Leaf blades mostly flat. Spikes or racemes slender, scattered along main axis. Spikelets numerous, in 2 rows on one side of the rachis, breaking apart above glumes and between florets. Glumes awnless (ours) or awned, the upper generally longer and broader. Florets few to ca. one dozen per spikelet, mostly crowded and overlapping, the uppermost floret reduced to a rudiment. Lemma 3-veined, short-awned or awnless; palea well developed and often prominent.

Worldwide, largely tropical but some in temperate and often dry regions; 40 species. At least in western North America, including the Sonoran Desert, many of the species are characteristic of moist or marshy places—often with alkaline or semi-saline soils. In the Sonoran Desert only one member of the genus, *L. panicea,* commonly extends into the open desert away from temporarily or permanently wet habitats. References: Parodi 1919, 1927; Snow 1997, 1998.

1. Perennials with hard, knotty bases and cleistogenes; lemma truncate with 2 broad lobes, the midrib often extending into a short bristle not exceeding the lobes. ________ **L. dubia**

1′ Annuals (sometimes stout), base not knotty and without cleistogenes; lemma not as in 1, not truncate.
  2. Lemmas short-awned. ________ **L. viscida**
  2′ Lemmas awnless.
    3. Glabrous or minutely scabrous; spikelets 4.2—6.4 mm, 4-8-flowered. ________ **L. fusca**
    3′ Glabrous or leaf sheaths papillose-hispid with sparse, large hairs; spikelets 2.2–3.2 mm, (1) 2- or 3-flowered. ________ **L. panicea**

Leptochloa dubia (Kunth) Nees [*Chloris dubia* Kunth. *Diplachne dubia* (Kunth) Scribner] GREEN SPRANGLETOP. Tufted perennials with a tough, knotty base and well-developed roots, 40–100 cm; plants, especially the spikelets, often purple-tinged. Larger leaf blades 25–36 cm × 5–7 mm, somewhat bluish green; ligule a membrane fringed with white hairs. Enlarged, cleistogamous spikelets often at stem bases and in some leaf axils. Inflorescences 18–24 cm, with several or more unbranched primary branches, the branches triangular in cross section, the bases of the branches with minute hairs. Immature and mature spikelets quite different in appearance: florets of young spikelets are closely spaced, but at maturity they are well separated from each other, and the rachis is readily visible. Spikelets with 5–12 florets. Glumes persistent, the upper longer and broader than the lower. Spikelets breaking apart above the glumes and between the florets, each floret falling with its segment of the rachis. Lemmas 2.1–5.0 mm, broadly truncate (blunt-ended), the tip with 2 broad lobes and minutely fringed, the midrib extending into a bristle within the notch.

Mearns found this grass at Sonoyta in 1894; apparently it no longer occurs there. It is scarce in the Pinacate region but is occasionally encountered in sandy-silty soils in washes and some of the more heavily vegetated playas. The two collections since Mearns' are from the ecological and geographic limits for this species. The nearest other collections are from the Ajo and Kofa Mountains in southern Arizona. There are few records for this species at lower elevations within the Sonoran Desert. In north-central and northeastern Sonora and Arizona it commonly occurs toward the upper elevational limits of the desert and in mountains above the desert. Cattle eagerly seek this highly palatable grass.

Arizona to Oklahoma and Texas, Florida, Mexico including Baja California to Central America, and Ecuador to Chile and Argentina, Brazil, and Paraguay.

6 km W of Los Vidrios, sandy-silty wash in lava, *F 92-966*. Edge of playa among mesquite shrubs, 5 km S of Tinajas de los Pápagos, *Ezcurra 5 Nov 1982*. Sonoyta, *Mearns 2715* in 1894 (US).

Leptochloa fusca (Linnaeus) Kunth subsp. **uninervia** (J. Presl) N. Snow [*Megastachya uninervia* J. Presl. *Leptochloa uninervia* (J. Presl) Hitchcock & Chase. *L. imbricata* Thurber. *Diplachne uninervia* (J. Presl) Parodi] *ZACATE SALADO MEXICANO;* MEXICAN SPRANGLETOP. Robust annuals, non-seasonal but responding best to hot weather. Often forming large clumps 0.5–1.5 m and appearing perennial although actually annual. Herbage and inflorescence axis glabrous or minutely scabrous. Spikelets generally grayish "lead-colored," 4.2–6.4 mm, with 4–8 florets. Lemmas awnless.

Agricultural and urban weeds in well-watered, disturbed habitats, e.g., near Sonoyta, Puerto Peñasco, and south of San Luis; also riparian habitats along the Río Sonoyta. Southern United States to Argentina, Chile, and the Caribbean.

Snow (1998) reduced *L. uninervia* and *L. fascicularis* to subspecies of *L. fusca*. Subsp. *fascicularis* (Lamarck) N. Snow is widespread in the Americas including Sonora and Arizona but does not occur in the flora area. As treated by Snow, the *L. fusca* complex consists of 4 subspecies, 2 of them in the Old World.

Puerto Peñasco, *F 85-800*. Río Sonoyta, Sonoyta, *F 85-708*. Sonoyta River near Quitobaquito, *Mearns 2741* in 1894 (US). Ciénega de Santa Clara, *F 92-524*. Lerdo, *Palmer 945 in 1889* (US not seen; Vasey & Rose 1890). 30 km S of San Luis, *F 85-1040*.

Leptochloa panicea (Retzius) Ohwi subsp. **brachiata** (Steudel) N. Snow [*L. brachiata* Steudel. *L. filiformis* (Lamarck) P. Beauvois] *ZACATE SALADO, DESPARRAMO ROJO;* RED SPRANGLETOP. Summer-fall ephemerals, sometimes 45–80 cm but usually much shorter, delicate and filmy, reddish maroon or all green. Roots often poorly developed. Glabrous or the leaf sheaths sometimes sparsely hispid with large papillose-based hairs. Spikelets often reddish purple, 2.2–3.2 mm, 2- or 3-flowered, or occasionally 1-flowered at the end of a season, the mature (grain-bearing) spikelets sometimes appearing 1-flowered. Glumes generally shorter than the rest of the spikelet; lemmas awnless, the veins, especially the midrib, often rather sparsely pubescent with small soft hairs, or sometimes glabrous or essentially so.

Often on fine-textured soils where water temporarily accumulates; playas including crater centers, brushy alluvial fans and depressions, and larger washes and arroyos. Also a weed in disturbed habitats including agricultural areas. Widespread in Sonora; southern half of the United States to Argentina and Peru, often weedy, and introduced and spreading elsewhere in the New World as well as the Old World.

Leptochloa panicea is a nearly worldwide complex of 3 subspecies (Snow 1998). Subsp. *brachiata* is the most widespread of the 2 New World subspecies. This subspecies has previously been known as *L. mucronata* (Michaux) Kunth, *L. panicea* subsp. *mucronata* (Michaux) Nowack (in part), and *L. filiformis*. Subsp. *mucronata* is restricted to the southeastern United States west to Iowa and Texas. Subsp. *panicea* is native to the Old World.

S of San Luis, *F 85-1034*. MacDougal Crater, *F 10638*. Sykes Crater, *F 20017*. Papago Tanks, *MacDougal 7 Nov 1907* (US). Pinacate Junction, *F 86-347*. Suvuk, *Nabhan 3*.

Leptochloa viscida (Scribner) Beal [*Diplachne viscida* Scribner] STICKY SPRANGLETOP. Warm-weather ephemerals; highly variable in size depending on soil moisture, frequently 10–30 cm but moisture-starved plants reproducing as small as 1.5 cm; often viscid. (The viscid-sticky feature is often used as a key character, but plants from western Sonora are often only slightly viscid. You can usually find a few sparkling droplets on the flowering axis and branches, and this exudate is sometimes thick on the lemmas.) Stems conspicuously geniculate-spreading, especially at the base, and sometimes forming roots at these nodes. Inflorescences at first often contracted and partially enclosed in the leaf sheath, later expanding and much spreading. Spikelets several-flowered. Glumes and lemmas thin, whitish, or sometimes reddish; glumes with a prominent green midvein, the lemmas with 3 prominent green veins. Lemmas 2.0–2.8 mm, with a rounded, notched apex, the midrib extending into a slender awn to 1 mm.

Often in disturbed habitats and poorly drained, low-lying areas such as playas with fine-textured soils; sometimes seasonally abundant at Playa Díaz and similar habitats. Through much of its geographic range it is characteristic of "heavy-soiled bottomlands, margins of drying swales, roadsides and waste places" (Gould 1951:137). Sonora, Sinaloa, Chihuahua, both Baja California states, and New Mexico to California.

Pinacate Junction, *F 85-997.* Playa Díaz, *F 10809B.* Tinajas de los Pápagos, *Ezcurra 5 Nov 1982.* Playa los Vidrios, *Ezcurra 5 Nov 1982.*

LOLIUM RYEGRASS

Annuals or perennials. Usually with tall, slender spikes. Spikelets evenly spaced and solitary at nodes, several-flowered, flat, and fitting edgewise and alternately in concavities on opposite sides of a slightly curving to zigzag, continuous rachis. Spikelets breaking apart above the glumes and between the florets. Lower glume absent from most spikelets of an inflorescence, the terminal spikelets with 2 glumes.

Native to Eurasia, mostly cool or temperate regions; 8 species, several naturalized in the New World. Reference: Terrell 1968.

1. Glumes shorter than spikelets. ______ **L. perenne**
1′ Glumes as long as or longer than spikelets. ______ **L. temulentum**

Lolium perenne Linnaeus [*L. multiflorum* Lamarck. *L. perenne* var. *multiflorum* (Lamarck) Parnell. *L. perenne* var. *aristatum* Willdenow. *L. perenne* var. *italicum* (A. Braun) Parnell] *BALLICO ITALIANO;* ITALIAN RYEGRASS, WINTER RYE. Winter-spring annuals. Spikes slender and erect, the spikelets 1–2 cm, strongly flattened. Glumes about 1/3 to 2/3 as long as the spikelet. Lemmas awnless or with more or less straight awns to 1.5 cm.

This is the lawn grass commonly planted in southern Arizona and northern Sonora as a winter-spring replacement for Bermuda grass (*Cynodon dactylon*), as the latter turns brown after the first frost. Occasional in sandy soil in disturbed, urban habitats.

Puerto Peñasco, *F 85-771.* San Luis, *Reina 97-536.*

Lolium temulentum Linnaeus. DARNEL. Winter-spring annuals. Glumes (1) 1.5 (2+) cm, as long as or longer than the spikelets. Spikelets flattened, awns 0.5–1.5 cm.

Weed of cultivated fields in Arizona and probably in Sonora. Native to Europe, now widespread in North America. One collection from the flora area: *Palmer 946 in 1889* (US, labeled "Lower California," although Vasey & Rose's 1890 report says it is from Lerdo, Sonora).

MONANTHOCHLOË

Perennials with creeping wiry stems. Leaves short and 2-ranked (distichous). Inflorescences reduced to inconspicuous, leaflike single spikelets nearly hidden among leaves at ends of short shoots, but becoming conspicuous on staminate plants when the anthers are exserted at anthesis. Dioecious. Spikelets falling very tardily, awnless, 3–5-flowered, the upper florets rudimentary; stigmas and stamens conspicuously exserted. Glumes none.

One species in North America and 1 in Argentina. *Monanthochloë* shows similarities to *Reederochloa,* a monotypic genus from interior desert basins of central Mexico, and natural hybrids with *Distichlis spicata* are known from the Pacific coast of Baja California. These grasses are dioecious and have bicellular hydathodes (specialized, salt-excreting, enlarged, and sunken epidermal hairs). *Monanthochloë* also appears vegetatively convergent with its Old World counterpart *Aeluropus lagopoides.* These 4 genera of saltgrasses are in the tribe Aeluropodeae. References: Soderstrom & Decker 1964; Stephenson 1971.

Monanthochloë littoralis Engelmann. *ZACATE PLAYERO;* SHORE GRASS. Mats of upright, creeping, or spreading stems from numerous rhizomes and stolons. Stems with long and short shoots. Leaves of the long shoots separated by long internodes, relatively broader and sometimes longer than those of the short shoots, the latter in compact clusters. Leaf blades (4) 5–9 mm, stiff, sharp-pointed (often spinose-tipped), flattened to subulate with inrolled margins, scabrous or with short hairs on veins and margins near base of blade, the veins 9. (A specimen from Isla Tiburón [*F 9069*] has some long-shoot leaf blades 15 mm.) Spikelets (6.0) 9.0–9.5 mm; lemma of lowermost floret often 7.4–7.6 mm, the palea 0.2–0.9 mm longer than the lemma; veins green, the lemma multiple-veined, the palea 2-veined. Anthers 4.2–4.8 mm. Simultaneous mass flowering in spring, often in March. (Farther south, such as along the gulf coast of Baja California Sur, mass flowering occurs as early as January.)

Dense and extensive colonies in muddy soils of esteros and tidal flats inundated by higher tides. Coastal marshes in southern California, the Pacific and Atlantic sides of the northern half of Mexico including both coasts of the Baja California peninsula, Texas to Florida, and Cuba. Also inland in southern Texas and at Cuatro Ciénegas, Coahuila. This is the shortest-leaved grass in the Sonoran Desert.

16 km NNW Gustavo Sotelo, *Ezcurra 2 Apr 1981* (MEXU). Bahía Adair, 31°35′ N, 113°52′ W, *López-Portillo 9 Jul 1984.* Puerto Peñasco, *F 20971A.* 6.2 mi NW of Cerro Prieto, *F 86-516.*

MUHLENBERGIA MUHLY

Annuals or perennials of diverse growth habits. Ligules membranous or sometimes firm, not ciliate. Spikelets small, almost always 1-flowered, breaking off above the glumes. Lemmas firm, 3-veined, and 1-awned or sometimes awnless. Grain usually slender, fusiform, not readily falling from the floret; pericarp closely covering the seed.

Mostly New World, the Andes to North America, and Japan to the Himalayas; 160 species. Often in open, dry grassland in subtropical and warm-temperate climates; greatest diversity in Mexico and nondesert southwestern United States. References: Hitchcock 1935c; Peterson & Annable 1991; Reeder 1981.

1\. Ephemerals, soft and herbaceous including the stems; roots weakly developed; cleistogenes in lower leaf axils; panicles longer than broad, the pedicels mostly shorter than lemmas; glumes blunt, 0.5–1.1 mm. ______ **M. microsperma**

1′ Perennials with hard, knotty, woody bases and hard, wiry stems; root system well developed; cleistogenes none; panicles about as wide as long, the pedicels longer than lemmas; glumes acute to acuminate, 2.0–3.2 mm. ______ **M. porteri**

Muhlenbergia microsperma (de Candolle) Trinius [*Trichochloa microsperma* de Candolle] *LIENDRILLA CHICA;* LITTLE-SEED MUHLY. Non-seasonal ephemerals, (7) 10–40 (50) cm, soft and delicate, often suffused with purple except during summer-fall rainy season, the stems often weak, growing through other plants or decumbent-spreading in shaded places. Roots weakly developed. Leaves scabrous to pilose with short hairs; blades 1–8 cm, soon drying, often reduced when stressed by cold or drought; ligules membranous, translucent white, often ca. 1 mm, decurrent (running down the side of the upper leaf sheath). Panicles terminal, longer than wide (branches appressed-ascending when young, spreading at maturity, or rarely the branches remaining appressed), filmy, open and loosely flowered, usually purplish except during summer. Pedicels mostly shorter than the lemmas, but the pedicels at ends of panicle branches mostly as long as to longer than the lemmas. Glumes broad (obtuse), covering only the base of the lemma, the upper (larger) glume 0.5–1.1 mm. Lemmas short-haired on lower 1/3 and on the callus; body of lemma narrow, 1.8–2.5 (3) mm, tapering into a slender awn 14–28 mm; rather easily opened to release the golden brown grain 1.5–1.8 × 0.1–0.25 (0.3) mm.

Lower leaf axils with cleistogamous spikelets (cleistogenes), these sometimes also in leaf axils to the base of terminal inflorescences on decumbent stems; cleistogenes single or clustered, narrowly conical, 4–10 (12) mm, sometimes 2- or 3-flowered, the grain 1.3–1.95 × (0.3) 0.5–0.65 mm.

At all elevations in the Sonoyta region, Pinacate volcanic complex, granitic ranges, and the intervening desert flats; absent from moving dunes. Most numerous in better-watered, protected microhabitats such as among rocks, along arroyos, in canyons, on north- and east-facing slopes, and on cinder slopes and flats; also sand flats and playas. Often eliminated by animals except at the base of spiny shrubs. This is the most widespread and common *Muhlenbergia* in the Sonoran Desert. Southwestern United States (including southern California) to Guatemala, and Colombia and Venezuela to Ecuador including the Galapagos Islands and Peru.

The cleistogene resembles a miniature cornucopia and is "without glumes, and much more turgid than is the chasmogene. . . . It is tightly folded in the swollen, spongy-indurate base of the sheath of a reduced leaf" (Chase 1918:257). The cleistogene, much more variable than the normal spikelet, is a contracted inflorescence. Its awn is greatly reduced or absent. Unlike the chasmogene grain, the cleistogene grain is difficult to extract; it is not shed and germinates within its hard protective sheaths. The cleistogene grain resembles the normal one but is larger (thicker and heavier). Cleistogamous spikelets seem to offer the seed a better chance of surviving drought than the smaller and relatively unprotected normal grain, and they may offer protection from seed predators. The cleistogenes—because they are heavier than normal grains, lack long filmy awns, and are situated at the base of the plant—should have a reduced range of dispersal with a greater probability of remaining at or near the place where their parent was biologically successful. The terminal panicles are often eaten by animals, giving additional advantage to the dual reproductive strategy.

The chasmogamous spikelets of specimens from the flora area seem unusually small even for this variable species, with lemmas mostly 2.5 mm or shorter, whereas in Baja California the lemmas are (2.5) 3.0–3.5 (4.0) mm (C. Reeder 1981). At least in the flora area, the lemma, leaves, and plants are smaller among drought-stressed plants.

The most closely related species seems to be *M. appressa* C. Goodding, known from Arizona (e.g., Alamo Canyon, Ajo Mountains), California, and both Baja California states. This species is distinguished by closely appressed inflorescence branches (the panicles conspicuously contracted or narrowed), larger and longer lemmas and glumes, and longer ligules. The plants otherwise generally resemble those of *M. microsperma* in habit and have similar cleistogenes.

Quitobaquito, *F 88-422.* Elegante Crater, *F 19694.* Cerro Colorado, *F 10783.* Pinacate Peak, *F 19438.* Sykes Crater, *F 18973.* Moon Crater, *F 19101.* Pinacate Junction, *F 86-355.* Sierra del Rosario, *F 19101.*

Muhlenbergia porteri Scribner ex Beal. *ZACATE APAREJO;* BUSH MUHLY; *KU:KPADAG.* Perennials from a hard, knotty, woody base. Stems numerous, slender and wiry, hard, woody, somewhat brittle, long, and branched at various places, interweaving or somewhat tangled to form broad, rather globose or clambering bushy plants often 0.5–1+ m. Leaf blades 1.9–6.0 (8) cm, readily drought deciduous. Panicles more or less globose in outline, often 6–10 cm, open, and rather sparsely flowered. (The panicle often breaks off as a unit and is blown in the wind like a miniature tumbleweed.) Spikelets purplish most of the year, but greenish-straw colored with prominent green veins on the lemmas during summer and early fall. Pedicels slender, 2–9 (17) mm. Glumes subequal, 2.0–3.2 mm. Body of lemma 3.2–3.9 mm, sparsely hairy, the spikelets otherwise glabrous. Awn on lemma (2.5) 4–9 mm. Flowering various seasons including spring and summer.

Granitic hills and mountains in the Sonoyta region, infrequent along rocky arroyos of upper bajadas and north-facing slopes, often beneath large shrubs. Bush muhly was nearly extirpated from the border region of Organ Pipe Monument by cattle grazing. After the cattle were removed in the late 1970s there was a substantial increase in *M. porteri* along the southern border (Warren & Anderson 1987). By the mid-1980s it was common on the Arizona side of the border (eastward from ca. 10 km east of Aguajita Spring), sometimes within a few meters of the border fence, but rare on the Sonora side of the fence. Also in the Sierra Pinacate mostly above ca. 700 m on the north and east sides of the mountain; scattered although often locally extensive on rocky soils, crevices, and cliffs.

Widespread to the north in Arizona and eastward across most of northern Sonora. Also northern Baja California, Chihuahua, Coahuila, Durango, Nuevo León, San Luis Potosí, Zacatecas, and southern portions of California, Nevada, Utah, and Colorado to Arizona, New Mexico, and western Texas.

Sonoyta, *F 89-3.* 10 km SW of Sonoyta, *F 88-149.* Sierra de los Tanques, *F 89-25.* Sierra Pinacate, near summit, *F 18663.*

PANICUM MILLET, PANIC GRASS

Annuals or perennials of diverse habit and size. Spikelets awnless, of 2 florets, the lower one staminate or reduced to a lemma and palea or lemma only (ours), the upper one bisexual (fertile); spikelets characteristically breaking off below the glumes. (Among domesticated panicums the grain itself may break off, leaving the glumes and sterile lemma.) Both glumes usually present, the lower one often short and encircling the spikelet base; upper glume and sterile lemma similar in texture. Lemma of fertile floret firm, usually smooth and shiny, with inrolled margins tightly clasping the palea and enclosed grain.

More than 500 species, one of the largest genera of grasses in the world; pantropical, extending to warm-temperate regions in North America. Economic members include *P. miliaceum* (proso, broomcorn, hog, or common millet), an Old World domesticated grain grown for human and animal food and as birdseed; and *P. sumatrense* (Indian millet), a cultivated grain crop of southern Asia. References: Fairbrothers 1953; Zuloaga 1987; Zuloaga & Morrone 1996.

1. Fertile floret sessile (not stalked), the fertile lemma without fleshy expansions. ____ **P. hirticaule**
1′ Fertile floret on a very short stalk, the fertile lemma with 2 ear-like fleshy expansions at its base. ____ **P. alatum**
  2. Fertile floret strongly papillose on the entire surface. ____ **P. alatum** var. **alatum**
  2′ Fertile floret smooth, shiny, with papillae near the apex of the palea. ____ **P. alatum** var. **minus**

Panicum alatum Zuloaga & Morrone. Resembling *P. hirticaule* except in having spikelets with the upper (fertile) floret very short-stalked (stipitate) and 2 ear-like small fleshy expansions at the base of the fertile lemma. Southwestern United States to South America. Two varieties are known from the flora area, and the third variety occurs elsewhere in the desert of western Sonora.

The distinctions between *P. alatum* and *P. hirticaule* and their varieties are easy enough to discern with magnification. However, they are sometimes sympatric (e.g., *F 86-339A, F 86-339B*). Furthermore, they do not show obvious differences in geographic or ecologic distribution, and there seems to be no morphological distinction other than the spikelet characters.

Panicum alatum var. **alatum.** Spikelets 2.4–3.0 (3.2) mm; fertile floret strongly papillose on the entire surface. Southwestern United States and northern Mexico to Michoacán.

Cerro Colorado, *F 10781.* Playa Díaz, *F 10809A.*

Panicum alatum var. **minus** (Andersson) Zuloaga & Morrone [*P. hirticaule* var. *minus* Andersson] Spikelets 2.4–3.0 (3.2) mm long; fertile floret smooth, shiny, with papillae near the apex of the palea. Southwestern United States to South America.

Sierra Blanca, "pastizal efímeras," *Equihua 24 Aug 1982.* Suvuk, *Nabhan 365, 385.* Tinajas de los Pápagos, *F 86-496.* Pinacate Junction, *F 86-339B.*

Panicum hirticaule J. Presl. Zuloaga & Morrone (1996) describe 2 varieties. In var. *hirticaule* the fertile lemma is smooth. Var. *verrucosum,* distinguished by its conspicuously papillate fertile lemma, is not known from northwestern Sonora—the nearest records are from the Baboquivari Mountains in southern Arizona. Following the work of Reeder & Reeder (1998) I am recognizing the domesticated Sonoran panic grass as a third variety.

Panicum hirticaule var. **hirticaule** [*P. capillare* Linnaeus var. *hirticaule* (J. Presl) Gould] Hot-weather ephemerals, (5) 20–60 cm; mostly erect, often with a single main axis. Roots often weakly developed. Stems and leaves sometimes glabrate, or with spreading coarse usually papillose-based hairs, especially on the leaf sheaths, lower leaf-blade surfaces, and stems; flowering branches and spikelets glabrous. Panicles erect, the branches mostly straight and not drooping. Spikelets 2.7–3.1 mm, lower glume 1.5–2.0 mm (ca. 1/2 to 3/4 as long as spikelet). Glumes and lemmas with conspicuous longitudinal (parallel) green veins. Fertile lemma smooth and shiny. Grain 1.6–2.0 × 0.7–1.0 mm, shiny cream-white, becoming dark brown with age.

Nearly throughout the lowlands in the vicinity of the Pinacate volcanic complex, the Sonoyta region, at least the eastern granitic ranges, the intervening desert flats, and as an agricultural weed. Temporarily wet soils such as arroyo beds, playas, and shallow watercourses, and in favorable years widespread on open sandy flats and plains or even rocky desert; sometimes seasonally abundant and forming extensive stands. Widespread in the Sonoran Desert; southwestern United States to South America and the West Indies.

Pinacate Junction, *F 86-339A*. 1 km NW of Tezontle cinder mine, *F 86-378*.

Panicum hirticaule var. **miliaceum** (Vasey) Beetle [*P. capillare* Linnaeus var. *miliaceum* Vasey, *Contr. U.S. Natl. Herb.* 1:28, 1890. Not *P. miliaceum* Linnaeus. *P. sonorum* Beal, *Grasses N. Amer.* 2:130, 1896] SONORAN PANIC GRASS; *SHIMCHA, HESHMICHA* (COCOPA). Robust plants distinguished from var. *hirticaule* by usually larger, more floriferous panicles with generally drooping branches and other characteristics discussed below.

Once common along the lower Río Colorado. Castetter & Bell (1951) reported that Sonoran panic grass was an important crop cultivated (they say "semicultivated") by the Cocopa and other Yuman people along the lower Río Colorado and harvested for the edible grain. It has not been cultivated there since the early twentieth century, following the demise of the Río Colorado delta ecosystem and the Cocopa agricultural system that depended upon the living river. Specimens of the extinct panic grass grown by the Cocopa appear indistinguishable from the panic grass grown by the Guarijío people, but the precise relationships remain unknown. Presently this grass is grown only by a few conservative Guarijío families in the remote mountain areas of the Río Mayo region in southeastern Sonora, and by the Native Seed/SEARCH organization in Tucson from seed obtained from the Guarijío in recent years.

Nabhan & de Wet (1984) report a "wild, weedy form of *P. sonorum*" for the lower Río Colorado, subtropical regions of southeastern Sonora and southwestern Chihuahua, and southward along the Sierra Madre Occidental into Nayarit and perhaps southward. Their "wild race" can only be one of the other two varieties of *P. hirticaule* or *P. alatum*. They list 14 characters separating the "wild race" from the domesticated one, and report that the wild race differs from the domesticated one in its smaller, darker grain, and longer and denser hairs on the leaf blade. The "domesticated race" is reported to have a softer and sweet rather than bitter fruit case, reduced seed dormancy, and generally larger grain. They discuss plants with characters intermediate between the wild and domesticated "forms," but this information needs critical evaluation. The Guarijío treat var. *hirticaule* and var. *miliaceum* as distinct ethnospecies (Felger & David Yetman, unpublished field notes, 1997).

Near Lerdo, 1889, *Palmer 947* (US 2903025, type of *P. capillare* Linnaeus var. *miliaceum* Vasey). *Palmer without date,* "seeds used for food by the Indians, Sonora" (US 742141; presumably from near Lerdo, 1889, probably an isotype, the herbarium sheet is stamped "herbarium of F. Lamson-Scribner purchased by A. S. Hitchcock Feb. 2, 1905)." *Palmer in 1889,* "seeds largely used as food by the Cocopas" (US 973378; presumably from near Lerdo and probably another isotype).

PENNISETUM

Annuals or perennials, sometimes large and rather woody. Leaf blades flat or folded. Inflorescences of dense, spikelike panicles. Spikelets surrounded by separate (distinct) bristles or the bristles scarcely united at their bases, these forming a fascicle that falls with the spikelets as a unit, the bristles often plumose. Spikelets with 1 staminate or reduced and sterile floret below, and 1 fertile floret above; glumes 1 or 2, unequal. Worldwide, mostly tropics and subtropics; 100 species.

Pennisetum and *Cenchrus* are closely allied. The *Pennisetum* fascicle is less specialized than the *Cenchrus* bur. In *Pennisetum* the bristles are distinct or nearly so, whereas in *Cenchrus* the bristles (spines) are clearly united for a substantial portion of their length and are usually hard and spinescent; however, the variation is continuous. Among *Pennisetum* $x = 9$ seems to be the base chromosome number, whereas among *Cenchrus* sensu stricto $x = 17$. Intermediates such as *P. ciliare* have fascicles with bristles scarcely fused at the base. Economically important Old World members include pearl millet (*P. glaucum,* = *P. americanum, P. typhoides*), kikuyu grass (*P. clandestinum*), and elephant or Napier grass (*P. purpureum*). References: Chase 1921; DeLisle 1963; Pohl 1980; Sohns 1955.

1. Panicles 4.5–12.5 cm; longer fascicle bristles 8–13 mm. ________ **P. ciliare**
1′ Panicles 13–30 cm; longer fascicle bristles 22–34 mm. ________ **P. setaceum**

***Pennisetum ciliare** (Linnaeus) Link [*Cenchrus ciliaris* Linnaeus. DeLisle lists 50 synonyms] *ZACATE BUFFEL;* BUFFELGRASS. Perennials often with a knotty base or sometimes annuals in the flora area, 50–80 (100) cm. Roots thick and often encased in a jacket of sand laced together by a feltlike mat of root hairs (see *Aristida californica*). Stems erect to geniculate spreading. Leaves often reddish during cool weather; sheaths laterally compressed and keeled, glabrate to sparsely pilose with long slender hairs; hairs near collar usually sparse, white, and straight, reaching 5 mm with enlarged bases, the ligule membranous, ciliate; blades glabrate, scabrous, or sparsely pilose, especially on lower surface, larger blades 23–34 cm × 7–10 mm, the midrib prominent.

Panicles 4.5–12.5 cm, dense and spikelike. Fascicles crowded, essentially sessile with a thick, nubbin-like pedicellate stipe 0.3–0.5 mm; bristles mostly united at the very base, flexible and somewhat sinuous, purplish brown, unequal, the inner ones largest, plumose basally with spreading silky white hairs, the longest bristles 8–13.5 mm and conspicuously flattened, the outer bristles in a loose ring around the fascicle base, much smaller, many, slender, terete, and scabrous. Spikelets 4.5–5.5 mm, sessile or subsessile, 2 or 3 per fascicle.

Native to the warmer parts of Africa, Madagascar, and India, and widely introduced in hot, semiarid regions of the world for forage and fodder. This is the most extensively planted forage grass in Sonora—vast tracts of natural vegetation have been cleared and planted with buffelgrass. Plantings in Sonora began in the 1950s. Since the late 1960s and early 1970s it has spread into many natural areas in Sonora and Arizona, has become well established as an urban weed, and is particularly abundant along roadsides. Although not planted in the flora area, by the late 1980s it was fairly well established as a roadside weed. By that time it had become a common weed in the vicinity of Sonoyta and was scattered along the roadside of Mex 2 across the Pinacate region and the western granitic mountains, along various well-traveled dirt roads, and around agricultural fields. I have not seen it in undisturbed, natural desert areas in the flora area except occasionally in areas adjacent to dense roadside stands. I did not see it in northwestern Sonora during the early 1970s.

Because buffelgrass is such a hardy, drought-resistant, and excellent forage grass, considerable effort has been expended in new selections and introductions. Unfortunately, it is also a seriously invasive weed, and if it reaches more favorable habitats such as higher elevations in the Pinacate volcanic region, it will probably displace local native plants. Much of the variation relates to the variability in reproduction and chromosome numbers ($2n = 32, 34, 36, 38, 40, 45, 52, 54$; see Crins 1991). Reproduction is largely apomictic but facultative sexual reproduction is common (e.g., Bray 1978).

Two errors in identification of historic collections could be misleading with regard to dates of introduction for *P. ciliare* in Sonora. DeLisle (1963:325) cites a 1910 specimen of *Cenchrus ciliaris* from Sonora (*Rose et al. 12866*). It is actually *P. karwinskyi* Schrader (*C. multiflorus* J. Presl; Chase 1921), a species of subtropical deciduous forest or thornscrub. Another collection of *P. karwinskyi* from Sonora (*Wiggins & Rollins 429* in 1941) was originally misidentified and distributed as *C. ciliaris* (DeLisle 1963).

4 mi E of La Joyita on Mex 2, 21 May 1987, *Prigge 7244* (UCR). 38 km W of Sonoyta on Mex 2, 14 May 1985, *F 85-714.* S of Pinacate Junction, 14 Sep 1986, *F 86-374.* N of Puerto Peñasco, 25 Jun 1985, *F 85-774.* Observation: 7 mi S of San Luis, 6 Oct 1985.

***†Pennisetum setaceum** (Forsskal) Chiovenda [*Phalaris setacea* Forsskal. *Pennisetum ruppelii* Steudel] FOUNTAIN GRASS. Robust, densely tufted perennials often to 1 m. Panicles 13–30 cm, purplish during drier or cooler seasons. Fascicles on slender, bristly peduncles 1.0–2.5 mm; fascicle bristles separate, flexible and somewhat sinuous, terete, unequal, the longer ones 22–34 mm, plumose basally. Spikelets 5.5–6.5 mm, mostly on slender pedicels to 1.0 mm, 2 or 3 per fascicle. Reproductive most of the year, especially during warmer months.

Native to Africa. Common landscape plant in southern Arizona; weedy and spreading into natural habitats. Cultivated and weedy in Sonoyta but so far not spreading or naturalized in northwestern Sonora, although since 1986 it has spread northward along Arizona Highway 86 in Organ Pipe Monument (Felger 1990). If it reaches the higher elevations in the Sierra Pinacate, it probably will displace many native plants.

PHALARIS CANARY GRASS

Annuals or perennials. Panicles mostly spikelike, ovoid to cylindrical, sometimes with a few short appressed branches below. Spikelets awnless, laterally compressed, densely crowded, overlapping, breaking off above the glumes, and sometimes the panicle branchlets breaking apart. Glumes equal or nearly so, often boat-shaped and keeled, the keels frequently winged. Spikelets with 1 bisexual floret and 1 or 2 small, scalelike sterile florets below, these falling with the fertile floret, or sterile florets absent. Fertile lemma glabrous or hairy, often shiny, tightly clasping the palea and ovary.

Mostly temperate North America, Europe, and Africa, a few in southern South America, the center of diversity in the Mediterranean region; 15 species. In modern times several have spread by human agency and are worldwide. *P. canariensis* furnishes the canary seed of commerce, and Harding grass (*P. tuberosa*) is cultivated for forage. *P. canariensis* has been cultivated as a minor crop for food and glue making since ancient times. References: Anderson 1961; Baldini 1995.

1. Sterile florets 2; keel of glumes scarcely winged, the wing neither notched nor toothed. ____________ **P. caroliniana**

1′ Sterile florets 1; keel of glumes conspicuously winged and often notched or toothed. ____________ **P. minor**

***Phalaris caroliniana** Walter. *ALPISTILLO, ALPISTE SILVESTRE;* CAROLINA CANARY GRASS; *BA:BKAM.* Winter-spring ephemerals, highly variable in size. Stems often 1 to several. Panicles spikelike, ovoid to subovoid, usually 2.5–4.2 cm. Glumes 4.5–6.0 mm, the keel with a narrow wing toward the summit, the wing green, relatively thick, and not notched, or the wing virtually absent. Sterile lemmas 2, one on each side at base of the fertile floret. Fertile lemmas ovate-lanceolate (teardrop-shaped), 3.5–4.0 mm, and densely hairy.

In the flora area known only from the heavily grazed playa near Rancho los Vidrios (*Ezcurra 7 Oct 1981,* MEXU); not native. Known from a number of localities in southwestern Arizona. Oregon to Colorado and Virginia to the northern 2/3 of Mexico including Sonora and Baja California.

***Phalaris minor** Retzius. *ALPISTILLO, ALPISTE SILVESTRE;* LITTLE-SEED CANARY GRASS. Winter-spring annuals resembling *P. caroliniana,* highly variable in size. Panicles spikelike, mostly several centimeters in length, sometimes reaching 8–10 cm. Glumes 4.5–5.5 mm, with longitudinal green and white stripes or bands, the keel expanded into a conspicuous wing near the tip, the wings often notched or with small irregular teeth. Sterile floret 1.

Disturbed wetland habitats, roadsides, and an urban and agricultural weed; seasonally common in the vicinity of Sonoyta and San Luis, especially as a weed in wheat fields. Native to the Mediterranean, now worldwide.

S of San Luis, *F 91-69.* Roadside, 7 mi E of San Luis, *F 91-63A.* Pinacate Junction, *F 93-215.* Sonoyta, *F 85-700.*

PHRAGMITES

Perennial reeds. Several species in the temperate and warm regions: one in tropical and temperate regions worldwide, the others only in the Old World. Reference: Conert 1961.

Phragmites australis (Cavanilles) Trinius ex Steudel [*Arundo australis* Cavanilles. *A. phragmites* Linnaeus. *Phragmites communis* Trinius] *CARRIZO;* REEDGRASS, COMMON REED; *WA:PK.* Bamboo-like reeds 2–5 m tall with strong rhizomes and tough roots. Stems stout, reaching 1–1.5 cm in diameter, and also often producing numerous smaller stems as slender as 1.7 mm in diameter. Leaves large, 2-ranked and evenly spaced, the blades (7) 30–40 cm. Panicles terminal, large, and plumelike, reaching 30–45 cm. Spikelets several-flowered, the lower one staminate or sterile, the others bisexual. Rachillas with long silky white hairs, the lemmas glabrous.

Emergent from shallow brackish water or alkaline wet soil at springs. The roots often deeply buried in semi-saline or alkaline mud. New shoots appear in spring and rapidly grow to full height. Most of the leaves and many of the stems die off during the winter. Banks of the lower Río Colorado, Ciénega de Santa Clara, Laguna Prieta, pozos at La Salina, Río Sonoyta, and Burro Spring near Quitobaquito. Great stands once grew along the lower Río Colorado and its delta (Castetter & Bell 1951). At the Ciénega de Santa Clara, where it is emergent from water often as deep as 1 m, it covered many thousands of hectares in the late 1980s and 1990s (Zengel et al. 1995), and it was locally common at Laguna Prieta.

On February 17, 1699, Padre Kino and his party camped "en un carrizal que hace el arroyo" ('in a stand of carrizo situated in the arroyo'), "about 25 miles" westward from Sonoyta (Burrus 1971:230, 391). Kino named this place Los Carrizales. Ronald Ives (in Burrus 1971) shows it along the Río Sonoyta about 10 km downstream from Quitobaquito. In the 1990s a band of several large patches, grew along the east bank of the Río Sonoyta at the base of an alkaline bajada of Sierra de los Tanques (to the east), apparently the site of an underground seep or spring. This population is about 5 km upstream from where Ives maps Kino's Los Carrizales. No other *Phragmites* is known from the Río Sonoyta. Several widely scattered, highly localized *Phragmites* populations occur at waterholes and wetland habitats elsewhere in western Sonora and on Isla Tiburón (Felger & Moser 1985).

The "fruits usually disarticulate beneath the long-bearded rachilla, which probably aids in wind dispersal. They are also dispersed to some extent by water currents" (Croat 1978:147). In arid regions such as the Sonoran Desert, birds flying between waterholes might be dispersal agents. It occurs on every vegetated continent on earth and is perhaps the world's most widespread species of flowering plant.

Reedgrass was an important native resource. The stems were used for housing, arrow shafts, musical instruments, cordage, pipes, containers, knives, eating utensils, and many other practical and aesthetic purposes (Bean & Saubel 1972; Castetter & Bell 1951; Felger & Moser 1985; Uphof 1968). The Seri used the stems to construct ocean-going reed boats, or *balsas,* and people living near the lower Río Colorado made similar craft. Although the young shoots are edible, there is no indication that reedgrass was so used in the flora area. The Cocopa and others collected a sweet manna-like "honeydew," the exudate of aphids, from the leaves as a highly esteemed food (Heizer 1945; Jones 1945).

One might confuse *P. australis* with *Arundo donax,* which is extensively cultivated in the Sonoran Desert. *Arundo,* native to the Old World, is widely naturalized in the New World. *Arundo* generally is more robust, produces only large stems, and the base of the blade, or collar, clasping the stem is often somewhat swollen.(*Phragmites* usually has both large and slender stems.) The two reedgrasses can also be distinguished by microscopic examination of the phytoliths (Ollendorf, Mulholland, & Rapp 1988). *Arundo* (at least in the Sonoran Desert region) has a darker area or band at the junction of the leaf blade and sheath which apparently is not present in *Phragmites.* Some authors indicate a difference in the hairs around the auricle at the base of the leaf blade, but this feature does not seem to separate the two. The sure way to distinguish them is by the spikelets: *Phragmites* has densely hairy rachillas and glabrous lemmas, whereas *Arundo* has glabrous rachillas and conspicuously hairy lemmas.

12 km S of San Luis, *F 93-03*. Río Colorado 20 mi N of El Golfo, *F 75-57*. Ciénega de Santa Clara, *F 92-992*. La Salina, *F 86-544*. Laguna Prieta, *F 85-738*. Río Sonoyta S of El Huérfano, *F 92-973*.

PLEURAPHIS

Perennials, often with rhizomes or stolons. Leaves usually relatively short. Inflorescence a slender, densely flowered spike. Spikelets in clusters of 3 at each node of a zigzag rachis, the clusters falling as a unit. Central spikelet with 1 bisexual floret, and each lateral spikelet with 2 (3) staminate flowers. Glumes relatively thin, papery, separate, usually variously awned. Lemmas thin, short-awned or awnless. Dry regions in southwestern United States and northern Mexico; 3 species.

Among most populations of *Pleuraphis* one can usually find some plants in flower at any time of year if there is "any moisture at all" (John Reeder, personal communication 1996). The basic chromosome number is probably $x = 9$. There are no documented diploids; the lowest chromosome counts are for tetraploids with $2n = 36$. In *P. jamesii* $2n = 36$. Both *P. mutica* and *P. rigida* show amazing geographic chromosome races with $2n$ ranging from 36 to 108 for the latter and to 180 for the former (Reeder 1971, 1984). The separation of *Pleuraphis* from *Hilaria* is based on anatomical, molecular, and morphological characteristics (Travis Columbus, personal communication 1999). *Hilaria* is distinguished by its tough, firm, and relatively thick glumes conspicuously fused toward the bases, and by pistillate-flowered central spikelet. Reference: Sohns 1956.

Pleuraphis rigida Thurber [*Hilaria rigida* (Thurber) Bentham ex Scribner] *TOBOSO;* BIG GALLETA. Large tufted perennials forming dense clumps, often 0.5–1.2 m, the bases tough and knotty with short, almost woody rhizomes; roots tough, wiry, and well developed. Stems rigid and stout, with feltlike white hairs readily rubbing off. Leaf sheaths glabrous; blades spreading, (4) 7–23 cm, the length apparently depending upon soil moisture—shorter blades commonly stiff and sharp-tipped. Plants dormant or with considerable dieback during extended drought.

Inflorescences spikelike, (4.5) 6–14 cm, the zigzag main axis remaining long after the spikelets fall; sometimes developing a large, swollen gall structure 5–6 × 2–3 cm. Spikelet clusters dense, overlapping, with a dense basal tuft of white hair. Spikelets (excluding awns) (6.2) 7.2–10.0 mm, fringed on their blunt tips. Glumes, lemmas, and paleas more or less equal in length. Number and size of awns varying with position of spikelet and from one cluster to the next; glumes and lemmas each with (3 or 4) 5–7 veins, all to some or none extending into awns of 0.7–7.0 mm. Lemma margins densely ciliate with silky hairs. Spikelets often tinged pink to purple-brown during cool, dry seasons; usually drying straw colored with prominent veins. Flowering facultatively at any season.

Widespread through most of the flora area; arroyos, washes, sand flats and dunes, and less common in rocky places in both the granitic ranges and volcanic complex, near sea level to about 1000 m near the base of Pinacate Peak. Especially abundant on sand flats and lower dunes. Southwestern Utah, southern Nevada, southeastern California, western Arizona, northwestern Sonora southward along the coast to the vicinity of Puerto Libertad, and northeastern Baja California southward to the vicinity of San Felipe.

This is the most common and widespread large perennial grass in the flora area. Its success in such dry environments seems to be due in part to the large underground system of roots, stems, and rhizomes, as well as facultative variation in size, morphology, and flowering time. The maximum CO_2 uptake rate is very high over a wide range of temperatures—hence productivity can be substantial during any season. The high photosynthetic rates and high water-use efficiency also help explain its success in desert environments (Nobel 1980c). It is the most arid-inhabiting species in the genus. The spikelets are unmistakable and seem to exhibit more variation than other grasses in the region. The intrepid Arizona botanical collector Leslie N. Goodding called it "the redemption of the desert" (Charlotte Goodding Reeder, personal communication 1996).

There are at least 2 chromosomal races: Reeder (1984) found $2n = 36$ for plants from Baja California and in 6 counts from the flora area and southwestern Arizona. Plants from Utah, Nevada, and northern Arizona were $2n = 108$, interpreted as duodecaploids (Reeder 1977). There are no apparent morphological differences between the southern diploids and the northern multi-polyploids. (The report of $2n = 18$ by Holmgren & Holmgren [1977:48] is viewed with suspicion, especially if it was obtained from plants in the region of coverage of the Intermountain Flora where Reeder documented $2n = 108$.) Because polyploidy is almost invariably a derived condition, it is presumed that the diploid is more primitive. *P. rigida* and *Distichlis palmeri* are the only grasses that show evidence of originating in the arid region surrounding the northern Gulf of California, well within the Sonoran Desert.

Bahía de la Cholla, *Gould 4133.* 15 km SE of López Collada, *Ezcurra 3 Apr 1981* (MEXU). Colonia Díaz, *Mearns 2848.* N of Cerro Colorado, *F 86-373.* Sykes Crater, *F 20030.* Moon Crater, *F 19014.* NE of Sierra del Rosario, *F 20423.* 19 mi E of San Luis on Mex 2, *F 86-390.* Arizona, Yuma County: 5.5 km W of Nameer's Grave, *Reeder & Reeder 7589* ($2n = 36$).

POA BLUEGRASS

Mostly perennials, of diverse growth habits. Leaf blades narrow, the tip boat-shaped; ligules membranous. Inflorescences of terminal panicles, the branches usually clustered. Spikelets awnless, (1) 2 to several-flowered, laterally compressed, breaking apart above the glumes and between the florets. Uppermost florets reduced or rudimentary. Glumes slightly unequal, usually shorter than the first lemma, the lower glume usually shorter than the upper glume. Lemmas 5-veined, the margins membranous (scarious), and the base or callus often with cottony, cobwebby tufts of hair (a useful diagnostic character).

Worldwide, mostly temperate, arctic, and montane at lower latitudes; perhaps 150 species. Very few in deserts; only 2 *Poa* species, both winter-spring annuals, occur within the Sonoran Desert. Reference: Marsh 1952.

1. Spikelets 3.6–4.6 mm; lemmas 2.1–2.7 mm, glabrous or with silky hairs on veins but without a basal basal cottony web; well-watered urban habitats. ____ **P. annua**

1′ Spikelets 4.5–8 mm; lemmas 3.2–4.4 mm with a cottony web at base; desert habitats and roadsides. ____ **P. bigelovii**

***Poa annua** Linnaeus. *ZACATE AZUL ANUAL, PASTITO DE INVIERNO;* ANNUAL BLUEGRASS, WINTERGRASS. Mostly 5–15 cm, soft, glabrous, bright green, erect to spreading, branching from the base. Panicles pyramidal, 2–5 cm. Spikelets clustered at ends of branches, 3.6–4.6 mm, often with (2) 3 or 4 florets plus rudiments. Lower glume smaller than the upper one; upper glume 1.8–2.3 mm. Lemmas 2.1–2.7 mm, ovate, blunt-tipped, the veins silky-haired basally or sometimes essentially glabrous. Palea with minute silky hairs on the 2 keels.

This dwarf grass is a common lawn and garden weed in Sonoyta (*F 87-13*), and was formerly at Quitobaquito (17 Mar 1945, *Darrow 2405*). In the Sonoran Desert it tends to be small and short-lived, and seldom becomes established away from well-watered settlements or agricultural areas. Native to Europe and Asia, naturalized worldwide.

Poa bigelovii Vasey & Scribner. Mostly erect, (8) 12–55 cm, with weakly developed roots and delicate, slender stems. Spikelets pale green, 4.5–8.0 mm, with 3–8 florets, the young spikelets spear-shaped. Florets overlapping and compressed (flattened) against each other, spreading apart at maturity with both margins of the florets becoming visible. Lemmas 3.2–4.4 mm, the margins white hairy and membranous, the base with a dense cottony tuft or web.

Arizona Upland near Sonoyta; during favorable years along roadsides but mostly in small arroyos or washes; often growing in the protection of spiny shrubs. Common elsewhere in Arizona Upland, to the north and east of the flora area in Sonora and Arizona including Organ Pipe Monument and Cabeza Prieta Refuge. Lowland western Sonora south to the Sierra del Viejo near Caborca and a surprising disjunct in the Río Mayo region in southeastern Sonora. Southern California, Baja California, and Sonora to southern Utah, southern Colorado, Oklahoma, western Texas, Chihuahua, Coahuila, and Nuevo León.

Aguajita, *F 92-111.* 10 and 18 km SW of Sonoyta, *F 88-151, F 88-209.* Sierra del Viejo SW of Caborca, *Sanders 3527* (UCR).

POLYPOGON BEARD GRASS

Annuals and perennials. Leaf blades flat, thin, bright green. Panicles dense and contracted. Spikelets small, breaking away below the glumes, 1-flowered, mostly awned. Glumes equal, longer than the floret, somewhat scabrous, often with a slender awn. Lemmas membranous, 5-veined. Grain widest above the middle.

Worldwide, mostly warm-temperate regions and montane tropics; 18 species. Often in moist habitats, many in alkaline or partially saline wetlands.

This genus is closely related to *Agrostis* and hybridizes with it. *Agrostis,* mostly in temperate and cold regions, occurs at higher elevations in Arizona and Sonora, does not extend into the deserts, and is not especially salt or alkaline tolerant. *P. monspeliensis* and *P. viridis* are tetraploids, $2n = 28$, $x = 7$. The report of $2n = 14$ for New World *P. viridis* (Brown 1950) is undoubtedly in error, as are some other chromosome numbers in that report (John Reeder, personal communication 1993). References: Björkman 1960; Tutin 1980.

1. Glumes long-awned; ephemerals. ____ **P. monspeliensis**
1′ Glumes awnless; perennials. ____ **P. viridis**

***Polypogon monspeliensis** (Linnaeus) Desfontaines [*Alopecurus monspeliensis* Linnaeus] *ZACATE COLA DE ZORRA;* RABBITFOOT GRASS. Non-seasonal ephemerals, 8–100 cm. Leaf blades 3.5–22 cm × 5–10 mm. Panicles (1.5) 3–15 cm, densely flowered with tawny awns, terminal on a long stem, spikelike and sometimes with short, dense branches below. Glumes each with a slender awn 4–7 mm; body of glume 1.5–2.2 mm. Lemmas and paleas thin, translucent, slightly exceeding the grain, the lemma with a delicate, deciduous awn ca. 1 mm long. Grain about half as long as body of glume.

Alkaline soils along the Río Sonoyta, lower Río Colorado including the Ciénega de Santa Clara, and weeds in agricultural and urban areas, sometimes at roadsides; occasional in washes and arroyos—as long as the soil is at least temporarily wet. Also at La Salina in partially saline or alkaline wet soils with cattails (*Typha*), and sporadically in semi-saline soils along the coast with saltgrass (*Distichlis spicata*). It seems to be spreading in the flora area with the advance of human disturbance.

Elsewhere in western Sonora, e.g., near Bahía Kino and Guaymas, also on alkaline and semi-saline soils, such as irrigation canals and brackish marshes beneath cattails (*Typha*). Native to Europe, widely naturalized in western North America.

Río Sonoyta, S of El Papalote, *F 86-200.* Quitobaquito, *F 7677.* Arroyo Batamote, *F 87-29.* La Salina, *F 86-545.* Wellton-Mohawk drainage canal, *F 90-196.*

***Polypogon viridis** (Gouan) Breistroffer [*Agrostis viridis* Gouan. *A. semiverticillata* (Forsskal) C. Christensen] WATER BENT. Stoloniferous perennials, the stems erect to decumbent, reaching 30–50 cm, rooting at nodes. Leaf blades flat, 5.5–14 cm × 3.5–13 mm. Panicles 5–8 (12) cm, sometimes interrupted and with a few branches. Spikelets awnless. Glumes 1.4–2.1 mm, minutely scabrous and appearing speckled, membranous and translucent with a broad green midstripe or sometimes partially or all purplish. Lemmas 0.8–1.0 mm. Grain slightly shorter than the lemma. Growing and flowering March through the warmer months.

Well established in wet, alkaline-muddy soil and shallow water at Quitobaquito Springs (*Darrow 2409*), nearby Burro Spring, and Quitovac. Native to Eurasia, now worldwide. In the Sonoran Desert region *P. viridis* grows with *P. monspeliensis,* although *P. viridis* has a much more restricted distribution.

SCHISMUS MEDITERRANEAN GRASS

Small, tufted annuals (ours) and perennials with contracted panicles. Spikelets several-flowered, awnless, breaking apart above glumes and between florets. Glumes about as long as the entire spikelet, with membranous white margins. Lemmas rounded on back and notched at tip.

Five species; native to Africa and Eurasia. During years of favorable winter rains *S. arabicus* and *S. barbatus* are abundant across much of the northern part of the Sonoran Desert, forming dense stands sometimes to the exclusion of native ephemerals. They germinate and grow rapidly with cool-weather rains. The first stems and leaves often spread out close to the ground, excluding or preventing other plants from sprouting. Both species are widely naturalized in arid and semiarid regions around the world. References: Conert & Türpe 1974; Hoover 1936.

1. Lemmas hairy on back and margin, the palea shorter than the lemma, usually not reaching the lemma notch. ______ **S. arabicus**

1′ Lemmas glabrous (occasionally with few hairs below) or with some hairs on margin, the palea about as long as the lemma. ______ **S. barbatus**

***Schismus arabicus** Nees [*S. barbatus* subsp. *arabicus* (Nees) Maire & Weiller] Winter-spring ephemerals. Stems (4) 10–20 (22) cm, erect to often spreading or semiprostrate. Leaves mostly basal, the blades soft, bright green, narrow. Panicles compact, many-flowered. Glumes (2.9) 3.5–5.5 (6.2) mm, often purple tinged; lemmas 1.5–2.4 mm, the margin and back hairy, the apex shallowly to deeply notched, the lobes often acute; palea shorter than the lemma, usually not reaching the notch. Grain 0.7–0.9 mm, shiny golden brown, partly transparent, the embryo visible.

Seasonally abundant throughout the flora area except apparently absent from higher elevations of the Sierra Pinacate and the granitic mountains. Most abundant on sandy soils of sand flats, arroyos, and washes, interdune troughs, and at the base of larger dunes; also an agricultural and urban weed. The southernmost record in Sonora is from near Puerto Libertad. Widespread in southwestern United States and northwestern Mexico. The earliest North American record is 1933 in Arizona (Felger 1990).

Differences between *S. arabicus* and *S. barbatus* are subtle. Gould (1951) believed that they broadly intergrade. Although a case could be made to treat them as subspecies, most modern agrostologists maintain them as separate species (John Reeder, personal communication 1991). Conert & Türpe (1974:70) report that "in material from areas where the distribution of the two annual species overlaps the morphological distinctions often become obscured. Nevertheless, each species can be recognized by applying the given diacritical characters."

The two taxa are geographically segregated in the Old World; *S. arabicus* from southwest Africa to the western and northern Sahara and the western Mediterranean region, and *S. barbatus* from Kashmir and southern Russia west to Greece. They also show some ecological or geographic segregation in northwestern Sonora.

N of Sierra del Rosario, *F 20776*. 6 mi W of Los Vidrios, *F 20586*. MacDougal Crater, *Turner 86-18*. Hornaday Mts, *Burgess 6810*. 4 km W of Sonoyta, *F 86-147*. Puerto Peñasco, *F 85-764*.

***Schismus barbatus** (Linnaeus) Thellung [*Festuca barbata* Linnaeus] Resembling *S. arabicus* but the glumes tending to be slightly shorter, the lemma usually glabrous on the back or with hairs on the margin or occasionally near the base, the apical notch often shallow or minute, and the palea about as long as the lemma.

Common in the Sonoyta region and adjacent Arizona. Widespread in arid regions of southwestern United States and Baja California. This species does not seem to be as widespread in the flora area as *S. arabicus,* and does not seem to extend southward in Sonora. The earliest record in North America seems to be 1926 from southern Arizona, and by 1928 it had become established (Felger 1990). It is a good forage grass although the plants are usually too small to be significant.

Río Sonoyta, 10 km W of Sonoyta, *F 86-189*. 4 km E of San Luis, *Van Devender 92-352*. Aguajita, *Bowers 1043*.

SETARIA Bristlegrass

Annuals or perennials. Panicles contracted, usually spikelike with many spikelets and bristly; bristles scabrous, 1 to several below each spikelet and remaining on the plant when the spikelet is shed. (The bristles represent reduced panicle branches.) Spikelets awnless, 2-flowered, the lower floret staminate or sterile, the upper one fertile; fertile lemma firm and tightly grasping the palea and grain.

Tropical to warm-temperate regions, few in deserts; 100 species. Not known from the flora area but occurring nearby. This absence may be due to habitat destruction in the Sonoyta region and elsewhere near the Arizona border, as well as the relative paucity of dependable summer rains. Reference: Rominger 1962.

1. Perennials, roots stout and coarse. ____ **S. leucopila**
1′ Summer-fall ephemerals, roots weak.
  2. Fertile lemma very finely rugulose (evenly textured with a fine, beadlike textured surface, the beads in rows); spikelet more or less oval in outline, widest at middle. ____ **S. grisebachii**
  2′ Fertile lemma coarsely rugose with transverse wrinkles; spikelet diamond-shaped, widest below middle. ____ **S. liebmannii**

†Setaria grisebachii E. Fournier. Summer-fall ephemerals. Central America to southwestern United States. In Arizona and Sonora mostly at elevations slightly above the desert. Nearest record: Ajo Mountains, *L. N. Goodding 25 Sept 1943.*

†Setaria leucopila (Scribner & Merrill) K. Schumann [*Chaetochloa leucopila* Scribner & Merrill] Tufted perennials. Small arroyos and washes, mostly among mesquite brush, and rocky slopes. Near the Sonora border in Organ Pipe Monument and Sonora eastward from the northeastern edge of Sonoyta. Perhaps elsewhere in the Sonoyta region before recent habitat destruction. Also just north of the flora area in the Cabeza Prieta Refuge. Central and northwestern Mexico and southwestern United States. Nearest records: 1 km E of U.S. border crossing station at Lukeville, 15–20 m N of border fence, 1390 ft, *F 87-313.* Agua Dulce Spring, 1400 ft, Cabeza Prieta Refuge, *Simmons 24 Jan 1965.*

†Setaria liebmannii E. Fournier. Summer-fall ephemerals. It should be sought along the Río Sonoyta and the larger, more densely vegetated arroyos and clay-soil playas on the north side of the Pinacate volcanic field. This is the only ephemeral, or annual, bristlegrass in the lowland desert and thornscrub of western Sonora and northwestern Sinaloa. Central Mexico to Arizona. Nearest record: Ajo Mountains, *L. N. Goodding 25 Sep 1943.*

SORGHUM

Annuals or perennials, often large and coarse with thick stems and large, flat leaves. Inflorescence usually a large open panicle, or compact in some domesticated forms. Spikelets in pairs (or 3s at branch ends), one of the pair (or trio) sessile and bisexual, the other(s) pedicellate and male or sterile. Each sessile spikelet falling with the adjoining piece of rachis and pedicel(s) of its companion spikelet(s). Pedicellate spikelet(s) deciduous from pedicel(s) at maturity. Sterile or male spikelets awnless. Fertile spikelets often awned, with awned and awnless ones sometimes on the same panicle. Awns stout, twisted, and readily deciduous.

Old World, mostly tropical Africa; 20 species. Sorghums have been cultivated since ancient times in the Old World, and in modern times have achieved global agronomic importance. References: de Wet 1978; Snowden 1936, 1955.

1. Annuals; often with very thick stems; inflorescence often dense, the branches often floriferous to the base; leaves usually broad and straplike, often more than 3–4 cm wide. ____ **S. bicolor**
1′ Perennials or facultative annuals or ephemerals; stems not especially thick; inflorescence open, the branches not floriferous to the base; leaves often less than 2.5 cm wide. ____ **S. halepense**

***Sorghum bicolor** (Linnaeus) Moench [*Holcus bicolor* Linnaeus] *SORGO FORRAJERO, MAIZMILO;* SORGHUM. Robust annuals with very thick stems, the inflorescences densely branched, the branches floriferous to the base. Leaves large and coarse, usually more than 2 cm wide. Pedicellate (fertile) spikelet tending to be persistent, the grain plump, filling the spikelet, and spreading the glumes.

Several dwarf varieties, members of the subsp. *bicolor* complex, usually less than 1 m, are grown in the region for grain and fodder. Occasionally encountered along roadsides, often to 50+ cm with a thick stem, unbranched or with a few closely appressed erect branches, cornlike strap-shaped leaves 3–4 cm wide, and compact, many-flowered panicles with awnless spikelets and relatively large grain. Narrow-leaved, non-dwarf sorghum varieties, rarely encountered in the flora area, might be difficult to distinguish from *S. halepense,* but the annual habit and characters of the inflorescence, spikelets, and grain should separate the two taxa. The grain-filled fertile spikelets are rounder, fatter, and tend to be more persistent than those of *S. halepense.*

Sudan grass, subsp. *drummondii* (Steudel) de Wet [*S. sudanense* (Piper) Stapf], is sometimes grown in the region as a fodder crop. It closely resembles *S. halepense* but is an obligate annual, does not form rhizomes, is not quite as robust, and has generally less succulent stems. Annual Sudan grass is often preferred because Johnson grass is perennial and difficult to eradicate.

20 mi E of San Luis, roadside, *F 16697A* (dwarf cultivar). Ejido Morelia, W of Sonoyta, cultivated crop, *F 87-308* (Sudan grass).

Sorghum halepense (Linnaeus) Persoon [*Holcus halepensis* Linnaeus] *ZACATE JOHNSON;* JOHNSON GRASS. Coarse perennials 0.5–2+ m, with strong rhizomes in better-watered habitats, or facultative annuals or ephemerals without rhizomes. Leaf blades 0.8–2.3 (5) cm wide. Inflorescences open, the branches not floriferous to the base. Pedicellate (fertile) spikelet elliptic (more slender than those of the grain sorghums) and deciduous at maturity.

Weakly established at a few roadside localities and as an urban and farm weed, especially in the vicinity of Sonoyta; apparently not persisting in natural areas of the Sonoran Desert. Grown as a forage and fodder crop, but also a noxious weed. Native to the Mediterranean, now in the warmer parts of the world.

Sonoyta, *F 85-704.* Suvuk, *Nabhan 378.* E of Puerto Peñasco, *F 85-759.* Pinacate Junction, *F 85-996.* 20 mi E of San Luis, *F 16697B.*

SPOROBOLUS DROPSEED

Annuals or perennials of diverse sizes and growth habits, mostly tufted, some with creeping rhizomes. Ligules a line of hairs (ciliate), sometimes also with large hairs at the collar and leaf sheath "behind the collar." Spikelets small, almost always 1-flowered, glabrous, awnless, breaking apart above the glumes. Glumes usually different in size, sometimes subequal. Lemmas almost always 1-veined. Grain readily falling from spikelets at maturity (hence the name "dropseed"); pericarp usually thin and closely enclosing the seed but free from it, readily soaking up water, becoming mucilaginous when wet and forcibly ejecting the seed, the "naked" seed often clinging to the tip of the palea and lemma.

Warm, especially temperate regions worldwide, most diverse in the New World; 160 species. Closely related to *Muhlenbergia,* and the generic boundaries with *Eragrostis* are blurred (Clayton & Renvoize 1986). *Sporobolus* can be distinguished by its lack of awns (some species of *Muhlenbergia* also lack awns), ciliate ligules, thin and 1-veined lemmas, and the obovoid and usually laterally compressed grain that readily falls from the spikelet. Among *Sporobolus* "the species intergrade to such an extent that their limits are seldom sharply defined, and they sometimes seem to take the form of noda in a continuum of variation" (Clayton & Renvoize 1986:225).

The young panicles are strongly contracted in *S. cryptandrus, S. flexuosus,* and *S. pyramidatus,* and the lower portion of the panicle is also sometimes contracted in *S. airoides.* As the panicle matures the branches usually spread: the contrast is striking and can be confusing for identification. "Those species of *Sporobolus* in which some or all of the inflorescences remain enclosed in the substending leaf sheaths tend to reproduce cleistogamously, that is, pollen cannot be dispersed and obligately pollinates the stigma of the same floret. The percentage of seed set in these inbred plants is usually quite high" (Yatskievych 1999:727). Among the Sonoran Desert species, growth and flowering occur during the warmer times of the year. References: Hitchcock 1935c; Sutherland 1986.

1. Ephemerals; branches at lowest node of panicle whorled with (4) 5 or more branches (concealed by leaf sheath in younger inflorescences); spikelets 1.5–1.8 mm. ____ **S. pyramidatus**

1′ Perennials; branches at lower nodes mostly solitary, very rarely with 3 branches (lower nodes of young inflorescences often concealed by the leaf sheath); spikelets 1.8–2.4 mm.

  2. Large, coarse grasses, usually 1–1.5+ m, the stems stout, straight, and erect; hairs at collar region of leaf, if present, usually straight; anthers 1.0–1.5 mm; grain 0.95–1.2 mm. ____ **S. airoides**

  2′ Medium-sized, relatively delicate grasses, the stems slender, if more than 1 m long then usually weak and growing through shrubs; leaf collar with a dense and often tangled mat of hairs; anthers 0.2–0.6 mm; grain 0.7–0.9 mm.

    3. Stems erect-ascending and free standing, usually not more than 75 cm; pulvini not hardened into hooks, not as in 3′. ____ **S. cryptandrus**

    3′ Stems often reaching 1–1.6 m, the panicles long, lax, often curved, tangled, and clambering into shrubs; pulvini at base of panicle branches hardened into persistent hooks and with a few rather stout hairs. ____ **S. flexuosus**

Sporobolus airoides (Torrey) Torrey var. **airoides** [*Agrostis airoides* Torrey] *ZACATÓN ALCALINO;* ALKALI SACATON. Coarse perennials often 1.2–1.6 m, forming large clumps with tough, knotty bases and very well-developed roots. Leaves mostly basal, the blades coarse and tough, often (30) 50–85 cm; sheaths pilose, the hairs sparse at the upper corners of the margin ("inside" the summit of the leaf sheath, adjacent to the ligule), the collar ("outside" the summit of the leaf sheath) glabrous; lower portion of leaf sheaths near base of plants persistent, rather firm, shiny and smooth. Panicles openly branched at maturity, (15) 25–53 cm; larger panicle branches (6) 7–16 (20) cm; branches mostly solitary at each node (rarely with 3 or 4 branches per node). Spikelets 1.8–2.2 mm. Anthers 1.0–1.5 mm. Grain 0.95–1.2 mm. Mostly growing April-November, dormant during the cooler months.

Often locally abundant in sandy soils and low stabilized dunes at margins of salt flats, especially near the coast. Also at Bahía Adair pozos and alkaline seeps and oases including Laguna Prieta, occasional along the Río Sonoyta riverbed, and common in alkaline soils at Quitobaquito. Western United States and northern Mexico including coastal Sonora to northwestern Sinaloa, often on alkaline or semi-saline soils.

Gould (1951) reports "numerous intermediate forms" between *S. airoides* var. *airoides* and *S. airoides* var. *wrightii* (Munro ex Scribner) Gould (*S. wrightii* Munro ex Scribner). In var. *airoides* the panicles are more open, the branches arise at near-right angles rather than at acute angles (or branches nearly appressed), and the leaf blades tend to be more involute and narrower. I find a single, habitat-specific, large alkali sacaton grass in the Sonoran Desert of western Sonora. It is a valuable forage grass and the grain is edible (Heizer & Elsasser 1980).

Río Sonoyta, S of El Huérfano, *F 92-983*. Laguna Prieta, *F 85-746*. La Salina, *Turner 60-67*. Quitobaquito, *Bowers 1805*.

Sporobolus cryptandrus (Torrey) A. Gray [*Vilfa cryptandra* Torrey] SAND DROPSEED. Tufted perennials often 30–75 cm. Leaf sheaths usually pilose with relatively dense long white and often tangled hairs at the summit (junction of sheath and blade—this in addition to the smaller hairs of the ciliate ligule), these hairs usually extending onto part of the collar. Base, or sometimes entire panicle, partially enclosed by an enlarged, inflated leaf sheath (often with a reduced blade), the upper part of the panicle often but not always becoming free and the branchlets spreading at maturity. Panicles (including enclosed portion) often 20–30 cm, erect to arched toward apex; panicle branches single at nodes of exposed upper branches, often 3–5 (6) cm. Spikelets 1.9–2.4 mm, tan to lead-gray. Anthers 0.15–0.6 mm. Grain 0.7–0.9 mm.

Not common, widely scattered colonies or local populations, often in disturbed habitats and mostly in sandy soils: roadsides, sandy plains, sand flats, low dunes, washes, dry riverbeds, and floodplains. Vicinity of Puerto Peñasco and Sonoyta, and rare elsewhere along the Río Sonoyta. Occasional at higher elevations in the Sierra Pinacate (875 m, *F 19300*), these approaching *S. flexuosus* in having swollen pulvini with hairs, but with relatively small and mostly contracted panicles. Some specimens from nearby Arizona (Tinajas Altas and Tule Well) also appear intermediate in character. *S. cryptandrus* extends across northern Sonora and is characteristic of coastal Sonora in sandy habitats from Cerro Tepopa (vicinity 29°22′ N) southward.

Southern Canada to central Mexico. *S. cryptandrus* forms a taxonomically difficult complex with *S. contractus* Hitchcock, *S. flexuosus, S. giganteus* Nash, and several others (Jones & Fassett 1950; Sutherland 1986). *S. cryptandrus,* which often produces the smallest-sized plants, is the most common and widespread member of this complex, especially in southwestern United States, where it is often weedy. Members of the *S. cryptandrus* complex are often cleistogamous with reduced anthers.

W of Sonoyta, *F 86-320.* Puerto Peñasco, *F 86-511.*

Sporobolus flexuosus (Thurber ex Vasey) Rydberg [*S. cryptandrus* var. *flexuosus* Thurber ex Vasey] MESA DROPSEED. Tufted perennials similar to *S. cryptandrus* except as follows: Often reaching 1.0–1.6 m and lacy in appearance, the panicles to 60 cm, lax, often curving or drooping and loosely clambering into shrubs, the branchlets spreading; exposed panicle branches reaching 12–15 cm, often becoming tangled. Pulvini at base of panicle branches swollen and with several stout hairs, hardening at maturity, the branches or branchlets falling away leaving the pulvini and a short portion of the branch as persistent hooks becoming tangled and often anchoring the panicles onto shrubs.

Occasional in natural creosote bush scrub in sandy plains in the Sonoyta region (19 mi SW of Sonoyta, *F 86-507*); often growing through *Hymenoclea salsola* shrubs. Isolated populations southward in western Sonora to the vicinity of Bahía San Pedro. Northern Mexico at least in Sonora, Chihuahua, and Coahuila, and southwestern United States from southern California to western Texas and Utah. Common in parts of northern Arizona, but the distribution within the Sonoran Desert is limited. The distinctive characters of the pulvini often are seen only on mature panicles.

Sporobolus pyramidatus (Lamarck) Hitchcock [*Agrostis pyramidata* Lamarck. *Sporobolus patens* Swallen. *S. pulvinatus* Swallen] *ZACATE PIRÁMIDE;* WHORLED DROPSEED. Warm-weather ephemerals, erect to spreading, ca. 10–25 cm. Upper margin of leaf sheaths with sparse long white hairs not in a dense tuft. Panicles at first contracted and narrow, often partly covered by the leaf sheath, the branchlets ultimately spreading and the panicles somewhat pyramidal. Lower branches whorled, (4) 5–11 branches per node; upper branches single at the nodes. Branchlets very slender with glandular areas appearing as lighter-colored, elliptic depressions. (These glands, seen with magnification, may be cryptic and their extent variable.) Spikelets 1.5–1.8 mm. Anthers 0.3–0.4 mm. Grain 0.8–0.9 mm.

Seasonally common in sandy silt along the Río Sonoyta floodplain and larger washes from Quitobaquito eastward. Tolerant of alkaline and at least mildly saline soil. Hot-weather ephemerals in the Sonoran Desert, elsewhere often perennial. Common through much of Sonora; Kansas and Colorado to South America.

Sporobolus patens, of Arizona and Sonora, and *S. pulvinatus,* of southwestern United States and Mexico, seem to be nothing more than smaller plants of the highly variable *S. pyramidatus* (Charlotte Reeder, personal communication 1994; also see Gould 1951).

Floodplain of Río Sonoyta, SSW of Quitobaquito, *F 89-35.* Aguajita, *F 88-420.*

STIPA NEEDLEGRASS, FEATHERGRASS

Mostly tufted perennials. Leaf sheaths open, the bases often persistent; blades usually long, narrow, and rolled inward but sometimes flat. Panicles often longer than wide. Spikelets 1-flowered, readily separating above the glumes. Glumes longer than body of lemma, usually elongated and tapering, persistent, membranous, and often papery. Lemma hard, awned, and tightly enclosing the palea. Awn stout, usually much longer than the lemma body, usually twisted below, sometimes hairy or feathery, bent 1 or 2 times, and deciduous although sometimes tardily so; awn and body of lemma of different textures or colors, the junction abrupt.

Worldwide, near arctic to montane tropics; most numerous in temperate regions, few in deserts; 300 species. Six species are listed for the Sonoran Desert, but only *S. speciosa* occurs within the desert. The others are at elevations above the desert or at its margins along the Pacific side of Baja California.

The awn is a conspicuous feature in the genus. *Stipa* might be confused with 1-awned *Aristida* species, e.g., *A. ternipes* var. *ternipes*. In these aristidas (1) the lemma is continuous with the awn or awn column, (2) the body of the lemma and awn are similar in texture and color, (3) the awns are not twisted and not hairy below, and (4) the spikelets and glumes are much smaller and more delicate than those of *Stipa* species from southwestern North America. Although their gross morphology can be similar and the chromosome numbers are often the same, there are profound anatomical differences (Metcalf 1960; Reeder & Decker 1961). *Stipa* has C_3 photosynthesis and shows a mixture of pooid (festucoid) and bambusoid characters, whereas *Aristida* has C_4 photosynthesis and affinities with the eragrostoid-chloridoid grasses. References: Barkworth 1993; Cabrera & Torres 1970; Hitchcock 1925a, b.

Stipa speciosa Trinius & Ruprecht var. **speciosa** [*Acantherum speciosum* (Trinius & Ruprecht) Barkworth] DESERT NEEDLEGRASS. Perennials with a dense basal clump of leaves. Leaf sheaths reddish brown, the blades tough and rolled inward, the longer blades often 45–60 cm, less than 1 mm wide when dry. Stems often 65–75 cm with a dense spikelike panicle 10–26 cm. Young, emerging inflorescences and lower part of mature panicles partially enclosed by a broad sheath. Glumes 13–20 mm, papery, and persistent. Lemma awn 3.3–4.5 cm, with 1 bend, the awn column strongly twisted and bearded below the bend with slender white hairs (3) 5–6 (8) mm. Dormant during the coldest and hottest months; growing and flowering March-May and fall.

North and northwest slopes of Pinacate Peak and immediately adjacent cinder hills above 1000 m (*F 86-446*). Locally common in rock crevices on cliffs and in cinder soils. Not known elsewhere in Sonora. The nearest populations are in the Ajo and Tinajas Altas Mountains. Also on mountaintops in the Baja California Peninsula. The Pinacate Peak population is a Pleistocene relict. Nine millennia ago it occurred at 60 m elevation in the Hornaday Mountains and undoubtedly elsewhere at lower elevations (Van Devender et al. 1990).

Both Baja California states, Isla Angel de la Guarda in the Gulf of California, northwestern Sonora, California, Arizona, and Nevada to Colorado, also Argentina, Bolivia, and Chile. Several other varieties occur in South America but only var. *speciosa* is in North America, pointing to a probable South American origin. This species is in the subgenus *Pappostipa,* which has considerable diversity in South America.

TRIDENS

Tufted perennials. Spikelets 3–15-flowered, small, slightly compressed, breaking apart above the glumes and between the florets. Glumes 1-veined (upper glume 3-veined in *T. muticus* var. *elongatus*). Lemmas prominently 3-veined, the apex rounded to often notched with short lobes, the midvein often projecting as a mucro or short awn (North America), or sometimes 3-toothed or 3-awned (South America). Anthers dark purple. North America, mostly northern Mexico and southwestern to eastern United States, South America, and 1 species in Africa; 18 species. References: Parodi 1937; Tateoka 1961.

Tridens muticus (Torrey) Nash var. **muticus** [*Tricuspis muticus* Torrey. *Triodia muticus* (Torrey) Scribner] SLIM TRIDENS. Tightly tufted perennials with stiff, slender, erect stems (22) 28–40 (50) cm; often bluish glaucous. Collar with a dense tuft of long white hairs, the sheaths sparsely to moderately pubescent with similar hairs; leaf blades usually tightly rolled. Panicles slender, spikelike, with closely appressed short branches, each branch with 1 to several spikelets. Spikelets (5.5) 6.5–13.5 mm. Glumes persistent, shorter than the spikelets. Florets strongly overlapping. Lemmas (3) 3.5–5.2 mm, the lower half pilose. Spikelets green during hot and often humid weather of summer and early fall, and purplish at other seasons. Growth and flowering response apparently non-seasonal.

Widespread, generally in rocky, hilly to mountainous habitats, mostly in canyons and arroyos; Sonoyta region, Pinacate volcanic field, and granitic ranges. In the driest habitats, such as the Sierra del Rosario, mostly toward higher elevations on north- and east-facing slopes. Western Sonora south at least to the Sierra del Viejo southwest of Caborca. Also southeastern California, southern Nevada, southern Utah, Baja California, northern Sonora, Chihuahua, Coahuila, and Nuevo León; deserts, grasslands, and oak woodlands.

Some authors recognize *T. muticus* and *T. elongatus* Buckley as distinct species, whereas others treat them as varieties or reduce them to synonyms. Var. *elongatus* (Buckley) Shinners does not occur in the Sonoran Desert, ranges into higher elevations, and extends farther eastward and northward in the United States than does var. *muticus*. Var. *elongatus* is distinguished by its generally broader leaf blades and a longer second glume with 2 or more lateral veins.

10 km SW of Sonoyta, *F 88-153*. Tinaja de los Chivos, *Simmons 9 Nov 1963*. Tinajas de los Pápagos, *Ezcurra 5 Nov 1982*. Tinajas de Emilia, *F 19740*. Sierra del Rosario, *F 75-14A*.

TRITICUM *TRIGO,* WHEAT

Annuals or biennials. Leaf blades mostly flat. Inflorescence a somewhat flattened, compact terminal spike. Spikelets sessile and solitary at each node, 2–5 (9)-flowered, separating above the glumes and between the florets. Glumes and lemmas firm, keeled, awned to awnless. Native to the Old World; 10 species.

***Triticum aestivum** Linnaeus. *TRIGO;* WHEAT; *PILKAÑ.* Winter-spring annuals. Dwarf, high-yield modern varieties with stout, scabrous awns 7+ cm. Occasional at roadsides and along railroad track west of Puerto Peñasco, growing from grain fallen from passing trucks and trains; producing full grain heads but not reproducing. Winter wheat is extensively cultivated in the region. Eusebio Kino brought wheat to the Sonoyta Valley in 1701 (Burrus 1971), and the Cocopa grew wheat along the lower Río Colorado (Castetter & Bell 1951).

W of Sonoyta, *F 85-153*. El Papalote, *F 88-20*. Mex 2 at rd to Rancho Guadalupe Victoria, *F 20405*. Observations: Mex 2 on N side of Sierra del Viejo; San Luis.

ZEA

Four species native to Mexico.

***Zea mays** Linnaeus subsp. **mays.** *MAÍZ;* CORN, MAIZE. Maize has been cultivated with summer-fall rains at a few special places in the Pinacate region. These fields are at the arid limits of agriculture. Best known is the Romero floodwater field at Suvuk on the northeast side of the lava fields (Nabhan 1985). In the 1980s this field was still being planted during years of sufficient rainfall. In the 1960s there was a small milpa (cornfield) in deep, sandy soil a few kilometers east and slightly north of MacDougal Pass. Perhaps only in the sandy fields west of Jaisalmer in Rajasthan, India, has dryland agriculture been practiced in such an arid environment.

The Cocopa grew large quantities of maize along the lower Río Colorado, and it was also cultivated by the O'odham along the Río Sonoyta at the time of contact (Burrus 1971; Castetter & Bell 1951; Lumholtz 1912). Maize was also grown at Quitobaquito (Felger et al. 1992) and is a modern crop in the region.

POTAMOGETONACEAE PONDWEED FAMILY

Annual or perennial aquatic herbs, ranging from fsresh water to seawater. Leaves alternate or subopposite, usually submerged to floating. Flowers on spikes or headlike inflorescences usually above water or at the water surface. Worldwide; 3 genera, 95 species.

1. Leaf tips entire (without microscopic teeth); flowers usually more than 4 along a slender spike not enclosed in a leaf sheath; perianth segments 4, the stamens 4; fruits asymmetric, sessile, the fruiting peduncles not elongating. ______ **Potamogeton**

1′ Leaf tips usually with microscopic teeth; flowers 2 on opposite sides of a spike at first enclosed in a leaf sheath; perianth none, the stamens 2 (each stamen with 2 prominent cells); fruits symmetric or nearly so, stalked, the peduncles greatly elongating as fruits develop. ________ **Ruppia**

POTAMOGETON PONDWEED

Perennials. Leaves 2-ranked, alternate to sometimes subopposite; leaf blades submerged, floating, or rarely emergent, slender to broad, the margins entire. Flowers and fruits on necklace-like spikes. Flowers with 4 brownish green tepals, 4 stamens, and 4 (rarely 1) separate pistils. Fruits asymmetric or not, ovoid to obovoid (about as wide as long), 1-seeded. Worldwide but mostly in temperate regions of the Northern Hemisphere; 90 species. Reference: González Gutierrez 1989.

Potamogeton pectinatus Linnaeus. SLENDER PONDWEED. Dense, filmy green masses with very slender stems to more than 50 cm. (This species is reported to have slender creeping rhizomes at the end of which is a white tuberous root 1.0–1.5 cm, but tuberous roots are generally absent from specimens from Mexico.) Leaves alternate, reaching 5–15 cm, the leaf sheaths membranous and nearly transparent, the blades 0.5 mm wide, the tips acute and entire. Inflorescences and peduncles very slender, the flowers and fruits in unevenly spaced whorls, the peduncles often 5.5–12.5 cm, the fruiting spikes 1.8–2.5 (3.0) cm. Fruits brown, 2.5–4.0 mm, ovoid, asymmetric, and slightly compressed. Growing and reproductive during warmer months.

Sonoyta in the Presa Derivadora reservoir, also in the ponds at Quitobaquito and Quitovac with *Zannichellia* in shallow water to a depth of 1 m. Nearly worldwide, and one of the most widespread pondweeds in western North America. Superficially resembling *Ruppia,* but the inflorescence, even in the early flowering stages, is not enclosed in a leaf sheath, there are usually more than 2 flowers per spike, the flowers are not on opposite sides of the flowering stalk, and the fruits are very different.

Presa Derivadora, *F 86-137*. Quitobaquito, *F 86-270*. Quitovac, *Nabhan & Felger 10 Jul 1982.*

RUPPIA

Probably a single highly variable species, although some authors claim 2 or more species. "Related to, but more advanced than *Potamogeton*" (Cronquist 1981:1067); *Ruppia* is sometimes placed in its own monotypic family. Reference: Setchell 1946.

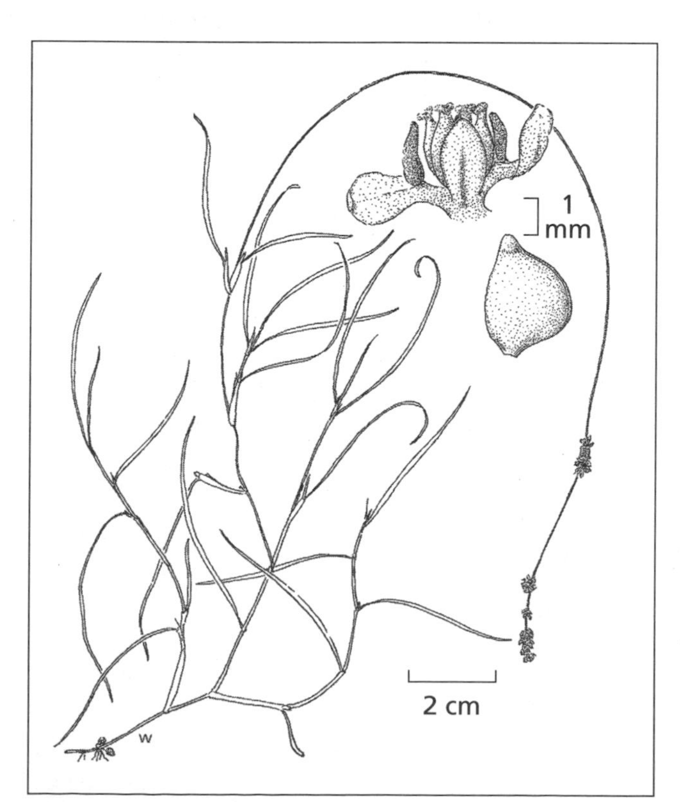

Potamogeton pectinatus. ILW (Wiggins 1980).

Ruppia maritima Linnaeus. DITCH GRASS. Apparently annuals. Stems very slender. Leaves alternate, 4–10 cm, expanded at the base into a prominent, membranous, nearly transparent sheath of fused stipules, the blades threadlike, less than 0.5 mm wide, the apex acute to obtuse with microscopic teeth. Inflorescences of 2 flowers, one above the other on opposite sides of the peduncle. Perianth none, the stamens 2, each with 2 prominent anthers, the pistils 4, separate, symmetrical or nearly so, ovoid to pear-shaped, sessile in young flowers, with stipes elongating in fruit. Inflorescence at first enclosed in a membranous spathe inside the expanded leaf sheath, the peduncles pushing through the sheath and greatly elongating, often spiraling to elevate the developing fruits to the surface of the water. Each flower

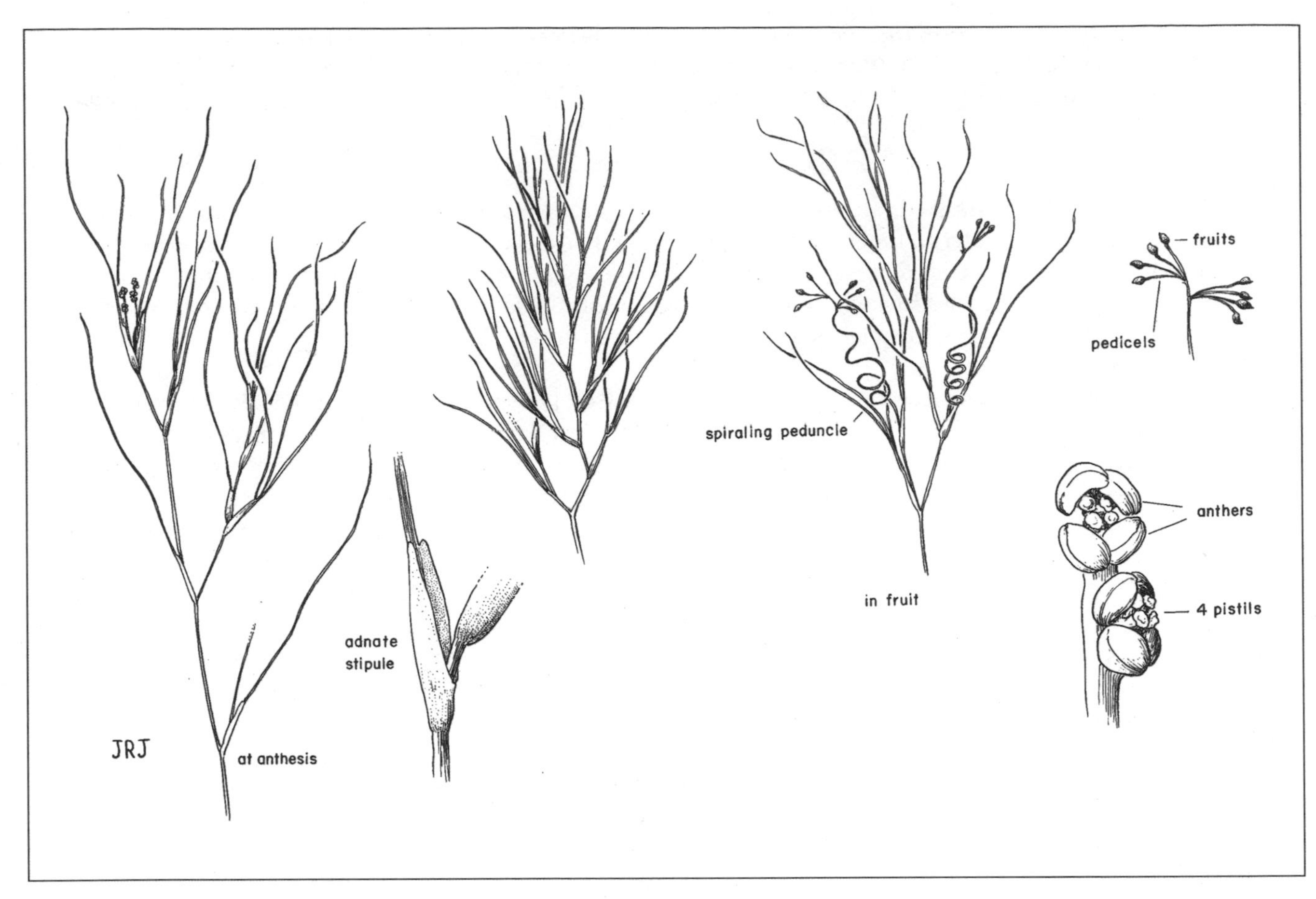

Ruppia maritima. Freshwater form with spiraling or coiled peduncles; note the inflated leaf sheath (adnate stipule). JRJ (Hitchcock, Cronquist, & Ownbey 1969).

producing 4 hard brown nutlets ca. 2 × 1 mm, each with a single seed 1 mm in diameter. (Caution: Do not confuse the 2 stamens for 2 separate male flowers, nor the 4 separate carpels or nutlets for 4 separate female flowers.) Actively growing, flowering, and fruiting April through summer.

In warm, clear, sometimes stagnant, fresh to somewhat brackish water often 10–50 cm deep, often rooted in soft mud and anaerobic muck; usually found during the hotter months. At several pozos at La Salina and wetlands in the Río Colorado delta region. The Quitobaquito plants have coiled peduncles, aligning them with *R. cirrhosa* (Petagna) Grande, the inland form in western North America and widespread elsewhere in the world.

Ruppia maritima sensu lato occurs worldwide in fresh and brackish water, and in seawater. In different regions of the Gulf of California, especially the Infiernillo Channel on the east side of Isla Tiburón, it grows during hot weather rooted in shallow subtidal seawater in protected bays, esteros, and channels. During the warmer months fragments of the plants are common in the beach drift (Felger & Moser 1985). It would not tolerate exposure to hot dry air, and its absence from the sea in the flora area is probably due to the extreme tidal fluctuations.

3.5 mi N of mouth of Santa Clara Slough, *F 90-194.* 1 km N of El Doctor, *F 92-249.* Lerdo, lagoon of brackish water, Apr 1889, *Palmer 935* (US). Ciénega de Santa Clara, *F 92-243.* La Salina, *F 86-542.* Rancho Guadalupe Victoria, water tank, apparently brought with water hauled from Caborca, *F 92-300.* Quitobaquito, *F 86-222.*

TYPHACEAE CATTAIL FAMILY

One genus, or 2 including *Sparganium* with 14 species.

TYPHA *TULE,* CATTAIL

Tall perennial herbs, glabrous and with starchy rhizomes; emergent from fresh to brackish water or in wet soil. Leaves erect and strap-shaped. Flowers unisexual, extremely numerous and very small, densely packed on tall cylindrical spikelike inflorescences, the staminate flowers above the pistillate flowers. Fruits minute, single-seeded, and achene-like (follicles). Worldwide, temperate to tropical; 10 species.

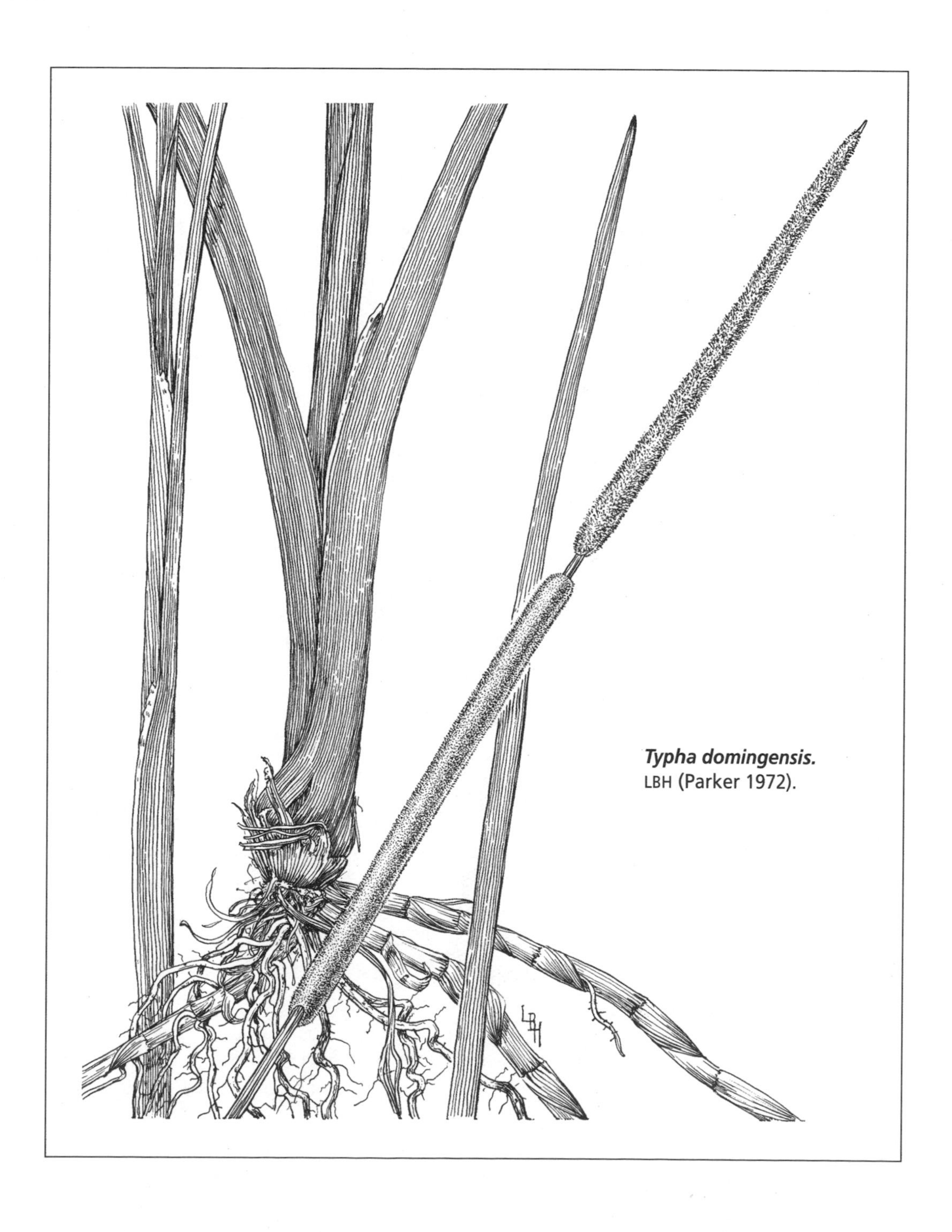

Typha domingensis.
LBH (Parker 1972).

Submerged seedlings have flaccid, ribbon-shaped leaves much like those of *Vallesneria* or the floating leaves of *Sparganium*. The tiny, lightweight fruits (0.02–0.03 mg) and associated hairs may be airborne over great distances. The seeds are reported to remain viable for many years, and a single spike of *T. domingensis* may produce 682,000 seeds. Cattails are well-known wild-food plants but should be cooked because raphides in the cells can be highly irritating. References: Hotchkiss & Dozier 1949; Mason 1957; Thieret & Luken 1996.

Typha domingensis Persoon. *TULE;* SOUTHERN CATTAIL; *'UDAWHAG.* Winter dormant, the above-water portions frost sensitive. Leaves 2–3 m × 6–14 mm, the lower portion spongy and thickened, the inner surface flat, the outer surface convex. Staminate portion of inflorescence separated from the pistillate portion by a barren gap (an interval without flowers).

Emergent from shallow water and also growing in wet soil; widely scattered in natural and disturbed wetlands. Cattails may come and go with the water supply: small, non-reproductive plants sometimes occur at Pinacate waterholes such as Tinaja del Tule but do not survive the summer drought. In 1961, Julian Hayden (personal communication 1985) saw a dense stand of tules at Tinaja del Bote, but they subsequently disappeared.

Well established along the Río Sonoyta, at La Salina pozos, and Laguna Prieta. The Laguna Prieta population seems to be receding due to lowering of the water table. Cattails were abundant along the lower Río Colorado and its delta and still grow thickly along irrigation canals in the lower Río Colorado valley and San Luis region, in the Ciénega de Santa Clara, and in local wetlands near El Doctor. *Typha* was the dominant plant across the vegetated portion of the Ciénega de Santa Clara in the late 1980s and the 1990s, covering about 4500 ha (Glenn et al. 1992; Glenn et al. 1996; Zengel et al. 1995). Here it grew emergent from water as deep as 1 m, in large pure-stand islands or intermixed or fringed with *carrizo* (*Phragmites australis*). According to local residents, the Ciénega de Santa Clara cattail population increased in the late 1980s and early 1990s and crowded out much of the *carrizo*. This is the largest stand of cattails in the state of Sonora.

Typha domingensis occurs at many wetland habitats within the Sonoran Desert, whereas the closely related *T. angustifolia* is more common in non-desert regions. As with most cattail species, *T. domingensis* is wide ranging, growing in brackish and fresh water across the southern 2/3 of the United States, through most of tropical America, and also in the Old World.

Specimens from the flora area have obovoid sterile ovaries characteristic of *T. domingensis,* but the leaves, often 6–8 mm wide, are rather narrow for this species. The gap between the male and female portions of the inflorescence is diagnostic. The O'odham at Quitovac used the leaves for basketmaking and the rootstock for food (Nabhan et al. 1982), and it was an important food plant for the Cocopa (Castetter & Bell 1951; Gifford 1933).

Río Sonoyta at Sonoyta, *F 85-698.* Laguna Prieta, *F 85-751.* Ciénega de Santa Clara, *F 92-526.* Aguajita Spring, *F 89-234.*

ZANNICHELLIACEAE HORNED PONDWEED FAMILY

Submerged herbaceous aquatics; monoecious or dioecious. Worldwide in fresh to brackish water; 4 genera, 10 species. Reference: Haynes & Holm-Nielsen 1987.

ZANNICHELLIA HORNED PONDWEED

Two species, one in Africa; or one highly variable species.

Zannichellia palustris Linnaeus. Probably perennials, or perhaps annuals or both. Leaves opposite, threadlike, entire. Monoecious. Male and female flowers usually in the same small cluster in leaf axils. Each male flower reduced to a single stamen. Each female flower with a cluster of often 4 separate green carpels (1 or more often breaking off), each carpel with an open, flared peltate stigma. Carpels (fruits) 3.0–4.2 mm (including stipe and beak), asymmetric, flattened, oblong, prominently beaked, often with small teeth or prickles on a crest and sometimes on the body, or smooth and entire, 1-seeded.

Quiet pools of the Río Sonoyta, ponds at Presa Derivadora, Quitobaquito, and Quitovac, and the El Doctor wetland. Worldwide in fresh and brackish water. Plants resembling *Potamogeton pectinatus* but distinguished by the opposite leaves and small clusters of nearly sessile, asymmetric fruits, each longer than wide.

Presa Derivadora, *F 86-135*. Río Sonoyta, 10 mi W of Sonoyta, *F 86-193*. Quitobaquito, *Peebles 14566*. N of El Doctor, *F 93-248*.

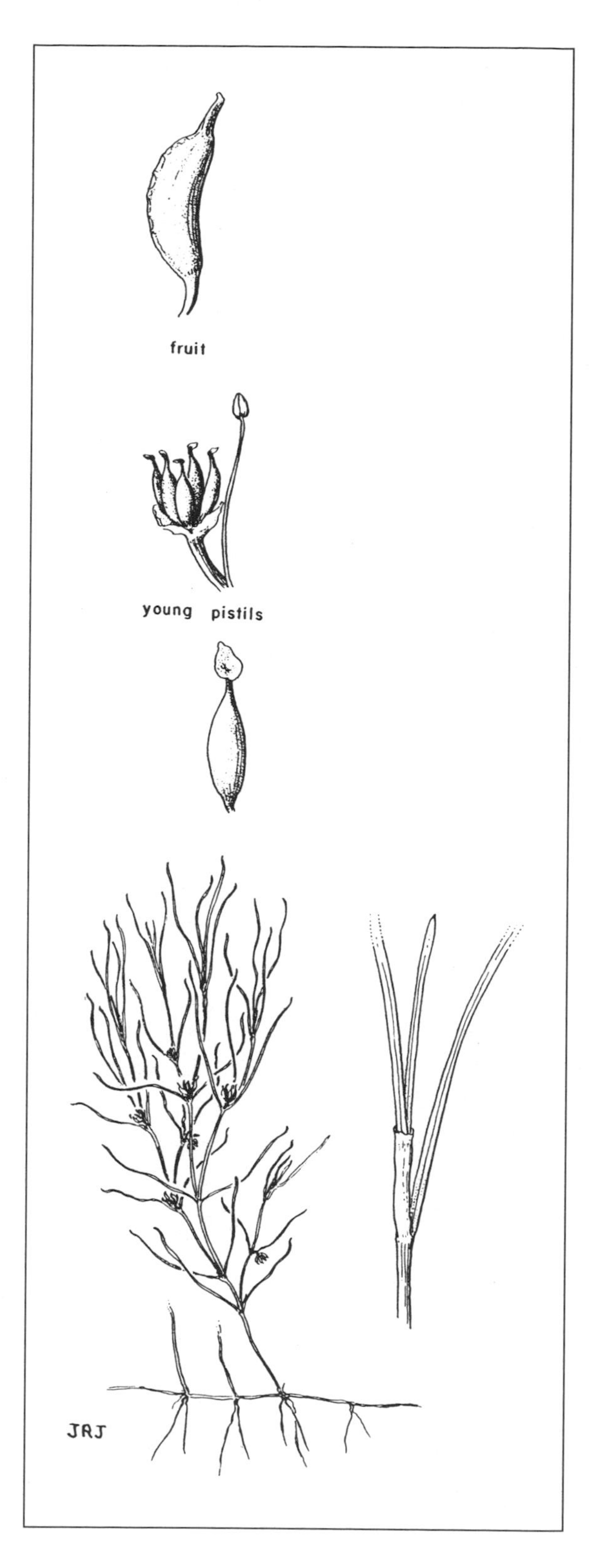

Zannichellia palustris. JRJ (Hitchcock, Cronquist, & Ownbey 1969).

Gazetteer

By Bill Broyles and Richard S. Felger

This gazetteer is designed to assist in locating and interpreting place names mentioned in this book. Most of the information is from the more extensive regional gazetteer by Broyles et al. (1997). Certain major areas or places are discussed in more detail in Part I of the text; for example, Ciénega de Santa Clara, Gran Desierto dunes, the Pinacate volcanic region, La Salina, Laguna Prieta, the Río Colorado, Quitobaquito. The names are listed alphabetically by key words; some variant, limited, or former names are given.

The information for each site or locality is organized in the following manner. Entry name (area), type of feature and its relationship to a larger geographic feature. Brief human or natural history is sometimes given. Synonyms or other names known to have been used are also given in select cases; the = sign means "also known as" or "the same as." Coordinates are given in degrees. In general metric values are given for places in Mexico and English values for places north of the international boundary. By convention, mapmakers locate expansive features (towns, bays, plains) by their approximate centers, valleys and streams by their mouths, and mountains by their summits. All localities are in Sonora unless otherwise noted. A few abbreviations are used for adjacent areas in Arizona: CP = Cabeza Prieta National Wildlife Refuge, GR = Barry M. Goldwater Air Force Range (excluding CP), and OP = Organ Pipe Cactus National Monument.

A

ABRA PLAIN, LA. (OP) Valley and watershed sloping SW from the Puerto Blanco Mountains to the Río Sonoyta. 31°57′ N, 112°56′ W. 1050–1700 ft.

ADAIR, BAHÍA. A broad bay running from Punta Cholla near Puerto Peñasco westward to Punta Borrascosa. Named by British Lieutenant Robert William Hale Hardy (1829:320) probably for John Adair (1655–1722). 31°28′ N, 113°49′ W.

AGUA DULCE. Shallow river pools on the Río Sonoyta 8 km WSW of Quitobaquito; Lumholtz (1912: 196) reported "mezquites at either side and streams of running water appearing at intervals, the last time at Agua Dulce." Vicinity of 31°56′ N, 113°06′ W. 270 m.

AGUA DULCE MOUNTAINS. (CP) Granitic mountain arc in SE corner of CP. 32°01′32″ N, 113°08′41″ W. 2850 ft.

AGUAJITA SPRING. (OP) Small, permanent spring in Aguajita Wash several meters N of the international border, N of El Papalote along the same wash, or 1.8 mi E of Quitobaquito. 31°56′23″ N, 113°00′37″ W. 1100 ft. In **AGUAJITA WASH,** the major drainage for La Abra Plain to the Río Sonoyta, E of Quitobaquito.

AGUA SALADA. Intermittent pools on Río Sonoyta about 5 km downstream (SW) of Agua Dulce. Because of evaporation and stagnation, the puddled water is brackish. 31°55′ N, 113°08′ W. 260 m.

AGUILA, PASO DEL. Pass where a prehistoric foot trail and now Mexico Highway 2 go through western gap between Sierra Nina and Sierra Tuseral. Site of El Aguila truck stop and restaurant. = El Puerto. 32°06′ N, 113°45′ W. 320 m.

AJO. (Pima Co., Ariz.) Copper-mining town north of OP. 32°22′18″ N, 112°51′36″ W. 1751 ft.

AJO MOUNTAINS. (OP) Major volcanic range straddling the eastern boundary of OP and the western boundary of Tohono O'odham Reservation. 32°01′36″ N, 112°41′24″ W. 4808 ft.

ALAMO CANYON. (OP) Large canyon on the western slope of the Ajo Mountains. Mouth at 32°04′06″ N, 112°43′27″ W. 2400 ft.

ALGODONES DUNES. Large dune field between Yuma and the Salton Sea in California. 32°57′ N, 115°06′ W. 567 ft.

ANDRADE, MESA DE. Sandy mesa above Río Colorado floodplain, 54 km S of San Luis and 5 km SW of Riito. Detached from Mesa de Sonora by a former channel of the river. 32°05′ N, 114°57′ W. 10 m.

ARENOSA, MESA. An arching mesa, on the N wall of the San Jacinto fault, between El Golfo and Bahía Adair. Composed of Tertiary and younger Río Colorado deltaic sediments, the uplifted mesa slopes inland to the NE, draining into the Gran Desierto dunes. The seaward bluffs are heavily eroded, creating "badlands" NE of El Golfo and exposing notable fossil beds. 31°40′ N, 114°25′ W. 180 m.

ARIZONA HIGHWAY 85. Road from Buckeye S through Gila Bend, Ajo, and OP to Lukeville; continuing as Mexico Highway 8 to Puerto Peñasco.

B

BAJA CALIFORNIA. The northern of the two Mexican states on the Baja California Peninsula. Mexicali is the capital.

BAJA CALIFORNIA SUR. The southern of the two Mexican states on the Baja California Peninsula. The capital is La Paz.

BATAMOTE. Hills, arroyo, arroyo crossing, and tinaja where the Río Sonoyta cuts through low hills midway between Sonoyta and Puerto Peñasco. Name also applied to the southern reach of the Río Sonoyta. Bridge where Mexico Highway 8 crosses Arroyo Batamote is at 31°40′35″ N, 113°18′00″ W. 125 m. Also **BATAMOTE HILLS,** of mid-Tertiary volcanic rock, 3–10 km NE of Batamote Crossing, 31°43′35″ N, 113°14′10″ W. 380 m. = Cerros Batamote, Colinas Batamote.

BLANCA, SIERRA. Granitic range 16 km SE of Pinacate Peak, surrounded by dunes and on the north by lava flows. 31°33′31″ N, 113°26′05″ W. 460 m.

BORRASCOSA, PUNTA. Sandy point with a lighthouse, SSE of La Salina at western lip of Bahía Adair. 31°29′32″ N, 114°01′42″ W.

BOTE, TINAJA DEL. Natural intermittent tinaja in arroyo bottom, 2 km N of Cráter de la Luna. Spanish for *tin can* or *bucket.* 31°46′55″ N, 113°40′45″ W. 90 m.

BULL PASTURE. (OP) Elevated basin on W slope of Ajo Mountains. 32°00′46″ N, 112°41′34″ W. ca. 3000–3400 ft.

BURRO SPRING. (OP) Small spring in Quitobaquito Hills, 0.6 mi N of Quitobaquito Springs. 31°57′12″ N, 113°01′08″ W. 1160 ft.

BUTLER MOUNTAINS. (GR) Small granitic range 4 mi W of Tinajas Altas Mountains, partially buried by drifting sands. 32°22′35″ N, 114°12′23″ W. 1169 ft.

C

CABEZA PRIETA MOUNTAINS. (CP) Major range of geologically complex basalts and granites in CP. 32°20′48″ N, 113°49′26″ W. 2830 ft.

CABEZA PRIETA NATIONAL WILDLIFE REFUGE. U.S. federal wildlife refuge and wilderness area covering 3480 km^2. In the SE corner of Yuma Co. and SW corner of Pima Co., Ariz. (Note: Abbreviated in text to Cabeza Prieta Refuge.)

CABEZA PRIETA TANKS. Usually perennial tinajas in the middle of the Cabeza Prieta Mountains, 1.2 mi N of Cabeza Prieta Peak. 32°18′27″ N, 113°48′09″ W. 1530 ft.

CACTUS FLAT. Sandy flat about 3 km E of Pinacate Junction; N of Playa Díaz and S of Mexico Highway 2, notable for extensive stands of cacti. 31°58′ N, 113°19′ W. 200 m.

CAMPO ROJO. Campsite 4 km E of Pinacate Peak. Located below a red volcanic butte, at the end of a rough road; the beginning of the easiest approach to Pinacate summit. = Emilia Camp; Hunter's Camp; Palo Verde Camp; Red Cone Camp. 31°46′30″ N, 113°26′50″ W. 420 m.

CARAVAJALES FLAT. Seasonally verdant flat 2 km SW of Tinajas de los Pápagos; named for Juan Caravajales, a Hia-ceḍ O'odham who farmed in the area. 31°54′ N, 113°36′ W. 210 m.

CARNEGIE PEAK. Second highest peak in the Sierra Pinacate, 1.2 km E of Pinacate Peak; a point on the rim of Volcán de Santa Clara. Named for Andrew Carnegie. 31°46′23″ N, 113°29′08.5″ W. 1130 m.

Carrizal, El. One of Padre Kino's campsites on the Río Sonoyta, about 10 km WSW of Quitobaquito, midway between Agua Dulce and Agua Salada (Ives 1989). Named for 'carrizo' (*Phragmites australis*). About 31°56′ N, 113°06′ W. 270 m.

Cerro Prieto, Estero. See Prieto, Estero Cerro.

Chivo, Cerro el. Cinder cone in the SW part of Pinacate volcanic field; 11 km W of Pinacate Peak. = Chivos Butte. 31°46′09″ N, 113°37′04″ W. 340 m.

Chivos, Tinaja de los. Deep, bedrock waterhole on the SW side of the Pinacate lava shield, 9 km WSW of Pinacate Peak. The water often lasts year-round. 31°44′54″ N, 113°36′21″ W. 180 m.

Cholla, Bahía de la. Bay 8 km WNW of Puerto Peñasco. Since the mid-twentieth century a tourist camp on the southern beach has grown into a sprawling vacation village called Cholla Bay. Named for an expansive stand of low-growing *Opuntia bigelovii*. Also **Estero la Cholla.** 31°21′40″ N, 113°36′40″ W.

Ciénega de Santa Clara. See Ciénega de **Santa Clara.**

Cipriano, Represo. Dam across arroyo near El Camino del Diablo, less than 1 km SW of U.S.-Mexico boundary marker 179. Apparently built to contain water draining to Las Playas to the NW, possibly by Cipriano Ortega and his brother Bartolo about 1887, and long since breached. 32°02′47″ N, 113°22′58″ W. 215 m.

Cipriano, Sierra. Granitic mountain 7 km SW of Sonoyta, on E side of Mexico Highway 8. This outlier of the 53 million-year-old Gunnery Range Granite is named for Cipriano Ortega (see **Santo Domingo**). 31°46′07″ N, 112°58′41″ W. 900 m.

Cloverleaf Crater. See **Molina Crater.**

Cocopah, Sierra. See Sierra **Cucapá.**

Colinas Batamote. See **Batamote.**

Colorado, Cerro. Large, isolated crater at NE side of Pinacate region, 25 km NE of Pinacate Peak. Named for the reddish rock, especially vivid at sunset. 750 m maximum diameter. 31°55′00″ N, 113°18′15″ W. 220 m at southern rim.

Colorado, Río. Mexican portion of the Colorado River. Draining lands as far N as Wyoming, the river traverses 2320 km (1450 m) to its mouth on the Gulf of California at 31°48′ N, 114°49′ W.

Cubabi, Sierra. Granitic mountain 8 km S of Sonoyta on the NE side of Mexico Highway 2. 31°43′15″ N, 112°49′36″ W. 1330 m.

Cucapá, Sierra. Range 16 km S of Mexicali, forming E wall of Laguna Salada basin. Named for the Cocopa who live in the Río Colorado delta region. = Sierra de los Cucapá, Sierra de los Cocopahs, Cocopa Mountains. 32°22′34″ N, 115°27′23″ W. 1080 m.

D

Derivadora, Presa. Dam 1.5 km E of Sonoyta on the Río Sonoyta. Built in 1951, it impounds a small lake, the largest freshwater pond in NW Sonora E of the Río Colorado. 31°51′33″ N, 112°49′59″ W. 390 m.

Desemboque San Ignacio, El. Seri Indian village on the shore of the Gulf of California. 29°30′ N, 112°24′ W.

Díaz, Colonia. Agricultural settlement in late nineteenth and early twentieth centuries on former meander of Río Colorado, 48 km SW of boundary marker 205 at San Luis. Possibly vicinity of 32°11′ N, 115°10′ W. 14 m (see Mearns 1907).

Díaz, Playa. Large playa (usually dry lakebed) immediately NW of Cerro Colorado. Pinacate Junction is at the NW margin of the playa. 31°56′21″ N, 113°19′08″ W. 200 m.

Doctor, Estación el. Settlement and railroad stop on Mesa de Sonora along Sonora Highway 40; 40 km NW of El Golfo de Santa Clara and near the Río Colorado. Freshwater pozos, well, and wetlands located immediately to W in bed of Río Colorado. 31°57′45″ N, 114°44′30″ W. 15 m.

Dos Playas. (CP) Internally drained twin playas in southern Mohawk Valley, 7 mi NW of Papago Well. About 250–300 acres each, barren when dry. 32°11′39″ N, 113°20′16″ W. 810 ft.

E

Elegante Crater. Maar crater 13 km NE of Pinacate Peak. Named by Lumholtz (1912) for its elegance; deepest of the Pinacate craters, 1460 m wide, 243 m deep. Cracks in the small playa in the lowest portion of the floor may indicate continuing subsidence. = Cráter Elegante, Volcán el Elegante. 31°50′46″ N, 113°23′26″ W. Rim 320 m, floor 80 m.

Emilia Camp. See **Campo Rojo.**

Emilia, Tinajas de. Intermittent to perennial tinajas 4 km ESE of Pinacate Peak and S of Campo Rojo. 31°45′43″ N, 113°27′06″ W. 420 m.

Enterrada, Sierra. Small granitic mountain SW of the Pinacate volcanic shield, 27 km WNW of Pinacate Peak. Surrounded and partially buried by wind-blown dunes. = Buried Range. 31°49′40″ N, 113°47′05″ W. 180 m.

Extraña, Sierra. Granitic mountain at western margin of the Pinacate lava shield, 20 km WNW of Pinacate Peak. 31°50′37″ N, 113°42′43″ W. 200 m.

G

Gila Bend. (Maricopa Co., Ariz.) Town 3 mi S of the once-perennial Gila River which bends westward here; junction of Arizona Highway 85 and Interstate 8. Gila Bend and the territory southward were part of Mexico until the Gadsden Purchase of 1853. 32°56′52″ N, 112°42′58″ W. 736 ft.

Gila Mountains. Large granite range mostly in western GR 15 mi E of Yuma, extending SE from Gila River to Tinajas Altas Mountains. 32°31′58″ N, 114°14′01″ W. 3156 ft.

Gila River. Major tributary of the Colorado River. It flows W from Gila Mountains of New Mexico to its mouth at Yuma. Confluence with the Colorado is at 32°43′11″ N, 114°33′16″ W. 120 ft.

(Barry M.) Goldwater Air Force Range. (Maricopa, Pima, and Yuma Cos., Ariz.) 2.6-million-acre U.S. Air Force and Marine Corps area established early in 1941. Formerly known as Luke Air Force Range and Williams Military Reservation, renamed for Arizona's long-time senator in 1986.

Golfo de Santa Clara, El. Beachside fishing and tourist village on Gulf of California near SE corner of the Río Colorado delta. Terminus of Sonora Highway 40 from San Luis R.C. Mesa Arenosa, to NE, rises more than 100 m above beach. 31°41′10″ N, 114°29′55″ W.

Gore, Isla. Mudflat island at mouth of Río Colorado, covered by spring tides. See Isla Montague. 31°40′40″ N, 114°39′25″ W.

Gran Desierto, El. The great sand desert of NW Sonora, especially the moving dunes; sometimes applied to entire region of NW Sonora. = Desierto del Altar. Altar Desert.

Grande, Cráter. See **Sykes Crater.**

Grijalva, Rancho. See Rancho **Guadalupe Victoria.**

Guadalupe Victoria, Rancho. Cattle ranch 10 km S of Mexico Highway 2 and 3 km S of La Lava microwave tower. Originally called Colonia Cuauhtémoc, established in the early 1950s. 32°00′36″ N, 113°34′23″ W. 225 m.

Gustavo Sotelo, Estación. Railroad maintenance station in the lonely vastness of the Gran Desierto, at km post 207. Named in honor of one of the four civil engineers who helped build the railroad across the Gran Desierto and died of thirst when their truck bogged in heavy sand in July 1937 (Barrios Matrecito 1977). They are Mexican heroes, memorialized in Niño Martini's 1964 novel *El Muro y la Trocha* and the movie *Viento Negro,* and remembered in four of these gandy dancer stations. See also Estaciones **López Collada** and **Sánchez Islas.** 31°33′34″ N, 113°42′25″ W. 8 m.

H

Halcón, Cerro el. Cinder cone 3.5 km N of Moon Crater. 31°47′30″ N, 113°41′10″ W. 160 m.

Hardy, Río. (Baja Calif.) Tributary of the Río Colorado on W side of the delta, drains valley SE of Mexicali. Mouth at 32°04′39″ N, 115°15′ W.

Hornaday Mountains. Small granitic range, 26 km NNW of Pinacate Peak, borders the NW margin of the Pinacate volcanic shield. Named for William T. Hornaday (1908). 31°59′55″ N, 113°37′38″ W. 440 m.

Horseshoe Bend. Name used by Edward Palmer in late nineteenth century for channel curve in delta region of Río Colorado. Vicinity of 31°57′30″ N, 115°07′30″ W. (See *Distichlis palmeri.*)

Hourglass Canyon. Canyon on W side of Sierra Pinacate, 3.2 km NW of Tinaja de los Chivos; named by Ann Woodin for the shape of a shadow cast on the canyon wall in springtime. 31°46′40″ N, 113°32′27″ W. 500 m.

Huarache, Tinaja. Bedrock waterhole, 6 km W of Pinacate Peak. = Pyramid Tank. Deepest of the Pinacate tinajas (Broyles 1996a). 31°45′39″ N, 113°34′42″ W. 380 m.

Huérfano, Cerro el. Butte 13 km W of Quitobaquito at bend of Río Sonoyta; 1.2 km SW of boundary marker 173. Named 'orphan' because it stands apart from its parent Sierra de la Salada. 31°57′05″ N, 113°04′55″ W. 340 m.

I–K

Isabel, Puerto. Steamboat landing, shipyard, and small port on Shipyard Slough, between the mouth of the Río Colorado and the Santa Clara Slough. This small American settlement on Mexican soil was active 1867–1877. Vicinity of 31°49′53″ N, 114°41′10″ W. 10 km W from E shore. Above reach of the tidal bore.

Ives Flow. Largest and southernmost lava flow in the Pinacate volcanic field, between Sierra Blanca and Pinacate Peak and covering more than 75 km^2. Its lava has a different composition than other Pinacate flows. Named for Ronald Lorenz Ives. Center ca. 31°41′ N, 113°28′ W. 100–700 m.

Johnson, Pico. See Sierra **Seri.**

Joyita, La. Truck stop and restaurant on Mexico Highway 2, 1.2 km NW of Cerro Pinto and 3.4 km W of El Sahuaro. Well dug in 1931 to serve travelers on the primitive Sonoyta–San Luis trail. 32°14′38″ N, 114°03′52″ W. 260 m.

Kino, Bahía. Fishing and resort settlement on the Sonora coast W of Hermosillo. 28°47′ N, 111°47′ W.

L

Lago Díaz. See Playa **Díaz.**

Lava, Microondas La or **Cerro la Lava.** Microwave tower 2 km S of Mexico Highway 2 and 2.5 km N of Rancho Guadalupe Victoria. The turnoff to La Lava is also the Biósfera El Pinacate entrance to MacDougal and Sykes Craters. 32°01′57″ N, 113°34′08″ W. 340 m.

Lechuguilla, Sierra. Granitic mountain N of Sierra del Viejo and Mexico Highway 2. Mostly in Sonora, the northern tip in Arizona. 32°12′26″ N, 113°57′11″ W. 720 m.

LERDO, COLONIA. Late nineteenth- and early twentieth-century farming settlement on the Sonora side of the Río Colorado delta. The main settlement was named Ciudad Lerdo.

In April 1889, Edward Palmer spent three days collecting specimens from the vicinity of Lerdo (see McVaugh 1956; Vasey & Rose 1890). Probably 9 km S of modern-day Riito, on SW corner of mesa de Andrade. Vicinity of 32°04′30″ N, 114°57′45″ W. 15 m.

LISAS, ESTERO LAS. Tidal marsh at Bahía Adair, ca. 40 km NW of Puerto Peñasco. 31°36′18″ N, 113°57′24″ W.

LOBOS, PUERTO. Fishing village and estero on the coast of Sonora. 30°16′ N, 112°15′ W.

LÓPEZ COLLADA, ESTACIÓN. Remote railroad maintenance station in the dunes of the Gran Desierto NW of Puerto Peñasco, at km post 174. See Estaciones **GUSTAVO SOTELO** and **SÁNCHEZ ISLAS.** 31°42′46″ N, 113°59′22″ W. 10 m.

LUIS ENCINAS JOHNSON, EJIDO. Farming settlement on Mesa Andrade in Río Colorado delta, 16 km S of Riito. 32°03′35″ N, 114°57′36″ W. 10 m.

LUKEVILLE. (Pima Co., Ariz.) Settlement and U.S. customs station on the international border N of Sonoyta. Founded by Charles Luke and named after himself or his son, World War I flying ace Frank Luke Jr., who was killed in action in 1918. 31°52′57″ N, 112°48′54″ W. 1410 ft.

LUNA, CRÁTER DE LA. See **MOON CRATER.**

M

MACDOUGAL CRATER. Large maar crater in the NW part of the Pinacate volcanic field, 26 km NW of Pinacate Peak; 1100 m wide, 130 m deep. Named for Dr. Daniel Trembly MacDougal by Hornaday (1908). 31°58′32″ N, 113°37′34″ W. 220 m at the crater rim.

MACDOUGAL PASS. Sandy gap through E side of Sierra Hornaday, 4 km SW of the present Rancho Guadalupe Victoria. The Hornaday (1908) expedition took its wagons through this long, sandy corridor between the lavas and the dunes from the border to MacDougal Crater. The present road to MacDougal Crater goes through another gap E of MacDougal Pass. 31°59′15″ N, 113°36′15″ W. 230 m.

MAYOR, EL. (Baja Calif.) Cocopa Indian settlement on the Río Hardy, 8 km above confluence with Río Colorado and near Sierra del Mayor. 32°05′27″ N, 115°18′31″ W. 10 m.

MENAGER'S DAM. (Tohono O'odham Nation, Pima Co., Ariz.) Earthen and rock dam across narrow gap between mountains in far SW corner of the reservation. 31°49′15″ N, 112°32′33″ W. 1742 ft.

MEXICO HIGHWAY 2. The northernmost highway across Mexico, and the only one connecting Baja California with the mainland. In the 1930s the road between Sonoyta and San Luis was a treacherous single-track trail through rough lava and shifting dunes. The Sonoyta–San Luis Highway portion was graded during World War II and paved between 1953 and 1955.

MEXICO HIGHWAY 8. Highway from Sonoyta to Puerto Peñasco. One of the first paved highways in Sonora, finished during World War II.

MINA DEL DESIERTO. Prospect and restaurant on Mexico Highway 2, 2.5 km W of Paso del Aguila in Sierra Nina. 32°06′58″ N, 113°48′ W. 285 m.

MOHAWK DUNES. (GR) Dunes covering 7,700 ha, extending 32 km along W side of Mohawk Mountains and 3.5 km W of the range. 32°35′ N, 113°43′ W. 680 ft.

MOLINA CRATER. Small, three-lobed maar crater 1 km SE of MacDougal Crater. The sections resemble a cloverleaf 150 m wide and 30 m deep. Each section has its own small playa. = Cráter Trébol, Cloverleaf Crater. 31°57′44″ N, 113°36′45″ W. 200 m at the crater rim.

MONTAGUE, ISLA. Large silt island in mouth of Río Colorado, inundated by extreme spring tides. 31°45′ N, 114°45′ W. Islas Montague, Gore, and Pelícano form the four mouths of the Río Colorado.

MOON CRATER. Maar crater 18 km W of Pinacate Peak at the SW side of the Pinacate volcanic field, bordered by dune fields; 26 m deep, 450 m wide. A volcanic cone rises from the central crater floor. Dune sand is filling the crater, and a small, ephemeral playa occupies the lowest portion of the crater floor. = Cráter de la Luna. 31°45′23″ N, 113°41′09″ W. 100 m.

MORÚA, ESTERO. Large estero (tidal marsh) ca. 10 km ESE of Puerto Peñasco. 31°17′ N, 113°26′ W at mouth.

N–O

NINA, SIERRA. Steep, granitic mountain S of Mexico Highway 2, 38 km NW of Pinacate Peak. Lumholtz (1912:237) named it for "my friend, Mrs. [Nina] John Gray, of Boston." = Sierra del Águila. 32°05′35″ N, 113°45′32″ W. 630 m.

ORGAN PIPE CACTUS NATIONAL MONUMENT. In western Pima Co., Ariz., it embraces 330,687 acres, much of it designated as wilderness (see Broyles 1996b). Part of the worldwide UNESCO Biosphere Reserve system. National Park Service headquarters and visitor center are 5 mi N of Sonoyta, at 31°57′18″ N, 112°48′00″ W. 1670 ft. (Note: Abbreviated in text to Organ Pipe Monument.)

P

PAPAGO WELL. (CP) Well at W end of Agua Dulce Mountains. 32°05′56″ N, 113°17′09″ W. 909 ft.

PÁPAGOS, TINAJAS DE LOS. Series of large bedrock tinajas, one of the few usually reliable waterholes in the Pinacate region, although in years of extreme drought it has gone dry (Julian Hayden, personal communication 1990). 31°54′59″ N, 113°36′17″ W. 220 m.

PAPALOTE, EL. Restaurant and truck stop along Mexico Highway 2, 18 km W of Sonoyta and 0.5 km S of Aguajita Spring. 31°56′16″ N, 113°00′44″ W. 325 m.

PASO DEL ÁGUILA. See Paso del **ÁGUILA.**

PASTIZAL DE EFÍMERAS. Silt flats NE of Sierra Blanca; barren and dry except in wettest of times. Similar flats S of Sierra Blanca. Thin soil and underlying salt pans (evaporite deposits) permit only shallow-rooted ephemeral plants and an occasional cholla. Vicinity of 31°33′ N, 113°27′ W. 70 m.

PELÍCANO, PUNTA. Small granitic peninsula at the S end of Bahía de la Cholla. 31°20′45″ N, 113°38′12″ W.

PEÑASCO, ESTERO. Tidal wetlands at Puerto Peñasco, destroyed when the modern harbor was built in the 1970s. 31°18′ N, 113°33′ W.

PEÑASCO, PUERTO. Fishing village founded in the early 1920s, now a rapidly growing town supported largely by tourism. Estimated 1996 population: 34,000. = Rocky Point. 31°18′05″ N, 113°32′55″ W. 10 m.

PEÑASCO, PUNTA. Volcanic hill at Puerto Peñasco. Originally named Rocky Point by Hardy (1829:310): "We passed Rocky Point, for so I named it." 31°17′55″ N, 113°32′47″ W. 50 m.

PINACATE, EL or **PINACATE VOLCANIC FIELD.** The dominating shield volcano in central Gran Desierto. Its lava derived from a mantle plume. The mountainous portion is often referred to as the **SIERRA PINACATE** or **VOLCÁN SANTA CLARA.**

PINACATE JUNCTION. A road entrance into the Pinacate region, 51 km W of Sonoyta; junction of Mexico Highway 2 and road S to Tezontle cinder mine and Cráter Elegante. = Los Vidrios entrance to Reserva de la Biósfera el Pinacate. 31°59′50″ N, 113°21′00″ W. 215 m.

PINACATE PEAK. Highest point in the Pinacate region. 31°46′21″ N, 113°29′54″ W, the summit variously reported at 1190–1291 m, probably ca. 1210 m.

PINTA, ESTERO LA. Tidal marsh ca. 25 km SSE of Puerto Peñasco. 31°15′ N, 113°14′ W.

Pinta Sands. (CP) Low dunes and sand flats along the Arizona-Sonora border at S end of Sierra Pinta; forming a collar around far northern end of Pinacate lava field. 32°09′08″ N, 113°32′54″ W. 750 ft.

Pinto, Cerro. Granitic mountain 2 km S of Mexico Highway 2; westernmost of the steep granitic mountains along Mexico Highway 2 west of the Pinacate volcanic field. A road leads to the microwave tower atop the N end. 32°13′06″ N, 114°03′15″ W. 680 m.

Playas, Las. (CP) Expansive, internally drained playa along El Camino del Diablo. 32°04′19″ N, 113°25′13″ W. 680 ft.

Pozo Nuevo, Rancho. Forlorn ranchito 1.3 km SW of the Sierra Extraña. Trapped between sand dunes and lava flows, it sits in a silt flat sometimes used as an illicit airstrip. 31°49′29″ N, 113°43′06″ W. 60 m.

Prieta, Laguna. Intermittent small lake 25 km SE of San Luis. 32°18′20″ N, 114°33′27″ W. 40 m.

Prieto, Cerro. Volcanic hill on the coast 12 km NW of Puerto Peñasco. 31°23′30″ N, 113°36′10″ W. 120 m.

Prieto, Estero Cerro. Tidal marsh 2–6 km NNW of Cerro Prieto; the mouth at 31°25′31″ N, 113°39′38″ W.

Puerto, El. See Paso del **Aguila.**

Puerto Blanco Mountains. (OP) Range 7 mi NNW of Lukeville. 32°00′16″ N, 112°51′26″ W. 3145 ft.

Pyramid Peak. Cinder cone surrounded by dunes 3.5 km NW of MacDougal Crater (see Ives 1964). 31°59′37″ N, 113°40′16″ W. 350 m.

Pyramid Tank. See Tinaja **Huarache.**

Q

Quitobaquito. (OP) Springs and oasis along international border 13 mi W of Sonoyta. 31°56′40″ N, 113°01′01″ W. 1100 ft.

Quitobaquito Hills. (OP) Low hills N of Quitobaquito Springs; geologically complex, largely granitic. 31°59′06″ N, 113°02′18″ W. 1844 ft.

Quitovac. Tohono O'odham and Hia-ceḍ O'odham settlement and oasis about 40 km SSE of Sonoyta, S of Mexico Highway 2. 31°30′59″ N, 112°45′11″ W. 420 m.

R

Red Cone Camp. See **Campo Rojo.**

Reserva de la Biósfera Alto Golfo y Delta del Río Colorado. (Sonora, Baja Calif.) Mexican national biosphere reserve protecting the upper Gulf of California and Río Colorado delta. Contains 942,270 ha, including a core or nuclear zone of 160,620 ha.

Reserva de la Biósfera El Pinacate y El Gran Desierto de Altar. (Sonora) Mexican national biosphere reserve W of Mexico Highway 8 covering 714,656 ha. Two core areas: one of 41,392 ha protecting most of the Sierra del Rosario, and another of 228,113 ha centered around the Pinacate volcanic region.

Riito. Agricultural settlement near Río Colorado at intersection of Sonora Highway 40 and the railroad, 42 km SSW of San Luis. 32°09′52″ N, 114°57′31″ W. 20 m.

Rocky Point. See Puerto **Peñasco** and Punta **Peñasco.**

Rosario, Sierra del. Isolated granitic range surrounded by dunes; summit at 32°05′25″ N, 114°12′20″ W. 562 m.

S

SAHUARO, EL. Truck stop on Mexico Highway 2, 2 km E of Cerro Pinto microwave tower. 32°13′45″ N, 114°01′50″ W. 290 m.

SALINA, LA. Grand salt basin 25 km SSW of Estación López Collada. This long, narrow salt flat near W edge of Bahía Adair is surrounded by dunes except for a former seaway to the south. A series of pozos, or waterholes, dots the margin of the salt flat. Lumholtz (1912:395) called it Salina Grande. 31°31′00″ N, 114°07′56″ W, near sea level.

SÁNCHEZ ISLAS, ESTACIÓN. Abandoned railroad maintenance station at km 129, 10 km NNE of El Golfo. See Estación **LÓPEZ COLLADA.** 31°45′41″ N, 114°27′29″ W. 123 m.

SAN FELIPE. (Baja Calif.) The northernmost town on the Baja California shore of the Gulf of California, 100 km S of Río Colorado delta and 193 km S of Mexicali. 31°02′30″ N, 114°50′ W.

SAN LUIS. (Yuma Co., Ariz.) Twin city of San Luis Río Colorado. Immediately north of the international border and port of entry. Population: ca. 5650. 32°29′13″ N, 114°46′53″ W. 136 ft.

SAN LUIS RÍO COLORADO. Fast-growing city in the NW corner of Sonora, adjacent to the E bank of the Río Colorado. Known as San Luis R.C., to distinguish it from younger sister city of San Luis, Ariz. A hub of transportation and agriculture, founded 1906. Estimated 1996 population: 123,090. 32°27′56″ N, 114°46′12″ W. 45 m. (Note: Throughout the text, San Luis is used to refer to San Luis R.C., and San Luis, Ariz. for the U.S. town.)

SANTA CLARA, CIÉNEGA DE. Major wetland in the Río Colorado delta. Fresh water from springs and the Río Colorado joins agricultural wastewater discharged from the Wellton-Mohawk Canal, forming the largest wetland vegetation habitat in the region (Glenn et al. 1992, 1996). 32°01′ N, 114°52′ W. Near sea level.

SANTA CLARA, SIERRA. See El **PINACATE.**

SANTA CLARA SLOUGH. Tidal slough on eastern edge of Río Colorado delta, 11 km NW of El Golfo de Santa Clara. Its upper end is the Rillito Salada, which ties into Ciénega de Santa Clara. Mouth at 31°45′ N, 114°35′ W.

SANTO DOMINGO. Former hacienda on the Río Sonoyta 12 km W (downstream) of Sonoyta. Developed by the infamous and later respected Cipriano Ortega, this feudal ranching and farming hacienda had its heyday 1870–1904 (Hornaday 1908; Hoy 1970, 1990). 31°53′55″ N, 112°57′40″ W. 360 m. The present-day Ejido Santo Domingo is 2 km downstream.

SERI, SIERRA. Large range on mainland opposite Isla Tiburón, north of Bahía Kino. 29°13′ N, 112°08′ W. 1063 m.

SODA, LA. Salt flats with several freshwater artesian pozos, which support a miniature, fragile wetland flora. About 20 km SE of López Collada near the shore. The name ‘soda’ refers to trona (a form of calcium carbonate) and halite (table salt) found here; “the absence of sulfate salts on these flats supports the hypothesis that the origin of the salts is not marine” but rather from the continental freshwater aquifer (Ezcurra et al. 1988:41). Vicinity of 31°37′ N, 113°49′ W. Near sea level.

SONORA HIGHWAY 40. Highway from San Luis R.C. to El Golfo.

SONORA, MESA DE. The eastern bluff above Río Colorado floodplain, from San Luis R.C. to El Golfo. It apparently includes Mesa de San Luis and is north of Mesa Arenosa.

SONOYTA. Agricultural and commercial community on the Río Sonoyta 3 km S of the international boundary at Lukeville. Includes Sonoyta Viejo and Sonoyta Nuevo. One of oldest settlements in the region. From prehistoric times until the 1980s the economy was primarily agricultural. Padre Kino arrived in 1698 and introduced cattle and new crops. In the aftermath of a flood on the night of August 6, 1891, the village was relocated downstream to its present location (Ives 1989; Lumholtz 1912).

Originally called Son Oidag or Sonoidac (O'odham for 'spring fields' or 'fields at the rock base'); later San Marcelo de Sonoydag (Ives 1989). Estimated 1990 population of 10,000. 31°51′45″ N, 112°50′45″ W. 400 m.

SONOYTA MOUNTAINS. (Sonora, OP) Low mountains 3 km W of Sonoyta; eroded remnant of an 8 mi^2 granite pluton. 31°55′08″ N, 112°50′57″ W. 2313 ft.

SONOYTA, RÍO. Small river, rising from springs E of Sonoyta. It flows westward from the town and then abruptly southward along the E side of the Pinacate region. Ultimately, when carrying rare surges of flooding, ends at Estero Morúa on the Gulf of California. Also **SONOYTA VALLEY,** the upper river valley of Río Sonoyta.

SUVUK. Hia-ceḍ O'odham camped here and cultivated tepary beans, maize, and squashes; Lumholtz (1912:330) reported it as their primary agricultural site. Later the Romero family dry-farmed (a *temporal*) nearby (Nabhan 1985). Vicinity of 31°44′45″ N, 113°22′50″ W. 180 m.

SYKES CRATER. Marvelous crater in NW part of Sierra Pinacate, 19 km NW of Pinacate Peak; 700 m in diameter, 220 m deep. Named for Godfrey Sykes (Hornaday 1908), it is the second deepest of the Pinacate craters. 31°56′24″ N, 113°34′06″ W. 400 m at rim.

T

TANQUES, SIERRA DE LOS. Low, sprawling granitic mountains and hills 19 km W of Sonoyta, S of Mexico Highway 2 and W of Mexico Highway 8. 31°49′56″ N, 113°05′27″ W. 595 m.

TECOLOTE, CERRO. Cinder cone 4.8 km NE of Cráter Elegante. 31°52′45″ N, 113°21′36″ W. 360 m.

TEZONTLE CINDER MINE. Cinder mine 18 km NE of Pinacate Peak. = La Morusa, Freeman Mine, La Laja, or Materiales Tezontle. 31°56′ N, 113°21′ W. 300 m.

TINAJAS ALTAS. (GR) The most significant tinajas in the region, with nine sets of intermittent and perennial pools. An arroyo from a hanging valley in these granitic mountains downcuts steeply through joint fractures to scour and pluck a staircase of pools. 32°18′42″ N, 114°03′03″ W. 1180 ft.

TINAJAS ALTAS MOUNTAINS. (GR) Large, rugged granitic range between Gila Mountains in Arizona and Cerro Pinto in Sonora. 32°16′24″ N, 114°02′45″ W. 2764 ft.

TORNILLAL, EL. Freshwater pozo on beach below bluffs of Mesa Arenosa, 12 km SE of El Golfo. The only natural waterhole between La Salina and El Golfo; Lumholtz (1912) reports that bitter, brackish water could be obtained by digging at the canyon's mouth. Now a tourist camp. Named for *tornillo* or screwbean (*Prosopis pubescens*) that grows near the waterhole. 31°33′40″ N, 114°17′51″ W.

TRÉBOL, CRÁTER. See **MOLINA** Crater.

TULE, TINAJA DEL. Large bedrock pool in a canyon bottom, 9 km W of Pinacate Peak; usually perennial but it has dried up during years of extreme drought. 31°46′13″ N, 113°35′51″ W. 260 m. Also **ARROYO TULE,** basalt-lined arroyo or canyon with a series of smaller tinajas extending 2 km S downstream from the large upper tinaja.

TULE WELL. (CP) Site of several historic wells at S end of Cabeza Prieta Mountains, now abandoned. It was a day's journey E of Tinajas Altas by foot or horseback. 32°13′34″ N, 113°44′55″ W. 1174 ft.

TUSERAL, SIERRA. Rugged granitic and gneiss mountains N of Mexico Highway 2 and extending into CP in Yuma Co. Arizona portion also called Tule Mountains. 32°07′11″ N, 113°42′25″ W. 820 m.

V

VIDRIOS, LOS. Truck stop and restaurant on Mexico Highway 2, 57 km W of Sonoyta. Established in 1953 by Don Juan Bermúdez, who ran the restaurant, gas station, and small motel from 1956 until his health failed in the mid-1980s. During this time Los Vidrios served as the unofficial cultural and information center for El Pinacate. For a number of years it was the only reliable place for meals and gas between Sonoyta and San Luis. Following Don Juan's death in 1985, the family sold Los Vidrios and the magic passed with its founder. 32°01′36″ N, 113°25′13″ W. 220 m.

VIDRIOS, RANCHO LOS. Ranch and represo 8 km SE of Cerro Colorado. = Los Vidrios Viejos. 31°51′43″ N, 113°14′03″ W. 215 m. Also **PLAYA LOS VIDRIOS,** a playa with vertisol soil at Rancho los Vidrios.

VIEJO, SIERRA DEL. Largest of the steep, granitic, or crystalline rock mountains W of the Pinacate volcanic field. Located along S side of Mexico Highway 2, 50 km NW of Pinacate Peak, it is about 20 km long and 3 km wide. The name Sierra del Viejo was used by Lumholtz (1912:300 & maps), and it may have been known by that name prior to his visit. Incorrectly labeled Sierra de los Alacranes on some maps. 32°07′51″ N, 113°54′27″ W. 790 m; base 400 m at the N side and 280 m on S side. Another, much larger mountain also called **SIERRA DEL VIEJO** is 40 km SSW of Caborca, Sonora.

W–Z

WELLTON. (Yuma Co., Ariz.) Settlement named for wells drilled when Southern Pacific Railroad came through in 1877. 1 mi N of U.S. Interstate Highway 8; 30 mi E of Yuma. 32°40′22″ N, 114°08′46″ W. 250 ft.

WELLTON-MOHAWK DRAINAGE CANAL. (Yuma Co., Ariz.) Wellton-Mohawk Main Outlet Drain Extension (MODE) drainage canal, discharges expended irrigation water from fields in lower Gila River Valley and carries away leached salts and field chemicals. During the 1980s and 1990s the Ciénega de Santa Clara became greatly enlarged from MODE canal water. = Canal de Descarga R. Sánchez Toboada. Terminus at N end of Ciénega de Santa Clara is about 10 km SW of Riito, at 32°03′26″ N, 114°53′56″ W.

WHY. (Pima Co., Ariz.) Settlement at Y-junction of Arizona Highways 85 and 86, 10 mi SE of Ajo. 32°16′07″ N, 112°44′17″ W. 1784 ft.

WILLIAMS SPRING. (OP) Seeps 1 mi NW of Quitobaquito. = Rincon Spring. 31°57′30″ N, 113°01′27″ W. 1095 ft.

YUMA. (Yuma Co., Ariz.) City on the Colorado River just below confluence with the Gila River. 32°43′31″ N, 114°37′25″ W. 213 ft.

Appendix A

Appendix A. Growth Forms and Distribution of Plants in Northwestern Sonora

GROWTH FORMS: MG = megaphanerophytes: trees more than 8 m tall **M** = microphanerophytes: trees or shrubs usually 2–8 m tall **N** = nanophanerophytes: shrubs or shrub-sized perennials, usually 0.5–2 m tall **C** = chamaephytes: perennials, the meristem (growth bud) above ground but usually less than 0.5 m tall **H** = hemicryptophytes: perennials with meristem at or near surface **G** = geophytes: the meristem below ground **T** = therophytes: annuals or ephemerals—**TN** = nonseasonal, **TS** = summer, **TW** = winter/spring **HY** = hydrophytes—**sub** = submerged, **em** = emergent. **AP** = joculative annuals or herbaceous perennials **p** = parasitic, **s** = succulent, **v** = vining.

HABITATS: SO = Sonoyta region **PI** = Pinacate volcanic complex (lower elevations) **UP** = Upper elevations of Sierra Pinacate **GR** = granitic ranges **SR** = Sierra del Rosario **DU** = dunes **F** = desert plains or flats (widespread, unless subregions are indicated: **e** = east, **s** = south, **w** = west, **n** = north) **CO** = coastal habitats **W** = wetlands (m = marine) **DI** = seriously disturbed ("weedy") habitats.

* = Non-native species
( ) = rare occurrence or locally extirpated

		SO	PI	UP	GR	SR	DU	F	CO	W	DI
PTERIDOPHYTES											
Marsileaceae											
Marsilea vestita	TN		PI					Fn		W	
Pteridaceae											
Astrolepis cochisensis	H			UP							
A. sinuata	H			UP							
Cheilanthes parryi	H			UP	GR	SR					
Notholaena californica	H	SO	PI	UP	GR	SR					
N. standleyi	H	SO									
Pellaea mucronata	H			UP							
Selaginellaceae											
Selaginella eremophila	H	SO			GR?						
GYMNOSPERMS											
Ephedraceae											
Ephedra aspera	N	SO	PI	UP	GR	SR					
E. trifurca	N						DU	Fs/w			
ANGIOSPERMS: DICOTYLEDONS											
Acanthaceae											
Carlowrightia arizonica	C	SO	PI		GR						
Justicia californica	N	SO	PI	UP	GR	SR					
Aizoaceae											
*Mesembryanthemum crystallinum	TWs	SO						Fs			
*M. nodiflorum	TWs	SO									
Sesuvium verrucosum	Hs									W	DI
Trianthema portulacastrum	TS	SO	PI		GR			F			DI
Amaranthaceae											
*Amaranthus albus	TS	SO									DI
A. crassipes	TS							Fe/n		W	
A. fimbriatus	TS	SO	PI	UP				F			
A. palmeri	TS	SO	PI		GR			F			DI

ANGIOSPERMS: DICOTYLEDONS (cont.)

		SO	PI	UP	GR	SR	DU	F	CO	W	DI
A. watsonii	TS										DI
Tidestromia lanuginosa	TS	SO	PI	UP?	GR	SR		F			DI
T. oblongifolia	C				GR						
Anacardiaceae											
Rhus aromatica	N			UP							
R. kearneyi	N				GR						
Apiaceae											
Bowlesia incana	TW	SO									
*Coriandrum sativum	TW	SO									
*Cyclospermum leptophyllum	TW	SO									
Daucus pusillus	TW	SO		UP							
Eryngium nasturtiifolium	TW							Fw		W	
Hydrocotyle verticillata										W	
Apocynaceae											
Apocynum cannabinum	C							Fw		W	
Aristolochiaceae											
Aristolochia watsonii	Gv	SO									
Asclepiadaceae											
Asclepias albicans	Ns		PI		GR	SR		(Fw)			
A. erosa	TW							Fw			
A. subulata	Ns	SO	PI			SR		F			
Metastelma arizonicum	Nv	SO									
Sarcostemma cynanchoides	Nv	SO	PI	UP	GR						
Asteraceae											
Acourtia wrightii	H	SO									
Adenophyllum porophylloides	C	SO									
Ambrosia acanthicarpa	TN	SO						Fs/e			
A. ambrosioides	N	SO	PI		GR						
A. confertiflora	H	SO	PI					F			DI
A. deltoidea	N	SO	PI	UP	GR		DU	F			
A. dumosa	N	SO	PI	UP	GR	SR	DU	F			
A. ilicifolia	N				GR	SR		Fw			
Artemisia ludoviciana	H			UP							
Baccharis emoryi	M							Fw		W	
B. salicifolia	N	SO	PI					Fs/e/n		W	DI
B. sarothroides	M	SO	PI					Fs/e			
Baileya multiradiata	TN	SO									
B. pauciradiata	TN		PI					Fe/s/w			
B. pleniradiata	TS		PI			SR	DU	F			
Bebbia juncea	N	SO	PI	UP	GR						
Brickellia atractyloides	N				GR						
B. coulteri	N	SO	PI	UP	GR						
Calycoseris parryi	TW			UP							
C. wrightii	TW	SO			GR						
*Carthamus tinctorius	TW	SO						Fn/w			DI
*Centaurea melitensis	TW	SO									DI

ANGIOSPERMS: DICOTYLEDONS (cont.)

		SO	PI	UP	GR	SR	DU	F	CO	W	DI
Chaenactis carphoclinia	TW	SO	PI		GR	SR		F			
C. stevioides	TW	?	PI					Fe/n			
Chloracantha spinosa	N	SO						Fw		W	DI
*Conyza canadensis	TS	SO						Fw			DI
*C. coulteri	TS	SO						(Fn)			DI
Dicoria canescens	TN						DU				
*Eclipta prostrata	TS	SO						Fw		W	DI
Encelia farinosa	N	SO	PI	UP	GR	SR		F			
E. frutescens	N						DU	Fs/n/w			
Erigeron divergens	TW	SO	PI					Fn		W	
E. oxyphyllus	H				GR						
Eriophyllum lanosum	TW	SO									
Filago arizonica	TW	SO	PI	UP	?						
F. californica	TW	SO	PI	UP	?						
F. depressa	TW	SO									
*Flaveria trinervia	TS							Fw			DI
Gaillardia arizonica	TW		PI					Fe			
Geraea cansescens	TW	SO	PI		GR	SR	DU	F			
Gnaphalium palustre	TW		PI							W	
Gutierrezia sarothrae	C	SO		UP							
Gymnosperma glutinosum	N	SO			GR						
*Helianthus annuus	TS										
H. niveus	TN						DU	Fs/w			
Heterotheca thiniicola	N						DU				
Hymenoclea monogyra	M	SO									
H. salsola	N	SO	PI		GR	SR		F			
Hymenothrix wislizenii	AP	SO									
Hymenoxys odorata	TW	SO									
Isocoma acradenia	N										
var. acradenia		SO						Fs			
var. eremophila							DU	Fs/w			
*Lactuca serriola	TW	SO									DI
Machaeranthera carnosa	C	SO						Fs		W	
M. coulteri	AP	SO	PI				DU	F			
M. pinnatifida	AP	SO	PI	UP	GR	SR					
Malacothrix glabrata	TW	SO	PI	UP	GR		DU	Fs/w/n			
M. sonorae	TW	SO		UP							
Monoptilon bellioides	TW	SO	PI	UP?	GR	SR		Fn			
Palafoxia arida	TW										
var. arida		SO	PI		GR	SR	DU	F			
var. gigantea							DU				
Pectis cylindrica	TS		PI					Fn			
P. papposa	TS	SO	PI	UP	GR	SR	DU	F			
Perityle emoryi	TW	SO	PI	UP	GR	SR	DU	F			
Peucephyllum schottii	N		PI	UP	GR	SR					
Pleurocoronis pluriseta	N	SO	PI	UP	GR	SR					
Pluchea odorata	AP	SO						Fs/w		W	
P. sericea	M	SO					(DU)	Fs/w		W	

ANGIOSPERMS: DICOTYLEDONS (cont.)

		SO	PI	UP	GR	SR	DU	F	CO	W	DI
Porophyllum gracile	C	SO	PI	UP	GR	SR					
Prenanthella exigua	TW	SO	PI		GR						
Psathyrotes ramosissima	TW		PI					Fn			
Psilostrophe cooperi	C	SO		UP							
Rafinesquia californica	TW			UP							
R. neomexicana	TW	SO	PI	UP	GR	SR		Fn			
Senecio lemmonii	TW	SO									
S. mohavensis	TW	SO	PI		GR?						
S. pinacatensis	N			UP							
*Sonchus asper	TW	SO	PI				(DU)	Fw/n		W	DI
*S. oleraceus	TW	SO	PI					Fs		W	DI
Stephanomeria pauciflora	N	SO	PI	UP	GR			F			
S. schottii	TW		PI			SR	DU	Fs/w			
Stylocline gnaphaloides	TW	SO									
S. micropoides	TW	SO	PI	UP	GR?						
Symphyotrichum subulatum	TS	SO								W	
Thymophylla concinna	TW	SO									
Trichoptilium incisum	TW	SO	PI		GR						
Trixis californica	N	SO	PI	UP	GR	SR					
Uropappus lindleyi	TW			UP							
*Verbesina encelioides	TN							Fs			DI
Viguiera parishii	N	SO	PI	UP	GR						
*Xanthium strumarium	TS	SO									DI
Bataceae											
Batis maritima	Hs								CO	Wm	
Berberidaceae											
Berberis haematocarpa	N			UP							
Boraginaceae											
Amsinckia intermedia	TN	SO			GR?						
A. tessellata	TW	SO	(PI)	UP							
Cryptantha angustifolia	TW	SO	PI	UP	GR	SR	DU	F			DI
C. barbigera	TW	SO	PI	UP	?						
C. costata	TW						DU	Fs/w			
C. ganderi	TW						DU	Fe/s/w			
C. holoptera	AP		PI	UP	GR						
C. maritima	TW	SO	PI	UP	GR	SR	DU	F			
C. micrantha	TW						DU	F			
C. pterocarya	TW	SO	PI	UP	GR						
C. racemosa	AP				GR	SR					
Heliotropium convolvulaceum	TS						DU	Fs/w			
H. curassavicum	Hs	SO						F		W	DI
Lappula occidentalis	TW	SO									
Pectocarya heterocarpa	TW	SO	PI		GR	SR		Fe/n			
P. platycarpa	TW	SO	PI								
P. recurvata	TW	SO	UP								
Tiquilia palmeri	H						(DU)	F			
T. plicata	H						DU	F			

ANGIOSPERMS: DICOTYLEDONS (cont.)

		SO	PI	UP	GR	SR	DU	F	CO	W	DI
Brassicaceae											
*Brassica nigra	TW	SO	PI					Fn			DI
*B. tournefortii	TW	SO	PI		?		DU	F			DI
Caulanthus lasiophyllus	TW	SO	PI	UP	GR	SR	?				
Descurainia pinnata	TW	SO	PI	UP	GR	SR		F			
Dimorphocarpa pinnatifida	TW						DU				
Dithyrea californica	TW		PI				DU	Fs/w/n			
Draba cuneifolia	TW	SO	PI	UP	GR						
*Eruca vesicaria	TW	SO									
Lepidium lasiocarpum	TW	SO	PI	UP	GR	SR		F			
Lesquerella tenella	TW	SO	PI					Fn			
Lyrocarpa coulteri	C	SO	PI	UP	GR	SR		F			
*Sisymbrium irio	TW	SO	PI					Fn/s			DI
Streptanthella longirostris	TW		PI			SR		F			
Thysanocarpus curvipes	TW	SO									
Burseraceae											
Bursera microphylla	M		PI	UP	GR	SR					
Cactaceae											
Carnegiea gigantea	Ms	SO	PI	UP	GR	SR		Fn			
Echinocactus polycephalus	Cs		PI		GR	SR		Fw			
Echinocereus engelmannii	Cs										
var. acicularis		SO									
var. chrysocentrus			PI	UP	GR			Fw/n			
E. nicholii	Cs	SO									
Echinomastus erectocentrus	Cs	SO									
Ferocactus cylindraceus	Ns	SO	PI	UP	GR						
F. emoryi	Ns	SO									
F. wislizeni	Ns	SO						Fn			
Lophocereus schottii	Ms	SO	PI	(UP)				Fe/n			
Mammillaria grahamii	Cs	SO	PI	UP	GR			Fe/n			
M. tetrancistra	Cs	?	PI	UP	GR	SR					
M. thornberi	Cs	SO									
Opuntia acanthocarpa	Ns	SO			GR						
O. arbuscula	Ns	SO						Fn			
O. basilaris	Hs				GR						
O. bigelovii	Ns	SO	PI	UP	GR			Fs			
O. chlorotica	Ns			UP							
O. echinocarpa	Ns		PI	UP	GR	SR		F			
O. engelmannii	Ns										
var. engelmannii		SO									
var. flavispina		SO									
O. fulgida	Ms	SO	PI		?			F			
O. kunzei	Cs		PI	(UP)				Fw/n			
O. leptocaulis	Ns	SO	PI								
O. ramosissima	Ns	SO	PI		GR	SR		Fw/n			
Peniocereus greggii	Ns	?	PI	UP							
P. striatus	Ns	SO									
Stenocereus thurberi	Ms	SO	PI								

ANGIOSPERMS: DICOTYLEDONS (cont.)

		SO	PI	UP	GR	SR	DU	F	CO	W	DI
Campanulaceae											
Nemacladus glanduliferus	TW	SO	PI	UP	?						
Capparaceae											
Capparis atamisquea	M	SO									
Cleome isomeris	N			UP							
*C. viscosa	TS							Fn			DI
Wislizenia refracta											
subsp. palmeri	N						DU	Fs/w			
subsp. refracta	TN	SO	PI					Fe		W	
Caryophyllaceae											
Achyronychia cooperi	TW	?	PI		GR	SR	DU	F			
Drymaria viscosa	TW						DU	Fs/n/w			
Loeflingia squarrosa	TW	SO									
Silene antirrhina	TW	SO									
Spergularia salina	TS	SO								W	
Chenopodiaceae											
Allenrolfea occidentalis	Ns							Fs/w	CO		
Atriplex barclayana	C								CO	Wm	
A. canescens	N						DU		CO		
A. elegans	TS	SO						Fn			DI
A. lentiformis	N	SO	(PI)					Fw			
A. linearis	N	SO	PI					Fw	CO		
A. pacifica	TW	SO	PI	UP				Fs			
A. polycarpa	N	SO	PI	UP	GR			F			
*A. wrightii	TS	SO									DI
*Bassia hyssopifolia	TS	SO						Fw			DI
*Chenopodium album	TW										DI
*C. berlandieri	TW	SO									
*C. murale	TW	SO	PI					Fs/w			DI
Monolepis nuttalliana	TW	SO									DI
Nitrophila occidentalis	Gs	SO						Fs/w		W	
Salicornia bigelovii	TNs								CO	Wm	
S. subterminalis	Ns								CO	Wm	
S. virginica	Hs								CO?	Wm	
*Salsola tragus	TS	SO	PI					Fs/w			DI
Sarcobatus vermiculatus	N						DU	Fs/w			
Suaeda esteroa	TNs								CO	Wm	
S. moquinii	Ns	SO							CO	Wm	DI
S. puertopenascoa	Cs									Wm	
Convolvulaceae											
Cressa truxillensis	H							Fs/w	CO	Wm	DI
Cuscuta salina	TSp/v	SO									
C. tuberculata	TSp/v	SO	PI								
C. umbellata	TSp/v	?	PI		?						
Evolvulus alsinoides	H	?	PI								
Ipomoea hederacea	TSv		PI								
*I. ×leucantha	TSv							Fw			DI

ANGIOSPERMS: DICOTYLEDONS (cont.)

		SO	PI	UP	GR	SR	DU	F	CO	W	DI
Crassulaceae											
Crassula connata	TWs	SO	PI	?	GR		DU	F			
Dudleya arizonica	Hs				GR						
Crossomataceae											
Crossosoma bigelovii	N	SO			GR						
Cucurbitaceae											
Brandegea bigelovii	TWv	SO	PI		GR			F			
*Cucumis melo	TS										DI
Cucurbita digitata	Gv	SO	PI	UP	GR			Fe/n/w			
Euphorbiaceae											
Acalypha californica	N	SO									
Croton californicus	C						DU	Fe/s			
C. sonorae	M	SO									
C. wigginsii	N						DU				
Ditaxis brandegeei	N		PI		GR						
D. lanceolata	C	SO	PI		GR	SR					
D. neomexicana	TN	SO	PI		GR	SR		F			
D. serrata var. californica	TW								CO		
var. serrata	TN		PI			SR	DU	Fs/n/w			
Euphorbia abramsiana	TN	SO						Fn			DI
E. albomarginata	G	?	PI					Fe/n			DI
E. arizonica	H	?	PI	UP							
E. eriantha	TN	SO	PI	UP	GR	SR		F			
*E. hyssopifolia	TN	SO									
E. micromera	AP	SO	PI		GR			F			
E. misera	Ns				GR						
E. pediculifera	AP	SO	PI	UP	GR			?			
E. petrina	TN								CO		
E. platysperma	TN						DU				
E. polycarpa	AP	SO	PI	UP	GR	SR	DU	F			
*E. prostrata	TS							Fn			
*E. serpens	TN							Fw?			DI?
E. setiloba	TN	SO	PI	?	GR	SR		F?			
E. trachysperma	TS							Fn			
Jatropha cinerea	N	SO		UP							
J. cuneata	Ns	SO	PI	UP	GR						
Sebastiania bilocularis	M	SO	(PI)								
Stillingia linearifolia	C		PI			SR	DU	F	CO		
S. spinulosa	TW							Fw			
Tragia nepetifolia	Cv	SO									
Fabaceae											
Acacia constricta	M	SO									
A. greggii	M	SO	PI	UP	GR						
Astragalus aridus	TW							Fw			
A. insularis	TW	SO					DU	Fe/s/w			
A. lentiginosus	TW							Fw			

ANGIOSPERMS: DICOTYLEDONS (cont.)

		SO	PI	UP	GR	SR	DU	F	CO	W	DI
A. magdalenae	TW										
var. magdalenae									CO		
var. peirsonii							DU				
A. nuttallianus	TW	SO	PI	UP				Fs/n			
A. sabulonum	TW							Fw			
Calliandra eriophylla	N	SO	PI	UP							
Dalea mollis	TN	SO	PI	UP	GR	SR	DU	F			
D. mollissima	TN							Fw			
Hoffmannseggia glauca	G	SO						Fs/w			DI
H. microphylla	N				GR	SR					
Lotus rigidus	C				GR						
L. salsuginosus	TW	SO	PI	UP							
L. strigosus	TW	SO	PI	UP	GR	SR	DU	F			
Lupinus arizonicus	TW	SO	PI	UP	GR	SR	DU	F			
L. sparsiflorus	TW	SO									
Marina parryi	TN	SO	PI	UP	GR						
*Melilotus indica	TW	SO								W	DI
Olneya tesota	M	SO	PI	UP	GR	SR	(DU)	F			
*Parkinsonia aculeata	M		(PI)					(Fs/w/n)			DI
P. florida	M	SO	PI			SR	(DU)	F			
P. microphylla	M	SO	PI	UP	GR						
*Pediomelum rhombifolium	H		PI					Fs			
Phaseolus filiformis	TNv	SO	PI	UP	GR	SR		F			
Prosopis glandulosa	M		PI		GR	SR	DU	F		W	
P. pubescens	M	SO						Fs/w		W	
P. velutina	M	SO									
Psorothamnus arborescens	N							Fe?			
P. emoryi	N						DU	Fs/w			
P. spinosus	M	SO					DU	F			
Senna covesii	H	SO	PI								
Sesbania herbacea	TS							Fw		W	DI
Fouquieriaceae											
Fouquieria splendens	M	SO	PI	UP	GR	SR		F			
Frankeniaceae											
Frankenia palmeri	C								CO		
F. salina	H								CO	Wm	
Gentianaceae											
Centaurium calycosum	TN	(SO)									
Eustoma exaltatum	H	(SO)									
Geraniaceae											
*Erodium cicutarium	TW	SO	PI	UP	GR			F			
E. texanum	TW	SO	PI		GR	SR	DU	F			
Hydrophyllaceae											
Eucrypta chrysanthemifolia	TW	SO	PI	UP	GR			Fe			
E. micrantha	TW	SO	PI	UP	GR	SR		Fe		W	
Nama demissum	TW		PI	UP				F			
N. hispidum	TW	SO	PI		GR	SR		F			

ANGIOSPERMS: DICOTYLEDONS (cont.)

		SO	PI	UP	GR	SR	DU	F	CO	W	DI
N. stenocarpum	TW		PI							W	
Phacelia affinis	TW		PI	UP							
P. ambigua	TW	SO	PI	UP	GR	SR		F			
P. cryptantha	TW			UP							
P. distans	TW	SO									
P. neglecta	TW		PI								
P. pedicellata	TW		PI	UP	GR	SR					
Pholistoma auritum	TW			UP							
Koeberliniaceae											
Koeberlinia spinosa	M							(Fn)			
Krameriaceae											
Krameria erecta	C	SO	PI		GR	SR					
K. grayi	N	SO	PI	UP	GR	SR					
Lamiaceae											
Hyptis emoryi	M	SO	PI	UP	GR	SR					
Salazaria mexicana	N			UP	GR						
Salvia columbariae	TW	SO	PI	UP	GR			F			
S. mohavensis	C			UP							
Teucrium cubense	TN		PI					Fn		W	
T. glandulosum	C			UP							
Lennoaceae											
Pholisma sonorae	Gp						DU	Fs/w			
Loasaceae											
Eucnide rupestris	TN		PI								
Mentzelia adhaerens	TN		PI	UP	(GR)						
M. affinis	TW	SO	PI								
M. albicaulis complex	TW	SO	PI	UP	GR	?	DU	F			
M. involucrata	TW	SO	PI	UP	GR	SR					
M. multiflora	TW		PI	UP	GR	SR	DU	F			
M. oreophila	AP				GR						
M. veatchiana	TW			UP							
Petalonyx linearis	C		PI	UP	GR						
P. thurberi	N	SO					DU				
Lythraceae											
Lythrum californicum	N									W	
Malpighiaceae											
Janusia gracilis	Cv	SO	PI		GR						
Malvaceae											
Abutilon californicum	N		PI	UP							
A. incanum	N	SO									
A. malacum	N	SO									
A. palmeri	C		PI								
Eremalche exilis	TW							Fn			
E. rotundifolia	TW		PI								
Herissantia crispa	AP	SO	PI		GR						
Hibiscus coulteri	C	SO									

ANGIOSPERMS: DICOTYLEDONS (cont.)

		SO	PI	UP	GR	SR	DU	F	CO	W	DI
H. denudatus	C	SO	PI	UP	GR	SR					
Horsfordia alata	M	SO	PI		GR	SR					
H. newberryi	M		PI		GR						
*Malva parviflora	TW	SO						F			DI
Malvella leprosa	G		PI					Fn/w			DI
M. sagittifolia	G		PI					Fe/n			
Sphaeralcea ambigua	C										
subsp. ambigua		SO	PI		GR	SR		(Fn)			
subsp. rosacea				UP							
S. coulteri	TW	SO	PI		GR			F			DI
*S. emoryi	AP	SO						Fn			
S. orcuttii	TW						DU	Fs/w			
Martyniaceae											
Proboscidea altheaefolia	G	SO	PI	UP				Fn			
P. parviflora	TS	SO	PI								
Molluginaceae											
*Mollugo cerviana	TS	SO	PI		GR		DU	F			
Nyctaginaceae											
Abronia maritima	Hs								CO?		
A. villosa	TW		PI				DU	F	CO		
Acleisanthes longiflora	H	SO									
Allionia incarnata	AP	SO	PI	UP	GR	SR					
*Boerhavia diffusa	TS	SO									DI
B. erecta	TS	SO	PI								DI
B. spicata	TS	SO	PI								DI
B. wrightii	TS	SO	PI		GR	SR		F			
Commicarpus scandens	N	(SO)									
Mirabilis bigelovii	H	SO	PI	UP	GR						
M. tenuiloba	H					SR					
Oleaceae											
Menodora scabra	N	SO			GR						
Onagraceae											
Camissonia arenaria	AP				GR	SR					
C. boothii	TW		PI		GR	SR	DU	Fs/w			
C. californica	TW	SO	PI	UP	GR	SR		F			
C. cardiophylla	AP		PI	UP	?						
C. chamaenerioides	TW	SO	PI	UP	GR	SR		F			
C. claviformis	TW										
subsp. peeblesii			PI		GR	SR		Fe/s/w			
subsp. rubescens			PI		GR		DU	F			
subsp. yumae						SR	DU	Fs/w			
Gaura parviflora	TN	SO	PI					Fn		W	
Oenothera arizonica	TW							Fn?			
O. deltoides	TW		PI				DU	F			
O. primiveris	TW	SO	PI	UP	GR	SR		F			
Orobanchaceae											
Orobanche cooperi	Gp	SO	PI	UP		SR	DU	F			

ANGIOSPERMS: DICOTYLEDONS (cont.)

		SO	PI	UP	GR	SR	DU	F	CO	W	DI
Papaveraceae											
Argemone gracilenta	H	?	PI	UP				F			
*A. ochroleuca	AP	SO						Fw			DI
Eschscholzia californica	TW	SO									
E. minutiflora	TW	SO	PI	UP	GR						
Pedaliaceae											
*Sesamum orientale	TS										DI
Phytolaccaceae											
Stegnosperma halimifolium	M		PI	UP							
Plantaginaceae											
Plantago ovata	TW	SO	PI	UP	GR	SR	DU	F			
P. patagonica	TW	SO									
Polemoniaceae											
Aliciella latifolia	TW		PI								
Eriastrum diffusum	TW	SO	PI	UP	GR			F			
Gilia minor	TW			UP							
G. stellata	TW	SO	PI	UP							
Langloisia setosissima	TW		PI								
Linanthus bigelovii	TW	SO	PI	UP	GR	SR	DU	F			
Loeseliastrum schottii	TW					SR	DU	Fs/w			
Polygonaceae											
Chorizanthe brevicornu	TW	SO	PI	UP	GR	SR		F			
C. corrugata	TW		PI		GR	SR					
C. rigida	TW	SO	PI	UP	GR	SR		F			
Eriogonum deflexum	TN	SO	PI	UP							
E. deserticola	N						DU				
E. fasciculatum	N										
var. fasciculatum									CO		
var. polifolium		?	PI	UP	GR	SR					
E. inflatum	H	SO	PI		GR	SR					
E. thomasii	TW	SO	PI	UP	GR	SR		F			
E. thurberi	TW	SO									
E. trichopes	TW	SO	PI		GR	SR	(DU)	F			
E. wrightii	N	SO	(PI)	UP	GR	SR					
Nemacaulis denudata	TW		PI	UP			DU	F			
*Polygonum argyrocoleon	TN	SO						Fw			DI
P. hydropiperoides	HYem	SO								W	
Rumex inconspicuus	TW	SO						Fs		W	
Portulacaceae											
Cistanthe ambigua	TWs					SR					
C. monandra	TWs		PI								
C. parryi	TWs		PI	UP							
Portulaca halimoides	TSs	SO	PI	UP				Fe			
*P. oleracea	TSs	SO									DI
P. retusa	TSs		PI								
Rafflesiaceae											
Pilostyles thurberi	Cp						DU	Fs/w			

ANGIOSPERMS: DICOTYLEDONS (cont.)

		SO	PI	UP	GR	SR	DU	F	CO	W	DI
Ranunculaceae											
Delphinium scaposum	G	SO									
Myosurus minimus	Hyem	(SO)									
Resedaceae											
Oligomeris linifolia	TW	SO	PI		GR	SR		F			
Rhamnaceae											
Colubrina californica	M	SO									
Condalia globosa	M	SO	PI								
Ziziphus obtusifolia	M	SO		UP				(Fe/w)			
Rubiaceae											
Galium stellatum	C			UP	GR	SR					
Rutaceae											
Thamnosma montana	N				GR						
Salicaceae											
Populus fremontii	MG	SO						Fw		W	
Salix exigua	M							Fw		W	DI
S. gooddingii	MG	SO						Fw		W	
Sapindaceae											
Dodonaea viscosa	N	SO									
Saururaceae											
Anemopsis californica	H	SO						Fw		W	
Scrophulariaceae											
Antirrhinum cyathiferum	TN	SO	PI	UP	GR	SR					
A. filipes	TWv	SO	PI	UP							
Keckiella antirrhinoides	N			UP							
Mimulus rubellus	TW		PI	UP							
Mohavea confertiflora	TW		PI	UP							
Penstemon parryi	TW	SO									
P. pseudospectabilis	H			UP	GR						
Veronica peregrina	TW	SO	PI					Fn		W	
Simaroubaceae											
Castela emoryi	M							(Fe)			
Solanaceae											
*Calibrachoa parviflora	TN	SO									
Chamaesaracha coronopus	H							Fn			
Datura discolor	TN	SO	PI		GR	SR	(DU)	F			
Lycium andersonii	N	SO	PI	UP	GR	SR					
L. brevipes	N	SO						Fw	CO		
L. californicum	N	SO						Fs			
L. exsertum	N	SO									
L. fremontii	N	SO						Fw			
L. macrodon	N	SO						Fe/w			
L. parishii	N	SO									
Nicotiana clevelandii	TW	SO	PI		GR	SR	DU	F			
*N. glauca	M	SO						Fs			DI
N. obtusifolia	C	SO	PI		GR						

ANGIOSPERMS: DICOTYLEDONS (cont.)

		SO	PI	UP	GR	SR	DU	F	CO	W	DI
Physalis acutifolia	TS	SO						Fe			DI
P. crassifolia	C	?	PI	UP	GR						
P. lobata	G							Fn			
P. pubescens	TS							Fw			?
*Solanum americanum	TN	SO									
*S. elaeagnifolium	H	SO						Fs			DI
S. hindsianum	N	SO	PI	UP	GR						
Sterculiaceae											
Ayenia compacta	C	SO			GR						
Tamaricaceae											
*Tamarix aphylla	MG						(DU)				
*T. ramosissima	M	SO						Fw		W	DI
Ulmaceae											
Celtis pallida	M	SO									
Urticaceae											
Parietaria floridana	TW	SO	PI	UP							
Verbenaceae											
Glandularia gooddingii	AP			UP							
*Phyla nodiflora	H	SO						Fs			DI
Verbena bracteata	TW		PI							W	
V. officinalis	AP	SO	PI							W	
Viscaceae											
Phoradendron californicum	p	SO	PI	UP	GR	SR	DU	F			
Zygophyllaceae											
Fagonia californica	C										
subsp. californica			PI	UP	GR	SR					
subsp. longipes		SO									
F. densa	C					SR					
F. pachyacantha	C		PI		GR						
Kallstroemia californica	TS	SO	PI		?			F			
K. grandiflora	TS	SO	PI		?			F			
Larrea divaricata											
subsp. tridentata	N	SO	PI	UP	GR	SR	DU	F			
var. arenaria	M						DU				
*Tribulus terrestris	TS	SO						F			DI
MONOCOTYLEDONS											
Agavaceae											
Agave deserti	Cs	SO			GR						
Hesperoyucca whipplei	Cs				GR						
Alismataceae											
Sagittaria longiloba	HYem									W	
Cyperaceae											
Cyperus esculentus	TS		PI							W	DI
C. laevigatus	HYem	SO						Fw		W	
C. odoratus	H	SO								W	

ANGIOSPERMS: MONOCOTYLEDONS (cont.)

		SO	PI	UP	GR	SR	DU	F	CO	W	DI
*C. rotundus	H							Fw			DI
C. squarrosus	TN	SO	PI							W	
Eleocharis geniculata	HYem	SO								W	
E. rostellata	H	SO						Fw		W	
Scirpus americanus	HYem	SO						Fw		W	
S. maritimus	HYem							Fw		W	
Hydrocharitaceae											
Najas marina	HYsub							Fw		W	
Juncaceae											
Juncus acutus	H							Fw		W	
J. articus	H	(SO)								W	
J. bufonius	TW	(SO)								W	
J. cooperi	H	(SO)						Fw		W	
Liliaceae											
Dichelostemma capitatum	G	SO		UP							
Hesperocallis undulata	G		PI				DU	F			
Triteleiopsis palmeri	G						DU	F			
Zephyranthes longifolia	G			UP							
Nolinaceae											
Nolina bigelovii	N				GR						
Poaceae											
Aristida adscensionis	TN	SO	PI	UP	GR	SR		F			
A. californica	H						DU	F			
A. parishii	H	SO?						Fn?			
A. purpurea	H	SO	PI	UP	GR	SR					
A. ternipes	H										
var. gentilis		SO									
var. ternipe		SO									
*Avena fatua	TW	SO						Fs			DI
Bothriochloa barbinodis	H			UP							
Bouteloua aristidoides	TS	SO	PI	UP	GR	SR	(DU)	F			
B. barbata	TS	SO	PI	UP	GR	SR	(DU)	F			
B. trifida	H	SO	PI								
Brachiaria arizonica	TS	SO?									
Bromus berterianus	TW			UP							
B. carinatus	TW	SO									
*B. catharticus	TW	SO						Fe			DI
*B. rubens	TW	SO	PI	UP				Fs			DI
*B. tectorum	TW	SO									
*Cenchrus echinatus	TS	SO						Fs			DI
*C. incertus	TN							(Fn)			DI?
C. palmeri	TN		PI		GR	SR	DU	F			
Chloris crinita	H	SO									
*C. virgata	TS	SO	PI					Fe/n			DI
*Cynodon dactylon	H	SO	PI					Fs/w		W	DI
*Dactyloctenium aegyptium	TS	SO								(W)	DI
Digitaria californica	H	SO		UP							

ANGIOSPERMS: MONOCOTYLEDONS (cont.)

		SO	PI	UP	GR	SR	DU	F	CO	W	DI
Distichlis palmeri	H								CO	Wm	
D. spicata	H	SO						Fs/w	CO	W	
*Echinochloa colonum	TS	SO	PI					Fw		W	DI
*E. crusgalli	TS	SO								W	DI
Enneapogon desvauxii	H	SO	PI	UP	GR						
*Eragrostis barrelieri	TN	SO									DI
*E. cilianensis	TS	SO	PI					Fe			DI
*E. lehmanniana	H		(PI)								
E. pectinacea	TS	SO	PI					Fn			
*Eriochloa acuminata	TS	SO						Fw			DI
E. aristata	TS		PI					Fn			
Erioneuron pulchellum	H	SO	PI	UP	GR	SR					
Festuca octoflora	TW	SO	PI	UP	GR						
Heteropogon contortus	H	SO	PI		GR	SR					
*Hordeum murinum	TW	SO	(PI)								DI
Leptochloa dubia	H	(SO)	(PI)								
L. fusca	TS	SO						Fs/w		W	DI
L. panicea	TS	SO	PI					Fn/e/w		W	DI
L. viscida	TS		PI					Fe/n			
*Lolium perenne	TW							Fs			DI
*L. temulentum	TW										DI
Monanthochloë littoralis	H								CO	Wm	
Muhlenbergia microsperma	TN	SO	PI	UP	GR	SR		F			
M. porteri	H	SO		UP							
Panicum alatum	TS										
var. alatum			PI								
var. minus			PI								
P. hirticaule	TS	SO	PI		GR?			Fe/w		W	DI
*Pennisetum ciliare	AP	SO	PI		GR						DI
*Phalaris caroliniana	TW							Fe			DI
*P. minor	TW	SO								W	DI
Phragmites australis	H	SO						Fw		W	
Pleuraphis rigida	H		PI	UP	GR		DU	F			
*Poa annua	TW	SO									DI
P. bigelovii	TW	SO									
*Polypogon monspeliensis	TN	SO						Fw	CO	W	DI
*P. viridis	H	?								W	
*Schismus arabicus	TW	SO	PI		GR	SR	DU	Fe/w			DI
*S. barbatus	TW	SO									DI
*Sorghum bicolor	TS	SO	PI								DI
*S. halepense	H	SO	PI					F			DI
Sporobolus airoides	H						(DU)	Fs/w	CO	(W)	
S. cryptandrus	H	SO		UP				Fs/e			
Sporobolus flexosus	H							Fe			
S. pyramidatus	TS	SO									
Stipa speciosa	H			UP							
Tridens muticus	H	SO	PI		GR	SR					
*Triticum aestivum	TW	SO						Fs/w			DI

ANGIOSPERMS: MONOCOTYLEDONS (cont.)

		SO	PI	UP	GR	SR	DU	F	CO	W	DI
Potamogetonaceae											
Potamogeton pectinatus	HYsub	SO								W	
Ruppia maritima	HYsub							Fw		W	
Typhaceae											
Typha domingensis	HYem	SO	PI					Fw		W	
Zannichelliaceae											
Zannichellia palustris	HYsub	SO								W	
TOTALS:		329	267	165	179	107	87	270	29	76	95

MG	M	N	H	TN	TS	TW	HY	AP	G
3	34	88	63	39	66	175	12	18	16

Appendix

B

Appendix B. Distribution of Plants in Sykes Crater (= Crater Grande)

BC = brown "cinders" **TB** = tuff breccia (o = outer slope; i = inner slope) **CL** = cliffs
TA = talus slopes **AL** = alluvium **PL** = playa

Plant name	BC	TB	CL	TA	AL	PL
Aliciella latifolia		TBo	CL			
Allionia incarnata		TBo			AL	
Amaranthus palmeri						PL
Antirrhinum cyathiferum		TBi				
Aristida adscensionis	BC	TB		TA		
Baileya pleniradiata					AL	
Bebbia juncea		TBi				
Boerhavia spp.	BC	TBi			AL	
Bouteloua aristidoides	BC	TBo		TA	AL	PL
B. barbata		TBi			AL	
Bursera microphylla		TBi	CL	TA		
Camissonia californica					AL	
C. cardiophylla		TB			AL	
C. chamaenerioides		TBi			AL	
C. claviformis		TBo		TA		
Carnegiea gigantea	BC	TBo			AL	
Chaenactis carphoclinia		TBo			AL	
Chorizanthe brevicornu	BC	TB			AL	
C. corrugata		TBo				
C. rigida		TBo			AL	
Cryptantha angustifolia		TB			AL	
C. barbigera					AL	
C. holoptera	BC	TBi	CL	TA	AL	
C. maritima		TBi		TA	AL	
C. pterocarya		TBi			AL	
Dalea mollis	BC	TB		TA	AL	
Descurainia pinnata					AL	
Ditaxis brandegeei	BC	TBi			AL	
D. lanceolata					AL	
D. neomexicana					AL	
Draba cuneifolia					AL	
Echinocactus polycephalus		TBo				
Encelia farinosa	BC	TB		TA	AL	
Eriastrum diffusum					AL	
Eriogonum deflexum		TBi				
E. inflatum		TBo				
E. thomasii		TB			AL	

Appendix B (cont.)

Plant name	BC	TB	CL	TA	AL	PL
E. trichopes		TBo				
Erioneuron pulchellum		TBo		TA		
Eschscholzia minutiflora		TB			AL	
Eucnide rupestris	BC	TB	CL	TA		
Eucrypta micrantha				TA	AL	
Euphorbia albomarginata						PL
E. pediculifera	BC	TBo				
E. polycarpa		TB		TA	AL	PL
E. setiloba		TB			AL	
Fagonia californica		TBo				
Fouquieria splendens	BC	TB		TA		
Gaura parviflora						PL
Geraea canescens					AL	PL
Gilia stellata						AL
Hibiscus denudatus				TA		AL
Hyptis emoryi		TBi	CL	TA	AL	
Jatropha cuneata	BC	TB		TA	AL	
Krameria grayi		TBi				
Larrea divaricata		TB		TA	AL	PL
Lepidium lasiocarpum					AL	PL
Linanthus bigelovii	BC	TBi		TA	AL	
Lotus strigosus	BC					
Lupinus arizonicus	BC	TB			AL	
Mammillaria grahamii		TBo				
Mentzelia involucrata	BC	TB			AL	
Mohavea confertiflora		TB			AL	
Monoptilon bellioides	BC			TA	AL	
Muhlenbergia microsperma	BC	TBi	CL		AL	PL
Nama demissum					AL	
N. hispidum	BC				AL	
Nemacladus glanduliferus	BC	TBi		TA	AL	
Nicotiana obtusifolia			CL	TA		
Notholaena californica			CL	TA		
Olneya tesota		TB			AL	
Opuntia bigelovii	BC	TBo		TA	AL	
O. fulgida					AL	
O. ramosissima		TBo				
Palafoxia arida					AL	
Parietaria floridana			CL	TA		

Appendix B (cont.)

Plant name	BC	TB	CL	TA	AL	PL
Parkinsonia microphylla		TB		TA	AL	
Pectis papposa					AL	PL
Pectocarya heterocarpa					AL	
P. platycarpa					AL	
Perityle emoryi	BC	TBi	CL	TA	AL	
Petalonyx linearis	BC	TBi				
Peucephyllum schottii		TB				
Phacelia ambigua		TB		TA	AL	
P. neglecta		TBo				
P. pedicellata		TBi		TA	AL	
Phaseolus filiformis		TBi			AL	
Physalis crassifolia		TB	CL	TA		
Plantago ovata					AL	PL
Pleuraphis rigida					AL	PL
Pleurocoronis pluriseta		TBi	CL			
Prosopis glandulosa						PL
Sphaeralcea ambigua		TBi		TA	AL	
S. coulteri						PL
Streptanthella longirostris					AL	
Teucrium cubense						PL
Tidestromia lanuginosa					AL	
Trichoptilium incisum		TBo				
Trixis californica				TA		
TOTALS (99 species):	24	63	12	31	63	15

Appendix

C

Appendix C. Commonly Cultivated Trees and Shrubs in Northwestern Sonora

SO = Sonoyta **SL** = San Luis **PP** = Puerto Peñasco

Plant name	SO	SL	PP
Agavaceae			
Agave americana: maguey; common agave	SO	SL	PP
Yucca spp.	SO	SL	PP
Anacardiaceae			
Mangifera indica: mango			PP
Schinus molle: pirúl; pepper tree	SO	SL	PP
S. terebinthifolius: Brazilian pepper tree	SO	SL	PP
Apocynaceae			
Carissa macrocarpa (*C. grandiflora*)			PP
Nerium oleander: laurel; oleander	SO	SL	PP
Plumeria rubra: súchil; plumeria, fragipani		SL	PP
Arecaceae			
Phoenix dactylifera: dátil; date palm	SO	SL	PP
P. roebelenii: pygmy date palm		SL	
Syagrus romanzoffiana: palma real; queen palm		SL	PP
Washingtonia filifera: palmera; fan palm	SO	SL	PP
W. robusta: palmera; fan palm	SO	SL	PP
Bignoniaceae			
Jacaranda mimosifolia: jacaranda		SL	PP
Tecoma stans var. *stans: lluvia de oro, palo de arco;* yellow trumpet	SO	SL	PP
Cactaceae			
Opuntia ficus-indica: nopal tunero; prickly pear	SO	SL	PP
Caprifoliaceae			
Sambucus mexicana: tápiro; elderberry	SO		
Caricaceae			
Carica papaya: papaya (marginal at SL and SO)			PP
Casuarinaceae			
Casuarina cunninghamiana: pino; river she-oak		SL	
Cupressaceae			
Cupressus sempervirens cv. *stricta, ciprés;* Italian cypress	SO	SL	PP
Thuja orientalis: tuya; oriental arborvitae	SO	SL	PP
Fabaceae			
Bauhinia variegata: orquidea, salacasuchil; purple orchid tree	SO	SL	PP
Caesalpinia gilliesii: tabachín; bird-of-paradise bush	SO	SL	PP
C. pulcherrima: tabachín; red bird-of-paradise bush	SO	SL	PP
Ceratonia siliqua: algarrobo; carob	SO	SL	PP
Delonix regia: árbol del fuego; royal poinciana	SO	SL	PP
Leucaena leucocephala: guaje; white lead tree	SO	SL	PP
Parkinsonia aculeata: palo verde, retama; Mexican palo verde	SO	SL	PP
Pithecellobium dulce: guamúchil	SO	SL	PP
Prosopis hybrids: *mesquite;* "Chilean" or South American mesquites; hybrids mostly involving *P. alba* and *P. nigra,* and perhaps *P. chilensis* and *P. flexuosa*	SO	SL	PP

Appendix C (cont.)

Plant name	SO	SL	PP
Malvaceae			
Hibiscus rosa-sinensis: obelisco; hibiscus	SO	SL	PP
Meliaceae			
Melia azedarach: paraiso, piocha; chinaberry	SO	SL	PP
Moraceae			
Ficus carica: higuiera; fig	SO	SL	PP
F. nitida: yucateco, laurel de la india		SL	PP
F. pandurina			PP
Morus alba: mora; white mulberry	SO	SL	PP
Myoporaceae			
Myoporum cf. *laetum*			PP
Myrtaceae			
Callistemon citrinus: lemon bottlebrush	SO	SL	PP
Eucalyptus camaldulensis: eucalipto; red gum	SO	SL	PP
Psidium guajava: guayaba; common guava	SO	SL	PP
Nyctaginaceae			
Bougainvillea sp. & hybrids: *bouganvilea;* bougainvillea	SO	SL	PP
Oleaceae			
Fraxinus velutina: fresno; Arizona ash	SO	SL	PP
Olea europaea: olivo; olive	SO	SL	PP
Poaceae			
Arundo donax: carrizo; giant reed or cane	SO	SL	PP
Saccharum officinale: caña azúcar; sugarcane	SO	SL	
Polygonaceae			
Antigonon leptopus: San Miguelito; queen's wreath	SO	SL	PP
Proteaceae			
Grevillea robusta: grevilea; silk oak	SO	SL	PP
Punicaceae			
Punica granatum: granada; pomegranate	SO	SL	PP
Rhamnaceae			
Ziziphus jujuba: pera cimarrón; jujube	SO		
Rosaceae			
Pyracantha coccinea: piracanta; pyracantha	SO	SL	PP
Rosa hybrids: *rosal;* rose	SO	SL	PP
Rutaceae			
Citrus cultivars: *toronja;* grapefruit; *limón;* lime; lemon; *naranja;* orange	SO	SL	PP
Ruta graveolens: ruda; rue	SO	SL	PP
Salicaceae			
Populus fremontii: álamo; cottonwood	SO	SL	PP
Tamaraceae			
Tamarix aphylla: pino salada; athel or salt cedar	SO	SL	PP
Verbenaceae			
Lantana camara and hybrids: *frutilla;* lantana	SO	SL	PP

Appendix C (cont.)

Plant name	SO	SL	PP
L. montevidensis: trailing lantana	SO	SL	PP
Vitex trifolia: carnavalito; pigeon berry		SL	PP
Vitaceae			
Vitis spp.: *uva;* grape	SO	SL	PP

Appendix

D

Appendix D. Non-Native Plants in Northwestern Sonora

	Habitat		
	Ruderal	Disturbed	Natural
Dicotyledons			
Aizoaceae			
Mesembryanthemum crystallinum		D	N
M. nodiflorum		D	
Amaranthaceae			
Amaranthus albus	R		
Apiaceae			
Coriandrum sativum		D	
Cyclospermum leptophyllum	R		
Asteraceae			
Carthamus tinctorius		D	
Centaurea melitensis		D	N
Conyza canadensis	R	D	
C. coulteri	R	D	N
Eclipta prostrata	R	D	
Flaveria trinervia	R		
Helianthus annuus	R	D	
Lactuca serriola	R	D	
Sonchus asper	R	D	N
S. oleraceus	R	D	N
Verbesina encelioides	R	D	
Xanthium strumarium	R	D	
Brassicaceae			
Brassica nigra	R	D	
B. tournefortii	R	D	N
Eruca vesicaria	R	D	
Sisymbrium irio	R	D	N
Capparaceae			
Cleome viscosa		D	
Chenopodiaceae			
Atriplex wrightii		D	
Bassia hyssopifolia	R	D	
Chenopodium album	R		
C. berlandieri	R?	D	
C. murale	R	D	N
Salsola tragus	R	D	
Convolvulaceae			
Ipomoea ×leucantha	R		
Cucurbitaceae			
Cucumis melo	R		
Euphorbiaceae			
Euphorbia hyssopifolia	R	D	
E. prostrata			N
E. serpens	R		

Appendix D (cont.)

	Habitat		
	Ruderal	Disturbed	Natural
Fabaceae			
Melilotus indica		D	
Parkinsonia aculeata		D	
Pediomelum rhombifolium	R		
Geraniaceae			
Erodium cicutarium	R	D	N
Malvaceae			
Malva parviflora	R	D	
Sphaeralcea emoryi		D	
Molluginaceae			
Mollugo cerviana			N
Nyctaginaceae			
Boerhavia diffusa	R	D	
Papaveraceae			
Argemone ochroleuca	R	D	
Pedaliaceae			
Sesamum orientale		D	
Polygonaceae			
Polygonum argyrocoleon	R	D	
†Rumex crispus	R		
Portulacaceae			
Portulaca oleracea	R	D	
Solanaceae			
Calibrachoa parviflora			N
Nicotiana glauca	R	D	
Solanum americanum	R		
S. elaeagnifolium	R	D	
Tamaricaceae			
Tamarix aphylla	R	D	
T. ramosissima	R	D	N
Verbenaceae			
Phyla nodiflora	R		
Zygophyllaceae			
Tribulus terrestris	R	D	
Monocotyledons			
Cyperaceae			
Cyperus rotundus	R		
Poaceae			
Avena fatua	R	D	
Bromus catharticus	R	D	
B. rubens	R	D	N
B. tectorum		D	

Appendix D (cont.)

	Habitat		
	Ruderal	Disturbed	Natural
Monocotyledons Poaceae (cont.)			
Cenchrus echinatus		D	
C. incertus		D	N
Chloris virgata		D	N
Cynodon dactylon	R	D	N
Dactyloctenium aegyptium	R		
Echinochloa colonum	R	D	N
E. crusgalli	R	D	
†Eleusine indica	R		
Eragrostis barrelieri	R		
E. cilianensis	R	D	N
E. lehmanniana			N
Eriochloa acuminata	R		
E. contracta	R		
Hordeum arizonicum	R		
H. murinum	R		
Lolium perenne	R		
L. temulentum	R		
Pennisetum ciliare	R	D	
†P. setaceum		D	
Phalaris caroliniana		D	
P. minor	R	D	
Poa annua	R		
Polypogon monspeliensis	R	D	N
P. viridis			N
Schismus arabicus		D	N
S. barbatus		D	N
Sorghum bicolor	R	D	
S. halepense	R	D	
Triticum aestivum	R	D	
TOTALS: 88	64	61	23

Appendix

E

Appendix E. Systematic Arrangement and Relative Abundance of the Grasses in Northwestern Sonora

Bold = species that are well established, thriving members of the natural ecosystem of the region; they would continue to survive in the region even if human influences were removed.

UPPER CASE BOLD = common or abundant grasses that are significant in the local vegetation.

( ) = Grasses that are not truly established in the flora; includes plants in disturbed habitats, or ones known from a single or few collections at the margin of the flora area, or perhaps were once present or cultivated by earlier people. These are plants which would probably not survive if the region were to revert to fully pristine, non-human-influenced conditions. Also included are grasses suspected of occurring in the flora area.

* = plants not native to the region; included are some that are native in nearby regions, such as in Arizona or elsewhere in Sonora, but ones which do not seem to be part of the native flora of northwestern Sonora.

Subfamily ARUNDINOIDEAE
Tribe Arundineae
Phragmites australis
Tribe Danthonieae
***SCHISMUS ARABICUS**
***S. BARBATUS**

Subfamily CHLORIDOIDEAE (ERAGROSTOIDEAE)
Tribe Aeluropodeae
DISTICHLIS PALMERI
D. spicata
MONANTHOCHLOË LITTORALIS
Tribe Aristideae
ARISTIDA ADSCENSIONIS
A. CALIFORNICA
(A. parishii)
A. PURPUREA
(A. ternipes var. gentilis)
A. ternipes var. ternipes
Tribe Chlorideae (includes Eragrostideae)
BOUTELOUA ARISTIDOIDES
B. BARBATA
B. trifida
Chloris crinita
***C. virgata**
***Cynodon dactylon**
*(Dactyloctenium aegyptium)
*(Eleusine indica)
*(Eragrostis barrelieri)
***E. cilianensis**
*(E. lehmanniana)
E. pectinacea
ERIONEURON PULCHELLUM
(Leptochloa dubia)
L. fusca
L. panicea
L. viscida
MUHLENBERGIA MICROSPERMA
M. porteri
PLEURAPHIS RIGIDA
SPOROBOLUS AIROIDES
S. cryptandrus
S. flexuosus
S. pyramidatus
Tridens muticus
Tribe Pappophoreae
ENNEAPOGON DESVAUXII

Subfamily PANICOIDEEAE
Tribe Andropogoneae
Bothriochloa barbinodis
Heteropogon contortus
*(Sorghum bicolor)
*(S. halepense)

Tribe Paniceae
(Brachiaria arizonica)
*(Cenchrus echinatus)
*(C. incertus)
C. PALMERI
Digitaria californica
***Echinochloa colonum**
*(E. crusgalli)
*(Eriochloa acuminata)
E. aristata
*(E. contracta)
Panicum alatum var. **alatum**
P. alatum var. **minus**
P. hirticaule var. **hirticaule**
(**P. hirticaule** var. **miliaceum**)
*(Pennisetum ciliare)
*(P. setaceum)
(Setaria grisebachii)
(S. leucopila)
(S. liebmannii)

Appendix E (cont.)

Subfamily POOIDEAE
Tribe Aveneae
*(Avena fatua)
*(Phalaris caroliniana)
*(P. minor)
***Polypogon monspeliensis**
*(P. viridis)
Tribe Poeae
Bromus berterianus
B. carinatus
*(B. catharticus)
***B. rubens**
*(B. tectorum)
FESTUCA OCTOFLORA
*(Lolium perenne)
*(L. temulentum)
*(Poa annua)
P. bigelovii

Tribe Stipeae
Stipa speciosa
Tribe Triticeae
(Hordeum arizonicum)
*(H. murinum)
*(Triticum aestivum)

Appendix

F

Appendix F. Geographic Distributions of Grasses in Northwestern Sonora

The geographic regions are defined as follows:

Sonoran Desert, Gulf of California: marine influenced, coastal desert portion of the Gulf of California in Sonora and the Baja California Peninsula, including the Río Colorado delta.

Sonoran Desert Region: arid region surrounding the Gulf of California, includes most of Baja California Peninsula, western Sonora, southeastern California, and western and southern Arizona. Closely approximating Sonoran Desert boundaries defined by Shreve (1951) with modifications by Brown (1982), Felger & Lowe (1976), and Búrquez et al. (1999).

Southwest North America: southwestern United States and northwestern Mexico; U.S. portion may include western Texas, Oklahoma, Arizona (mostly southern, western, or southwestern), southern Utah, sometimes southern Colorado, and southeastern California. Mexico portion includes Sonora and may include the Baja California Peninsula, Sinaloa (especially northwestern portion), Chihuahua, Coahuila, and neighboring states.

Megamexico: commonly includes part or most of the southwestern United States, much or most of Mexico, and sometimes northern Central America such as Guatemala. This pattern is sometimes listed as "Latin American" by other authors, but Latin American may also include South America.

North America: includes a major portion of North America; may extend from Alaska or Canada to Mexico and the West Indies, but few range north of the southern half of the United States. May be western and/or southern United States and northern to central Mexico.

New World: North and South America, and often also in Central America and the West Indies. Some are amphitropical—north of the tropics in North America and south of the tropics in South America and absent in between.

Old World: Native to the Old World and introduced, accidentally or intentionally, into the New World in post-Columbian times; most are widespread weeds.

Cosmopolitan: Old World and New World, on at least three continents.

* = Grasses not native to the floral area.

Sonoran Desert, Gulf of California

Distichlis palmeri

Sonoran Desert Region

Aristida californica var. *californica*
Cenchrus palmeri

Southwest North America

Aristida parishii
Bouteloua trifida
Erioneuron pulchellum
Hordeum arizonicum
Leptochloa viscida
Muhlenbergia porteri
Panicum hirticaule var. *miliaceum*
Pleuraphis rigida
Poa bigelovii
Sporobolus flexuosus
Tridens muticus var. *muticus*

Appendix F (cont.)

Megamexico

Aristida purpurea var. *nealleyi*
A. ternipes var. *gentilis*
Brachiaria arizonica
**Eriochloa acuminata* var. *acuminata*
E. aristata var. *aristata*
Panicum alatum var. *alatum*
Setaria grisebachii
S. leucopila
S. liebmannii

North America

Bromus carinatus
Eragrostis pectinacea
**Eriochloa contracta*
Festuca octoflora
Monanthochloë littoralis
**Phalaris caroliniana*
Sporobolus airoides var. *airoides*
S. cryptandrus

New World

Aristida ternipes var. *ternipes*
Bothriochloa barbinodis (amphitropical)
Bouteloua aristidoides (amphitropical)
B. barbata (amphitropical)
Bromus berterianus (amphitropical)
**Bromus catharticus* var. *catharticus* (South America)
**Cenchrus echinatus*
**C. incertus*
Chloris crinita (amphitropical)
**C. virgata*
Digitaria californica var. *californica*
Distichlis spicata
Leptochloa dubia (amphitropical)
L. fusca subsp. *uninervia*
L. panicea subsp. *brachiata*
Muhlenbergia microsperma
Panicum alatum var. *minus*
P. hirticaule var. *hirticaule*
Sporobolus pyramidatus
Stipa speciosa var. *speciosa* (amphitropical)

Old World

**Avena fatua*
**Bromus rubens*
**B. tectorum*
**Cynodon dactylon* var. *dactylon*
**Dactyloctenium aegyptium*
**Echinochloa colonum* var. *colonum*
**E. crusgalli* var. *crusgalli*

Appendix F (cont.)

Old World (cont.)

*Eleusine indica
*Eragrostis barrelieri
*E. cilianensis
*E. lehmanniana
*Hordeum murinum subsp. glaucum
*Lolium perenne
*L. temulentum
*Pennisetum ciliare
*P. setaceum
*Phalaris minor
*Poa annua
*Polypogon monspeliensis
*P. viridis
*Schismus arabicus
*S. barbatus
*Sorghum bicolor
*S. halepense
*Triticum aestivum

Cosmopolitan

Aristida adscensionis
Enneapogon desvauxii
Heteropogon contortus
Phragmites australis

Literature Cited

Abdallah, M. S. & **H. C. D. de Wit.** 1967–68. The Resedaceae. Mededelingen Landbouwhogeschol Wageningen (The Netherlands) 67:1–98; 68:99–416.

Abel, W. E. & **D. F. Austin.** 1981. Introgressive hybridization between *Ipomoea trichocarpa* and *Ipomoea lacunosa* (Convolvulaceae). Bulletin of the Torrey Botanical Club 108:231–239.

Abrams, L. 1923. Illustrated flora of the Pacific states. Vol. 1. Stanford: Stanford University Press.

———. 1944. Illustrated flora of the Pacific states. Vol. 2. Stanford: Stanford University Press.

———. 1951. Illustrated flora of the Pacific states. Vol. 3. Stanford: Stanford University Press.

Abrams, L. & **Ferris, R. S.** 1960. Illustrated flora of the Pacific states. Vol. 4. Stanford: Stanford University Press.

Adams, C. D. 1972. Flowering plants of Jamaica. Jamaica: University of the West Indies.

Adams, K. R. 1987. Little barley (*Hordeum pusillum* Nutt.) as a possible New World domesticate. In part 3 of La Ciudad: Specialized studies in the economy, environment, and culture of La Ciudad, J. E. Kissenburg, G. E. Rice & B. L. Shears, eds., 203–237. Arizona State University Anthropological Field Studies 20. Tempe: Arizona State University.

Aellen, P. & **T. Just.** 1943. Key and synopsis of the American species of the genus *Chenopodium* L. American Midland Naturalist 30:47–76.

Alcorn, S. M., S. E. McGregor, & **G. Olin.** 1962. Pollination requirements of the organpipe cactus. Cactus and Succulent Journal (U.S.) 34:134–138.

Allred, K. W. 1984. Morphologic variation and classification of the North American *Aristida purpurea* complex (Gramineae). Brittonia 36:382–395.

———. 1992. The genus *Aristida* (Gramineae) in California. Great Basin Naturalist 52:41–52.

———. 1994. A new name for *Aristida hamulosa* (Gramineae). Phytologia 77:411–413.

Allred, K. W. & **F. W. Gould.** 1983. Systematics of the *Bothriochloa saccharoides* complex (Poaceae: Andropogoneae). Systematic Botany 8:168–184.

Allred, K. W. & **J. Valdés-Reyna.** 1997. The *Aristida pansa* complex and a key to the Divaricatae group of North America (Gramineae: Aristideae). Brittonia 49:54–66.

Al-Shehbaz, I. A. 1986. The genera of Lepidieae (Cruciferae; Brassicaceae) in the southeastern United States. Journal of the Arnold Arboretum 67:265–311.

Alvarez de Williams, A. 1983. Cocopa. In vol. 10 of Handbook of North American Indians, A. Ortiz, ed, 99–112. Washington, D.C.: Smithsonian Institution.

———. 1987. Environment and edible flora of the Cocopa. Environment Southwest 519:22–27 (San Diego Museum of Natural History).

Anderson, C. 1972. A monograph of the Mexican and Central American species of *Trixis* (Compositae). Memoirs of the New York Botanical Garden 22:1–68.

Anderson, D. E. 1961. Taxonomy and distribution of the genus *Phalaris*. Iowa State Journal of Science 36:1–96.

———. 1974. Taxonomy of the genus *Chloris* (Gramineae). Brigham Young University Science Bulletin, Biological Series 19:1–133.

Anderson, E. F. 1986. A revision of the genus *Neolloydia* B. & R. (Cactaceae). Bradleya 4:1–28.

Anderson, W. R. 1982. Notes on neotropical Malpighiaceae—I. Contributions from the University of Michigan Herbarium 15:93–136.

Argus, G. W. 1986. The genus *Salix* (Salicaceae) in the southeastern United States. Systematic Botany Monographs 9:1–121.

———. 1995. Salicaceae, Willow Family, part 2: *Salix* L., Willow [Vascular Plants of Arizona]. Journal of the Arizona-Nevada Academy of Sciences 29:39–62.

Arnow, L. A. 1987. Gramineae A.L. Juss. Grass family. In A Utah flora, S. L. Welsh, N. D. Atwood, S. Goodrich, & L. C. Higgins, eds., 684–788. Great Basin Naturalist Memoirs 9. Provo: Brigham Young University.

Aronson, J. A., D. Pasternak, & **A. Danon.** 1988. Introduction and first evaluation of 120 halophytes under seawater irrigation. In Arid lands today and tomorrow, E. E. Whitehead, C. F. Hutchinson, B. N. Timmermann, & R. G. Varady, eds., 737–746. Boulder, Colo.: Westview Press.

Arriagada, J. E. 1998. The genera of Inuleae (Compositae; Asteraceae) in the southeastern United States. Harvard Papers in Botany vol. 3:1–48.

Ashri, A. & **P. F. Knowles.** 1959. Further notes on *Carthamus* in California. Leaflets of Western Botany 9:5–8.

Atwood, N. D. 1975. A revision of the *Phacelia crenulatae* group (Hydrophyllaceae) for North America. The Great Basin Naturalist 35:127–190.

Austin, D. F. 1978. The *Ipomoea batatas* complex—I: Taxonomy. Bulletin of the Torrey Botanical Club 105: 114–129.

———. 1990. Comments on southwestern United States *Evolvulus* L. and *Ipomoea* L. Madroño 37:124–132.

———. 1992. Rare Convolvulaceae in the southwestern United States. Annals of the Missouri Botanical Garden 79:8–16.

———. 1998. Convolvulaceae, Morning Glory Family [Vascular Plants of Arizona]. Arizona-Nevada Academy of Science 30:61–83.

Austin, D. F. & **R. A. Pedraza.** 1983. Los géneros de Convolvulaceae en México. Boletín de la Sociedad Botánica de México 44:3–16.

Averett, J. E. 1973. Biosystematic study of *Chamaesaracha*. Rhodora 75:325–365.

———. 1979. Biosystematics of the physaloid genera of the Solanaceae in North America. In The biology and taxonomy of the Solanaceae, J. G. Hawkes, R. N. Lester, & A. D. Skelding, eds., 493–503. London: Academic Press.

Bacon, J. D. 1984. Chromosome numbers and taxonomic notes in the genus *Nama* (Hydrophyllaceae)—II. Sida 10:269–275.

Bailowitz, R. A. 1988. Systematics of *Ascia* (*Ganyra*) (Pieridae) populations in the Sonoran Desert. Journal of Research on the Lepidoptera 26:73–81.

Baker, H. G. 1986. Yuccas and yucca moths—A historical commentary. Annals of the Missouri Botanical Garden 73:556–564.

Baker, M. A. & **D. J. Pinkava.** 1987. A cytological and morphological analysis of a triploid apomict, *Opuntia* × *kelvinensis* (subgenus *Cylindropuntia,* Cactaceae). Brittonia 39:387–401.

Bakshi, T. S. & **R. N. Kapil.** 1954. The morphology and ecology of *Mollugo cerviana* Ser. Journal of the Indian Botanical Society 33:309–328.

Baldini, R. M. 1995. Revision of the genus *Phalarais* L. (Gramineae). Webbia 49:265–329.

Ball, P. W. & **T. G. Tutin.** 1959. Notes on annual species of *Salicornia* in Britain. Watsonia 4:193–205.

Balslev, H. 1996. Juncaceae. Flora Neotropica Monograph 68:1–167.

Barber, S. 1982. Taxonomic studies in the *Verbena stricta* complex (Verbenaceae). Systematic Botany 7:433–456.

Barbour, M. G. 1968. Germination requirements of the desert shrub *Larrea divaricata*. Ecology 49:915–923.

———. 1969. Patterns of genetic similarity between *Larrea divaricata* of North and South America. American Midland Naturalist 81:54–67.

Barkley, F. A. 1937. A monographic study of *Rhus* and its immediate allies in North and Central America, including the West Indies. Annals of the Missouri Botanical Garden 24:265–498.

———. 1940. *Schmaltzia*. American Midland Naturalist 24:647–665.

Barkley, T. M. 1978. *Senecio*. North American Flora (series 2) 10:50–139.

———. 1999. The Segregates of *Senecio,* s.l., and *Cacalia,* s.l., in the flora of North America north of Mexico. Sida 18:659–670.

Barkley, T. M., B. L. Clark, & **A. M. Funston.** 1996. The segregate Genera of *Senecio,* s.l., and *Cacalia,* s.l. (Asteraceae: Senecioneae) in Mexico and Central America. In Compositae: Systematics. Proceedings of the International Compositae Conference, Kew 1994, D.J.N. Hind & H. J. Beentje, eds., 613–620. Kew: Royal Botanic Gardens.

Barkworth, M. E. 1993. North American Stipeae (Gramineae): Taxonomic changes and other comments. Phytologia 74:1–25.

Barneby, R. C. 1964. Atlas of North American Astragalus. Memoirs of the New York Botanical Garden 13:1–1188.

———. 1977. *Dalea* imagines. Memoirs of the New York Botanical Garden 27:1–891.

———. 1989. Intermountain flora: Vascular plants of the Intermountain West, U.S.A., vol. 3, part B. New York: New York Botanical Garden.

———. 1998. Silk tree, guanacaste, monkey's earring: A generic system for the synandrous Mimosaceae of the Americas—III: *Calliandra.* Memoirs of the New York Botanical Garden 74.

Barneby, R. C. & **E. C. Twisselmann.** 1970. Notes on *Loeflingia* (Caryophyllaceae). Madroño 20:398–408.

Barringer, K. 1997. *Aristolochia.* In vol. 3 of Flora of North America north of Mexico, 44–50. New York: Oxford University Press.

Barrios Matrecito, V. 1977. Por las rutas del desierto. Patronato de ediciones culturales. San Luis Río Colorado, Sonora. 3rd ed. Hermosillo, Sonora: Gobierno del Estado del Sonora, Secretaría del Fomento Educativo y Cultura.

Bassett, I. J. & **B. R. Baum.** 1969. Conspecificity of *Plantago fastigiata* of North America with *P. ovata* of the Old World. Canadian Journal of Botany 47:1865–1868.

Bassett, I. J. & **C. W. Crompton.** 1978. The genus *Suaeda* (Chenopodiaceae) in Canada. Canadian Journal of Botany 56:581–591.

Bates, D. M. 1968. Generic relationships in the Malvaceae, tribe Malveae. Gentes Herbarum 10:117–135.

Baum, B. R. 1967. Introduced and naturalized tamarisks in the United States and Canada (Tamaricaceae). Baileya 15:19–25.

———. 1977. Oats: Wild and cultivated, a monograph of the genus *Avena* L. (Poaceae). Canada Department of Agriculture, Research Branch, Biosystematics Research Institute, Monograph 14. Ottawa, Canada.

———. 1978. The genus *Tamarix*. Jerusalem: Israel Academy of Science and Humanities.

Bean, L. J. & **K. S. Saubel.** 1972. Temalpakh. Banning, Calif.: Malki Museum Press.

Becerra, J. X. & **D. L. Venable.** 1999. Nuclear ribosomal DNA phylogeny and its implications for evolutionary trends in Mexican *Bursera* (Burseraceae). American Journal of Botany 86:1047–1057.

Beetle, A. A. 1943. The North American variations of *Distichlis spicata.* Bulletin of the Torrey Botanical Club 70:638–650.

———. 1955. The genus *Distichlis*. Revista Argentina Agronomía 22:86–94.

———. 1974. Noteworthy grasses from Mexico, II. Phytologia 28:313–318.

———. 1983, 1987. Las gramíneas de México. 2 vols. México, D.F.: COTECOCA, Secretaría de Agricultura y Recursos Hidraulicos.

———. 1987. Noteworthy grasses from Mexico 13. Phytologia 63:209–297.

Beetle, A. A. & **D. Johnson.** 1991. Gramíneas de Sonora. Hermosillo: Gobierno del Estado de Sonora, Secretaría de Fomento Gandadero.

Beliz, T. 1986. A revision of *Cuscuta* sect. *Cleistogrammica* using phenic and cladistic analysis with a comparison of reproductive mechanisms and host preferences in species from California, Mexico, and Central America. Ph.D. diss., University of California, Berkeley.

Bell, F., K. M. Anderson, & **Y. G. Stewart.** 1980. The Quitobaquito cemetery and its history. Tucson: National Parks Service, Western Archeological Center.

Bell, W. H. & **E. F. Castetter.** 1937. The utilization of mesquite and screwbean by the aborigines in the American Southwest. University of New Mexico Bulletin 314:1–55.

Bemis, W. P. & **T. W. Whitaker.** 1965. Natural hybridization between *Cucurbita digitata* and *C. palmata.* Madroño 18:39–47.

———. 1969. The xerophytic *Cucurbita* of northwestern Mexico and southwestern United States. Madroño 20:33–41.

Benham, D. M. & **M. D. Windham.** 1993. *Astrolepis*. In vol. 2 of Flora of North America north of Mexico, 140–143. New York: Oxford University Press.

Benson, L. 1941. The mesquites and screw-beans of the United States. American Journal of Botany 28:748–754.

———. 1969a. The cacti of Arizona. 3rd ed. Tucson: University of Arizona Press.

———. 1969b. The native cacti of California. Stanford: Stanford University Press.

———. 1982. The cacti of the United States and Canada. Stanford: Stanford University Press. Stanford.

Benson, L. & **R. A. Darrow.** 1945. A manual of southwestern desert trees and shrubs. University of Arizona Bulletin 15, no. 2. Tucson: University of Arizona.

———. 1981. Trees and shrubs of the southwestern deserts. 3rd ed. Tucson: University of Arizona Press.

Betancourt, J. L., T. R. Van Devender, & **P. S. Martin,** eds. 1990. Packrat middens: The last 40,000 years of biotic change. Tucson: University of Arizona Press.

Bierner, M. W. & **R. K. Jansen.** 1998. Systematic implications of DNA restriction site variation in *Hymenoxys* and *Tetraneuris* (Asteraceae, Helineae, Gallardinae). Lundellia 1:17–26

Björkman, S. O. 1960. Studies in *Agrostis* and related genera. Symbolae Botanica Upsaliensis 17:1–112.

Blake, S. F. 1913. A revision of *Encelia* and related genera. Proceedings of the American Academy of Arts and Sciences 49:346–396.

———. 1926. *Baccharis.* In Trees and shrubs of Mexico, P. C. Standley. Contributions from the United States National Herbarium 23 (part 5):1499–1507.

Bloom, W. W. 1955. Comparative viability of sporocarps of *Marsilea quadrifolia* L. in relation to age. Illinois Academy of Science Transactions 47:72–76.

Bogin, C. 1955. Revision of the genus *Sagittaria* (Alismataceae). Memoirs of the New York Botanical Garden 9:179–233.

Bogle, A. L. 1969. The genera of Portulacaceae and Basellaceae in the southeastern United States. Journal of the Arnold Arboretum 50:566–598.

———. 1970. The genera of Molluginaceae and Aizoaceae in the southeastern United States. Journal of the Arnold Arboretum 51:431–462.

Boldt, P. E. 1989. *Baccharis* (Asteraceae): A review of its taxonomy, phytochemistry, ecology, economic status, natural enemies, and the potential for its biological control in the United States. Temple, Tex.: U.S.D.A., Agricultural Research Service, Grassland, Soil and Water Research Laboratory.

Bolger, D. J., J. L. Neff, & **B. B. Simpson.** 1995. Multiple origins of the yucca–yucca moth association. Proceedings of the National Academy of Sciences of the United States of America 92:6864–6867.

Bolton, H. E. 1936. Coronado. Albuquerque: University of New Mexico Press.

Bothmer, R. von, N. Jacobsen, C. Baden, R. B. Jorgensen, & **I. Linde-Laursen.** 1991. An ecogeographical study of the genus *Hordeum.* Systematics & Ecogeographic Studies on Crop Genepools 7. Rome: International Board for Plant Genetic Resources.

Boufford, D. E. 1992. Urticaceae, Nettle Family [Vascular Plants of Arizona]. Journal of the Arizona-Nevada Academy of Science 26:43–49.

———. 1997. Urticaceae. In vol. 3 of Flora of North America north of Mexico, 400–413. New York: Oxford University Press.

Boulos, L. 1972–74. Rèvision systèmatique du genre *Sonchus* L. s.l.—I–IV. *Sonchus.* Särtryck ur Botaniska Notiser 125:287–319; 126:155–196; 127:7–37, 402–451.

Bourell, M. & **T. F. Daniel.** 1988. Noteworthy collections: California. Madroño 35:279–280.

Bowden, W. M. & **H. A. Senn.** 1962. Chromosome numbers in 28 grass genera from South America. Canadian Journal of Botany 40:1115–1124.

Bowers, J. E. 1980. Flora of Organ Pipe Cactus National Monument. Journal of the Arizona-Nevada Academy of Science 15:1–11, 33–47.

Boyd, J. W., D. S. Murray, & **R. J. Tyrl.** 1984. Silverleaf nightshade, *Solanum elaeagnifolium*, origin, distribution, and relation to man. Economic Botany 38:210–217.

Bradley, C. E. & **A. J. Haagen-Smit.** 1949. The essential oil of *Pectis papposa.* Economic Botany 3:407–412.

Brako, L. & **J. L. Zarucchi.** 1993. Catalogue of the flowering plants and gymnosperms of Peru. St. Louis: Missouri Botanical Garden.

Bravo-Hollis, H. 1978. Las cactaceas de México. Vol. 1. 2nd ed. México, D.F.: Universidad Nacional Autónoma de México.

Bravo-Hollis, H. & **H. Sánchez-Mejorada R.** 1989. Claves para la identificación de las cactaceas de México. Cactaceas y Succulentas Mexicanas, número especial. México, D.F.

———. 1991a. Las cactaceas de México. Vol. 2. México, D.F.: Universidad Nacional Autónoma de México.

———. 1991b. Las cactaceas de México. Vol. 3. México, D.F.: Universidad Nacional Autónoma de México.

Bray, R. A. 1978. Evidence for facultative apomixis in *Cenchrus ciliaris.* Euphytica 27:801–804.

Bray, W. L. 1898. On the relation of the flora of the Lower Sonoran Zone in North America to the flora of the arid zones of Chile and Argentina. Botanical Gazette 26:121–147.

Bretting, P. K. 1982. Morphological differentiation of *Proboscidea parviflora* ssp. *parviflora* (Martyniaceae) under domestication. American Journal of Botany 69:1531–1537.

———. 1985. Nomenclatural changes in *Proboscidea* (Martyniaceae). The Southwestern Naturalist 30:150–151.

Bretting, P. K. & **S. Nilsson.** 1988. Pollen morphology of the Martyniaceae and its systematic implications. Systematic Botany 13:51–59.

Britton, N. L. & **J. N. Rose.** 1919–23. The Cactaceae. 4 vols. Carnegie Institute of Washington Publication 248. Washington, D.C.: Carnegie Institute.

Brizicky, G. K. 1963. Taxonomic and nomenclatural notes on the genus *Rhus* (Anacardiaceae). Journal of the Arnold Arboretum 44:60–80.

———. 1964. The genera of Rhamnaceae in the southeastern United States. Journal of the Arnold Arboretum 45:439–463.

———. 1968. *Herissantia, Bogenhardia,* and *Gayoides* (Malvaceae). Journal of the Arnold Arboretum 49:278–279.

Brown, D. E. 1982. Biotic communities of the American Southwest—United States and Mexico. Desert Plants 4:3–341.

Brown, J. H. & **A. C. Gibson.** 1983. Biogeography. St. Louis: Mosby.

Brown, R. C. 1978. Biosystematics of *Psilostrophe* (Compositae: Helenieae). Part 2: Artificial hybridization and systematic treatment. Madroño 25:187–201.

Brown, W. V. 1950. A cytological study of some Texas Gramineae. Bulletin of the Torrey Botanical Club 77: 63–76.

Broyles, B. 1996a. Surface water resources for prehistoric peoples in western Papaguería of the North American Southwest. Journal of Arid Environments 33:483–495.

———. 1996b. Organ Pipe Cactus National Monument: A Sonoran Desert Sanctuary. Tucson: Southwest Parks and Monuments Association.

Broyles, B., R. S. Felger, G. P. Nabhan, & **L. Evans.** 1997. Our grand desert: A gazetteer for northwestern Sonora, southwestern Arizona, and northeastern Baja California. Journal of the Southwest 39:703–855.

Bruhl, J. J. 1995. Sedge genera of the world: relationships and a new classification of the Cyperaceae. Australian Journal of Systematic Botany 8:125–305.

Brummitt, R. K. & **C. E. Powell,** eds. 1992. Authors of plant names. Kew: Royal Botanic Gardens.

Buchmann, S. L. 1987. The ecology of oil flowers and their bees. Annual Review of Ecology and Systematics 18:343–369.

Buddell, G. F. & **J. W. Thieret.** 1997. Saururaceae. In vol. 3 of Flora of North America north of Mexico, 37–38. New York: Oxford University Press.

Burgess, T. L. 1988. The relationship between climate and leaf shape in the *Agave cerulata* complex. Ph.D. diss., University of Arizona, Tucson.

Burgess, T. L., J. E. Bowers, & **R. M. Turner.** 1991. Exotic plants at the Desert Laboratory, Tucson, Arizona. Madroño 38:96–114.

Burkart, A. E. 1952. Las Leguminosas argentinas. Buenos Aires: Acme Agency.

———. 1969. Gramíneas. In part 2 of Flora ilustrada de entre ríos (Argentina). Buenos Aires: Collección Científica del I.N.T.A.

———. 1976. A monograph of the genus *Prosopis* (Leguminosae subfam. Mimosoideae). Journal of the Arnold Arboretum 57:219–525.

Burkart, A., N. S. Troncoso de Burkart, & **N. M. Bacigalupo.** 1987. Flora illustrada de entre ríos (Argentina), part 3. Buenos Aires: Collección Científica del I.N.T.A.

Búrquez, A. 1998. Historical summary of the formation of the biosphere reserve El Pinacate y el Gran Desierto de Altar. In The Sierra Pinacate, J. Hayden, 74–77. Tucson: University of Arizona Press.

Búrquez, A. & **C. Castillo.** 1994. Reserve de la biósfera El Pinacate y Gran Desierto de Altar: entomo biológico y social. Estudios Sociales 5(9):9–64 (Hermosillo, Sonora).

Búrquez, A. & **A. Martínez-Yrízar.** 1997. Conservation and landscape transformation in Sonora, México. Journal of the Southwest 39:371–378.

Búrquez, A., A. Martínez-Yrízar, R. S. Felger, & **D. Yetman.** 1999. Vegetation and habitat diversity at the southern edge of the Sonoran Desert. In Ecology of Sonoran Desert plants and plant communities, R. H. Robichaux, ed., 36–67. Tucson: University of Arizona Press.

Búrquez, A. & **M. A. Quintana.** 1994. Islands of diversity: Ironwood ecology and richness of perennials in a Sonoran Desert biological reserve. In Ironwood: An ecological and cultural keystone of the Sonoran Desert, G. P. Nabhan & J. L. Carr, eds., 9–28. Conservation International Occasional Paper No. 1. Washington, D.C.

Burrus, E. J. 1971. Kino and Manje: Explorers of Sonora and Arizona. Rome and St. Louis: Jesuit Historical Institute.

Burtt, B. L. 1991. Umbelliferae of southern Africa: An introduction and annotated check-list. Edinburgh Journal of Botany 48:133–282.

Buxbaum, F. 1975a. Die Gattung *Echinocereus.* In Die Kakteen, Lieferung 60, CVIIC, H. Krainz, ed. Stuttgart.

———. 1975b. Gattung *Peniocereus.* In Die Kakteen, CIIa, H. Krainz, ed. Stuttgart.

Bywater, M. & **G. E. Wickens.** 1984. New World species of the genus *Crassula.* Kew Bulletin 39:699–728.

Cabrera, A. L. 1970. Parte 2: Gramineae. In Flora de la provincia de Buenos Aires. Buenos Aires: Colección Científica del I.N.T.A.

Cabrera, A. L. & **M. A. Torres.** 1970. *Stipa* L. In Flora de la provincia de Buenos Aires, A. L. Cabrera, ed., 255–290. Buenos Aires: Colección Científica del I.N.T.A.

Carlquist, S. 1956. On the generic limits of *Eriophyllum* (Compositae) and related genera. Madroño 13:226–239.

Carolin, R. C. 1987. A review of the family Portulacaceae. Australian Journal of Botany 35:383–412.

Carr, B. L., J. V. Crisci, & **P. C. Hoch.** 1990. A cladistic analysis of the genus *Gaura* (Onagraceae). Systematic Botany 15:454–461.

Carter, A. M. 1974. The genus *Cercidium* (Leguminosae: Caesalpinioideae) in the Sonoran Desert of Mexico and the United States. Proceedings of the California Academy of Sciences (series 4) 40:17–57.

Castetter, E. F. & **W. H. Bell.** 1937. The aboriginal utilization of the tall cacti in the American Southwest. University of New Mexico Bulletin 307:3–48.

———. 1942. Pima and Papago Indian agriculture. Albuquerque: University of New Mexico Press.

———. 1951. Yuman Indian agriculture. Albuquerque: University of New Mexico Press.

Chamberland, M. 1991. Biosystematics of the *Echinocactus polycephalus* complex (Cactaceae). Master's thesis, Arizona State University, Tempe.

———. 1997. Systematics of the *Echinocactus polycephalus* complex (Cactaceae). Systematic Botany 22:303–313.

Chase, A. 1918. Axillary cleistogenes in some American grasses. American Journal of Botany 5:254–258.

———. 1920. The North American species of *Cenchrus.* Contributions from the United States National Herbarium 22:45–77.

———. 1921. The North American species of *Pennisetum.* Contributions from the United States National Herbarium 22:209–234.

———. 1946. *Enneapogon desvauxii* and *Pappophorum wrightii:* an agrostological detective story. Madroño 8:187–189.

———. 1958. Nota sobre la presencia de *Distichlis* en Africa. Revista Argentina de Agronomía 25:195.

Chiang-Cabrera, F. 1981. A taxonomic study of the North American species of *Lycium* (Solanaceae). Ph.D. diss., University of Texas, Austin.

———. 1983. Nomenclatural changes for new sectional delimitation in *Lycium* (Solanaceae) in the New World. Taxon 32:456–458.

Childs, T. 1954. Sketch of the "Sand Papago." The Kiva 19:27–39.

Christian, J. A. & **D. B. Dunn.** 1970. Nomenclature of the *Lupinus arizonicus* complex. Transactions of the Missouri Academy of Science 4:95–98.

Christy, C. M. 1998. Loasaceae: Stickleaf or blazing-star family [Vascular Plants of Arizona]. Journal of Arizona-Nevada Academy of Science 30:96–111.

Clark, C. 1997. *Eschscholzia.* In vol. 3 of Flora of North America north of Mexico, 308–312. New York: Oxford University Press.

———. 1998. Phylogeny and adaptation in the *Encelia* alliance (Asteraceae: Heliantheae). Aliso 17:89–98.

Clary, K. H. & **B. B. Simpson.** 1995. Systematics and character evolution of the genus *Yucca* L. (Agavaceae): Evidence from morphology and molecular analyses. Boletín de la Sociedad Botánica de México 56:77–88.

Clayton, W. D. 1972. Gramineae. In vol. 3, part 2 of Flora of West Tropical Africa, J. Hutchinson & J. M. Dalzielp, eds., 349–512. 2d. ed. London: Crown Agents for Overseas Governments and Administrations.

Clayton, W. D. & **J. R. Harlan.** 1970. The genus *Cynodon* L. C. Rich. in tropical Africa. Kew Bulletin 24:185–189.

Clayton, W. D. & **S. A. Renvoize.** 1986. Genera graminum, grasses of the world. Kew Bulletin Additional Series XIII. London: Her Majesty's Stationery Office.

Cody, M. L. 1984. Branching patterns in columnar cacti. In vol. 13 of Being alive on land: Tasks for vegetation studies, eds. N. S. Margaris, M. Arianoustou-Farragitako, & W. C. Oechel, 201–236. The Hague: W. Junk.

Cole, G. A. & **M. C. Whiteside.** 1965. An ecological reconnaissance of Quitobaquito spring, Arizona. Journal of the Arizona Academy of Sciences 3:159–163.

Cole, K. L. 1986. The lower Colorado River valley: A Pleistocene desert. Quaternary Research 25:392–400.

Collins, S. L. & **W. H. Blackwell.** 1979. *Bassia* (Chenopodiaceae) in North America. Sida 8:57–64.

Columbus, J. T. 1999. An expanded circumscription of *Bouteloua* (Gramineae: Chloridoideae). Aliso 18:61–65.

Columbus, J. T., M. S. Kinney, R. Pant, & **M. E. Siqueiros Delgado.** 1998. Cladistic parsimony analysis of internal transcribed spacer region (nrDNA) sequences of *Bouteloua* and relatives (Gramineae: Chlorideae). Aliso 17:99–130.

Conert, H. J. 1961. Die Systematik und Anatomie der Arundineae. Weinheim: Cramer.

Conert, H. J. & **A. M. Türpe.** 1974. Revision der Gattung *Schismus*. Abhandlungen der Senckenbergischen naturforschenden Gesellschaft (Frankfurt). 532:1–81.

Constance, L. 1937. A systematic study of the genus *Eriophyllum* Lag. University of California Publications in Botany 18:69–135.

———. 1938. The genus *Eucrypta* Nutt. Lloydia 1:143–152.

———. 1939. The genus *Pholistoma* Lilja. Bulletin of the Torrey Botanical Club 66:341–352.

———. 1990. Tardy transfers from *Apium* to *Ciclospermum* (Apiaceae). Brittonia 42:276–278.

Correll, D. S. & **M. C. Johnston.** 1970. Manual of the vascular plants of Texas. Renner, Tex.: Texas Research Foundation.

Cortés, E. A., M. A. Fernández, E. M. Franco, & **E. Vera.** 1976. Geología del área volcánica del Pinacate en el Desierto de Altar, Sonora, Mexico. Ph.D. diss., Instituto Politécnico Nacional, México, D.F.

Cothrun, J. D. 1969. Some aspects of the germination and attachment of *Ammobroma sonorae,* a root parasite of desert shrubs. Ph.D. diss., Oklahoma State University.

Cozzo, D. 1946. Relación anatómica entre la estructura del leño de las especias argentinas de *Capparis* y *Atamisquea*. Lilloa 12:29–37.

Craig, R. T. 1945. The *Mammillaria* handbook. Pasadena, Calif.: Abbey Garden Press.

Crins, W. J. 1989. The Tamaricaceae in the southeastern United States. Journal of the Arnold Arboretum 70: 403–425.

———. 1991. The genera of Paniceae (Gramineae: Panicoideae) in the Southeastern United States. Journal of the Arnold Arboretum (Supplementary Series) 1:171–312.

Cristobal, C. L. 1960. Revisión del género *Ayenia* (Sterculiaceae). Opera Lilloana 4:1–230.

Croat, T. B. 1978. Flora of Barro Colorado Island. Stanford: Stanford University Press.

Cronquist, A. 1943. The separation of *Erigeron* from *Conyza,* Bulletin of the Torrey Botanical Club 70:629–632.

———. 1947. Revision of the North American species of *Erigeron,* north of Mexico. Brittonia 6:121–300.

———. 1981. An integrated system of classification of the flowering plants. New York: Columbia University Press.

———. 1994. Intermountain flora: Vascular plants of the Intermountain West, U.S.A., vol. 5. New York: New York Botanical Garden.

Cronquist, A., A. H. Holmgren, N. H. Holmgren, & J. L. Reveal. 1972. Intermountain flora: Vascular plants of the Intermountain West, U.S.A., vol. 1. New York: Haffner Publishing Co.

Cronquist, A., A. H. Holmgren, N. H. Holmgren, J. L. Reveal, & P. K. Holmgren. 1977. Intermountain flora: Vascular plants of the Intermountain West, U.S.A., vol 6. New York: Columbia University Press.

———. 1984. Intermountain flora: Vascular plants of the Intermountain West, U.S.A., vol. 4. New York: New York Botanical Garden.

Cronquist, A., N. H. Holmgren, & P. K. Holmgren. 1997. Intermountain flora: Vascular plants of the Intermountain West, U.S.A., vol. 3, part A. New York: New York Botanical Garden.

Cronquist, A. & D. D. Keck. 1957. A reconstitution of the genus *Machaeranthera*. Brittonia 9:231–239.

Crosswhite, F. S. 1980. The annual saguaro harvest and crop cycle of the Papago, with reference to ecology and symbolism. Desert Plants 2:3–61.

Cuatrecasas, J. 1968a. Revisión de las especies columbianas del género *Baccharis*. Revista de la Academia Colombiana de Ciencias Exactas, Físicas y Naturales 13:5–102.

———. 1968b. Notas adicionales, taxonomicas y corologicas, sobre *Baccharis*. Revista de la Academia Colombiana de Ciencias Exactas, Físicas y Naturales 13:201–226.

Daniel, T. F. 1983. *Carlowrightia* (Acanthaceae). Flora Neotropica Monograph 34:1–116.

———. 1984. The Acanthaceae of the southwestern United States. Desert Plants 5:162–179.

———. 1988. Taxonomic, nomenclatural, and reproductive notes on *Carlowrightia* (Acanthaceae). Brittonia 40:245–255.

Danin, A. 1996. Plants of desert dunes. Berlin: Springer-Verlag.

Danin, A., I. Baker, & H. G. Baker. 1978. Cytogeography and taxonomy of the *Portulaca oleracea* L. polyploid complex. Israel Journal of Botany 27:177–211.

D'Antoni, H. L. & O. T. Solbrig. 1977. Algarrobos in South American cultures past and present. In Mesquite, B. B. Simpson, ed., 189–199. Stroudsburg, Pa.: Dowden, Hutchinson, & Ross.

D'Arcy, W. G. 1973. Solanaceae. Flora of Panama. Annals of the Missouri Botanical Garden 60:573–780.

———, ed. 1986. Solanaceae: Biology and systematics. New York Columbia University Press.

———. 1989. Solanaceae. In Nomenclatural notes for the North American flora I, J. T. Kartesz & K. N. Gandhi. Phytologia 67:464–465.

Darlington, J. 1934. A monograph of the genus *Mentzelia*. Annals of the Missouri Botanical Garden 21:103–226.

Davidse, G., M. Sousa S., & S. Knapp, eds. 1995. Flora Mesoamericana. Vol. I, Psilotaceae a Salviniaceae. México, D.F.: Universidad Autónoma de México, Instituto de Biología.

Davis, O. K., ed. 1990. Quaternary geology of Bahía Adair and the Gran Desierto region. Deserts, past and future evolution. International Geological Correlation Program no. 252. Tucson: Arizona Geological Survey. An abridged version published as "Quaternary and environmental geology of the northeastern Gulf of California." In Geologic excursions through the Sonoran Desert Region, Arizona and Sonora, G. E. Gehrels & J. E. Spencer, eds. 136–153. Arizona Geological Survey Special Paper 7. Tucson: Arizona Geological Survey.

Davis, W. S. & P. H. Raven. 1962. Three new species related to *Malacothrix clevelandii*. Madroño 16:258–266.

Davis, W. S. & H. J. Thompson. 1967. A revision of *Petalonyx* (Loasaceae) with a consideration of affinities in subfamily Gronovioideae. Madroño 19:1–18.

Day, A. G. 1993. *Gilia*. In The Jepson manual, J. C. Hickman, ed., 828–839. Berkeley: University of California Press.

Dehgan, B. & G. L. Webster. 1978. Three new species of *Jatropha* (Euphorbiaceae) from western Mexico. Madroño 25:30–39.

———. 1979. Morphology and infrageneric relationships of the genus *Jatropha* (Euphorbiaceae). University of California Publications in Botany 74:1–73.

Delgado-Salinas, A., T. Turley, A. Richman, & M. Lavin. 1999. Phylogenetic analysis of the cultivated and wild species of *Phaseolus* (Fabaceae). Systematic Botany 24:438–460.

DeLisle, D. G. 1963. Taxonomy and distribution of the genus *Cenchrus*. Iowa State Journal of Science 37:259–351.

De Mason, D. A. 1984. Offshoot variability in *Yucca whipplei* subsp. *percursa* (Agavaceae). Madroño 31:197–202.

Dempster, L. T. 1979. Rubiaceae. In vol. 4, part 2 of A flora of California, W. L. Jepson, ed., 1–47. Berkeley: University of California.

Dempster, L. T. & **F. Ehrendorfer.** 1965. Evolution of the *Galium multiflorum* complex in western North America. Part 2: Critical taxonomic revision. Brittonia 17:289–334.

Desvaux, E. 1853. Gramíneas. In Historia fisica y política de Chile, C. Gay, ed., 233–469. Botánica 6. Paris: H. Bossange.

Detling, L. E. 1939. A revision of the North American species of *Descurainia*. American Midland Naturalist 22:481–520.

de Wet, J.M.J. 1968. Biosystematics of the *Bothriochloa barbinodis* complex (Gramineae). American Journal of Botany 55:1246–1250

———. 1978. Systematics and evolution of *Sorghum* sect. *Sorghum* (Gramineae). American Journal of Botany 65:477–484.

de Wet, J.M.J. & **J. R. Harlan.** 1970. Biosystematics of Cynodon L. C. Rich. (Gramineae). *Taxon* 19:565–569.

de Wet, J.M.J., K. E. Prasada Rao, M. H. Mengesha, & **D. E. Brink.** 1983. Domestication of sawa millet (*Echinochloa colona*). Economic Botany 37:283–291.

Diaz, H. B. & **V. Markgraf,** eds. 1992. El Niño: Historical and paleoclimatic aspects of the southern oscillation. Cambridge: Cambridge University Press.

Dice, J. C. 1988. Systematic studies in the *Nolina bigelovii–N. parryi* (Nolinaceae) complex. Master's thesis, San Diego State University, San Diego.

Diggs, G. M., Jr., B. L. Lipsomb, & **R. J. O'Kennon.** 1999. Shinners & Mahler's Illustrated Flora of North Central Texas. Fort Worth: Botanical Research Institute of Texas.

Dimmitt, M. A. 1987. The hybrid palo verde 'Desert Museum': A new, superior tree for desert landscape. Desert Plants 8:99–103.

Dorn, R. D. 1976. A synopsis of American *Salix*. Canadian Journal of Botany 54:2769–2789.

———. 1977. Willows of the Rocky Mountain states. Rhodora 79:390–429.

Dressler, R. L. & **J. Kuijt.** 1968. A second species of *Ammobroma* (Lennoaceae) in Sinaloa, Mexico. Madroño 19:179–182.

Duke, J. A. 1961. Preliminary revision of the genus *Drymaria*. Annals of the Missouri Botanical Garden 48: 173–268.

Dunford, M. P. 1984. Cytotype distribution of *Atriplex canescens* (Chenopodiaceae) of southern New Mexico and adjacent Texas. Southwestern Naturalist 29:223–228.

Dunn, D. B., J. A. Christian, & **C. T. Dziekanowski.** 1966. Nomenclature of the California *Lupinus concinnus–L. sparsiflorus* complex. Aliso 6:45–49.

Durrenberger, R. W. & **X. Murrieta.** 1978. Clima del estado de Sonora, Mexico. Climatological Publications, Mexican Climatology Series 3. Tempe: Laboratory of Climatology, Arizona State University.

Eckenwalder, J. E. 1977. North American cottonwoods (*Populus,* Salicaceae) of sections *Abaso* and *Aigeiros*. Journal of the Arnold Arboretum 58:193–208.

———. 1992. Salicaceae, Willow Family, part one: *Populus* [Vascular Plants of Arizona]. Journal of the Arizona-Nevada Academy of Science 26:29–33.

Eddlemann, W. R. 1989. Biology of the Yuma clapper rail in the southwestern United States and northwestern Mexico. Yuma: U.S. Bureau of Reclamation.

Ediger, R. I. 1970. Revision of section Suffruticosi of the genus *Senecio* (Compositae). Sida 3:504–524.

Ehrenfeld, J. G. 1976. Reproductive biology of three species of *Euphorbia* subgenus *Chamaesyce* (Euphorbiaceae). American Journal of Botany 63:406–413.

———. 1979. Pollination of three species of *Euphorbia* subgenus *Chamaesyce,* with special reference to bees. American Midland Naturalist 101:87–98.

El Hadidi, M. N. 1966. The genus *Fagonia* L. in Egypt. Candollea 21:20–40.

———. 1974. Weitere Beobachtungen an der Gattung *Fagonia* L. Mitteilungen der Botanischen Staatssammlung München 11:379–404.

Emory, W.H.P. & **W. V. Brown.** 1958. Apomixis in the Gramineae tribe Andropogoneae: *Heteropogon contortus*. Madroño 14:238–246.

Engelmann, G. 1859. Cactaceae of the boundary. In Report on the United States and Mexican Boundary Survey. Vol 2, part 1, W. H. Emory, ed. Washington, D.C.: C. Wendell.

Engelmann, G. & **J. M. Bigelow.** 1856 (1857). Description of the Cactaceae. In Reports of explorations and surveys for a railroad from the Mississippi River to the Pacific Ocean, A. W. Whipple, vol. 4. Washington, D.C.

Epling, C. 1938. The California Salvias: A review of *Salvia* section *Audibertia*. Annals of the Missouri Botanical Garden 25:95–188.

———. 1939. A revision of *Salvia* subgenus *Calosphace*. Repertorium Specierum Novarum Regni Vegetabilis Beih (Berlin) 100:1–383.

———. 1949. Revisión del género *Hyptis* (Labiatae). Revista Museo La Plata 2, Sec. 7, Botanica 30:153–497. La Plata, Argentina: Museo de la Plata.

Ernst, W. R. & **H. J. Thompson.** 1963. The Loasaceae in the southeastern United States. Journal of the Arnold Arboretum 44:138–142.

Espinosa-Garcia, F. J. 1985. El género *Gnaphalium* L. (Compositae: Inuleae) en el valle de México. Master's thesis, Universidad Autónoma de México, México, D.F.

Euler, R. C. & **V. H. Jones.** 1956. Hermetic sealing as a technique of food preservation among the Indians of the American Southwest. Proceedings of the American Philosophical Society 100:87–99.

Everly, M. L. 1947. A taxonomic study of the genus *Perityle* and related genera. Contributions from the Dudley Herbarium 3:377–396.

Ezcurra, E. 1984. The vegetation of El Pinacate, Sonora: A quantitative study. Ph.D. diss., University College of North Wales, Bangor.

Ezcurra, E., M. Equihua, & **J. López-Portillo.** 1987. The desert vegetation of El Pinacate, Sonora, Mexico. Vegetatio 71:49–60.

Ezcurra, E., R. S. Felger, A. Russell, & **M. Equihua.** 1988. Freshwater islands in a desert sand sea: The hydrology, flora, and phytogeography of the Gran Desierto oases of northwestern Mexico. Desert Plants 9:35–44, 55–63.

Ezcurra, E. & **V. Rodrígues.** 1986. Rainfall patterns in the Gran Desierto, Sonora, Mexico. Journal of Arid Environments 10:13–28.

Fairbrothers, D. E. 1953. Relationships in the *Capillaria* group of *Panicum* in Arizona and New Mexico. American Journal of Botany 40:708–714.

Feinbrun-Dothan, N. 1986. Flora Palestina. Part 4, Alismataceae to Orchidaceae. 2 vols. Jerusalem: Israel Academy of Sciences and Humanities.

Felger, R. S. 1977. Mesquite in Indian cultures of southwestern North America. In Mesquite, B. B. Simpson, ed., 150–176. Stroudsburg, Pa.: Dowden, Hutchinson, & Ross.

———. 1979. Ancient crops for the twenty-first century. In New agricultural crops, G. A. Ritchie, ed., 5–20. Boulder: Westview Press.

———. 1980. Vegetation and flora of the Gran Desierto, Sonora, Mexico. Desert Plants 2:87–114.

———. 1990. Non-native plants of Organ Pipe Cactus National Monument. University of Arizona, Cooperative National Park Resource Studies Unit, Technical Report 31:1–93.

———. 1991. *Senecio pinacatensis* (Asteraceae), a new species from the Pinacate Region of Sonora, Mexico. Phytologia 71:326–332.

———. 1992. Synopsis of the vascular plants of northwestern Sonora, Mexico. Ecologica 2:11–44.

———. 1993. *Mirabilis tenuiloba* S. Watson (Nyctaginaceae): New for Arizona. Madroño 40:178.

———. 1999. The flora of Cañón del Nacapule: A desert-bounded tropical canyon near Guaymas, Sonora, Mexico. Proceedings of the San Diego Society of Natural History 35:1–42.

Felger, R. S. & **B. Broyles,** eds. 1997. Dry Borders. Journal of the Southwest 39:303–860.

Felger, R. S. & **J. Henrickson.** 1997. Convergent adaptive morphology of a Sonoran Desert cactus (*Peniocereus striatus*) and an African spurge (*Euphorbia cryptospinosa*). Haseltonia 5:77–85.

Felger, R. S., M. B. Johnson, & **M. F. Wilson.** 2000. Trees of Sonora, Mexico. New York: Oxford University Press.

Felger, R. S. & **C. H. Lowe.** 1967. Clinal variation in the surface-volume relationships of the columnar cactus *Lophocereus schottii* in northwestern Mexico. Ecology 48:530–536.

———. 1970. New combinations for plant taxa in northwestern Mexico and southwestern United States. Journal of the Arizona Academy of Sciences 6:82–84.

———. 1976. The island and coastal vegetation and flora of the northern part of the Gulf of California, Mexico. Natural History Museum of Los Angeles County, Contributions in Science 285:1–59.

Felger, R. S. & **M. B. Moser.** 1985. People of the desert and sea: Ethnobotany of the Seri Indians. Tucson: University of Arizona Press.

Felger, R. S., P. L. Warren, S. A. Anderson, & **G. P. Nabhan.** 1992. Vascular plants of a desert oasis: Flora and ethnobotany of Quitobaquito, Organ Pipe Cactus National Monument, Arizona. Proceedings of the San Diego Society of Natural History 8:1–39.

Felger R. S., M. Wilson, B. Broyles, & **G. P. Nabhan.** 1997. The Binational Sonoran Desert Biosphere Network and its plant life. Journal of the Southwest 39:411–560.

Ferguson, D. J. 1988. *Opuntia macrocentra* Eng. and *Opuntia chlorotica* Eng. & Big. Cactus and Succulent Journal (U.S.) 60:155–160.

Fernald, M. L. 1900. A synopsis of the Mexican and Central American species of *Salvia*. Proceedings of the American Academy of Arts and Sciences 35:489–573.

Fishbein, M. 1996. Phylogenetic relationships of *Asclepias* L. and the importance of pollinators to the evolution of the milkweed inflorescence. Ph.D. diss., University of Arizona, Tucson.

Fleming, T. H. & **J. N. Holland.** 1998. The evolution of obligate pollination mutualism: Senita cactus and senita moth. Oecologia 114:368–375.

Fleming, T. H., M. D. Tuttle, & **M. A. Horner.** 1996. Pollination biology and relative importance of nocturnal and diurnal pollinators in three species of Sonoran Desert columnar cacti. Southwestern Naturalist 41: 257–269.

Flora of North America Editorial Committee. 1993. Flora of North America north of Mexico. Vol. 2. New York: Oxford University Press.

Flores, E. M. & **E. M. Engleman.** 1976. Apuntas sobre anatomía y morfología de las semillas de cactaceas—I: Desarrollo y estructura. Revista de Biología Tropical 24:199–227.

Fontana, B. L. 1980. Ethnobotany of the saguaro: An annotated bibliography. Desert Plants 2:63–78.

Ford, K. C. 1975. Las yerbas de la gente: A study of Hispano-American medicinal plants. Museum of Anthropology, University of Michigan, Anthropological Papers 60:1–431.

Fosberg, F. R. 1978. Studies in the genus *Boerhavia* L. (Nyctaginaceae). Smithsonian Contributions to Botany 39:1–20.

Fosberg, F. R. & **S. A. Renvoize.** 1977. The flora of Aldabra and neighbouring islands. Kew Bulletin (additional series) 7.

Freeman, D. C., E. D. McArthur, & **K. T. Harper.** 1984. The adaptive significance of sexual lability in plants using *Atriplex canescens* as a principal example. Annals of the Missouri Botanical Garden 71:265–277.

Frenkel, R. E. 1977. Ruderal vegetation along some California roadsides. Berkeley: University of California Press.

Fryxell, P. A. 1974. The North American malvellas (Malvaceae). Southwestern Naturalist 19:97–103.

———. 1980. A revision of the American species of *Hibiscus* section *Bombicella* (Malvaceae). U.S.D.A. Technical Bulletin 1624:1–53.

———. 1983. A revision of *Abutilon* sect. *Oligocarpae* (Malvaceae) including a new species from Mexico. Madroño 30:84–92.

———. 1985. Additional novelties in Mexican Malvaceae. Systematic Botany 10:268–272.

———. 1988. Malvaceae of Mexico. Systematic Botany Monographs 25:1–522.

Galloway, L. A. 1975. Systematics of the North American desert species of *Abronia* and *Tripterocalyx* (Nyctaginaceae). Brittonia 27:328–347.

García-Mendoza, A. & **R. Galván V.** 1995. Riqueza de las familias Agavaceae y Nolinaceae en México. Boletín de la Sociedad Botánica de México 56:7–24.

Gastony, G. J. & **D. R. Rollo.** 1998. Cheilanthoid ferns (Pteridaceae: Cheilanthoideae) in the southwestern United States and adjacent Mexico—A molecular phylogenetic reassessment of generic lines. Aliso 17: 131–144.

Gentry, H. S. 1942. Rio Mayo plants: A study of the flora and vegetation of the valley of the Rio Mayo, Sonora. Carnegie Institution of Washington Publication 527:1–328.

———. 1972. The agave family in Sonora. U.S.D.A. Agricultural Handbook 399:1–195.

———. 1978. *Nolina* in and around the Sonoran Desert. Saguaroland Bulletin 32:112–116.

———. 1982. Agaves of continental North America. Tucson: University of Arizona Press.

Gibson, A. C. 1990. The systematics and evolution of subtribe Stenocereinae, 8. Organ Pipe Cactus and its closest relatives. Cactus and Succulent Journal (U.S.) 62:13–24.

———. 1991. The systematics and evolution of subtribe Stenocereinae, 11. *Stenocereus dumortieri* versus *Isolatocereus dumortieri.* Cactus and Succulent Journal (U.S.) 63:184–190.

Gibson, A. C. & **K. E. Horak.** 1978. Systematic anatomy and phylogeny of Mexican columnar cacti. Annals of the Missouri Botanical Garden 65:999–1057.

Gibson, A. C. & **P. S. Nobel.** 1986. The cactus primer. Cambridge: Harvard University Press.

Gibson, A. C., K. C. Spencer, R. Bajaj, & **J. L. McLaughlin.** 1986. The ever-changing landscape of cactus systematics. Annals of the Missouri Botanical Garden 73:532–555.

Gifford, E. W. 1933. The Cocopa. University of California Publications in American Archaeology and Ethnology 31:257–334.

Gillis, W. T. 1977. *Pluchea* revisited. Taxon 26:587–591.

Glad, J. B. 1976. Taxonomy of *Mentzelia mollis* and allied species (Loasaceae). Madroño 23:283–292.

Glass, C. & **R. Foster.** 1975. The genus *Echinomastus* in the Chihuahuan Desert. Cactus and Succulent Journal (U.S.) 47:218–223.

Glenn, E. P. 1987. Relationship between cation accumulation and water content of salt-tolerant grasses and a sedge. Plant Cell and Environment 10:205–212.

Glenn, E. P., R. S. Felger, A. Búrquez, & **D. S. Turner.** 1992. Ciénega de Santa Clara: Endangered wetland in the Colorado River delta, Sonora, Mexico. Natural Resources Journal 32:817–824.

Glenn, E. P., C. Lee, R. Felger & **S. Zengel.** 1996. Effects of Water Management on the wetlands of the Colorado River delta, Mexico. Conservation Biology 1175–1186.

Glenn, E. P., J. W. O'Leary, M. C. Watson, T. L. Thompson, & **R. O. Kuehl.** 1991. *Salicornia bigelovii* Torr.: An oilseed halophyte for seawater irrigation. Science 251:1065–1067.

Godfrey, R. K. 1952. *Pluchea* section *Stylimnus,* in North America. Journal of the Elisha Mitchell Scientific Society 68:238–271.

Gómez, S. A. 1953. Caparidáceas argentinas. Lilloa 26:279–341.

González-Elizondo, M. S. & **D. Gómez-Sánchez.** 1992. Notes on *Helianthus* (Compositae-Heliantheae) from Mexico. Phytologia 72:63–70.

González Gutierrez, M. S. 1989. El género *Potamogeton* (Potamogetonaceae) en México. Acta Botánica Mexicana 6:1–43.

Goodman, G. J. 1934. A revision of the North American species of the genus *Chorizanthe.* Annals of the Missouri Botanical Garden 21:1–102.

Goodspeed, T. H. 1954. The genus *Nicotiana.* Waltham, Mass.: Chronica Botanica.

Gottlieb, L. D. 1972. A proposal for classification of the annual species of *Stephanomeria* (Compositae). Madroño 21:463–481.

Gould, F. W. 1951. Grasses of the southwestern United States. Tucson: University of Arizona Press.

———. 1953. A cytotaxonomic study in the genus *Andropogon.* American Journal of Botany 40:297–306.

———. 1957. New North American andropogons of subgenus *Amphilophis* and a key to those species occurring in the United States. Madroño 14:18–29.

———. 1979. The genus Bouteloua (Poaceae). Annals of the Missouri Botanical Garden 66:348–416.

Gould, F. W., M. A. Ali, & **D. E. Fairbrothers.** 1972. A revision of *Echinochloa* in the United States. American Midland Naturalist 87:36–59.

Gould, F. W. & **T. W. Box.** 1965. Grasses of the Texas Coastal Bend. College Station: Texas A & M University Press.

Gould, F. W. & **R. Moran.** 1981. The grasses of Baja California, Mexico. San Diego Society of Natural History Memoir 12:1–140.

Gould, F. W. & **R. B. Shaw.** 1968. Grass systematics. New York: McGraw-Hill.

Graham, S. A. 1964. The genera of Lythraceae in the southeastern United States. Journal of the Arnold Arboretum 45:235–250.

———1985. A revision of *Ammannia* (Lythraceae) in the Western Hemisphere. Journal of the Arnold Arboretum 66:395–420.

Grant, A. L. 1924. A monograph of the genus *Mimulus*. Annals of the Missouri Botanical Garden 11:99–388.

Grant, V. 1959. Natural history of the phlox family. The Hague: Martinus Nijhoff.

Grant, V. & **A. Grant.** 1956. Generic and taxonomic studies in *Gilia*—VIII. The cobwebby gilias. Aliso 3:203–287.

———. 1979a. The pollination spectrum in the southwestern American cactus flora. Plant Systematics and Evolution 133:29–37.

———. 1979b. Pollination of *Echinocereus fasciculatus* and *Ferocactus wislizenii*. Plant Systematics and Evolution 132:85–90.

Grant, V. & **P. D. Hurd.** 1979. Pollination of the southwestern opuntias. Plant Systematics and Evolution 133:15–28.

Gray, A. 1852. Plantae Wrightianae Texano-Neo-Mexicanae, part 1. Smithsonian Contributions to Knowledge 3 (V):1–146 & plates 1–10. Washington, D.C.: Smithsonian Institution.

———. 1876. Botany of California. Vol 1. Cambridge: John Wilson and Son, University Press.

Gregory, D. P. 1963. Hawkmoth pollination in the genus *Oenothera*. Aliso 5:357–419.

Griffiths, D. 1912. The grama grasses: *Bouteloua* and related genera. Contributions from the United States National Herbarium 14:343–428.

Grimes, J. W. 1990. A revision of the New World species of Psoraleeae (Leguminosae: Papilionoideae). Memoirs of the New York Botanical Garden 61:1–114.

Guittonneau, G. 1972. Contribution a l'étude biosystématique du genre *Erodium* L'Hér. dans le bassin méditérranéen occidental. Boissiera 20:1–154.

Hadac, E. & **J. Chrteck.** 1970. Notes on the taxonomy of Cuscutaceae. Folia Geobotanica Phytotaxonomica 5:443–445.

———. 1973. Some further notes on the taxonomy and nomenclature of Cuscutaceae. Folia Geobotanica Phytotaxonomica 8:219–221.

Hall, H. M. & **F. E. Clements.** 1923. The phylogenetic method in taxonomy. The North American species of *Artemisia*, *Chrysothamnus*, and *Atriplex*. Carnegie Institute of Washington Publication 326:1–355.

Hammer, K., N. Romeike, & **C. Tittel.** 1983. Vorarbeiten zur monographischen Darstellung von Wildpfanzensortimenten: *Datura* L., sectiones *Dutra* Bernh., *Ceratocaulis* Bernh., et *Datura*. Kulturpflanze 31:13–75.

Hammond, G. P. & **A. Rey.** 1940. Narratives of the Coronado Expedition 1540–1542. Coronado Cuarto Centennial Publications, 1540–1940, vol. 2. Albuquerque: University of New Mexico Press.

Hardy, R.W.H. 1829. Travels in the interior of Mexico in 1825, 1826, 1927, and 1828. London: Colburn and Beatley. Reprint, Glorieta, N.Mex.: Rio Grande Press, 1977.

Harlan, J. R. & **J.M.J. de Wet.** 1969. Sources of variation in *Cynodon dactylon* (L.) Pers. Crop Science 9:774–778.

Harrison, H. K. 1972. Contributions to the study of the genus *Eriastrum*—II: Notes concerning the type specimens and descriptions of the species. Brigham Young University Science Bulletin, Biological Series 16(4):1–26.

Hartman, R. L. 1986. Apocynaceae. In Flora of the Great Plains, Great Plains Flora Association, eds., 610–613. Lawrence: University of Kansas.

———. 1990. A conspectus of *Machaeranthera* (Asteraceae: Astereae). Phytologia 68:439–465.

———. 1993. Caryophyllaceae (in part). In The Jepson manual, J. C. Hickman, ed., 475–497. Berkeley: University of California Press.

Hartman, R. L., J. D. Bacon, & **C. F. Bohnsteadt.** 1975. Biosystematics of *Draba cuneifolia* and *D. platycarpa* (Cruciferae) with emphasis on volatile oil and flavonoid constituents. Brittonia 27:317–327.

Hartmann, W. K. 1989. Desert heart. Tucson: Fisher Books.

Hastings, J. R. 1964. Climatological data for Sonora and northern Sinaloa. Technical Reports on the Meteorology and Climatology of Arid Regions 15. Tucson: University of Arizona Institute of Atmospheric Physics.

Hastings, J. R. & **R. R. Humphrey.** 1969. Climatological data and statistics for Sonora and northern Sinaloa. Technical Reports on the Meteorology and Climatology of Arid Regions 19. Tucson: University of Arizona Institute of Atmospheric Physics.

Hawkes, J. G., R. N. Lester, & **A. D. Skelding,** eds. 1979. The biology and taxonomy of the Solanaceae. London: Academic Press.

Hawkins, J. A. 1996. Systematics of *Parkinsonia* L. and *Cercidium* Tul. (Leguminosae, Caesalpinoideae). Ph.D. thesis, University of Oxford, U.K.

Hawkins, J. A., L. White Olascoaga, C. E. Hughes, J. R. Contreras Jiménez, & **P. Mercado Ruaro.** 1999. Investigation and documentation of hybridization between *Parkinsonia aculeata* and *Cercidium praecox* (Leguminosae: Caesalpinoidea). Plant Systematics and Evolution 216:49–68.

Hayden, J. D. 1967. Summary of prehistory and history of the Sierra Pinacate, Sonora. American Antiquity 32:335–344.

———. 1969. Gyratory crushers of the Sierra Pinacate, Sonora. American Antiquity 34:154–161.

———. 1972. Hohokam petroglyphs of the Sierra Pinacate, Sonora, and the Hohokam shell expeditions. The Kiva 37:74–83.

———. 1976. Pre-altithermal archaeology in the Sierra Pinacate, Sonora, Mexico. American Antiquity 41: 274–289.

———. 1982. Ground figures of the Sierra Pinacate, Sonora, Mexico, In Hohokam and Patayan: Prehistory of southwest Arizona, R. H. McGuire & M. B. Schiffer, eds., 581–588. New York: Academic Press.

———. 1997. Laguna Prieta. Journal of the Southwest 39:321–330.

———. 1998. The Sierra Pinacate. Tucson: University of Arizona Press.

Haynes, R. R. 1979. Revision of North and Central American *Najas* (Najadaceae). Sida 8:34–56.

Haynes, R. R. & **L. B. Holm-Nielsen.** 1987. The Zannichelliaceae in the southeastern United States. Journal of the Arnold Arboretum 68:259–268.

———. 1994. The Alismataceae. Flora Neotropica Monograph 64:1–112.

Heiser, C. B. 1944. Monograph of *Psilostrophe.* Annals of the Missouri Botanical Garden 31:279–300.

———. 1961. Morphological and cytological variation in *Helianthus petiolaris* with notes on related species. Evolution 15:247–258.

———. 1963. Artificial intergeneric hybrids of *Helianthus* and *Viguiera*. Madroño 17:118–127.

Heiser, C. B., D. M. Smith, S. B. Clevenger, & **W. C. Martin Jr.** 1969. The North American sunflowers (*Helianthus*). Memoirs of the Torrey Botanical Club 22:(3)1–218.

Heizer, R. 1945. Honey-dew "sugar" in western North America. The Masterkey 14:140–145.

Heizer, R. F. & **A. B. Elsasser.** 1980. The natural world of the California Indians. Berkeley: University of California Press.

Henrard, J. T. 1926–33. A critical revision of the genus Aristida. Leiden: Mededeelingen's Rijks Herbarium.

———. 1929–33. A monograph of the genus Aristida. Leiden: Mededeelingen's Rijks Herbarium.

———. 1937. A study of the genus *Vulpia*. Blumea 2:299–326.

———. 1950. Monograph of the genus Digitaria. Leiden: Universitare Pers Leiden.

Henrickson, J. 1972. A taxonomic revision of the Fouquieriaceae. Aliso 7:439–537.

———. 1977a. Leaf production and flowering in ocotillos. Cactus and Succulent Journal (U.S.) 49:133–137.

———. 1977b. Saline habitats and halophytic vegetation of the Chihuahuan Desert region, In Transactions of the symposium on the biological resources of the Chihuahuan Desert Region, R. H. Wauer & D. H. Riskind, eds., 289–314. U.S. Department of the Interior, National Park Service Transactions and Proceedings Series 3. Washington, D.C.: Government Printing Office.

———. 1999. Studies in New World *Amaranthus* (Amaranthaceae). Sida 18:783–807.

Henrickson, J. & **M. C. Johnston.** n.d. A flora of the Chihuahuan Desert region. Unpublished manuscript.

Hernández Sandoval, L. 1995. Análisis cladístico de la familia Agavaceae. Boletín de la Sociedad Botánica de México 56:57–68.

Hershkovitz, M. A. 1990. Nomenclatural changes in Portulacaceae. Phytologia 68:267–270.

———. 1991a. Taxonomic notes on *Cistanthe, Calandrinia,* and *Talinum* (Portulacaceae). Phytologia 70:209–225.

———. 1991b. Phylogenetic assessment and revised circumscription of *Cistanthe* Spach (Portulacaceae). Annals of the Missouri Botanical Garden 78:1009–1021.

———. 1993. Revised circumscriptions and subgeneric taxonomies of *Calandrinia* and *Montiopsis* (Portulacaceae) with notes on the phylogeny of the portulacaceous alliance. Annals of the Missouri Botanical Garden 80:333–365.

Hickman, J. C., ed. 1993. The Jepson manual. Berkeley: University of California Press.

Higgins, L. C. 1979. Boraginaceae of the southwestern United States. The Great Basin Naturalist 39:293–350.

Hill, R. J. 1976. Taxonomic and phylogenetic significance of seed coat microsculpturing in *Mentzelia* (Loasaceae) in Wyoming and adjacent western states. Brittonia 28:86–112.

Hinton, B. D. 1969. *Parietaria hespera* (Urticaceae), a new species of the southwestern United States. Sida 3:293–297.

Hitchcock, A. S. 1913. Mexican grasses in the United States National Herbarium. Contributions from the United States National Herbarium 17:181–389.

———. 1925a. The North American species of *Stipa*. Contributions from the United States National Herbarium 24:215–262.

———. 1925b. Synopsis of the South American species of *Stipa*. Contributions from the United States National Herbarium 24:263–289.

———. 1935a. *Aristida*. North American Flora 17:376–406.

———. 1935b. *Muhlenbergia*. North American Flora 17:431–476.

———. 1935c. *Sporobolus*. North American Flora 17:481–496.

———. 1951. Manual of the grasses of the United States. 2nd ed. revised by A. Chase. U.S.D.A. Miscellaneous Publication 200. Washington, D.C.: Government Printing Office.

Hitchcock, C. L. 1932. A monograph of the genus *Lycium* of the western hemisphere. Annals of the Missouri Botanical Garden 19:179–374.

———. 1933. A taxonomic study of the genus *Nama* I. American Journal of Botany 20:415–431, 518–534.

———. 1936. The genus *Lepidium* in the United States. Madroño 3:265–320.

———. 1945. The Mexican, Central American, and West Indian Lepidia. Madroño 8:118–143.

Hitchcock, C. L. & A. Cronquist. 1964. Vascular plants of the Pacific Northwest, part 2. Seattle: University of Washington Press.

Hitchcock, C. L., A. Cronquist, & M. Ownbey. 1969. Vascular plants of the Pacific Northwest, part 1. Seattle: University of Washington Press.

Hitchcock, C. L., A. Cronquist, M. Ownbey, & J. W. Thompson. 1959. Vascular plants of the Pacific Northwest, part 4. Seattle: University of Washington Press.

Hitchcock, C. L. & B. Maguire. 1947. A revision of the North American species of *Silene*. University of Washington Publications in Biology 13:1–73.

Holm, L., D. L. Plucknett, J. V. Pancho, & J. P. Herberger. 1977. The world's worst weeds. Honolulu: University Press of Hawaii.

Holm, R. W. 1950. The American species of *Sarcostemma* R. Br. (Asclepiadaceae). Annals of the Missouri Botanical Garden 37:477–560.

Holmes, W. C. 1981. *Cleome viscosa* L. (Capparidaceae)—New to Louisiana. Sida 9:187.

Holmgren, A. & N. Holmgren. 1977. Poaceae. In A. Cronquist et al., eds., 175–584. Intermountain Flora, vol. 6. New York: Columbia University Press.

Holmgren, N. H. 1984. Scrophulariaceae. In A. Cronquist et al., eds., 344–506. Intermountain Flora, vol. 4. New York: New York Botanical Garden.

Holmgren, P. K., N. H. Holmgren, & L. C. Barnett. 1990. Index herbariorum. Part I: The herbaria of the world. 8th ed. New York: New York Botanical Garden.

Hoover, R. F. 1936. Notes on California grasses. Madroño 3:227–230.

———. 1940. The genus *Dichelostemma*. American Midland Naturalist 24:463–476.

———. 1941. A systematic study of *Triteleia*. American Midland Naturalist 25:73–100.

Hopkins, C. O. & W. H. Blackwell Jr. 1977. Synopsis of *Suaeda* (Chenopodiaceae) in North America. Sida 7:147–173.

Horak, M. J. & **J. S. Holt.** 1986. Isozyme variability and breeding systems in populations of yellow nutsedge (*Cyperus esculentus*). Weed Science 34:538–543.

Hornaday, W. T. 1908. Camp-fires on desert and lava. New York: Charles Scribner's Sons. Reprint, Tucson: University of Arizona Press, 1983.

Hotchkiss, N. & **H. L. Dozier.** 1949. Taxonomy and distribution of North American cat-tails. American Midland Naturalist 41:237–254.

Howell, J. T. 1946. A revision of *Phacelia* sect. *Euglypta.* The American Midland Naturalist 36:381–411.

———. 1956. A review of *Calyptridium parryi.* Leaflets of Western Botany 8:9–11.

Hoy, W. E. 1970. Organ pipe cactus historical research. Unpublished manuscript on file at Organ Pipe Cactus National Monument.

———. 1990. Sonoyta and Santo Domingo. Journal of Arizona History 31:117–140.

Huey, L. M. 1942. A vertebrate faunal survey of the Organ Pipe Cactus National Monument, Arizona. Transactions of the San Diego Society of Natural History 9:353–376.

Huft, M. 1984. A review of *Euphorbia* (Euphorbiaceae) in Baja California. Annals of the Missouri Botanical Garden 71:1021–1027.

Humphrey, R. R. 1960. Forage production on Arizona ranges: Pima, Pinal, and Santa Cruz Counties. University of Arizona, Agricultural Experiment Station Bulletin 302:1–138.

Humphrey, R. R., A. L. Brown, & **A. C. Everson.** 1958. Arizona range grasses. University of Arizona, Agricultural Experiment Station Bulletin 298:1–104.

Hunt, D. R. 1971. Schumann and Buxbaum reconciled, the Schumann system of *Mammillaria* classification brought provisionally up-to-date. Cactus and Succulent Journal (U.K.) 33:53–72.

———. 1978. The classification of *Mammillaria.* National Cactus and Succulent Journal 32:75–81.

———. 1983–87. A new review of *Mammillaria* names. Bradleya 1:105–128; 2:65–96; 3:53–66; 4:39–64; 5:17–48.

Hunter, K. 1998. Molecular systematics, polyploidy, and paleoecology of *Larrea.* In The Desert Tortoise Council: Abstracts, 23rd annual meeting and symposium, 18–19. Ridgecrest, Calif.: Desert Tortoise Council.

Hunziker, A. T. 1984. Capparaceae. In Los géneros de fanerogamas de Argentina, L. Hauman, ed., 94–95. Boletín de la Sociedad Argentina de Botanica, vol. 23. Córdoba, Argentina: Sociedad Argentina de Botanica.

Hunziker, J. H., R. A. Palacios, A. G. de Valesi, & **L. Poggio.** 1972. Species disjunctions in *Larrea:* Evidence from morphology, cytogenetics, phenolic compounds, and seed albumins. Annals of the Missouri Botanical Garden 59:224–233.

Ihlenfeldt, H. & **U. Grabow-Seidensticker.** 1979. The genus *Sesamum* L. and the origin of the cultivated sesame, In Taxonomic aspects of African economic botany, G. Kunkel, ed., 53–60. Proceedings of the 11th Plenary Meeting of A.E.T.F.A.T. Las Palmas, Canary Islands. Las Palmas: Excelentísimo Ayuntamiento de las Palmas de Gran Canaria.

Iljin, M. M. 1936. Contribution à la systétique du genre *Suaeda* et de la tribu Suaedeae. Sovetskaja Botanik (Leningrad) 5:39–49.

Iltis, H. H. 1957. Studies in the Capparidaceae)—III, Evolution and phylogeny of the Western American Cleomoideae. Annals of the Missouri Botanical Garden 44:77–119.

———. 1960. Studies in the Capparidaceae VII, Old World cleomes adventive in the New World. Brittonia 12:279–294.

Imbrie, J. & **K. P. Imbrie.** 1979. Ice ages: Solving the mystery. Hillside, N.J.: Enslow.

Ingram, J. W. 1970. *Argythamnia.* In Manual of the vascular plants of Texas, D. S. Correll & M. C. Johnston, eds., 939–942. Renner, Tex.: Texas Research Foundation.

———. 1980. The generic limits of *Argythamnia* (Euphorbiaceae) defined. Gentes Herbarum 11:427–436.

Inouye, R. S. 1991. Population biology of desert annual plants. In The ecology of desert communities, G. A. Polis, eds., 27–54. Tucson: University of Arizona Press.

I.O.S. [International Organization for Succulent Plant Study] Working Party. 1986. The genera of the Cactaceae: Towards a new consensus. Bradleya 4:65–78.

Irwin, H. S. & **R. C. Barneby.** 1982. The American Cassiinae. Memoirs of the New York Botanical Garden 35:1–918.

Isely, D. 1969. Legumes of the United States. Native *Acacia.* Sida 3:365–386.

———. 1990. Leguminosae (Fabaceae). Vascular flora of the southeastern United States 3 (2):1–258. Chapel Hill: The University of North Carolina Press.

———. 1998. Native and naturalized Leguminosae (Fabaceae) of the United States. Provo, Utah: Monte L. Bean Life Science Museum, Brigham Young University.

Ives, R. L. 1950a. The Sonoyta oasis. Journal of Geography 49:1–15.

———. 1950b. Puerto Peñasco, Sonora. Journal of Geography 49:349–361.

———. 1964. The Pinacate region, Sonora, Mexico. Occasional Papers of the California Academy of Sciences 47:1–43.

———. 1966. Kino's exploration of the Pinacate region. Journal of Arizona History 7:59–75.

———. 1989. Land of lava, ash, and sand. Tucson: Arizona Historical Society.

Jahns, R. H. 1959. Collapse depressions of the Pinacate volcanic field, Sonora, Mexico. In Arizona Geological Society Guidebook 2, 165–184. Tucson: Arizona Geological Society.

James, L. E. & **D. W. Kyhos.** 1961. The nature of the fleshy shoot of *Allenrolfea* and allied genera. American Journal of Botany 48:101–108.

Jansen, R. & **B. Parfitt.** 1977. *Allenrolfea mexicana* Lundell (Chenopodiaceae): Its conspecificity with *A. occidentalis* Kuntze. Rhodora 79:130–132.

Jansen, R. K., R. S. Wallace, K. J. Kim, & **K. L. Chambers.** 1991. Systematic implications of chloroplast DNA variation in the subtribe Microseridinae (Asteraceae: Lactuceae). American Journal of Botany 78:1015–1027.

Jeffrey, C. 1980. A review of the Cucurbitaceae. Botanical Journal of the Linnean Society 81:233–247.

Johnson, A. F. 1978. A new subspecies of *Abronia maritima* from Baja California, Mexico. Madroño 25:224–227.

Johnson, D. M. 1985. New records of longevity of *Marsilea* sporocarps. American Fern Journal 75:30–31.

———. 1986. Systematics of the New World species of *Marsilea* (Marsileaceae). Systematic Botany Monographs 11:1–87.

Johnson, M. B. 1992. The genus *Bursera* (Burseraceae) in Sonora, Mexico, and Arizona, U.S.A. Desert Plants 10:126–144.

Johnson, R. R. 1969. Monograph of the plant genus *Porophyllum* (Compositae: Helenieae). University of Kansas Science Bulletin 48:225–267.

Johnson, R. R., B. T. Brown & **S. Goldwasser.** 1983. Avian use of Quitobaquito Springs oasis, Organ Pipe Cactus National Monument, Arizona. University of Arizona, Cooperative National Park Resource Studies Unit, Technical Report 13:1–16.

Johnston, I. M. 1924a. Expedition of the California Academy of Sciences to the Gulf of California in 1921: The botany (vascular plants). Proceedings of the California Academy of Sciences (series 4) 12:951–1218.

———. 1924b. Taxonomic records concerning American Spermatophytes: *Parkinsonia* and *Cerdidium.* Contributions from the Gray Herbarium 70:61–68.

———. 1925. Studies in the Boraginaceae—IV. The North American species of *Cryptantha.* Contributions from the Gray Herbarium 74:1–114.

———. 1939. Studies in the Boraginaceae, XIII. New or otherwise noteworthy species, chiefly from western United States. Journal of the Arnold Arboretum 20:375–402.

Johnston, M. C. 1962a. The North American mesquites *Prosopis* sect. *Algarobia* (Leguminosae). Brittonia 14: 72–90.

———. 1962b. Revision of *Condalia* including *Microrhamnus* (Rhamnaceae). Brittonia 14:332–368.

———. 1963. The species of *Ziziphus* indigenous to the United States and Mexico. American Journal of Botany 50:1020–1027.

———. 1971. Revision of *Colubrina* (Rhamnaceae). Brittonia 23:2–53.

———. 1990. The vascular plants of Texas. Austin: Marshall C. Johnston.

Jones, C. E., L. J. Colin, T. R. Ericson, & **D. K. Dorsett.** 1998. Hybridization between *Cercidium floridum* and *C. microphyllum* (Fabaceae) in California. Madroño 45:110–118.

Jones, E. K. & **N. C. Fassett.** 1950. Subspecific variation in *Sporobolus cryptandrus*. Rhodora 52:125–126.

Jones, V. H. 1945. The use of honey-dew as food by Indians. The Masterkey 14:145–149.

Kahn, R. & **C. E. Jarvis.** 1989. The correct name for the plant known as *Pluchea symphytifolia* (Miller) Gillis (Asteraceae). Taxon 38:659–662.

Karis, P. O. 1995. Cladistics of the subtribe Ambrosiinae (Asteraceae). Systematic Botany 20:40–54.

Kearney, T. H. 1935. The North American species of *Sphaeralcea* subgenus *Eusphaeralcea*. University of California Publications in Botany 19:1–127.

———. 1951. The American genera of Malvaceae. American Midland Naturalist 46:93–131.

———. 1956. Notes on Malvaceae. VIII. *Eremalche*. Madroño 13:241–243.

Kearney, T. H. & **R. H. Peebles.** 1939. Arizona plants: New species, varieties and combinations. Journal of the Washington Academy of Sciences 29:474–492.

———. 1960. Arizona flora. 2nd ed. Berkeley: University of California Press.

Keator, G. 1992. Studies in the genus *Dichelostemma*. Four Seasons 9(1):24–39 (Berkeley, Calif.: Regional Parks Botanical Garden).

Keck, D. D. 1936. Studies in *Penstemon*—II. The section *Hesperothamnus*. Madroño 3:200–219.

———. 1946. A revision of the *Artemisia vulgaris* complex in North America. Proceedings of the California Academy of Sciences (series 4) 25:421–468.

Keil, D. J. 1975. *Pectis cylindrica* (Compositae) established as a member of the Texas flora and confirmed as a distinct species. Southwestern Naturalist 20:286–287.

———. 1977. A revision of *Pectis* section *Pectothrix* (Compositae: Tageteae). Rhodora 79:32–78.

———. 1996. *Pectis*. In The Comps of Mexico, 6: Tagetaea and Anthemideae, B. L. Turner, ed., 22–43. Phytologia Memoirs 10.

Keller, S. 1979. A revision of the genus *Wislizenia* (Capparidaceae) based on population studies. Brittonia 31: 333–351.

Kelley, W. A. 1990. Comments on *Portulaca* in California. Madroño 36:281–282.

Kelly, W. H. 1977. *Cocopa ethnography*. Anthropological Papers of the University of Arizona 29:1–150.

Kemp, P. R. 1983. Phenological patterns of Chihuahuan Desert plants in relation to the timing of water availability. Journal of Ecology 71:427–436.

Kiger, R. W. 1997. Papaveraceae. In vol. 3 of Flora of North America north of Mexico, 300–302. New York: Oxford University Press.

King, R. M. 1967. Studies in the Eupatorieae (Compositae). I. *Pleurocoronis*. Rhodora 69:35–45.

———. 1987. The genera of the Eupatorieae (Asteraceae). Monographs in Systematic Botany from the Missouri Botanical Garden 22:1–581.

Kingsley, K. J. & **R. A. Bailowitz.** 1987. Grasshoppers and butterflies of the Quitobaquito management area, Organ Pipe Cactus National Monument, Arizona. University of Arizona, Cooperative National Park Resource Studies Unit, Technical Report 21:1–25.

Kingsley, K. J., R. A. Bailowitz, & **R. L. Smith.** 1987. A preliminary investigation of the arthropod fauna of Quitobaquito Springs area, Organ Pipe Cactus National Monument, Arizona. University of Arizona, Cooperative National Park Resource Studies Unit, Technical Report 23:1–24.

Kladiwa, L. & **H. W. Fittkau.** 1971. Genus *Neolloydia*. In Die Kakteen, CVIIIBH, H. Krainz, ed. Stuttgart.

Klein, W. M. 1970. The evolution of three diploid species of *Oenothera* subgenus *Anogra* (Onagraceae). Evolution 24:578–597.

Knowles, D. F. 1944. Interspecific hybridizations of *Bromus*. Genetics 29:128–140.

Kohlmann, B. & **S. Sánchez-Colón.** 1984. Estudio areográfico del género *Bursera* en México. Cactaceas y Suculentas Mexicanas 29:27–32.

Koutnik, D. 1984. *Chamaesyce* (Euphorbiaceae)—A newly recognized genus in southern Africa. South African Journal of Botany 3:262–264.

———. 1985. New combinations in California *Chamaesyce* (Euphorbiaceae). Madroño 32:187–189.

Koyama, T. 1958. Taxonomic study of the genus *Scirpus* Linné. Journal, Faculty of Science, Tokyo University, Section 3, Botany 7:271–366.

———. 1962. The genus *Scirpus* Linn.: Some North American Aphylloid species. Canadian Journal of Botany 40:913–37.

———. 1963. The genus *Scirpus* Linn.: Critical species of the section *Pterolepis*. Canadian Journal of Botany 41:1107–1131.

Kresan, P. L. 1997. A geologic tour of the lower Colorado River region of Arizona and Sonora. Journal of the Southwest 39:567–612.

Krombein, K. V. 1961. Some insect visitors of mat *Euphorbia* in southeastern Arizona (Hymenoptera, Diptera). Entomological News 72:80–83.

Kuijt, J. 1969. The biology of parasitic flowering plants. Berkeley: University of California Press.

———. 1982. The Viscaceae in the southeastern United States. Journal of the Arnold Arboretum 63:401–410.

———. 1997. *Phoradendron olae* Kuijt: A new species from Mexico pivotal in the taxonomy of the genus, with comments on *P. californicum* Nutt. Brittonia 49:181–188.

Kyhos, D. W. 1967. Natural hybridization between *Encelia* and *Geraea* (Compositae) and some related experimental investigations. Madroño 19:33–43.

———. 1971. Evidence of different adaptations of flower color variants of *Encelia farinosa* (Compositae). Madroño 21:49–61.

Lafèrriere, J. E. 1992. Berberidaceae, Barberry Family [Vascular Plants of Arizona]. Journal of the Arizona-Nevada Academy of Science 26:2–4.

Lakey, J. 1984. A review of generic concepts in American Phaseolinae (Fabaceae, Faboideae). Iselya 2:21–64.

Lamson-Scribner, F. 1897, 1899. American grasses (illustrated). U.S.D.A. Division of Agrostology Bulletin 7 & 17. Washington, D.C.: Government Printing Office.

Lancaster, N., R. Greeley, & **P. R. Christensen.** 1987. Dunes of the Gran Desierto sand-sea, Sonora, Mexico. Earth Surface Processes and Landforms 12:277–288.

Lane, M. A. 1982. Generic limits of *Xanthocephalum, Gutierrezia, Amphiachyris, Gymnosperma, Greenella,* and *Thurovia* (Compositae: Astereae). Systematic Botany 7:405–416.

———. 1985. Taxonomy of *Gutierrezia* (Compositae: Astereae). Systematic Botany 10:7–28.

Lavin, M. 1988. Systematics of *Coursetia* (Leguminosae-Papilionoideae). Systematic Botany Monographs 21: 1–167.

Lavin, M. & M. Sousa S. 1995. Phylogenetic systematics and biogeography of the tribe Robinieae (Leguminosae). Systematic Botany Monographs 45:1–165.

Legrand, C. D. 1962. Las especies Americanas de *Portulaca*. Anales del Museo Nacional de Montevideo 7:1–147.

Lehto, E. 1979. "Extinct" wire-lettuce, *Stephanomeria schottii* (Compositae), rediscovered in Arizona after more than one hundred years. Desert Plants 1:22.

Leister, O. A. 1970. Resedaceae. Flora of Southern Africa 13:177–184. Pretoria: Republic of South Africa, Department of Agriculture Technical Services.

Lellinger, D. B. 1985. A field manual of ferns and fern allies of United States and Canada. Washington, D.C.: Smithsonian Institution.

Leopold, A. 1949. A sand county almanac. New York: Oxford University Press.

Levin, G. 1994. Systematics of the *Acalypha californica* complex (Euphorbiaceae). Madroño 41:254–265.

Liede, S. 1996. *Sarcostemma* (Asclepiadaceae)—A controversial generic circumscription reconsidered: Morphological evidence. Systematic Botany 21:31–44.

Lindsay, E. G., J. W. MacSwain, & **P. H. Raven.** 1963. Comparative behavior of bees and Onagraceae—I: Oenothera bees of the Colorado Desert. University of California Publications in Entomology 33:1–24.

Lindsay, G. E. 1963. The genus *Lophocereus*. Cactus and Succulent Journal (U.S.) 35:176–192.

———. 1996. The Taxonomy and ecology of the genus *Ferocactus*: Explorations in the U.S.A. and Mexico. Tireless Termites Press.

Ling, Y. R. 1995. The New World *Artemisia* L. In Advances in Compositae systematics, D. J. N. Hind, C. Jeffrey, & G. V. Pope, eds., 255–281. Kew: Royal Botanic Gardens.

Lingenfelter, R. E. 1978. Steamboats on the Colorado River. Tucson: University of Arizona Press.

Lippold, H. 1978. Die Gattung *Dodonaea* Miller (Sapindaceae) in Amerika. Wissenschaftliche Zeitschrift der Friedrich-Schiller Universität Jena. Beiträga zur Phytotaxonomie 6:79–126.

Liston, A., L. H. Rieseberg, & **T. S. Elias.** 1989. Genetic similarity is high between intercontinental disjunct species of *Senecio* (Asteraceae). American Journal of Botany 76:383–388.

Little, E. L., Jr. 1950. Southwestern trees, a guide to the native species of New Mexico and Arizona. U.S.D.A. Agricultural Handbook 9:1–109.

Lonard, R. I. & **F. W. Gould.** 1974. The North American species of *Vulpia* (Gramineae). Madroño 22:217–230.

Lord, E. M. 1981. Cleistogamy: A tool for the study of floral morphogenesis, function, and evolution. The Botanical Review 47:421–449.

Löve, D. & P. Dansereau. 1959. Biosystematic studies on *Xanthium:* Taxonomic appraisal and ecological status. Canadian Journal of Botany 37:173–208.

Lowden, R. M. 1986. Taxonomy of the genus *Najas* L. (Najadaceae) in the neotropics. Aquatic Botany 24: 147–184.

Lumholtz, C. S. 1912. New trails in Mexico. New York: Scribner.

Lynch, D. J. 1981. Genesis and geochronology of alkaline volcanism in the Pinacate volcanic field, northwestern Sonora, Mexico. Ph.D. diss., University of Arizona, Tucson.

Mabberley, D. J. 1997. The plant-book. 2d ed. Cambridge: Cambridge University Press.

Mabry, T. J., J. H. Hunziker, & **D. R. DiFeo.** 1977. Creosote bush: Biology and chemistry of *Larrea* in New World deserts. Stroudsburg, Pa.: Dowden, Hutchinson, & Ross.

Malone, C. R. & **V. W. Proctor.** 1965. Dispersal of *Marsilea mucronata* by water birds. American Fern Journal 55:167–170.

Manning, S. D. 1991. The genera of Pedaliaceae in the southeastern United States. Journal of the Arnold Arboretum (supplementary series) 1:313–347.

Maréchal, R., J. M. Mascherpa, & **F. Stainier.** 1978. Etude taxonométrique d'un groupe complexe d'especes des genres *Phaseolus* et *Vigna* (Papilionoideae). Boissiera 28:1–273.

Marsh, V. L. 1952. A taxonomic revision of the genus *Poa* of the United States and southern Canada. American Midland Naturalist 47:202–250.

Martin, P. S., D. Yetman, M. Fishbein, P. Jenkins, T. R. Van Devender, & **R. K. Wilson,** eds. 1998. Gentry's Río Mayo Plants. Tucson: University of Arizona Press.

Martin, S. C. 1948. Mesquite seeds remain viable after 44 years. Ecology 29:393.

Mason, C. T., Jr. 1960. Notes on the flora of Arizona II. Leaflets of Western Botany 9:87–88.

———. 1992. Crossosomataceae, Crossosoma Family [Vascular Plants of Arizona]. Journal of the Arizona-Nevada Academy of Science 26:7–9.

Mason, H. L. 1945. The genus *Eriastrum* and the influence of Bentham and Gray upon the problem of generic confusion in Polemoniaceae. Madroño 8:65–91.

———. 1957. A flora of the marshes of California. Berkeley: University of California Press.

Matthei, O. 1986. El género *Bromus* L. (Poaceae) en Chile. Gayana (Botánica) 43:47–110.

Matthews, J. F. & **D. W. Ketron.** 1991. Two new combinations in *Portulaca* (Portulaceae). Castanea 56:304–305.

Matthews, J. F., D. W. Ketron, & **S. F. Zane.** 1993. The biology and taxonomy of the *Portulaca oleracea* L. (Portulaceae) complex in North America. Rhodora 95:166–183.

Matthews, J. F. & **P. A. Levins.** 1985. *Portulaca pilosa* L., *P. mundula* I. M. Johnst., and *P. parvula* Gray in the Southwest. Sida 11:45–61.

———. 1986. The systematic significance of seed morphology in *Portulaca* (Portulacaceae) under scanning electron microscopy. Systematic Botany 11:302–308.

Mathias, M. E. & **L. Constance.** 1944–45. Umbelliferae. North American Flora 28B:43–295.

———. 1965. A revision of the genus *Bowlesia* Ruiz & Pav. (Umbelliferae-Hydrocotyloideae) and its relatives. University of California Publications in Botany 38:1–73.

May, L. A. 1973. Resource reconnaissance of the Gran Desierto region, northwestern Sonora, Mexico. Master's thesis, University of Arizona, Tucson.

McAuliffe, J. R. 1984. Saguaro–nurse tree associations in the Sonoran Desert: Competitive effects of saguaros. Oecologia 64:319–321.

———. 1999. The Sonoran Desert: Landscape complexity and ecological diversity. In Ecology of Sonoran Desert Plants and Plant Communities, R. H. Robichaux, ed., 68–114. Tucson: University of Arizona Press.

McClintock, E. & **C. Epling.** 1946. A revision of *Teucrium* in the New World, with observations on its variation, geographical distribution and history. Brittonia 5:491–510.

McDonough, W. T. 1964. Germination responses of *Carnegiea gigantea* and *Stenocereus thurberi*. Ecology 45: 155–159.

McGregor, R. L., J. L. Gentry, & **R. E. Brooks.** 1986. Solanaceae. In Flora of the Great Plains. Great Plains Flora Association, eds., 637–651. Lawrence: University of Kansas.

McGregor, S. E., S. M. Alcorn, & **G. Olin.** 1962. Pollination and pollinating agents of the saguaro. Ecology 43:259–267.

McKelvey, S. D. 1938. Yuccas of the southwestern United States, part 1. Jamaica Plain, Mass.: Arnold Arboretum of Harvard University.

———. 1947. Yuccas of the southwestern United States, part 2. Jamaica Plain, Mass.: Arnold Arboretum of Harvard University.

McLaughlin, S. P. & **J. E. Bowers.** 1999. Diversity and affinities of the flora of the Sonoran floristic province. In Ecology of Sonoran Desert Plants and Plant Communities, R. H. Robichaux, ed., 12–35. Tucson: University of Arizona Press.

McLaughlin, S. P., J. E. Bowers, & **K.R.F. Hall.** 1987. Vascular plants of eastern Imperial County, California. Madroño 34:359–378.

McVaugh, R. 1939. Some realignments in the genus *Nemacladus.* American Midland Naturalist 22:521–550.

———. 1945. The genus *Jatropha* in America: Principal intrageneric groups. Bulletin of the Torrey Botanical Club 72:271–294.

———. 1956. Edward Palmer, plant explorer of the American West. Norman: University of Oklahoma Press.

———. 1983. Flora Novo-Galiciana. Vol. 14, Gramineae. Ann Arbor: University of Michigan Press.

———. 1984. Flora Novo-Galiciana. Vol. 12, Compositae. Ann Arbor: University of Michigan Press.

———. 1987. Flora Novo-Galiciana. Vol. 5, Leguminosae. Ann Arbor: University of Michigan Press.

———. 1989. Flora Novo-Galiciana. Vol. 15, Bromeliaceae to Dioscoreaceae. Ann Arbor: University of Michigan Press.

McVaugh, R. & **J. Rzedowski.** 1965. Synopsis of the genus *Bursera* L. in western Mexico, with notes on the material of *Bursera* collected by Sessé and Moçiño. Kew Bulletin 18:317–382.

Mearns, E. A. 1892–93. Field books, Mexican boundary survey, vol. 3. On file, United States National Herbarium, library, Natural History Museum, Smithsonian Institution. Washington, D.C.

———. 1907. Mammals of the Mexican boundary of the United States. United States National Museum Bulletin 56:1–530.

Meine, C. 1988. Aldo Leopold. Madison: University of Wisconsin Press.

Merriam, R. 1972. Reconnaissance geologic map of the Sonoyta quadrangle, northwest Sonora, Mexico. Geological Society of America Bulletin 83:3533–3536.

Metcalf, C. R. 1960. Anatomy of the monocotyledons. Vol. 1, Gramineae. Oxford: Clarendon Press.

Meyer, T. 1957. Las especies *Menodora* (Oleaceae) de Argentina, Bolivia, Paraguay, y Uruguay. Lilloa 28:209–245.

Mickel, J. T. 1979a. How to know the ferns and fern allies. Dubuque, Iowa: William C. Brown.

———. 1979b. The fern genus *Cheilanthes* in continental United States. Phytologia 41:431–437.

Miller, K. I. & **G. L. Webster.** 1967. A preliminary revision of *Tragia* (Euphorbiaceae) in the United States. Rhodora 69:241–305.

Miller, R. R. & **L. A. Fuiman.** 1987. Description and conservation status of *Cyprinodon macularius eremus,* a new subspecies of pupfish from Organ Pipe Cactus National Monument. Copeia 1987:593–609.

Mooney, H. A. & **J. A. Drake,** eds. 1986. Ecology of biological invasions of North America and Hawaii. New York: Springer-Verlag.

Mooney, H. A., J. Ehleringer, & **J. A. Berry.** 1976. High photosynthetic capacity of a winter annual in Death Valley. Science 194:322–324.

Mooney, H. A. & **W. A. Emboden.** 1968. The relationship of terpene composition, morphology, and distribution of populations of *Bursera microphylla* (Burseraceae). Brittonia 20:44–51.

Mooney, H. A., B. B. Simpson, & **O. T. Solbrig.** 1977. Phenology, morphology, physiology. In Mesquite, B. B. Simpson, ed., 26–43. Stroudsburg, Pa.: Dowden, Hutchinson, & Ross.

Mooney, H. A. & **B. R. Strain.** 1964. Bark photosynthesis in ocotillo. Madroño 17:230–233.

Moran, R. 1969. Twelve new dicots from Baja California, Mexico. Transactions of the San Diego Society of Natural History 15:273–274.

———. 1992. Pygmy weed (*Crassula connata*) etc. in western North America. Cactus and Succulent Journal (U.S.) 64:223–231.

Moran, R. & **R. S. Felger.** 1968. *Castela polyandra,* a new species in a new section; union of *Holacantha* with *Castela* (Simaroubaceae). Transactions of the San Diego Society of Natural History 15:31–40.

Morefield, J. D. 1993a. *Chaenactis,* In The Jepson manual, J. C. Hickman, ed., 223–226. Berkeley: University of California Press.

———. 1993b. *Filago.* In The Jepson manual, J. C. Hickman, ed., 266–268. Berkeley: University of California Press.

———. 1993c. *Stylocline.* In The Jepson manual, J. C. Hickman, ed., 348–349. Berkeley: University of California Press.

Morgan, D. R. 1997. Reticulate evolution in *Machaeranthera* (Asteraceae). Systematic Botany 22:599–615.

Moss, C. E. 1954. The species of *Arthrocnemum* and *Salicornia* in southern Africa. Journal of South African Botany 20:1–22.

Mosyakin, S. L. 1996. A taxonomic synopsis of the genus *Salsola* (Chenopodiaceae) in North America. Annals of the Missouri Botanical Garden 83:387–395.

Mulroy, T. W. & **P. W. Rundel.** 1977. Annual plants: Adaptations to desert environments. BioScience 27:109–114.

Munz, P. A. 1965. Onagraceae. North American Flora (series 2) 5:1–278.

———. 1968. Supplement to A California flora. Berkeley: University of California Press.

———. 1974. A flora of southern California. Berkeley: University of California Press.

Munz, P. A. & **D. D. Keck.** 1963. A California flora. Berkeley: University of California Press.

Munz, P. A. & **J. C. Roos.** 1950. California Miscellany 2. Aliso 2:217–238.

Nabhan, G. P. 1985. Gathering the desert. Tucson: University of Arizona Press.

Nabhan, G. P. & **J. L. Carr,** eds. 1994. Ironwood: An Ecological and cultural keystone of the Sonoran Desert. Conservation International Occasional Paper No. 1. Washington, D.C.

Nabhan, G. P. & **J. M. J. de Wet.** 1984. *Panicum sonorum* in Sonoran Desert agriculture. Economic Botany 38: 65–82.

Nabhan, G. P. & **R. S. Felger.** 1978. Teparies in southwestern North America. Economic Botany 32:2–19.

Nabhan, G. P., A. M. Rea, K. L. Reichhardt, E. Mellink, & **C. F. Hutchinson.** 1982. Papago influences on habitat and biotic diversity: Quitovac oasis ethnoecology. Journal of Ethnobiology 2:124–143.

Nabhan, G. P., A. Whiting, H. Dobyns, R. Hevly, & **R. Euler.** 1981. Devil's claw domestication: Evidence from southwestern Indian fields. Journal of Ethnobiology 1:135–164.

Navas Bustamante, L. E. 1973. Flora de la Cuenca de Santiago de Chile. Vol 1, Gramineae. Santiago: Universidad de Chile.

Nee, M. 1993. Solanaceae. In The Jepson manual, J. C. Hickman, ed., 1068–1077. Berkeley: University of California Press.

Nelson, A. & **P. B. Kennedy.** 1908. New plants from the Great Basin. Muhlenbergia 3:138.

Nesom, G. L. 1982. Nomenclatural changes and clarifications in Mexican *Erigeron* (Asteraceae). Sida 9:223–229.

———. 1988. *Baccharis monoica* (Compositae: Astereae), a monoecious species of the *B. salicifolia* complex from Mexico and Central America. Phytologia 65:160–164.

———. 1989a. Infrageneric taxonomy of New World *Erigeron* (Compositae: Astereae). Phytologia 67:67–93.

———. 1989b. *Aster intricatus* (Asteraceae: Astereae) transferred to *Machaeranthera.* Phytologia 67:438–440.

———. 1989c. New species, new sections, and a taxonomic overview of *Pluchea* (Compositae: Inuleae). Phytologia 67:158–167.

———. 1990a. Taxonomy of the genus *Laennecia* (Asteraceae: Astereae). Phytologia 68:205–228.

———. 1990b. Further definition of *Conyza* (Asteraceae: Astereae). Phytologia 68:229–233.

———. 1991. Taxonomy of *Isocoma* (Compositae: Astereae). Phytologia 70:69–114.

———. 1992. Revision of *Erigeron* sect. *Linearifolii* (Asteraceae: Astereae). Phytologia 72:157–208.

———. 1994a. Subtribal classification of the Astereae (Asteraceae). Phytologia 76:193–274.

———. 1994b. Review of the taxonomy of *Aster* sensu lato (Asteraceae: Astereae), emphasizing the New World species. Phytologia 77:141–297.

———. 1997. Review: "A revision of *Heterotheca* sect. *Phyllotheca* (Nutt.) Harms (Compositae: Astereae)" by J. C. Semple. Phytologia 83:7–21.

Nesom, G. L., Y. Suh, D. R. Morgan, S. D. Sundberg, & **B. B. Simpson.** 1991. *Chloracantha:* A new genus of North American Astereae (Asteraceae). Phytologia 70:371–381.

Nicora, E. G. 1978. Flora Patagonica. Parte 3, Gramineae. Buenos Aires: Colección Científica del Instituto Nacional de Tecnología Agropecuaria.

Nobel, P. S. 1977. Water relations and photosynthesis of a barrel cactus, *Ferocactus acanthodes,* in the Colorado Desert. Oecologia 27:117–133.

———. 1978. Microhabitat, water relations, and photosynthesis of a desert fern, *Notholaena parryi.* Oecologia 31:293–309.

———. 1980a. Morphology, nurse plants, and minimum apical temperatures for young *Carnegiea gigantea.* Botanical Gazette 41:181–191.

———. 1980b. Morphology, surface temperatures, and northern limits of columnar cacti in the Sonoran Desert. Ecology 61:1–7.

———. 1980c. Water vapor conductance and CO_2 uptake for leaves of a C_4 desert grass, *Hilaria rigida.* Ecology 61:252–258.

———. 1982. Low-temperature tolerance and cold hardening of cacti. Ecology 63:1650–1656.

———. 1988. Environmental biology of agaves and cacti. New York: Cambridge University Press.

Nordlindh, T. 1972. Notes on the variation and taxonomy in the *Scirpus maritimus* complex. Botaniska Notiser 125:397–405.

O'Kane, S. L., Jr., I. Al-Shehbaz, & **N. J. Turland.** 1999. Proposal to conserve the name *Lesquerella* against *Physaria* (Cruciferae). Taxon 48:163–164.

Ollendorf, A. L., S. C. Mulholland, & **G. Rapp Jr.** 1988. Phytolith analysis as a means of plant identification: *Arundo donax* and *Phragmites communis.* Annals of Botany 61:209–214.

Ooststroom, S. J. van. 1934. A monograph of the genus *Evolvulus.* Mededeelingen Botanisch Museum en Herbarium (Rijks Universiteit Utrecht) 14:1–267.

Ortlieb, L. & **J. Roldan Q.** 1981. Geología del noroeste de México y del sur de Arizona. Hermosillo: Estación Regional del Noroeste, Instituto de Geología, Universidad Nacional Autónoma de México.

Ownbey, G. B. 1958. Monograph of the genus *Argemone* for North America and the West Indies. Memoirs of the Torrey Botanical Club 21:1–159.

———. 1997. *Argemone.* In vol. 3 of Flora of North America north of Mexico, 314–322. New York: Oxford University Press.

Parfitt, B. D. 1987. *Echinocereus nicholii* (L. D. Benson) Parfitt, Stat. Nov. (Cactaceae). Phytologia 63:157–158.

Parfitt, B. D. & **M. A. Baker.** 1993. *Opuntia.* In The Jepson manual, J. C. Hickman, ed., 452–456. Berkeley: University of California Press.

Parfitt, B. D. & **C. H. Pickett.** 1980. Insect pollination of prickly-pears (*Opuntia:* Cactaceae). Southwestern Naturalist 25:104–107.

Parfitt, B. D. & **D. J. Pinkava.** 1988. Nomenclatural and systematic reassessment of *Opuntia engelmannii* and *O. lindheimeri* (Cactaceae). Madroño 35:342–349.

Parish, S. B. 1890. Notes on the naturalized plants of southern California. VII. Zoe 1:300–303.

Parker, K. C. 1987a. Site-related demographic patterns of organ pipe cactus populations in southern Arizona. Bulletin of the Torrey Botanical Club 114:149–155.

———. 1987b. Seedcrop characteristics and minimum reproductive size of organ pipe (*Stenocereus thurberi*) in southern Arizona. Madroño 34:294–303.

Parker, K. F. 1958. Arizona ranch, farm, and garden weeds. Agricultural Extension Service circular 265. Tucson: University of Arizona.

———. 1972. An illustrated guide to Arizona weeds. Tucson: University of Arizona Press.

Parodi, L. R. 1919. Las chlorideas de la república Argentina. Revista Facultad de Agronomía y Veterinario (Buenos Aires) 2:233–339.

———. 1927. Revisión de las gramíneas argentinas del género *Diplachne.* Revista Facultad de Agronomía y Veterinario (Buenos Aires) 6:21–43.

———. 1937. Revisión de las gramíneas del género *Tridens* de la flora Argentina. Revista Argentina de Agronomía 4:241–257.

Patterson, R. W. 1993. *Eriastrum*. In The Jepson manual, J. C. Hickman, ed., 826–828. Berkeley: University of California Press.

Pavlick, L. 1995. *Bromus* L. of North America. Victoria, Canada: Royal British Columbia Museum.

Payne, W. W. 1964. A re-evaluation of the genus *Ambrosia* (Compositae). Journal of the Arnold Arboretum 45:401–438.

Pennell, F. W. 1921. *Veronica* in North and South America. Rhodora 23:1–22, 29–41.

Peterson, K. M. 1974. On the correct name for the appressed-winged variety of *Hymenoclea salsola* (Compositae: Ambrosieae). Brittonia 26:397.

Peterson, K. M. & **W. W. Payne.** 1973. The genus *Hymenoclea* (Compositae: Ambrosieae). Brittonia 25:243–256.

Peterson, P. M. & **C. R. Annable.** 1991. Systematics of the annual species of *Muhlenbergia* (Poaceae-Eragrostideae). Systematic Botany Monographs 31:1–109.

Pfeifer, H. W. 1970. A taxonomic revision of the Pentandrous species of *Aristolochia*. University of Connecticut Publication Series 1:1–134.

Phillips, A. R., J. T. Marshall, & **G. Monson.** 1964. The birds of Arizona. Tucson: University of Arizona Press.

Philips, S. M. 1972. A survey of *Eleusine* Gaertn. (Gramineae) in Africa. Kew Bulletin 27:251–270.

Pillay, M. & **K. W. Hilu.** 1995. Chloroplast-DNA restriction site analysis in the genus *Bromus* (Poaceae). American Journal of Botany 82:239–249.

Pinkava, D. J. 1995. Cactaceae, part 1. The Cereoid Cacti [Vascular Plants of Arizona]. Journal of the Arizona-Nevada Academy of Science 29:6–12.

Pinkava, D. J. & **H. S. Gentry,** eds. 1985. Symposium on the genus *Agave*. Desert Plants 7:34–116.

Pinkava, D. J., E. Lehto, T. Reeves, & **L. McGill.** 1978. Plants new to Arizona—IV, and new distributional records of noteworthy species. Journal of the Arizona-Nevada Academy of Science 13:84.

Pinkava, D. J., E. Lehto, T. Reeves, & **E. Sundell.** 1975. Plants new to Arizona flora. Journal of the Arizona-Nevada Academy of Science 10:146.

Pinkava, D. J. & **. D. Parfitt.** 1988. Nomenclatural changes in Chihuahuan Desert *Opuntia* (Cactaceae). Sida 13:125–130.

Pinkava, D. J., B. D. Parfitt, M. A. Baker, & **R. D. Worthington.** 1992. Chromosome numbers in some cacti of western North America— VI, with nomenclatural changes. Madroño 39:98–113.

Pinto-Escobar, P. 1986. El género *Bromus* en los Andes centrales de Sudamerica. Caldasia 15:15–34.

Polhill, R. M. & **P. H. Raven,** eds. 1981. Advances in legume systematics. Parts 1 and 2. Kew: Royal Botanic Gardens.

Pohl, R. W. 1980. Family #15, Gramineae. Flora Costaricensis. Fieldiana Botany, New Series 4:1–595.

Porter, D. M. 1963. The taxonomy and distribution of the Zygophyllaceae of Baja California, Mexico. Contributions from the Gray Herbarium 192: 99–135.

———. 1969. The genus *Kallstroemia* (Zygophyllaceae). Contributions from the Gray Herbarium 198: 41–153.

———. 1972. The genera of Zygophyllaceae in the southeastern United States. Journal of the Arnold Arboretum 53:531–552.

Porter, J. M. 1997. Phylogeny of Polemoniaceae based on nuclear ribosomal internal transcribed spacer DNA sequences. Aliso 15:57–77.

———.1998. Aliciella: A recircumscribed genus of Polemoniaceae. Aliso 17:23–46.

Porter, S. C. 1989. Some geological implications of average Quaternary glacial conditions. Quaternary Research 32:245–261.

Powell, A. M. 1972. Taxonomy of *Amauria* (Compositae-Peritylinae). Madroño 21:516–525.

———. 1974. Taxonomy of *Perityle* section *Perityle* (Compositae—Peritylinae). Rhodora 76:229–306.

———. 1978. Systematics of *Flaveria* (Flaveriinae-Asteraceae). Annals of the Missouri Botanical Garden 65: 590–636.

Rahn, K. 1979a. *Plantago* ser. *Gnaphaloides* Rahn. A taxonomic revision. Botanisk Tidsskrift 73:137–154.

———. 1979b. *Plantago* ser. *Ovata:* A taxonomic revision. Botanisk Tidsskrift 74:13–20.

Raunkiaer, C. 1934. The life-forms of plants and statistical plant geography. Oxford: Clarendon Press.

Raven, P. H. 1962. The systematics of *Oenothera,* subgenus *Chylismia.* University of California Publications in Botany 34:1–122.

———. 1964. The generic subdivision of Onagraceae, tribe Onagreae. Brittonia 16:276–288.

———. 1969. A revision of the genus *Camissonia* (Onagraceae). Contributions from the United States National Herbarium 37:161–396.

———. 1970. *Oenothera brandegeei* from Baja California, Mexico and a review of subgenus *Pachylophus* (Onagraceae). Madroño 20:350–354.

Raven, P. H. & **D. P. Gregory.** 1972. A revision of the genus *Gaura* (Onagraceae). Memoirs of the Torrey Botanical Club 23:1–96.

Raven, P. H., D. W. Kyhos, D. E. Breedlove, & **W. W. Payne.** 1968. Polyploidy in *Ambrosia dumosa* (Compositae: Ambrosieae). Brittonia 20:205–211.

Ray, P. M. & **H. F. Chisaki.** 1957. Studies on *Amsinckia.* American Journal of Botany 44:529–554.

Rea, A. 1983. Once a river. Tucson: University of Arizona Press.

———. 1997. At the desert's green edge: An ethnobotany of the Gila River Pima. Tucson: University of Arizona Press.

Rebman, J. P. 1995. Biosystematics of *Opuntia* subgenus *Cylindropuntia* (Cactaceae): The chollas of Lower California, Mexico. Ph.D diss, Arizona State University, Tempe.

———. 1999. A new cholla (Cactaceae) from Baja California, Mexico. Haseltonia 6:17–22.

Rechinger, K. H., Jr. 1937. The North American species of *Rumex.* Field Museum of Natural History, Botanical Series 17:1–151.

Record, S. J. & **R. W. Hess.** 1943. Timbers of the New World. New Haven, Conn.: Yale University Press.

Reeder, C. 1981. *Muhlenbergia.* In The grasses of Baja California, Mexico, F. W. Gould & R. Moran, 67–78. San Diego Society of Natural History Memoir 12.

Reeder, J. R. 1971. Notes on Mexican grasses IX: Miscellaneous chromosome numbers—3. Brittonia 23:105–117.

———. 1977. Chromosome numbers in western grasses. American Journal of Botany 64:102–110.

———. 1984. Chromosome number reports 82. Taxon 33:132–133.

———. 1986. Another look at *Eragrostis tephrosanthos* (Gramineae). Phytologia 60:153–154.

Reeder, J. R., & **D. J. Crawford.** 1970. Affinities of *Erioneuron* and *Munroa* (Gramineae). Abstract in American Journal of Botany 57:752.

Reeder, J. R. & **H. F. Decker.** 1961. Affinities of *Stipa* and *Aristida.* Abstract in American Journal of Botany 48:549.

Reeder, J. R. & **R. S. Felger.** 1989. The *Aristida californica-glabrata* complex (Gramineae). Madroño 36:187–197.

Reeder, J. R. & **C. G. Reeder.** 1998. Poaceae. In Gentry's Río Mayo Plants, P. S. Martin et al., eds., 498–520. Tucson: University of Arizona Press.

Reisner, M. 1986. Cadillac desert. New York: Viking.

Renvoize, S. A. 1968. The Afro-Asian species of *Enneapogon* P. Beauv. Kew Bulletin 22:393–401.

Reveal, J. L. 1968. Notes on *Eriogonum*—IV. A revision of the *Eriogonum deflexum* complex. Brittonia 20:13–33.

———. 1969. The subgeneric concept in *Eriogonum* (Polygonaceae), In Current topics in plant science, J. Gunckel Jr., eds., 229–249. New York: Academic Press.

———. 1970. Additional notes on the California buckwheats (*Eriogonum,* Polygonaceae). Aliso 7:217–230.

———. 1976. *Eriogonum* (Polygonaceae) of Arizona and New Mexico. Phytologia 34:409–484.

———. 1978. Distribution and phylogeny of Eriogonoideae (Polygonaceae). Great Basin Naturalist Memoirs 2:169–190.

———. 1989a. A checklist of the Eriogonoideae (Polygonaceae). Phytologia 66:266–294.

———. 1989b. The eriogonoid flora of California (Polygonaceae: Eriogonoideae). Phytologia 66:295–414.

Reveal, J. L. & **B. J. Ertter.** 1980. The genus *Nemacaulis* (Polygonaceae). Madroño 27:101–109.

Reveal, J. L. & **C. B. Hardham.** 1989. A revision of the annual species of *Chorizanthe* (Polygonaceae: Eriogonoideae). Phytologia 66:98–198.

Reynolds, J. F. & **D. J. Crawford.** 1980. A quantitative study of variation in the *Chenopodium atrovirens-desiccatum-pratericola* complex. American Journal of Botany 67:1380–1390.

Richardson, A. T. 1977. Monograph of the genus *Tiquilia* (*Coldenia, sensu lato*), Boraginaceae: Ehretioideae. Rhodora 79:467–572.

Rieseberg, L. H., S. M. Beckstrom-Sternberg, A. Liston, & **D. M. Arias.** 1991. Phylogenetic and systematic inference from chloroplast DNA and isozyme variation in *Helianthus* sect. *Helianthus* (Asteraceae). Systematic Botany 16:50–76.

Robbins, W. W, M. K. Bellue, & **W. S. Ball.** 1951. Weeds of California. State of California, Documents Section. Sacramento.

Robertson, K. R. 1972. The Malpighiaceae in the southeastern United States. Journal of the Arnold Arboretum 53:101–112.

Robinson, B. L. 1917. A monograph of the genus *Brickellia.* Memoirs of the Gray Herbarium 1:1–151.

Robinson, H. & **J. Cuatrecasas.** 1973. The generic limits of *Pluchea* and *Tessaria* (Inuleae, Asteraceae). Phytologia 27:277–285.

Robinson, K. R. 1981. The genera of Amaranthaceae in the southeastern United States. Journal of the Arnold Arboretum 62:267–314.

Rogers, C. E., T. E. Thompson, & **G. J. Seiler.** 1982. Sunflower species in the United States. Bismark, N. Dak.: National Sunflower Association.

Rogers, D. J. 1951. A revision of *Stillingia* in the New World. Annals of the Missouri Botanical Garden 38: 207–259.

Rollins, R. C. 1979. *Dithyrea* and a related genus (Cruciferae). The Bussey Institution, Harvard University, 3–32.

———. 1981. Weeds of the Cruciferae (Brassicaceae) in North America. Journal of the Arnold Arboretum 62:517–540.

———. 1984. *Draba* (Cruciferae) in Mexico and Guatemala. Contributions from the Gray Herbarium 213: 1–10.

———. 1993. The Cruciferae of continental North America. Stanford: Stanford University Press.

Rollins, R. C. & **U. C. Banerjee.** 1975. Atlas of the trichomes of *Lesquerella* (Cruciferae). Cambridge, Mass.: The Bussey Institution, Harvard University.

Rollins, R. C. & **E. A. Shaw.** 1973. The genus *Lesquerella* (Cruciferae) in North America. Cambridge, Mass.: Harvard University Press.

Romanczuk, C. & **M. A. del Pero.** 1978. Las especies del género *Celtis* (Ulmaceae) de la Argentina. Darwiniana 21:541–577.

Rominger, J. M. 1962. Taxonomy of *Setaria* (Gramineae) in North America. Illinois Biological Monographs 29:1–132.

Rosatti, T. J. 1989. The genera of suborder Apocynineae (Apocynaceae and Asclepiadaceae) in the southeastern United States. Journal of the Arnold Arboretum 70:307–401, 443–514.

Rose, J. N. & **P. C. Standley.** 1912. Report on a collection of plants from the Pinacate region of Sonora. Contributions from the United States National Herbarium 16:5–20.

Rossbach, R. B. 1940. *Spergularia* in North and South America. Rhodora 42:57–83, 105–143, 158–193, 203–213.

Rutherford, R. J. 1970. The anatomy and cytology of *Pilostyles thurberi* Gray (Rafflesiaceae). Aliso 7:263–288.

Rydberg, P. A. 1922. *Dicoria.* North American Flora 33:11–13.

———. 1924. *Sesban.* North American Flora 24:202–205.

Rzedowski, J. 1978. Claves para la identificación de los géneros de la familia Compositae en México. Acta Científica Potosina 7:1–145.

Rzedowski, J. & **G. Calderón de Rzedowski.** 1988. Análisis de la distribución geográfica del complejo *Prosopis* (Leguminosae, Mimosoideae) en Norteamérica. Acta Botánica Mexicana 3:7–19.

Rzedowski, J. & **E. Ezcurra.** 1986. Una nueva especie de *Haplopappus* (Compositae: Astereae) de las dunas del noroeste de Sonora, México. Ciencia Interamericana 26:16–18.

Rzedowski, J. & **H. Kruse.** 1979. Algunas tendencias evolutivas en *Bursera* (Burseraceae). Taxon 28:103–116.

Sánchez-Mejorada R., H. 1973. El género *Neoevansia* Marshall, historia y revisión. Cactaceas y Suculentas Mexicanas 18:13–27.

———. 1974. Revisión del género *Peniocereus* (las cactaceas). Toluca: Gobierno del Estado de México, Dirección de Agricultura y Ganadería.

Sanders, A. C. 1998. Polygonaceae. In Gentry's Río Mayo Plants, P. S. Martin et al., eds., 415–417. Tucson: University of Arizona Press.

Sanderson, M. J. & **M. F. Wojchiechowski.** 1996. Diversification rates in a temperate legume clade: Are there "so many species" of *Astragalus* (Fabaceae)? American Journal of Botany 83:1488–1502.

Sauer, J. D. 1955. Revision of the dioecious amaranths. Madroño 13:5–46.

———. 1967. The grain amaranths and their relatives: A revised taxonomic and geographic survey. Annals of the Missouri Botanical Garden 54:103–137.

Schilling, E. E. 1981. Systematics of *Solanum* sect. *Solanum* (Solanaceae) in North America. Systematic Botany 6:172–185.

———. 1990a. Taxonomic revision of *Viguiera* subgenus *Bahiopsis* (Asteraceae: Heliantheae). Madroño 37: 149–170.

———. 1990b. The black nightshades (*Solanum* section *Solanum*) of the Indian subcontinent. Botanical Journal of the Linnean Society 102:253–259.

Schilling, E. E. & **C. B. Heiser.** 1981. Infrageneric classification of *Helianthus* (Compositae). Taxon 30:393–403.

Schilling, E. E., C. R. Linder, R. D. Noyes, & **L. H. Rieseberg.** 1998. Phylogenetic relationships in *Helianthus* (Asteraceae) based on nuclear ribosomal DNA internal transcribed space region sequence data. Systematic Botany 23:177–187.

Schmutz, E. M., B. H. Freeman, & **R. E. Reed.** 1968. Livestock-poisoning plants of Arizona. Tucson: University of Arizona Press.

Schmutz, E. M. & **L. B. Hamilton.** 1979. Plants that poison. Flagstaff: Northland Press.

Scholz, H. 1964. Zygophyllaceae, In vol. 2. of A. Engler's Syllabus der Pflanzenfamilien, 12th ed., H. Melchior, ed., 51–252. Berlin: Gebrüder Borntraeger.

Schreiber, A. 1974. Die Gattung *Fagonia* (Zygophyllaceae) in Südwestafrika. Mitteilungen der Botanischen Staatssammlung München 11:571–578.

Schuyler, A. E. 1967. A taxonomic revision of North American leafy species of *Scirpus*. Proceedings of the Academy of Natural Sciences of Philadelphia 119:295–323.

———. 1974. Typification and application of the names *Scirpus americanus* Pers., *S. olneyi* Gray, and *S. pungens* Vahl. Rhodora 76:51–52.

Seegeler, C. J. P. 1989. *Sesamum orientale* L. (Pedaliaceae): Sesame's correct name. Taxon 38:656–659.

Seigler, D. S. & **E. Wollenweber.** 1983. Chemical variation in *Notholaena standleyi*. American Journal of Botany 70:790–798.

Semple, J. C. 1996. A revision of *Heterotheca* sect. *Phyllotheca* (Nutt.) Harms (Compositae: Astereae). University of Waterloo Biological Series 37:1–164.

Semple, J. C., S. Heard, & **C. Xiang.** 1996. The asters of Ontario (Compositae: Astereae): *Diplactis* Raf., *Oclemena* Greene, *Doellingeria* Nees and *Aster* L. (including *Canadanthus* Nesom, *Symphyotrichum* Nees and *Virgulus* Raf.). University of Waterloo Biological Series 38:1–94.

Setchell, W. A. 1946. The genus *Ruppia* L. Proceedings of the California Academy of Sciences (series 4) 25:469–478.

Shaw, R. B. & **R. D. Webster.** 1987. The genus *Eriochloa* (Poaceae: Paniceae) in North and Central America. Sida 12:165–207.

Shear, C. L. 1900. Studies on American grasses: A revision of the North American species of *Bromus* occurring north of Mexico. U.S.D.A. Division of Agrostology Bulletin 23:1–66.

Sheldon, C. 1979. The wilderness of desert bighorns & Seri Indians, D. B. Brown, P. M. Webb, & N. B. Carmony, eds. Phoenix: Arizona Bighorn Sheep Society.

Sherman-Broyles, S. L., W. T. Barker, & **L. M. Schulz.** 1997. *Celtis*. In vol. 3 of Flora of North America north of Mexico, 376–379. New York: Oxford University Press.

Shinners, L. H. 1957. Synopsis of the genus *Eustoma* (Gentianaceae). Southwestern Naturalist 2:38–43.

Shishkin, B. K., ed. 1970. Centrospermae. In vol. 6 of Flora of the U.S.S.R. trans. N. Landau. Washington, D.C.: Smithsonian Institution.

Shmida, A. & **T. L. Burgess.** 1988. Plant growth-form strategies and vegetation types in arid environments. In Plant form and vegetation structure, M. J. A. Werger et al., eds., 211–241. The Hague: SPB Academic Publishing.

Shreve, F. 1951. Vegetation of the Sonoran Desert. Carnegie Institution of Washington Publication 591:1–192.

Simmons, N. M. 1966. Flora of the Cabeza Prieta Game Range. Journal of the Arizona Academy of Science 4:93–104.

Simpson, B. B. 1977a. Breeding systems of dominant perennial plants of two disjunct warm desert ecosystems. Oecologia 27:203–226.

———, ed. 1977b. Mesquite. Stroudsburg, Pa.: Dowden, Hutchinson, & Ross.

———. 1978. Compositae tribe Mutiseae. North American Flora (series 2) 10:7–10.

———. 1989. Krameriaceae. Flora Neotropica Monograph 49:1–108.

Simpson, B. B. & **B. M. Miao.** 1997. The circumscription of *Hoffmannseggia* (Fabaceae, Caesalpinioideae, Caesalpinieae) and its allies using morphological and cpDNA restriction site data. Plant Systematics and Evolution 205:157–178.

Small, J. K. 1895. A monograph of the North American species of the genus *Polygonum*. Memoirs from the Department of Botany of Columbia College, vol. 1. Lancaster, Pa: The New Era Print.

Smith, J. M. 1976. A taxonomic study of *Acleisanthes* (Nyctaginaceae). Wrightia 5:261–276.

Snow, N. 1997. Phylogeny and systematics of *Leptochloa* P. Beauvois sensu lato (Poaceae, Chloridoideae). Ph.D. diss., Washington University, St. Louis.

———. 1998. Nomenclatural changes in *Leptochloa* P. Beauvois sensu lato (Poaceae, Chloridoideae). Novon 8:77–80.

Snowden, J. D. 1936. The cultivated races of *Sorghum*. London: Adlard & Sons.

———. 1955. The wild fodder sorghums of the section *Eu-Sorghum*. Botanical Journal of the Linnean Society 55:191–260.

Soderstrom, T. R. & **J. H. Beaman.** 1968. The genus *Bromus* (Gramineae) in Mexico and Central America. Michigan State University, Biological Series 3:465–520.

Soderstrom, T. R. & **H. F. Decker.** 1964. *Reederochloa,* a new genus of dioecious grass from Mexico. Brittonia 16:334–339.

Soderstrom, T. R., K. W. Hilu, C. S. Campbell, & **M. E. Barkworth,** eds. 1987. Grass systematics and evolution. Washington, D.C.: Smithsonian Institution Press.

Sohns, E. R. 1955. *Cenchrus* and *Pennisetum:* Fascicle morphology. Journal of the Washington Academy of Science 45:135–143.

———. 1956. The genus *Hilaria* (Gramineae). Journal of the Washington Academy of Sciences 46:311–321.

Solbrig, O. T. 1961. Note on *Gymnosperma glutinosum* (Compositae-Astereae). Leaflets of Western Botany 9: 147–150.

———. 1972. The floristic disjunctions between the "monte" in Argentina and the "Sonoran Desert" in Mexico and the United States. Annals of the Missouri Botanical Garden 59:218–223.

Spellenberg, R. & **R. K. Delson.** 1977. Aspects of reproduction in Chihuahuan Desert Nyctaginaceae. In Transactions of the symposium on the biological resources of the Chihuahuan Desert region, United States and Mexico, R. H. Wauer & D. H. Riskind, eds., 273–287. U.S. Department of the Interior, National Park Service Transactions and Proceedings Series 3. Washington, D.C.: Government Printing Office.

Spira, T. P. & **L. K. Wagner.** 1983. Viability of seeds up to 211 years old extracted from adobe brick buildings of California and northern Mexico. American Journal of Botany 70:303–307.

Standley, P. C. 1911. The American species of *Fagonia*. Proceedings of the Biological Society of Washington 24:243–250.

———. 1916. Chenopodiaceae. North American Flora 21:3–93.

———. 1917. Amaranthaceae. North American Flora 21:95–169.

———. 1918. Allioniaceae. North American Flora 21:171–254.

Starr, G. D. 1985. New World salvias cultivated in the southwestern United States. Master's thesis, University of Arizona, Tucson.

Stebbins, G. L. 1981. Chromosomes and evolution in the genus *Bromus* (Gramineae). Botanishe Jahrbücher für Sytematik und Planzengeschichte 102:359–379.

Steenbergh, W. F. & **C. H. Lowe.** 1976. Ecology of the saguaro—I: The role of freezing weather in a warm-desert plant population. In Research in the Parks, 49–92. U.S. Department of the Interior, National Parks Symposium Series 1.

———. 1977. Ecology of the saguaro—II: Reproduction, germination, establishment, growth, and survival of the young plant. National Park Service Scientific Monograph Series 8:1–242.

———. 1983. Ecology of the saguaro—III: Growth and demography. National Park Service Scientific Monograph Series 17:1–128.

Steinmann, V. W. & **R. S. Felger.** 1995. New combinations for *Euphorbia* and *Ditaxis* (Euphorbiaceae) in northwestern Mexico and southwestern United States. Madroño 42:455–457.

———. 1997. The Euphorbiaceae in Sonora, Mexico. Aliso 16:1–71.

Stephenson, S. N. 1971. A putative *Distichlis* × *Monanthochloë* (Poaceae) hybrid from Baja California, Mexico. Madroño 21:125–127.

Stevenson, D. W. 1993. Ephedraceae. In vol. 2 of Flora of North America north of Mexico, 428–434. New York: Oxford University Press.

Steyermark, J. A. 1932. A revision of the genus *Menodora*. Annals of the Missouri Botanical Garden 19:87–176.

Stockwell, P. 1940. A revision of the genus *Chaenactis*. Contributions from the Dudley Herbarium 3:89–167.

Straw, R. M. 1966. A redefinition of *Penstemon* (Scrophulariaceae). Brittonia 18:80–95.

———. 1967. *Keckiella:* New name for *Keckia* Straw (Scrophulariaceae). Brittonia 19:203–204.

Strong, S. K. 1977. A study of *Abutilon* (Malvaceae) in the southwestern United States and Mexico. Ph.D. diss., University of Texas, Austin.

Strother, J. L. 1969. Systematics of *Dyssodia* Cavanilles (Compositae: Tageteae). University of California Publications in Botany 48:1–88.

———. 1978. *Cacaliopsis, Luina, Rainiera, Tetradymia, Peucephyllum, Lepidospartum*. North American Flora (series 2) 10:160–173.

———. 1982. *Dicoria argentea* (Compositae: Ambrosiinae), a new species from Sonora, Mexico. Madroño 29: 101–103.

———. 1986. Renovation of *Dyssodia* (Compositae: Tageteae). Sida 11:371–378.

Strother, J. L. & **G. Pilz.** 1975. Taxonomy of *Psathyrotes* (Compositae: Senecioneae). Madroño 23:24–40.

Stutz, H. C., J. M. Melby, & **G. K. Livingston.** 1975. Evolutionary studies of *Atriplex canescens*. American Journal of Botany 62:236–245.

Stutz, H. C. & **S. C. Sanderson.** 1979. The role of polyploidy in the evolution of *Atriplex:* A relic *gigas* diploid population of *Atriplex canescens*. In Arid land resources, J. R. Goodin & D. K. Northington, eds., 615–621. Lubbock: International Center for Arid and Semi-Arid Land Studies, Texas Tech University.

Suh, Y. & **B. B. Simpson.** 1990. Phylogenetic analysis of chloroplast DNA in North American *Gutierrezia* and related genera. Sytematic Botany 15:660–670.

Suksdorf, W. 1931. Untersüchungen in der Gattung *Amsinckia*. Werdenda 1:47–113.

Sundberg, S. D. 1986. The systematics of *Aster* subg. *Oxytripolium* (Compositae) and historically allied species. Ph.D. diss., University of Texas, Austin.

———. 1991. Infraspecific classification of *Chloracantha spinosa* (Benth.) Nesom (Asteraceae) Astereae. Phytologia 70:382–391.

Sundell, E. 1981. The New World species of *Cynanchum* subgenus *Mellichampa* (Asclepiadaceae). Evolutionary Monographs 5:1–63.

Sutherland, D. 1986. Poaceae, In Flora of the Great Plains, Great Plains Flora Association, eds., 1113–1235. Lawrence: University of Kansas.

Svenson, H. K. 1957. Scirpeae. Flora North America 18:505–556.

Swallen, J. R. 1964. Gramineae. In vol. 1 of Flora and vegetation of the Sonoran Desert, by F. Shreve & I. L. Wiggins, 237–301. Stanford: Stanford University Press.

Swallen, J. R. & **E. Hernández X.** 1961. Clave de los géneros mexicanos de Gramíneas. Boletín de la Sociedad Botánico de México 26:52–118.

Sykes, G. 1937. The Colorado delta. American Geographical Society Special Publication 19. Washington, D.C.: Carnegie Institution of Washington and the American Geographical Society of New York.

Tanaka, T. 1976. Tanaka's cyclopedia of edible plants of the world. Tokyo: Keigaku.

Tateoka, T. 1961. A biosystematic study of *Tridens* (Gramineae). American Journal of Botany 48:565–573.

Taylor, N. P. 1979. Notes on *Ferocactus* B. & R. Cactus and Succulent Journal (U.K.) 41:88–94.

———. 1984. A review of *Ferocactus* Britton & Rose. Bradleya 2:19–38.

———. 1985. The genus *Echinocereus*. Portland, Ore.: Timber Press.

Taylor, N. P. & **J. Y. Clark.** 1983. Seed-morphology and classification in *Ferocactus* subg. *Ferocactus*. Bradleya 1: 3–16.

Terrell, E. E. 1968. A taxonomic revision of the genus *Lolium,* U.S.D.A. Technical Bulletin 1392:1–65.

Thieret, J. W. 1971. The genera of Orobanchaceae in the southeastern United States. Journal of the Arnold Arboretum 52:404–434.

Thieret, J. W. & **J. O. Luken.** 1996. The Typhaceae in the southeastern United States. Harvard Papers in Botany no. 8:27–56.

Thomas, R. K. 1991. Papago land use west of the Papago Indian Reservation, south of the Gila River, and the problem of Sand Papago identity. In Ethnology of northwestern Mexico: A Sourcebook, R. H. McGuire, ed., 357–399. Spanish Borderlands Sourcebook 6. New York: Garland Publishing.

Thompson, D. M. 1988. Systematics of *Antirrhinum* (Scrophulariaceae) in the New World. Systematic Botany Monographs 22:1–142.

Thompson, H. J. & **W. R. Ernst.** 1967. Floral biology and systematics of *Eucnide* (Loasaceae). Journal of the Arnold Arboretum 48:56–88.

Thompson, H. J. & **J. E. Roberts.** 1971. Observations on *Mentzelia* in southern California. Phytologia 21:279–288.

Tiedemann, A. R. 1984. Proceedings—Symposium on the biology of *Atriplex* and related Chenopods. May 2–6, 1983, Provo, UT. U.S.D.A. Forest Service General Technical Report INT-172. Ogden: U.S. Forest Service.

Tillet, S. S. 1967. The maritime species of *Abronia* (Nyctaginaceae). Brittonia 19:299–327.

Timbrook, S. 1986. Segregation of *Loeseliastrum* from *Langloisia* (Polemoniaceae). Madroño 33:157–174.

Tomb, A. S. 1972. Re-establishment of the genus *Prenanthella* Rydb. (Compositae: Cichorieae). Brittonia 24: 223–228.

———. 1974. Chromosome numbers and generic relationships in the subtribe Stephanomeriinae (Compositae: Cichorieae). Brittonia 26:203–216.

Torrey, J. 1855. Letter addressed to Dr. John Torrey, on *Ammobroma sonorae,* by Mr. A. B. Gray. Proceedings of the American Academy of Arts 9:223–236.

———. 1856. In Reports of explorations and surveys for a railroad from the Mississippi River to the Pacific Ocean, A. W. Whipple, vol. 4. Washington, D.C.

———. 1858. Botany of the Boundary. In vol. 2, part 1 of Report on the United States and Mexican Boundary Survey, W. H. Emory, ed., 29–270. Washington, D.C.: C. Wendell.

Toursarkissian, M. 1975. Las nictagináceas Argentinas. Revista del Museo Argentino de Ciencias Naturales (Botánica) 5:27–83.

Trent, J. S. & **K. W. Allred.** 1990. A taxonomic comparison of *Aristida ternipes* and *Aristida hamulosa* (Gramineae). Sida 14:251–261.

Tryon, A. F. 1957. A revision of the fern genus *Pellaea* section *Pellaea.* Annals of the Missouri Botanical Garden 44:125–193.

Tryon, R. M. 1955. *Selaginella rupestris* and its allies. Annals of the Missouri Botanical Garden 42:1–99.

———. 1956. A revision of the American species of *Notholaena*. Contributions from the Gray Herbarium 179: 1–106.

Tryon, R. M. & **A. F. Tryon.** 1973. Geography, spores, and evolutionary relations in the cheilanthoid ferns. Botanical Journal of the Linnean Society 67:145–153.

———. 1982. Ferns and allied plants: With special reference to tropical America. New York: Springer-Verlag.

Tucker, G. C. 1983. The taxonomy of *Cyperus* (Cyperaceae) in Costa Rica and Panama. Systematic Botany Monographs 2:1–85.

———. 1987. The genera of Cyperaceae in the southeastern United States. Journal of the Arnold Arboretum 68:361–445.

———. 1994. Revision of the Mexican species of *Cyperus* (Cyperaceae). Systematic Botany Monographs 43: 1–213.

———. 1996. The genera of Pooideae (Gramineae) in the southeastern United States. Harvard Papers in Botany no. 9:11–90.

Turner, B. L. 1962. Taxonomy of *Hymenothrix* (Helenieae, Compositae). Brittonia 14:101–120.

———. 1978. A new species of *Brickellia,* subgenus *Phanerostylis* (Asteraceae). Brittonia 30:342–344.

———. 1986. Reduction of *Machaeranthera arida* to varietal status under *M. coulteri* (Asteraceae: Astereae). Phytologia 61:143–146.

———. 1990. Taxonomic overview of *Brickellia coulteri* (Asteraceae-Eupatorieae), including *B. brachiata* and *B. megalodonta.* Phytologia 68:234–238.

———. 1991. An overview of the North American species of *Menodora* (Oleaceae). Phytologia 71:340–356.

———. 1994. Revisionary study of the genus *Allionia* (Nyctaginaceae). Phytologia 77:45–55.

———. 1996. The Comps of Mexico, 6: Tagetaea and Anthemideae. Phytologia Memoirs 10:1–93.

———. 1997. The Comps of Mexico, 1: Eupatorieae. Phytologia Memoirs 11:1–272.

Turner, B. L. & **T. Barkley.** 1990. Taxonomic overview of the *Senecio flaccidus* complex in North America, including *S. douglasii.* Phytologia 69:51–55.

Turner, B. L. & **R. Hartman.** 1976. Infraspecific categories of *Machaeranthera pinnatifida* (Compositae). Wrightia 5:308–315.

Turner, B. L. & **M. I. Morris.** 1975. New taxa of *Palafoxia* (Asteraceae: Helenieae). Madroño 23:79–80.

———. 1976. Systematics of *Palafoxia* (Asteraceae: Helenieae). Rhodora 78:567–628.

Turner, M. W. 1993. Systematic study of the genus *Baileya* (Asteraceae: Helenieae). Sida 15:491–508.

Turner, R. M. 1990. Long-term vegetation change at a fully protected Sonoran Desert site. Ecology 71:464–477.

Turner, R. M., J. E. Bowers, & **T. L. Burgess.** 1995. Sonoran Desert plants: An ecological atlas. Tucson: University of Arizona Press.

Tutin, T. G. 1980. *Polypogon.* Flora Europaea 5:236.

Tzvelev, N. N. 1968. Notae de gramineis flora U.R.S.S. Novosti Systematiki Vysshikh Rastenii 5:15–30.

——— [Tsvelev]. 1983. Grasses of the Soviet Union. Academy of Sciences of the U.S.S.R. English ed. in 2 vols. Translated by B. R. Sharma. New Delhi: Amerind Publishing Co.

Ulibarri, E. A. 1979. Las especies argentinas del género *Hoffmanseggia.* Darwiniana 22:135–158.

Umber, R. E. 1979. The genus *Glandularia* (Verbenaceae) in North America. Systematic Botany 4:72–102.

Unger, G. 1992. Die grossen Kugelkakteen Nordamerikas. Graz: Gottfried Unger.

Uphof, J.C.T. 1968. Dictionary of economic plants, 2nd ed. Lehre, Germany: J. Cramer.

Urtecho, R. J. 1996. A taxonomic study of the Mexican species of *Tragia* (Euphorbiaceae). Ph.D. diss., University of California, Davis.

Valdés-Reyna, J. & **S. L. Hatch.** 1997. A revision of *Erioneuron* and *Dasyochloa* (Poaceae: Eragrostideae). Sida 17:645–666.

Van Devender, T. R. 1987. Holocene vegetation and climate in the Puerto Blanco Mountains, southwestern Arizona. Quaternary Research 27:51–72.

———. 1990. Late Quaternary vegetation and climate of the Sonoran Desert, United States and Mexico. In Packrat middens: The last 40,000 years of biotic change, J. L. Betancourt, T. R. Van Devender, & P. S. Martin, eds., 134–165. Tucson: University of Arizona Press.

Van Devender, T. R., T. L. Burgess, R. S. Felger, & **R. M. Turner.** 1990. Holocene vegetation of the Hornaday Mountains of northwestern Sonora, Mexico. Proceedings of the San Diego Society of Natural History 2:1–19.

Van Devender, T. R., L. J. Toolin, & **T. L. Burgess.** 1990. The ecology and paleoecology of grasses in selected Sonoran Desert plant communities. In Packrat middens: The last 40,000 years of biotic change, J. L. Betancourt, T. R. Van Devender, & P. S. Martin, eds., 326–349. Tucson: University of Arizona Press.

Van Valkenburgh, R. 1945. Tom Childs (interview notes). On file, Amerind Foundation. Dragoon, Ariz.

Vasek, F. C. 1980. Creosote bush: Long-lived clones in the Mojave Desert. American Journal of Botany 67: 246–255.

Vasey, G. 1889. New or little known plants. Garden and Forest 2:401–403.

Vasey, G. & **J. N. Rose.** 1890. List of plants collected by Dr. Edward Palmer in Lower California in 1889. Contributions from the United States National Herbarium 1:9–28.

Vassal, J. 1981. Acacieae. In Advances in legume systematics, R. M. Polhill & P. H. Raven, eds., 169–171. Kew: Royal Botanic Gardens.

Venable, D. L. & **C. E. Pake.** 1999. Population ecology of Sonoran Desert annual plants. In Ecology of Sonoran Desert Plants and Plant Communities, R. H. Robichaux, ed., 115–142. Tucson: University of Arizona Press.

Veno, B. A. 1979. A revision of the genus *Pectocarya* (Boraginaceae) including reduction to synonymy of the genus *Harpagonella* (Boraginaceae). Ph.D. diss., University of California, Los Angeles.

Wagenitz, G. 1976. Two species of the "*Filago germanica*" group (Compositae-Inuleae) in the United States. Sida 6:221–223.

Wagner, W. L. 1998. Species status for a Sonoran Desert annual member of *Oenothera* sect. *Anogra*. Novon 8:307–310.

Wagner, W. L., R. Stockhouse, & **W. M. Klein.** 1985. The systematics and evolution of the *Oenothera caespitosa* species complex Onagraceae). Monographs in Systematic Botany from the Missouri Botanical Garden 12:1–103.

Wahl, H. A. 1954. A preliminary study of the genus *Chenopodium* in North America. Bartonia 27:1–46.

Waller, S. S. & **J. K. Lewis.** 1979. Occurrence of C_3 and C_4 photosynthetic pathways in North American grasses. Journal of Range Management 32:12–28.

Warnock, M. J. 1997. *Delphinium.* In vol. 3 of Flora of North America north of Mexico, 196–240. New York: Oxford University Press.

Warren, P. L. & **L. S. Anderson.** 1987. Vegetation recovery following livestock removal near Quitobaquito Spring, Organ Pipe Cactus National Monument, Arizona. University of Arizona, Cooperative National Park Resource Studies Unit, Technical Report 20:1–40.

Warrick, G. D. & **P. R. Krausman.** 1989. Barrel cacti consumption by desert bighorn sheep. The Southwestern Naturalist 34:483–486.

Waterfall, U. T. 1958. A taxonomic study of the genus *Physalis* in North America north of Mexico. Rhodora 60:107–114, 128–142, 152–173.

———. 1967. *Physalis* in Mexico, Central America, and the West Indies. Rhodora 69:82–120, 203–239, 319–329.

Watson, M. C. & W. R. Ferren Jr. 1991. A new species of *Suaeda* (Chenopodiaceae) from coastal northwestern Sonora, Mexico. Madroño 38:30–36.

Watson, S. 1889. Upon a collection of plants made by Dr. E. Palmer in 1887. Proceedings of the American Academy of Arts and Sciences 24:36–82.

Webber, J. M. 1953. Yuccas of the southwest. U.S.D.A. Monograph 17:1–97.

Webster, G. L. 1975. Conspectus of a new classification of the Euphorbiaceae. Taxon 24:593–601.

———. 1994. Synopsis of the genera and suprageneric taxa of Euphorbiaceae. Annals of the Missouri Botanical Garden 81:33–144.

Webster, G. L., W. V. Brown, & **B. N. Smith.** 1975. Systematics of photosynthetic carbon fixation pathways in *Euphorbia*. Taxon 24:27–33.

Webster, R. D. & **S. L. Hatch.** 1981. Taxonomic relationships of Texas specimens of *Digitaria ciliaris* and *Digitaria bicornis* (Poaceae). Sida 9:34–42.

Wells, P. V. & **J. H. Hunziker.** 1976. Origin of the creosotebush (*Larrea*) deserts of southwestern North America. Annals of the Missouri Botanical Garden 63:843–861.

West, J. G. 1984. A revision of *Dodonaea* Miller (Sapindaceae) in Australia. Brunonia 7:1–194.

Whalen, M. 1977. Taxonomy of *Bebbia* (Compositae: Heliantheae). Madroño 24:112–123.

Whalen, M. A. 1987. Systematics of *Frankenia* (Frankeniaceae) in North and South America. Systematic Botany Monographs 17:1–93.

Whalen, M. D. 1984. Conspectus of species groups in *Solanum* subgenus *Leptostemonum*. Gentes Herbarum 12:179–282.

Wheeler, L. C. 1936. Revision of the *Euphorbia polycarpa* group of the southwestern United States and adjacent Mexico: A preliminary treatment. Bulletin of the Torrey Botanical Club 63:397–416, 429–450.

———. 1941. *Euphorbia* subgenus *Chamaesyce* in Canada and the United States exclusive of southern Florida. Rhodora 43:97–154, 168–205, 223–286.

Wherry, E. T. 1961. Remarks on the genus *Linanthus*. Aliso 5:9–10.

Whetstone, R. D., T. A. Atkinson, & **D. D. Spaulding.** 1997. Berberidaceae. In vol. 3 of Flora of North America north of Mexico, 272–273. New York: Oxford University Press.

Whitaker T. W. & **Bemis, W. P.** 1964. Evolution in the genus *Cucurbita*. Evolution 18:553–559.

Whitaker, T. W. & **G. N. Davis.** 1962. Cucurbits: Botany, cultivation, and utilization. New York: Interscience Publishers.

Whittemore, A. T. 1997a. *Berberis*. In vol. 3 of Flora of North America north of Mexico, 276–287. New York: Oxford University Press.

———. 1997b. *Myosurus*. In vol. 3 of Flora of North America north of Mexico, 135–138. New York: Oxford University Press.

Whittemore, A. T. & **B. D. Parfitt.** 1997. Ranunculaceae. In vol. 3 of Flora of North America north of Mexico, 85–87. New York: Oxford University Press.

Wiens, D. 1964. Revision of the acataphyllous species of *Phoradendron*. Brittonia 16:11–54.

Wiggins, I. L. 1937. Effects of the January freeze upon the pitahaya in Arizona. Cactus and Succulent Journal (U.S.) 8:171.

———. 1964. Flora of the Sonoran Desert. In Flora and vegetation of the Sonoran Desert, by F. Shreve & I. L. Wiggins, 189–1740. 2 vols. Stanford: Stanford University Press.

———. 1971. The genus *Selaginella* in Baja California, Mexico. American Fern Journal 61:149–160.

———. 1980. Flora of Baja California. Stanford: Stanford University Press.

Wiggins, I. L. & **R. C. Rollins.** 1943. New and noteworthy plants from Sonora, Mexico. Contributions from the Dudley Herbarium 3:266–284.

Williams, E. W. 1957. The genus *Malacothrix* (Compositae) American Midland Naturalist 58:494–512.

Willson, J. & **R. Spellenberg.** 1977. Observation on anthocarp anatomy in the subtribe *Mirabilinae* (Nyctaginaceae). Madroño 24:104–111.

Wilson, H. D. 1980. Artificial hybridization among species of *Chenopodium* sect. *Chenopodium*. Systematic Botany 5:253–263.

Wilson, K. L. 1990. Typification and application of *Scirpus geniculatus* L. Cyperaceae Newsletter 7:6–7.

Wilson, R. C. 1972. *Abronia:* I. Distribution, ecology, and habit of nine species of *Abronia* found in California. Aliso 7:421–437.

Windham, M. D. 1993a. *Notholaena*. In vol. 2 of Flora of North America north of Mexico, 143–149. New York: Oxford University Press.

———. 1993b. *Pellaea*. In vol. 2 of Flora of North America north of Mexico, 175–186. New York: Oxford University Press.

Windham, M. D. & **E. W. Rabe.** 1993. *Cheilanthes*. In vol. 2 of Flora of North America north of Mexico, 152–169. New York: Oxford University Press.

Winograd, I. J., J. M. Landwehr, K. R. Ludwig, T. B. Coplen, & **A. C. Riggs.** 1997. Duration and structure of the past four interglaciations. Quaternary Research 48:141–154.

Wood, C. E. 1971. The Saururaceae in the southeastern United States. Journal of the Arnold Arboretum 52: 479–485.

Woodcock, D. 1986. The late Pleistocene of Death Valley: A climatic reconstruction based on macrofossil data. Phytogeography, Paleoclimatology, Paleoecology 57:273–283.

Woodson, R. E. 1930. Studies in the Apocynaceae—1: A critical study of the Apocynoideae (with special reference to the genus *Apocynum*). Annals of the Missouri Botanical Garden 17:1–212.

———. 1954. The North American species of *Asclepias*. Annals of the Missouri Botanical Garden 41:1–211.

Woodson, R. E., R. W. Schery, & **H. J. Kidd.** 1961. Nyctaginaceae. Flora of Panama. Annals of the Missouri Botanical Garden 48:51–65.

Wyatt, R. & **S. B. Broyles.** 1990. Reproductive biology of milkweeds (*Asclepias*): Recent advances. In Biological approaches and evolutionary trends in plants, S. Kawano, ed., 255–272. London: Academic Press.

Yabuno, T. 1966. Biosystematic study of the genus *Echinochloa*. Japanese Journal of Botany 19:277–323.

Yatskievych, G. 1982. A conspectus of the Lennoaceae. Phytologia 52:73–74.

———. 1985. Notes on the biology of the Lennoaceae. Cactus and Succulent Journal (U.S.) 57:73–79.

———. 1999. Styermark's Flora of Missouri, vol. 1. Jefferson City, Mo: The Missouri Department of Conservation.

Yatskievych, G. & **C. T. Mason Jr.** 1986. A revision of the Lennoaceae. Systematic Botany 11:531–548.

Yatskievych, G. & **M. C. Windham.** 1986. Notes on Arizona Pteridophyta. Journal of the Arizona-Nevada Academy of Science 21:19–21.

Yensen, S. B. & **C. W. Weber.** 1986. Composition of *Distichlis palmeri* grain, a saltgrass. Journal of Food Science 51:1089–1090.

———. 1987. Protein quality of *Distichlis palmeri* grain, a saltgrass. Nutrition Reports International 35:963–972.

Young, D. A. 1975. Systematics of *Rhus* subgenus *Lobadium* section *Styphonia* (Anacardiaceae). Ph.D. diss., Claremont Graduate School, Claremont.

Yuncker, T. G. 1932. The genus *Cuscuta*. Memoirs of the Torrey Botanical Club 18:109–331.

———. 1965. *Cuscuta*. North American Flora (series 2) 4:1–40.

Zavortink, J. E. 1966. A revision of *Mentzelia,* section *Trachyphytum* (Loasaceae). Ph.D. diss., University of California, Los Angeles.

Zengel, S. A. & **E. P. Glenn.** 1996. Presence of the endangered desert pupfish (*Cyprinodon macularius,* Cyprinidontidae) in Cienega de Santa Clara, Mexico, following an extensive marsh dry-down. Southwestern Naturalist 41:73–78.

Zengel, S. A., V. Meretsky, E. P. Glenn, R. S. Felger, & **D. Ortiz.** 1995. Ciénega de Santa Clara, a remnant wetland in the Río Colorado delta (Mexico): Vegetation distribution and the effects of water flow reduction. Ecological Engineering 4:19–36.

Zigmond, M. L. 1981. Kawaiisu ethnobotany. Salt Lake City: University of Utah Press.

Zimmerman, A. D. 1985. Systematics of the genus *Coryphantha* (Cactaceae). Ph.D. diss., University of Texas, Austin.

Zohary, M. 1966. Flora Palestina. Part 1. Jerusalem: The Israel Academy of Sciences and Humanities.

Zuloaga, F. O. 1987. Systematics of New World species of *Panicum* (Poaceae: Paniceae). In Grass systematics and evolution, T. R. Soderstrom et al., eds., 287–306. Washington, D.C.: Smithsonian Institution Press.

Zuloaga, F. O. & **O. Morrone.** 1996. Revisión de las especies americanas de *Panicum* subgénero *Panicum* sección *Panicum* (Poaceae: Panicoideae: Paniceae). Annals of the Missouri Botanical Garden 83:200–280.

Zuloaga, F. O., E. G. Nicora, Z. E. Rúgolo de Agrasar, O. Morrone, J. Pensiero, & **A. M. Cialdella.** 1994. Catálogo de la Familia Poaceae en la República Argentina. Monographs in Systematic Botany from Missouri Botanical Garden 47:1–178.

Index

Main page entries for plant names (common and scientific) are in **boldface**; pages where plants are illustrated are in *italic*; synonym (if same genus) is in [square brackets] following plant name, otherwise see by synonym's generic name. See gazetteer for place names not found in index.

B

C

D

E

F

G

M

P

Q

R

S

T

U

V

W

XYZ